# SECOND IEEE INTERNATIONAL CONFERENCE ON FUZZY SYSTEMS

## SAN FRANCISCO, CALIFORNIA
## MARCH 28 - APRIL 1, 1993
## VOLUME I

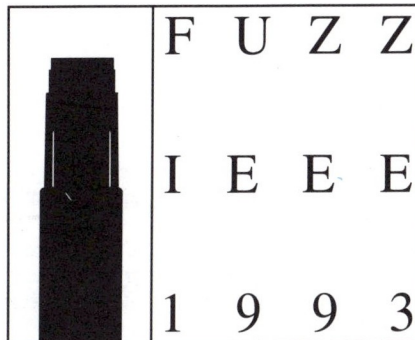

FUZZ IEEE 1993

IEEE

IEEE NEURAL NETWORKS COUNCIL

**1993 IEEE International Conference on Fuzzy Systems**



IEEE Catalog Number: 93CH3136-9
ISBN Softbound: 0-7803-0614-7
Casebound: 0-7803-0615-5
Microfiche: 0-7803-0616-3
Library of Congress Number: 92-52722

Additional copies of this publication are available from

IEEE Service Center
445 Hoes Lane
P.O. Box 1331
Piscataway, NJ 08855-1331
1-800-678-IEEE

**COVER BACKGROUND:**

UNC/MCNC Fuzzy Logic Inference Chip by Hiroyuki Watanabe (University of North Carolina, Chapel Hill), Wayne D. Dettroff and Kathy E. Yount.

The chip, fabricated in MCNC, is used to control navigation of an autonomous robot at Oak Ridge National Laboratory. It includes 688,131 transistors and runs with 10 MHz clock cycle. It performs about 150,000 fuzzy logic inferences per second by executing up to 51 fuzzy if-then rules in parallel.

## A Message from Dr. Russell C. Eberhart,
*President, IEEE Neural Networks Council*

Welcome to the 1993 International Conference on Neural Networks and to the Second IEEE International Conference on Fuzzy Systems, sponsored by the IEEE Neural Networks Council. General Chairman Enrique Ruspini and the Program Committee have assembled fine technical programs in both fuzzy systems and neural networks. This is the first year these conferences have been held concurrently. The synergism between the two fields will, I'm sure, make this an exciting week.

I live in North Carolina, but I have been to San Francisco so often over the past 25 years that I consider it my "second hometown." Overall, I'd say that San Francisco has the best selection of restaurants in the United States. And it has cultural and scenic attractions to match. From Chinatown to Fisherman's Wharf to the Golden Gate, there is more than enough to keep you busy every minute you're not attending the conference sessions.

I encourage you to attend the International Joint Conference on Neural Networks (IJCNN) being held this October in Nagoya, Japan. This will be the Neural Network Council's first major conference in Japan, and we are excited about the opportunity to learn more about what is happening in neural networks in Asia.

In 1994, the Neural Networks Council is sponsoring a landmark event, the World Congress on Computational Intelligence. This exciting congress will include three major conferences: The International Conference on Neural Networks (ICNN)'94, the 1994 International Conference on Fuzzy Systems (FUZZ-IEEE'94), and the 1994 Conference on Evolutionary Computation (CEC). The event will be held at the Walt Disney World Dolphin Hotel in Florida, from June 26 until July 1, 1994. It will occur during the World Soccer Cup, which is being held in Orlando and other U.S. cities (the first time it has ever been held in the United States). You will be able to attend all three conferences for one registration fee.

Thank you for coming to San Francisco! I hope that the conferences meet or exceed your expectations. Please see me or any of the IEEE Neural Networks Council officers or AdCom members if you have suggestions on how we can better serve you.

**Dr. Russell C. Eberhart**

## A Message from the Program Chair

When I was asked to organize the scientific program of the Second IEEE International Conference on Fuzzy Systems (FUZZ-IEEE'93), I accepted the challenge fully aware of the difficulties in achieving a balanced conference program. I wanted the conference to include the most notable and recent theoretical results in fuzzy sets, but still reflect the versatility of its maturing fuzzy-logic based technology.

As a result, the conference is organized along four tracks, addressing the theory of approximate reasoning, the role of fuzzy logic in information processing, the application of fuzzy systems to engineering problems, and the design, synthesis, and analysis of fuzzy logic controllers. The boundaries of these tracks are obviously fuzzy (and sometimes artificial), given the unavoidable overlap between categories such as decision making and optimal control, machine learning and adaptive neural controllers, to name a few. However, the tracks provide a basic structure with sufficient coherence to guide the conference participant throughout its content.

This year, FUZZ-IEEE'93 is held in conjunction with the 1993 International Conference on Neural Networks (ICNN93). We hope that this spatial and temporal concurrence will help highlight the strong connection and synergism between these two emerging technologies. Their interaction and complementarity will be illustrated throughout some of the plenary sessions and in the special NN/FL integration sessions.

An indication of the growth of the field of fuzzy sets is a 25% increase in the number of contributed papers submitted to FUZZ-IEEE'93. This year, we received 375 papers (336 submitted and 39 invited). Of these, 157 papers will be presented in the regular sessions, 34 in the invited sessions, and 73 in the poster sessions. Furthermore, five additional invited papers will be presented in the conference plenary sessions.

I would like to thank the large number of people who have made this conference possible. I want to extend my appreciation to the authors, who enthusiastically responded to our call for papers; the reviewers, who provided guidance and suggestions to enhance the papers' scientific content; the IEEE Neural Network Council, which gave us the opportunity and the financial support to organize this conference; and the management at General Electric Corporate Research and Development (with special thanks to Dr. Norman Sondheimer) who gave me the time and resources required to perform this task.

There is, however, one person who deserves special recognition. I want to thank my friend and colleague, Dr. Enrique Ruspini, the FUZZ-IEEE'93 Conference Chair, whose dedication to this conference has been a constant source of help and inspiration for me.

**Technical Program Chair, FUZZ-IEEE'93**

**Piero P. Bonissone**

## Greetings from the General Chairman,

On behalf of the Institute of Electrical and Electronic Engineers, its Neural Networks Council, and the Organizing Committees of the Second IEEEInternational Conference on Fuzzy Systems and the 1993 IEEE-International Conference on Neural Networks, it is my pleasure to welcome the participants of FUZZ-IEEE'93 and ICNN'93 to its beautiful venue. We trust that the natural beauty of the Bay Area and the cosmopolitan atmosphere of San Francisco will provide a most congenial background to your professional activities.

I felt most honored when Jim Bezdek, in the Spring of 1991,recommended my appointment to head a major conference in the field where both of us have spent most of our professional careers. About one year later, when I was entrusted, once again, with the responsibility to organize another major meeting I must confess that I felt rather elated but also very worried at the magnitude of the task that lay ahead.

Now that both FUZZ-IEEE'93 and ICNN'93 have been successfully planned and organized, I am most happy to have disregarded those apprehensions and to have engaged, for the past year, in the coordination of a major technological event that brings together two major disciplines thatare redefining our notions of modeling, learning, and system analysis. Beyond sharing common objectives related to the understanding and controlof complex systems, fuzzy logic and neural networks have been successfully applied in combination to treat many important problems—-a fact that should be obvious to the readers of these Proceedings. I cannot find words to express my satisfaction to have contributed to this first combined gathering of specialists of bothfields.

As I surmised last year, the secret to this success rested on the selection of a competent team to coordinate the multiple details ofsuch a complex event. The ability to organize such a pioneering meeting, to prepare its records both in conventional and compact-diskformats, to prepare a videotape detailing the state of the art, to secure the participation of interested researchers and exhibitors attests to their ability and dedication.

I am most thankful to all members of the Advisory Boards, Organizing Committees, and Program Committees of FUZZ-IEEE'93 and ICNN'93 fortheir continued help and support. Special recognition goes to AvivBergman, who helped to conceive and supported the production of our video program, Wei Xu for his timely advice and dedicated effort onbehalf of our exhibits program, Jim Bezdek for his trust and for his single-handed organization of the tutorial program, Richard Tong, who kept and keeps tracking our finances, Camerone Welch for her dedication and superb communication skills, and Andy Worth, who coordinated our student volunteers. I am also most thankful to Nomi Feldman and her team at Meeting Management, who gave us indispensable logistic and organizational support and who handled significant crises in a most competent manner. I am particularly proud of the support provided by my family, specially by my wife Susana, who helped tirelessly to assure that the data would be duly recorded and that the mail would leave on time, and to my son Gabriel, who took care of the art work production. Ann Burgmeyer and Janet Romano of IEEE Publishing Services, Reed Carlstrom and John Sands of Young Minds, Inc. made sure that Conference Proceedings would be prepared in the short time that our schedule allowed and, by efficient performance and judicious advice, greatly simplified my task. I amalso thankful to Peter-Wiesner and Beth Murray of the IEEE Educational Activities Board andto Ossama Khatib, of Stanford University, Paul Marca and his staff at the Stanford Instructional Television Network for their help with our video documentary and to Perry Sensi of IEEE Conference Services for his help and advice.

I want to acknowledge Russ Eberhart, Bob Marks, Jim Bezdek and all members of the Neural Networks Council for their trust and encouragement. I am indebted to John Lowrance, Ray Perrault, and the management of SRI International, who have strongly supported my research endeavorsand, particularly, my efforts to organize these meetings. I am most grateful to Shiro Usui and Elie Sanchez, who were invaluable, as ICNN'93Program Cochairs, to coordinate the details of this worldwide project.

I cannot find words to tell of my deep enduring gratitude to Piero Bonissone and Hamid Berenji, who worked closely with me to assure the quality of our program, and with whom, I am happy to say, I will-continue to collaborate in upcoming IEEE events. And, finally, to Lotfi Zadeh, who gave us so much guidance and encouragement through the years.

**Enrique H. Ruspini**
***General Chairman, FUZZ-IEEE'93 and ICNN'93***

## SECOND IEEE INTERNATIONAL CONFERENCE ON FUZZY SYSTEMS (FUZZ-IEEE'93)

**San Francisco Hotel, San Francisco, California**
**March 28 – April 1, 1993**

**GENERAL CHAIRMAN:** **Enrique H. Ruspini**
**Artificial Intelligence Center**
**SRI International**

**TECHNICAL PROGRAM COMMITTEE:**

**Technical Program Chairman: Piero P. Bonissone**
**General Electric Corporate Research and Development**

**Sponsored by the IEEE Neural Networks Council**
**in cooperation with**

IEEE Circuits and Systems Society
IEEE Communications Society
IEEE Control Systems Society
IEEE Systems, Man, and Cybernetics Society
International Fuzzy Systems Association (IFSA)
North American Fuzzy Information Processing Society (NAFIPS)
Japan Society for Fuzzy Theory and Systems (SOFT)
European Laboratory for Intelligent Techniques Engineering (ELITE)


**ADVISORY BOARD:**

J. Bezdek
D. Dubois
G. Klir
H. Prade
E. Sanchez
Ph. Smets
M. Sugeno
T. Terano
E. Trillas
T. Yamakawa
L.A. Zadeh
H.J. Zimmermann

**COMMITEE:**

P. Adlassnig
C. Alsina
K. Asai
J. Baldwin
W. Bandler
H. Berenji
J. Bezdek
P. Bonissone
B. Bouchon-Meunier
C. Brown
J. Buckley
C. Carlsson
S.S. Chen
B. D'Ambrosio
M. Delgado
D. Driankov
D. Dubois
F. Esteva
M. Fedrizzi
C. Freksa
T. Fukuda
M.A. Gil
A. Giordana
M. Gupta
K. Hirota
J. Kacprzyk
A. Kandel
J. Keller
P. Klement
G. Klir
L. Kóczy
D. Kraft
R. Kruse
R. Lea
X. Liu
R. López de Mántaras
E. H. Mamdani
I. Masaki
J. Mendel
S. Miyamoto
M. Mizumoto
M. Mukaidono
C. Negoita
V. Novak
S. Orlovski
S. Ovchinnikov
S. Pal
W. Pedrycz

F. Petry
H. Prade
T. Radecki
A. Ralescu
D. Ralescu
A. Ramer
M. Roubens
E. Ruspini
A. Saffiotti
L. Saitta
E. Sanchez
Ph. Smets
M. Sugeno
T. Takagi
H. Tanaka
T. Terano
M. Togai
R. Tong
E. Trillas
I. Turksen
M. Umano
L. Valverde
J. L. Verdegay
P. Wang
P.-Z. Wang
H. Watanabe
T. Whalen
W. Xu
R. Yager
T. Yamakawa
J. Yao
S. Yasunobu
J. Yen
L. Zadeh
M. Zemankova
H.J. Zimmermann

## ORGANIZING COMMITTEE:

**FINANCE:** R. Tong (Chair), R. Nutter

**PUBLICITY:** H. Berenji (Chair), B. D'Ambrosio, R. López de Mántaras, T. Takagi

**LOCAL ARRANGEMENTS:** S. Ovchinnikov

**TUTORIALS:** J. Bezdek (Chair), H. Berenji, H. Watanabe

**EXHIBITS:** W. Xu (Chair), A. Ralescu, M. Togai, L. Valverde, T. Yamakawa, H.J. Zimmermann

**PRESS/PUBLIC RELATIONS:** C. Welch

**VOLUNTEERS:** A. Worth

# SECOND IEEE INTERNATIONAL CONFERENCE ON FUZZY SYSTEMS

**San Francisco Hilton Hotel, San Francisco, California**

**March 28 - April 1, 1993**

## Table Of Contents

## VOLUME I





** Not available at time of printing*

## MONDAY, MARCH 29, 1993, 2:00-3:30 PM

### MACHINE LEARNING II: FUZZY LOGIC INFERENCE

Chairman: Hamid R. Berenji, NASA Ames Research Center



### FUZZY CONTROL DESIGN I



### ROBOTICS APPLICATIONS



### FUZZY REASONING II: THEORY



## MONDAY, MARCH 29, 1993, 4:30-6:00 PM

### FUZZY CONTROL II: NEURAL NETWORKS/MODEL-BASED METHODS



### FUZZY CONTROL DESIGN II



## INVITED SESSION:

### INDUSTRIAL APPLICATIONS OF FUZZY LOGIC IN GERMANY

Chairman and Organizer: Rudolf Felix, Fuzzy Demonstration Centre, Dortmund





* *Not available at time of printing*

FUZZY REASONING III: CONNECTIVES



**MONDAY, MARCH 29, 1993**

**POSTER PAPERS**

FUZZY CONTROL



ROBOTICS



**TUESDAY, MARCH 30, 1993, 8:30-10:00 AM**

**PLENARY SESSION**





** Not available at time of printing*

## FUZZY LOGIC AND NEURAL NETWORKS I



## FUZZY CONTROL III: STABILITY/ROBUSTNESS



## INDUSTRIAL APPLICATIONS OF FUZZY LOGIC IN EUROPE

Chairman: Rudolf Felix, Fuzzy Demonstration Center, Dortmund



## FUZZY REASONING IV: STRUCTURES



## TUESDAY, MARCH 30, 1993, 2:00-3:30 PM

## FUZZY LOGIC AND NEURAL NETWORKS II



## FUZZY CONTROL IV: STABILITY



## HARDWARE I

Chair: H. Watanabe, University of North Carolina, Chapel Hill





** Not available at time of printing*

* *Not available at time of printing*

GENETIC ALGORITHMS



NEURAL NETWORKS AND FUZZY SETS



# VOLUME II

## WEDNESDAY, MARCH 31, 1993, 8:30-10:00 AM

## JOINT FUZZ-IEEE '93 - ICNN '93 PLENARY

Prof. Carver Mead, California Institute of Technology
Prof. Teuvo Kohonen, Helsinki University of Technology *(See ICNN Program)*

## WEDNESDAY, MARCH 31, 1993, 11:00-12:30 AM

MODELING I (APPLICATIONS)



FUZZY CONTROL VI: TUNING

Chair: P. Wang, Duke University



IMAGE UNDERSTANDING I (SEGMENTATION)





** Not available at time of printing*

**INVITED SESSION:**

FUZZY EXPERT SYSTEMS I

Chairman: Ramón López de Mántaras, Center of Advanced Studies CSIC, Blanes, Spain



**WEDNESDAY, MARCH 31, 1993, 2:00-3:30 PM**

MODELING II



FUZZY CONTROL APPLICATIONS I



IMAGE UNDERSTANDING II



FUZZY EXPERT SYSTEMS II



**WEDNESDAY, MARCH 31, 1993, 4:30-6:00 PM**

MODELING III: MODELING STRUCTURES



FUZZY CONTROL APPLICATIONS II





* *Not available at time of printing*

SIGNAL PROCESSING I



**INVITED SESSION:**

APPROXIMATE REASONING
Chairman: Llorenç Valverde, University of the Balearic Islands, Spain



**WEDNESDAY, MARCH 31, 1993**

**POSTER PAPERS**

FUZZY REASONING, KNOWLEDGE REPRESENTATION



INFORMATION AND EXPERT SYSTEMS



DECISION ANALYSIS, PLANNING, SCHEDULING



MODELING





** Not available at time of printing*

THEORY



**THURSDAY, APRIL 1, 1993, 8:30-10:00 AM**

**PLENARY SESSION**



**THURSDAY, APRIL 1, 1993, 11:00 AM-12:30 PM**

MODELING IV



FUZZY CONTROL APPLICATIONS III



SIGNAL PROCESSING II



**INVITED SESSION:**

FUZZY REASONING VII

Chairman: Francesc Esteva, Center of Advanced Studies CSIC, Blanes, Spain



**THURSDAY, APRIL 1, 1993, 2:00-3:30 PM**

DECISION ANALYSIS I





* *Not available at time of printing*

## FUZZY CONTROL DESIGN III



## PATTERN RECOGNITION AND CLUSTER ANALYSIS



## KNOWLEDGE REPRESENTATION



# THURSDAY, APRIL 1, 1993, 4:30-6:00 PM

## DECISION ANALYSIS II



## FUZZY CONTROL VII: OPTIMIZATION



## PATTERN RECOGNITION AND CLUSTERING ANALYSIS II



## FUZZY ARITHMETIC





** Not available at time of printing*

# Supervised Learning in Fuzzy Systems: Algorithms and Computational Capabilities *

Chi-Cheng Jou
Department of Control Engineering
National Chiao Tung University
Hsinchu, Taiwan 30050, R.O.C.

*Abstract*—**This paper presents model structures for fuzzy systems and accompanies these model structures with learning algorithms. The emphasis of the paper is on basic principles of the design, operating characteristics, and adaptation of fuzzy systems. Several supervised learning algorithms for the adjustment of parameters are discussed. Results of simulations of function approximation and system identification demonstrate that the model structures and supervised learning algorithms suggested for fuzzy systems are practically feasible.**

## I. Introduction

The first fuzzy system models go back to the 1970s. Around this time, Zadeh [1] suggested the description of a system as a set of fuzzy logical rules with fuzzy sets. In particular, the compositional rule of inference was considered to be the backbone of the system. The first application of a fuzzy system was by Mamdani [2], who used a fuzzy system in the control of a laboratory model steam engine. Since then, almost all fuzzy system designs have been based on the original construct suggested by Mamdani. Early approaches to synthesizing fuzzy systems usually presumed that fuzzy logical rules were to be obtained from the laws of physics, expert knowledge, and so on. However, in many respects such fuzzy logical rules are not accurate enough for engineering purposes, and thus fuzzy systems have not met with widespread application in the past two decades.

From a system-theoretic viewpoint, a fuzzy system represents merely a versatile nonlinear static map. Functional values are stored in a distributed rule-based fashion so that the value of the function at any point in the input space is derived by aggregating the consequences of the fuzzy logical rules. Since fuzzy systems can, in principle, be constructed to approximate relations between variables regardless of their analytical dependency, they can be thought of as model-free estimators [3, 4]. Thus, a fuzzy system may be capable of altering or adjusting the fuzzy logical rules in such a way that its behavior or performance improves through contact with its environment. In other words, adaptation may be incorporated into fuzzy systems. Several characteristics of adaptive fuzzy systems make development of such systems worthwhile. First, they can be trained to perform specific filtering and decision-making tasks, and they do not require the elaborate synthesis procedures usually needed for conventional fuzzy systems. Second, they can adapt automatically to changing environments, and they can generalize their behavior to deal with new situations after being trained on a small set of training patterns. Finally, in some model formats they can be described as nonlinear systems with time-varying parameters.

The aims of this paper are to provide appropriate model structures for fuzzy systems, to accompany these model structures with supervised learning algorithms, and to demonstrate the computational capabilities of these learning algorithms. In seeking to reach these objectives, we will present certain basic principles of the design, operating characteristics, and adaptation of fuzzy systems. Results of simulations of function approximation and system identification demonstrate that model structures and supervised learning algorithms we suggest for fuzzy systems are practically feasible.

## II. Model Structures for Fuzzy Systems

In this context, by the term *fuzzy system*, we mean that the model gives a complete description of the system response using fuzzy logical rules. An appropriately chosen model structure can greatly simplify the learning procedure and facilitate system design. A fuzzy system with $m$ fuzzy logical rules can be described by

$$f(\mathbf{x}) = \frac{\sum_{j=1}^{m} \prod_{i=1}^{n} \mu_{A_{ij}}(x_i) w_j}{\sum_{l=1}^{m} \prod_{k=1}^{n} \mu_{A_{kl}}(x_k)} = \frac{\sum_{j=1}^{m} \xi_j(\mathbf{x}) w_j}{\sum_{l=1}^{m} \xi_l(\mathbf{x})} \quad (1)$$

where the $n$-dimensional vector $\mathbf{x} = (x_1, \ldots, x_n)$ denotes the input, function $\mu_{A_{ij}}$ denotes the membership function

*This work was supported by the R.O.C. National Science Council under grant NSC81-0422-E009-03.

0-7803-0614-7/93/$03.00 ©1993 IEEE

of the $j$th input fuzzy set for the $i$th input variable, and scalar $w_j$ denotes the $j$th output fuzzy singleton. For convenience, we define the notion $\xi_j(\mathbf{x}) = \prod_{i=1}^{n} \mu_{A_{ij}}(x_i)$. To derive such an algebraic form for fuzzy systems, we have made the following assumptions: (1) the max-product inference scheme is used for evaluating the overall output fuzzy set; (2) the output fuzzy sets are distinct singletons; and (3) the centroid defuzzification scheme is used to produce a single numerical output from the resulting output fuzzy set. This algebraic form brings fuzzy systems closer to the objective of automatic adjustment of system behavior. To simplify our notation, we consider the normalized activation of each fuzzy logical rule as $\phi_j(\mathbf{x}) = \xi_j(\mathbf{x}) / \sum_{l=1}^{m} \xi_l(\mathbf{x})$. Thus, the fuzzy system defined in (1) can be expressed by $f(\mathbf{x}) = \sum_{j=1}^{m} \phi_j(\mathbf{x}) w_j$.

To obtain a direct application of the above fuzzy systems, we shall define a function $\mu_{A_{ij}}$, which is characterized by a set of parameters. We will consider $\mu_{A_{ij}}$ to be a radially-symmetric function, with a single maximum at the center, that drops off rapidly to zero at large radii. For a radially-symmetric function $\mu_{A_{ij}}$, $m_{ij}$ and $\sigma_{ij}$ represent the center and the effective width in the space of the $i$th input variable, respectively. Thus, this model makes use of input fuzzy sets each having a local receptive field with a sensitivity curve that changes as a function of the distance from a particular point in the input space. The use of membership functions of input fuzzy sets with overlapping receptive fields clearly provides a kind of interpolation and extrapolation.

## III. Learning Algorithms

In this section we are concerned primarily with on-line algorithms for estimation of the parameters within a given fuzzy system. In particular, we emphasize prediction error methods because of their direct applicability to adaptive filtering, prediction, pattern classification, and control.

### *A. The LMS Algorithm*

The least-mean-square algorithm, or LMS algorithm [5], uses a special estimate of the gradient descent. Because the class of fuzzy systems presented takes the form of linear models, we may take $\{\phi_j\}$ to be a set of *basis functions* and $\{w_j\}$ to be a set of adjustable weights. Since the $\phi_j$ functions determining the basic representation of the problem do not have to belong to a specific restricted class of functions in order for parameter estimation methods to apply, we can exercise wide latitude in selecting these functions. They are given a concrete interpretation where a method is applied to a specific problem, and this interpretation constitutes the basic representation of the problem. Some choices of $\phi_j$ functions are better than others, but basically selecting $\phi_j$ functions is an art. Our aim here is to demonstrate how the system behavior evolves while *fixing* $\phi_j$ functions. Note that the evolution is of the output fuzzy singletons, $w_j$, and not of the normalized activations of the rules themselves, $\phi_j$. We summarize the LMS algorithm for the adaptation of $w_j$ as follows:

1. Specify the $\phi_j$ functions according to a priori information and initialize the parameters $w_j$ at random.
2. Choose an input-output pair $(\mathbf{x}, \varsigma)$.

For the current input vector $\mathbf{x}$, compute the normalized activations $\phi_j$ and system output $y = f(\mathbf{x})$.

3. Carry out a modification step by changing the parameters $w_j$ according to

$$w_j^{new} = w_j^{old} + \eta[\varsigma - y]\phi_j \qquad (2)$$

and continue with step 2.

The term $\eta (0 < \eta < 1)$ represents the learning rate. Even in the case of a linear model, the fuzzy system output is a nonlinear function of the input patterns when the $\phi_j$ functions are nonlinear functions of the patterns. However, such a fuzzy system is a linear function of the parameters, since the $\phi_j$ functions are not parameterized and are not involved in the learning process. In this case, the $\phi_j$ functions do not complicate the parameter estimation procedure. Thus, well-behaved learning methods for linear models can be applied. Despite the flexibility that nonlinear $\phi_j$ functions add to linear models, fuzzy systems defined by nonlinear functions of the parameters are of considerable interest because they provide wider latitude for adjusting behavior.

### *B. The Back-Propagation Algorithm*

The back-propagation algorithm is central to much current work on learning in neural networks [6]. The algorithm gives a prescription for changing the weights in any feed-forward network to learn a set of training patterns. The basis is simply gradient descent. Back-propagation can be applied to fuzzy systems, since fuzzy systems can be represented in feed-forward model format. We consider each membership grade $\mu_{A_{ij}}$ to be described by the membership function $\mu_{A_{ij}}(m_{ij}, \sigma_{ij}; x_i)$, where $m_{ij}$ and $\sigma_{ij}$ denote the center and effective width that define $\mu_{A_{ij}}$. Further, we assume that $\mu_{A_{ij}}$ is a continuous differentiable function of $m_{ij}$ and $\sigma_{ij}$. We now seek $m_{ij}$, $\sigma_{ij}$ and $w_j$ such that the fuzzy system maps a given set of training patterns. In practice, the error measure is given by the sum of the squared errors over all training pairs. Thus, determining the parameters is equivalent to the problem of minimizing the error measure. The gradient descent procedure offers the simplest way of solving this problem.

The back-propagation algorithm carrying out an approximate gradient descent is summarized as follows:

1. Initialize the parameters $m_{ij}$, $\sigma_{ij}$ and $w_j$. In the absence of any a priori knowledge, they can be chosen ran-

domly.
2. Choose an input-output pair $(\mathbf{x}, \varsigma)$.
3. Apply $\mathbf{x}$ to the fuzzy system and compute the output $y$.
4. Carry out a modification step by changing the parameters according to

$$w_j^{new} = w_j^{old} + \eta_w [\varsigma - y] \phi_j \tag{3}$$

$$m_{ij}^{new} = m_{ij}^{old} + \eta_m [\varsigma - y] \frac{\partial y}{\partial m_{ij}} \tag{4}$$

$$\sigma_{ij}^{new} = \sigma_{ij}^{old} + \eta_\sigma [\varsigma - y] \frac{\partial y}{\partial \sigma_{ij}} \tag{5}$$

and return to step 2.

The terms $\eta_w$, $\eta_m$, and $\eta_\sigma$ represent the learning rates for parameters $w_j$, $m_{ij}$, and $\sigma_{ij}$, respectively. The derivatives $\partial y / \partial m_{ij}$ and $\partial y / \partial \sigma_{ij}$ can be obtained using the chain rule. Thus, for any input vector $\mathbf{x}$, membership functions $\mu_{A_{ij}}$ are required to be differentiable with respect to parameters $m_{ij}$ and $\sigma_{ij}$. The back-propagation algorithm can become stuck in a local minimum of the error measure. Since the error measure is an extremely complicated function of all the parameters $w_j$, $m_{ij}$, and $\sigma_{ij}$, it can have numerous local minima. Depending on the initial parameter estimates, the back-propagation algorithm always leads to the nearest minimum. Thus, whether a good minimum is obtained depends on the initial values for the parameters and on the form of the error measure. In spite of these problems, the back-propagation algorithm is so general that it can deal with a given problem without any a priori information.

### *C. The Hybrid Learning Algorithm*

The determination of the $\phi_j$ functions can be regarded as a procedure for selecting good features. Feature selection itself is a widely studied problem [7]. Concepts from competitive learning techniques can be used to select suitable features. Competitive learning methods encode pattern information according to coding principles that are built into the learning rule. The aim of competitive learning is to cluster the input data so that the receptive fields of the fuzzy logical rules are found by the system itself from the correlations of the input data. The hybrid learning algorithm for fuzzy systems comprises two parts: the first is concerned with determining the normalized activation functions $\phi_j$ using competitive learning, and the second with adjusting the output singletons $w_j$ based on output performance. In the following we consider a hybrid learning algorithm based on fuzzy competitive learning [8]. The algorithm is summarized as follows:

1) *Determination of centers*:
1. Initialize the centers of the fuzzy logical rules, $\mathbf{m}_j = (m_{1j}, \ldots, m_{nj})$, randomly.
2. Choose an input pattern $\mathbf{x}$.
3. For the current input pattern $\mathbf{x}$, compute the cluster-membership grades $\mu_{C_j}$ by

$$\mu_{C_j} = \frac{(\|\mathbf{x} - \mathbf{m}_j\|^2)^{-\alpha}}{\sum_{l=1}^{m} (\|\mathbf{x} - \mathbf{m}_l\|^2)^{-\alpha}} \tag{6}$$

where $\alpha > 0$ indicates the weighting exponent.
4. Carry out a modification step by changing the parameters $m_{ij}$ according to

$$m_{ij}^{new} = m_{ij}^{old} + \eta' \mu_{C_j} [x_i - m_{ij}^{old}] \tag{7}$$

and continue with step 2.

2) *Determination of effective widths*: Specify the effective widths of the fuzzy logical rules, $\sigma_{ij}$, by

$$\sigma_{ij} = \sigma = \frac{\gamma}{m} \sum_{l=1}^{m} \min_{k \neq l} \|\mathbf{m}_l - \mathbf{m}_k\| \tag{8}$$

where $\gamma$ is a width scaling factor.

3) *Adaptation of output singletons*: Carry out the LMS algorithm.

The positive factor $\gamma$ adjusting the effective widths determines the total activations of the fuzzy logical rules, $\xi_j$. The appropriate value of $\gamma$ depends on the number of fuzzy logical rules and the distribution of the input data. In practice, the value of $\gamma$ is chosen so that the total activations of the fuzzy logical rules for any input approximates unity. The difference between the LMS algorithm and the hybrid learning algorithm is in the determination of the $\phi_j$ functions. In the LMS algorithm, the choice of the basic problem representation embodied in the $\phi_j$ functions is based on how well we understand the problem, and thus strong assumptions are made about the problem. In contrast, the hybrid learning algorithm discovers the $\phi_j$ functions under weak assumptions about the problem, and this approach calls for a more flexible representation. In both learning algorithms, the input membership functions $\mu_{A_{ij}}$ are not necessarily differentiable with respect to their centers and effective widths, because the structures of the $\phi_j$ functions do not directly enter into the derivation of both parameter estimation algorithms.

### *D. The Two-Stage Back-Propagation Algorithm*

Since fuzzy systems are nonlinear models, the final parameter estimates strongly depend on initial estimates and on the vagaries of training experience. Incorporating prior knowledge into the selection of initial parameter values can greatly accelerate, or otherwise improve, the learning process. To take advantage of this property, we present a two-stage back-propagation algorithm [9], which combines the fuzzy competitive learning algorithm and

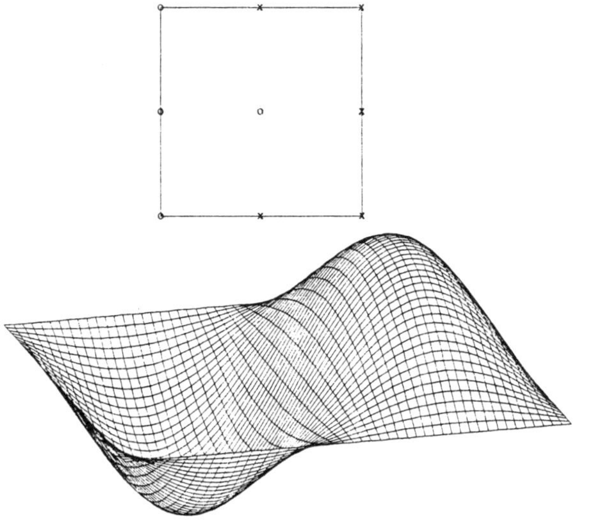

Fig. 1. Two sets of training patterns presented to fuzzy systems for learning. For the set of bipolar patterns, the symbols o and $\times$ represent the output values 1 and $-1$, respectively. For the set of continuous training patterns, the input-output relationship is described by $y = \sin(\pi x_1)\cos(0.5\pi x_2)$. For both sets of training patterns, the inputs $x_1$ and $x_2$ are bounded in $[-1, 1]$.

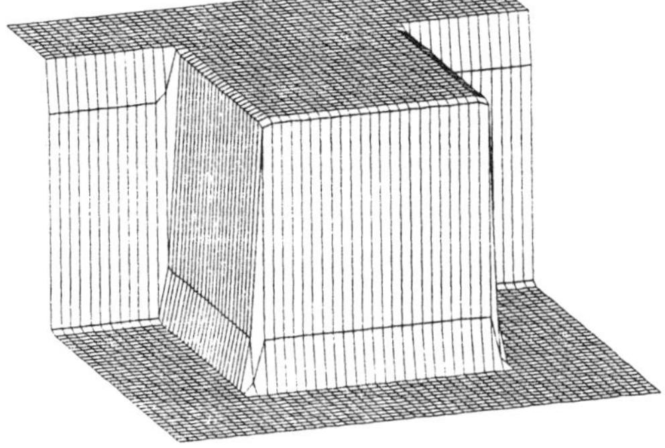

(a)

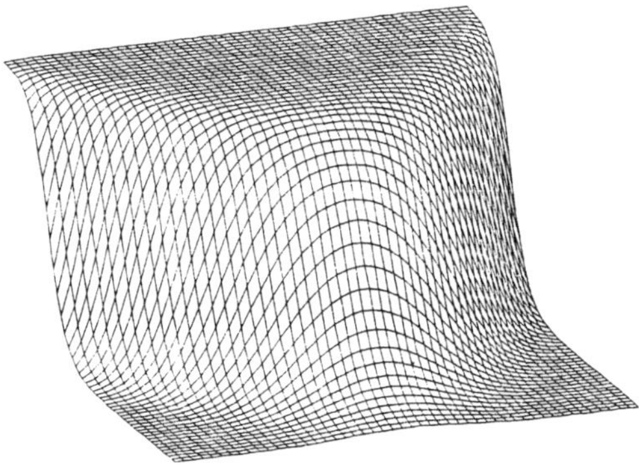

(b)

Fig. 2. The input-output relationships, for the set of bipolar patterns, learned by fuzzy systems using (a) the LMS algorithm and (b) the back-propagation algorithm.

the back-propagation algorithm. In the two-stage back-propagation algorithm, parameter estimation consists of determining the initial estimates of the fuzzy logical rules and the adjustment of the rules based on an error function.

In the first stage, the initial estimates of the fuzzy logical rules are determined by fuzzy competitive learning. In the second stage, all the parameters that define the fuzzy logical rules are modified by applying the back-propagation algorithm. We summarize the two-stage back-propagation algorithm as follows:

1) *Initialization of the fuzzy logical rules*: Determine the initial parameters $m_{ij}$ and $\sigma_{ij}$ according to I and II in the hybrid learning algorithm and initialize $w_j$ at random.

2) *Adaptation of the fuzzy logical rules*: Carry out the back-propagation algorithm.

## IV. Simulation Examples

### *A. Function Approximation*

Our simulations are conducted on two sets of training patterns, as shown in Fig. 1. The training patterns are in two-dimensional input space and one-dimensional output space. Gaussian functions are selected for the construction of the input membership functions, since Gaussian functions are highly nonlinear but have many well-defined features. A Gaussian membership function $\mu_{A_{ij}}(x_i)$ is defined by $\mu_{A_{ij}}(x_i) = \exp(-|x_i - m_{ij}|^2/2\sigma_{ij}^2)$, where $m_{ij}$ and $\sigma_{ij}$ represent, respectively, the mean and the marginal standard deviation of the $j$th fuzzy set with respect to the $i$th input variable. For the set of bipolar training patterns, the LMS algorithm is used to train a fuzzy system with nine fuzzy logical rules, in which the centers $\mathbf{m}_j$ are distributed regularly in the input space, and the back-propagation algorithm is used to train a fuzzy system with two fuzzy logical rules. The LMS algorithm employs the learning rate $\eta = 0.1$ and continues for 200 training steps, while the back-propagation algorithm employs the learning rates $\eta_m = 0.005$, $\eta_\sigma = 0.001$, and $\eta_w = 0.05$, and continues for 3000 training steps. Fig. 2 depicts the three-dimensional representations learned by the fuzzy systems using the LMS algorithm and the back-propagation algorithm. For the LMS algorithm, each stimulus maximally activates a particular fuzzy logical rule, and thus a sharp three-dimensional representation is obtained rapidly. The back-propagation algorithm, on the other hand, generates a rather smooth input-output relationship. Fig. 3 shows the decision boundaries obtained from learning the set of patterns with bipolar outputs.

For the set of continuous training patterns, we train

a fuzzy system with 25 fuzzy logical rules using the LMS algorithm, and another fuzzy system with three fuzzy logical rules using the back-propagation algorithm. The LMS algorithm employs the learning rate $\eta = 0.01$ and continues for 1000 training steps, while the back-propagation algorithm employs the learning rates $\eta_m = \eta_\sigma = \eta_w = 0.05$ and continues for 500 training steps. The resulting input-output relationships learned by the fuzzy systems are shown in Fig. 4. The rms errors for the LMS algorithm and the back-propagation algorithm are 0.0390 and 0.0413, respectively. The degree of success is evident from the similarity to the desired surface of the surfaces generated by the fuzzy systems.

*B. System Identification*

In the following, results of simulated identification of a nonlinear dynamic system using fuzzy systems are presented. The plant to be identified is described by

$$y(k+1) = \frac{(y(k)+2.5)y(k)y(k-1)}{1+y^2(k)+y^2(k-1)} + u(k) \qquad (9)$$

where $[u(k), y(k)]$ represents the input-output pair of the plant at time $k$. The identification task consists of setting up a parameterized fuzzy system and adjusting the parameters of the system to optimize a performance function based on the error between the plant and the fuzzy system output. In the process of modeling, the plant output is fed back into the fuzzy system. This implies that in this case the identification model has the form $\hat{y}(k+1) = f(y(k), y(k-1)) + u(k)$, where the fuzzy system has two inputs, $y(k)$ and $y(k-1)$, and generates a scalar output $f(y(k), y(k-1))$, and $\hat{y}(k+1)$ denotes the output of the identification model. Since both the unknown plant and the fuzzy system are driven by the same input, the fuzzy system adjusts itself with the goal of causing the output of the identification model to match that of the unknown plant. Upon convergence, the input-output response relationships should match. Note that because no feedback loop exists in the identification model, the gradient descent method can be used to adjust the parameters effectively.

In the identification model, the fuzzy system contains ten fuzzy logical rules. Again, Gaussian functions are used to construct the input membership functions. In our simulations, the hybrid learning algorithm and the two-stage back-propagation algorithm are applied because of the inhomogeneous stimulus distributions of $y(k)$ and $y(k-1)$. The training input stimuli $u(k)$ are randomly generated in the interval $[-2, 2]$. Fig. 5 shows the clustering of the input data using fuzzy competitive learning. The resulting trajectories for the ten cluster centers are shown for a run of 1000 input stimuli. The initial positions are indicated by $\circ$. For each fuzzy logical rule, the circle represents

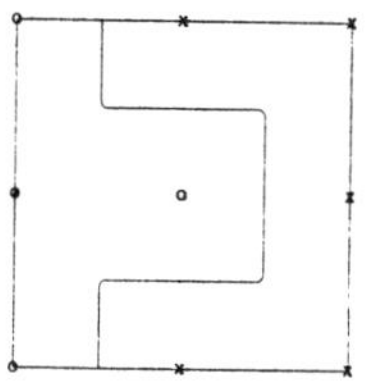

(a)

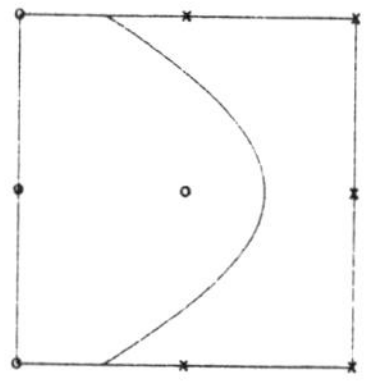

(b)

Fig. 3. The decision boundaries of the input-output relationships learned by the fuzzy systems using (a) the LMS algorithm and (b) the back-propagation.

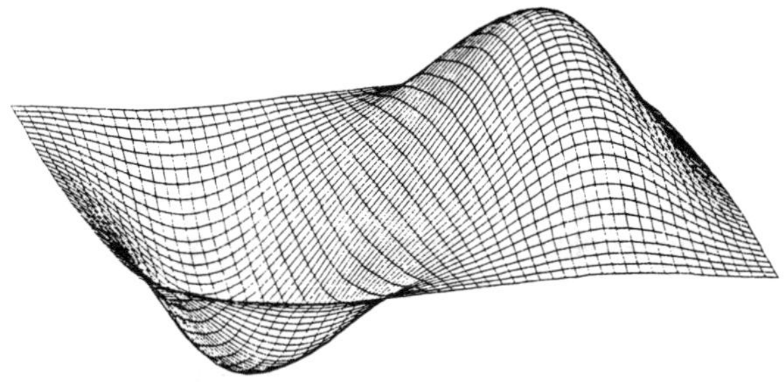

(a)

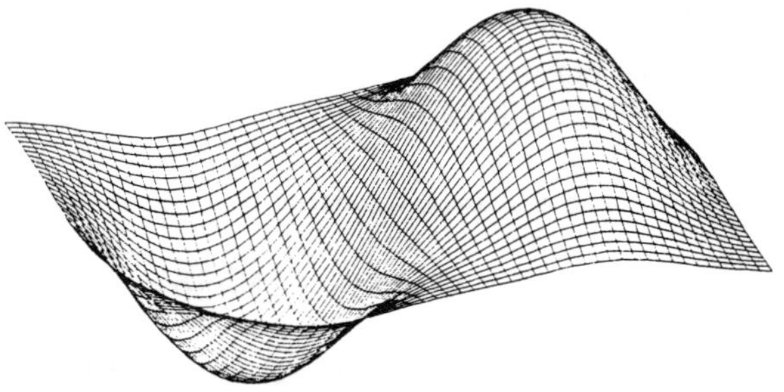

(b)

Fig. 4. The input-output relationships, for the set of continuous training patterns, learned by fuzzy systems using (a) the LMS algorithm and (b) the back-propagation algorithm.

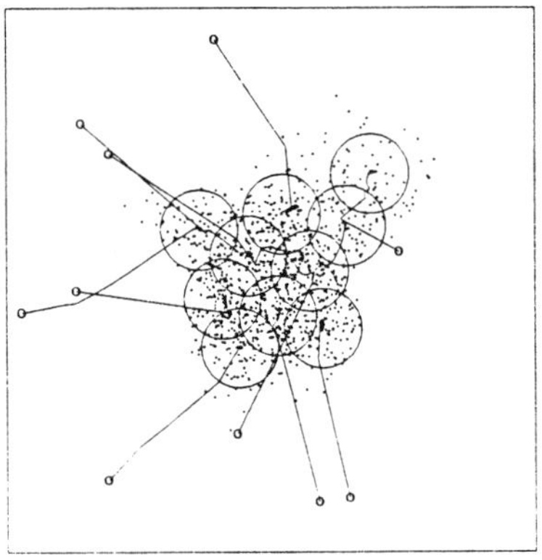

Fig. 5. The clustering of the input data using fuzzy competitive learning.

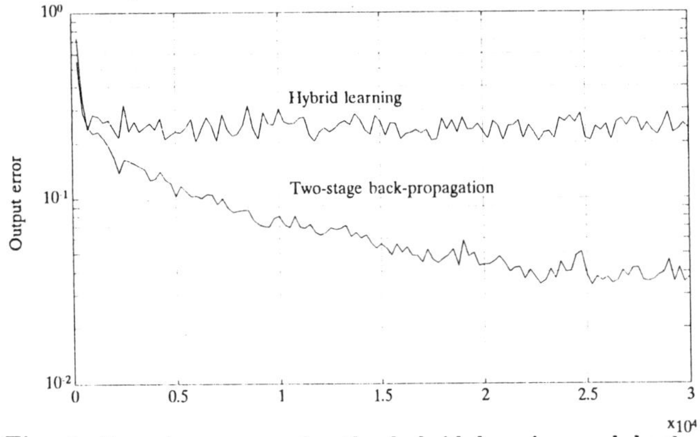


Fig. 6. Learning curves for the hybrid learning and back-propagation algorithms.

the locations at which the membership grades are 0.5. After the input membership functions are determined by fuzzy competitive learning, the training process is continued for 30,000 time steps using the hybrid learning algorithm with the learning rate $\eta = 0.05$ and the two-stage back-propagation algorithm with the learning rates $\eta_m = 0.1$, $\eta_\sigma = 0.01$, and $\eta_w = 0.4$. Fig. 6 shows the rms errors during learning. Each curve is the average of 250 training steps. As can be seen from the figure, with the same number of rules and training steps, the fuzzy system using the two-stage back-propagation algorithm can identify the nonlinear plant more accurately than the system using the hybrid learning algorithm.

## V. Concluding Remarks

In this paper we have considered a class of fuzzy systems and presented a number of supervised learning algorithms for the adjustment of the fuzzy logical rules in fuzzy systems. When the basic problem representation is known, we can use the LMS algorithm. However, this algorithm usually requires a large number of fuzzy logical rules and is not optimal, since the input membership functions are not optimized with respect to output performance. In addition to modifying the output singletons, the back-propagation algorithm varies the parameters of the input fuzzy sets during learning, and thus only a small number of fuzzy logical rules are needed. Both the LMS and back-propagation algorithms are applicable in the case of a sufficiently homogeneous set of input stimuli. When the distribution of input stimuli is not known, the concept of fuzzy competitive learning can be incorporated into the learning process before supervised learning, resulting in the hybrid learning and two-stage back-propagation algorithms, which are based on the LMS and back-propagation algorithms, respectively. The great appeal of fuzzy systems using supervised learning lies in their computational efficiency and the small number of adjustable parameters they require. Simulations performed on sets of bipolar and continuous patterns and a low-order nonlinear dynamic system show that function approximation and system identification by fuzzy systems using the back-propagation and two-stage back-propagation algorithms, respectively, can be very effective. The present exercise provides adequate justification for current efforts to characterize unknown systems using the fuzzy system approach.

## References


[1] L. A. Zadeh, "Outline of a new approach to the analysis of complex systems and decision process," *IEEE Trans. Syst. Man Cybern.*, vol. 3, pp. 28-44, 1993.

[2] E. H. Mamdani, "Application of fuzzy algorithms for control of simple dynamic plant," *Inst. Electr. Eng.*, vol. 121, pp. 1585-1588, 1974.

[3] B. Kosko, "Fuzzy systems as universal approximators," *IEEE Int. Conf. on Fuzzy Systems*, pp. 1153-1162, 1992.

[4] L. Wang, "Fuzzy systems are universal approximators," *IEEE Int. Conf. on Fuzzy Systems*, pp. 1163-1170, 1992.

[5] B. Widrow, and M. E. Hoff, "Adaptive switching circuits," *1960 IRE WESCON Convention Record*, Part 4, pp. 96-104, 1960.

[6] D. E. Rumelhart, G. E. Hinton, and R. J. Williams, "Learning representations by back-propagation errors," *Nature*, vol. 323, pp. 533-536, 1986.

[7] R. O. Duda, and P. E. Hart, *Pattern Classification and Scene Analysis*. New York: Wiley, 1973.

[8] C. C. Jou, "A fuzzy competitive learning algorithm for clustering," *Int. Joint Conf. on Neural Networks*, vol. III, pp. 631-636, 1992.

[9] C. C. Jou, "Learning in fuzzy systems: the two-stage algorithm," *Int. Joint Conf. on Neural Networks*, vol. III, pp. 535-540, 1992.

# Uncertain Reasoning in an ID3 Machine Learning Framework

Peter E. Maher
Department of Mathematics and Computer Science
University of Missouri - St. Louis
St. Louis, MO 63121, USA.

email : maher@arch.umsl.edu

Daniel St. Clair[1]
University if Missouri - Rolla
Engineering Education Center
St. Louis, MO 63121, USA.

email : stclair@umrgec.eec.umr.edu

***Abstract* - Quinlan's ID3 is a symbolic machine learning algorithm which uses training examples as input and constructs a decision tree as output. The algorithm's popularity comes from its ease of application and the comprehensibility of the decision trees (rule sets) it constructs. While ID3 often produces excellent results when precise training and testing data are available, it usually fails miserably in the presence of uncertain training and/or testing data.**

**The UR-ID3 algorithm described combines uncertain reasoning with the rule set produced by ID3 to create a machine learning algorithm which is robust in the presence of uncertain training and testing data. Experimental results are presented which compare the new algorithm's performance with that of ID3 and backpropagation neural networks.**

## I. Introduction

Decision-tree classifier systems, such as Quinlan's ID3 [4], take training examples as input and produce a classification tree as output. Each training instance contains a set of attribute values and an associated classification. Fig. 1 shows a typical decision tree produced by ID3. Each interior node represents an attribute $A_i$ while each branch, $a_{ij}$, represents the $j^{th}$ value of attribute $A_i$. Leaf nodes, $L_j$, correspond to sets, each of which contain one or more classifications, $C_q$. The highlighted path in the decision tree in Fig. 1 represents the rule :

$$A_1(a_{11}) \wedge A_4(a_{43}) \wedge A_2(a_{22}) \Rightarrow C(L_4)$$

or equivalently :

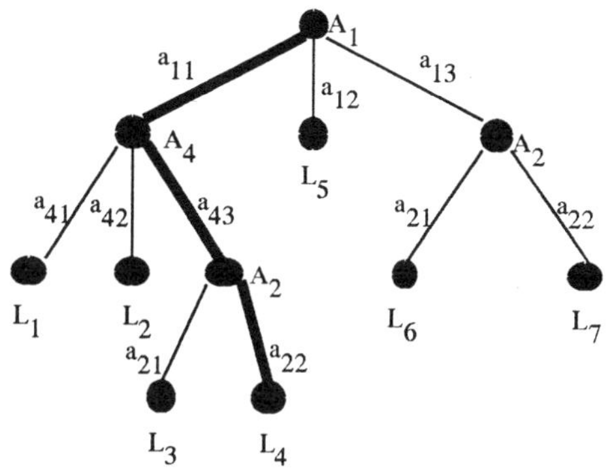


Fig. 1. Typical Decision Tree

**If** the value of attribute $A_1$ is $a_{11}$, and
the value of attribute $A_4$ is $a_{43}$, and
the value of attribute $A_2$ is $a_{22}$
**then** all classifications $C_i \in C(L_4)$ may be assumed.

During learning, ID3 and its variations [6,11] use entropy as an information metric to select the attribute to be used at each node. By using entropy, ID3 tries to construct shallow, wide trees that represent a set of general rules. Ideally, each path in the tree terminates with a leaf node containing a single conclusion ($|C(L_j)| = 1$, for all j). Such a classification tree would be able to completely differentiate each of the concepts it has learned [9]. In application areas such as aircraft fault diagnostics, it is not unusual to have multiple conclusions appearing at a leaf node [8].

One problem with the standard decision-tree approach to machine learning is that uncertain data, either in training and/or testing, often produces poor classification accuracies.

[1] Dr. St. Clair is also Visiting Principal Scientist at McDonnell Douglas Research Laboratories. This work is partially supported by the McDonnell Douglas Independent Research and Development program.

0-7803-0614-7/93$03.00 

## II. Goals of the UR_ID3 Algorithm

The major aim of the UR-ID3 algorithm is to permit the use of uncertain attribute values within the ID3 framework. This technique allows noisy data to be handled in an effective manner, thus eliminating the difficulties presented by the standard ID3 algorithm in such situations. The representation of uncertain, or noisy, attribute values, in both the training and testing examples, will first be presented. An algorithm will then be described that permits these uncertain values to be processed and a classification to be derived.

Uncertain, or approximate, values are denoted by ~N, and are defined by means of a triangular membership function [12] as shown in Fig. 2, in which 'mem(x)' denotes the membership of a value x to the fuzzy set ~N. The value 5 has been used in the definition of 'approximate.' However, this can be parameterized and treated as a *certainty factor*.

To facilitate such quantities being associated with the attribute values in the ID3 decision tree, a method of determining the support for the similarity of any two values must be developed. This support will be represented as an interval $[S_n,S_p] \subseteq [0,1]$, in which the lower bound, $S_n$, is the *necessary support*, and the upper bound, $S_p$, is the *possible support* for the similarity [10]. Consider the case of comparing two approximate values, say ~M, and ~N. Diagrammatically, their membership functions are as shown in Fig. 3.

The notion of *possibility theory* [13] is utilized in the calculations. The notation $\pi(\sim M \mid \sim N)$ represents the maximum membership value at the points of intersection of the membership functions for ~M and ~N. The interval representing the support for the similarity of the two fuzzy numbers ~M and ~N can then be defined as follows :

$$[S_n,S_p] = [\min(1 - \pi(\neg \sim M \mid \sim N),\ 1 - \pi(\neg \sim N \mid \sim M)),\ \pi(\sim M \mid \sim N)].$$

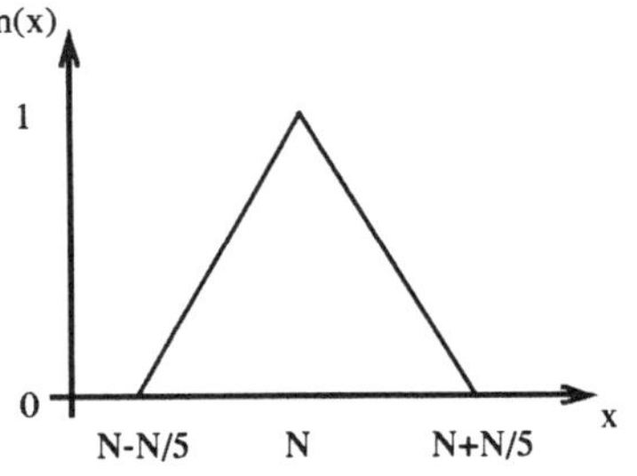


Fig. 2. Diagrammatic Representation of ~N

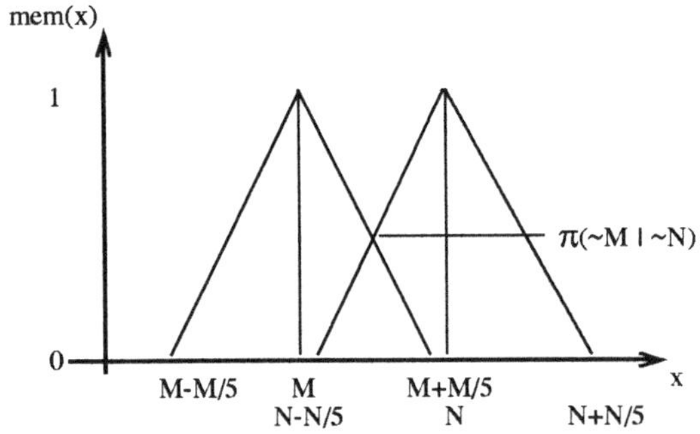


Fig. 3. Diagrammatic Representation of ~M and ~N

The minimum must be taken since, in general, the values $\pi(\neg \sim M \mid \sim N)$ and $\pi(\neg \sim N \mid \sim M)$ will be different.

The comparison of two fuzzy numbers M and N is taken to be the support interval for the numbers and is denoted by $M \oplus N$. Consider, for example, the support intervals derived by comparison of the following fuzzy numbers :

(i) $\sim 10 \oplus \sim 12$ : [0,0.545]
(ii) $\sim 10 \oplus \sim 14$ : [0,0.167]
(iii) $\sim 10 \oplus \sim 100$ : [0,0].

As a special case of the technique described above, it can be noted that when comparing two precise values (where the base of the triangle is a single point) the support interval will be [1,1] if the two values are identical, and [0,0] otherwise. The computation reduces to standard ID3 whenever there are no uncertain attribute values.

The result of the UR-ID3 algorithm will be to produce a support interval for *each* possible classification of a test sample. Full support for a classification is given by [1,1], and full support against a classification is represented by [0,0]. In general, UR-ID3 will produce a set of support intervals that are not extreme cases and a decision as to the correct classification must be made based on this information.

## III. Description of the UR-ID3 Algorithm

The UR-ID3 algorithm begins by using ID3 to construct a decision tree from training values that are assumed to be certain. Following tree construction, the value associated with each branch of the tree is considered to be approximate. Attribute values from the data being compared against the tree are also considered to be uncertain. As a result, a set of support intervals giving likelihoods for each of the possible classifications can be produced by traversing the tree as described below. It is assumed, throughout this paper that each leaf node, $L_j$ ($j = 1$ to $t$), is associated with at most one classification, $C_i$ ($i = 1$ to $s$). Note that a classification, $C_i$, may appear at more than one leaf node. However, the UR-ID3 algorithm is easily extended to handle the more general case.

The first phase of the process is to derive a support interval at each leaf, $L_j$, of the tree. This interval, denoted by $\theta_j$, will then represent the likelihood of the attribute values on the path from the root of the tree down to the leaf node matching those of the new object. The path from the root to the leaf node $L_j$ will be denoted by $P_j$, and the corresponding set of attribute values by $V_{Pj}$. A set of support intervals, $\theta_{Pj}$, is calculated, each element of which results from the comparison of a member of $V_{Pj}$ with the corresponding attribute value in the new object.

The logical AND operator, denoted by $\wedge$, is defined on two intervals as :

$$[a_1,b_1] \wedge [a_2,b_2] = [a_1a_2, b_1b_2].$$

This is used to combine intervals along path $P_j$ into a single support interval, $\theta_j$. This interval represents the support for the classification that appears at leaf node $L_j$. Associated with each leaf node there is an integer which indicates the number of instances of this leaf node occurring during the training period. If the number of instances at leaf node $L_j$ is denoted by $N_j$, then the support interval can be weighted by the weighting factor $N_j/\Sigma N_k$. Thus,

$$\theta_j' = \theta_j * N_j/\Sigma N_k, \text{ for each } j,$$

where $\Sigma N_k$ is the total number of training samples, and the operator "*" denotes the product of the scaler $N_j/\Sigma N_k$ with each of the endpoints of the interval $\theta_j$. The same process is carried out for each leaf in the tree, and a set of support intervals, $\theta' = \{\theta_1', \dots, \theta_t'\}$ is derived.

The next phase of the process determines a single support interval for each possible classification. For each classification $C_i$ there is a subset $L_{Ci} \subseteq L$, the set of all leaf nodes, such that the classification at each node in this subset is $C_i$. This is the set of leaf nodes that provide support for classification $C_i$. There is a corresponding set of support intervals associated with each $L_{Ci}$ and these intervals must now be combined to produce a single interval, denoted by $\theta_{Ci}$, giving the likelihood of classification $C_i$ being correct. The logical OR operator is used to perform this combination. This operation is defined on two intervals as :

$$[a_1, b_1] \vee [a_2, b_2] = [a_1 + a_2 - a_1a_2, b_1 + b_2 - b_1b_2].$$

The classification of a test sample can now be accomplished by considering the corresponding set of support intervals for each possible classification. The lower bound of the interval represents the degree of evidence present that supports the classification, while the upper bound represents the possible support given more evidence. There are many ways in which the classification decision can be made, and this will, in general, be dependent upon the application. A conservative approach would be to choose the lower bound, or $S_n$ value of the interval. This is the technique used in all experiments described in this paper.

## IV. A Simple Example

As an example to illustrate the UR-ID3 method, the classical Iris data will be considered. The Iris flower dataset was developed by R.A. Fisher in 1936 [2]. It contains measurements of petal length, petal width, sepal length, and sepal width from species Virginica, Versicolor, and Setosa. According to Mingers [3] there is little noise in the data set; however, its attributes contain many unique values. In the tree shown in Fig. 4, the labels on the branches correspond to the attribute values. Classifications, and their number of occurrences, are noted at each leaf.

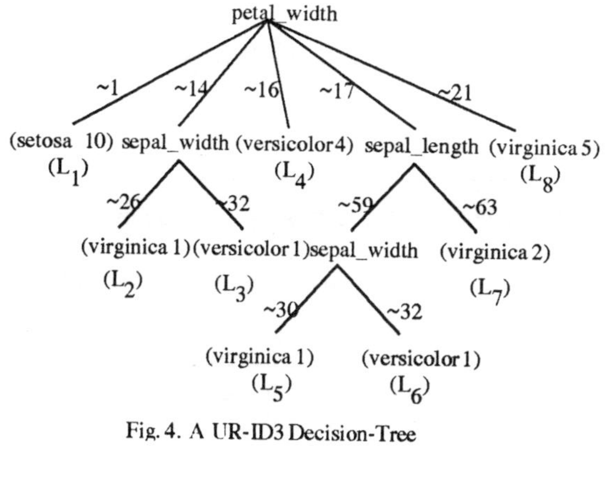


Fig. 4. A UR-ID3 Decision-Tree

Uncertainty has been introduced into the tree by defining each of the values as "approximate." The tree was formed, however, by considering all attribute values as exact; that is, the standard ID3 technique was used.

Suppose that the object to be classified has the following attribute values :

(petal_width, petal_length,
sepal_width, sepal_length)
= (~ 18, ~ 45, ~ 29, ~ 60).

The results of comparing the petal_width of ~18 with all petal_width values at the first level of the tree are as follows :

$\sim 18 \oplus \sim 1$ : [0,0];
$\sim 18 \oplus \sim 14$ : [0,0.375];
$\sim 18 \oplus \sim 16$ : [0.176,0.706]
$\sim 18 \oplus \sim 17$ : [0.343,0.857];
$\sim 18 \oplus \sim 21$ : [0.077,0.615]

Moving down to the next level of the tree, the attributes of sepal_width and sepal_length are considered. The computation of the corresponding support intervals is similar to that shown above, as it is for the subsequent levels of the decision tree.

The support intervals for each leaf node of the tree can now be computed by using the logical AND operator. The total number of training samples is 25, and the results are as follows :

$\theta'_1$ : [0,0]; $\theta'_2$ : [0,0.01]; $\theta'_3$ : [0,0.011];
$\theta'_4$ : [0.028,0.113]; $\theta'_5$ : [0.003,0.03];
$\theta'_6$ : [0.002,0.025]; $\theta'_7$ : [0.01,0.06];
$\theta'_8$ : [0.015,0.123]

The support for each possible classification can now be determined, and the results are :

$\theta_{setosa}$ : [0,0]; $\theta_{versicolor}$ : [0.030,0.145];
$\theta_{virginica}$ : [0.035,0.211]

Thus, it would appear that Virginica is the most likely classification since its necessary support is the highest.

## V. Some Results and Comparisons With Other Methods

Several experiments were conducted to evaluate UR-ID3 learning accuracies against those produced by ID3, a "nearest neighbor" variant of ID3 (ID3nn), and the well-known backpropagation (BP) neural network. The ID3nn algorithm is identical to ID3 except during testing. If a path in the ID3 tree cannot be found because the attribute value being tested does not appear in the tree, then the path is chosen which has the value closest to the one in the test example [7]. The BP algorithm is the only neural network algorithm in the group. It was included since it is known to produce good results on data having continuous attributes [5].

Three different domains were chosen on which to test the algorithms. The first two domains represent real data, while the third was simulated data. These domains are characterized in Table I.

1) The Iris flower dataset is described in Section 3 of this paper. Fisher used it to develop statistical methods of classification. This dataset was selected due to its wide use in the literature.

2) The Thyroid dataset, obtained from the Garvan Institute of Medical Research, Sydney, contains diagnoses of thyroid conditions based upon a number of boolean attributes and several continuous valued attributes. This dataset was included because it contains both discrete and continuous-valued attributes.

3) Breiman et al's waveform recognition dataset is simulated data representing the problem of recognizing three waveforms [1]. Each waveform is generated by a function which randomly combines two of three given triangular waveforms and adds noise. The classification of the resulting waveform is determined by the two input waveforms chosen. Random Gaussian noise added to each value makes this problem difficult.

Training and test datasets were generated from each of the datasets by randomly selecting 75% of the data for training and using the remaining 25% for testing.

TABLE I
Dataset Summary

| Dataset | Number Instances | No. Input Attributes | Number Classes |
|---|---|---|---|
| Iris | 150 | 4 | 3 |
| Thyroid | 80 | 20 | 23 |
| Breiman | 400 | 21 | 3 |

TABLE II
ID3 Trees

| Dataset | Number Internal Nodes | Number Rule (Leaf Nodes) |
| --- | --- | --- |
| Iris1 | 4 | 46 |
| Iris2 | 4 | 52 |
| Iris3 | 6 | 38 |
| Thyroid1 | 4 | 37 |
| Thyroid2 | 4 | 35 |
| Breiman | 1 | 201 |

Six train/test sets of data were generated; three for Iris, two for Thyroid, and one for Breiman. The ID3 algorithm was then used to generate decision trees for each set of data (see Table II).

Table III shows the percentage of correct classifications produced by each of the algorithms. The UR-ID3 algorithm consistently performs significantly better than both the standard ID3, and the ID3nn algorithms, except for the Breiman data in which ID3nn and UR-ID3 give the same degree of accuracy. This is due to the nature of the tree produced by ID3 and to the dataset. This tree has only one level and contains 201 leaf nodes.

The ID3-based algorithms, which are symbolic in nature, are known to be worse in modelling domains in which there are a large number of continuous-value attributes than the backpropagation (analog) algorithms.

TABLE III
Percent of Correct Classifications by Algorithm

| Dataset | ID3 | ID3nn | UR-ID3 | BP |
| --- | --- | --- | --- | --- |
| Iris1 | 75.7 | 89.2 | 94.6 | 91.9 |
| Iris2 | 71.1 | 92.1 | 94.7 | 94.7 |
| Iris3 | 78.9 | 94.7 | 94.7 | 92.1 |
| Thyroid1 | 65.0 | 85.0 | 90.0 | 90.0 |
| Thyroid2 | 50.0 | 70.0 | 80.0 | 100.0 |
| Breiman | 31.3 | 67.7 | 67.7 | 89.9 |

However, with the exception of one thyroid dataset and the Breiman dataset, the UR-ID3 algorithm performs at least as well if not better than BP.

The second thyroid dataset contains only 5 different classifications out of a possible 23; this is believed to be the primary cause of the UR-ID3 algorithm not performing well. The Breiman dataset contains purely continuous data. Each data item is specified to 6 decimal places. This, together with the structure of the tree, makes it much more suited to a neural network approach.

## VI. Conclusions

The ID3 algorithm coupled with fuzzy logic produces an algorithm whose results on domains with continuous-valued attributes are generally better than those produced by ID3 and ID3nn. The algorithm is equally applicable to domains which contain a combination of discrete and continuous-valued attributes. The discrete attributes need not be ordered. Not only does UR-ID3 produce results competitive with those produced by backpropagation, but UR-ID3 rules are easier to interpret than the backpropagation network.

## References


[1] Breiman, L., Friedman, J.H., Olshen, R.A., and Stone, C.J., *Classification and Regression Trees*, Wadsworth & Brooks/Cole, 1984.

[2] Fisher, R.A., "The use of uultiple measurements in taxonomic problems," *Annals of Eugenics*, vol. 7, 1936, pp. 179-188.

[3] Migers, J., "An empirical comparison of selection measures for decision-tree induction," *Machine Learning*, vol. 3, 1989, pp 319-342.

[4] Quinlan, J.R., "Induction of Decision Trees", Machine Learning, Vol. 1, pp 81-106, 1986.

[5] Rumelhart, D.E., Hinton, G.E., and Williams, R.J., "Learning interior representation by error propagation," *Parallel Distributed Processing*, vol. 1, Ch. 8, D.E. Rumelhart and J.L. McClelland eds., MIT Press Cambridge, Mass., 1986.

[6] Schlimmer, J.C., and Fisher, D., "A Case Study of Incremental Concept Induction," *Proceedings of the Fifth National Conference on Artificial Intelligence*, vol. 1, Morgan Kaufmann Publishers, Inc., pp. 496-501, August 1986.

[7] St. Clair, D.C., Bond, W.E., Rigler, A.K., and Aylward, S., "An evaluation of learning performance in backpropagation neural networks and decision-tree

classifier systems," *Proceedings of the 1992 ACM/SIGAPP Symposium on Applied Computing*, vol. II, ACM Press, pp. 636-642, March 1992.

[8] St. Clair, D.C., Bond, W.E., Flachsbart, B.B., and Vigland, A.R., "An architecture for adaptive learning in rule-based diagnostic expert systems," *AI in Armament Workshop-Diagnostics*, American Institute of Aeronautics & Astronautics, March 1988.

[9] St. Clair, D.C., Bond, W.E., and Sabharwal, C.L., "Measuring concept strength in classification trees," *Fifth International Conference on Methodologies for Intelligent Systems* : Selected Papers, M.L. Emrich, M.S. Phifer, B. Huber, M. Zermankova, and Z. Ras eds., International Center for the Application of Information Technology, University of Tennessee, Knoxville, pp. 68-179, 1990.

[10] Shafer, G., *A Mathematical Theory of Evidence*, Princeton, New Jersey, Princeton University Press, 1976.

[11] Utgoff, P.E., "Incremental Induction of Decision Trees," *Machine Learning*, vol. 4, no. 2, pp. 81-106, 1989.

[12] Zadeh, L.A., "Fuzzy sets," *Information and Control*, 8, pp. 338-353, 1965.

[13] Zadeh, L.A., "Fuzzy sets as a basis for a theory of possibility," *Fuzzy Sets and Systems*, vol. 1, 1978.

# TRAINING OF FUZZY LOGIC SYSTEMS USING NEAREST NEIGHBORHOOD CLUSTERING

Li-Xin Wang
Computer Science Division
Department of Electrical Engineering and Computer Science
University of California at Berkeley
Berkeley, CA 94720

***Abstract—***

**In this paper, we first construct an optimal fuzzy logic system which is capable of matching all the input-output pairs in the training set to arbitrary accuracy. Then, we develop an adaptive version of the optimal fuzzy logic system using the nearest neighborhood clustering algorithm. To do this we first determine clusters of the sample data using the nearest neighborhood clustering algorithm, and then view the clusters as sample data and use the optimal fuzzy logic system to match them. We use this adaptive fuzzy system as adaptive controllers for nonlinear dynamic systems.**

## I. INTRODUCTION

A number of training algorithms for fuzzy logic systems have been proposed in the literature [3-5]. In [3], the fuzzy logic systems are represented as three-layer feedforward networks and a back-propagation training algorithm is developed to adjust the parameters of the fuzzy logic systems to make them match the desired input-output pairs. Although this back-propagation fuzzy system worked very well for a number of examples, in principle it may be trapped at a local minimum and converge slowly. To overcome this problem, another training algorithm for fuzzy logic systems was developed in [4] using the Gram-Schmidt orthogonal least squares algorithm. Because this algorithm performs a linear search, it will not be trapped at a local minimum. However, this algorithm is computationally expensive and may not be practical for large-scale problems. To overcome this disadvantage, a third training algorithm was developed [5] which was based on a very simple table look-up scheme.

Although the above three training algorithms for fuzzy logic systems have their advantages in particular situations, they are not optimal in the sense that all of them cannot guarantee that the resulting fuzzy logic system can match all the desired input-output pairs in the training set to arbitrary accuracy. For some practical problems, sample data (i.e., input-output pairs) may be expensive to obtain. For example, a test flight of a new aircraft is very expensive. For these small-sample problems, we may want a fuzzy logic system that is capable of matching all the input-output pairs to any given accuracy.

In this paper, we first develop such an optimal fuzzy logic system (Section 2). Then, we extend the optimal fuzzy logic system to large-sample problems (Section 3). To do this we first determine clusters of the sample data using the nearest neighborhood clustering algorithm, and then view the clusters as sample data and use the optimal fuzzy logic system to match them. We call this approach an adaptive version of the optimal fuzzy logic system, or an adaptive fuzzy system. We also show how to combine linguistic fuzzy IF-THEN rules and numerical input-output pairs using this adaptive fuzzy system. In Section 3, we use this adaptive fuzzy system as controllers for nonlinear dynamic systems, and simulate the approach for the same examples used in [1] for testing the neural controllers. Section 4 concludes this paper.

## II. AN OPTIMAL FUZZY LOGIC SYSTEM

Suppose that we are given $N$ input-output pairs $(\underline{x}^l, y^l), l = 1,2,...,N$, and $N$ is small, say, $N = 20$. Our task is to construct a fuzzy logic system $f(\underline{x})$ which can match all the $N$ pairs to any given accuracy, i.e., for any given $\epsilon > 0$, we require that $|f(\underline{x}^l) - y^l| < \epsilon$ for all $l = 1,2,...,N$.

This optimal fuzzy logic system is constructed as

$$f(\underline{x}) = \frac{\sum_{l=1}^{N} y^l exp(-\frac{|\underline{x}-\underline{x}^l|^2}{\sigma^2})}{\sum_{l=1}^{N} exp(-\frac{|\underline{x}-\underline{x}^l|^2}{\sigma^2})}. \quad (1)$$

As shown in [3,4], this fuzzy logic system is constructed

0-7803-0614-7/93$03.00 ©1993IEEE

using centroid defuzzifier, singleton fuzzifier, product inference, and Gaussian membership function. The following theorem shows that by properly choosing the parameter $\sigma$, the fuzzy logic system (1) can match all the $N$ input-output pairs to any given accuracy.

Theorem 1: For arbitrary $\epsilon > 0$, there exists $\sigma^* > 0$ such that the fuzzy logic system (1) with $\sigma = \sigma^*$ has the property that

$$|f(\underline{x}^l) - y^l| < \epsilon \tag{2}$$

for all $l = 1, 2, ..., N$.

Proof: Evaluating $f(\underline{x})$ of (1) at $\underline{x} = \underline{x}^l$, we have

$$f(\underline{x}^l) = \frac{y^l + \sum_{j \neq l=1}^{N} y^j exp(-\frac{|\underline{x}^l - \underline{x}^j|^2}{\sigma^2})}{1 + \sum_{j \neq l=1}^{N} exp(-\frac{|\underline{x}^l - \underline{x}^j|^2}{\sigma^2})}, \tag{3}$$

where $l = 1, 2, ..., N$. First, assume that $\underline{x}^l \neq \underline{x}^j$ for $l \neq j$; thus, for arbitrary $\epsilon_1 > 0$ and any $l, j = 1, 2, ..., N$, $l \neq j$, we can make $exp[-(\underline{x}^l - \underline{x}^j)^2/\sigma^{*2}] < \epsilon_1$ by properly choosing $\sigma^*$, because $exp[-(\underline{x}^l - \underline{x}^j)^2/\sigma^{*2}] \to 0$ as $\sigma^* \to 0$ if $\underline{x}^l \neq \underline{x}^j$. Therefore we have (2). If $\underline{x}^l = \underline{x}^{j_0}$ for some $j_0 \neq l$, and there are $r - 1$ such $j_0$; then (3) can be written as

$$f(\underline{x}^l) = \frac{ry^l + \sum_j y^j exp(-\frac{|\underline{x}^l - \underline{x}^j|^2}{\sigma^2})}{r + \sum_j exp(-\frac{|\underline{x}^l - \underline{x}^j|^2}{\sigma^2})}, \tag{4}$$

where the $\sum_j$ is over all $j$'s in $\{1, 2, ..., N\}$ except $l$ and the $j_0$'s. Using the same arguments as above, we can prove the truth of (2). Q.E.D.

Remark 1: The $\sigma$ is a smoothing parameter: the smaller the $\sigma$, the smaller the matching error $|f(\underline{x}^l) - y^l|$, but the less smooth the $f(\underline{x})$ becomes. We know that if $f(\underline{x})$ is not smooth, it may not generalize well for the data points not in the training set. Thus, the $\sigma$ should be properly chosen to provide a balance between matching and generalization. Because the $\sigma$ is a one-dimensional parameter, it is usually not difficult to determine an appropriate $\sigma$ for a practical problem. Sometimes, a few trial and error may determine a good $\sigma$. As a general rule, large $\sigma$ can smooth out noisy data, while small $\sigma$ can make $f(\underline{x})$ as nonlinear as is required to approximate closely the training data.

Remark 2: The $f(\underline{x})$ is a general nonlinear regression which provides a smooth interpolation between the observed points $(\underline{x}^l, y^l)$. It is well behaved even for very small $\sigma$.

## III. An Adaptive Version of the Optimal Fuzzy Logic System

The optimal fuzzy logic system (1) uses one rule for one input-output pair in the training set, thus it is no longer a practical system if the number of input-output pairs in the training set is large. For these large-sample problems, various clustering techniques can be used to group the samples so that a group can be represented by only one rule in the fuzzy logic system. Here, we use the simple nearest neighborhood clustering scheme.

*An Adaptive Version of the Optimal Fuzzy Logic System*:

- Starting with the first input-output pair $(\underline{x}^1, y^1)$, establish a cluster center $\underline{x}_0^1$ at $\underline{x}^1$, and set $A^1(1) = y^1$, $B^1(1) = 1$. Select a radius $r$.
- Suppose that when we consider the $k$'th input-output pair $(\underline{x}^k, y^k)$, $k = 2, 3, ...,$ there are $M$ clusters with centers at $\underline{x}_0^1, \underline{x}_0^2, ..., \underline{x}_0^M$. Compute the distances of $\underline{x}^k$ to these $M$ cluster centers, $|\underline{x}^k - \underline{x}_0^l|$, $l = 1, 2, ..., M$, and let the smallest distances be $|\underline{x}^k - \underline{x}_0^{l_k}|$, i.e., the nearest cluster to $\underline{x}^k$ is $\underline{x}_0^{l_k}$. Then:

  a) If $|\underline{x}^k - \underline{x}_0^{l_k}| > r$, establish $\underline{x}^k$ as a new cluster center $\underline{x}_0^{M+1} = \underline{x}^k$, set $A^{M+1}(k) = y^k$, $B^{M+1}(k) = 1$, and keep $A^l(k) = A^l(k-1)$, $B^l(k) = B^l(k-1)$ for $l = 1, 2, ..., M$.

  b) If $|\underline{x}^k - \underline{x}_0^{l_k}| \leq r$, do the following:

$$A^{l_k}(k) = A^{l_k}(k-1) + y^k, \tag{5}$$
$$B^{l_k}(k) = B^{l_k}(k-1) + 1, \tag{6}$$

  and set

$$A^l(k) = A^l(k-1), \tag{7}$$
$$B^l(k) = B^l(k-1), \tag{8}$$

  for $l = 1, 2, ..., M$ with $l \neq l_k$.

- The adaptive fuzzy system at the $k$'th step is computed as

$$f_k(\underline{x}) = \frac{\sum_{l=1}^{M} A^l(k) exp(-\frac{|\underline{x} - \underline{x}_0^l|^2}{\sigma^2})}{\sum_{l=1}^{M} B^l(k) exp(-\frac{|\underline{x} - \underline{x}_0^l|^2}{\sigma^2})} \tag{9}$$

  if $\underline{x}^k$ does not establish a new cluster; and, if $\underline{x}^k$ establishes a new cluster, change the $M$ in (9) to $M + 1$.

Remark 3: Comparing (9) with (1) and noticing (5)-(8), we see that if we replace the $\underline{x}^l$ in (1) by the center of the cluster to which the $\underline{x}^l$ belongs, then (1) becomes (9). This is why we call (9) an adaptive version of the optimal fuzzy logic system.

Remark 4: The radius $r$ determines the complexity of the adaptive fuzzy system. For smaller radius $r$, we have more clusters which result in a more sophisticated nonlinear regression at the price of more computation to evaluate it. Because $r$ is a one-dimensional parameter (like $\sigma$), we may find an appropriate $r$ for a specific problem by trial and error .

Remark 5: Because for each input-output pair a new cluster may be formed, this adaptive fuzzy system performs parameter as well as structure adaptations in a uniform fashion.

Since the $A$ and $B$ coefficients in (9) are determined using recursive equations, it is easy to add a forgetting factor to (5)-(8). This is desirable if the adaptive fuzzy system is being used to model a system with changing characteristics. For these cases, we replace (5), (6) by

$$A^{l_k}(k) = \frac{\tau-1}{\tau}A^{l_k}(k-1)+\frac{1}{\tau}y^k, \quad (10)$$

$$B^{l_k}(k) = \frac{\tau-1}{\tau}B^{l_k}(k-1)+\frac{1}{\tau}, \quad (11)$$

and replace (7), (8) by

$$A^{l}(k) = \frac{\tau-1}{\tau}A^{l}(k-1), \quad (12)$$

$$B^{l}(k) = \frac{\tau-1}{\tau}B^{l}(k-1), \quad (13)$$

where $\tau$ can be considered as a time constant of an exponential decay function. For practical considerations, there should be a lower threshold for $B^l$ so that when sufficient time has elapsed without update for a particular cluster (which results in the $B^l$ to be smaller than the threshold), that cluster would be eliminated.

Finally, we see how to combine linguistic fuzzy IF-THEN rules and numerical input-output pairs using this adaptive fuzzy system. Here, we use linguistic rules and input-output pairs to construct two separate fuzzy logic systems, and the final adaptive fuzzy system is obtained by combining them through weighted average. Specifically, let $f^L(\underline{x})$ be an fuzzy logic system constructed from linguistic fuzzy IF-THEN rules, then the final adaptive fuzzy system is

$$f(\underline{x}) = \alpha f^L(\underline{x}) + (1-\alpha)f_k(\underline{x}), \quad (14)$$

where $f_k(\underline{x})$ is given by (9), and $\alpha \in [0,1]$ is a weighting factor. If there are no linguistic rules, set $\alpha = 0$.

## IV. Application to Adaptive Control of Nonlinear Dynamic Systems

In this section, we use the adaptive fuzzy system (9) as a basic building block of adaptive controllers for nonlinear dynamic systems. We use the same examples as in [1].

Example 1: We consider here the problem of controlling the plant which is described by the difference equation

$$y(k+1) = g[y(k), y(k-1)] + u(k) \quad (15)$$

where the function

$$g[y(k), y(k-1)] = \frac{y(k)y(k-1)[y(k)+2.5]}{1+y^2(k)+y^2(k-1)} \quad (16)$$

is assumed to be unknown. The aim of control is to determine a controller $u(k)$ (based on the adaptive fuzzy system) such that the output $y(k)$ of the closed-loop system follows the output $y_m(k)$ of the following reference model:

$$y_m(k+1) = 0.6y_m(k) + 0.2y_m(k-1) + r(k). \quad (17)$$

where $r(k) = sin(2\pi k/25)$, i.e., we want $e(k) = y(k) - y_m(k)$ converges to zero as $k$ goes to infinite.

If the function $g[*]$ of (16) is known, we can construct a controller as follow:

$$u(k) = -g[y(k), y(k-1)] + 0.6y(k) + 0.2y(k-1) + r(k), \quad (18)$$

which, when applied to (15), results in

$$y(k+1) = 0.6y(k) + 0.2y(k-1) + r(k). \quad (19)$$

Combining (17) and (19), we have

$$e(k+1) = 0.6e(k) + 0.2e(k-1), \quad (20)$$

from which it follows that $lim_{k\to\infty}e(k) = 0$. However, since $g[*]$ is unknown, the controller (18) cannot be implemented. To solve this problem, we replace the $g[*]$ in (18) by the adaptive fuzzy system (9), i.e., we use the following controller:

$$u(k) = -f_k[y(k), y(k-1)] + 0.6y(k) + 0.2y(k-1) + r(k). \quad (21)$$

where $f_k[*]$ is in the form of (9) with $\underline{x} = (y(k), y(k-1))^T$. This results in the nonlinear difference equation

$$y(k+1) = g[y(k), y(k-1)] - f[y(k), y(k-1)] + 0.6y(k) + 0.2y(k-1) + r(k) \quad (22)$$

governing the behavior of the closed-loop system. The overall control system is shown in Fig. 1. From Fig. 1 we see that the adaptive fuzzy controller consists of two parts: an identifier and a controller. The identifier uses the adaptive fuzzy system $f$ to approximate the unknown nonlinear function $g$ in the plant, and this $f$ is then copied to the controller.

We simulated the following two cases for this example:

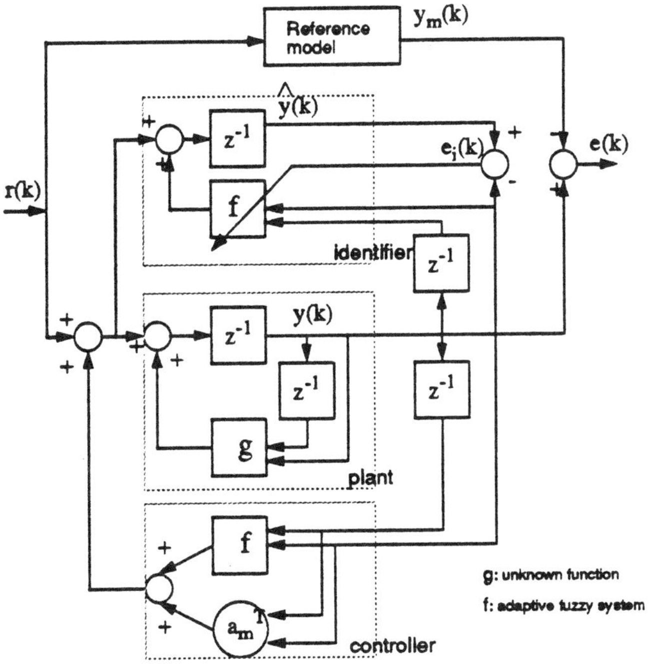


Figure 1: Overall adaptive fuzzy control system for Example 1.

- *Case 1*: The controller in Fig. 1 was first disconnceted and only the identifier was operating to identify the unknown plant. In this identification phase, we chose the input $u(k)$ to be an i.i.d. random signal uniformly distributed in the interval $[-3, 3]$. After the identification procedure was terminated, (21) was used to generate the control input, i.e., the controller in Fig. 1 began operating with $f$ copied from the final $f$ in the identifier. Figures 2 and 3 show the output $y(k)$ of the closed-loop system with this controller together with the reference model output $y_m(k)$ for the cases where the identification procedure was terminated at $k = 100$ and $k = 500$, respectively. In these simulations, we chose $\sigma = 0.3$ and $r = 0.3$. From these simulation results we see that: 1) with only 100 steps of training the identifier could produce an accurate model which results in a very good tracking control, and 2) with more steps of training the control performance was improved. In [1], the neural controller achieved similar performance when the identification procedure was carried out for $10^5$ steps.

- *Case 2*: The identifier and the controller operated simultaneously (as shown in Fig. 1) from $k = 0$. We still chose $\sigma = 0.3$ and $r = 0.3$. Figure 4 shows the $y(k)$ and $y_m(k)$ for this simulation. We see that the control was almost perfect.

Example 2: In this example we consider the plant

$$y(k+1) = \frac{5y(k)y(k-1)}{1+y^2(k)+y^2(k-1)+y^2(k-2)} + u(k) + 0.8u(k-1), \quad (23)$$

where the nonlinear function is assumed to be unknown. The aim is to design a controller $u(k)$ such that $y(k)$ will follow the reference model

$$y_m(k+1) = 0.32y_m(k) + 0.64y_m(k-1) - 0.5y_m(k-2) + sin(2\pi k/25). \quad (24)$$

Using the same idea as in Example 1, we choose

$$u(k) = -f_k[y(k), y(k-1), y(k-2)] - 0.8u(k-1) + 0.32y(k) + 0.64y(k-1) - 0.5y(k-2) + sin(2\pi k/25), \quad (25)$$

where $f_k[*]$ is in the form of (9). The basic configuration of the overall control scheme is the same as Fig. 1. Figure 5 shows the $y(k)$ and $y_m(k)$ when both the identifier and the controller began operating from $k = 0$. We chose $\sigma = 0.3$ and $r = 0.3$ in this simulation. We see, again, that the performance was very good.

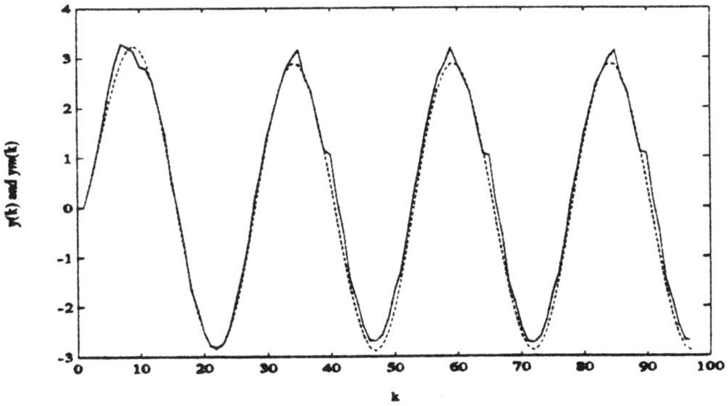


Figure 2: The output $y(k)$ (solid line) of the closed-loop system and the reference trajectory $y_m(k)$ (dashed line) for the Case 1 in Example 1 when the identification proceudre was terminated at $k = 100$.

## V. Conclusions

In this paper, we first constructed an optimal fuzzy logic system which is capable of matching all the input-output pairs in the training set to arbitrary accuracy. Then, we developed an adaptive version of the optimal fuzzy logic

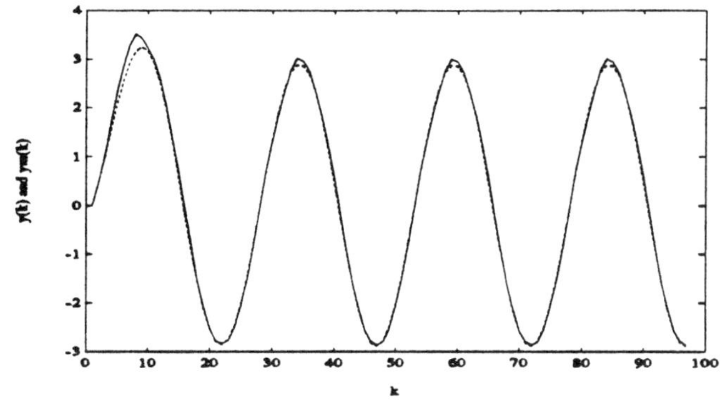


Figure 3: The output $y(k)$ (solid line) of the closed-loop system and the reference trajectory $y_m(k)$ (dashed line) for the Case 1 in Example 1 when the identification proceudre was terminated at $k = 500$.

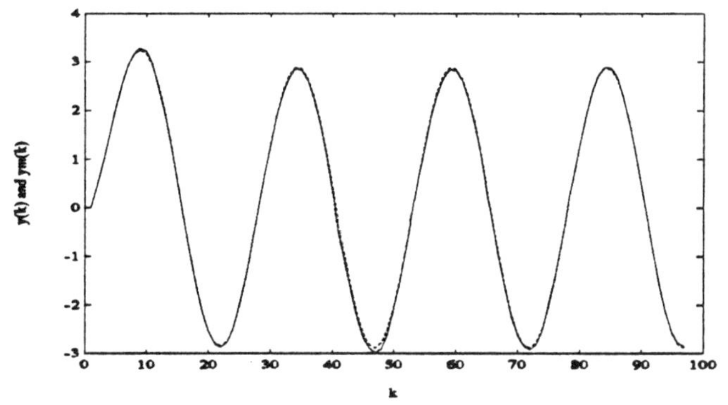


Figure 4: The output $y(k)$ (solid line) of the closed-loop system and the reference trajectory $y_m(k)$ (dashed line) for the Case 2 in Example 1.

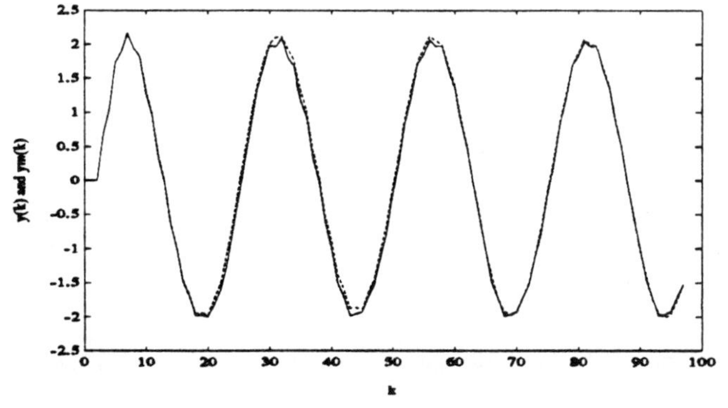


Figure 5: The output $y(k)$ (solid line) of the closed-loop system and the reference trajectory $y_m(k)$ (dashed line) for Example 2.

system using the nearest neighborhood clustering algorithm. We used this adaptive fuzzy system to construct adaptive controllers for nonlinear dynamic systems, and the simulation results showed that the adaptive fuzzy controller could produce very good tracking control. The advantages of this method of constructing fuzzy logic systems are: 1) it performs parameter and structure identifications simultaneously, 2) it performs a one-pass operation over the training data and is therefore computationally efficient, and 3) it presents a uniform framework for combining numerical data and fuzzy IF-THEN rules.

## VI. Acknowledgements

This work was supported by the Rockwell International Science Center. The author would like to thank Prof. Lotfi Zadeh for his constant encouragement.

## VII. References

[1] Narendra, K. S. and K. Parthasarathy, "Identification and control of dynamical systems using neural networks," *IEEE Trans. on Neural Networks*, Vol.1, No.1, pp.4-27, 1990.

[2] Specht, D. F., "A general regression neural network," *IEEE Trans. on Neural Networks*, Vol. 2, No. 6, pp. 568-576, 1991.

[3] Wang, L. X. and J. M. Mendel, "Back-propagation fuzzy systems as nonlinear dynamic system identifiers," *Proc. IEEE International Conf. on Fuzzy Systems*, pp. 1409-1418, San Diego, 1992.

[4] Wang, L. X. and J. M. Mendel, "Fuzzy basis functions, universal approximation, and orthogonal least squares learning," *IEEE Trans. on Neural Networks*, Vol. 3, No. 5, pp. 807-814, 1992.

[5] Wang, L. X. and J. M. Mendel, "Generating fuzzy rules by learning from examples," *IEEE Trans. on Systems, Man, and Cybern.*, Vol. 22, No. 6, 1992.

# Learning Fuzzy Concept Definitions

M. Botta, A. Giordana and L. Saitta
Dipartimento di Informatica, Universita` di Torino
Corso Svizzera 185, 10149 Torino, Italy
Tel. (+39) 11 771-2002, Fax. (+39) 11 751-603
E-mail: attilio@di.unito.it

***Abstract*--The symbolic approach to machine learning has developed algorithms for learning First Order Logic concept definitions. Nevertheless, most of them are limited because of their impossibility to cope with numeric features, typical of real-world data. In this paper, a method to face this problem is proposed. In particular, an extended version of the system ML-SMART is described, which is capable to automatically adjust the values of fuzzy sets used to define the semantics of the predicates in the concept description language. The learning strategy works in two separate phases: in the first one, the structure of the concept definition is learned by choosing tentative values for the fuzzy sets; in the second phase, the values are refined using a simple genetic algorithm, trying to get closer to an optimum assignment. The system is evaluated on a complex artificial domain, that shows the good potentialities of this approach.**

## I. INTRODUCTION

In the recent literature, inducing descriptions of structured concepts from examples is receiving a growing interest. After the first proposed approaches [1,2,3,4], a fundamental methodological contribution came from Michalski, with the system INDUCE [5,6]. Then, the topic has been further developed in other systems, such as ML-SMART [7], RIGEL [8], GOLEM [9], FOIL [10], CLINT [11], the system developed by Kodratoff [12] and FOCL [13]. In developing FOIL, Quinlan identified the task of learning concept descriptions with the task of learning the intensional definition of a relation.

However, most systems for learning structured concepts (or relations) are not able to deal with numerical features. The only exception is represented by INDUCE and by others derived from it, such as RIGEL [8] and INDUBI [14,15] that have the capacity of learning the value of numerical constants occurring in predicates as terms. The ability to deal with numerical features is fundamental in order to apply machine learning to classification problems in real world, where data are noisy and features are inherently fuzzy.

In this paper, we describe an improved version of ML-SMART, called SMART+, that is able, at least in part, to face such difficulty. In particular, the semantics of logical predicates is defined by means of trapezoidal fuzzy sets, specified by means of parametric constants automatically induced from a set of learning events.

The learning process take place in two consecutive phases. In the first one, the structure of the logical formulas is learned together with tentative values for the fuzzy sets. In the second one, this values are optimised using a Genetic Algorithm.

A brief overview of SMART+ is given in Section II, whereas Section III describes the method for inducing fuzzy set values. Section IV contains an example of application of the method on a complex artificial domain simulating a problem of pattern recognition, and some conclusive remarks follows in Section V.

## II. SMART+ OVERVIEW

SMART+ is a problem solver explicitly designed for the task of learning concept descriptions from examples supplied by a teacher. Given a set $H_0$ of concepts and a set $F_0$ of labelled instances, SMART+ generates as output a classification theory described in a first order logic language **L** which is basically an Horn clause language extended with functions, negation and quantifiers. In particular, a well formed formula (wff) in the language **L** takes the form:

$$H_i \wedge \varphi(t_1,t_2,...,t_n) \rightarrow H_j \quad (1)$$

being $H_i \subset H_j \subset H_0$ predicates denoting sets of concepts and $\varphi$ a logical formula stating a condition over the terms $t_1$, $t_2$, ..., $t_n$. Expression (1) actually means that if an event f, to be classified, is an instance of a concept h belonging to the set $H_i$ and the condition $\varphi(t_1,t_2,...,t_n)$ is true of f, then h is included in $H_j$. As the set of concepts $H_j$ implied by a rule is, in general, in the body of another rule, the knowledge learned by SMART+ defines a structured classification theory which can be described as a discrimination graph.

The formula $\varphi(t_1,t_2,...,t_n)$ is built up using predicates in a set **P**, connectives $\wedge$ and $\neg$ and quantifiers ATM, ATL and EX. As described in [16], these quantifiers stand for ATMost, ATLeast and EXactly, respectively, and can be considered as an extension of the standard existential quantifier (similar to the numeric quantifiers used in the system INDUCE [5]).

In order to cope with the fuzziness inherent to real-world data, a continuous-valued semantics is adopted for the predicates. For each predicate p∈**P** a corresponding semantic function $f_p$, mapping a numerical base variable to the interval [0,1], must be defined by the teacher in order to specify how to evaluate the truth of p on the learning events. As a matter of fact, $f_p$ is a function that evaluates the membership $\mu_p$ of a numerical feature $V_p$ with respect to a trapezoidal fuzzy set $S_p$. The value of $S_p$ can be defined by means of parameters that will occur in the predicate p as variable terms. In particular, terms in a predicate p are subdivided into two categories: objects and parameters. The first ones, denoted by the symbols $x_1, x_2, \ldots, x_n$, can be bound only to items corresponding to learning instances or components of them, whereas the second ones, denoted by the symbols $k_1, k_2, \ldots, k_m$, correspond to fuzzy set parameters and can be bound only to numerical constants. In particular, trapezoidal fuzzy sets are considered and defined as in Fig. 1(a), in a way that allows to identify the fuzzy set by a pair of real parameters, $k_1$ and $k_2$. Special cases are reported in Fig. 1(b) and 1(c), in which open intervals are represented: in this case, only one numerical parameter is necessary to define the fuzzy set.

For each parameter k, the teacher must specify the range in which k has to be searched for and the approximation requested for tuning its value. A default value can also be specified; this will be used by the system in the case that this learning capability is explicitly disabled by the teacher.

As well, the semantics for the logical connectives can be chosen by the teacher, in order to realise a specific evidential calculus. By default, SMART+ adopts a pair of t-norm and t-conorm functions [17,18] for the AND and OR connectives, respectively. The semantics of the quantifiers can be expressed by means of this connectives [16,18].

## III. OPTIMIZING NUMERICAL PREDICATES

In order to produce the structured classification theory described in the previous section, SMART+ adopts a reduction to subproblems technique. A subproblem is defined by a pair <H, F> where H denotes a set of concepts and F denotes a set of learning events. The initial problem is defined by the pair $\langle H_0, F_0 \rangle$. Given a subproblem <H, F>, the system searches new classification rules by developing a tree of formulas related by the *more-specific-than* relation. Given a formula φ in the frontier of the tree, a more specific formula ψ can be obtained either by introducing some quantifier or by adding to φ a new literal corresponding to an instance of a predicate p∈**P**. In this second case, the values that minimise the entropy are chosen for the numerical parameters occurring in p.

When a formula φ covering only instances of a subset H' of H is found, the classification rule $H \wedge \varphi \rightarrow H'$ is proposed. However, the parameters occurring in φ has been defined one at a time, and thus, the found assignment might be far from an optimal one. Then, each newly proposed rule undergoes a refinement step, in which a better assignment for the parameters is searched for by using a Genetic Algorithm.

The search stops in a subproblem when all the instances F are classified by some rule or when a maximum amount of computational resources has been exhausted. In this case, rules slightly inconsistent can also be accepted, trading off completeness and consistency criteria. At last, it is possible that some instance remains unclassified in some subproblem.

New subproblems are defined by merging together all the learning events covered by rules implying the same set H of classes. The learning process continues until all the subproblems containing instances of more than one class have been solved.

In order to account for the inconsistencies, a weight $w = \frac{|E(\varphi)|}{|\varphi^*|}$ is associated to each rule, being $|E(\varphi)|$, the number of learning events correctly classified by the rule, and $|\varphi^*|$ the total number of learning events matched by φ. The weight represents an estimate of the probability that the classification generated by φ is correct and it is accounted for in the evidential calculus of the performance system of SMART+, when the learned knowledge is validated on a test set.

### A. Local Optimisation Algorithm

Two kinds of statistical information, extracted from the extension φ* of a formula φ, are exploited to guide the learning process inside a subproblem and, in particular, to select the values for the numeric parameters. This information consists of the *instance distribution histogram* (denoted by

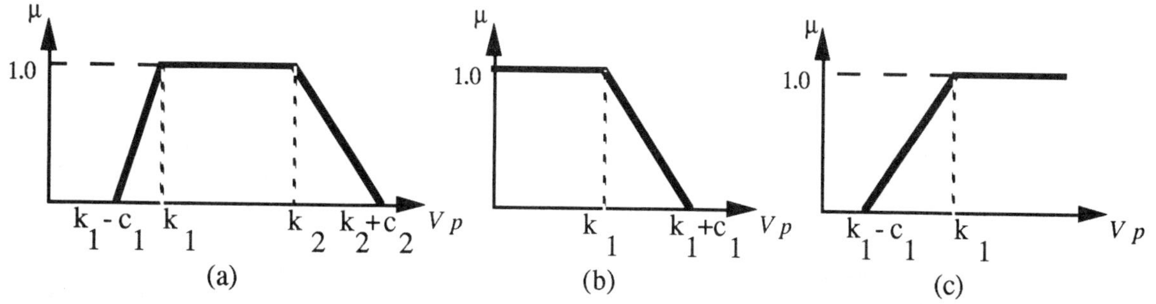


Fig. 1 - Example of trapezoidal fuzzy sets used to define the semantics of predicates. The ordinate μ represents the truth value, whereas the abscissa $V_p$ represents the values of the base variable.

$\mathbf{m}(\varphi)$) and of the *object distribution histogram* (denoted by $\mathbf{m^*}(\varphi)$) over the concepts in the set $H_0$. Given a relation $\varphi^*$, $\mathbf{m}(\varphi)$ is a n-dimensional vector with $n=|H_0|$; each component $m_i$ ($1\leq i \leq n$) is an integer corresponding to the number of events $f \in F_0$ which are instances of the concept $h_i \in H_0$ and are covered by $\varphi$. Analogously, $\mathbf{m^*}(\varphi)$ is an n-dimensional vector, whose generic component $m^*_i$ corresponds to the global number of different bindings in $\varphi^*$, between variables in $\varphi$ and objects of instances of class $h_i$. Moreover, for each subproblem sp = <H,F>, an analogous pair of vectors $\mathbf{M}$(sp) and $\mathbf{M^*}$(sp), reporting a distribution similar to $\mathbf{m}$ and $\mathbf{m^*}$ evaluated on all the instances F, are also defined.

When a literal p is proposed for specialising a formula $\varphi$, the vector $\mathbf{m}(\varphi\wedge p)$ is evaluated. The new formula $\psi = \varphi\wedge p$ is created only if it covers a still significant amount of instances for at least one class. If p contains some parameter, a search is performed in order to find values defining formulas with a "good" statistical evaluation, i.e., that cover many examples and are discriminant.

Before selecting values for parameters, the system partitions the classes having a non null component in $\mathbf{m}(\varphi)$ into two groups: the ones which are likely to be classified by the new formula, and the ones which are likely to be excluded. Then, values optimising the discrimination capability with respect to this partition are searched for.

Partition in the two groups of classes is described by a Boolean vector $\mathbf{t}(\varphi)$, whose component $t_i$ ($1\leq i\leq |H_0|$) is set to 1 if $\frac{m_i(\varphi)}{M_i(sp)}$ is greater than a user defined threshold $v_0$, and 0 otherwise. Vector $\mathbf{t}(\varphi)$ is said the *target* of $\varphi$.

The evaluation criteria used to select parameter values are based on a kind of entropy measure trading off completeness and consistency. Given a formula $\varphi$, the vector $\mathbf{m}(\varphi)$ and the corresponding target $\mathbf{t}(\varphi)$, formula $\varphi$ is scored by using the following heuristic rule:

$$\left(\frac{1}{n}\sum_{t_i\neq 0}\frac{m_i(\varphi)}{M_i(sp)}\right) * \left(1-\frac{1}{\tilde{n}}\sum_{t_i=0}\frac{m_i(\varphi)}{M_i(sp)}\right) \qquad (2)$$

where n is the number of elements of $\mathbf{t}(\varphi)$ different from 0 and ñ is the number of elements of $\mathbf{t}(\varphi)$ equal to 0. The first part of (2) represents the degree of completeness of formula $\varphi$, whereas the second part is a measure of consistency with respect to the classes discriminated by $\mathbf{t}(\varphi)$. The higher the completeness degree and the consistency level, the more discriminating the formula $\varphi$ is.

In structured domains, various settings for the parameters of a predicate p can produce formulas having the same instance distribution histogram but different object distribution histogram; in this case, given a formula $\varphi(x_1,...,x_n) = \psi(x_1,...,x_n) \wedge p(x_1,...,x_n,k_1,...,k_m)$, the object distribution histograms $\mathbf{m^*}(\varphi)$ and $\mathbf{m^*}(\psi)$ and a target $\mathbf{t}(\varphi)$, the following rule is also used to score formula $\varphi$:

$$\left(\frac{1}{n}\sum_{t_i\neq 0}\frac{m^*_i(\varphi)}{m^*_i(\psi)}\right) * \left(1-\frac{1}{\tilde{n}}\sum_{t_i=0}\frac{m^*_i(\varphi)}{m^*_i(\psi)}\right) \qquad (3)$$

In (3) n and ñ are the same as in (2). Whereas rule (2) considers the number of instances matched by $\varphi$, rule (3) takes into account the number of models existing for $\varphi$; empirically, this criterion is in the spirit of minimising the entropy of $\varphi$ with respect to the models.

The algorithm searching the values for the parameters of a predicate p can now be described. It receives as input arguments a formula $\varphi$, a predicate p and, for each parameter k occurring in p, a triple <min, max, step> describing the range of variability and the granularity of k. A set of suitable values, to be used in order to specialise formula $\varphi$ with predicate p, is returned as the result.

```
Algorithm FLOP (φ, p, {(min_i, max_i, step_i) : ∀ i ∈ [1..n]})
  ∀ i ∈ [1..n] k_i = min_i;
  i = 1;
  Let BEST be a table whose entries are sets of triples (m,
     m*, (k1,...,kn)) indexed by a target mask t
  while i ≤ n do
     m = instance distribution histogram for φ∧p(k1,...,kn);
     m* = object distribution histogram for φ ∧ p(k1,...,kn);
     t = target mask for m
     if m and m* are "relevant"
        then BEST[t] = BEST[t] ∪ {(m, m*, (k1,...,kn))}
     endif
     k_i = k_i + step_i;
     while i ≤ n and k_i > max_i do
        k_i = min_i;
        i = i + 1;
        if i ≤ n then k_i = k_i + step_i endif
     enddo
     if i ≤ n then i = 1  endif
  enddo
  BESTPARS = {}
  for each entry in BEST do
     evaluate every triple (m, m*, (k1,...,kn)) using
     expressions (2) and (3) and add the highest scored
     triple to BESTPARS
  endfor
  return (BESTPARS)
```

Informally, vectors **m** and **m*** are evaluated for all the possible assignments of values to the parameters of p, according to the range and the granularity. Then, the promising **m**'s are identified and the corresponding targets are determined. Finally, for every target, the most suited setting of the parameters is selected. The FLOP algorithm has an exponential complexity in the number of parameters of the predicate p, therefore a coarse granularity is recommended in order to limit the size of the number of **m** and **m*** to be evaluated.

However, the use of a coarse granularity can lead to inaccuracy in this learning phase. Furthermore, the specialisation process leads to learn parameter values for a given predicate when the ones of other predicates have been already chosen. In many cases this strategy prevent from learning optimal values. On the other hand a contextual optimisation of all the parameters in the multidimensional space of numerical features would be impossible in this learning phase. Therefore the refinement step described in the following can be useful in order to increase the accuracy of the classification theory.

### *B. Global Optimisation Using a Genetic Algorithm*

Genetic Algorithms [19] can be an excellent tool for optimising multimodal numerical functions, such as the semantic functions implicitly associated to the classification rules. In the present case a simple Genetic Algorithm of the type proposed by DeJong [20] has been used and implemented as described in [19].

Given a formula $\varphi$, all the parameters $k_i$ occurring in $\varphi$ define a string of reals $rs = \#:[k_1, k_2, k_3, ..., k_n]$; as a predicate p can occur more than once in $\varphi$, for each occurrence of p in $\varphi$ the parameters of p will have a different occurrence in *rs*. A string *rs* is then converted into a bit string representation in order to use the standard Genetic Algorithm.

In order to prevent the crossover operator technique from generating values outside the range assigned by the teacher to the parameters, the following conversion algorithm is adopted to map a string of reals *rs* into the corresponding bit string *bs*:

a) Given a parameter $k_i$ and the range $[min_i, max_i]$ assigned by the teacher, $k_i$ is represented as:
$$k_i = min_i + \delta_i / \Delta_i \quad (4)$$
being $0 \le \delta_i \le 2^{N_i}$, where $N_i$ is the number of bits chosen by the teacher for representing the increment $\delta_i$. The value of $\Delta_i$ is obtained from $N_i$ according to the following equation:
$$\Delta_i = \frac{2^{N_i}}{(max_i - min_i)} \quad (5)$$
b) Given a string $rs= \#:[k_1, k_2, k_3, .... k_n]$, for each parameter $k_i$ in *rs* a field of $N_i$ bits will be defined in *bs* where the value of the corresponding $\delta_i$ is represented.

Given a formula $\varphi$, the Genetic Algorithm is then applied trying to increase the completeness and the consistency of $\varphi$. An initial population $A(\varphi)$ of bit strings is generated at random and the fitness of each individual *bs* is evaluated by computing the extension $\varphi^*(bs)$ and by ranking the corresponding **m** and **m*** vectors according to the rules (2) and (3). Afterwards the usual genetic evolution begins:

a) A subset $A_r(\varphi)$ of $A(\varphi)$ is chosen for reproduction,
b) Pairs of individuals from $A_r(\varphi)$ are chosen for mating and then a new population $A'(\varphi)$ is obtained by replacing $A_r(\varphi)$ in $A(\varphi)$ with the new individuals obtained by applying crossover and mutation in $A_r(\varphi)$. The fitness of each new individual is evaluated again, as described above.

Actually, two types of crossover $c_1$ and $c_2$ are used. Crossover $c_1$ is the standard *single point crossover*, whereas crossover $c_2$ is defined as follows: given two parents $bs_1$ and $bs_2$, a parameter $k_c$ is selected at random and two offsprings $bs_1'$ and $bs_2'$ are generated in three steps:

a) The parameters to the left of $k_c$ in $bs_1'$ and $bs_2'$ are obtained by copying the parameters in $bs_1$ and $bs_2$, respectively.
b) The parameter $k_c$ in both offsprings is obtained by averaging the two corresponding values in the parents.
c) The parameters to the right of $k_c$ are obtained in $bs_1'$ by copying the corresponding ones in $bs_1$ and viceversa in $bs_2'$.

The choice between $c_1$ and $c_2$ is controlled by two probabilities $P_1$ and $P_2$, respectively, such that:
$$P_1 + P_2 \le 1$$
The standard mutation operator is applied with a probability $P_m = 0.001$.

## IV. EVALUATION ON A TEST CASE

The application domain is quite complex, even though artificial, and bears many features of a real-world one. Ten capital letters of the English alphabet have been chosen, a Horn clause theory has been invented and used by a random choice theorem prover in order to generate instances of these letters, some of which are shown in Fig. 2. Each letter is represented as a set of segments that are described by initial and final xy coordinates in a Cartesian plane. From these attributes, other features can be extracted, such as the length of a segment, its orientation, its preceding and following segments, and so on.

Fig. 2 – Some instances taken from the artificial character domain.

Afterwards, SMART+ has been run under three different conditions: (a) using the default values for fuzzy sets, as assigned by the teacher; (b) performing only a local optimisation according to the algorithm described in Section 3.1; (c) using local optimisation plus the global optimisation performed by Genetic Algorithm. Table I reports the results obtained in the three cases on a learning set of 1000 instances (100 per class). The first two columns refer to the number of formulas in the learned knowledge and the relative computational time spent by SMART+, respectively. The second three columns refer to the recognition rate, the error rate and the ambiguity rate, respectively, evaluated on an independent test set of 5000 events (500 per class).

TABLE I
RESULTS OBTAINED BY SMART+ ON THE TEST EXAMPLE

| | Number of learned formulas | Computation Time | Recognition Rate | Error Rate | Ambiguity Rate |
|---|---|---|---|---|---|
| No OPT | 81 | 1 | 41.48 | 3.82 | 54.70 |
| Local OPT | 47 | 2.5 | 98.68 | 0.12 | 1.20 |
| Local+GA OPT | 28 | 4.5 | 99.70 | 0.0 | 0.30 |

From Table I, the capability of learning fuzzy sets appears determinant to the success of the application. In fact, using the default values the performances achieved were very poor, whereas by only using local optimisation the recognition rate raise up to 98%. A further significant improvement is then added by the global optimisation algorithm that brings the recognition rate to 99.7%. On the other hand the simplicity and compactness of the knowledge base increases according to the performances as it is possible to grasp from the values reported in the first column, at the cost of a greater computation time.

## V. CONCLUSION

We have presented a method for learning fuzzy set values in first order logic environment, which combines an extension of the methodology developed for learning relations with genetic algorithms. In particular, genetic algorithms can be an effective tool for refining numerical parameters such as thresholds, weights and coefficients which control the flexible matching of a symbolic expression against a real world instance. The applicability of the method has been demonstrated on a non trivial learning problem of pattern recognition.

However, the conclusion we can take from this experiment is that the symbolic approach proposed by Artificial Intelligence can be extended in order to be effective to deal with patterns of real world such as complex signals. In particular this approach can compete quite well with other methods, such as Neural Networks, in domains where structural knowledge is relevant and where there exists background knowledge which can be exploited.

## REFERENCES


[1] P.H. Winston, "Learning Structural Descriptions from Examples", in P.H. Winston (Ed.),*The Psychology of Computer Vision*, McGraw Hill, New York, 1975, pp. 157-209.

[2] S.A. Vere, "Induction of Concepts in the Predicate Calculus", *Proc. 4th IJCAI*, (Tbilisi, USSR, 1975), pp. 281-287.

[3] G.D. Plotkin, "A Note on Inductive generalisation", *Machine Intelligence, 6*, 1970, 101-124.

[4] F. Hayes-Roth, J. Mc Dermott, "An Interfence Matching Technique for Inducing Abstractions", *Communications of ACM, 21*, 1978, 401-410.

[5] R.S. Michalski, "Pattern Recognition as Rule-Guided Inductive Inference", *IEEE Trans. on Pattern Analysis and Machine Intelligence, PAMI-2*, 1980, 349-361.

[6] R.S. Michalski, "A Theory and Methodology of Inductive Learning", *Artificial Intelligence, 20*, 1983, pp. 111-161.

[7] F. Bergadano, A. Giordana, L. Saitta, "Automated Concept Acquisition in Noisy Environment", *IEEE Transaction on Pattern Analysis and Machine Intelligence, PAMI-10*, 1988, pp. 555-578.

[8] R. Gemello, F. Mana, L. Saitta, "RIGEL: An Inductive Learning System", *Machine Learning, 6*, 1991, 7-36.

[9] S. Muggleton, C. Feng, "Efficient Induction of Logic Programs", *Proc. First Conference on Algorithmic Learning Theory*, (Tokyo, Japan, 1990).

[10] R. Quinlan, "Learning logical definitions from relations", *Machine Learning, 5*, 1990, 239-266.

[11] L. De Raedt, M. Bruynooghe, "Towards Friendly Concept-Learners", *Proc. IJCAI-91*, (Sidney, Australia, 1991), pp. 849-854.

[12] Y. Kodratoff, J-G. Ganascia, "Improving the Generalization Step in Learning", in R. Michalski, T. Mitchell and J. Carbonell (Eds.), *Machine Learning: An Artificial Intelligence Approach, Vol. II*, Morgan Kaufmann, (Los Altos, CA), 1986, 215-244.

[13] M. Pazzani, D. Kibler, "The Utility of Knowledge in Inductive Learning", *Machine Learning, 9*, 1992, 57-94.

[14] F. Esposito, D. Malerba, G. Semeraro, "Flexible Matching for Noisy Structural Descriptions", *Proc. IJCAI-91*, (Sidney, Australia, 1991), pp. 658-664.

[15] F. Esposito, D. Malerba, G. Semeraro, "Classification in Noisy Environments Using a Distance Measure Between Structural Symbolic Descriptions", *IEEE Transaction on Pattern Analysis and Machine Intelligence, PAMI-14*, 1992, 390-402.

[16] R. Yager, "Quantified Propositions in a Linguistic Logic", *Int. J. of Man-Machine Studies, vol. 19*, 1983, pp. 195-227.

[17] S. Weber, "A General Concept of Fuzzy Connectives, Negations and Implications Based on t-norms and t-conorms", *Fuzzy Sets and Systems, vol. 11*, 1983, pp. 115-134.

[18] F. Bergadano, A. Giordana, and L. Saitta, "Learning from Examples in Presence of Uncertainty," in Approximate Reasoning in Intelligent Systems Decision and Control, ed. E. Sanchez and L. Zadeh, pp. 105-124, Pergamon Press , 1986.

[19] D.E. Goldberg,*Genetic Algorithms*, Addison Wesley Publishing Company, Reading, MA 1989.

[20] K.A. De Jong, "Analysis of the Behaviour of a Class of Genetic Adaptive Systems", *Doctoral dissertation*, Department of Computer and Communication Sciences, University of Michigan (Ann Arbor, MI, 1975).

# Synthesis of Nonlinear Controllers via Fuzzy Logic

**Reza Langari**

Department of Mechanical Engineering
Texas A&M University
College Station, TX 77843-3123

*Abstract*—In this paper an analytic formulation of a class of fuzzy logic control algorithms is presented. This formulation makes use of the idea that characteristic functions of fuzzy sets may be considered to be algebraic functions of their arguments. Placing common constraints such as continuity and boundedness on these functions, and making use of a variation of the so called centroid of area defuzzification rule, results in an analytic formulation of the control law that (*i*) can be used to explain how fuzzy control algorithms function as they do, and (*ii*) makes formal analysis of fuzzy control systems, at least in principle, possible.

## I Introduction

Fuzzy Logic Control(FLC) is fast emerging as an alternative to conventional control techniques in situations where it may be infeasible to formulate an analytic model of the process or, alternatively, when performance measures are not readily transformed into standard control theoretic design objectives– as steady state or transient characteristics of the closed loop system. Industrial processes such as Gas Metal Arc Welding(GMAW), whose complexity exceeds the limitations of analytic modeling tools are prime candidate for application of FLC[6].

In spite of the success that FLC continues to demonstrate, however, precisely what function fuzzy logic control helps accomplish in a given situation remains to be fully understood. Attempts to deal with this issue have taken the form of empirically based comparative studies, such as comparison with PID control[8], or analytic studies, as for instance by Kickert and Mamdani[4], Buckley and Ying[1], and Langari[5]. The present paper extends the results presented in the latter work and develops a more general analytic description that is applicable to a broader class of fuzzy logic control algorithms.

The remainder of the paper is as follows. First an overview of fuzzy control, which is primarily meant to establish the notation for the presentation of our main results, is presented. Next, we derive the aforementioned analytic description and discuss its implications. We conclude the paper by a summary of its major points.

## II Fuzzy Control Systems

In this paper we consider fuzzy linguistic control system, tak-

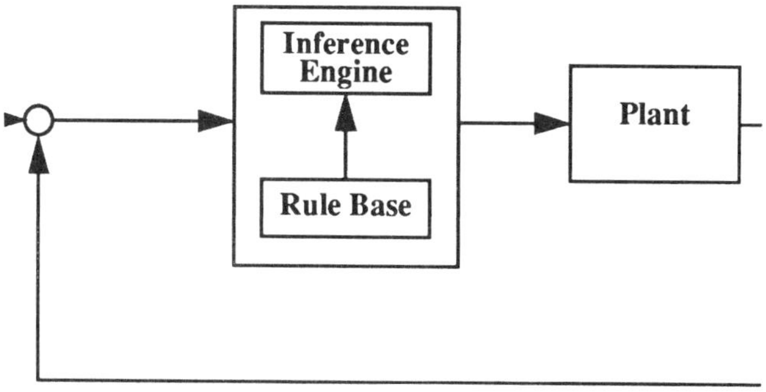


**Figure 1 Architecture for Fuzzy Control**

ing the form shown in Figure 1, where the knowledge of operation of the plant is explicitly represented as *condition* → *action* rules of the form:

$R_{j,l}$: if $e(t)$ is $\tilde{A}_j$ and $de(t)$ is $\tilde{B}_l$ then $u(t)$ is $\tilde{C}_{j,l}$,

where $e(t)$ denotes the instantaneous value[1] of the process error at time $t$ and $de(t)$ is short for $\mathcal{D}(e;t)$, which stands for $de/dt$ or $\int^t e d\tau$. Further, $\tilde{A}_j$, $\tilde{B}_l$ and $\tilde{C}_{j,l}$ belong to collections $\tilde{\mathcal{A}}$, $\tilde{\mathcal{B}}$, and $\tilde{\mathcal{C}}$ of fuzzy subsets defined over the domains of definition of the relevant variables(Figure 2.) Further, in above definitions, $R_{j,l}$ denotes the $j,l^{th}$ rule in the rule set $\mathcal{R}$. In particular, $R_{j,l}$ may be viewed as associating elements, $\tilde{A}_j$ of $\tilde{\mathcal{A}}$ and $\tilde{B}_l$ of $\tilde{\mathcal{B}}$ with element $\tilde{C}_{j,l}$ of $\tilde{\mathcal{C}}$.

### 2.1 Control Computation

Suppose, at some instance $t$, the error $e(t)$ has positive grades of membership, $\mu_{\tilde{A}_j}(e(t))$ and $\mu_{\tilde{A}_{j+1}}(e(t))$ to some pair

1. For simplicity, we shall use notation consistent with continuous time representation of fuzzy control systems.

$\tilde{A}_j$ and $\tilde{A}_{j+1}$ of $\tilde{\mathcal{A}}$. Similarly, suppose *de(t)* belongs to some pair $\tilde{B}_l$ and $\tilde{B}_{l+1}$ of $\tilde{\mathcal{B}}$. Accordingly, at this instant, the following control rules apply[1]:

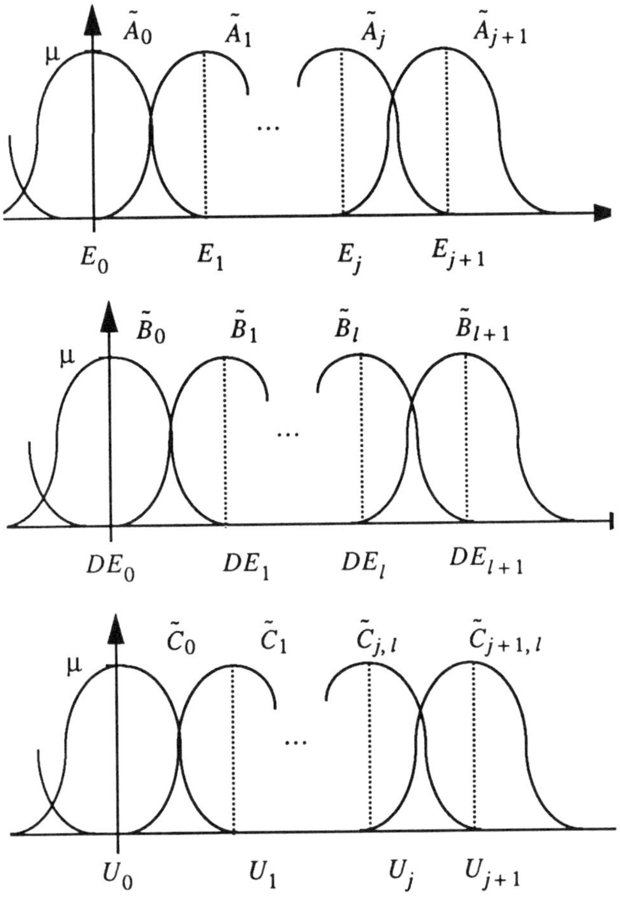


**Figure 2 Collections $\mathcal{A}$ , $\mathcal{B}$ , and $\mathcal{C}$ of fuzzy subsets partitioning *E*, *DE*, and *U*.**

$R_{j,l}$: if $e(t)$ is $\tilde{A}_j$ and $de(t)$ is $\tilde{B}_l$ then $u(t)$ is $\tilde{C}_{j,l}$,

$R_{j+1,l}$: if $e(t)$ is $\tilde{A}_{j+1}$ and $de(t)$ is $\tilde{B}_l$ then $u(t)$ is $\tilde{C}_{j+1,l}$,

$R_{j+1,l+1}$ : if $e(t)$ is $\tilde{A}_{j+1}$ and $de(t)$ is then $u(t)$ is $\tilde{C}_{j+1,l+1}$ ,

$R_{j,l+1}$ : if $e(t)$ is $\tilde{A}_j$ and $de(t)$ is $\tilde{B}_{l+1}$ then $u(t)$ is $\tilde{C}_{j,l+1}$ ,

with each rule satisfied to some degree. The corresponding *truth value* is defined, for instance for the first rule, by

$$\mu_{j,l} = min(\mu_{\tilde{A}_j}(e(t)), \mu_{\tilde{B}_l}(de(t))), \quad (1)$$

or, by

$$\mu_{j,l} = \mu_{\tilde{A}_j}(e(t))\mu_{\tilde{B}_l}(de(t)). \quad (2)$$

The truth values of other rules in the above set are similarly defined. Note that the *product* instead of *min* results in interactivity between the truth values of the components of the antecedent clause. This fact is essential to our analytic treatment, and will be made use of later in the paper.

Now, representing the consequent clause of each $R_{j,l}$ rule, that is $\tilde{C}_{j,l}$, by its single representative, or *defuzzified*, value that is $U_{j,l}$, defined as

$$U_{j,l} = \frac{\int u\mu_{\tilde{C}_{j,l}}(u)}{\int \mu_{\tilde{C}_{j,l}}(u)} , \quad (3)$$

the control action, $u(t)$ , is computed as

$$u(t) = \frac{\sum_{j,l}\mu_{j,l}U_{j,l}}{\sum_{j,l}\mu_{j,l}} , \quad (4)$$

where *j* and *l* range over the indices of all applicable rules.

Note that this approach is based on a variation of the Centroid of Area(COA) defuzzification rule[2], but as we shall see later has better analytical properties. For instance $U_{j,l}$ is the COA of $\tilde{C}_{j,l}$ and *would be* exactly equal to $u(t)$ if the *i*th rule were the *only* applicable rule. In effect the above approach superimposes the action of various rules and is exactly equivalent to the COA rule if triangular shaped membership functions were used. In other instances, the equivalence is approximate, but as we shall see shortly this distinction is irrelevant in so far as the presented results are concerned.

It is interesting to observe the fuzzy control action as a function of controller inputs. In particular, when the *e* and *de* are exactly at the center values of the relevant fuzzy

1. In principle it is possible for more than four rules to be applicable at any given time. The treatment as presented in this section, however, remains valid with the appropriate changes to the number of applicable rules where necessary.

subsets, for instance $\tilde{A}_j$ and $\tilde{B}_l$, or $\tilde{A}_{j+1}$ and $\tilde{B}_{l+1}$, then the control action is precisely that which is given by the defuzzified value of the rule, that is $U_{j,l}$. In other instances the fuzzy inference engine computes the control action, $u(t)$ by interpolation, that $u(t)$ is a weighted average of $U_{j,l}$, $U_{j+1,l}$, ... and the associated weights are the truth values of the rules. This is the point that is rigorously established later on.

# III Problem Statement

We consider the single input, single output fuzzy logic control systems where the control rules take the form described in Section 2. The objective here is to develop an analytic representation of the control law, $u = FLC(e, de)$, which (*i*) remains faithful to the spirit in which the original rules were stated and (*ii*) offers a description of what the control law does and also what it does not do.

## 3.1 Assumptions and Definitions

Let us denote the domains of definition of *e*, *de* and *u* by *E*, *DE*, and *U* respectively. Then as shown in Figure 2, collections $\tilde{\mathcal{A}} = \{\tilde{A}_j\}$ of *unimodal, convex*, and *normal* fuzzy subsets[2] effectively partition *E*, *DE*, and *U*, respectively, as follows. Each element $\tilde{A}_j$, a fuzzy subset, is centered at some $E_j \in E$ and is characterized by some $\mu_{\tilde{A}_j}$. Similarly, each element $\tilde{B}_l \in \tilde{\mathcal{B}}$ is centered at some $DE_l \in DE$, and is characterized by $\mu_{\tilde{B}_l}$. In addition, each element $\tilde{C}_{j,l} \in \tilde{C}$ is effectively represented by its *defuzzified value*, $U_{j,l}$. As discussed in Section 2.1, $U_{j,l}$ is the Centroid of Area(COA) representation of $\tilde{C}_{j,l}$. We shall further place the following constraint on $U_{j,l}$s.

**Assumption 1** For each $U_{j,l}$, there exist a pair $K_j$ and $K'_l$ of real constants such that the following relationship holds.

$$U_{j,l} = K_jE_j + K'_lDE_l.$$

We then define, for each *j* and likewise for each *l*, $\Delta K_j$ and $\Delta K'_l$ as follows:

$$\Delta K_j = K_{j+1} - K_j, \tag{5}$$

$$\Delta K'_l = K'_{l+1} - K'_l. \tag{6}$$

Next we place some constraints on $\tilde{\mathcal{A}}$ and $\tilde{\mathcal{B}}$ as follows. First, we require that $\tilde{\mathcal{A}}$ and $\tilde{\mathcal{B}}$ form *approximate fuzzy partitions* of *E* and *DE*, respectively.

**Assumption 2** Let $\tilde{\mathcal{A}} = \{\tilde{A}_j\}$ (and $\tilde{\mathcal{B}} = \{\tilde{B}_l\}$ ) be collection(s) of fuzzy subsets define over *E* (and *DE*). Then, for each element $e \in E$,

$$\sum_j \mu_{\tilde{A}_j}(e) \cong 1.$$

(A similar condition holds for $\tilde{\mathcal{B}}$.)

In other words, the sum total of membership of each element $e \in E$, for instance, to the elements of $\tilde{\mathcal{A}}$, should be nearly unity.

The interpretation of Assumption 2 is that, *externally, fuzzy classification must be compatible with feature based classification in terms of classical sets, where each element is categorized under one and only one class.* Note that in view of the statement of Assumption 2 , compatibility is not strict. Indeed, if we were to interpret this definition in a strict sense, we would have to require that $\sum_j \mu_{\tilde{A}_j}(e) = 1$, as its is done in Langari[5]. We do, however, generally require that the deviation of $\sum_j \mu_{\tilde{A}_j}(e)$ from its nominal value of one be small. We may formalize this as

$$\delta\mu = \sup_e \left|\sum_j \mu_{\tilde{A}_j}(e) - 1\right| \ll 1 \tag{7}$$

(A similar condition holds for $\tilde{\mathcal{B}}$.)

We shall also establish some notation as follows. Let $\bar{\mu}_{\tilde{A}_j}(e)$ denote the value of $\mu_{\tilde{A}_j}(e)$ that would satisfy the *ideal* true fuzzy partitioning condition. We then denote the difference between these two values by $\delta\mu_{\tilde{A}_j}(e)$:

$$\delta\mu_{\tilde{A}_j}(e) = \mu_{\tilde{A}_j}(e) - \bar{\mu}_{\tilde{A}_j}(e). \tag{8}$$

(A similar definition holds for $\tilde{\mathcal{B}}$. With this in mind, it is easy to see that $\delta\mu$ given by (7) can be written as

$$\delta\mu = \sup_e \left|\sum_j \delta\mu_{\tilde{A}_j}(e)\right|, \quad (9)$$

which of course as before must be small compared to one.

## 3.2 Formulation of the control law

Let us recall that any given instance the control action, as described in Section 2.1, can be written as

$$u(t) = \frac{\mu_{j,l}U_{j,l} + \mu_{j+1,l}U_{j+1,l} + \mu_{j+1,l+1}U_{j+1,l+1} + \mu_{j,l}}{\mu_{j,l} + \mu_{j+1,l} + \mu_{j,l} + \mu_{j,l+1}} \quad (10)$$

Now considering the fact that, following Section 2.1,

$$\mu_{j,l} = \mu_{\tilde{A}_j}(e)\,\mu_{\tilde{B}_l}(de), \quad (11)$$

which can be approximated as,

$$\mu_{j,l} \cong \bar{\mu}_{j,l} + \delta\mu_{j,l}, \quad (12)$$

with

$$\bar{\mu}_{j,l} = \bar{\mu}_{\tilde{A}_j}(e)\,\bar{\mu}_{\tilde{B}_l}(de), \quad (13)$$

and

$$\delta\mu_{j,l} = \bar{\mu}_{\tilde{A}_j}(e)\,\delta\mu_{\tilde{B}_l}(de) + \delta\mu_{\tilde{A}_j}(e)\,\bar{\mu}_{\tilde{B}_l}(de) \quad (14)$$

Accordingly, the expression on the right side of (10) can be written as

$$u(t) = \frac{\sum_{j,l}\bar{\mu}_{j,l}U_{j,l} + \sum_{j,l}\delta\mu_{j,l}U_{j,l}}{\sum_{j,l}\bar{\mu}_{j,l} + \sum_{j,l}\delta\mu_{j,l}}. \quad (15)$$

Further, the sum,

$$\sum_{j,l}\bar{\mu}_{j,l} = \bar{\mu}_{j,l} + \bar{\mu}_{j+1,l} + \bar{\mu}_{j+1,l+1} + \bar{\mu}_{j,l+1},$$

is 1, as shown in Langari[5], and $\sum_{j,l}\bar{\mu}_{j,l}U_{j,l}$ is the control action if the true fuzzy partitioning assumption discussed above were to hold. With this in mind, the expression for $u(t)$ can be written as

$$u(t) = \bar{u}(t) + \delta u(t), \quad (16)$$

where

$$\bar{u}(t) = \sum_{j,l}\bar{\mu}_{j,l}U_{j,l}, \quad (17)$$

$$\delta u(t) = \sum_{j,l}\delta\mu_{j,l}U_{j,l} - \Big(\sum_{j,l}\bar{\mu}_{j,l}\Big)\bar{u}(t). \quad (18)$$

Now substituting for $U_{j,l}$, $U_{j+1,l}$, ... their equivalent values from Assumption 1, we have:

$$U_{j+1,l} = U_{j,l} + \Delta, \quad (19)$$

$$U_{j+1,l+1} = U_{j,l} + \Delta + \Delta', \quad (20)$$

$$U_{j,l+1} = U_{j,l} + \Delta', \quad (21)$$

where, denoting $E_{j+1} - E_j$ by $\Delta E_j$ and $DE_{l+1} - DE_l$ by $\Delta DE_l$,

$$\Delta = \Delta K_j E_j + K_j \Delta E_j + \Delta K_j \Delta E_j, \quad (22)$$

$$\Delta' = \Delta K'_l DE_l + K'_l \Delta DE_l + \Delta K'_l \Delta DE_l. \quad (23)$$

Further, note that we can write $\bar{\mu}_{\tilde{A}_{j+1}}(e)$ as

$$\bar{\mu}_{\tilde{A}_{j+1}}(e(t)) = \left[\frac{1}{\Delta E_j} + F_j(e(t) - E_j)\right](e(t) - E_j) \quad (24)$$

which, as shown in Figure 3, reflects the use of $F_j$ to denote the deviation of $\bar{\mu}_{\tilde{A}_{j+1}}$ from its *extreme* linear form. (Similar approach is taken regarding $\bar{\mu}_{\tilde{B}_{l+1}}$.)

Substituting from above in the expression for $\bar{u}(t)$, given by (17), we have[1]:

$$\bar{u} = \bar{u}_1 + \bar{u}_2, \quad (25)$$

where,

$$\bar{u}_1 = K_j e + \left[\frac{\Delta K_j}{\Delta E_j}E_{j+1} + F_j(e - E_j)\Delta\right](e - E_j), \quad (26)$$

and,

$$\bar{u}_2 = K'_l de + \left[\frac{\Delta K'_l}{\Delta DE_l}DE_{l+1} + F'_l(de - DE_l)\Delta'\right](de - DE_l) \quad (27)$$

1. For simplicity of expression, and where no ambiguity may arise, we shall drop $t$ from the algebraic expressions involving functions of time.

Likewise, after some algebraic manipulations, we can

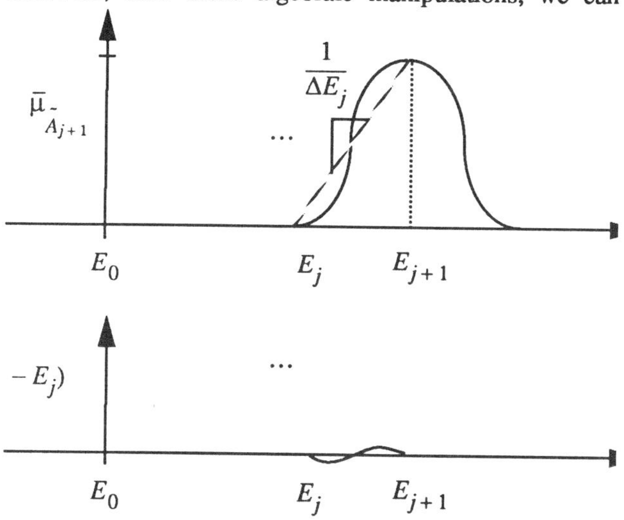


**Figure 3 Deviation of $\bar{\mu}_{\tilde{A}_{j+1}}$ p from its extreme linear**

show that the expression for $\delta u$, given by (18) yields

$$\delta u = \delta\mu_j \bar{u}_1 + \delta\mu'_l \bar{u}_2, \quad (28)$$

where

$$\delta\mu_j = \delta\mu_{\tilde{A}_j}(e) + \delta\mu_{\tilde{A}_{j+1}}(e), \quad (29)$$

$$\delta\mu'_l = \delta\mu_{\tilde{B}_l}(de) + \delta\mu_{\tilde{B}_{l+1}}(de). \quad (30)$$

# IV Discussion

The expression for the control law derived above can be interpreted as follows. First, $u$ can be thought of as the sum of the contributions from $\bar{u}$ and $\delta u$. Now $\bar{u}$ is itself sum of the linear terms $K_j e$ and $K'_l de$ and the two additional terms,

$$\left[\frac{\Delta K_j}{\Delta E_j} E_{j+1} + F_j (e - E_j) \Delta\right] (e - E_j),$$

and,

$$\left[\frac{\Delta K'_l}{\Delta DE_l} DE_{l+1} + F'_l (de - DE_l) \Delta'\right] (de - DE_l).$$

These terms approach, $\frac{\Delta K_j}{\Delta E_j} E_{j+1} (e - E_j)$, and $\frac{\Delta K'_l}{\Delta DE_l} DE_{l+1} (de - DE_l)$, respectively, in the limit[1], and thus $\bar{u}$ approaches its limiting form, which is *piecewise linear* in each of $e$ and $de$. The terms $\delta u$ is a perturbation on $\bar{u}$ and approaches zero under the assumption of true fuzzy partitioning, which is the limiting case of the constraint stated under Assumption 2 earlier.

The key idea here is that $u$ is in the limit a piecewise linear function of each of its arguments. Such function may be thought of as being made up of quadrangular tiles forming a surface in the three dimensional space of the Cartesian coordinate frame, $e \times de \times u$. Such a function, can approximate *any* continuous function of its arguments over a finite interval to an arbitrary degree of accuracy[7]. The implication of this statement is that the class of fuzzy control algorithms we have considered may be used as nonlinear compensators for instance in relatively simple but common situations where the plant may be described by $\dot{x} = Ax + g(x)u$ and where $u$ can be designed to compensate for the nonlinearity in the input channel stated in terms of $g$. The resulting system would be considered *feedback linearized*[3].

A simple instance of this situation is in process control where the plant may be described by

$$\ddot{y} + a_1(y, \dot{y})\dot{y} + a_2(y, \dot{y})y = u. \quad (31)$$

Then $u$ defined as

$$u = u_c + (\hat{a}_1(y, \dot{y})\dot{y} + \hat{a}_2(y, \dot{y})y), \quad (32)$$

where $\hat{a}_1$ and $\hat{a}$ are estimates of $a_1$ and $a_2$ respectively, renders the following system

$$\ddot{y} = u_c, \quad (33)$$

provided $\|a_1 - \hat{a}_1\|$ and $\|a_2 - \hat{a}_2\|$ are negligible. Now given the earlier argument that a fuzzy control algorithm of the type we have considered can *emulate* any continuous function of its arguments over a finite interval, it is easy to see that $u$ given by (32) can indeed be implemented in terms of some fuzzy control algorithm. In the same vein there are numerous situations where fuzzy control can be and is used to compensate for some dominant nonlinearity in the plant. There are limits though as can be expected. For instance, if only the plant output and its derivative are directly accessible in the implementation of the fuzzy control algorithm, then correspondingly limited set of functions can be emulated and as such limited class of nonlinearities in the plant may be compensated for. This is not a limitation of fuzzy control per se but only of the manner of spec-

1. It is not difficult to see that as the characteristic functions approach their limiting triangular form, $F_j$ and $F'_l$ approach zero and in fact become identically zero in the limit.

ification of the rules and, at least in principle, can be dealt with using the framework described above.

## V Conclusion

This paper presents a framework for analyzing the functional behavior of a common class of fuzzy control algorithms. In particular, it shows that in principle fuzzy control may be used to compensate for nonlinear plant behavior and as such can be viewed as a feedback linearization strategy, albeit using rules rather than differential-algebraic relationships. The consequence of this statement is that fuzzy control does indeed accomplish more than its counterparts such as PID in that it not only compensates for dynamic deficiencies in the plant as PID does, but it also linearizes some nonlinear plants and as such provides uniform dynamic behavior, which PID for instance can not.

## VI References


[1] J. J. Buckley and H. Ying. Fuzzy controller theory: Limit theorems for linear fuzzy control rules. *Automatica*, 25(3), 1989.

[2] D. Dubois and H. Prade. *Fuzzy Sets and Systems*. Academic Press, 1980.

[3] A. Isidori. *Nonlinear Control Systems*. Springer Verlag, 1989.

[4] W. J. M. Kickert and E. H. Mamdani. Analysis of a fuzzy logic controller. *Fuzzy Sets and Systems*, 1(1), 1978.

[5] G. Langari. A nonlinear formulation of a class of fuzzy linguistic control systems. In *Proceedings of the 1992 American Control Conference, Chicago, Illinois*, June 1992.

[6] G. Langari and M. Tomizuka. Fuzzy linguistic control of arc welding, in *Sensors and Controls for Manufacturing*. ASME, 1988.

[7] H. L. Royden. *Real Analysis*. MacMillan, 1988.

[8] K. L. Tang and R. J. Mulholland. Comparing fuzzy logic with classical controllers, *IEEE Transactions on Systems, Man, and Cybernetics*, 17(6), November/December 1987.

[9] L. A. Zadeh. Outline of an approach to the analysis of complex systems and decision processes. *IEEE Transactions on Systems, Man, and Cybernetics*, 1973.

# Concept of Stability Margin for Fuzzy Systems and Design of Robust Fuzzy Controllers

Kazuo TANAKA and Manabu SANO

Department of Mechanical Systems Engineering
Kanazawa University
2-40-20 Kodatsuno Kanazawa 920 Japan
Tel. +81-762-61-2101 (ext.405)
Fax. +81-762-63-3849

***Abstract*** *- This paper discusses robust stability for fuzzy systems with premise parameter uncertainty and design problem of robust fuzzy controllers. We derive four conditions for ensuring stability of fuzzy control systems by weakening a stability condition (Basic Stability Condition) proposed by Tanaka and Sugeno. We introduce concept of stability margin and consider a design problem of robust controllers for fuzzy systems with premise parameter uncertainty. Simulation results show the robustness of the fuzzy controller.*

## 1. Introduction

One of the most important concepts concerning the properties of control systems is stability. It is, however, difficult to analyze stability of fuzzy control system because it is essentially nonlinear. Recently, some useful stability analysis methods [1]-[3], which are based on nonlinear stability theory, have been reported. In order to develop fuzzy control theory in the future, we will have to seek more advanced stability theory. An approach to construct more advanced stability theory is to develop an analysis technique for robust stability.

This paper discusses robust stability for fuzzy systems with premise parameter uncertainty and design problem of robust fuzzy controllers. We derive four conditions for ensuring stability of fuzzy control systems by weakening a stability condition (Basic Stability Condition) proposed by Tanaka and Sugeno [4], [5]:non-robust condition (NC), weak non-robust condition (WNC), robust condition (RC) and weak robust condition (WRC). Concept of stability margin is introduced from the WRC and is applied to a robust controller design. Simulation results show the robustness of the fuzzy controller.

## 2 Basic stability condition and weak stability condition

The fuzzy system, proposed by Takagi and Sugeno [6], is described by fuzzy IF-THEN rules which locally represent linear input-output relations of a system. This fuzzy system is of the following form:

Rule i:IF x(k) is $\mathcal{A}_{1i}$ and $\cdots$ and x(k-n+1) is $\mathcal{A}_{ni}$
THEN $\mathbf{x}_i(k+1) = \mathbf{A}_i\,\mathbf{x}(k) + \mathbf{B}_i\,\mathbf{u}(k)$,

where

$\mathbf{x}^T(k) = [x(k), x(k-1), \cdots, x(k-n+1)]$,
$\mathbf{u}^T(k) = [u(k), u(k-1), \cdots, u(k-m+1)]$,

i=1,2, $\cdots$, r and r is the number of IF-THEN rules. $\mathbf{x}_i(k+1)$ is the output from the i-th IF-THEN rule, and $\mathcal{A}_{ij}$ is a fuzzy set. Given a pair of ($\mathbf{x}(k)$, $\mathbf{u}(k)$), the final output of the fuzzy system is inferred as follows.

$$\mathbf{x}(k+1) = \frac{\sum_{i=1}^{r} w_i(k)\{\mathbf{A}_i\mathbf{x}(k) + \mathbf{B}_i\mathbf{u}(k)\}}{\sum_{i=1}^{r} w_i(k)}, \tag{1}$$

where

$$w_i(k) = \prod_{j=1}^{n} \mathcal{A}_{ij}(x(k-j+1)),$$

$\mathcal{A}_{ij}(x(k-j+1))$ is the grade of membership of x((k-j+1)) in $\mathcal{A}_{ij}$.

The free system of Eq.(1) is defined as

$$\mathbf{x}(k+1) = \frac{\sum_{i=1}^{r} w_i(k)\mathbf{A}_i\mathbf{x}(k)}{\sum_{i=1}^{r} w_i(k)}, \tag{2}$$

where it is assume that

$$\sum_{i=1}^{r} w_i(k) > 0, \qquad w_i(k) \geq 0, \quad i=1, 2, \sim, r \tag{3}$$

for all k. Each linear consequent equation represented by $\mathbf{A}_i\,\mathbf{x}(k)$ is called "subsystem".

Tanaka and Sugeno [4], [5] derived a stability condition for ensuring stability of Eq.(2) in accordance with the definition of stability in the sense of Lyapunov. Theorem 2.1 gives the stability condition.

*Theorem 2.1*

*The equilibrium of a fuzzy system described by Eq.(2) is asymptotically stable in the large if there exists a common positive definite matrix* **P** *such that*

$$\mathbf{A}_i^T\,\mathbf{P}\,\mathbf{A}_i - \mathbf{P} < \mathbf{0} \tag{4}$$

*for* i=1, 2, $\cdots$, r, *that is, for all the subsystems.*

The proof of this theorem is given in the literature [5]. This theorem is reduced to the Lyapunov stability theorem for linear discrete systems when r=1. Theorem 2.1 gives, of course, a sufficient condition for ensuring stability of Eq.(2). We may intuitively guess that a fuzzy system is globally stable if all subsystems are stable. However, this is not the case in general: we shall discuss it in Example 2.1.

0-7803-0614-7/93$03.00 ©1993IEEE

We should notice that the stability condition of Eq.(4) depends only on $\mathbf{A}_i$. In other words, the stability condition of Eq.(4) does not depend on $w_i(k)$.

[Example 2.1]

Let us consider the following fuzzy free system.

Rule 1:IF x(k-1) is $\mathcal{A}_1$ THEN $\mathbf{x}_1(k+1) = \mathbf{A}_1\mathbf{x}(k)$,

Rule 2:IF x(k-1) is $\mathcal{A}_2$ THEN $\mathbf{x}_2(k+1) = \mathbf{A}_2\mathbf{x}(k)$,

where

$$\mathbf{x}(k)^T = [x(k) \quad x(k-1)],$$

$$\mathbf{A}_1 = \begin{bmatrix} 1.0 & -0.5 \\ 1.0 & 0 \end{bmatrix}, \qquad \mathbf{A}_2 = \begin{bmatrix} -1.0 & -0.5 \\ 1.0 & 0 \end{bmatrix}.$$

Fig.1 shows membership functions of $\mathcal{A}_1$ and $\mathcal{A}_2$. Fig.2 (a) and (b) illustrate the behavior of the following linear systems for the initial condition x(0)=-0.70 and x(1)=0.90, respectively : $\mathbf{x}(k+1) = \mathbf{A}_1\mathbf{x}(k)$ and $\mathbf{x}(k+1) = \mathbf{A}_2\mathbf{x}(k)$. These linear systems are stable since $\mathbf{A}_1$ and $\mathbf{A}_2$ are stable matrices. However, the fuzzy system, which consists of these stable linear systems, is unstable as shown in Fig.2 (c).

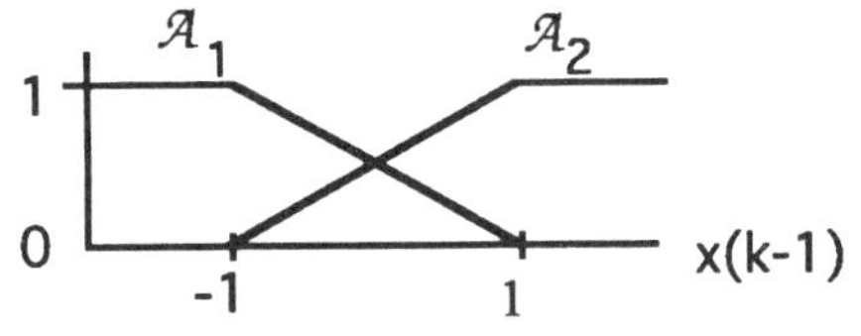


Fig.1 Membership functions.

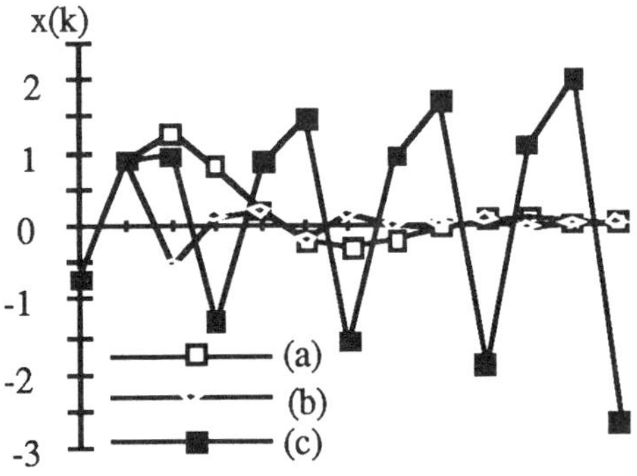


Fig.2 (a)Behavior of $\mathbf{A}_1\mathbf{x}(k)$,
(b)Behavior of $\mathbf{A}_2\mathbf{x}(k)$,
(c)Behavior of fuzzy system.

Next, we weaken the stability condition of Eq.(4). It is possible to find a Lyapunov function more effectively if we can weaken the stability condition of Eq.(4). Theorem 2.2 gives a weakened condition for Eq.(4).

*Theorem 2.2*

*The equilibrium of a fuzzy system described by Eq.(2) is asymptotically stable in the large if there exists a common positive definite matrix* **P** *such that*

$$\mathbf{S} = \sum_{i=1}^{r} w_i(k)\, w_i(k)\, \{\mathbf{A}_i^T\mathbf{P}\mathbf{A}_i - \mathbf{P}\}$$
$$+ \sum_{i<j} w_i(k)\, w_j(k) \{\mathbf{A}_i^T\mathbf{P}\mathbf{A}_i - \mathbf{P} + \mathbf{A}_j^T\mathbf{P}\mathbf{A}_j - \mathbf{P}$$
$$- (\mathbf{A}_i-\mathbf{A}_j)^T\mathbf{P}(\mathbf{A}_i-\mathbf{A}_j) \;-\mathbf{P}\} < \mathbf{0}. \qquad (5)$$

*(proof)It follows directly from the proof of Theorem 2.1.*

As mentioned above, the stability condition of Eq.(4) depends only on $\mathbf{A}_i$. The stability condition of Eq.(5) depends not only on $\mathbf{A}_i$ but also on $w_i(k)$. This means that it is possible to find a positive definite matrix **P** such that Eq.(5) is satisfied even if there does not exist a common positive definite matrix **P** such that Eq.(4) is satisfied. We shall shown it in Example 3.1. However, we should notice that Theorem 2.2 gives, of course, a sufficient condition for ensuring stability of Eq.(2).

## 3 Design problems of fuzzy controller and robust fuzzy controller

We have shown the stability conditions for fuzzy systems in the previous section. We apply the stability conditions to design problems of fuzzy controllers and robust fuzzy controllers in this section.

Let us consider the following fuzzy system again.

Rule i:IF x(k) is $\mathcal{A}_{1i}$ and · · · and x(k-n+1) is $\mathcal{A}_{ni}$
THEN $\mathbf{x}_i(k+1) = \mathbf{A}_i\, \mathbf{x}(k) + \mathbf{B}_i\, \mathbf{u}(k)$,

Assume that this fuzzy system has premise parameter uncertainty. Therefore, the final output of the fuzzy system is calculated as follows.

$$\mathbf{x}(k+1) = \frac{\sum_{i=1}^{r} (w_i(k)+\Delta w_i(k))\, \{A_i\mathbf{x}(k)+B_i\mathbf{u}(k)\}}{\sum_{i=1}^{r} (w_i(k) + \Delta w_i(k))}, \qquad (6)$$

where $\Delta w_i(k)$ denotes unknown premise parameter uncertainty. It is assumed that

$-1 \le \Delta w_i(k) \le 1$, for all i

$0 \le w_i(k) + \Delta w_i(k) \le 1$, for all i

$$\sum_{i=1}^{r} (w_i(k) + \Delta w_i(k)) > 0,$$

for all k.

We attempt to stabilize the fuzzy system with premise parameter uncertainty, that is, Eq.(6), by the following controller.

Rule i:IF x(k) is $\mathcal{A}_{1i}$ and · · · and x(k-n+1) is $\mathcal{A}_{ni}$
THEN $\mathbf{u}_i(k) = \mathbf{F}_i\, \mathbf{x}(k)$

The final output of this fuzzy controller is calculated by Eq.(7).

$$\mathbf{u}(k) = \frac{\sum_{i=1}^{r} w_i(k)\mathbf{F}_i\mathbf{x}(k)}{\sum_{i=1}^{r} w_i(k)}, \qquad (7)$$

where we must use $w_i(k)$ instead of $w_i(k)+\Delta w_i(k)$ as a weight of i-th rule since $\Delta w_i(k)$'s are unknown premise parameters. In this case, fuzzy controller design is to determine the consequent matrices $\mathbf{F}_i$.

We derive four conditions for ensuring stability of fuzzy control systems:non-robust condition (NC), weak non-robust condition (WNC), robust condition (RC) and weak robust condition (WRC). NC and WNC are conditions for ensuring stability of fuzzy control systems with no premise parameter uncertainty, that is, $\Delta w_i(k)=0$ for all i and k. RC and WRC are conditions for ensuring stability of fuzzy control systems with premise parameter uncertainty.

### 3.1 NC and WNC

By substituting Eq.(7) into Eq.(1), we obtain

$$\mathbf{x}(k+1) = \frac{\sum_{i=1}^{r}\sum_{j=1}^{r} w_i(k)w_j(k)\,\{\mathbf{A}_i+\mathbf{B}_i\mathbf{F}_j\}\mathbf{x}(k)}{\sum_{i=1}^{r}\sum_{j=1}^{r} w_i(k)w_j(k)} \quad (8)$$

From Eq.(8),

$$\mathbf{x}(k+1) = \frac{\sum_{i=1}^{r}\sum_{j=1}^{r} w_i(k)w_j(k)\,\{\mathbf{A}_i+\mathbf{B}_i\mathbf{F}_j\}\mathbf{x}(k)}{\sum_{i=1}^{r}\sum_{j=1}^{r} w_i(k)w_j(k)}$$

$$= \frac{1}{R}\Big[\sum_{i=1}^{r} w_i(k)w_i(k)\,\mathbf{G}_{ii}\mathbf{x}(k) + 2\sum_{i<j} w_i(k)w_j(k)\frac{\mathbf{G}_{ij}+\mathbf{G}_{ji}}{2}\mathbf{x}(k)\Big], \quad (9)$$

where

$$\mathbf{G}_{ij} = \mathbf{A}_i + \mathbf{B}_i\mathbf{F}_j,$$

$$R = \sum_{i=1}^{r}\sum_{j=1}^{r} w_i(k)w_j(k).$$

Without loss of generality, Eq.(9) can be rewritten as follows.

$$\mathbf{x}(k+1) = \frac{\sum_{i=1}^{r(r+1)/2} v_i(k)\mathbf{H}_i\mathbf{x}(k)}{\sum_{i=1}^{r(r+1)/2} v_i(k)}, \quad (10)$$

where

$\mathbf{H}_{\sum_{t=1}^{j}(t-1)+i} = \mathbf{G}_{ij}$, $\quad i = j$

$\mathbf{H}_{\sum_{t=1}^{j}(t-1)+i} = (\mathbf{G}_{ij} + \mathbf{G}_{ji})/2$, $\quad i < j$

$v_{\sum_{t=1}^{j}(t-1)+i}(k) = w_i(k)w_j(k)$, $\quad i = j$

$v_{\sum_{t=1}^{j}(t-1)+i}(k) = 2w_i(k)w_j(k)$. $\quad i < j$

By applying Theorem 2.1 to Eq.(10), we derive a NC for fuzzy control systems, Eq.(10), with no premise parameter uncertainty, that is, $\Delta w_i(k)=0$ for all i and k. Theorem 3.1 gives the NC for Eq.(10).

*Theorem 3.1 : non-robust condition (NC)*

*The equilibrium of a fuzzy control system described by Eq.(10) is asymptotically stable in the large if there exists a common positive definite matrix* **P** *such that*

$$\mathbf{H}_i^T \mathbf{P} \mathbf{H}_i - \mathbf{P} < \mathbf{0} \quad (11)$$

*for* $i = 1, 2, \cdots, r(r+1)/2$.

*(proof)It follows directly from Theorem 2.1.*

The non-robust design problem for Theorem 3.1 is to select $\mathbf{F}_j$ ($j = 1, 2, \cdots, r$) such that

$$\mathbf{H}_i^T \mathbf{P} \mathbf{H}_i - \mathbf{P} < \mathbf{0} \qquad i = 1,2,\cdots, r(r+1)/2$$

for a common positive definite matrix **P** when $\mathbf{A}_j$ and $\mathbf{B}_j$ are given.

Next, we weaken the NC of Theorem 3.1 by applying Theorem 2.2 to Eq.(10). Theorem 3.2 gives the weakened NC, that is, WNC.

*Theorem 3.2 : weak non-robust condition (WNC)*

*The equilibrium of a fuzzy control system described by Eq.(10) is asymptotically stable in the large if there exists a common positive definite matrix* **P** *such that*

$$\mathbf{S} = \sum_{i=1}^{r(r+1)/2} v_i(k)\,v_i(k)\,\{\mathbf{H}_i^T\mathbf{P}\mathbf{H}_i - \mathbf{P}\} + \sum_{i<j} v_i(k)\,v_j(k)\{\mathbf{H}_i^T\mathbf{P}\mathbf{H}_i - \mathbf{P} + \mathbf{H}_j^T\mathbf{P}\mathbf{H}_j - \mathbf{P} - (\mathbf{H}_i-\mathbf{H}_j)^T\mathbf{P}(\mathbf{H}_i-\mathbf{H}_j) - \mathbf{P}\} < \mathbf{0}. \quad (12)$$

*(proof) It follows directly from Theorem 2.2.*

The non-robust design problem for Theorem 3.2 is to select $\mathbf{F}_j$ ($j = 1, 2, \cdots, r$) such that Eq.(12) is satisfied for a common positive definite matrix **P** when $\mathbf{A}_j$ and $\mathbf{B}_j$ are given.

It will be shown in Example 3.1 that we can find a positive definite matrix **P** such that Eq.(12) is satisfied even if there does not exist a common positive definite matrix **P** such that Eq.(11) is satisfied.

[Example 3.1]

Let us consider the following fuzzy system and fuzzy controller.

Rule 1:IF x(k) is $\mathcal{A}_1$ THEN $\mathbf{x}_1(k+1) = \mathbf{A}_1\mathbf{x}(k) + \mathbf{B}_1 u(k)$,

Rule 2:IF x(k) is $\mathcal{A}_2$ THEN $\mathbf{x}_2(k+1) = \mathbf{A}_2\mathbf{x}(k) + \mathbf{B}_2 u(k)$,

where

$$\mathbf{A}_1 = \begin{bmatrix} 1.0 & 0.5 \\ 1.0 & 0 \end{bmatrix}, \quad \mathbf{B}_1 = \begin{bmatrix} 0.2 \\ 0 \end{bmatrix}, \quad \mathbf{A}_2 = \begin{bmatrix} 1.0 & -0.5 \\ 1.0 & 0 \end{bmatrix}, \quad \mathbf{B}_2 = \begin{bmatrix} 0.58 \\ 0 \end{bmatrix}.$$

Rule 1:IF x(k) is $\mathcal{A}_1$ THEN $\mathbf{x}_1(k+1) = \mathbf{F}_1\mathbf{x}(k)$,

Rule 2:IF x(k) is $\mathcal{A}_2$ THEN $\mathbf{x}_2(k+1) = \mathbf{F}_2\mathbf{x}(k)$,

where

$$\mathbf{F}_1 = [-5.0 \quad -3.75], \quad \mathbf{F}_2 = [-1.72 \quad 0.43].$$

We obtain

$$\mathbf{H}_1 = \mathbf{H}_3 = \begin{bmatrix} 0 & -0.25 \\ 1 & 0 \end{bmatrix}, \quad \mathbf{H}_2 = \begin{bmatrix} -0.62 & -1.04 \\ 1 & 0 \end{bmatrix}.$$

It is found that there does not exist a common positive definite matrix **P** such that

$$\mathbf{H}_i^T \mathbf{P} \mathbf{H}_i - \mathbf{P} < \mathbf{0} \qquad i=1, 2, 3$$

since $\mathbf{H}_2$ is an unstable matrix. Therefore, the controller parameter matrices $\mathbf{F}_1$ and $\mathbf{F}_2$ do not satisfy the NC of Eq.(11).

Next, we show that the WNC of Eq.(12) is satisfied if we select

$$\mathbf{P} = \begin{bmatrix} 2.2 & 0 \\ 0 & 1.2 \end{bmatrix}$$

as a common positive definite matrix **P**. By substituting $\mathbf{H}_1 \sim \mathbf{H}_3$ and **P** into Eq.(12), we obtain

$$\mathbf{S} = \begin{bmatrix} \alpha & \gamma \\ \gamma & \beta \end{bmatrix},$$

where

$$\alpha = -v_1^2(k) - 0.15v_2^2(k) - v_3^2(k) - 2v_1(k)v_2(k) - 2v_1(k)v_3(k) - 2v_2(k)v_3(k) - 2v_2(k)v_4(k),$$

$$\beta = -1.06v_1^2(k) - 1.25v_1(k)v_2(k) + 1.20v_2^2(k) - 2.13v_1(k)v_3(k) - 1.25v_2(k)v_3(k) - 1.06v_3^2(k),$$

$$\gamma = 0.34v_1(k)v_2(k) + 1.43v_2^2(k) + 0.34v_2(k)v_3(k).$$

In order that **S** is a negative definite matrix, it must be satisfied that

$$\alpha < 0 \text{ and } \alpha \times \beta + \gamma \times \gamma > 0.$$

It is clear that $\alpha < 0$. On the other hand,

$$\begin{aligned}\alpha\times\beta + \gamma\times\gamma = {} & 1.06v_1^4(k) + 2.22v_2^4(k) + 1.06v_3^4(k) \\ & + 3.38v_1^3(k)v_2(k) + 1.34v_1^2(k)v_2^2(k) - 3.19v_1(k)v_2^3(k) \\ & + 4.25v_1^3(k)v_3(k) + 10.1v_1^2(k)v_2(k)v_3(k) \\ & + 2.68v_1(k)v_2^2(k)v_3(k) - 3.19v_2^3(k)v_3(k) + 6.38v_1^2(k)v_3^2(k) \\ & + 10.1v_1(k)v_2(k)v_3^2(k) + 1.34v_2^2(k)v_3^2(k) + 4.25v_1(k)v_3^3(k) \\ & + 3.38v_2(k)v_3^3(k).\end{aligned}$$

By substituting

$$v_1(k) = w_1(k)w_1(k),$$
$$v_2(k) = 2w_1(k)w_2(k),$$
$$v_3(k) = w_2(k)w_2(k)$$

into the above equation, we obtain

$$\begin{aligned} f_1(w_1(k),w_2(k)) = {} & 1.06w_1^8(k) + 6.75w_1^7(k)w_2(k) \\ & + 9.62w_1^6(k)w_2^2(k) - 5.29w_1^5(k)w_2^3(k) \\ & - 18.4w_1^4(k)w_2^4(k) - 5.29w_1^3(k)w_2^5(k) \\ & + 9.62w_1^2(k)w_2^6(k) + 6.75w_1(k)w_2^7(k) \\ & + 1.06w_2^8(k), \end{aligned}$$

where

$$f_1(w_1(k),w_2(k)) = \alpha \times \beta + \gamma \times \gamma.$$

It is found that

$$f_1(w_1(k),w_2(k)) > 0$$

when $0 \le w_1(k) \le 1$ and $0 \le w_2(k) \le 1$. This means that the WNC of Eq.(12) is satisfied for $0 \le w_1(k) \le 1$ and $0 \le w_2(k) \le 1$, that is, for any fuzzy sets $\mathcal{A}_1$ and $\mathcal{A}_2$. Therefore, $\mathbf{F}_1$ and $\mathbf{F}_2$ guarantee stability of the fuzzy control system. Fig.3 shows surface of the $f_1(w_1(k),w_2(k))$ over $0 \le w_1(k) \le 1$ and $0 \le w_2(k) \le 1$.

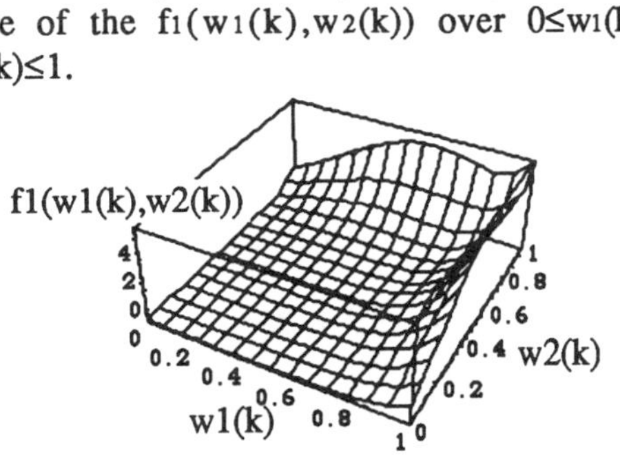


Fig.3 Surface of $f_1(w_1(k),w_2(k))$

It is seen from Fig.3 that

$$f_1(w_1(k),w_2(k)) > 0$$

when $0 \le w_1(k) \le 1$ and $0 \le w_2(k) \le 1$.

### 3.2 RC and WRC

By substituting Eq.(7) into Eq.(6), we obtain

$$\mathbf{x}(k+1) = \frac{\sum_{i=1}^{r}\sum_{j=1}^{r}(w_i(k)+\Delta w_i(k))w_j(k)\mathbf{G}_{ij}\mathbf{x}(k)}{\sum_{i=1}^{r}\sum_{j=1}^{r}(w_i(k)+\Delta w_i(k))w_j(k)} \quad (13)$$

where

$$\mathbf{G}_{ij} = \mathbf{A}_i + \mathbf{B}_i\mathbf{F}_j.$$

Without loss of generality, Eq.(13) can be rewritten as follows.

$$\mathbf{x}(k+1) = \frac{\sum_{i=1}^{r\times r} v_i(k)\mathbf{H}_i\mathbf{x}(k)}{\sum_{i=1}^{r\times r} v_i(k)} \quad (14)$$

where

$$\mathbf{H}_{(i-1)\times r+j} = \mathbf{G}_{ij}, \quad (15)$$

$$v_{(i-1)\times r+j}(k) = (w_i(k)+\Delta w_i(k))w_j(k), \quad (16)$$

for all i and j. By applying Theorem 2.1 to Eq.(14), we derive a RC for fuzzy control system, Eq.(14), with premise parameter uncertainty. Theorem 3.3 gives the RC for Eq.(14).

*Theorem 3.3 : robust condition (RC)*

*The equilibrium of a fuzzy control system described by Eq.(14) is asymptotically stable in the large if there exists a common positive definite matrix* **P** *such that*

$$\mathbf{H}_i^T \mathbf{P} \mathbf{H}_i - \mathbf{P} < 0 \quad (17)$$

*for* $i = 1, 2, \cdots, r \times r$.

*(proof)It follows directly from Theorem 2.1.*

The robust design problem for Theorem 3.3 is to select $\mathbf{F}_j$ ($j = 1, 2, \cdots, r$) such that

$$\mathbf{H}_i^T \mathbf{P} \mathbf{H}_i - \mathbf{P} < 0 \qquad i = 1, 2, \cdots, r \times r$$

for a common positive definite matrix **P** when $\mathbf{A}_j$ and $\mathbf{B}_j$ are given. It is clear that Eq.(17) implies Eq.(11).

Next, we weaken the RC of Theorem 3.3 by applying Theorem 2.2 to Eq.(14). Theorem 3.4 gives the weakened RC, that is, WRC.

*Theorem 3.4 : weak robust condition (WRC)*

*The equilibrium of a fuzzy control system described by Eq.(14) is asymptotically stable in the large if there exists a common positive definite matrix* **P** *such that*

$$\mathbf{S} = \sum_{i=1}^{r\times r} v_i(k)\,v_i(k)\,\{\mathbf{H}_i^T\mathbf{P}\mathbf{H}_i - \mathbf{P}\}$$
$$+ \sum_{i<j} v_i(k)\,v_j(k)\{\mathbf{H}_i^T\mathbf{P}\mathbf{H}_i - \mathbf{P} + \mathbf{H}_j^T\mathbf{P}\mathbf{H}_j - \mathbf{P}$$

$$- (\mathbf{H}_i - \mathbf{H}_j)^T \mathbf{P} (\mathbf{H}_i - \mathbf{H}_j) \ - \mathbf{P} \} < \mathbf{0}. \qquad (18)$$

*(proof)It follows directly from Theorem 2.2.*

The robust design problem for Theorem 3.4 is to select $\mathbf{F}_j$ $(j = 1, 2, \cdots, r)$ such that Eq.(18) is satisfied for a common positive definite matrix $\mathbf{P}$ when $\mathbf{A}_j$ and $\mathbf{B}_j$ are given. It is clear that Eq.(18) implies Eq.(12).

### 3.3 Stability margin

We introduce concept of stability margin for fuzzy control systems with premise parameter uncertainty from the WRC in Theorem 3.4. By substituting Eqs.(15) and (16) into Eq.(18) and by solving Eq.(18) for $\Delta w_i(k)$'s, we can find an admissible region of $\Delta w_i(k)$'s such that Eq.(18) is satisfied. This admissible region can be regarded as an index which represents stability margin for fuzzy control systems with premise parameter uncertainty. We define an Index of Stability Margin (ISM) as follows.

[Definition 3.1]

$$\text{ISM} = S_{adm}/S_{all},$$

where $S_{all}$ is area (or volume) of all changeable region in premise parameters space and $S_{adm}$ is area (or volume) of admissible region in all the changeable region.

If Eq.(18) is satisfied for $-1 \leq \Delta w_i(k) \leq 1$, $\forall i$ and $\forall k$, it is clear that ISM=1. This means that robust control can be perfectly realized. Conversely, when ISM=0, stability of fuzzy systems with premise parameter uncertainty may not be guaranteed. Thus, ISM can be regarded as an index of stability margin. In Example 3.2, we concretely illustrate a calculation way of ISM.

[Example 3.2]

Let us consider the following fuzzy system and fuzzy controller.

Rule 1:IF $x(k)$ is $\mathcal{A}_1$ THEN $x_1(k+1) = \mathbf{A}_1 x(k) + \mathbf{B}_1 u(k)$,

Rule 2:IF $x(k)$ is $\mathcal{A}_2$ THEN $x_2(k+1) = \mathbf{A}_2 x(k) + \mathbf{B}_2 u(k)$,

where

$$\mathbf{A}_1 = \begin{bmatrix} 1.0 & 0.5 \\ 1.0 & 0 \end{bmatrix}, \quad \mathbf{B}_1 = \begin{bmatrix} 0.2 \\ 0 \end{bmatrix}, \quad \mathbf{A}_2 = \begin{bmatrix} 1.0 & -0.5 \\ 1.0 & 0 \end{bmatrix}, \quad \mathbf{B}_2 = \begin{bmatrix} 0.4 \\ 0 \end{bmatrix}.$$

Rule 1:IF $x(k)$ is $\mathcal{A}_1$ THEN $x_1(k+1) = \mathbf{F}_1 x(k)$,

Rule 2:IF $x(k)$ is $\mathcal{A}_2$ THEN $x_2(k+1) = \mathbf{F}_2 x(k)$,

where

$$\mathbf{F}_1 = [-5.0 \quad -3.75], \qquad \mathbf{F}_2 = [-2.5 \quad 0.63].$$

Fig.4 shows membership functions of the fuzzy sets $\mathcal{A}_1$ and $\mathcal{A}_2$.

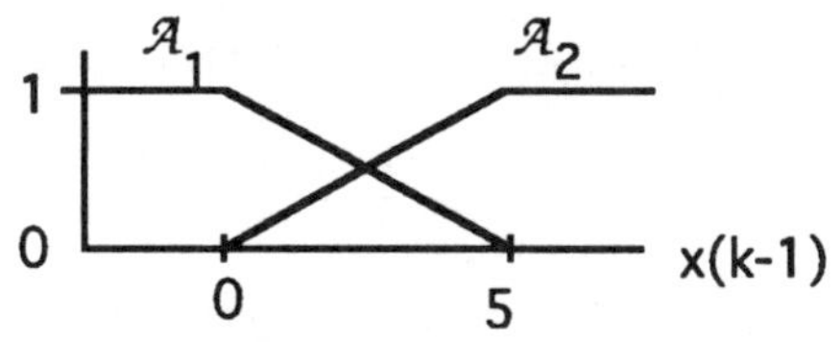


Fig.4 Membership functions

We obtain

$$\mathbf{H}_1 = \mathbf{H}_4 = \begin{bmatrix} 0 & -0.25 \\ 1 & 0 \end{bmatrix},$$

$$\mathbf{H}_2 = \begin{bmatrix} 0.5 & 0.63 \\ 1 & 0 \end{bmatrix}, \quad \mathbf{H}_3 = \begin{bmatrix} -1 & -2 \\ 1 & 0 \end{bmatrix}.$$

It is found that there does not exist a common positive definite matrix $\mathbf{P}$ such that

$$\mathbf{H}_i^T \mathbf{P} \mathbf{H}_i - \mathbf{P} < \mathbf{0} \qquad i=1, 2, 3, 4$$

since $\mathbf{H}_2$ and $\mathbf{H}_3$ are unstable matrices. Therefore, the controller parameter matrices $\mathbf{F}_1$ and $\mathbf{F}_2$ do not satisfy the RC of Eq.(17).

Let us find an admissible region of premise parameter uncertainty $\Delta w_i(k)$'s which satisfy the WRC of Eq.(18). Assume that we select

$$\mathbf{P} = \begin{bmatrix} 2.2 & 0 \\ 0 & 1.2 \end{bmatrix}$$

as a common positive definite matrix $\mathbf{P}$. By substituting $\mathbf{H}_1 \sim \mathbf{H}_4$ and $\mathbf{P}$ into Eq.(16), we obtain

$$\mathbf{S} = \begin{bmatrix} \alpha & \gamma \\ \gamma & \beta \end{bmatrix},$$

where

$$\begin{aligned} \alpha = & -v_1^2(k) - 0.45v_2^2(k) + 1.2v_3^2(k) - v_4^2(k) \\ & -2v_1(k)v_2(k) - 2v_1(k)v_3(k) - 4.2v_2(k)v_3(k) \\ & -2v_1(k)v_4(k) - 2v_2(k)v_4(k) - 2v_3(k)v_4(k), \end{aligned}$$

$$\begin{aligned} \beta = & -1.06v_1^2(k) - 0.34v_2^2(k) - 7.6v_3^2(k) - 1.06v_4^2(k) \\ & -3.09v_1(k)v_2(k) - 0.20v_1(k)v_3(k) - 2.13v_1(k)v_4(k) \\ & -7.90v_2(k)v_3(k) - 3.09v_2(k)v_4(k) - 0.20v_3(k)v_4(k), \end{aligned}$$

$$\begin{aligned} \gamma = & \ 0.69v_2^2(k) + 4.40v_3^2(k) - 0.28v_1(k)v_2(k) \\ & +0.55v_1(k)v_3(k) - 3.58v_2(k)v_3(k) \\ & -0.28v_2(k)v_4(k) + 0.55v_3(k)v_4(k). \end{aligned}$$

In order that $\mathbf{S}$ is a negative definite matrix, it must be satisfied that

$$\alpha < 0 \text{ and } \alpha \times \beta + \gamma \times \gamma > 0.$$

To obtain a graphical image of an admissible region which guarantees robust stability, we assume that

$$w_1(k) + w_2(k) = 1,$$

$$(w_1(k) + \Delta w_1(k)) + (w_2(k) + \Delta w_2(k)) = 1.$$

By substituting

$$v_1(k) = (w_1(k) + \Delta w_1(k))w_1(k),$$

$$v_2(k) = (w_1(k) + \Delta w_1(k))w_2(k),$$

$$v_3(k) = (w_2(k) + \Delta w_2(k))w_1(k),$$

$$v_4(k) = (w_2(k) + \Delta w_2(k))w_2(k),$$

into the above equation and by eliminating $w_2(k)$ and $\Delta w_2(k)$, we can derive

$$\begin{aligned} g_1(w_1(k), \Delta w_1(k)) = & -1.0 + 0.55\Delta w_1^2(k) - 1.1w_1(k)\Delta w_1(k) \\ & +1.1w_1(k)\Delta w_1^2(k) + 0.55w_1^2(k) \\ & +0.55w_1^2(k)\Delta w_1^2(k) - 1.1w_1^3(k) \\ & +1.1w_1^3(k)\Delta w_1(k) + 0.55w_1^4(k), \end{aligned}$$

$$\begin{aligned} g_2(w_1(k), \Delta w_1(k)) = & \ 1.06 + 0.96\Delta w_1(k) - 2.34\Delta w_1^2(k) \\ & -0.96w_1(k) + 5.65w_1(k)\Delta w_1(k)) \\ & -4.69w_1(k)\Delta w_1^2(k) - 1.38w_1^2(k) \\ & -2.34w_1^2(k)\Delta w_1^2(k) + 4.69w_1^3(k) \\ & -4.69w_1^3(k)\Delta w_1(k) - 2.34w_1^4(k), \end{aligned}$$

where

$$g_1(w_1(k),\Delta w_1(k)) = \alpha,$$
$$g_2(w_1(k),\Delta w_1(k)) = \alpha\times\beta + \gamma\times\gamma.$$

Fig.5 shows negative region such that $g_1(w_1(k),\Delta w_1(k)) < 0$ and positive region such that $g_2(w_1(k),\Delta w_1(k)) > 0$. The stability region, which guarantees robust stability, is intersection region of the negative region and the positive region. On the other hand, all the changeable region shown in Fig.6 (a) can be obtained since $0 \leq w_1(k)+\Delta w_1(k) \leq 1$. Fig.6 (a) shows the stability region and all the changeable region. The admissible region is intersection region of the stability region and the changeable region. Fig.6 (b) shows the admissible region. From Definition 3.1 and Fig.6 (b), we can approximately calculate stability margin of this control system.

$$S_{all} = 1\times1=1,$$
$$S_{adm} \cong S_{all}-S_{tra}= 1-(1-0.25)(1-0.25)/2 \cong 1-0.28=0.72,$$
$$ISM = S_{adm}/S_{all} \cong 0.72,$$

where $S_{tra}$ denotes the area of the triangle which consists of points A, B and C in Fig.6 (b).

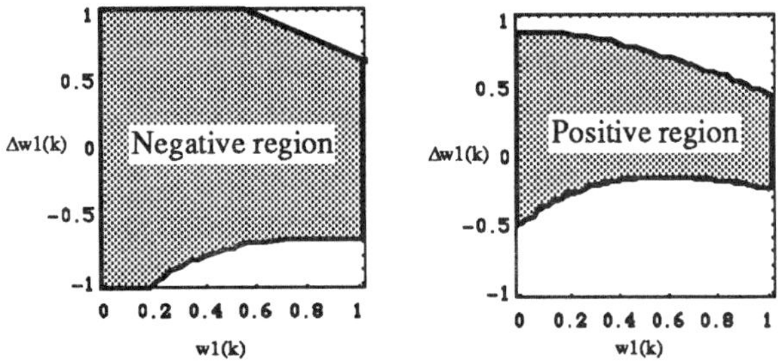


(a) (b)

Fig.5 (a) Negative region of $g_1(w_1(k),\Delta w_1(k))$.
(b) Positive region of $g_2(w_1(k),\Delta w_1(k))$.

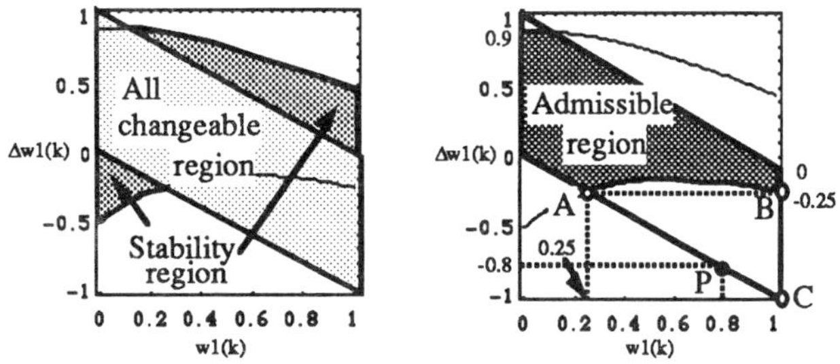


(a) (b)

Fig.6 (a) Stability region and changeable region.
(b) Admissible region for robust stability.

Let us consider robust stability for three cases of premise parameters uncertainty.

case (a) $\Delta w_1(k) = 0,$
$\Delta w_2(k) = 0,$

case (b) $w_1(k) + \Delta w_1(k) = \mathcal{A}_1( x(k)+4 ),$
$w_2(k) + \Delta w_2(k) = \mathcal{A}_2( x(k)+4 ),$

case (c) $w_1(k) + \Delta w_1(k) = \mathcal{A}_1( x(k)-4 ),$
$w_2(k) + \Delta w_2(k) = \mathcal{A}_2( x(k)-4 ).$

Case (a) means that the fuzzy system has no premise parameter uncertainty. In cases (a) and (b), all pairs of $(w_1(k),\Delta w_1(k))$ for all x(k) perfectly belong to the admissible region. Therefore, robust stability is guaranteed in these cases, that is, the WRC is satisfied. In case (c), some pairs of $(w_1(k),\Delta w_1(k))$ for all x(k) does not belong to the admissible region. For example, when x(k)=1, $w_1(k)$=0.8 and $\Delta w_1(k)$=-0.8, that is, (0.8,-0.8) which is the point "P" in Fig.6 (b). This point "P" does not belong to the admissible region. Therefore, robust stability may not be guaranteed in this case. Fig.7 shows control results for x(0) = x(1) = 0.5. It is found from Fig.8 that the control system of case (c) is not asymptotically stable in the large.

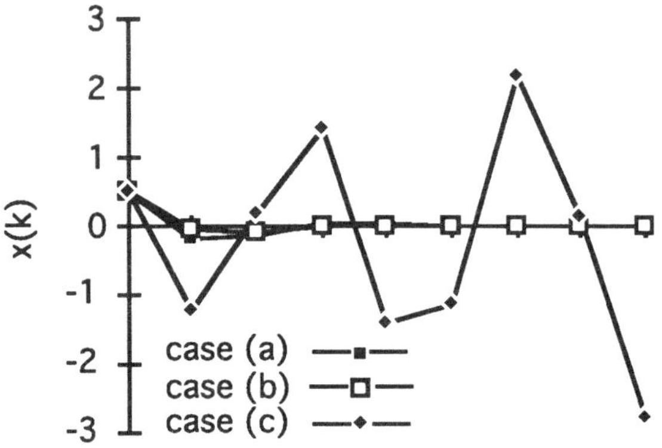


Fig.7 Control results.

## 4. Conclusion

We have discussed robust stability for fuzzy systems and design problem of robust fuzzy controllers. We have weakened a stability condition proposed by Tanaka and Sugeno and derived four conditions for ensuring stability of fuzzy control systems:non-robust condition, weak non-robust condition, robust condition and weak robust condition. Concept of stability margin has been introduced from the weak robust condition. We have considered a design problem of robust controllers for fuzzy systems with premise parameter uncertainty.

## References


[1]R. Lamgari and M. Tomizuka:Analysis and Synthesis of Fuzzy Lingustic Control Systems, 1990 ASME Winter Anual Meeting, pp.35-42 (1990).

[2]S. Kitamura and T. Kurozumi:Extended Cricle Criterion, and Stability Analysis of fuzzy Control Systems, Proc. of the International Fuzzy Eng. Symposium 91, vol.2 pp.634-643 (1991).

[3]F. Hara and M. Ishibe:Simulation Study on the Existence of Limit Cycle Oscillation in a Fuzzy Control System, procedings of the Korea-Japan Joint Conference on Fuzzy Systems and Engineering, pp.25-28 (1992).

[4]K.Tanaka and M.Sugeno:Stability Analysis of Fuzzy Systems Using Lyapunov's Direct Method, Proc. of NAFIPS'90, pp.133-136, (1990).

[5]K.Tanaka and M.Sugeno:Stability Analysis and Design of Fuzzy Control Systems, FUZZY SETS AND SYSTEMS 45, no.2, pp.135-156 (1992).

[6]T.Takagi and M.Sugeno:Fuzzy Identification of Systems and Its Applications to Modeling and Control, IEEE SMC-15 no.1, pp.116-132, (1985).

# A Two-Input Two-Output Fuzzy Controller Is the Sum of Two Nonlinear PI controllers with Variable Gains

Hao Ying

1) Department of Physiology and Biophysics
2) Biomedical Engineering Center
3) Office of Academic Computing
University of Texas Medical Branch
Galveston, TX 77555, USA.

**Abstract — Analytical structure of a two-input two-output fuzzy controller is proven to be the sum of two nonlinear PI controllers with variable gains continuously varying with process outputs. The fuzzy controller consists of the following components: (1) two-member input fuzzy sets with linear membership functions; (2) five-member output fuzzy sets with singleton membership functions; (3) 16 fuzzy control rules; (4) probabilistic AND fuzzy logic; and (5) the center of gravity defuzzification algorithm.**

## I. INTRODUCTION

There have been a number of multiple-input multiple-output (MIMO) fuzzy control applications, which include the flexible wing aircraft control [1] and the steam generating unit regulation [5]. Some theoretical aspects of the "traditional" fuzzy-relation-based MIMO fuzzy controllers have also been probed. The structure of a MIMO fuzzy control system was studied [2]. A MIMO fuzzy controller under Godel's implication was investigated [3]. Research on linguistic decoupling control of fuzzy multivariable processes was conducted [6][7]. For detailed information on the status of fuzzy control theory, readers are referred to the recent survey [4].

To advance fuzzy control theory, an analytical framework in relation to nonfuzzy control theory is necessary. In this paper, we will use the previously developed methodology [8] to study a two-input two-output fuzzy controller. Analytical structure of the fuzzy controller will be derived in terms of proportional-integral (PI) controller.

## II. COMPONENTS OF A FUZZY CONTROLLER

The following state variables $x_i(k)$ ($i = 1, 2, 3, 4$) serve as inputs of the two-input two-output fuzzy controller:

0-7803-0614-7/93$03.00 

$$x_1(k) = k_1 \cdot e_1(k) = k_1(SP_1 - y_1(k)), \tag{2.1}$$
$$x_2(k) = k_2 \cdot \Delta e_1(k) = k_2(y_1(k-1) - y_1(k)), \tag{2.2}$$
$$x_3(k) = k_3 \cdot e_2(k) = k_3(SP_2 - y_2(k)), \tag{2.3}$$
$$x_4(k) = k_4 \cdot \Delta e_2(k) = k_4(y_2(k-1) - y_2(k)), \tag{2.4}$$

where $k_i$'s are input scalars. $y_1(k)$ and $y_2(k)$ are coupling outputs of the two-input two-output process with respective setpoints, $SP_1$ and $SP_2$. $e_1(k)$ and $e_2(k)$ denote process output errors while $\Delta e_1(k)$ and $\Delta e_2(k)$ designate rate change of errors. k and k-1 represent current and previous sampling time, respectively. In this study, we select such $k_i$'s that $-L \le x_i(k) \le L$ $(L > 0)$.

$x_i(k)$'s are fuzzified by input fuzzy sets, each of which has two members. The members are denoted by $X_i^1$ and $X_i^{-1}$, respectively, which may be verbally called "positive" and "negative." The membership functions of $X_i^1$'s ($X_i^{-1}$'s) are identical and are linear functions:

$$\mu_1(x_i) = \frac{L + x_i(k)}{2L}, \tag{2.5}$$

$$\mu_{-1}(x_i) = \frac{L - x_i(k)}{2L}, \tag{2.6}$$

where

$$\mu_1(x_i) + \mu_{-1}(x_i) = 1. \tag{2.7}$$

$\mu_1(x_i)$ is the membership of $x_i$ for $X_i^1$ while $\mu_{-1}(x_i)$ is the membership of $x_i$ for $X_i^{-1}$.

There are two incremental outputs of the fuzzy controller, namely $\Delta u_1$ and $\Delta u_2$, defined in [-H, H]. There are five members, $\Delta U_m^{-2}$, $\Delta U_m^{-1}$, $\Delta U_m^{0}$, $\Delta U_m^{1}$ and $\Delta U_m^{2}$ ($m = 1, 2$), which may be named as "very negative," "negative" "zero," "positive," and "very positive," respectively. The membership function of the j-th member ($j = -2, -1, 0, 1, 2$)

is one at j·V, where V = H/2, and is zero anywhere else, which is called a singleton membership function.

Fuzzy control rules in this study are:

IF $x_1(k)=X_1^{I_1}$ AND $x_2(k)=X_2^{I_2}$ AND $x_3(k)=X_3^{I_3}$ AND

$$x_4(k)=X_4^{I_4} \text{ THEN } \Delta u_1(k)=\Delta U_1^{J_1},\ \Delta u_2(k)=\Delta U_2^{J_2} \tag{2.8}$$

where

$$J_1=\frac{I_1+I_2}{2}+\delta_1\frac{I_3+I_4}{2}, \tag{2.9}$$

$$J_2=\delta_2\frac{I_1+I_2}{2}+\frac{I_3+I_4}{2}. \tag{2.10}$$

$\delta_2$ ($\delta_1$) should be -1 if increasing $y_1(k)$ ($y_2(k)$) causes increase of $y_2(k)$ ($y_1(k)$). Otherwise, $\delta_2$ ($\delta_1$) should be +1. $\delta_1$ ($\delta_2$) should be zero if $y_1(k)$ ($y_2(k)$) does not couple with $y_2(k)$ ($y_1(k)$). It should be noted that $J_m$ = -2, -1, 0, 1 or 2. Because there are total $2^4$ different combinations for the four scaled inputs, 16 control rules are needed, as listed in Table 1.

To evaluate AND's in the antecedents of the fuzzy control rules, probabilistic AND fuzzy logic is used. The resulting memberships, $\mu_{J_m}$'s, are assigned to $\Delta U_m^{J_m}$'s in the consequences of the fuzzy control rules. That is,

$$\mu_{J_m}=\mu_{I_1}(x_1)\cdot\mu_{I_2}(x_2)\cdot\mu_{I_3}(x_3)\cdot\mu_{I_4}(x_4). \tag{2.11}$$

The crisp incremental outputs, $\Delta u_m(k)$'s, are calculated by the center of gravity defuzzification algorithm:

$$\Delta u_m(k)=\frac{\sum\left(\mu_{J_m}\cdot J_m\cdot V\right)}{\sum\mu_{J_m}}, \qquad m=1,2. \tag{2.12}$$

New outputs of the fuzzy controller are computed by adding $\Delta u_m(k)$'s to the previous outputs, $u_m(k-1)$'s.

## III. ANALYTICAL STRUCTURE OF THE FUZZY CONTROLLER IN RELATION TO LINEAR PI CONTROLLER

Theorem.

The analytical structure of the two-input two-output fuzzy controller is the sum of two nonlinear PI controllers with variable gains:

$$\Delta u_1(k)=K_i^1e_1(k)+K_p^1\Delta e_1(k)+\delta_1\left(K_i^2e_2(k)+K_p^2\Delta e_2(k)\right) \tag{3.1}$$

$$\Delta u_2(k)=\delta_2\left(K_i^3e_1(k)+K_p^3\Delta e_1(k)\right)+K_i^4e_2(k)+K_p^4\Delta e_2(k) \tag{3.2}$$

where the variable proportional-gains ($K_p^i$'s) and integral-gains ($K_i^i$'s) are

$$K_p^1=\frac{\beta_1k_2H}{2L},\ K_p^2=\frac{\beta_2k_4H}{2L},\ K_p^3=\frac{\beta_1k_2H}{2L},\ K_p^4=\frac{\beta_2k_4H}{2L} \tag{3.3}$$

$$K_i^1=\frac{\beta_1k_1H}{2L},\ K_i^2=\frac{\beta_2k_3H}{2L},\ K_i^3=\frac{\beta_1k_1H}{2L},\ K_i^4=\frac{\beta_2k_3H}{2L} \tag{3.4}$$

and

$$\beta_1=\frac{16L^8-L^2(x_3^2L^2+x_3^2x_4^2+x_4^2L^2+L^4)(x_1x_2+L^2)}{32L^8-(x_1^2L^2+x_1^2x_2^2+x_2^2L^2+L^4)(x_3^2L^2+x_3^2x_4^2+x_4^2L^2+L^4)}$$

$$\beta_2=\frac{16L^8-L^2(x_1^2L^2+x_1^2x_2^2+x_2^2L^2+L^4)(x_3x_4+L^2)}{32L^8-(x_1^2L^2+x_1^2x_2^2+x_2^2L^2+L^4)(x_3^2L^2+x_3^2x_4^2+x_4^2L^2+L^4)}. \tag{3.5}$$

Proof. Based on Table 1 and the expressions (2.5), (2.6) and (2.11), 16 memberships for $\Delta U_1^{J_1}$ and $\Delta U_2^{J_2}$ can be obtained. Substitute the memberships into the defuzzification algorithm (2.12) and simplify the results. The structure of the fuzzy controller shown above can be attained. Obviously the gains change with $x_i(k)$'s. $\blacksquare$

There is a structural duality: exchanging $x_1(k)$ with $x_2(k)$ or $x_3(k)$ with $x_4(k)$ does not affect the expressions of $\beta_1$ and $\beta_2$ and therefore does not affect the expressions of $\Delta u_m(k)$'s.

## IV. CONCLUSIONS

In this paper, we have proven that the analytical structure of a two-input two-output fuzzy controller is the sum of two nonlinear PI controllers whose proportional-gains and integral-gains continuously change with process outputs. Based on the methodology developed in this paper, analytical structures of other MIMO fuzzy controllers with more inputs and outputs can also be derived. However, the task becomes more difficult as the number of inputs and outputs involved increases. For a MIMO fuzzy controller using errors and rate change of errors of κ process outputs as inputs, at least 4

$^{\kappa}$ control rules are needed. As a result, analytical derivation of the structure become very hard.

## REFERENCES


[1] S. Chiu, S. Chand, D. Moore, and A. Chaudhary, "Fuzzy logic for control of roll and moment for a flexible wing aircraft," *IEEE Control Systems Magazine*, 11, 42-48, 1991.

[2] M. M. Gupta, J. B. Kiszka, and G. M. Trojan, "Multivariable structure of fuzzy control systems," *IEEE Transactions on Systems, Man and Cybernetics*, 16, 638-656, 1986.

[3] J. B. Kiszka, M. M. Gupta, and G. M. Trojan, "Multivariable fuzzy controller under Godel's implication," *Fuzzy Sets and Systems*, 34, 301-321, 1990.

[4] C. C. Lee, "Fuzzy logic in control systems: Fuzzy logic controller," *IEEE Transactions on Systems, Man and Cybernetics*, 20, 404-435, 1990.

[5] K. S. Ray, "Application of fuzzy logic controller to a block-decoupled nonlinear steam generating unit (210 [MW])," *Control Theory and Advanced Technology*, 3, 343-374, 1987.

[6] C. W. Xu and Y. Z. Lu, "Decoupling in fuzzy systems: a cascade compensation approach," *Fuzzy Sets and Systems*, 29, 177-185, 1989.

[7] C. W. Xu, "Linguistic decoupling control of fuzzy multivariable processes," *Fuzzy Sets and Systems*, 44, 209-217, 1991.

[8] H. Ying, W. Siler and J. J. Buckley, "Fuzzy control theory: a nonlinear case," *Automatica*, 26, 513-520, 1990.


Table 1. Sixteen control rules. The antecedents are $X_i^{I_i}$ ($i = 1, 2, 3, 4$) and the consequences are $\Delta U_m^{J_m}$ ($m = 1, 2$).

| | IF $x_1(k)=X_1^{I_1}$ AND $x_2(k)= X_2^{I_2}$ AND $x_3(k)= X_3^{I_3}$ AND $x_4(k)=X_4^{I_4}$ THEN $\Delta u_1(k)=\Delta U_1^{J_1}$, $\Delta u_2(k)=\Delta U_2^{J_2}$, where $J_1 = \frac{I_1+I_2}{2}+\delta_1\frac{I_3+I_4}{2}$, $J_2 = \delta_2\frac{I_1+I_2}{2}+\frac{I_3+I_4}{2}$. | | | | | |
|---|---|---|---|---|---|---|
| rule No. | $I_1$ | $I_2$ | $I_3$ | $I_4$ | $J_1$ | $J_2$ |
| 1 | 1 | 1 | 1 | 1 | $1+\delta_1$ | $1+\delta_2$ |
| 2 | 1 | -1 | 1 | 1 | $\delta_1$ | 1 |
| 3 | -1 | 1 | 1 | 1 | $\delta_1$ | 1 |
| 4 | -1 | -1 | 1 | 1 | $-1+\delta_1$ | $1-\delta_2$ |
| 5 | 1 | 1 | -1 | -1 | $1-\delta_1$ | $-1+\delta_2$ |
| 6 | 1 | -1 | -1 | -1 | $-\delta_1$ | -1 |
| 7 | -1 | 1 | -1 | -1 | $-\delta_1$ | -1 |
| 8 | -1 | -1 | -1 | -1 | $-1-\delta_1$ | $-1-\delta_2$ |
| 9 | 1 | 1 | 1 | -1 | 1 | $\delta_2$ |
| 10 | 1 | -1 | 1 | -1 | 0 | 0 |
| 11 | -1 | 1 | 1 | -1 | 0 | 0 |
| 12 | -1 | -1 | 1 | -1 | -1 | $-\delta_2$ |
| 13 | 1 | 1 | -1 | 1 | 1 | $\delta_2$ |
| 14 | 1 | -1 | -1 | 1 | 0 | 0 |
| 15 | -1 | 1 | -1 | 1 | 0 | 0 |
| 16 | -1 | -1 | -1 | 1 | -1 | $-\delta_2$ |

# Minimizing Rules of Fuzzy Logic System by Using a Systematic Approach

Chuan-Chang Hung*and Benito Fernández R.†
NeuroEngineering Research & Development Laboratory
Mechanical Engineering Department
The University of Texas at Austin

*Abstract*— **Fuzzy logic systems have been extensively applied for control and decision systems. The control strategy may be viewed as a rule based design. All fuzzy rules contribute to some degree to the final inference or decision, however, some rules fired weakly do not contribute significantly to the final decision and may be "eliminated" (reduced). It is desired to minimize the rules in order to reduce the computation time to make a faster decision. Karnaugh maps have provided systematic methods for simplifying switching functions in logic design of binary digital systems. Based on this idea, this paper will present a novel method to help us reduce fuzzy rules. Comparisons will be made between systems utilizing reduced rules and original rules to verify the outputs. As a practical example of a nonlinear system, an inverted-pendulum will be controlled by minimum rules to illustrate the performance and applicability of this proposed method.**

## I. Introduction

During the past years, rule-based controllers combined with fuzzy logic and approximate reasoning have emerged as one of the most promising application of fuzzy set theory [1]. The rule-based system has been used to emulate and even surpass the decision-making of human reasoning in control processes. However, the design of a rule set of a rule-based system is initially developed to form an approximate or incomplete knowledge base for control strategy. Therefore, it is possible to design more rules than that required by the actual process, but still cannot simulate the entire control process. As a result, it will decrease the effectiveness of control processes due to the time required for firing all the rules and incomplete knowledge of the rules. It is desired to use the most effective rule set instead of implementing all possible rules, and provide a compensation for incomplete knowledge. Then, the computation time will be reduced and a control decision will be made more precisely.

Karnaugh maps have provided systematic methods for simplifying switching functions in logic design of binary digital systems [3]. Given a word description of the desired behavior of a logic network, we can write the output of the network as function of the input variables. This function can be specified as an algebraic expression. Then, we can facilitate algebraic manipulation by using the Minimum maps with boolean algebra. By doing this, we can obtain the minimum sum-of-products or minimum product-of-sums form of a function. Although fuzzy rule-based systems have been used to solve problems that involve ill-defined entities between true and false, we intend to investigate the similarity of fuzzy logic systems and binary digital systems and present a novel method to help us reduce the fuzzy rules based on Minimum maps. The minimum rule base being reduced is expected to have a similar performance as that of the original rule base, even in the case of incomplete knowledge. Therefore, the time required for computing the control signal will be deceased. We will begin the presentation of this systematic approach for reducing the fuzzy rules. Next, the comparison is made between systems utilizing the reduced rules and the original rule-base to verify the outputs. Finally, an inverted-pendulum is controlled by the minimum set of rules to illustrate the performance and applicability of the proposed method.

## II. A Systematic Approach

In order to derive the algebraic simplification of switching functions, we need to translate word descriptors of the rules into like algebraic equations. Fuzzy logic allows premises and conclusions of the rules to be fuzzy propositions. The truth values of propositions are expressed as linguistic variables with varying degrees of truth. Therefore, the linguistic value of fuzzy sets are converted into a binary representation in order to apply Minimum maps. After the above transformation, given a set of fuzzy rules (completely or incompletely specified), we can display

*Graduate Student
†Assistant Professor

0-7803-0614-7/93/$03.00 

them on a *minimum fuzzy map*. The fuzzy rules may be represented as a minterm expression (standard sum of products), a maxterm expression (standard product of sums), or as an algebraic form. As a result, the minimum sum-of-products or minimum product-of-sums form rules that can be derived from the map.

In the following section, we will elaborate the procedure of reducing the fuzzy rules by our systematic approach based on minimum fuzzy maps.

*A. The Procedure of Reducing Rules*

1. In order to apply Karnaugh maps to represent fuzzy rules, the linguistic values (fuzzy sets) of input and output variables in [0,1] must be regarded as crisp values (0 or 1) without considering the overlap between each fuzzy set.

2. Discretize the input and output variables as n-bit integers, having $2^n$ possible values. Hence, each linguistic values can be represented as one binary value.

3. While displaying the rules on the map, we can minimize the n-variable functions using Karnaugh maps based on different values of only one variable $(Y(True) + Y^{'}(False) = 1)$. For instance, the minimum sum-of-products expressions can be derived by using the theorem

$$XY^{'} + XY = X \qquad (1)$$

However, each variable in a fuzzy rule has multiple linguistic values in terms of the multiple fuzzy sets. In addition, the simplified form is indicated by looping the corresponding output value on the map, not just the corresponding $1's$ on a conventional map that has only a true or false value. Therefore, the effects of the variables are deleted according to the changes of all linguistic values in the loop. Here, the eq.(1) is modified as follows for simplifying the fuzzy rules:

$$X_1Y_1 + X_1Y_2 + .... + X_1Y_n = X_1 \qquad (2)$$

where, $X_1$ is the one of linguistic values of $X$ variable. $Y_1, Y_2, ..., Y_n$ are n linguistic values of $Y$ variable

Meanwhile, outputs with the same binary value can be reduced into sets formed by the minimum sum-of-products or minimum product-of-sum. In what follows, there are two cases that show this simplification:

(a) Completely-specified rules:

| B | | A: $A_1$ 00 | $A_2$ 01 | $A_3$ 10 | $A_4$ 11 |
|---|---|---|---|---|---|
| $B_1$ | 00 | | 10 | | |
| $B_2$ | 01 | 01 | 01 | 01 | 01 |
| $B_3$ | 10 | | 00 | | |
| $B_4$ | 11 | | 1 0 | | |

$$\begin{aligned} C &= A_2B_1 + A_1B_2 + A_2B_2 + A_3B_2 \\ &\quad + A_4B_2 + A_2B_3 + A_2B_4 \qquad (3) \\ &= A_2B_1 + B_2 + A_2B_3 + A_2B_4. \end{aligned}$$

where, $A$ and $B$ are input variables, $C$ is the output variable. $A_1, A_2, A_3$, and $A_4$ are linguistic values of variable $A$, they are represented as binary numbers such as $A_1 = 00, A_2 = 01, A_3 = 10$, and $A_4 = 11$. $B_1, B_2, B_3$, and $B_4$ are linguistic values of variable $B$, they are represented as binary numbers such as $B_1 = 00, B_2 = 01, B_3 = 10$, and $B_4 = 11$. In the entity of the map, $00, 01, 10$, and $11$ are output linguistic values for $C_1, C_2, C_3$, and $C_4$ respectively.

(b) Incompletely-specified rules:

| B | | A: $A_1$ 00 | $A_2$ 01 | $A_3$ 10 | $A_4$ 11 |
|---|---|---|---|---|---|
| $B_1$ | 00 | | 10 | | |
| $B_2$ | 01 | 01 | × | 01 | 01 |
| $B_3$ | 10 | | 10 | | |
| $B_4$ | 11 | | 10 | | |

$$\begin{aligned} C &= A_1B_2 + A_3B_2 + A_4B_2 + A_2B_2(\times) \\ &\quad + A_2B_1 + A_2B_3 + A_2B_4 \qquad (4) \\ &= B_2 + A_2. \end{aligned}$$

where, the required minterm are indicated by a binary number on the map,and the don't care minterms are indicated by ×'s. When choosing terms to form the minimum sum of products, all the same binary number must be covered, but the ×'s are only used if they will simplify the resulting expression.

4. Define the membership function to fuzzify the linguistic values of input and output variables for new expression.

5. Apply the concept of fuzzy logic [4] for new rules to execute the inference procedure and to defuzzify the output in order to obtain the output value.

## III. Examples

### *A. One bit problem: Two fuzzy sets*

The fuzzy system has two inputs $A$ and $B$, and one output $C$. We discretize each input and output spaces into two regions in terms of two linguistic values which are defined as Positive Small (PS) and Negative Small (NS). The fuzzy rules are listed as followed:

If A is PS and B is NS, then C is PS;
If A is PS and B is PS, then C is PS;
If A is NS and B is PS, then C is NS;
If A is NS and B is NS, then C is NS.

**First step:**
The linguistic value of input and output variables are approximated as crisp values without considering overlapping. Each variable is 1 bit code, so the value of the variable can be represented as binary number: 1 is NS and 0 is PS.

**Second Step:**
Translate the word description of the *fuzzy rules* into *algebraic expression* and display them in the Karnaugh map, e.g.

$$C = A_1B_2 + A_1B_1 + A_2B_1 + A_2B_2 \tag{5}$$

| | | | (PS) $A_1$ 0 | (NS) $A_2$ 1 |
|---|---|---|---|---|
| (PS) | $B_1$ | 0 | 0 | 1 |
| (NS) | $B_2$ | 1 | 0 | 1 |

where, $A$ and $B$ are input variables, $C$ is the output variable. $A_1$ and $A_2$ are linguistic values of variable $A$, they are represented as binary numbers: $A_1$ = 0= PS, $A_2$ = 1= NS. $B_1$ and $B_2$ are linguistic values of variable $B$, they are represented as binary numbers: $B_1$ = 0= PS, $B_2$ = 1= NS. In the entity of the map 0 and 1 are output linguistic values for $C_1$ and $C_2$

**Third Step:**
A 0 in column $A_1$ row $B_1$ of above Figure indicates that $A_1B_1$ is a minterm of $C$. Similarly, a 1 in column $A_1$ row $B_2$ indicates that $A_1B_2$ is a minterm. Minterms in adjacent squares of the map can be combined since they differ in only one variable. Thus, $A_1B_1$ and $A_1B_2$ combine to form $A_1$. Likewise, the $A_2B_1$ and $A_2B_2$ can be combined to form $A_2$ by looping the corresponding $0's$ on the map. Therefore, the algebraic simplification of a completely-specified set of rules can be derived as:

$$C = A_1 + A_2 \tag{6}$$

**Fourth Step:**
Assign the triangle membership function in Fig. 1 to the linguistic values (PS, NS) of $A$ variable and transfer the algebraic expression to word description of fuzzy rules. They are described as:

If A is PS, then C is PS; If A is NS, then C is NS

**Fifth Step:**
Apply the concept of fuzzy logic to get output value

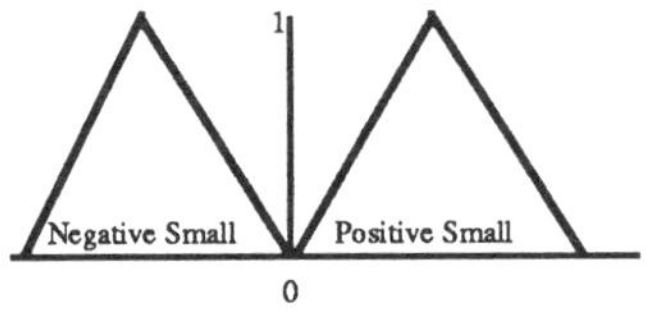


Fig. 1 The Membership Function of Input Variables

**Validation:**
Suppose the input $A$ and $B$ and $C$ output spaces are the same, the real interval [-1,1]. The $A$, $B$ and $C$ are partitioned into the same two fuzzy sets:

PS = [0,1]
NS = [-1,0]

Consider the input data pair $A = 0.75$, $B = 0.5$ activates the original fuzzy rules. We combine the antecedent membership value with minimum with the conjunctive AND. Then, the correlation-minimum inference procedure [2] activates the output fuzzy set to some degree. We can then compute the fuzzy centroid with the following equation. To determine the specific output value of the fuzzy set, the control actions are weighted by their membership values and averaged. This weighted control action is known to provide a sluggish, yet more robust control.

$$\bar{C} = \frac{\sum_{j=1}^{p} w_j C_j}{\sum_{j=1}^{p} w_j} \tag{7}$$

where, $\bar{C}$ is the specific output value, and $w_j$ is the scalar activation value of the $j^{th}$ fuzzy rule's consequence which is the minimum of the n antecedent membership values, and $C_j$ is the $j^{th}$ value in the discretized output universe of discourse $C=\{C_1, C_2, ...C_j, ...C_n\}$.
The specific $C$ value is calculated from the original rules to be 0.5. Similarly, the value of output is 0.5 from the new minimum rules. Hence, we can conclude that we can apply the minimum rules and obtain the same output values as when applying the original set of rules in complete knowledge base . However, the minimum set of rules provides less computation time than the original ones.

*B. Two bits problem: Three fuzzy sets*

The fuzzy system has the same input and output as the previous example. However, each input and output spaces are discretized into three regions in terms of three linguistic values which are defined as Zero (ZE), Positive Small (PS) and Negative Small (NS). The fuzzy rules are listed as follows:

If A is ZE and B is ZE, then C is ZE;
If A is ZE and B is PS, then C is NS;
If A is ZE and B is NS, then C is PS;
If A is PS and B is NS, then C is PS;
If A is NS and B is PS, then C is PS.

**First step:**
The linguistic value of input and output variables as crisp values without considering overlapping. Each variable is a two bit code, so the value of the variable can be represented as binary number: 00 is ZE, 01 is PS, 10 is NS and 11 are defined as don't care values (NIL) (NIL: not specified linguistic value in variable domain)

**Second Step:**
Translate the word description of the *fuzzy rule* into *algebraic expression* and display in the Karnaugh map, e. g.

$$C = A_1B_1 + A_1B_2 + A_1B_3 + A_2B_1 + A_3B_2 \quad (8)$$

| | | | (ZE) $A_1$ 00 | (PS) $A_2$ 01 | (NS) $A_3$ 10 | NIL $A_4$ 11 |
|---|---|---|---|---|---|---|
| (ZE) | $B_1$ | 00 | 00 | | | |
| (PS) | $B_2$ | 01 | 10 | ×× | 10 | ×× |
| (NS) | $B_3$ | 10 | 01 | 01 | ×× | ×× |
| NIL | $B_4$ | 11 | | | | |

where, $A$ and $B$ are input variables, $C$ is the output variable. $A_1$, $A_2$, $A_3$ and $A_4$ are linguistic values of variable $A$, they are represented as binary numbers such as $A_1$ = 00= ZE, $A_2$ = 01= PS, $A_3$ = 01= NS, $A_4$ = 11 = NIL. $B_1$, $B_2$, $B_3$ and $B_4$ are linguistic values of variable $B$, they are represented as binary numbers such as $B_1$ = 00= ZE, $B_2$ = 01= PS, $B_3$ = 10= NS, $B_4$ = 11= NIL. In the entity of the map 00, 01, 10 and ×× are output linguistic values for $C_1$, $C_2$, $C_3$ and $C_4$ respectively.

**Third Step:**
When choosing terms to form the minimum sum-of-products, all the outputs with similar values must be converted, but the ××'s are only used if they simply the resulting expression. $A_1B_2$ and $A_3B_2$ with don't-cares combining to form $B_2$ by looping the corresponding $10's$ on the map. Likewise, the $A_1B_3$ and $A_2B_3$ can be combined to form $B_3$ by looping the corresponding $01's$ on the map. Therefore, the algebraic simplification of an incompletely-specified rule can be derived as:

$$C = A_1B_1 + B_2 + B_3 \quad (9)$$

**Fourth Step:**
Assign the triangle membership function with overlapping between fuzzy sets in Fig. 2 to the linguistic values (ZE, PS, NS) of variable $A$ and transfer the algebraic expression to word description of fuzzy rules. They are described as:

If A is ZE and B is ZE, then C is ZE;
If B is PS, then C is NS; If B is NS, then C is PS.

**Fifth Step:**
Apply the concept of fuzzy logic to get output value.

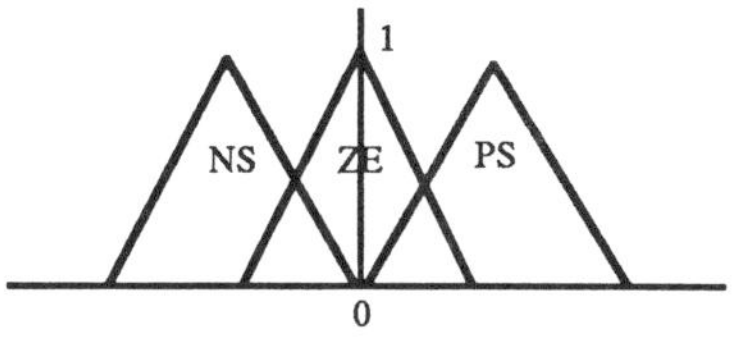


Fig. 2 The Membership Function of Input Variables

**Validation**
Suppose the input $A$ and $B$ and $C$ output spaces are the same, the real interval [-1,1]. The $A$, $B$ and $C$ are partitioned into the same two fuzzy sets:

ZE = [-0.2, 0.2]
PS = [0,1]
NS = [-1,0]

Consider the input data pair $A = 0.1$, $B = 0.01$ activating the original fuzzy rules. The specific $C$ value is calculated at -0.0833 from Eq. 7. Similarly, the value of output is -0.0833 from the new minimum rules. Moreover, consider the values of input data to $A = 0.5$, $B = 0.3$. The specific $C$ value cannot be obtained from the original fuzzy rules, because total weights of all activated rules are zero in the denominator of Eq. 7. However, the value of output can be calculated to be -0.5 from the new minimum rules. Hence, we conclude that we can apply the minimum rules in getting more complete results while the original rules lack complete information to compute the output value.

## IV. Control Case

In this section, the control problem for the inverted-pendulum is used to interpret how the 3-bits concept can be used to reduce the fuzzy rules by forming a Karnaugh map. This control system has two inputs $e$(the error of the pendulum's angle from the balanced position) and $\dot{e}$ (the error of the pendulum's angular velocity) and one output $u$(torque). We discretize each input and output spaces into five regions in terms of five linguistic values which is defined as ZERO (ZE), Positive Small (PS), Positive Large (PL), Negative Small (NS) and Negative Large (NL). The fuzzy rules are listed as follows:

If $e$ is ZE and $\dot{e}$ is NS, then $u$ is PS;
If $e$ is ZE and $\dot{e}$ is NL, then $u$ is PL;
If $e$ is ZE and $\dot{e}$ is PS, then $u$ is NS;
If $e$ is ZE and $\dot{e}$ is PL, then $u$ is NL;
If $e$ is ZE and $\dot{e}$ is ZE, then $u$ is ZE;
If $e$ is NS and $\dot{e}$ is ZE, then $u$ is PS;
If $e$ is NL and $\dot{e}$ is ZE, then $u$ is PL;
If $e$ is PS and $\dot{e}$ is ZE, then $u$ is NS;
If $e$ is PL and $\dot{e}$ is ZE, then $u$ is NL;
If $e$ is PS and $\dot{e}$ is NS, then $u$ is NS;
If $e$ is PL and $\dot{e}$ is NS, then $u$ is NL;
If $e$ is PL and $\dot{e}$ is PL, then $u$ is NL;
If $e$ is NS and $\dot{e}$ is PS, then $u$ is PS;
If $e$ is NL and $\dot{e}$ is PS, then $u$ is PL;
If $e$ is NL and $\dot{e}$ is PL, then $u$ is PL.

**First step:**
The linguistic value of input and output variables as crisp value without considering overlapping. Each variable is three bit code, so the value of the variable can be represented as binary numbers: 010 is ZE, 011 is PS, 100 is PL, 001 is NS and 000 is NL. The other binary numbers 101, 110 and 111 are defined as don't care values (NIL) of the linguistic variable.

**Second Step:**
Translate the word description of the *fuzzy rule* into and *algebraic expression* display in the Karnaugh map, e.g.

$$\begin{aligned} u &= e_1\dot{e}_3 + e_1\dot{e}_4 + e_1\dot{e}_5 + e_2\dot{e}_3 + e_2\dot{e}_4 + e_3\dot{e}_1 \\ &\quad + e_3\dot{e}_2 + e_3\dot{e}_3 + e_3\dot{e}_4 + e_3\dot{e}_5 + e_4\dot{e}_2 + e_4\dot{e}_3 \\ &\quad + e_5\dot{e}_1 + e_5\dot{e}_2 + e_5\dot{e}_3 \end{aligned} \tag{10}$$

| | (NL) $e_1$ 000 | (NS) $e_2$ 001 | (ZE) $e_3$ 010 | (PS) $e_4$ 011 | (PL) $e_5$ 100 | nil $e_6$ 101 | nil $e_7$ 110 | nil $e_8$ 111 |
|---|---|---|---|---|---|---|---|---|
| (NL) $\dot{e}_1$ 000 | x x x | x x x | 100 | x x x | 000 | | | |
| (NS) $\dot{e}_2$ 001 | x x x | x x x | 011 | 001 | 000 | | | |
| (ZE) $\dot{e}_3$ 010 | 100 | 011 | 010 | 001 | 000 | | | |
| (PS) $\dot{e}_4$ 011 | 100 | 011 | 001 | x x x | x x x | | | |
| (PL) $\dot{e}_5$ 100 | 100 | x x x | 000 | x x x | x x x | | | |
| nil $\dot{e}_6$ 101 | | | | | | | | |
| nil $\dot{e}_7$ 110 | | | | | | | | |
| nil $\dot{e}_8$ 111 | | | | | | | | |

where, $e$ and $\dot{e}$ are input variables, $u$ is output variable. $e_1$, $e_2$, $e_3$, $e_4$, $e_5$, $e_6$, $e_7$, $e_8$ are linguistic values of variable $e$, they are represented as binary numbers such as $e_1 = 000 =$ NL, $e_2 = 001$=NS, $e_3 = 010 =$ ZE, $e_4 = 011 =$ PS, $e_5 = 100 =$ PL, $e_6 = 100 =$ NIL, $e_7 = 110 =$ NIL and $e_8 = 111 =$ NIL. $\dot{e}_1$, $\dot{e}_2$, $\dot{e}_3$, $\dot{e}_4$, $\dot{e}_5$, $\dot{e}_6$, $\dot{e}_7$, $\dot{e}_8$ are linguistic values of variable $\dot{e}$, they are represented as binary numbers such as $\dot{e}_1 = 000$=NL, $\dot{e}_2 = 001$=NS, $\dot{e}_3 = 010 =$ ZE, $\dot{e}_4 = 011 =$ PS, $\dot{e}_5 = 100$= PL, $\dot{e}_6 = 100 =$ NIL (don't care value), $\dot{e}_7 = 110 =$ NIL and $\dot{e}_8 = 111 =$ NIL. In the entity of the map 000, 001, 010, 011, 100, and 101 are output linguistic values for $u_1$= NL, $u_2$= NS, $u_3$ = ZE, $u_4$ = PS and $u_5$ = PL respectively and $\times\times\times$ is the don't care value.

**Third Step:**
To simplify the expression, $e_1\dot{e}_3$, $e_1\dot{e}_4$ and $e_1\dot{e}_5$ with don't cares combining to form $e_1$ by looping the corresponding same output value $100's$ on the map. Likewise, the $e_2\dot{e}_3$ and $e_2\dot{e}_4$ can be combined to form $e_2$, $e_4\dot{e}_2$ and $e_4\dot{e}_3$ can be combined to form $e_4$ and the $e_5\dot{e}_1$, $e_5\dot{e}_2$ and $e_5\dot{e}_3$ can be combined to form $e_5$ by looping the corresponding $010's$, $001's$ and $000's$ on the map respectively. Therefore, the algebraic simplification of an incompletely-specified rule can be derived as:

$$\begin{aligned} u &= e_1\dot{e}_3 + e_2\dot{e}_3 + e_3\dot{e}_1 + e_3\dot{e}_2 + e_3\dot{e}_3 + e_3\dot{e}_4 \\ &\quad + e_3\dot{e}_5 + e_4\dot{e}_3 + e_5\dot{e}_3 \end{aligned} \tag{11}$$

**Fourth Step:**
Assign the triangle membership function with overlapping between fuzzy sets in Fig. 3 to the linguistic values (ZE, PS, NS) of variable $A$ and transfer the algebraic expression to word description of fuzzy rules. They are described as:

If $e$ is ZE and $\dot{e}$ is NS, then $u$ is PS;
If $e$ is ZE and $\dot{e}$ is NL, then $u$ is PL;
If $e$ is ZE and $\dot{e}$ is PS, then $u$ is NS;
If $e$ is ZE and $\dot{e}$ is PL, then $u$ is NL;
If $e$ is ZE and $\dot{e}$ is ZE, then $u$ is ZE;
If $e$ is NS, then $u$ is PS; If $e$ is NL, then $u$ is PL;
If $e$ is PS, then $u$ is NS; If $e$ is PL, then $u$ is NL.

**Fifth Step:**
Apply the concept of fuzzy logic to get output value.

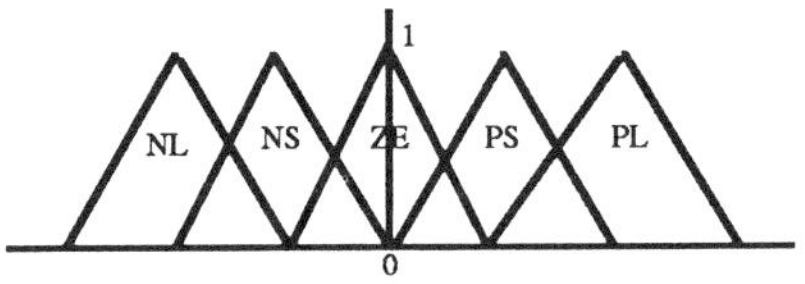


Fig. 3 The Membership Function of Input Variables

**Comparison:**
This nonlinear system is simulated to compare the performance of using the original rules and the minimum rules in controlling the balance of the inverted pendulum. The initial condition for the pendulum is set on $e = 0.2$ rad and $\dot{e} = 0.0$ rad/sec. The response is shown in Fig. 4 and Fig. 5 respectively. The figures show that the controller of the pendulum from the original 15 rules and the reduced 9 rules have similar response. Next, we change the initial condition to $e = 0.5$ rad and $\dot{e} = 0.0$ rad/sec. The results show that the response of the reduced rules become slightly faster than the original rules. Based on these results, we can see that the reducing rules can provide similar performance as that of the original rules in the control process. Moreover, the computation time is obviously shorter by using the reduced rules and it alleviates the complexity of implementation for the rule base.

We plotted the 3-D control surface to visualize the controller of the two set of rules.The control surfaces show the control u (vertical axes) corresponding to all combinations of values of the two input state variables error and error rate (horizon plane). We find the control surface of the reduced set of rules smoother, that is, it reflects that fact that the rule-based controller uses fewer rules than the original rule-based controller.

## V. Conclusion

In this paper, we interpret and demonstrate the applicability of using Karnaugh maps to reduce rules base controller. We compare the output between the reduced set and the original set rules. We conclude that the expensive computation time will be reduced by using the minimum rules. Meanwhile, the reduced rules have similar performance as the original rules. As a result, this approach can provide a low-cost and robust means of design of the fuzzy rule-based controller. We would like to cautious that the results presented here are preliminary and a formal treatise of the matter is forthcoming. Limitations are obvious in the case of incomplete knowledge of the rules. The main goal of the study was to venture into "Conceptual" generalization using fuzzy logic.

## References


[1] Gupta, M. M. and T. Yamakawa, "Fuzzy Logic in Knowledge-Based Systems, Decision and Control", North-Holland, 1988.

[2] Kosko, B.,"Neural Networks and Fuzzy Systems ", Printice Hall, Englewood Cliffs, NI07632, 1992.

[3] Roth, Charles H.,"Fundamentals of Logic Design", Third Edition, West Publishing Co. 1985.

[4] Zadeh, L. A. "Outline of a new approach to the analysis complex systems and decision processes',*IEEE Transactions on Systems, Man, and Cybernetics*, pp. 28-44, 1973.


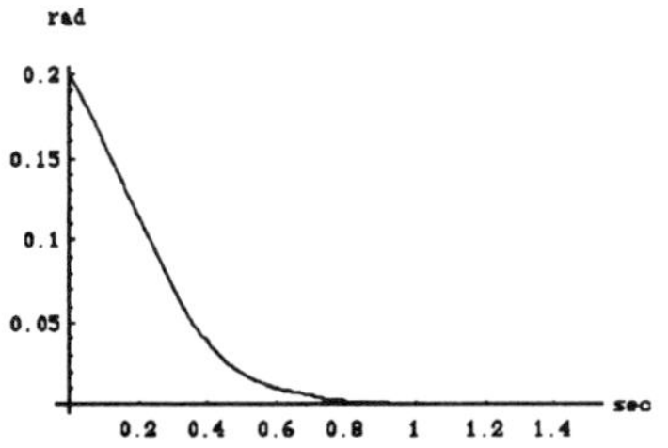


Fig. 4 The angle response from the original rules

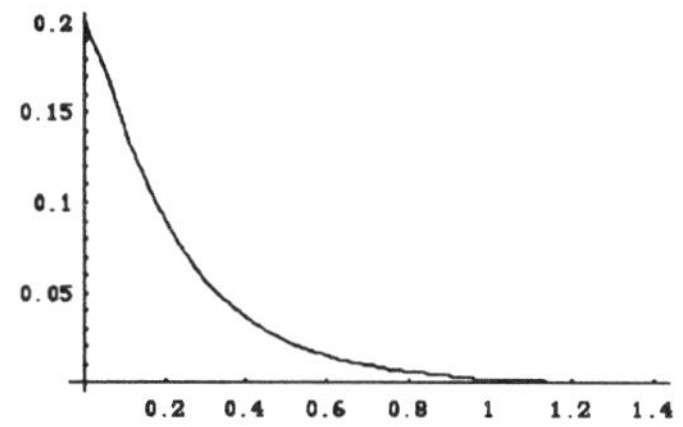


Fig. 5 The angle response from the reduced rules

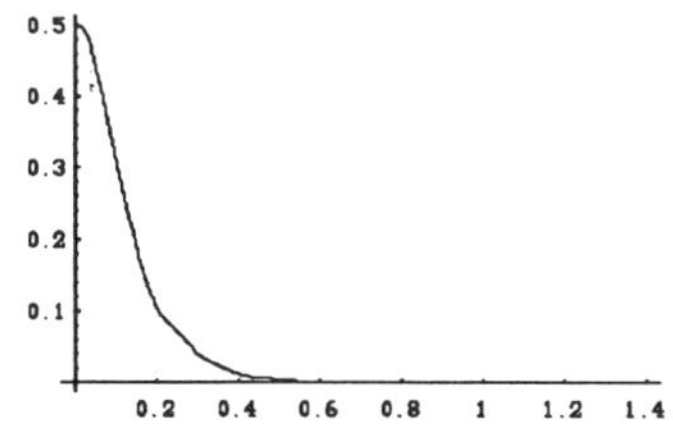


Fig. 6 The angle response from the original rules

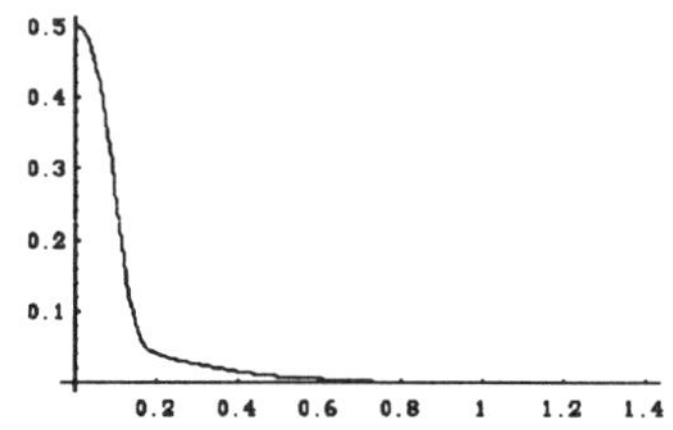


Fig. 7 The angle response from the reduced rules

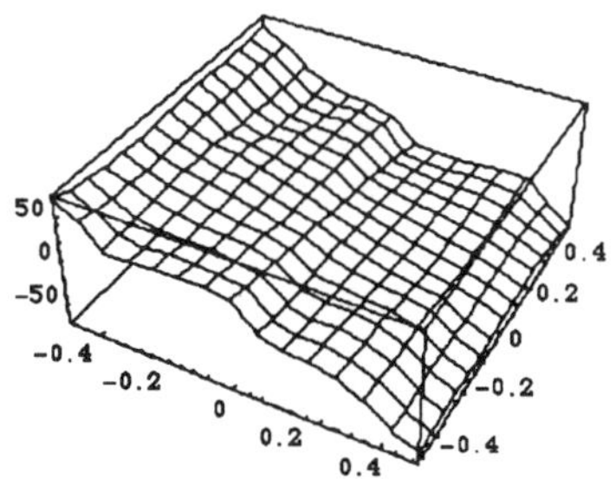


Fig. 8 The control surface from the original rules

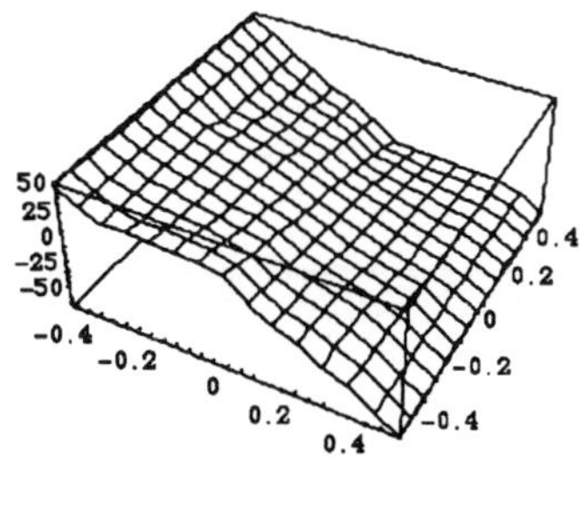


Fig. 9 The control surface from the reduced rules

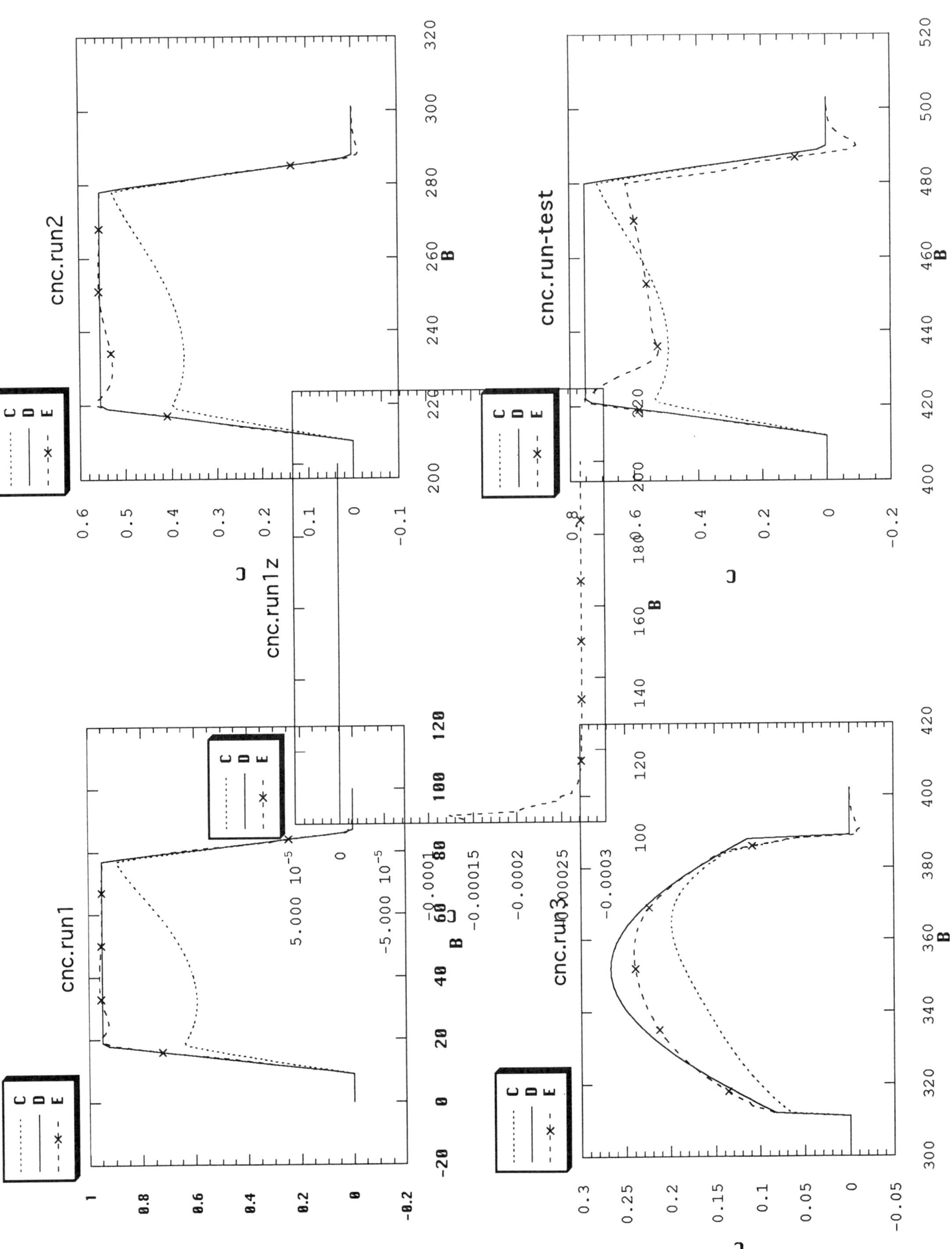

cnc.run1
cnc.run2
cnc.run1z
cnc.run3
cnc.run-test
C
D
E
B

# Trainable Fuzzy and Neural-Fuzzy Systems for Idle-Speed Control

L. A. Feldkamp and G. V. Puskorius
Research Laboratory, Ford Motor Company
Suite 1100, Village Plaza
23400 Michigan Avenue
Dearborn, Michigan 48124
lfeldkam@smail.srl.ford.com

***Abstract*— We describe the use of a neural-network based procedure to train fuzzy or hybrid neural-fuzzy systems as vehicle idle-speed controllers. Simulation with a nonlinear model containing a significant delay is used, and we attempt to simulate the effects of realistic sampling and controller update frequencies. The present treatment may be regarded as a step toward on-line training with an actual system. The fuzzy system has a parameterized form similar to that described previously, allowing us to use methods identical to those we use for training neural networks. We illustrate the results of training by imposing various torque disturbances and showing the controller actions and the response of the system.**

## I. Introduction

The control of the speed of an automobile's engine under idle conditions is an old problem, but one that continues to be interesting and important. Modern passenger-car engines are usually controlled by computer to yield a combination of performance, economy, and emissions that is far superior to that of engines controlled by earlier methods. The whole of engine control is fertile ground for the application of new techniques employing neural networks and fuzzy logic. The idle-speed control (ISC) problem seems a logical first application of these methods, since ISC is comparatively isolated from other aspects of the overall engine control problem. At the same time, idle-speed control is a challenging problem, as it represents an attempt to regulate, against both anticipated and unanticipated disturbances, a highly nonlinear system operating well away from its optimal region. Because of time delays, the system state vector is incompletely known. Furthermore, the length of such delays can vary with time, making an analytical treatment very difficult. Finally, though the system itself may be regarded as continuous, information from it is available only in sampled form and control may be exercised only at discrete times.

Traditional approaches to the problem involve building system models whose parameters can be determined by a series of measurements. Even if the model is chosen to have nonlinearities, the controller design has usually been based on linearization of the model about an appropriate operating point. Delay elements in the model can either be explicitly ignored, as in PID control, or can be treated by posing the controller in state-variable form, in terms of both externally observed system states and appropriately defined internal states. Since the latter are not directly observable, it is asserted that they can be reconstructed by an *observer*, such as a Kalman filter. It is probably accurate to say that most production idle-speed controllers do not employ this last sophistication.

Recent papers [1] present a different approach using a fuzzy proportional-integral (PI) controller. The authors made use of simulation studies for a large part of the design process, using experimentation with an actual system largely for verification. The two-dimensional input space was divided into 64 regions, each of which corresponds to a rule of a max-min fuzzy system. Adequate performance was observed when all but 16 rules had been eliminated. Comparison to an LQ optimal controller derived by one of the authors [2] suggested that the fuzzy controller was slightly inferior. The difference between these should not be overinterpreted, however, since the LQ controller had the benefit of more information (the manifold pressure in addition to the speed error) and was subjected to optimization in terms of a performance function, a process that does not seem to have been applied in the design of the fuzzy controller.

A full approach to the ISC problem would involve simulation studies followed by on-line training of both fuzzy and neural controllers. In the present paper we describe simulation-based training of fuzzy controllers. We make use of a representative model that incorporates nonlinearity and time delay. Because we view the model as a convenient means of preparing to deal with the physical system, we attempt to simulate some of the harsh real-

0-7803-0614-7/93$03.00 ©1993IEEE

ity that may be anticipated. For example, we impose restrictions on the quality and timeliness of information supplied to the controller and on when control actions can be computed and supplied to the system. We carry out training on the fuzzy system using the same procedure that we have applied to the training of both neural and fuzzy systems for other problems, such as vehicle active suspension [3] and anti-lock braking [4]. During training, we minimize a quadratic performance function, as in conventional optimal control. We discuss the controller performance that is obtained when such fuzzy systems are provided inputs from fixed or trained preprocessing layers. Whether fixed or trained, the preprocessing layer is recurrent, i.e., it uses both current and past information from the system being controlled. Elsewhere [5] we consider in greater detail the efficacy of recurrent neural architectures for the ISC problem.

We have organized the remainder of this paper as follows. Section 2 contains a description of the engine model under idle conditions and details the way we interact with the model during the training process. In Section 3 we describe various controller architectures and in Section 4 consider the process of training. In Section 5 we discuss the results of training and make comparisons to results for a conventional control architecture and relate this work to that on recurrent neural networks. In Section 6 we provide some final comments.

## II. Engine Model

The dynamic engine model employed in this study was derived from steady-state engine map data and empirical information by Powell and Cook [6], with revisions as described by Vachtsevanos et al. [7]. The engine model parameters are for a 4-cylinder fuel injected engine. The model is a two-state, two-input system. The states are manifold pressure $P$ in kPa and engine speed $N$ in rpm; the control inputs are throttle angle $\theta$ and spark advance $\delta$ in degrees. Disturbances act on the engine in the form of load torques $T_d$ in N-m.

The evolution of the system is described by the following set of coupled equations:

$$
\begin{aligned}
\dot{P} &= k_P(\dot{m}_{ai} - \dot{m}_{ao}), \; k_P \;=\; 42.40 \\
\dot{N} &= k_N(T_i - T_L), \; k_N \;=\; 54.26 \\
\dot{m}_{ai} &= (1 + 0.907\theta + 0.0998\theta^2)g(P) \\
g(P) &= \begin{cases} 1 & P < 50.6625 \\ 0.0197(101.325P - P^2)^{\frac{1}{2}} & P \geq 50.6625 \end{cases} \\
\dot{m}_{ao} &= -0.0005968N - 0.1336P \\
&\quad + 0.0005341NP + 0.000001757NP^2 \\
m_{ao} &= \frac{\dot{m}_{ao}(t-\tau)}{120N}, \; \tau \;=\; \frac{45}{N} \\
T_i &= -39.22 + 325024m_{ao} - 0.0112\delta^2 \\
&\quad + 0.000675\delta N(2\pi/60) + 0.635\delta \\
&\quad + 0.0216N(2\pi/60) - 0.000102N^2(2\pi/60)^2 \\
T_L &= (N/263.17)^2 + T_d \, .
\end{aligned}
$$

For notational simplicity, we have suppressed explicit dependence on time in these equations, except in the case of the delayed $\dot{m}_{ao}$. The time delay $\tau$ is treated here as a lumped quantity that represents the effect of the time between induction of the fuel mixture into a particular cylinder and the corresponding power stroke. The value of $\tau$ given above corresponds to an induction-power lag of 270 degrees.

Controller design with the neural network methods used here depends almost entirely on observing the output behavior of the system for actively determined input patterns. Since the above differential equations exhibit "stiff" behavior in certain regions of operation, a backward Euler scheme with a step size of 1 ms is used. Control commands are throttle angle $\theta$ in the range 5–25 degrees and spark advance $\delta$ in the range 10–45 degrees. Measurements of manifold pressure $P$ and engine speed $N$ are assumed to be available 4 times per engine revolution, with the current engine position being calculated from the engine speed. Control commands are computed at 20 ms intervals but are applied twice per engine revolution at fixed values of crank angle. With the assumed timing, crank rotation of at least 90 degrees (20 ms at 750 rpm) occurs before a measurement can be reflected in an applied control command. Measured manifold pressure $P_m$ is corrupted by Gaussian noise of zero mean and variance 1 kPa$^2$, while measured engine speed $N_m$ is corrupted by Gaussian noise of zero mean and variance of 6.25 rpm$^2$. Note that while control actions are computed at uniform time intervals, state variables are measured and control is asserted uniformly in engine position. Disturbances, ranging from 0 to 61 N-m, may begin and end at any time within the resolution implied by the 1 ms update interval for the differential equations. This means that changes in disturbance input occur asynchronously with the application of control and the measurement of system outputs. The range of applied disturbances is relatively large, apparently about twice as large as considered in [1].

The primary goal of ISC is to maintain a constant engine speed while the system experiences unobserved disturbances. For the engine model described above, a set point of 750 rpm is desired. The two controls affect engine speed rather differently. The throttle command $\theta$ has a large dynamic range (thousands of rpm), but its effect is delayed by a time inversely proportional to engine speed (for a steady state engine speed of 750 rpm, the delay is 60 ms). On the other hand, the spark advance $\delta$ has an immediate effect on the engine speed, but its dynamic range is roughly an order of magnitude smaller than that of the throttle command. In addition to maintaining engine speed at 750 rpm, we must respect secondary criteria. For example, we would like to maximize fuel efficiency while simultaneously minimizing vehicle vibrations. A means of increasing fuel efficiency is to operate the engine with spark angle advanced. For the engine model described above, the timing that results in maximum brake torque (MBT) for a steady-state speed of 750 rpm is found to be 30.9 degrees of spark advance. However, the spark advance must be retarded relative to this value to leave room to respond to torque load disturbances. We choose, somewhat arbitrarily, the nominal value $\delta = 22.9$ degrees (8 degrees retarded from MBT) to allow rapid system response to torque disturbances while maintaining a reasonable level of fuel efficiency.

## III. Controller Architectures

We have previously described [8] our procedure for tuning or training max-min fuzzy and hybrid neural-fuzzy systems. In brief, we parameterize membership functions in terms of analytic functions and replace MAX and MIN operations with soft approximations. This results in a computational structure that is completely differentiable and hence can be optimized (trained) in the same way as a neural network. For convenience, we made the following minor changes: 1) the analytic form for soft minimum given in [8] was replaced by the form used by Berenji [9], $smin(x, y) = \frac{xe^{-kx}+ye^{-ky}}{e^{-kx}+e^{-ky}}$, which maintains order invariance when generalized to more than two arguments. 2) The output membership functions were taken to be singletons with weighted average defuzzification. This modification simplifies and speeds up the training process.

We experimented with several possibilities for inputs to the fuzzy system. In the first we constructed estimates of speed error proportional and integral terms. This is the approach taken by Abate and Dosio [1]. After training several different fuzzy systems, we concluded that these inputs contain insufficient information for satisfactory control of the present system. We next supplied the fuzzy system with five inputs: speed error, integral of speed error, difference in speed error between consecutive time steps, manifold pressure, and difference in pressure between time steps. We term this PID/PD information. We derived these five inputs, as described in [5], by means of a preprocessing layer of seven recurrent nodes with a prespecified connection pattern and weights. This fixed preprocessing layer is also used for the training of a conventional PID/PD linear controller. Finally, we allowed the weights of the preprocessing layer, initialized to PID/PD values, to be trained along with the parameters of the fuzzy system.

The consequences of dealing with more than two or three inputs to a fuzzy system, whether trained or not, poses some difficulties that have nowhere been adequately explored. In our view, at least three approaches can be taken. The first is to construct a "fuzzy lookup table" by fully tiling the relevant portion of the input space using, for example, Gaussian input membership functions. The latter could be held fixed and only the output membership functions, perhaps taken as singletons, allowed to be trained. This approach could require hundreds or thousands of rules (here $n^5$, where $n$ determines the resolution along each dimension) and a correspondingly large number of trainable parameters. Assuming a limited range on the input functions, indexing could be used so as to involve only a small fraction of the rules in calculation of each control action, but even this number would be significant, of order $2^5$ or $3^5$.

A second approach would be to use intuitive understanding of the problem to create a much smaller number of rules, many of which might involve a subset of the input variables. This procedure has been shown to work well in many cases, mostly simpler problems possessing limited dynamics or involving a single control action.

Our present approach is to train a system with relatively few rules (here 24), counting on the flexibility of the input membership functions to make up for the sparse coverage of the input space. Each rule of the fuzzy system has two (singleton) outputs. After defuzzification, the outputs of the fuzzy system feed sigmoidal nodes which explicitly limit the controller outputs $\theta$ and $\delta$ to their allowed ranges.

## IV. Training

We employ the same two-step procedure that we use to train neural controllers. In the first step, we train a neural network to identify the input-output characteristics of the unknown dynamical system. Then the trained identification network is used to provide estimates of the dynamic derivatives of plant outputs with respect to the trainable controller parameters. A training algorithm based upon a decoupled extended Kalman filter (DEKF) is employed for training both the identification network and the fuzzy controller, as described in [3, 10, 11, 12]. On the basis of considerable experience, we feel that the training of both feedforward and recurrent neural networks as well as fuzzy systems by DEKF algorithms generally leads to superior results with less total training time and fewer presentations of training data than does training by pure gradient descent.

Training of controllers is performed in an indirect fashion by using the identification network to model the input-output behavior of the system. In this scheme, the desired control signals are not known, but rather must be inferred indirectly through a specification of the system's desired behavior. This is provided by a subjectively and empirically determined cost function. For the ISC problem, we choose a quadratic cost function consisting of four terms. The first component penalizes deviations of engine speed from the desired value, here 750 rpm. The remaining

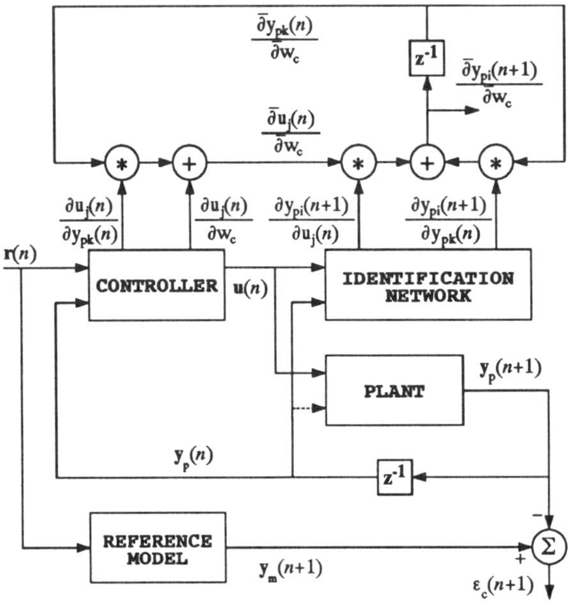


Figure 1: Recurrent derivative structure for controller training. This figure assumes that there are no internal feedback connections within either the identification or control networks, although there are external recurrent connections. The vertical lines emanating from the identification and control networks denote derivatives that are computed by backpropagation. The *total* partial derivatives that this recurrent derivative structure produces are used with the error vector $\varepsilon_c(n+1)$ to update the weights of the controller.

terms discourage certain behaviors of the control signals. We penalize deviations of the spark advance from 22.9 degrees for reasons discussed above. The third and fourth terms penalize large changes in throttle and spark angle between two successive control time steps. In this way we implement a "smoothness" constraint [13] that tends to inhibit oscillatory behavior in the controls for dynamical systems with significant internal time delays. The contribution to the cost function at time step $n$ is given by

$$\begin{aligned} C(n) &= \frac{1}{2}\left(\beta_1(750 - N(n+1))^2 + \beta_2(22.9 - \delta(n))^2 \right. \\ &\quad \left. + \beta_3(\theta(n-1) - \theta(n))^2 + \beta_4(\delta(n-1) - \delta(n))^2\right), \end{aligned}$$

where the empirical weighting factors are $\beta_1 = 1.6 \times 10^{-7}$ for $N_m(n+1) > 750$ and $\beta_1 = 2.4^{-7}$ for $N_m(n+1) \le 750$; $\beta_2 = 3.25 \times 10^{-7}$; $\beta_3 = 10^{-3}$; and $\beta_4 = 1.63 \times 10^{-5}$.

A critical step in training of controllers by gradient methods is the proper computation of derivatives of plant and controller outputs with respect to the trainable controller parameters. The computation of these derivatives is guided by two observations: 1) the evolution of the system state is defined recursively in terms of the previous state; 2) the computed control signals are defined recursively as a function of the measured system state and/or output, which is itself a function of the previous state and previously applied control signals. Hence the derivatives of the system state and controller outputs with respect to the trainable controller parameters should likewise be defined recursively in terms of the derivatives from previous time steps. The temporal evolution and computation of the *total* partial derivative of a component of plant output with respect to a controller parameter is illustrated in the sensitivity circuit of Figure 1. The derivatives computed can then be used by gradient methods such as dynamic backpropagation [14] or Kalman-filter-based algorithms to update the controller's trainable parameters.

A neural identification network consisting of a single layer of 8 completely interconnected (recurrent) nodes with linear activation functions was trained to model the input-output characteristics of the plant. The network has four inputs, consisting of the measured system state, $P_m$ and $N_m$, and the control inputs, $\theta$ and $\delta$. The output of the network is a prediction of the system state 20 ms into the future. The feedback weights of linear recurrent nodes are constrained to be less than unity in magnitude by treating them as the outputs of bipolar sigmoidal functions. This tends to make the training process computationally stable for linear recurrent nodes. The control inputs and torque disturbances were varied through their entire ranges during training of the identification networks. Since the torque disturbances are not observable, the prediction of system state by the identification network will be in error, particularly during large transitions in torque disturbance. However, we only require that on average the derivatives of system state with respect to controller parameters be correct.

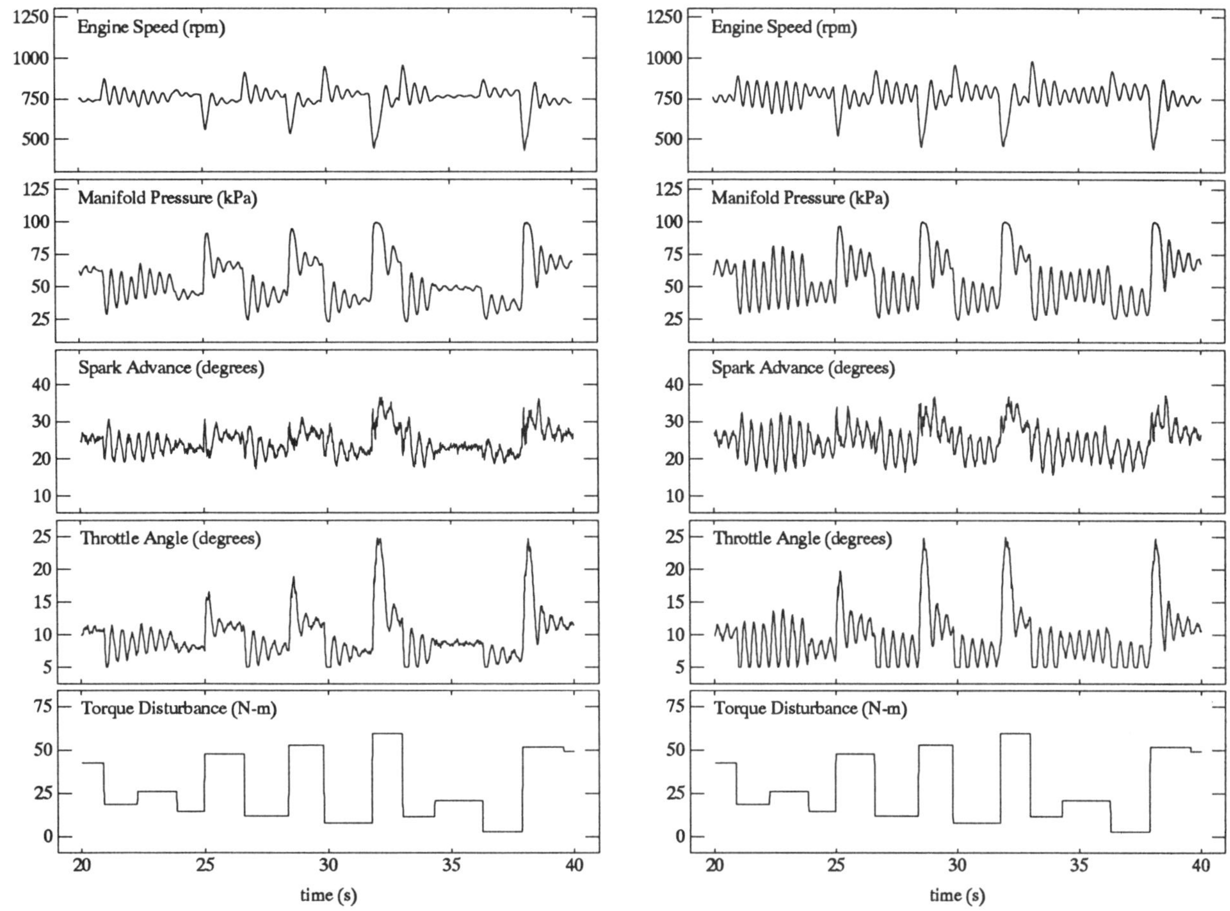


Figure 2: On the left is shown the behavior of the trained PID/PD controller for a test sequence of load disturbances, shown in the bottom panel. On the right is shown the performance of the (unmodified) controller for an altered plant with an offset in the measurement of manifold pressure, as described in the text.

## V. Results

Results for a trained PID/PD linear controller are given in Figure 2 for a sequence of load disturbances independent from but generated in the same way as the disturbance pattern used in training. The response to both small and large disturbances reflects the unavoidable induction-power delay as well as delays in data acquisition and application of control. With this controller, the engine exhibits considerable oscillation in its response to torque disturbances.

The results obtained for a fuzzy system fed by the fixed preprocessing layer are superior to that of the PID/PD controller, but are not nearly as good as that obtained when the weights of the preprocessing layer are trained along with the fuzzy system. Figure 3 shows the performance of this hybrid controller. After the engine speed recovers from each disturbance, the response is fairly well damped. Additional testing with closely spaced disturbances disclosed no tendency to instability. In response to long periods of constant disturbance, the speed remains nearly constant (with very small fluctuations due to simulated measurement noise) at approximately the target value of 750 rpm.

As a measure of controller robustness, we altered the plant equations and introduced a systematic offset (-10 kPa) in the measurement of manifold pressure. The constants $k_P$ and $k_N$ are set to 38.16 and 59.69, respectively. The intake mass air equation is changed to $\dot{m}_{ai} = (1.2 + 0.907\theta + 0.12\theta^2)g(P)$. The expression for load torque is given by $T_L = (N/236.85)^2 + T_d$. The performance of the PID/PI controller on this altered plant, shown on the right side of Figure 2, is not very satisfactory. Though the performance of the hybrid controller on the altered plant (right side of Figure 3) is not as good as on the unaltered plant (greater ringing), it is clearly superior to that of the PID/PD controller.

As a further test, we repeated the test of the hybrid controller with a different altered plant (offset of +10 kPa, $k_P$ = 59.69, $k_N$ = 38.16, $\dot{m}_{ai} = (0.8 + 0.907\theta + 0.08\theta^2)g(P)$, and $T_L = (N/300)^2 + T_d$). The response in

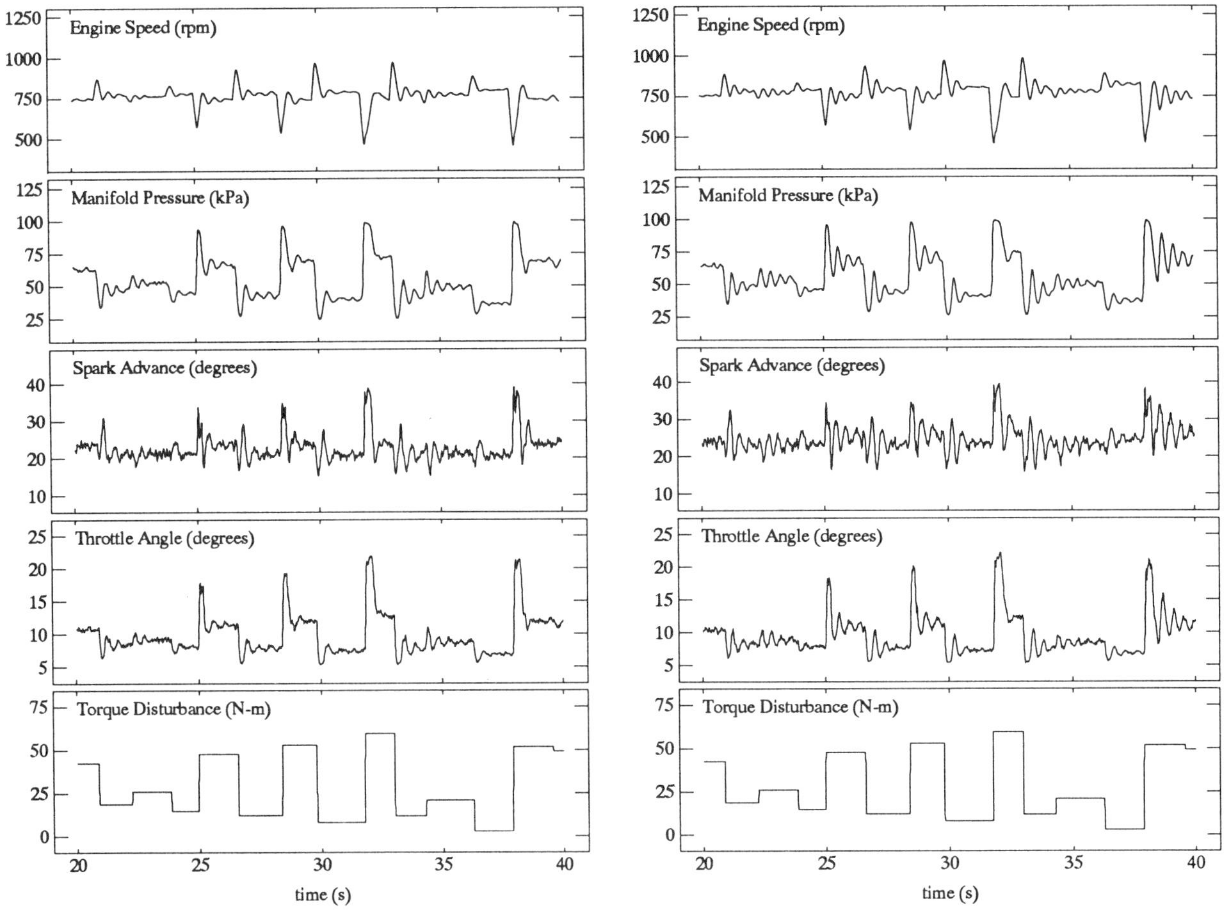


Figure 3: Illustration (left) of the behavior of the neural-fuzzy hybrid controller for the original plant and (right) for the altered plant and measurement offset.

engine speed was almost as good as for the original plant, i.e., better than for the first altered plant.

## VI. Discussion

Control of the plant defined here seems to be a difficult problem. We found that the training process proceeded more slowly than similarly executed training for simpler plants. We doubt that a purely intuitive approach to fuzzy control would have much success in dealing with the model presented here. (Relaxing the key stipulations of our problem statement, as by neglecting the induction-power delay, makes the problem **much** easier.) To the extent that the model and supplementary conditions used in this exercise reflect the complexity of an actual system, it is not surprising that production ISC systems, which tend to contain relatively simple control architectures, sometimes exhibit interesting behavior.

This work confirms the evident principles that a good controller architecture must 1) be able to extract the required information from the available data and 2) have sufficient representational power to form good control actions when this information has been extracted. With the same fixed PID/PD preprocessor, a fuzzy system performed better than a linear controller, presumably due to greater representational power. Similarly, when preceding the same fuzzy system architecture, a trainable preprocessing layer gave rise to a better controller than did a fixed preprocessor, undoubtedly because it was providing better information. In parallel work, we have used a more extensive recurrent layer to produce even better control for this same plant. In [5] we present results obtained with a recurrent hidden layer of 8 nodes feeding 2 recurrent output nodes. Nearly equivalent performance can be obtained without recurrence in the output layer, either by single sigmoidal nodes or by a simple fuzzy system. This supports our view that a properly trained recurrent preprocessing layer can recover useful state information that may not be present in temporal combinations, such as PID, that are constructed *a priori*.

**Acknowledgments:** We thank our colleague B. K. Powell for helpful discussions of idle-speed control and other aspects of engine dynamics. We are grateful to S. S. Farinwata and Prof. G. Vachtsevanos of Georgia Institute of Technology for stimulating discussions and for providing details of the revised engine model.

## References

[1] M. Abate and N. Dosio (1990). Use of Fuzzy Logic for Engine Idle Speed Control. *SAE Paper 900594*, 107–114; M. Abate (1991). An Application of Fuzzy Logic to Engine Control. In *Proceedings of the Fuzzy and Neural Systems and Vehicle Applications '91 Conference*, November 8–9, 1991, Tokyo, Japan.

[2] M. Abate and V. Di Nunzio (1990). Idle Speed Control Using Optimal Regulation. *SAE Paper 905008*, 87–96.

[3] L. A. Feldkamp, G. V. Puskorius, L. I. Davis, Jr., and F. Yuan (1991). Decoupled Kalman Training of Neural and Fuzzy Controllers for Automotive Systems. *Proceedings of the Fuzzy and Neural Systems and Vehicle Applications '91 Conference*, November 8–9, 1991, Tokyo, Japan; L. A. Feldkamp, G. V. Puskorius, L. I. Davis, Jr., and F. Yuan (1992). Neural Control Systems Trained by Dynamic Gradient Methods for Automotive Applications. In *Proceedings of the 1992 International Joint Conference on Neural Networks* (Baltimore 1992) vol. II 798–804.

[4] L. I. Davis, Jr., G. V. Puskorius, F. Yuan, and L. A. Feldkamp (1992). Neural Network Modeling and Control of an Anti-Lock Brake System. *Proceedings of Intelligent Vehicles '92* (Detroit 1992) 179–184.

[5] G. V. Puskorius and L. A. Feldkamp (1993). Automotive Engine Idle Speed Control with Recurrent Neural Networks. Submitted to the *American Control Conference, 1993.*

[6] B. K. Powell and J. A. Cook (1987). Nonlinear Low Frequency Phenomenological Engine Modeling and Analysis. *Proceedings of the 1987 American Control Conference*, vol. 1, 336–340; J. A. Cook and B. K. Powell (1988). Modeling of an Internal Combustion Engine for Control Analysis. *IEEE Control Systems Magazine*, vol. 8, No. 4, 20–26.

[7] G. Vachtsevanos, S. S. Farinwata and H. Kang (1992). A Systematic Design Method for Fuzzy Logic Control With Application to Automotive Idle Speed Control. To appear in *Proceedings of the 31st IEEE Conference on Decision and Control* (Tuscon, AZ 1992).

[8] F. Yuan, L. A. Feldkamp, L. I. Davis, Jr., and G. V. Puskorius (1992). Training a Hybrid Neural-Fuzzy System. In *Proceedings of the 1992 International Joint Conference on Neural Networks* (Baltimore 1992) II-739–II-744; L. A. Feldkamp, G. V. Puskorius, F. Yuan, and L. I. Davis, Jr. (1992). Architecture and Training of a Hybrid Neural-Fuzzy System. In *Proceedings of the 2nd International Conference on Fuzzy Logic and Neural Networks* (Iizuka, Japan 1992) 131–134.

[9] H. Berenji (1992). Basic Concepts of Fuzzy Control. Tutorial in the *IEEE International Conference on Fuzzy Systems* (San Diego 1992).

[10] G. V. Puskorius and L. A. Feldkamp (1991). Decoupled Extended Kalman Filter Training of Feedforward Layered Networks. In *International Joint Conference on Neural Networks* (Seattle 1991), vol. I, 771–777. New York: IEEE.

[11] G. V. Puskorius and L. A. Feldkamp (1992). Recurrent Network Training with the Decoupled Extended Kalman Filter Algorithm. In *Proceedings of the 1992 SPIE Conference on the Science of Artificial Neural Networks* (Orlando 1992).

[12] G. V. Puskorius and L. A. Feldkamp (1992). Model Reference Adaptive Control with Recurrent Networks Trained by the Dynamic DEKF Algorithm. In *Proceedings of the 1992 International Joint Conference on Neural Networks* (Baltimore 1992).

[13] M. I. Jordan (1989) Generic Constraints on Underspecified Target Trajectories. In *International Joint Conference on Neural Networks* (Washington D.C. 1989), vol. I, 217–225. New York: IEEE.

[14] K. S. Narendra and K. Parthasarathy (1990). Identification and Control of Dynamical Systems Using Neural Networks. *IEEE Transactions on Neural Networks* **1**, no. 1, 4–27; ibid. (1991). Gradient Methods for the Optimization of Dynamical Systems Containing Neural Networks. *IEEE Transactions on Neural Networks* **2**, no. 2, 252–262.

# Application of Fuzzy Logic to Shift Scheduling Method for Automatic Transmission

S.Sakaguchi, I.Sakai, T.Haga
HONDA R&D Co., Ltd.
1-4-1 Chuo Wako-shi
Saitama 351-01 , JAPAN

*Abstract*-**Fuzzy control, which is now widely used in a varied number of fields, has also proved its usefulness in automobile engineering, where it can be used to automatically achieve constant speeds and control anti-lock brake systems. As seen in the case of automatic control of plants, where the operator's strategy is compiled for the purpose, these systems can only be realized after a knowledge base has been established. Knowledge bases have been based on interviews and questionnaires given to operators and their knowledge in how control devices work. This approach was tended to obtain only the most characteristic knowledge and failed to provide satisfactory control results. In view of this problem, we are proposing a way to construct a knowledge base that has the potential for a wide range of applications.**

**The behavior of human beings is generally regarded as a cycle of recognition of their environment, judgement, and action. This model of behavior consists of two processes: a responsive action devoid of intentional judgement and action based on knowledge and restrictions. Using this model, the authors have constructed a three-layer knowledge base that draws on common sense, knowledge based on experience, and restriction In this way a satisfactory knowledge base is constructed so that the reliability of control devices can be significantly enhanced.**

**This paper describes how the layers of knowledge have been constructed by applying the fuzzy control to the shift scheduling of automatic transmission cars(AT vehicles), and introduces the field test results of the scheduling system by a fuzzy control device to verify the effectiveness of the aforementioned approach .**

## I. INTRODUCTION

In recent years, automobiles are designed to have higher performances and more sophisticated functions. As the vehicle performances and functions became higher and more sophisticated, it has become difficult to explain its function without human elements because of the qualitative advancement achieved and the amount of information processed in a modern automobile. In other words, today's automobiles are required to have more sophisticated operational functions and features, in addition to the basic function of being mere by a transportation device in order to respond to more demanding human perceptions. There is also a stronger demand for qualitative improvement of controls that have been considered not so simple to materialize.

On the other hand, for automating complicated controls performed manually by humans, engineers are now placing emphasis on the fuzzy control method which controls objective system by first focusing on human knowledge and then materializing its controlling methods. The fuzzy control attempts to achieve control by incorporating human decision-making abilities into the controller as demonstrated by operators skilled in operating highly complicated systems. In short, the method aims to simulate human behavior using a computer.

For an example, a human operator's control strategy needs to be first established on knowledge base for automating a manufacturing plant.

A conventional approach for composing the knowledge is "questionnaire and interview" methods. This approach, however, tends to extract knowledge of only very specific features and not of overall circumstance. As a result, it is often unsuccessful to materialize the control by merely describing the knowledge, obtained through question and interview methods, in accordance of the rules of "IF...THEN..." which in turn poses a serious problem in designing control systems.

This paper proposes, in consideration of above mentioned problem, a method to construct highly flexible knowledge base and introduces the field test results of the fuzzy control applied to shift scheduling system of vehicle automatic transmission.

## II. HUMAN THINKING PROCESS

When in the process of taking action, humans are not conscious of the mathematical model based on which the intended control is implemented. Rather, they roughly recognize the circumstance, then make a judgement and take action. The process can be broken down into four steps : recognition->judgement->decision->action.(Fig. 1) It is easy to describe this series of actions by the production rule of "IF...THEN...". Since this rule itself is regarded as the know-how for the control obtained through experience, practical application of the rule can be considered to be an effective approach to automate the complicated operations performed by humans.

This process reveals a strong linkage of the human action with knowledge they process, the information obtained from the outside world, and object consciousness of how to reflect the intention in their mind to the action. Thus, once the following problems are solved, automatic control by a computer will begin to approximate the flexible operation by humans : how to extract and compile the information in the brains; and how to introduce useful information into the computer.

0-7803-0614-7/93$03.00 ©1993IEEE

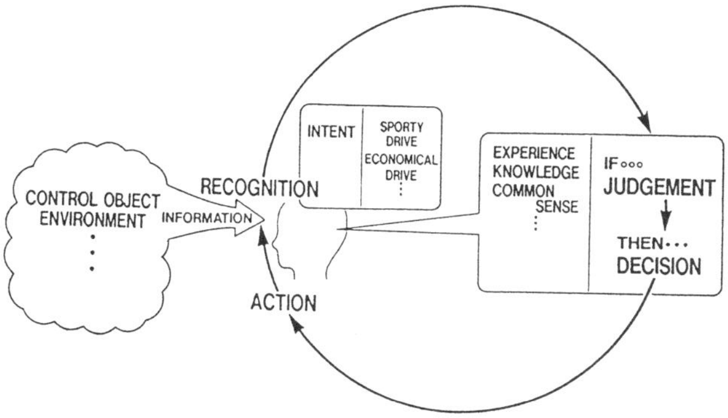


Fig. 1 Human Thinking Process

Another problem is finding how to feed the information about the environment into the control system. This issue relates closely to human's conscious awareness of objectives, as seen in the aforementioned process.

The authors are proposing a way to extract knowledge of human subject and construct a knowledge base through analyzing the human thinking process from recognition to action.

## III. DESIGNING CONTOROLLER

### A. Compiling knowledge

Human actions are not based on the same levels of decision-making, but are believed to take place at different levels. Fig. 2 depicts the basis of this relationship. In other words, the human actions are divided into two categories, one is the action by reflection and another is the action based on knowledge and restriction. The former is what is called common sense, intuitive, or learned through repeated decision making, by which the action reflectively takes place. The latter corresponds to experiential knowledge used to select a better approach to achieve the intended result. Therefore, the former is a decision making process without going through the intentional judgement process and the latter a decision-making accompanied by the intentional judgement process.

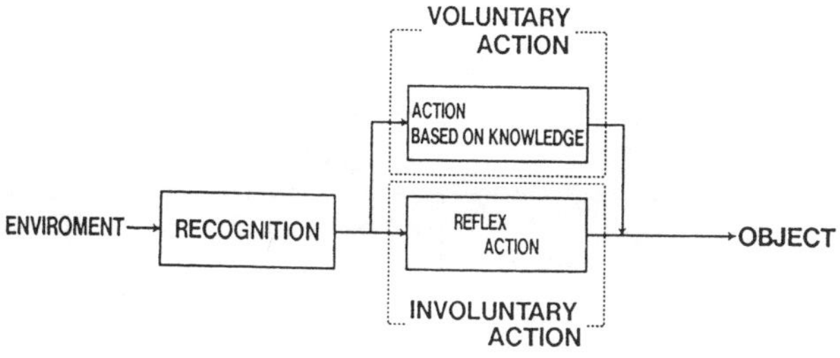


Fig. 2 Human Actions

The authors propose to construct a knowledge base using three categories of knowledge : (1) knowledge of common sense that oversees the whole; (2) knowledge of experience that is locally effective or characteristic; and (3) restriction. This will prevent an evasion of knowledge and make the controller to work on the basis of common sense in the normal situations and activate the experience when necessary, thereby improving their reliability to a large extent.

### B. Extracting knowledge

For extracting knowledge, knowledge based on experiences can be extracted rather easily from human subject since a human himself can consciously extract the knowledge whenever it is needed, while common sense knowledge is difficult to extract since humans are unconscious of the knowledge. In other words, humans are not normally aware of what they consider common sense. For this reason, questionnaires and interviews method are not appropriate in extracting this knowledge.

*1) Abstracting common sense knowledge* : Recognizing the problems as described above, the authors decided, instead of extracting knowledge equivalent to common sense knowledge through questionnaires and interviews method, to extract common sense knowledge by asking, "What is the most fundamental requirement of the object of control ?" Consider the water tank model in Fig. 4. A valve on the tank is operated to keep the surface of the water as a target level. In this case, the most fundamental requirement of the object of control is to increase the volume of water when the surface is below the target level and to decrease the volume when the surface is over, or simply:

(1)To open the valve when the surface is below the target level

(2)To close the valve when the surface is over the target level

This knowledge is so basic that it alone cannot control the water level with high precision. And yet, the water level cannot be controlled adequately without this knowledge, though humans are not conscious of it and usually take it for granted.

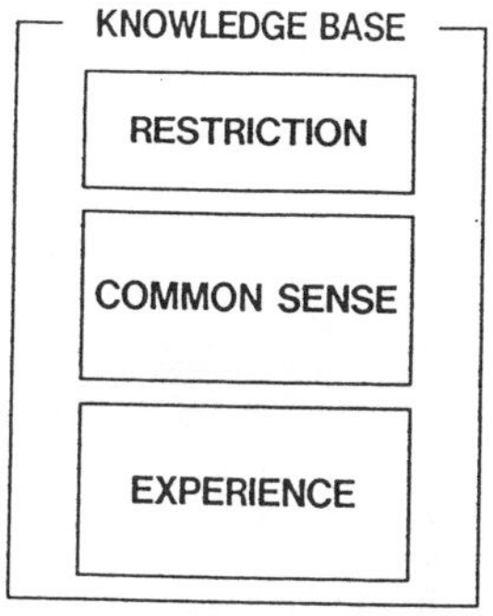


Fig. 3 Configuration of Knowledge Base

In questionnaires and interviews, when asked to explain how this water level is controlled, in most cases, only the knowledge of how to control the level is obtainable in such a characteristic situations.

(1)Adjust the valve opening when the water surface is going to overshoot or undershoot, and

(2)make the valve opening larger if the water surface is below the target level and is still reading.

.

.

.

As this example shows, knowledge obtained through questionnaires and interviews often already contains common sense as a basic component. That is, knowledge is normally extracted not as an overall knowledge but as localized effective knowledge. However, sets of knowledge specific to certain circumstances can hardly control systems satisfactorily. The approach suggested in this paper is effective in extracting the common sense of humans.

*2) Abstracting experimental knowledge* : As already explained, it is quite easy to extract knowledge that corresponds to experience through questionnaires and interviews. In the water tank model, humans often open or close the valve with the firm intention of bringing the water level to a particular level. Humans would promptly give the following knowledge when asked in a questionnaire or interviews, how to control the water level. Namely, if the level is below the target but is approaching it, they will:

(1)Close the valve if the water surface is rising and approaching with rapid speed to the target level,

(2)close the valve slightly if the water surface is rising slowly to the target level.

.

.

.

In comparison with the common knowledge described above, this knowledge is effective and characteristic in an extremely limited situation. Because of its characteristics, the knowledge can remain with the human consciousness and can easily be extracted as well.

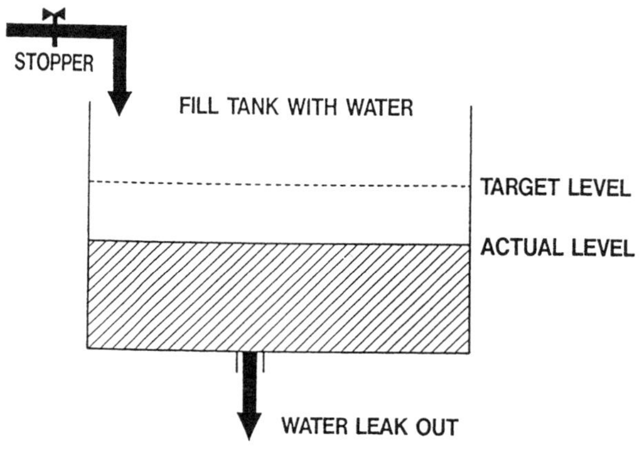


Fig. 4 Water Tank Model

*C. Inferring intention*

The information that are treated when humans make decision on their action can be classified into two categories: information obtained from external environment and intention closely linked to one's objective. To replace humans with an automatic control system, therefore, one key issue besides the acquisition of knowledge is learning how to incorporate human intentions into the control system. In order to resolve the subject, the authors have first formed human intent based on the information obtained from the human action and environment, and then adopted a multistage fuzzy inference by incorporating its result as input information for the controller.(Fig. 5) This makes it possible to express an antecedent of the knowledge base as a simple description and to express the control rule in short simple phrases, thus leading to improved maintenance and simpler settings. In addition, the separate input information provides a benefit that the control rule does not require rewriting even when the precision of inference is to be upgraded.

*D. Overall architecture*

As a result of the above study, the fuzzy control system has been composed as an overall architecture as shown in Fig. 6, in which the knowledge base is divided into common sense, experience, and restriction for assessment, and the rules of common sense and experience are referred to as fuzzy inference. The restriction is written as an algorithm

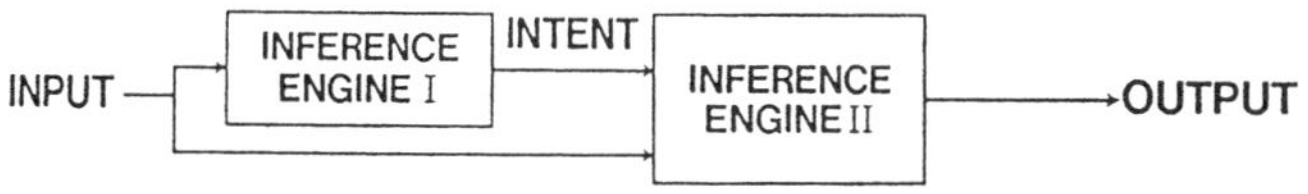


Fig. 5 Multistage Inference

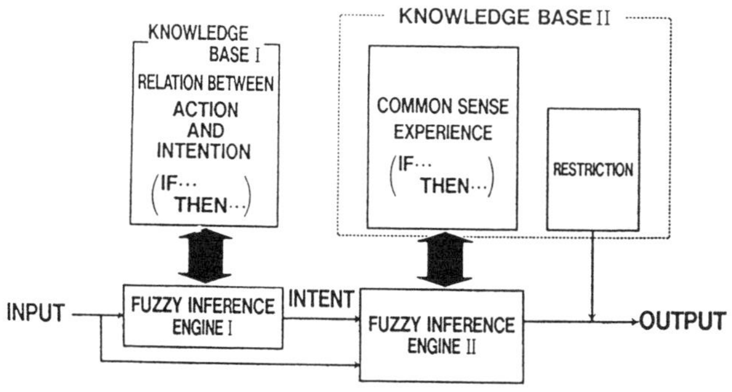


Fig. 6 Overall Architecture

to monitor fuzzy inference output because it is an item that cannot be compromised. Given that the input data contains both the intention and environmental information, a module is added to infer the intention. This architecture seems to have potential for a wide range of applications. This paper adopts the shift scheduling system for AT vehicles to verify the appropriateness of an overall architecture of the system.

## IV. FUZZY AT SHIFT SCHEDULING SYSTEM

### *A. Conventional AT shift scheduling method*

As shown in Fig. 7, shift scheduling is usually based on two parameters, vehicle speed and throttle opening. Presently, microprocessors are increasingly used to determine the shift scheduling. In some of the shifting scheduling, command values retrieved from gear shift mapping are automatically adjusted and corrected in detail.

### *B. Fuzzy AT shift scheduling system*

*1) Block composition* : Fig. 8 shows a fuzzy AT shift scheduling system that consists of an input block, an inference block, a knowledge base and an output block.

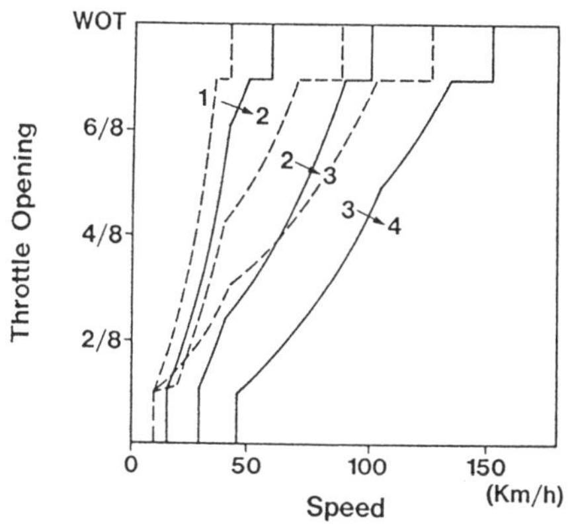


Fig. 7 Gear Shift Map

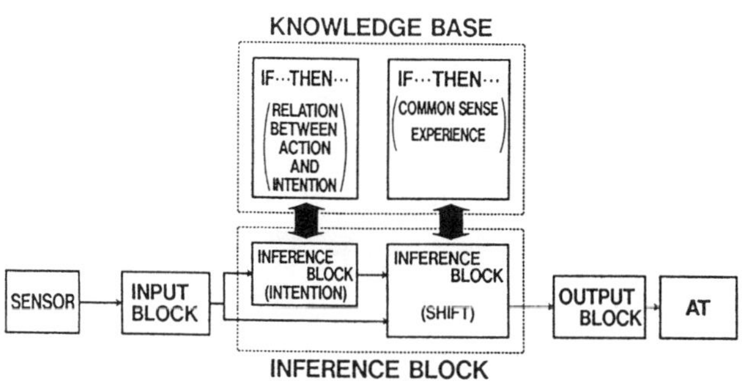


Fig. 8 Fuzzy AT Shift Scheduling System

Input block --- This block receives six input variables --- vehicle speed, throttle opening, engine speed, intake manifold pressure, shift positions and brake signals -- and calculates information necessary for controlling the shifting schedule by performing pre-processing including such statistical processing as averaging and filtering calculations.

Inference block --- This block can be divided into two blocks; intentional inference block and shift position inference block. The intentional inference block decides how much the driver intends to slow down through fuzzy inference using the information from the input block and sets of rules based on the knowledge base. The shift position inference block determines desirable shift positions through fuzzy inference using the information from the input block, results of the intentional inference obtained and sets of rules based on the knowledge base.

Knowledge base --- This block is composed of two sets of knowledge; one used for intentional inference and another used for shift scheduling. In the former, the relationship of the driver's intention to slow down and status of the vehicle is stored as a proposition of the "IF...THEN..." statements. In the latter, the common sense mentioned earlier and driver's control expertise represented as experience are stored as a number of propositions made using "IF...THEN..." statements, where the ambiguity of recognition is expressed by fuzzy sets.

Output block --- This block performs engine over-rev checking of a shift position which is inferred by the inference block and decides the eventual shift position before issuing a command to the actuator. Known restrictions in the human knowledge are thereby considered.

*2) Knowledge base*

*2)-1 Knowledge base for shift scheduling* : As explained before, the knowledge for the shift scheduling consists of common sense and experience. For the common sense, it would be most appropriate to adopt the most fundamental characteristic required by the object to be controlled. The authors consider the following two points as the most essential requirements for shift scheduling of automatic transmissions.

(1) Selection of shifting gears according to the vehicle speed.

(2) Shiftings in accordance with the vehicle speeds are variable in accordance with the throttle opening.

These points can be expressed in the form of "IF...THEN..." statements as follows:

(1) If the vehicle speed is high, then use a higher gear, and if the vehicle speed is low, then use a lower gear.

(2) If the throttle opening is large, then use a lower gear, and if the throttle opening is small, then use a higher gear.

These two sets of knowledge have been extracted through a study of characteristics that AT vehicles should have. Together they form an overall knowledge base that works in all driving conditions and agrees with human instinct.

Experiences, under special circumstance where common sense could not be utilized, are responsible for correcting the output power produced by common sense. During uphill driving, for an example, a common sense of "shifting into a higher gear as the vehicle accelerates" is not necessary applicable. In such a case, the driver's experience leads him to use a lower gear. Experience is the basis of this decision.

A list of the concrete knowledge base in the AT shift scheduling system and its membership functions are described in Fig. 9.

*2)-2 Knowledge base for intentional inference*: Human intentions can be determined only through their actions, since there is no sensor which can read intentions. To predict the driver's intention, it is necessary therefore to clarify the situation (i.e., what action he takes)in which he has an intention to slow down or decides not to slow down. For this reason, the authors created a state transition diagram, Fig. 10 which shows, for an example, that the state changes from the non-existence to the existence of the intention to decrease speed when the throttle is closed, the braking pedal is pressed and the vehicle slows down(arrow A). It is not, however, clear from this diagram whether the intention intensifies even when the pedal is released(arrow B). On the basis of this diagram, the relationship between these actions and changing direction of the intention with the "IF...THEN..." statements

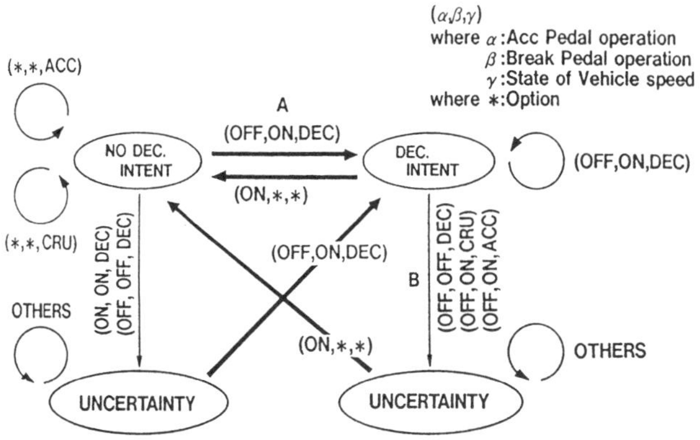

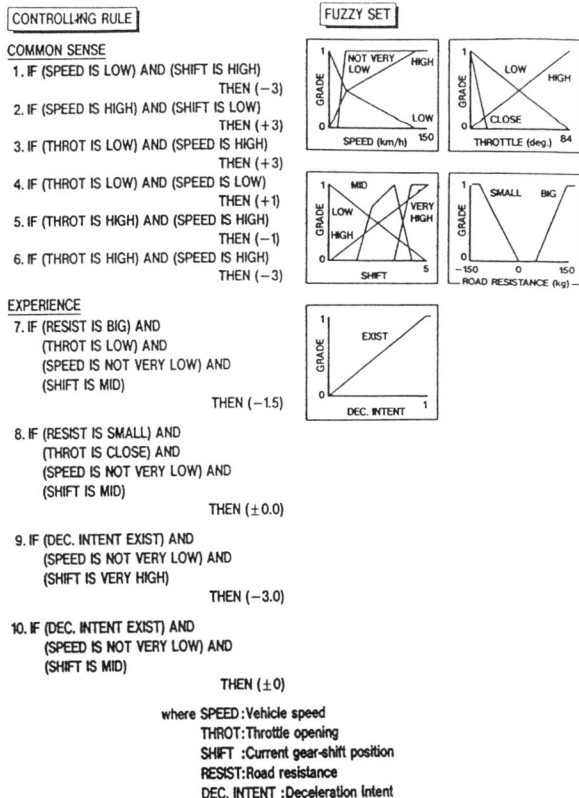

Fig. 10 State Transition Diagram

to make the knowledge base to infer the intention to decrease speed are described. It is basically composed of the following two rules:

(1) Intensifies the intent to decrease the speed when the throttle is closed, the braking pedal is pressed down and the vehicle slows down, and

(2) weakens when the throttle is opened.

In actuality, rules have been written for each of the different situations in which a vehicle is used to better estimate the driver's intention. A list of concrete knowledge base and membership functions are described in the following figure.

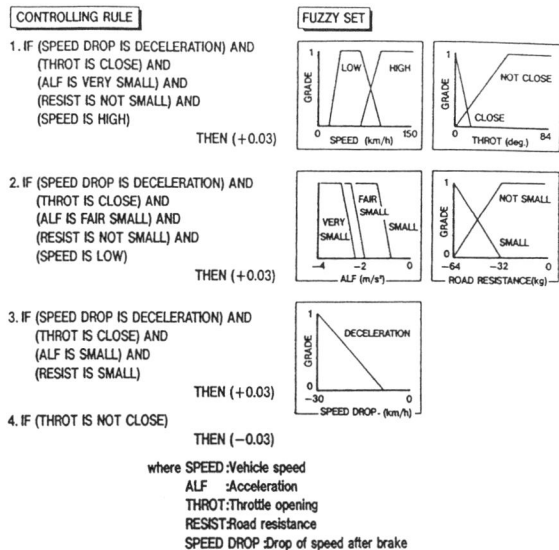

CONTROLLING RULE

COMMON SENSE

1. IF (SPEED IS LOW) AND (SHIFT IS HIGH) THEN (−3)
2. IF (SPEED IS HIGH) AND (SHIFT IS LOW) THEN (+3)
3. IF (THROT IS LOW) AND (SPEED IS HIGH) THEN (+3)
4. IF (THROT IS LOW) AND (SPEED IS LOW) THEN (+1)
5. IF (THROT IS HIGH) AND (SPEED IS HIGH) THEN (−1)
6. IF (THROT IS HIGH) AND (SPEED IS HIGH) THEN (−3)

EXPERIENCE

7. IF (RESIST IS BIG) AND (THROT IS LOW) AND (SPEED IS NOT VERY LOW) AND (SHIFT IS MID) THEN (−1.5)
8. IF (RESIST IS SMALL) AND (THROT IS CLOSE) AND (SPEED IS NOT VERY LOW) AND (SHIFT IS MID) THEN (±0.0)
9. IF (DEC. INTENT EXIST) AND (SPEED IS NOT VERY LOW) AND (SHIFT IS VERY HIGH) THEN (−3.0)
10. IF (DEC. INTENT EXIST) AND (SPEED IS NOT VERY LOW) AND (SHIFT IS MID) THEN (±0)

FUZZY SET

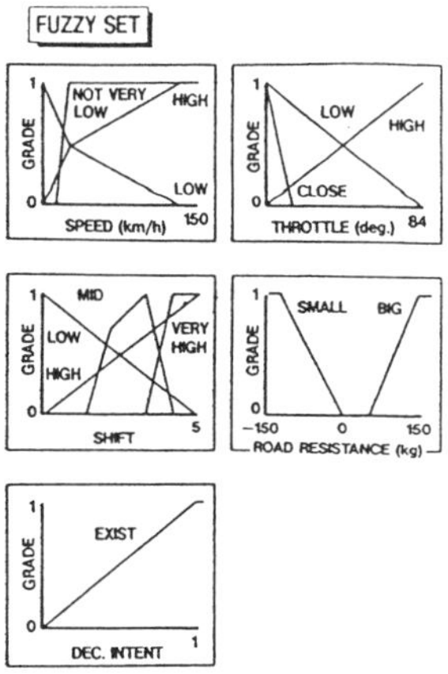


where SPEED:Vehicle speed
THROT:Throttle opening
SHIFT :Current gear-shift position
RESIST:Road resistance
DEC. INTENT :Deceleration Intent

Fig. 9 "IF...THEN..." Rules and Fuzzy Sets(SHIFT POSITION)

CONTROLLING RULE

1. IF (SPEED DROP IS DECELERATION) AND (THROT IS CLOSE) AND (ALF IS VERY SMALL) AND (RESIST IS NOT SMALL) AND (SPEED IS HIGH) THEN (+0.03)
2. IF (SPEED DROP IS DECELERATION) AND (THROT IS CLOSE) AND (ALF IS FAIR SMALL) AND (RESIST IS NOT SMALL) AND (SPEED IS LOW) THEN (+0.03)
3. IF (SPEED DROP IS DECELERATION) AND (THROT IS CLOSE) AND (ALF IS SMALL) AND (RESIST IS SMALL) THEN (+0.03)
4. IF (THROT IS NOT CLOSE) THEN (−0.03)

FUZZY SET

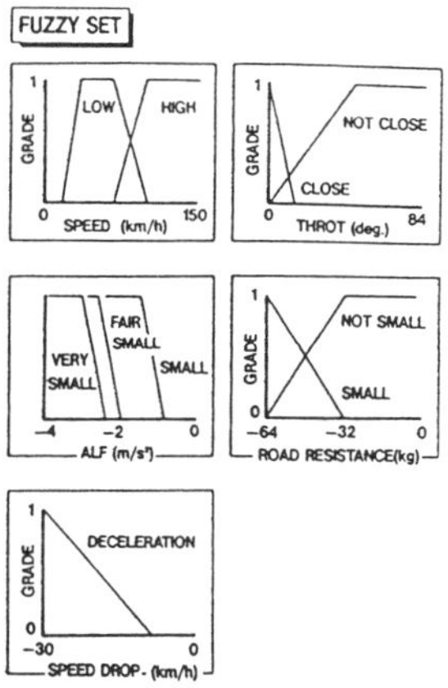


where SPEED:Vehicle speed
ALF :Acceleration
THROT:Throttle opening
RESIST:Road resistance
SPEED DROP :Drop of speed after brake

Fig. 11 "IF...THEN..." Rules and Fuzzy Sets (INTENTION)

## V. FIELD TEST RESULTS

Field tests on the fuzzy shift scheduling system were conducted on mountainous roads and city streets.

*A.Mountainous road test*

A mountain road in California was selected as a typical mountainous road and the fuzzy control system (FUZZY) was compared with a conventional automatic drive system (DRIVE).

*1) Uphill driving* : Fig. 12 shows the uphill driving performances of the two systems, while Fig. 13 shows the frequencies of the shifts used and the number of gear changes. As the figures indicate, when the driver releases the throttle of the conventional system during the uphill driving, the transmission unintentionally shift up creating power insufficiency which in turn generates shift-down. On the other hand, the fuzzy system under the same circumstance will hold the gear appropriately at the third gear, eliminating busy shiftings and improving driveability.

Table I compares the fuel consumption of the two systems to further show the advantage of the fuzzy control system, which seems to be the result of the smaller slip loss in the torque converter arising from the effective utilization of lower gear during uphill driving.

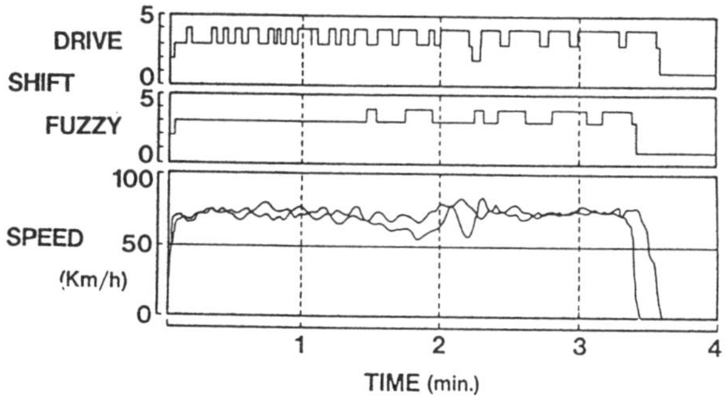


Fig. 12 Comparison of Fuzzy and D-Range I (UPHILL DRIVING)

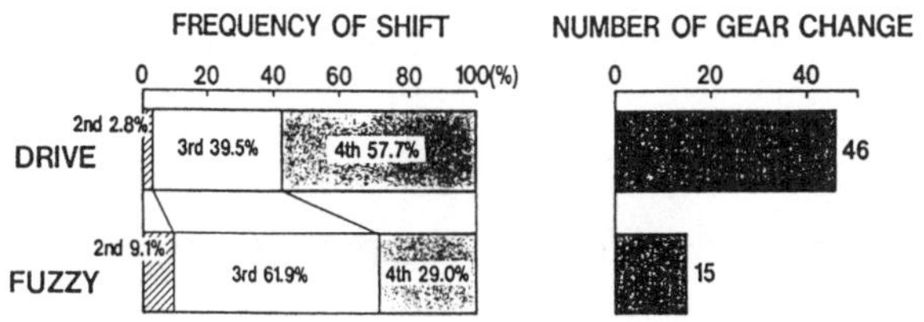


Fig. 13 Comparison of Fuzzy and D-Range II (UPHILL DRIVING)

Table I Fuel Consumption

| | |
|---|---|
| DRIVE | 4.43km/ℓ (10.42MPG) |
| FUZZY | 5.00km/ℓ (11.76MPG) |

*2) Downhill driving* : Fig. 14 shows the downhill driving performances of the two systems, while Fig. 15 shows the frequencies of shifting and the number of braking. These figures indicate that the fourth gear is almost always maintained without the effective application of engine braking under the conventional system, while the fuzzy system makes proper down shiftings as if it is responding to the driver's intention to decrease the vehicle speed. Subsequently, the gear is held at the third to apply engine braking, thereby substantially improving the driving stability.

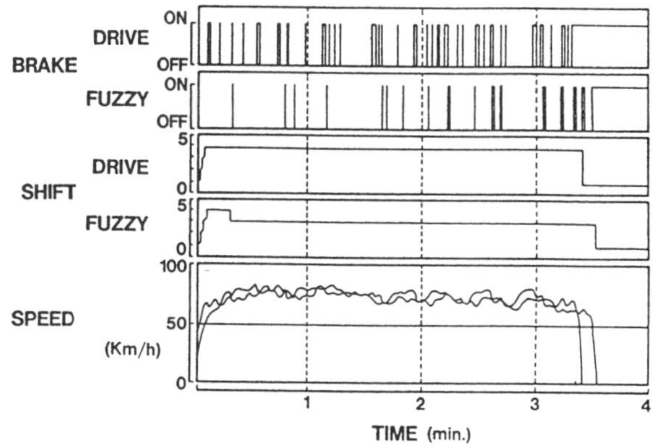


Fig. 14 Comparison of Fuzzy and D-Range III (DOWNHILL DRIVING)

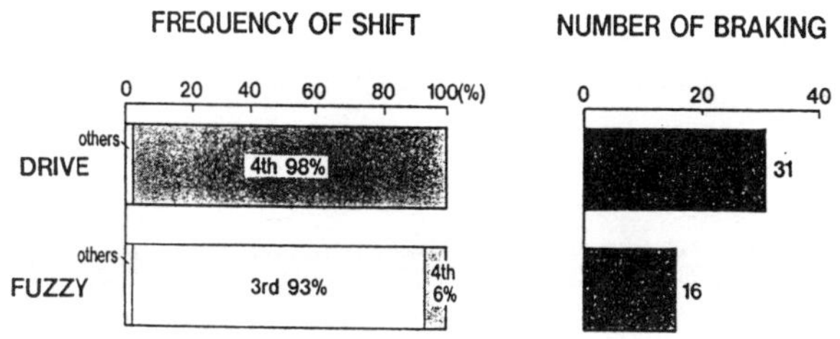


Fig. 15 Comparison of Fuzzy and D-Range IV (DOWNHILL DRIVING)

*B. City street test*

To re-create typical city street conditions, the authors selected a road in the suburbs of Tokyo and compared the driving performance of the fuzzy control system with that of the conventional AT system. On city streets where there are no extreme slopes, the fuzzy control system makes scheduling, which relies on common sense in the knowledge base, in large part in the same way as the conventional system. This means that the proposed system requires the driver to conduct mostly the same throttle pedal operations as those conducted under the conventional system.

## VI. CONCLUSION

Since human behavior can be assessed as a three-layer construction, this paper has proposed to construct a knowledge base for the fuzzy control system by considering common sense, experience and restriction, and to incorporate the human intention into system control as input information.

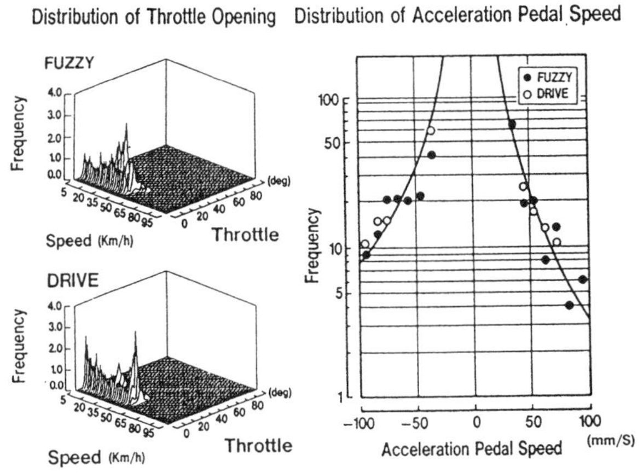


Fig. 16 Accelerator Operation for City Driving

On the basis of this concept, the authors have developed a system to determine shift scheduling of an automatic transmission.

Field tests have proved that the developed system can realize a shift scheduling which satisfactorily reflects human experiences and intentions. In uphill and downhill driving in particular, the fuzzy system, in comparison with the map method employed by the conventional AT system, shows a significant improvement in the driveability particularly in such areas as riding comfort and firm driving feeling.

In conclusion, this fuzzy control system was actually applied to the production model, 3.2-liter V-6 engine introduced in September of 1992 and has received excellent evaluations in the Japanese market.

## REFERENCES


[1] T. Yamazaki and M.Sugeno, "Fuzzy Control",*System and Control* Vol. 28 No. 10 PP442-446,(July, 1984)

[2] O. Itoh, " Application of Fuzzy Control to Activated Sludge Process" *Preprints of Second IFSA Congress,* PP282-285, (1987)

[3] E. H. Mamdani, "application of Fuzzy Algorithms for Control of Simple Dynamic Plant", *Proc. IEEE,* 121, PP1585-1588, (1974)

[4] Lotfi. A. Zadeh, "Fuzzy Sets", *Information and Control,* Vol. 8, PP338-358 (1965)

[5] S. Yasunobu, "Predictive Fuzzy Control and Application for Automatic Container Crane Operation System", *Preprints of Second IFSA Congress,* PP349-352, (1987)

[6] S. Yasunobu, "A Predictive Fuzzy Control for Automatic Train Operation", *System and Control* Vol. 28 No. 10 PP605-613, (October 1984)

[7] H. Takahashi, "Subjective Evaluation Modeling Using Fuzzy Logic and Neural Network", *Third IFSA Congress,* PP520-523, (1989)

[8] I. Sakai, S. Sakaguchi, T. Haga, "Shift Scheduling Method of Automatic Transmission Vehicles with Application of Fuzzy Logic", *23rd FISITA Congress* Vol. 1, PP343-347, (1990)

[9] I. Sakai,S. Sakaguchi, T. Haga, M. Togai, "Shift Scheduling Method of Automatic Transmission Vehicles Using Fuzzy Logic", *IEEE Roundtable Discussion on "Fuzzy and Neural Systems, and Vehicles Applications",* paper #27, (1991)

# Sensor Integration System utilizing Fuzzy Inference with LED Displacement Sensor and Vision System

Koji SHIMOJIMA, Toshio FUKUDA, Fumihito ARAI, and Hideo MATSUURA
Dept. of Mech. Eng., Nagoya University
Furo-cho, Chikusa-ku, Nagoya 464-01, Japan

***Abstract*-This paper deals with a 3-D measurement system applied to a curved metal surface carving system, and a multi-sensor integration system based on fuzzy inference. The measurement system consists of two different sensors, a LED displacement sensor and a vision system. The LED displacement sensor is used as a part of the vision system based on the active stereo sensing method. In addition, the LED displacement sensor's outputs are used for calibrating camera parameters. Therefore, the system can calibrate the camera parameters easily. Neural networks are used to compensate the output of the image processing for some errors, such as camera parameter's error and lens distortion. By utilizing the neural network, we can use a vision system as accurately as possible. We use a sensor integration method based on the fuzzy inference. Fuzzy inference's input consists of information on the change in the sensor output and the position change of the sensor system, together with the environmental data of measurement. For this integration system, we can use the sensory system accurately. The proposed system is shown to be effective through extensive experiments.**

## I. INTRODUCTION

Automated carving systems with robots are not widely applied to make a work repaired and reusable from a used one which has distortion and defects. Because the works' surface's shapes, which were once used, are complicated and different from each other, it requires more skillful experiences, so that most of these works have been carried out by human experts. In recent years, however, the number of the experts is decreasing, even it takes a long time to become an expert. Therefore, automated manufacturing systems for repair and reusable production have been desired so far from these industrial sectors.

Most of the present automated manufacturing systems need the CAD data of works or sensory data of a force-torque sensor to carve works [1,2]. For the manufacturing of the reusable products which have complicated surfaces and profiles different from the original, the present systems are not suitable. Because the CAD data do not express the profile of used works but the original one, therefore the systems using the CAD data cannot carve the used one. While the systems with force-torque sensor cannot recognize the desirable shape from the sensory data. Besides, it is very difficult to analyze force or torque accurately. Therefore, they cannot carve precisely the object.

To solve these problems, we proposed an automated carving system with an integrated measurement system [3]. The system measured each target's surface shapes with the measuring system and made the carving path from the measurement data. We also proposed a sensor integration system for the measuring system [3, 4]. In this measuring system, we used some eddy current sensors. This system had two problems. One was that this system could not measure the rough surface, because the sensors had to be very close to the target surface to measure. The other was that the system could not recognize the position on the target surface where the system was measuring. Recognition of the measuring position is very important for precise carving, because the carving path is made from the measuring path.

For these problems, we proposed a new measurement system and its integration method. The system consists of two LED displacement sensors (this sensor can measure displacement at a distance of 40 mm), and two CCD cameras (for the measuring point's recognition on the target surface). The LED displacement sensor is not only used as a displacement sensor but also as a marker of the measuring point for vision system. Therefore, the measurement system can measure the surface shape of the target at a distance and recognize the measuring point.

The vision system applied the active stereo method consists of a CCD camera and a LED displacement sensor. This vision system has two characteristics: One is that the calibration of the camera parameters is very easy because the system calibrates the camera parameters by the measuring data of the LED displacement sensor. The other is that a neural network (**NN**) is utilized as a compensator of the vision system's output for errors, such as camera parameters' error and lens distortion. By utilizing the neural network, we can use the vision system as accurately as possible.

In order to integrate various sensory data, the sensor integration system (**SIS**) has been studied [5]. Multiple sensory systems have a major problem, which is how to integrate sensory data to produce more reliable and accurate output. Various SIS methods have been reported so far: hypothesis testing by sensor models and Bayesian approach [6], confidence distance matrix [7], estimation by performance and cost criteria [8], estimation of cost by Bayesian method [9], and using Kalman filter [10]. Most of these systems are based on statistical techniques, therefore they can use sensors effectively under a stable measuring environment. Without having the knowledge of the sensors' specification, they do not have the flexibility against changes of sensor's environments.

In the previous system [3,4], we used a **SIS** using the **fuzzy inference** to integrate the measurement data. These SISs evaluate the suitability for the environment of each sensor, e.g., the intensity of a spot-light's reflection for the LED displacement sensor, and then used the appropriate sensor for sensors' environments. Though this SIS could not evaluate internal conditions of sensors, e.g., the sensitivity for the intensity of the spot-light's reflection for the LED displacement sensor.

The measurement system proposed here is mounted at the tip of the manipulator, then changes in outputs of a sensor are caused by the position changes of the manipulator. Therefore, new SIS's inputs consist of informations on both the change in sensor outputs and the changes of the manipulator's position , together with the environmental data of a sensor. Then the new **SIS** can evaluate both the suitability for sensors' environments and the internal condition of a sensor in order to integrate the measurement data.

For this sensor integration system, the measurement system can measure the object as accurately as possible under any sensors' environments and internal conditions. The effect of this **SIS** is shown through extensive experiments.

## II. 3-D MEASUREMENT SYSTEM

The previous system [3] used plural eddy current sensors. An eddy current sensor cannot measure displacement at a distance, therefore the measurement system had to be close to the work's surface. As a result, the system could not measure a rough surface. The system also could not recognize the measurement point on the target surface where the system was measuring because the system did not have a vision system. The measurement point on the target's surface is very important for precise carving, because the position of the sensor system is related with that of the carving tool in carving.

For these problems, we propose a new measurement system, which can measure displacement from a distance and recognize the point where the system measures. The fundamental measurement system consists of a set of a CCD camera and a LED displacement sensor. In this paper, we use a pair of the measurement system in order to measure the angle between the sensors and the object's surface. The specifications of the measurement system are as follows:

LED displacement sensor:
- measurable region — 35mm to 45mm.
- measurement accuracy — 20 μm.

CCD camera 1:
- the focal distance — 7.5mm.

CCD camera 2:
- the focal distance — 15mm.

Image processing board:
- 512 pixel x 512 pixel at 256 gray scale.

For the combination of these sensors, the measurement system can measure the rough surface shape and recognize the height of the measurement system along the z axis accurately. Accuracy of the LED displacement sensor is greatly effected by a reflection of an object surface. Accuracy of the vision system does not depend on the reflection of object surface but the shape of the LED displacement sensor's spot-light. In order to integrate these sensor outputs in consideration of each sensors' specification, we use a sensor integration system based on the fuzzy inference. Fig. 1 shows the outline of the sensor integration system.

### *A. LED Displacement Sensor*

The measurement system has 2 LED displacement sensors. The sensors measure distance to the object surface along the x axis and the angle between the sensor and the object surface around the y axis. The spot-light of the LED displacement sensor is also utilized in the vision system based on the triangulation method as a marker of the measurement point on the object surface. The output of the LED displacement sensor is utilized for calibrating the camera parameters of the vision system.

### *B. Vision system*

The image data obtained through the CCD camera is utilized for measuring distance to the object surface both along the x axis and the z axis. A large number of studies have been made on the vision system based on the triangulation method. These vision systems are classified into two types, such as a passive stereo vision and an active stereo vision. The passive stereo method [11], e.g., the binocular vision and the photometric stereo, has the stereo matching problem. Accuracy of the measurement is related with that of the correspondences among different images. Furthermore, it takes much time to compute the distance. The active stereo method, e.g., using spot-light, slit-light and pattern-light [12,13], can compute the distance from single image, but this algorithm needs artificial source of light such as a laser.

In this paper, considering the image processing speed and accuracy in measurement, we utilize the active stereo method to the 3D measurement system and the source of light is the LED displacement sensor. The position of the sensor system along the z axis is determined by recognition of both spot-lights' positions and the border line between the carving part and the no-carving part on the object. Fig. 2 shows the spot-light's detection flow chart and Fig. 3 shows the border line detection flow chart.

### *C. Active stereo method*

Distance between the sensor and the object is measured from the spot-light's position of the LED displacement sensor based on the triangulation (see Figs.4 and 5). Distances along the x axis (Li) and the z axis (Zi) between the sensor and the object are calculated as follows:

$$Li = Z0 \cdot \tan(\theta 0 + \theta i) \qquad (1)$$

$$Zi = Li / \tan(\theta 0 + \theta i) \qquad (2)$$

$$\tan\theta i = dFx \cdot Fxi / fi \qquad (3)$$

where Fxi, fi, dFx, θ0, and Z0 means the gravity center

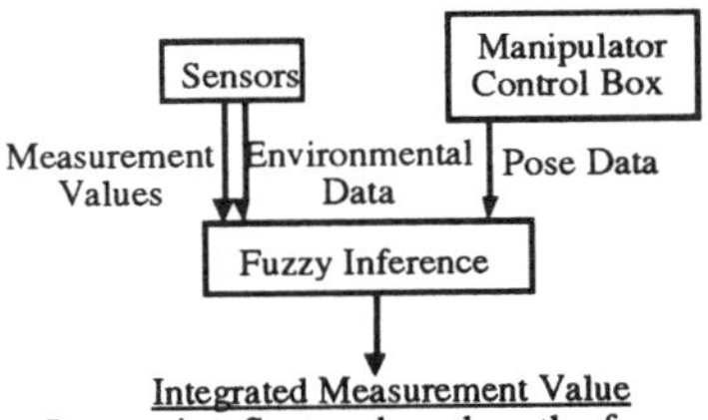

Fig.1 Sensor Integration System based on the fuzzy inference.

position of the spot-light, the focal distance of the CCD camera, a size of a pixel, the angle and the distance between the CCD camera and the LED displacement sensor respectively.

### *D. Camera calibration*

In order to use the vision system accurately, camera parameters have to be calibrated accurately. In this vision system, camera calibration is very easy and accurate because the source of light is the LED displacement sensor and outputs of the LED displacement sensor are utilized for the camera calibration data. In this system, we calibrate θ0, Z0 and dFx.

The sensor system is mounted at the tip of the manipulator. Data for the camera calibration is obtained by moving the manipulator. Data consist of both outputs of the LED displacement sensor and the gravity center position of the spot-light at each measurement point. We defined the camera parameters, which make the total error minimum between LED displacement sensor's outputs and vision system's outputs (which are calculated by eqs.1 to 3) minimum.

### *E. Neural network compensator*

Vision systems have some problems. The lens distortion is one of the problems. In addition, there is some error in camera calibration. In this system, we utilize a **NN** for compensation of the lens distortion and camera calibration's error, since NNs have a function of interpolation of the learning data. The structure of the NN is 3 layers (input layer 1 unit, hidden layer 30 units, and output layer 1 unit) and we use NN for each spot-light of the LED displacement sensor.

An input of the NN is the gravity center position of the LED displacement sensor and an output of the NN is compensation value of the gravity center position. We use the back propagation algorithm [14] for leaning. Equations concerning the input/output of NN are as follows:

$$In = Gn/512 \quad (4)$$

$$Tn = (Rn - Gn + 40)/80 \quad (5)$$

where In, Gn, Tn, and Rn means the input of the NN, the gravity center position of the LED displacement sensor, the training data, and the gravity center position calculated from the distance between sensors and the object respectively. In this paper, we determine the maximum of compensation as 40 pixels. Fig.6 shows the construction of the vision system.

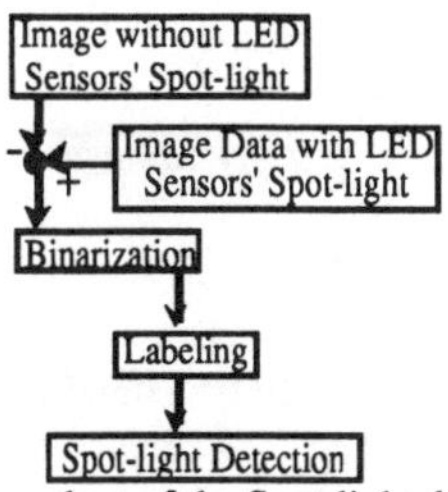

Fig.2 Flow chart of the Spot-light detection.

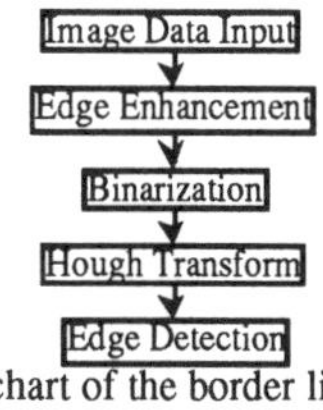

Fig.3 Flow chart of the border line detection.

### *F. Integration of measurement values by the fuzzy inference*

In this measurement system, we can obtain 3 measurement values at the same measurement point. One is a LED displacement sensor, while the others are two CCD cameras. The measurement system is mounted at the tip of the industrial 5 axis manipulator. Therefore, if the manipulator can move accurately, we can select or integrate the measurement values by comparing the variation of each sensor output with the variation of manipulator's positions. In ordinary industrial manipulator systems, because of characteristics of a actuator, accuracy of linkage setting, friction, and backlash, there is a difference between desired trajectory and actual trajectory, and we cannot measure a magnitude of the difference.

In order to integrate the measurement values, we propose the **SIS** based on the **fuzzy inference**. We consider the rate of the variation of manipulator's positions to the variation of sensors' outputs as one of the inputs of the fuzzy inference. The rate is expressed in eq.6 as X1.

If $P(t)-P(t-1)\neq 0$:

$$X1=\frac{|S(t)-Si(t-1)-(P(t)-P(t-1))|}{|P(t)-P(t-1)|}$$

If $P(t)-P(t-1)=0$:

$$X1=P(t)-P(t-1) \quad (6)$$

where P(t) means the position of the sensor system at time t, Si(t) means the output of the SIS.

The LED displacement sensor's output is affected by the intensity of a spot-light's reflection, while the output of the vision system is affected by the aspect value of the spot-light on the image and accuracy of camera calibration not the intensity. Another fuzzy inference's input X2 is the affected factor of a sensor, such as the intensity of the spot-light of the LED displacement sensor and the aspect value of the vision sensor. The normalized value of X2 means the environmental data of measurement and if X2 is nearly equal to

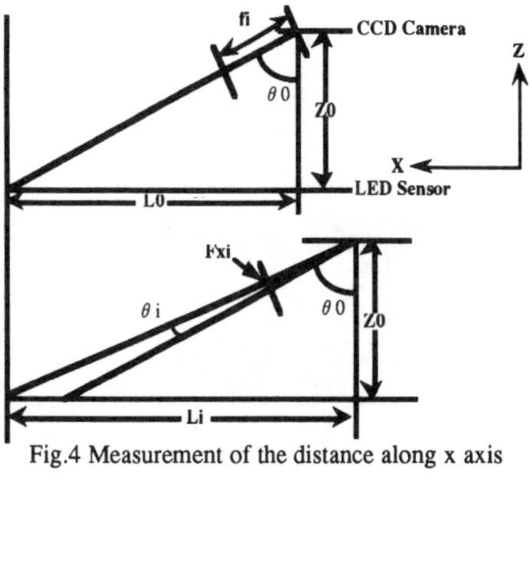

Fig.4 Measurement of the distance along x axis

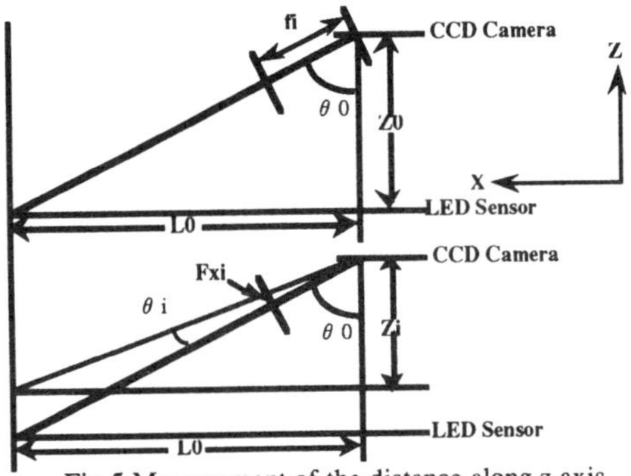

Fig.5 Measurement of the distance along z axis.

1, the condition is suitable for a sensor and in the case of X2 nearly equal to 0, the condition is almost prohibit for a sensor. Table 1 expresses the fuzzy rules. In the table, the suitability of a sensor for measuring is higher in order of S,MS,M,MB, and B. We determined the membership functions and rules by operator's experience.

The fuzzy inference's output is used as the suitability of each sensor for measurement and the SIS's output is determined by comparing with the suitabilities of all sensors.

## III. EXPERIMENTS AND RESULTS

Experiments were carried out by using the industrial 5 axis manipulator and the sensor system (see Fig. 7) set at the tip of the manipulator. A measurement object was set on the X-Z table whose accuracy was 10μm.

Three types of experiments were performed: 1) camera calibration, 2) compensation for the vision system utilizing NN and 3) sensor integration system. The reference data were obtained from the X-Z table.

### *A. Calibration and compensation*

Parameters of the vision system were calibrated by 2 ways. One was that the manipulator was moved to several different points in order to move the sensor system and then the camera parameters calibrated by the outputs of the LED displacement sensor of each point. The other was that the X-Z table was moved to several different points in order to move the object, and the parameters calibrated by the output of the X-Z table.

TABLE 1: FUZZY RULES

| | | X1 | | | | |
|---|---|---|---|---|---|---|
| | | S | MS | M | MB | B |
| X2 | S | M | MS | MS | S | S |
| | MS | M | M | MS | MS | S |
| | M | MB | M | M | MS | MS |
| | MB | MB | MB | M | M | MS |
| | B | B | MB | MB | M | M |

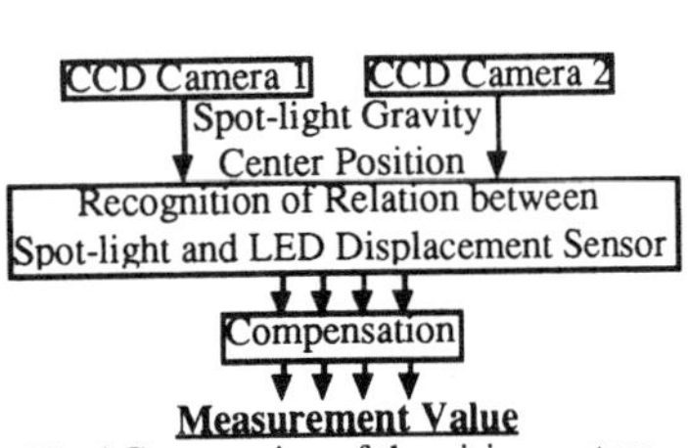

Fig.6 Construction of the vision system.

Fig.7 Experimental system

Two NNs were trained to compensate for the measurement error of the vision system. One was used for the vision system calibrated by the LED displacement sensor. The NN was trained by the output of the LED displacement sensor when the object was in the area of the LED displacement sensor's measurable region and by the manipulator's pose data when the object was out of the measurable range. The other was used for the vision system calibrated by the X-Z table and trained by the outputs of the X-Z table.

Experiments were carried out by moving the X-Z table to move the object. Fig.8 shows results of measurement of the vision system calibrated by the LED displacement sensor with/without NNs' compensation. Fig.9 shows the error between the reference data obtained from the X-Z table and the measurement values of Fig.8. Table 2 shows the average error between the measurement value and the reference.

Experimental results showed that accuracy of calibration by the X-Z table was about 2 times as high as that by the LED displacement sensor. Utilizing NN's compensator, accuracy of the vision system calibrated by the LED displacement sensor was about 5 times as high as that without NN. In the case of the X-Z table, accuracy with NN was about 2.5 times as high as that without NN. The reason of improvement of accuracy was that the vision system can measure distance between the sensors and the object accurately in all measurable area by NN's compensation for the lens distortion and camera calibration's error.

The measurement errors were increased above 70mm because the vision system was based on the triangulation method and accuracy fell with the distance. Furthermore, the intensity of the spot-light was low and the system could not detect the gravity center position of the spot-light accurately.

### *B. Experiments of sensor integration system*

We experienced the **SIS** with the environmental data of measurement and the pose data of the manipulator to compare it with other **SIS**s. Experimental **SIS**s were 3 types:

**SIS1**: Integrated by the fuzzy inference with sensors' environments and the pose data of manipulator.

TABLE 2: AVERAGE MEASUREMENT ERROR (MM)
MEASUREMENT RANGE (30MM TO 100MM) / THE CALIBRATION AREA

| | With NN | Without NN |
|---|---|---|
| LED Sensor | 0.20/0.08 | 1.04/0.13 |
| X-Z Table | 0.22/0.02 | 0.50/0.17 |

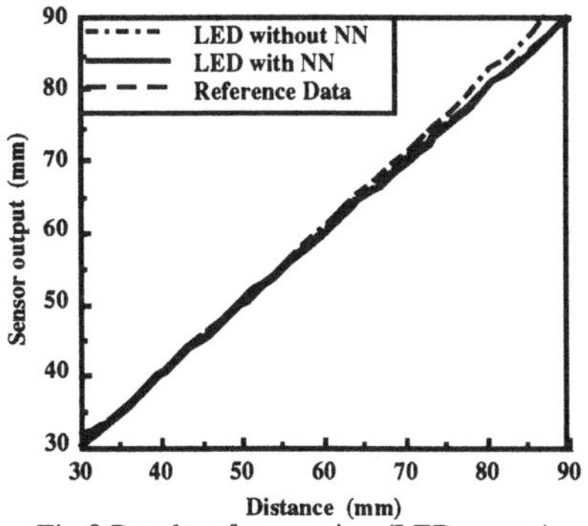

Fig.8 Results of measuring (LED sensor)

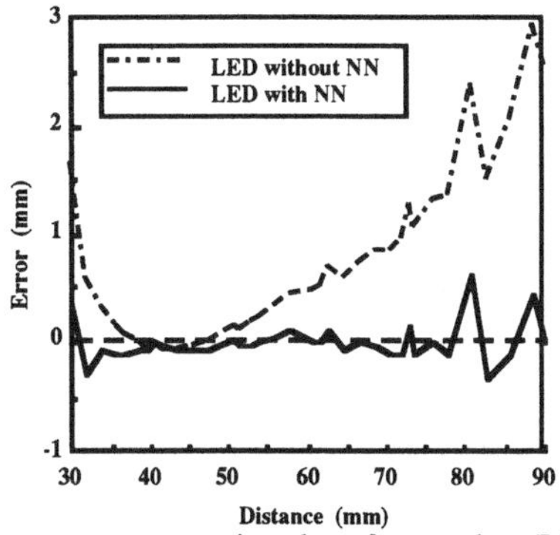

Fig.9 Measurement error against the reference data (LED sensor)

**SIS2**: Integrated by the fuzzy inference with the pose data of manipulator.

**SIS3**: Integrated by the fuzzy inference with sensors' environments.

The X-Z table was moved from 45mm to 35 mm apart from the sensory system at an interval of 1mm, in order to move the object. The sensory system measured the distance at each point in several cases of sensors' environments and sensors' internal conditions such as color of the object, a sensitivity of the LED displacement sensor, the camera parameters and errors between desired positions and actual positions of the manipulator.

Figs. 10 and 11 show experimental results and table 3 shows average errors in each condition. In the table and figures., O means color of the object, (B/black, DG/dark gray, LG/light gray, and W/white), L means the sensitivity of the LED displacement sensor, (B/for a black object and W/for a white), C means accuracy of camera parameters, (G/camera parameters with little errors and B/camera parameters with some errors), and +/0/- means errors between desired positions and actual positions of the manipulator, (+/the actual position is larger than the desired positions by 200μm, 0/the actual positions are equal to the desired positions, and -/the actual positions is smaller than the desired positions by 200μm). Figure (a), (b), and (c) in Figs.10 and 11 shows the measurement error of each sensor, **SIS1** and **SIS3**, and **SIS2** respectively.

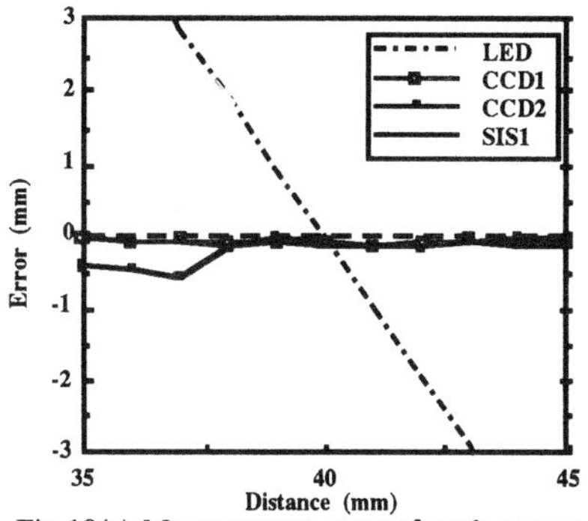

Fig.10(a) Measurement error of each sensor

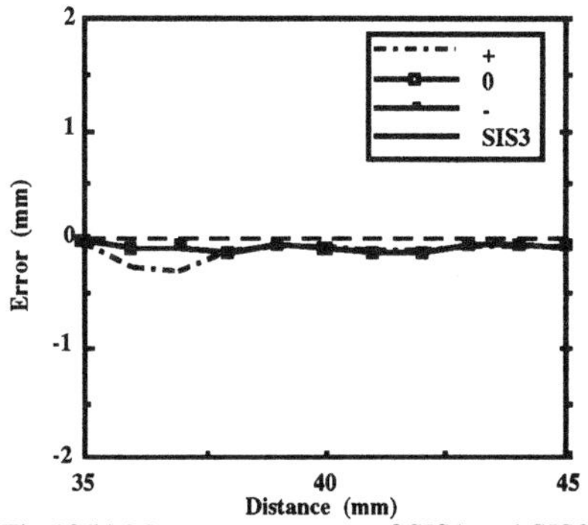

Fig.10(b) Measurement error of SIS1 and SIS3

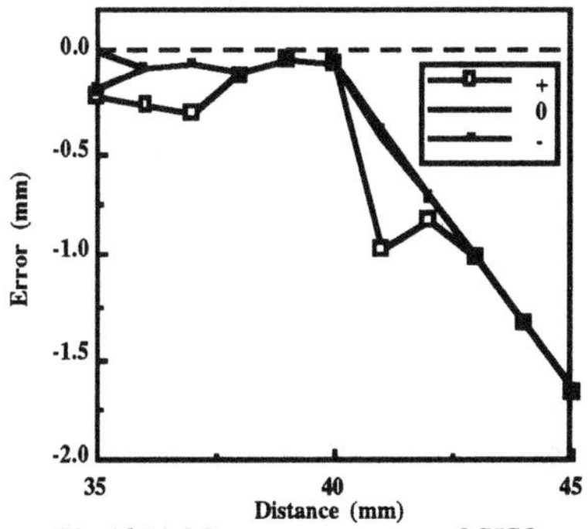

Fig.10(c) Measurement error of SIS2
Fig.10 Experimental results(O/B, L/W, C/G)

Figs. 10 show the experimental results under the following conditions: the object was black, the sensitivity of the LED parameter has little error. The LED displacement sensor could not sense the spot-light because of the unsuitable sensitivity, the sensor could not measure distance. **SIS1** and **SIS3** could measure the distance accurately, because both **SIS1** and **SIS3** had the information of environmental data of measurement for integration of measurement data and integrated the data with the appropriate sensors. Because of a lack of the information, the **SIS2** suffered measurement error from the LED displacement sensor's outputs. However, since **SIS2** integrated measurement values at each measurement with the information of the manipulator's motion, the error was decreased at each step.

Figs. 11 show the results in the case of the white object, the sensitivity for black, and camera parameters with some errors. Because of the unsuitable sensitivity, outputs of the LED displacement sensor were over-sensitive to the spot-light and had measurement errors. **SIS1** and **SIS2** could measure

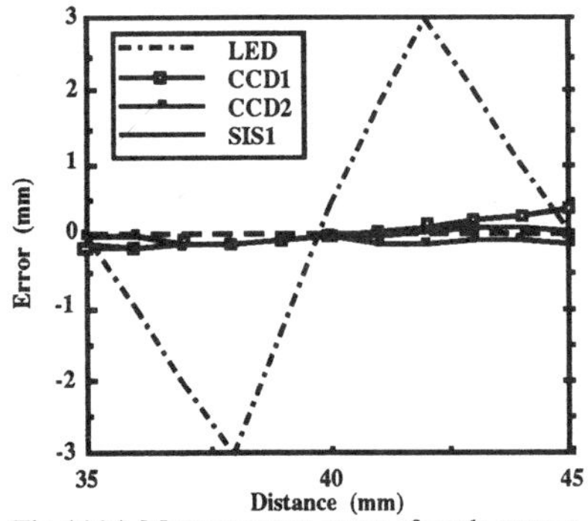

Fig.11(a) Measurement error of each sensor

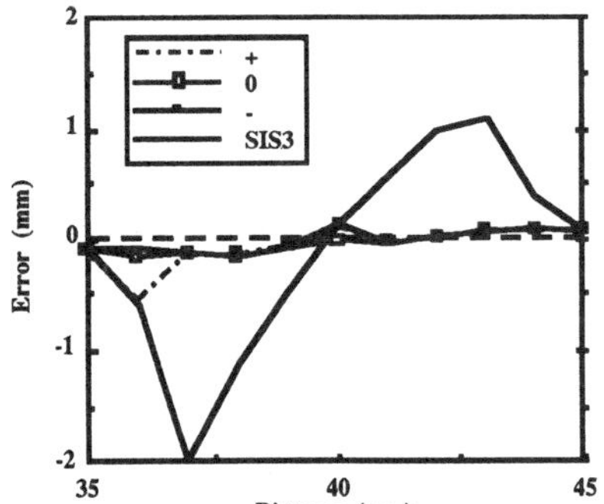

Fig.11(b) Measurement error of SIS1 and SIS3

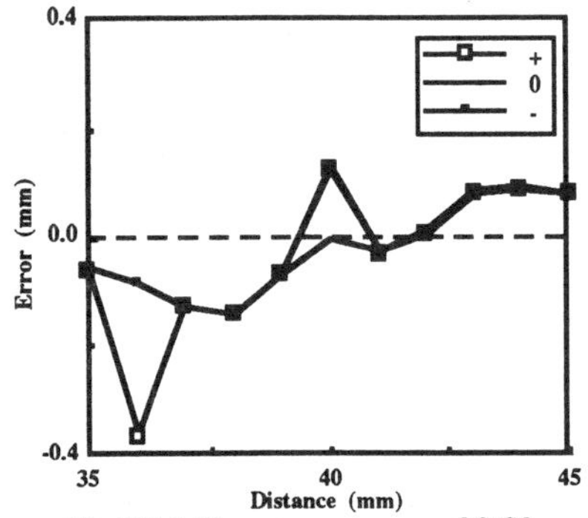

Fig.11(c) Measurement error of SIS2

Fig.11 Experimental results(O/W, L/B, C/B)

distance accurately, because of information of the manipulator's motion. Both SIS1 and SIS2 could recognize the accurate sensors and integrate the measurement values. **SIS3** could not recognize the error caused by the sensor's internal parameters and thus did not measure accurately.

From table 3, **SIS1** was the most accurate of three SISs in any sensors' external and internal conditions. **SIS1** also had a robustness against the error of the manipulator's motion.

## IV. Concluding Remarks

In this paper, we presented a 3D measurement system and its integration method as follows:

1) Camera calibration method,
2) NN's compensation for a vision system,
3) Sensor integration system based on the fuzzy inference with information of sensors' environments and the manipulator's pose.

We also showed effectiveness of the proposed SIS through some experiments as follows:

1) Utilizing outputs of the LED displacement sensor, camera parameters can be calibrated easily.
2) Using NN's compensator for the lens distortion and camera calibration's errors, the vision system can measure accurately.
3) Utilizing the fuzzy inference with moving data of the sensor system and environmental data of measurement, the measurement system can measure accurately and stable against variation of environments and internal parameters of sensors.

Future works are: 1) improvement of the image processing time, 2) parameter calibration of the manipulator for accurate sensor data integration, 3) optimization of the membership functions and rules of the fuzzy inference to optimize sensor data integration.

## REFERENCE

[1] F.M.Proctor, R.J.Norcross, K.N.Murphy: Automating Robot Programming in the Cleaning and Deburring Workstation of the AMRF, SME, Technical Paper, pp.1/11, (1989)

[2] K. Kashiwagi, K. Ono, E. Izumi, T. Kurenuma, K. Yamada: Force Controlled Robot for Grinding, IEEE Int'l Workshop on Intelligent Robots and Systems(IROS '90), pp.1001/1006, (1990).

[3] T.Fukuda, K.Shimojima, F.Arai, H.Matsuura: Multi-Sensor Integration System based on Fuzzy Inference and Neural Network for Industrial Application, IEEE Int'l Conf. on Fuzzy Systems 1992(Fuzz-IEEE '92), pp. 907/914, (1992)

[4] T.Fukuda, K.Shimojima, F.Arai, H.Matsuura: A Multi-Sensor Integration System with Fuzzy Inference and Neural Network, Pacific Rim Int'l Conf. on Artificial Intelligence 90(PRICAI '90), pp. 859/864, (1990)

[5] R.C. Luo, M.G. Kay: Multi-sensor integration and fusion in intelligent systems, IEEE Trans. on System, Man, and Cybernetics, Vol. 19, No. 5, pp. 901/931, (1989).

[6] H. F. Durrant-Whyte: Sensor models and multi-sensor integration, Int'l J. Robot. Res., Vol. 7, No.6, Dec., pp. 97/113,(1988).

[7] R. C. Luo, M. Lin: Dynamic multi-sensor data fusion system for intelligent robots, IEEE J. Robot. Automat., Vol. 4, No. 4, pp. 386/396, (1988).

[8] Y. F. Zheng: Integration of multiple sensors into a robotic system and its performance evaluation, IEEE Trans. Robotics and Automat., Vol. 5, No. 5, pp. 658/669 (1989).

[9] J. M. Richardson, K. A. Marsh: Fusion of multi-sensor data, Int'l. J. Robot. Res., Vol. 7, No. 6, pp 78/96,(1988).

[10] Y. Nakamura, Y. Xu: Geometrical fusion method for multi-sensor robotics system, Prof. IEEE Int'l. Conf. Robotics and Automat., pp. 668/673, (1989).

[11] T. Kanade, M. Okutomi: A Stereo Matching Algorithm with an Adaptive Window: Theory and Experiment, IEEE Int'l Conf. on Robotics and Automation, pp.1088/1095, (1991).

[12] K. Sato, S. Inokuchi: Three-dimensional surface measurement by space encoding range imaging, J. of Robotic System, vol.2, no.1, pp.27/39, (1985).

[13] R. Gutsche, T. Stahs, F.M. Wahl: Path Generation with a Universal 3d Sensor, IEEE Int'l Conf. on Robotics and Automation, pp. 838/843, (1991).

[14] D.E. Rumelhart, J.L. McClelland, and The PDP Research Group: Parallel Distributed Processing, The MIT Press, USA, (1986)

TABLE 3: AVERAGE OF MEASUREMENT ERROR (mm)

| Condition | Sensor | | | SIS1 | | | SIS2 | | | SIS3 |
|---|---|---|---|---|---|---|---|---|---|---|
| | LED | CCD1 | CCD2 | + | 0 | - | + | 0 | - | |
| O/W, L/W, C/G | 0.06 | 0.02 | 0.06 | 0.03 | 0.03 | 0.03 | 0.03 | 0.03 | 0.03 | 0.04 |
| O/W, L/W, C/B | 0.06 | 0.15 | 0.06 | 0.05 | 0.04 | 0.04 | 0.04 | 0.04 | 0.05 | 0.05 |
| O/W, L/B, C/G | 1.42 | 0.02 | 0.07 | 0.09 | 0.04 | 0.05 | 0.05 | 0.04 | 0.05 | 0.64 |
| O/W, L/B, C/B | 1.42 | 0.15 | 0.07 | 0.13 | 0.08 | 0.08 | 0.10 | 0.07 | 0.08 | 0.69 |
| O/LG, L/W, C/G | 0.06 | 0.07 | 0.10 | 0.05 | 0.05 | 0.05 | 0.06 | 0.04 | 0.05 | 0.06 |
| O/DG, L/W, C/B | 0.14 | 0.14 | 0.12 | 0.12 | 0.11 | 0.07 | 0.08 | 0.10 | 0.07 | 0.12 |
| O/B, L/W, C/G | 2.63 | 0.07 | 0.18 | 0.11 | 0.07 | 0.07 | 0.62 | 0.50 | 0.51 | 0.07 |
| O/B, L/W, C/B | 2.63 | 0.16 | 0.18 | 0.15 | 0.13 | 0.13 | 0.62 | 0.55 | 0.54 | 0.13 |
| O/B, L/B, C/G | 0.12 | 0.09 | 0.25 | 0.15 | 0.11 | 0.11 | 0.16 | 0.11 | 0.11 | 0.11 |
| O/B, L/B, C/B | 0.12 | 0.16 | 0.25 | 0.16 | 0.13 | 0.15 | 0.15 | 0.12 | 0.15 | 0.13 |

# Fuzzy Logic Technology & the Intelligent Highway System (IHS)

Bruno Bosacchi
AT&T Bell Laboratories - Princeton NJ 08542-0900
&
Ichiro Masaki
General Motors Research Laboratories - Warren MI 48090

*Abstract* - **This paper presents an overview of the applications of Fuzzy Logic technology (FLT) to the Intelligent Highway System and, more specifically, to one of its essential components, the Intelligent Vehicle (IV). In particular, status and trends of the effort are highlighted by reviewing and discussing some representative papers from two recent conferences on the subject (Intelligent Vehicle '92 [IV'92], Tokyo, 1991, and Intelligent Vehicle '93 [IV'93], Detroit, 1992).**

## I. Introduction

Fuzzy Logic Technology (FLT) has been enjoying a remarkable popularity over the past few years, thanks to an abundance of commercial applications which have fueled the FLT development with technical challenges and financial incentives. However, with the exception of a few major projects, most of the commercial successes so far concern relatively simple problems in the home appliances and consumer electronics markets. To maintain its momentum, FLT may need to scale up its effort to more challenging tasks, in markets lucrative enough to justify and support the required R&D effort.

In this respect, that complex of efforts and technologies which in this paper we group under the names of Intelligent Highway System (IHS) and Intelligent Vehicle (IV) appears as a promising field. The IV/IHS technical challenges will be discussed later on. As for its financial reward, we will just mention a few projections on two of its most important sectors: the automotive electronics and the communication products and services associated with the flow of information in the intelligent highway network. The automotive electronics is the fastest growing segment of the electronics technology. According to forecasts by BPA, its volume in the year 2000 will be higher than the volume of the entire electronics industry in the year 1980! And according to Leading Edge Reports, the world automotive electronic market, which was $2 billion in 1980 and $27 billion in 1990, is projected to reach $61 billion in 1995! As for the IHS communication products and services, AT&T estimates that the US market alone may reach a total volume of $5 million in the next 5 years, and $212 billion over the next 20 years! [1]

In this paper we attempt to provide a picture of the FLT potential in the IV/IHS market by discussing status and trends of the field in the light of papers presented at two recent conferences specifically devoted to the subject: Intelligent Vehicle '91 [IV'91], held in Tokyo, in Nov.1991, and Intelligent Vehicle '92 [IV'92], held in Detroit in June 1992. We first present an overview of the IV/SHS effort (Sect.2), and then discuss the potential role of FLT (Sect.3). Finally, in Sect. 4, we highlight some specific FLT applications by reviewing representative papers from IV'91 and IV'92.

## II. Overview of the IV/IHS effort [2]

We denote under the umbrella name of IV/IHS a wide range of electronic-based systems and technologies aimed at assisting the driver and providing new and better ways to improve traffic efficiency and safety. This entails the realization of a well managed system of "intelligent drivers in intelligent vehicles on intelligent highways". This ubiquitous request for "intelligence" translates into concepts such as human-like behavior in the car automatic operations, and human-friendly interface between the driver on one side, and the vehicle and highway system on the other. Indeed these concepts, as shown, for example, by the Trilby plan of General Motors or the Delphi prediction of the University of Michigan,[3] are a powerful driver of the IV/IHS effort.

The pursuit of the IV/IHS objectives requires the complex interaction of a high number of players (drivers, highway authorities, car manufacturers, communication companies, commercial fleets, etc.). As a consequence, the effort is most effectively managed at the national level, with national programs such as the IVHS (Intelligent Vehicle Highway System) in the US, PROMETHEUS in Europe, and AMTICS in Japan.

For the sake of discussion, we can distinguish three main components in the IV/SHS effort: vehicle control, information, and overall system management. (In the IVHS initiative of the US, these components closely pattern the Advanced Vehicle Control System, the Advanced Driver Information System, and the Advanced Traffic Management System, respectively).

0-7803-0614-7/93/$03.00 

The vehicle control component is concerned with the use of advanced control engineering techniques and "intelligent" control technologies to assure the passengers a safe and comfortable ride. The task also includes the communication of signals across the interface vehicle/highway, to provide the vehicle with the information needed by the implementation of intelligent control functions and autonomous navigation. Problems such as gear shift control, cruise control, automatic system control, collision avoidance, autonomous navigation, etc. belong to this effort. Of course, some of these functions already exist in some form aboard to-day cars. However, the IV/IHS thrust is towards the introduction of a considerable level of "intelligence", which translates mainly in the requirement of "human-like" features in the control operations.

The information component aims at using a broad spectrum of wireless communication technologies (cellular, satellite, short range beacon, etc.) to establish a complete two way-communication (audio, video and data) between the passengers inside the car and the world outside. It is basically the implementation at the car level of the Personal Communication Network concept). An important role is played by information delivery technologies, such as image processing, voice recognition, display systems, etc., and the drive towards more "intelligence", in this effort, translates mainly in the requirement of "human-friendly" interfaces.

Finally, the overall management component aims at the centralized system management of the highway operations, gathering, in traffic management centers, information from the vehicles and from a highway network of video motion detectors, image processing systems, character recognition devices, etc., in interaction with highway dispatch centers (police, ambulances, tow-trucks, etc.). It includes the monitoring of traffic conditions, the analysis of flow patterns and traffic composition to map out traffic strategies, the redirection of vehicles to reduce traffic congestion, the rapid response to emergencies and highways accidents, the identification of vehicles, the charging of tolls, the detection of traffic rule violations, etc. Here again, though many of the above task already exist in an elementary form, the drive is towards more "intelligence". This translates mainly in the need of system optimization and quick decisions in complex and ill-defined situations.

## III. THE ROLE OF FLT IN THE IV/IHS EFFORT

In this Section we present a general discussion of the role of FLT in the IV/IHS effort, as it emerges by our reading of the 27 papers presented at IV'91 and IV'92 which, on a total of 102 contributions, were somehow related to FLT. (These papers are listed in Appendix A) [4]

The discussion of Sect.2 has shown that the IV/IHS effort encompasses tasks ranging from straightforward control engineering problems, to that complex interplay of control, communication and decision which characterizes the information and management systems in which the driver-vehicle system is embedded. This wide range of tasks offers an ideal testground for FLT, which, in the management of complex control and decision problem has not only its main objective, but even its original motivation. [5]

Vehicle control engineering includes many functions which are characterized by complex and nonlinear behavior and cannot be completely described by mathematical models. The potential of FLT for dealing with such cases is well known. Therefore, vehicle motion control is a natural target for FLT, with its ability to capture the heuristic "rules of thumb" ordinarily used by human operators. In particular, FLT is intended to provide more human-like behavior and better performance than conventional PID controllers in terms of frequency and smoothness of change in the controlling variables. This point is well documented by the commercial success of FLT in the home appliances and consumer electronics applications, and is object of strong interest in the automotive industry, as can be seen from the affiliation of the authors of the papers of App.A.

On the other hand, automotive control problems tend to be more complex than those of home appliances and consumer electronics, and are further complicated by considerations of safety, reliability, etc. Therefore more fundamental work is probably needed to compensate for the ad-hoc nature of fuzzy rule design and establish procedures to estimate the completeness and consistency of the rules and the stability of the solutions. The move of FLT towards control problems of higher complexity is also accompanied by some interesting trends, such as the emergence of VLSI hardware implementations of fuzzy logic microcontrollers and the integration of FLT with other "intelligent" technologies such as Neural Networks, Genetic Algorithms, Object Oriented Programming, etc. For example, several IV'91 & IV'92 papers integrate FLT and Neural Networks to elicit membership functions, build knowledge bases, implement fuzzy operations, etc.

At a more subtle level of control engineering, FLT looks promising in dealing with the vaguer field of human-vehicle relationship through the modeling of the human recognition, judgement and decision processes which occur during driving operations. Due to the trend towards human-like behavior, which conventional engineering concepts cannot realistically model, this is a field of increasing importance. A few papers address the possibility of using FLT to introduce in the control process variables such as driver's feelings, predictions and intentions, rather than simple reliance on the "mechanical" input from sensors.

Traditionally, the IV conferences have been mostly focused on the Intelligent Vehicle component of the IHS effort, i.e., on the vehicle control and the management of the control signals across the vehicle/highway interface. So it may not be surprising that only a few among the IV'91 & IV'92 papers address the application of FLT to the information and management systems in which the driver-vehicle system is embedded, leaving the impression of a strong unbalance in favour of the control component. This unbalance, however, is probably real, and mirrors the general situation in which FLT finds itself. In spite of the extensive and impressive literature on the unique effectiveness of FLT for decision and information processing problems in ill-defined situations, when processing of heuristic information is required, or when human interpretation plays a relevant role, significant practical applications still wait for development. Perhaps, given its commercial importance, the IV/IHS field may provide the decisive push.

## IV. Specific IV/IHS Applications of FLT

Since space requirements preclude a detailed review of all the papers of Appendix A, we have grouped them in Table I according to specific IV/IHS tasks, with the tasks approximately listed in order of increasing complexity, from straightforward control engineering to autonomous navigation and overall system management. In the review which follows, we limit our attention to a few representative contributions which illustrate and support the discussion of Sect.3, with a certain emphasis on contributions from IV'92, since a review of the papers of IV'91 has already been given.[6]

**Table 1.** Automotive Applications of FLT

| IV/IHS Application | FLT Papers |
|---|---|
| Engine Diagnostics & Control | #17 |
| Anti-Lock Braking System | #4 |
| Active Suspension System | #5, #19 |
| Integrated Chassis Control | #1 |
| Cruise Control | #13, #24 |
| Gear Shift Control | #2, #14, #15 |
| Automatic Steering Control | #7, #21- #23 |
| Electronic Brake System | #18 |
| Vision Enhancement System | #25 |
| Collision Avoidance System | #11, #12 |
| Automobile Tracking Problem | #6 |
| Autonomous Navigation System | #8- #11, #18, #21- #23, #25 |
| Traffic Signal Control | #20 |

As an example of advantageous application of FLT in a straightforward control problem, we consider paper #17, which describes the implementation and testing of a fuzzy controller (FC) for idle speed, an important parameter to optimize stability, fuel consumption, power output, drivability and exhaust emission of an engine. This paper reports that, when compared to a conventional PID controller, and to an Optimal Linear Quadratic (OLQ) regulator, both of which use mathematical models, the FC performs much better than the first, and only slightly below the second. It is further claimed that FC should outperform the OLQ in the long run, due to its superior robustness in adjusting to the unavoidable drift in engine parameters, due to wear, manufacturing tolerances, nonlinearities, operating conditions, etc., which affect the optimality of the tuning.

At a more sophisticated level of control engineering, it is claimed that FLT can outperform conventional control not only in terms of reduced overshoot and undershoot, and less frequent and smoother changes in the variables under control, but also in terms of adaptive behavior and "intelligent" operational mode. These features, for example, are observed in the cruise control systems discussed in paper #13 and #24. The first reports on an FLT constant speed cruise controller which since 1987 has been used as standard equipment on certain Japanese models. Since the traffic congestion on Japanese road does not easily allow constant speed, the controller has been being modified to allow adaptive control, i.e., the vehicle speed is automatically adjusted by a fuzzy estimate of the optimal distance from the vehicle running ahead. The second describes an intelligent human-like cruise control which optimizes speed and distance with respect to preceding vehicles, taking also into account specific driving and weather conditions.

Increasing further the sophistication level, various attempts are being made to introduce intelligent "human-like" behavior in the implementation of the control operation. A good example is provided by the problem of automatic gear shift control, discussed in several papers (#2, #14, #15). The engineering of automatic transmission is by now a mature field. Yet, it is a common experience of automatic transmission car drivers that gears do not always shift properly under certain driving conditions. For example, when the driver lets up on the accelerator to navigate a slight downgrade while driving an uphill slope, the gear may automatically shift, even when, seeing the uphill ahead, a "human" driver would avoid needless shifting. The same situation is encountered on a winding road.

To introduce human-like behavior in the automatic transmission problem one needs to "input" the driving environment in the control system. In paper #14 this is achieved using a multiplicity of inputs (car speed, throttle opening, engine speed, engine negative pressure, shift position and brake signals) to generate IF...THEN rules which take into account common sense experience and driver intentions. Papers #2 and #15 take a different approach inferring the driving environment from the pattern of accelerating behavior of the driver. In particular, they assume that car speed, displacement and rate of displacement of the accelerator pedal sufficiently characterize the driving environment. In paper #2, data related to the driving pattern (accelerator inputs, speed changes, brake and steering input, and road conditions) of a large number of drivers on a large variety of roads are recorded and statistically analyzed to extract distinctive features which are then used to create fuzzy logic (FL) rules and membership functions. During operation, the vehicle speed, the accelerator displacement and displacement rate are used as inputs of the fuzzy inference scheme, whereas the output is computed as the minimum value (logical product) of the grade obtained from their respective membership functions. This result is compared with the value of the corresponding parameter predetermined by the statistical analysis and, with the help of the FL rules, a decision is made as to whether the vehicle is traveling in conditions requiring the switch to a more appropriate gear shifting pattern. In paper #15, the control knowledge of experience drivers is used to create a Fuzzy Controller (FC) with a desirable gear shifting behavior. This FC is realized with a neural network chip which is used to represent membership functions, implement rules and, through a novel learning algorithm, to adjust the weight of each rule, thus achieving self-tuning capabilities.

The role of FLT in the management of signals across the interface vehicle/highway, necessary to implement "intelligent" and "human-like" control operation is abundantly exemplified in the large majority of papers. The unique ability of FLT to deal with the interpretation of fuzzy signals is used in a variety of tasks, from automatic steering control, cruise control, collision avoidance, etc. up to the complex interplay of information and control of system operations such as autonomous navigation.

A typical example is provided by paper #7, which reports on a fuzzy controller for human-like automatic steering control, using fuzzy inferences through a fuzzy LSI chip and real-time self-adjustment of the membership functions. Human like operation in automatic steering control is also pursued in papers 21, 22, and 23. Paper 21 even attempts to improve on human-like behavior by eliminating human fatigue! The system has been designed to substitute human drivers and increase duration and difficulty of durability tests during the development of new models.

A second example for dealing with fuzzy signals is provided by paper #27 on the vision system of an autonomous vehicle. This system analyzes 2D images to derive scene descriptions which are used to control the vehicle as it moves in a 3D environment. Deriving scene description requires searching and matching the extracted features against the features of the model base. This is a very complex task due to the feature instabilities caused by changing viewpoints. Paper #27 proposes using viewpoint invariance relations (VIRs) on line segments, such as parallelism, collinearity, proximity, as the primitives on which to base the analysis of the segmentation results. Whereas in a noise-free image the extraction of VIR's is relatively straightforward, in the actual practice the task is strongly complicated by the uncertainties of the image formation process and by the imprecision of the information. For example, a pair of parallel segments characterizing a scene may only appear as "nearly" parallel. The paper proposes to use FLT to model these uncertainties. For example, the parallelism of a pair of "approximately parallel" segments is quantitatively characterized by the angle 0 between their directions, and the fuzzy set "approximate parallelism" (AP) is represented by the membership function u(AP)=0. A method ("anchoring & adjusting") is then introduced to aggregate the various pieces of fuzzy information.

A final example on the management of fuzzy signals, which also anticipates a prototype situation in any future traffic information and management system, is the development of a fuzzy map-matching algorithm for autonomous navigation, discussed in paper #16. In general, the vehicle location is obtained by processing inputs from positioning sensors and is then related to a geographic database. Due to sensor errors and database imperfections, the map-matching process involves many ambiguities and a resulting vagueness in the determination of the vehicle location in the road network. The fuzzy algorithm proposed in #16 uses FL to manage this kind of problems.

We conclude the review with two other examples of FLT application to information and traffic management problems. Paper #20 develops an adaptive distributed approach to traffic signal control, in which the signal timing parameters at a given intersection are adjusted as functions of the local traffic conditions and of the signal timing parameters at adjacent intersections. The signal timing at an intersection is defined by three parameters: cycle time, phase split, and offset. Fuzzy decision rules are employed to adjust these parameters based only on local information. According to the author, this approach provides a fault-tolerant, highly responsive traffic management system, whose effectiveness has been demonstrated through a simulation of the traffic flow in a network of controlled intersections.

Finally, in paper #18, a discussion is given of an exploratory study on the application of fuzzy control to emergency braking in the framework of ISIS, a French interactive road information display system. ISIS is meant to supplement roadside signs with IR beacons which broadcast data, instructions and warnings to passing vehicles. In particular, ISIS aims at ensuring observance of speed limits and stop signs, even "dismissing" the driver who does not collaborate to slow down the vehicle or bring it to a complete stop. Preliminary results on experimental vehicles are reported as encouraging.

### V. Conclusions

We conclude this review with some broad summarizing remarks. The papers of IV'91 and IV'92 provide a good overview of the challenges and prospects, and a reasonably updated picture of status and trends of the FLT application to the the IVHS field (For more recent developments, the readers are invited to attend the 1993 edition of the conference, Intelligent Vehicles '93, which will be held Tokyo, in July '93). IVHS is a challenging and potentially rewarding field for FLT, for both control and management/decision aspects. For the latter, interesting application are beginning to appear, but its enormous potential is still virtually untapped. As for the control aspects, the volume itself of the effort, well documented by the IV'91 & IV'92 contributions, clearly shows that the FLT community has the challenge of scaling up its effort from the relatively simple successes in the consumer electronics and home appliances markets to the more demanding applications in the automotive field, where issues of safety and reliability are of paramount importance. An area where fundamental work is urgently needed is the development of convenient criteria to assess stability, robustness, and reliability of the solutions, to avoid the extensive simulations presently used to deal with this problem.


### Acknowledgements

The authors are indebted to several people at AT&T Bell Labs and General Motors Research and Environment Staff, in particular S.R.Nagel and G.G.Dodd, for their interest and encouragement.


### References


[1] AT&T Focus, Nov. 1992, p.18

[2] For a review, see, for example, the Special Issue of IEEE Trans. on Vehicular Technology, **40**, #1, (Feb. 1991).

[3] The University of Michigan: Delphi IV, Vol.2, Technology (TECH-63A), p.172-173

[4] Information on the Proceedings of IV'91 and IV'92 can be obtained from Ichiro Masaki. See also "Vision-Based Vehicle Guidance", ed by I.Masaki (Springer-Verlag, 1991)

[5] L.A.Zadeh Inf. and Control **8**, 338-352 (1965)

[6] B.Bosacchi & I.Masaki in Science of Artificial Neural Networks, ed. by D.W.Ruck, SPIE Proc. #**1710**, p. 686, (1992)


## APPENDIX A

*List of FLT papers presented at IV'91 & IV'92*

### IV'91

1. Y.Uyeda (OMRON, Japan) *Improvement of Chassis Control by Fuzzy Controller*
2. H.Takahashi (Nissan Motor Co., Japan) *A Method of Detecting the Driving Environment Using Fuzzy Reasoning*
3. M.J.Patyra (Carnegie Mellon University, USA) *Implementation of Fuzzy Operations with Neural Networks*
4. P.M.Basehore, J.T.Yestrebsky, K.A.Tucker (American NeuraLogix, Inc., USA) *Innovative Architecture and Theoretical Considerations Yield Efficient Fuzzy Logic Controller VLSI Design*
5. L.A.Feldkamp, G.V.Puskorius, L.I.Davis, F.Yuan (Ford Motor Co., USA) *Decoupled Kalman Training of Neural and Fuzzy Controllers for Automotive Systems*
6. J.R.Jang (University of California, Berkeley, USA) *A Self-Learning Fuzzy Controller with Application to Automobile Tracking Problems*
7. Y.Hashimoto, T.Shigematsu, K.Ohnishi, N.Ohta (Toyota Motor Co., Toyota Central R&D Labs., Inc., Japan) *Application of Fuzzy Algorithm to Automatic Steering Control*
8. T.Hessburg, M.Tomizuka (University of California, Berkeley, USA) *A Fuzzy Rule Based Controller for Automotive Vehicle Guidance*
9. E.H.Ruspini, D.C.Ruspini (SRI International, & NASA Ames Research Center, USA) *Autonomous Vehicle Motion Planning Using Fuzzy Logic*
10. L.Huang, W.Kao, H.Oshizawa, M.Tomizuka (ZEXEL USA Co. & University of California at Berkeley, USA) *A Fuzzy Map-Matching Algorithm for Automotive Navigation Systems*
11. R.N.Lea, Y.Jani, M.G.Murphy, M.Togai (NASA, LinCom Co., University of Houston, Togai InfraLogic Inc., USA.) *Design and Performance of a Fuzzy Logic Based Vehicle Controller for Autonomous Collision Avoidance*
12. R.Fujioka (OMRON, Japan) *An Anticollision System Concept*
13. A.Takayama, A.Hirako (Isuzu Advanced Engineering Center, Ltd., Japan) *Adaptive Cruise Control According to Optimal Distance*

14. I.Sakai, S.Sakaguchi, T.Haga, M.Togai (Honda R&D & Togai InfraLogic,Inc., Japan) *Shift Scheduling Method of Automatic Transmission Vehicles Using Fuzzy Logic*

15. Y.Dote, K.Hayashi, J.Nasu, M.Strefezza, A.Takayama, A.Hirako (Muroran Institute of Technology & Isuzu Advanced Engineering Center, Ltd., Japan) *Neuro Fuzzy Transmission Control for Automobile*

16. B.S.Widmann, G.R.Widmann (GM Hughes Aircraft Co., USA) *Implementation of Object Oriented Constructs for Use in Fuzzy Logic Classifiers*

17. M.Abate (Centro Ricerche FIAT - Orbassano, Italy) *An Application of Fuzzy Logic to Engine Control*

18. J.P.Aurrand-Lions, L.Fournier, P.Jarri, M. de Saint Blancard, E.Sanchez (University of Aix-Marseille II & IMT, Peugeot S.A., France) *Application of Fuzzy Control for ISIS Vehicle Braking*

### IV'92

19. E.C.Yeh, H.J.Tsao (National Tsing Hua University, R.O.C.) *Fuzzy Control for Active Suspension Design*

20. S.Chiu (Rockwell International, U.S.A.) *Adaptive Traffic Signal Control Using Fuzzy Logic*

21. K.Ohnishi, J.Komura, T.Ishibashi (Toyota Motor Corporation, Japan) *Development of Automatic Driving System on Rough Road - Realization of Highly Reliable Automatic Driving System*

22. T.Shigematu, Y.Hashimoto, T.Watanabe (Toyota Motor Corporation, Japan) *Development of Automatic Driving System on Rough Road - Automatic Steering Control by Fuzzy Algorithm*

23. N.Ooka, T.Tsuboi, H.Oka (Nippondenso Co., Ltd., Japan) *Development of Automatic Driving System on Rough Road - Fault Tolerant Structure for Electronic Controller*

24. R.Muller, G.Nockler (Daimler-Benz AG, Germany) *Intelligent Cruise Control with Fuzzy Logic*

25. B.S.Widmann, G.R.Widmann (GM Hughes Aircraft Co., USA) *Fuzzy Logic Object Oriented Classes*

26. S.Baocheng, L.Xihui, Z.F.Zhang (Chinese Academies of E.I. & of SINICA) *A New Approach to Fuzzy Control by Using Neural Networks*

27. S.Toh (Arizona State U., U.S.A.) *Extracting Viewpoint Invariance Relations Using Fuzzy Sets*

# Toward a Logic for Fuzzy Syllogisms *

Daniel G. Schwartz
Department of Computer Science
Florida State University
Tallahassee, Florida 32306, U.S.A.
schwartz@cs.fsu.edu

***Abstract*— This paper describes a formal logic that encodes the reasoning embodied in syllogisms of the form "*Most* birds can fly; Tweety is a bird; therefore, it is *likely* that Tweety can fly" and of the form "*Usually,* if something is a bird, it can fly; Tweety is a bird; therefore, it is *likely* that Tweety can fly." More exactly, we define a formal language in which one can express syllogisms of these and related forms, and then this language is provided with a rigorously defined semantics in which such syllogisms are valid. In process we establish a general approach to formalizing the notions of fuzzy quantification (*all, most, many, few,* etc.), usuality (*always, usually, often, seldom,* etc.), and likelihood (*certain, likely, uncertain, unlikely,* etc.). Some of the interrelations between these concepts that are expressible in this formalism are also explored.**

## Introduction

The subject of fuzzy quantification, dealing with modifiers like *all, most, many, few,* etc., was first discussed by Zadeh and Bellman [27] and subsequently was developed at length by Zadeh [22–25]. Zadeh [25] also introduced the notion of usuality, dealing with modifiers like *always, usually, often, seldom,* etc., and those studies were continued in [26]. The latter two papers additionally mention, but do not explore, an apparent connection between usuality and quantification.

Usuality logic is also termed "dispositional logic" by Zadeh, in reference to the fact that when one knows that a certain phenomenon "usually" occurs, then one is "disposed" to expect that it will continue to occur in the same way in the future.[1] This tacitly suggests a connection between usuality and likelihood, which deals with terms like *certain, likely, uncertain, unlikely,* etc. No explicit mention of such a connection appears in Zadeh's papers, but fuzzy likelihood (or "fuzzy probability") was studied by him as early as [21], and was taken up again in [24].

In developing a semantics for assertions like "*Most* students are *young*," where *young* is given as a fuzzy subset of a collection of ages, Zadeh interprets *most* as a fuzzy subset $S$ of the unit interval, computes a real number referred to as the *relative sigma count* of the number of students which are young, and then reinterprets the original assertion as being equivalent with the assertion "$\Sigma$count(*young/student*) is $S$." The *sigma count* of a fuzzy set, e.g., $\Sigma$count(*young*), is a generalization of the notion of cardinality for crisp sets, so that the relative sigma count is the corresponding generalization of the relative cardinalities of crisp sets, i.e., it is the sigma count of the number of students which are *young* divided by the total number of students. Thus the semantics based on sigma counts is a natural and straightforward generalization to fuzzy sets of the simple probability measure that is based on relative cardinalities of crisp sets. In fact, if one applies the formula for the relative sigma count to only crisp sets, one gets back the standard probability measure. Zadeh's later papers point out that essentially the same semantics works as well for usuality.

In this approach, the degree of compatibility that the computed relative sigma count has with, say, the term *most* (i.e., the degree of membership of the sigma count with the fuzzy set $S$) is implicitly interpreted as a degree of truth for the overall assertion. Thus propositions like the one given above are not categorically either *true* or *false*, but have intermediary gradations. This feature has intuitive appeal, but it also introduces an added level of complexity. The semantics becomes bivalent, and correspondingly simpler, if one interprets the quantifiers as crisp subintervals rather than as fuzzy subsets of $[0, 1]$.

Another approach to fuzzy quantification has been explored by Yager [14–20]. Yager's semantics begins with

*Supported by the U.S. Office of Naval Research, grant number N00014-92-J-1749.

[1] Private communication during IPMU-92 (Information Processing and the Management of Uncertainty in Knowledge-Based Systems), Palma de Mallorca, Spain, July 6-10.

0-7803-0614-7/93$03.00 ©1993IEEE

the observation that the classical "for all" as in "*All* students are young," is logically equivalent with the conjunction "student-1 is young AND student-2 is young AND etc.," where "student-1," "student-2," etc. are all the individual students in the underlying domain of discourse. Similarly the classical "there exists" as in "*There exists* a student that is young" is logically equivalent with the disjunction "student-1 is young OR student-2 is young OR etc." Taking the fuzzy quantifiers like *most* and *few* as being intermediary between *for all* and *there exists*, Yager represents these as operators which are weakenings of the logical AND and/or strengthenings of the logical OR.[2] These operators, known as *ordered weighted averages*, or OWA operators, are shown to have many of the algebraic properties one would desire of fuzzy quantifiers.

Further exploration of the probabilistic interpretation of fuzzy quantifiers has been conducted by Dubois and Prade [5], Amarger, Epenoy, and Grihon [1], and Amarger, Dubois, and Prade [3–5]. These works mostly consider the case of fuzzy quantifiers applied to propositions involving only crisp predicates and where the quantifiers are given as subintervals of $[0, 1]$. The authors limit themselves to this simpler case for the purpose of studying the problem of inference chaining, but clearly reserve the option of generalizing to the case of fuzzy predicates and fuzzy quantifiers in the original sense of Zadeh. Some preliminary efforts in the latter direction have appeared in [5].

The main problem addressed in those studies is to answer the following, for arbitrary fuzzy quantifiers $Q_1, Q_2$: Given that $Q_1$ $A$'s are $B$'s, and that $Q_2$ $B$'s are $C$'s, what can be said about how many $A$'s might be $C$'s? It is shown that when $Q_1$ and $Q_2$ are interpreted as subintervals of probabilities, then an algorithm can be applied to compute upper and lower bounds for the probability that an $A$ will be a $C$. The interval determined by these upper and lower bounds is normally larger, and hence less precise, than the ones started with. This correlates with the normal intuition that imprecision accumulates as the reasoning chains increase in length.

The present paper concerns a related but somewhat different problem. This is to lay out precisely the relation between the concepts of fuzzy quantification, usuality, and likelihood. The key ideas originated several years ago with the desire to devise a logic which captures the reasoning embodied in syllogisms such as

> *Most* birds can fly.
> Tweety is a bird.
> ―――――――――――
> It is *likely* that Tweety can fly.

and

> *Usually*, if something is a bird, it can fly.
> Tweety is a bird.
> ―――――――――――
> It is *likely* that Tweety can fly.

Schwartz [10] laid down the intuitive underpinnings for a possible semantics which would tie the three concepts together. It was observed that the kind of relations expressed in the above syllogisms apply as well across all five lines of the following table.

| *Quantification* | *Usuality* | *Likelihood* |
|---|---|---|
| all | always | certainly |
| most | usually | likely |
| many/about half | frequently/often | uncertain |
| few/some | occasionally/seldom | unlikely |
| no | never | certainly not |

The foregoing syllogisms show the implicit connection between the three modifiers appearing on the second line. As another example, the syllogism "*Few* cats have no tails; Felix is a cat; therefore it is *unlikely* that Felix has no tail" similarly illustrates the fourth line. The present work considers only the above five modifiers in each category, inasmuch as this serves to illustrate the general methodology. The table may of course be expanded by adding further modifiers, like *almost all, very many, almost never, very unlikely*, and so on. The psychological studies summarized by Mosteller and Youtz [8] list 52 different modifiers related to either likelihood or usuality.

After [10] there followed [11] and [12], which sought to develop a two-leveled formal language suitable for expressing the above relationships in a semantics that employs min-max interpretations of the logical AND and OR. This approach was adopted because it yields rather pleasant algebraic properties, establishing in particular a convenient interrelation between the two linguistic levels. Shortly after writing the latter paper, however, this approach was abandoned when it became clear that it does not lead to an intuitively acceptable interpretation of the logical IMPLIES. It was decided at this point to return to the probabilistic approach originally proposed by Zadeh. Subsequently, Dubois and Prade published a note [6] which gives an even more compelling reason to abandon that version of the min-max approach, this being that it is far too restrictive to have any practical applications.

The ideas presented here represent the more recent change of direction. A logic dubbed QUAL 1.0 (Quantification, Usuality, And Likelihood, Version 1.0) is presented, which addresses the problem of developing a formal language and semantics adequate for expressing such syllogisms as the foregoing, as well as other intuitively plausible interrelationships between the three concepts. As in the works by Amarger, et. al., this considers only the simplest case of interval-based modifiers applied to

[2]These remarks are confirmed by private communication with Yager during IPMU-92.

crisp predicates. Generalization to fuzzy predicates will be considered at a later time.

The only other works known to this author that attempt to formalize this sort reasoning are those of Nilsson [9] and Halpern and Rabin [7], which focus exclusively on likelihood and employ a possible-worlds semantics. The possible-worlds approach has an established appeal, but it it does not lend itself conveniently to implementation in automated reasoning systems. It is believed that the current approach will prove to be more practical in this respect.

## LANGUAGES

Let us agree to represent the modifiers in the foregoing table, in top-down then left-right order, by $\mathcal{Q}_2$, $\mathcal{Q}_1$, $\mathcal{Q}_0$, $\mathcal{Q}_{-1}$, $\mathcal{Q}_{-2}$, $\mathcal{U}_2$, $\ldots$, $\mathcal{U}_{-2}$, $\mathcal{L}_2$, $\ldots$, $\mathcal{L}_{-2}$. In this treatment, $\mathcal{L}_0$, representing "uncertain whether," expresses uncertainty in the sense of there being positive information to the effect that the proposition in question is about 50% likely to be true, as opposed to uncertainty due to a lack of information in this regard. Furthermore, the usuality modifiers here express only those aspects by which the English language modifiers correlate with their corresponding quantifiers and likelihood modifiers—e.g, $\mathcal{U}_1$, representing "usually," represents the sense in which this means "in most instances," but not the sense it which it may mean "for most *times*." Formulating the latter would require a much more complicated system, amounting to a full-fledged temporal logic in the sense of Shoham [13].

As *symbols* let us employ: exactly one *(individual) variable*, denoted by $x$; some *(individual) constants*, denoted generically by $a, b, \ldots$; some unary *relation symbols*, denoted generically by $\alpha, \beta, \ldots$; some *logical connectives*, denoted by $\neg, \vee, \wedge, \rightarrow, \dot{\rightarrow}, \ddot{\neg}$, and $\ddot{\vee}$; the abovementioned *quantifiers*, *usuality modifiers*, and *likelihood modifiers*; and *parentheses* and *comma*, denoted as usual. As *formulas* we shall have the members of the sets

$$\mathrm{F}_1 = \{\alpha(x) | \alpha \text{ is a relation symbol}\}$$

$$\mathrm{F}_2 = \mathrm{F}_1 \cup \{\neg P, (P \vee Q), (P \wedge Q) | P, Q \in \mathrm{F}_1 \cup \mathrm{F}_2\}$$

$$\mathrm{F}_3 = \{(P \rightarrow Q) | P, Q \in \mathrm{F}_2\}$$

$$\mathrm{F}_4 = \{\mathcal{Q}_2\mathcal{L}_i P, \mathcal{Q}_i\mathcal{L}_2 P, \mathcal{Q}_2(\mathcal{L}_2 P \dot{\rightarrow} \mathcal{L}_i Q), \mathcal{U}_2\mathcal{L}_i P, \mathcal{U}_i\mathcal{L}_2 P, \mathcal{U}_2(\mathcal{L}_2 P \dot{\rightarrow} \mathcal{L}_i Q) | P, Q \in \mathrm{F}_2 \cup \mathrm{F}_3\}$$

$$\mathrm{F}_5 = \mathrm{F}_4 \cup \{\ddot{\neg} P, (P \ddot{\vee} Q) | P, Q \in \mathrm{F}_4 \cup \mathrm{F}_5\}$$

$$\mathrm{F}'_1 = \{P(a/x) | P \in \mathrm{F}_1 \text{ and } a \text{ is an individual constant}\}$$

$$\mathrm{F}'_2 = \{P(a/x) | P \in \mathrm{F}_2 \text{ and } a \text{ is an individual constant}\}$$

$$\mathrm{F}'_3 = \{P(a/x) | P \in \mathrm{F}_3 \text{ and } a \text{ is an individual constant}\}$$

$$\mathrm{F}'_4 = \{\mathcal{L}_i P(a/x), (\mathcal{L}_2 P \dot{\rightarrow} \mathcal{L}_i Q)(a/x) | P, Q \in \mathrm{F}_2 \cup \mathrm{F}_3 \text{ and } a \text{ is an individual constant}\}$$

$$\mathrm{F}'_5 = \mathrm{F}'_4 \cup \{\ddot{\neg} P, (P \ddot{\vee} Q) | P, Q \in \mathrm{F}'_4 \cup \mathrm{F}'_5\}$$

where $P(a/x)$ denotes the formula obtained from $P$ by replacing every occurrence of the variable $x$ by an occurrence of the constant $a$. As abbreviations let us employ

$$(P \ddot{\wedge} Q) \quad \text{for} \quad \ddot{\neg}(\ddot{\neg} P \ddot{\vee} \ddot{\neg} Q)$$

$$(P \ddot{\rightarrow} Q) \quad \text{for} \quad (\ddot{\neg} P \ddot{\vee} Q)$$

$$(P \ddot{\leftrightarrow} Q) \quad \text{for} \quad ((P \ddot{\rightarrow} Q) \ddot{\wedge} (Q \ddot{\rightarrow} P))$$

By a *language* $\Lambda$ will be meant any collection of symbols and formulas as describe above. Languages thus differ from one another essentially only in their choice of individual constants and relation symbols. Depending on how these symbol sets are chosen, the languages may be either finite or infinite.

To illustrate, the first of the foregoing syllogisms may be represented in a language employing the individual constant Tweety and the unary relations Bird and CanFly as

$$\begin{array}{l} \mathcal{Q}_1\mathcal{L}_2(\mathrm{Bird}(x) \rightarrow \mathrm{CanFly}(x)) \\ \underline{\mathcal{L}_2\mathrm{Bird}(\mathrm{Tweety}) \qquad\qquad\qquad} \\ \mathcal{L}_1\mathrm{CanFly}(\mathrm{Tweety}) \end{array}$$

In words: For *most* $x$ it is *certain* that, if $x$ is a Bird then $x$ CanFly; it is *certain* that Tweety is a Bird; therefore it is *likely* that Tweety CanFly.

## SEMANTICS

An interpretation $I$ for a language $\Lambda$ will consist of: a *universe* $U_I$ of *individuals* (here assume $U_I$ is finite); assignment of a unique individual $a_I \in U_I$ to each individual constant $a$ of $\Lambda$; assignment of a unique unary relation $\alpha_I$ on $U_I$ to each relation symbol $\alpha$ of $\Lambda$; a *likelihood mapping* $l_I$ which associates each formula in $\mathrm{F}_2 \cup \mathrm{F}_3 \cup \mathrm{F}'_2 \cup \mathrm{F}'_3$ with a number in $[0,1]$ and a *truth valuation* $v_I$ which associates each formula in $\mathrm{F}_4 \cup \mathrm{F}_5 \cup \mathrm{F}'_4 \cup \mathrm{F}'_5$ with a *truth value* $T$ or $F$. The subscript $I$ will be dropped when the intended meaning is clear.

Given assignments for the individual constants and unary relation symbols, the mappings $l$ and $v$ are induced in the following way. Observe that the assignments $\alpha_I$ induce the assignment of a unique subset $P_I$ of $U_I$ to each formula $P \in \mathrm{F}_2$ by

$$(\neg P)_I = (P_I)^c$$

$$(P \vee Q)_I = P_I \cup Q_I$$

$$(P \wedge Q)_I = P_I \cap Q_I$$

For subsets $X \subset U$ let a *proportional size* $\sigma$ be defined by

$$\sigma(X) = |X|/|U|$$

where $|\cdot|$ denotes cardinality. Then $l$ is defined by

for $P \in \mathrm{F}_2$, $l(P) = \sigma(P_I)$

for $P \in \mathrm{F}_3$, $l(P \rightarrow Q) = \sigma(P_I \cap Q_I)/\sigma(P_I)$

for $P(a/x) \in \mathrm{F}'_2 \cup \mathrm{F}'_3$, $l(P(a/x)) = l(P)$

[*Note*: The latter is a very strong condition which forbids any form of likelihood maintenance and will be modified in future extensions.]

Next we select for each $i = -2, \ldots, 2$ a *likelihood interval* $\iota_i \subseteq [0,1]$ according to a scheme such as

$\iota_2 = [1,1]$ (singleton 1)

$\iota_1 = [\frac{2}{3}, 1)$

$\iota_0 = (\frac{1}{3}, \frac{2}{3})$

$\iota_{-1} = (0, \frac{1}{3}]$

$\iota_{-2} = [0,0]$ (singleton 0)

These intervals are associated below with the corresponding likelihood modifiers. It should be clear that the selection of such intervals is to a certain extent arbitrary. There have been numerous studies to determine what probabilities human subjects tend to associate with these and other modifiers (cf. [8]), and one may if one wishes choose intervals which more closely match the empirical results. Given such intervals, the valuation mapping $v$ is defined by

$v(\mathcal{Q}_2\mathcal{L}_i P) = T$ iff $l(P) \in \iota_i$

$v(\mathcal{U}_2\mathcal{L}_i P) = T$ iff $l(P) \in \iota_i$

$v(\mathcal{Q}_i\mathcal{L}_2 P) = T$ iff $l(P) \in \iota_i$

$v(\mathcal{U}_i\mathcal{L}_2 P) = T$ iff $l(P) \in \iota_i$

$v(\mathcal{Q}_2(\mathcal{L}_2 P \dot{\rightarrow} \mathcal{L}_i Q)) = T$ iff $l(P \rightarrow Q) \in \iota_i$

$v(\mathcal{U}_2(\mathcal{L}_2 P \dot{\rightarrow} \mathcal{L}_i Q)) = T$ iff $l(P \rightarrow Q) \in \iota_i$

$v(\mathcal{L}_i P) = T$ iff $l(P) \in \iota_i$

$v(\mathcal{L}_2 P \dot{\rightarrow} \mathcal{L}_i Q) = T$ iff $l(P \rightarrow Q) \in \iota_i$

$v(\ddot{\neg} P) = T$ iff $V(P) = F$

$v(P \ddot{\vee} Q) = T$ iff either $v(P) = T$ or $v(Q) = T$

## Key Properties

It is easy to see that the foregoing syllogism is validated by any interpretation $I$ of the kind just described, i.e., if the premises of the syllogism are both $T$ in $I$, then so also will be the conclusion. More generally, it can be seen that this semantics validates classical modus ponens

$$\frac{\begin{array}{c} P \ddot{\rightarrow} Q \\ P \end{array}}{Q}$$

for all $P, Q \in \mathrm{F}_5 \cup \mathrm{F}'_5$, and the substitution rule

$$\frac{P}{P(a/x)}$$

for all $P \in \mathrm{F}_5$. In addition, it validates all formulas of the forms

$\mathcal{Q}_i\mathcal{L}_2(P \rightarrow Q) \ddot{\leftrightarrow} \mathcal{Q}_2(\mathcal{L}_2 P \dot{\rightarrow} \mathcal{L}_i Q)$

$\mathcal{U}_i\mathcal{L}_2(P \rightarrow Q) \ddot{\leftrightarrow} \mathcal{U}_2(\mathcal{L}_2 P \dot{\rightarrow} \mathcal{L}_i Q)$

$\mathcal{Q}_i P \ddot{\leftrightarrow} \mathcal{U}_i P$

which express the interrelations, respectively, between quantification and likelihood, between usuality and likelihood, and between quantification and usuality. The desire to have this feature underlies the design of the logic's rather unorthodox three-leveled linguistic structure. It may also be seen that negations behave in intuitively plausible ways. These and other properties of the logic will be discussed more fully in future works.

## References


[1] Amarger, S., Epenoy, R., and Grihon, S., Reasoning with conditional probabilities – a linear programming based method. *Proceedings of the DRUMS (Defeasible Reasoning and Uncertainty Management Systems) Esprit Project, RP2 Workshop*, Albi, France, April 26–28, 1990, pp. 154–167.

[2] Amarger, S., Dubois, D., and Prade, H., Handling imprecisely known conditional probabilities. *Proceedings of UNICOM, "AI and Computer Power: The Impact on Statistics,"* London, UK, March 13–14, 1991, 32 pp.

[3] Amarger, S., Dubois, D., and Prade, H., Constraint propagation with imprecise conditional probabilities. *Proceedings of the Seventh Conference on Uncertainty in Artificial Intelligence (UAI-91)*, UCLA, Los Angeles, CA, Morgan Kaufmann, Palo Alto, CA, July 13–15, 1991, pp. 26–34.

[4] Amarger, S., Dubois, D., and Prade, H., Imprecise quantifiers and conditional probabilities. In R. Kruse

and P. Siegel (eds.), *Symbolic and Quantitative Approaches to Uncertainty*, Proceedings of the European Conference ECSQAU, Marseille, France, October 1991, Springer-Verlag, NY, 1991, pp. 31–37.

[5] Dubois, D. and Prade, H., On fuzzy syllogisms. *Computational Intelligence*, 4 (1988) 171–179.

[6] Dubois, D. and Prade, H., About D.G. Schwartz's likelihood logic. *BUSEFAL: Bulletin for Studies and Exchanges on Fuzziness and its Applications*, 51 (1992) 49–52.

[7] Halpern, J.Y. and Rabin, M.O., A logic to reason about likelihood. *Artificial Intelligence*, 32 (1987) 379–405.

[8] Mosteller, F. and Youtz, C., Quantifying probabilistic expressions. *Statistical Science*, 5 (1990) 2–34.

[9] Nilsson, N.J., Probabilistic logic. *Artificial Intelligence*, 28 (1986) 71–87.

[10] Schwartz, D.G., On the relation between fuzzy quantification, usuality, and linguistic likelihood. *Proceedings of the Third International Conference on Information Processing and the Management of Uncertainty in Knowledge Based Systems, IPMU-90*, Paris, France, July 2–6, 1990, pp. 263–265.

[11] Schwartz, D.G., A min-max semantics for fuzzy likelihood, *Proceedings of the IEEE International Conference on Fuzzy Systems (FUZZ-IEEE)*, San Diego, California, March 8–12, 1992, pp. 1393-1398.

[12] Schwartz, D.G., Representing fuzzy quantifiers as modal operators, *Proceedings of the International Conference on Information Processing and Management of Uncertainty in Knowledge Based Systems, IPMU-92*, Palma de Mallorca, Spain, July 6–10, 1992, pp. 663–668.

[13] Shoham, Y., *Reasoning About Change: Time and Causation from the Standpoint of Artificial Intelligence*, MIT Press, Cambridge, MA, 1988.

[14] Yager, R.R., Quantified propositions in linguistic logic. *International Journal of Man-Machine Studies*, 19 (1983) 195–227.

[15] Yager, R.R., Quantifiers in the formulation of multiple objective decision functions. *Information Sciences*, 31 (1983) 107–139.

[16] Yager, R.R., Reasoning with fuzzy quantified statements: Part I. *Kybernetes*, 14 (1985) 233–240.

[17] Yager, R.R., Reasoning with fuzzy quantified statements: Part II. *Kybernetes*, 15 (1986) 111–120.

[18] Yager, R.R., On ordered weighted averaging operators in multi-criteria decision making. *IEEE Transactions on Systems, Man and Cybernetics*, 18 (1988) 183–190.

[19] Yager, R.R., Connectives and quantifiers in fuzzy sets. *Fuzzy Sets and Systems*, 40 (1991) 39–76.

[20] Yager, R.R., Interpreting linguistically quantified propositions. Technical Report MII-1117, Machine Intelligence Institute, Iona College, New Rochelle, NY 10801, 1992.

[21] Zadeh, L.A., The concept of a linguistic variable and its application to approximate reasoning, Part I. *Information Sciences*, 8 (1975) 199–249; Part II, 8, 301–357; Part III, 9, 43–80.

[22] Zadeh, L.A., PRUF—a meaning representation language for natural languages. *International Journal of Man-Machine Studies*, 10 (1978) 395–460.

[23] Zadeh, L.A., A computational approach to fuzzy quantifiers in natural languages. *Computers and Mathematics*, 9 (1983) 149–184.

[24] Zadeh, L.A., Fuzzy probabilities. *Information Processing and Management*, 20 (1984) 363–372.

[25] Zadeh, L.A., Syllogistic reasoning in fuzzy logic and its application to usuality and reasoning with dispositions. *IEEE Transactions on Systems, Man, and Cybernetics*, 15 (1985) 754–763.

[26] Zadeh, L.A., Outline of a theory of usuality based on fuzzy logic. In A. Jones, A. Kaufmann, and H.-J. Zimmerman (eds.), *Fuzzy Sets Theory and Applications*, Reidel, Dordrecht, Holland, 1986, pp. 79–97.

[27] Zadeh, L.A. and Bellman, R.E., Local and fuzzy logics. In J.M. Dunn and G. Epstein (eds.), *Modern Uses of Multiple-Valued Logic*, Reidel, Dordrecht, Holland, 1977, pp. 103–165.

# A Decision-Making Approach to A Syllogistic Reasoning Problem with Consistency Maintenance in a Knowledge Base*

Hiroshi Narazaki
Electronics Research Laboratory
Kobe Steel, Ltd.
5-5, Takatsukadai 1-chome, Nishi-ku, Kobe, 651-22 Japan
Ismail Burhan Turksen
Department of Industrial Engineering,
University of Toronto
Toronto, Ontario, M5S1A4, Canada

***Abstract***— **We discuss a syllogistic reasoning problem with the knowledge $Q_1(A \to B)$, $Q_2(B \to C)$ where $Q_i$, $i = 1, 2$, are imprecise quantifiers such as "Most" and "Usually" and $A$, $B$, and $C$ are fuzzy sets such as "Big cars." We propose a new approach that integrates syllogistic reasoning and knowledge-base consistency level maintenance based on the framework of multi-objective decision-making. The syllogistic reasoning problem is formulated as a default estimation problem of missing information in the observed data using the prior knowledge. The empirical knowledge is a collection of observations and it is used for the evaluation of the consistency levels of the prior knowledge.**

## I. Introduction

Let's start with a simple example as follows: Suppose that we know (1)"Most American cars are big", (2)"Most big cars are expensive", and (3)"John's car is an American car." Then the question is how expensive John's car is likely to be. This problem can be put in a more general form as follows:

**R1:** $Q_1(A \to B)$
**R2:** $Q_2(B \to C)$
**R3:** $Q_3(A \to C)$

where $A$, $B$ and $C$ are fuzzy sets and $Q_1$ and $Q_2$ are imprecise quantifiers such as "Most", and the question is what quantifier is suitable for $Q_3$, i.e., an estimation of how expensive John's car is likely to be. This is a typical problem appearing in our commonsense inference. Thus, it is an important task to establish a computational model for such syllogistic reasoning cases. A notable approach was put forward by Zadeh[1][2]. Before discussing our approach, we review Zadeh's approach briefly.

*This work was done while the authors were with Department of Systems Science, Tokyo Institute of Technology, 4259 Nagatsuta, Midori-ku, Yokohama, 227 Japan

## II. Zadeh's Approach

Zadeh defines the *sigma-count* and the *relative sigma-count* which are generalizations of the cardinality of a set and the conditional probability, respectively.

Let's $U = \{u_i; i = 1, 2, ...\}$ be a set of objects in the domain. In the above example, $u_i$ is a car that we observe on the street. $u_i$ is associated with membership degrees in fuzzy sets $A$, $B$, and $C$("American cars", "Big cars", and "Expensive cars" in our case). We denote those membership degrees as $\mu_A(u_i)$, $\mu_B(u_i)$, and $\mu_C(u_i)$, respectively.

The sigma-count of $A$, $\Sigma count(A)$, is defined as $\Sigma count(A) = \sum_{u_i \in U} \mu_A(u_i)$. The relative sigma count, $\Sigma count(B|A)$, is defined as $\Sigma count(B|A) = \Sigma count(A \cap B)/\Sigma count(A)$ where the sigma count of the intersection $A \cap B$ is calculated by $\Sigma count(A \cap B) = \sum_{u_i \in U} \mu_A(u_i) \wedge \mu_B(u_i)$, using a minimum operator $x \wedge y = Min(x, y)$.

An imprecise quantifier such as "Most" is associated with a fuzzy set over the domain of the relative sigma count(Fig.1). $Q_1(A \to B)$ is interpreted as an *elastic constraint* for the relative sigma count $\Sigma count(B|A)$, or a possibility distribution assignment to the relative sigma count.

Zadeh discussed two problems; (1) *compositionality*, and (2) *robustness*. The first problem is concerned with the representation of the resulting quantifier $Q_3$ in terms of $Q_1$ and $Q_2$. The second problem is related to the sensitivity: Consider a crisp case where $Q_1$ and $Q_2$ are given as "All." Then we will find that $Q_3$ should also be "All." However, even a slight uncertainty in the second quantifier

0-7803-0614-7/93$03.00 ©1993IEEE

$Q2$ (e.g. "Almost all") makes the resulting quantifier $Q_3$ a vacuous one, i.e., "from none to all." To avoid this, Zadeh introduced an additional assumption, *Major Premise Reversibility(MPR)*, which assumes that $Q_1(B \rightarrow A)$ is also the case. This results in assuming the equal cardinality of $A$ and $B$, i.e., $\Sigma count(A) = \Sigma count(B)$.

## III. Our Problem

We first distinguish *prior knowledge* and *empirical knowledge*. We assume that the prior knowledge consists of two rules, $Q_1(A \rightarrow B)$ and $Q_2(B \rightarrow C)$. The empirical knowledge is a collection of observations. Suppose that we have made $N$ observations so far. Then we represent our empirical knowledge as

$$E = \{D(u_i), i = 1, 2, ..., N\}$$

where $u_i$ is an object and $D(u_i)$ represents a triplet $(a_i, b_i, c_i)$. $a_i$ and $b_i$ and $c_i$ represent membership degrees of $u_i$ in fuzzy sets $A$, $B$, and $C$, i.e., $\mu_A(u_i)$, $\mu_B(u_i)$, and $\mu_C(u_i)$, respectively.

Given $\{D(u_i), i = 1, 2, ..., N\}$, we can calculate the sigma counts and relative sigma counts. In this paper, we use slightly different notations for the sigma counts and relative sigma counts, i.e., we use $S_{N,A}$ for $\Sigma count(A)$ and $S_{N,B|A}$ for $\Sigma count(B|A)$. The suffix $N$ explicitly shows that the counts are calculated from $\{u_1, ..., u_N\}$. The second suffix, $A$ or $B|A$, shows what the sigma count is concerned with.

Now consider a case where we are receiving a new object $u_{N+1}$ in the $N+1$st observation. Further, suppose that we are not completely sure about $u_{N+1}$, i.e., some of the values in the triplet $D(u_{N+1}) = (a_{N+1}, b_{N+1}, c_{N+1})$ are missing. However, let us assume that at least one of them is given.

We define two problems; (1) *estimation problem* and (2) *evaluation problem*. The estimation problem is to *guess* the missing values in $D(u_{N+1})$. The evaluation problem is concerned with the *consistency* of our prior knowledge with the empirical knowledge where the *consistency* is defined as follows: We say that our prior knowledge, say $Q_1(A \rightarrow B)$, is consistent with our empirical knowledge $E$ with (at least) degree $\gamma$ when the relative sigma count $S_{N,B|A}$ is in the $\gamma$ level set of the fuzzy set associated with the quantifier $Q_1$. More precisely, $R1$ and $R2$ are consistent with the empirical knowledge up to the $N$th observation with degree $\gamma_1$ and $\gamma_2$, respectively, when

$$S_{N,B|A} \in [l_{Q1}(\gamma_1), u_{Q1}(\gamma_1)] \qquad (1.1)$$

$$S_{N,C|B} \in [l_{Q2}(\gamma_2), u_{Q2}(\gamma_2)] \qquad (1.2)$$

where $[l_{Q1}(\gamma_1), u_{Q1}(\gamma_1)]$ and $[l_{Q2}(\gamma_2), u_{Q2}(\gamma_2)]$ are $\gamma_1$ and $\gamma_2$ level sets of the fuzzy sets associated with the quantifiers $Q_1$ and $Q_2$, respectively. For computational simplicity, we assume, throughout this paper, that a consistency degree can be mapped to a unique closed level set. This assumption means that the fuzzy set for a quantifier should be convex.

We solve the estimation problem as follows: "*Find default values for the missing values in the triplet $D(u_{N+1}) = (a_{N+1}, b_{N+1}, c_{N+1})$ so that the consistency degrees for $R1$ and $R2$ can be maximized.*"

In general, the uncertain reasoning gives interval solution as discussed in Turksen's CNF-DNF interval logic[3] and Baldwin's support logic[4]. Unfortunately, if we want to maintain the interval during the course of inference, in addition to the computational complexity, the interval tends to become too broad like "*from none to all*" as mentioned above. The point-valued approach selects a particular solution from the admissible interval using some preference criterion such as maximum entropy. In our *default estimation*, the consistency maximization is used as the preference criterion.

Compared with the conventional inference methods, our approach has the following unique features: First we are no more working on individual objects, but our interest is focussed on the *meta level*, i.e., the consistency between prior and empirical knowledge. Secondly, the estimation and evaluation problems are integrated as two aspects of the *same* inference process. The conventional inference method concentrates on the estimation problem, leaving the evaluation problem for a separate learning or adaptive mechanisms. However, there is no need for this separation in our method. Thirdly, our approach does not rely on the MPR assumption.

Our method is formulated as a multi-objective optimization problem: In general, the *separate* feasibility of the goals in (1.1) and (1.2) does not guarantee their *simultaneous* feasibility. In this paper we concentrate our efforts on clarifying the feasibility conditions.

It should be also noted that our method treats the inference as a *multi agent problem*. The conventional inference method treats knowledge pieces merely as data to be processed by a central inference engine. In contrast, our method treats each piece as an agent pursuing his own objective, i.e., the consistency maximization, and the inference is formulated as a process to maximally achieve those agents' objectives by resolving competition and conflicts.

## IV. Feasibility Conditions

We need to consider two levels of feasibility conditions; (1) *separate feasibility conditions* and (2) *simultaneous feasibility condition*.

The interaction between two rules, i.e., $R1$ and $R2$, occurs due to the common variable $b_{N+1}$ shared by them. For example, consider a case where $a_{N+1}$ and $c_{N+1}$ are given and $b_{N+1}$ is unknown. Then we need to *estimate* the unknown variables using both $R1$ and $R2$ so that their individual consistency degrees can be maximized. However, there is no guarantee that $R1$ and $R2$ agree on the same value of $b_{N+1}$. Now let $B_{R1}$ be an interval of $b_{N+1}$ such that $b_{N+1} \in B_{R1}$ achieves the goal of $R1$ in (1.1). Similarly, let $B_{R2}$ be an interval of $b_{N+1}$ such that $b_{N+1} \in B_{R2}$ achieves the goal of $R2$ in (1.2). First, for the separate feasibility, $B_{R1}$ and $B_{R2}$ should not be empty. Further, for the simultaneous feasibility of the goals in (1.1) and (1.2), $B_{R1} \cap B_{R2}$ should not be empty. If $B_{R1} \cap B_{R2} \neq \phi$, we can choose the *default value* of $b_{N+1}$ using the following criterion:

$$Maximize\ Min\{\gamma_1 - s_1, \gamma_2 - s_2\} \quad (2.1)$$

$$\text{subject to } B_{R1}, B_{R2}, B_{R1} \cap B_{R2} \neq \phi \quad (2.2)$$

where $s_1$ and $s_2$ are parameters to be specified by the user which represent the minimum requirements for the consistency degrees. Intuitively speaking, the consistency degrees can be maximized in the multi-objective sense until $B_{R1} \cap B_{R2}$ collapses to a point, $b^{op}_{N+1}$. This $b^{op}_{N+1}$ gives the default estimation of $b_{N+1}$.

In general, we have to consider the following 6 cases.

**Case 1:** Given $a_{N+1}$ and $b_{N+1}$, estimate $c_{N+1}$.

**Case 2:** Given $b_{N+1}$ and $c_{N+1}$, estimate $a_{N+1}$.

**Case 3 (Interpolative Chaining):** Given $a_{N+1}$ and $c_{N+1}$, estimate $b_{N+1}$.

**Case 4 (Forward Chaining):** Given $a_{N+1}$, estimate $b_{N+1}$ and $c_{N+1}$.

**Case 5 (Backward Chaining):** Given $c_{N+1}$, estimate $b_{N+1}$ and $a_{N+1}$.

**Case 6:** Given $b_{N+1}$, estimate $a_{N+1}$ and $c_{N+1}$.

In Cases 1, 2, and 6, we have only to consider the separate feasibility condition. Case 3 is exactly what we considered and discussed above as the interaction between the two rules. In the following discussions we consider only Case 4 and Case 5. In [5], we discussed Case 5 in a simplified manner where only the lower bounds of the relative sigma counts are considered. Here we generalize the discussion, considering also the upper bounds of the relative sigma counts as in (1.1) and (1.2). A more complete and detailed discussions could be found in [6].

### *A. Preliminaries*

First we discuss how $S_{N+1,B|A}$ and $S_{N+1,C|B}$ varies depending on $a_{N+1}$, $b_{N+1}$, and $c_{N+1}$.

Let's consider $S_{N+1,B|A}$ with given $a_{N+1}$. See Fig.2(a) for illustration. $S_{N+1,B|A}$ is monotone increasing and the possible interval of $S_{N+1,B|A}$, written as $I_{B|A,a_{N+1}}$, is given as

$$I_{B|A,a_{N+1}} = [\frac{S_{N,B|A}S_{N,A}}{S_{N,A}+a_{N+1}}, \frac{S_{N,B|A}S_{N,A}+a_{N+1}}{S_{N,A}+a_{N+1}}] \quad (3)$$

where the lower suffix $a_{N+1}$ in $I_{B|A,a_{N+1}}$ means that $a_{N+1}$ is given.

Next, consider $S_{N+1,C|B}$ with given $c_{N+1}$. See Fig.2(b) for illustration. $S_{N+1,C|B}$ increases as $b_{N+1}$ increases and reaches the maximum when $b_{N+1} = c_{N+1}$. For $b_{N+1} > c_{N+1}$, $S_{N+1,C|B}$ is a monotone decreasing function of $b_{N+1}$. Thus, the possible interval of $S_{N+1,C|B}$, written as $I_{C|B,c_{N+1}}$, is as follows:

$$I_{C|B,c_{N+1}} = [Min(S_{N,C|B}, \frac{S_{N,C|B}S_{N,B}+c_{N+1}}{S_{N,B}+1}), \frac{S_{N,C|B}S_{N,B}+c_{N+1}}{S_{N,B}+c_{N+1}}] \quad (4)$$

where the lower suffix $c_{N+1}$ in $I_{C|B,c_{N+1}}$ means that $c_{N+1}$ is given.

Let's define the following two intervals of the sigma-counts of $R1$ and $R2$:

$$I_{R1} = [l_{Q_1}(\gamma_1), u_{Q_1}(\gamma_1)] \cap I_{B|A,a_{N+1}}$$

$$I_{R2} = [l_{Q_2}(\gamma_2), u_{Q_2}(\gamma_2)] \cap I_{C|B,c_{N+1}}$$

$I_{R1}$ and $I_{R2}$ are intervals for $S_{N+1,B|A}$ and $S_{N+1,C|B}$ that are both achievable and desirable with respect to the consistency requirements (1.1) and (1.2) stated above when $a_{N+1}$ and $c_{N+1}$ are given. Using (3) and (4), we can express $I_{R1}$ and $I_{R2}$ as follows:

$$I_{R1} = [Max\{l_{Q1}(\gamma_1), \frac{S_{N,B|A}S_{N,A}}{S_{N,A}+a_{N+1}}\}, Min\{u_{Q1}(\gamma_1), \frac{S_{N,B|A}S_{N,A}+a_{N+1}}{S_{N,A}+a_{N+1}}\}] \quad (5.1)$$

$$I_{R2} = [Max\{l_{Q2}, Min(S_{N,C|B}, \frac{S_{N,C|B}S_{N,B}+c_{N+1}}{S_{N,B}+1})\}, Min\{u_{Q2}, \frac{S_{N,C|B}S_{N,B}+c_{N+1}}{S_{N,B}+c_{N+1}}\}] \quad (5.2)$$

Let $B_{R1}$ be an interval of $b_{N+1}$ that realizes $S_{N+1,B|A} \in I_{R1}$, and it will be seen from Fig.2(a) that $B_{R1} = [B^l_{R1}, B^u_{R1}]$ can be obtained by solving $S_{N+1,B|A} = I^l_{R1}$ and $S_{N+1,B|A} = I^u_{R1}$ where we abbreviate $I_{R1}$ in (5.1) as $I_{R1} = [I^l_{R1}, I^u_{R1}]$. Direct calculation yields

$$B^l_{R1} = I^l_{R1}(S_{N,A}+a_{N+1}) - S_{N,B|A}S_{N,A} \quad (6.1)$$

When $I^u_{R1} < I^u_{B|A,a_{N+1}}$,

$$B^u_{R1} = I^u_{R1}(S_{N,A}+a_{N+1}) - S_{N,B|A}S_{N,A} \quad (6.2)$$

Otherwise, when $I^u_{R1} = I^u_{B|A,a_{N+1}}$, we have

$$B^u_{R1} = 1 \quad (6.3)$$

Given $S_{N,A}$, $S_{N,B|A}$, $\gamma_1$, $[l_{Q1}(\gamma_1), u_{Q1}(\gamma_1)]$, and $a_{N+1}$, *the separate feasibility condition for R1, $B_{R1} \neq \phi$, is equivalent to $I^l_{R1} \leq I^u_{R1}$* which, from (5.1), can be expressed by the following two inequality constraints:

$$l_{Q1}(\gamma_1) - \frac{S_{N,B|A}S_{N,A} + a_{N+1}}{S_{N,A} + a_{N+1}} \leq 0$$

$$-u_{Q1}(\gamma_1) + \frac{S_{N,B|A}S_{N,A}}{S_{N,A} + a_{N+1}} \leq 0 \qquad (7)$$

Let $B_{R2}$ be an interval of $b_{N+1}$ that realizes $S_{N+1,C|B} \in I_{R2}$. See Fig.2(b) for illustration. Similarly, we want to solve $S_{N+1,C|B} = I^l_{R2}$ and $S_{N+1,C|B} = I^l_{R2}$. However, the result becomes more complicated because $B_{R2}$ may be split into two disjoint intervals, $B^1_{R2}$ and $B^2_{R2}$. Let's denote $B^1_{R2} = [L(B^1_{R2}), U(B^1_{R2})]$ and $B^2_{R2} = [L(B^2_{R2}), U(B^2_{R2})]$. Then direct calculation yields

$$L(B^1_{R2}) = \frac{S_{N,B}(I^l_{R2} - S_{N,C|B})}{1 - I^l_{R2}} \qquad (8.1)$$

$$U(B^1_{R2}) = \frac{S_{N,B}(I^u_{R2} - S_{N,C|B})}{1 - I^u_{R2}} \qquad (8.2)$$

$$L(B^2_{R2}) = \frac{S_{N,B}S_{N,C|B} + c_{N+1}}{I^u_{R2}} - S_{N,B} \qquad (8.3)$$

$$U(B^2_{R2}) = \frac{S_{N,B}S_{N,C|B} + c_{N+1}}{I^l_{R2}} - S_{N,B} \qquad (8.4)$$

Notice that, though not explicitly written, the upper and lower bounds should be *saturated* at 1 and 0, respectively. However, we do not include the saturation, because the lower bound in the negative region and the upper bound in the area beyond 1 are just trivially satisfied.

Given $S_{N,B}$, $S_{N,C|B}$, $\gamma_2$, $[l_{Q2}(\gamma_2), u_{Q2}(\gamma_2)]$, and $c_{N+1}$, *the separate feasibility condition for $R_2$, $B_{R2} \neq \phi$, is equivalent to $I^l_{R2} \leq I^u_{R2}$* which, from (5.2), can be expressed by the following two inequality constraints:

$$l_{Q2}(\gamma_2) - \frac{S_{N,C|B}S_{N,B} + c_{N+1}}{S_{N,B} + c_{N+1}} \leq 0$$

$$-u_{Q2}(\gamma_2) + Min\{S_{N,C|B}, \frac{S_{N,C|B}S_{N,B} + c_{N+1}}{S_{N,B} + 1}\} \leq 0 \quad (9)$$

Finally, *the simultaneous feasibility condition, $B_{R1} \cap B_{R2} \neq \phi$, is stated as follows:*

$$L(B^1_{R2}) \leq B^u_{R1}, \quad U(B^2_{R2}) \geq B^l_{R1}, \text{ and}$$

$$Min\{B^l_{R1} - U(B^1_{R2}),\ L(B^2_{R2}) - B^u_{R1}\} \leq 0 \qquad (10)$$

(The third inequality can be read as "$U(B^1_{R2}) \geq B^l_{R1}$ or $L(B^2_{R2}) \leq B^u_{R1}$"). The first two conditions guarantee $B_{R1} \cap B_{R2} \neq \phi$ when the "gap" between $B^1_{R1}$ and $B^2_{R2}$ is filled to make $[L(B^1_{R2}), U(B^2_{R2})]$ a connected interval. The third condition excludes the case where $B_{R1}$ falls in the gap between $B^1_{R2}$ and $B^2_{R2}$, which results in $B_{R1} \cap B_{R2} = \phi$.

## V. Problem Formulation

### A. Inequality Constraints

Let's consider Case 4. The feasibility conditions are expanded into the following inequality constraints:

$$g_1(\gamma_1, \gamma_2) = l_{Q1}(\gamma_1) - \frac{S_{N,B|A}S_{N,A} + a_{N+1}}{S_{N,A} + a_{N+1}} \leq 0 \quad (11.1)$$

$$g_2(\gamma_1, \gamma_2) = -u_{Q1}(\gamma_1) + \frac{S_{N,B|A}S_{N,A}}{S_{N,A} + a_{N+1}} \leq 0 \quad (11.2)$$

$$g_3(\gamma_1, \gamma_2) = l_{Q2}(\gamma_2) - \frac{B^u_{R1} + S_{N,B}S_{N,C|B}}{B^u_{R1} + S_{N,B}} \leq 0 \quad (11.3)$$

$$g_4(\gamma_1, \gamma_2) = \frac{B^l_{R1} + S_{N,B}S_{N,C|B}}{B^l_{R1} + S_{N,B}} - u_{Q2}(\gamma_2) \leq 0 \quad (11.4)$$

$$g_5(\gamma_1, \gamma_2) = \frac{S_{N,B}(l_{Q2}(\gamma_2) - S_{N,C|B})}{1 - l_{Q2}(\gamma_2)} -$$

$$Min\{B^u_{R1}, u_{Q2}(\gamma_2)(B^u_{R1} + S_{N,B}) - S_{N,B}S_{N,C|B}\} \leq 0 \quad (11.5)$$

In reality, it is sufficient to choose

$$c_{N+1} \in [Max\{l_{Q2}(\gamma_2)(B^l_{R1} + S_{N,B}) - S_{N,B}S_{N,C|B},$$

$$B^l_{R1}, \frac{S_{N,B}(l_{Q2}(\gamma_2) - S_{N,C|B})}{1 - l_{Q2}(\gamma_2)}\},$$

$$Min\{u_{Q2}(\gamma_2)(B^u_{R1} + S_{N,B}) - S_{N,B}S_{N,C|B}, B^u_{R1}\}] \quad (11.6)$$

$$b_{N+1} \in B_{R1} \cap B_{R2} (\neq \phi) \qquad (11.7)$$

Though we omit the complete proof, the outline for deriving above constraints is as follows: With $a_{N+1}$ given, (11.1) and (11.2) guarantee $B_{R1} \neq 0$. Because $c_{N+1}$ is a free variable, after determining $B_{R2} \neq \phi$ and $B_{R1} \cap B_{R2} \neq \phi$, the constraints containing $c_{N+1}$ are modified as the constraints for $c_{N+1}$ in (11.6). The others that do not contain $c_{N+1}$, together with the non-emptiness condition of the interval (11.6), are added to the inequality constraints in (11.3)-(11.5). In performing the inference with given $\gamma_1$ and $\gamma_2$, we need to first choose $c_{N+1}$ from the interval (11.6) to determine $B_{R2}$, and then we can compute $B_{R1} \cap B_{R2}$ from which we can choose $b_{N+1}$.

Similarly, the feasibility conditions for Case 5 are expanded into the following inequality constraints:

$$g_1(\gamma_1, \gamma_2) = l_{Q2}(\gamma_2) - \frac{S_{N,C|B}S_{N,B} + c_{N+1}}{S_{N,B} + c_{N+1}} \leq 0 \quad (12.1)$$

$$g_2(\gamma_1, \gamma_2) = -u_{Q2}(\gamma_2) +$$

$$Min\{S_{N,C|B}, \frac{S_{N,C|B}S_{N,B} + c_{N+1}}{S_{N,B} + 1}\} \leq 0 \quad (12.2)$$

$$g_3(\gamma_1, \gamma_2) = Max\{0, \frac{S_{N,A}(l_{Q1}(\gamma_1) - S_{N,B|A})}{1 - l_{Q1}(\gamma_1)}\} -$$

$$Min\{1, \frac{U(B_{R2}^2) + S_{N,A}S_{N,B|A}}{l_{Q1}(\gamma_1)} - S_{N,A}\} \leq 0 \qquad (12.3)$$

In reality, it is sufficient to choose

$$a_{N+1} \in [Max\{0, \frac{S_{N,A}(l_{Q1}(\gamma_1) - S_{N,B|A})}{1 - l_{Q1}(\gamma_1)},$$

$$\frac{L(B_{R2}^2) + S_{N,B|A}S_{N,A}}{u_{Q1}(\gamma_1)} - S_{N,A}\},$$

$$Min\{1, \frac{U(B_{R2}^2) + S_{N,A}S_{N,B|A}}{l_{Q1}(\gamma_1)} - S_{N,A}\}] \qquad (12.4)$$

$$b_{N+1} \in B_{R1} \cap B_{R2} (\neq \phi) \qquad (12.5)$$

The above inequalities are derived in a similar manner to Case 4(Proof omitted). In performing the inference, we need to choose $a_{N+1}$ first from (12.4) to compute $B_{R1} \cap B_{R2}$ from which we can choose $b_{N+1}$.

## VI. Search Problem

The discussions so far formulate the syllogistic reasoning problem for $R1$ to $R2$ as a traditional optimization problem with inequality constraints. There are already many methods available for the above optimization calculation. Unfortunately, in our method some of the $g_j(\gamma_1, \gamma_2)$'s in (11.1)-(11.5), or in (12.1)-(12.3) have discontinuous derivatives due to $Max$ and $Min$ operations in the formulae. However, owing to the *monotonicity property* of the constraints for $g_j(\gamma_1, \gamma_2) \leq 0$(Proof omitted), i.e.,$\partial g_j(\gamma_1, \gamma_2)/\partial \gamma_i \geq 0$, $i = 1, 2$, we can use the following simplified search method.

**Step 0:** Initial value setting: Let $\gamma_1^c$ and $\gamma_2^c$ be the current values of $\gamma_1$ and $\gamma_2$ respectively. First, $\gamma_1^c \leftarrow \gamma_1^f$, $\gamma_2^c \leftarrow \gamma_2^f$ where $\gamma_1^f$ and $\gamma_2^f$ constitute an initial feasible solution.

**Step 1:** For $\gamma_1^c$ and $\gamma_2^c$, determine which of $\gamma_1$ and $\gamma_2$ we should improve based on (2.1). For example, if $\gamma_1 - s_1 < \gamma_2 - s_2$, we should improve $\gamma_1$ because $\gamma_1$ is crucial for (2.1). Let $\gamma_i$ be either $\gamma_1$ or $\gamma_2$ which we want to improve.

**Step 2:** Increase $\gamma_i$ to $\gamma_i^c + \delta$ where $\delta$ is a positive improvement step such as 0.05. Evaluate the constraints. If no constraint is violated, go back to Step 1 for further improvement. Otherwise go to the next step.

**Step 3:** Find a maximum value of $\gamma_i$ in $[\gamma_i^c, \gamma_i^c + \delta)$ so that all constraints are satisfied. One heuristics is to increase the value of $\gamma_i$ as $\gamma_i = \gamma_i^c + \delta - \delta'$ by decreasing $\delta'$ as $\delta/2$, $\delta/4$, ..., while all constraints are satisfied.

**Step 4:** Increase the other of $\gamma_i$, written as $\gamma_{i'}$(e.g., if $\gamma_i$ is $\gamma_1$, then $\gamma_{i'}$ is $\gamma_2$). Increase $\gamma_{i'}$ as much as possible with keeping all constraints satisfied. The search for $\gamma_{i'}$ may be done in a similar manner as Step 2 and 3. We increase $\gamma_{i'}$ as $\gamma_{i'}^{New} \leftarrow \gamma_{i'}^{Old} + \delta$ until constraint violation occurs. Then we find a smaller value of $\gamma_{i'}$ as in the case for $\gamma_i$ in Step 3 to remedy the constraint violation(s).

Step 1, 2, and 3 improve the criterion (2.1). However, notice that the solution so obtained may be only a *weak Pareto optimum* in that the other consistency level may be further improved without deteriorating the criterion or violating the constraints in Step 4. This simplified search method is used in the example below.

## VII. An Illustrative Example

Suppose that we are given the following two rules; (1) $Most(A \rightarrow B)$, and (2) $Most(B \rightarrow C)$ and the three observed data are shown in Table 1. For simplicity, we give the membership function for "Most" by a linear shape as in Fig.3. For an initial solution we use $\gamma_1 = \gamma_2 = 0.5$ and for the specification parameters $s_1$ and $s_2$ in (2.1), we use $s_1 = s_2 = 0$ which means we maximize $\gamma_1$ and $\gamma_2$.

We demonstrate our method by using the following two problems:

**Problem 1:** Suppose $a_4$ is given as 0.7. Estimate $b_4$ and $c_4$, and evaluate the credibility of the prior knowledge, i.e., a forward chaining in Case4.

**Problem 2:** Suppose $c_4$ is given as 0.8. Estimate $b_4$ and $a_4$, and evaluate the credibility of the prior knowledge, i.e., a backward chaining in Case 5.

We start with Problem 1: With the iteration of Step 1-3, it is found that $\gamma_1 = 0.75$ and $\gamma_2 = 0.725$. At this point, $\gamma_2$ determines $L$, and therefore, we try further to improve $\gamma_1$ in Step 4, and we obtain the result $\gamma_1 = 0.7625$. Using $\gamma_1 = 0.7625$ and $\gamma_2 = 0.725$, first we identify the admissible interval for $c_4$, which turns out to be [0.909,1]. We choose any value for $c_4$ from this interval. Here we choose $c_4 = 0.95$. Now we can identify $B_{R1} \cap B_{R2}$ as [0.909, 0.957], and we choose $b_4 = 0.95$. Thus, our final inference result is $(a_4, b_4, c_4) = (0.7, 0.95, 0.95)$. Actually, this result achieves $\gamma_1 = 0.769$ and $\gamma_2 = 0.729$.

Next we discuss Problem 2: With the iteration of Step 1-3 it is found that $\gamma_1 = 0.75$ and $\gamma_2 = 0.7125$. In this case $\gamma_2$ determines $L$, and therefore, we try further to improve $\gamma_1$ by Step 4, and we obtain $\gamma_1 = 0.775$. Using $\gamma_1 = 0.775$ and $\gamma_2 = 0.7125$, we first identify the admissible interval for $a_4$ as [0.767, 0.808]. We choose $a_4 = 0.8$. Then $B_{R1} \cap B_{R2}$ is given as [0.796, 0.803]. We choose $b_{N+1} = 0.80$. Thus our inference result is $(a_4, b_4, c_4) = (0.80, 0.80, 0.80)$. Actually, by using these results, we obtain $\gamma_1 = 0.778$ and $\gamma_2 = 0.714$.

## VIII. Conclusion

It is shown that a syllogistic reasoning problem can be formulated as a multi-objective consistency level maximization problem.

Our method integrates the estimation and evaluation problems in a unified manner. The estimation problem is a reasoning problem in the traditional sense. The evaluation problem is a feedback of our empirical knowledge to the prior knowledge.

In future research, we can generalize our discussions in the following directions: First, the current discussions critically depends on the particular T-norm, i.e., Min operator. We can generalize the discussion for a wide variety of T-norms. Secondly, in this paper we discussed a reasoning pattern identified as $\{Q_1(A \rightarrow B),\ Q_2(B \rightarrow C)\} \Longrightarrow Q_3(A \rightarrow C)$. We can extend our discussion to a wide class of reasoning patterns such as a *Consequent Conjunction Syllogism* discussed by Zadeh[5] as follows: $\{(Q_1$A's are B's), $(Q_2$A's are C's)$\} \Longrightarrow \{Q_3$A's are (B and C)'s$\}$. Thirdly, our chaining problem simply consists of two rules. We can generalize our method to a problem of longer chains. Finally, the definition of sigma count may be generalized to interval valued sigma count which gives rise to a new set of syllogistic default reasoning problems in the sense of the interval-valued fuzzy sets introduced by Turksen[8]. All these need further investigation.


References

[1] L.A.Zadeh, "A Theory of Commonsense Knowledge," in *Issues of Vagueness*, H.J.Skala, S.Termini, and E. Trillas, Eds., Doedrecht:Reidel, pp.257-296, 1984.

[2] L.A.Zadeh, "Syllogistic Reasoning in Fuzzy Logic and its Application to Usuality and Reasoning with Dispositions," IEEE Trans.SMC, Vol.SMC-15, pp.754-763, 1985.

[3] I.B. Turksen, "Interval Valued Fuzzy Sets Based On Normal Forms," Fuzzy Sets and Systems, Vol.20, pp.191-210, 1986.

[4] J.F. Baldwin, "Evidential Support Logic Programming," Fuzzy Sets and Systems, Vol.24, pp.1-26, 1987.

[5] H.Narazaki and I.B. Turksen, "A Syllogistic Reasoning as a Multi-Objective Default Expectation Process," Proc. IIZUKA'92, Vol.1, pp.291-294, IIZUKA, JAPAN, 1992.

[6] H.Narazaki and I.B. Turksen, "An Integrated Approach for a Syllogistic Reasoning and Knowledge Consistency Level Maintenance," submitted.


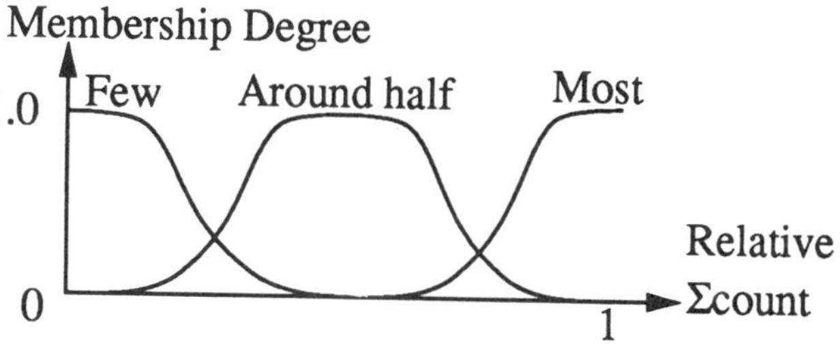


**Fig.1 Fuzzy sets for imprecise quantifiers**

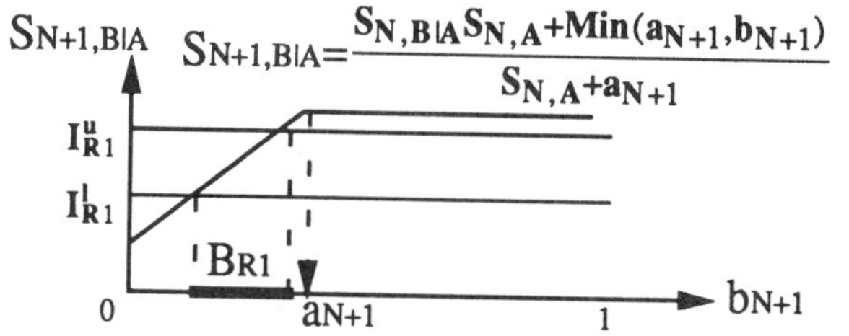


**Fig.2(a) $S_{N+1,B|A}$ and $B_{R1}$**

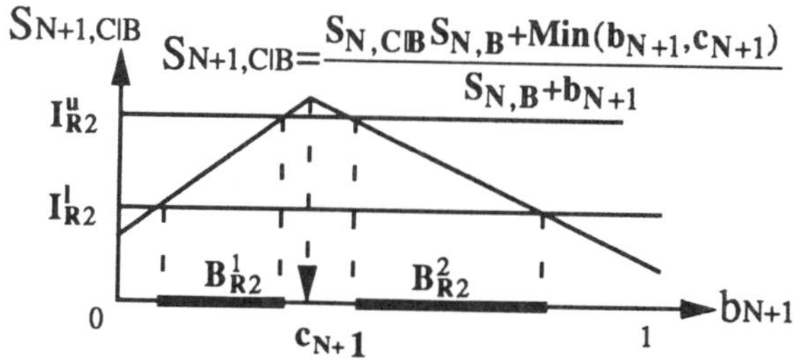


**Fig.2(b) $S_{N+1,C|B}$ and $B_{R2}$**

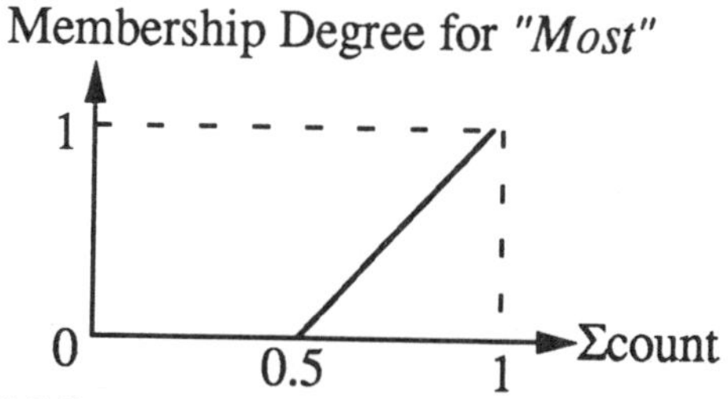


**Fig.3 Membership Function for "Most"**

**Table 1 Empirical Knowledge**

| Data No. \ Fuzzy Sets | A | B | C |
|---|---|---|---|
| 1 | 0.9 | 0.6 | 0.8 |
| 2 | 0.3 | 0.5 | 0.1 |
| 3 | 0.7 | 0.9 | 1.0 |

# Parallel Fuzzy Resolution Inference on Fuzzy Neural Logic Network

Liya Ding, Zuliang Shen

Institute of Systems Science, National University of Singapore
Heng Mui Keng Terrace, Kent Ridge, Singapore 0511
E-mail: liya@iss.nus.sg; zuliang@iss.nus.sg.

***Abstract*—A form of parallel fuzzy resolution inference on fuzzy neural logic network (FPRIN for short) is introduced. In continuance of research on parallel resolution inference on neural logic network (PRIN), PRIN is extended for fuzzy logic by the fuzzy resolution principle, fuzzy linear resolution and the fuzzy perceptron in neural logic network. By using FPRIN, fuzzy resolution inference can be executed in parallel without loss of its logical properties.**

## 1. INTRODUCTION

Flexible information processing realized by massively parallel processing is a main target of future real world computing technology [17]. Fuzzy logic has established a foundation for the processing of uncertain or incomplete information. On the other hand, neural networks technology has been widely used in recent years. One of the important features of neural networks is the possibility of parallel processing. Hence it is an interesting topic to fuse fuzzy logic and neural network to realize fuzzy parallel logic inference.

In this paper, a parallel fuzzy resolution inference on fuzzy neural logic network (FPRIN) is put forward, by extending PRIN from binary logic to fuzzy logic. The essential idea and theoretical basis of FPRIN are from several research results: PRIN [1], fuzzy resolution principle [3] and exponential form(EF) in fuzzy logic [7]. In FPRIN, fuzzy resolution inference is processed by two levels: (1) symbolic level represented by the FPRIN network itself; (2) fuzzy inference level expressed as the access of fuzzy truth values and confidence values. Representing all logical relationships given by fuzzy rules and facts as a well-constructed neural network, it is possible to execute logical inference by the access in parallel of the truth values and the confidence values on the neural network.

We employ three neural unit models to implement FPRIN. The fuzzy neural logic unit model (FNLU) expresses fuzzy logical operations, the fuzzy neural inference unit model (FNIU) expresses the procedure of fuzzy refutation based on fuzzy linear resolution, and the fuzzy neural knowledge unit (FNKU) transmits the information from the premise to the conclusion of a rule.

According to the transformation laws, a FPRIN network can be built by transforming a given set of fuzzy rules and facts to a corresponding neural network combined by FNLUs, FNIUs and FNKUs. All truth values of fuzzy rules, facts and query are dealt with as the inputs of FNIUs, and outputs are used to represent the truth values as well as the confidences of conclusions.

The discussion is divided into several stages. Section 2 gives a brief introduction to the theoretical basis of parallel fuzzy resolution. Three basic units of FPRIN are introduced in Section 3. In Section 4, the construction of FPRIN is described by transformation laws. The inference on FPRIN and its logical properties are discussed in Section 5 with some examples. Section 6 gives out conclusions.

## 2. PARALLEL FUZZY RESOLUTION INFERENCE

The theoretical basis of FPRIN can be found from: the fuzzy resolution principle [3], the exponential form of fuzzy logic [7] and the parallel resolution inference on neural logic network(PRIN) [1].

### 2.1 Fuzzy Resolution Principle

The resolution principle on binary logic [12] is a major breakthrough of computer science and artificial intelligence. Since fuzzy set theory and fuzzy logic were established by L.A.Zadeh [13], many scientists have tried to find an equally effective resolving method in fuzzy logic [14]. The fuzzy resolution principle has been put forward to extend the resolution principle from binary logic to fuzzy logic [4].

In binary logic, there are the complementary laws that $T_I(x)\vee T_I(\neg x)\equiv 1$, $T_I(x)\wedge T_I(\neg x)\equiv 0$ for any interpretation I. This means that any variable x and its negation $\neg x$ constitute a complete contradictory pair. The two complementary laws made up the deep base of resolution principle.

However, in fuzzy logic, these laws do not hold. In fuzzy resolution principle, some important concepts such as *fuzzy contradiction* and *contradictory degree*, *fuzzy resolvent* and *confidence of resolvent* have been introduced to solve this difficulty [3]. By these concepts, a fuzzy refutation procedure based on fuzzy resolution principle can be established.

*Definition 2.1* (Contradiction) [3]: Let a pair of complementary variables x and $\neg x$ be under any interpretation I. Then $x \wedge \neg x$ is said to be a *contradiction* under the given interpretation I. When $T_I(x)\wedge T_I(\neg x) = 0$, then it is said to be *complete contradictory*. When $T_I(x)\wedge T_I(\neg x) = 0.5$, then it is said to be *nor-contradictory*. When $T_I(x)\wedge T_I(\neg x) \in (0, 0.5)$, then it is said to be *incomplete contradictory*.

In fuzzy logic, instead of the complementary laws of binary logic, we have the relations: $T_I(x)\vee T_I(\neg x) \geq 0.5$ and $T_I(x)\wedge T_I(\neg x) \leq 0.5$ for any interpretation I. Except the point 0.5 (usually considered as 'unknown'), the intervals [0, 0.5) and (0.5, 1] can be considered as fuzzy false and fuzzy true [6] as the extensions of false (0) and true (1) of binary logic,

0-7803-0614-7/93$03.00 ©1993IEEE

respectively. Therefore, Definition 2.1 gives the important basis for extending resolution principle to fuzzy logic.

*Definition 2.2* (Truth confidence, Contradictory degree)[3,5]: Let the *truth confidence* of a variable x be

$$\mathbf{c}(x) = 2 \times T_I(x) - 1 \qquad (2.1)$$

under a given interpretation I. Let the *contradictory degree* of a contradiction $x \wedge \neg x$ be

$$\mathbf{cd}(x) = \max(T_I(x), T_I(\neg x)) - \min(T_I(x), T_I(\neg x)) \qquad (2.2)$$

under a given interpretation I.

It is evident that $\mathbf{c}(x) \in [-1, 1]$ and $\mathbf{cd}(x) = |\mathbf{c}(x)| \in [0, 1]$ hold for any interpretation.

*Definition 2.3* (Fuzzy resolvent, Confidence of resolvent) [3]: Consider two clauses $C_1$, $C_2$,

$$C_1 = x \vee L_1, \quad C_2 = \neg x \vee L_2$$

where $L_1$ and $L_2$ do not contain the literal x or $\neg x$ as a factor and have no pair of complementary variables. Then the clause $L_1 \vee L_2$ is said to be a *resolvent* of $C_1$ and $C_2$ whose keyword is x and the contradictory degree of the keyword is $\mathbf{cd}(x)$. A *resolvent* of $C_1$ and $C_2$ is written as $R(C_1, C_2)$, and a *fuzzy resolvent* of $C_1$ and $C_2$ is written as $R(C_1, C_2)_{\mathbf{cd}}$, where $\mathbf{cd} = \mathbf{cd}(x)$ is the contradictory degree of the keyword or is said to be the *confidence of resolvent* of $R(C_1, C_2)$.

It is important that as a special case of the fuzzy resolvent, a given fact can be always looked as a resolvent with $\mathbf{cd} \equiv 1$. Further, any resolvent with $\mathbf{cd} = 0$ is always *unknown* (or *meaningless*) in the inference.

*Definition 2.4* (Mixed truth value of fuzzy resolvent) [5]: Let $(L_1 \vee L_2)_{\mathbf{cd}}$ be the fuzzy resolvent of $C_1 = x \vee L_1$ and $C_2 = \neg x \vee L_2$. The *mixed truth value* of $(L_1 \vee L_2)_{\mathbf{cd}}$ is defined as

$$\mathbf{MT}((L_1 \vee L_2)_{\mathbf{cd}}) = \mathbf{c}(L_1 \vee L_2) \times \mathbf{cd}(x) \times 0.5 + 0.5 \qquad (2.3)$$

where, $\mathbf{c}(L_1 \vee L_2)$ is the truth confidence of $L_1 \vee L_2$, $\mathbf{cd}(x)$ is the confidence of the resolvent or the contradictory degree of the contradiction $x \wedge \neg x$.

The fuzzy refutational procedure based on the fuzzy resolution principle is complete, even though the premises and consequences, as well as the inference results may be all incomplete (or fuzzy) in the procedure. The logical completeness of the fuzzy resolution principle was proved in 1988 [3].

### *2.2 Exponential form of fuzzy logic*

The fuzzy resolution principle also established the theoretical basis of processing fuzzy inference on several different levels. In other words, a fuzzy inference is possible to deal with at different levels of symbolic logic and numeric truth values. The *exponential form* of fuzzy logic (EF for short) provides one kind of representation form of fuzzy truth values.

By the EF, any fuzzy truth value (in fuzzy valued logic or fuzzy linguistic valued logic [16]) is defined as a truth base with an exponential confidence, denoted by $B^{\mathbf{c}}$. When $\mathbf{c}$ is a membership function, EF represents fuzzy linguistic truth value. As a sub-case of linguistic valued logic, when the exponential confidence function is equal to a constant, the EF represents a truth value on fuzzy valued logic.

*Definition 2.5* (Basic Expression of EF) [7]: Let $a \in [0, 1]$ be a truth value in fuzzy valued logic [16], then a is equivalent to an exponential form $B^{\mathbf{c}}$ if and only if

$$a = (B - \mathbf{U}) \times \mathbf{c} + \mathbf{U}, \qquad (2.4)$$

where $B \in [0, 1]$ is called the *fuzzy truth base*, $\mathbf{c} \in (-\infty, \infty)$ is called the *confidence exponent*, $\mathbf{U}$ = unknown is called an *unknown* or *meaningless point* for inference, which is equal to 0(truth I) or 0.5(truth II) [16].

Representing all truth values in a common truth base, the calculations on truth values given in different truth bases can easily be accomplished from the exponential confidences.

A fuzzy inference can be divided to both of the truth base level and the exponential confidence level using EF. Obviously, when B = 1 is used as the common base, the inference on truth base level can be looked as a symbolic inference. Therefore, if we can represent the inference relation of binary logic level by the structure of neural network and let the numerical flow in the neural network reflect the inference relation of exponential confidence level, some kind of fuzzy inference on neural networks should be possible.

### *2.3 Parallel Resolution Inference on Neural Logic Network*

Parallel resolution inference on neural logic network (PRIN) has been proposed for resolution inference on binary propositional logic. The basic idea of PRIN is based on the understanding of refutation procedure and linear resolution [12]. The given knowledge base represented by rules and facts is used for every step of inference. Here, the knowledge base is supposed to be unchanged during one inference procedure. In other words, a refutation procedure can be considered as a dynamic query flow on a static data structure.

In FPRIN, the main neural network structure of PRIN that represents inference relation of binary logic level will be kept. The knowledge output function, conclusion output function, as well as all weights in neural inference unit will be defined based on the fuzzy resolution principle and the exponential form of fuzzy logic. The fuzzy neural logic unit will be defined instead of NLU in PRIN. The logical properties of *fuzzy neural AND* and *fuzzy neural OR* will be given in Section 3. In binary logic, the weight of any rule is considered as 1 [15], so it can be omitted. In fuzzy logic, a rule may have a weight that represents the influence or importance from its premise to its consequence. A set of n given rules that have the same consequence proposition is dealt with as one fuzzy rule with n possible premises that may have different weights linking them to the consequence. In this case, a fuzzy neural knowledge unit is used to transmit the weighted information of each premise to the consequence.

## 3. BASIC UNITS OF FPRIN

### *3.1 The Representation of Fuzzy Truth Value in FPRIN*

By exponential form, t = T(A), the fuzzy truth value of a fact A, is represented by $B^{\mathbf{c}}$ as defined in Definition 2.5, where B is the truth base, and $\mathbf{c}$ is the confidence exponent. In our discussion, we will always let B = 1, and the unknown point $\mathbf{U}$ = 0.5, so by Definition 2.5, $\mathbf{c}$ can be calculated by

$$c = (t - 0.5) \times 2 = 2t - 1 \qquad t \in [0, 1] \tag{3.1}$$

Comparing with Definition 2.2, when a truth base B = 1 and an unknown point **U** = 0.5 are given, obviously the confidence exponent just represents the truth confidence of the A.

In FPRIN, any fuzzy truth value is represented by a pair structure called the *truth pair*. The truth value of a fuzzy resolvent $R_{cd}$ is represented as:

$$(\mathbf{c1}, \mathbf{c2}) \tag{3.2}$$

where **c1** = c(R) is the truth confidence of R, **c2** = **cd** is the confidence of resolvent of R. As a special case of a fuzzy resolvent, the truth value of a given fact can be always represented as a pair (2t-1, 1), where 2t-1 is the truth confidence, the confidence of resolvent is 1. The *mixed truth value* of a truth pair (c1, c2) is defined as:

$$\mathbf{MT}((\mathbf{c1}, \mathbf{c2})) = \mathbf{c1} \times \mathbf{c2} \times 0.5 + 0.5. \tag{3.3}$$

The truth value of each element of a query is assigned to be (-1, 1), and of an unknown to be (0, 0). When the truth value of a fact is not given, it will be considered unknown. In FPRIN, the status of no-query is assigned by a special pair (2, 2) called *undefined.*

Corresponding the pair structure of truth values, the weight of a fuzzy rule is also represented by a pair in FPRIN and denoted by:

$$(w_1, w_2), \tag{3.4}$$

where $w_1$ and $w_2$ are applied for the calculations with the corresponding truth confidence and confidence of resolvent respectively, and $w_2$ is always assigned the value 1.

### *3.2 Fuzzy Neural Logic Unit(FNLU)*

A FNLU(Fig. 1) is used to express fuzzy logical operation. It has n (integer n≥1) inputs, noted as $I_j = (i_{j1}, i_{j2})$ $(1 \le j \le n)$ and one output $O = (o_1, o_2)$. Each corresponding $W_j = (w_{j1}, w_{j2})$ is the weight for the branch linking the value of input $I_j$ to the output. O is calculated by

$$o_1 = \mathcal{L}_j(i_{j1} \times i_{j2}, w_{j1}) / \mathcal{L}_j(i_{j2}, w_{j2}) \tag{3.5}$$

$$o_2 = \mathcal{L}_j(i_{j2}, w_{j2}) \tag{3.6}$$

The function $\mathcal{L}$ is a neural gate using propagation algorithm of fuzzy perceptron [8]. Suppose n pairs

$$(i_1, w_1), (i_2, w_2), \ldots, (i_n, w_n)$$

have been given. The function $\mathcal{L}$ is calculated by following steps:

(1) Rearranging the sequence of all pairs as $i_j \le i_{j+1}$ $(1 \le j \le n\text{-}1)$:

$$(i'_1, w'_1), (i'_2, w'_2), \ldots, (i'_n, w'_n) \tag{3.7}$$

(2) Calculating all differences: $d_1, d_2, \ldots, d_n$

where,

$$d_j = i'_j \qquad j = 1 \tag{3.8}$$

$$d_j = i'_j - i'_{j-1} \qquad j > 1 \tag{3.9}$$

(3) Getting qualified weights: $w^*_1, w^*_2, \ldots, w^*_n$

where,

$$w^*_j = \begin{cases} 1 & \sum_{h=j}^{n} (w'_h) \ge 1 \\ 0 & \text{otherwise} \end{cases} \tag{3.10}$$

(4)

$$\mathcal{L}_j(i_j, w_j) = \sum_j (d_j \times w^*_j) \tag{3.11}$$

*Definition 3.1* (Fuzzy Neural AND, Fuzzy Neural OR): Suppose fuzzy neural inputs pairs and the corresponding weights pairs of a FNLU be given as:

$$(i_{11}, i_{12}), (i_{21}, i_{22}), \ldots, (i_{n1}, i_{n2})$$

$$(w_{11}, w_{12}), (w_{21}, w_{22}), \ldots, (w_{n1}, w_{n2})$$

The *fuzzy neural AND* is defined as

$$\underset{j}{\text{AND}}((i_{j1}, i_{j2})) = (\min_j (i_{j1} \times i_{j2}) \div \min_j (i_{j2}), \min_j (i_{j2})) \tag{3.12}$$

and *the fuzzy neural OR* is defined as

$$\underset{j}{\text{OR}}((i_{j1}, i_{j2})) = (\max_j (i_{j1} \times i_{j2}) \div \max_j (i_{j2}), \max_j (i_{j2})) \tag{3.13}$$

By (3.7) – (3.11), it is evident that if let $w_j = 1/n$ (j=1, ..., n) then $\mathcal{L}(i_j, w_j) = \min(i_j)$ will be held and if let $w_j = 1$ then $\mathcal{L}(i_j\ w_j) = \max(i_j)$ will be held. In other words, when $w_{j1} = 1/n$ and $w_{j2} = 1/n$, for all $1 \le j \le n$, a FNLU realizes a fuzzy neural AND operator; and when $w_{j1} = 1$, $w_{j2} = 1$ for all $1 \le j \le n$, a FNLU realizes a fuzzy neural OR operator.

### *3.3 Fuzzy Neural Inference Unit(FNIU)*

A FNIU(Fig.2) is used to express the procedure of fuzzy refutation based on the fuzzy resolution principle and fuzzy linear resolution. Each fuzzy proposition, as a fact or a part of a rule in a given set, is transformed to a FNIU node. It has two inputs: a *knowledge input* $I_k = (i_{k1}, i_{k2})$ and a *query input* $I_q = (i_{q1}, i_{q2})$ as well as two outputs: a *knowledge output* $O_k = (o_{k1}, o_{k2})$ and a *conclusion output* $O_c = (o_{c1}, o_{c2})$. $W_k = (w_{k1}, w_{k2})$ is *knowledge weight.* The $w_{k1}$ depends on the weight of fuzzy rule, but the $w_{k2}$ is always assigned the value 1. The *query weight* $W_q = (w_{q1}, w_{q2})$ is always assigned by (1, 1). The output $O_k$ is calculated by:

$$\begin{aligned} O_k &= \mathcal{F}(I_k, W_k) \\ &= (o_{k1}, o_{k2}) \\ &= (f1(i_{k1} \times w_{k1}), f2(f1(i_{k1} \times w_{k1}), i_{k2} \times w_{k2})) \end{aligned} \tag{3.14}$$

where $\mathcal{F}$ is the *knowledge function.* It is realized by $f1$ and $f2$. $f1$ is called the *truth knowledge function* and defined as:

$$f1(x) = \begin{cases} x & x \in [-1, 1] \\ -1 & x < -1 \\ 1 & x > 1 \end{cases} \tag{3.15}$$

$f2$ is called the *confidence knowledge function* and defined as:

$$f2(x, y) = \min(x, y) \tag{3.16}$$

$O_c$ is calculated by:

$$\begin{aligned} O_c &= \mathcal{G}(I_q, W_q, I_k, W_k) \\ &= (o_{c1}, c_{c2}) \\ &= (g1(i_{q1} \times w_{q1}, i_{q2} \times w_{q2}, f1(i_{k1} \times w_{k1})), \\ &\quad g2(g1(i_{q1} \times w_{q1}, i_{q2} \times w_{q2}, o_{k1}), i_{k2} \times w_{k2})) \end{aligned} \tag{3.17}$$

where $\mathcal{G}$ is the *conclusion function.* It is realized by $g1$ and $g2$, where $g1$ is called the *truth function* and defined as:

$$g1(x, y, z) = \begin{cases} (x + y) + z & (x+y)+z \in [-1, 1] \\ 2 & \text{otherwise} \end{cases} \tag{3.18}$$

while $g2$ is called the *confidence function* and defined as:

$$g2(x, y) = \begin{cases} y & x \in [-1, 1] \\ x & \text{otherwise} \end{cases} \quad (3.19)$$

The *confidence of conclusion* $c_c$ is defined as:

$$c_c = \min(o_{c1}, o_{c2}) \quad (3.20)$$

The *mixed truth value of conclusion* is defined as:

$$\mathbf{MT}((o_{c1}, o_{c2})) = o_{c1} \times o_{c2} \times 0.5 + 0.5. \quad (3.21)$$

*3.4 Fuzzy Neural Knowledge Unit (FNKU)*

For transmitting the information from premises to consequences of fuzzy rules, we have another kind of unit, called the fuzzy neural knowledge unit (FNKU) (Fig. 3). A FNKU has one input $I = (i_1, i_2)$ and one output $O = (o_1, o_2)$. It can be looked as a special FNIU that has not query input and conclusion output. $W = (w_1, w_2)$ is the weight. The output is calculated by

$$O = (o_1, o_2) \quad (3.22)$$
$$= ( h(i_1 \times w_1), h(i_2 \times w_2) )$$

where $h$ is the *transmitting function* defined as:

$$h(x) = \begin{cases} x & x \in [-1, 1] \\ -1 & x < -1 \\ 1 & x > 1 \end{cases} \quad (3.23)$$

## 4. CONSTRUCTION OF FPRIN

For transforming a given set of facts and rules to a corresponding FPRIN, transformation laws are used. All rules in our discussion will be focused on Horn-set [14]. A proposition that may appear in a rule or a fact is transformed to a FNIU node. The relations among propositions are represented by FNLU nodes, FNKU nodes as well as the linkage of corresponding nodes (FNIU, FNLU, FNKU). Hereafter, a FNIU node corresponding to proposition A is also denoted by A and called *node A in FPRIN* or *node* A for short.

*Transformation law 1 (fact law)*:

Fact A with T(A) = t

is transformed to a FNIU as shown in Fig. 4, where $W_k = (1, 1)$, $W_q = (1, 1)$, $I_k = (2t-1, 1)$.

*Transformation law 2 (AND conditions law)*:

Rule $P_1, P_2, ..., P_n \rightarrow Q \ \{w_1\}$

is transformed to a FPRIN neural logic network as shown in Fig.5, where $W_k = (w_1, 1)$, $W_q = (1, 1)$.

*Transformation law 3 (AND-OR conditions law)*:

A set of rules: $P_1 \rightarrow Q \ \{w_1\}$

$P_{j1}, P_{j2}, ..., P_{jm} \rightarrow Q \ \{w_j\}$

$P_n \rightarrow Q \ \{w_n\}$

is transformed to a FPRIN neural logic network as shown in Fig.6, where $W_k = (1, 1)$, $W_q = (1, 1)$, $W_{pj} = (w_j, 1)$, $(j = 1, 2, ..., n)$.

*Transformation law 4 (rule-fact law)*:

Rule $P_1, P_2, ..., P_n \rightarrow Q \ \{w_1\}$

Fact Q with T(Q) = t

are transformed to a FPRIN neural logic network as shown in Fig.7, where $W_k = (1,1)$, $W_q = (1,1)$, $W_P = (w_1,1)$, $W_k = (1, 1)$. It can be seen as a special case of Transformation Law 3.

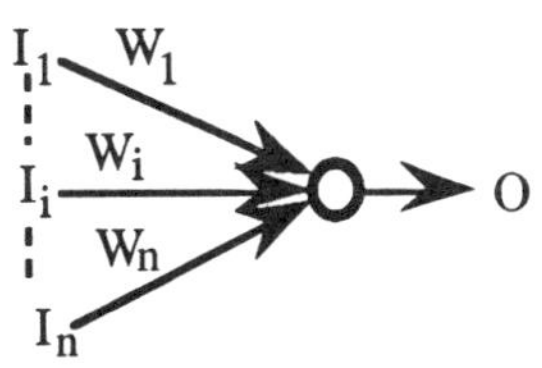


Fig.1. FNLU

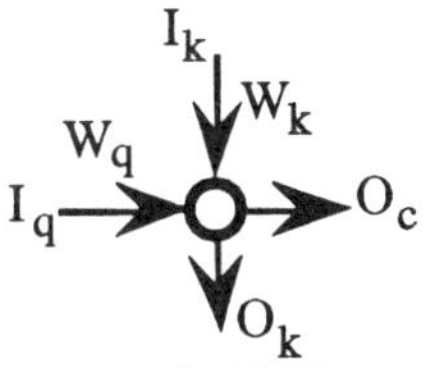


Fig.2. FNIU

Fig. 3 FNKU

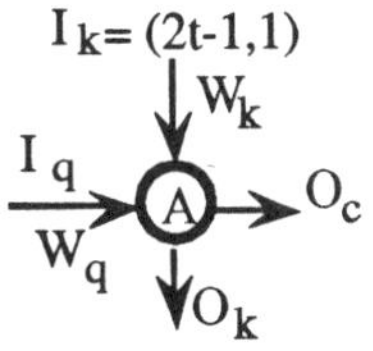


Fig. 4 *fact law*

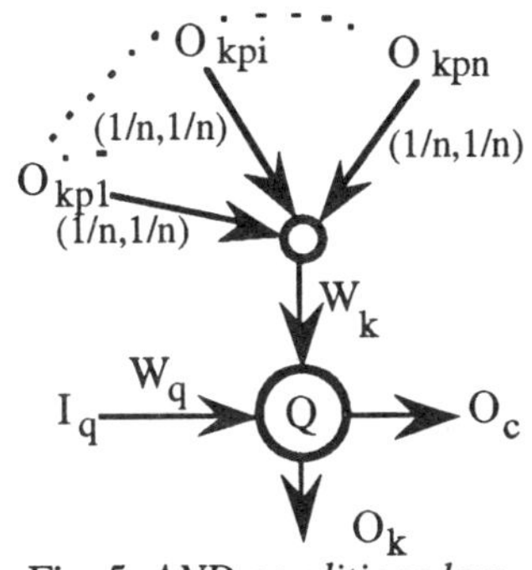


Fig. 5 *AND conditions law*

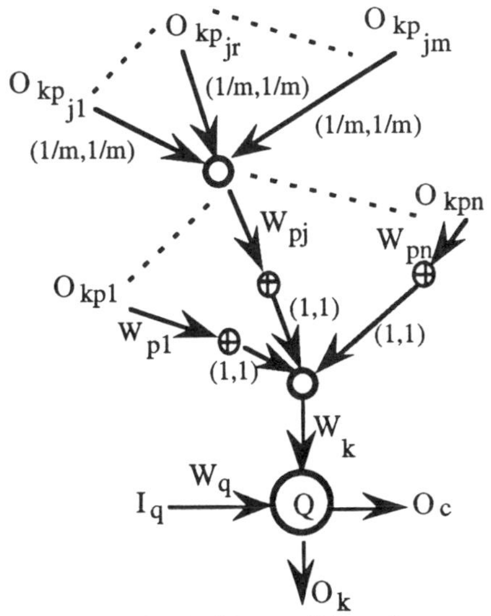


Fig. 6 *AND-OR conditions law*

*Transformation law 5 (simplification law)*:

Using Transformation law 2 or 3, when n = 1 we can get the FPRINs as shown in Fig. 8(a), 8(b), respectively. These FPRINs can be simplified to an equivalence as Fig. 8(c). In other words, we can directly transform a given rule $P_1 \rightarrow Q \ \{w_1\}$ to a FPRIN as shown in Fig. 8(c), where $W_k = (w_1, 1)$, $W_q = (1, 1)$.

*Transformation law 6 (knowledge-copy law)*:

Rules: $A \rightarrow S \ \{w_1\}$

$A \rightarrow R \ \{w_2\}$

are transformed to a FPRIN neural logic network as shown in Fig. 9, where $W_{k1} = (w_1, 1)$, $W_{k2} = (w_2, 1)$, $W_{q1} = (1, 1)$, $W_{q2} = (1, 1)$.

*Example 4.1:* Suppose the rules and facts

$r_1$: $P \rightarrow Q$ {0.8} $r_2$: $Q, R \rightarrow V$ {1}
$r_3$: $W \rightarrow V$ {0.9} $r_4$: P with T(P) = 1
$r_5$: Q with T(Q) = 0.7 $r_6$: R with T(R) = 0.6
$r_7$: W with T(W) = 0.7

are given. The corresponding FPRIN(Fig.10) is constructed by using the transformation laws step by step.

(step 1): Transform $r_4$ by transformation law 1;
(step 2): Transform $r_6$ by transformation law 1;
(step 3): Transform $r_7$ by transformation law 1;
(step 4): Transform $r_1$, $r_5$ and result of step 1 by transformation law 4;
(step 5): Transform $r_2$, $r_3$ and results of step 2-4 by transformation law 3.

## 5. INFERENCE IN FPRIN

### *5.1 The Inference on FPRIN*

For a given FPRIN network constructed by transformation laws 1~6, suppose that there are m FNIU nodes that represent m propositions. Let a sequence of these FNIU nodes be decided and noted as:

$$A_1, A_2, ..., A_m. \qquad (5.1)$$

By this sequence, the knowledge input vector $\mathbf{I_k}$, the query vector $\mathbf{I_q}$, the knowledge output vector $\mathbf{O_k}$ and the conclusion output vector $\mathbf{O_c}$ are represented as follows:

$$\mathbf{I_k} = [I_{k1}, I_{k2}, ..., I_{km}] \qquad (5.2)$$
$$\mathbf{I_q} = [I_{q1}, I_{q2}, ..., I_{qm}] \qquad (5.3)$$
$$\mathbf{O_k} = [O_{k1}, O_{k2}, ..., O_{km}] \qquad (5.4)$$
$$\mathbf{O_c} = [O_{c1}, O_{c2}, ..., O_{cm}] \qquad (5.5)$$

where $I_{kj}$, $I_{qj}$, $O_{kj}$, $O_{cj}$ (j=1, 2, ..., m) are the knowledge input, query input, knowledge output and the conclusion output of $A_j$ respectively. The knowledge weight vector $\mathbf{W_k}$ and the query weight vector $\mathbf{W_q}$ are represented as:

$$\mathbf{W_k} = [W_{k1}, W_{k2}, ..., W_{km}] \qquad (5.6)$$
$$\mathbf{W_q} = [W_{q1}, W_{q2}, ..., W_{qm}]. \qquad (5.7)$$

where $W_{kj}$, $W_{qj}$ (j=1, 2, ..., m) are the knowledge weight and the query weight of $A_j$, respectively. The knowledge output vector and the conclusion output vector are calculated by:

$$\mathbf{O_k} = \mathcal{F}(\mathbf{I_k}, \mathbf{W_k}) \qquad (5.8)$$
$$\mathbf{O_c} = \mathcal{G}(\mathbf{I_q}, \mathbf{W_q}, \mathbf{O_k}), \qquad (5.9)$$

where $\mathcal{F}$ and $\mathcal{G}$ are the knowledge function and the conclusion function defined in section 3.

Obviously, when a PFRIN network and a sequence of the FNIU nodes in this FPRIN have been given, it means that the $\mathbf{I_k}$ is also known. Thus, if any query by $\mathbf{I_q}$ corresponding to this FPRIN is asked, we can get the conclusion by only executing the truth function and the confidence function to the knowledge input vector and the query input vector, no matter how many propositions are included in the query or how many rules need to be used for the inference. Thus, a parallel fuzzy inference can be executed.

*Example 5.1*: See the same FPRIN network in Example 4.1 (Fig. 10). Suppose the consequence of nodes is decided as: P, Q, R, W, V,

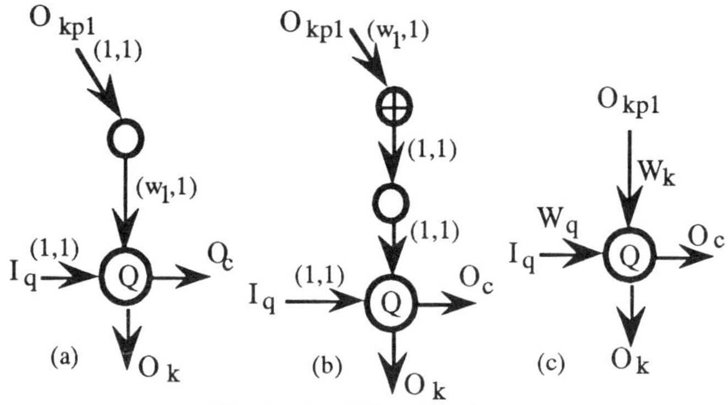

Fig. 8 *simplification law*

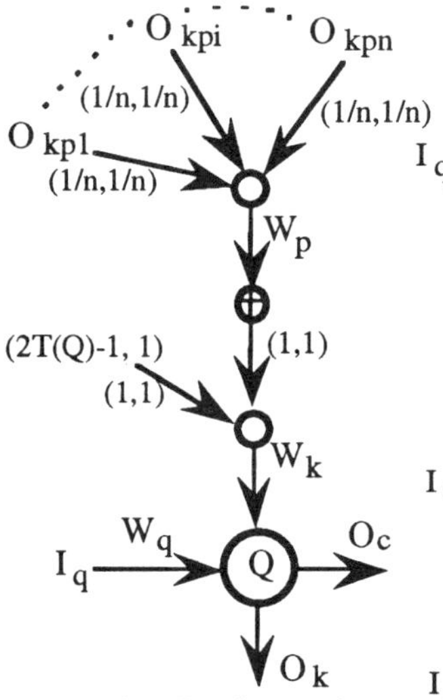
Fig. 7 *rule-fact law*

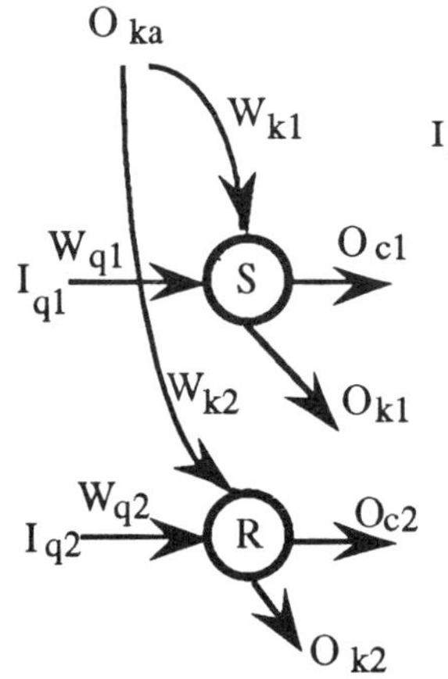
Fig. 9 *knowledge-copy law*

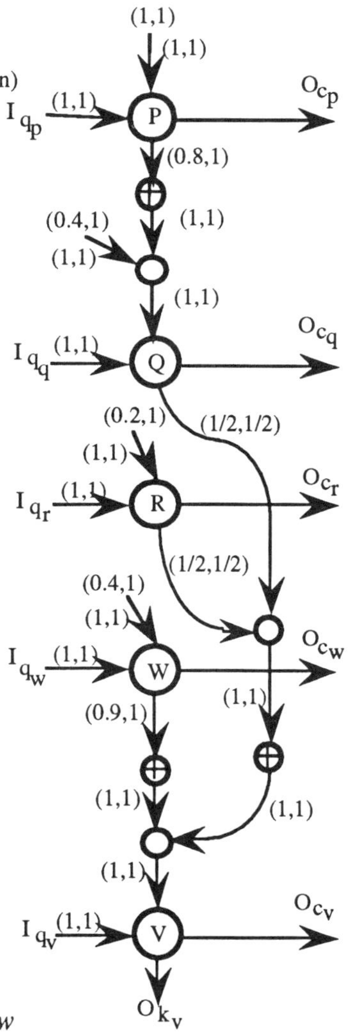
Fig. 10 An Example

then we have:

$\mathbf{I_k}$ = [(1,1), (0.8,1), (0.2,1), (0.4,1), (0.36,0.4)]
$\mathbf{O_k}$ = [(1,1), (0.8,0.8), (0.2,0.2), (0.4,0.4), (0.36,0.36)]
$\mathbf{W_k}$ = [(1, 1), (1, 1), (1, 1), (1, 1), (1, 1)]
$\mathbf{W_q}$ = [(1, 1), (1, 1), (1, 1), (1, 1), (1, 1)].

(I) If a query Q

is posed, it can be expressed as

$$I_q = [(2, 2), \mathbf{(1, -1)}, (2, 2), (2, 2), (2, 2)].$$

Thus, we get the conclusion output vector:

$$O_c = [(2, 2), \mathbf{(0.8, 1)}, (2, 2), (2, 2), (2, 2)].$$

This means the conclusion Q is proved with the confidence of conclusion

$$c_c = \min(0.8, 1) = 0.8$$

or in other words, Q is proved to be fuzzy true, and its mixed truth value is

$$MT(Q) = 1 \times 0.8 \times 0.5 + 0.5 = 0.9.$$

(II) If a query P and V is asked, then we have

$$I_q = [\mathbf{(-1, 1)}, (2, 2), (2, 2), (2, 2), \mathbf{(-1, 1)}].$$

Similarly, we get

$$O_c = [\mathbf{(1, 1)}, (2, 2), (2, 2), (2, 2), \mathbf{(0.36, 0.4)}].$$

In other words, we get the conclusion: P and V is proved with the confidence of conclusion

$$c_c = \min(\min(1, 1), \min(0.36, 0.4)) = 0.36$$

and the mixed truth value

$$MT(P \text{ and } V) = \min(MT(P), MT(V)) = 0.36 \times 0.4 \times 0.5 + 0.5 = 0.572.$$

*5.2 Logical Properties*

From the basic units and the definitions defined in Section 3, there follow several important properties.

*Property 5.1*: Suppose the truth value of fact A is $t \in [0,1]$, then by the conclusion function $\mathcal{G}$, when A is queried the conclusion output $O_c = (o_{c1}, o_{c2})$ should be

$$o_{c1} = o_{k1} = i_{k1} = 2t - 1 \quad (5.10)$$

$$o_{c2} = o_{k2} = i_{k2} = 1 \quad (5.11)$$

The mixed truth value of the conclusion of A should be

$$MT(O_c) = t \quad (5.12)$$

It means that $o_{c1}$ represents the truth confidence of A and $o_{c2}$ represents the confidence of resolvent of A. Further, the mixed truth value of the conclusion is just the truth value of A. On the other hand, when no query is addressed to A, then the conclusion output will always be the special pair (2, 2) that represents *undefined*.

*Property 5.2*: In a *fuzzy neural AND*, the mixed truth value of the output will always be the minimum of the mixed truth values of the inputs, and the confidence of resolvent of output will always be the minimum of the confidences of resolvent of the inputs. That is:

$$MT(O) = \min_j (MT(I_j)) \quad (5.13)$$

$$o_2 = \min_j (i_{j2}) \quad (5.14)$$

*Property 5.3*: In a *fuzzy neural OR*, the mixed truth value of output will always be the maximum of the mixed truth values of the inputs, and the confidence of resolvent of output will always be the maximum of the confidences of resolvent of the inputs. That is:

$$MT(O) = \max_j (MT(I_j)) \quad (5.15)$$

$$o_2 = \max_j (i_{j2}) \quad (5.16)$$

## 6. CONCLUSION

The parallel resolution inference on neural logic network(PRIN) has been extended from binary logic to fuzzy logic based on the fuzzy resolution principle. The FPRIN also realized an application of exponential form of fuzzy logic. Six transformation laws have been introduced. By those laws, any Horn-set can be transferred to a corresponding FPRIN network. Parallel fuzzy resolution inference can be done in such a neural network without loss of its logical completeness. Our next effort will be to extend FPRIN from fuzzy propositional logic to fuzzy predicate logic.

## REFERENCES


[1] L.Ding, "A proposal of parallel resolution inference on neural logic network", Proc. 2nd International Conference on Fuzzy Logic and Neural Networks(Iizuka'92), pp.237-240, 1992.

[2] T.J.Reynolds, H.H.Teh, and B.T.Low, "Programming in neural logic", Institute of Systems Science, National University of Singapore, 1990.

[3] Z.Shen, L.Ding, and M.Mukaidono, "Fuzzy resolution principle", Proc. 18th International Symposium on Multiple-valued Logic, IEEE, pp.210-215, 1988.

[4] Z.Shen, L.Ding, M.Mukaidono: "A theoretical framework of fuzzy prolog machine", *Fuzzy Computing* (M.M.Gupta & T.Yamakawa, Ed.), Elsevier Science Publishers B.V., North-Holland, pp.89-100, 1988.

[5] M.Mukaidono, Z.Shen, and L.Ding, "Fundamentals of fuzzy prolog", *Int. J. Approximate Reasoning*, Vol.3, No.2, pp.179-193, 1989.

[6] L.Ding, Z.Shen, and M.Mukaidono, "Fuzzy linear resolution as the inference engine of intelligent systems", *Methodologies for Intelligent Systems*, 4(Z.W.Ras, Ed.), Elsevier Science Publishing Co., Inc., New York, pp.1-8, 1989.

[7] Z.Shen, L.Ding, "Exponential form of fuzzy logic", Proc. XII/TIMS XXXI Joint International Conference, 1992.

[8] T.H.Goh, P.Z.Wang, H.C.Lui, "Learning algorithm for the enhanced fuzzy perceptron", Proc. IJCNN'92, III, pp.435-440, 1992.

[9] S.C.Chan, L.S.Hsu, K.F.Loe and H.H.Teh, "Neural logic networks", *Progress in Neural Networks*, Vol. II, (Ed. by Omid M. Omidvar), Ablex Publishing Co., 1991.

[10] L.S.Hsu, H.H.Teh, S.C.Chan and K.F.Loe, "Imprecise reasoning using neural networks", Proc. 23rd Annual Hawaii Int. Conf. on System Science, IEEE, pp.363-368, 1990.

[11] S.C.Chan, L.S.Hsu, S.Brody and H.H.Teh, "Neural three-valued logic networks", Abstracts, Proc. IJCNN, Washington, USA, pp.549, 1989.

[12] J.A.Robinson, "A machine oriented logic based on the resolution principle", *J. ACM*, Vol.12, No.1, pp.23-41, 1965.

[13] L.A.Zadeh, *FUZZY SETS AND APPLICATIONS: Selected papers by L.A.Zadeh*, (R.R.Yager, et cl. Ed.), John Wiley & Sons, 1987.

[14] C.L.Chang, R.C.T.Lee, *Symbolic Logic and Mechanical Theorem Proving*, Academic Press, Inc., 1973.

[15] L.Ding, Z.Shen, and H.C.Lui, "Weight of fuzzy rule in approximate case-based reasoning", Proc. ISKIT'92, pp.75-78, 1992.

[16] L. Ding, Z. Shen, and M. Mukaidono, "A new method of approximate reasoning", Proc. 19th International Symposium on Multiple-valued Logic, IEEE, pp.179-185, 1989.

[17] Feasibility Study Committee of the Real-World Computing Program, *The Master Plan for the Real-World Computing Program(DRAFT)*, MITI, Japan, 1992.

# REINFORCEMENT STRUCTURE/PARAMETER LEARNING FOR NEURAL-NETWORK-BASED FUZZY LOGIC CONTROL SYSTEMS

*C. T. Lin*
Department of Computer and Information Science
National Chiao-Tung University
Hsinchu, Taiwan, R.O.C.

*C. S. George Lee*
School of Electrical Engineering
Purdue University
West Lafayette, Indiana 47907

## ABSTRACT

This paper proposes a Reinforcement Neural-Network-Based Fuzzy Logic Control System (RNN-FLCS) for solving various reinforcement learning problems. The proposed RNN-FLCS is best applied to learning environments where obtaining exact training data is expensive. It is constructed by integrating two Neural-Network-Based Fuzzy Logic Controllers (NN-FLCs), each of which is a connectionist model with a feedforward multi-layered network developed for the realization of a fuzzy logic controller. One NN-FLC functions as a fuzzy predictor and the other as a fuzzy controller. Using the temporal difference prediction method, the fuzzy predictor can predict the external reinforcement signal and provide a more informative internal reinforcement signal to the fuzzy controller. The fuzzy controller performs a stochastic exploratory algorithm to adapt itself according to the internal reinforcement signal. During the learning process, the proposed RNN-FLCS can construct a fuzzy logic control system automatically and dynamically through a reward-penalty signal or through very simple fuzzy information feedback; both structure learning and parameter learning are performed simultaneously in the two NN-FLCs using the fuzzy similarity measure. Simulation results are presented to illustrate the performance and applicability of the proposed RNN-FLCS.

## 1. Introduction

Most of the supervised and unsupervised learning algorithms for neural networks require precise training data sets for setting the weights and the connectivity of the links for various applications [1]. For some real-world applications, precise data for training/learning are usually difficult and expensive, if not impossible, to obtain. For this reason, there has been a growing interest in reinforcement learning algorithms for neural networks [2-4]. In this paper, we are extending our previous work on neural-network-based fuzzy logic control systems (NN-FLC) [5,6] to the reinforcement learning problem, And we apply the technique of associative reinforcement learning to our proposed reinforcement NN-FLC learning system. The proposed NN-based learning system can construct a fuzzy logic control and decision system automatically and dynamically through a reward-penalty signal (i.e., good/bad signal) or through very simple fuzzy feedback information such as "high," "too high," "low," and "too low." Moreover, there is a possibility of a long time delay between an action and the resulting reinforcement feedback information. To achieve the goal of solving reinforcement learning problems in fuzzy logic systems, a Reinforcement Neural-Network-Based Fuzzy Logic Control System (RNN-FLCS) is proposed which consists of two closely integrated NN-FLCs. Structurally, these two NN-FLCs share the first two layers of the original NN-FLC in [5]; that is, they use the same distributed representation of input patterns. This representation is the overlapping type and is dynamically adjustable through the learning process. Associated with the proposed RNN-FLCS is the reinforcement structure/parameter learning algorithm which dynamically determine the proper network size, connections, and parameters of the RNN-FLCS through an external reinforcement signal. Furthermore, learning can proceed even in the period without any external reinforcement feedback.

This work was supported in part by the National Science Foundation under Grant CDR 8803017 to the Engineering Research Center for Intelligent Manufacturing Systems and a grant from the Ford Foundation.

## 2. Neural-Network-Based Fuzzy Logic Controller

This section introduces the structure and functions of our previously proposed Neural-Network-Based Fuzzy Logic Controller (NN-FLC) [5,6], which is a basic component of the proposed RNN-FLCS. Figure 1 shows the connectionist structure of our NN-FLC. The system has five layers. We shall next describe the functions of the nodes in each of the five layers of the proposed connectionist model. In the following, $f$ is an integration function of a node, which combines activation from other nodes to provide net input for this node, and $a$ is an activation function of a node, which outputs an activation value as a function of net input. In the following equations, superscript is used to indicate the layer number.

- **Layer 1:** The nodes in this layer transmit input values directly to the next layer. That is,

$$f = u_i^1 \quad \text{and} \quad a = f. \tag{1}$$

From (1), the link weight at layer one ( $w_i^1$ ) is unity.

- **Layer 2:** If we use a single node to perform a simple membership function, then the output function of this node should be this membership function. For example, for a bell-shaped function,

$$f = M_{x_i}^j(m_{ij}, \sigma_{ij}) = \frac{-(u_i^2 - m_{ij})^2}{\sigma_{ij}^2} \quad \text{and} \quad a = e^f, \tag{2}$$

where $m_{ij}$ and $\sigma_{ij}$ are, respectively, the center (or mean)

0-7803-0614-7/93$03.00 ©1993IEEE

and the width (or variance) of the bell-shaped function of the $j$th term of the $i$th input linguistic variable $x_i$.

- **Layer 3:** The links in this layer are used to perform precondition matching of fuzzy logic rules. Hence, the rule nodes should perform the fuzzy AND operation,

$$f = \min(u_1^3, u_2^3, \cdots, u_p^3) \text{ and } a = f. \tag{3}$$

The link weight in layer three ($w_i^3$) is then unity.

- **Layer 4:** The links at layer four should perform the fuzzy OR operation to integrate the fired rules which have the same consequence:

$$f = \sum_{i=1}^{p} u_i^4 \text{ and } a = \min(1, f). \tag{4}$$

Hence, the link weight $w_i^4 = 1$.

- **Layer 5:** The nodes in this layer transmit the decision signal out of the network. These nodes and the layer-five links attached to them act as the defuzzifier. If $m_{ij}^5$'s and $\sigma_{ij}^5$'s are the centers and the widths of the membership functions, respectively, then the following functions can be used to simulate the *center of area* defuzzification method [7]:

$$f = \sum w_{ij}^5 u_i^5 = \sum (m_{ij}\sigma_{ij}) u_i^5 \text{ and } a = \frac{f}{\sum \sigma_{ij} u_i^5}. \tag{5}$$

Here the link weight at layer five ( $w_{ij}^5$ ) is $m_{ij}\sigma_{ij}$.

## 3. Structure/Parameter Learning Algorithm for the RNN-FLCS with a Single-Step Fuzzy Predictor

Unlike the supervised learning problem, the reinforcement learning problem has only very simple "evaluative" information called reinforcement signal available for learning. In this paper, the reinforcement signal $r(t)$ is defined as a value between -1 and 1 corresponding to various degrees of failure or success. We also assume that $r(t)$ is the reinforcement signal available at time step $t$ and is caused by the input and actions chosen at time step $t-1$ or even affected by earlier inputs and actions. The objective of learning is to maximize the reinforcement signal. To resolve the reinforcement learning problems, a new structure, called the Reinforcement Neural-Network-Based Fuzzy Logic Control System (RNN-FLCS), is proposed. The proposed RNN-FLCS, as shown in Fig. 2, integrates two NN-FLCs into a learning system: one NN-FLC for the fuzzy controller and the other for the fuzzy predictor. These two NN-FLCs share the same layers 1 and 2 and have individual layer 3 to layer 5. In this section, a reinforcement learning algorithm is proposed for the RNN-FLCS with a single-step fuzzy predictor to solve simpler reinforcement learning problems in which a reinforcement signal is only one time step behind its corresponding action. For the case that there is a long time delay between an action and the resulting reinforcement signal, a more powerful multi-step fuzzy predictor is necessary for the RNN-FLCS. This will be discussed in Section 4.

### 3.1. Stochastic Exploration

In this subsection, we first develop the learning algorithm for the action network. The goal of the reinforcement structure/parameter learning algorithm is to adjust the parameters (e.g., $m_i$'s) of the action network or to change the connectionist structure or even to add new nodes, if necessary, such that the reinforcement signal is maximum. That is, $\Delta m_i \propto \frac{\partial r}{\partial m_i}$. To know $\frac{\partial r}{\partial m_i}$, we need to know $\frac{\partial r}{\partial y}$, where $y$ is the output of the action network. In our learning algorithm, the gradient information, $\frac{\partial r}{\partial y}$, is estimated by the stochastic exploratory method [8]. In estimating the gradient information, the output $y$ of the action network does not act on the environment directly. Instead, it is treated as a mean (expected) action. The actual action, $\hat{y}$, is chosen by exploring a range around this mean point. This range of exploration corresponds to the variance of a probability function, which is the normal distribution in our design. This amount of exploration, $\sigma(t)$, is chosen as

$$\sigma(t) = \frac{k}{2}[1 - \tanh(p(t))] = \frac{k}{1 + e^{2p(t)}}, \tag{6}$$

where $k$ is a search-range scaling constant which can be simply set to 1, and $p(t)$ is the predicted (expected) reinforcement signal used to predict $r(t)$. Once the variance has been decided, the actual output of the stochastic node can be set as $\hat{y}(t) = N(y(t), \sigma(t))$. The gradient information is estimated as

$$\frac{\partial r}{\partial y} \approx [r(t) - p(t)]\left[\frac{\hat{y}(t-1) - y(t-1)}{\sigma(t-1)}\right] \equiv [r(t) - p(t)]\left[\frac{\hat{y} - y}{\sigma}\right]_{t-1} \tag{7}$$

where the subscript, $t-1$, represents the time displacement. Assuming that $w$ is an adjustable parameter in a node (e.g., the center of a membership function), the general parameter learning rule used is

$$w(t+1) = w(t) + \eta\left(\frac{\partial r}{\partial w}\right), \text{ and } \frac{\partial r}{\partial w} = \frac{\partial r}{\partial a}\frac{\partial a}{\partial f}\frac{\partial f}{\partial w}, \tag{8}$$

where $\eta$ is the learning rate.

- **Layer 5:** Using (5), (7), and (8), the expected updated amount of the center parameter is

$$\Delta m_i(t) = \eta[r(t) - p(t)]\left[\frac{\hat{y} - y}{\sigma}\right]_{t-1}\left[\frac{\sigma_i u_i}{\sum \sigma_i u_i}\right]_{t-1}. \tag{9}$$

Similarly, the expected updated amount of the width parameter is

$$\Delta\sigma_i(t) = \eta[r(t) - p(t)]\left[\frac{\hat{y} - y}{\sigma}\right]_{t-1}\left[\frac{m_i u_i(\sum \sigma_i u_i) - (\sum m_i \sigma_i u_i) u_i}{(\sum \sigma_i u_i)^2}\right]_{t-1} \tag{10}$$

The error to be propagated to the preceding layer is

$$\delta^5(t) = \frac{\partial r}{\partial f^5} = \frac{\partial r}{\partial a}\frac{\partial a}{\partial f^5} = [r(t) - p(t)]\left[\frac{\hat{y} - y}{\sigma}\right]_{t-1}. \tag{11}$$

**Fuzzy Similarity Measure:** In this step, the system will decide whether the current structure should be changed or not according to the expected updated amount of the center and width parameters (in (9) and (10)). To do this, the expected center and width are, respectively, computed as

$$m_{i-new} = m_i(t) + \Delta m_i(t) \text{ and } \sigma_{i-new} = \sigma_i(t) + \Delta\sigma_i(t). \quad (12)$$

From the current membership functions of output linguistic variables, we want to find the one which is the most similar to the expected membership function by measuring their fuzzy similarity. The fuzzy similarity measure is proposed in [6] to measure the similarity between two fuzzy sets.

Let $M(m_i, \sigma_i)$ represent the bell-shaped membership function with center $m_i$ and width $\sigma_i$. Let

$$degree(i, t) = E[M(m_{i-new}, \sigma_{i-new}), M(m_{i-closest}, \sigma_{i-closest})]$$
$$= \max_{1 \le j \le k} E[M(m_{i-new}, \sigma_{i-new}), M(m_j, \sigma_j)], \quad (13)$$

where $k = |T(y)|$, and $E(\cdot,\cdot)$ is the fuzzy similarity measure defined in [6]. If $A$ and $B$ are two fuzzy sets with bell-shaped membership functions with centers $m_1$, $m_2$ and widths $\sigma_1, \sigma_2$, then the approximate fuzzy similarity measure of $A$ and $B$, $E(A, B)$, can be computed as follow [6]: Assuming $m_1 \ge m_2$,

$$E(A, B) = \frac{M(A \cap B)}{M(A \cup B)} = \frac{M(A \cap B)}{\sigma_1\sqrt{\pi} + \sigma_2\sqrt{\pi} - M(A \cap B)}. \quad (14)$$

Here $M(A \cap B)$ is defined in [6]. After the most similar membership function $M(m_{i-closest}, \sigma_{i-closest})$ to the expected membership function $M(m_{i-new}, \sigma_{i-new})$ has been found, the following adjustment is made:

IF $degree(i, t) < \alpha(t)$,
  THEN
    create a new node $M(m_{i-new}, \sigma_{i-new})$ in layer 4
      and denote this new node as the $i-closest$ node;
    do the structure learning process;
  ELSE IF $M(m_{i-closest}, \sigma_{i-closest}) \ne M(m_i, \sigma_i)$
    THEN
      do the structure learning process;
    ELSE

$$m_i(t+1) = m_{i-new} \text{ and } \sigma_i(t+1) = \sigma_{i-new}.$$

$\alpha(t)$ in the above adjustment is a monotonically increasing scalar similarity criterion.

**Structure Learning:** To find the rules whose consequences should be changed, we set a *firing strength threshold*, $\beta$. Only the rules whose firing strengths are higher than this threshold are treated as *really firing* rules. Assuming that the term node $M(m_i, \sigma_i)$ in layer 4 has inputs from rule nodes $1 \cdots l$ in layer 3, whose corresponding firing strengths are $a_i^3$'s, $i = 1 \cdots l$, then

IF $a_i^3(t) \ge \beta$, THEN change the consequence of the $i$th rule node from $M(m_i, \sigma_i)$ to $M(m_{i-new}, \sigma_{i-new})$.

- **Layer 4:** There is no parameter to be adjusted in this layer. Only the error signals ($\delta_i^4$'s) need to be computed and propagated. From (5) and (8), the error signal $\delta_i^4$ is derived as

$$\delta_i^4(t) = [r(t) - p(t)]\left[\frac{\hat{y} - y}{\sigma}\right]_{t-1}\left[\frac{m_i\sigma_i(\sum\sigma_i u_i) - (\sum m_i\sigma_i u_i)\sigma_i}{(\sum\sigma_i u_i)^2}\right]_{t-1}. \quad (15)$$

In the multi-output case, the computations in layers five and four are exactly the same as the above using the same internal reinforcement signals and proceeding independently for each output linguistic variable.

- **Layer 3:** As in layer four, only the error signals need to be computed. According to (4) and (8), this error signal can be derived as $\delta_i^3(t) = \delta_i^4(t)$. If there are multiple outputs, then the error signal becomes $\delta_i^3(t) = \sum_k \delta_k^4(t)$, where the summation is performed over the consequences of a rule node; that is, the error of a rule node is the summation of the errors of its consequences.

- **Layer 2:** Using (2) and (8), the adaptive rule of $m_{ij}$ and $\sigma_{ij}$ are

$$m_{ij}(t+1) = m_{ij}(t) - \eta\left[\frac{\partial r}{\partial a_i}\right]_t\left[e^{f_i}\frac{2(u_i - m_{ij})}{\sigma_{ij}^2}\right]_{t-1}, \quad (16)$$

$$\sigma_{ij}(t+1) = \sigma_{ij}(t) - \eta\left[\frac{\partial r}{\partial a_i}\right]_t\left[e^{f_i}\frac{2(u_i - m_{ij})^2}{\sigma_{ij}^3}\right]_{t-1},$$

where $\frac{\partial r}{\partial a_i} = \sum_k q_k(t)$ (see [5] for details).

### 3.2. Single-Step Fuzzy Predictor

We shall use an NN-FLC to develop a single-step fuzzy predictor (evaluation network) as shown in Fig. 2. The function of the single-step fuzzy predictor is to predict the external reinforcement signal, $r(t)$, one time step ahead, that is, at time $t-1$. Here, $r(t)$ is the real reinforcement signal resulting from the inputs and actions chosen at time step $t-1$, but it can only be known at time step $t$. If the fuzzy predictor can produce a signal, $p(t)$, which is the prediction of $r(t)$ but is available at time step $t-1$, then the time delay problem can be solved. With a correct predicted signal, $p(t)$, a better action can be chosen by the action network at time step $t-1$, and the corresponding learning can be performed on the action network at time step $t$ upon receiving the external reinforcement signal $r(t)$. As indicated in the last subsection, $p(t)$ is necessary for the stochastic exploration with a multiparameter probability distribution (in (6)). The other internal reinforcement signal, $\hat{r}(t)$, in Fig. 2 is set as $\hat{r}(t) = r(t) - p(t)$, which is the prediction error for computing (7) by the action network. The single-step prediction is the extreme case of the multi-step prediction which will be presented in the next section. The goal to train the single-step fuzzy predictor is to minimize the squared error prediction:

$$E = \frac{1}{2}[r(t) - p(t)]^2, \quad (17)$$

where $r(t)$ represents the desired output (real external reinforcement signal), and $p(t)$ is the current output (predicted reinforcement signal). Then the gradient information can be easily derived as $\frac{\partial E}{\partial p} = p(t) - r(t)$. Similar to the learning rule developed in the last section, we can derive the structure/parameter learning algorithm for the single-step fuzzy predictor using the general parameter learning rule: $w(t+1) = w(t) + \eta(-\frac{\partial E}{\partial w})$, where $w$ is the adjustable parameters in the fuzzy predictor. The learning equations are the same as (8)-(16) if $\frac{\partial r}{\partial y}$ is replaced by $(-\frac{\partial E}{\partial p})$ and the effects caused by this replacement are properly updated, that is, all the terms $[r(t)-p(t)]\left[\frac{\hat{y}-y}{\sigma}\right]_{t-1}$ in (8)-(16) are replaced with the term $[r(t)-p(t)]$.

## 4. Multi-Step Fuzzy Predictor

When both the reinforcement signal and input patterns from the environment may depend arbitrarily on the past history of the network output and the network may only receive a reinforcement signal after a long sequence of outputs, the credit assignment problem becomes severe. This *temporal credit assignment* problem results because we need to assign credit or blame to each step individually in such a long sequence for an eventual success or failure. To solve the temporal credit assignment problem, the technique based on the temporal-difference methods, which are often closely related to the dynamic programming techniques [4], is used [2,9]. Unlike the single-step prediction or the supervised learning method which assigns credit according to the difference between the predicted and actual output, the temporal-difference methods assign credit according to the difference between temporally successive predictions. Some important temporal-difference equations of three different cases are summarized below.

- **Case 1:** *Prediction of final outcome.* Given the observation-outcome sequences of the form $x_1, x_2, \cdots, x_m, z$, where each $x_t$ is an input vector available at time step $t$ from the environment, and $z$ is the external reinforcement signal available at time step $m+1$. For each observation-outcome sequence, the fuzzy predictor produces a corresponding sequence of predictions $p_1, p_2, \cdots, p_m$, each of which is an estimate of $z$. For this prediction problem, the learning rule is $\Delta w_t = \eta(p_t - p_{t-1})\sum_{k=1}^{t-1}\lambda^{t-k-1}\nabla_w p_k$, where $p_{m+1} \equiv z$, $0 \le \lambda \le 1$, and $\eta$ is the learning rate. $\lambda$ is the recency weighting factor with which alternations to the predictions of observation vectors occurring $k$ steps in the past are weighted by $\lambda^k$.
- **Case 2:** *Prediction of finite cumulative outcomes.* In this case, $p_t$ predicts the remaining cumulative cost given the $t$th observation, $x_t$, rather than the overall cost for the sequence. Let $r_t$ be the actual cost incurred between time steps $t-1$ and $t$. Then $p_{t-1}$ is to predict $z_{t-1} = \sum_{k=t}^{m+1} r_k$. Thus, the learning rule is $\Delta w_t = \eta(r_t + p_t - p_{t-1})\sum_{k=1}^{t-1}\lambda^{t-k-1}\nabla_w p_k$.
- **Case 3:** *Prediction of infinite discounted cumulative outcomes.* In this case, $p_{t-1}$ predicts $z_{t-1} = \sum_{k=0}^{\infty}\gamma^k r_{t+k} = r_t + \gamma p_t$, where the discount-rate parameter $\gamma$, $0 \le \gamma < 1$, determines the extent to which we are concerned with short- or long-range prediction. This is used for prediction problems in which exact success or failure may never become completely known. In this case, the prediction error is $(r_t + \gamma p_t) - p_{t-1}$, and the learning rule is $\Delta w_t = \eta(r_t + \gamma p_t - p_{t-1})\sum_{k=1}^{t-1}\lambda^{t-k-1}\nabla_w p_k$.

In applying the temporal difference procedures to the proposed RNN-FLCS, we let $\lambda = 0$ due to its efficiency and accuracy [9]. A general learning rule used for the above three cases is

$$\Delta w_t = \eta(r_t + \gamma p_t - p_{t-1})\nabla_w p_{t-1}, \tag{18}$$

where $\gamma$, $0 \le \gamma < 1$, is a discount-rate parameter, and $\eta$ is the learning rate.

We shall next derive the learning rule of the multi-step fuzzy predictor according to (18). In this case, $p(t)$ is the single output of the fuzzy predictor (evaluation network) for the network's current parameter, $w(t)$, and current given input vector, $x(t)$, at time step $t$. Here, $p(t)$ can be any kind of prediction output in the various cases of the multi-step prediction problem stated above. According to (18), let $\hat{r}(t) = r(t) + \gamma p(t) - p(t-1)$, $0 \le \gamma < 1$. Then $\hat{r}(t)$ is the error signal of the output node of the multi-step fuzzy predictor. The general parameter learning rule then is

$$\Delta w(t) = \eta\hat{r}(t)\left[\frac{\partial p}{\partial w}\right]_{t-1}, \tag{19}$$

where $w$ is the network parameter (i.e., $m_i$ or $\sigma_i$). The learning rule for each layer in the fuzzy predictor can be computed as in (8)-(16). The only exception is that the error signal is different. Thus, the learning equations for the multi-step fuzzy predictor are the same as in (8)-(16) but with the term $[r(t)-p(t)]\left[\frac{\hat{y}-y}{\sigma}\right]_{t-1}$ replaced by the term $\hat{r}(t)$. Also the multi-step fuzzy predictor will provide two internal reinforcement signals, the prediction output, $p(t)$, and the prediction error, $\hat{r}(t)$, to the action network for its learning (see Fig. 2).

The learning algorithm for the action network is the same as that derived in Subsection 3.1 above. However, due to the different nature of the internal reinforcement signal, $\hat{r}(t)$, the learning algorithm of the action network with the multi-step fuzzy predictor will be different. The goal of the action network is to maximize the external reinforcement signal, $r(t)$. Thus, we need to estimate the gradient information, $\frac{\partial r}{\partial y}$, as we did above. With the internal reinforcement signals, $p(t)$ and $\hat{r}(t)$, from the evaluation network, the action network can perform the stochastic exploration and learning. The prediction signal $p(t)$ is used to decide the variance of the normal distribution function in the stochastic exploration in (6). Then the actual output, $\hat{y}(t)$, can be determined according to Subsection 3.1. Since $\hat{r}(t)$ is the prediction error, the gradient information is estimated as

$$\frac{\partial r}{\partial y} = \hat{r}(t)\left[\frac{\hat{y}-y}{\sigma}\right]_{t-1} = [r(t) - [p(t-1) - \gamma p(t)]]\left[\frac{\hat{y}-y}{\sigma}\right]_{t-1} . \quad (20)$$

From the above equation, we can observe that if $r(t) > [p(t-1) - \gamma p(t)]$, the actual action, $\hat{y}(t-1)$, is better than the expected action, $y(t-1)$. So $y(t-1)$ should be moved closer to $\hat{y}(t-1)$. On the other side, $y(t-1)$ should be moved further away from $\hat{y}(t-1)$.

Having the gradient information, $\frac{\partial r}{\partial y}$ (in (20)), the learning algorithm of the action network can be determined in the same way as in the previous section. The exact learning equations are the same as in (8)-(16) except that $[r(t) - p(t)]\left[\frac{\hat{y}-y}{\sigma}\right]_{t-1}$ is replaced by the new error term $[r(t) + \gamma p(t) - p(t-1)]\left[\frac{\hat{y}-y}{\sigma}\right]_{t-1}$.

## 5. An Illustrative Example

The proposed RNN-FLCS with multi-step fuzzy predictor has been simulated on a Sun SPARCstation for the cart-pole balancing problem. As shown in Fig. 3, there are four input state variables in this system: $\theta$, angle of the pole from an upright position (in degrees); $\dot{\theta}$, angular velocity of the pole (in degrees/sec); $x$, horizontal position of the cart's center (in meter); and $\dot{x}$, velocity of the cart (in m/s). The only control action is $f$, which is the amount of force (in N) applied to the cart to move it toward the left or the right. The system fails and receives a penalty signal, -1, when the pole falls past a certain angle ($\pm$ 12° is used here) or the cart runs into the bounds of its track (the distance is $2.4m$ from the center to both ends of the track). The goal of this control problem is to train the RNN-FLCS such that it can decide a sequence of forces with proper magnitudes to apply to the cart to balance the pole without failure for as long as possible.

In the computer simulation, similar to [2], the learning system was tested for 10 runs by trying to use the same learning parameter values in [2]. Each run consisted of a sequence of trials, where each trial began with the same initial condition: $\theta[0] = \dot{\theta}[0] = x[0] = \dot{x}[0] = 0$, and ended with a failure signal indicating that either $|\theta[t]| > 12°$ or $|x[t]| > 2.4m$. The input fuzzy partitions were set as $|T(x)| = 3$, $|T(\dot{x})| = 3$, $|T(\theta)| = 6$, and $|T(\dot{\theta})| = 3$ for all runs. For each run, the input (output) membership functions were initialized in the way that they covered the whole input (output) space evenly, and the output fuzzy partition was initialized as $|T(f)| = 7$. Runs were consisted of at most 50 trials unless the duration of each run exceeded 500,000 time steps. A run was called "success" and terminated after 500,000 time steps before all 50 trials took place; otherwise it was called "failure" and terminated at the end of its 50th trial. The simulation results (see Fig. 4) showed that the RNN-FLCS can learn to balance the pole within 20 trials. Most of the 10 runs even can complete the learning before 10 trials. This performance is better than that presented in [2,10] and compatible to that in [11]. In most runs, the final number of learned output membership functions is less than 15 as compared to 189 output membership functions that were used in [11] for each run. That is, one output membership function for each (overlapping) grid of input space.

## 6. Conclusion

This paper described the development of integrating two NN-FLCs into an integrated Reinforcement Neural-Network-Based Fuzzy Logic Control System (RNN-FLCS) for solving various reinforcement learning problems. Furthermore, by combining the techniques of temporal difference, stochastic exploration, and the previously proposed on-line supervised structure/parameter learning algorithm, a reinforcement structure/parameter learning algorithm was derived for the RNN-FLCS. Using the proposed structure and learning algorithm, a fuzzy logic controller to control a plant and a fuzzy predictor to model the plant can be set up dynamically through simultaneous structure/parameter learning for various classes of reinforcement learning problems. The proposed RNN-FLCS makes the design of fuzzy logic controllers more practical for real-world applications since it greatly lessens the quality and quantity requirements of the feedback training signals.

## 7. References


[1] G. E. Hinton, "Connectionist learning procedures," *Artificial Intelligence*, Vol. 40, No. 1, 1989, pp. 143-150.

[2] A. G. Barto, R. S. Sutton, and C. W. Anderson, "Neuronlike adaptive elements that can solve difficult learning control problems," *IEEE Trans. Syst. Man Cybern.*, Vol. 13, No. 5, pp. 834-847, 1983.

[3] R. J. Williams, "A class of gradient-estimating algorithms for reinforcement learning in neural networks," *Proc. of 1987 Int'l Joint Conf. Neural Networks*, San Diego, pp. II601-608, 1987.

[4] P. J. Werbos, "A menu of design for reinforcement learning over time," in *Neural Networks for Control*, Chapter 3, W. T. Miller, III, R. S. Sutton, and P. J. Werbos, eds, Cambridge: MIT Press, 1990.

[5] C. T. Lin and C. S. G. Lee, "Neural-network-based fuzzy logic control and decision system," *IEEE Trans. on Computers*, Vol. 40, No. 12, pp. 1320-1336, Dec. 1991.

[6] C. T. Lin and C. S. G. Lee, "Real-time supervised structure/parameter learning for Fuzzy Neural Network," *Proc. of 1992 IEEE Int'l Conf. on Fuzzy Systems*, San Diego, CA, pp. 1283-1290, 1992.

[7] L. A. Zadeh, "Fuzzy logic," *IEEE Computer*, pp. 83-93, April 1988.

[8] J. A. Franklin, "Input space representation for reinforcement learning control," *Proc. of IEEE Int'l Conf. Intelligent Machine*, pp. 115-122, 1989.

[9] R. S. Sutton, "Learning to predict by the methods of temporal difference," *Machine Learning*, Vol. 3, pp. 9-44, 1988.

[10] C. W. Anderson, "Strategy learning with multilayer connectionist representations," *Proc. of Fourth Int'l Workshop on Machine Learning*, pp. 103-114, Irvine, CA, June 1987.

[11] C. C. Lee and H. R. Berenji, "An intelligent controller based on approximate reasoning and reinforcement learning," *Proc. of IEEE Int'l Conf. Intelligent Machine*, pp. 200-205, 1989.

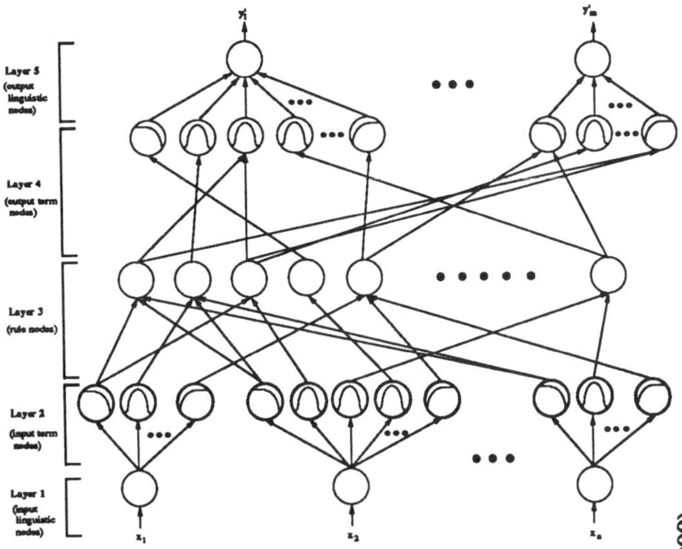

**Fig. 1.** Proposed neural-network-based fuzzy logic controller (NN-FLC).

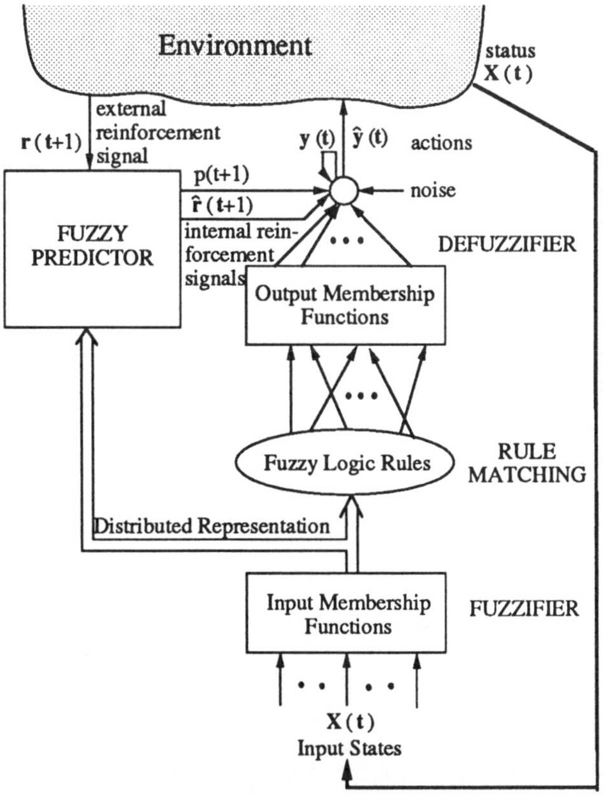


**Fig. 2.** Proposed reinforcement neural-network-based fuzzy logic control system (RNN-FLCS).

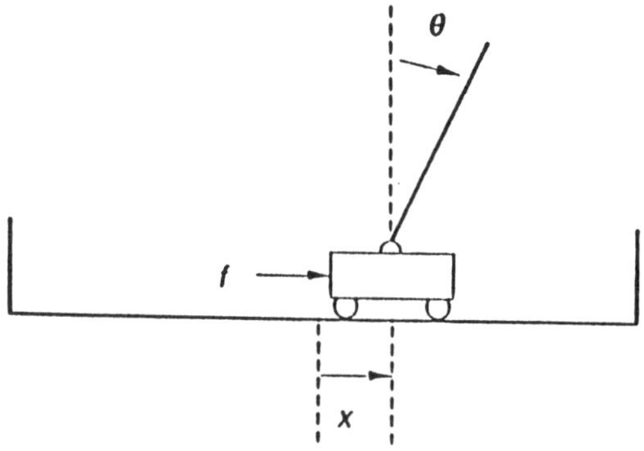


**Fig. 3.** The cart-pole balancing problem.

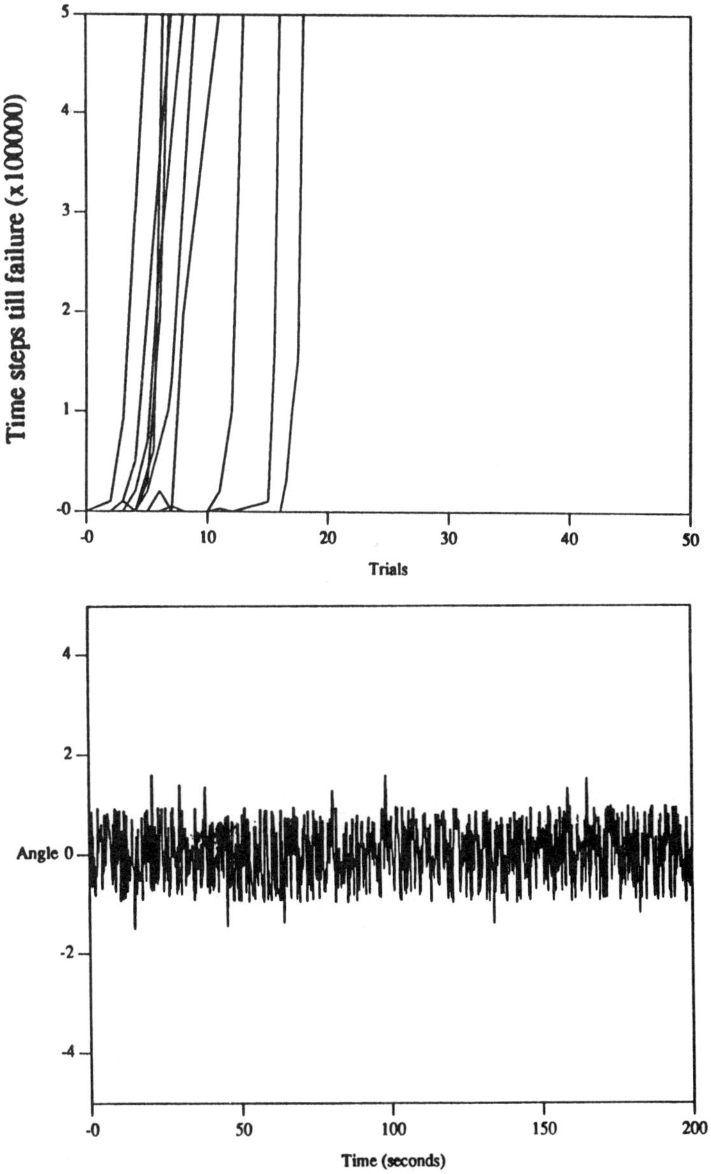


**Fig. 4.** Performance of the RNN-FLCS on the the cart-pole balancing problem.

# A Neuro-Fuzzy Classifier and Its Applications

Chuen-Tsai Sun
Department of Computer and Information Science
National Chiao Tung University
Hsinchu, Taiwan 30050
E-mail: ctsun@weber.cis.nctu.edu.tw

Jyh-Shing Jang
Department of Electrical Engineering and Computer Sciences
University of California, Berkeley, CA 94720
E-mail: jang@diva.berkeley.edu

## Abstract

*Fuzzy classification* is the task of partitioning a feature space into fuzzy classes. A learn-by-example mechanism is desirable to automate the construction process of a fuzzy classifier. In this paper we introduce a method of employing *adaptive networks* to solve a fuzzy classification problem. System parameters, such as the *membership functions* defined for each feature and the *parameterized t-norms* used to combine conjunctive conditions are calibrated with backpropagation.

To explain this new approach, first we introduce the concept of adaptive networks and derive a supervised learning procedure based on a gradient descent algorithm to update the parameters in an adaptive network. Next, we apply the proposed architecture to two problems: two-spiral classification and Iris categorization. From the experiment results, it is summarized that the adaptively adjusted classifier performs well on an Iris classification problem. The results are discussed from the viewpoint of feature selection.

## I. Introduction

Conventional approaches of pattern classification involve clustering training samples and associating clusters to given categories. The complexity and limitations of previous mechanisms are largely due to the lacking of an effective way of defining the boundaries among clusters. This problem becomes more intractable when the number of features used for classification increases.

On the contrary, *fuzzy classification* [9, 14] assumes the boundary between two neighboring classes as a continuous, overlapping area within which an object has partial membership in each class. This viewpoint not only reflects the reality of many applications in which categories have fuzzy boundaries, but also provides a simple representation of the potentially complex partition of the feature space. In brief, we use *fuzzy if-then rules* to describe a classifier. A typical fuzzy classification rule is like:

$$\text{if } X_1 \text{ is } A \text{ and } X_2 \text{ is } B \text{ then } Z \text{ is } C,$$

where $X_1$ and $X_2$ are features or input variables; $A$, $B$ are *linguistic terms* [13] characterized by appropriate *membership functions* [12], which describe the features of an object $Z$. The *firing strength* or the degree of appropriateness of this rule with respect to a given object is the degree of belonging of this object to the class $C$.

As such, a fuzzy rule gives a meaningful expression of the qualitative aspects of human recognition. Based on the result of pattern matching between rule antecedents and input signals, a number of fuzzy rules are triggered in parallel with various values of firing strength. Individually invoked actions are considered together with a combination logic.

Further, we want the system to have learning ability of updating and fine-tuning itself based on newly coming information. Researchers have been trying to automate the classifier construction process based on a training data set. We propose a method of using adaptive networks for this purpose. We use experimental data to verify the effectiveness of this approach.

## II. Learning with Adaptive Networks

An adaptive network is a multi-layer feed-forward network in which each node performs a particular function (*node*

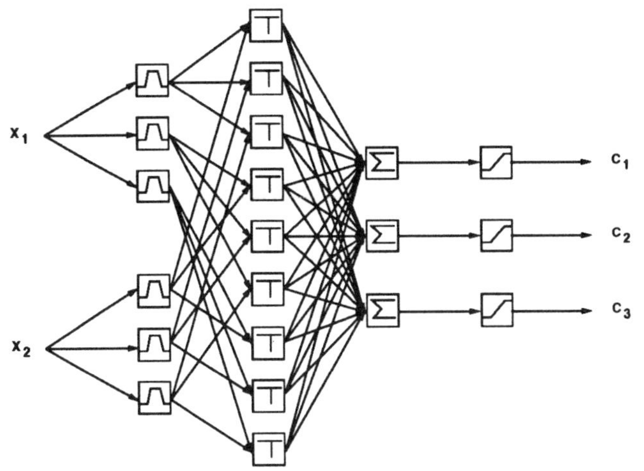


Figure 1: *An adaptive-network-based fuzzy classifier.*

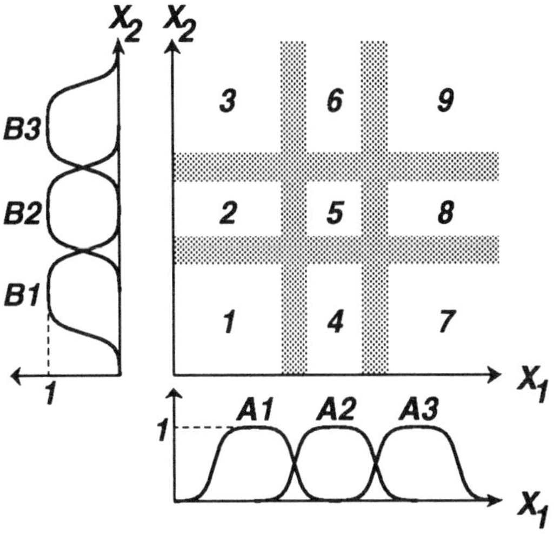


Figure 2: *Partition of feature space.*

*function*) based on incoming signals and a set of parameters pertaining to this node. The type of node function may vary from node to node; and the choice of node function depends on the overall function that the network is designed to carry out.

Figure 1 demonstrates the adaptive-network-based classifier architecture with two input variables, $X_1$ and $X_2$. The training data are categorized by three classes, $C_1, C_2$ and $C_3$. Each input is represented as three linguistic terms, thus we have nine fuzzy rules. In our model the nodes in the same layer have the same type of node function.

Each node in *Layer 1* is associated with a parameterized bell-shaped membership function represented as

$$\mu_A(X_i) = \frac{1}{1 + [(\frac{x_i - c_i}{a_i})^2]^{b_i}}, \quad (1)$$

where $X_i$ is one of the input variables, $A$ is the linguistic term associated with this node function, and $\{a_i,\ b_i,\ c_i\}$ is the parameter set.

The initial values of the parameters are set in such a way that the membership functions along each axis satisfy $\epsilon$ *completeness* [6] ($\epsilon = 0.5$ in our case), *normality* and *convexity* [5]. Figure 2 illustrates the concept. Although these initial membership functions are set heuristically and subjectively, they do provide an easy interpretation parallel to human thinking. The parameters are then tuned with *backpropagation,* a gradient descent method, in the learning process based on a given training data set.

Each node in *Layer 2* generates a signal corresponding to the conjunctive combination of individual degrees of match. The output signal corresponds to the firing strength of a fuzzy rule with respect to an object to be categorized. In most pattern classification and query-retrieval systems, the conjunction operator plays an important role and its interpretation changes across contexts. Since there does not exist a single operator that is suitable for all applications, we can use *parameterized t-norms* at *Layer 2* to cope with this dynamic property of classifier design. Bonissone provided a detailed discussion on t-norms and their parameterized versions, see [2]. For example, we can use Hamacher's t-norm:

$$T_H(x_1, x_2, \gamma) = \frac{x_1 x_2}{\gamma + (1-\gamma)(x_1 + x_2 - x_1 x_2)}, \quad (2)$$

where $x_i$'s are the operands and $\gamma$ is a non-negative parameter.

In some other applications, e.g., see [15], features are combined in a compensatory way. For these situations, *mean operators* [11] are more appropriate than conjunctive operators. To find a good mean operator for a certain system, we can also implement a parameterized operator and use training data to calibrate it. For instance, we can use the one proposed by Dyckhoff and Pedrycz:

$$M_{DP}(x_1, x_2, \gamma) = \frac{(x_1^\gamma + x_2^\gamma)^{1/\gamma}}{2}, \quad (3)$$

where $\gamma \geq 1$. Note that we can either use a parameterized operator for each node in *Layer 2* or employ a single one for the whole layer. Whether an operator is local or global depends on applications. Moreover, a *parameterized fuzzy quantifier* [3] can also be introduced into this picture based on the same concept. The combinational parameters are also fine-tuned by backpropagation.

We take the linear combination of the firing strengths of the rules at *Layer 3* and apply a sigmoidal function at *Layer 4* to calculate a degree of belonging to a certain

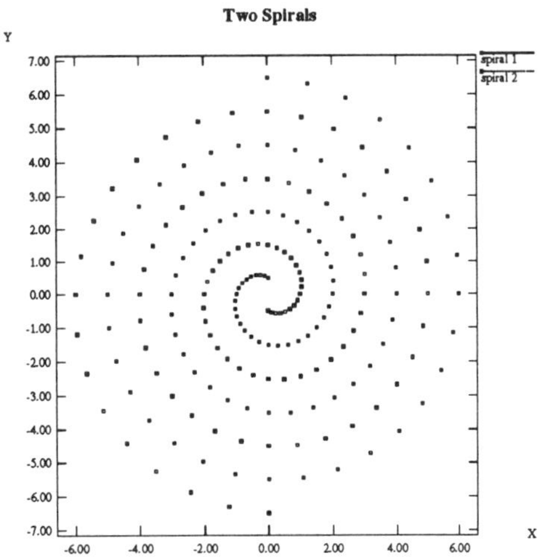


Figure 3: *Training data for the two-spiral problem.*

class. Through experience we found the following definition of error measure useful in classification problems. As before, assume we have three classes, the error measure $E$ can be formulated as:

$$\begin{aligned} E \;=\; & do_1(1-do_2)(1-do_3)\{(sig[m(co_2-co_1)] \\ & + sig[m(co_3-co_1)]\} \\ & + (1-do_1)do_2(1-do_3)\{(sig[m(co_1-co_2)] \\ & + sig[m(co_3-co_2)]\} \\ & + (1-do_1)(1-do_2)do_3\{(sig[m(co_1-co_3)] \\ & + sig[m(co_2-co_3)]\} \end{aligned} \tag{4}$$

where $do_i$'s are desired outputs and $co_i$'s are calculated outputs. The first, second and third terms account for the conditions when the desired classes are class 1, 2 and 3, respectively. Since this error measure is reasonable when the maximum selector is used, it is therefore referred to as *maximum-type error measure*.

The maximum-type error measure introduced above can increase the degrees of freedom and therefore is suitable for crisp-output neuro-fuzzy classifiers. Meanwhile, the slow convergence of the gradient descent is compensated for by the proper choice of the maximum-type error measure, so the learning process will not suffer from the drawbacks of the gradient descent. In the following we present two application examples which employ neuro-fuzzy classifiers with maximum-type error measures to do crisp pattern classification. with maximum-type error measures to do crisp pattern classification.

## III. Two-Spiral Problem

The two-spiral problem was proposed by Alexis P. Wieland on the connectionist mailing list as an interesting benchmark task for neural networks. The task requires a neural network classifier with two inputs and one output to learning a mapping that distinguishes between points of two intertwined spirals. The two sets of spiral data consist of 194 points, with 97 points for each spiral. One spiral is generated as a mirror image of the other, making the problem highly nonlinear-separable.

As pointed out by Wieland, this task has several features that makes it an interesting test for neural network's learning algorithms. First of all, it requires the neural networks to learn the highly nonlinear separation of the input space, which is difficult for most current learning algorithms. Secondly, its 2-dimensional input space makes it easy to plot the overall input-output relations as a 3-dimensional surface or 2-dimensional image for visual inspection and analysis.

To proceed with the simulation, first the rule number has to be decided. Since the input partition is checkerboard-like, we expect that the partition number (or equivalently, the number of membership functions) on either input $x$ and $y$ should be equal to the maximum number of alternations between classes along one dimension when the other is fixed. In the two-spiral problem, the maximum number of alternations on $y$ is approximately 14, which occurs on the straight line $x = 0$; the maximum number of alternations on $x$ is approximately 13, which occurs on $y = 0$.

Using the maximum-type error measure defined above, we perform four runs of simulation; the number of membership functions on both inputs is varied from 10 to 13 sequentially. It is found that 13 is the minimum number for the network to classify the two spirals correctly. This agrees with our observation of the maximum number of alternations along each dimension.

As mentioned earlier, this problem is suitable for visual inspection or analysis on the classifier's input-output behavior through data visualization techniques such as 3-D surface or 2-D image. Figure 4 depicts the classifier's input-output behavior; each of the four images is composed of 22,500 (150×150) pixels which are 2 bit deep.

Although we can always employ a large number of membership to achieve a perfect classification, this kind of over-parameterized structure is not recommended since it not only slows down the learning but also degrades the generalization power for unseen data sets; this is just like the case in over-parameterized neural networks. Therefore, the ability to determine the number of membership functions from visual inspection is a very practical and useful technique that enables us to find roughly a minimum structure of a neuro-fuzzy classifier to do the job. For neural networks, we do not have similar quick and easy techniques to determine the minimum structure (node numbers and layer numbers) simply due to the uniformity in the node function.

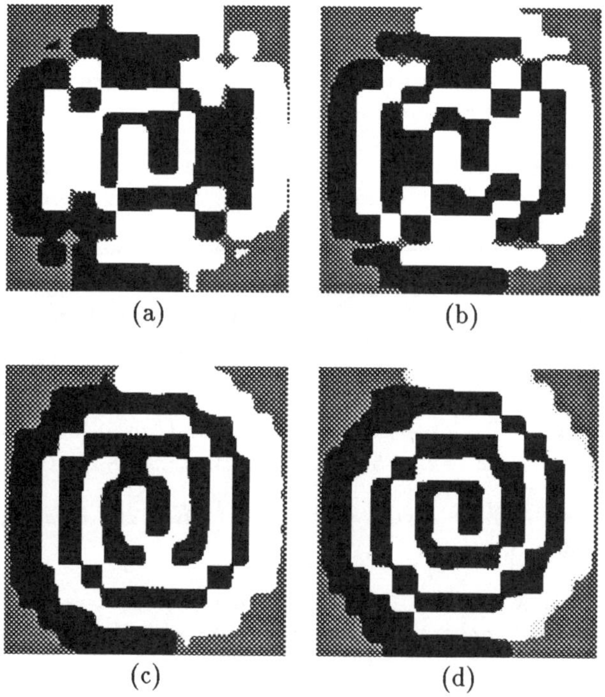


Figure 4: *Image representation of classifier's input-output behavior.* The number of membership functions on $x$ and $y$ equals to (a) 10; (b) 11; (c) 12 and (d) 13.

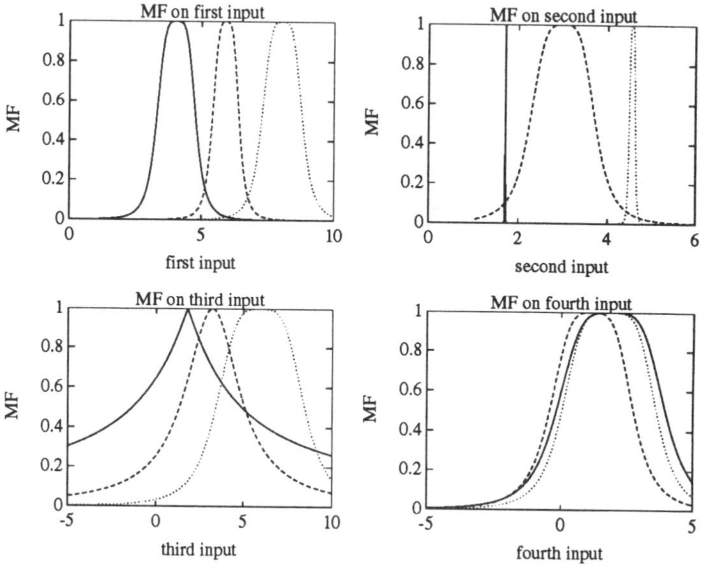


Figure 5: *Membership functions after 20 training epochs in Iris classification.*

## IV. Iris Categorization

Next we apply the proposed scheme on an Iris classification problem of finding the mapping between four input variables (sepal length, sepal width, petal length, and petal width) and three classes (Setosa, Versicolor, and Virginica).

There are 150 samples in the data set and we use 120 of them as training data and the other 30 for testing. Initially, each feature dimension is partitioned into 3 homogeneously distributed overlapping regions. We constructed a network and trained it with the training data set for 20 epochs. The adjusted network was then evaluated by the testing data set. The desired and calculated outputs matched for all 30 testing data. The discriminating power of the classifier was well validated.

Another advantage of our fuzzy approach is that the resulted model gives us insights of the data characteristics. We analyzed the data set with statistical methods and found that, individually speaking, both inputs 3 and 4 have stronger (but incomplete) discriminating ability than inputs 1 or 2. However, if we choose input 3 as the primary salient feature, we need input 1 to be the secondary feature to complete the feature space partition. This analytical conclusion was predicted by the adjusted parameters in our adaptive network model. Figure 5 shows the final membership functions.

In this figure, the membership functions of the third input clearly gives the ranges of the salient feature. On the other hand, inputs 2 and 4 are neglected by different ways. For input 2, two of the initial functions shrank to peaks and left the remaining one to cover the entire dimension. For input 4, the three functions adjusted themselves to overlap each other and redundantly covered the dimension. The various behavior remains an interesting research topic to be further studied in the future.

## V. Concluding remarks

An adaptive classifier partitions the feature space based on labeled training data. In the context of fuzzy classification, classes are overlapping and each training data item is associated with numbers in the unit interval representing degrees of belonging, one value for each class. The overlapping among regions provides the natural smoothness for the input-output mapping. This characteristic makes this model suitable for classification problems, especially for those with overlapping categories.

We proposed a general fuzzy classification scheme with learning ability using an adaptive network, which can be used in pattern recognition, decision analysis, and many other fields. Membership parameters were identified with the model. Parameterized t-norms and mean operators were brought into this picture to make the classification scheme more flexible. The resulted membership function served the need of feature selection.

The proposed neuro-fuzzy approach is better than neural network classifiers in the sense that prior knowledge about the training data set can be encoded into the parameters of the neuro-fuzzy classifier. This encoded knowledge, usually acquired from human experts or data visualization techniques, can almost always allow the learning process to begin from a good initial point not far away from the optimal one in the parameter space, thus speeding up the convergence to the optimal or a near-optimal point. Moreover, the parameters obtained after the learning process can be easily transformed into structure knowledge in the form of fuzzy if-then rules.

## References


[1] K. J. Astrom and B. Wittenmark. *Computer Controller Systems: Theory and Design.* Prentice-Hall, Inc., 1984.

[2] Piero P. Bonissone. Summarizing and propagating uncertain information with triangular norms. *International Journal of Approximate Reasoning*, 1:71–101, 1987.

[3] J. Kacprzyk and R. R. Yager. "Softer" optimization and control models via fuzzy linguistic quantifiers. *Information Sciences*, 34:157–178, 1984.

[4] R. E. Kalman. A new approach to linear filtering and prediction problems. *Transactions of the ASME. Journal of Basic Engineering*, pages 35–45, March 1960.

[5] Arnold Kaufmann and Madan M. Gupta. *Introduction to Fuzzy Arithmetic: Theory and Applications.* Van Nostrand Reinhold Co., 1985.

[6] C.C. Lee. Fuzzy logic in control systems: Fuzzy logic controller. *IEEE Trans. on Systems, Man, and Cybernetics*, 20(2):404–435, 1990.

[7] John Moody and Christian Darken. Learning with localized receptive fields. Technical Report YALEU/DCS/RR-649, Department of Computer Science, Yale University, 1988.

[8] Stephen M. Omohundro. Geometric learning algorithms. Technical Report TR-89-041, International Computer Science Institute, 1989.

[9] Sergei Ovchinnikov. Similarity relations, fuzzy partitions, and fuzzy orderings. *Fuzzy Sets and Systems*, 40:107–126, 1991.

[10] G.M. Reaven and R.G. Miller. An inquiry into the nature of diabetes mellitus using a multidimensional analysis. Technical Report 33, Division of Biostatistics, Stanford University, 1979.

[11] Ronald R. Yager. Connectives and quantifiers in fuzzy logic. *Fuzzy Sets and Systems*, 40:39–75, 1991.

[12] Lotfi A. Zadeh. Fuzzy sets. *Information and Control*, 8:338–353, 1965.

[13] Lotfi A. Zadeh. The concept of a linguistic variable and its application to approximate reasoning. *Information Sciences*, 8:199–249, 1975.

[14] Lotfi A. Zadeh. Fuzzy sets and their application to pattern classification and clustering analysis. In J. van Ryzin, editor, *Classification and clustering*, pages 251–299. Academic Press, New York, 1978.

[15] H. J. Zimmermann and P. Zysno. Latent connectives in human decision making. *Fuzzy Sets and Systems*, 4:37–51, 1980.

# A NEW CLASS OF FUZZY LOGIC CONTROLLERS : THE CT-FLC

Labib Sultan*

Director of R&D
MentaLogic Systems Inc.
7500 Woodbine avenue
Markham, Ontario L3R 1A8
Canada

Talib Janabi

Director of Technology
MentaLogic Systems Inc.
7500 Woodbine avenue
Markham, Ontario L3R 1A8
Canada

## ABSTRACT

This paper outlines the cognitive mechanisms employed in the development of a new class of fuzzy logic controllers called the Clearness Transformation Fuzzy Logic Controller "CT-FLC". This controller employs a new scheme for approximate reasoning which gives better time performance and yield high control quality than the currently existing fuzzy controllers. The paper reports on the performance of the CT-FLC using results of simulation studies on the control of complex nonlinear systems.

## 1. INTRODUCTION

A fuzzy logic controller can be considered as a control expert system which simulates the human thinking in the interpretation of real world data using fuzzy set labels and performs an approximate reasoning using the compositional rule of inference (CRI) introduced by Zadeh [7]. The CRI represents the core of the deduction mechanism of the controller. It performs a composition of the fuzzy sets and the matrices of fuzzy rules (knowledge on the input-output relationship) using Max-Min operations. Fuzzy logic controllers propagate the numerical measurement of the process variables into fuzzy sets (fuzzification), deduce the control actions in the form of fuzzy sets using the CRI(pattern matching and conflict resolution), and translates the deduced fuzzy actions into crisp actions to be applied to the controlled process (defuzzifiaction). Hence, a fuzzy controller is an expert system consisting of an inference engine which employs the CRI method for approximate reasoning and two convertors: numerical to linguistic (fuzzifier) and linguistic to numerical (defuzzifier) to facilitate its communication with real world processes.

* The author is an Associate professor at the Department of Computer Science, Glendon Campus, York University, 2275 Bayview Av., Toronto, Ontario M4N 3M6, Canada.
The author would like to acknowledge the support of NSERC.

Many successful applications have been reported on the application of fuzzy controllers employing the CRI reasoning scheme [3]. However, the following are some of the difficulties which limit their wide applications:

- difficulties to cope with real time requirements for the control of fast processes or multivariable applications.
- Difficulties in tuning and obtaining stable controllers (the trial and error method is still the basic method in improving the expert knowledge toward developing a tuned and stable fuzzy controller).

These difficulties stem from the computational scheme employed by the CRI. The tuning problem is also related together with other factors to the complexity of this scheme. It is obvious that the solution to these problems is not technologically driven (i.e. by developing high speed fuzzy controller processor chips and advanced circuitry design ). The goal is to find more flexible scheme for approximate reasoning which can operate in real time and provides more insight to the behaviour of the controller so that the tuning and stability problems can be solved efficiently.

In this paper the operation of fuzzy logic controllers is analysed in a framework based on the cognitive tasks analysis for decision making in supervisory control environment [1, 2]. In this framework a new class of fuzzy controllers, called the CT-FLC, is developed [4,5,6]. The CT-FLC is characterized as a device which simulates the human approximate reasoning at the level of fuzzy pattern processing. It employs pattern processing operations such as: recognition, assessment, association and approximation. The paper discuses the cognitive schemes employed by the CT-FLC, and presents simulation results on the time, control performance and robustness of this controller.

## 2. THE FRAMEWORK OF THE COGNITIVE TASKS ANALYSIS OF FUZZY LOGIC CONTROLLERS

## 2.1 The Control Tasks of the Operator

The cognitive task analysis proposed by Rassmussen [2] is utilized to establish the operation of fuzzy controllers at three levels: skill-based, rule-based and knowledge-based. These levels have been elaborated in a step-ladder diagram to reflect the complexity of the tasks handled by operators. We consider that fuzzy logic controllers cover the tasks of skill-based and rule-based levels. The tasks of the knowledge based level, where decisions are elaborated as a compromise between purposive policies such as safety and production policies, etc. , fall beyond the operation of fuzzy controllers as parameter driven control systems. At these levels, the cognitive tasks achieved by skilled operators are :

(i ) Observation, and detection the patterns of the process current status.
(ii) Assessment and evaluation of the detected patterns;
(iii) Actions planning; and
(iv) Actions execution

## 2.2. The Tasks of the Fuzzy Controller

The tasks of operators (i) and (iv) correspond to the fuzzification and defuzzification phases of the fuzzy controller, respectively. The second and third tasks are related to the approximate reasoning procedure employed by fuzzy controllers. Further, we use the concept of fuzzy patterns to formalize each task of the fuzzy controller. A "fuzzy pattern" is a cognitive entity and hence it is more flexible than its mathematical synonym the "fuzzy set" in describing the cognitive tasks.

The fuzzification task of the fuzzy controller covers tasks (i) and (ii), whereby the observed numerical measurement of the process variables (e.g., the value of temperature = 30 c°) would be mapped to fuzzy patterns such as *NORMAL, SLIGHTLY HIGH*, etc. The next task is the action planning. This task is currently performed by the fuzzy controller using the CRI. Generally, this task is convenient to be referenced as "the associative pattern matching" activity, whereby the pattern(s) generated by the fuzzification phase are used to activate patterns of the control action(s). The translation of these patterns into precise actions is performed by the approximate reasoning and defuzzification tasks . Hence, the fuzzification, pattern matching, approximate reasoning and defuzzification are the major tasks performed by the fuzzy controller. These are the same tasks which are performed by operators in their normal practice in the supervisory control environment and are consistent with Rasmussen cognitive task analysis diagram [2]. However, we can notice that there is no distinction between the pattern matching task and the approximate reasoning task in the operation of fuzzy controllers. The CRI performs both tasks using the Max-Min operation.

### 2.3 Interpretation of the Approximate Reasoning of the Fuzzy Controller

### (i) The CRI Scheme.

The approximate reasoning mechanism of the CRI can be given the following cognitive interpretation. The output of the controller is generated by firing all the possible rules in the controller knowledge base. These deductions are performed by Max-Min operation to produce the action of each fired rule. The final action generated by the defuzzifier is the *averaging* of all the deduced actions.

### (ii) The CTFLC Approximate Reasoning Scheme.

One of the alternative schemes for approximate reasoning is called the clearness transformation mechanism for approximate reasoning [4,6]. The cognitive interpretation of this scheme is as follows: it is proposed that the human performs an assessment of the clearness of the perceived fuzzy patterns of the process and activates the relevant rules on how to react, rather than calling all the rules (knowledge) about the process. He/she then qualifies and quantifies the action(s) to be taken based on his/her assessment to the detected patterns. The clearer the detected patterns of the process (less fuzziness) the more confident and relevant actions will be taken by the operator to recover the process to its normal operation. The contrary is also true. This approximation has been formalized mathematically [6] as the clearness transformation mechanism for approximate reasoning and employed as a basis for the design of the CT-FLC.

## 3. THE CONCEPTS OF THE CT-FLC FUZZY CONTROLLERS

The CT-FLC is a device which operates and makes its decisions at the level of fuzzy pattern processing. Two concepts have been used in the development of the CT-FLC: the concept of a fuzzy pattern and the formulation of the clearness transformation mechanism for approximate reasoning [4,6].

### (i) The Concept of a Fuzzy Pattern

***A fuzzy pattern (FP)*** is defined by a triple **< *S, D, A*>**, where:

S - is the syntactical description of a fuzzy pattern;
D - is the domain to which a fuzzy pattern is attached; and

A- is the clearness assessment of a fuzzy pattern.

### (ii) The Clearness Transformation Mechanism for Approximate Reasoning

The approximate reasoning module of the CT-FLC employs pattern operations such as: clearness assessment, pattern association, clearness transformation between fuzzy patterns and other cognitive operation related to the processing of the fuzzy patterns generated by the fuzzifier and the pattern matching modules. The approximate reasoning mechanism employed in the CT-FLC does not require the use of fuzzy matrices and the application of Max-Min operations to produce the fuzzy patterns of the control actions. The replacement of the cumbersome computation of Max-Min by simple pattern processing operations improves the time and functional performance of the CT-FLC considerably compared to the conventional fuzzy controller.

### (iii) The Operational modules of the CT-FLC

The CT-FLC is a device which has a modular structure consisting of four modules: The fuzzifier, The fuzzy pattern matching, The approximate reasoning, and The Defuzzifier. The CT-FLC has been implemented with conventional micro-controllers and recently designed as an Application Specific Integrated Circuit (ASIC) by MSI. This ASIC is a complete self- contained fuzzy controller system which implements the functionality of the four modules of the controller on one chip.

## 4. EVALUATION OF THE CTFLC

### 4.1 The Control Quality and Robustness of The CT-FLC

For the simulation study presented here the variables used at the controller input are the error and change of error. The system chosen is a single-input single output nonlinear control system representing a difficult system from control point of view. Figure (1) shows the block diagram of the closed loop system. The linear element of the system is described by the following transfer function:

$$G(s) = \frac{1}{s(s+0.2)}$$

The nonlinear element is on-off plus dead-zone as shown in figure (2). The membership functions used for the error, change of error and output are the same and shown in figure (3). The number of control rules which are used for the control of this system is 64 as shown in figure (4). The system responses before and after compensation using the CT-FLC controller are shown in figure (5). As can be seen from this figure, the controller was capable of eliminating the steady-state error caused by the dead-zone, as well as, avoiding limit cycle.

Further, the robustness of the CT-FLC controller was tested using white noise with variance equal to 0.01 at the input of the nonlinear element simulating process noise, and at the output of the linear element simulating sensor noise (see figure 1). This noise is applied to the system together with disturbances of magnitudes equal to the set-point (100% disturbance). Again these disturbances are applied to the input and output of the linear element simulating process and load disturbances. The white noise was added to the input of the linear element from the beginning of the simulation (time zero). The process disturbance at the input of the linear element was thrown onto the system after 20 seconds and was allowed to remain for one second simulating a pulse disturbance, while the disturbance at the output which was also thrown onto the system after 20 seconds (in a separate test) but allowed to remain permanently simulating a shut-down of a downstream equipment. The system was subjected to a unit step change for the purposes of these tests.

Figure (6a) shows the closed loop system response after adding the process noise but before compensation. Figure (6b) shows the closed loop response of the system after compensation with process noise and process disturbance (at the input to the nonlinear element). Figure (7a) shows the closed loop system response after adding the sensor noise but before compensation. Figure (7b) shows the closed loop response of the system after compensation with sensor noise and load disturbance (at the output of the linear element). In figure (7) the noise was imposed on the actual output (not only on the feedback signal) to illustrate the presence of noise. It is clear that the controller was capable of keeping the stability of the system and was able to bring it to the set-point. The performance and robustness of the controller were remarkable in guiding the system to its set-point despite the existence of the dead zone and on-off characteristics of the nonlinearity.

### 4.2. The Real Time Performance of the CT-FLC

The time performance of the CT-FLC was evaluated using two types of industrial micro-controllers: ZILOG-Z/8 (C11 model) and Motorola (M68HC11). These two industrial micro-controllers were selected to estimate the time performance of the CT-FLC in handling the control of the nonlinear system described in Sec. (4.1). The time was estimated for a control cycle of the controller operation. The results are shown in table-2.

Table 1.

| Micro controller | CT-FLC *time, ms* | CRI Controller *time, ms* |
|---|---|---|
| Z-8C11 | 3-4 | 700 |
| M68HC11 | 2 | 550 |

These results have a considerable impact on the applicability of the standard micro-controllers. The CT-FLC algorithm will open a new avenue toward the use of these general purpose controllers to implement the functionality of fuzzy logic controllers in handling various industrial applications.

## 4.3 The Tunability of the CT-FLC

Mentalogic Systems Inc. (MSI) has developed an intelligent fuzzy autotuning system for single-input single-output control systems. The system optimizes the performance of the controller by improving the initial knowledge-base loaded in the controller. This system is also developed as part of an integrated system for fuzzy logic controller applications development outlined in [4,6]. Further work is under way to develop this fuzzy tuner for multivariable control systems.

## REFERENCES


[1] P.C. Cacciabue, G. Mancini and G. Guida , A Knowledge Based Approach to Modelling the Mental processes of Operators. Proc. IEAEA/CEC/OECD/NEA Int. Conf. on Man-Machine Interface, Tokyo, Japan, , pp.71-78, 1988.

[2] J. Rasmussen, Information processing and Human-Machine Interaction. An Approach to Cognitive Engineering. North- Holland, (1986).

[3] M. Sugeno, The Industrial Applications of Fuzzy Controllers, Elsevier Science Publishers B.V., (1985).

[4] L. Sulatn and T. Janabi , " The Cognitive bases for the design of a new class of fuzzy logic controllers", Proc. of NAFIPS-92 Intl. Conf., pp.

[5] L.Sultan and T. Janabi, " Intelligent Fuzzy Controller Development System For Handling Complex Control Problems". The Canadian Conference on Electrical and Computer Engineering, CCECE' 92.

[6] L. Sultan and T. Janabi, " The Clearness Transformation Fuzzy Logic Controller". A patent pending in the USA.

[7] L.A. Zadeh, "Outline of a New Approach to the Analysis of Complex Systems and Decision Processes," IEEE Trans Systems, Man and Cyber.,vol. SMC-3, pp.28-44,1973.


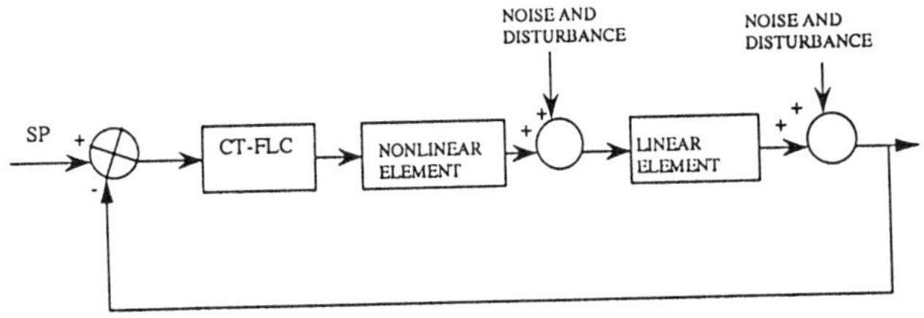


Figure1. Block Diagram of the Closed Loop Control System

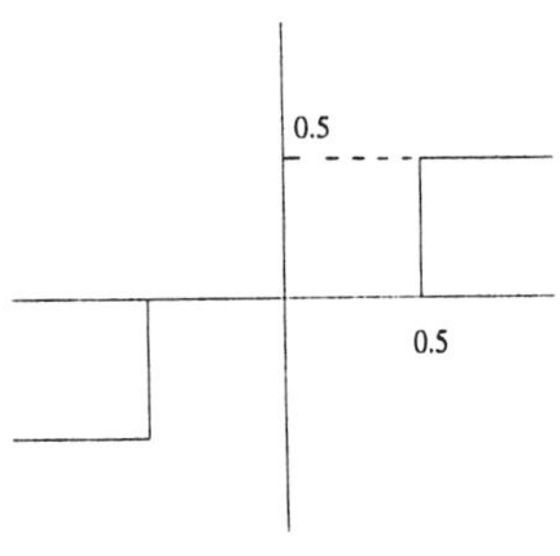


Figure 2. On-Off Plus Dead-Zone

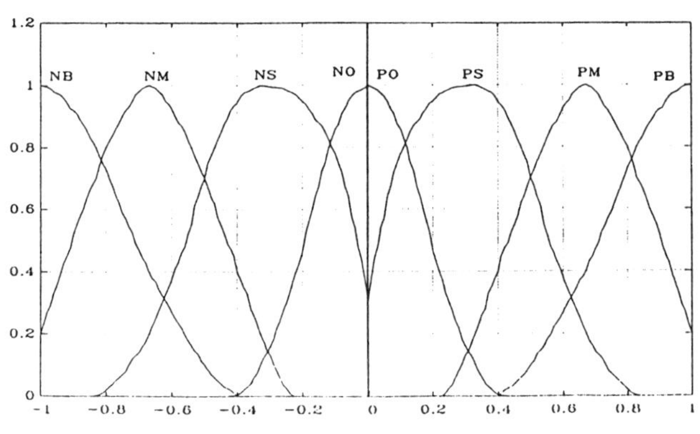


Figure 3. Fuzzy Sets

| IF | AND | THEN |
|---|---|---|
| E = NB | CE = NB or CE = NM | CA = NB |
| E = NB | CE = NS or CE = NZ or CE = PZ or CE = PS | CA = NS |
| E = NB | CE = PM or CE = PB | CA = NZ |
| E = NM | CE = NB or CE = NM or CE = NS | CA = NB |
| E = NM | CE = NZ or CE = PS | CA = NS |
| E = NM | CE = PZ | CA = NZ |
| E = NM | CE = PM or CE = PB | CA = NM |
| E = NS | CE = NB or CE = NM | CA = NS |
| E = NS | CE = NS or CE = NZ | CA = NB |
| E = NS | CE = PZ or CE = PS | CA = NS |
| E = NS | CE = PM or CE = PB | CA = NM |
| E = NZ | CE = NB or CE = NM or CE = NS | CA = NS |
| E = NZ | CE = PZ | CA = NZ |
| E = NZ | CE = NZ | CA = NB |
| E = NZ | CE = PS or CE = PM or CE = PB | CA = NM |
| E = PZ | CE = NB | CA = NZ |
| E = PZ | CE = NM or CE = NS | CA = PZ |
| E = PZ | CE = NZ | CA = PS |
| E = PZ | CE = PZ or CE = PS or CE = PM or CE = PB | CA = PB |
| E = PS | CE = NB or CE = NM or CE = NS or CE = NZ | CA = PZ |
| E = PS | CE = PZ or CE = PS or CE = PM or CE = PB | CA = PB |
| E = PM | CE = NB or CE = NM or CE = NS | CA = PZ |
| E = PM | CE = NZ | CA = PS |
| E = PM | CE = PZ | CA = PB |
| E = PM | CE = PS or CE = PM or CE = PB | CA = PM |
| E = PB | CE = NB or CE = NM or CE = NS | CA = PZ |
| E = PB | CE = NZ | CA = PB |
| E = PB | CE = PZ or CE = PS | CA = PS |
| E = PB | CE = PM or CE = PB | CA = PM |

**Figure 4. Table of Rules**

The abreviations used are:
E = Error
CE = Change in Error
CA = Control Action
NB = Negative Big
NM = Negative Medium
NS = Negative
NZ = Negative Zero
PB = Positive Big
PM = Positive Medium
PS = Positive Small
PZ = Positive Zero

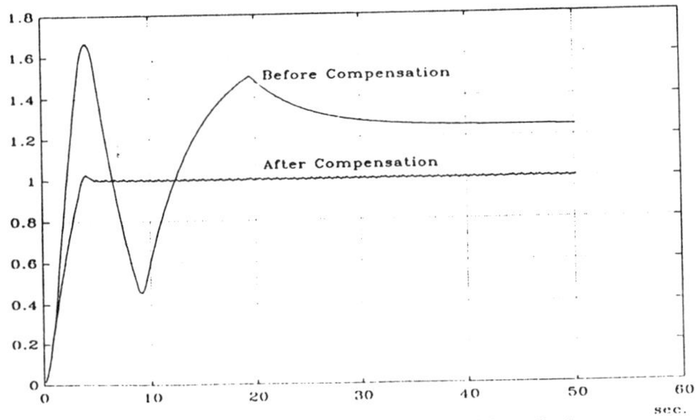


Figure 5. Unit Step Response of Closed loop System

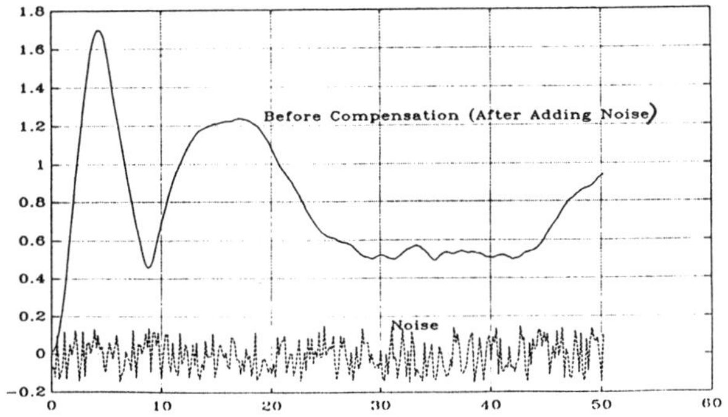


Figure 6a. Unit Step Response of Closed Loop System

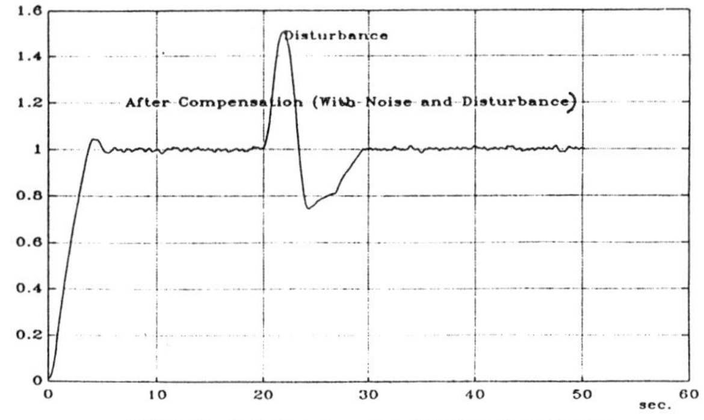


Figure 6b. Unit Step Response of Closed Loop System

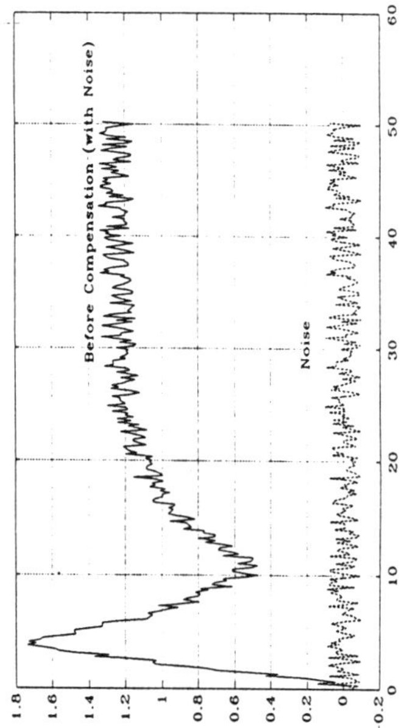


Figure 7a. Step Response of Closed Loop System

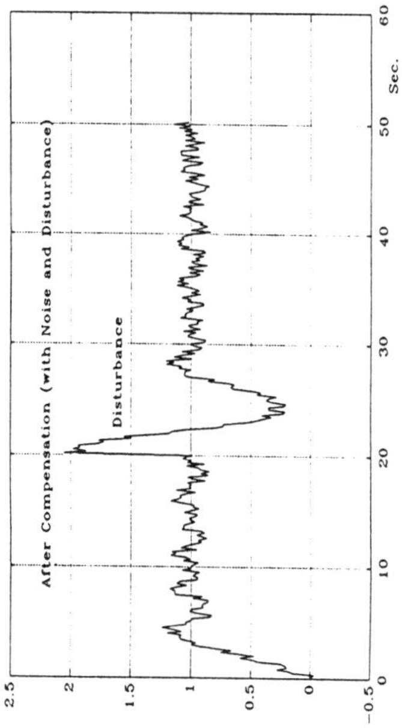


Figure 7b. Step Response of Closed Loop System

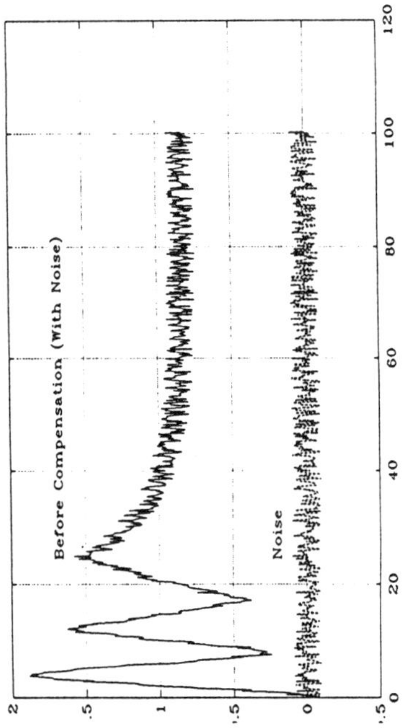


Figure 8a. Step Response of Closed Loop System

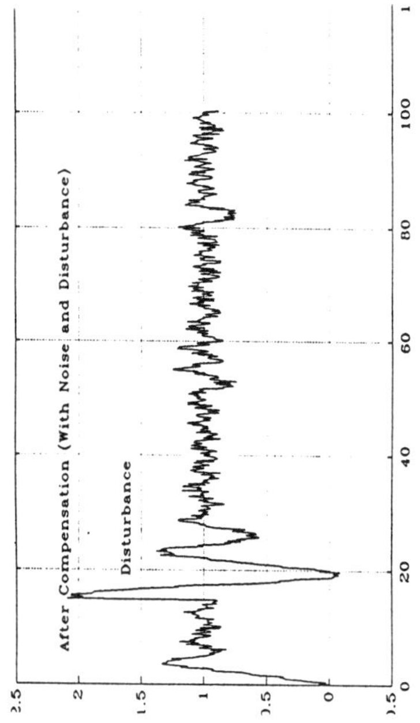


Figure 8b. Step Response of Closed Loop System

# AUTOMATIC SYNTHESIS, ANALYSIS AND IMPLEMENTATION OF A FUZZY CONTROLLER

Andrea Pagni, Rinaldo Poluzzi, Gianguido Rizzotto
Corporate Advanced System Architectures
SGS-THOMSON Microelectronics
Via C. Olivetti 2
20041 Agrate Brianza (MI) ITALY

Matteo Lo Presti
Fuzzy Logic Research Group, Co.Ri.M.Me.
Strada Statale 114, Torre Galiera
95121 Catania ITALY

***Abstract* - Many applications of fuzzy controllers have been successfully performed since Prof. Zadeh introduced the concept of fuzzy sets. They have been used to control ill defined or complex nonlinear systems. A large diffusion of such control structures has been prevented by the quite total absence of design methods and hardware device. A method based on Cell-to-Cell approach to extract the control rules for a fuzzy controller is proposed in this paper. This procedure was implemented to control the position of a DC motor. In order to obtain the high performances, in terms of FIPS, that complex control requires, the rules and the membership functions has been implemented via an hardware solution: WARP. The global architecture alongside with some generalities on its particular approach are here presented.***

## I. CELL-TO-CELL MAPPING APPROACH

The growing interest towards the dynamics of nonlinear systems has driven the development of research method like Poincarre' maps. This approach allows to study complex nonlinear systems using a point-to-point technique through a great number of integrations. Poincarre' maps allows to show typical aspect of nonlinear systems, like chaotic phenomena, periodic motions, etc., carrying out their importance in the analysis on local ambit. The major limitation of these technique is the impossibility to be used in a global analysis of the systems, because of the great computational burden resulting from a point-to-point analysis. Moreover, the knowledge of the system model and the inaccuracy of computational method makes this approach redundant. For these reasons it's far more convenient an analysis approach like Cell-to-Cell Mapping: as realizing a subdivision of the state space in a finite numbers of cells carrying out a system structural analysis for each cell. In this way, each state variable is considered like a collection of a finite numbers of cells. Each point belonging to a cell is associated to the variations of the same cell. Let:

$$\dot{x}(t) = F(t, x(t)) \tag{1}$$

the mathematical model of the system, and **x** the state vector. The coordinate axis of a state variable $x_i$ is divided into a large number of intervals, with interval size $\mathbf{h_i}$. Let $\mathbf{Z_i}$ an integer that identifies the interval **i**. The interval $Z_i$ of the $x_i$-axis is defined to be the one which contains all $x_i$ satisfying the relation:

$$(Z_i - 1/2)h_i \le x_i \le (Z_i + 1/2)h_i \tag{2}$$

The N-tuple $Z_i = 1,2,....,N$, denoted **Z**, is called "cell vector" of the state space. The state space is now considered as a collection of cells, and the mapping from x(n) to x(n+1) of the point-to-point approach is here substituted by cell mapping **C** from **Z**(n) to **Z**(n+1):

$$\mathbf{Z}(n+1) = \mathbf{C}(\mathbf{Z}(n)) \quad Z_i(n+1) = C_i(\mathbf{Z}(n)) \tag{3}$$

## II. FUZZY CONTROL AUTOMATIC SYNTHESIS

The fuzzy controller synthesis consists in the determination of the rules allowing to control a system. This task, generally, requires an expert that, on the basis of heuristic knowledge of the system, establishes and describes a control algorithm. In this paragraph it's described a method based on cell-to-cell mapping to perform the automatic synthesis of a fuzzy controller. The procedure of controller synthesis is divided into two different phases. In the first phase it's calculated the "optimum" control variable for each cell of the space state. This is obtained by using an optimization procedure of a particular performances index, obtaining a map of control inputs for the system. In the second phase, this map, is changed into a fuzzy control algorithm.
The fuzzy system is of Mamdani type with symmetric membership function for the consequent variable; in this



0-7803-0614-7/93$03.00 ©1993IEEE

case the centroid of the consequents membership function is fixed and independent from the activation value of the antecedent part of the rule if Max-dot inference method is adopted.In the same phase the number of fuzzy sets and the shape of the membership functions for each input variables of the controller must be fixed a priori. The consequent parameters are computed by using a particular procedure that will be described hereafter.

*A. Optimization phase*

Let's suppose to have a discrete model of the system to be controlled:

$$x_1(k+1) = f_1(x(k), u, k) \tag{4}$$

$$x_2(k+1) = f_2(x(k), u, k) \tag{5}$$

$$x_n(k+1) = f_n(x(k), u, k) \tag{6}$$

where $x \in R^n$, is the system state vector, u is the input signal, and k is the discrete time. This procedure is available also for those systems that cannot be mathematically modelized, but a neural or linguistic model can be obtained. Once identified the state variables of the system to be controlled it is fixed the maximum range of variation for each of these variables. These variables represent the fuzzy controller inputs not yet fuzzificate. The space state is divided into an adequate number of cells, each cell representing a possible initial condition, from which starting we want to control the system. The space state discretization strongly influences the control system performances: an approximate discretization allows an approximate control law. For each cell the optimum control law $u_o(.)$ is computed by using the following optimization procedure:

1) it is fixed a functional F(.,u), resuming the desired system performances, for example: F can be equal to the weighted sum of the control variable quadratic errors towards the reference values.

2) for each cell the functional initial value $F_{in}$ is computed: the controller input variable values are those calculated starting from the initial conditions represented by the cell and under the action of $u_{nom}$;

3) it is calculated $u_o$ minimizing the functional.

This procedure is a gradient descent based algorithm, for this reason the problem of local minima is present; to overcome this problem a simulated annealing has been implemented. In the case of high dimensional state spaces, cell-to-cell mapping leads to exponential computational burden; due to intrinsic parallelism of the cell-to-cell approach this time can be strongly optimized adopting a Transputer network. At the end of the optimization procedure we obtain for each cell an "optimum" value,$u_o$, of the control variable to be imposed to the system in order to obtain the desired performances in a finite number of sampling period. This map represents with more details part of the information that, otherwise, should be developed by an expert.

*B. Controller synthesis*

Once finished the optimization procedure a n-dimension matrix is obtained, with n system order, whose element represents the optimum control law. The next step is to translate this control law in a fuzzy algorithm. The number of control variables is equal to the system state variables. Fixed the number and the shape of the fuzzy sets for each state variable in heuristic way, we get by consequence the rules number: this is equal to the number of combinations obtainable from n set each containing $n_1,..,n_n$ elements:

$$n_R = n_1 * .... * n_n$$

where $n_i$ represents the number of fuzzy set for the i-th variable. At this point the only action is to calculate the consequent membership functions shape of each rules. For this task it is possible to use the control map obtained by using cell-to-cell mapping. Fixed the antecedent membership functions shape, the activation grade of each rules for each cell (determinate by the minimum between the activation grade corresponding to each fuzzy set) remain fixed. Fig. 1 reports the case of a second order system with two fuzzy sets definite for each state variable.The defuzzification process it is performed by using the following rule:

$$u_o = \frac{\sum_i^{rules} \mu(R_i) \cdot y(R_i)}{\sum_i^{rules} \mu(R_i)} \tag{7}$$

where $\mu(R_i)$ is the activation grade for the i-th rule and $y(R_i)$ is the corresponding centroid. If symmetric membership functions for the consequents are used, the centroids of these fuzzy sets are independent from the activation grade of the antecedents.From these premises, the precedent equation can be rewritten for each cell as:

$$\sum_1^{rules} \mu(R_i) \cdot y(R_i) = u_o \sum_1^{rules} \mu(R_i) \tag{8}$$

In this equation the only unknown values are the $y(R_i)$, centroids of the $n_R$ rules; in fact $y_o$ has been computed for each cell by using cell-to-cell mapping optimizing the functional F and $\mu(R_i)$ are the activation grades of the antecedent of each rules in correspondence of the considered cell.

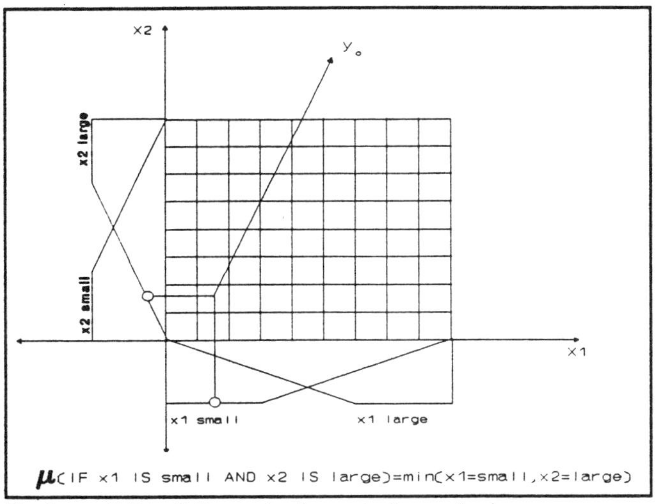

Fig. 1. Cell-to-Cell mapping for controller synthesis

Once the precedent equation for each cell is rewritten and a system of m equation in $n_R$ unknown is obtained, with m equal to the number of cell and bigger than $n_R$ of the kind:

$$A \cdot y(R_i) = b \tag{9}$$

A value for the vector $y(R_i)$ can be obtained by using the least square method:

$$y(R_i) = (A^t \cdot A)^{-1} \cdot A^t \cdot b \tag{10}$$

As this equation fixes the fuzzy sets centroids associated with each rules, the fuzzy controller remains univocally fixed.

## III. CELL-TO-CELL APPROACH LIKE ANALYSIS METHOD

Observing that the fuzzy controller is a non linear system, the traditional analysis methods, useful for the class of linear system, are not applicable to the processes controlled by the fuzzy strategy. It has been observed that a global analysis of non linear system is possible only by using simulation. A simulation algorithm requires a long time to determine the system behavior inside the interest domain.
Cell-to-cell represents a method for producing in reasonable time an approximate but complete characterization of the system inside the interest region of the state space.
This can be obtained starting from a system model that can be analytic, linguistic or mixed, that is by analytic and linguistic parts (this is the case of a system controlled by a fuzzy controller). Since the number of cells is finite, the cell obtained by using the mapping C must verify one of the three sequent conditions:

1) the cell falls into a motion of periodicity k; i.e. starting from a cell the system state get back to the same cell after k steps;

2) the cell falls into a periodic motion after r step; i.e. it is far r step from a periodic motion;

3) the cell diverges in a finite number of steps going out from the interest domain.

From the above considerations comes out that in order to describe the system behavior, each cell must be characterized by three numbers assigned by the analysis algorithm during its evolution. The set of these terns shows the approximate behavior of the system; the elements comprising these terns are:

- a group number G(z) equal for all cells belonging to the same periodic motion;

- a step number S(z) showing how far is the cell from the periodic motion characterizing the group to which the cell belongs to;

- the periodicity P(G(z)) showing the number of cells that constitute the periodic motion.

A global analysis can be realized analyzing the system behavior starting from all cells. The cells are divided in "virgin cells" , "under process cells" and "processed cells"; all virgin cells have a group number equal to 0. When a virgin cell came to be considerate, its group number it is settled to -1; the final cell number will be equal to the group number to which it belong. The algorithm start considering the first cell and using the system model to compute both the system state after a time t and the cell to which it belongs to. The procedure is iterate for all the cells. At each step a cell can show three different conditions:

1) the generated cell has not yet been processed, i.e. the analysis algorithm in its evolution has not generated this cell; in this case the cell group number is fixed to -1 pointing out that it is under process;

2) the generated cell has a positive group number, this means that it was processed; in this case the generated sequence is mapped into a cell with known properties; the group number and the periodicity of the found cell is assigned to the sequence;

3) the founded cell is the one under process; this means that was found a new periodic motion, therefore to all the cells of the sequence a new group number is assigned.

Once finished this phase, the algorithm considers a new virgin cell repeating the previous procedure until all cells belonging to the interest domain has been examined.

### *A. Position control of a DC motor*

The procedure of automatic fuzzy controller synthesis was used to design a fuzzy control system for the control of the rotor position of a DC motor; the DC motor controlled via statoric voltage represents an example of electromechanical system particularly interesting for its frequent application in the servo systems. The motor is supposed to be connected to a load with a friction f and inertia J. Writing the equations that express the equilibrium between the electromagnetic forces in the statoric circuit and the equilibrium between mechanical torques applied to the motor shaft, discretizing the system equation it is possible to obtain the following discrete model of the system:

$$\theta(k+1)=\theta(k)+a_{12}\dot{\theta}(k)+b_1\eta u(k)$$

$$\dot{\theta}(k+1)=a_{22}\dot{\theta}(k)+b_2\eta u(k)$$

where: $a_{12}=\tau(1-\exp(-T/\tau))$ $\quad a_{22}=\exp(-T/\tau)$

$b_1=T+\tau(\exp(-T/\tau)-1)$ $\quad b_2=1-\exp(-T/\tau)$

$$\tau=\frac{R\cdot J}{Rf+K_b\cdot K_m}=0.283 \qquad \eta=\frac{K_m}{Rf+K_b\cdot K_m}=0.90566$$

The control task is to track a variable trajectory with a zero steady state error, minimum settling time and minimum overshoot. The automatic fuzzy controller synthesis algorithm previously described was implemented on a transputer network, by the great computational effort coming from the optimization of the functional F for all the cells, and for the intrinsic parallelism of the computation algorithm. The system is of a second order one. The input variables to the controller are the error between rotor position and setpoint, and the rotor angular speed. Has been observed that the angular speed is connected to the speed variation error in fact:

$$\Delta e \approx \dot{\theta}(k)\cdot T$$

The fuzzy controller output is the statoric voltage. The ranges of these variables are imposed by the system physic limits; in our case they are:

$$e=setpoint-\theta\rightarrow[-0.2,0.2]rad$$

$$\dot{\theta}\rightarrow[-5,5]rad/\sec \qquad u\rightarrow[-25,25]Volt$$

Following the previous procedure, it is necessary fix the functional F that characterizes the wanted system performances; in this case for the control task we have:

$$F=(setpoint-\theta)^2$$

i.e. for each cell the functional is equal to the quadratic error. The space state was divided considering 101 cells for each variable obtaining, consequently, 10201 cells.

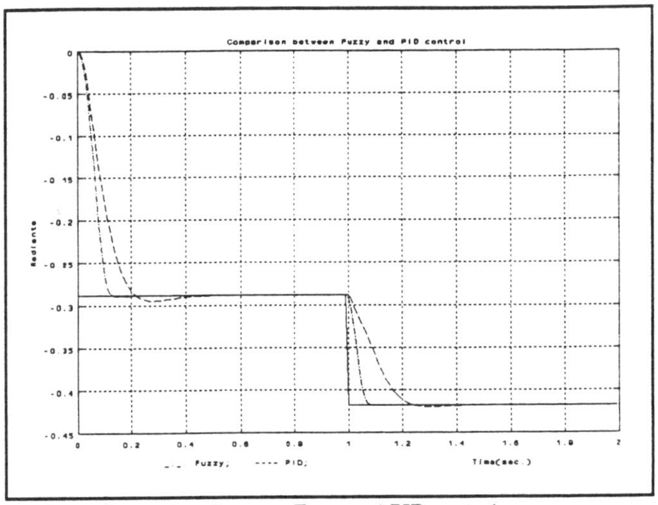

Fig. 2. Comparison between Fuzzy and PID control

By optimizing the functional F, the "Cell-to-Cell controller" was computed, i.e. a two dimensional matrix, whose elements represent the optimum control law, starting from the initial conditions represented from the considerate cell. Now it is necessary to translate this control map into a fuzzy controller. For this task three triangular membership functions for each control variables have been used.
By using the formalism previously introduced we get: $\mathbf{n_R=9}$; it is necessary to determinate the centroid of the 9 rules obtained considering all the possible combinations of the fuzzy sets of the two control variables. According to previously procedure a system with 10201 equations and 9 unknown values has been obtained; solving this system by using a least square algorithm the centroid of each rule has been computed. Fig. 2 shows the result of the control realized by using the developed procedure; this results are compared to those obtained using a PID; it was noted that the proposed fuzzy controller has better performances than PID both for the control tasks and for the control robustness; it was noted in fact that fuzzy controller is more insensitive than PID for either the system parametric variations and external noises. The global performances and the stability of the controlled system was tested by using the analysis algorithm previously described.

## IV. WARP: Weight Associative Rule Processor

WARP is a dedicated VLSI machine whose architecture has been designed in order to efficiently exploit data provided by the cell-to-cell approach . The major innovation with respect of the traditional Fuzzy machines has been the adoption of **different data structures** for the various phases of the computational cycle. In fact one of the greatest limiting factor in the traditional fuzzy architectures is the use of the same data representation for both the computation connected to the left and to the right side of a rule.

WARP architecture is based on the storage of the information in two main memories devices: one for the M.F.s of the left side of the rules and one for those connected to the right side of the rules. In order to represent the membership functions connected to the Fuzzy variables of the **left side** of the rules we adopted a **vectorial representation** of the Membership Functions based on 64 ($2^6$) or 128 ($2^7$) elements, each possessing 16 ($2^4$) truth levels. The utilization of vectors for this phase of the Fuzzy calculus has the great advantage that in the case of a controller, for each rule the data involved in the computing are one or more M.F's (representing the knowledge of the system) and one or more crisp values (representing the input from the "external" world). With this data representation, in order to find the match level (hereafter called $\alpha_i$) between the input and the stored M.F.s it's sufficient to get the various $\alpha$ corresponding to the truth level of the element located by the projection of the input in the universe of discourse. Storing in succession all the $\alpha$ values of a term set, representing the membership functions connected to left side of the rule, as showed by Fig. 3, it's possible to retrieve all the $\alpha$ value of a term set using the crisp input value for calculate the address in the fuzzy memory device utilized.

The number of memory access is function of the M.F.'s comprising a term set and inversely proportional of the size of the memory word, obtaining a significative reduction of the number of access in comparison with more traditional information storage methods. Although the illustrated method for storing and retrieving the various $\alpha$ values connected to a fuzzy variable it's highly efficiently, once calculated the related $\Theta$ value (the truth level for modifying a variable of the right side) of a the vectorial computation become slow due to the huge time consuming process of modifying the M.F.'s of the right side of the rule with the threshold value provided and assembling all the M.F.'s that will form the M.F. furnished as output.

Taking in account some limiting factors like:

- the number of parallel computational elements that realistically can comprise such a device

- the linear increase in memory size when trying to augment the number of element which characterize a Membership Function.

- the necessity to cycle over all the elements of the Membership Function provided as output in order to carry out the defuzzification phase

It clearly appears how such a management result inefficient.

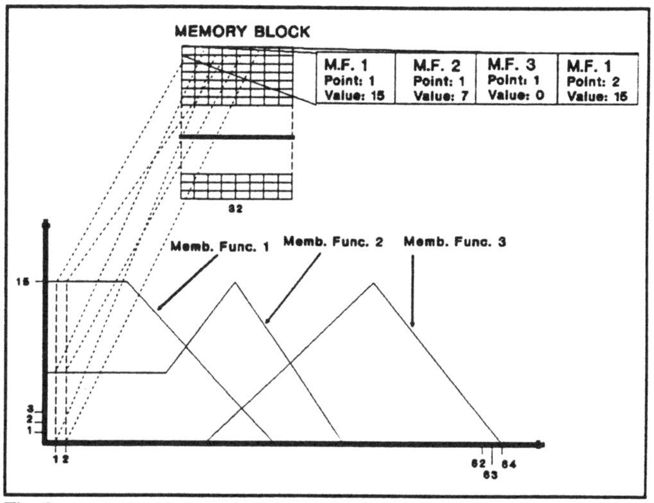

Fig. 3. Antecedent memory organization

**WARP allows to overcame those limitations.** The main consideration has been made on analyzing what **effectively** are the computational requirements of the M.F.s of the right side of the rules. Making the realistic assumption that in the case of a relatively simple controller the output of a rule isn't used as input for another rule what is necessary is the value of the area underlined by a M.F. for each truth level and the point of application (barycentre) of this weight. Having a limited number of possible truth values (16) coming from the left side of a rule, it's possible to represent a membership function utilizing 15 word of memory, each containing both the value of the area and the point of application. It appears clear that utilizing such a method for storing information, the inferencing method adopted (Max-Min or Max-Dot) it's perfectly transparent with respect of the computational architecture in fact the only difference between those methods lies in the different value of the area of the resulting M.F. Moreover a great computational advantage of that approach is that part of the fuzzy calculus can in effect be performed off the line.

## V. WARP ARCHITECTURE

The WARP architecture has been defined with the purpose to efficiently manage the data structures above illustrate.The global architecture of the machine is illustrated in Fig. 4. The **Fuzzifier** section is devoted to the calculus of the memory location and to the retrieving of the $\alpha$ values. Inside the **Memory Blocks**, the data representing the membership functions are stored according to the scheme of section IV. To obtain higher performances the memory device devoted to the M.F.s of the **left** side of the rules has been divided in 4 independent blocks. Each of these blocks contains all the value of one or more fuzzy variables, allowing the parallel retrieval of the $\alpha$ values.

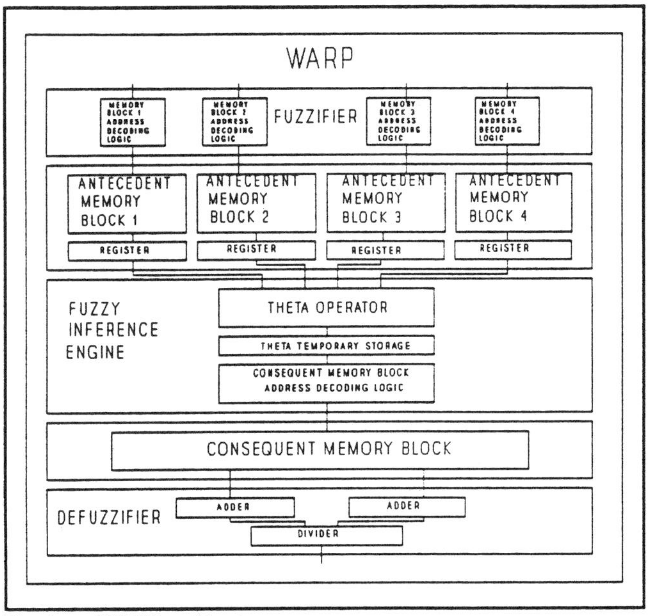


Fig. 4. WARP architecture

The α values founded are memorized in a set of devoted register and then opportunely processed in the **Fuzzy Inference Engine** to calculate the Θ value of each rule.
The adoption of the vectorial data representation for the M.F. of the left side of the rules, allows this operation to be performed in an highly efficiently way via the Theta-operator. The Θ values are used to calculate the address of the memory word, containing the values inside the **Consequent Memory Block** where are stored the membership functions related to the fuzzy variables of the right side of the rules. The assembling of all the M.F.s comprising an output and the defuzzification process are carried out in the **Defuzzifier** block. In the case of WARP the defuzzification adopted is the one defined "method of the centroyds". This particular architecture makes unfeasible the splitting of the memory devoted to the M.F.s of the right side of the rules as this would also require an increase of the number of defuzzification blocks.

## VI. CONCLUSIONS

Cell-to-Cell mapping technique for synthesis and analysis of a fuzzy controller provides a certain number of advantages. The synthesis algorithm in addition to being completely automatic it is also wholly independent from the dynamic characteristics of the controlled plant. As showed this control technique allows to perform a stability analysis of the controlled system. Beside the above explained advantages, must be outlined that the synthesis algorithm above described it is characterized by a considerable computational burden although being simple at structural level. This is mostly due to the number of cells rather than on the number/type of operations performed on the single cell. It is worth to note that this phase, performed off-the-line, can be usefully performed on a transputer network (due to being characterized by an high level of parallelism), so greatly reducing the computational time.
Since many different control applications require high performance (in term of FIPS) and reliable controller, we present a dedicated Hardware structure: WARP.
In order to provide an answer to a wide number of application requests, WARP design relies on concepts of flexibility and modularity. The innovative approach of WARP it is represented by the adoption of different data structures for representing the membership functions characterizing the fuzzy variables of the antecedent and consequent side of the rules.The careful analysis of the computational requirements during the various stage of the fuzzy processing and the subsequent mapping in adequate hardware structures has lead to the achievement of high level of computational efficiency permitting performance in the order of 10 MFIPS to be obtained while reducing the number of parallel computational elements. Moreover the architecture is totally transparent with respect of the types of memory utilized (EEPROM, FLASH, ...) and technology (Sub-μ CMOS, ...) so allowing the device to be used for a wide range of applications.

## VII. REFERENCES


[1] C.S. Hsu and R.S. Guttalu
An unravelling algorithm for global analysis of dynamical system: an application of Cell-to-Cell Mapping
Journal of App. Mechanics, Vol. 47 pag. 931

[2] Y.Y. Chen, T.C. Tsao
A new approach for the global analysis of a fuzzy dynamical systems
Proceedings of 27th conference on Decision and Control. Dec. 1988

[3] S.M. Smith, D.J. Comer
Automated calibration of a fuzzy logic controller using cell state space algorithm
IEEE Transaction on Control Systems. Aug. 91

[4] T. Yamakawa
"Intrinsic Fuzzy electronic circuits for sixth generation Computer"
Fuzzy Computing, M.M. Gupta and T. Yamakawa, Elseiver Publisher B.V. (North Holland), 1988

[5] H. Watanabe, W.D. Dettloff and K. E. Yount
"A VLSI Fuzzy Logic controller with Reconfigurable Cascadable Architecture"
IEEE Journal of solid-state circuits. Vol 25, No. 2, April 1990

[6] A. Pagni, R. Poluzzi, G.G. Rizzotto
"WARP: Weight Associative Rule Processor. An Innovative Fuzzy Logic Controller"
IIZUKA'92 - 2ND International Conference on Fuzzy Logic and Neural Networks

# Design of Fuzzy Controllers Based on Frequency and Transient Characteristics

Kazuo TANAKA and Manabu SANO

Department of Mechanical Systems Engineering
Kanazawa University
2-40-20 Kodatsuno Kanazawa 920 Japan
Tel. +81-762-61-2101 (ext. 405)
Fax. +81-762-63-3849

***Abstract*** *- This paper proposes design methods of fuzzy phase-lead compensators based on frequency and transient characteristics. The main feature of fuzzy phase-lead compensators proposed in this paper is to have parameters for effectively compensating phase characteristics in control systems. We derive two important theorems by introducing concept of frequency and transient characteristics. One is a theorem for judging whether a fuzzy phase-lead compensator should be used or not. The other is a theorem for realizing a phase-lead compensation. A design method of fuzzy phase-lead compensators for linear controlled objects is constructed using these theorems. Furthermore, it is extended to a design method for unknown or nonlinear controlled objects. Simulation results show validity of these design methods.*

## 1. Introduction

Fuzzy control was first introduced in the early 1970's by Mamdani [1]. However, we lack at present theoretical controller design methods although fuzzy control has been applied to many real industrial processes. This paper presents theoretical compensation methods based on frequency and transient characteristics in fuzzy control systems.

The main purpose of controller design is to realize control system such that transient characteristics such as speed of response and damping characteristics are satisfied. The best way of realizing such a design is to introduce concepts of frequency characteristics such as gain crossover frequency and phase margin. Because transient characteristics are strongly related to frequency characteristics. For example, gain crossover frequency is related to speed of response, and phase margin is related to damping characteristics.

Generally speaking, it is necessary to compensate phase characteristics of control systems in order to improve damping characteristics. It is, however, known that compensation of phase characteristics is not generally easy. It has been said in the field of fuzzy control that a phase-lead compensation can be achieved if we use a coordinate transformation of e-ė phase plane, exactly speaking, rotation of e-ė phase plane [2]-[4], [6]. However, this paper will show that the coordinate transformation does not always realize an effective phase-lead compensation. We propose a new coordinate transformation for effectively realizing it.

We derive two important theorems by introducing concepts of frequency and transient characteristics. One is a theorem for judging whether a fuzzy phase-lead compensator should be used or not. This theorem is useful in the viewpoint of practical applications. Because, fuzzy control has been applied to some controlled objects which can be easily compensated by a simple linear controller. The other is a theorem for realizing a phase-lead compensation. A design method of fuzzy phase-lead compensators for linear controlled objects is constructed using these theorems. Furthermore, it is extended to a design method for unknown or nonlinear controlled objects. Let us define symbols which will be used in this paper.

$G(s)$:A transfer function of a controlled object.
$Gc(S)$:A transfer function of a linear PI controller $(a+b\cdot s)/s$.
$Gc^*(S)$:A transfer function of a linear PI controller $(a^*+b^*\cdot s)/s$.
$w_{CG}$:A gain crossover frequency of open loop transfer function $Gc(s)G(s)$.
$\theta m$:A phase margin of open loop transfer function $Gc(s)G(s)$.
$w_{0CG}$:A desired gain crossover frequency of open loop transfer function.
$\theta 0m$:A desired phase margin of open loop transfer function.
$\varepsilon p$:A overshoot of transient response of control system.
$Tp$:A time to peak of transient response of control system.
$\varepsilon 0p$:A overshoot of desired transient response of control system.
$T0p$:A time to peak of desired transient response of control system.
$g(w)$:A gain in the frequency $w$ of $Gc(s)G(s)$.
$\Psi(w)$:A phase in the frequency $w$ of $Gc(s)G(s)$.
$g^*(w)$:A gain in the frequency $w$ of $Gc^*(s)G(s)$.
$\Psi^*(w)$:A phase in the frequency $w$ of $Gc^*(s)G(s)$.

## 2 Frequency and Transient Characteristics

Fig.1 shows an example of Bode diagram. Generally speaking, $w_{CG}$ and $\theta m$ are related to speed of response and damping characteristics, respectively.

Let us consider the following second order lag system.

$$G(s) = \frac{\omega_n^2}{s^2 + 2\zeta\omega_n s + \omega_n^2},$$

0-7803-0614-7/93$03.00 ©1993IEEE

where $\zeta$ is damping ratio and $\omega n$ is undamping natural frequency. It is known that $\omega_{CG}$ and θm can be represented by $\zeta$ and $\omega n$.

$$\theta_m = 90 - \tan^{-1}\sqrt{0.25\sqrt{4+\frac{1}{\zeta^4}} - 0.5}, \quad (1)$$

$$\omega_{CG} = \sqrt{\sqrt{4\zeta^4+1} - 2\zeta^2}\,\omega_n. \quad (2)$$

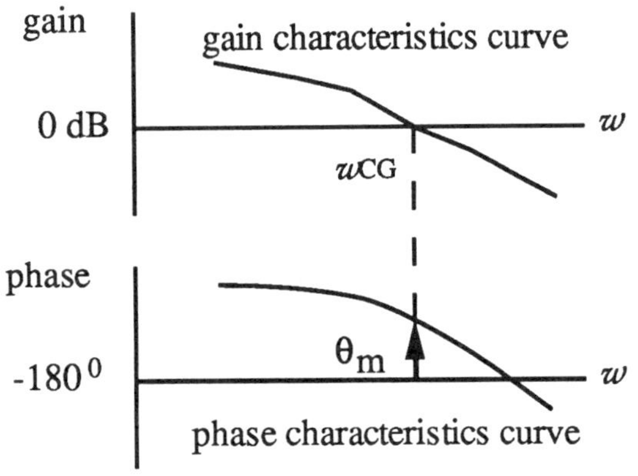


Fig.1 Bode diagram

Fig.2 shows a transient response of second order lag system. $\zeta$ and $\omega n$ can be represented by εp and Tp as follows.

$$\zeta = \frac{-\frac{1}{\pi}\ln\frac{\varepsilon_p}{100}}{\sqrt{1+(-\frac{1}{\pi}\ln\frac{\varepsilon_p}{100})^2}} \quad (3)$$

$$\omega_n = \frac{\pi}{T_p\sqrt{1-\zeta^2}} \quad (4)$$

The transient characteristics of a high order lag system with a overshoot εp and a time to peak Tp can be approximated by a second order lag system with $\zeta$ and $\omega n$ calculated by substituting εp and Tp into Eqs.(3) and (4).

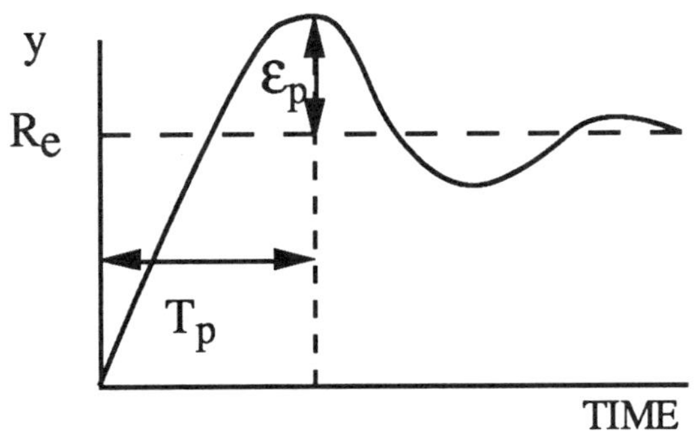


Fig.2 Transient response

## 3. Phase-lead Compensation

The easiest way of improving speed of response is to increase gain, that is, to perform gain compensation. When applying the ordinary fuzzy controller whose parameters are scaling factors for premise variables and a consequent variable, we can easily realize it by increasing the value of scaling factor for the consequent variable. However, overshoot is caused by increasing gain. To avoid overshoot, that is, to improving damping characteristics without losing speed of response, phase compensation is necessary. There is no clear relation between scaling factors of the ordinary fuzzy controller and phase characteristics. It is, therefore, difficult to effectively realize phase compensation in the ordinary fuzzy controller with the scaling factors.

Fujii [6] has shown a transformation matrix of realizing phase-lead compensation. We will show that this does not always realize an effective phase-lead compensation. We derive an important theorem by introducing phase margin and gain crossover frequency in the frequency domain. This theorem gives a new transformation matrix which can realize an effective phase-lead compensation. Furthermore, we construct a fuzzy phase-lead compensator and propose two design procedures of the compensator. The feature of the compensator is to have parameters for effectively improving phase characteristics.

### 3.1 Outline of Fuzzy Phase-lead Compensation

We derive two important theorems by introducing concepts of frequency and transient characteristics. One is a theorem for judging whether a fuzzy phase-lead compensator should be used or not. The other is a theorem for realizing a phase-lead compensation.

[Theorem 3.1]

*Assume that a controlled object is represented by*

$G(s)=k/(s^2+p1s+p2)$.

*Furthermore, assume that* $\zeta 0$ *and* $\omega 0n$ *are calculated by substituting* ε0p *and* T0p *into* Eqs.(3) *and* (4) *and that we use a linear PI controller*

$Gc(S)=(a+b\cdot s)/s$,

*where*

$a = \omega 0n^2(p1 - 2\,\zeta 0\,\omega 0n)/k$, (5)

$b = (\omega 0n^2 + 2\,p1\zeta 0\omega 0n - 4\,\zeta 0^2\omega 0n^2 - p2)/k$. (6)

*If*

$\zeta 0\ \omega 0n < p1/(2+h)$, (7)

*where* h *is a safe coefficient and* h > 1, *then the control system ( the closed loop transfer function* Gc(s)G(s)/(1+Gc(s)G(s)) *) can be approximated by the second order lag system*

$$\frac{\omega_n^2}{s^2+2\zeta\omega_n s+\omega_n^2}.$$

The proof is omitted due to lack of space. The value of h must be theoretically a large number. It is, however, practically sufficient that h=2~3. Theorem 3.1 shows that it is sufficient to use a linear PI controller $Gc(S)=(a+b\cdot s)/s$, where

$$a = w0n^2(p1 - 2\,\zeta 0 w0n)/k,$$

$$b = (w0n^2 + 2\,p1\zeta 0 w0n - 4\,\zeta 0^2 w0n^2 - p2)/k,$$

if Eq.(7) is satisfied. In other words, we do not need to design a fuzzy phase-lead compensator if Eq.(7) is satisfied. We should attempt to design a fuzzy phase-lead compensator only if Eq.(7) is not satisfied. However, we must notice that a desired transient response with $\varepsilon 0p$ and T0p may be realized by a linear controller even if Eq.(7) is not satisfied. Because, Eq.(7) of Theorem 3.1 gives a sufficient condition. It is, however, generally difficult to realize a desired transient response by a PI controller in this case.

[Theorem 3.2]

*If we use*

$$G^*c(s)=(a^*+b^*\cdot s)/s$$

*instead of*

$$Gc(s)=(a+b\cdot s)/s,$$

*where*

$$[a^*\ b^*] = [a\ b]\cdot \mathbf{T}(\theta_c, w_{CG}), \tag{8}$$

$$\mathbf{T}(\theta_c, w_{CG}) = \begin{bmatrix} \cos(-\theta_c) & -\dfrac{1}{w_{CG}}\sin(-\theta_c) \\ w_{CG}\sin(-\theta_c) & \cos(-\theta_c) \end{bmatrix}, \tag{9}$$

*then the gain crossover frequency and the phase margin of open loop transfer function of* G*c(s)G(s) *become* $w_{CG}$ *and* $\theta m+\theta c$, *respectively.*

The proof is omitted due to lack of space. Fig.3 shows an example of Bode diagrams of Gc(s)G(s) and G*c(s)G(s).

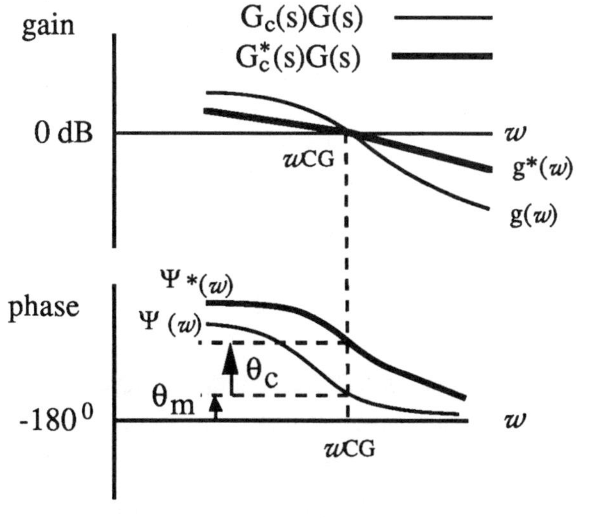


Fig.3 Phase -lead compensation

It is found from Fig.3 that a phase-lead compensation with the lead angle $\theta c$ can be realized without changing the gain crossover frequency $w_{CG}$, that is,

$$\Psi^*(w_{CG})=\Psi(w_{CG})+\theta c = \theta m+\theta c,$$

if we use

$$G^*c(s)=(a^*+b^*\cdot s)/s$$

instead of

$$Gc(s)=(a+b\cdot s)/s.$$

However, we should notice that the gain characteristics change in the frequency range excepting the gain crossover frequency $w_{CG}$. Exactly speaking,

$$g^*(w) < g(w), \qquad w < w_{CG}$$

$$g^*(w) = g(w), \qquad w = w_{CG}$$

$$g^*(w) > g(w). \qquad w > w_{CG}$$

As mentioned above, it is known that $w_{CG}$ and $\theta m$ are related to speed of response and damping characteristics, respectively. The main purpose of phase-lead compensation is to design a control system such that damping characteristics are satisfied without losing speed of response, that is, to increase phase margin without changing $w_{CG}$. By using the transformation matrix of Eq.(9), we can easily increase phase margin without changing $w_{CG}$.

Fujii [6] has reported that a phase-lead compensation in fuzzy control systems can be realized if we use the following transformation matrix.

$$\begin{bmatrix} \cos(-\theta_c) & -\sin(-\theta_c) \\ \sin(-\theta_c) & \cos(-\theta_c) \end{bmatrix} \tag{10}$$

A coordinate transformation achieved by the transformation matrix Eq.(10) means rotation of e-$\dot{e}$ coordinate system. Eq.(10) is a special case of Eq.(9). In other words, Eq.(10) is equivalent to Eq.(9) when $w_{CG}=1$. The phase-lead compensation by Eq.(10) is useful only if $w_{CG}=1$. However, required design performance with respect to speed of response is not always $w_{CG}=1$. We should use the transformation matrix Eq.(9) instead of Eq.(10) in order to effectively realize phase-lead compensation.

We construct a fuzzy phase-lead compensator by introducing the transformation matrix of Eq.(9) shown in Theorem 3.2.

Rule 1: IF $\Phi$ is about "$-\pi$ or 0 or $\pi$"
THEN

$$\dot{u}_1 = [a\ b]\cdot\begin{bmatrix} e \\ \dot{e} \end{bmatrix},$$

(11)

Rule 2: IF $\Phi$ is about "$-\pi/2$ or $\pi/2$"
THEN

$$\dot{u}_2 = [a\ b]\cdot \mathbf{T}(\theta_c, w_{CG})\cdot\begin{bmatrix} e \\ \dot{e} \end{bmatrix} = [a^*\ b^*]\cdot\begin{bmatrix} e \\ \dot{e} \end{bmatrix},$$

where $\Phi = \tan^{-1}(e/\dot{e})$. The final output of this controller is calculated as

$$\dot{u} = \frac{w_1\dot{u}_1 + w_2\dot{u}_2}{w_1 + w_2}\ , \tag{12}$$

where w1 is a membership value of the fuzzy set, about "$-\pi$ or 0 or $\pi$", of Rule 1 and w2 is a membership value of the fuzzy set, about "$-\pi/2$ or $\pi/2$", of Rule 2. Fig.4 shows these fuzzy sets, where q is a premise parameter of these fuzzy sets and $-\pi/2 < q < \pi/2$.

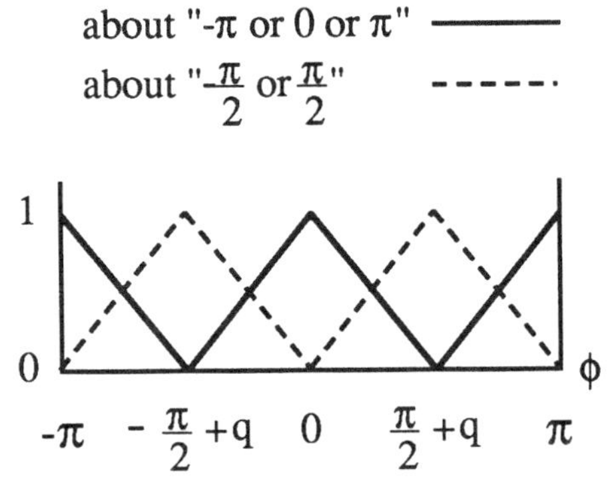


Fig.4 Fuzzy sets

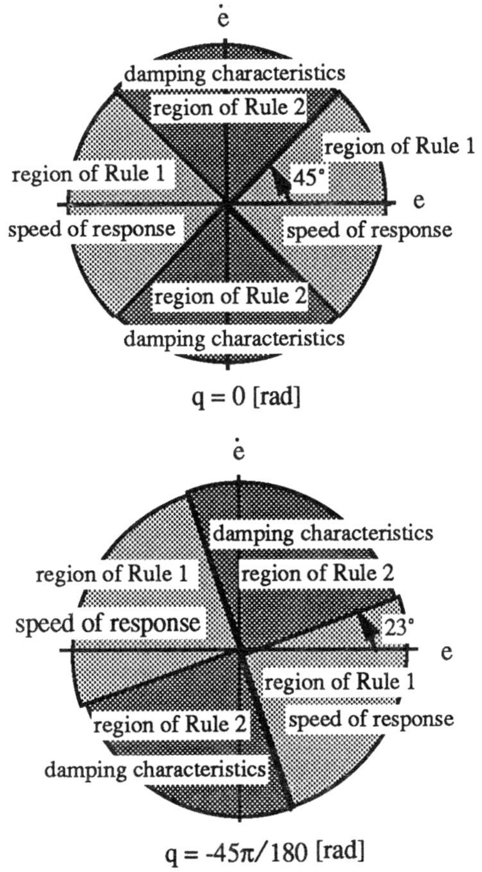


Fig.5 Fuzzy partition

The parameters of the fuzzy compensator are a, b, $\omega_{CG}$ and $\theta c$ in Eq.(11) and q in Fig.4. The controller design is to determine these parameters. Since Rule 1 and Rule 2 are related to speed of response and damping characteristics respectively, we should determine a and b such that speed of response is satisfied, that is, $Tp \leq Top$ and $\omega_{CG}$ and $\theta c$ such that damping characteristics is satisfied, that is $\varepsilon p \leq \varepsilon op$. If $Top<Tp$ or $\varepsilon op < \varepsilon p$ after determining a, b, $\omega_{CG}$ and $\theta c$, we can adjust the value of q. If the value of q is decreased, speed of response is improved. Conversely, if the value of q is increased, damping characteristics is improved. Therefore, we should decrease it if $Top<Tp$. We should increase it if $\varepsilon op < \varepsilon p$. Fig.5 shows fuzzy partitions for each value of the premise parameter q. Each region such that membership value of each rule is more than 0.5 is shown. As shown in Fig.5, we can adjust the regions by changing the premise parameter q.

The above points will be considered in design procedures. The design procedures will be given in Sections 3.2 and 3.3. The former is a design procedure for linear controlled objects. The latter is a design procedure for unknown nonlinear controlled objects.

### 3.2 Design Procedure 1
(Case of linear controlled objects)

Assume that a controlled object G(s) is linear. The procedure consists of five parts.

[Step 1]

Select a desired transient response of control system with a overshoot $\varepsilon 0p$ and a time to peak T0p. From the values of $\varepsilon 0p$ and T0p, calculate the values of

$$\zeta 0,\ \omega 0n,\ \theta 0m \text{ and } \omega 0_{CG}$$

by Eqs.(1) ~ (4). From Theorem 3.1, judge whether a fuzzy phase-lead compensator should be used or not. If a fuzzy compensator should be used, then go to [Step 2].

[Step 2]

Assume that $\theta c=0$. Then, since a=a* and b=b*, the fuzzy controller is reduced to a linear controller $Gc(s)=(a+b\cdot s)/s$. The purpose of this step is to determine a and b in Rule 1. The speed of response is not satisfied if we use a and b calculated from Eqs.(5) and (6), because the value of a real root is close to those of real parts of complex roots. This means that speed of response is affected by the real root. So, calculate a and b by substituting the following $\zeta 0'$ instead of $\zeta 0$ into Eqs. (5) and (6).

$$\zeta 0' = p1/(\ (3+h)\omega 0n\ )$$

The above equation can be derived by solving

$$\zeta 0'\omega 0n = p1/(2+(h+1))$$

for $\zeta 0'$.

[Step 3]

If the linear controller satisfies damping characteristics, that is, $\varepsilon p \leq \varepsilon op$, then end, else calculate the phase margin $\theta m$ and the gain crossover frequency $\omega_{CG}$ of Gc(s)G(s). Next, derive a* and b* from the transformation matrix $\mathbf{T}(\theta c,\omega_{CG})$ of Eq.(9), where

$$\theta c = \theta 0m - \theta m.$$

[Step 4]

Assume that p=0. Investigate whether $T_p \leq T_{0p}$ and $\varepsilon_p \leq \varepsilon_{0p}$ or not. If both of them are satisfied, then end, else go to [Step 5].

[Step 5]

Adjust the premise parameter q. Decreases the value of q if $T_p > T_{0p}$ and $\varepsilon_p \leq \varepsilon_{0p}$. Conversely, increases it if $T_p \leq T_{0p}$ and $\varepsilon_p > \varepsilon_{0p}$. If the desired transient response can not be realized by adjusting the premise parameter q, then h=h+1 and go back to [Step 2].

### 3.3 Design Procedure 2

(Case of unknown nonlinear controlled objects)

The above design procedure cannot be applied to unknown or nonlinear systems since frequency characteristics cannot be used in this case. However, without loss of generality, the above design procedure can be extended to a design method for unknown nonlinear controlled objects, because there are clear relations between frequency characteristics and transient characteristics.

[Step 1]

Select a desired transient response of control system with a overshoot $\varepsilon_{0p}$ and a time to peak $T_{0p}$.

[Step 2]

Assume that $\theta_c=0$. Then, since a=a* and b=b*, the fuzzy controller is reduced to a linear controller $G_c(s)=(a+b\cdot s)/s$. Determine a and b of Rule 1 such that $T_p \leq T_{0p}$.

[Step 3]

If the linear controller satisfies damping characteristics, that is, $\varepsilon_p \leq \varepsilon_{op}$, then end, else go to [Step 4].

[Step 4]

Assume that $\omega_{0CG}=1$ and p=0. Increase gradually the value of $\theta_c$ of Rule 2 until $\varepsilon_p \leq \varepsilon_{0p}$. This means that a* and b* of Rule 2 are adjusted. Investigate whether $T_p \leq T_{0p}$ and $\varepsilon_p \leq \varepsilon_{0p}$ or not. If both of them are satisfied , then end, else go to [Step 5].

[Step 5]

Adjust the premise parameter q. The value of q decreases if $T_p > T_{0p}$ and $\varepsilon_p \leq \varepsilon_{0p}$. Conversely, it increases if $T_p \leq T_{0p}$ and $\varepsilon_p > \varepsilon_{0p}$. If the desired transient response can not be realized by adjusting the premise parameter q, select other values of a and b such that the value of $T_p$ is smaller, where $\theta_c=0$, that is, a*=a and b*=b, and go back to [Step 3] .

A self-learning fuzzy control can be realized if the above procedure is automatically performed. This is a subject for future study.

## 4. Examples

We illustrate some examples of the above design methods.

### 4.1 Example 1

Let us consider the following controlled object.

$$\ddot{y} = -10\dot{y} - 16y + 2.45u \qquad (13)$$

We design a fuzzy compensator by the design procedure 1. Fujii [6] reported that a desired transient response is realized by a fuzzy controller. However, we show that the desired transient response is easily realized by a simple linear control. In other words, it is sufficient to use a linear PI controller in this case.

[Step 1]

In [6], $\varepsilon_{0p}$=6.0[%] and $T_{0p}$ =1.2[sec.]. From Eqs.(1)~(4), we obtain

$\zeta_0$=0.68, $\omega_{0n}$=3.51,
$\theta_{0m}$=63.34[deg],
$\omega_{0CG}$=2.35[rad/sec].

Then,

$\zeta_0\ \omega_{0n}$ = 2.39 < p1/(2+h) = 2.5,

where h=2. Eq.(7) of Theorem 3.1 is satisfied. Therefore, we can realize the desired transient response by using a linear controller

$G_c(s)=(a+b\cdot s)/s$,

where a=25.62 and b=8.36. Fig.7 shows the simulation result. It is found from Fig.7 that the PI controller realizes the desired transient characteristics.

### 4.2 Example 2

Let us consider the following controlled object.

$$\ddot{y} = -4\dot{y} - 4y + 3u \qquad (14)$$

We design a fuzzy compensator by the design procedure 1.

[Step 1]

We select a desired transient response as follows: $\varepsilon_{0p}$ = 5.0[%] and $T_{0p}$ =1.5[sec.]. From Eqs.(1)~(4), we obtain that

$\zeta_0$=0.69, $\omega_{0n}$=2.89,
$\theta_{0m}$=64.63[deg],
$\omega_{0CG}$ = 1.89[rad/sec].

Then,

$\zeta_0\ \omega_{0n}$ = 1.85 > p1/(2+h) = 1.00,

where h=2. Eq.(7) of Theorem 3.1 is not satisfied. It is, therefore, difficult to realize the desired transient response by a linear PI controller.

[Step 2]

We obtain $\zeta_0'$=0.277. Therefore, a=6.70 and b=2.74.

[Step 3]

$\omega_{CG}$=1.89.
$\theta_m$ = 39.94[deg].
$\theta_c = \theta_{0m} - \theta_m$ = 64.63 - 39.94 = 24.69.
Therefore, a*=3.92 and b*=3.97.

[Step 4]~[Step 5]

Tp=6.9[sec.] and εp=0.02[%] when q=0[rad]. So, we adjust the premise parameter q. Tp=1.5[sec.] and ε0p=4[%] when q=-60π/180[rad]. Fig.8 shows the simulation result.

### 4.3 Example 3

Let us consider the following nonlinear controlled object.

$$\ddot{y} = -4\dot{y} - 4y^2 + 3u \cdot (1 - \sin(0.1 \cdot \pi \cdot \dot{u})) \quad (15)$$

We design a fuzzy compensator by the design procedure 2.

[Step 1]

We select a desired transient response as follows: ε0p = 10[%] and T0p = 2[sec.].

[Step 2]

a =2 and b = 0.5.

[Step 3]~[Step 4]

θc = 40[deg].

Therefore, a*=1.21 and b*= 1.67.

[Step 5]

Tp=2.0[sec.] and εp=10[%] when q=0[rad]. Fig.9 shows simulation result.

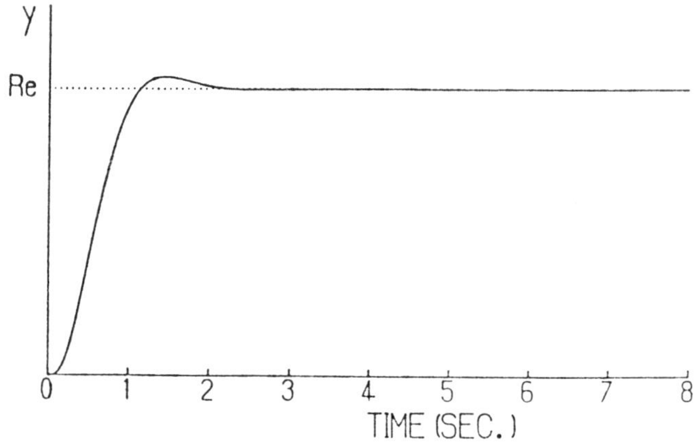


Fig.7 Control result (Example 1)

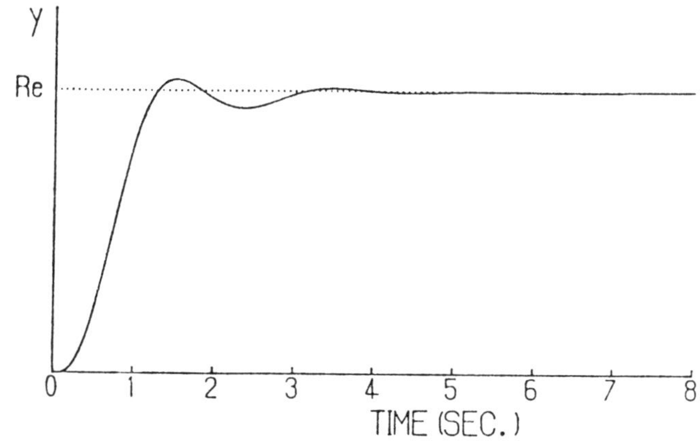

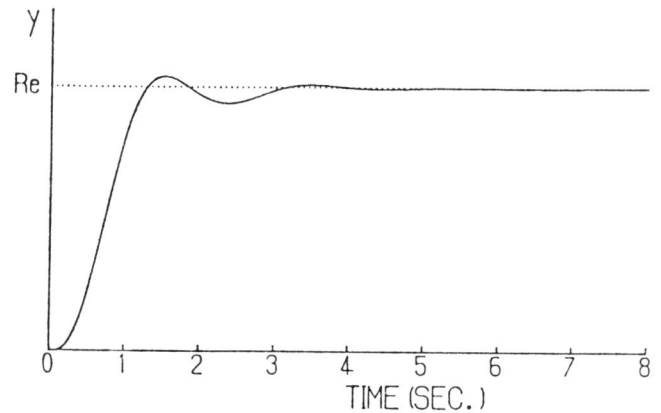

Fig.8 Control result (Example 2)

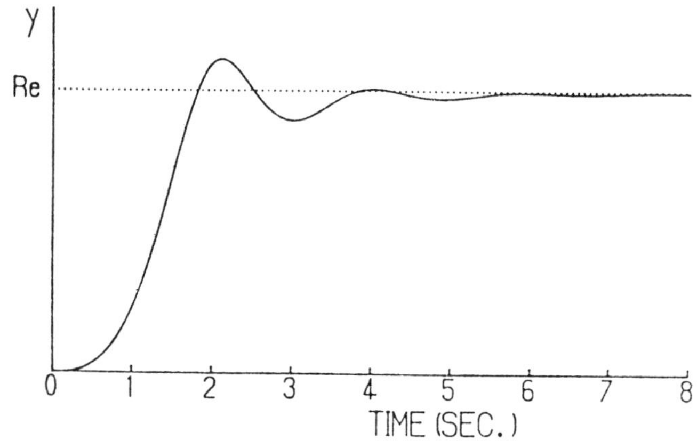


Fig.9 Control result (Example 3)

## 5. Conclusion

The compensation methods of fuzzy control systems based on frequency and transient characteristics have been discussed. We have derived two important theorems. One is a theorem for judging whether a fuzzy phase-lead compensator should be used or not. The other is a theorem for realizing a phase-lead compensation in fuzzy control systems. A design method of fuzzy phase-lead compensators for linear controlled objects has been proposed using these theorems. Moreover, it has been extended to a design method for unknown or nonlinear controlled objects.

## References

[1]E.H.Mamdani:Applications of Fuzzy Algorithms for Control of Simple Dynamic Plant, Proc. IEE, vol.121, no.12, pp.1585-1588 (1974).

[2]M.Sugeno:Fuzzy Control, Nikkankogyou Publ. Co., (1988).

[3]K.Tanaka and M.Sano:A New Tuning Method of Fuzzy Control, Proceeding of 4th IFSA World Congress, vol.1, pp.207-210 (1991)

[4]A.Ishigame et al.:Design of Electric Power System Stabilizer Based on Fuzzy Control Theory, IEEE International Conference on Fuzzy Systems, pp.973-980 (1992).

[5]K.Tanaka at el.:A Phase-lead Compensation of Fuzzy Control Systems, Journal of Japan Society for Fuzzy Theory and Systems, vol.4, no.3, pp.163-170 ( in Japanese).

[6]A.Fujii, T.Ueyama and N.Yoshitani:Design of Fuzzy Controller using Frequency Response, Proceeding of 5th Fuzzy System Symposium, pp.115-120 (1989) ( in Japanese).

# Fuzzy Logic Control for High Order Systems

Serge BOVERIE(1)(2), Bernard DEMAYA(2), Jean–Michel LEQUELLEC(1)(2), André TITLI(2)(3)(4)

(1) SIEMENS AUTOMOTIVE SA
Avenue du MIrail – BP 1149
31036 Toulouse Cedex France

(2) SIEMENS AUTOMOTIVE SA
Laboratoire MIRGAS
Avenue du Mirail – BP 1149
31036 Toulouse Cedex France

(3) INSA TOULOUSE
Avenue de Rangueil
31077 Toulouse Cedex
France

(4) LAAS–CNRS
7, Avenue du Colonel Roche
31077 Toulouse Cedex
France

***Abstract***– **In this paper we introduce a fuzzy control algorithm for high order processes. This algorithm is based on a serie association between a simple fuzzy controller and an incremental block. Its application to several linear systems submitted to severe external and internal disturbances have shown the robustness and the intrinsic performances of such a controller.**

***Keywords:*** **Fuzzy sets, fuzzy controllers, process control, benchmark examples.**

## I. INTRODUCTION

Previous studies have shown the great advantages of the fuzzy control technique in the control of low order processes ([2], [3], [4] & [5] ). These advantages can be expressed in terms of:

– Ease of the synthesis of the fuzzy controller (no requirement of any mathematical model of the process to be controlled).

– Intrinsic performance.

– Robustness with respect to internal and external disturbances.

But, conventional fuzzy controllers, based on the use as inputs of the error $\varepsilon$ and the change of the error $\Delta\varepsilon$, are not able to control processes where the order is higher than two. So, in order to solve these problems, it could be possible to design a multi–input fuzzy controller and to use it in the control of high order processes. For example, to control a third order system, we could define a controller with the three following inputs : $\varepsilon$, $\Delta\varepsilon$, $\Delta^2\varepsilon$. However, this method would lead to a loss of interest of Fuzzy Control all the more so as the control rules would become more complex, or even impossible to describe and their number would then reach a discriminatory number.This is why, in our study, we have made sure that we found a compromise solution that allows on one hand to conserve the advantages of the fuzzy controller, and on the other hand to solve the control problem of high order processes.

Several methods can be considered.

The first one consists of making a series association of a fuzzy controller and a non–linear controller: the output of the first controller is used to drive the second controller ( Fig. 1).

We can also make the hypothesis that the association of the fuzzy controller with the process can be considered as a non–linear system ( Fig. 1).

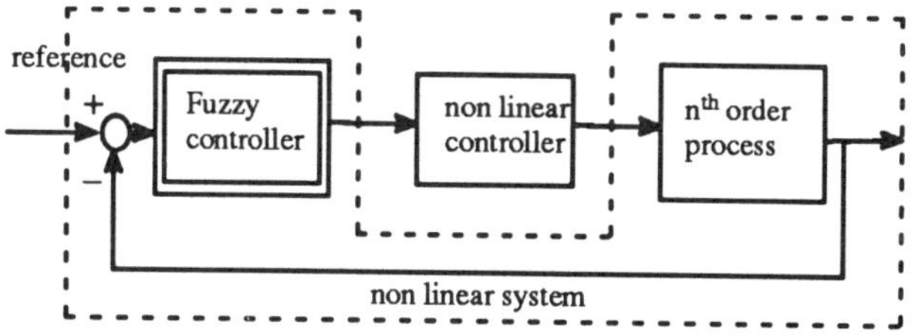


Fig.1 –diagram of a series association between a fuzzy controller and a non linear controller

However, the design of the associated non–linear controller is very tricky and needs a precise mathematical model of the process. Because of this contradictory design element with fuzzy techniques, this method was dismissed. The final chosen control structure, which is given by a second method, can be described by the Fig. 2 :

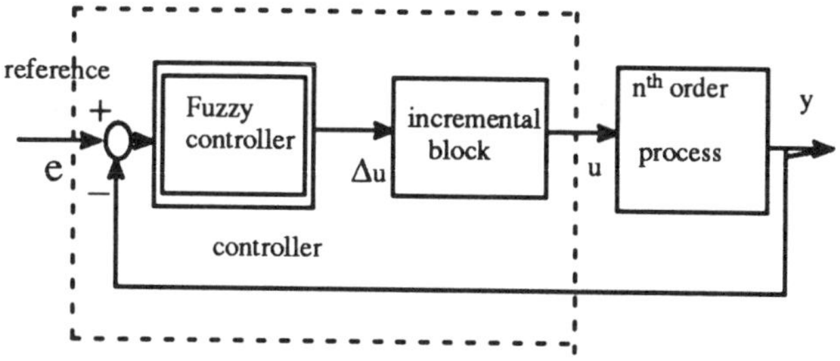


Fig.2 – Fuzzy Controller associated with an incremental block

This structure is based on the association between a conventional fuzzy controller and an incremental block (F.I.A.), and allows the adaptation of the overall controller to the order of the process to be controlled.In order to explain this structure in detail, our study can be divided into three parts:

– The first part is dedicated to the design of the basic fuzzy controller,

– The second part is dedicated to the design of the incremental structure,

– The third part is dedicated to simulation results. In this last part, we describe the results obtained for different classes of processes and we try to bring out the surprising robustness of these controllers.

## II. Synthesis of the Fuzzy Controller

### *A. Synthesis of the basic fuzzy controller*

The basic fuzzy controller is designed with a qualitative technique applied in the phase plane, based on the analysis of the evolution of the error between the reference signal and the output of the process, and the change of this error ( Fig. 3).

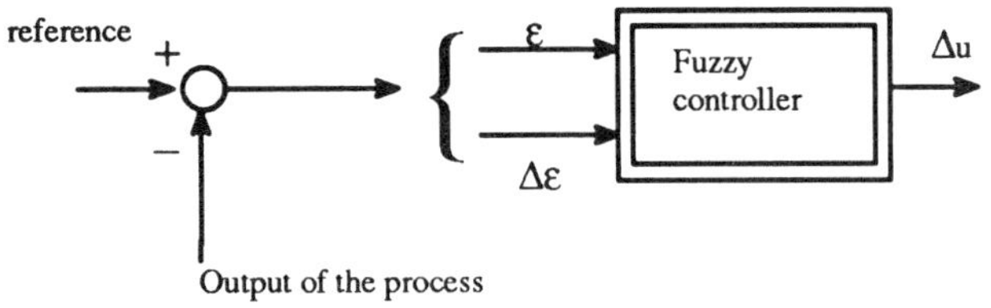


Fig.3 –Skeleton diagram of the fuzzy controller

By referring to the behavior of the open loop process in the qualitative phase plane [6], the control rules of the fuzzy controller can be defined as follows.

One defines local behavior in phase plane (ε, Δε) divided into four quadrants, according to the sign of $\Delta^2\varepsilon$.
The goal of the control strategies is to reach the equilibrium point (ε=0, Δε=0) , by generating control actions as ε and Δε move in the phase plane on trajectories in the clockwise direction (this is a characteristic behavior of closed loop systems).

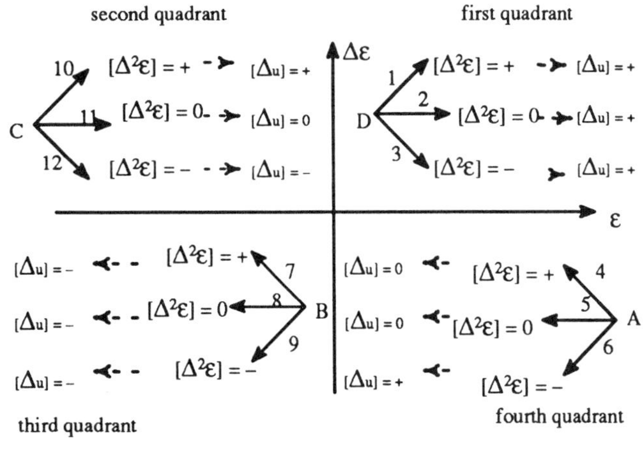


−: $\Delta^2\varepsilon$ negative; + : $\Delta^2\varepsilon$ positive; 0 : $\Delta^2\varepsilon$ Null

[Δu] = deduced control value

Fig.4 –Analysis of the open loop process trajectories in the phase plane

In each point of the phase plane we determine the sign of the control which insures the convergence of the closed loop system toward the equilibrium point. This control action is related with the local behavior of the open loop process. In Fig. 4 we have represented the possible behavior of the process and the deduced control action.

For example, in the fourth quadrant, the trajectory number 4 leads the process to the equilibrium. So, in order to conserve the natural convergence, the control action must be held or increased if we want to improve the convergence, that is Δu ≥ 0. The trajectory number 6 favours the natural crossing of the Δε axis, but, if ε is big and Δε small, in order to accelerate the convergence, it could be possible to generate a positive increment, Δu>0. When ε is small and Δε big to force the process to converge to the equilibrium point, the controller must generate a negative increment, that is Δu ≤ 0. For the trajectory number 5, the control increment can be chosen equal to Δu = 0 .

In the quadrant 1, the trajectories 1 and 2 don't allow the natural crossing of ε axis. Then, it is necessary to generate a positive incremental signal Δu in order to insure this crossing. On the other hand, the trajectory 3 allows the natural crossing of ε axis, so a control action Δu = 0 is enough; but, we can apply a positive increment to accelerate the convergence.

We apply a similar reasoning to deduce the control actions that the controller will have to generate in the two other quadrants ( Fig. 5).

Finally when we apply the control values the closed loop process behavior becomes as Fig. 5, where the straight line δ represent the natural convergence line (on this line the control value is equal to zero).
Above this line the control value is positive.
Below this line the control value is negative.
The double arrow represents the global behavior of the closed loop system. The quadrant 2 and 4 are called "convergence quadrants"

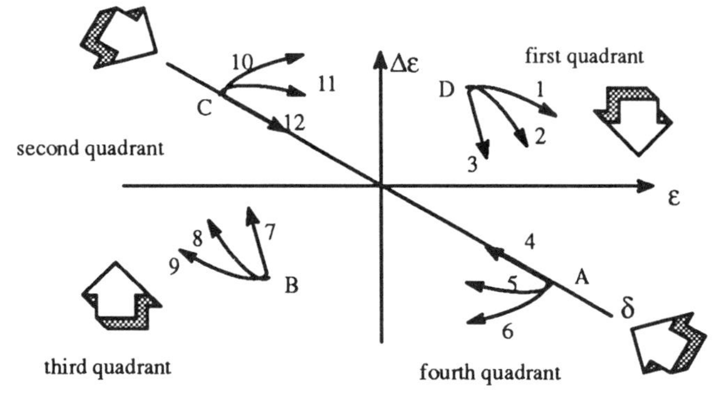


Fig.5 –Global behavior of the closedloop system

The qualitative reasoning method only allows to determine the sign of the control action.
In order to modulate these control actions, it is interesting to divide the universe of discourse of ε and Δε into fuzzy sets, and to define some fuzzy control decision related with the membership degree of ε and Δε to these fuzzy sets. We can distinguish several

fuzzy sets for the variables ε, Δε, and the control Δu, for example Negative **B**ig, Negative **M**edium, Negative **S**mall, . . .

From this analysis, we can deduce the control rules defined in the phase plane. These rules are given in table 1. In this table we find again the anti–diagonal where the control value is equal to zero and which is equivalent to the qualitative natural convergence line:

TABLE I
RULE BASE

| X\X | NB | NM | NS | ZE | PS | PM | PB |
|---|---|---|---|---|---|---|---|
| PB | ZE | PS | PM | PB | PB | PB | PB |
| PM | NS | ZE | PS | PM | PB | PB | PB |
| PS | NM | NS | ZE | PS | PM | PB | PB |
| ZE | NB | NM | NS | ZE | PS | PM | PB |
| NS | NB | NB | NM | NS | ZE | PS | PM |
| NM | NB | NB | NB | NM | NS | ZE | PS |
| NB | NB | NB | NB | NB | NM | NS | ZE |

*B. Synthesis of the incremental block*

For a $n^{th}$ order process, it is well known that the order of the control system must be (n–1) (using for example a pole compensation method) , and, in this way, the structure of the controller can be described by the following symbolic form: $PID^{(n-1)}$.

In our case, the fuzzy controller is a PD type, so, in order to respect the preceding condition, the incremental bloc must have a $PID^{(n-2)}$ form. From this, we can deduce the main shape of this control block :

$$U = K\int_0^t \Delta u(\tau)d\tau + k_0\Delta u + k_2\Delta u^{(1)} + ... + k_{n-2}\Delta u^{(n-2)} \qquad (1)$$

We can discretize this equation writing:

$$\Delta u_j^{(1)} = \frac{\Delta u_j - \Delta u_{j-1}}{\Delta t} \quad ,$$

$$\Delta u_j^{(2)} = \frac{\Delta u_j - 2.\Delta u_{j-1} + \Delta u_{j-2}}{\Delta t^2} \qquad (3)$$

$$\cdots$$

$$\Delta u_j^{(n-2)} = \frac{f(\Delta u_j\, , \, \ldots . , \Delta u_{j-n+2})}{\Delta t^{n-2}}$$

Where $\Delta u_j$ is the value of control increment at instant j calculated by the fuzzy controller at the sampling period Δt. Replacing in (2) the different terms by their expression, one gets:

$$U_j = K\sum_{i=0}^{i=j}\Delta u_i \; + \; k_1\Delta u_j \; + \; k_2\frac{\Delta u_j - \Delta u_{j-1}}{\Delta t} \; + \ldots\ldots.. \qquad (4)$$

Finally, we obtain the control structure:

$$U_j = K\sum_{i=0}^{i=j}\Delta u_i + k'_0\Delta u_j + k'_1\Delta u_{j-1} + \ldots + (-1)^{(n-2)}k'_{n-2}\Delta u_{j-n+2} \quad (5)$$

with:

$$\left\{\begin{aligned} k'_0 &= k_0 + \frac{k_1}{\Delta t} + \; . \; . \; . \; . + \frac{k_{n-2}}{\Delta t^{n-2}} \\ k'_1 &= \frac{k_1}{\Delta t} + \; . \; . \; . \; . + (n-2)\frac{k_{n-2}}{\Delta t^{n-2}} \\ &\vdots \\ k'_{n-2} &= \frac{k_{n-2}}{\Delta t^{n-2}} \end{aligned}\right. \qquad (6)$$

We notice that the incremental block has n degrees of freedom for a $n^{th}$ order process and therefore, we can suppose that this structure make the tuning of the fuzzy controller more complex. But, we have shown, experimentally, that the sensitivity of tuning parameters is very small, which confirms the great robustness of fuzzy controller and justify the choice of the incremental fuzzy controller.

## III. SIMULATION - RESULTS

The principles developed in this paper have been tested on different processes defined in the table below ([7], [8], [9] & [10]):

For each process, the nominal values of the $a_i$ terms and the gain K are given below :

- first order : a = 1 ; K = 1
- second order : $a_1$ = 1.4 , $a_2$ = 1 ; K = 1
- third order : $a_1$ = 1.75, $a_2$ = 2.15, $a_3$ = 1 ; K = 1

TABLE II
THE DIFFERENT MODELS

| | |
|---|---|
| First order process : | $G_1(s) = \frac{K}{s + a}$ |
| Second order process : | $G_2(s) = \frac{K}{s^2 + a_1 s + a_2}$ |
| Third order process : | $G_3(s) = \frac{K}{s^3 + a_1 s^2 + a_2 s + a_3}$ |

Different disturbances were then applied to the process (TABLE III) :– external disturbances : bias on the control, white noise on the measurement.– modification of the model parameters : modification of the plant poles and of the static gain.

TABLE III

THE DIFFERENT DISTURBANCES AND MODIFICATIONS OF THE MODEL PARAMETERS

| COMPLEXITY LEVEL | 0 | 1 | 2 |
|---|---|---|---|
| Disturbance level in % of the nominal value | 100 % | 10 % | 60 % |
| Sensor nois (R M S) | 0 | 0.02 | 0.16 |
| Unmodelled dynamic. (sec) | 0 | 0.10 | 0.25 |
| Time delay (sec) | 0 | 0.05 | 0.1 |
| Unmodelled plant zero (sec) (only for second and third order plants) | 0 | 0.10 | 0.2 |
| Pole variation max value | Δa=±2 for first order Δa=±3 for 2nd, 3rd order | IDEM | IDEM |
| Gain variation % of the nominal gain | -50 to 200 for first order -50 to 300 for 2nd to 3rd order | IDEM | IDEM |

Plant perturbations (last three rows)

The overall process can be described by Fig. 6:

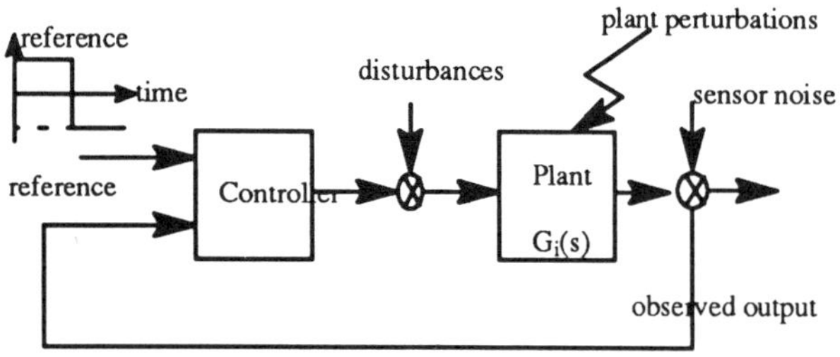


Fig.6 –Description of the process

First order process: The F.I.A. have performed sucessfully till the second complexity level.
These results show a good robustness of the system even when its becomes unstable. f Fig. 7–8)

Second order process: The F.I.A. have performed sucessfully till the first complexity level.
However, when the system is held stable the controller can insure a good behavior of the process for the complexity level 2 ( Fig. 9–10).

Third order process: The F.I.A. have performed sucessfully till the 0 complexity level.
For the complexity level 1 the controller insure a good behavior as far as $-2 < \Delta a < +3$.
For the complexity level 3 the controller insure a good behavior till the process is held stable ( Fig. 11–12).

## IV. CONCLUSION

In this paper, devoted to the control of high order systems by the use of fuzzy control, we have developed the design of a basic fuzzy controller, mainly by the definition of the set of rules (a rule table) with a qualitative reasonning in the phase plane.
In order to conserve the simplicity of the basic fuzzy controller and to take profit of the well known robustness of this kind of controller we have assiociated this basic fuzzy controller to an incremental controller with the goal to have a controller with an order corresponding with the order of the process.
Of course, the number of parameters to be tuned to get good performances has increased, but, on the other hand the number of rules to be written has been limited.

Simulation results have proven the efficiency of this approach.

## ACKNOLEDGMENT

The authors would thanks ADME organization which has partially supported this work.

## REFERENCES

[1] – L.A. Zadeh, " Fuzzy logic ", IEEE computer, 1988, pp. 83–93.

[2] – S.Boverie, B.Demaya, A.Titli, "Fuzzy logic control compared with other automatic control approaches", Proceedings of the 30th IEEE conference on decision control, december 1991, pp. 1212–1216.

[3] – S.Boverie, B.Demaya, R. Ketata, A.Titli, "Performance evaluation of a fuzzy controller", Proceedings of the IFAC SICICA in Malaga, May 1992.

[4] – L. P. Holmblad, J.J. Østergaard , " Control of a Cement Kiln by Fuzzy Logic", North Holland, 1982, Fuzzy Information and Decision Processes, pp. 389–399

[5] – W.J. Kickert, R.Van Nauta Lemke, "Application of a Fuzzy Controller in a Warm Water Plant", Automatica, Vol. 12, pp. 301–308

[6] – L. Foulloy, "Projet commande symbolique et neuromimétique", Pôle Automatisation Intégrée, GR Automatique, Centre national de la Recherche Scientifique, Rapport 91–2, novembre 91.

[7] – M. K. Masten, H. E. Cohen , "A Showcase of Adaptive Controller Designs", International Journal of Adaptive Control and Signal Processing, 1989, Vol. 3, pp. 95–101

[8] – M. K. Masten, H. E. Cohen , "An advanced Showcase of Adaptive Controller Designs", International Journal of Adaptive Control and Signal Processing, 1990, Vol. 4, pp. 89–98

[9] – M'Saad, I.D. Landau, M. Samaan, " Further Evaluation of Partial State Model Reference Adaptive Design", International Journal of Adaptive Control and Signal Processing, 1990, Vol. 4. pp. 133–148

[10] – M'Saad, I.D. Landau, M. Duque, M. Samaan, "Example applications of the partial state model reference adaptive design technique", International Journal of Adaptive Control and Signal Processing, 1989, Vol. 3. pp. 155–165

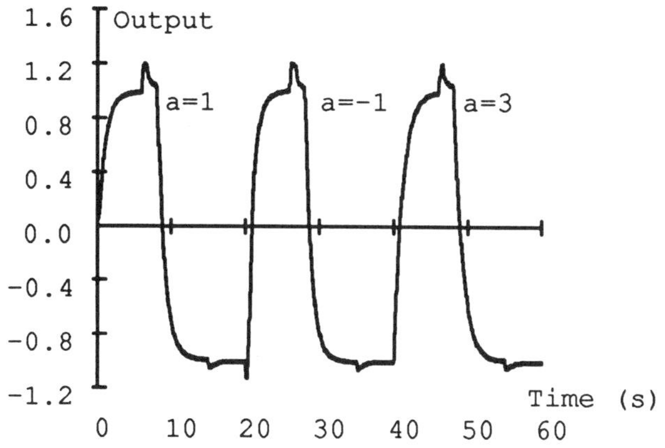


Fig.7 –First order with K=1 (level 0)

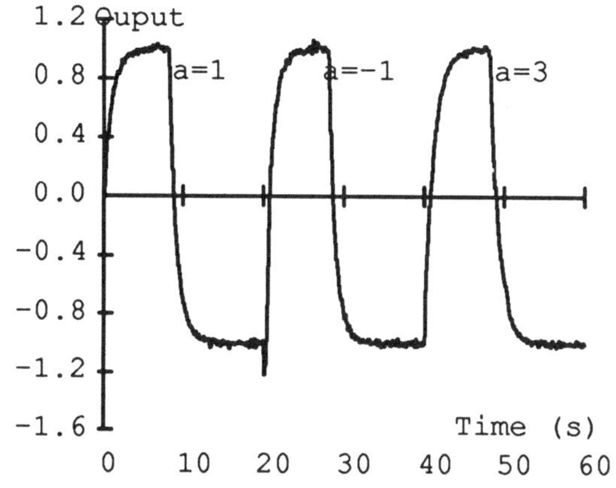


Fig.8 –First order with K=1 (level 1)

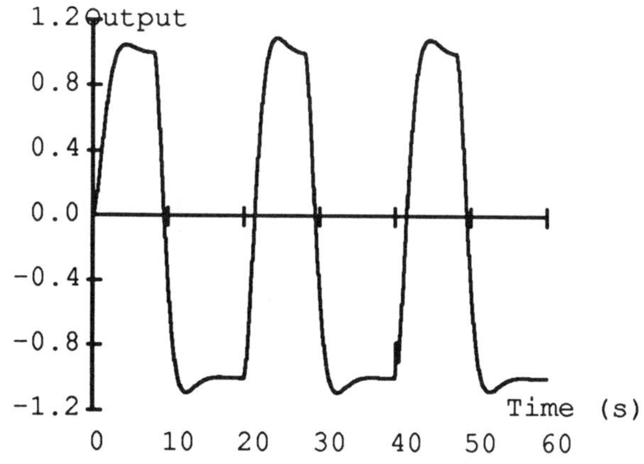


Fig.9 –Second order, with K=1, (level 0)

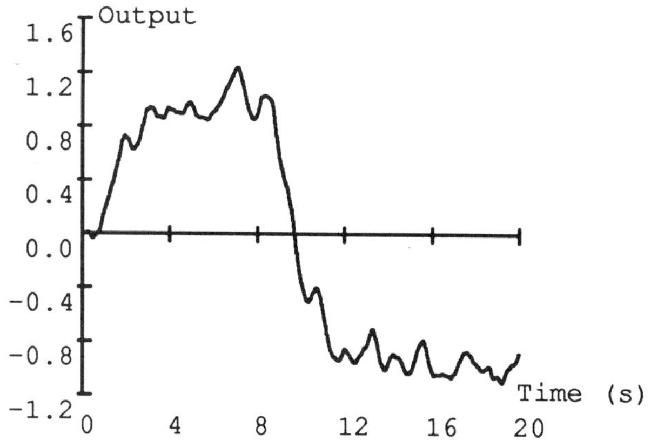


Fig.10 –Second order , with K=1, $a_1$=4.4, $a_2$=3 (level 2)

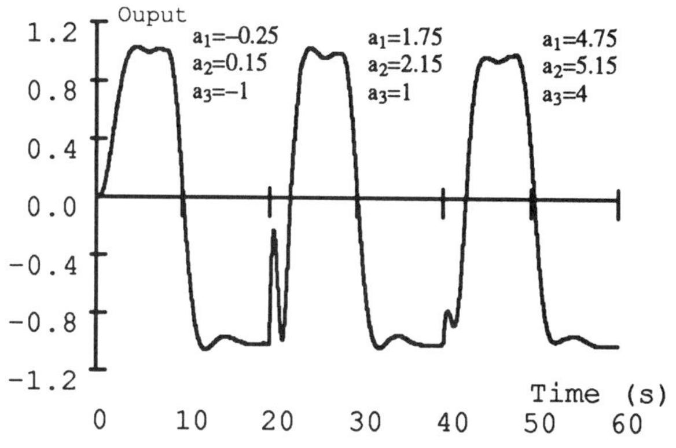


Fig.11 –Third order with K=1, 3, 0.5; level (0)

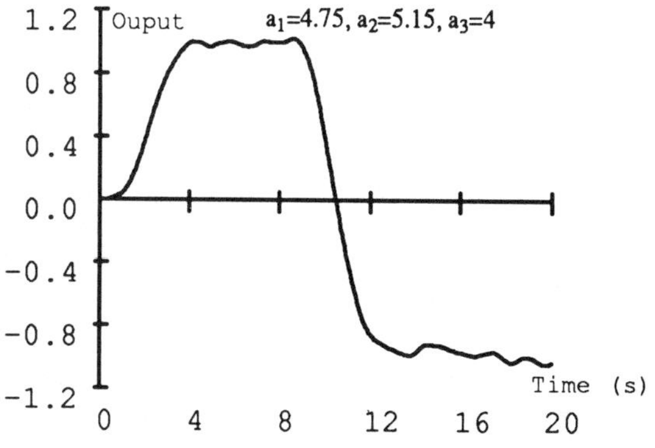


Fig.12 –Third order with K=1, level 1

# Adaptive Fuzzy–Control for Flexible–Link Manipulators: a Hybrid Frequency–Time Domain Scheme *

Anthony Tzes Kyriakos Kyriakides
Department of Mechanical Engineering
Polytechnic University
333 Jay Street
Brooklyn, NY 11201

***Abstract*— This paper addresses the implementation of an adaptive fuzzy controller for flexible link robot arms. The design technique is a hybrid scheme involving both frequency and time domain techniques. The eigenvalues of the open loop plant can be estimated through application of a frequency domain based identification algorithm. The region of the eigenvalue space, within which the system operates, is partitioned into fuzzy cells. Membership functions are assigned to the fuzzy sets of the eigenvalue universe of discourse. The degree of uncertainty on the estimated eigenvalues is encountered through these membership functions. The knowledge data base consists of feedback gains required to place the closed loop poles at predefined locations. A rule based controller infers the control input variable weighting each gain with the value of the membership functions at the identified eigenvalue. The aforementioned controller is demonstrated in simulation studies.**

## I. Introduction

The area of Fuzzy Logic Controllers (FLCs) has attracted much attention in recent years, and the number of relevant works is far too large for citation here; recent review works include [1, 2]. The FLC operates in an uncertain fuzzy environment and consist of a fuzzification interface, a knowledge base, the decision making logic, and a defuzzification interface. Several heuristic guidelines have been compiled for the implementation of the individual modules of a FLC. The associated functions of the FLC components can be summarized in the following. The fuzzification interface transforms a crisp measurement corrupted with noise into a fuzzy singleton within a certain universe of discourse. The knowledge base characterizes the manipulation of fuzzy data (data base) [3] and the control policy (linguistic control rules) in a FLC environment. The decision making logic infers fuzzy logic control actions from a rule base [4] based either on the fuzzy model of the process or on expert experience. The defuzzification interface yields a deterministic control input from an inferred control action.

In this paper, a self–tuning fuzzy controller for the one–link flexible manipulator is considered. Self–tuning fuzzy control schemes attempt to decrease the effects of plant uncertainty by identifying the system on–line, thereby yielding a closed loop with reduced sensitivity and improved performance over non-adaptive algorithms. Such schemes therefore consist of two components–the controller and the estimator.

The estimator must provide a reliable system estimate in the presence of unmodeled dynamics and unmeasurable disturbances. The Time-varying Transfer Function Estimation identification technique [5] is used to provide the system eigenvalues consistent with the measurement data and the inherent model structure. Based on the knowledge of the operating constraints in the identified system, the estimate can be restricted within a region of the eigenvalue space. Specifically, the unknown eigenvalue vector will be contained within an interval of the eigenvalue space, which ranges from the eigenvalue vectors of the unloaded to those of the fully loaded operating mode.

Fuzzy controllers inquire the uncertain information provided by the identifier to infer the appropriate control action. In fuzzy logic, the associated universe of discourse encloses the estimated eigenvalue vector. Fuzzy membership functions are assigned to layers partitioning this area. The knowledge data base consists of feedback gains required to place the closed loop poles at predefined locations, and a rule based controller infers the output variable from the values of the membership functions at the estimated eigenvalue vector.

*This material is based upon work supported by the NSF under Grant No 91-12362

## II. Problem Statement

In this paper we address the issue of adaptive feedback control on the single flexible link robot manipulator. The primary objective of the design is the synthesis of a closed-loop system with certain dynamic characteristics, which will enable it to follow a given reference signal. A secondary engineering task is the achievement of an adaptive behavior throughout the operating modes of the system (e.g. fully loaded, partially loaded). The design approach consists of two separate parts. The first part is involved with the implementation of an on-line identification of the unknown plant. At this point, the particular characteristics of the ligthly damped structure are taken under consideration. The second part is the design of a suitable feedback controller, whose structure depends on that of the identifier. The filter is designed in the frequency domain and it is based on the TTFE algorithm, while the controller is a hybrid unit whose components are a fuzzy-rulebase and a cell control technique with proportional plus derivative $(P+D)$ output feedback gains.

The system under consideration has the following mathematical description,

$$\begin{aligned} \dot{x} &= Ax + bu \\ y &= Cx + du + v \end{aligned} \tag{1}$$

where, $x \in R^n$ is the state vector, $u \in R$ is the input, $y \in R^4$ is the output measurement vector consisting of the shaft position, the shaft velocity, the tip position and the tip acceleration, while $v \in R^4$ is a random processes of measurement noise. The derivation of the state equations and the constant matrices $A$, $b$, $C$ and $d$ for the sinlge flexible link are given in [6].

### *A. The Identification Issue*

Since the system dynamics are unknown, the design of a proper identification algorithm is an important topic of the problem. In this paper we employ identification techniques developed in the frequency–domain. The simplest frequency–domain adaptive filter is one in which the input signal $u(t)$ and output $y(t)$ are accumulated into buffer memories to form $N$-point data blocks. These blocks are then transformed by $N$-point Fast Fourier Transform to their equivalent frequency–domain blocks, $U$, $Y$, at the $t$ time instant. The Empirical Transfer Function (ETFE) [7], estimates the assumed linear transfer function $H(s) = \frac{Y(s)}{U(s)}$ in the frequency domain via a non-recursive updating scheme at the time $t$ as

$$\hat{H}_i(t) = \frac{Y_i(t)}{U_i(t)}, \; \hat{H}_{N-i}(t) = \bar{\hat{H}}_i(t), \; i \in [0 \le i \le \frac{N}{2}, U(i) \neq 0]$$

where $i$ corresponds to the $i$th bin in the frequency domain, or $\frac{2\pi i}{NT}\ \frac{rad}{sec}$, and $\bar{\hat{H}}_i(t)$ is the complex conjugate to $\hat{H}_i(t)$. Notice that $\hat{H}_i(t)$ can be updated every $\gamma$, $1 \le \gamma < N$ samples.

### *B. The controller issue*

As we mentioned before, the design of the controller depends on the choice of the identification scheme. In our case, the output of the estimator filter is the current information about the flexible modes of the robot arm. In order to take advantage of this information we engage a type of cell $P+D$ control law. The general idea is to assign certain feedback gains to several cells inside the interval of interest of the eigenfrequency space. Every cell infers the feedback gain calculated at its midpoint open-loop eigenfrequency, whenever the updated eigenfrequency vector lies within it. Selection of overlapping cells defines a fuzzy-partition of the eigenfrequency space. This kind of partition requires a fuzzy-logic decision making, for a given point of the space to be a member of any one of the participating cells. In the case of a non–fuzzy partition each sub–set of the eigenfrequency space is associated with one and only one gain. Therefore, the resulting feedback gain function is discontinuous on the boundaries of the non–overlapping cells. On the other hand, the choice of overlapping cells serves merely to avoid these discontinuities of the feedback gain values along any path on the area covered by the cells. We point out that the selection of a cell control strategy [8] associates a small number of pre-defined feedback gains with a large area of the eigenfrequency space. A descriptive presentation of the proposed adaptive-fuzzy control scheme is shown in Figure 1.

Let us now assume a simple proportional feedback, $k_p$, which is closing the loop from the shaft position output port. To simplify our design, we approximate the flexible link system with its first eigenfrequency. As the value of the gain $k_p$ increases the poles move to the left (away from their open-loop location). For a certain value of the proportional gain the roots reach their leftmost placement. Further increase in the gain will drive them towards the zeros of the system, Figure 2.

## III. Transfer Function Identification Method

In this section we introduce an identification scheme based in the frequency domain. Frequency response information of finite duration is used to update the filter parameters. The RLS method is employed to adjust the frequency weights of the filter. This scheme, called Time-varying Transfer Function Estimation (TTFE) reduces the variance of the estimated frequency response derived from ETFE since it assumes that:

> The adjacent frequency bins $H_i(k)$, $H_j(k)$ are correlated through the following relation
>
> $$H_i(k) = \frac{\sum_{j=(i-\Delta_i) \bmod N}^{(i+\Delta_i) \bmod N} \epsilon_j{}^i H_j(k)}{\sum_{j=(i-\Delta_i) \bmod N}^{(i+\Delta_i) \bmod N} \epsilon_j{}^i}$$
>
> indicating that the estimate $H_i(k)$ is related to all the adjacent frequencies within a modulus $\Delta_i$ with a corre-

sponding weight $\epsilon_j^i$ for the frequency point (bin) $\omega_j$.

The frequency bin $H_i(k)$ is related with the $H_i(k-1)$ bin of the hybrid time–frequency domain through a finite impulse response model $Y_i(k) = H_i(k-1)U_i(k-1)$.

Compared with the time-domain techniques, TTFE does not suffer from the severe problem of nonpersistent excitation of the system, since the algorithm can be modified in order not to update the values of the transfer function in the frequency bands where the spectrum of the input $U(e^{j\omega})$ takes on values close to zero.

## IV. The Design of a Fuzzy $P+D$ Control Law

The central idea in the fuzzy–controller design methodology is the creation of the decision rulebase. The inference engine induced by that rulebase is not governed by the conventional (single-valued) logic. Instead, the decision making is based on fuzzy (multi-valued) logic. The necessity of this reasoning is due to a lack of ability in identifying–with certain possibility–the membership of an input-element to the rulebase. In the literature, a large class of controllers, which are based on fuzzy-logic, possesses an expert inference rule. This approach has some disadvantages, among them the fact that the controller is very specialized and depends on the ideal operator, whose performance attempts to imitate in an automated manner. A smaller class of fuzzy controllers is designed similarly to the well known PD and PID schemes [9]. These techniques are often associated with an expertise. This is done either directly, by simply assigning an ideal control action to a pair of error/change-of-error measurements, or indirectly by adjusting the gains of the individual control parts so that certain specifications are satisfied. In all of these cases, the tuning of the controller is based on the input–output system behavior.

The control law suggested in this paper relies on a $P+D$ technique. Having in mind the behavior of the branch of the root–locus, which departs from the dominant conjugate poles as the proportional gain $k_p$ increases, a reasonable control policy could have been to place the closed–loop eigenvalues at the admissible location with the smallest real part. Let us denote this pair of eigenvalues by $(\lambda_d, \bar{\lambda}_d)$. This task can be easily accomplished invoking the method of [10], in which no expertise is required. The advantage of this control policy is the direct modification of the system dynamics to the desirable dynamic behavior. In addition the $P+D$ technique is easily implemented through a cell control strategy [8]. This combined approach, emanating from gain scheduling, assigns to each cell the feedback gains sufficient to locate the closed-loop poles in a neighborhood of the design point.

Consider the system of (1) with the following feedback

$$u = k_p y_1 + k_d \dot{y}_1$$

where $k_p$ and $k_d$ are constant gains that can be evaluated as follows, [11]. The time description of the closed-loop system is

$$\dot{x} = (I - bk_d C_{12})^{-1}(A + bk_p C_{12})x$$

where $C_{12}$ consists of the first two rows of the $C$ matrix and $I$ is the unit matrix. It is required that the system behaves as,

$$\dot{x} = Fx$$

where the eigenvalues of $F$ have negative real parts and are chosen among the admissible locations of the closed-loop poles. In our case the complex eigenvalues of $F$ have the smallest possible real part. Thus

$$F = (I - bk_d C)^{-1}(A + bk_p C)$$

which can be written as,

$$Z = F - A = b[k_p\ k_d]S = bhS;\ S = [C^T\ (CF)^T]^T. \qquad (2)$$

The solution of equation (2) is obtained from:

$$h = (b^T b)^{-1} b^T Z S^T (SS^T)^{-1}.$$

The gains $k_p$ and $k_d$ resulting from the departitioning of $h$ have the following form,

$$k_p \equiv p(\lambda_o);\ k_d \equiv d(\lambda_o)$$

where $\lambda_o$ is the representative of the dominant complex pair of the open-loop roots and $p$, $d$ are real valued functions of a complex variable.

### *A. Fuzzy Partition of the Eigenfrequency Space*

In order to understand the need for a fuzzy partition of the eigenfrequency space the designer should think in a linguistic manner. Since there is uncertainty about the proximity of the estimated eigenfrequency vector to its true value, a linguistic-type description of this distance can be engaged. Suppose that we select several points of the eigenfrequency space for which the feedback gains place the closed loop poles in their specific location. For a dominant pole design, where the eigenfrequency space is one–dimensional, the gains are evaluated on $s_1$ eigenfrequencies, i.e.

$$k_{p,i} = p(\lambda_{1,i});\ k_{d,i} = d(\lambda_{1,i});\ i = 1, \ldots, s_1.$$

If the same gains are inferred for eigenfrequency vectors "away" from these points the closed loop eigenvalues will deviate from their desirable placement. In the sequel, we introduce the fuzzy-set description of the term "away" (or "close"), in order to deal with a quantity whose exact value is unknown. We point out that, the eigenfrequencies do not have to coincide with frequency bins of the TTFE identifying algorithm. Later on, we will show this alternative choice to be an advantage of the proposed method versus a simple gain scheduling based on the quantification of the frequency domain.

The grade of membership of the estimated eigenfrequency vector in a particular fuzzy-set decreases (increases) as the vector moves "away" ("close") from (to) the point of full grade. In this sense we program the controller to weight the gain assigned to the identified frequency with the value of the membership function, associated with the corresponding fuzzy-set. This weighting process continues until all of the fuzzy-sets, which share the particular element, are encountered. We denote by $A_{i,j}$ the $j$-th fuzzy-set of the partition in the $\lambda_i$ eigenfrequency direction. As an example, we consider the fuzzy-reasoning for the membership of the first eigenfrequency, $\hat{\lambda}_1$, identified by the TTFE scheme in a two-dimensional modal approximation. Assume that the eigenfrequency $\hat{\lambda}_1$ is related to $s_1$ fuzzy-sets, namely $A_{1,j}$ ($j = 1, 2, \ldots, s_1$), while the second eigenfrequency $\hat{\lambda}_2$ belongs to the $m$-th fuzzy-set, $A_{2,m}$. The rulebase of the decision on the gain $k_p$ takes the following form,

if $\hat{\lambda}_1$ is $A_{1,j}$ and $\hat{\lambda}_2$ is $A_{2,m}$, then the gain is $k_{p,j,m}$

The next step is to weight the inferred gains with the value of the corresponding membership function. The weighted average of these gains is the defuzzified output of the rulebase.

### *B. Control Defuzzification*

The rulebase, which operates with fuzzy-logic, infers several outcomes for a single event. Therefore, there is a need for a defuzzification interface between the controller and the actuator. As we mentioned before the final output of the inference engine is in a form of a weighted sum over the set of inferred gains. Assuming a partition of $s_1$ fuzzy-sets in the $\lambda_1$ direction and $s_2$ in the $\lambda_2$ direction, the sets $A_{i,j}$ and $A_{i,m}$ form the fuzzy-cell $C_{j,m}$. The usual defuzzification average [9], for a two-dimensional eigenfrequency space, is given as follows

$$u_f = \frac{\sum_{j=1}^{s_1} \sum_{m=1}^{s_2} \tilde{u}_{j,m} w_{j,m}}{\sum_{j=1}^{s_1} \sum_{m=1}^{s_2} w_{j,m}}, \tag{3}$$

where $w_{j,m}$ the weight associated with the fuzzy-cell $C_{j,m}$. The control action, $\tilde{u}_{j,m}$, is in the following feedback form,

$$\tilde{u}_{j,m} = k_{p,j,m} y_1 + k_{d,j,m} \dot{y}_1. \tag{4}$$

Combining the linear equations (3) and (4), we get the inferred control in a compact form,

$$u_f = k_p y_1 + k_d \dot{y}_1$$

where $k_p$ and $k_d$ are expressed as a weighted average

$$k_{p(d)} = \frac{\sum_{j=1}^{s_1} \sum_{m=1}^{s_2} k_{p(d),j,m} w_{j,m}}{\sum_{j=1}^{s_1} \sum_{m=1}^{s_2} w_{j,m}}.$$

### *C. Selection of the Membership Functions*

The choice of the membership functions is important for the adequate behavior of the controller along any path of the fuzzy-partition. The usage of certain types of membership functions is standard in the literature of fuzzy-logic applications. In this section our main concern is the shape of these functions. Linear segments are in most cases a convenient choice. Because of the feedback structure of the controller, discontinuities in the first derivative of the defuzzified gain are likely to cause impulsive responses in the output, similar to the case of a non-fuzzy cell control action. Parabolic segments [12], have several advantages drawn from the fuzzy-set theory. Among their advantages is the smooth behavior of the first derivative. In addition these functions have already been incorporated in applications of feedback control based on fuzzy-logic [9]. In the sequel we denote by $f_{i,j}$ the membership function corresponding to the $j$-th fuzzy-set of the $\lambda_i$ eigenfrequency partition, shown in Figure 3.

### *D. Interface between the TTFE Filter and the Fuzzy-Partition*

In the previous section we saw that the TTFE identification algorithm estimates a point in the eigenspace space, which converges to the true eigenfrequency vector. This is the information that should be incorporated by the controller. For this case of a single point estimation output, the evaluation of the membership function at the specific point is sufficient to grade the membership of the point to the corresponding fuzzy-set. Therefore, the quantity

$$f_{i,j_i} = f_{i,j_i}(\hat{\lambda}_i); \ i = 1, \ldots, n; \ j_i = 1, \ldots, s_i$$

will be used for the selection of a weighting factor to the corresponding gain, $k_{p(d),j_1,\ldots,j_n}$, in the defuzzification process. Recalling the two-dimensional modal approximation, the pair of $A_{1,j}$ and $A_{2,m}$ fuzzy-sets corresponds to the $C_{j,m}$ fuzzy-cell. Using the minimum operator [13], the weighting factor of the gain inferred on the $C_{j,m}$ fuzzy-cell is given as follows,

$$w_{j,m} = \min(f_{1,j}, f_{2,m}); \ j = 1, \ldots, s_1; \ m = 1, \ldots, s_2$$

where $s_1$ ($s_2$) is the number of fuzzy-sets in the direction of $\lambda_1$ ($\lambda_2$) eigenfrequency. Considering our earlier example, the following scheme gives an idea for the scanning of the partitioned space using $\lambda_1$,

if $\lambda_1$ is $A_{1,j}$, then $w_{j,m}$ weights the gain $k_{p(d),j,m}$.

This ruling process can be easily generalized to the case of the $n$-dimensional modal approximation.

## V. Simulation Results

An experimental flexible-link model [6] is chosen to show the performance of the proposed fuzzy controller. This model contains only the first and the second bending modes. The values of the first eigenfrequencies for the unloaded ($u$) and the fully loaded ($l$) cases have as follows,

$$\lambda_{u,1} = 6.85 \; \frac{rad}{sec}; \; \lambda_{l,1} = 3.05 \; \frac{rad}{sec}.$$

The shifting of the open-loop poles due to payload variations is shown in Figure 4. In order to demonstrate the effectiveness of the TTFE identification algorithm we perform a discontinuous change in the system dynamics, by switching the payload. In terms of its first eigenfrequency, the system changes rapidly from 3.62 to 6.28 $\frac{rad}{sec}$. Results from an open-loop identification of (1) are shown in Figure 5. The transient stage of the identification, which has a length of 20 *sec*, is expected to cause a delay in the adaptation of the controller to the new system dynamics. The pseudo-random noise sequence, $\tilde{v}$, which corrupts the output measurements has a level of 0.01. The sampling time is $T = 20$ *msec* and the length of the FFT is $N = 1024$ bins. Therefore, the resolution in the frequency domain is $\frac{2\pi}{NT} = 0.3068 \; \frac{rad}{sec}$.

The desired closed-loop system in the particular application has its dominant complex poles placed on their leftmost admissible placement. The reference input signal, $r(n)$, is a step of unit amplitude (set-point command). To increase the speed of convergence of the identification, we perform the initial FFT of the TTFE algorithm off-line. In this example the control action starts at the beginning of the 20th second.

The fuzzy-rulebase, which is one-dimensional, consists of 11 fuzzy-cells. An ad hoc procedure enabled as to choose appropriate lengths for its fuzzy-components. Augmentation of the fuzzy-partition results in the improvement of the system response. As a trade-off, the number of rules and gains needed to be stored in the data base of the controller should increase respectively. These gains form the data base of the controller.

The closed loop responses are depicted in Figure 6. The system undergoes the same rapid change in its dynamics, as in the open-loop simulation. The response with the adaptive fuzzy controller (solid line) matches closely that of an ideal pole placement controller.

## VI. Conclusion

An adaptive fuzzy approach for the control of a flexible-link robot arm has been presented. The proposed scheme consists of an estimator and a fuzzy controller. The TTFE identification technique is used to provide an estimate of the flexible eigenfrequencies of the system, consistent with the measured data. The fuzzy controller utilizes this information to infer the control action. Fuzzy membership functions are assigned to layers partitioning the universe of discourse which encloses the aforementioned set. A rule based controller infers the output variable, by weighting each gain (from the knowledge base) with the value of the membership functions at the identified eigenfrequency.

Application of the proposed scheme to an experimental flexible-link model is demonstrated and discussed. This technique was proven to be very promising for control design in systems with lightly damped poles. Research is currently undergoing on the development of an adaptive partitioning scheme for the fuzzy rulebase, while the improvement of the identification algorithm is investigated through the introduction of a chirp $z$-transform approach.

## References


[1] C. C. Lee, "Fuzzy Logic in Control Systems: Fuzzy Logic Controller -Part I," *IEEE Transactions on Systems, Man, and Cybernetics*, vol. 20, pp. 404–418, March 1990.

[2] C. C. Lee, "Fuzzy Logic in Control Systems: Fuzzy Logic Controller -Part II," *IEEE Transactions on Systems, Man, and Cybernetics*, vol. 20, pp. 419–435, March 1990.

[3] D. A. Rutherford and G. C. Bloore, "The implementation of fuzzy algorithms for control," *Proceedings of the IEEE*, vol. 64, no. 4, pp. 572–573, 1976.

[4] M. Braae and D. A. Rutherford, "Theoretical and linguistic aspects of the fuzzy logic controller," *Automatica*, vol. 15, pp. 553–577, 1979.

[5] A. Tzes and S. Yurkovich, "A Frequency Domain Identification Scheme for Flexible Structure Control," *Trans. ASME, J. Dyn., Meas., and Control*, vol. 112, pp. 427–434, Sep. 1990.

[6] A. Tzes and S. Yurkovich, "Adaptive Precompensators for Flexible Link Manipulator Control," in *Proceedings of the IEEE Conference on Decision and Control*, (Tampa, FL), pp. 2083–2088, Dec. 1989.

[7] L. Ljung, *System Identification Theory For The User.* Englewood Cliffs, NJ: Prentice–Hall, 1987.

[8] Y. Chen, "Rules extraction for fuzzy control systems," in *IEEE International Conference on System, Man and Cybernetics*, (Cambridge, MA), pp. 526–527, Nov. 1989.

[9] H. Kang and G. Vachtsevanos, "Model reference fuzzy control," in *Proceedings of the IEEE Conference on Decision and Control*, (Tampa, FL), pp. 751–756, Dec. 1989.

[10] P. N. Paraskevopoulos, "On the pole assignment by proportional plus derivative output feedback," *Electronic Letters*, vol. 14, pp. 34–35, 1976.

[11] S. Daley and K. F. Gill, "Comparison of a fuzzy logic controller with a P+D control law," *Transactions of the ASME, Dynamic Sys., Meas. and Control*, vol. 111, pp. 128–137, June 1989.

[12] L. A. Zadeh, "Fuzzy sets as a basis for a theory of possibility," *Fuzzy Sets Systems*, vol. 1, no. 1, pp. 3–28, 1978.

[13] W. Pedrycz, "An approach to the analysis of fuzzy systems," *International Journal of Control*, vol. 34, no. 3, pp. 403–421, 1981.

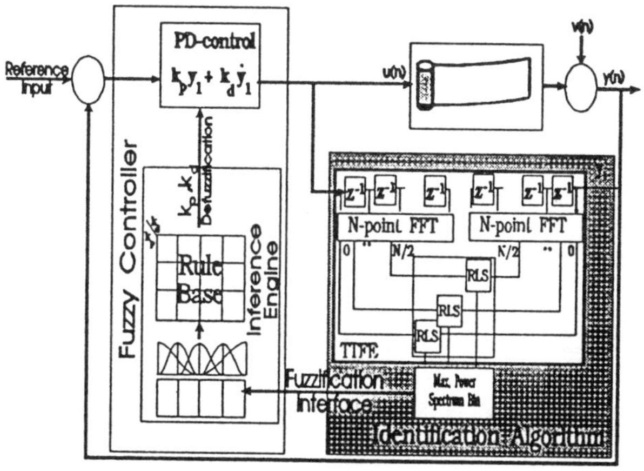


Figure 1: Adaptive Fuzzy P+D Control

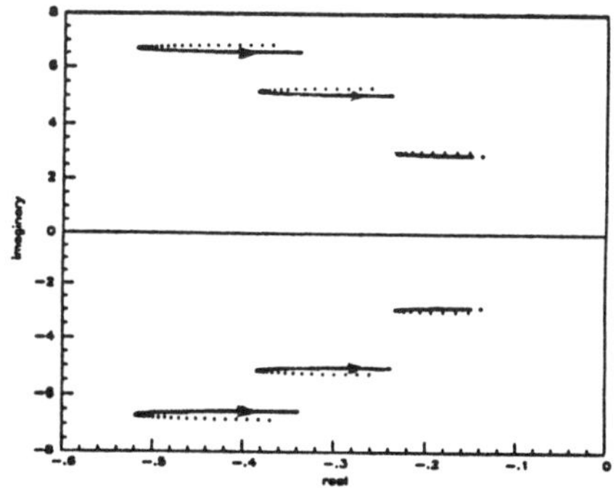


Figure 2: Root Loci associated with the First Eigenfrequency for the Unloaded (left), Half-loaded (center) and Fully-loaded (right) Operating Modes

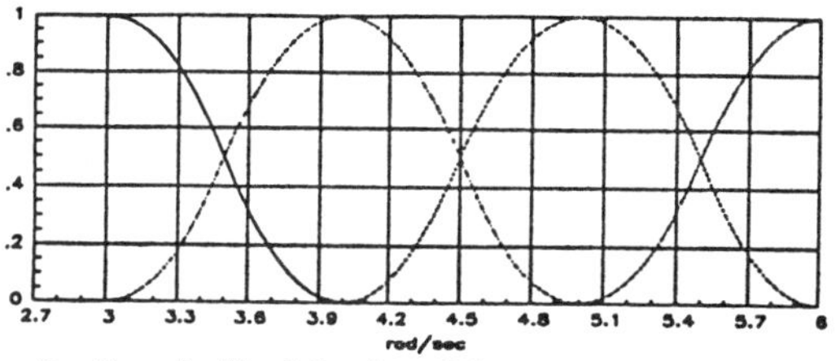


Figure 3: Parabolic Membership Functions for a sample Fuzzy Partition

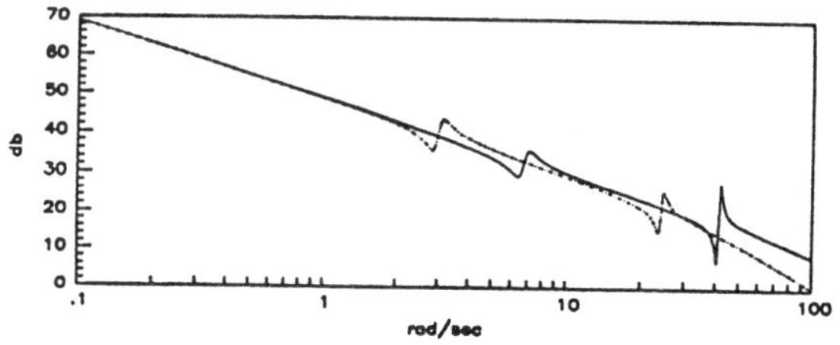


Figure 4: Bode Plots for the Unloaded (solid line) and the Fully-loaded Operating Modes

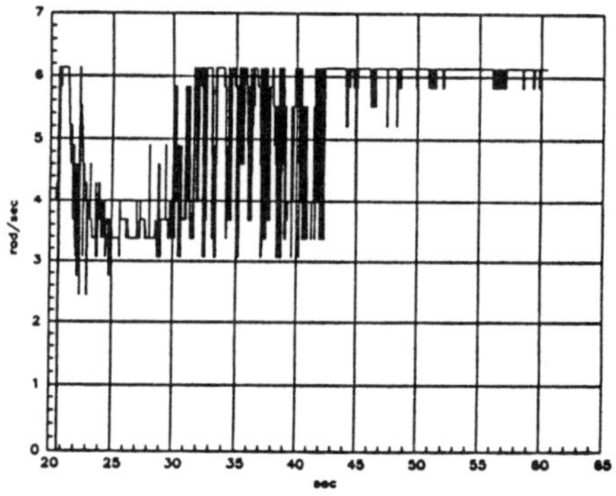


Figure 5: Open Loop TTFE Identification

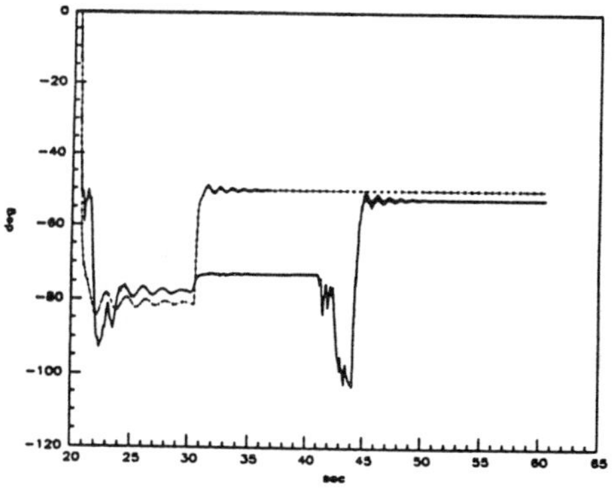


Figure 6: System Response (Shaft Position) of the Presented Method versus the Exact Pole Placement

# *Fuzzy Logic Based Robotic Arm Control*

Robert N. Lea, Ph.D.
NASA/ Johnson Space Center
Houston, Texas 77058

Jeffrey Hoblit
LinCom Corporation
Houston, Texas 77058

Yashvant Jani, Ph.D.
Togai InfraLogic Inc.
Houston, Texas, 77058

***Abstract*** **- Fuzzy logic and neural networks have emerged as new methods for decision making and control within the last decade. Non-linear complex systems can be easily controlled utilizing fuzzy logic principles. The dynamical behavior of robotic systems is complex, especially in presence of loads. Robotic joints experience and exhibit friction, stiction and gear backlash effects. Due to lack of proper linearization of these effects, modern control theory based on state space methods can not provide adequate control for robotic systems. Furthermore, inversion of Jacobian matrices, particularly when they are near singular, provide another computational problem especially for redundant joint robotic systems.**

**At the Johnson Space Center (JSC), we are investigating the feasibility of applying fuzzy logic based control for robotic systems. Functions such as tracking, approach and grapple are typically performed by a robotic arm in a manual mode or semiautomatic mode where a point of resolution is driven to a desired point by kinematic inversion. We are developing fuzzy logic based algorithms for semiautomatic mode so that the computational problem of Jacobian inversion can be eliminated. The difference between the desired location and current location is an input vector to the controller that generates joint rate commands. A six degree-of-freedom robotic arm simulation is used to evaluate the performance of the fuzzy logic based controller. In this paper, development of the control algorithm, simulation testing, results and the performance of the controller are reported.**

## I. INTRODUCTION

Industrial applications of fuzzy logic and neural networks have increased significantly within the last five years in the U.S. as evidenced by the Industrial Conferences [1, 2] as well as the recent FUZZ-IEEE '92 held at San Diego by the Institute of Electrical and Electronics Engineers [3]. Many of the applications are related to the control of non-linear and complex systems. Of particular interest are the control applications for a robotic arm. The control of planner arms [4,5] with redundant degrees of freedom has been demonstrated in simulation as well as hardware implementations. The control rules used in maintaining the trajectory are simple and do not require the inversion of a Jacobian matrix as in the traditional approach [6]. Advantages in computational speed and accuracy are realized.

Research activities in the Software Technology Laboratory at the JSC include the development of robust algorithms for decision making using fuzzy logic, neural networks, genetic algorithms, Dempster-Shafer theory and fuzzy clustering. We have demonstrated a robust control of the space shuttle relative trajectory during proximity operations [7], tether length control during deployment, on-station and retrieval phases [8], and control of Mars Rover trajectories and collision avoidance [9]. Fuzzy control algorithms utilized in these applications are primarily based on reducing the error and error rate which are related through the systems dynamics and are perceived via proper sensor measurements. This technique of reducing error and error rate was first applied to the attitude control of the space shuttle [10] where angle error and its rate error were reduced to within an acceptable range. Later the same technique was applied to reducing length and length rate errors for the tethered satellite system. Application of the same algorithm to the space shuttle robotic arm is presented here. We first describe the configuration of the Remote Manipulator System (RMS) of the shuttle in section 2 and then the development of the control algorithm for the point of resolution (POR) mode in section 3. The kinematic simulation test cases are then described in section 4 and test results and preliminary conclusions are discussed in section 5. Finally our future work in this area is summarized in section 6. References are provided at the end.

## II. ROBOTIC ARM CONFIGURATION

The shuttle RMS is a six rotational joint arm as shown conceptually in fig. 1. The first joint is known as a yaw joint as it rotates the arm around the z-axis of the base frame. The next two joints are called pitch joints because the rotation is through the y-axis of the base frame when the arm is in its rest position. The next three joints are pitch, yaw and roll and correspond to the rotational sequence used in defining the shuttle orientation in on-orbit operations.

The first joint is about 13 inches from the base. The lengths of the six segments between the joints are approximately 23, 251, 278, 18, 22, and 13 inches, where the last length corresponds to the distance between the joint and POR. The state of the robotic arm can be specified in two different ways: 1) six joint angles will accurately describe the orientation of the arm, or 2) position and orientation of the POR in the base frame. The RMS base frame is related to the shuttle vehicle frame through a fixed coordinate transformation matrix. The relationship between the POR frame and the base frame is determined through the six

0-7803-0614-7/93$03.00 

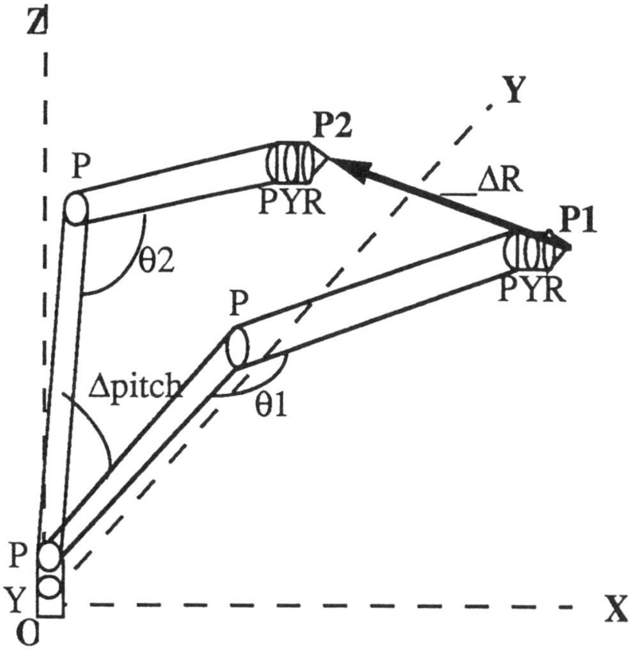


Fig. 1. The Shuttle RMS configuration and motion from point P1 to point P2.

rotations of the joints as well as the distance between the joints. Each joint angle providing the rotation is described in the coordinate frame at the joint. A kinematic forward transformation can be performed on the first state to obtain the position and orientation of the POR in the base frame. The position and orientation of the POR are related to joint angles via a Jacobian as follows.

$$[\,x, \theta\,]^T = J(\gamma)\,[\,\gamma\,]^T \tag{1}$$

where, x = position of the POR in base frame, $\theta$ = orientation of the POR, $\gamma$ = joint angles, and $J(\gamma)$ = Jacobian which is dependent on the joint angles only. Typically, sensors measure the joint angles and rates to provide proper feedback to the controller. Once the measurements are received, the motion is easily specified in terms of position and orientation of the POR in the base frame using forward kinematics. A camera mounted at the end of the arm can provide some inputs for the POR position in the base frame.

The objective of this study is to specify the desired position and orientation of the POR in the base frame, and let the controller manipulate the joint angles and rates to achieve that state. A path planner can easily work in the base frame, while the controller can work on the joint angle and rate. For a robotic arm, the Cartesian velocities of the POR are related to the joint velocities via the Jacobian matrix which is a non-linear trigonometric function of joint angles and lengths of arm segments.

$$[\,v, \omega\,]^T = J(\gamma, \gamma\text{-dot})\,[\,\gamma\text{-dot}\,]^T \tag{2}$$

where, $\gamma$, $\gamma$-dot = joint angles and rates, v = POR velocity, $\omega$ = angular velocity and J ($\gamma$, $\gamma$-dot ) is approximately equal to the Jacobian J ($\gamma$) when $\gamma$-dot is very small. Note that this is an instantaneous relationship and the Jacobian changes as the angle changes, even if the rates remain constant. If the joint rates are not very small, then, the Jacobian becomes a much more complex function of rates also.

## III. DEVELOPMENT OF CONTROL ALGORITHMS

A typical control strategy is to difference the actual state vector from the desired state vector and use the deltas as the desired velocity and angular velocity in equation (2). Desired (or commanded) joint rates can then be derived by inverting the Jacobian if it is non-singular. Otherwise the inversion problems will prohibit the solution and control can not be achieved.

Our strategy is to use the fuzzy logic based attitude controller and avoid the Jacobian inversion. We noticed that if the deltas are transformed from the Cartesian coordinate frame [Δx, Δy, Δz] to spherical coordinate frame [ΔR, Δelev, Δazim], equivalently, range error, elevation error and azimuth error, then, each delta can be one to one related to each joint angle. For example, the azimuth error can be related to the first RMS joint which is a yaw joint. Thus, by moving the first joint, any azimuth error Δazim can be corrected. As the first joint moves, the rate of azimuth correction can be computed for the controller's use. Now, there are two parameters, azimuth error and its rate error that can be used in the fuzzy logic based attitude controller. The output of the controller is interpreted as the commanded joint rate instead of a jet firing command.

Similarly the elevation and range errors can be corrected by moving the second and third joint. The last three joints relate to the POR orientation angles directly because we use the same Shuttle rotational sequence to define POR orientation. In general, this in not true for any robotic arm, and one must clearly identify a correct correspondence. Thus, all six errors [ Δazim, Δelev, ΔR, Δpitch, Δyaw, Δroll ] can be corrected by moving the six joints in order. As soon as these errors are nulled, the state vector is at the desired value. The controller outputs the joint rate commands and the servo motors provide these commanded rates for each corresponding joint. Algorithmic steps as described below are simple : 1) compute deltas, 2) transform them into the spherical frame, and 3) apply the attitude controller to correct each error.

Inputs to the fuzzy guidance-controller are current and desired state vectors whose first three components are the position x, y, and z, and last three components are roll, pitch and yaw angles. The current state vector is derived by applying forward kinematics (equation. 1) to the current joint angles. The next step is to compute the deltas between the desired state and the current state by subtracting the two. These deltas [DEL[1] through DEL[6]] are then transformed into delta range, delta elevation, delta azimuth and delta angles as follows.

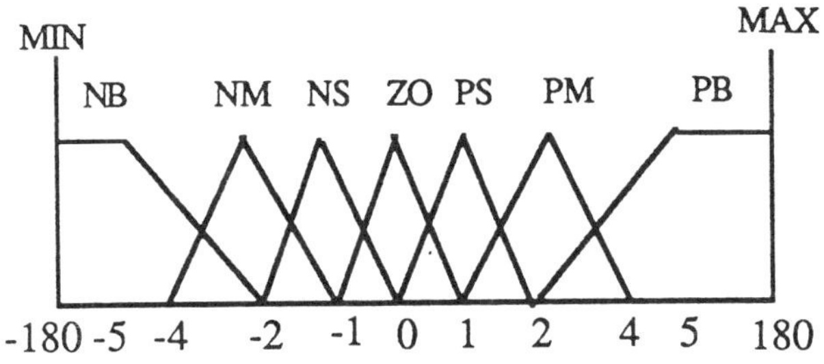


Fig. 2a - Fuzzy membership functions for input angle and angle rate.

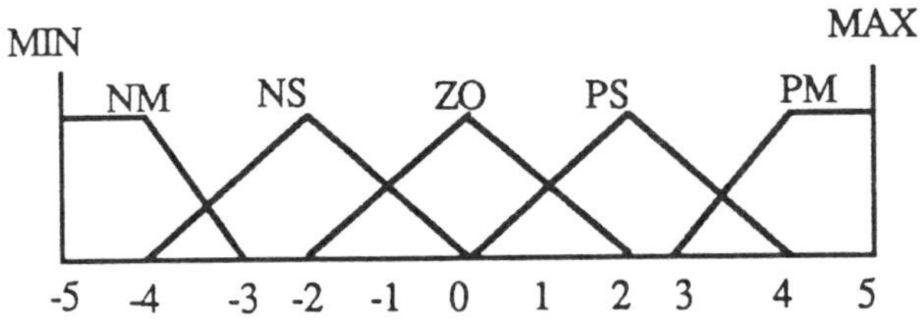


Fig. 2b - Fuzzy membership functions for output commanded joint rates

ΔR = sqrt (DEL[1]*DEL[1] + DEL[2]*DEL[2] + DEL[3]*DEL[3] );
Δelev = ATAN2 (DEL[3],sqrt(DEL[2]*DEL[2] + DEL[1]*DEL[1]);
Δazim = ATAN2 (DEL[1],DEL[2]);
Δpitch = DEL[5];
Δyaw = DEL[6];
Δroll = DEL[4];

Also the rates of these parameters are computed by using previous values.

ΔΔR = ΔR - ΔR_previous ;
ΔΔelev = Δelev - Δelev_previous ;
ΔΔazim = Δazim - Δazim_previous ;
ΔΔpitch = Δpitch - Δpitch_previous ;
ΔΔyaw = Δyaw - Δyaw_previous ;
ΔΔroll = Δroll - Δroll_previous ;

An attitude controller implementation [10] based on the phase plane is used six times to generate the commanded joint rates for the robotic arm. The membership functions for angle and angle rate are shown in fig. 2a and the membership functions for commanded (desired) joint rates are shown in fig. 2b. Derivation of the membership functions is based on the deadband values used in the phase plane, and the commanded values required to null these errors. Originally, these membership functions were derived for the Shuttle attitude control and no change has been made. The rulebase used to compute the commanded joint rate is shown in Table I.

This attitude controller has been tested for the shuttle operations in a very high fidelity simulation and has shown excellent results in comparison with the existing conventional controller [10]. This controller has also been integrated with the translational controller for relative trajectory and attitude control and has shown robust control as well as fuel savings [7]. A tether length controller has been derived based on this concept using length error and length rate error and has been tested in a high fidelity simulation with bead model to show advantages [8]. For the robotic arm control, the commanded joint rates are achieved by the servo motors to move the POR to the desired position and orientation.

For joint_1, Phase_plane (Δazim, ΔΔazim, Gamma_1);
For joint_2, Phase_plane (Δelev, ΔΔelev, Gamma_2);
For joint_3, Phase_plane (ΔR, ΔΔR, Gamma_3);
For joint_4, Phase_plane (Δpitch, ΔΔpitch, Gamma_4);
For joint_5, Phase_plane (Δyaw, ΔΔyaw, Gamma_5);
For joint_6, Phase_plane (Δroll, ΔΔroll, Gamma_6);

## IV. SIMULATION AND TESTING

The RMS forward kinematics implementation based on equation (1) in the orbital operations simulation is used to test our approach. The simulation is initialized with an initial arm position and orientation and the POR is commanded to achieve a desired position and orientation. Based on the starting and final POR states, the algorithm generates the rate commands for each joint. Perfect servo motors with first order lag are used to achieve the commanded rate for each joint. The cycle time is 80 ms which is consistent with the shuttle cycle time. The simulation flow is shown in fig. 3 with our algorithm in the loop. Four tests are performed to verify the operation of our controller. Initial and final states of all four test cases are given in Table II. In all test cases, the new POR

TABLE I
Fuzzy Rulebase For The Robotic Joint Rate Control

| Rate Error \ Angle Error | NB | NM | NS | ZO | PS | PM | PB |
|---|---|---|---|---|---|---|---|
| NB | PM | PM | PS | PM | | | |
| NM | PM | PM | PS | PM | | | |
| NS | PS | PS | PS | PS | | | |
| ZO | PS | PS | PS | ZO | NS | NS | NS |
| PS | | | | NS | NS | NS | NS |
| PM | | | | NM | NS | NM | NM |
| PB | | | | NM | NS | NM | NM |

KEY:
NB - Negative Big
NM - Negative Medium
NS - Negative Small
ZO - Zero
PS - Positive Small
PM - Positive Medium
PB - Positive Big

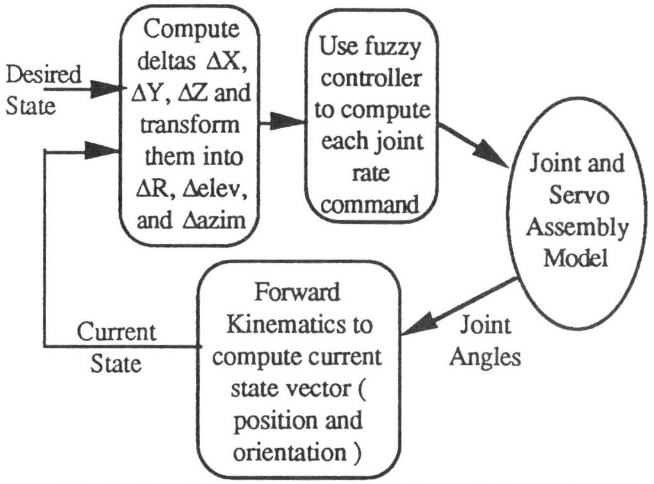


Fig. 3. Simulation flow with RMS forward kinematics and fuzzy controller

state was commanded 10 seconds into the run. Total run time of the simulation is 300 seconds.

The first test is a simple POR translation. The arm end point is commanded to go from its current position and attitude to a new position with no change in attitude. Motion was commanded on all three axes, with a large displacement on the X axis. The second test case commanded motion on the X axis as well as an attitude change. For the third test case, the arm is commanded to perform a three axes translation and three axes rotation. The fourth test case is a combination of maneuvers, simulating the "handling" of an object. The arm begins straight out and moves with a combined translation and rotation to line up on a location. It then moves straight in on the Z axis to grab the object, backs straight up on Z to lift the object, and then proceeds to a new location to drop the object. Thus, it has four desired points to reach in a given delta time.

## V. RESULTS

In all cases the hybrid controller successfully achieved and held the commanded POR states. Approximately 60 seconds of run time was used in the maneuvering. The remaining run time demonstrated the ability of the control system to hold the desired position and orientation. Plots showing the POR translations and rotations as well as the joint angles and rates were generated for analysis. Observations from these plots show a few interesting characteristics of the controller.

First, the commanded POR states (x, y, z, pitch, yaw, roll) are not all reached at precisely the same time, and the path followed by each joint is not precisely smooth. The implication is that the POR did not traverse a straight line in the work space when going from its initial to commanded state. As the controller is currently implemented, each joint takes out its error as quickly as possible. Due to the configuration of the arm, some joints are able to eliminate error faster than other joints resulting in the arm tip moving "up and over" instead of straight to the desired position.

Second, there is some over shoot of the desired position along the Z axis in the first test case. The control of the arm is actually done in a spherical coordinate system. The work space, and the plots, are assumed to be represented by a Cartesian system. These different representations cause this over shoot. The control system is not controlling the shoulder pitch joint to drive the arm to a specific Z position, but to a specific elevation angle from the base of the arm to the arm tip. When the length (radial) of the arm is correct, as controlled by the elbow pitch joint, then the correct elevation angle will directly correspond to the desired Cartesian Z position. However, if the length of the arm is too large then the correct elevation angle will correspond to a larger than desired Cartesian Z position. So, when the over all length of the arm is commanded to decrease, there is a risk of over shooting the desired Z position when the elevation angle control (shoulder pitch joint) obtains its desired angle before the arm length control (elbow pitch joint) obtains its desired angle.

Finally, some error may occur in end tip orientation during large maneuvers, even when no orientation movement is commanded. The pitch plot of the first test case in fig. 4 shows this clearly around 15 seconds into the run. This is

Table II
Initial and final states of the end position (POR) for the test cases for RMS motion.

| | Starting POR | | | | | | Final (Cmd) POR | | | | | |
|---|---|---|---|---|---|---|---|---|---|---|---|---|
| TC # | X (Ft) | Y (Ft) | Z (Ft) | Pitch (Deg) | Yaw (Deg) | Roll (Deg) | X (Ft) | Y (Ft) | Z (Ft) | Pitch (Deg) | Yaw (Deg) | Roll (Deg) |
| 1 | 50 | 0 | 1.2 | 0 | 0 | 0 | 20 | 5 | 5 | 0 | 0 | 0 |
| 2 | 50 | 0 | 1.2 | 0 | 0 | 0 | 30 | 0 | 0 | 30 | 40 | 50 |
| 3 | 50 | 0 | 1.2 | 0 | 0 | 0 | 30 | 10 | 10 | 30 | 40 | 50 |
| 4 | 50 | 0 | 1.2 | 0 | 0 | 0 | 30 | 0 | -20 | 90 | 0 | 0 |
| | | | | | | | 30 | 0 | -23 | 90 | 0 | 0 |
| | | | | | | | 30 | 0 | -20 | 90 | 0 | 0 |
| | | | | | | | 30 | 0 | 0 | 0 | 0 | 0 |

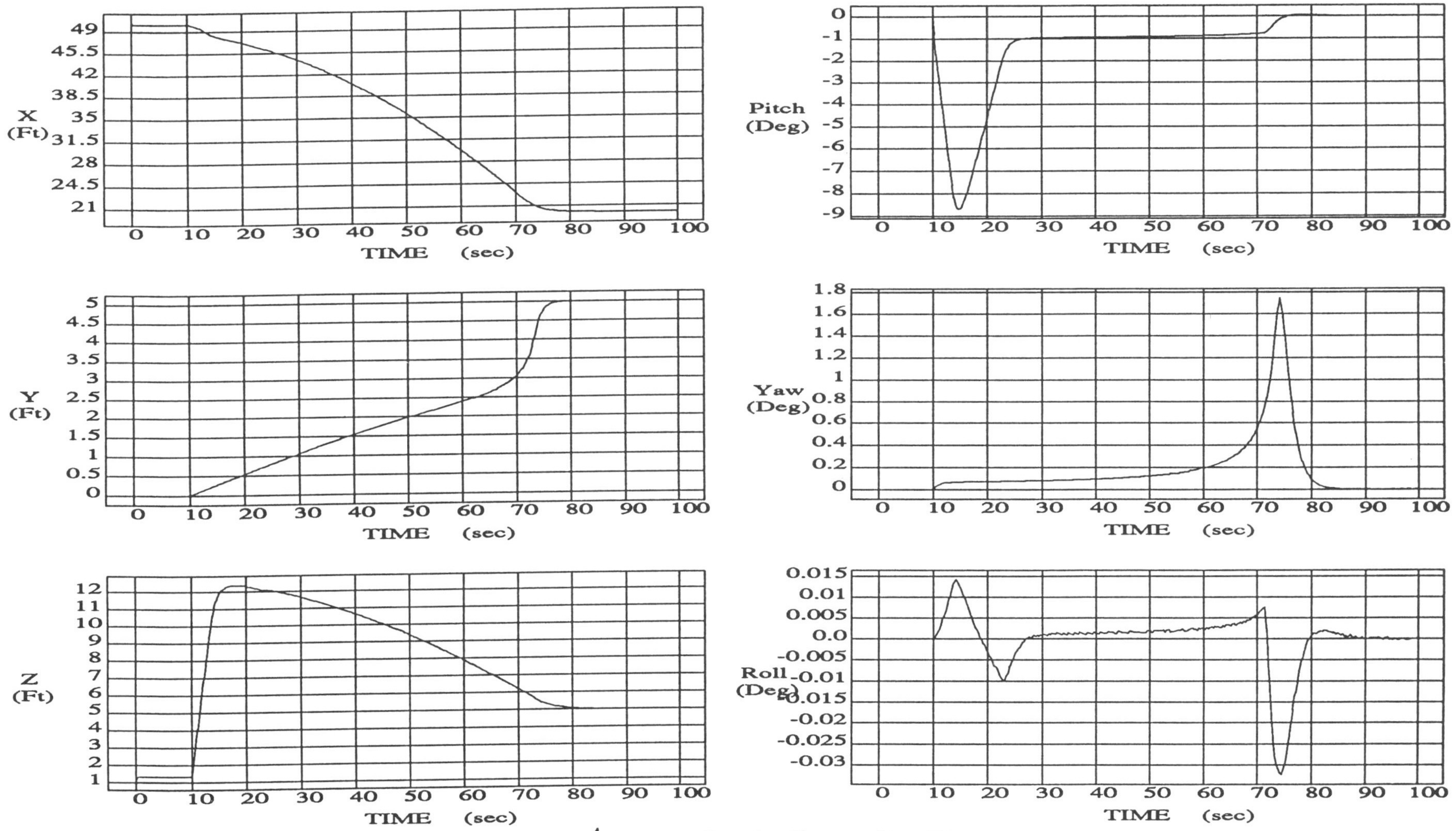


Fig. 4 Fuzzy Logic Controller Test #1
Translation And Rotation Of POR

caused by a brief period of time when both the shoulder pitch and the elbow pitch joints are driving negative. The controller commands are limited to 2 deg/sec in magnitude. Since two joints are pitching down at almost 2 deg/sec the combined change in pitch orientation is almost 4 deg/sec in the downward direction. With the controller limited to 2 deg/sec in command authority it is not possible for the wrist pitch joint to keep up with the accumulating error.

All of the behavior of the controller commented on above is caused to some degree by each joint controlling its own portion of the system independent of the other joints. Although the over all goal is clearly reached and held in each test case the exact path taken to the goal is not under complete control. If it is desired to exactly control the path taken to the desired POR this could be accomplished in several ways. For example, intermediate points along the desired path could be specified or the rate at which each joint moves towards its specified goal could be controlled by some set of overseeing rules. In either event the benefits of rule based control are still maintained.

## VI. SUMMARY

It has been shown that there is a generic fuzzy algorithm that can be applied to the control of manipulator joints for maintaining a POR trajectory. Future work in this project will involve additional testing for many different cases, and tuning of rulebase and membership functions if necessary. As noted earlier, the motion of the joints needs to be correlated, especially when it is known that a particular joint has a large effect on a certain error and an adverse effect on another error. To minimize these adverse effects, the commanded joint rates should be properly related to each other. The relations among joint rates and their ratios need to be derived based on simulation results and hardware information. The fuzzy membership functions can be tuned for each joint to achieve rate matching. However, developing additional rules may offer advantages especially in light of collision avoidance and redundant degrees of freedom.

After thorough testing of the algorithm using a six degree of freedom arm, it will be extended to handle more than six degrees of freedom. Handling redundant degrees of freedom is very important, especially in light of obstacle avoidance and path planning. The rulebase for collision avoidance will also be developed, particularly for the objects within the work space which need to be avoided by each joint.

## REFERENCES

[1] Proceeding of the Industrial Conference on Fuzzy Systems ICFS-91, sponsored by The Microelectronics and Computer Technology Corporation, Austin, Texas, June 27-28, 1991.

[2] Proceeding of the 1st International Workshop on Industrial Applications of Fuzzy Control and Intelligent Systems, sponsored by Center for Fuzzy Logic and Intelligent Systems Research, Texas A&M University, College Station, Texas, Nov. 21-22, 1991.

[3] Proceeding of the IEEE International Conference on Fuzzy Systems 1992, (FUZZ-IEEE 92), held at San Diego, California, March 8-12, 1992.

[4] A. Nedungadi : "A Fuzzy Robot Controller - Hardware Implementation", Proceedings of FUZZ-IEEE 92 held at San Diego, California, pp. 1325, March 8-12, 1992.

[5] G.V.S. Raju, & J. Zhou : "Fuzzy Rule Based Approach for Robot Motion Control", Proceedings of FUZZ-IEEE 92 held at San Diego, California, pp. 1349, March 8-12, 1992.

[6] J.J. Craig : Introduction to Robotics Mechanics & Control, Addison-Wesley Publishing Company, 1985. (Chapters 5 and 6)

[7] R.N. Lea, J. Hoblit and Y. Jani : "A Fuzzy Logic Based Spacecraft Controller for Six Degree of Freedom Control and Performance Results", Proceedings of AIAA Guidance, Navigation and Control Conference, New Orleans, August 12-14, 1991.

[8] R.N. Lea, J. Villarreal, Y. Jani, & C. Copeland : "Tether Operations Using Fuzzy Logic Based Length Control", Proceedings of FUZZ-IEEE 92 held at San Diego, California, pp. 1335, 1992.

[9] R.N. Lea, Y. Jani, M.G. Murphy & M. Togai : "Design and Performance of A Fuzzy Logic Based Vehicle Controller for Autonomous Collision avoidance", Proceedings of Fuzzy Neural Systems : Applications to Vehicle Control, Tokyo, Japan, November 1991.

[10] R.N. Lea, J. Hoblit & Y. Jani : "Performance Comparison of A Fuzzy Logic Based Attitude Controller with the Shuttle On-orbit Digital Auto Pilot", Proceedings of North American Fuzzy Information Processing Society (NAFIPS - '91) Workshop, Columbia, May 14-17, 1991.

# Blending Reactivity and Goal-Directedness in a Fuzzy Controller

**Alessandro Saffiotti*** **Enrique H. Ruspini** **Kurt Konolige**
Artificial Intelligence Center, SRI International
Menlo Park, CA 94025, U.S.A.
saffiotti@ai.sri.com

***Abstract*— Controlling the movement of an autonomous mobile robot requires the ability to pursue strategic goals in a highly reactive way. We describe a fuzzy controller for such a mobile robot that can take abstract goals into consideration. Through the use of fuzzy logic, reactive behavior (e.g., avoiding obstacles on the way) and goal-oriented behavior (e.g., trying to reach a given location) are smoothly blended into one sequence of control actions. The fuzzy controller has been implemented on the SRI robot Flakey.**

## I. Introduction

Autonomous operation of a mobile robot in a real environment poses a series of problems. In the general case, knowledge of the environment is partial and approximate; sensing is noisy; the dynamics of the environment can only be partially predicted; and robot's hardware execution is not completely reliable. Though, the robot must take decisions and execute actions at the time-scale of the environment. Classical planning approaches have been criticized for not being able to adequately cope with this situation, and a number of reactive approaches to robot control have been proposed (e.g., [Firby, 1987; Kaelbling, 1987; Gat, 1991]), including the use of fuzzy control techniques (e.g., [Sugeno and Nishida, 1985; Yen and Pfluger, 1992]). Reactivity provides immediate response to unpredicted environmental situations by giving up the idea of reasoning about future consequences of actions. Reasoning about future consequences (sometimes called "strategic planning"), however, is still needed in order to intelligently solve complex tasks (e.g., by deciding not to carry an oil lantern downstairs to look for a gas leak [Firby, 1987].)

One solution to the dual need for strategic planning and reactivity is to adopt a two-level model: at the upper level, a planner decides a sequence of abstract goals to be achieved, based on the available knowledge; at the lower level, a reactive controller achieves these goals while dealing with the environmental contingencies. This solution requires that the reactive controller be able to simultaneously satisfy strategic goals coming from the planner (e.g., going to the end of the corridor), and low-level "innate" goals (e.g., avoiding obstacles on the way). A major problem in the design of such a controller is how to resolve conflicts between simultaneous goals.

In this paper, we describe a reactive controller for an autonomous mobile robot that uses fuzzy logic for trading off conflicting goals. This controller has been implemented on the SRI robot Flakey, and its performance demonstrated at the first AAAI robot competition, where Flakey finished second [Congdon *et al.*, 1993]. The formal bases for the proposed controller have been set forth by Ruspini [Ruspini, 1990; Ruspini and Ruspini, 1991; Ruspini, 1991a] after the seminal works by Zadeh (e.g., [Zadeh, 1978]). In a nutshell, each goal is associated with a function that maps each perceived situation to a measure of desirability of possible actions from the point of view of that goal. The notion of a *control structure* is used for introducing high-level goals into the fuzzy controller. Intuitively, a control structure is an object in the robot's workspace, together with a desirability relation: typical control structures are locations to reach, walls to follow, doors to enter, and so on. Each desirability function induces a particular behavior — one obtained by executing the actions with higher desirability. Behaviors induced by many simultaneous goals can be smoothly blended by using the mechanisms of fuzzy logic. In particular, reactive and goal-oriented behaviors are blended in this way into one sequence of control actions.

The next section gives a brief overview of Flakey. Section III sketches the architecture of the controller, and describes the way behaviors are implemented, and how they are blended together. Section IV deals with the introduction of high-level goals into the reactive controller. Section V discusses the results, and concludes.

*On leave from Iridia, Université Libre de Bruxelles, Brussels, Belgium.

0-7803-0614-7/93$03.00 ©1993IEEE

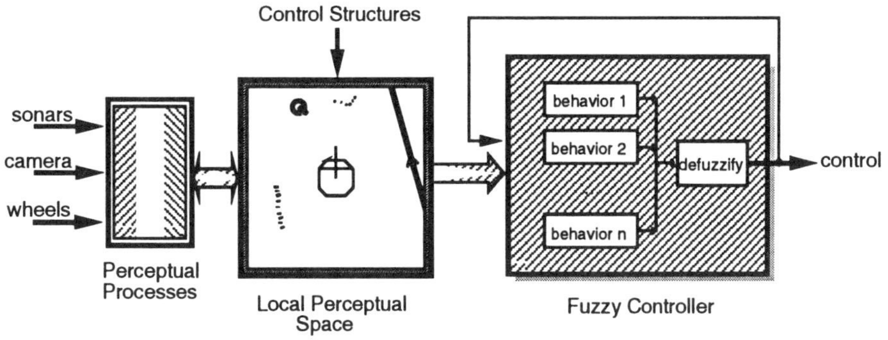


Figure 1: Architecture of Flakey (partial)

## II. The mobile robot test-bed

Flakey is a custom-built mobile robot platform approximately 1 meter high and .6 meter in diameter for use in an indoor environment. There are two independently-driven wheels, one on each side, giving a maximum linear velocity of about .5 meters/sec. Flakey sensors include a ring of 12 sonars, giving information about distances of objects up to about 2 meters; wheel encoders, providing information about current linear and rotational velocity; and a video camera, currently used in combination with a laser to provide dense depth information over a small area in front of Flakey. On-board computers are dedicated to low-level sensor interpretation, motor control, and radio communication with an off-board Sparc station. Though it is possible to run the high level interpretation and control processes on board, they are normally run remotely for programming convenience.

Figure 1 illustrates the part of Flakey's architecture that is relevant to the controller. The sensorial input is processed by a number of interpretation processes at different levels of abstraction and complexity, and the results of interpretation are stored in the *local perceptual space* (LPS). The LPS represents a Cartesian plane centered on Flakey where all the information about the vicinity of Flakey is registered. In Figure 1, points corresponding to surfaces identified by the sonars and the camera are visible in the LPS — Flakey is the the octagon in the middle of the LPS, in top-view. The other objects in the LPS are "artifacts" associated to *control structures*, and are discussed in Section V. The content of the LPS constitutes the input to the controller: this checks its input and generates a control action every 100 milliseconds.

## III. Reactive fuzzy controller

The fuzzy controller is centered on the notion of *behavior*. Intuitively, a behavior is one particular control regime that focuses on achieving one specific, predetermined goal (e.g., avoiding obstacles). Hence, we can think of a behavior as a mapping from configurations in the LPS to actions to perform. More precisely, and following [Ruspini, 1991b], we say that each behavior $B$ is associated with a desirability function

$$Des_B : \text{LPS} \times \text{Control} \rightarrow [0,1]$$

that measures, for each configuration $s$ of the LPS and value $c$ of a control variable, the desirability $Des_B(s,c)$ of applying control values $c$ in the situation $s$ *from the point of view of $B$*. Equivalently, we can say that $Des_B$ associates each situation $s$ with the fuzzy set $\widetilde{C}$ of control values characterized by the membership function $\mu_{\widetilde{C}}(c) = Des_B(s,c)$. Notice that in general, $c$ is a n-dimensional vector of values for all the control variables; in the case of Flakey, the control variables include linear acceleration and turning angle.

In practice, each behavior is implemented by a fuzzy machine structured as shown in Figure 2. The *fuzzy state* is a vector of fuzzy variables (each having a value in $[0,1]$) representing the truth values of a set of fuzzy propositions of interest (e.g., "obstacle-close-on-left"). At every

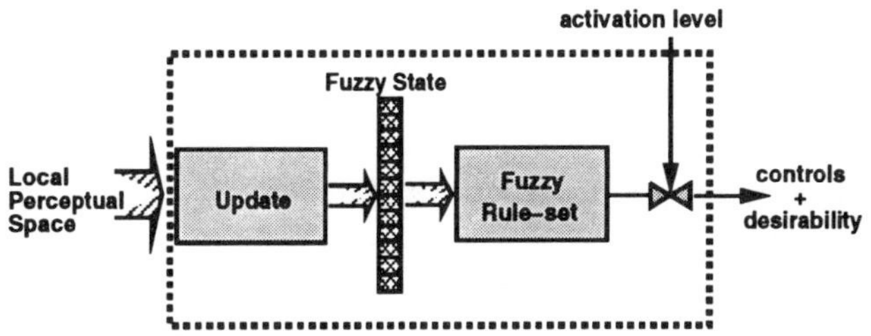


Figure 2: Implementation of a behavior.

cycle, the **Update** module look at the (partially) interpreted perceptual input stored in LPS, and produces a new fuzzy state. The **Fuzzy Rule-Set** module contains a set of fuzzy rules of the form "If $A$ then $c$" where $A$ is a fuzzy expression composed by predicates in the fuzzy states plus the fuzzy connectives AND, OR and NOT; and $c$ is a vector of values for the control variables. Max, min, and complement to 1 are used to compute the truth value of disjunction, conjunction and negation, respectively. An example of a control rule is:

```
IF obstacle-close-in-front
AND NOT obstacle-close-on-left
THEN turn -6 degrees
```

Each "If $A$ then $c$" rule computes the degree of desirability of applying control value $c$ as a function of the degree at which the current state happens to be similar to $A$. The outputs of all the rules in a rule-set are unioned using the max T-conorm: the function computed in this way is meant to provide an approximation of the $Des_B$ function above.[1] This desirability function is fed to the **Defuzzify** module for computing one single control value. We presently do defuzzification according to the centroid approach: the resulting control value is given by

$$\frac{\int c\, Des_B(c)\, dc}{\int Des_B(c)\, dc}.$$

As shown in Figure 1 above, many behaviors can be simultaneously active in the controller, each aiming at one particular goal — e.g., one for avoiding obstacles; one for keeping a constant speed; one for heading toward a beacon; etc. Correspondingly, many instances of the fuzzy machine depicted in Figure 2 simultaneously run in the controller, each one implementing one behavior's desirability function. All these desirability functions are merged into a composite one by the max T-norm; the defuzzification module converts the resulting tradeoff desirabilities into one crisp control decision. Care must be taken, however, of possible conflicts among behaviors aiming at different, incompatible goals. These conflicts would result in desirability functions that assign high values to opposite actions: simple T-norm composition should not be applied in these cases. The key observation here is that each behavior has in general its own *context* of applicability. Correspondingly, we would like that the impact of the control actions suggested by each behavior be weighted according to that behavior's degree of applicability to the current situation. For instance, the actions proposed by the obstacle avoidance behavior should receive higher priority when there is a danger of collision, at the expense of the other, concurrent behaviors. In order to do this, the output of each rule-set is discarded by the value of the corresponding *activation level*: typically, the activation level is represented by some variable in the fuzzy state. This corresponds to arbitrate the relative dominance of different behaviors by a set of meta-rules of the form

$$\text{IF } A' \text{ THEN activate\_behavior } B \qquad (1)$$

where $A'$ is a LPS configuration. Notice that this solution is formally equivalent to transforming each rule "If $A$ then $c$" in $B$ into a rule "If $A'$ and $A$ then $c$" (see [Berenji *et al.*, 1990] for a similar approach to conflict resolution.)

As an example, consider the way Flakey "wanders" around. In the wandering mode, three behaviors coexist in the controller: AVOID-OBSTACLES, AVOID-COLLISIONS and GO-FORWARD. GO-FORWARD just keeps Flakey going at a fixed velocity, given as a parameter. AVOID-OBSTACLES looks at the last 5 seconds' sonar readings in the LPS, and guides Flakey away form occupied areas. AVOID-COLLISIONS looks at the nearest sonar readings and proposes drastic actions (immediate stop and turn) when a serious risk of collision is detected. The activation levels of AVOID-OBSTACLES and AVOID-COLLISIONS are given by the fuzzy state variable "approaching-obstacle"; the complement of this value gives the activation level for GO-FORWARD. The visual result for an external observer is that Flakey "follows its nose", while smoothly turning away from obstacles as it approaches them.

## IV. BEYOND PURE REACTIVITY

The behaviors discussed in the previous section are purely reactive: at each cycle, Flakey selects an action solely on the basis of the current state of the world as perceived by its sensors and represented in the local perceptual space. Engaging into more purposeful activities than just wandering around requires more than pure reactivity: we need to take explicit goals into consideration. For example, we may want Flakey to reach a given position at a given velocity, and still (reactively) avoid the obstacles on the way.[2]

In our approach, a goal is represented by a *control structure*. Intuitively, a control structure is virtual object (an *artifact*) that we put in the LPS, associated with a behavior that encodes the way to react to the presence of this object. For example, a "control-point" is a marker for a $(x, y)$ location, together with a heading and a velocity: the associated behavior GO-TO-CP reacts to the presence of a control point in the LPS by generating the commands to reach that position, heading and velocity. In Figure 1 there are two artifacts: a control point to reach (left), and a wall to follow (right).

[1] See [Ruspini, 1991a; Ruspini, 1991b] for an account of fuzzy logic and fuzzy control in terms of similarity and desirability measures, and the use of T-norms and T-conorms in this context.

[2] Reactive behaviors are also associated with (innate) goals, hardwired in the definition of the behavior. We are now interested in dynamically assigning specific strategic goals to Flakey.

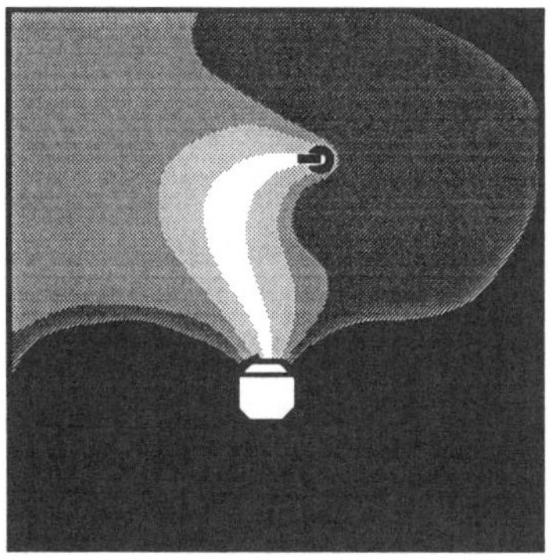

Figure 3: Path families generated by actions with increasing values of $Des_{FS}$.

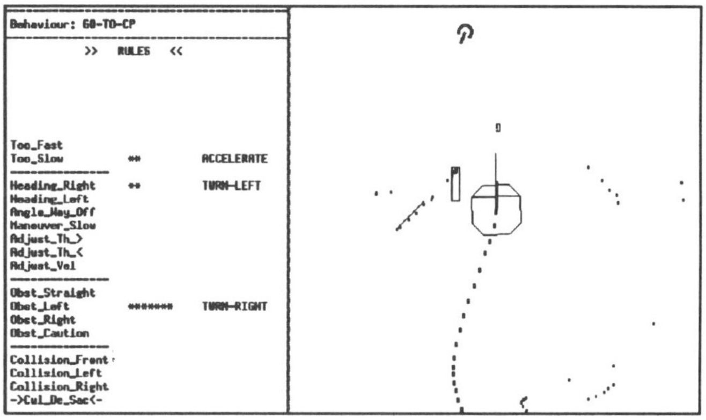



Figure 4: A snapshot of Flakey's control window while achieving a control point.

More precisely, a control structure is a pair

$$S = \langle Q_S, R_S \rangle,$$

where $Q_S$ is an artifact, and $R_S$ is a fuzzy relation between the position of Flakey and that of the artifact.[3] Such a control structure implicitly defines a goal: the goal to achieve, and maintain, the given relation between Flakey and the artifact $Q_S$. Intuitively, if $Q_S$ is at position $q$, $R_S(q,p)$ says how much a position $p$ of Flakey satisfies this goal. If the position of Flakey is such that $R_S(q,p) = \alpha$, we say that the control structure $S$ is satisfied to the degree $\alpha$.

The $R_S$ relation induces a desirability function $Des_S$ in the following way. Given the set $P$ of possible positions of Flakey, and the set $C$ of possible control values, let $\text{Exec}(p,c)$ denote the new position reached by applying control $c$ from position $p$. Then, the desirability *from the viewpoint of the control structure* $S$ of executing $c$ when Flakey is at position $p$ (and $Q_S$ is at $q$) is given by

$$Des_S(q,p,c) = R_S(q, \text{Exec}(p,c))$$

However, not all positions are equally reachable by Flakey: moving to certain positions will require more effort (changes in velocity and/or direction, time, etc.) than moving to others. To account for this, we consider a second desirability function $Des_F$: $Des_F(p,c)$ measures the desirability *from the viewpoint of Flakey's motion capabilities* of executing control action $c$ when Flakey is at position $p$. The desirability of control actions from the joint viewpoints of feasibility for Flakey, and effectiveness with respect to the control structure $S$, is measured by the combination $Des_S \otimes Des_F$ (where $\otimes$ is a T-norm). Figure 3 illustrates one such combined desirability function: here, $S$ is a control point, represented by the semi-circle near the top (the "tail" indicates the desired entry orientation.) The fading from black to white illustrates the increase in the value of $Des_S \otimes Des_F$ for some families of possible paths.

We have already seen how $Des_S$ can induce, for each LPS configuration, a fuzzy set of possible controls. As we did in the case of reactive behaviors, we approximate this fuzzy set using rules on the form "If $A$ then $c$". The only difference is that $A$ now refers to artifacts rather than to sensorial input.[4] We have designed sets of rules for many "purposeful" behaviors, including going to a $x, y$ position; achieving a control point; following a wall; crossing a door; and so forth. Each ruleset consists of a small number (four to eight) of rules. Purposeful behaviors can coexist with other behaviors, either purposeful and reactive: the context-dependent blending of behaviors explained above provides arbitration and guarantees the smooth integration of directed activities and reactivity.

Figure 4 exemplifies the performance of the integration. The picture shows Flakey's control window during an actual run: on the right is Flakey's local perceptual space. Flakey sits in the middle of the window, pointing upwards; the small points all around mark sonar readings, indicating the possible presence of some object; the rectangle on the left of Flakey highlight a dangerously close object. The window on the left lists all the currently active rules, grouped into rule-sets: topmost, the rules for the GO-FORWARD behavior; below, those for GO-TO-CP, for AVOID-OBSTACLES, and for AVOID-COLLISIONS. In the shown situation, Flakey is going too slow and heading right of the CP: hence, some desirability is given to the

[3] Positions are actually points in a $(x, y, \theta, v)$ 4-D space.

[4] Alternatively, these rules can be thought as responding to input from a "virtual sensor" that senses the position of an artifact.

*accelerate* and the *turn-left* actions. However, the close obstacle on the left causes the activation level of the AVOID-OBSTACLES behavior to be high, at the expenses of the other behaviors; hence, the *turn-right* action suggested by AVOID-OBSTACLES receives high total desirability (as indicated by the 7 stars). The small box in front of Flakey indicates the resulting turning control — some degrees on the right. The overall result of the blending is that Flakey makes its way among obstacles while *en route* to achieving the position and bearing of the given control point. The smoothness of the movement in evident in the wake of small boxes that Flakey left behind it (one box per second). Flakey's speed was between 200 and 300 mm/sec.

One word is worth spent on the problem of local minima, ubiquitous in approaches to robot navigation based on local combination of behaviors [Latombe, 1991]. The problem is illustrated in Figure 5 (top): the robot needs to mediate the tendency to move toward the goal, and the tendency to stay away from the obstacle. A straightforward combination of these two opposite tendencies (whether they are described by desirability measures, potential fields [Khatib, 1986], motor schemas [Arkin, 1990], or other) may result in the production of a zone of local equilibrium (local minimum): when coming from the left edge, the robot would be first attracted and then trapped into this zone. By using meta-rules like the 1 above to reason about the relative importance of goals, our context-dependent blending of behaviors provides a way around this problem. Figure 5 shows the path followed by Flakey in a simulated run (top), and the corresponding activation levels of the KEEP-OFF and REACH behaviors (bottom). In (a), Flakey has perceived the obstacle; as the obstacle becomes nearer, the KEEP-OFF behavior becomes more active, at the expenses of the REACH behavior. In this way, the "attractive power" of the goal is gradually shaded away by the obstacle, and Flakey responds more and more to the obstacle-avoidance suggestions alone. The REACH behavior re-gains importance, however, as soon as Flakey is out of danger (b).

## V. CONCLUSIONS

We have defined a mechanism based on fuzzy logic for blending multiple behaviors aimed at achieving different, possibly conflicting goals. Goals are either built-in, as in most fuzzy controllers, or dynamically set from outside the controller. Typically, the built-in goals correspond to reactive behaviors (like avoiding collisions), while the dynamic ones are strategic goals communicated by a planner. Context-dependent blending of behaviors ensures that strategic goals be achieved as much as possible, while maintaining a high reactivity.

Our behavior blending mechanism has been originally inspired to the technique proposed by Berenji et al. [Berenji *et al.*, 1990] for dealing with multiple goals in fuzzy control. There are however two important differences: first, our context mechanism dynamically modifies the degrees of importance of each goal; second, we allow the introduction of high-level, situation-specific goals in the controller.

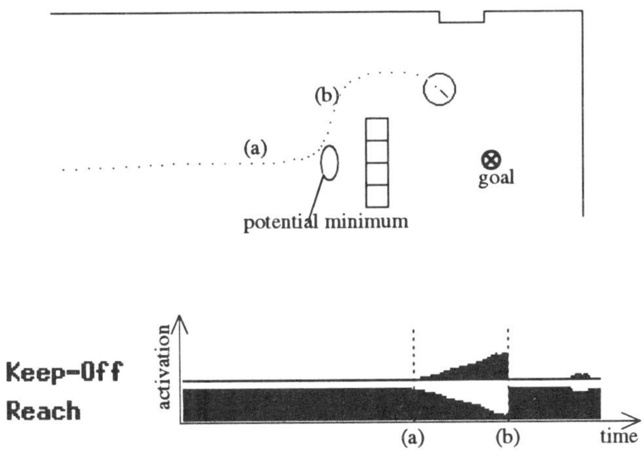


Figure 5: How context-dependent blending of behaviors avoids potential local minima.

From another perspective, the work presented here fits in the tradition of the "two level" approaches to robot control, where a strategic planner is used to generate guidelines to a reactive controller (e.g., [Arkin, 1990; Payton *et al.*, 1990; Gat, 1991]). In our case, a plan consists in a sequence of control structures. For example, a plan to exit building E could consist in three successive *corridors* to follow, one *control point* in the entrance hall close to the door, and the exit *door* itself. The context of applicability of each control structure is used to decide when each control structure becomes relevant. (see [Saffiotti *et al.*, 1993; Saffiotti, 1993] for more on this issue). We believe that having based our architecture on fuzzy logic results in improved robustness (e.g., more tolerance to sensor noise and knowledge imprecision), while granting a better understanding of the underlying mechanisms.

Finally, many current approaches to robot control deal with multiple goals using the so-called "potential fields" method [Khatib, 1986]: goals are represented by pseudo-forces, which may be thought of as representatives of most desirable behavior from that goal's viewpoint. These optimal forces are then combined, as physical vectors, to produce a resultant force that summarizes their joint effect. In our approach, by contrast, the goals' desirability functions, rather than a summary description, are combined into a joint desirability function, from which a most desired tradeoff control is extracted. Moreover, this combination takes behaviors' context of applicability into account; this provides a key to eliminate the local minima arising from the combination of conflicting goals.

The technique proposed in this paper has been implemented in the SRI mobile robot Flakey, resulting in extremely smooth and reliable movement. The performance of Flakey's controller has been demonstrated at the first AAAI robot competition in San Jose, CA [Congdon *et al.*, 1993]. Flakey accomplished all the given tasks while smoothly getting around obstacles (whose positions were not known beforehand) and people, and placed second behind Michigan University's CARMEL. Flakey's reliable reactivity is best summarized in one judge's comment: "Only robot I felt I could sit or lie down in front of." (What he actually did!)

**Acknowledgments** John Lowrance, Daniela Musto, Karen Myers and Leonard Wesley contributed to the development of the ideas presented in this paper. Nicolas Helft implemented a first version of Flakey's controller.

The first author has been supported by a grant from the National Council of Research of Italy. Research performed by the second author leading to the conceptual structures used in the autonomous mobile vehicle controller was supported by the U.S. Air Force Office of Scientific Research under Contract No. F49620–91–C–0060. Support for Kurt Konolige came partially from ONR Contract No. N00014–89–C–0095. Additional support was provided by SRI International.

## References

[Arkin, 1990] Arkin, Ronald C. 1990. The impact of cybernetics on the design of a mobile robot system: a case study. *IEEE Trans. on Systems, Man, and Cybernetics* 20(6):1245–1257.

[Berenji *et al.*, 1990] Berenji, H.; Chen, Y-Y.; Lee, C-C.; Jang, J-S.; and Murugesan, S. 1990. A hierarchical approach to designing approximate reasoning-based controllers for dynamic physical systems. In *Procs. of the 6th Conf. on Uncertainty in Artificial Intelligence*, Cambridge, MA.

[Congdon *et al.*, 1993] Congdon, C.; Huber, M.; Kortenkamp, D.; Konolige, K.; Myers, K.; and Saffiotti, A. 1993. CARMEL vs. Flakey: A comparison of two winners. *AI Magazine.* To appear.

[Firby, 1987] Firby, J. R. 1987. An investigation into reactive planning in complex domains. In *Procs. of the AAAI Conf.*

[Gat, 1991] Gat, E. 1991. *Reliable Goal-Directed Reactive Control for Real-World Autonomous Mobile Robots.* Ph.D. Dissertation, Virginia Polytechnic Institute and State University.

[Kaelbling, 1987] Kaelbling, L. P. 1987. An architecture for intelligent reactive systems. In Georgeff, M.P. and Lansky, A.L., editors 1987, *Reasoning about Actions and Plans.* Morgan Kaufmann.

[Khatib, 1986] Khatib, O. 1986. Real-time obstacle avoidance for manipulators and mobile robots. *The International Journal of Robotics Research* 5(1):90–98.

[Latombe, 1991] Latombe, J.C. 1991. *Robot Motion Planning.* Kluver Academic Publishers, Boston, MA.

[Payton *et al.*, 1990] Payton, D. W.; Rosenblatt, J. K.; and Keirsey, D. M. 1990. Plan guided reaction. *IEEE Trans. on Systems, Man, and Cybernetics* 20(6).

[Ruspini and Ruspini, 1991] Ruspini, E. H. and Ruspini, D. 1991. Autonomous vehicle motion planning using fuzzy logic. In *Procs. of the IEEE Round Table on Fuzzy and Neural Sys. and Vehicle Appl.*, Tokyo, Japan.

[Ruspini, 1990] Ruspini, E. H. 1990. Fuzzy logic in the Flakey robot. In *Procs. of the Int. Conf. on Fuzzy Logic and Neural Networks (IIZUKA)*, Japan. 767–770.

[Ruspini, 1991a] Ruspini, E. H. 1991a. On the semantics of fuzzy logic. *Int. J. of Approximate Reasoning* 5.

[Ruspini, 1991b] Ruspini, E. H. 1991b. Truth as utility: A conceptual systesis. In *Procs. of the 7th Conf. on Uncertainty in Artificial Intelligence*, Los Angeles, CA.

[Saffiotti *et al.*, 1993] Saffiotti, A.; Konolige, K.; and Ruspini, E. H. 1993. Now, do it! Technical report, SRI Artificial Intelligence Center, Menlo Park, California. Forthcoming.

[Saffiotti, 1993] Saffiotti, A. 1993. Some notes on the integration of planning and reactivity in autonomous mobile robots. In *Procs. of the AAAI Spring Symposium on Foundations of Automatic Planning*, Stanford, CA.

[Sugeno and Nishida, 1985] Sugeno, M. and Nishida, M. 1985. Fuzzy control of a model car. *Fuzzy Sets and Systems* 16:103–113.

[Yen and Pfluger, 1992] Yen, J. and Pfluger, N. 1992. A fuzzy logic based robot navigation system. In *Procs. of the AAAI Fall Symposium on Mobile Robot Navigation*, Boston, MA. 195–199.

[Zadeh, 1978] Zadeh, L.A. 1978. Fuzzy sets as a basis for a theory of possibility. *Fuzzy Sets and Systems* 1:3–28.

# On the Distributivity between t-Norms and t-Conorms

Carlo Bertoluzza
Dipartimento di Informatica e Sistemistica
Universitá di Pavia
Via Abbiategrasso 209 - 27100 Pavia - Italy

**Abstract. The notions of t-conorm and of t-norm are very important in the study of many problems, as for example the analysis of the decomposable measures in the sense of Dubois-Prade and in the sense of Forte-Kampé de Fériet ([1,2]). In this paper we determine the couples (t-conorm,t-norm) which are distributive one to the other. We shall also present two problems, related with belief and information measures, where the distributivity plays an important role.**

## I. PRELIMINARIES

The definition of t-norm and t-conorm which we give here, is a slight generalization of the one given for example in [1] and [4]; the modification has been made in order to apply the results obtained here to the problems presented in the paragraphs 5 and 6.

Let $[a,b]$ be a closed interval of the extended positive real line $\mathbb{R}^+$.

**Definition 1.1** A *t-conorm* on $[a,b]$ is a map $\top : [a,b]^2 \to [a,b]$ such that

(1.1) $x \top y = y \top x$

(1.2) $(x \top y) \top z = x \top (y \top z)$

(1.3) $x_1 < x_2 \Longrightarrow x_1 \top y \leq x_2 \top y$

(1.4.a) $x \top a = x$

A *t-norm* on $[a,b]$ is a map $\circ : [a,b]^2 \to [a,b]$ with the same properties of a t-conorm, except for (1.4.a) which is substituted by

(1.4.b) $x \circ b = x$

It is easy to recognize that

(1.5) $[a]$ $x \top y \geq \sup(x,y)$ $\quad$ $[b]$ $x \circ y \leq \inf(x,y)$

(1.6) $[a]$ $x \top b = b$ $\quad$ $[b]$ $x \circ a = a$

In the representation theorems which we state below, we suppose that the laws $\top$ and $\circ$ are continuous. This is the only restriction we make.

Let $\Delta_\top$ be the set of the *t-idempotent* elements, that is

(1.7) $\Delta_\top = \{x \in [a,b] \mid x \top x = x\}$

It may be recognized that

(1.8) $\alpha \in \Delta_\top \Longrightarrow x \top \alpha = \sup(x,\alpha)$

In fact

(a) if $\alpha > x$, then $\alpha = \alpha \top \alpha \leq x \top \alpha \leq \alpha \top \alpha = \alpha$

(b) if $x > \alpha$ then let us pose $\theta(x) = \alpha \top x$ . $\theta$ is a continuous function which takes all the values between its minimum $\theta(\alpha) = \alpha$ and its maximum $\theta(b) = b$ , and thereby $\forall x \in [\alpha, b]$ , there exists $\xi \in [\alpha,b]$ such that $x = \theta(\xi)$ . Then we have $\alpha \top x = \alpha \top \theta(\xi) = \alpha \top (\alpha \top \xi) = (\alpha \top \alpha) \top \xi = \alpha \top \xi = x$ .

Clearly $\Delta_\top$ is closed; therefore its complemetary set

$$\overline{\Delta}_\top = \{x \in [a,b] \mid x \top x > x\}$$

is the union of a finite or countable family of open disjoint intervals

$$\overline{\Delta}_\top = \bigcup_{i \in I \subset \mathbb{N}} ]a_i, b_i[ \quad , \quad a_i, b_i \in \Delta_\top$$

The restriction $\top_i$ of the law $\top$ to the subset $[a_i,b_i]^2$ fulfills all the conditions of the main theorem 3.3 in [3]. Then there exists a strictly increasing function $g_i : [a_i,b_i] \to \mathbb{R}^+$ with $g_i(a_i) = 0$ such that

$$x \top y = \overline{g}_i\{g_i(x) + g_i(y)\} \qquad \forall (x,y) \in [a_i,b_i]^2$$

where $\overline{g}_i(\xi) = g_i^{-1}[\sup(\xi, g_i(b_i))]$ is the pseudo inverse of $g_i$ . We may then state the following

**Theorem 1.1.a** Any continuous t-conorm is given by

$$(1.9.a) \qquad x \top y = \begin{cases} [1] \quad \overline{g}_i[g_i(x) + g_i(y)] \\ \quad\;\; \text{if } (x,y) \in [a_i,b_i]^2 \\ [2] \quad \sup(x,y) \\ \quad\;\; \text{otherwise} \end{cases}$$

0-7803-0614-7/93$03.00 ©1993IEEE

Part [1] has just been proved. (1.8) proves part [2] whenever $x$ or $y$ are T−idempotent. To complete the proof suppose that $x \in [a_i, b_i]$ , $y \in [a_j, b_j]$ , $i \neq j$ , $a_j \geq b_i$ (obviously $y \geq x$). Then we have $y = a_i \top y \leq x \top y \leq b_i \top y = y$ .

In a symmetric way we can establish a representation theorem for t-norms. In fact it may be recognized that, if $\alpha$ belong to the $\circ$−idempotent set

$$(1.7.a) \qquad \Delta_\circ = \{x \in [a, b] \mid x \circ x = x\}$$

then we have

$$(1.8.b) \qquad \alpha \circ x = \inf(\alpha, x)$$

and moreover

$$\overline{\Delta}_\circ = \{x \in [a, b] \mid x \circ x < x\} = \cup_{r \in R \subset \mathbf{N}} ]\alpha_r, \beta_r[$$
$$\text{(with } \alpha_r, \beta_r \in \Delta_\circ)$$
$$(x, y) \in [\alpha_r, \beta_r] \Longrightarrow x \circ y = \overline{\gamma}_r[\gamma_r(x) + \gamma_r(y)]$$

where $\gamma_r$ is a strictly decreasing function with $\gamma_r(\beta_r) = 0$ , and $\overline{\gamma}_r(\xi)$ is the pseudo-inverse of $\gamma_r$ defined by $\overline{\gamma}_r(\xi) = \gamma_r^{-1}[\sup(\xi, \gamma_r(\alpha_r))]$ . Finally we have

**Theorem 1.1.b** Any continuous t-norm is given by

$$(1.9.b) \qquad x \circ y = \begin{cases} [1] \quad \overline{\gamma}_r[\gamma_r(x) + \gamma_r(y)] \\ \qquad \text{if } (x, y) \in [\alpha_r, \beta_r]^2 \\ [2] \quad \inf(x, y) \\ \qquad \text{otherwise} \end{cases}$$

In this paper we deal with the problem of the distributivity of a t-norm with respect to a t-conorm. In other words we search for all the couples $(\top, \circ)$ fulfilling the property

$$(1.10.a) \qquad x \circ (y \top z) = (x \circ y) \top (x \circ z)$$

We bind ourselves to study the problem under the hypotesis that equation (1.10.a) holds whenever $x < b$ and $y \top z < b$ because this limitation allows us to apply the results obtained here to many cases of interest, as we will see later .

We should consider also the dual problem, that is the distributivity of a t-conorm with respect to a t-norm

$$(1.10.b) \qquad x \top (y \circ z) = (x \top y) \circ (x \top z)$$

but it is evident that the solution of the second problem is the dual form of the solution of the first one.

## II. The sets $\Delta_\top$ and $\Delta_\circ$

In the representation theorems 1.1.a and 1.1.b, the sets of the idempotent elements $\Delta_\top$ and $\Delta_\circ$ are two arbitrary closed subsets of $[a, b]$ , containig the end points $a$ and $b$. The distributivity equation (1.10.a) imposes some limitations on the structure of the set $\Delta_\top$. In fact we shall prove the following

**Theorem 2.1.a** Let $\alpha_\circ = \sup\{x | x \in \Delta_\circ - \{b\}\}$ . Then either

$$(2.1) \qquad \Delta_\top = [a, b]$$

or

$$(2.2) \qquad \Delta_\top = [a, \alpha_\circ] \cup \{b\}$$

Obviously $\alpha_\circ \in \Delta_\circ$ , because $\Delta_\circ$ is closed, and $\overline{\Delta}_\top$ is either the empty set or the open interval $]\alpha_\circ, b[$ .

Whe prove this theorem by means of the following lemmas.

**Lemma 2.1** $\Delta_\circ \subset \Delta_\top$
In fact, if $x \circ x = x$ , then by using (1.5.b)(1.10.a) and (1.5.a) we have

$$x \geq x \circ (x \top x) = (x \circ x) \top (x \circ x) = x \top x \geq x$$

Therefore $x \top x = x$ and the lemma has been proved.

**Lemma 2.2** If $x \in \Delta_\circ$ and $y < x$ , then $y \in \Delta_\top$
In fact, if $x \in \Delta_\circ$ and $y < x$ , then we have $x \circ y = \inf(x, y) = y$ , and, by using lemma 2.1, we obtain

$$y \top y = (y \circ x) \top (y \circ x) = y \circ (x \top x) = y \circ x = y$$

which proves the lemma.

Thus we proved that $[a, \alpha_\circ] \subset \Delta_\top$ . If $\alpha_\circ = b$ , then theorem 2.1 is proved. If $\alpha_\circ < b$ , then $]\alpha_\circ, b[ \subset \overline{\Delta}_\circ$ , and from (1.9.b) we obtain

$$(2.3) \qquad (x, y) \in ]\alpha_\circ, b[^2 \Longrightarrow x \circ y = \overline{\gamma}[\gamma(x) + \gamma(y)]$$

where $\gamma$ is a strictly decreasing function defined on $[\alpha_\circ, b]$ , with $\gamma(b) = 0$ .

**Lemma 2.3** If $x \in \Delta_\top - \{b\}$ and $y < x$ , then $y \in \Delta_\top$ .
This fact has just been proved by lemma 2.2 in the case where $x \leq \alpha_\circ$ . If $x > \alpha_\circ$ we can consider the function $h(\xi) = x \circ \xi$ . It assumes all the values between its minimum $h(a) = a$ and its maximum $h(b) = x$ . Then $\forall y \in [a, x]$ there exists a value $\eta$ such that $y = h(\eta) = x \circ \eta$ . Because $x \in \Delta_\top$ we have

$$y = x \circ \eta = (x \top x) \circ \eta = (x \circ \eta) \top (x \circ \eta) = y \top y$$

that is $y \in \Delta_\top$ .

Lemma 2.2 proves that the set $\Delta_\top$ of the T−idempotent elements has the form

$$\Delta_\top = [a, \beta]$$

where $\beta = \sup\{x \in [a,b[ \mid xTx = x\}$ . If $\beta = b$ then the theorem 2.1.a is proved. If $\beta < b$ , then we shall complete the proof of this theorem by proving that $\beta = \alpha_\circ$ .

Since $\beta \circ y$ is a T−idempotent element we have always

$$(2.4) \qquad \beta \circ (yTy) = (\beta \circ y)T(\beta \circ y) = \beta \circ y$$

Let $y$ be a value such that $\beta < y < b$ ; then $yTy \neq y$ and the three elements $\beta, y, yTy$ belong to the closed interval $[\alpha_\circ, b]$ . But in this interval the operation $\circ$ has the form (2.3), and by posing $yTy = t$ , from relation (2.4) we have

$$(2.5) \qquad \overline{\gamma}[\gamma(\beta) + \gamma(t)] = \overline{\gamma}[\gamma(\beta) + \gamma(y)]$$

where obviously $t \neq y$ . It is evident that (2.5) may hold if and only if each of its member equals to $\alpha_\circ$. Thus for all $y \in ]\beta, b[$ we have $\beta \circ y = \alpha_\circ$ and, from the continuity of the t-norm $\circ$, we obtain

$$\alpha_\circ = \lim_{y \to b} \beta \circ y = \beta$$

and theorem 2.1.a is proved.

If, instead of considering the distributivity in the form (1.10.a), we study the distributivity of a t-conorm with respect to a t-norm [that is the distributivity in the form (1.10.b)], then we can prove, in a completely symmetric way, the following

**Theorem 2.1.b** Let $\alpha_T = \inf\{x | x \in \Delta_T - \{a\}\}$ . Then either

$$\Delta_\circ = [a, b]$$

or

$$\Delta_\circ = \{a\} \cup [\alpha_T, b]$$

By using the theorems 2.1.a and 2.1.b, we can recognize that the following theorem holds:

**Theorem 2.2** If both (1.10.a) and (1.10.b) hold, then either

$$\Delta_\circ = \Delta_T = [a, b]$$

or

$$\Delta_\circ = \Delta_T = \{a, b\}$$

In fact, if an element $x \in ]a.b[$ is $\circ$−idempotent, then by theorem 2.1.a $[a, x] \subset \Delta_T$ , and consequently by theorem 2.1.b $\Delta_\circ = [a, b]$ . By using theorem 2.1.a again we conclude that also $\Delta_T = [a, b]$ . We obtain the same result, by using theorems 2.1.a and 2.1.b in an inverted order, even in the case where an element $x \in ]a, b[$ is T−idempotent, and this complete the proof of the theorem.

## III. The form of the distributive pairs $(\circ, T)$

**Theorem 3.1.a** If $\alpha_\circ = b$ , that is if $\Delta_T = [a, b]$ , then it follows from (1.9.a) that $xTy = \sup(x, y)$ everywhere. In this case the distributivity equation (1.10.a) is satisfied whatever be the t-norm $\circ$. In fact, because of the monotonicity of the law $\circ$, it is obvious that

$$x \circ \sup(y, z) = \sup(x \circ y, x \circ z)$$

**Theorem 3.2.a** If $\alpha_\circ < b$ , then the distributivity equation (1.10.a) is fulfilled if and only if the the restrictions of the laws T and $\circ$ to the domain $]\alpha_\circ, b[^2$ have the form

$$(3.1) \qquad xTy = \overline{g}[g(x) + g(y)]$$
$$x \circ y = g^{-1}[g(x) \cdot g(y)]$$

In other words let $g(\cdot)$ [resp. $\gamma(\cdot)$ ] be the function which defines the restriction of the law T (resp. $\circ$) to the set $[\alpha_\circ, b]$ . Then (3.1) means that the following relation between the functions $g(\cdot)$ and $\gamma(\cdot)$ holds

$$(3.2) \qquad g(x) = e^{-k \cdot \gamma(x)} \qquad k > 0$$
$$\gamma(x) = -\frac{1}{k} \log[g(x)]$$

We prove this result by means of the following three lemmas:

**Lemma 3.1** For each $x \in ]\alpha_\circ, b[$ we have

$$(3.3) \qquad \sup\{y \mid y \circ x = \alpha_\circ\} = \alpha_\circ$$

It is evident that the left hand side of (3.3) (which we shall indicate by $\tilde{x}$) is strictly less than $b$, because $b \circ x = x > \alpha_\circ$ . On the other hand it can not be less than $\alpha_\circ$ because $\alpha_\circ \circ x = \alpha_\circ$ . Let us suppose that $\tilde{x} > \alpha_\circ$ ; then we can choose $y$ and $z$ such that $\alpha_\circ < y < \tilde{x} < z < b$ , and we have obviously $x \circ y = \alpha_\circ$ (by the monotonicity of $\circ$) and $x \circ z > \alpha_\circ$ (by the definition of $\tilde{x}$). Using the distributivity equation we obtain

$$x \circ (yTz) = (x \circ y)T(x \circ z) = \alpha_\circ T(y \circ z) = x \circ z$$

But $yTz = \xi > z$ because both the elements $y$ and $z$ belong to the same open interval $]\alpha_\circ, b[ \subset \overline{\Delta}_T$ . The three relations

$$\xi \neq z$$
$$\xi \circ x = z \circ x$$
$$z \circ x > \alpha_\circ$$

are not compatible, because relation (2.3) ensures that

$$z \neq \xi \, , \; z \circ x = z \circ \xi \Longrightarrow z \circ x = \alpha_\circ$$

Thus it is absurd to suppose that $\tilde{x} > \alpha_\circ$ , and the lemma is proved.

**Lemma 3.2** The function $\gamma(x)$ , which determines [via (2.3)] the restriction of $\circ$ to the interval $[\alpha_\circ, b]$ , fulfills the property

$$(3.3) \qquad \lim_{x \to \alpha_\circ} \gamma(x) = +\infty$$

In fact, if $\lim_{x\to\alpha_o}\gamma(x) = a < +\infty$, there exists an element $\tilde{x} > \alpha_o$ with $\gamma(\tilde{x}) > \frac{a}{2}$. Then by using (2.3) we obtain $\tilde{x} \circ \tilde{x} = \alpha_o$, which contradicts lemma 3.1. Then (3.3) is proved.

Property (3.3) implies that $\overline{\gamma} = \gamma^{-1}$ everywere, and then

$$(3.4) \qquad x \circ y = \gamma^{-1}[\gamma(x) + \gamma(y)] \qquad \forall(x,y) \in [\alpha_o, b]^2$$

Now let us consider all the values $x, y, z$ in $]\alpha_o, b[$ such that $yTz < b$. As $x \circ y < y$ and $x \circ z < z$ we also have $(x \circ y)T(x \circ z) < b$. Then for these values we have

$$yTz = g^{-1}[g(y) + g(z)]$$
$$(x \circ y)T(x \circ z) = g^{-1}[g(x \circ y) + g(x \circ z)]$$

Thus the distributivity equation (1.10.a) becomes

$$(3.5) \qquad \gamma^{-1}[\gamma(x) + \gamma * g^{-1}\{g(y) + g(z)\}] = \\ = g^{-1}[g * \gamma^{-1}\{\gamma(x) + \gamma(y)\} + \\ + g * \gamma^{-1}\{\gamma(x) + \gamma(z)\}]$$

where $*$ represents the function composition operation.

**Lemma 3.3** The functions $g(x)$ and $\gamma(x)$ fulfill (3.5) if and only if

$$(3.6) \qquad \gamma^{-1}[\gamma(x) + \gamma(y)] = g^{-1}[g(x) \cdot g(y)]$$

In fact we can write (3.5) in the form

$$g * \gamma^{-1}[\gamma(x) + \gamma * g^{-1}\{g(y) + g(z)\}] = \\ = g * \gamma^{-1}[\gamma(x) + \gamma(y)] + g * \gamma^{-1}[\gamma(x) + \gamma(z)]$$

Then by posing $h(\xi) = g * \gamma^{-1}(\xi)$, $u = g(x)$, $v = g(y)$, we have

$$h[\gamma(x) + h^{-1}(u+v)] = \\ = h[\gamma(x) + h^{-1}(u)] + h[\gamma(x) + h^{-1}(v)]$$

This is the Cauchy equation in the unknown function

$$m_x(\xi) = h[\gamma(x) + h^{-1}(\xi)]$$

The function $m_x(\xi)$ is continuous, because it is obtained as composition of continuous functions, and therefore $m_x(v) = c(x) \cdot v$, that is

$$h[\gamma(x) + h^{-1}(v)] = g * \gamma^{-1}[\gamma(x) + \gamma * g^{-1}(v)] = c(x) \cdot v$$

which may be written in the form

$$\gamma^{-1}[\gamma(x) + \gamma(y)] = g^{-1}[c(x) \cdot g(y)]$$

The left hand side of this relation is symmetric in $x$ and $y$, and the same must hold for the right one. This implies $c(x) = \lambda \cdot g(x)$ ; without loss of generality we can pose $\lambda = 1$, and so the part "only if" of the lemma is proved. The part "if" is a straightforward computation.

The relations (3.6) and (3.4) prove the point (3.1) of the theorem. In order to prove (3.2), we pose $u = g(x)$, $v = g(y)$, $h(\xi) = g * \gamma^{-1}(\xi)$ in (3.6), thus obtaining

$$h(u+v) = h(u) \cdot h(v)$$

This is a Cauchy equation whose continuous solutions are the functions

$$h(u) = g * \gamma^{-1}(u) = e^{c \cdot u}$$

from which we obtain

$$(3.7) \qquad g(t) = e^{c \cdot \gamma(t)} \quad , \quad \gamma(t) = \frac{1}{c} \log g(t)$$

As pointed out in theorems 1.1.a and 1.1.b the functions $g(t)$ and $\gamma(t)$ are respectively strictly increasing and strictly decreasing; it follows that the constant $c$ in (3.7) is negative, and the theorem 3.2.a is completely proved.

In a dual way we can characterise the distributivity of a t-conorm with respect to a t-norm, by means of the following two theorems.

**Theorem 3.1.b** All the t-conorms T are distributive with respect to the extremal t-norm

$$x \circ y = \inf(x, y)$$

In fact, because of the monotonicity of T, $xT\inf(y,z) = \inf[xTy, xTz]$

**Theorem 3.2.b** If the constant $\alpha_T$ of theorem 2.1.b is strictly greater than $a$, then the distributivity equation (1.10.b) is fulfilled if and only if the restiction of the laws T and $\circ$ to the domain $[a, \alpha_T]^2$ have the form

$$(3.8) \qquad x \circ y = \overline{\gamma}[\gamma(x) + \gamma(y)] \\ xTy = \gamma^{-1}[\gamma(x) \cdot \gamma(y)]$$

which is equivalent to

$$(3.9) \qquad \gamma(x) = e^{-\bar{k} \cdot g(x)} \qquad \bar{k} > 0 \\ g(x) = -\frac{1}{\bar{k}} \log[\gamma(x)]$$

We can complete this paragraph by using theorems 3.2 to prove a well known result: the only pair (t-norm, t-conorm) which are distributive one to the other is the pair (inf, sup).

In fact we know, from theorem 2.2, that the bidistributivity implies either

$$\Delta_o = \Delta_T = [a, b]$$

or

$$\Delta_o = \Delta_T = \{a, b\}$$

In the first case we have $xTy = \sup(x,y)$, $x \circ y = \inf(x,y)$, whereas in the second case we have, everywere in $[a,b]^2$

$$xTy = \overline{g}[g(x) + g(y)]$$
$$x \circ y = \overline{\gamma}[\gamma(x) + \gamma(y)]$$

and by theorems 3.2 (a and b), the following relation

$$(3.10) \qquad \gamma(x) = e^{-\tilde{k}\cdot[e^{-k\cdot\gamma(x)}]}$$

necessarily holds for all values of $x$. But no function $\gamma(t)$ does satisfy this identity, and thereby no one pair of bidistributive (t-norm, t-conorm) corresponds to the second case.

## IV. Examples

In [4] S.Weber presents some examples of t-conorms (and of t-norms) on [0, 1]. Using the theorems 3.2.a (and 3.2.b) we associate to each one its distributive t-norm (and respectively its distributive t-conorm). In each of the formulas (4.1) the first line shows a t-conorm and the second one represents its distributive t-norm, whereas in each of (4.2) the first line is a t-norm and the second is its distributive t-conorm. It may be remarked, by comparing (4.1.a) and (4.1.b), that the distributivity of a particular t-norm with respect to a particular t-conorm does not imply the distributivity of the those t-conorm with respect to those t-norm.

(4.1.a) $\min(x+y, 1)$
$x \cdot y$

(4.1.b) $x + y - xy$
$1 - \exp[-\, log(1-x) \cdot \log(1-y)]$

(4.1.c) $(x^p + y^p - x^p y^p)^{\frac{1}{p}}$
$[1 - \exp\{-\frac{1}{p}\log(1-x^p) \cdot \log(1-y^p)\}]^{\frac{1}{p}}$

(4.1.d) $\min[(x^{-p} + y-^{p})^{-\frac{1}{p}}, 1]$
$[-\frac{1}{p} x^{-p} y^{-p}]^{-\frac{1}{p}}$

(4.1.e) $\min[x + y + \lambda xy, 1]$
$\frac{1}{\lambda}[\exp\{\frac{1}{\lambda}\log(1+\lambda x)\log(1+\lambda y)\} - 1]$

(4.1.f) $[\max[(1-x)^{-p} + (1-y)^{-p} - 1, 0]]^{-\frac{1}{p}}$
$1 - [\frac{1}{p}\{1 - (1-x)^{-p}\} \cdot$
$\cdot \{1 - (1-y)\}^{-p} + 1]^{-\frac{1}{p}}$

(4.2.a) $x \cdot y$
$\exp[-\log x \cdot \log y$

(4.2.b) $(x^{-p} + y-^{p} - 1)^{-\frac{1}{p}}$
$[\frac{1}{p}(x^{-p} - 1)(y^{-p} - 1) + 1]^{-\frac{1}{p}}$

(4.2.c) $\max(x + y - 1, 0)$
$1 - (1-x)(1-y)$

(4.2.d) $[\max(x^{-p} + y^{-p} - 1, 0)]^{-\frac{1}{p}}$
$[1 - \frac{1}{p}(1 - x^{-p})(1 - y^{-p})]^{-\frac{1}{p}}$

(4.2.e) $\max[x + y - 1 - \lambda(1-x) \cdot (1-y), 0]$
$1 - \log(1 + \lambda(1-y))\} - 1]$

(4.2.f) $\max[\frac{x + y - 1 + \lambda xy}{1+\lambda}, 0]$
$\frac{1}{\lambda}[(1+\lambda)\exp\{\frac{1}{\lambda}\log\frac{1+\lambda}{1+\lambda\cdot x} \cdot$
$\cdot \log\frac{1+\lambda}{1+\lambda\cdot y}\} - 1]$

In the next paragraph we present two problems where the distributivity is the fundamental property which we must require in order to reach the objective. In the first problem we need the distributivity of a t-norm with respect to a t-conorm, while in the second one a t-conorm should be distributive with respect to a t-norm.

## V. Product of decomposable measures

Let $(\Omega_1, S_1, m_1)$ and $(\Omega_2, S_2, m_2)$ be two decomposable Sugeno's spaces with t-conorms indicated by $\top_1$ and $\top_2$ (obviously $\top_1$ and $\top_2$ are defined on [0, 1] ). Then let $(\Omega, S)$ be the cartesian product of the structures $(\Omega 1, S1)$ and $(\Omega_2, S_2)$ , that is $\Omega = \Omega_1 \times \Omega_2$ , and $S$ is the algebra generated by $S_1 \times S_2$ .
A natural problem consists in constructing a decomposable measure $m$ (with t-conorm $\top$) over the structure $(\Omega, S)$ such that its marginal measures $m^{(1)}(A) = m(A \times \Omega_2)$ and $m^{(2)}(B) = m(\Omega_1 \times B)$ concide respectively with $m_1(A)$ and $m_2(B)$ .
A natural way to construct the measure $m$ starts with the determination of the measure of the sets of the form $A = A_1 \times A_2$ ; then the measure of a finite union of sets of this kind may be determined by using the t-conorm $\top$. If $S_1$ and $S_2$ are finitely generated (as well as in the case

where the measure $m$ results to be $\sigma$-decomposable with respect to the law $\top$), then the construction just described determines completely the measure $m$.
In order to construct $m(A) = m(A_1 \times A_2)$ we suppose that there exists a function $f : [a,b]^2 \to [a,b]$ such that

$$(5.1) \qquad m(A_1 \times A_2) = f[m_1(A_1), m_2(A_2)] = m_1(A_1) \circ m_2(A_2)$$

The following properties hold

**Property 5.1** $x \circ y = y \circ x$
In fact we may write $A_1 \times A_2 = (A_1 \times \Omega_2) \cap (\Omega_1 \times A_2)$, and from (5.1) we have

$$m[(A_1 \times \Omega_2) \cap (\Omega \times A_2)] = m(A_1 \times \Omega_2) \circ m(\Omega_1 \times A_2)$$

But $(A_1 \times \Omega_2) \cap (\Omega_1 \times A_2) = (\Omega_1 \times A_2) \cap (A_1 \times \Omega_2)$ ; thus

$$m[(A_1 \times \Omega_2) \cap (\Omega_1 \times A_2)] = m[(\Omega_1 \times A_2) \cap (A_1 \times \Omega_2)] = m(\Omega_1 \times A_2) \circ m(A_1 \times \Omega_2)$$

and the property is proved.

**Property 5.2** $x \circ 1 = x$. In fact $m_1(A_1) = m(A_1 \times \Omega_2) = m_1(A_1) \circ 1$.

**Property 5.3** $x_1 < x_2 \Longrightarrow x_1 \circ y \leq x_2 \circ y$.
In fact if $A_1 \subset A_2$, then $A_1 \times B \subset A_2 \times B$, and consequently $m_1(A_1) < m_1(A_2) \Longrightarrow m(A_1 \times B) \leq m(A_2 \times B)$, that is property 5.3 holds.

Thus the function $\circ$ satisfies three of the four characteristic properties of a t-norm. The missing one (the associativity) can not be derived directly from (5.1). But it is a good property if we want to construct, in a canonical form, the measure over the product of three (or more) Sugeno's structures $(\Omega_1, S_1) \times (\Omega_2, S_2) \times (\Omega_3, S_3)$. In fact the set $A_1 \times A_2 \times A_3$ may be written as $(A_1 \times A_2) \times A_3$ or as $A_1 \times (A_2 \times A_3)$. Thus $m(A_1 \times A_2 \times A_3$ may be written, using (5.1), either in the form

$$m(A_1 \times A_2 \times A_3 = f_1[m_{12}(A_1 \times A_2), m_3(A_3)] = f_1[f_2[m_1(A_1), m_2(A_2)], m_3(A_3)]$$

or in the form

$$m(A_1 \times A_2 \times A_3 = f_3[m_1(A_1), m_{23}(A_2 \times A_3)] = f_3[m_1(A_1), f_4[m_2(A_2), m_3(A_3)]]$$

where we used different forms of the function $f$ appearing in (5.1) corresponding to different cartesian products. It is easy to prove that $f_1 = f_2 = f_3 = f_4 = f$, so, by posing $m_1(A_1) = x$, $m_2(A_2) = y$, $m_3(A_3) = z$, $f(x,y) = x \circ y$, we obtain

$$(x \circ y) \circ z = x \circ (y \circ z)$$

and in this case the law $f = \circ$ is associative

Thus the class of the t-norms is the most natural family of functions which may be used to construct, via (5.1), product measures. A kind of compatibility between the laws $\circ, \top_1, \top_2, \top$ arises from the well known relations

$$(5.2) \qquad (A_1 \cup B_1) \times A_2 = (A_1 \times A_2) \cup (B_1 \times A_2)$$
$$A_1 \times (A_2 \cup B_2) = (A_1 \times A_2) \cup (A_1 \times B_2)$$

If $A_1 \cap B_1 = A_2 \cap B_2 = \emptyset$, then from (5.2) we have

$$(5.3.a) \quad [m_1(A_1) \top_1 m_1(B_1)] \circ m_2(A_2) = [m_1(A_1) \circ m_2(A_2)] \top [m_1(B_1) \circ m_2(A_2)]$$

$$(5.3.b) \quad m_1(A_1) \circ [m_2(A_2) \top_2 m_2(B_2)] = [m_1(A_1) \circ m_2(A_2)] \top [m_1(A_1) \circ m_2(B_2)]$$

We can easily recognize from these relations, that the construction of the measure $m$ is possible only if $\top_1 = \top_2 = \top$. In fact if we pose in (5.3.a) $A_2 = \Omega_2$, that is $m_2(A_2) = m_2(\Omega_2) = 1$, and use property 5.2, then we obtain

$$m_1(A_1) \top_1 m_1(B_1) = m_1(A_1) \top m_1(B_1)$$

that is $x \top_1 y = x \top y$, whenever $A_1 \cap B_1 = \emptyset$ and, "a fortiori" whenever $x \top y < 1$. Because of the monotonicity of $\top$ and $\top_1$ the same relation holds even if $x \top y = 1$. Then we proved that $\top = \top_1$. Using (5.3.b) we prove in the same way that $\top = \top_2$.

Now the compatibility relations (5.3) can be written in the form

$$(5.4) \qquad (x \top y) \circ z = (x \circ z) \top (y \circ z)$$

Thus we may conclude that the cartesian product of two decomposable measures ia also decomposable if and only if

(a) the t-conorm $\top_1$ coincides with the conorm $\top_2$ [$\top_1 = \top_2 = \top$],
(b) the t-norm $\circ$ is distributive with respect to the t-conorm $\top$ as soon as $x$ and $y$ represent measures of disjoint subsets.

(It may be remarked that $x$ and $y$ represent measures of disjoint setsif (but not only if) $x \top y < b$. This is the reason which motivates the hypotesis made in paragraph 1.)

By using the results just established, we can write explicitly the measure $m$ of the sets of $S$ which are of the form $X = \cup_{i \in I} A_i \times B_i$ with $I \in \mathbb{N}$, card(I) $< +\infty$, $A_r \cap A_s = \emptyset$ or $B_r \cap B_s = \emptyset$

Let $\mu = \sup\{g^{-1} g[m_1(A_i)] \cdot g[m_2(B_i)] \mid i \in I\}$. If $\mu \leq \alpha_\top$, then

$$(5.5) \qquad m(X) = \mu$$

If $\mu > \alpha_\top$, then

$$(5.6) \qquad m(X) = g^{-1}[\sum_{j \in J} g\{m_1(A_j)\} \cdot g\{m_2(B_j)\}]$$

where $J \subset I$ is the set of the indices $j$ for which

$$g^{-1}[g\{m_1(A_j)\} \cdot g\{m_2(B_j)\}] > \alpha_T$$

The same expressions (5.5)(5.6) hold even in the case where $card(I) = +\infty$, provided that the measure $m$ is $\sigma$-decomposable with respect to the law T.

It must be remarked that in the expression (5.6) we wrote $g^{-1}$ instead of $\bar{g}$. This is not a mistake because, as $X \subset \Omega$ we have $m(X) \leq m(\Omega) = 1$. Then

$$\sum g[m_1(A_j)] \cdot g[m_2(B_j)] =$$
$$= \sum g[m(A_j \times m(B_j)] \leq g[m(\Omega)] = g(b)$$

and in this case $\bar{g} = g^{-1}$.

As an example we consider the t-conorm (4.2), where $\alpha_T = 0$, $g(x) = -\log(1-x)$, $g^{-1}(t) = 1 - e^{-1}$. Then $m(X) = 1 - \exp\{\sum \log[1 - m_1(A_j)] \cdot \log[1 - m_2(B_j)]\}$

## VI. An application to information measures

In the axiomatic theory proposed by J.Kampé de Fériet and B.Forte ([1,5]), an information space is a structure $(\Omega, \mathcal{S}, \mathcal{K}, J)$, where

- $\Omega$ is the set of the elementary events,
- $\mathcal{S} \subset \mathcal{P}(\Omega)$ is the algebra of the observable events,
- $\mathcal{K}$ is a family of pairs of m-independent sub-algebras,
- $J : \mathcal{S} \to \mathbb{R}^+$ is a map with the following properties

$$(6.1) \quad J(\Omega) = 0 \quad , \quad J(\emptyset) = +\infty$$

$$(6.2) \quad A \subset B \Longrightarrow J(A) \geq J(B)$$

$$(6.3.a) \quad (\mathcal{A}, \mathcal{B}) \in \mathcal{K} \ , \ A \in \mathcal{A} \ . \ B \in \mathcal{B} \Longrightarrow$$
$$\Longrightarrow J(A \cap B) = J(A) + J(B)$$

(Two algebras $\mathcal{A}$ and $\mathcal{B}$ contained in $\mathcal{S}$ are m-independent if $A \cap B \neq \emptyset$ for all non-empty sets $A \in \mathcal{A}$ and $B \in \mathcal{B}$. It has to be remarked that the family $\mathcal{K}$ does not contain all the m-independent pairs $(\mathcal{A}, \mathcal{B})$. It contains only the couples which are supposed to be independent with respect to the information measure (J-independent). For a complete analysis of this subject see e.g. [6])

In the actual example we refer to a particular subclass of information spaces, that is the ones for which the following supplementary condition holds:

$$(6.4) \quad A \cap B = \emptyset \Longrightarrow J(A \cup B) = F[J(A), J(B)]$$

They are indicated as composable spaces, and the function $F : \mathbb{R}^{+2} \to \mathbb{R}^+$ is called *the composition law*. It is easy to recognize that $F$ is a t-norm on $\mathbb{R}^+$.

Now let us consider three subset $A, B_1, B_2$ with $A \in \mathcal{A}$, $B \in \mathcal{B}$ (and consequently $B_1 \cup B_2 \in \mathcal{B}$), $(\mathcal{A}, \mathcal{B}) \in \mathcal{K}$. Then, by appling (6.3.1)(6.4) to the relation $A \cap (B_1 \cup B_2) = (A \cap B_1) \cup (A \cap B_2)$ we obtain

$$(6.5) \quad J(A) + F[J(B_1), J(B_2)] =$$
$$= F[J(A) + J(B_1), J(A) + J(B_2)]$$

This means that the t-conorm "+" is distributive with respect to the t-norm $F$ at least in correspondence to the values $x, y_1, y_2$ which are information measures of sets $A, B_1, B_2$ belonging to J-independent algebras.

If the class $\mathcal{K}$ contains a suitable number of elements, or equivalently if we are looking for "universal" t-norms (see e.g. [7]), then we can use theorems (3.1.b) and (3.2.b) to recognize that only the t-norms

$$(6.6) \quad F(x,y) = \inf(x,y)$$

$$(6.7) \quad F(x,y) = \sup\{0, -k \cdot \log[e^{-\frac{x}{k}} + e^{-\frac{y}{k}}]\}$$

with $k > 0$, are compatible with the relation (1.10.b).

This result imposes a strong limitation to the choice of the information measures which may be associate to an information structure $(\Omega, \mathcal{S}, \mathcal{K})$, whenever the class $\mathcal{K}$ is quite "rich". This fact motivates Kampé de Fériet and others [8], to generalise the definition of information measure. They propose to substitute axiom (6.3.1) by the following one

$$(6.3.b) \quad (\mathcal{A}, \mathcal{B}) \in \mathcal{K} \ , \ A \in \mathcal{A} \ , \ B \in \mathcal{B} \Longrightarrow$$
$$\Longrightarrow J(A \cap B) = G[J(A), J(B)]$$

It has been recognized, without difficulties, that the function $G$ (independence law) must be a t-conorm on $\mathbb{R}^+$. Thus the compatibility equation (6.5) becomes

$$(6.8) \quad G[J(A), F\{J(B_1), J(B_2)\}] =$$
$$= F[G\{J(A), J(B_1)\}, G\{J(A), J(B_2)\}]$$

that is a distributivity equation of the t-conorm $G$ with respect to the t-norm $F$. Under the same condition specified for the classical case, we may use theorems (3.1.b) and (3.2.b) to obtain all the universal pairs ($G = t-conorm$, $F = t-norm$) of compatible independence and composition laws.

We wish to remark that, while the t-conorms $G$ may have all the possible expressions specified in theorem 1.1.a, the t-norms $F$ must be choosen among those for which $\Delta_o = \Delta_F = \{0\} \cup [\alpha_o, +\infty]$, including the two limit cases $\alpha_o = 0$, $\alpha_o = +\infty$.

## VII. References


[1] D.Dubois, H.Prade : *"Théorie des possibilités"*, Masson, Paris 1988

[2] J.Kampé de Fériet : *"Note di teoria dell'informazione"* Quad. Gruppi di ricerca del CNR, Roma 1972

[3] C.H.Ling : *"Representation of associative function"*, Publ.Math. 12 1965

[4] S.Weber : *"$\perp$ −decomposable measures and integrals for archimedean t-conorms"* J.Math.Anal. and Appl, 101 1984

[5] B.Forte, J.Kampé de Fériet : *"Information et Probabilité"*, CRAS Paris, 265-A 1965

[6] C.Bertoluzza, S.De Simoni : *"Elementi di teoria dell'-informazione"*, Atti del convegno sull' Inferenza Statistica Pura, Dipartimento Statistico, Firenze 1991

[7] P.Benvenuti : *"Sulle misure d'informazione compositive con traccia compositiva universale"*, Rend.Mat. 3-4, vol II, serie IV 1969

[8] F.Barbaini, C.Bertoluzza : *"Sul concetto d'indipendenza in teoria dell'informazione"* Statistica, Anno XLV, n.2 1985

# Alternating Projection onto Fuzzy Convex Sets

Seho Oh and Robert J. Marks II
Department of Electrical Engr., FT-10
University of Washington
Seattle, WA 98195

***Abstract***— Alternating projections onto convex sets (POCS) is powerful tool for signal and image restoration. However, if POCS is among three or more nonintersecting convex sets, the result is not unique and POCS is generally not useful. This, however, can be overcome by allowing solutions that are in some sense, 'close' to each convex set. Such relaxation can be achieved through fuzzification of the sets into fuzzy convex sets. By performing POCS among the $\alpha$-cuts of fuzzified sets, good solutions can be obtained. We propose morphological dilation as a fuzzification procedure. Fuzzy POCS is illustrated through application to the problems of time-bandwidth product minimization, signal extrapolation and solution of simultaneous equations.

## Introduction

*Alternating projections onto convex sets* (POCS) [1] is a remarkably powerful method of signal recovery and synthesis. A (crisp) set, $A$, is convex if $\vec{x}_1 \in A$ and $\vec{x}_2 \in A$ implies that $\lambda\vec{x}_1 + (1-\lambda)\vec{x}_2 \in A$ for all $0 \leq \lambda \leq 1$. In other words, the line segment connecting $\vec{x}_1$ and $\vec{x}_2$ is totally subsumed in $A$. Examples of sets of signals that are convex are the sets of bandlimited signals, duration limited signals, bounded signals, signals with energy less than one, signals with unit area, and complex signals with a specified phase.

The projection onto a convex set is illustrate in Figure 1. For a given $\vec{y} \notin A$, the projection onto $A$ is the unique vector $\vec{x} \in A$ such that the mean square distance between $\vec{x}$ and $\vec{y}$ is minimum. If $\vec{y} \in A$, then the projection onto $A$ is $\vec{y}$.

Here is the remarkable result of POCS. Given two or more convex sets with nonempty intersection, alternately projecting among the sets will converge to a point included in the intersection [2]. This is illustrated in Figure 2. If two convex sets do not intersect, then convergence is to a limit cycle that is a mean square solution to the problem. Specifically, the cycle is between points in each set that are closest in the mean square sense to the other set [3]. This is illustrated in Figure 3.

POCS breaks down in the important case where three or more convex sets do not intersect [4]. POCS converges to greedy limit cycles that are dependent on the ordering of the projections and do not display any desirable optimality properties. This is illustrated in Figure 4.

This third case, however, can be successfully addressed by fuzzy POCS. The problem becomes one of finding a solution that is, in some sense, equally close to each of the convex sets. The concept of 'close' suggests a fuzzification of the nonintersecting convex sets to fuzzy convex sets [5]. Even if three or more crisp sets do not intersect, $\alpha$-cuts of their fuzzification can. In some cases, there exists an $\alpha$ such that intersection occurs at a single point. This is illustrated in Figure 5.

## Fuzzy Convex Sets

The fuzzy set $A_f$ ($f$ for fuzzy) on the universal set $E$ is defined by the membership function $\mu_A(\cdot)$ which maps $E$ to the real value $[0,1]$. The set $A_f$ can be written as

$$A_f = \{\vec{x}/\mu_A(\vec{x}) \mid \vec{x} \in E\}$$

Let $A_f^\alpha$ denote the crisp set corresponding to an $\alpha$-cut of $A_f$.

$$A_f^\alpha = \begin{cases} \{\vec{x} \mid \mu_A(\vec{x}) \geq \alpha, \vec{x} \in E\} & ; \quad \alpha \neq 0 \\ E & ; \quad \alpha = 0 \end{cases} \tag{1}$$

The fuzzy set $A_f$ is convex if all of its $\alpha$-cuts ($0 \leq \alpha \leq 1$) are convex. Equivalently [6], the fuzzy set $A_f$ is convex if and only if for every $0 \leq \lambda \leq 1$,

$$\mu_A[\lambda\vec{x}_1 + (1-\lambda)\vec{x}_2] \geq \min\{\mu_A(\vec{x}_1), \mu_A(\vec{x}_2)\}$$

## Fuzzified Convex Sets and Their Projections

Two methods of fuzzification of crisp convex sets of signals to fuzzy convex sets are useful in fuzzy POCS. If the crisp convex set is parameterized, the fuzzy convex set, in many cases, can be generated by fuzzification of parameter set. There is a homomorphism between the signal and

0-7803-0614-7/93/$03.00 

parameter sets. The parameter set typically exists on an interval (*e.g.* $0 \le$ Bandwidth $\le \Omega$ for a set of bandlimited functions and $0 \le$ Energy $\le E$ for a set of signals with energy less than or equal to $E$) and is therefore typically convex. Fuzzification is achieved by dilation of this set with a convex dilation kernel. If the dilation kernel is convex, then the result is an $\alpha$-cut of a fuzzy convex set. The degree of membership of a signal in the fuzzy signal set is equal to that of the membership of the parameter in the fuzzified parameter set. If, on the other hand, the crisp set of functions is not parameterized, fuzzification can be achieved through the direct morphological dilation of the signal in the set. By choosing convex dilation kernels of increasing dimension, $\alpha$-cuts of the fuzzified convex set can be generated.

We now illustrate with some specific examples.

*Bandlimited Signal*

The set of the bandlimited signals with bandwidth $\Omega$ is

$$A_1 = \{x(t) \mid X(\omega) = 0 \quad \text{for} \quad |\omega| > \Omega\}$$

where the Fourier transform is

$$X(\omega) = \int_{-\infty}^{\infty} x(t)e^{-j\omega t}dt$$

Clearly, $A_1$ is convex. Let $\Omega_\alpha$ be a nondecreasing function of $\alpha$ for $0 \le \alpha \le 1$ and $\Omega_1 = 0$. The dilation kernel [1] [7, 8] used to generate the $\alpha$-cut of the fuzzification is

$$H_1^\alpha = \{\omega \mid |\omega| \le \Omega_\alpha\}$$

The $\alpha$-cut of the fuzzified set is

$$A_{1_f}^\alpha = \{x(t) \mid X(\omega) = 0 \quad \text{for} \quad |\omega| > \Omega + \Omega_\alpha\}$$

and the projection onto the convex $\alpha$-cut is

$$\mathcal{P}_1^\alpha x \Leftrightarrow \begin{cases} 0 & ; \quad |\omega| > \Omega + \Omega_\alpha \\ X(\omega) & ; \quad \text{otherwise} \end{cases}$$

where $\Leftrightarrow$ denotes a Fourier transform pair.

*Timelimited Signal*

The convex set of timelimited signals is

$$A_2 = \{x(t) \mid x(t) = 0 \quad \text{for} \quad |t| > \tau\}$$

where $2\tau$ is the centered interval over which the signal can be nonzero. The dilation kernel for the $\alpha$-cut of the fuzzification operator is

$$H_2^\alpha = \{t \mid |t| \le \tau_\alpha\}$$

[1]The dilation of the set $C \subset E$ with dilation kernel $D \subset E$ is $C \oplus D = \{\vec{x} \mid \text{there exist } \vec{y} \text{ such that } \vec{y} \in C \text{ and } \vec{x} - \vec{y} \in D\}$

where $\tau_\alpha$ is a nondecreasing function $\alpha$ for $0 \le \alpha \le 1$ and $\tau_1 = 0$. The $\alpha$-cut of the fuzzified set and the corresponding projection are

$$A_{2_f}^\alpha = \{x(t) \mid x(t) = 0 \quad \text{for} \quad |t| > \tau + \tau_\alpha\}$$

and

$$\mathcal{P}_2^\alpha x = \begin{cases} 0 & ; \quad |t| > \tau + \tau_\alpha \\ x(t) & ; \quad \text{otherwise} \end{cases}$$

*Signals with Bounded Error*

For a given signal $p(t)$, the convex set of signals with a bound of $KR(t)$ is

$$A_3 = \{x(t) \mid |x(t) - p(t)| \le KR(t)\}$$

Using the dilation kernel

$$H_3^\alpha = \{k \mid |k| \le K_\alpha\}$$

where $K_\alpha$ is a nondecreasing function $\alpha$ for $0 \le \alpha \le 1$ and $K_1 = 0$. The convex $\alpha$-cut of the fuzzified set is

$$A_{3_f}^\alpha = \{x(t) \mid |x(t) - p(t)| \le R_\alpha(t)\}$$

where $R_\alpha(t) = (K + K_\alpha)R(t)$. The projection onto $\alpha$-cut of the fuzzy set is

$$\mathcal{P}_3^\alpha x = \begin{cases} p(t) + q(t) & ; \quad |x(t) - p(t)| > R_\alpha(t) \\ x(t) & ; \quad \text{otherwise} \end{cases}$$

where $q(t) = R_\alpha(t)[x(t) - p(t)]/|x(t) - p(t)|$.

*Fuzzification by Signal Dilation*

When the constraint set is not specified by a parameter or parameter set, then $\alpha$-cuts of the fuzzification can be performed by direct dilation of each signal in the set. Let $g(\vec{x})$ be the fuzzification dilation kernel which maps $E$ to the real value $[0,1]$. Then the fuzzification of the crisp set, $A$, to the fuzzy set $A_f$ is defined by

$$A_f = \{\vec{x}/\mu(\vec{x}) \mid \vec{x} \in E\} \tag{2}$$

where

$$\mu(\vec{x}) = \max_{\vec{y}}\{g(\vec{x} - \vec{y}) \mid \vec{y} \in A\}$$

Then we have the following theorem.

**Theorem 1** *Let the crisp set $A$ be convex and let $G = \{\vec{x}/g(\vec{x}) \mid \vec{x} \in E\}$ be a convex dilation kernel. Then $A_f$ in equation (2) is a fuzzy convex set.*

The proof of the above theorem is in Appendix A.

We specify

$$g(\vec{x}) = m(\| \vec{x} \|)$$

Let $m(0) = 1$ and $m(z)$ be a monotonic decreasing function for $z \geq 0$. Then the $\alpha$-cut of the fuzzification kernel, $g(\vec{x})$, is always a (convex) sphere. Let $m(R_\alpha) = \alpha$. We then have the following theorem for the projection onto $A_f^\alpha$

**Theorem 2** *Let* $\vec{x}_0 \notin A_f^\alpha$ *and* $\vec{x}_p^\alpha = \mathcal{P}^\alpha \vec{x}_0$. *Then*

$$\vec{x}_p^\alpha = \vec{x}_p^1 + \frac{R_\alpha}{\| \vec{x}_0 - \vec{x}_p^1 \|}(\vec{x}_0 - \vec{x}_p^1)$$

where $\vec{x}_p^1$ is the projection onto the crisp set. The proof of is in Appendix B.

## Method of Alternating Projections onto Fuzzy Convex Sets

The optimal POCS solution is achieved by the maximum value of $\alpha$ that results in a non empty intersection of all signal sets. In certain cases, this intersection can be at a single point.

To find the optimal solution, we start at a large value of $\alpha$ and iterate. If the iteration does not converge, $\alpha$ is decreased and the iteration is repeated. If convergence does occur, a search can be performed between the current and previous values of $\alpha$ for the optimal solution.

## Examples

In this section, we will give the examples of signal synthesis and restoration based on fuzzy POCS.

*Example 1 : Time-Bandwidth Product*

Our problem is to find a one dimensional signal $x[n]$ which is positive, bandlimited, timelimited signal and has a specified area $V = \sum_n x[n]$. There exists no signal that satisfies all of these constraints. To apply fuzzy POCS, we will keep constant area and positivity sets crisp. The sets of bandlimited and timelimited signals, though, will be fuzzified, The convex crisp sets are

$$\begin{aligned} A_1 &= \{x[n] \mid X[k] = 0 \text{ for } k \neq 0\} \\ A_2 &= \{x[n] \mid x[n] = 0 \text{ for } n \neq 0\} \\ A_4 &= \{x[n] \mid x[n] \geq 0\} \\ A_5 &= \left\{ x[n] \mid \sum_{n=-L/2}^{L/2-1} x[n] = V \right\} \end{aligned}$$

where $L$ is the length of the $x[n]$ and $X[k]$ is the discrete Fourier transform of $x[n]$. The $\alpha$-cuts of the fuzzified sets are

$$A_{1_f}^\alpha = \left\{ x[n] \mid X[k] = 0 \text{ for } |k| > -\sqrt{\xi}\Theta_\alpha \right\}$$

$$A_{2_f}^\alpha = \left\{ x[n] \mid x[n] = 0 \text{ for } |n| > -\frac{\Theta_\alpha}{\sqrt{\xi}} \right\}$$

where $\Theta_\alpha = \frac{L}{4}\log\alpha$ and $\xi$ parameterize the relative importance between bandlimitedness and timelimitedness. Figure 6 using $V = L = 1024$ shows the results for various values of $\xi$. Figure 6a, 6b and 6c result when $\xi$ is 4, 1 and 0.25, respectively. In Figure 6, the solid lines show $x[n]$ and the broken lines show the Gaussian function which has the same peak value, same mean, and the same variance as the signal, $x[n]$. When $\xi$ is large, then timelimitedness is more important than bandlimitedness as shown in Figure 6. Not surprisingly, fuzzy POCS yields a result quite close to the Gaussian curve. The Gaussian is the function which displays the minimum time-bandwidth product [9].

*Example 2 : Bandlimited Signal Extrapolation*

This example is motivated by the celebrated Papoulis Gerchberg algorithm [10, 11]. The problem is estimation of a high bandwidth signal, $p[n]$, with a signal of lower bandwidth. Assume the signal is $p[n] = \text{sinc}(2Bn)$. Let $A_1$ be the set of signals with frequency components no greater than $\frac{B}{2}$. Note that $x[n] \notin A_1$. In our simulation, $B = 1/64$. The crisp convex sets are

$$\begin{aligned} A_1 &= \{x[n] \mid X[k] = 0 \text{ for } |k| > L/128\} \\ A_3 &= \{x[n] \mid x[n] = p[n]\} \end{aligned}$$

where $p[n] = \text{sinc}(n/32)$. The $\alpha$-cuts of the fuzzified set are

$$A_{1_f}^\alpha = \{x[n] \mid X[k] = 0 \text{ for } |k| \leq \Phi_\alpha\}$$

$$A_{3_f}^\alpha = \left\{ x[n] \mid |x[n] - p[n]| \leq \frac{\Psi[n]}{\sqrt{\xi}}\log\alpha \right\}$$

where

$$\Phi_\alpha = -\frac{\sqrt{\xi}L}{4}\log\alpha + \frac{L}{128},$$

$$\Psi[n] = \frac{1 - \exp(-|2n|/L)}{\log\alpha_{min}},$$

and $\xi$ parameterized the relative importance of the two constraints. $L$ is the length of $x[n]$, and $\alpha$ is varied from 1 to $\alpha_{min} = 1/3$. Figure 7a, 7b and 7c correspond to values of $\xi = 4$, 1 and 0.25 respectively. In our case, if $\xi$ is large, then the known signal is more important than the signal being bandlimited. Figure 8 shows $|X[k]|$ for each case.

*Example 3 : Solution of a Set of Overdetermined Linear Equations*

This example outlines solution of the overdetermined linear equation, $\underline{Q}\vec{x} = \vec{y}$, by POCS, assuming that $\underline{Q}^T\underline{Q}$ is nonsingular. The minimum mean square error solution is

$$\vec{x}_a^* = (\underline{Q}^T\underline{Q})^{-1}\underline{Q}^T\vec{y}$$

In other words, for a given $\underline{Q}$ and $\vec{y}$,

$$\min_{\vec{x}}\{\| \underline{Q}\vec{x} - \vec{y} \|\} = \| \underline{Q}\vec{x}_a^* - \vec{y} \| \quad (3)$$

where $\| \cdot \|$ is $l_2$ norm.

For the POCS solution, however, the result is a quite different. Let

$$\underline{Q} = [\vec{q}_1, \vec{q}_2, \cdots, \vec{q}_N]^T$$

and $\vec{y} = [y_1, y_2, \cdots, y_N]^T$. Then the linear equation can be written as [12]

$$\vec{q}_i^T \vec{x} = y_i \quad \text{for} \quad i = 1, 2, \cdots, N$$

The crisp solution set for the $i^{th}$ equation is

$$B_i = \{\vec{x} \mid |\vec{q}_i^T \vec{x} - y_i| \leq 0\}$$

Using the fuzzification with a dilation kernel $h(\alpha) = e^{-\alpha}$ on the singleton value of $\vec{q}_i^T \vec{x} - y_i$, the $\alpha$-cut of the fuzzification is

$$B_{i_f}^{\alpha} = \{\vec{x} \mid |\vec{q}_i^T \vec{x} - y_i| \leq -\log\alpha\} \quad (4)$$

We now seek the maximum value of $\alpha$ (minimum of $-\log\alpha$) which satisfies (4). In other words,

$$\min_{\vec{x}}\{\max_i[|\vec{q}_i^T \vec{x}_p^* - y_i|]\}$$

$$= \min_{\vec{x}}\{\| \underline{Q}\vec{x}_p^* - \vec{y} \|_\infty\}$$

where $\| \cdot \|_\infty$ is $l_\infty$ norm of the metric space. Thus, in contrast to minimum mean square ($l_2$ norm) solution in (3), we obtain a minimum $L_\infty$ norm solution using fuzzy POCS.

When we use the nonparameterized method for fuzzification with $g(\vec{x}) = m(\| \vec{x} \|)$ and $m(z) = e^{-z}$, the $\alpha$-cut of the fuzzification is

$$C_{i_f}^{\alpha} = \{\vec{x} \mid |\vec{q}_i^T \vec{x} - y_i| \leq - \| \vec{q}_i \| \log\alpha\} \quad (5)$$

We now seek the maximum value of $\alpha$ which satisfies (5). In other words,

$$\min_{\vec{x}}\left\{\max_i\left[\frac{|\vec{q}_i^T \vec{x}_q^* - y_i|}{\| \vec{q}_i \|}\right]\right\}$$

$$= \min_{\vec{x}}\{\| \underline{D}_Q^{-1}\underline{Q}\vec{x}_q^* - \underline{D}_Q^{-1}\vec{y} \|_\infty\}$$

where

$$\underline{D}_Q = diag[\| \vec{q}_1 \|, \| \vec{q}_2 \|, \cdots, \| \vec{q}_N \|]$$

## Notes

1. Consider the case, illustrated in Figure 3, where two crisp convex sets do not intersect. Let the limit cycle be between points $\vec{y}_A$ and $\vec{y}_B$. If both sets are fuzzified using the same convex dilation kernel, then an optimal solution using fuzzy POCS is the point $(\vec{y}_A + \vec{y}_B)/2$.

2. Our procedure using fuzzy POCS begins with small convex sets. The sets grow until intersection occurs. Alternately, initialization can be initiated with large $\alpha$-cuts of convex sets. The sets are reduced in size until iteration breaks into a small limit cycle.


## Acknowledgements

The authors acknowledge the financial support provided by Boeing Computer Services, Seattle WA and the Washington Technology Center, the University of Washington, Seattle, WA.


## References


[1] H. Stark, editor, **Image Recovery: Theory and Application**, (Academic Press, Orlando, 1987).

[2] D.C. Youla and H. Webb, "Image restoration by method of convex set projections: Part I - Theory", *IEEE Trans. Med. Imaging*, vol MI-1, pp.81-94, 1982.

[3] M.H. Goldburg and R.J. Marks II "Signal synthesis in the presence of an inconsistent set of constraints", *IEEE Transactions on Circuits and Systems*, vol. CAS-32 pp. 647-663 (1985).

[4] D.C. Youla and V. Velasco, "Extensions of a result on the synthesis of signals in the presence of inconsistent constraints", *IEEE Transactions on Circuits and Systems*, vol. CAS-33, pp.465-468 (1986).

[5] M.R. Civanlar and H.J. Trussel, "Digital signal restoration using fuzzy sets", *IEEE Transactions on Acoustics, Speech and Signal Processing*, vol. ASSP-34, p.919 (1986).

[6] L. Zadeh, "Fuzzy Sets", **Information Control**, vol.8, pp.338-353, 1965. Reprinted in J.C. Bezdek & S.K. Pal, **Fuzzy Models for Pattern Recognition**, (IEEE Press, 1992).

[7] J. Serra, **Image analysis and mathematical morphology**, (Academic Press, New York, 1982).

[8] P. Maragos and R. W. Schafer, "Morphological filter - Part I: Their set-theoretic analysis and relations to linear shift-invariant filters", *IEEE Trans. on Acoustic, Speech, and Signal Processing*, vol.ASSP-35, no.8, 1987. pp.1153-1169.

[9] J.B. Thomas, **An Introduction to Statistical Communication Theory**, (Wiley, NY, 1969).

[10] A. Papoulis, "A new algorithm in spectral analysis and bandlimited signal extrapolation", *IEEE Transactions on Circuits and Systems*, vol.CAS-22, pp.735-742 (1975).

[11] R.J. Marks II, **An Introduction to Shannon Sampling and Interpolation Theory**, (Springer-Verlag, 1991).

[12] S. Kuo and R.J. Mammon, "Resolution enhancement of tomographic images using the row action projection method", *IEEE Transactions on Medical Imaging*, vol.10, no, pp.593-601 (1992).

# Appendices

## PROOF OF THEOREM 1

Let $G^\alpha$ denote an $\alpha$-cut of $G$

$$G^\alpha = \{\vec{x} \mid g(\vec{x}) \geq \alpha\}$$

Let $\vec{x}_1 \in A^\alpha_f$ and $\vec{x}_2 \in A^\alpha_f$. Then there exist $\vec{y}_1$ and $\vec{y}_2$ such that

$$\vec{x}_1 - \vec{y}_1 \in G^\alpha, \quad \vec{x}_2 - \vec{y}_2 \in G^\alpha$$

Now, we examine $\gamma\vec{x}_1 + (1-\gamma)\vec{x}_2$,

$$\gamma\vec{x}_1 + (1-\gamma)\vec{x}_2$$

$$= \gamma(\vec{y}_1 + \vec{x}_1 - \vec{y}_1) + (1-\gamma)(\vec{y}_2 + \vec{x}_2 - \vec{y}_2)$$

$$= [\gamma(\vec{x}_1 - \vec{y}_1) + (1-\gamma)(\vec{x}_2 - \vec{y}_2)]$$

$$+[\gamma\vec{y}_1 + (1-\gamma)\vec{y}_2]$$

Because $A$ and $G^\alpha$ are convex sets, we have

$$\gamma\vec{y}_1 + (1-\gamma)\vec{y}_2 \in A$$

$$\gamma(\vec{x}_1 - \vec{y}_1) + (1-\gamma)(\vec{x}_2 - \vec{y}_2) \in G^\alpha$$

So,

$$\gamma\vec{x}_1 + (1-\gamma)\vec{x}_2 \in A^\alpha_f$$

Therefore our proof is complete.
Q. E. D.

## PROOF OF THEOREM 2

Before we prove the Theorem, we show the following Lemma

**Lemma 1** *For any* $\vec{y} \in A$ *and* $\vec{x} \notin A^\alpha_f$, *then* $\| \vec{x} - \vec{y} \| > R_\alpha$

**Proof** : Assume that $\| \vec{x} - \vec{y} \| \leq R_\alpha$. Then $\vec{y} \in A$ and $\| \vec{x} - \vec{y} \| \leq R_\alpha$ implies $\vec{x} \in A^\alpha_f$. This contradicts the assumption and our proof is complete. Q.E.D.

**Proof of Theorem 2**

$$\vec{x}^\alpha_p - \vec{x}^1_p = \frac{R_\alpha}{\| \vec{x}_0 - \vec{x}^1_p \|}(\vec{x}_0 - \vec{x}^1_p)$$

Here, $\| \vec{x}^\alpha_p - \vec{x}^1_p \| = R_\alpha$ and $\vec{x}^1_p \in A$. Therefore $\vec{x}^\alpha_p \in A^\alpha_f$. Also,

$$\begin{aligned} \vec{x}_0 - \vec{x}^\alpha_p &= (\vec{x}_0 - \vec{x}^1_p) + (\vec{x}^1_p - \vec{x}^\alpha_p) \\ &= (\vec{x}_0 - \vec{x}^1_p)\left[1 - \frac{R_\alpha}{\| \vec{x}_0 - \vec{x}^1_p \|}\right] \end{aligned}$$

$$\| \vec{x}_0 - \vec{x}^\alpha_p \| = \| \vec{x}_0 - \vec{x}^1_p \| - R_\alpha$$

For any $\vec{y} \in A^\alpha_f$, let $\vec{y}^1_p = \mathcal{P}^1\vec{y}$. Then $\| \vec{y} - \vec{y}^1_p \| \leq R_\alpha$.

$$\| \vec{y} - \vec{x}_0 \| \geq \| \vec{y}^1_p - \vec{x}_0 \| - \| \vec{y} - \vec{y}^1_p \| \geq \| \vec{y}^1_p - \vec{x}_0 \| - R_\alpha$$

If $\vec{y}^1_p \neq \vec{x}^1_p$, then

$$\| \vec{y} - \vec{x}_0 \| \geq \| \vec{y}^1_p - \vec{x}_0 \| - R_\alpha$$

$$> \| \vec{x}^1_p - \vec{x}_0 \| - R_\alpha = \| \vec{x}_0 - \vec{x}^\alpha_p \|$$

Therefore

$$\| \vec{y} - \vec{x}_0 \| > \| \vec{x}_0 - \vec{x}^\alpha_p \|$$

If $\vec{y}^1_p = \vec{x}^1_p$, then

$$\| \vec{y} - \vec{x}_0 \| \geq \| \vec{x}^1_p - \vec{x}_0 \| - \| \vec{y} - \vec{x}^1_p \|$$

The equality holds for only the case that $\vec{y} = \vec{x}^\alpha_p$.

$$\| \vec{y} - \vec{x}_0 \| > \| \vec{x}_0 - \vec{x}^\alpha_p \|$$

Therefore our proof is complete.
Q. E. D.

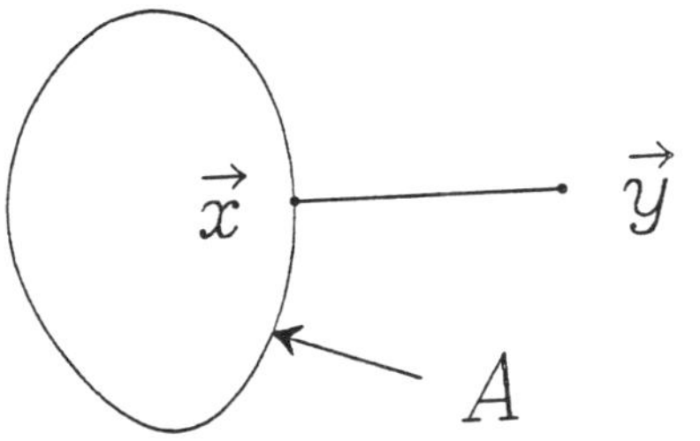


**Figure 1**: The set, $A$, is convex. The projection of $\vec{y}$ onto $A$ is the unique element in $A$ closest to $\vec{y}$ in the mean square sense.

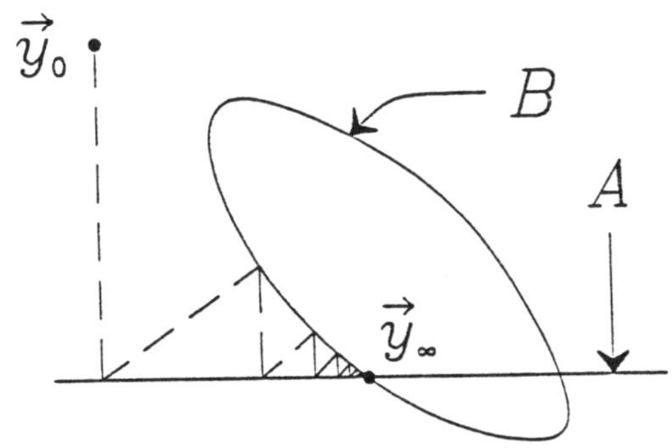


**Figure 2**: Alternating projection between two or more convex sets with nonempty intersection results in convergence to a fixed point, in that intersection. Here, sets $A$ (a line segment) and $B$ are convex. Initializing the iteration at $\vec{y}_0$, convergence is to $\vec{y}_\infty \in A \cap B$.

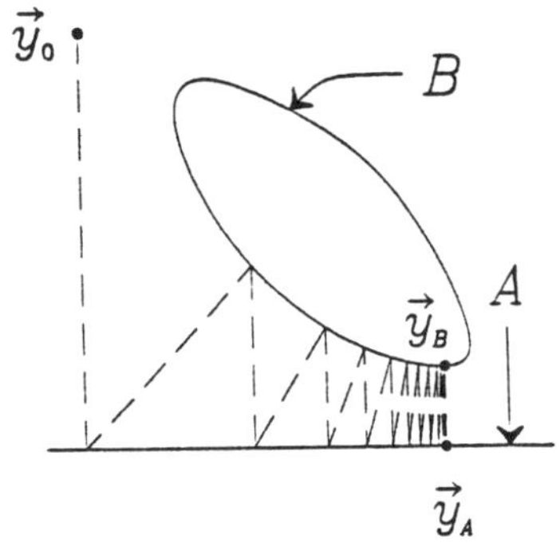


**Figure 3**: If two convex sets, $A$ and $B$, do not intersect, POCS converges to a limit cycle, here between the points $\vec{y}_A$ and $\vec{y}_B$. The point $\vec{y}_B \in B$ is the point in $B$ closest to the set $A$. A similar property is true for $\vec{y}_A$.

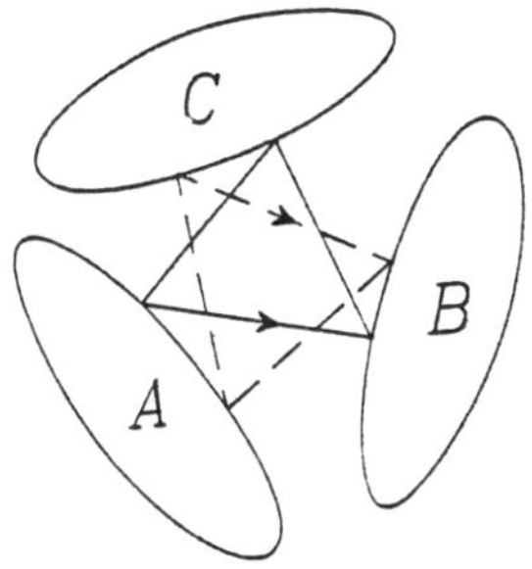


**Figure 4**: If three or more convex sets do not intersect, the POCS convergences to greedy limit cycles with no particularly useful properties. As illustrated here, the limit cycles can differ for different choices of set ordering.

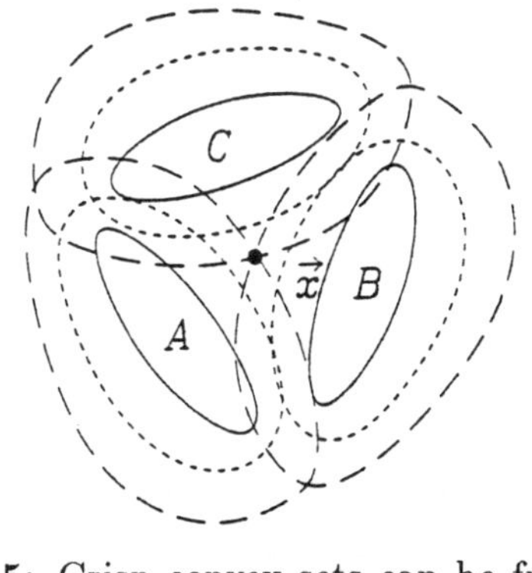


**Figure 5**: Crisp convex sets can be fuzzified into fuzzy convex sets. The $\alpha$-cuts of the fuzzifications are convex. As illustrated here, there can exist an $\alpha$-cut of each of the convex sets such that the resulting intersection is nonempty. Application of POCS to these $\alpha$-cuts will result to convergence to a point in this intersection. The solution, for large $\alpha$, is then 'close' to each of the underlying crisp sets.

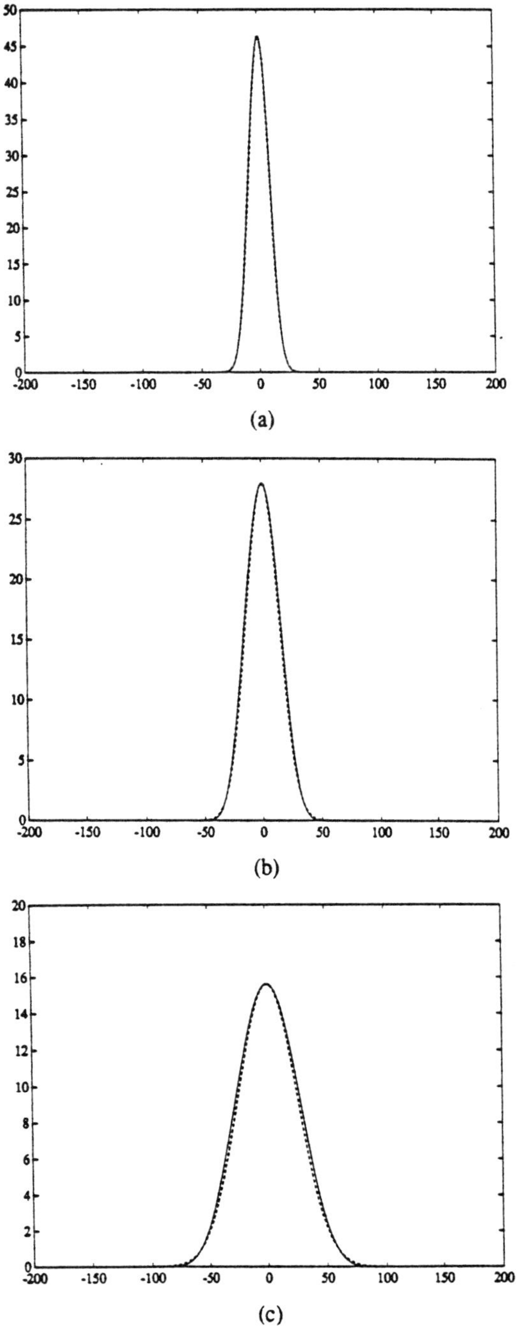


Figure 6: Fuzzy POCS solutions of a signal that is both time limited and bandlimited. The importance of being bandlimited increases from (a) to (c). The result is compared to a Gaussian curve fit (broken line) in each case.

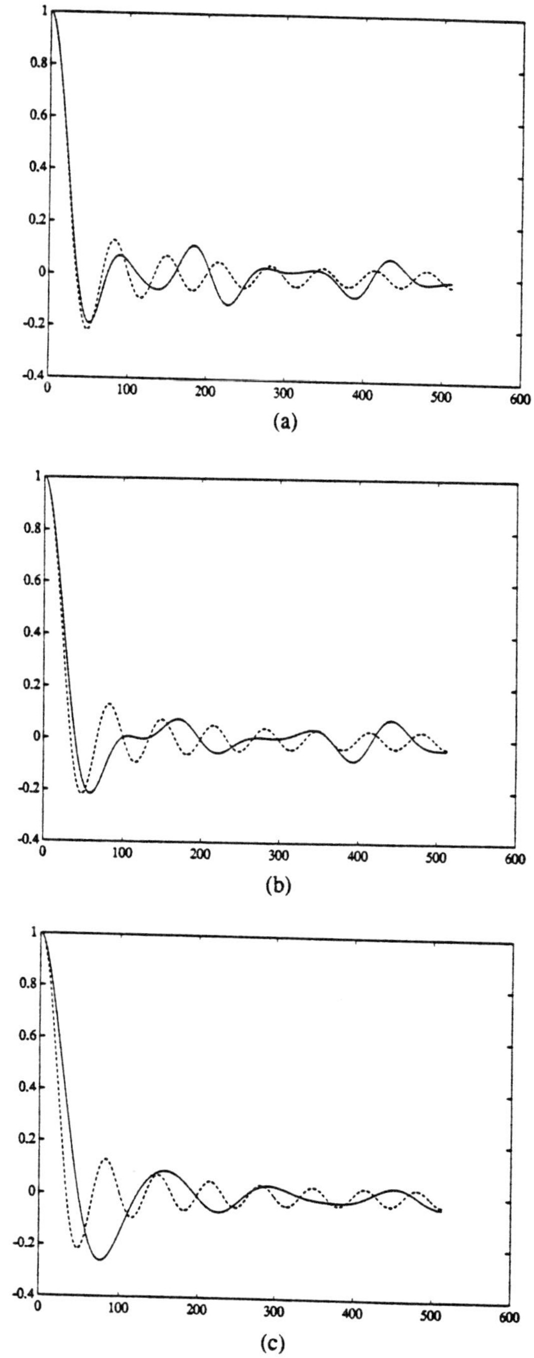


Figure 7: A bandlimited signal is plotted here with a broken line. We attempt to fit a signal with lower bandwidth to the known signal while simultaneously keeping the error within a specified boundary. The fuzzy POCS results are shown, from (a) to (c), as the allowable bandwidth increases and the bound constraint is relaxed.

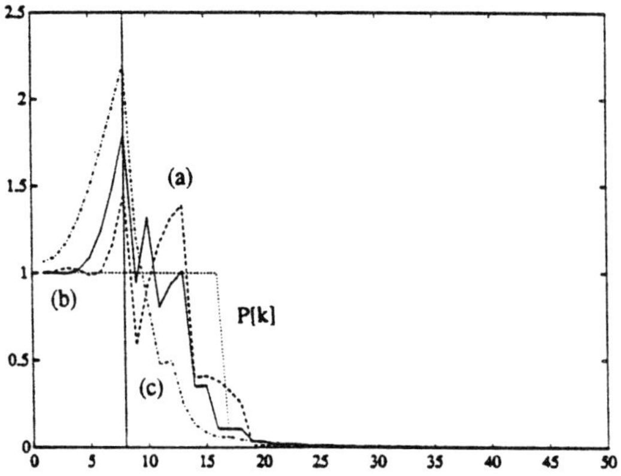


**Figure 8**: The magnitudes of the discrete Fourier transform of the signals in Figure 7(a) through Figure 7(c).

# Comparing Hasse Diagrams

Bienvenido Jose A. Juliano, Jr. and Wyllis Bandler
Department of Computer Science B–173
The Florida State University
Tallahassee, Florida 32306–4019, U.S.A.

***Abstract*— In this paper, we investigate the applicability of some relational strength measures when comparing and analyzing Hasse diagrams. The structures used were derived from a particular line of research at our Department, ongoing as of this writing, that deals with human perception of an urban environment.**

## I. Introduction

A methodology for comparing Hasse diagrams is presented in this paper. The Hasse diagrams used are derived from data collected in an Urbanistics project, described in [1], that investigates human perception of an urban environment. Knowledge was elicited by allowing interviewees to "relate" bipolar constructs, containing both positive and negative poles, with tangible landmarks in the vicinity of their neighborhood. The triangular products of these *fuzzified* relations were used to derive, say, construct–construct relations from which Hasse diagrams are generated. These diagrams represent certain knowledge structures that the interviewee possesses.

## II. Hasse Diagrams

Relations are analyzed in order to extract deep structure from them. One way to do this is to derive certain closures and interiors depending on the relational properties under investigation. Perhaps one of the most interesting group of properties is that exhibited by *orders* and *preorders.* Normally, Hasse diagrams are used to graphically illustrate the different categories hidden by such relations.

### A. Processing Methodology

The data used in this investigation were processed using a conceptual procedure presented in [2].Given some fuzzy relation $T \subseteq X \times Y$, fuzzy local preorders (say, $R \subseteq X \times X$) were derived by taking the local preorder closure of the triangle product[1] $(T \triangleleft T^{-1})$.

The following *Hassefication procedure*, given in [2], is then used

1. Take an $\alpha$–cut $R_\alpha$.
2. Form $S = \text{sym int} R_\alpha$. This is a *local equivalence.*
3. Remove the zero–class $C_0$, consisting of all $x$'s unrelated by $S$ to any elements.
4. Let $E$ be the *factor set* of $X \setminus C_0$ according to $S$.
5. Denote by $\preceq$ the *factor relation* $R_\alpha/S$, an *order.*
6. Let $P = (E, \preceq)$. Draw the Hasse diagram $H(P)$.

### B. Graph Theoretic Concepts

A pair $P = (E, \preceq)$ is a (partially) ordered set if for all $x, y, z \in E$

1. $x \preceq x$
2. $x \preceq y$, $y \preceq x$ implies $x = y$
3. $x \preceq y$, $y \preceq z$ implies $x \preceq z$

$P = (E, \preceq)$ can also be interpreted as a *directed graph* $D = D(P)$ defined on a set $E$ of vertices and with edges of the form $(x, y)$ whenever $x \preceq y$. The *comparability graph* $G = G(P)$ is the undirected version of $D(P)$. Hence, $G(P)$ has vertex set $E$ and edges $\{x, y\}$ whenever $x \preceq y$. Note that in cases when the relation is a *strict order,* $D(P)$ and $G(P)$ may be defined without loops [3].

We say that $y$ covers $x$ if $x \prec y$ and there is no $z \in E$ such that $x \prec z \prec y$. Hence, the complete information about $P$ is contained in the Hasse diagram $H = H(P)$, namely the subgraph of $D(P)$ retaining just the covering edges. Usually $H(P)$ is drawn as an undirected graph with the understanding that the orientation is from "bottom" to "top".

### C. Choosing Alpha–Cuts

By taking various alpha–cuts on the relations considered, one can note the relationship between the cardinality of the factor set $|X|$, the *mean fuzzy cardinality*[2] of

[1] This study was restricted to the use of the Kleene–Dienes fuzzy implication operator.

[2] We denote a formulation for the mean fuzzy cardinality of a fuzzy relation $R$ by

$$\dot{R} = \frac{\Sigma\text{Count}(R)}{|X|^2}$$

0-7803-0614-7/93$03.00 ©1993IEEE

TABLE I
NUMBER OF STRATA AND CARDINALITIES FOR VARIOUS ALPHA–CUTS CONSIDERED

| *Original Relation* | *α–cut Value* | *No. of Strata* | *Factor Set* $\|X\|$ | *Factor Relation* $\dot{R}$ |
|---|---|---|---|---|
| | 0.40 | 2 | 2 | 0.750 |
| | 0.50 | 3 | 4 | 0.562 |
| **gr17** | 0.60 | 5 | 12 | 0.312 |
| | 0.80 | 3 | 10 | 0.280 |
| | 1.00 | 2 | 6 | 0.277 |
| | 0.20 | 3 | 4 | 0.562 |
| | 0.40 | 2 | 4 | 0.375 |
| **gr18** | 0.60 | 2 | 6 | 0.277 |
| | 0.80 | 2 | 10 | 0.180 |
| | 1.00 | 3 | 14 | 0.183 |
| | 0.40 | 1 | 1 | 1.000 |
| **gr21** | 0.50 | 4 | 6 | 0.472 |
| | 0.60 | 4 | 7 | 0.428 |
| | 0.80 | 2 | 3 | 0.555 |

the factor relation, and the number of strata in the corresponding Hasse diagram. Notice from Table I that the cardinality of the factor set, which is the number of nodes in the Hasse diagram, peaks at about the same α–cut that the number of strata peaks. We based our choice of values for α–cuts on this observation. The α values considered are:[3]

- $\alpha = 0.60$ and $\alpha = 0.80$ for **gr17**
- $\alpha = 0.60$ and $\alpha = 0.80$ for **gr18**
- $\alpha = 0.50$ and $\alpha = 0.60$ for **gr21**

The corresponding Hasse diagrams for the α–values chosen are illustrated in Figures 1 to 6. To identify the Hasse diagrams considered, "$H_{i,j}$" will denote the Hasse diagram for **gr*i*** at an alpha–cut of $\alpha = j$. So, $H_{18,0.60}$ denotes a Hasse diagram for **gr18** at $\alpha = 0.60$.

## III. A Method for Mapping Hasse Diagrams

In [5], fuzzy relational products were used in comparing and verifying the correctness of medical knowledge structures. Several alpha–cuts and fuzzy implication operators were considered in that study.

Hasse diagrams in this study represent deep structure: a relation between (equivalence) classes. Let $P_A = (A, \preceq_A)$ and $P_B = (B, \preceq_B)$ be two (partially) ordered sets obtained through the Hassefication procedure in the previous section. Then, $A$ and $B$ are subsets of $2^X$ for some set $X$; *i.e.* $A, B \subseteq 2^X$. We can define a homomorphism between the Hasse diagrams $H_A = H(P_A)$ and $H_B = H(P_B)$ to be a function $\mathcal{F} : H_A \rightarrow H_B$ that maps from $H_A$ to $H_B$ consisting of a pair $\mathcal{F} = \langle f, g \rangle$ where

- $f : A \rightarrow B$ is a mapping between the factor sets of $H_A$ and $H_B$ ($f \subseteq 2^X \times 2^X$); and
- $g$ is a partial mapping from paths in $H_A$ into paths in $H_B$ (based on the *factor relations* $\preceq_A$ and $\preceq_B$).

Defining the mappings $f$ and $g$ are discussed in the next two sections. The notations are a modification of that used in [4].

### A. Mapping Nodes Between Hasse Diagrams

There are many ways to define a mapping $f$ between the factor sets of two Hasse diagrams under consideration. For example, one may use the notion of containment. In our case, the mapping $f$ was based on the notion of overlap.

We wish to form $\mathcal{F} : H_A \rightarrow H_B$ to be some mapping between the Hasse diagrams $H_A$ and $H_B$. Define $f$ as follows: match a node $\bar{x} \in A$ with a node $\bar{y} \in B$, which are of course (equivalence) classes[4], if the two nodes overlap.

For simplicity, we can use the following function to describe $f$:

$$f(\bar{x}, \bar{y}) = \begin{cases} 1.00, & \text{if } \bar{x} \cap \bar{y} \neq \emptyset; \text{ and} \\ 0.00, & \text{otherwise} \end{cases} \quad (1)$$

From Figure 7, since $\bar{x} \cap \bar{y} = \{d\} \neq \emptyset$, then $f(\bar{x}, \bar{y}) = 1.00$. Similarly, $\bar{w} \cap \bar{z} = \{a\} \neq \emptyset$, and so $f(\bar{w}, \bar{z}) = 1.00$. Of course, $f(\bar{x}, \bar{z}) = f(\bar{w}, \bar{y}) = 0.00$ since both these pairs of nodes do not overlap.

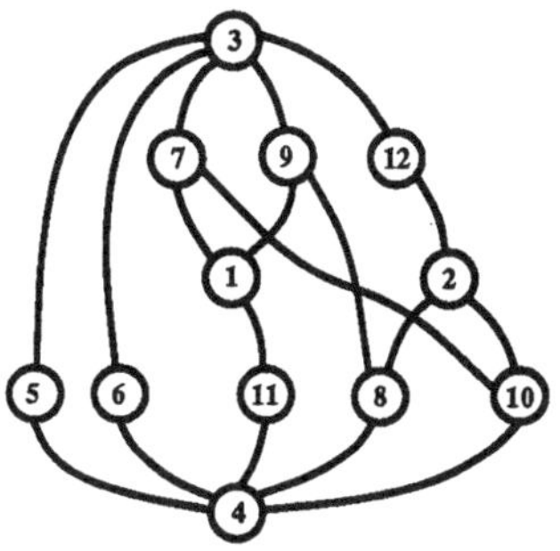


Figure 1: $H_{17,0.60}$, Hasse diagram for **gr17** at $\alpha = 0.60$

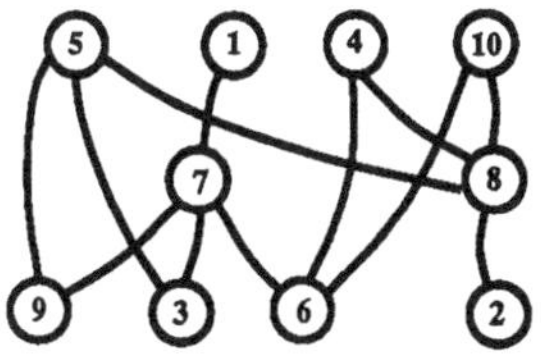


Figure 2: $H_{17,0.80}$, Hasse diagram for **gr17** at $\alpha = 0.80$

[3]The **grx**'s indicate data from the URBS project representing knowledge structures elicited from interviewees. Any other detailed specification is irrelevant to the present study.

[4]The use of the overhead bar was to emphasize that $\bar{x}$ and $\bar{y}$ are actually sets; $\bar{x}, \bar{y} \in 2^X$.

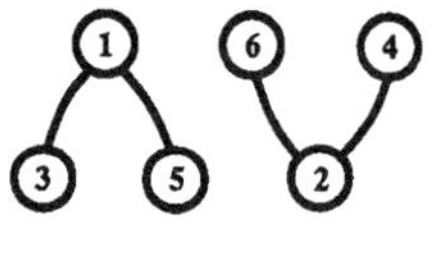

Figure 3: $H_{18,0.60}$, Hasse diagram for **gr18** at $\alpha = 0.60$

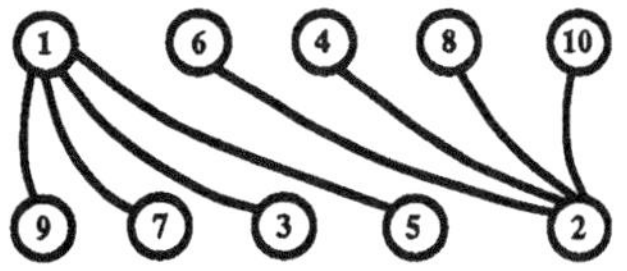

Figure 4: $H_{18,0.80}$, Hasse diagram for **gr18** at $\alpha = 0.80$

A more restrictive mapping may be derived by fuzzifying the above function further. For example, we can use the notion of cardinalities to define f as

$$f(\bar{x},\bar{y}) \;=\; \frac{|\bar{x}\cap\bar{y}|}{|\bar{x}\cup\bar{y}|} \qquad (2)$$

In this case then

$$f(\bar{x},\bar{y}) \;=\; \frac{|\{d\}\cap\{d\}|}{|\{d\}|} \;=\; 1.00 \quad \text{and}$$
$$f(\bar{w},\bar{z}) \;=\; \frac{|\{a\}\cap\{a,b,c,e,f\}|}{|\{a,b,c,e,f\}|} \;=\; 0.20.$$

Equation (1) was used in defining $f$ for the analyses presented in this paper.

*B. Mapping Paths Between Hasse Diagrams*

Similarly, there are a number of ways to define the mapping $g$. This mapping will depend on the node mapping $f$. Consider all $\bar{x},\bar{w} \in A$ such that $f(\bar{x}) = \bar{y}$, $f(\bar{w}) = \bar{z}$ for $\bar{y},\bar{z} \in B$. If $(\bar{x},\bar{w}) \in P_A$ and $(\bar{y},\bar{z}) \in P_B$ then we can match these two paths. An implication operator may be used in order to derive the degree to which these paths are mapped:

$$g\left((\bar{x},\bar{w}),(\bar{y},\bar{z})\right) \;=\; \mathsf{deg}\left(\bar{x}P_A\bar{w} \;\rightarrow\; \bar{y}P_B\bar{z}\right) \qquad (3)$$

From Figure 7, $P_A$ and $P_B$ are crisp relations. This results from the Hassefication procedure used in the Urbanistics project: the factor sets and factor relations are crisp.

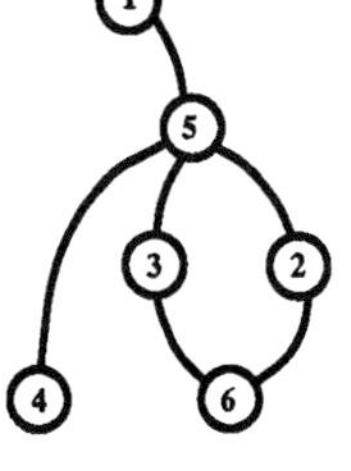

Figure 5: $H_{21,0.50}$, Hasse diagram for **gr21** at $\alpha = 0.50$

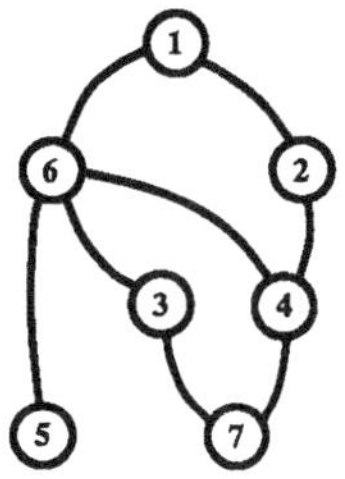

Figure 6: $H_{21,0.60}$, Hasse diagram for **gr21** at $\alpha = 0.60$

Hence, $g\left((\bar{x},\bar{w}),(\bar{y},\bar{z})\right) = \mathsf{deg}\left(\bar{x}P_A\bar{w} \rightarrow \bar{y}P_B\bar{z}\right) = 1.00$. In this case, the choice of fuzzy implication operator to use becomes irrelevant.

A fuzzier value for $g$ may be computed if instead of using the factor relations, say $\bar{x}P_A\bar{w}$, we use the local preorder closure $R_A$ to derive $\bigvee xR_Aw$ for all $x \in \bar{x}$ and $w \in \bar{w}$. Referring to Figures 8 and 9

$$\begin{aligned}
\bar{x}P_A\bar{w} &= \bigvee_{x\in\bar{x},w\in\bar{w}} xR_Aw \;=\; dR_Aa \;=\; 0.50;\\
\bar{y}P_B\bar{z} &= \bigvee_{y\in\bar{y},z\in\bar{z}} yR_Bz\\
&= \bigvee\{dR_Aa, dR_Ab, dR_Ac, dR_Ae, dR_Af\}\\
&= \bigvee\{1.0, 0.8, 0.6, 0.6, 1.0\} \;=\; 1.00
\end{aligned}$$

And so

$$\begin{aligned}
g\left((\bar{x},\bar{w}),(\bar{y},\bar{z})\right) &= \mathsf{deg}\left(0.50 \rightarrow 1.00\right)\\
&= \left((1-0.50)\vee 1.00\right) \;=\; 1.00.
\end{aligned}$$

Equation (3) was used for the mappings derived in this study. Although there are other ways to make the path mappings more restrictive, the comparison and investigation of these and other approaches are beyond the scope of the current investigation.

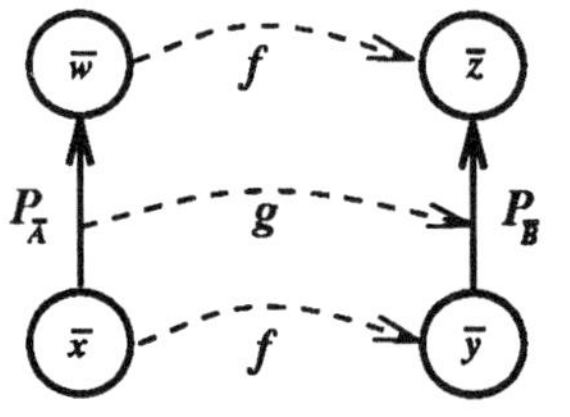

Figure 7: Sample portion of two Hasse diagrams

**NOTE:** This figure was derived from the Hasse diagram mapping $\mathcal{F}: H_{17,0.80} \rightarrow H_{18,0.60}$ (refer to Figures 2 and 3) where $\bar{x} = \{d\}$, $\bar{w} = \{a\}$, $\bar{y} = \{d\}$, and $\bar{z} = \{a,b,c,e,f\}$. Nodes $\bar{x}$ and $\bar{w}$ are actually nodes 3 and 5, respectively, in $H_{17,0.80}$. While $\bar{y}$ and $\bar{z}$ are nodes 3 and 1, respectively, in $H_{18,0.60}$.

| | b | | c | | d | | | | a | | e | | f | |
|---|---|---|---|---|---|---|---|---|---|---|---|---|---|---|
| b | .6 | 0 | .6 | 0 | .6 | 0 | .5 | 0 | .6 | 0 | .6 | 0 | .4 | 0 |
| | .5 | .6 | .6 | .5 | .5 | .6 | .5 | .5 | .5 | .5 | .5 | .5 | .5 | .6 |
| c | .5 | 0 | 1 | 0 | .5 | 0 | .5 | 0 | .5 | 0 | .5 | 0 | .4 | 0 |
| | .6 | .6 | .6 | 1 | .6 | .8 | .6 | .6 | 1 | .6 | .6 | .8 | .6 | .8 |
| d | .6 | 0 | .8 | 0 | .8 | 0 | .5 | 0 | .8 | 0 | .8 | 0 | .4 | 0 |
| | .5 | .6 | .6 | .5 | .5 | .8 | .5 | .5 | .5 | .5 | .5 | .5 | .5 | .6 |
| | .5 | 0 | .6 | 0 | .5 | 0 | .6 | 0 | .5 | 0 | .5 | 0 | .4 | 0 |
| | .5 | .5 | .6 | .5 | .5 | .5 | .5 | .6 | .5 | .5 | .5 | .5 | .5 | .5 |
| a | .5 | 0 | .6 | 0 | .5 | 0 | .5 | 0 | 1 | 0 | .5 | 0 | .4 | 0 |
| | .5 | .6 | 1 | .5 | .5 | .8 | .5 | .5 | .5 | 1 | 1 | .5 | .5 | .8 |
| e | .5 | 0 | .8 | 0 | .5 | 0 | .5 | 0 | .5 | 0 | 1 | 0 | .4 | 0 |
| | .5 | .6 | .6 | .5 | .5 | .8 | .5 | .5 | 1 | .5 | .5 | 1 | .5 | .8 |
| f | .6 | 0 | .8 | 0 | .6 | 0 | .5 | 0 | .8 | 0 | .8 | 0 | .8 | 0 |
| | .5 | .4 | .6 | .4 | .5 | .4 | .5 | .4 | .5 | .4 | .5 | .4 | .4 | .8 |

Figure 8: Local preorder closure of **gr17**.

| | b | | c | | d | | | | a | | e | | f | |
|---|---|---|---|---|---|---|---|---|---|---|---|---|---|---|
| b | 1 | 0 | .6 | 0 | 0 | 0 | .4 | 0 | 1 | 0 | .6 | 0 | 1 | 0 |
| | .2 | 1 | .2 | .8 | .2 | .8 | .2 | .8 | .2 | .8 | .2 | .8 | .2 | .8 |
| c | .8 | 0 | 1 | 0 | 0 | 0 | .4 | 0 | 1 | 0 | .6 | 0 | 1 | 0 |
| | .2 | .6 | .2 | 1 | .2 | .6 | .2 | .6 | .2 | .6 | .2 | .6 | .2 | .6 |
| d | .8 | 0 | .6 | 0 | 1 | 0 | .4 | 0 | 1 | 0 | .6 | 0 | 1 | 0 |
| | .2 | 0 | .2 | 0 | 0 | 1 | .2 | 0 | .2 | 0 | .2 | 0 | .2 | 0 |
| | .8 | 0 | .6 | 0 | 0 | 0 | 1 | 0 | 1 | 0 | .6 | 0 | 1 | 0 |
| | .2 | .4 | .2 | .4 | .2 | .4 | .2 | 1 | .2 | .4 | .2 | .4 | .2 | .4 |
| a | .8 | 0 | .6 | 0 | 0 | 0 | .4 | 0 | 1 | 0 | .6 | 0 | .8 | 0 |
| | .2 | 1 | .2 | 1 | .2 | 1 | .2 | 1 | .2 | 1 | .2 | 1 | .2 | 1 |
| e | .8 | 0 | .6 | 0 | 0 | 0 | .4 | 0 | 1 | 0 | 1 | 0 | 1 | 0 |
| | .2 | .6 | .2 | .6 | .2 | .6 | .2 | .6 | .2 | .6 | .2 | 1 | .2 | .6 |
| f | .8 | 0 | .6 | 0 | 0 | 0 | .4 | 0 | 1 | 0 | .6 | 0 | 1 | 0 |
| | .2 | 1 | .2 | 1 | .2 | 1 | .2 | 1 | .2 | .8 | .2 | 1 | .2 | 1 |

Figure 9: Local preorder closure of **gr18**.

## IV. Summary of Results

Mappings were derived both *within Hasse diagrams* (the diagrams considered come from the same data source, e.g. both from **gr17**) and *between Hasse diagrams* (coming from different sources). In Table II, the cardinalities of both the original factor sets and factor relations are summarized, along with the values for the corresponding *induced directed subgraphs* for each mapping considered in this paper. Recall that the factor sets are denoted by $A$ and $B$ and the resultant factor relations are $P_A$ and $P_B$. We can deduce the following about the mappings used if we focus on the values for the induced subgraphs rather than the original graphs:

1. $|f| \geq \max(|induced\ A|, |induced\ B|)$; and
2. $|g| \geq \max(|induced\ P_A|, |induced\ P_B|)$.

This pattern is observed when considering the induced subgraphs. Whereas no observable patterns are observed when relying on the original graphs.

In order to derive an estimate of the "goodness of fit" of the mappings presented in this paper, the following methodology was used: obtain a measure for both the node maps and the path maps for each graph pair by dividing the scalar cardinality of the induced graph by the scalar cardinality of the original graph. Hence,

- for the node mapping, $f$, use

$$\max\left(\frac{|induced\ A|}{|original\ A|}, \frac{|induced\ B|}{|original\ B|}\right)$$

- for the path mapping, $g$, use

$$\max\left(\frac{|induced\ P_A|}{|original\ P_A|}, \frac{|induced\ P_B|}{|original\ P_B|}\right)$$

These values are listed in Table III. Notice that the values obtained for the mappings *within Hasse diagrams* indicate excellent fit. This supports the plausibility of the mapping method introduced in this paper. The table also seems to indicate that of all the mappings *between Hasse diagrams* considered, the one with the best fit is $H_{17,0.60} \rightarrow H_{18,0.60}$.

TABLE II
Scalar Cardinalities for Original and Induced Graphs

| *Mapping* | $H_A$ | | $H_B$ | |
|---|---|---|---|---|
| $\mathcal{F}: H_A \rightarrow H_B$ | *original* | *induced* | *original* | *induced* |
| $H_{17,0.60} \rightarrow H_{17,0.80}$ | 12,45 | 10,28 | 10,28 | 10,28 |
| $H_{18,0.60} \rightarrow H_{18,0.80}$ | 6,10 | 6,10 | 10,18 | 10,18 |
| $H_{21,0.50} \rightarrow H_{21,0.60}$ | 6,17 | 6,17 | 7,21 | 7,21 |
| $H_{17,0.60} \rightarrow H_{18,0.60}$ | 12,45 | 12,32 | 6,10 | 6,10 |
| $H_{17,0.80} \rightarrow H_{18,0.60}$ | 10,28 | 10,24 | 6,10 | 4,6 |
| $H_{18,0.60} \rightarrow H_{21,0.50}$ | 6,10 | 3,5 | 6,17 | 6,16 |
| $H_{18,0.80} \rightarrow H_{17,0.60}$ | 10,18 | 10,12 | 12,45 | 12,18 |
| $H_{21,0.50} \rightarrow H_{17,0.60}$ | 6,17 | 6,12 | 12,45 | 6,11 |
| $H_{21,0.60} \rightarrow H_{18,0.60}$ | 7,21 | 7,19 | 6,10 | 3,5 |

**NOTE:** The numbers $n, m$ indicate cardinalities for the factor set and for the factor relation, respectively.

TABLE III
Estimates of "goodness of fit" for Hasse mappings

| *Mapping* | *Nodes Map* | *Links Map* |
|---|---|---|
| $\mathcal{F}: H_{i,j} \rightarrow H_{k,l}$ | $f$ | $g$ |
| $H_{17,0.60} \rightarrow H_{17,0.80}$ | 1.000 | 1.000 |
| $H_{18,0.60} \rightarrow H_{18,0.80}$ | 1.000 | 1.000 |
| $H_{21,0.50} \rightarrow H_{21,0.60}$ | 1.000 | 1.000 |
| $H_{17,0.60} \rightarrow H_{18,0.60}$ | 1.000 | 1.000 |
| $H_{17,0.80} \rightarrow H_{18,0.60}$ | 1.000 | 0.857 |
| $H_{18,0.60} \rightarrow H_{21,0.50}$ | 1.000 | 0.941 |
| $H_{18,0.80} \rightarrow H_{17,0.60}$ | 1.000 | 0.666 |
| $H_{21,0.50} \rightarrow H_{17,0.60}$ | 1.000 | 0.705 |
| $H_{21,0.60} \rightarrow H_{18,0.60}$ | 1.000 | 0.904 |

## V. Conclusions and Recommendations

The kind of mappings presented in this paper may be used to compute certain *congruences* between structures. In [6] and [7] the significance of the congruence between the *aspatial* and *spatial* structures that constitute a hypersystem for urban knowledge representation is discussed. Congruences between *actual* and *normative* structures are also mentioned. In particular, the *Structure Comparator* of the proposed General Meta-Knowledge Base in [7] may use these formulations to "assess the overall degree of congruence between two given structures".

Congruence may be used to determine which groups of people share the same view of their urban surroundings. This may also be used to approximate degrees of satisfaction with the urban environment. Urban planners could use such information to determine which part of the city needs more immediate attention. It is hoped that the methodology presented in this paper will eventually be embodied in a working system that will automate the analysis and diagnosis of data from certain urban studies.

## References

[1] W. Bandler and V. Mancini, "Internal representation of the built environment," in *Proceedings of NAFIPS '88*, North American Fuzzy Information Processing Society, San Francisco State University, San Francisco, California, June 1988, pp. 1–6.

[2] W. Bandler and L. J. Kohout, "Special properties, closures and interiors of crisp and fuzzy relations," *Fuzzy Sets and Systems*, 26:317–331, 1988.

[3] U. Faigle and R. S. Bonn, "Orders and graphs," in *Computational Graph Theory*, G. Tinhofer, E. Mayr, H. Noltemeier, M. M. Syslo, and R. Albrecht, Eds. Wien: Springer–Verlag, 1990, pp. 109–124.

[4] W. Bandler and L. J. Kohout, "Mathematical relations," in *International Encyclopedia of Systems and Control*, M. G. Singh *et al.*, Eds. New York: Pergamon Press, 1986, pp. 4000–4008.

[5] M. Ben–Ahmeida, L. J. Kohout, and W. Bandler, "The use of fuzzy relational products in comparison and verification of correctness of knowledge structures," in *Knowledge–Based Systems for Multiple Environments*, L. J. Kohout, J. Anderson, and W. Bandler, Eds. England: Ashgate Publishing, 1992, pp. 283–333.

[6] V. Mancini and W. Bandler, "Congruence of structures in urban knowledge representation," in *Uncertainty and Intelligent Systems*, Lecture Notes in Computer Science, 313, B. Bouchon, L. Saitta, and R. R. Yager, Eds. Berlin: Springer–Verlag, 1988, pp. 219–225.

[7] W. Bandler, V. Mancini, and E. Stiller, "Knowledge representation for a consultative system on urban problems," in *Expert Systems World Congress Proceedings*, vol. 2, J. Liebowitz, Ed. New York: Pergamon Press, 1991, pp. 828–834.

# CONTROL OF METAL-CUTTING PROCESS USING NEURAL FUZZY CONTROLLER

M. Balazinski
École Polytechnique de Montréal, Mechanical Engineering,
C.P. 6079, Succ. «A», Montréal (Québec) Canada, H3C 3A7

E. Czogala and T. Sadowski
Institute of Electronics, Technical University of Silesia
Pstrowskiego 16, 44-101 Gliwice, Poland

*Abstract*—This paper recalls the idea of a neural controller with application to the control of an industrial machining process. Two structures of a neural fuzzy controller, which links the ideas of a fuzzy controller and a neural network, are suggested. Results of simulations comparing performance of such a controller to that of a conventional fuzzy logic controller are shown. The experiment indicates that the performance of the proposed neural fuzzy controller is satisfactory. Hence, using a neural network, it is possible to build a controller that performs equally well as a fuzzy logic controller; moreover, it is more flexible and faster.

*Keywords*: Neural networks, neural fuzzy controller, fuzzy control, metal cutting.

## I. Introduction

Numerous applications of the fuzzy logic controller to the control of ill-defined complex processes have been reported since Mamdani's first paper [9]. Conventional and modern control theories need a precise knowledge of the model of the process to be controlled and exact measurements of input and output parameters. However, due to the complexity and vagueness of practical processes, the application of these theories is still limited.

In many real processes, control relies heavily upon human experience. Skilled human operators can control such processes quite successfully without any quantitative models in mind. The control strategy of the human operator is mainly based on linguistic qualitative knowledge concerning the behavior of an ill-defined process. Since most machining processes are stochastic, nonlinear and ill-defined, also metal-cutting processes fall into such a category of complex processes which are attractive to be controlled by means of fuzzy logic [10].

Considering the applications of fuzzy logic to control machining processes, the work of Zhu, Shumsheruddin and Bollinger [15] should be mentioned as the first. A simple loop, based on error and change in error was employed to control the surface finish in grinding by changing the feed rate. The knowledge base of the controller consists of thirty one rules based on the operator expertise. Sakai and Ohkusa [12,13] considered the application of the idea of fuzzy logic controller to turning. They obtained a preliminary framework for computing fuzzy control rules. Ralston and Ward have examined fuzzy logic control of turning within the context of computer numerical control and adaptive control [10,11] and, together with Dressman and Karwowski, they have considered a related expert system for machining. In his significant work Dubois [6] introduced fuzzy arithmetics to optimization of cutting conditions. Several knowledge-based systems, including expert systems, have been applied to simulate the fuzzy control of machining as well (cf. [10]).

Some approaches to the concept of a neural-network-based controllers employed to the control of various ill-defined, complex processes have been reported recently [1,2,3,4]. The aim of this paper is to recall this useful concept by pointing out its potential application to the control of machining processes, such as turning, milling, grinding etc.

## II. Structures of the Neural Fuzzy Controller

### A. The idea of a fuzzy logic controller

Fuzzy logic controllers may by considered as a special case of generalized decision tables with finite sets of conditions and actions in the rules [8]. Such controllers can be also viewed as models of a human operator determining the appropriate values of the control signal or its increment based on observation of process variables (e.g. *Error*, *Change in*

0-7803-0614-7/93$03.00 ©1993IEEE

*Error*, *Sum of Errors* etc.). The design of a fuzzy logic controller includes, after process recognition, a specification of the collection of control rules consisting of linguistic statements linking the inputs of the controller with the appropriate outputs.

The imprecise knowledge delivered by a human operator is usually expressed by fuzzy rules written in the form:

$$R_r\text{: If A is } A_i^{(r)} \text{ and B is } B_j^{(r)} \text{ and } \ldots \text{ and C is } C_p^{(r)} \text{ then U is } U_k^{(r)} \text{ and V is } V_l^{(r)} \text{ and } \ldots \text{ and W is } W_q^{(r)} \quad (1)$$

where $A_i^{(r)}$, $B_j^{(r)}, \ldots,$ $C_p^{(r)}$ denote linguistic values of the condition variables defined in universes of discourse: **X**, **Y**, ..., **Z** and $U_k^{(r)}$, $V_l^{(r)}, \ldots,$ $W_q^{(r)}$ stand for linguistic values of the conclusion variables defined in universes of discourse **U**, **V**, ..., **W** respectively; finally, *r* denotes the number of the rule. Such a rule corresponds to a relation which may be represented by a fuzzy implication [8].

Approximate reasoning is performed by means of the compositional rule of inference [14] which may be written in the form:

$$(U', V', \ldots, W') = (A', B', \ldots, C') \circ R \quad (2)$$

**R** represents the global relation aggregating all the rules and may be expressed as

$$R = \text{also}_r\,(R_r) \quad (3)$$

where the sentence-connective "also" may denote any *t*-, *s*-norms [8] (e.g. **min**, **max** operators) or average operators. The symbol $\circ$ stands for the compositional rule of inference operation (e.g. **sup-min**, **sup-product** etc.). (A', B',..., C') denote inputs, and (U',V',...,W') stand for outputs.

Taking into account the fact that fuzzy rules are usually formulated for each output separately we may consider the simplified rules (with one conclusion only). In this case the output may be obtained as follows:

$$U' = (A', B', \ldots, C') \circ R \quad (4)$$

Let us assume that the simplest fuzzy controller contains a knowledge base consisting of a collection of fuzzy control rules which have the form

$R_r$: **If** Error = $A_i^{(r)}$
**and** Change in Error = $B_j^{(r)}$
**then** Control Action = $U_k^{(r)}$ (5)

where *r* stands for the rule index. $A_i^{(r)}$, $B_j^{(r)}$, $U_k^{(r)}$ are linguistic values (fuzzy sets) for the linguistic variables *Error*, *Change in Error* and *Control Action* defined in universes of discourse **X**, **Y**, **U**, respectively.

We should mention here the explicit connective 'and' between the variables *Error* and *Change in Error* and the implicit sentence connective 'also' which links all the rules in the knowledge base.

A fuzzy control rule is usually implemented by a fuzzy implication (a fuzzy relation in **X** × **Y** × **U**):

$$R_r = (A_i^{(r)} \text{ and } B_j^{(r)}) \rightarrow U_k^{(r)} \quad (6)$$

where $(A_i^{(r)}$ and $B_j^{(r)})$ may be interpreted as a fuzzy set $A_i^{(r)} \times B_j^{(r)}$ in **X** × **Y**.

The input information: A' (error) and B' (change in error) being given, the control action U' can be deduced employing the compositional rule of inference, the definitions of fuzzy implication and connectives 'and' and 'also'. Even if we choose a particular compositional rule of inference, fuzzy implication and both connectives 'and' and 'also', the inference process can still be realized in different ways. Namely, if we consider input information (error and change in error) as vectors, we shall write the compositional rule of inference in the form:

$$U' = B' \circ (A' \circ R) \quad (7)$$

where R is the global relation obtained by connecting all the rules.

We can also use another notation and apply the following formula:

$$U' = (B' \times A') \circ R \quad (8)$$

Taking into account, for example, **sup-prod** as composition operator, **prod** for implication, **prod** for 'and' and **sum** for 'also' connectives, we get the same inference result from both formulas (7) and (8) respectively.

Using the membership function representation [8], we can write

$$U'(u) = \sum_r \sup_{x\in X, y\in Y} \left[(B'(y)\cdot A'(x))\cdot(A_i^{(r)}(x)\cdot B_j^{(r)}(y)\cdot U_k^{(r)}(u))\right] \quad (9)$$

Taking singletons (Kronecker delta) for A'(x), B'(y), when measurements are available, formula (9) can be simplified:

$$U'(u) = \sum_r \left[A_i^{(r)}(x_0)\cdot B_j^{(r)}(y_0)\cdot U_k^{(r)}(u))\right] \quad (10)$$

As a defuzzification method, center of gravity can be used.

It should be noted here that a different selection of operators may produce different inference results.

Two possible structures of a neural fuzzy controller will be described below.

*B. Discrete (m-h-n) neural controller (neural fuzzy controller)*

Taking into account the input information, two versions of the neural fuzzy controller may be considered [2]. The difference between them lies in the shape of the input layer, which can be linear ('vector' version of the controller) or rectangular ('matrix' version).

Let Card(**X**), Card(**Y**), ..., Card(**Z**), Card(**U**), Card(**V**), ..., Card(**W**) denote the respective cardinal numbers of the aforesaid discretized universes of discourse. The number of input neurons for the 'vector' version of a neural network can be determined as [5]

$$m = \mathrm{Card}(\mathbf{X}) + \mathrm{Card}(\mathbf{Y}) + \ldots + \mathrm{Card}(\mathbf{Z}) \tag{11}$$

while the number of output neurons

$$n = \mathrm{Card}(\mathbf{U}) + \mathrm{Card}(\mathbf{V}) + \ldots + \mathrm{Card}(\mathbf{W}) \tag{12}$$

For the 'matrix' version of a neural network the number of input neurons can be determined as

$$m = \mathrm{Card}(\mathbf{X}) * \mathrm{Card}(\mathbf{Y}) * \ldots * \mathrm{Card}(\mathbf{Z}) \tag{13}$$

and the number of output neurons as

$$n = \mathrm{Card}(\mathbf{U}) * \mathrm{Card}(\mathbf{V}) * \ldots * \mathrm{Card}(\mathbf{W}) \tag{14}$$

It is easy to show that the approach to the inference process in a fuzzy controller mentioned in the previous subsection leads to the above-described construction of two versions of the neural fuzzy controller which can be trained using the same input information and the control results can be compared.

Considering the discretization of the universes of discourse for error - **X**, change in error - **Y**, and control action - **U**, we can now construct two versions of the neural fuzzy controller. The structure of multilayer perceptron seems to be sufficient for the discussed task [1,5]. The structure of the input layer is considered to be linear for the 'vector' version and rectangular for the 'matrix' version (see Fig. 1)

According to formula (11) and taking into account the discretization of universes of discourse, the vector version will have $m_1$ = Card(**X**) + Card(**Y**) input neurons. The number of output neurons is given by n = Card(**U**). Denoting the number of units in the hidden layers as $h_1$, $h_2$,... , we can annotate the structure of the 'vector' network as ($m_1$ - $h_1$ - $h_2$ - ... - n).

For the 'matrix' version according to formula (13) and

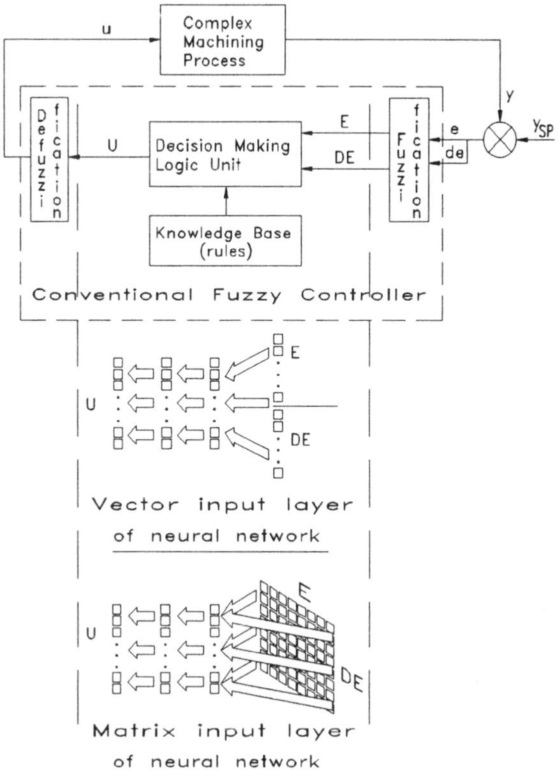


**Fig. 1** Control loop employing a fuzzy logic controller and both versions of neural fuzzy controller

using the same discretization as in the vector version, we will have $m_2$ = Card(**X**) * Card(**Y**) input neurons. Assuming the same number of output neurons i.e. n = Card(**U**) and denoting the number of units in the hidden layers as $h_1$, $h_2$, ... , we can also annotate the structure of the network matrix version as ($m_2$ - $h_1$ - $h_2$ - ... - n)

For instance, let us consider a 'vector' neural fuzzy controller using 10 input neurons for error and change in error, and 10 output neurons for the control value. Assuming that all the values lie within the intervals (0, 1), for the triple (0.1, 0.4, 0.6) we may write the following input and output vectors:

input: 0100000000 0001000000

output: 0000010000

Note that discretization enforces rounding input and output data to several values, corresponding to input or output neurons. Such a network does not exhibit interpolative properties; however, its advantage is that it may accept 'sampled' fuzzy information, not only crisp (singleton) data.

*C. Training and process control*

The neural fuzzy controller presented above may be trained

off-line by means of quantitative measurements expressed by triples (*Error*, *Change in Error*, *Control Action*) obtained during the observation of the process (sampling its parameters). The structure of the controller forces rescaling and discretization of the input-output data. It should be mentioned here that neural nets may also be initially trained using information obtained from the control rules (qualitative knowledge) [5]. As a learning scheme, the widely used backpropagation algorithm can be applied.

After training, the network can be used to control the process. This is accomplished by feeding process data (error and change in error) to the input layer of the network, which then recalls an appropriate action.

## III. NUMERICAL RESULTS

We will present here some numerical results obtained by simulating the control of a machining process.

In order to obtain comparable results we have used a slightly modified knowledge base originating from Zhu et al. [15]. The fuzzy controller used in our experiments employed **sup-prod** for compositional operation, **prod** for the 'and' connective between rule premises, **sum** for the sentence connective 'also'.

As an example let us mention a turning process, in which a constant cutting force (static case) should be assumed to assure the proper wear of the cutting tool. The changeable depth of cutting is compensated by the change of the feed rate. Basing on [7], the relation between the cutting depth, feed rate and the cutting force in the y-direction can be approximated by the following formula:

$$F_y = C_y \cdot d^{e_y} \cdot f^{u_y} \quad (15)$$

where $d$ denotes the cutting depth, $f$ stands for the respective feed rate and $C_y$, $e_y$, $u_y$ are constant coefficients.

Under chosen cutting conditions [7], formula (15) takes the form

$$F_y = 876 d^{0.9} f^{0.75} \quad (16)$$

Assuming a constant force $F_{yo}$ = 3050.4 [N], the range of the cutting depth $d \in [3..5]$ [mm] corresponds to the feed rate $f \in [0.75..1.4]$ [mm/s]. In this case the cutting depth was the value controlled, the feed rate being the control value.

In the first stage of our experiment we used a fuzzy controller represented by equations (7) and (10) to simulate a human operator. The set point was preprogrammed to change within the interval [3..5] (see Fig. 2 and 3). The results of control, shown in Fig. 2, were then used to train the neural fuzzy controller. For modeling the controller ('vector' version only) we employed three-layer feedforward

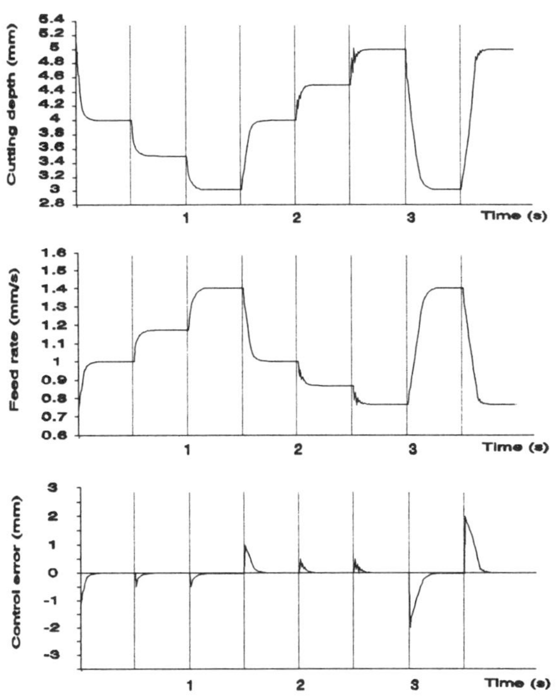


**Fig. 2** Results of control using fuzzy logic controller

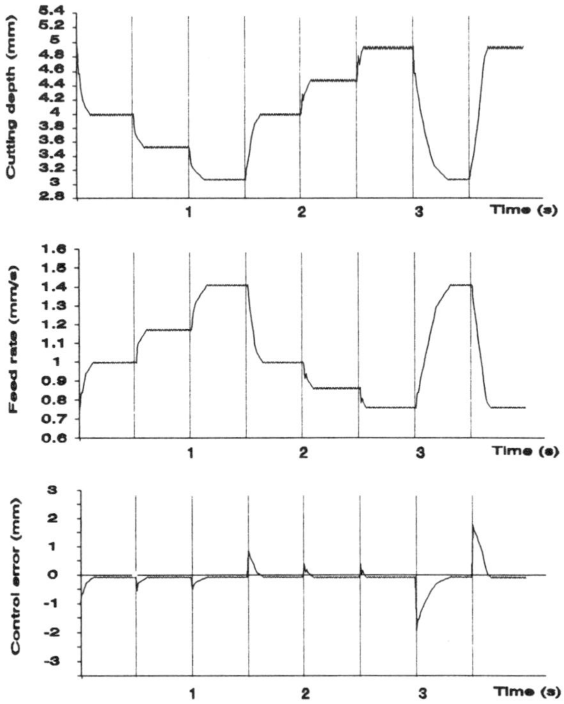


**Fig. 3** Results of control using neural fuzzy controller

network with sigmoidal elements arranged in a 20-20-10 structure. A backpropagation algorithm was used for training, with learning rate of 0.6 and momentum factor of 0.3. A random pattern presentation scheme was used for training; 200 training rounds were performed. The connectivity matrices were initially randomized with values from within the interval $[-0.5; 0.5]$. The ranges of error, change in error and drive were rescaled to 'fill' the whole range covered by the input and output neurons; rescaling was performed on the basis of previous observation of the process (i.e. operator's experience). The error and change in error values were clamped to the interval $[-0.5; 0.5]$, while drive values lay within the interval of $[0.75; 1.4]$.

In the second stage the same control program was performed using previously trained neural fuzzy controller. The results are shown in Fig. 3.

For the purpose of comparative study a quality index was defined as below:

$$QI = \sum_{i=0}^{N} \frac{(CD_i - SP_i)^2}{N+1} \tag{17}$$

where $CD_i$ denotes the controlled value (cutting depth), SP is the set point and N is the total number of observation points.

Comparing both stages we can note that both the fuzzy and the neural fuzzy controller behave similarly. Control performed by the neural fuzzy controller was minimally worse (quality index of 12.546 versus 11.797 of the fuzzy logic controller), and oscillations, resulting from relatively rough discretization, can be observed in steady states. It should be noted that the speed of a neural fuzzy controller is greater than that of a classical fuzzy controller, even though the parallel structure of the neural network is simulated.

## IV. Concluding remarks

The results of numerical experiments show that the neural controller performs equally well as a conventional fuzzy logic controller. Moreover, it is much more flexible (adaptive) and faster than the latter. The accuracy of control is sufficient, as it results from the performed experiments.

As an objective for future research, the input and output discretization problem should be considered: the larger the number of input (output) neurons, the better the accuracy of the controller; however, the larger the network itself, the longer the training time. Also the number of hidden neurons and learning parameters should be examined deeper.

It should also be noted that the structure of the neural fuzzy controller allows introduction of data expressed as fuzzy sets (which are 'sampled' at its inputs and outputs).

## References


[1] M. Balazinski, E. Czogala and T. Sadowski, Neural controllers and their application to the control of a machining process, Archives of Theoretical and Applied Computer Science, Polish Academy of Sciences, in press.

[2] U. Brunsmann, E. Czogala, and H. von Koch, On modelling of a fuzzy controller by means of two versions of multilayer feedforward neural networks, BUSEFAL 50, March 1992.

[3] J.J. Buckley and E. Czogala, Fuzzy models, fuzzy controllers, and neural nets, Archives of Theoretical and Applied Computer Science, Polish Academy of Sciences, in press.

[4] J.J. Buckley, Y. Hayashi and E. Czogala, On the equivalence of neural nets and fuzzy expert systems, Fuzzy Sets and Systems, in press.

[5] Czogala E., Sadowski T., A Method of Conversion of Fuzzy Decision Tables into Neural Networks, Biocybernetics and Biomedical Engineering, Inst. of Biocybernetics and Biomedical Engineering, Polish Academy of Sciences, Warsaw, in press.

[6] Dubois D., An Application of Fuzzy Arithmetic to the Optimization of Industrial Machining Processes, Mathematical Modelling, vol. 9, no. 6. 1987, 461-475.

[7] Kaczmarek J., *Principles of Machining by Cutting, Abrasion and Erosion*, Peter Peregrinus Ltd., 1976.

[8] C.C. Lee, Fuzzy logic in control systems: fuzzy logic controller - Part I/Part II, IEEE Trans. on Systems, Man and Cybernetics, vol. 20, No. 2, March/April 1990, 404-435.

[9] E.H. Mamdani, Applications of fuzzy algorithms for simple dynamic plant, Proc. IEE, vol. 121, No. 12, 1974, 1585-1588.

[10] P.A.S. Ralston and T.L. Ward, Fuzzy logic control of machining, Manufacturing Review, vol. 3, No. 3, 1990, 147-154.

[11] P.A.S. Ralston, T.L. Ward, L.M. Lambert, Simulation of Fuzzy Logic Control of a Lathe, Symposium on Advanced Manufacturing, Lexington, Kentucky, September 26-28, 1988, 21-25.

[12] Y. Sakai and K. Ohkusa, A Fuzzy Controller in

Turning Process Automation, Industrial Applications od Fuzzy Control, Elsevier Science Publishers B.V. (North-Holland), 1985, 139-151.

[13] Y. Sakai, K. Ohkusa, On a Control System for Cutting Process, 16th CIRP International Seminar on Manufacturing Systems, Tokyo, 1984, 188-195.

[14] L.A. Zadeh, Outline of a new approach to the analysis of complex systems and decision processes, IEEE Trans. on Systems, Man and Cybernetics, vol. 3, 1973, 28-44.

[15] J.Y. Zhu, A.A. Shumsheruddin, J.G. Bollinger, Control of Machine Tools Using the Fuzzy Control Technique, Annals of CIRP vol. 31, 1982, 347-352.

# Design and Testing of a Fuzzy Logic/Neural Network Hybrid Controller for Three-Pump Liquid Level/Temperature Control

Gail A. Cordes, Herschel B. Smartt, John A. Johnson,
Denis E. Clark, Keith L. Wickham
Idaho National Engineering Laboratory, EG&G Idaho, Inc.
Idaho Falls, Idaho 83415, U.S.A.

*Abstract*—**This paper describes multivariable, adaptive controllers that incorporate both fuzzy logic rules and neural networks. The controllers were implemented in laboratory trials and were robust, producing smooth temperature and water level response curves with short time constants. The paper also presents the projected role of these intelligent controllers in high level control applications and the reasons why intelligent controllers can advantageously replace classical controllers in many situations.**

## INTRODUCTION

In the future, intelligent systems for control will be a necessity for optimal operation of autonomous or semiautonomous systems located in space and other hazardous environments. What is really needed is a control system that adapts to the changing environment and process. Hybrid intelligent systems promise to provide this adaptive capability [1]. The Idaho National Engineering Laboratory (INEL) has an ongoing research project to study hybrid intelligent systems that combine conventional procedural techniques with intelligent technologies, including fuzzy logic, expert systems, neural networks, genetic algorithms, and intelligent databases, and to apply these hybrid systems in controllers that adapt to the changing process.

In Fiscal Year (FY) 1991 and FY 1992, the research focused on fuzzy logic applications and the potential for combining neural networks and fuzzy logic in one control application. During the research, we concluded that within the decade, the cutting edge of intelligent systems research will be the development of hybrid systems that incorporate the several intelligent technologies and human intelligence to reach cooperative decisions. This is not just a routine file transfer among systems but a utilization of the strengths of these technologies along with the human to creatively deal with a changing environment in real time.

Much research is currently underway at other institutions, as well as the INEL, to design intelligent control systems and to define the interactions of the human, the computer, and the world system in the control system. A distilled design of these human and computer communications has been constructed and is shown in Fig. 1. This figure shows that the reality of the world/system is perceived by the human senses as well as the sensor hardware. The state of the world as perceived by the human and the immediate condition of the human are transmitted to the computer. In return, the state of the world as sensed by the hardware is transmitted to the human. The human might immediately and directly interact with the world/system, and the computer might implement automatic action to affect the world/system. Drawing upon information from the human, the hardware sensors, and the existing database of past actions and descriptive information of the human and the world/system, the computer evaluates the state of the world/system in terms of immediate situation assessment and a prioritized threat assessment for the future. This synthesized state of the world/system is used to initiate a control action for the world/system, with the human retaining final decision-making power to intervene with the action and specify an alternate action.

The communication requirements of such a system demand the development of an advanced information processing technology and a complex human and computer

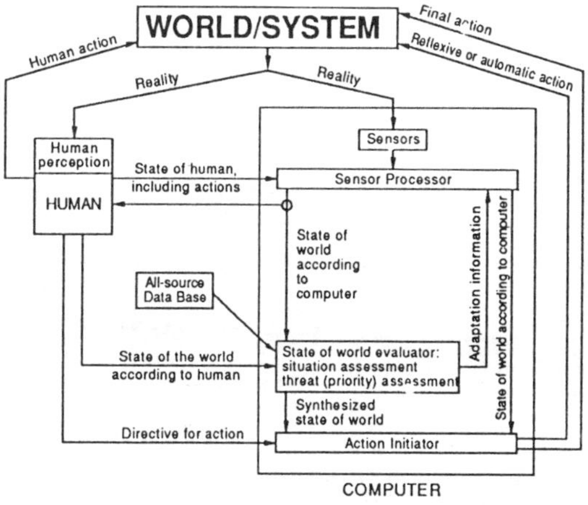


Fig. 1. Human/computer communications for control.

Work supported through the EG&G Idaho Laboratory Directed Research and Development Program under Department of Energy Idaho Field Office Contract DE-AC07-76ID01570.

0-7803-0614-7/93$03.00 ©1993IEEE

partnership in decisionmaking. The research described below explored the first part of this challenge, that of developing hybrid intelligent systems that could provide autonomous intelligent control, without including the human in the loop.

## LABORATORY TEST BED

The research tasks were intended to result in actual implementation as well as the theoretical design of the intelligent controller designs. For this purpose, a simple, environmentally benign laboratory test bed was constructed. As shown in Fig. 2, the test bed is a water tank with temperature and water level sensors. The water level is calculated using the automatic scale readout. Three separate peristaltic pumps designed for high volume flow add hot and cold water to the tank from reservoirs and remove mixed water. The controller thus is multiple-input and multiple-output. Controllers are linked to the hardware through the control computer.

We felt that a controller which simultaneously controlled temperature and level would be an interesting demonstration of the capabilities of intelligent control, since writing multiple-input, multiple-output conventional controllers can be a time-consuming process requiring substantial control expertise. Deliberate system disturbances used to challenge the controller can include, for example, the sudden addition of hot or cold water, and the addition of heat via an immersion heater. Controller response can also be gauged by the system's response to step changes in the temperature and level setpoints. In addition, the water tank volume and the apparent water level can also be modified to further test the adaptability of the controller.

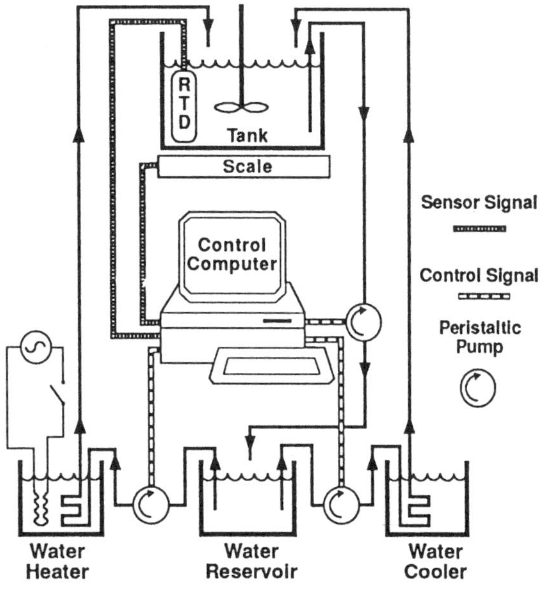


Fig. 2. Diagram of laboratory test bed.

## CONTROLLERS

Previous work [2,3] suggested a fuzzy logic based, neural network tuned, hybrid controller for this research. Conceptually, we thought that the continuous learning hybrid system can be categorized along two continuum which represent (1) the amount of preknowledge of the system being controlled, and (2) the time constant for the controller. A long time constant would be on the order of several minutes. The initial controllers developed in this work could be categorized as having high preknowledge and a long time constant.

The purpose of the research was to investigate various approaches to intelligent control using the test-bed demonstration system in the laboratory.

Four controller designs evolved during this research. The first controller consisted of two classical proportional integral derivative (PID) controllers. The second controller was based on fuzzy logic alone. For the third controller, two neural networks trained by back propagation were added to the design to modify the gain on the fuzzy logic output. The fourth controller design incorporates self-adapting neural networks.

These controllers are discussed below along with the experimental results from the first three controllers. The self-adapting controller design is discussed, but it was not implemented.

The experimental tests consisted of a sequence of trials for each of the three controllers that were actually implemented: classical control, fuzzy logic rules alone, and fuzzy logic with outputs modified by a nonadaptable neural network trained by back propagation. Because of space limitations, only a few typical response curves are given in this paper.

The purpose of the research was to investigate various approaches to intelligent control using the test-bed demonstration system in the laboratory.

## EXPERIMENTAL RESULTS

### *Classical Controller*

To provide some insight into the techniques of building classical controllers, two independent single input, single output, PID controllers were applied to the system.

A PID controller consists of three parts. The proportional part produces a proportional band within which the control signal is proportional to the error. A high gain (narrow band) forces the control value closer to the setpoint, but if the gain is too high, it will oscillate. If the gain is too low, control response is sluggish. The integral part of the controller governs control at the setpoint by slowly changing the output based on the past history of the error. The derivative part of the controller assists in responding to sudden changes in control inputs. Both integral and derivative terms can cause oscillations if their gains are too high.

This controller was linked to the test bed and tuned using the common sense approach. That is, in sequential order, the gains for the proportional, integral, and derivative terms were increased as much as possible until oscillations occurred, and then the gain was decreased slightly. This yielded a PID controller that was as responsive as possible without going into oscillation. The resulting test curves showed good system response with no oscillations.

*Fuzzy Logic Controller*

This controller was based on fuzzy logic alone. For this fuzzy logic controller, the input variables of tank water temperature and level were linguistically described by the variable values "cold, cool, about_right, warm, and hot" and "empty, low, about_right, high, and full", respectively. The output variables of cold/hot pump speed ratio and input/drain pump speed ratio were described as "cold_input, cool-input, medium_input, warm_input, and hot_input" and "way_more-in, more_in, in_and_out_even, more_out, and way_more_out", respectively. The triangular membership functions are shown in Fig. 3.

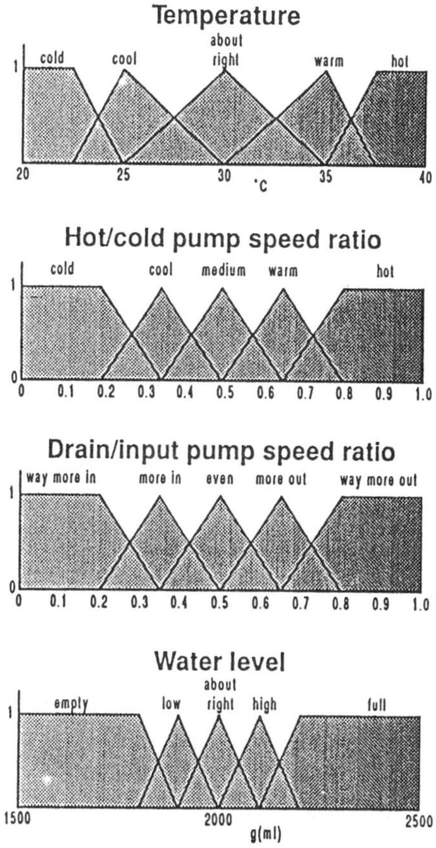

Fig. 3. Fuzzy rule membership functions.

Several approaches to formulating the fuzzy logic rules were tried. All shared the triangular, five-level membership functions for temperature, level, and pump speed shown in Fig. 3.

The most successful fuzzy rule base in the preliminary trials took an input/output and hot/cold ratio approach and treated level and temperature as independent variables. The rule base contained ten fuzzy rules that related the input and output variables, for example,

If Current_temperature IS Cold THEN
    Cold_to_Hot_pump_speed_ratio =
        Hot_input
    Input_to_Drain_pump_speed_ratio =
        In_and_out_even

The major difference between this and other approaches was that the ratio approach gives incremental pump speeds rather than actual pump output speeds. This rule base was investigated more extensively than the others and was used as the research continued. Please note that all versions of the fuzzy logic rule base easily controlled the hardware system, but the response curves varied as the gain was changed and different challenges perturbed the water temperature and level.

Even though these controllers performed well in the trials, the gain had to be adjusted manually and was constant during the entire run. With a small gain, the response curves smoothly approached the setpoint values, but the time constant was long. A large gain reduced the time constant significantly, but it induced oscillations in the measured variables as the setpoints were reached.

For example, in one run, approximately 300 mL of hot (~40°C) tap water was added to the control volume. At a gain of 10 for temperature and level, level and temperature returned to the control values. However, the level oscillated, increasing for about a minute, and then suddenly dropping back. Increasing the temperature gain to 20 and reducing the level gain to 4 considerably slowed the response time but eliminated the oscillation in the level control at the control value.

## *Fuzzy Logic/Neural Network Controller, Without Adaptation*

A mechanism for automatically varying the gains as the trial progressed would allow use of large gains at the onset of the trial and smaller gains as the variables approached the setpoints. Two neural networks were added to the fuzzy logic controller to achieve real time adaptability based on system performance as the trial ran. In designing the neural networks, the temperature and water level response curves were studied as various perturbations challenged the system and the controller gain was varied manually. Using about forty data points from each of these observations as a training set, two neural networks were trained, one for temperature and one for level. Each neural network consists of two input nodes, five hidden layer nodes, and one output node. The neural networks were feed-forward networks trained for 5000 cycles, about 10 minutes, with backpropagation using numerical values for error (difference between system variable and setpoint), slope of the error, and best-estimates of appropriate gain. The neural network design is shown in Fig. 4.

With the neural network adaptors, the response curves improved. The neural networks provided more rapid response times to the system without inducing oscillations about the setpoints. For example, Fig. 5 shows the response curves from a run in which 300 ml of hot tap water was added to the control volume. The figure shows the resulting level and temperature in detail, and the gains for each. Please note that the neural network-selected gains more or less follow the amount of error, although they are also influenced by the slope, or change in error. The total time shown on the plots is about 10 minutes.

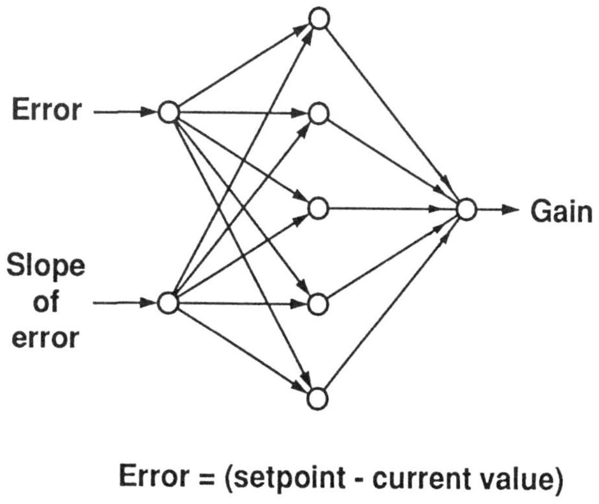


Fig. 4. Neural network design, no adaptation.

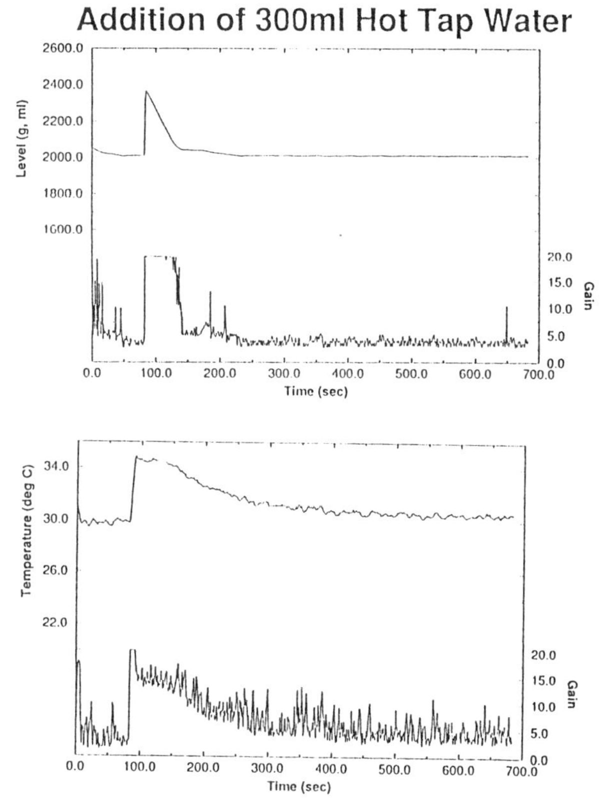


Fig. 5. Typical response curves, fuzzy logic/neural network control.

## *Fuzzy Logic/Neural Network Controller, With Adaptation*

A truly autonomous controller will learn the characteristics of the system as trials are run. Therefore, for follow-on work, a fourth controller is being designed with a self-training, fully adaptable neural network that could learn the optimum gain characteristics from the operating test bed. This controller design is intended to eliminate the need for a training set and allow learning to proceed throughout the operation of the system, from trial to trial. The proposed self-adapting neural network schematic is shown in Fig. 6. This controller is the subject of continuing research.

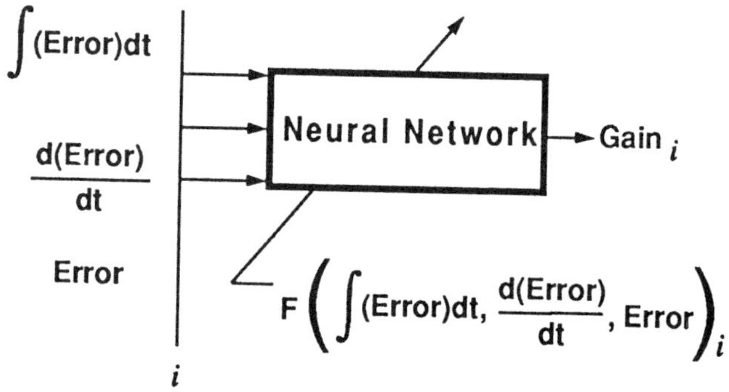


Fig. 6. Neural network schematic, with adaptation.

## OBSERVATIONS ON INTELLIGENT VERSUS CLASSICAL CONTROL

The question of comparing classical controllers with intelligent controllers is often raised. In this research, we explored the theoretical differences [4] and demonstrated some strengths and weaknesses of both categories of controllers.

Classical controllers make a good product based on observations of the quality of the finished product. When the quality deteriorates, faulty products are discarded, and the production system is modified to make a better product. The role of intelligent control is to provide the machine with the capability of making a good product once the state of the process is available to the machine through intelligent sensors. Intelligent controllers thus control the attributes desired in the product, making a good product the first time in the face of uncertainties and disturbances.

Intelligent control should be considered in applications for which a dynamic process model (process transfer function) does not exist, but expert knowledge or experimental data are available. Other candidates for intelligent control are processes where product properties must be controlled as opposed to process parameters. Intelligent control uses the tools of classical control theory, process modeling and artificial intelligence. Techniques such as artificial neural networks and fuzzy logic are not substitutes for control theory, but may be used as transfer function generators to be incorporated into controllers. They may also be used as generators of maps between input and output parameters, given a set of discrete data points for training. Both fuzzy logic and neural networks provide a mechanism for generating a continuous input-output map. Input-to-output mapping functions are, in fact, the same as transfer functions. Transfer function based analysis and design methods form a major portion of dynamic systems analysis [5], signal analysis [6], and control theory.

One difference between neural networks and fuzzy logic is that neural networks require numerical data for training and application. Fuzzy logic requires non-numerical information in the form of conditional logic statements for input. A companion area of research, not dealt with in this paper, is intelligent sensors that can find and interpret features in the input data even without exact prior knowledge of the spatial or temporal location of such features. These sensors also carry out self-calibration and validation checks.

Other researchers confirm that fuzzy logic is a convenient method for implementing controllers [7], and they report that a fuzzy logic controller has some desirable noise rejection properties that a classical controller does not have [8].

## SUMMARY

The significance of these results to United States industry lies in the simplicity of the controller formulation and the adaptability offered by the neural network. Only a short time is required to write the fuzzy logic rules and to demonstrate successful control. Identifying the best measurement functions and writing a minimal set of fuzzy rules takes longer, but a mathematical description of the system is not necessary. Thus, a real projected savings in development costs can be realized.

Incorporation of the neural network adaptability promises an approach to controllers that will function accurately despite a changing environment or degradation in a production plant, turning out a product that is good the first time.

## REFERENCES


[1] J. C. Bezdek and Sankar K. Pal, ed., *Fuzzy Models For Pattern Recognition*, IEEE Press, Piscataway, NJ, 1992.

[2] H. B. Smartt, J. A. Johnson, C. J. Einerson, and G. A. Cordes, "Application of a connectionist fuzzy logic system to control of gas metal arc welding," *Proceedings of ANNIE '91: Artificial Neural Networks in Engineering*, St. Louis, MO, November 1991.

[3] G. A. Cordes, S. R. Bryan, R. H. Powell, and D. R. Chick, "Applications of neural networks technology to setpoint control of a simulated reactor experiment loop," *Proceedings of the American Nuclear Society Topical Meeting AI91: Frontiers in Innovative Computing for the Nuclear Industry*, Jackson, WY, September 1991.

[4] H. B. Smartt, *Intelligent sensing and control of processes*, unpublished report, EG&G, Idaho, Idaho National Engineering Laboratory, August 1992.

[5] R. H. Cannon, *Dynamics of physical systems*, McGraw-Hill Publishing Co. (1967).

[6] K. Beauchamp and C. Yuen, *Digital methods for signal analysis*, George Allen and Unwin, London (1979).

[7] J. Clymer, P. Corey, and J. Gardner, "Discrete event fuzzy airport control," *IEEE Transactions on Systems, Man and Cybernetics*, vol. 22, no. 2, March/April 1992.

[8] R. Sutton and D. R. Towill, "An introduction to the use of fuzzy sets on the implementation of control algorithms," *J. Instit. Electron. Radio Eng.*, vol. 55, no. 10, pp. 357-367, October 1985.

# Fuzzy Integral Action in Model Based Control Systems*

M. De Neyer, R. Gorez and J. Barreto

***Abstract*— Fuzzy model based control is presented as well as several fuzzy model based control schemes. More specially implementation of fuzzy integral actions introduced by means of these control structures with a view of rejecting disturbances is discussed. Tracking and robustness capabilities of these control systems is analyzed and appraised by the simulation of the control of a mechanical system.**

## I. Introduction

Most fuzzy control systems developed up to now, use simple control structures: Proportional-Derivative, Proportional-Integral, combination of the two previous and incremental controllers. On the other hand, fuzzy control applications often handle only tracking problems neglecting a very important issue in process control, disturbance rejection. From classical control theory, it is well known that a feedback control loop must include an integral action for ensuring suppression of constant disturbances. There exist several possibilities of introducing an integral action in a control system. The use of PI, PID or incremental controllers is the simplest way. That way of introducing an integral action may lead to stability problems if the actual process to control contains itself integrators. Another completely different approach has been developed since the fifties using inside the control system a model of the plant to be controlled. These model based control schemes exhibit nice disturbance rejection and robustness properties and may avoid the aforementioned stability problems. Since that approach had proven its usefulness in quantitative control, an extension to fuzzy control was tried and tested by simulation [1], [2], [3]. The goal was twofold: to study the feasibility of more complex fuzzy control structures and their disturbance rejection capabilities. This study is focussed on difficulties in the implementation of fuzzy integral actions in model based control. A brief description of model based control and fuzzy control is given in section II and III. Section IV deals with implementation of some fuzzy model based control systems which are applied to the control of a mechanical system. Some simulations results are presented in section V and finally conclusions are derived.

## II. Model based control

In model based control, the basic signal processed by the control system is the difference signal between the output signals of the actual plant and of a model of the latter. The difference signal is fed back to modify the reference signals entering either the plant and/or the model. Three types of model based control system can be distinguished by the way the difference signal is fed back. In the well known Internal model Control scheme [7], the difference signal is returned to a common reference signal entering both the plant and the process (Fig.1, connection ①). In others the return signal modifies either plant inputs either model inputs. If the difference signal changes the model input signal, it yields a Luenberger observer (simplified configuration) [6] as in Fig.1 (connection ②); the model output is constrained to track the plant output . Whereas if it modifies the plant input signal, a dual scheme of the previous one is set up (Fig.1, connection ③) and the plant output is forced to follow the model output. This control scheme is known as Model Following Control [5].

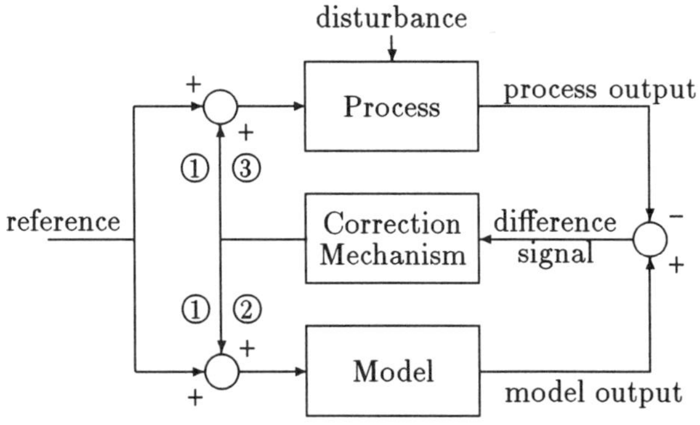


Figure 1: Model based control structures

The properties of these control systems are similar [4]. First, if the model perfectly matches the plant to be controlled, in absence of disturbances the return signal is null,

*This paper presents research results of the Belgian Program on Interuniversity Poles of attraction initiated by the Belgian State, Prime Minister's Office, Science Policy Programming. The scientific responsibility rests with its authors.

0-7803-0614-7/93$03.00 ©1993IEEE

resulting in an open-loop control. Then the tracking dynamics of the system is that of the plant. Second, in presence of external disturbances or model mismatchings (internal disturbances), the difference signal is not null and the correction mechanism tries to suppress or at least to reduce the effects of these disturbances, regulation dynamics of the system being fixed by means of the correction mechanism.

Although model based control has nice properties, its application must be limited to the control of stable process. As mentioned above in ideal conditions and without disturbance, the control is open-loop implying that unstable processes must be stabilized by a supplementary controller with a view to be included in a model based control system. All these control structures are defined independently of the type of knowledge representation used to describe the components of the control system e.g. transfer functions, fuzzy relations, neural networks.

## III. Fuzzy control

Classical fuzzy control structure is presented in Fig.2. It involves three parts: the process to be controlled, a fuzzy controller and interfaces. An important feature of fuzzy control is that two types of representation for values are used. While a fuzzy control system internally handles fuzzy values described by membership functions, its receives from measuring devices quantitative numerical values and must provide actuators with quantitative numerical values. Two dual interfaces perform conversions from one type to another in an analog way to AD-DA converters. Inside the fuzzy controller, the relation between the controller inputs and outputs is described by a set of linguistic decision rules, expressed as *if... then...* statements. Each rule triggers a control decision with a grade depending on the fulfillment of its conditional part by the actual input values. A global output value is got through the union of all the decisions with appropriate weights. Before its application to the process, the fuzzy control value is defuzzified.

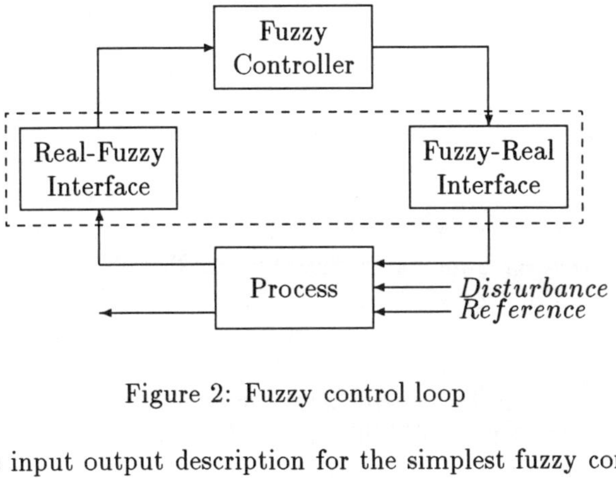


Figure 2: Fuzzy control loop

The input output description for the simplest fuzzy controller may be easily extended to more complex systems including several fuzzy components (compensator, filter, adder,...). Some major differences can be pointed out: component input signals may be intrinsically fuzzy coming from other fuzzy components, component outputs are not defuzzified if in turn they enter a fuzzy component, fuzzy states are allowed.

## IV. Fuzzy model based control

For illustrating the properties of model based control, we consider a simple process the nominal model of which is a double integrator. It can be viewed as a simplified model of an unidirectional motion of a robot link. Indeed models of these processes reduce to a double integrator if the viscous effects and the gravity force are neglected and if the inertia is normalized to one. The actual process output is related to its inputs by the following relation:

$$J\ddot{q} + B\dot{q} + Gq = \tau_m + \tau_l, \qquad (1)$$

$\ddot{q}$, $\dot{q}$, $q$ are respectively the acceleration, velocity, position, $\tau_m$, the control torque and $\tau_l$, the disturbance torque, $J$, the moment of inertia, $B$, the viscosity coefficient and $G$, the gravity coefficient. In nominal conditions, $J = 1$ and $B = 0 = G$. This process which is marginally unstable in open loop, is stabilized by a fuzzy PD controller (Fig.3).

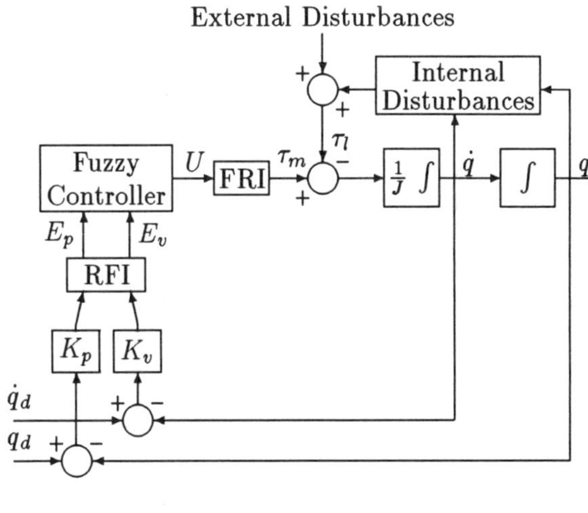


Figure 3: Fuzzy PD control of a double integrator

$q$, $q_d$: actual and desired positions
$\dot{q}$, $\dot{q}_d$: actual and desired velocities
$\tau_m$, $\tau_l$: control and disturbance torques
RFI, FRI: real to fuzzy, fuzzy to real interfaces

If the tracking dynamics is satisfactory in nominal conditions with a PD control, i.e position and velocity errors are null in steady state, in presence of constant disturbance (load torque) or unmodelled dynamics ($B \neq 0 \neq G$) the error doesn't vanish asymptotically. Adding an integral action in the control system allows at least the suppression of the static error and moreover since viscous friction and gravity may be seen as internal disturbances (constant in steady state), the system can be robust with respect to these unmodelled dynamics.

Two model based control structures, internal model control and model following control structures, were selected in order to introduce an integral action so that the static and velocity errors should be zero even in presence of model mismatchings. Depending on the place of the model based system with respect to the stabilizing loop (bold line in Fig.4, 5, 6), several combinations are possible. Three were analyzed: an internal model

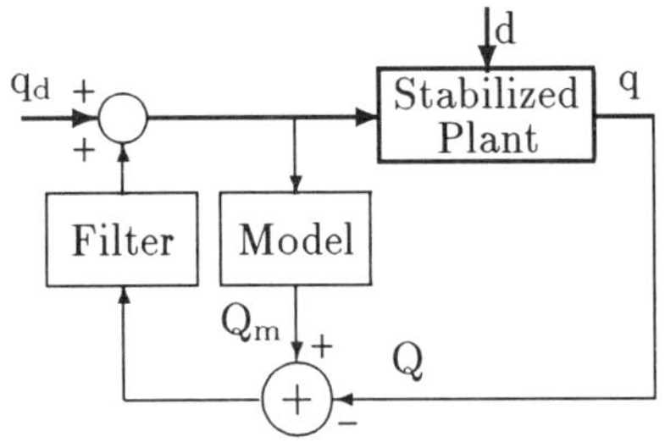


Figure 4: Internal model control system outside the main control loop

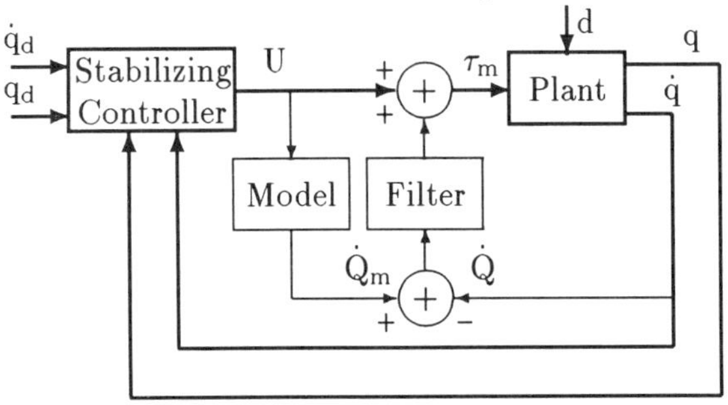


Figure 5: Model following control system inside the main control loop

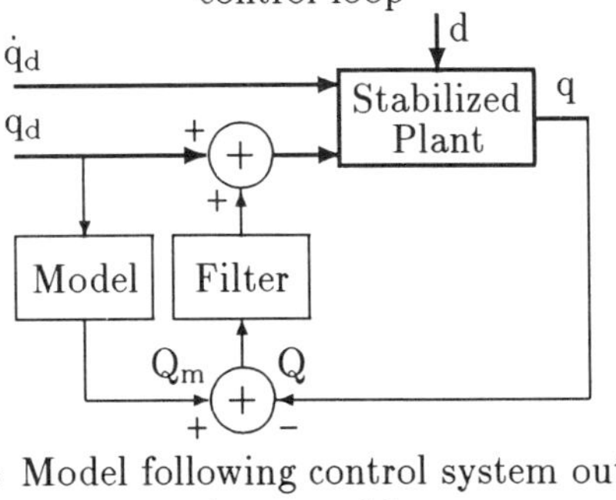


Figure 6: Model following control system outside the main control loop

$d$: disturbances  $\oplus$ : fuzzy adder

$Q_m$, $\dot{Q}_m$: fuzzy position and velocity estimates

$Q$, $\dot{Q}$: fuzzy position and velocity

control system outside the main control loop (IMC, Fig.4) and a model following control system outside the main control loop (MFC, Fig.6) and inside the main control loop (DO, Fig.5). The latter scheme is also called disturbance observer. Note that when using a disturbance observer (Fig.5), the model is concerned with the relationship between the control variable $\tau_m$ and a secondary output variable, the velocity $\dot{q}$. This choice allows a model order reduction; indeed under ideal and nominal conditions, the relation linking torque to velocity is simply an integration. In the IMC system (Fig.4) and the MFC system of figure 6), the stabilized plant is a second order system. This could be modelled by different systems with an increasing order from a simple delay to a second order system, depending on the relative values of the stabilizing loop settling time and the sampling interval used for modelling. However in order to preserve the good asymptotical tracking properties of the stabilized plant, the models in the IMC (Fig.4) and the MFC (Fig.6) systems must have an identical asymptotical response to that of the stabilized plant so that the difference signal is asymptotically null in tracking. In these two model based control systems, tracking of step or ramp-wise reference leads to a null steady state error iff the models are respectively at least first and second order models. Such a constraint does not exist for the DO system (Fig.5).

On the other hand in feedback control, constant disturbances are rejected if the control loop includes an integrator between the plant output and the disturbance entry. It is achieved within the DO system (Fig.5) if the model includes an integrator, and within the IMC scheme (Fig.4) if the model is a least first order system. The MFC system (Fig.6) will suppress the effect of disturbances if the correction mechanism (Filter) contains an integrator.

### *A. Fuzzy Implementation*

All the fuzzy components involved in the subsequent control systems have two inputs and one output. Their input output behaviour is described by a set of linguistic rules the generic form of which is

$$\textit{If}\ \mathrm{In}_1\ \textit{is}\ \mathrm{A_i}\ \textit{and}\ \mathrm{In}_2\ \textit{is}\ \mathrm{A_j}\ \textit{then set}\ \mathrm{Out}\ \textit{equal to}\ \mathrm{A_{ij}}, \qquad (2)$$

and can be abbreviated by the functional expression $Out = F(In_1, In_2)$ where $In_1$, $In_2$ are the fuzzy input variables and $Out$ the fuzzy output. $A_i$, $A_j$ and $A_{ij}$ are reference fuzzy values of these variables. Seven reference fuzzy values are defined by triangular membership functions on the universe of discourse [-1,1]: negative big (NB), negative medium (NM), negative small (NS), negligible (Z) and the corresponding positive values (PS, PM, PB). The rule base of a fuzzy component can be summarized in a array that must be read as follows: values of $In_1$ and $In_2$ respectively sit in the first column and in the first row. For a given pair of input values, the output of the corresponding rule is defined by the fuzzy value at the cross of the row corresponding to the $In_1$ value and the column to the $In_2$ value (see Table 1). A max-min inference law is used in the fuzzy calculations. Non null rule contributions are gathered in one global fuzzy value through a union. Before its application to a non fuzzy component, fuzzy values are defuzzified by a center of gravity method.

*A.1. Stabilizing control loop*

The rule base of the fuzzy PD controller $U = F_a(E_p, E_v)$ used for stabilizing is summarized in Table.1: $E_p$ and $E_v$ are respectively the position and velocity errors and $U$, the control variable .

Table 1: Rule-Base: Out = $F_a(In_1, In_2)$

| *Out* | | $In_2$ | | | | | | |
|---|---|---|---|---|---|---|---|---|
| | | **NB** | **NM** | **NS** | **Z** | **PS** | **PM** | **PB** |
| $In_1$ | **PB** | Z | PS | PM | PB | PB | PB | PB |
| | **PM** | NS | Z | PS | PM | PB | PB | PB |
| | **PS** | NM | NS | Z | PS | PM | PB | PB |
| | **Z** | NB | NM | NS | Z | PS | PM | PB |
| | **NS** | NB | NB | NM | NS | Z | PS | PM |
| | **NM** | NB | NB | NB | NM | NS | Z | PS |
| | **NB** | NB | NB | NB | NB | NM | NS | Z |

*A.2. IMC system*

Different models with increasing complexity representing more and more properly the process behaviour may be used. If the sampling period is larger than the process rise-time, the process transient response may be skipped so that the simplest model is a one step ahead shift operator. The model output $Q_m(k+1)$ predicts the expected steady state value $U(k)$. A finer prediction by a first order model $Q_m(k+1) = F_m(Q_m(k), U(k))$ allows a smaller sampling period and should yield a faster response. A first order model rule base is given in Table 2. While both proposed models motivate a control system with integral action affording the full rejection of constant disturbances, a successful tracking of ramp wise signals is not achieved. A second order model $Q_m(k+1) = 2 \cdot F_a(U(k), -\frac{1}{2}U(k-1))$ might afford good tracking performances (with a rule base as in Table 1).

The fuzzy estimate of the process output and the actual process output are compared trough a fuzzy comparator $C = F_a(Q_m, -Q)$ characterized by the rules of Table 1. By sake of simplicity, the filter was drop out as well as in the following implementation.

*A.3. DO system*

The integration relating torque to velocity can be quite naturally modelled by a fuzzy accumulator that estimates the process output velocity $\dot{Q}_m$ from the main control variable $U(k)$: $\dot{Q}_m(k+1) = F_a(\dot{Q}_m(k), U(k))$ . The fuzzy accumulator, the adder and the comparator obey to the same rules given in Table 1.

Table 2: Fuzzy first-order predictor

| $Q_m(k+1)$ | | $Q_m(k)$ | | | | | | |
|---|---|---|---|---|---|---|---|---|
| | | **NB** | **NM** | **NS** | **Z** | **PS** | **PM** | **PB** |
| U(k) | **PB** | NS | Z | PS | PS | PM | PB | PB |
| | **PM** | NM | NS | Z | PS | PM | PM | PM |
| | **PS** | NM | NS | Z | PS | PS | PS | PM |
| | **Z** | NM | NS | Z | Z | Z | PS | PM |
| | **NS** | NM | NS | NS | NS | Z | PS | PM |
| | **NM** | NM | NM | NM | NS | Z | PS | PM |
| | **NB** | NB | NB | NM | NS | NS | Z | PS |

*A.4. MFC system*

Only one model was selected, the simplest: a static gain (unitary) with a sampling period at least equal to the rise-time of the stabilized plant ($T = 5sec$). In this particular case, the control structure reduces to a low rate PID controller. The two fuzzy components of this control system have already been defined in the previous control systems; the filter is an accumulator (Table 1). Note the accumulator uses the latest error value :$E_i(k+1) = F_a(E_i(k), E(k))$ where $E$ is the position error and $E_i$ the accumulator output.

*B. Comments on fuzzy integral actions*

The key point in all the mentioned control systems is that they introduce an integral action, either directly through an accumulator (DO and MFC) or through a feedback loop such that the equivalent effect is an integral action (IMC). Fuzzy accumulator suffers a major problem: an accumulator is instable in the BIBO sense and its fuzzy feature is another source of instability. When the time goes up, at each iteration, the support set of the accumulator output becomes bigger until it reaches its maximum size (the support set is then equal to the universe of discourse) [2]. The support set should be infinitively large to not loose information at the bounds and to properly accumulate. This explosion of fuzziness due to the fuzzy closed loop (fuzzy accumulator, Fig.7) can be avoided by arbitrarily limiting the fuzziness of the state input. We choose to limit the state input to its two most important linguistic values (using a defuzzification followed by a fuzzification procedures). Even with that simplification, some undesirable behaviour (saturation of the accumulator output below its maximal value) may be still present but this time due to the max-min procedure used for inference. Most of time fuzzy explosion is present when there exists a fuzzy loop in the system or several fuzzy components are cascaded without defuzzification between them. Nevertheless the first order model used in this study is a counter example of fuzzy explosion, the state support set breadth being constant along the time. It must be noted that the model is stable.

The integration loops in fuzzy IMC also suffer stability

problems. The integration effect is obtained by looping through a fuzzy model in the feedback loop (integration loop, Fig.7). But for first and second order models, the open loop tracking responses to ramp wise signals don't reach a constant steady state value, oscillating around the expected steady state value. The max and min operators used in fuzzy logic are responsible for that behaviour. And when the loop is closed, sustained oscillations are observed with a first order model and the system becomes instable with a second order model. A more linear rule of inference (a sum-product instead of a max min operator) allows avoiding oscillations in the open loop responses of the first and second order models but in that case, the input output function of the rule base of the second order model is exactly equivalent to an quantitative addition.

## V. Simulation results

Many simulation runs were performed for the three control systems described above ([1], [3]), in both cases where the model is identical to the nominal process and when there are some model mismatchings ($J \neq 1$, $B \neq 0$, sampling period variation). Some typical simulation results are presented in Fig.8 to 11. For comparison purposes, a quantitative implementation of the same control systems was performed (the fuzzy implementation results are in solid line and the quantitative results in dashed line). All the graphics represent position errors in function of time. The first column gives results concerning the DO system(Fig.5) and the second results of the MFC system (Fig.6). The effect of viscous friction is shown in Fig.10 and 11, each plot consisting in three successive simulation runs for different values of the viscous friction coefficient $B = 0, 0.25, 1$. Simulation results of fuzzy IMC systems (Fig.4) are not presented here. As explained in the previous paragraph, tracking of ramp wise signals is not successful; nevertheless disturbance rejection capability is very good [1].

## VI. Conclusions

The conclusions of this study are multifold. Feasibility of more complex fuzzy control systems, in particular fuzzy model based control systems, were successfully tested although a fully fuzzy implementation is not always achievable. A direct fuzzy integral action can be properly implemented only if the accumulator state is defuzzified. More generally fuzzy loops do not work. Fuzzy integral action by looping through a fuzzy model is not achievable in tracking.

On the other hand the simulation study shows that the performances of fuzzy model based control systems in tracking and regulation are very good. Moreover these control systems are robust with respect to parameters variations and unmodelled dynamics. Nevertheless the behaviour of the quantitative and fuzzy versions of these control systems being similar, in particular for the model following control system outside the main control loop, the quantitative implementation remains a better choice for the studied case due to a greater complexity of the fuzzy implementation. Fuzzy control will reveal its interest in control of complex and ill-defined processes where quantitative techniques are not always appropriate, e.g. where an accurate model of the plant is not available or the plant is strongly nonlinear.

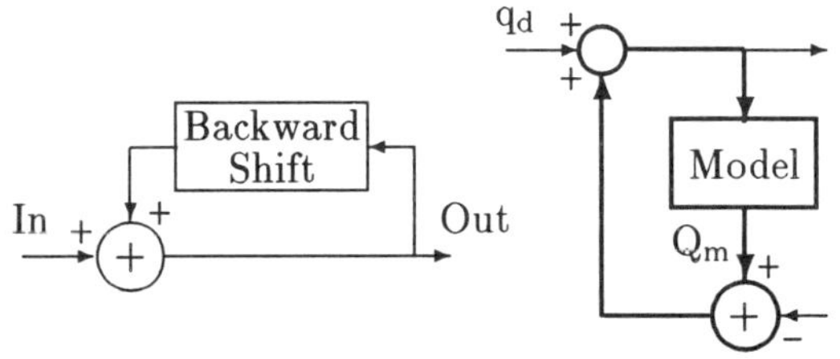


Figure 7: Fuzzy accumulator    Integration loop

## References

[1] M. De Neyer, R. Gorez, and J. Barreto. Comparative analysis of control systems with fuzzy or quantitative internal models. In *Europ. Sim. Symp.: Intelligent Process Control and Scheduling*, pages 145–150, Gent, November 1991.

[2] M. De Neyer, R. Gorez, and J. Barreto. Disturbance rejection based on fuzzy models. In M.G.Singh and L.Travé-Massuyès, editor, *Decision Support Systems and Qualitative Reasoning*, pages 215–220. Elseviers Sience Pub., Amsterdam, 1991.

[3] R. Gorez, M. De Neyer, D. Galardini, and J. Barreto. Model based control systems: Fuzzy and qualitative realizations. In *Symp. in Intelligent Components and Instruments for Control Applications*, pages 681–686, Malaga, May 1992.

[4] R. Gorez, D. Galardini, and K. Y. Zhu. Model-based control systems. In P. Borne, S.G. Tzafestas, and N.E. Radhy, editors, *Mathematics of the Analysis and Design of Process Control*, pages 223–232. Elsevier Science Publishers, Amsterdam, The Netherlands, 1992.

[5] Y. Landau. *Adaptive control: The model reference approach.* M. Dekker Inc., New York, 1979.

[6] D.G. Luenberger. An introduction to observers. *IEEE Trans. Automatic Control*, 10(6):596–602, 1971.

[7] M. Morari and E. Zafiriou. *Robust Process Control.* Prentice-Hall, Englewod Cliffs, N.J., 1989.

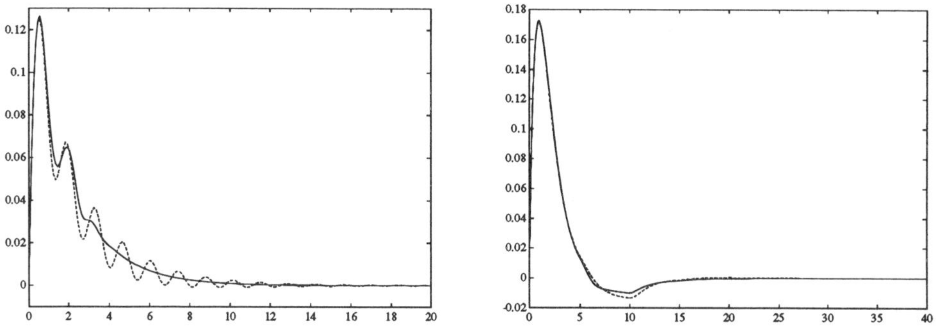


Figure 8: Nominal responses to a ramp-wise reference signal ($Velocity = 0.5$)

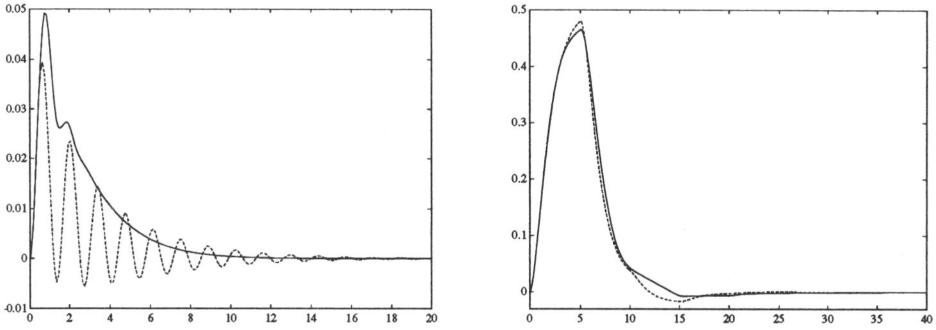


Figure 9: Nominal responses to a constant disturbance ($Disturbance = 0.5$)

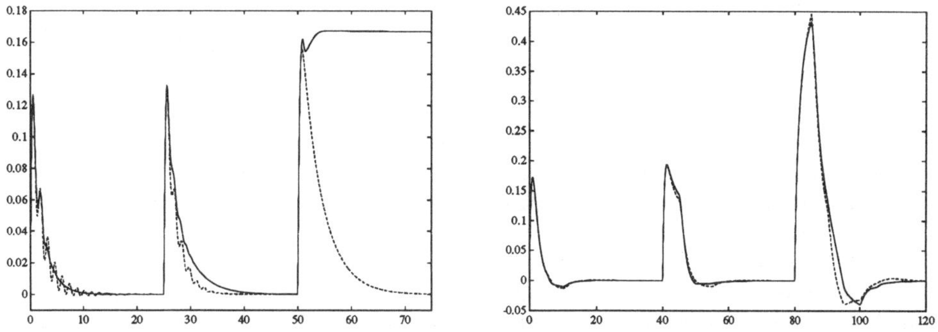


Figure 10: Responses to a ramp-wise reference signal ($V = 0.5$; $B = 0,\ 0.25,\ 1$)

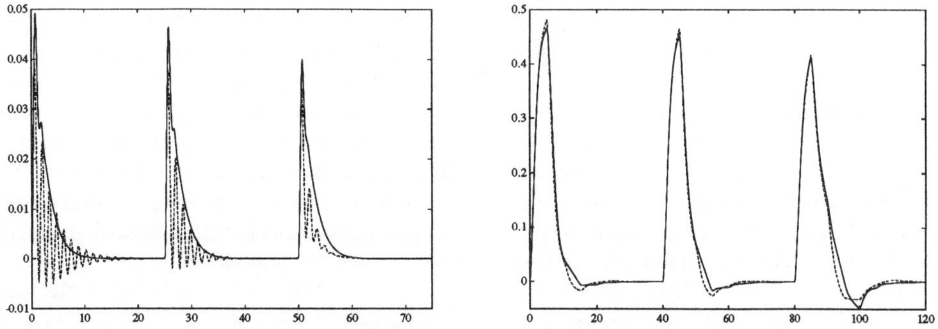


Figure 11: Responses to a constant disturbance ($D = 0.5$; $B = 0,\ 0.25,\ 1$)

**Disturbance Observer** **Model Following Control**

# A Fuzzy Adaptive Controller Using Reinforcement Learning Neural Networks

Augustine O. Esogbue
James A. Murrell
School of Industrial and Systems Engineering
Georgia Institute of Technology
Atlanta, Georgia 30332-0205

*Abstract*—This paper describes an adaptive controller for complex processes which is capable of learning effective control using process data and improving its control through on-line adaptation. The controller is applicable to processes with multivariable states and with uncertain or nonlinear dynamics for which analytical models or standard control algorithms are either impractical or cannot be derived. This controller performs a fuzzy discretization of the process state and control variable spaces and implements fuzzy logic control rules as a fuzzy relation. The membership functions of the fuzzy discretization are adjusted on-line and the fuzzy control rules are learned using a performance measure as feedback reinforcement. The fuzzy discretization procedure employs a data compression technique permitting multivariable state vector inputs. The controller is implemented with neural networks. Simulation results with the controller applied to a simple dynamical system demonstrate the effectiveness of the controller.

## I. INTRODUCTION

Fuzzy control has proven effective for complex, nonlinear, imprecisely-defined processes for which standard models and controls are impractical or cannot be derived. However, deriving fuzzy control rules is often difficult and time-consuming. Furthermore, problems of high-dimensionality are incurred in the implementation of controls for systems with multiple inputs and outputs. More efficient and systematic methods for knowledge acquisition and fuzzy controller synthesis are needed, such as adaptive fuzzy controllers capable of learning from process data to automatically generate a set of fuzzy control rules and improve on them over time.

Considerable work has been done in this area [1], [2], [3], [4], [5], [6], [7]. The controller proposed here has a unique combination of features and capabilities. It is an adaptive fuzzy controller capable of learning from process data on-line. This controller performs a fuzzy discretization of the process state and control variable spaces and implements fuzzy logic control rules as a fuzzy relation. The membership functions of the fuzzy discretization are adjusted on-line and the fuzzy relation is learned using a performance measure as feedback reinforcement, requiring little prior knowledge about the process; no training data sets nor any error signal derived from knowledge of the desired plant trajectory are needed. The fuzzy discretization procedure employs a statistical data compression technique permitting multivariable state vector inputs. Additional plant variables can be added without a geometric increase in the complexity of the controller structure. This procedure extracts the essential information from each variable needed to form fuzzy subsets of the process state space. While it adapts both the membership functions and the control rule state-control association, the controller primarily learns the control rule associations, unlike many other methods which fix the rule relationships and adjust the membership functions. The controller is implmented with neural networks, featuring a self-organizing neural network, a reinforcement learning neural network, and an associative memory network.

## II. THE CONTROLLER OPERATION

The operation of the controller is summarized here. At each interval of a discrete time sequence, the current process state vector is input to the controller. Its membership in each of several reference fuzzy subsets of the input space is calculated in terms of its similarity to the ideal, prototype member of each fuzzy set. Initially the locations of the prototype vectors in the state space are uniformly distributed. Throughout the time sequence, an adaptive algorithm adjusts these locations to reflect the actual clustering of the state vectors into fuzzy sets. The dispersion of the corresponding membership functions are also adapted to the state vector inputs by a similar algorithm. Once an input state has been given its fuzzy characterization in terms of the reference fuzzy sets, the appropriate control fuzzy set is selected. Initially, the selection is arbitrary, but a learning algorithm based on the reinforcement of a performance measure is used to increase the frequency with which good controls are selected. In the process, the controller learns a fuzzy relation between the input state vector and output control vector which embodies the fuzzy control rules. After the learning phase, the fuzzy relation is used to calculate fuzzy control in terms of the reference fuzzy sets of the control space. From this, a crisp control vector is computed.

The controller has five subsystems: the Statistical Fuzzy Discretization Network (SFDN), Fuzzy Correlation Network (FCN), Stochastic Learning Correlation Network (SLCN), Control Activation Network (CAN), and the Performance Evaluation System (PES). A block diagram of these and the plant is shown in Fig. 1.

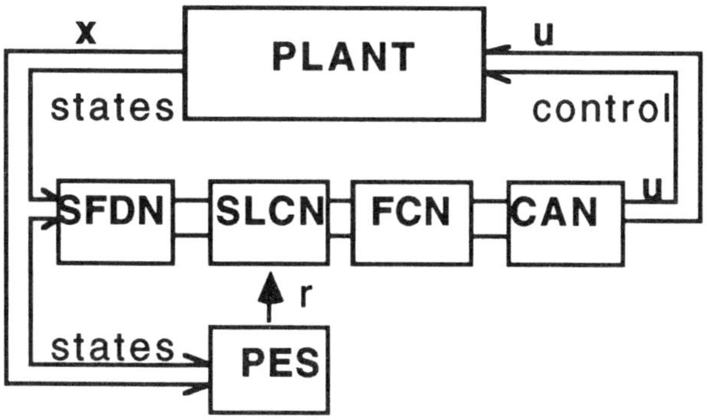


Fig. 1. Controller subsystems and plant

*A. Statistical Fuzzy Discretization Network (SFDN)*

This subsystem consists of a network of automata nodes ("neurons") arranged in a grid (see Fig. 2). Each node receives as its input the current process state vector. Every time a vector is input, each node computes an output that represents the degree of membership in the fuzzy subset of the input space that corresponds to that node. This output, called the node activation, is a measure of the degree of similarity of the input state vector to the ideal or prototype member of that fuzzy set. It is computed as some combination of the state vector and a location parameter vector associated with that node which represents the prototype process state for the corresponding fuzzy set. Also associated with each node is a parameter that encodes the degree of dispersion or spread of its fuzzy set membership function used to calculate the node activation. For a particular membership function form, the location parameters and spread parameter together define a fuzzy set. The SFDN thus performs a fuzzy discretization, inducing a fuzzy partition of the state space $X$ into reference fuzzy subsets $X_1, X_2, \dots, X_r$, each represented by a node in the grid (see [8]).

If the measure of similarity is closeness in terms of Euclidean distance, the location of any vector in the state space should result in the corresponding location on the node map being most activated. Similar inputs should result in

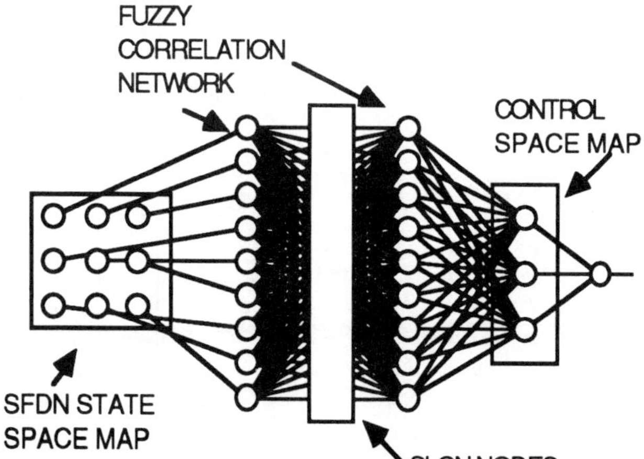


Fig. 2. Configuration of Controller Neural Networks

large activations of nodes which are close to each other in the grid. In the case of a 2-dimensional state space and 2-dimensional grid, the node grid forms a fuzzy map of the input space. The topological ordering and distances among vectors in the state space are preserved in the ordering and proportional distances among the location vectors of the nodes in the map. The process state is then characterized by its fuzzy location on this 2-dimensional fuzzy map based on its similarity to the prototypes of the fuzzy sets defined by the nodes of the map. This can be generalized for a state space of more than two dimensions and for similarity measures other than the Euclidean metric. If the state space were more than 2-dimensional, order and distance could not both be preserved in a 2-dimensional map, but the relative order and closeness (i.e., the topological ordering) could be preserved. (For discussion, see [9].) The choice of similarity measure depends on the nature of the particular process to be controlled. For example, measures such as the normalized inner product (correlation) emphasize the relationships among the components, i.e., the "shape" or pattern in the vector rather than the actual magnitudes of each component. While other fuzzy discretization methods often rely on clustering algorithms that use only a distance metric as a measure, the network described may use alternative similarity measures.

The state space must be restricted by setting upper and lower bounds in each dimension. The precision of the fuzzy discretization is determined by the number of nodes in the grid relative to the number of state space dimensions and the size of the bounded intervals in each dimension. Other fuzzy discretization schemes generate a complete set of fuzzy subset terms for each dimension of the space, thereby increasing geometrically the potential number of fuzzy rules as the dimension increases. In this method, there is one set of fuzzy terms whatever the dimension of the state space. Whether the number of nodes needs to be increased as the dimension increases is a matter of discretion involving a trade-off between the precision of the fuzzification and a polynomial (rather than geometric) increase in the number of nodes.

It is assumed that little is known a priori about the process, so a meaningful aggregation of the process states into fuzzy sets is not known. Thus, the location parameter vectors are initially distributed throughout the state space so that they are in order and uniformly spaced according to the similarity measure. With the Euclidean distance similarity measure, the initial distribution would be the usual uniform distribution.

The network described here is an extension of Kohonen's self-organizing feature map to fuzzy characterization of dynamic plant states [9]. The output of the $i^{th}$ node in the map is

$$a_i = \exp(-\|\boldsymbol{x} - \boldsymbol{m}_i\| / s_i) \tag{9}$$

where $\boldsymbol{x}$ is the state vector input, $\boldsymbol{m}_i$ is the vector of location parameters and $s_i$ the spread parameter for the $i^{th}$ node, and it is assumed that the choice of similarity measure is the Euclidean metric and the functional form of the membership functions is a gaussian function. Thus, the vector $\boldsymbol{a}$ of node activations is the fuzzy hyperstate due to input state vector $\boldsymbol{x}$:

$$\boldsymbol{a} = (\pi_{X_1}(\boldsymbol{x}), ..., \pi_{X_r}(\boldsymbol{x}))^T \qquad (10)$$

where $\pi_{X_i}(\bullet)$ is the membership function for fuzzy set $X_i$ given by (9).

A sequence of state vectors is input to the network over time, and an adjustment algorithm adapts the location parameters to reflect the actual clustering of the state vectors by which the aggregation into fuzzy subsets is determined. A simplified version of the update rule of the $j^{th}$ component of $\boldsymbol{m}_i$ for this example is

$$m_{ij,k+1} = \begin{cases} m_{ij,k} + \alpha_k(x_{j,k} - m_{ij,k}) & \text{for } i \in T_{c_k} \\ m_{ij,k} & \text{otherwise} \end{cases} \qquad (11)$$

where $k$ indexes the time step of the algorithm, $T_{c_k}$ is a small neighborhood of nodes in the grid within a radius $c_k$ around the node most activated by $\mathbf{x}_k$, and $\alpha_k$ and $c_k$ are decreasing functions of $k$. The basic concept for updating the spread parameter is given by

$$s_{i,k+1} = \begin{cases} s_{i,k} + \eta_k(|x_{j,k} - m_{ij,k}|^2 - s_{i,k}) & \text{for } i \in T_{c_k} \\ s_{i,k} & \text{otherwise} \end{cases} \qquad (12)$$

where $\eta_k$ is also a decreasing function of $k$.

The usual Kohonen algorithm requires an uncorrelated time sequence of inputs, whereas the sequence of states from a dynamical process is usually highly correlated. To employ the above algorithm, it is necessary to avoid the correlated sequence by one of several possible methods. For example, the sequence could be infrequently sampled, i.e., the algorithm time step $k$ is larger than the time step $t$ of the input sequence. Alternatively, the data could be collected together in a batch which is then randomized before it is applied to the inputs.

The SFDN provides a means of aggregating similar plant states, thus permitting implementation of the control as a discrete relation. The adaptation update equations are of the simple delta-rule type, which is more easily implemented in real time than clustering algorithms and permits the parallel distributed computation of neural networks.

### *B. Fuzzy Correlation Network (FCN)*

The Fuzzy Correlation Network implements the fuzzy control rules as a fuzzy relation $\boldsymbol{G}$ (learned by the SLCN) which associates the collection of fuzzy sets $X_1, ..., X_r$ for input vectors $x \in X$ with the fuzzy sets $U_1, ..., U_s$ for the controls $u \in U$. This is accomplished with a fuzzy associative memory (FAM) or correlation network [10], [11]. This can be illustrated as two parallel strings of nodes (see Fig. 2). The $i^{th}$ node on one side represents the degree to which $X_i$ has been selected, given by the SFDN node output $a_i$ for state $x$. Each of these is linked to every node on the other side. The output $b_j$ of the $j^{th}$ node on the other side indicates the degree to which fuzzy output set $U_j$ is the correct choice given the activations of the $X_i$'s, or the "firing strength" of each rule for which that fuzzy control is the consequent. The connection weight parameter $g_{ij}$ indicates the degree to which $X_i$ relates to $U_j$. The control rule "If $\boldsymbol{x}$ is $X_i$ then $\boldsymbol{u}$ is $U_j$" is represented by a strong link $g_{ij}$ between node $X_i$ and node $U_j$. The network connection weight matrix $\boldsymbol{G} = \{g_{ij}\}$ specfies a fuzzy relation on $\{X_1,...,X_r\} \times \{U_1,...,U_s\}$. Thus, given input activation $\boldsymbol{a}$, the fuzzy hyperstate, the fuzzy control vector is given by

$$\boldsymbol{b}^T = \sigma(\boldsymbol{a}^T\boldsymbol{G}) \qquad (13)$$

where $\sigma$ is a vector-valued function whose components are $b_j = \sigma_j(\boldsymbol{a}^T\boldsymbol{G}_{\bullet j})$, $\boldsymbol{G}_{\bullet j}$ is the $j^{th}$ column of $\boldsymbol{G}$, and each $\sigma_j$ is some type of limiting function, (i.e., $\sigma_i$ satisfies $\sigma_i(\alpha) \rightarrow 0$ as $\alpha \rightarrow -\infty$ and $\sigma_i(\alpha) \rightarrow 1$ as $\alpha \rightarrow \infty$), such as a sigmoid function. Thus, this implementation uses the product and limited-sum logic operators instead of the common min-max logic (see [12]). The product/limited-sum logic is easily implemented with a neural network associative memory.

### *C. Stochastic Learning Correlation Network (SLCN)*

The purpose of this subsystem is to test and learn the effectiveness of pairing a particular control vector fuzzy set with each given state vector fuzzy set, using the performance evaluation provided by the Performance Evaluation System (PES), then use this knowledge base to generate the fuzzy control relation used by the FCN. Both the SLCN and the FCN receive as input the fuzzy hyperstate $\boldsymbol{a}$ output by the state map grid of the SFDN. The first phase of operation of the controller is a learning phase in which the fuzzy control vector $\boldsymbol{b}$ is generated by the SLCN. In the second phase of operation, the fuzzy relation learned by the SLCN is used by the FCN to generate $\boldsymbol{b}$.

Initially, nothing is known about what control vector gives the best response when the process is in a given state. So, a fuzzy output is just picked and the performance measure indicates how good the selection was. If it was not that good, the controller will be less inclined to pick that control again the next time the system enters that fuzzy state. If the selection was good, that control action is reinforced, so that it is more likely to be selected next time. The network that implements this is akin to stochastic learning automata described by Narendra [13], [14].

The Stochastic Learning Correlation Network consists of a matrix of nodes where each row corresponds to a particular fuzzy input state and each column to a particular fuzzy control action. The degree of activation of a node indicates the fuzzy degree to which it selects the control fuzzy subset to which it is assigned. Each node has a spread parameter $h_{ij}$ for the $i^{th}$ fuzzy state and the $j^{th}$ fuzzy control. The location parameters $\lambda_{ij}$ (scalar) are adjusted so that they always fall at the center of the spread function. In this case, a box-shaped function defined on a bounded interval is used. Thus, the output of the node for the $i^{th}$ fuzzy state and the $j^{th}$ fuzzy control is given by

$$c_{ij,t} = \begin{cases} 1 & \text{if } \lambda_{ij,t} - h_{ij,t} < \xi_t < \lambda_{ij,t} + h_{ij,t} \\ 0 & \text{otherwise} \end{cases} \tag{14}$$

$$\lambda_{ij,t} = \sum_{k=1}^{j-1} h_{ij,t} + \frac{1}{2} h_{ij,t} \tag{15}$$

where $h_{ij,t}$ and $\lambda_{ij,t}$ are location and spread parameters, respectively, at time $t$ for node $(i,j)$, and $\xi_t \in [0,1]$ is generated by a chaotic or pseudo-random process and serves as the input to the node. The node with the largest spread parameter is then the one that will have the maximum activation most often. The fuzzy control vector **b** is given by

$$b_{j,t} = \begin{cases} c_{ij,t} & \text{for } i = \text{argmax}(a_i) \\ 0 & \text{otherwise} \end{cases} \tag{16}$$

When $i$ is the index of the most activated input fuzzy set, then the most activated node in the $i^{th}$ row of the SLCN node matrix selects the control fuzzy set.

The update algorithm for $h_{ij}$ is given as

$$h_{ij,t+1} = (h_{ij,t} + r_t\, a_{i,t}\, b_{j,t}\, \gamma_{ij,t}) \,/\, (1 + \Sigma_{j} r_t\, a_{i,t}\, b_{j,t}\, \gamma_{ij,t}) \tag{17}$$

$$\gamma_{ij,t} = \begin{cases} h_{ij,t} & \text{if } r_t < 0 \\ 1 - h_{ij,t} & \text{if } r_t \geq 0 \end{cases} \tag{18}$$

where $r_t$ is the reinforcement which is a function of the performance measure $p_t$, and $a_{i,t}$ and $b_{i,t}$ are the input and output activations, respectively, for the $i^{th}$ fuzzy state and $j^{th}$ fuzzy control at time $t$. The product $a_{i,t}\, b_{j,t}$ is the association or correlation between the state and control fuzzy sets. The quantities $p_t$ and $r_t$ are computed by the PES described in Sct. II.E. Initially $h_{ij} = 1/|U|$ for $j = 1$ to $|U|$, i.e., the spread parameters of the nodes for fuzzy state $i$ are set equal to each other, so one control would be picked as often as any other. Then this procedure maintains the sets $C_{i,t}=\{g_{ij,t}\,; j=1, 2, ..., |U|\}$ and $D_{i,t}=\{h_{ij,t}\,; j=1, 2, ..., |U|\}$ each as a unit simplex. Thus, the control is stochastically selected by using the set $D_{i,t}=\{h_{ij,t}\,; j=1, 2, ..., |U|\}$ as a probability distribution over the fuzzy discretization of control set $U$.

The update algorithm of (17)-(18) weights most heavily the connection between a given (fuzzy discretized) state and the applied control which resulted in the best performance measure. This is essentially the same heuristic that is often used when fuzzy control rules are developed off-line by human experts rather than by on-line adaptive control. Each spread parameter indicates the level of performance for the corresponding fuzzy action when at a particular fuzzy state. The best action is the prototype of good performance and the other parameters indicate how close to that the other choices are. As the SLCN matrix learns which fuzzy control gives the best results for each fuzzy state, the parameters of the fuzzy relation can be computed using the spread parameters. The fuzzy relation parameters $g_{ij}$ are given as follows:

$$g_{ij,t+1} = \phi(h_{i1,t}, h_{i2,t}, \ldots, h_{is,t}) \tag{19}$$

where $\phi(\cdot)$ is a high pass filter function; a simple and adequate implementation would be

$$g_{ij,t+1} = \frac{\sigma(h_{ij,t})}{1 + \Sigma_j \sigma(h_{ij,t})} \tag{20}$$

$$\sigma(\alpha) = \begin{cases} \alpha - 1/|U| & \text{if } \alpha > 1/|U| \\ 0 & \text{otherwise} \end{cases} \tag{21}$$

Thus, in the second phase of operation, rather than stochastically selecting the fuzzy control using using the SLCN nodes, the fuzzy control $\boldsymbol{b}$ is calculated by the FCN as $\boldsymbol{b}^T = \sigma(\boldsymbol{a}^T \boldsymbol{G})$. The second phase can begin as the learning process of the spread parameters becomes mature. If the process is time-varying and has shifted enough so that the existing control policy of the fuzzy relation is no longer adequate, then the learning mode can be shifted back to the initial phase to learn a new control policy. This second phase is a forcing of a fixed policy as opposed to the random policy given by the SLCN. The two phases can be interspersed, so that the fixed policy phase increases in frequency as the uncertainty decreases. Continued occasional use of the learning phase allows the controller to continue to improve. This mixing of randomized policies with the forcing of fixed policies is useful for those plants which have only a small random component. The level of learning or certainty in the relation parameters can be measured with an entropy measure on $C_{i,t}$, $D_{i,t}$ or the components of $\boldsymbol{b}$. For example,

$$e_i = -\sum_{k=1}^{|U|} g_{ik} \ln(g_{ik}) \quad \text{or} \quad e = -\sum_{k=1}^{|U|} b_k \ln(b_k) \tag{22}$$

and the forcing frequency can be set to be a decreasing function of $e$ or of all the $e_i$.

The adaptive control described here is dual control in which the dual objectives are achieving good control and obtaining sufficiently variable inputs to make good statistical estimates. The stochastic selection of the controls provides the persistent excitation and randomization needed for consistent estimation of the relation $\boldsymbol{G}$ via the estimated quantities $p_t$, the performance.

### *D. Control Activation Network (CAN)*

The input to the Control Activation Network is the fuzzy control vector $\boldsymbol{b}$. This fuzzy control is defuzzified to produce a crisp output quantity $\boldsymbol{u}$, which is a vector for multivariable control processes. Each CAN node has its location parameter vector set to the desired control vector prototype levels. Its input is the vector $\boldsymbol{b}$, and its output is a crisp control vector

$u$. The nodes can be set up as a map network to adapt the fuzzy sets according to the control vectors that are actually being output from the controller.

Using the max criterion defuzzification method, the $B_i$ node with the largest activation (degree of truth) triggers the activation of the CAN node whose output is the prototype value $\overline{u}_j$ corresponding to the fuzzy set $B_j$. Alternatively, in the center of area method, $u$ is calculated as the normalized weighted sum over the fuzzy sets $B_i$ in which the weights are the activations $b_j$, given as

$$u = \left( \sum_{k=1}^{|U|} b_k s_k \overline{u}_k \right) \Big/ \sum_{k=1}^{|U|} b_k \qquad (23)$$

where $s_k$ are the spread parameters for the output fuzzy set membership functions. If the membership function form is symmetrical, then the effect of the spread is trivial.

### *E. Performance Evaluation System (PES)*

The particular nature of the plant or process to be controlled and the characterization of the desired performance dictate the details of how the performance evaluation network is configured. When a performance measure is available which is an analytical function of the plant states or output, then the reinforcment signal $r_t$ is simply a normalization of the performance measure $p_t$ to lie in the interval [—1,1] or [0,1]. It is often the case, however, that complex processes which have no known well-defined plant model also do not have a well-defined formula for computing performance. Rather, there is a certain qualitative goal or objective to be reached, but it is not known what the values of the plant variables should be when that goal is reached. Even when a formula in terms of the variables is known, there is often an unknown delay between the control action taken and the effect on the plant variables, so that the result of the current control is not known until some future time. In such cases, various methods of estimating the performance evaluation function must be used. Our investigation into performance function estimation methods will be reported elsewhere.

The most straight-forward approach to the situation in which a qualitative determination of reaching the goal is given only after a period of many time intervals is described here. Reaching the goal is indicated by $p_t = 1$ (success) and reaching forbidden states (such as plant shutdown due to variables out-of-bounds) is indicated by —1 (failure). For each state entered at each time, a control action is taken. The average performance over time of this state-control pair is computed and updated whenever there is a success or failure. The reinforcement signal $r_t$ can then simply be the current value of this average for the current state-control pair that just occurred.

## III. RESULTS AND PROJECT STATUS

The ability of the controller to learn effective control has been tested through simulations in which the controller is applied to the inverted pendulum problem [15] with two state variables and one control variable. The dynamic equations and parameters are the same as those used in [2]. In these experiments, the controller uses a 5 by 5 node state map, a 1 by 5 node control map, the Euclidean metric and gaussian membership functions for similarity, and adapts the state space fuzzy sets but not the control fuzzy sets. The performance measure is the cumulated average method described in Sct. II.E., where failure is the pole angle exeeeding +12 or —12 degrees from vertical, or angular velocity exceeding +25 or —25, and success is achieving a membership degree of at least 0.9 in a gaussian membership function centered at the 0 degree, 0 angular velocity point (for the set "near zero"). A new trial begins after every failure with randomly selected initial states within the above ranges.

Each run uses a different inital seed for the pseudo-random generator. There is a high degree of variability in the learning phase, so rather than test how many trials it takes to balance the pole, the test is to determine how well the controller learned the control surface after fixed numbers of trials. The controller was run in the learning phase for 25, 50 and 100 trials. At these intervals, the controller was switched to the second phase and its response tested for initial states (10,0), (—10,0), (7,7), (—7,7), (7,—7), (—7,—7), and (0,0). If the controller did not drive the state directly to a point near (0,0), with small final acceleration, then the second phase trial was stopped. The points (0,0), (—7,7) and (7,—7) are easy, and the controller was sometimes able to drive the state to (0,0) after 25 trials. At 50 trials, the controller was sometimes able to drive the controller to near (0,0) for inital state (—10,0) or (10,0). In all but a few cases, 100 trials were required to learn a control surface that could drive all of the inital points to near (0,0) within about 1000 time steps (1 second of real time). Fig. 3 shows the time response of the angle and angular velocity, and Fig. 4 shows the phase space trajectory for initial state (10,0) using the relation weights learned after 100 trials. The distorted grid in Fig. 4 shows the locations of the state space map node vectors.

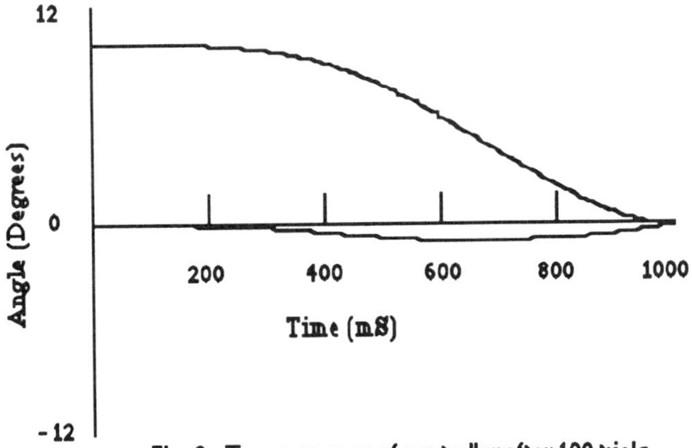


Fig. 3. Time response of controller after 100 trials

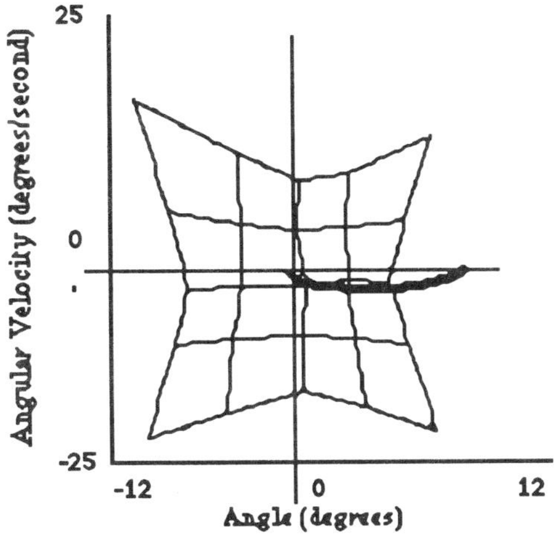


Fig. 4. Phase space plot of control response

IV. CONCLUDING REMARKS AND FUTURE WORK

A new fuzzy adaptive controller using reinforcement learning neural networks has been described. The basic capability to learn effective control has been demonstrated for a simple dynamical system. It learns this control on-line using only the input of the process state and a scalar feedback signal that gives a measure of the performance by whatever criteria can be determined from observation of the process. It does not need to be trained with examples of correct state-control pairs. It does not need an explicit error signal which either requires a priori knowledge of what the correct control vectors are or some type of plant model that is used to estimate them. It holds the promise of handling multi-dimensional systems without a proliferation of fuzzy subsets or discretized values. Also, the relation itself is learned rather than simply adapting the fuzzy sets to fixed rules.

Software test beds are currently under development which will permit studies of the system performance with larger numbers of state and control variables, testing of the data compression properties, and testing the effects of using different alternative similarity measures and various performance measures. Additional study of random/fixed policy schedules will be made also.

The type of control that this controller can provide is potentially applicable in many settings. It can be used in control situations in which there are multiple sensor measurements which may be noisy or imprecise or which require sensor fusion to generate a coherent picture of the process state. It can be used for high-level decision-making or control of data processing, intelligent system reconfiguration in response to changing conditions, or to direct flow in networks. Control of highly nonlinear dynamical systems (e.g., robot arms, etc.) for which it has been difficult to apply standard control theory methods is another application area where adaptive fuzzy controllers have proven effective. It can be used for failure detection and diagnosis or in a statistical process monitoring and control mode. Wherever intelligent decision-making in real-time is required for coping with an uncertain, noisy and/or changing environment, this type of automatic controller may be useful. In future work, we plan to explore the capabilities of this controller in some of these application areas.

ACKNOWLEDGEMENT

This research was sponsored in part by the National Science Foundation under Grant ECS-9216004.

REFERENCES

[1] Lin, C.T. and Lee, C.S.G., "Reinforcement structure/parameter learning for neural-network-based fuzzy logic control systems," this conference.

[2] Jang, J.S. R., 1992, "Fuzzy controller design without domain experts," *Proc. of IEEE Int. Conf. on Fuzzy Systems 1992*, pp. 289-296.

[3] Berenji, H.R., "A reinforcement learning-based architecture for fuzzy logic control," *Int. J. of Approximate Reasoning*, vol. 6, pp. 267-292, 1992.

[4] Lee, C.C., "Intelligent Control Based on Fuzzy Logic and Neural Net Theory," *Proc. Int. Conf. Fuzzy Logic and Neural Networks*, Iizuka, Japan 1990, pp. 759-764.

[5] Patrikar, A. and Provence, J., "A self-organizing controller for dynamic processes using neural networks," *Int. Joint Conf. on Neural Networks IJCNN 1990*, vol. 3, pp. 359-364.

[6] Procyk, T.J. and Mamdani, E.H., "A linguistic self-organizing process controller," *Automatica*, vol. 15, pp. 15-30, 1979.

[7] Xu, Chen-Wei and Lu, Yong-zai,"Fuzzy model identification and self-learning for dynamic systems," *IEEE Tran. Syst., Man, Cybern.*, vol. 17, pp. 683-689, 1987.

[8] Pedrycz, W., "An identification algorithm in fuzzy relational systems," *Fuzzy Sets and Systems*, vol. 13, pp. 153-167, 1984.

[9] Kohonen, T., *Self-Organization and Associative Memory*, Springer-Verlag, 1988.

[10] Pedrycz, W., Hirota, K. and Takagi, T., "Fuzzy associative memories: concepts, architectures and algorithms," *Proc. of IFES Conf.* 1991, pp. 163-174.

[11] Kosko, B., *Neural Networks and Fuzzy Systems*, Englewood Cliffs, NJ: Prentice Hall, 1992.

[12] Pedrycz, W., *Fuzzy Control and Fuzzy Systems*, Taunton, England: Research Studies Press, Ltd., 1979.

[13] Narendra, Kumpati and Thathachar, M. A. L., *Learning Automata; An Introduction*, Englewood Cliffs, NJ: Prentice-Hall, 1989.

[14] Murrell, J.A., "Stochastic Learning Automata in Adaptive Control Problems," Master's expository, Operations Research Dept., University of N. Carolina at Chapel Hill, 1988.

[15] Barto, A. G., Sutton, R. S., Anderson, C. W., "Neuron-like adaptive elements that can solve difficult learning problems", *IEEE Trans. on Systems, Man and Cybern.*, vol. 13, pp. 834-846, 1983.

# An Outline of the Intuitive Design of Fuzzy Logic and its Efficient Implementation.

D J Ostrowski, P Y K Cheung & K Roubaud

Dept. Of Elec. Eng., Imperial College, London. (UK)

LSI Logic Europe plc., Bracknell, Berks. (UK)

(fax +44 71-581-4419, email dostr@ic.ac.uk)

***Abstract*—From intuitive design with graphical tuning of independent fuzzy rules, to a very fast and low area implementation, automatically, using VHDL throughout. The suggested design approach allows the rulebase to be represented in the format that the user prefers and simulated and refined, but to be translated to a single format for implementation. Features of this include 1) only processing active rules 2) independent rules that allow one rule to be tuned without affecting the others 3) rules of different strengths allowing the relative importance of a rule to be expressed and 4) precalculation of the centre of area (COA) of fuzzy consequents considerably speeding defuzzification.**

## I. INTRODUCTION

The paper is presented in three main sections, the first discusses a range of design elements useful in fuzzy processor or controller design. Three representations of fuzziness are presented along with their origins, the need for each and their suitability for implementation. The use of truly independent rules, allowing the relative strengths and generality of a rule to be expressed, is discussed. A precalculation of fuzzy consequents, that speeds defuzzification, is shown. Finally in this section a special type of consequent is proposed that may be useful in defining fuzzy rules that discourage a certain action.

The second part is concerned with the efficient implementation of designs, and discusses taking designs from the form most intuitively useful to the designer and translating them to a common representation resulting in implementations for high speed and small silicon area, or efficient software. All the design elements presented in the first section may be automatically translated to this format. The method of translation is described, followed by an efficient method of storing rule antecedents. A suitable rule processing array, that finds the active consequents from the active antecedents, is explained. A method of precalculating and storing rule consequents is described and the simplified defuzzification is explained. Finally in this section the timing and hardware resources of the implementation are reviewed.

The third and final part of this paper looks at the use of VHDL code as a tool for developing fuzzy logic and as a readily available route through design and simulation. The features of VHDL that make it particularly suited to fuzzy logic design are also described.

## II. FUZZY DESIGN

It is a major requirement of fuzzy logic design that an intuitive understanding of the rulebase is retained throughout design / simulation / modification. The design process of a fuzzy controller or processor must start with a knowledge of the open loop behaviour of the system, the control or processing goals and the sensors and transducers linked to it, from this alone it is possible to develop a simple rulebase and assess its performance through simulation. The first task is to develop a rulebase, that is a list of antecedents (for each input) linked to a consequent (for a particular output). The rulebase though may be represented in several different ways, and an initial choice facing the designer is the best representation of fuzziness for the application. Fuzziness may be encapsulated entirely in a fuzzification of the input as seen in fuzzy associative matrices (FAM's) (Fig 1(a)), as described by Kosko[1]. Here each input is divided into a set of linguistic fuzzy groups and the rules are all composed of these linguistic terms. Fuzziness may be expressed in both input and rule antecedent (Fig 1(b)) as described by Dettloff[2], here there is much greater flexibility as each rule may have its own unique fuzzy set as may each input value. Thirdly fuzziness may be contained entirely within the rule antecedent (Fig 1(c)), here the crisp input value is treated as a fuzzy singleton and each rule is represented by a unique fuzzy set.

This work was supported by a Science and Engineering Research Council (UK) grant, thanks also to LSI Logic Europe plc. for advice and support.

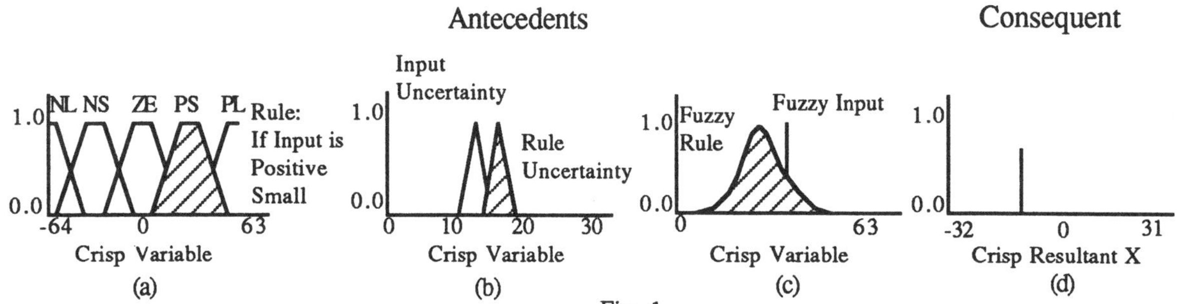


Fig. 1

0-7803-0614-7/93$03.00 ©1993IEEE

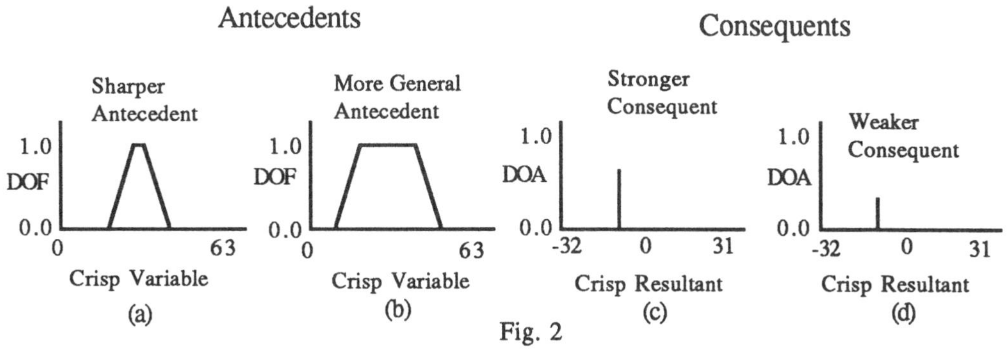

Fig. 2

These different representations lend themselves to different types of problems, the designer should be free to choose between them, because any format may be translated to that shown in Fig 1(c) and as described in the following section. This selection may be all in one format, or individual inputs may be assigned a particular format, whichever provides the most intuitive representation. At this stage in the design process, implementation may be disregarded as the target format will be efficient in both speed and area unless the rulebase is so simple to be entirely represented as a simple FAM. Dettloff's format (Fig 1(b)) clearly provides the greatest flexibility, but the implementation suggested in [2] has been criticised by Chiueh[3] for its inefficiency. FAM's as in Fig 1(a) that use the same input fuzzification for all rule antecedents and the same set of fuzzy groups for all consequents (as in Eichfeld [4]) offer more efficient implementations than that shown here, but at the expense of independent rules (where one rule may be tuned without affecting the others) and relative rule strengths (where rules may have similar consequents, but greater or lesser influence). Designs that may be effectively defined and refined as FAM's will suffer a loss of efficiency in implementation using the approach described in this paper. All other designs may use a mixture of representations which will be translated to 1(c) for implementation, while retaining efficiency and flexibility.

There is also flexibility in the representations of consequents. Rule consequents are usually represented as fuzzy groups, and for FAM's it is necessary, for efficiency, to use only one set of groups for all rules. But it should be noted that rule consequents do not necessarily need to be stored (or represented) as fuzzy sets (Fig 1(d)). If the fuzzy set for the consequent has a symmetrical shape, its centre of area (COA) does not change with the rule degree of fulfilment (dof), allowing the simpler singleton representation. For more complex fuzzy consequents of non-symmetrical shape, the complex representation is allowed and precalculation used. If the defuzzification used finds the COA of each fuzzy consequent, and then the COA of all consequents, precalculation is possible. The COA of each rule consequent may be pre-calculated and stored, for every possible degree of action (doa), resulting in a unique singleton for each rule and doa. This is equivalent to the summation of consequent fuzzy sets (sand mountain), rather than union (sand dunes) in both the case of simple and complex consequents.

In all these examples the antecedents are always shown as having a maximum degree of fulfilment dof = 1, it is suggested that the antecedent alone is used to define the fuzziness of a rule (Fig 2(a & b)), and the consequent is used to define its relative strength (Fig 2(c & d)). Note the preferred use of simplified (fuzzy singleton) consequents. As each rule is defined in terms of its own unique antecedents and consequent, these groups may be manipulated without affecting other rules. A major advantage of this form is that it is possible to adjust an individual rules broadness or narrowness of its antecedents and the relative strength of its consequents. That is to set one rule above another in precedence or generality, essential to expressing the importance or ubiquity of an individual rule. This is unlike the case of fuzzifying inputs into named groups where all rules manipulate the same set of named groups. The design elements described here are intended for use with a graphical/menu approach to human refining of rules after their initial definition in the rulebase. A suggested approach to rule tuning (where individual rules have been identified or are suspected of causing an undesirable response) are the four operations listed by Schmucker[5] (concentration, dilation, intensification and deintensification) along with altering the ranges of the antecedents and the relative strength of the consequents.

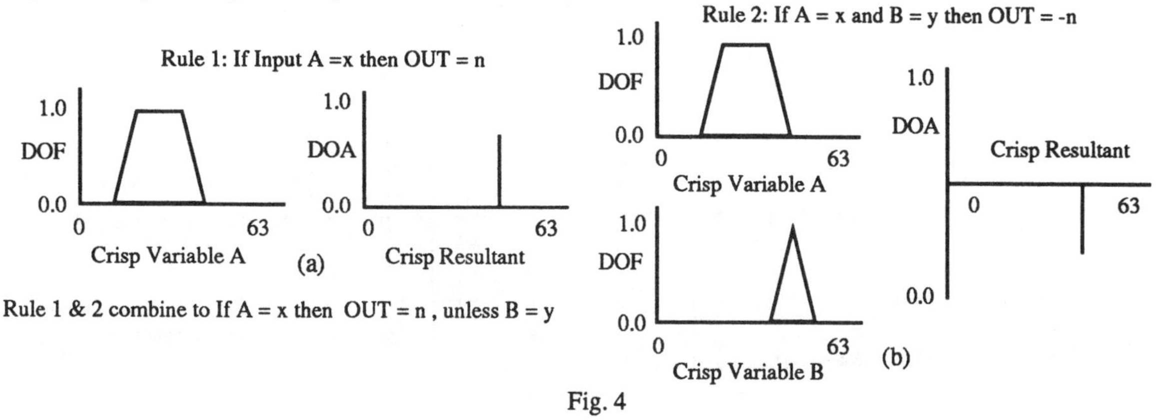

Fig. 4

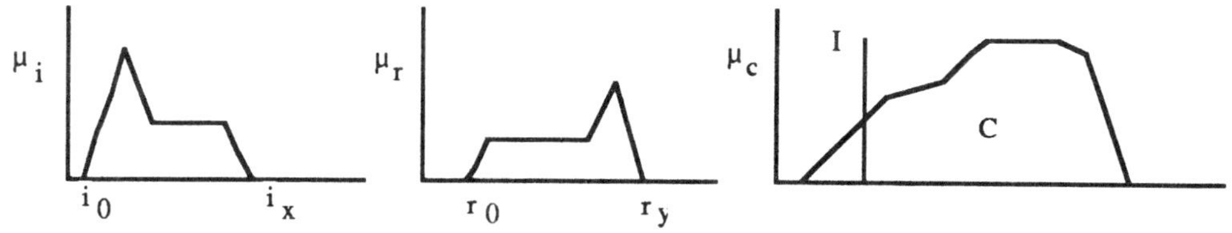


$I = \{\mu_i(i_0), \ldots \mu_i(i_x)\}$ $R = \{\mu_r(r_0), \ldots \mu_r(r_y)\}$ $C = \{\mu_c(i_x = r_0), \ldots \mu_c(i_0 = r_y)\}$

(a) (b) (c)

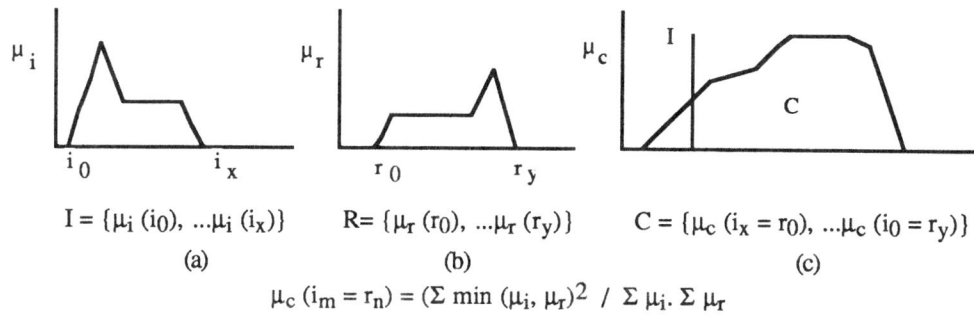

$$\mu_c(i_m = r_n) = (\Sigma \min(\mu_i, \mu_r)^2 \;/\; \Sigma \mu_i . \Sigma \mu_r$$

Fig. 5

Finally in this section a special type of consequent is proposed that adds a new feature in rule definition. Allowing consequents of 'negative weight' provides **If...then forbid...** and **If...then ...unless...** rules.

Fig. 3

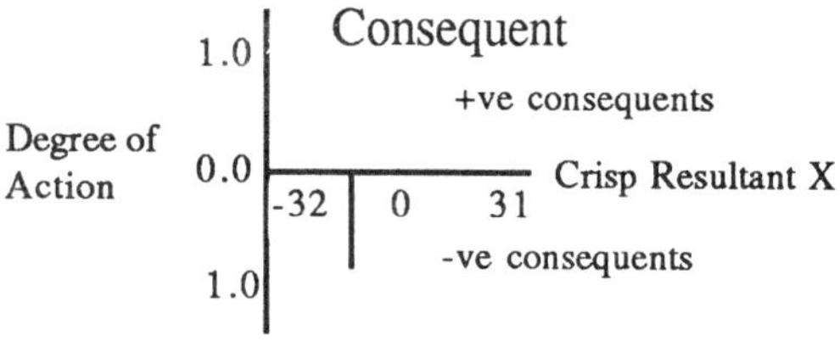


If the consequent has 'negative weight' it will push the defuzzified resultant away from the specified consequent, this is the If...then forbid... rule (Fig 3). Note that the degree of action is scaled to +1.0 on both sides, the area below the resultant axis represents consequents of 'negative weight' to discourage actions and that above of 'positive weight' to encourage a particular action. This type of rule may be useful for specifying safe operating area rules or modifying an existing rulebase, by specifying the inverse consequent to negate an existing rule. Negative consequents are also used to build If...then...unless... rules, where two rules have similar antecedents and exactly inverse consequents. Firing one rule allows the consequent, firing both negates it. Rule 1 in Fig 4(a) has a single rule antecedent relating to input A and a same simple positive consequent. Rule 2 in Fig 4(b) has exactly the antecedent for input A, but is also dependent on input B, and has exactly the inverse consequent of rule 1.

These two rules combine to If A =x then OUT =n unless B = y, so that $\mu_{doa} = \mu_{dofa} - \mu_{dofb}$ for $a > b$

$= 0$ for $a < b$

This is dependent on both A and B, unlike the rule If A = x and not B = y then OUT = n, which is dependent on the smaller variable as $\mu_{doa} = \min(\mu_{dofa}, (1 - \mu_{dofb}))$.

## III. EFFICIENT IMPLEMENTATION

The rulebase has been optimised, and the designer allowed to do so without constraint, through cycles of design / simulation / refinement, now it must be implemented. The approach described here is equally suited to hardware or software. It also ensures that only **active** rules are processed and as much precalculation as possible is carried out. In the first section a variety of representations of antecedents were described (Fig 1(a, b & c)), the first task is to translate the antecedent rulebase into a common format. The first two (Fig 1(a & b)) can be translated into Fig 5(c), the third (Fig 1(c)) is identical to this, where an input fuzzy singleton intersects a fuzzy rule set. Fig 5(a) shows a fuzzy set for an input, Fig 5(b) shows a set for a rule. The combined rule set in Fig 5(c) is found by using the formula below them to find the normalised ratio of the area of overlap to the area of the sets, for every possible degree of intersection. Any fuzzy input set and rule set may be translated this way, and can be reduced to a rule set indexed by a crisp input value. These antecedent rule sets are stored as a rom

Rulebase ROM

| Rule | Degree | Rule | Degree | Rule | Degree | Rule | Degree |
|---|---|---|---|---|---|---|---|
| R1 | 0.9 | R2 | 0.1 | R0 | 0.0 | R0 | 0.0 |
| R1 | 0.8 | R2 | 0.2 | R0 | 0.0 | R0 | 0.0 |
| R1 | 0.4 | R2 | 0.4 | R0 | 0.0 | R0 | 0.0 |
| R1 | 0.2 | R2 | 0.4 | R3 | 0.2 | R0 | 0.0 |
| R2 | 0.2 | R3 | 0.6 | R0 | 0.0 | R0 | 0.0 |
| R3 | 1.0 | R0 | 0.0 | R0 | 0.0 | R0 | 0.0 |
| R3 | 0.5 | R4 | 0.1 | R0 | 0.0 | R0 | 0.0 |
| R3 | 0.5 | R4 | 0.2 | R5 | 0.1 | R6 | 0.1 |
| R3 | 0.1 | R5 | 0.8 | R6 | 0.3 | R0 | 0.0 |
| R5 | 0.3 | R6 | 0.8 | R0 | 0.0 | R0 | 0.0 |
| R6 | 0.3 | R0 | 0.0 | R0 | 0.0 | R0 | 0.0 |

Fig. 6

Crisp Input Value (Fuzzy singleton) used as address

Line of data at address is rule antecedents for that input

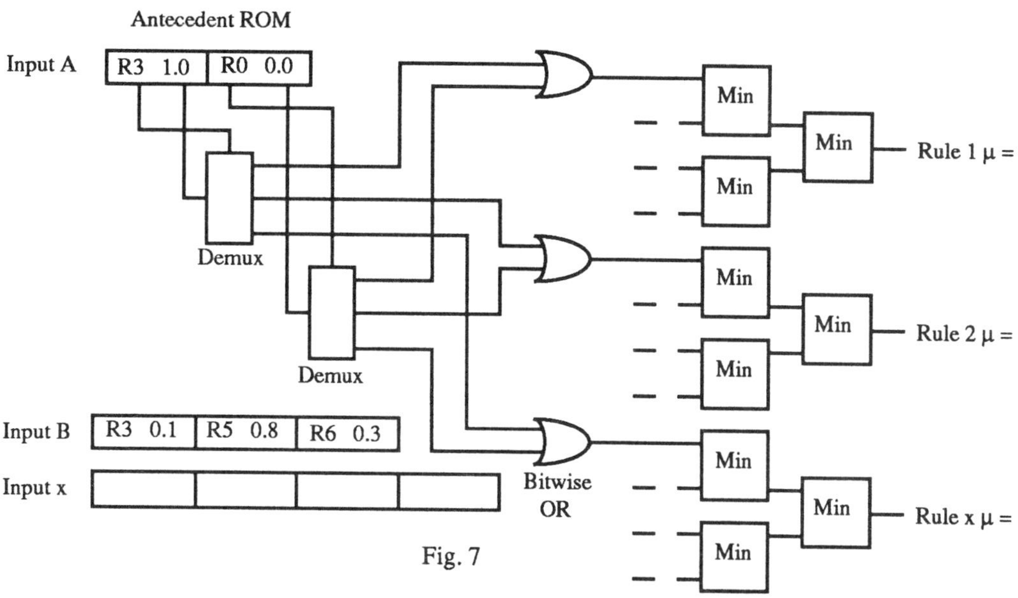

Fig. 7

or ram tables, one table for each input. The value from the input is used to address the table (Fig 6) and the address range corresponds to the sensor precision (or the required precision, if less). Each line in the table holds the degree of fulfilment of the rules activated by that input value.

Note that as most rules have dof = 0 for most of their input range, the table may be further reduced by storing rule/degree pairs for their active range only. The address range remains the same but the word length is much reduced to only the rules activated by that particular value of that input. Empty locations are assigned the non-existent 'rule 0', always dof = 0. Each table outputs a list of potentially active rule antecedents that must be compared with the outputs of the other input tables (the antecedent rulebases). From this the active rule list is built. It should be noted that each input, over most of its range, activates less than the maximum number of potentially simultaneously active rules.

Rule processing starts from the potentially active rule lists (which are symbolic in the case of hardware, not an actual entity) output by each antecedent table and consists of finding the smallest dof for every rule on the potentially active rule lists, e.g. 2 inputs may activate a rule and a third does not, so dof = 0. For software implementations rule processing is instruction intensive, mostly memory read/write and min operations. For hardware an array of demultiplexes and min units are used (Fig 7).

There is a demultiplex for every potentially active rule, the rule number is used as the selector on the demux and assigns the degree to the correct min unit. There is a binary tree of min units for every rule, and each tree is dedicated to a particular rule (but min units are simple). The output of all the min unit trees (called the active rule list, also symbolic in hardware) activate particular consequent roms or rams, this is the consequent rulebase. Here also, between the demuxes and min units, a set of OR gates are encountered. These are necessary as only active rules are stored, rules that have dof = 0 are not stored in the antecedent rulebase and must be regenerated where a rule is not activated by all its inputs. All unselected outputs of the demultiplexes outputs all 0's. If any demux for a particular output sends a dof ≠ 0, this is bitwise OR'd with the 0's sent by the others, and passes through to the min unit input.

The consequent rulebase (Fig 8) has a similar format to the antecedent rulebase, the consequent of each rule is held in a separate table and the degree acts as the rom or ram address. At each address is held another data pair for the magnitude and moment, (position * magnitude) of the rule resultant. This is the is the precalculated area and position * area of the fuzzy consequent. One line in the table is accessed by the dof of each active rule, and this data used in the defuzzification calculation.

The process of defuzzification then is finding the COA of each of these spot magnitudes and moments, (some of which may be negative if they stem from If...then forbid... rules).

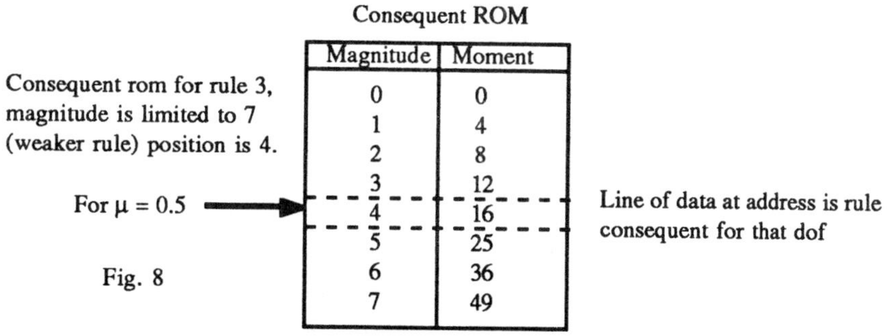
Consequent ROM
Magnitude
Moment
0 0
1 4
2 8
3 12
4 16
5 25
6 36
7 49
Consequent rom for rule 3, magnitude is limited to 7 (weaker rule) position is 4.
For μ = 0.5
Line of data at address is rule consequent for that dof
Fig. 8

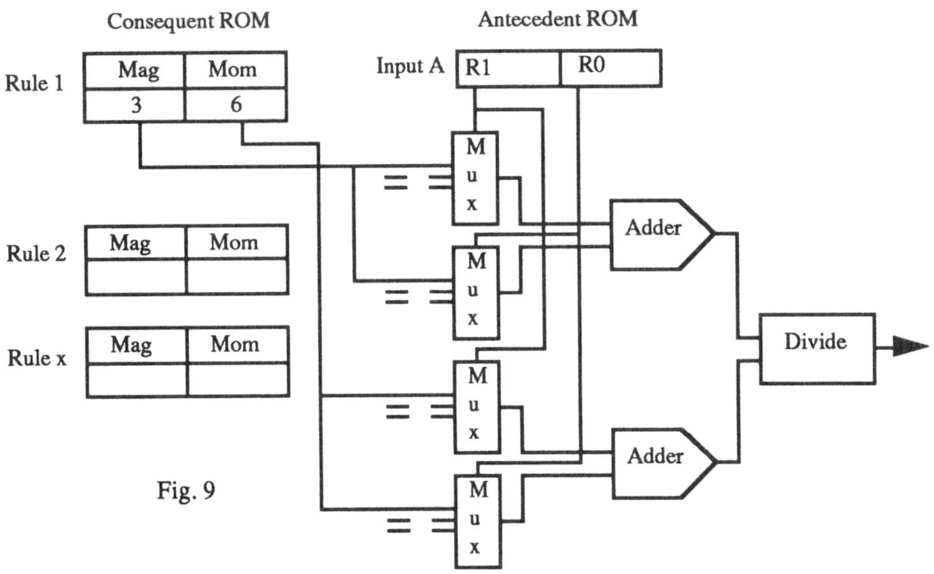


Fig. 9

A conventional COA method of finding the crisp resultant of these consequents is used, but as this is sparse (only spot values for each active rule) the processing time is much reduced. As the moment and magnitude of each resultant is known, the sum of all magnitudes and the sum of all moments is found, and the moment divided by the magnitude to find the crisp resultant. For hardware two binary trees of adders and one divider are used (Fig 9), in software this is mostly memory read and add operations.

A novel approach is taken to allocating adders to **active** consequents. The output of the antecedent rulebase that has the smallest maximum potentially simultaneously active rules is used (Input A in the above example). This steers the outputs of the consequent rulebases that **maybe** active through the multiplexes to the adder trees. A rule must be activated by all inputs to be active, so no more than those activated by the input described above can be active. The non-existent 'rule 0' is required for unassigned adder inputs, this forces the muxes to output all 0's. Two potential trouble spots should be noted in the defuzz circuit, firstly that no rules are fired and a divide by 0 error is encountered. Secondly, if negative weight consequents are used (and twos complement arithmetic), and only a negative consequent rule is fired. These must both be avoided by checks on the rule set. Negative consequents may also attempt to push the defuzzified consequent out of its normal range, these must be trapped and the range limit substituted.

Hardware implementations feature four main blocks 1) antecedent rulebase rom (one for each input) 2) processing array of demuxes OR gates and min units 3) consequent rulebase rom (one for each rule) 4) defuzz built from muxes, adders and a divider. Describing the hardware resources and timing with:

I inputs of B bits of resolution (crisp input values)
R rules of which
A are potentially active rules (antecedents)
of D bits of resolution (degrees of fulfilment)
C are active rules (consequents)
of M bits of resolution (magnitude & moment)

Hardware:

| | | | |
|---|---|---|---|
| Antecedent roms: | | I (one for each input) | |
| | Size: | B (addresses) | |
| | | A * D (bits) | (+ rule no. tag) |
| Rule proc. array: | | Demultiplexes: | I * A |
| | | OR gates: | R * I * D |
| | | Min units: | R (I - 1) |
| Consequent roms: | | R (one for each rule) | |
| | Size: | D (addresses) | |
| | | M (bits) | |
| Defuzz: | | Multiplexes: | 2C |
| | | Adders: | 2 (C - 1) |
| | | Divider: | 1 |

Timing: (input to output)

Antecedent ROM access time +
Demultiplex delay +
OR gate delay +
$lg_2$ I * min unit delay
Consequent ROM access time +
Multiplex delay +
$lg_2$ C * adder delay +
Divider delay

The hardware outlined here shows a linear increase in area with complexity (inputs and rules), a linear increase in rule processing resources with inputs and active rules only, and only a **$lg_2$ (active rules)** decrease in speed with a linear increase in **active** rules or a **$lg_2$ (inputs)** decrease in speed with a linear increase in inputs. (All of these remain true whether all processing is concurrent or inputs are evaluated sequentially).

## IV. VHDL FOR FUZZY LOGIC DESIGN

Our experience of VHDL is described in this section to encourage its use in fuzzy logic design, a recommended introduction is Perry [6]. VHDL is a hardware description language that allows designs to be defined at all levels from a behavioural description down to gate level and for these designs to be simulated. It has been used throughout the design described in this paper. It was found to be an excellent route from defining the rulebase, through timestep simulation and refinement, and is suited to automatic translation and synthesis of fuzzy logic designs.

The key steps in using VHDL to develop a fuzzy controller are writing a model of the controlled system and developing a rulebase. Rulebases require models of rom's, a feature of the language is file type operators that allow access to any file (which should be organised as data word per line) to emulate rom's. Tools written in 'c' can be used to modify these 'rulebase roms'. VHDL has a defined test strategy where every entity in a design should have a corresponding test driver, this is behavioural VHDL code written to fully exercise it. At the top level of a fuzzy design (which may or may not contain sub-entities, depending on whether it is a structural or behavioural description) the test driver may be a model of the target application. Public domain maths libraries of floating point functions are available for writing such models that are suited to time step simulation. This enables a rapid route through design / simulation / modification needed to refine the rulebase. Re-use of design entities is one of VHDL's main advantages along with its compatibility with many vendors through the IEEE standard and its suitability for high level design and simulation.

## CONCLUSION

The key idea in the design process of fuzzy logic is the ability to represent the rulebase in the most intuitive format and that it may be readily simulated and modified while retaining an understanding of the effects of individual rules. This should be followed by automatic translation to efficient hardware or software, to complete the design process and to produce small area and high speed designs. Further work will look at more complex designs with larger rulebases and multiple outputs.

## REFERENCES

[1] Bart Kosko, *Neural Networks and Fuzzy Systems*, Prentice-Hall. ISBN 0-13-612334-1

[2] W. Dettloff, H. Watanabe & K. Yount, "A fuzzy logic controller with reconfigurable, cascadable architecture," *IEEE J. Solid State Circuits*, Vol. 25, No. 2, Apr 1990, pp. 376-82.

[3] T. Chiueh, "Optimization of fuzzy logic implementation," *Proc. 21st Int. Sypm. Multiple-valued logic*, 1991.

[4] H. Eichfeld, M. Lohner & M. Muller, "Architecture of a CMOS fuzzy logic controller with optimised memory organisation and operator design," *FUZZ-IEEE '92*.

[5] Kurt. J. Schmucker, *Fuzzy Sets, Natural language computations, and risk analysis*, 1984.

[6] Douglas L. Perry, *VHDL*, McGraw-Hill. ISBN 0-07-049433-9

# Fuzzy System Design Through Fuzzy Clustering and Optimal Predefuzzification*

Sam-Kit Sin *and* Rui J.P. deFigueiredo
Department of Electrical and Computer Engineering
University of California, Irvine, CA 92717
Tel: (714) 856 7043 FAX: (714) 725 3408 Email: rui@uci.edu

***Abstract*— A new approach to the design of fuzzy systems is presented, assuming that the system specification is given in terms of a large number of sample I/O (input/output) pairs. The proposed method consists of two stages of processing. First, $K$ fuzzy relation patches are obtained by using a fuzzy clustering technique (such as the Fuzzy $K$-Means algorithm) in the input-output joint universe of discourse. The number $K$ of fuzzy clusters, is selected and justified based on some cluster validity measure. Each fuzzy relation patch thus discovered then constitutes a fuzzy rule in the proposed system. Second, as in the case of the Takagi-Sugeno (TS) fuzzy model, a (nonfuzzy) function is associated with each rule, which can be regarded as a *predefuzzifier* for that rule. Each of these functions is obtained in an optimal way, such that an appropriately defined objective (or loss) function is minimized. An example is included to illustrate the approach.**

## I. Introduction

Recently, the theory of fuzzy sets has been applied in many fields such as engineering, social sciences, and medical sciences. Smart fuzzy control systems have already been built into transportation systems, home appliances, etc. The structural components in these systems are a number of fuzzy rules expressed in terms of fuzzy sets. These rules are usually in the form of if-then statements, which have a straightforward natural language interpretation.

In this paper, a new method for the design of fuzzy systems is presented from a function approximation point of view. In this approach, it is assumed that the system specification is given in terms of a number of given crisp I/O pairs. Fuzzy rules are generated by fuzzy clustering in the input-output product space. The concept of *optimal predefuzzification* is introduced in which neural-network-based techniques may be incorporated. Thus the proposed approach will provide a new framework for (i) optimal fuzzy system design, (ii) merger of fuzzy set and neural network approaches to intelligent system design, and (iii) applications to intelligent control and signal processing.

*Supported by the ONR under Contract N00014-91-J-1072.

## II. Background on Fuzzy Systems

The basic theory of fuzzy sets and fuzzy systems can be found in many texts (such as [1, 2, 3]). A *fuzzy set* $\tilde{X}$ is an "uncertain" set defined on an *universe of discourse* $U$, which can be continuous or discrete. $\tilde{X}$ is specified by a *membership function*, usually denoted as $\mu_{\tilde{X}} : U \to [0,1]$, such that $\mu_{\tilde{X}}(x)$, $x \in U$, is interpreted as *the extent of certainty* to which $x$ belongs to $\tilde{X}$.

For concreteness we shall only consider fuzzy systems with single input and single output, in which the input to the system belongs to an input universe of discourse denoted by $X$ and the output to an output universe of discourse denoted by $Y$. Extension of our discussion to higher input and/or output dimension will be straightforward.

### A. A Conventional Fuzzy Model

The structural elements in a fuzzy system are a number of fuzzy rules specified in terms of fuzzy sets defined on the input and output universes of discourse. For example, let $\tilde{X}_i$ for $i = 1, 2, \ldots, n$ be $n$ fuzzy sets defined on $X$ and $\tilde{Y}_i$ for $i = 1, 2, \ldots, m$ be $m$ fuzzy sets defined on $Y$. For convenience, let $P$ and $Q$ be two (nonfuzzy) sets of fuzzy sets, such that $P \stackrel{\text{def}}{=} \left\{\tilde{X}_1, \tilde{X}_2, \cdots, \tilde{X}_n\right\}$ and $Q \stackrel{\text{def}}{=} \left\{\tilde{Y}_1, \tilde{Y}_2, \cdots, \tilde{Y}_m\right\}$. Suppose the system is specified with $K$ rules, then the $k$th rule, denoted as $r_k$, can be expressed as "If $\tilde{A}_k$ then $\tilde{B}_k$", or $\tilde{A}_k \to \tilde{B}_k$, or $\left(\tilde{A}_k, \tilde{B}_k\right)$, etc., where $\tilde{A}_k \in P$ and $\tilde{B}_k \in Q$ for $k = 1, 2, \ldots, K$. Clearly, $K \leq mn$.

For each given rule $r_k$, a fuzzy relation $\tilde{R}_k$ can be defined on $X \times Y$, such that

$$\mu_{\tilde{R}_k}(x, y) = \min\left[\mu_{\tilde{A}_k}(x), \mu_{\tilde{B}_k}(y)\right], \quad x \in X,\ y \in Y. \quad (1)$$

Then, given a crisp input $x_i \in X$ to $r_k$, the conclusion

0-7803-0614-7/93$03.00 ©1993IEEE

will be a fuzzy set $\tilde{B}'_k(x_i)$ whose membership function is given by,

$$\mu_{\tilde{B}'_k(x_i)} = \mu_{\tilde{R}_k}(x_i, y), \quad y \in Y \ . \tag{2}$$

The combined conclusion $\tilde{Y}(x_i)$ from all the $K$ rules specified for the system is then given by

$$\tilde{Y}(x_i) = \bigcup_{k=1}^{K} \tilde{B}'_k(x_i) \ . \tag{3}$$

Usually, a crisp output is derived from the fuzzy conclusion $\tilde{Y}(x_i)$ through an operation known as defuzzification. A common approach to such an operation is the centroid-defuzzification method [3]. Assume $Y$ continuous, then the centroid-defuzzified output is given by

$$f(x_i) = \text{centroid-defuzz}(\tilde{Y}(x_i)) = \frac{\int_Y y\mu_{\tilde{Y}(x_i)}(y)dy}{\int_Y \mu_{\tilde{Y}(x_i)}(y)dy} \ , \tag{4}$$

where $f$ can be regarded as the (nonlinear) mapping realized by the fuzzy system (with $K$ rules and a centroid-defuzzifier). A similar operation can also be defined if $Y$ is discrete [3].

To design a fuzzy system with certain desirable behavior, the specification is usually given in terms of a, usually large, number of sample I/O pairs, say $(x_i, y_i)$ for $i = 1, 2, \ldots, N$. It is required that, to create the fuzzy system, the selected rules be the ones that are most consistent with the given sample I/O pairs. In design methods proposed in the past, a number of fuzzy sets are first defined on the input and output universes of discourse, usually based on experience, and then the most consistent rules are selected from the candidate rules as derived from the predefined fuzzy sets. A number of measures for the consistency of a candidate rule with respect to the training data have been proposed, such as those reported in [3, 4]. If desired, the shapes of the fuzzy sets can then be "tuned" using subjective judgement and the rule-selection process repeated until the performance of the system becomes satisfactory. Nevertheless, it seems that a systematic approach to the automation of the fuzzy system design process is still lacking.

### *B. The Takagi-Sugeno (TS) Fuzzy Model*

In [5], Takagi and Sugeno proposed a rather interesting fuzzy model for function approximation, in which each rule $r_k$ in the system, where $k = 1, 2, \ldots, K$, is associated with a linear function $g_k$, such that

"If $\tilde{A}_k$, then $g_k(x) = a_{k0} + a_{k1}x$ ",

where $x$ is a crisp input and $a_{k0}$ and $a_{k1}$ are constant parameters estimated from the given data. Instead of providing a fuzzy conclusion, each rule in the TS model is associated with a linear functional relation between the input and output variables corresponding to that rule. The combined defuzzified output $y$ from this system is a weighted average of the individual output $g_k(x)$ from each constituent rule $r_k$, and is given by

$$y = f(x) = \sum_{i=1}^{K} \mu_{\tilde{A}_k}(x) g_k(x) \Big/ \sum_{i=1}^{K} \mu_{\tilde{A}_k}(x) \ . \tag{5}$$

In [6], a weighted recursive least squares (WRLS) algorithm was proposed to obtain the parameters $a_{0k}$, $a_{1k}$, $k = 1, 2, \ldots, K$, in association with a set of rather complex rules for adjusting the shapes of the required fuzzy set membership functions.

An immediate extension of the TS fuzzy model is to let $g_k$ be a general nonlinear function for $k = 1, 2, \ldots, K$. The case when all the output functions $g_k$ are linear will be referred to as the *linear TS fuzzy model*.

## III. The Proposed New Fuzzy Model

From (2) we notice that, in a conventional fuzzy system, inference with a given rule can be made via the fuzzy relation implied by that rule. In fact, the fuzzy relation derived from each given rule defines a patch in the input-output product space. The conclusion from a fuzzy system can be obtained once the fuzzy relation patches are specified, without referring again to the fuzzy sets from which they come from. Therefore, instead of specifying fuzzy sets in each universe of discourse and selecting the most consistent rules from the candidate rules derived from these fuzzy sets, one can directly estimate the appropriate fuzzy relation patches in order to construct the desired fuzzy system.

On the other hand, using the TS fuzzy model, it is needed to specify fuzzy sets only in the input universe of discourse; the conclusion part of each rule is replaced by a precise (nonfuzzy) functional description. In fact, since the input fuzzy sets partition the input universe of discourse into a number of patches, the linear TS model can be regarded as a form of *fuzzy piecewise linear model*. The main advantage of this approach is that it has the potential of generating a better (local) fit of the given data; the disadvantage being that the rules thus specified do not have the same natural language appeal (as in a conventional fuzzy system) since each output function is, in general, a "black box" (in other words, these rules are only "partially fuzzy").

In the sequel, a new method is proposed to obtain a fuzzy model which maintains a fuzzy language description, while sharing the advantage of a TS model. It consists of the two stages as described below.

### A. Stage 1: Estimation of Fuzzy Relation Patches by Fuzzy Clustering

With the given I/O pairs $(x_i, y_i)$, where $x_i \in X$, $y_i \in Y$, and $i = 1, 2, \ldots, N$, a fuzzy clustering is performed in the input-output product space $X \times Y$, which will group together I/O pairs that are geometrically close to each other in the joint universe of discourse $X \times Y$. Such clustering can be achieved, for instance, by using the Fuzzy $K$-Means algorithm (also known as the Fuzzy ISODATA algorithm), where $K$ denotes the number of fuzzy clusters into which the data are to be partitioned. Detailed derivation of the Fuzzy $K$-Means algorithm can be found in [7, 8], and proofs of its convergence are given in [9, 10]. More fuzzy clustering techniques are collected in [11].

Suppose Fuzzy $K$-Means is used to generate $K$ cluster centers denoted by $(x_k^*, y_k^*)$ for $k = 1, 2, \ldots, K$. The fuzzy relation $\tilde{R}_k$ corresponding to the $k$th cluster is such that

$$\mu_{\tilde{R}_k}(x, y) = \left[\sum_{i=1}^{K}\left(\frac{(x - x_k^*)^2 + (y - y_k^*)^2}{(x - x_i^*)^2 + (y - y_i^*)^2}\right)^{(1-m)}\right]^{-1}, \quad \left\{\begin{array}{l} x \in X,\ y \in Y, \\ k = 1, 2, \ldots, K, \end{array}\right. \tag{6}$$

where $m > 1$. Notice that $\mu_{\tilde{R}_k}(x, y) = 1$ if $(x, y) = (x_k^*, y_k^*)$, and that $0 \leq \mu_{\tilde{R}_k}(x, y) < 1$ otherwise.

Fig. 1 shows, for example, 100 I/O pairs $(x_i, y_i)$ (each data point marked by a "+") which are generated according to the following equation:

$$y_i = (1 - e^{-x_i/2}) \sin(3\pi x_i/2), \quad x_i \in [0, 1].$$

Based on these data, 6 fuzzy clusters are found by running the Fuzzy $K$-means algorithm with $K = 6$ and $m = 2$. The corresponding fuzzy patches are also shown in the same figure.

The $K$ fuzzy relations obtained by fuzzy clustering will then constitute the $K$ fuzzy rules in our proposed fuzzy system. Several measures of validity of the clustering results have been proposed, such as those reported in [12] and the references therein. The usual strategy is to perform the clustering for several values of $K$ and then choose the $K$ with the highest validity.

### B. Stage 2: Construction of Optimal Predefuzzifiers for Each Fuzzy Relation Patch

As soon as the fuzzy rule patches are found, a local fit of the given data, $(x_i, y_i)$, $i = 1, 2, \ldots, N$, can be obtained similar to the TS fuzzy model. Each fuzzy rule patch defined by $\tilde{R}_k$ is associated with a function $g_k$, which could be linear or nonlinear, and is expressed in terms of a number of parameters. These parameters are computed to minimize an objective function $J_k$ which is defined as

$$J_k(g_k) = \sum_{i=1}^{N} \mu_{\tilde{R}_k}(x_i, y_i)(g_k(x_i) - y_i)^2, \quad k = 1, 2, \ldots, K. \tag{7}$$

In particular, we can model $g_k$ as a *multilayer feedforward artificial neural network* whose synaptic weights are trained such that $J_k$ is minimized.

Given a crisp input, the $k$th rule will produce a fuzzy conclusion via the fuzzy relation patch $\tilde{R}_k$. Therefore, $g_k(x)$ can be interpreted as the *predefuzzified output* derived from the fuzzy conclusion produced by the $k$th rule. Also, $g_k$ can be regarded as a *predefuzzifier* for the $k$th rule, whose parameters are computed optimally in the sense that $J_k$ is minimized.

Furthermore, the $K$ output values $g_k(x)$ for $k = 1, 2, \ldots, K$ from the $K$ constituent rules (discovered during Stage 1), can be combined in a way similar to that used in the TS model. Following the TS model, the defuzzified output $y$ from the $K$ rules is given by

$$y = \frac{\sum_{k=1}^{K} g_k(x)\mu_{\tilde{R}_k}(x, g_k(x))}{\sum_{k=1}^{K} \mu_{\tilde{R}_k}(x, g_k(x))}. \tag{8}$$

**Remarks:**
It can be noticed that in the new method proposed above, there is no need to estimate the fuzzy membership functions explicitly. The fuzzy relation patches (each patch represents a fuzzy rule) are obtained directly by fuzzy clustering and the number of rules needed are decided by cluster validation. If a natural language description of each fuzzy rule in the form of if-then statement is desired, a fuzzy rule can be "projected" on each universe of discourse. For example, for the relation $\tilde{R}_k$, we can obtain fuzzy sets $\hat{A}_k$ and $\hat{B}_k$, defined on $X$ and $Y$ respectively, by a projection operation, such that

$$\mu_{\hat{A}_k}(x) = \max_{y \in Y} \mu_{\tilde{R}_k}(x, y), \quad x \in X, \tag{9}$$

$$\mu_{\hat{B}_k}(y) = \max_{x \in X} \mu_{\tilde{R}_k}(x, y), \quad y \in Y. \tag{10}$$

Appropriate qualitative labels may then be attached to these projections and the fuzzy rule may be approximated by "If $\hat{A}$ then $\hat{B}$". However, it should be emphasized that the "if-then" composition of $\hat{A}_k$ and $\hat{B}_k$ is normally different from $\tilde{R}_k$. Nevertheless, $\tilde{R}_k$, $k = 1, 2, \ldots, K$, are a more accurate representation of the system.

## IV. An Example

Below we present an example to illustrate the proposed approach. A (discrete) test signal $s(n)$ is generated ac-

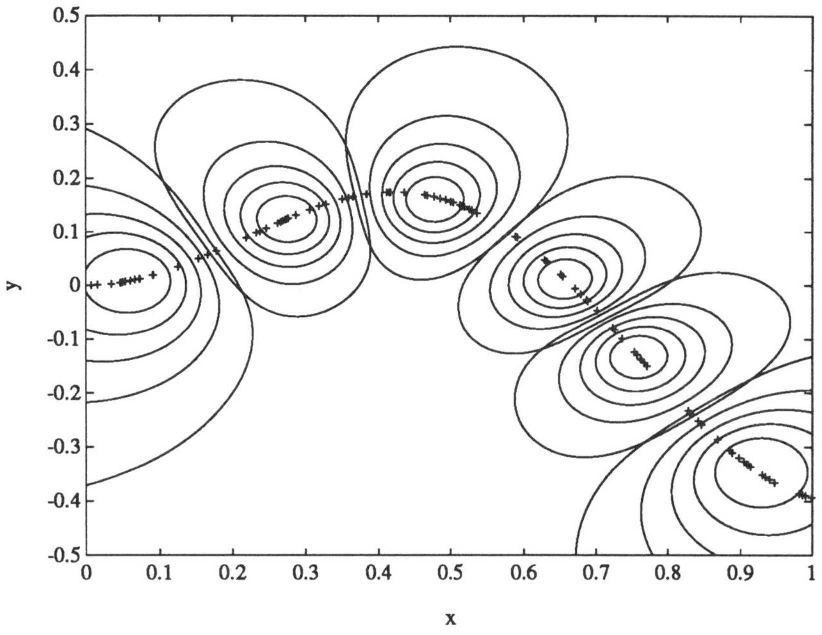


Figure 1: An example of fuzzy clustering, using the Fuzzy $K$-means algorithm, based on 100 I/O pairs (each data point marked by a "+"), with $K = 6$ and $m = 2$.

cording to the following model:

$$s(n) = e^{-|n-50|/10}\sin(n\pi/10), \quad n = 0, 1, 2, \ldots, 100 .$$

This signal is shown in Fig. 2. It is required in this example to obtain a second order fuzzy predictive model for $s(n)$, in which we let

$$s(n) = f(s(n-1), s(n-2)) + \epsilon(n), \quad n = 2, 3, \ldots, 100,$$

where $f$ is a fuzzy system to be determined and $\epsilon(n)$ denotes the modeling error.

Let $\mathbf{x}_n = [s(n-1),\ s(n-2)]^{\mathrm{T}}$, $y_n = s(n)$, and perform a fuzzy clustering on the I/O pairs $(\mathbf{x}_n, y_n)$ for $n = 2, 3, \ldots, 100$, using the Fuzzy $K$-means algorithm with $K = m = 2$. That is, only two rules are used to specify the fuzzy system $f$. Denote the two fuzzy cluster centers as $(\mathbf{x}_1^*, y_1^*)$ and $(\mathbf{x}_2^*, y_2^*)$, and the corresponding fuzzy patches as $\tilde{R}_1$ and $\tilde{R}_2$, respectively. Then associate each fuzzy patch with a linear function $g_k : \Re^2 \rightarrow \Re$ such that

$$g_k(\mathbf{x}) = \mathbf{a}_k^{\mathrm{T}}\hat{\mathbf{x}}\ , \quad \mathbf{x} \in \Re^2, \quad \hat{\mathbf{x}} = \begin{bmatrix} 1 \\ \mathbf{x} \end{bmatrix} ,$$

where $\mathbf{a}_k = [a_{k0},\ a_{k1},\ a_{k2}]^{\mathrm{T}}$ and $k = 1, 2$.

The parameters $a_{ki}$, $k = 1, 2$, $i = 0, 1, 2$, are found such that $J_k$ in (7) is minimized. The solution to this problem is well known and is given by

$$\mathbf{a}_k = \mathbf{R}_k^{-1}\hat{\mathbf{X}}\mathbf{D}_k^2\mathbf{y}\ , \quad k = 1, 2\ ,$$

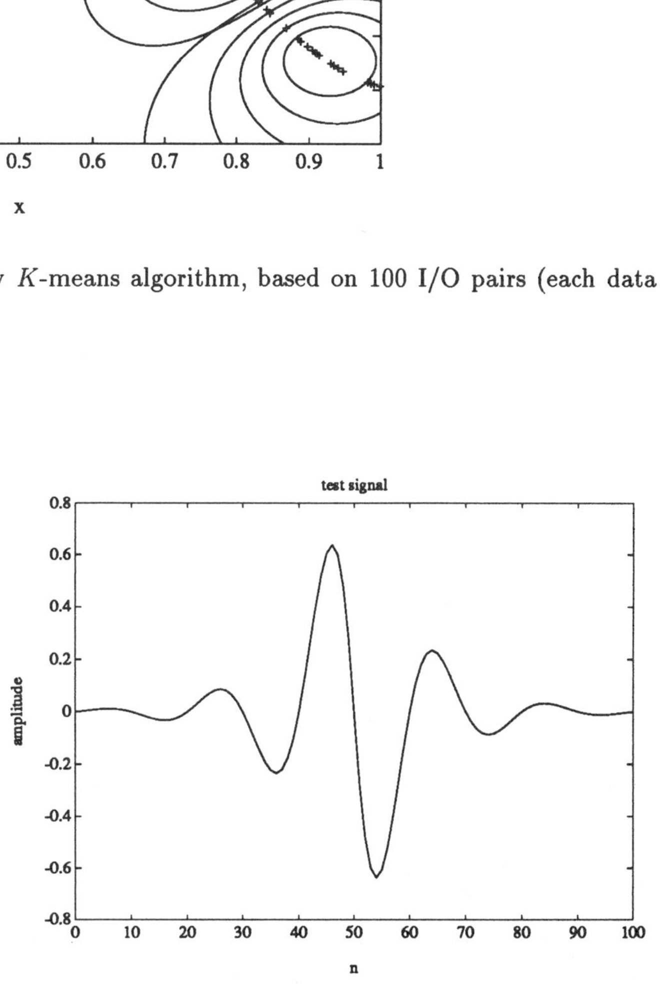


Figure 2: A test signal $s(n)$ used to illustrate the proposed fuzzy modeling, where $s(n) = e^{-|n-50|/10}\sin(n\pi/10)$ for $n = 0, 1, 2, \ldots, 100$.

where

$$\mathbf{y} = [y_2, y_3, \cdots, y_{100}]^{\mathrm{T}} ,$$

$$\hat{\mathbf{X}} = [\ \hat{\mathbf{x}}_2 \ \ \hat{\mathbf{x}}_3 \ \ \cdots \ \ \hat{\mathbf{x}}_{100}\ ] = \begin{bmatrix} 1 & 1 & \cdots & 1 \\ \mathbf{x}_2 & \mathbf{x}_3 & \cdots & \mathbf{x}_{100} \end{bmatrix} ,$$

$$\mathbf{D}_k = \mathrm{diag}([\mu_{\tilde{R}_k}(\mathbf{x}_2, y_2),\ \mu_{\tilde{R}_k}(\mathbf{x}_3, y_3), \cdots, \mu_{\tilde{R}_k}(\mathbf{x}_{100}, y_{100})]) ,$$

and

$$\mathbf{R}_k = \hat{\mathbf{X}}\mathbf{D}_k^2\hat{\mathbf{X}}^{\mathrm{T}} .$$

The reconstructed signal $\hat{s}(n)$ and the residual error sequence $\epsilon(n)$ are plotted in Figs. 3 and 4 respectively, where

$$\hat{s}(n) = f(\mathbf{x}_n) ,$$

and

$$\epsilon(n) = s(n) - \hat{s}(n) .$$

As shown in these results, even with only two fuzzy rules and linear predefuzzifiers, the fuzzy system obtained with the proposed technique models the given data very well.

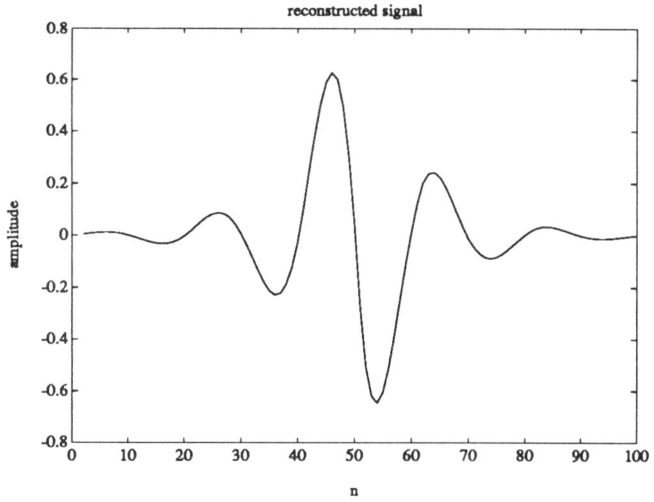


Figure 3: Reconstructed signal $\hat{s}(n)$, where $\hat{s}(n) = f(\mathbf{x}_n)$, $n = 2, 3, \ldots, 100$.

## V. Conclusions

A systematic method for the design of fuzzy systems is presented, assuming that the system specification is given in terms of a number of sample I/O pairs. In this new approach, there is no need to guess the numbers and shapes of fuzzy sets in the input and output universes of discourse; the fuzzy relation patches that are required to specify a fuzzy system are directly obtained by fuzzy clustering in the input-output joint universe of discourse and the number of clusters can be determined by using an appropriate measure of cluster validity. Also, the proposed model maintains a natural language appeal similar to that in a conventional fuzzy system, while having the advantage of providing a close local fit of the given data through the use of an optimal predefuzzifier in association with each fuzzy rule in the system. The viability of this model has been demonstrated with an illustrative example.

## References


[1] G. Klir and T. Folger, *Fuzzy sets, uncertainty, and information*, Englewood Cliffs, N.J.: Prentice Hall, 1988.

[2] H.J. Zimmermann, *Fuzzy set theory — and its applications*, Boston: Klumer-Nijhoff, 1984.

[3] B. Kosko, *Neural networks and fuzzy systems*, Prentice Hall, 1992.

[4] C.M. Higgins and R.M. Goodman, "Learning fuzzy rule-based neural networks for function approximation", *Proceedings of the International Joint Conference on Neural Networks, Baltimore, June 7-11, 1992*, pp. I-251–I-256.

[5] T. Takagi and M. Sugeno, "Fuzzy identification of systems and its applications to modeling and control", *IEEE Trans. Systems, Man, and Cybernetics*, Vol. SMC-15, No. 1, 1987, pp. 116-132.

[6] M. Sugeno and Kazuo Tanaka, "Successive identification of a fuzzy model and its applications to prediction of a complex system", *Fuzzy Sets and Systems*, Vol. 42, 1991, pp. 335–344.

[7] J.C. Dunn, "A fuzzy relative of the ISODATA Process and its use in detecting compact well-separated clusters", *J. Cybernetics*, Vol. 3, No. 3, 1973, pp. 32–57.

[8] J.C. Bezdek, "Fuzzy mathematics in pattern classification", Ph.D dissertation, Appl. Math., Cornell University, Ithaca, NY, 1973.

[9] J.C. Bezdek, "A convergence theorem for the fuzzy ISODATA clustering algorithms", *IEEE Transactions on Pattern Analysis and Machine Intelligence*, Vol. PAMI-2, No. 1, 1980, pp. 1–8.

[10] J.C. Bezdek, R.J. Hathaway, M.J. Sabin, and W.T. Tucker, "Convergence theory for fuzzy c-Means: counterexamples and repairs", *IEEE Transactions on Systems, Man, and Cybernetics*, Vol. SMC-17, No. 5, 1987, pp. 873–877.

[11] J.C. Bezdek and S.K. Pal (eds.), *Fuzzy Models for Pattern Recognition*, 1992, New York, NY: IEEE Press.

[12] X.L. Xie and G. Beni, "A validity measure for fuzzy clustering", *IEEE Transactions on Pattern Analysis and Machine Intelligence*, Vol. PAMI-13, No. 8, 1991, pp. 841–847.

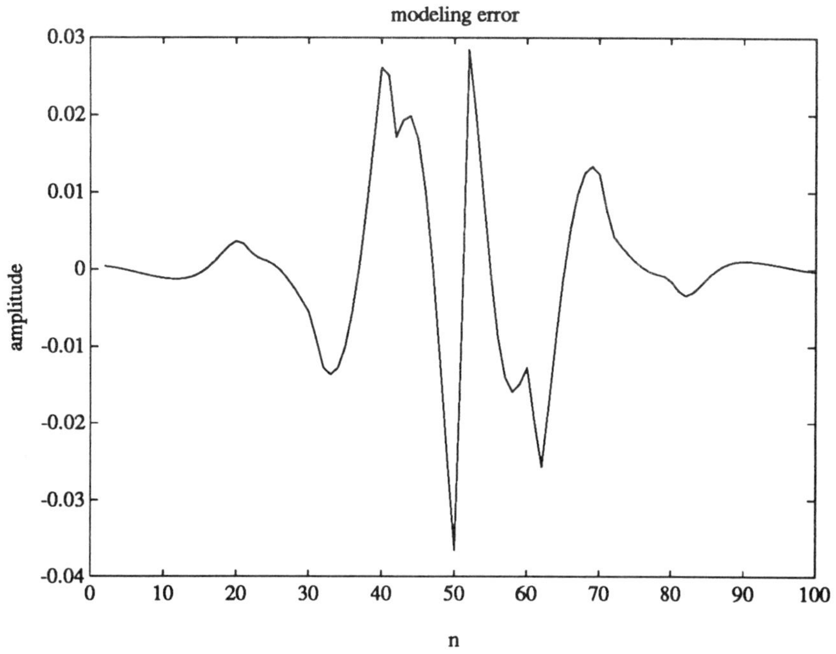


Figure 4: Modeling error $\epsilon(n) = s(n) - \hat{s}(n)$, $n = 2, 3, \ldots, 100$.

# A DESIGN METHOD FOR A CLASS OF FUZZY HIERARCHICAL CONTROLLERS

A. RUEDA and W. PEDRYCZ
DEPt. OF ELECTRICAL AND COMPUTER ENG.
THE UNIVERSITY OF MANITOBA
WINNIPEG, MANITOBA, CANADA R3T 2N2

***Abstract*: A design method for a hierarchical control structure is introduced in this paper. The structure involves two different levels of control, namely a coordination level and an execution level. At the coordination level a fuzzy controller is used to identify the status of the system under control and to activate appropriate local control modules, located at the execution level. The design of the fuzzy system is based on the control requirements and on the response of the local control elements. Simulations show that the controller can satisfy the control objective.**

## 1 Introduction

As originally proposed by Zadeh [14], and experimentally applyed by Mamdani et al [4,7,8], a fuzzy controller is a simple–level structure. The idea of having a hierarchically organized control structure has been studied in [1,9–11,13]. The concept can also be found in optimizing learning in neural networks [1]. In this hierarchical approach a coordination level is used to identify the status of a nonlinear system and to select one among specialized local control algorithms satisfying a control objective.

The coordination level is an evident domain for a fuzzy controller, where it is required to decide which local control algorithm to apply, on the basis of general knowledge about control policies. The coordination signals activate local control algorithms at the execution level and coordinate their impact on the final control action. This local control algorithms can be simple regulators commonly used in industrial applications, for instance, PID controllers (proportional–integral–derivative). Each of them is tuned to perform well under certain operating conditions of the nonlinear system.

Some experimental results have been presented in [9], using this architecture. It was shown that an excellent performance can be achieved. However, the fuzzy system was designed by intuition and experimentation, no method was presented. In this paper, a design method is presented so that the good performance reported in [9] can be achieved in an structured manner.

The paper is organized as follows: in Section 2, we introduce the overall structure of the control architecture; in Section 3, we present a method for the design of the fuzzy system; afterwards, in Section 4, simulations are presented; and, finally, conclusions are included in Section 5.

Supported by NSERC and MICRONET, Canada and CONACYT, Mexico.

## 2 Structure of the System

It is considered a single–input/single–output system, shown in Fig. 1, that has a fuzzy system in the coordination level and PID controllers in the execution level, as in [9]. The fuzzy controller is driven by error and change of error, and it infers a value of a selection variable called λ. Depending on the values of λ different PID controllers become active. All the transitions are smooth, guided by the membership functions of the fuzzy sets.

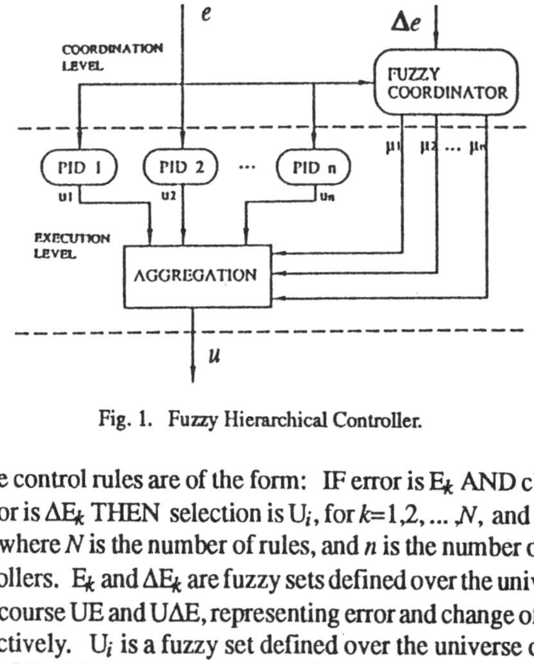


Fig. 1. Fuzzy Hierarchical Controller.

The control rules are of the form: IF error is $E_k$ AND change of error is $\Delta E_k$ THEN selection is $U_i$, for $k=1,2,\ldots,N$, and $i=1,2,\ldots,n$, where $N$ is the number of rules, and $n$ is the number of PID controllers. $E_k$ and $\Delta E_k$ are fuzzy sets defined over the universes of discourse UE and UΔE, representing error and change of error respectively. $U_i$ is a fuzzy set defined over the universe of discourse UL of the selection variable. There are $n$ fuzzy sets representing each of the PID controllers, as shown in Fig. 2.

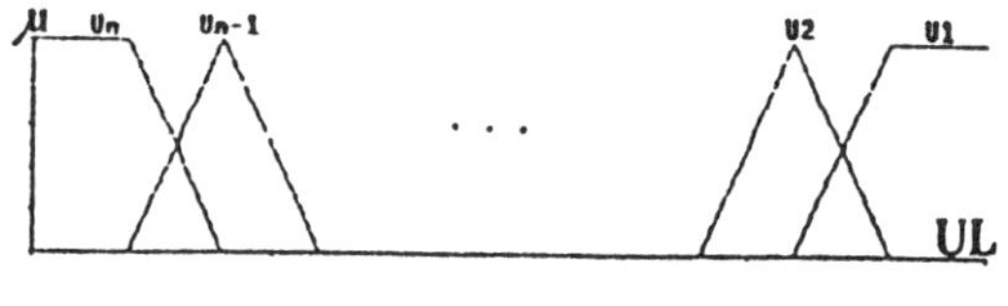


Fig. 2. Degrees of Activation According to λ.

The rules are combined into a three–dimensional fuzzy relation R, so that the inference machine can infer infer the degrees of preference of the PID controllers, given fuzzy sets of error and its change. The final control signal according to the levels of activation of the individual PID controllers, is produced by an aggregation block (Fig. 1), using a center of gravity method given by:

$$u = \frac{\sum_{i=1}^{n} u_i \, \mu_i(\lambda)}{\sum_{i=1}^{n} \mu_i(\lambda)} \tag{1}$$

where $u_i$ is the control output of the $i$th PID controller, and $\mu_i(\lambda)$ characterizes its degree of preference.

## 3 SELECTION OF THE MEMBERSHIP FUNCTIONS

The membership functions for error, change of error and selection variable are parameterized by the scalars α, β and γ that are determined from the step response of the PID controllers. Let $S_E(e,\alpha) = \{\mu_{EP}, \mu_{EZ}, \mu_{EN}\}$, $S_{\Delta E}(\Delta e,\beta) = \{\mu_{\Delta EP}, \mu_{\Delta EZ}, \mu_{\Delta EN}\}$, and $S_{UL}(\lambda,\gamma) = \{\mu_{U1}, \mu_{U2}\}$ be the sets of membership functions be defined by:

$$\begin{aligned} \mu_{PE} &= \mu_P(\alpha, e) & \mu_{ZE} &= \mu_Z(\alpha, e) & \mu_{NE} &= \mu_N(\alpha, e) \\ \mu_{\Delta PE} &= \mu_P(\beta, \Delta e) & \mu_{\Delta ZE} &= \mu_Z(\beta, \Delta e) & \mu_{\Delta NE} &= \mu_N(\beta, \Delta e) \end{aligned} \tag{2}$$

where

$$\begin{aligned} \mu_P(p,x) &= \begin{cases} \frac{1}{p}x & 0 \le x \le p \\ 1 & p < x \end{cases} \\ \mu_Z(p,x) &= \begin{cases} 1 - \frac{1}{p}x & 0 \le x \le p \\ 1 + \frac{1}{p}x & -p \le x < 0 \end{cases} \\ \mu_N(p,x) &= \begin{cases} -\frac{1}{p}x & -p \le x \le 0 \\ 1 & x < p \end{cases} \\ \mu_{U1} &= \begin{cases} \frac{7}{4\gamma}\lambda & 0 \le \lambda \le \frac{4}{7}\gamma \\ 1 & \frac{4}{7}\gamma < \lambda \end{cases} \\ \mu_{U2} &= \begin{cases} 1 & 0 \le \lambda \le \frac{1}{7}\gamma \\ \frac{7}{6\gamma}(\gamma - \lambda) & \frac{1}{7}\gamma < \lambda \end{cases} \end{aligned} \tag{3}$$

The parameters are selected from the combined steady–state step–response state–plane of the two PID controllers, according to the following criteria:

$$\alpha = \frac{e_1 + e_2}{2} \qquad \beta = \frac{\Delta e_1 + \Delta e_2}{2} \qquad \gamma = \sqrt{\alpha^2 + \beta^2} \tag{4}$$

where

$$\begin{aligned} e_1 &= \max |e_{PID1}| & \Delta e_1 &= \max |\Delta e_{PID1}| \\ e_2 &= \max |e_{PID2}| & \Delta e_2 &= \max |\Delta e_{PID2}| \end{aligned} \tag{5}$$

In the next section, experimental results are presented, where the fuzzy coordinator is designed using above criteria.

## 4 SIMULATIONS

We present simulation results of the proposed control architecture applied to the nonlinear water tank. The control objective is to get the system to follow a reference input, with fast response free of oscillations. Two discrete–time PID controllers were tuned to perform in two different ways, namely one gives a fast response and the other a good regulation. We show the response of each PID individual controller, and the hierarchical structure with the fuzzy coordinator.

*Dynamical model of the plant*

The plant used in the experiments is a water tank as shown in Fig. 3. The input is $u$, operating the inlet valve, that has saturation and time delay, and the output is the water level $h$. The system is modelled as in [2,9] by:

$$\frac{d}{dt} h = (q_{in} - q_{out})/area$$

$$area = (h+1)/7 \qquad q_{in} = q_{max}\ cval \tag{6}$$

$$cval = \begin{cases} 0 & u < 0 \\ u & 0 \le u \le 1 \\ 1 & u > 1 \end{cases} \qquad q_{out} = a_{out}\sqrt{2g\ max(h,0)}$$

where $q_{max}$=1, $g$=9.81m/sec$^2$, and $a_{out}$=RECT[0,0.125] (random noise with a rectangular distribution. There is a time delay in the inlet valve, modelled as part of the controller for simplicity.

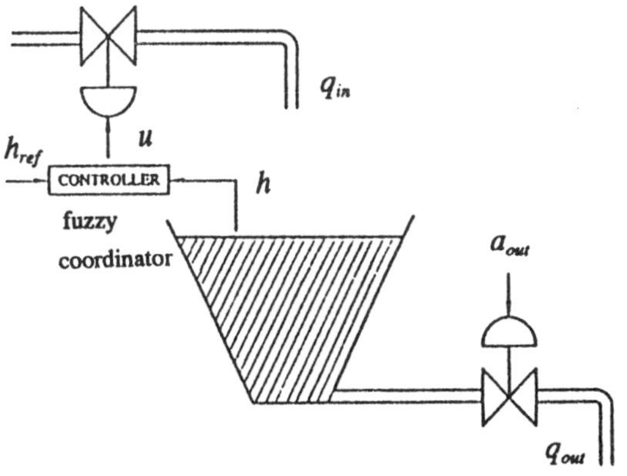


Fig. 3. Model of the plant.

*Control objective*

The control objective is to minimize the error and change of error of the closed–loop system. We define them as:

$$e = h_{ref} - h \qquad \Delta e = h_t - h_{t-1} \tag{7}$$

In the next section, we present a design procedure based on the step response of the individual PID controllers, that allows us to define the membership functions of the fuzzy sets.

*PID controllers*

Two discrete-time PID controllers with anti-reset windup are used [2]:

$$w = K\left[b\ h_{ref} - h\right] + \left[I_{t-1} + \frac{K\Delta t}{Ti}e\right] + \left(\frac{Td}{Td + M\Delta t}\right)\left[D_{t-1} - \frac{K}{M}\Delta e\right] \qquad (8)$$

$$I_t = I_{t-1} + (u - w)\frac{\Delta t}{Tt} \qquad D_t = \frac{T_d}{T_d + Nh}D_{t-1} - \frac{KT_dN}{T_d + Nh}e$$

$$z = \begin{Bmatrix} u_{\min} & | & w < u_{\min} \\ w & | & u_{\min} \le w \le u_{\max} \\ u_{\max} & | & w > u_{\max} \end{Bmatrix} \qquad u = z_{t-2}$$

where $K$, $Ti$ and $Td$ are the proportional gain, the integration and the derivation times respectively, $b$ is the set point weight factor, $M$ is the maximum derivative gain, $Tt$ is the tracking constant, $u_{max}$ and $u_{min}$ are the maximum and minimum values of the control output, and $\Delta t$ is the sampling period. The time delay of the valve of the tank has been set to 2 sampling periods. The following parameters are taken as in [9] for comparison:

| PID 1: | PID 2: | Both: |
|---|---|---|
| $K_1$=15 | $K_2$=1 | $u_{max}$=1 |
| $b_1$=1 | $b_2$=0 | $u_{min}$=0 |
| $Ti_1$=0.1 | $Ti_2$=15 | $\Delta t$=0.1 |
| $Tt_1$=0.1 | $Tt_2$=1 | |
| $Td_1$=10 | $Td_2$=10 | |
| $N_1$=10 | $N_2$=0 | |

Applying the procedure presented in Section 3, we find the following values for the parameters of the membership functions: $\alpha = 0.38$, $\beta = 0.1$, and $\gamma = 0.4$. In Fig. 4, the membership functions are displayed (the universes of discourse of E and ΔE have been normalized). The next step is to define the fuzzy relation, using the rules mentioned in Section 2. The performance index is based in observation of the response of the system, and it forms a combination of velocity of response and regulation properties.

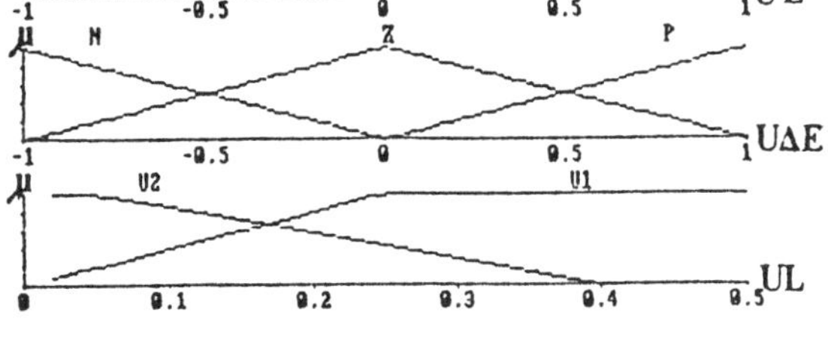

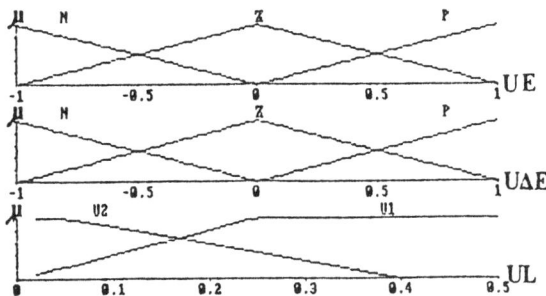

Fig. 4. Membership functions for E, ΔE and U.

*Experiment 1*

The reference level $h_{ref}$ is changed 3 times and the response is observed for the PID controllers and for the fuzzy coordinator. In Fig. 5a, the response of the system is shown, when PID 1 is applied only. Similarly, in Fig. 5b, it can be observed that PID 2 is slower but it has better regulation. The best features of both are combined by the fuzzy coordinator, as shown in Fig. 5c.

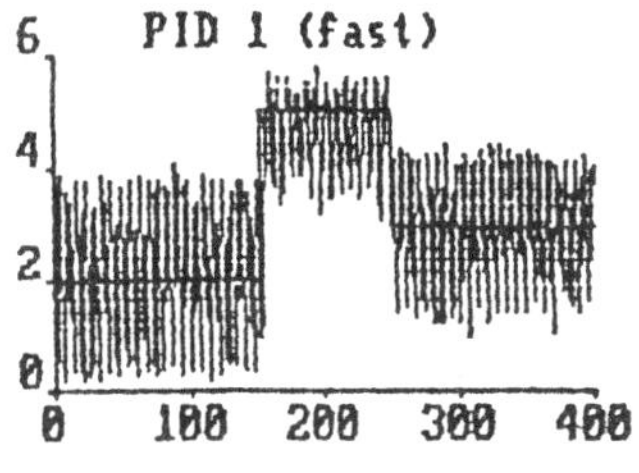


Fig. 5a. Response with PID 1.

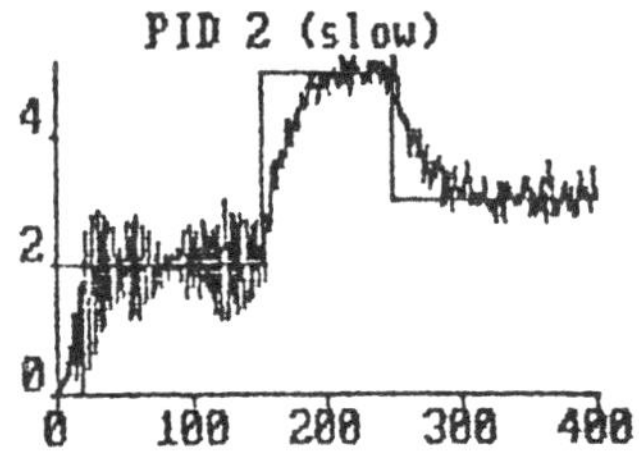


Fig. 5b. Response with PID 2.

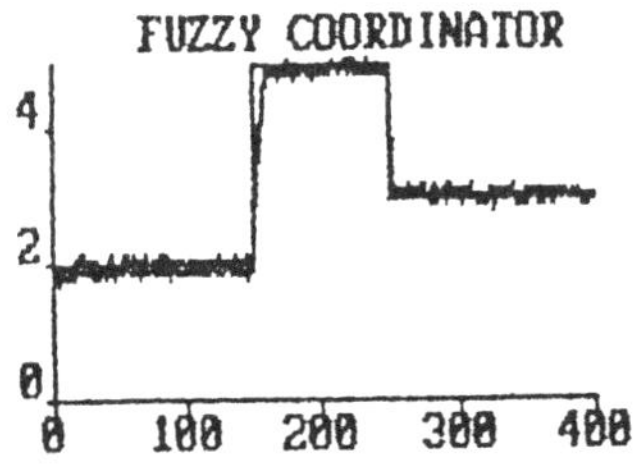


Fig. 5c. Response with fuzzy coordinator.

*Experiment 2*

The reference is set to be a sine wave. As in Experiment 1, it can be observed from Fig. 6a, 6b, and 6c that the response of the system is fast with good regulation.

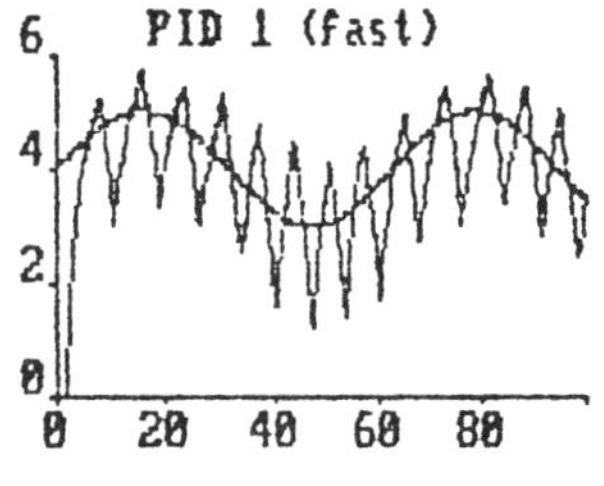


Fig. 6a. Response with PID 1.

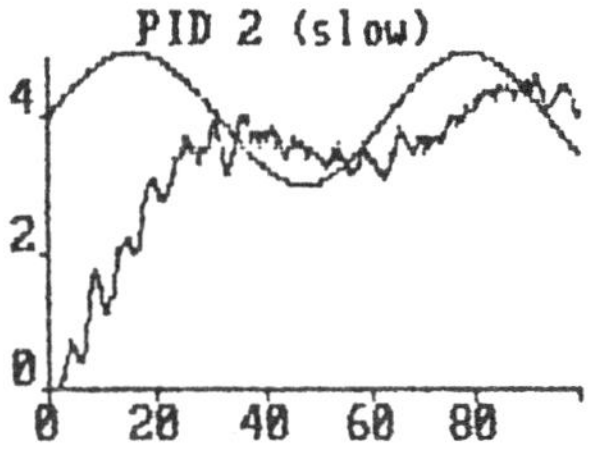

Fig. 6b. Response with PID 2.

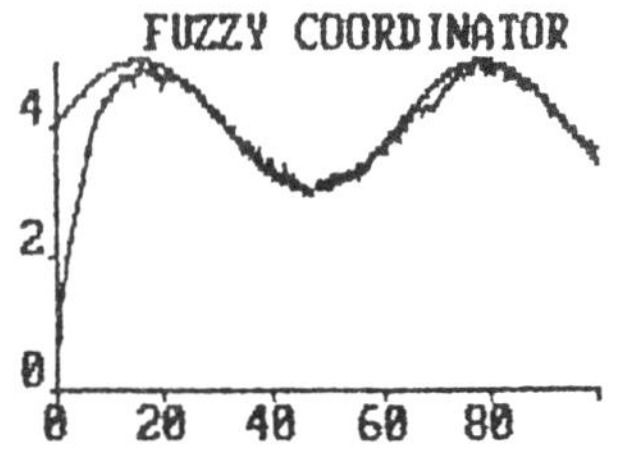

Fig. 6c. Response with fuzzy coordinator.

## 5 CONCLUSIONS

A design method for a hierarchical controller using a fuzzy system has been presented. The architecture was applied to the control of a nonlinear system, and it was shown in simulations that an excellent performance can be achieved despite of the nonlinearities and uncertainty. It is remarkable that the model of the plant is not required in the design of the controller, as opposed to classical nonlinear control techniques.

This architecture can be easily implemented, since the designer does not have to fine tune the PID controllers, or do mathematical analysis. Simply, obtain a couple of PID controllers roughly tuned to satisfy different requirements, and apply the design method. The architecture is a good example that a combination of rule–based and conventional systems can simplify a controller design and achieve and excellent performance

## REFERENCES

[1] Arabshahi, P., Choi, J. J., Mark II, J. J., Caudell, T. P., "Fuzzy Control of Backpropagation", IEEE Int. Conf. on Fuzzy Systems, San Diego 1992, 967–972.

[2] Astrom, K. J., Wittenmark, B., Computer Controlled Systems, Prentice Hall, Englewood Cliffs, N. J. 1984.

[3] Chow, K. H. F., A Self–Organizing Fuzzy Controller, M. Sc. Thesis, Dept. of Electrical Engineering, The University of Manitoba, Canada, 1984.

[4] Kickert, W. J. M., Mamdani, E. H., "Analysis of a Fuzzy Logic Controller", Fuzzy Sets and Systems, 1, 1978, 29–44.

[5] Koivo, H. (ed), Int. Journal of Adaptive Control and Signal Processing, Special Issue: Expert Systems in Control Design and Tuning, vol. 5, no. 1, 1991.

[6] Lee, C. C., "Fuzzy Logic in Control Systems: Fuzzy Logic Controller", Parts I and II, IEEE Trans. on Syst. Man and Cybern., vol. 20, no. 2, 1990, 404–435.

[7] Mamdani, E. H., "Applications of Fuzzy Algorithms for Control of Simple Dynamic Plants", Proc. IEEE, 121, 1976, 1585–1588.

[8] Mamdani, E. H., King, P. J., "The Application of Fuzzy Control Systems to Industrial Processes", Automatica, 13, 1977, 235–242.

[9] Rueda, A., Pedrycz, W., "Fuzzy Coordinator in Control Problems", Proc. of the NAFIPS Conf., Puerto Vallarta, Mexico 1992, vol. I, 322–329.

[10] Sugeno, M. (ed.), Industrial Applications of Fuzzy Control, North Holland, Amsterdam 1985.

[11] Pedrycz, W., Fuzzy Control and Fuzzy Systems (2nd edition), Research Studies Press, John Wiley, Taunton 1992.

[12] Tong, R. M., "A Control Engineering Review of Fuzzy Systems", Automatica, 13, 1977, 559–569.

[13] Van der Veen, J. C. T., Fuzzy Sets, Theoretical Reflections, Application to Ship Steering, M. Sc. Thesis, Dept. of Electrical Eng., Delft University of Technology, The Netherlands, 1976.

[14] Zadeh, L. A., "Outline of a New Approach to the Analysis of Complex Syxtems and Decision Processes", IEEE Trans. on Syst. Man and Cybern., vol. 3, no. 1, 1973, 28–44.

# Industrial Applications of Fuzzy Logic in North-Rhine Westphalia

Prof. Dr. Bernd Reusch
Fuzzy Initiative North-Rhine Westphalia
Martin–Schmeißer–Weg 18
W-4600 Dortmund 50

***Abstract*— This article reports on field research done for the Fuzzy Initiative in North-Rhine Westphalia.**

## I. Introduction

In 1991 the author proposed what is now called "Fuzzy Initiative North-Rhine Westphalia (NRW)" to the Department of Economy of this particular state NRW of the Federal Republic of Germany. The main phase of this initiative started at the beginning of 1992 and will terminate at the end of 1994. At this moment, five groups are (partially) funded by the aforementioned ministry: one in Aachen, three in Dortmund and one in Düsseldorf [1]. The initiative's primary goal is not research (consider the funding ministry!), but to translate results into industrial reality by various means including courses, seminars and joint projects with partners in industry. It is also set that we should mainly address medium and small scale enterprises, not exclusively from the state NRW. Another task, taken by the "Fuzzy Demonstrations-Zentrum Dortmund", was to study state of the art, trends and potentials of Fuzzy Technology in NRW. It is this study we want to report on here.

## II. Results

In order to make the study useful to a broader audience, results of other studies are summarised [3, 5, 4, 2]. Also included is information on available hard- and software as well as indications on "who is who" in Europe and available textbooks. All this was taken from the usual sources and will not be reported here.

In addition we sent out questionnaires to companies to get their view on Fuzzy Logic and also visited a number of selected companies and tried to find out in (sometimes) lengthy discussions what are the potentials of Fuzzy Logic for them in our view, being conservative in interpreting the interviews.

The second part of the study was interviewing partners in selected companies of different branches. It turned out that personal contact and some more time gave more insight into what are the actual possibilities of Fuzzy Logic. It turned out that a lot of people still associated Fuzzy Logic ( mistakingly ) with process control or electronics.

### *A. The Questionnaire*

A questionnaire was prepared and sent to more then thousand companies in NRW. The target group was divided into two subgroups: one which was known to have interest in Fuzzy Logic (group A) and another one which was randomly selected (group B). The questionnaire concentrated on the current reception of fuzzy methods as well as on the potentials seen in the future. Most of the answers were given by executive staff.

Of the companies written to, 135 replied (88 of group A and 47 of group B). In 34 of all electronic data processing is limited to administration; 98 use CAD in construction, production planning etc. The distribution of all addressees over branches is as follows: steel and metal-processing industry 13, chemical industry 14, mechanical engineering 37, automotive 6, electronics 35, civil engineering 11, textile industry 4, and others 14. This more or less reflects the industrial structure of NRW.

The questionnaire is divided into several sections. The first section addresses the companies' competitive situation. One third of all has a higher opinion of the competitive power of their company compared to the whole branch. In group B, there was a balance between those who ranked their company higher than the whole branch and those who ranked their company lower.

Another section asked about areas or problems that needed more attention. Some of them seemed relevant across different branches: waste management, production control, development and quality assurance were mentioned most. Further problem areas were: recording of factory data, production in small volume, flow of information, sales planning, complex control tasks, technical optimisation, recycling, project management and running, marketing, cost evaluation and packaging.

In a further section the way of introducing technical



0-7803-0614-7/93/$03.00 ©1993 IEEE

know-how and modernisation into the company is examined. About 77% of all replies prefer continuing their employees' education. Some 50% use cooperations and external consulting for this purpose. 65% believe that independent consulting would contribute to the company's competitive power.

| aspect | product | process |
|---|---|---|
| control (CO) | 80 | 67 |
| pattern recognition (PR) | 19 | 16 |
| data analysis (DA) | 29 | 23 |
| decision making (DM) | 25 | 23 |
| optimisation (OP) | 65 | 72 |
| production planning (PP) | 47 | 89 |

Table 1: Aspects with Fuzzy potential

To get information about possible areas of Fuzzy Logic application, we asked about certain aspects of products and production processes. Results (in absolute numbers) are listed in table 1.

Differences between branches appear as follows: steel and metal-processing industry ranks optimisation problems highest, mechanical, electrical and automotive industries see the highest demand in Fuzzy Control.

| industry | CO | PR | DA | DM | OP | PP |
|---|---|---|---|---|---|---|
| metal | 30 | 7 | 7 | 7 | 46 | 38 |
| chemistry | 14 | 7 | | 14 | 14 | |
| mechanical | 46 | 21 | 24 | 27 | 49 | 35 |
| automotive | 83 | 16 | | 16 | 50 | |
| electronic | 77 | 17 | 28 | 25 | 48 | 8 |
| civil engineering | 9 | | 18 | 9 | 18 | 18 |
| textile | | | | 25 | | |
| other | 50 | 42 | 35 | | 50 | 21 |

Table 2: Possible fields of application (percentages)

The last section discusses knowledge on Fuzzy Set theory and Fuzzy Technology in industry. It turned out that 30% never heard of Fuzzy Logic before. The others were informed mainly by technical publications. In 5 out of 135 companies Fuzzy Logic is in practical use. In total 23 discuss, plan or test the use of Fuzzy Logic. We also asked about possible applications of Fuzzy Logic in the addressed companies. Mechanical, automotive and electronic industries put the most emphasis on control applications, whereas metal-processing industry rates optimisation higher. In textile industry only decision making seemed to be relevant. Details of the distribution over branches are shown in Table 2. The areas of control (CO), pattern recognition and image processing (PR), data analysis (DA), decision making (DM), optimisation (OP) and production planning (PP) were explicitly taken into consideration.

Conclusions:

- Fuzzy logic is not exotic anymore.
- The Fuzzy Initiative NRW is known to 20% of the people written to.
- In 17% of the companies Fuzzy Logic is currently used or observed.
- Training is very important for further education of employees.
- Applications in control and optimisation problems are highly needed and suitable for Fuzzy Logic.
- Especially in mechanical industry production planning presents some potential for Fuzzy Logic.
- Data analysis and decision support is needed mainly in mechanical and electronical industry.

### *B. Interviews*

More than 30 companies in different branches were visited.

#### *B.1 Banks*

The area of *credit-worthiness* seems very promising for Fuzzy Logic. Current methods for scoring of borrowers are unsatisfactory because they use sharp bounds. It is felt that a lot of solvent persons are excluded from obtaining a loan. Intuitively, the decisions around the cut-off should be improvable. Furthermore, banks seem to be very interested in predictions of social developments for their planning of reserves. Tendency statements on the prospects of succeeding of investment strategies are very desirable too. For *customer consultation service* a decision support system on a local computer could be very helpful.

#### *B.2 Chemical industry*

Processes are characterised as control processes with long dead times (4 to 8 hours are not unusual). Since *exceptional situations* are mostly handled by human operators, there is a potential for Fuzzy Logic in augmenting a conventional controller with rules for exception handling or plausibility checks. Especially avoidance of severe faults seems to be a good field of application. Evaluation of process data is another topic that offers chances to Fuzzy Logic, since deeper insight into the behaviour of processes as indicated by measured data could be used to foresee problematic situations in the future. A classification of situations or tendencies could improve process performance.

Another interesting point is the exploration of new products or variations of older ones. Fuzzy expert systems and fuzzy mathematics seem to be appropriate at least as supplement to traditional models. The goal is to get simulation models more rapidly.

A number of sensors for more intelligent process control is still to be developed.

### *B.3 Electronics*

As commonly known, electronical controllers offer a wide range of application for Fuzzy Control. However, it is not adequate to replace good solutions, if there is no advantage in using Fuzzy Logic. We expect that future extensions of existing systems will take Fuzzy Logic into account. The point is that many more features will become possible by its use. Typical examples are heating and air conditioning systems. Improved control of the start phase could be designed with Fuzzy Logic.

In mass products Fuzzy Logic may be used in improving the user interface in future product generations.

Some situations were reported to us where Fuzzy Logic will not be implemented in the end product but could be used for simulation in the development process, i. e. for rapid prototyping.

In the design of electronic circuits a lot of decisions and optimisations have to be done. Some of them use fuzzy knowledge that is currently part of the human designer. For the choice of a suitable concept, circuit classes can be described qualitatively. For the decomposition of a design into components, a fuzzy formulation of requirements would be acceptable. Optimisation of individual components can be done using fuzzy methods.

### *B.4 Energy industry*

Similar to applications in chemical industry, process control can be enhanced with Fuzzy Logic. Typically, exact models are not available. There is an ongoing progress in distributing controllers all over the plant. A common evaluation of all signals could include Fuzzy Logic for the classification of situations.

### *B.5 Mechanical industry*

If machines are manufactured singly, the composition of a product must be supported by knowledge stored from earlier projects. A fuzzy rule base could provide increased confidence in defining a product.

### *B.6 Steel and metal-processing industry*

In steelworks a lot of processes are controlled via human knowledge. It seems to be very difficult to identify this knowledge and the signals that lead to an action. It is also quite difficult to get the right data for decision making. Sensors are expensive and sensitive to operating errors. Alternative solutions with cheaper sensors and compensation with fuzzy expert systems should be considered.

Again, exploration of the product space is an interesting aspect. Today, more than 2000 kinds of steel are known. An advantage of European industry is to be seen in the skill to develop new products with desired properties. Since the effect of parameters is widely unmodelled, experience is an important factor.

Production planning systems are likely to be improved. In many situations too many data has to be provided, a good deal of which is non-essential. Systems dealing with incomplete information would be very much appreciated.

### *B.7 Paper mills*

Most important is the maintenance of the fully continuous process. Potential problems must be recognised as early as possible. Information systems that are able to locate an error with high probability would be very useful. Today, only a portion of the gathered data is evaluated and knowledge about fault conditions is stored only in paper form.

### *B.8 Textile industry*

In textile industry, the degree of automation is usually very high. However, especially in refining cloths much work is still done manually, including quality assurance. In clothing industry the sewing machines have to be adjusted very often because of differences in cloth quality.

## *C. Support of Fuzzy Technology*

A number of companies sited in North-Rhine Westphalia offer tools for the development of fuzzy systems and services such as training, seminars and consulting. From Inform, Aachen, a tool called fuzzyTECH is available for design and implementation of Fuzzy Controllers. Its main feature is an on-line editing of the working system. Under the name *FUNNYLAB* the company MacroTek, Dortmund, offers another tool for developing Fuzzy Control applications. Membership functions can be learned through an integrated neural network. Ingenieurbüro Schulz, Dortmund, has integrated a fuzzy component in its AGO-Software-System which provides also conventional control techniques. The FC-LAB, another Fuzzy Controller development programme, offered by a well-known publisher, has also been developed in Dortmund.

Other companies, such as MIT GmbH, Aachen, amira GmbH, Duisburg, FLS and ZeTec, both Dortmund, offer many kinds of support for Fuzzy Technology. A number of projects accompanied by these companies or institutions like ELITE in Aachen or FDZ in Dortmund is currently in progress.

### *D. Education*

Fuzzy Logic plays an important role at many universities and Fachhochschulen (polytechnics). Great efforts in teaching principles of Fuzzy Logic and supervision of diploma and Ph. D. theses are made in Aachen, Dortmund, Duisburg and Paderborn, also in Bochum, Bonn, Hagen, Münster, Siegen to a lesser degree.

## III. Résumé

It should be pointed out that the importance of Fuzzy Technology is well understood in Germany, not only in North-Rhine Westphalia. A vast majority of the people written to has at least a basic understanding of what is meant by Fuzzy Logic. Further dissemination should be achieved by courses and training. Especially regarding fuzzy methods other than Fuzzy Control more instructional work has to be done.

The fields of Fuzzy Control and Fuzzy Optimisation gained much attention in our surveys. In Fuzzy Control the future will most likely be in the enhancement of existing applications, in order to cover non-linear behaviour and to obtain higher security. From our interviews the following insights may be derived:

- *Product definition* is a potential field of application. The exploration of new materials as well as the combination of many components to a new product lent themselves to experiments using fuzzy methods.
- In the area of *production planning systems* existing tools may be improved by the integration with fuzzy modules. Handling of incomplete information would be of particular interest.
- *Optimisation methods* and *decision support systems* based on Fuzzy Logic are still not widely known. There are many fields of application, e. g. in logistics, electronic circuit design, yield optimisation.
- In many cases fuzzy expert systems would be much simpler and yet yield better results than conventional ones.
- In some branches of industry better forecasting would be very much appreciated. We guess that a combination of data analysis and Fuzzy Expert systems would contribute to improved prognosis.

Some of our interview partners didn't recognise application areas of Fuzzy Technology until the discussion. The need for more thorough informing became evident.

Because of economic problems in steel industry, mining and textile industry, many companies in NRW have to turn in new directions. The great variety of possible applications makes Fuzzy Technology interesting especially for those companies.

## IV. Conclusion

We reported on parts of the study "Potential der Fuzzy-Technologie in Nordrhein-Westfalen" (in German). If you have more questions, please contact me under the following address. You may also obtain the study as a whole for a moderate price.

Address:
Prof. Dr. Bernd Reusch
Fuzzy Demonstrations-Zentrum
Martin-Schmeißer-Weg 18
W-4600 Dortmund 50

## References

[1] a) Lehrstuhl für Elektrische Steuerung und Regelung, Universität Dortmund

b) MIT–Management intelligenter Technologien GmbH, Aachen

c) Fuzzy Demonstrations-Zentrum Dortmund

d) ZeTec–Zentrum für FuzzyInformationsTechnik GmbH, Dortmund

e) Siemens Entwicklungszentrum für Mikroelektronik, Düsseldorf

[2] Fuzzy Logic in Japan. Deutsche Industrie- und Handelskammer in Japan, Institut für Marktforschung, 1992.

[3] Was der Fuzzy-Markt verspricht. *Elektronik*, (23):58, 1992.

[4] Th. Kille. Marktentwicklung und strategische Aspekte von Fuzzy Logic. In *Schweizer Fuzzy Logic Fachkonferenz*. Prognos AG, Basel, 1992.

[5] R. K. Miller and T. C. Walker. *Fuzzy Controllers*. SEAI Technical Publications, 1992.

# Qualitative Pattern Analysis for Industrial Quality Assurance

Rudolf Felix, Stefan Reddig
Fuzzy Demonstrations-Zentrum Dortmund
Martin-Schmeißer-Weg 18, D-4600 Dortmund 50, Germany,
Tel. +49 231 9 744 751, Fax. +49 231 9 744 777

**Abstract—A successful application of a fuzzy notion of similarity called analogy is presented. In contrast to other methods both information supporting and distracting characteristics of images is considered in parallel. The method has been successfully applied in the process of industrial quality assurance in the domain of material tests based on the investigation of microscope images of material cuts.**

## I. INTRODUCTION

Industrial quality assurance requires an increasing amount of automation.

A wide-spread part of quality assurance refers to the analysis of surfaces of materials and cuts through it. The surfaces or cuts are treated using optical devices like microscope or video cameras and the characteristics of the obtained images are used as a fundamental quality information.

Since a variety of such quality tests is performed by human personal and the quality criteria cannot be measured with absolute precision, qualitative pattern analysis gains importance for the automation of industrial quality assurance tests.

This paper presents the concept of a tool for qualitative pattern analysis. The tool is based on a fuzzy similarity notion already applied successfully in the domain of qualitative decision making [1, 2, 3, 4].

In contrast to other approaches [5, 6, 7, 8, 9] of similarity the presented notions consider both a similarity estimation and a non-similarity estimation of the patterns analyzed.

## II. ANALOGY OF PATTERNS AS NOTION OF SIMILARITY

The concept of a relation of similarity is essentially a generalization of the concept of an relation of equivalence , i.e. a relation which is reflexive, symmetric and transitive [1]. Industrial quality assurance and the performed quality checks usually require a given number of quality levels. Each quality level represents a class of objects (let us say products) of a certain quality. Provided that the quality levels are defined appropriately, they define a relation of similarity on all the products of which the quality has to be tested.

In case of quality tests based on pattern or image analysis the quality levels are represented by reference images and the problem is to investigate the test image of the actual test candidate and determine its most similar reference image (see Fig. 1 on the next page). Then, of course, the appropriateness of the quality check depends on the appropriateness of the notion of similarity used.

It is interesting to observe that the notions of similarity known in literature, for example [5, 6, 7, 8, 9], refer to characteristics of a pattern as an image (or a part of it) supporting its recognition and seemingly ignoring the information on the non-presence of aspects or the information more or less contradictory to the concerned characteristics of the pattern.

Subsequently the notion of analogy of images is presented. This notion refers to both the positive and the negative aspects of the characteristics of similarity . The notion of analogy is based on the following characterisation of an object (for example image).

Assume that there is a non-empty, finite set of characteristics $C = \{c_1, \ldots, c_n\}$ by which the objects of the domain under consideration can be described. Then, for each object two fuzzy sets, the support set $S$ and the distraction set $D$ are defined. Let $i = \{1, \ldots, n\}$ and let

$$\forall_i \quad \mu_S(c_i) = \begin{cases} \delta, & \text{the characteristic i supported by the degree } \delta \\ 0, & \text{else} \end{cases} \qquad (1)$$

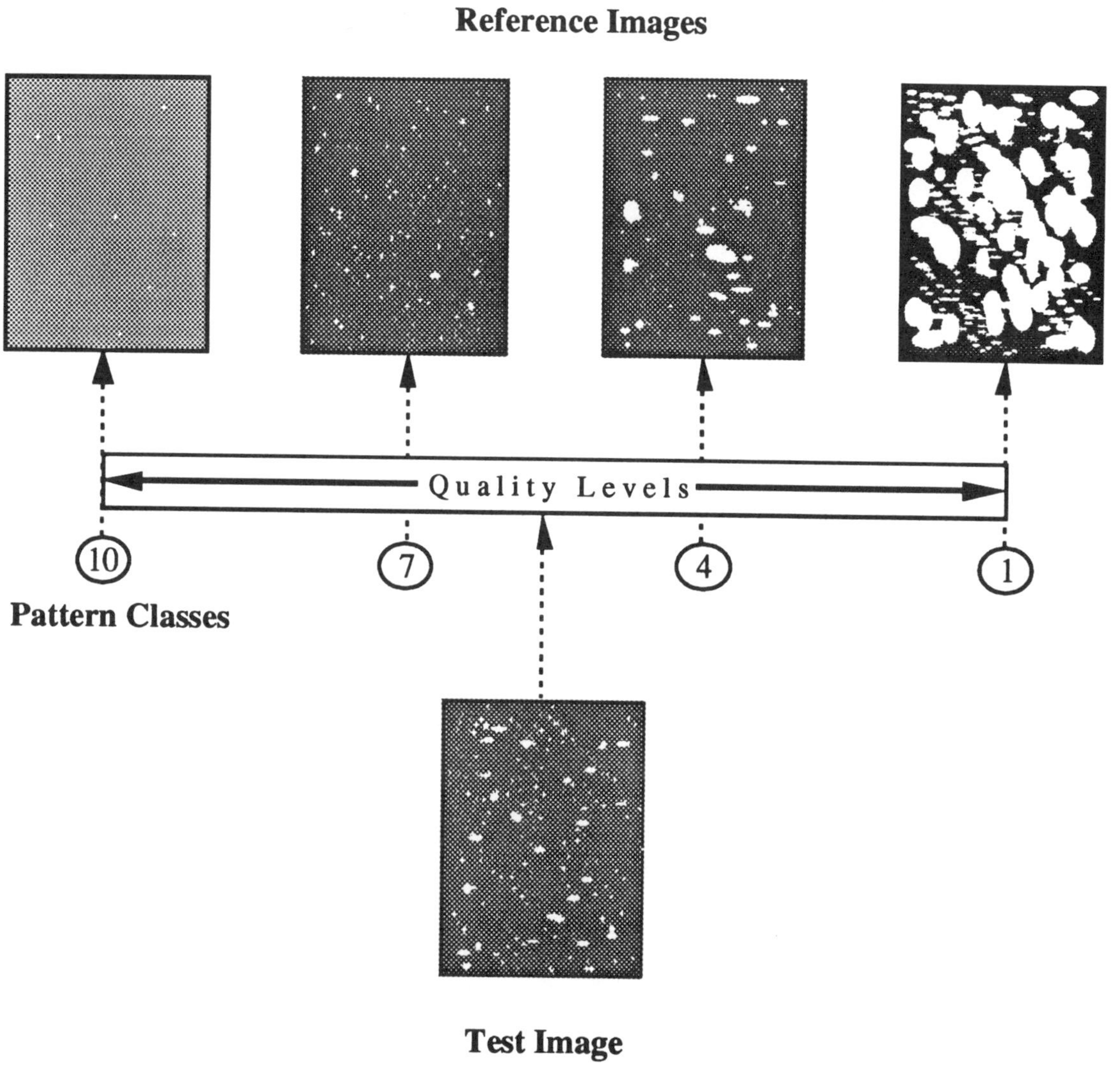


Fig. 1. Quality check by comparing the test image with the reference image.

and

$$\forall_i \quad \mu_D(c_i) = \begin{cases} \delta, & \text{the characteristic i distracted by the degree } \delta \\ 0, & \text{else} \end{cases} \qquad (2)$$

Thus, each object $O$ of the domain can be described by the corresponding support and distraction sets $\mu_S^O$ and $\mu_D^O$ respectively.

In order to measure the similarity of two objects (images), the fuzzy intersections and non-intersections (or in some cases inclusions or non-inclusions) are considered. Let $\cap$ be a fuzzy intersection relation defined by a membership function

$$\mu_\cap: \quad O_1 \times O_2 \to [0,1] \qquad (3)$$

where $O_1, O_2$ are two fuzzy subsets of the universe of discourse. Let $\pitchfork$ be a non-intersection defined by $\mu_\pitchfork := 1 - \mu_\cap$. Then the following relation called *analogy* can be considered as a notion of similarity between the objects $O_1$ and $O_2$:

$$\mu_{\text{analogy}}(O_1, O_2) :=$$
$$\min\left(\mu_s^{O_1} \cap \mu_s^{O_2}, \mu_s^{O_1} \pitchfork \mu_D^{O_2}, \mu_s^{O_2} \pitchfork \mu_D^{O_1}, \mu_D^{O_1} \cap \mu_D^{O_2}\right) \qquad (4)$$

where $\mu_s^{O_i}$ are support sets and $\mu_D^{O_i}$ distraction sets.

If the objects under consideration are gray level images, for example, then the set of characteristics $C = \{c_1, \ldots, c_n\}$ can be obtained by a sectorization of the image and an additional gray level estimation of each sector. Since the possible gray levels range between white and black, an estimation of the degree of white and the degree of black is only a matter of normalization.

Now, in order to perform the qualitative pattern analysis, the reference images are represented by their support and distraction sets and for each candidate to be analyzed the actual support and distraction sets are calculated. Subsequently, the analogy between the candidate and each reference image is determined. Since the reference images represent quality levels, the reference image with the highest degree of analogy to the candidate implies its quality.

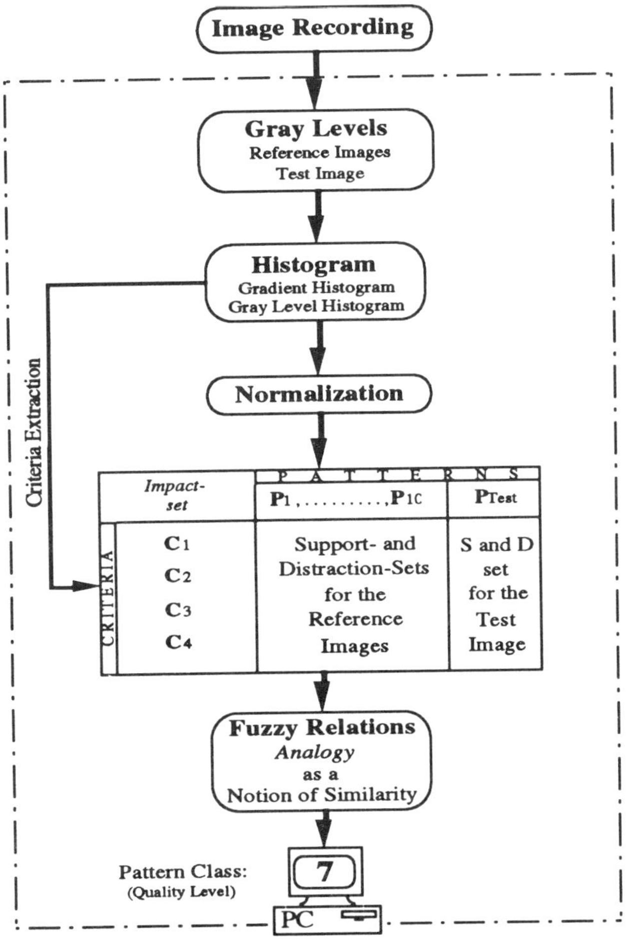


Fig. 2. An overview of the automated analysis process.

## III. EXAMPLES

The method presented above has been successfully applied in industrial quality assurance checking for cut images in the domain of material quality. Compared with tests performed by human testers, the number of successful cases reaches the 95% - 99% mark. Fig. 2 gives an overview of the automated analysis process.

## IV. SUMMARY

A successful application of a fuzzy notion of similarity called analogy has been presented. In contrast to other methods both information supporting and distracting characteristics of images is considered in parallel. The method has been successfully applied in the process of industrial quality assurance in the domain of material tests based on the investigation of microscope images of material cuts.

## V. REFERENCES


[1] R. E. Bellmann and L. A. Zadeh, "Decision Making in a Fuzzy Environment," Management Sciences 17, 1970

[2] R. Felix, "Entscheidungen bei qualitativen Zielen," Universität Dortmund, Fachbereich Informatik, PhD Dissertation (in German) Dortmund, Germany, 1992

[3] R. Felix, "Multiple Attribute Decision Making based on Fuzzy Relationships between Objectives," Proceedings of the 2nd International Conferenca on Fuzzy Logic and Neural Networks, Iizuka, Japan, 1992

[4] R. Felix, "Goal-oriented Selection of Alternatives Based on Fuzzy Sets," IPMU Information Processing and Management of Uncertainty in Knowledge-based Systems, Paris, France, 1990

[5] L. A. Zadeh, "Similarity Relations and Fuzzy Orderings," Yager, Ovchinnikov, Tong, Nguyen, FUZZY SETS AND APPLICATIONS, Selected Papers by L.A. Zadeh, Wiley-Interscience, 1987

[6] S. Ovchinnikov, "Similarity relations, fuzzy partitions, and fuzzy orderings," Fuzzy Sets and Systems vol. 40, 1991, pp. 107-126, North Holland

[7] E. H. Ruspini, "On the Semantics of Fuzzy Logic," International Journal of Approximate Reasoning, vol. 5, Number1, 1991, pp. 45 - 88

[8] E. H. Ruspini, "ON TRUTH, UTILITY, and SIMILARITY," Fuzzy Engineering toward Human Friendly Systems, vol. 1, Nov. 1991, pp. 42 - 50

[9] T. Calvo, "On fuzzy similarity relations," Fuzzy Sets and Systems vol. 47, 1992, pp. 121-123, North Holland

# Fuzzy-Controller for Flow and Mixture Regulation in an Aircraft Deicing Vehicle

Heiko Knappe
SCHRÖDER airporttechnik, deicer division
Mittenheimerstr. 74, D-8042 Oberschleissheim
GERMANY

***Abstract* - The results of traditional control theory often are unsatifactory when dealing with real, non-linear systems of high order that show parameter and structure uncertainties. This is mainly due to the necessary reduction in order and simplification of the system. They are necessary for designing conventional controllers, however, jeopardize exact mathematical modells. A fuzzy controller on the other hand does not need any mathematical description of the system. Therefore it ist especially suitable to control technical, non-linear processes for which verbal rules are known.**

**This paper introduces the *fuzzy flow and mixture regulation of a mobile deicer*, manufactured by *SCHRÖDER airporttechnik* company in Germany. In comparison to the originally used structure-variable linear controller, the fuzzy control system proves to be considerably better in situations of guidance and disturbances. It clearly shows higher robustness in practical operation. On the basis of the positive results a study was performed to research the robustness of a fuzzy (hybrid) system, concerning parameter and structure uncertainties. First, efforts were made to find a way to systematically design fuzzy controllers. A simulation of a two-mass-system was used.**

## I. Introduction

### *A. The Task of Aircraft Deicing*

In winter months freezing temperatures, cold rain and snow cause ice to form on aircraft wings and impair the safety during take-off. In order to avoid this risk, special de-/anti-icing vehicles have to be available at every airport. They free the aircraft wings, tail unit and landing gear from the ice and protect them from icing up again. This has to be performed shortly before take-off. A mixture of *Anti-/Deicing-Fluid* (*ADF*) and hot water is used. In case of incorrect deicing the aerodynamic attributes might be disturbed, causing danger of an aircraft crash.

In Germany, either airports or airlines (e.g. *Lufthansa, Swissair* ...) are responsible for deicing. For all users of deicers safety has to be the most important factor. Consequently especially trained personnel and reliable vehicles (mainly for meeting the nominal mixture of ADF and water) are a must. The merely seasonal use of the equipment constitutes an additional problem.

Fig. 1. *Mobile Deicing Unit* using Fuzzy-Technology. Vehicle manufactured by *SCHRÖDER airporttechnik* company and in operation at Düsseldorf airport (Germany)

For the choice of the appropriate proportion, users face a conflict. In order to provide the necessary safety, the mixture should contain a high concentration of ADF. Due to economical (extremly high price of ADF) and environmental reasons (ADF damages the environment and should be recycled), however, only lower concentrations are wanted. In order to keep the amount of ADF to a minimum and still provide the necessary safety, the demanded proportion and output have to be reached in almost no time and continousely be kept stable.

### *B. De-/Anti-icing Fluids*

Two different de-/anti-icing fluids are used. *ADF 1* is mainly used in the United States and Canada. It has the disadvantage to provide only an extremly short protection time of about 5 minutes. On the other hand it ist chemically robust and can be recycled. In Germany and other European countries the use of *ADF 2* is demanded by law, since this fluid provides a 20 minute protection and therefore higher safety. Its disadvantage is its high sensibility. Inadequate treatment causes damage to the molecular structure of the fluid an jeopardizes the deicing result. Moreover the low viscosity at low temperatures and the minimal ability to heat exchange constitutes problems. The long structures of molecules ADF 2 consits of, are destroyed when the fuild passes narrow paths at high speed and under high pressure or temperatures above 90⁰ C. Whereas ADF 1 can only be used for deicing processes next to the run-way, it ist possible to perform deicing at the gate with ADF 2.

*C. Deicing Operation*

Figure 1 shows a mobile deicing unit, manufactured by *SCHRÖDER airporttechnik*, Germany. It is in use at Düsseldorf airport. The vehicle contains two tanks for water and ADF and a heating and mixing system. With the help of a 14 m telescopic boom the operator reaches the optimal positions to spray the heated water-ADF-mixture (about 80⁰ C) from the monitor onto the aircraft wings.

The first step is to remove ice and snow with a hot mixture and under high pressure. Next, a protective cover of pure ADF is sprayed onto the aircraft. The weather and aircraft conditions determine the proportion to be used. Practical experience shows that in everyday operation the operator just opens and closes the spray monitor quickly in order to reduce the time and medium needed for the process and to allow maneuvering around the aircraft. Despite this kind of procedure the nominal proportion has to be guarantied. Consequently it is necessary that the controller reaches the demanded mixture in almost no time.

## II. Technical Realisation of the Proportioning System

### *A. The Proportioning System*

Due to the sensitivity of ADF 2 the tubing must be without any edges and narrow paths. In addition worm gear pumps have to be used. These pumps are driven hydraulically by the p.t.o. of the vehicle. The proportioning system can be designed in two ways:

Either ADF and water are pumped from the different tanks by the same pump. Proportioning happens before the medium reaches the pump. The output flow shows constant pressure (Figure 2). The ADF-water proportions are determined by two valves that are located in the suction pipe of the tanks. In this case it is easy to control the output pressure by the volumetric worm gear pump. This is achieved by controlling the oil pressure in the hydraulic system. In addition it is possible to control the ADF and water flow easily by the two valves. However, the ADF might be damaged by passing the valve. This is especially true when the ADF valve is almost closed. In this case the correct throughput and proportion cannot be guarantied due to the low ADF viscosity.

Two seperate pumps (Figure 3), on the other hand, have the advantage to only minimally stress the ADF. Moreover, it is possible to exactly keep the proportion. This, however, makes a more difficult control system necessary. The throughput of the pumps can only be regulated indirectly by the openings of the proportional valves in the hydraulic system. The actual water and ADF flow is measured by two inductive flow meters before the two fluids are combined. Additionally the pressure of the mixture is measured at the output.

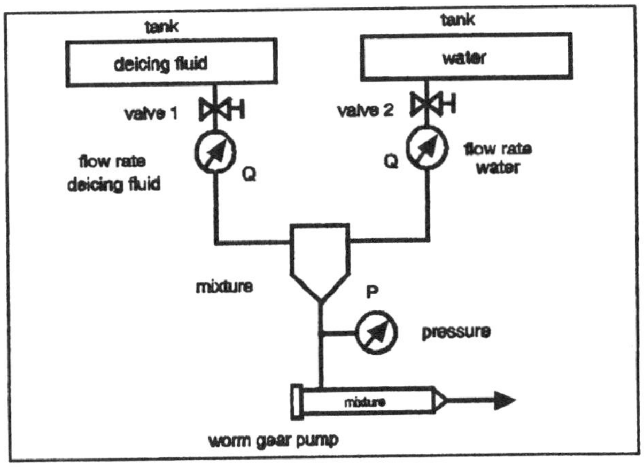


Fig. 2. Proportioning System with one pump

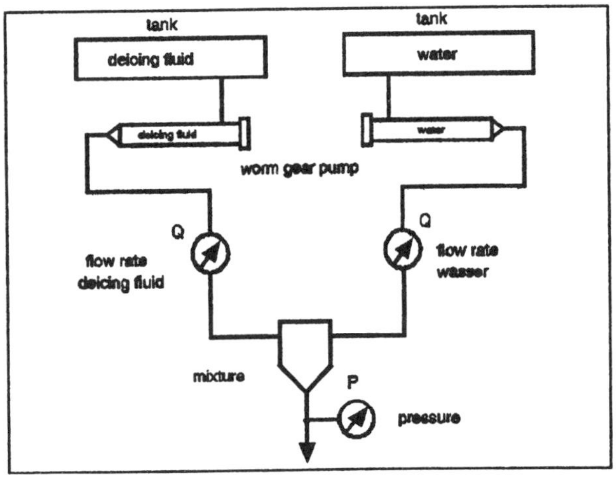


Fig. 3. Poportioning System with two pumps

The efficiency of the worm gear pumps shows a non-linear dependance on temperature an pressure. This makes an optimal regulation of the throughput difficult. The exact control is also impeded through dead-times and non-linearities in the positioning and regulating units. The inductive flow meter for example, has a high dead-time. Consequently the result of a correction can only be measured with some delay. In addition the signals of throughput and output pressure are degraded by high noise.

It is the goal of the control system to reach the nominal throughput as soon as possible and regulate it with deviations of less than 1 litre. This is the only way to guarantee the correct proportion. Since the opening of the spray monitor can be adjusted during operation, the control system not only has to keep the nominal proportion stable but also adjust it to changed througputs. Time again is a crucial factor. If the monitor is closed a high pressure with no throughput results. In this case the pump has to be shut off immediatelly. When the pressure in the pipe system sinks again the pumps have to be restarted.

*B. Parameter Switching PI-Controller System*

In the *SCHRÖDER deicers* a PI control system for the regulation of proportion and throughput was used so far. Here the parameters of the controller are switched, depending on the deviation of the real throughput from the nominal one. For fine tuning only small values are chosen for the integrator (I-part) in order to be able to keep the target amout stable within one litre. In case of larger deviations (more than 3 litres) the controller switches to coarse tuning, so that corrections on the pump can be made in shorter time. A control system for the system pressure is superimposed over the throughput regulation. It is activated as soon as there is a pressure of at least 7 bar in the tubing and limits the maximal pressure to 10 bar. In order to avoid constant switching from one controller to the other, a hysteresis is implemented. Moreover, the parameters for each controller have different values depending on the amount of throughput.

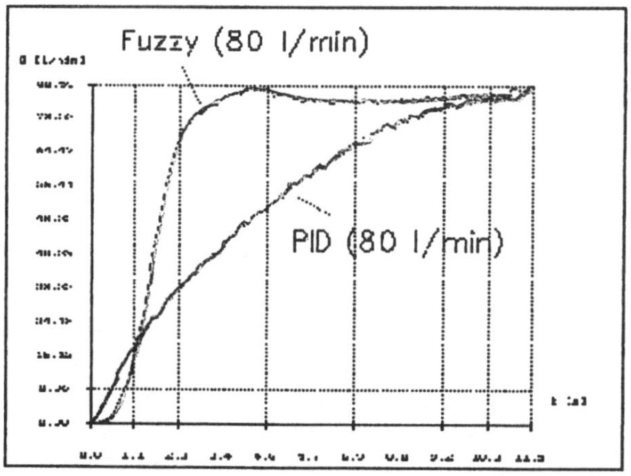


Figur 4: Vergleich der Regelergebnisse vor und nach dem Einsatz

Since it was not economical to develop a suitable modell of the system, the diffenrent paramters of the PI concept had to be determined by trial and error. Not all inner states of the system can be measured. Therefore it is impossible to create the optimal controller and the results of the PI concept are not fully satisfactory. In order to avoid oscillations in the controller it is important to choose a slow design. This causes the nominal throughput of ADF and water to be reached only after a period of 10 seconds. In case of disturbances, (e.g. through changing the opening of the monitor in size) the time for the adjustment of the controller is about 5 seconds with deviations from the nominal value of 10%. This, however, is not acceptable. Obviously the effort to compensate the non-linearities in the system by switching between differently dimensioned PI controllers is not sufficient. During the development of the PI controller several rules could be ovserved that determined the characteristics of the throughput. They constituted the basis for the definition of in what situation the system should switch from one controller to the other. Unfortunately, the implementation of a parameter or structure switching system is difficult and timeconsuming. In addition not every approach is successful and not all possibilities can be tested.

*C. The Fuzzy-Control System*

The PI experience provided a number of *IF-THEN*-rules for the different states of the system. A verbal description for a suitable strategy was available. Consequently fuzzy technology was used to implement this knowledge in the controller. The development was done with a personal computer, that was connected to the microprocessor system of the vehicle through the serial RS232 interface for the exchange of data. After the controller was optimized, it was implemented on the PLC in form of a look-up table.

*D: Comparison of the Fuzzy-Control-System with Conventional Control-Systems*

The results of the fuzzy controller were clearly better than those of the conventional controller. One of the biggest advantages was the short time needed for development and implementation. Moreover, a complicated system or switching between different controllers could be avoided. The non-linearity of the fuzzy controller allowed to reach good characteristics in coarse and fine tuning and a high robustness (changes in throughput, wear). It was not prone to oscillation. In addition, with a fuzzy controller several different measurements could be considered at a time. In this case pressure and throughput were regulated within one controller. Due to the better dynamics of the controller even the complicated system of switching the pump on and off when the monitor was closed or opened was not necessary anymore. The pump stops and starts fast enough.

The rules of a fuzzy system determine the coarse characteristics of the controller. Most of the time proper rules are found without difficulty. There are, however, a number of additional parameters that have to be defined, before getting suitable results. The correct determination of these parameters (position and form of membership functions, operators, methods of inference and defuzzification) is a crucial factor for the quality of the controller. They determine the interpolation between the different rules. Although this fine tuning is quite laborious, the implementation of the multi-layer PI- system is not any less timeconsuming. With the help of Fuzzy-PLC systems that are available meanwhile, it is possible to reach the goal quickly through the possibility to comfortably optimize the controller on-line.

TABLE I

Advantages of the Fuzzy-PI-Controller and Disadvantages of the Parameter Variable PI-Control-System

| **Disadvantages of the PI-Control-System** | **Advantages of the Fuzzy-PI-Control-System** |
|---|---|
| • Timeconsuming software development<br>• PLC programming prone to mistake<br>• PI-Parameters cannot be calculated - they have to be determined by trial and error (no guarantee for a optimal solution)<br>• Unsatisfactory guidance and disturbance attributes despite of structure variable controlling<br>• Unsatisfactory robustness for the large range of flow rates of 0 l/min to 400 l/min and wear | • Extremelly short development time<br>• Control rules can be found through observation only and be implemented and tested quickly<br>• Improved guidance and disturbance attributes through non-linear Control<br>• Higher robustness even within large ranges of flow rates and with the occurance of wear<br>• Parameter corrections can be avoided through higher robustness<br>• Only one controller is needed for pressure and throughput<br>• No additional logic necessary for starting and stopping the pumps |

Figure 4 shows the comparison of the measured throughput in the PI system versus the fuzzy system at 80 l/min nominal value. The fuzzy controller is considerably faster and reaches the target throughput already after a period of 3 seconds. As can be seen, fuzzy is also suitable for safety relevant processes. Practical use of the new fuzzy proportioning system shows its efficiency and the clear improvements. The disadvantages of the linear multi-layer PI-concept and the advantages of the non-linear fuzzy system are summarized in Table I.

## III. Control of an Elastic 2-Mass-System with Slack

Since the results of the fuzzy controlled proportioning system were convincing, a systematic approach to the design of fuzzy and fuzzy- hybrid controllers was researched. The modell chosen was an elastic 2-mass system with slack. It consisted of a DC engine with analog current controller. It was connected by slack transmission and an elastic shaft to a mass. In most cases this modell is an acceptable approach to industrial multi-mass-systems.

The mass was controlled in revolutions per minute (r.p.m.) and in position. The results of the fuzzy controller were compared to optimized linear state controllers. The goal of the project was to find a systematic way of defining a fuzzy and fuzzy-hybrid controller and to work out advantages and disadvantages of non-linear fuzzy controllers in comparison to linear approaches. Moreover, the robustness towards parameter and structure uncertainties was analyzed.

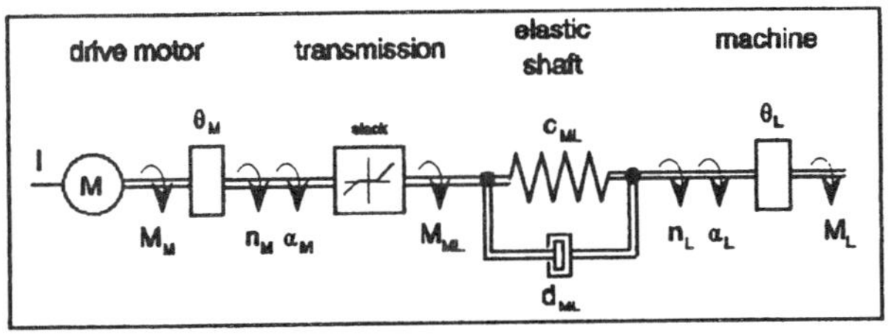


Figur 5: Elastic 2-Mass-System with Slack

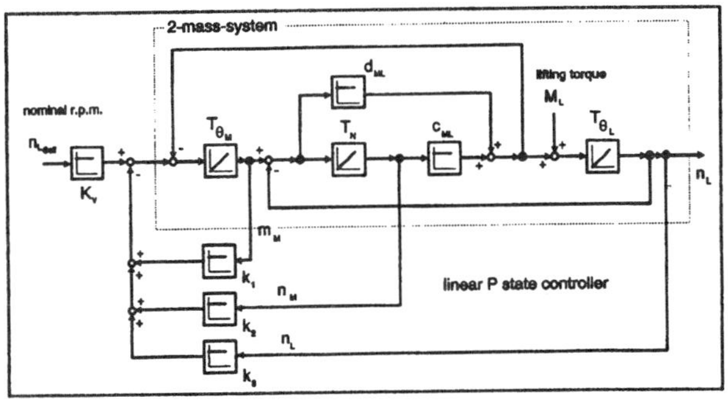


Fig. 6. r.p.m. Control - Signal Flow Chart of the System

### A. Control-System for r.p.m. Regulation without Slack

In a first step the slack was neglected. For the linear system (frequency 100Hz) a linear state P-(I) controller and a non-linear PI-Fuzzy controller were defined and optimized.

#### 1) Linear State P-(I)-Controller

Figure 6 shows the P- state controller for speed regulation of the machine. The following three states of the system are measured: $M_M$ (torque of the engine), $n_M$ (speed of the engine) and $n_L$ (speed of the second mass). They are linearly fed back as input into the control system. From the equation of state the characteristic polynominal can be determined. Moreover the linear coefficients of feedback $k_1$, $k_2$, $k_3$ and $k_V$ can be calculated for an ideal polynominal (e.g. asymptotic characteristic - polynominal of third order). In order to aviod stationary vagueness an additional integrator is used. Its task is to reduce the deviations between nominal and real values. It was necessary to limit the output of the controller to the maximally accepted current of the DC motor in order to be able to compare the two controllers.

This was achieved by choosing a suitable polynominal function.

*2) PI-Fuzzy-Controller*

In a first step the fuzzy controller was defined as a P-controller. The fuzzy system calculated the appropriate current for the DC motor from the three states of the system and the deviation from the target value. Even after optimizing the contröller for the distubance torque $M_L$ some deviation from the r.p.m. of the second mass was left. The characteristics of the controller were clearly improved by a parallel non-linear integrator. In this case the fuzzy system determined the suitable time constant of the integrator $T_I$ in addition to the motor current. This was necessary in order to guarantee stationary exactness and stability. Figure 9 shows the results of the simulation for the speed control (for the designed system frequency of 100 Hz). The fuzzy system was optimized for a jump stimulation of fixed height. Since the controller is non-linear the system is only optimized for specific situations and the results cannot necessarily be projected onto other activation functions.

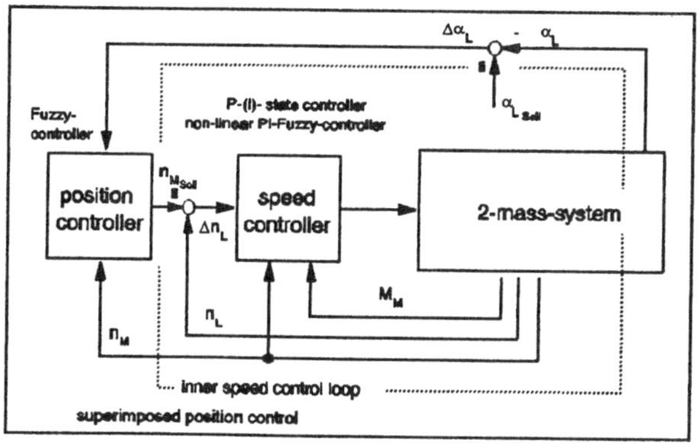


Fig. 7. Cascade connected Fuzzy-Position-Control

## B. Position Control without Slack

For the position control the scheme in figure 6 has to be completed with an integrator. Thus, the order of the system and therefore the number of states increases from 3 to 4. A PI state controller for the regulation of the position was developped parallel to the speed controller. In the r.p.m-control system all 3 states could be taken into consideration in each rule. Rules with 4 conditions, however, are much more difficult to survey and not suitable for a rule base. Consequently, the fuzzy controller was designed as a cascade. (figure 7). A superimposed position controller defined the nominal value for the revolutions per minute. The real angular velocity was adjusted correspondingly by the subordinate speed controller. This way, both controllers have only 3 inputs.

For the speed controller the previously described state controller and the optimized fuzzy-speed controller were used. Since the superimposed controller determines different stimulation functions, the non-linear fuzzy controller is not very suitable for this task due to its limited optimization for jump activations (fix heights). Despite this disadvantage the results of simulation were very close.

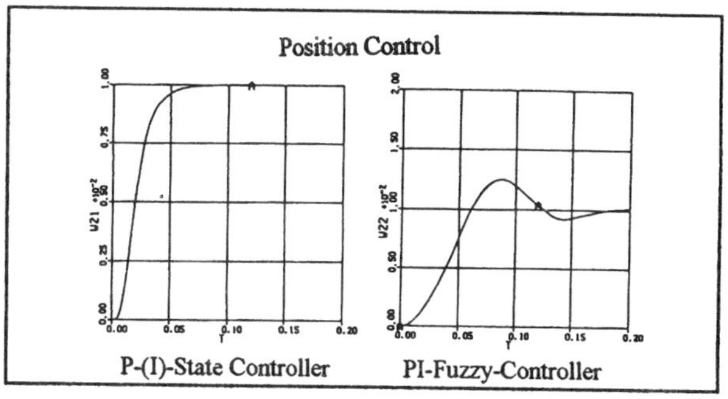


Fig. 8. Guidance Attributes of the P-(I)-State- and the PI-Fuzzy-Controllers for Position Control (Neglecting Slack)

Figure 8 compares the simulation results of the position control system of the linear PI state controller and the non-linear cascade connected fuzzy controller. It can be observed, that the state controller shows better guidance. This phenomenon can be explained. Due to cascade- connecting the control system, not all dependencies of the different states are taken into consideration, causing difficulties in regulating it properly.

## C. Vague Parameters

In order to be able to compare the robustness of the linear state controller and the fuzzy controller the designed system frequency was rised from 50 Hz to 200 Hz. (figure 9). The controllers were not adjusted to the change in parameters.

A harder system with 150 Hz is easier to be regulated than a softer system with 50 Hz. Consequently, the results of both controllers improved for the harder system. For the 50 Hz system the state controller, that was dimensioned for 100 Hz, clearly showed a higher overshoot and a longer transient time. The transient time and the amplitude of the overshoot, however, remained almost unchanged with the fuzzy controller.

## D. Consideration of Slack

So far the slack of the gearance was neglected. In order to judge the robustness of the fuzzy controller and the state controller concerning a non-linear vagueness in structure, the slack was introduced with different values and not considered when designing the controllers. The results of the simulation are found in figure 10. The slack caused the state controller to oscillate with high frequencies that are dependent on the designed system frequency and the slack chosen. The system performed continuous undesireable oscillations around the target value. The non-linear fuzzy-state- controller on the other hand proved to be much more robust than the closed linear control loop. It remained stable for the position control and did not oscillate even with a slack of 10°.

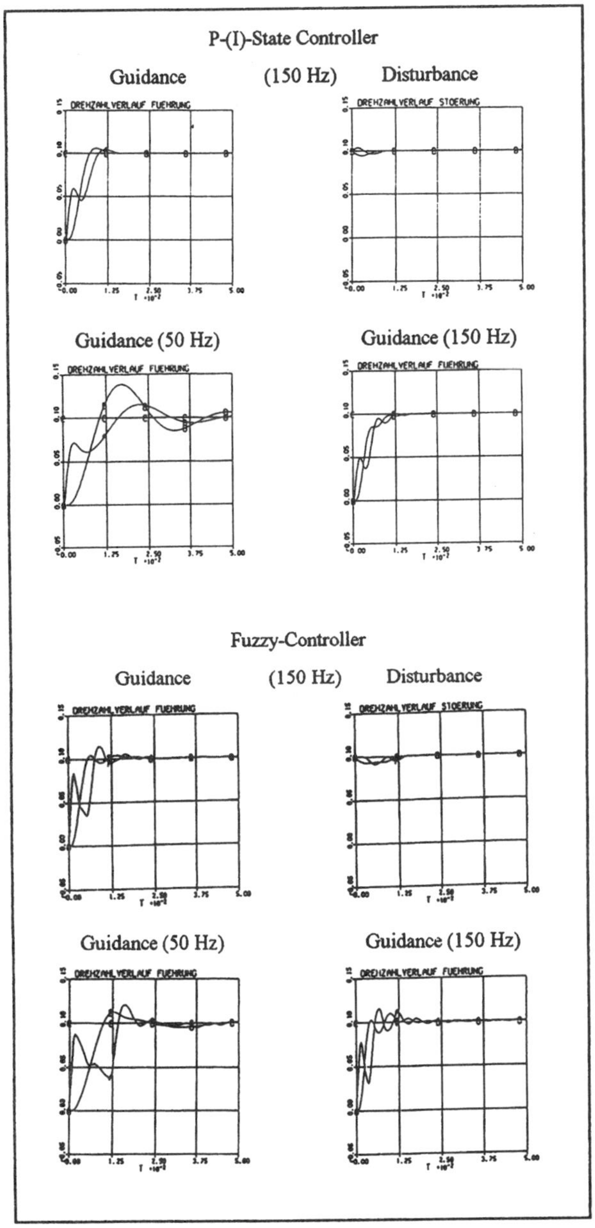


Fig. 9. Guidance and Disturbance Attributes of the P-(I)-State- and the PI-Fuzzy-Controller for Different Designed System Frequencies (50 Hz, 100 Hz, 150 Hz) (Speed Control) (A - Motor Mass $m_M$ / B - Maschine Mass $m_L$)

## IV. Conclusion

In principle a fuzzy controller constitutes a non-linear state controller. Due to the non-linearities it is impossible to determine the attributes of the controller for arbitrary stimulation functions, like this is true in linear control theory. The state space, therefore, is a helpful tool in combination with the trajectories of the system, to determine the attributes of the closed control loop. It aids to optimize the fuzzy controller. Unfortunately, this is only possible for systems with merely two inputs.

The (crisp) rules define fix points in the state space. The other parameters of the fuzzy system (e.g. the membership functions) are responsible for the interpolation between these points. Consequently the rules are used for coarse tuning, whereas the other fuzzy parameters are responsible for fine tunig. Since the interpolation can only be influenced to a certain limit, it is often necessary to increase the rulebase even during the phase of optimizing the controller. Through the newly created fix points in the system the attributes of the controller are described more accurately.

As could be seen from the simulations and the practical implementation, fuzzy logic bears a high potential. The optimization of the contoller, however, is difficult (experimental controller design) and should only be seen as a supplement to the conventional methods of control theory. Fuzzy control should be mainly used (by itself or in combination with classical controllers) in situations, where either no mathematical modell is available, it is too expensive to be developped or the states of the system cannot be measured properly.

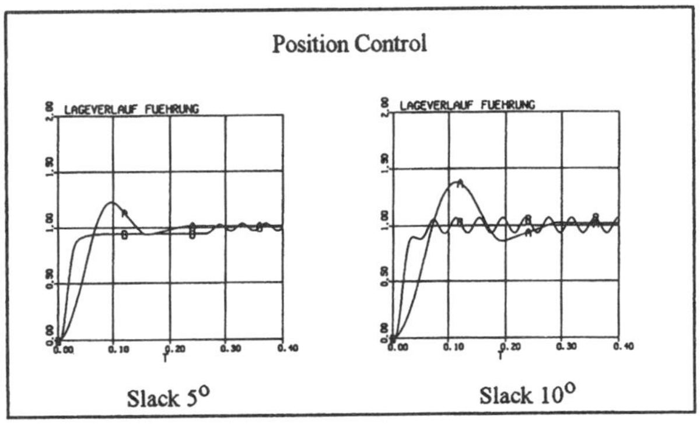


Fig. 10. Guidance Attributes of the P-(I)-State- (B) and the PI-Fuzzy-Controller (A) for Position Control (considering Slack) (5° und 10°)

## References

[1] H.-J.Zimmermann, *"Fuzzy Set Theory and its Applications"*, Kluwer Dordrecht, 2. Edition, 1991

[2] C. C. Lee, *"Fuzzy Logic in Control Systems: Fuzzy Logic Controller"*, Part I/II, IEEE Transactions on Systems, Man, and Cybernetics, Vol. 20, No. 2, March/April 1990

[3] M. Sugeno, *"An Introdutory Survey of Fuzzy Control"*, Information and Sciences 36, Page 59-83, Elsevier Science Publishing Co., Inc., 1985

[4] L. Zadeh, "Fuzzy Logic", IEEEComputer, Page83 - 93, April 1988

[5] T. Takagi, M. Sugeno, *"Fuzzy Identification of Systems and Its Applications to Modeling and Control"*, IEEE Transactions on Systems, Man, and Cybernetics, Vol. SMC-15, No. 1, January/February 1985

[6] H. Takagi, I. Hayashi, *"NN-Driven Fuzzy Reasoning"*, International Journal of Approximate Reasoning, Page 191 - 212, Elsevier Science Publishing Co., Inc., 1991

[7] H. Knappe, *"Fuzzy-Enteisungstechnik" ("Fuzzy De-/Anti-Icing Technique")*, Spektrum der Wissenschaft (Scientific American - German Edition), 1993

# On the Use of Fuzzy Integral as a Fuzzy Connective

Michel GRABISCH
Thomson-Sintra ASM
1, avenue Aristide Briand
94117 Arcueil, France
email grabisch@thomson-lcr.fr

**Abstract** Fuzzy integrals of discrete functions can be considered as a class of fuzzy connectives, of which several properties are already known: continuity, monotonicity, idempotency, etc... thus making them very near the concept of averaging operator. In a first part, relations with known classes of connectives are established, then in a second part, equivalence classes of fuzzy integrals are investigated. Lastly, approximation problems are examined.

## 1 Motivations

The concept of fuzzy integral with respect to a fuzzy measure has been introduced in 1974 by Sugeno [14, 15], and was further refined by several authors, including Weber [19], Kruse [11], Ichihashi *et al.* [7], and finally Murofushi and Sugeno [12, 13], who first introduced the Choquet integral, and secondly the concept of fuzzy t-conorm integral, which is perhaps the most general definition, encompassing original definitions of Sugeno, Weber, Kruse and the Choquet integral.

Unfortunately, fuzzy integrals and fuzzy measures have not received very much attention from people of applications fields, perhaps due to their rather abstract mathematical flavour. Nevertheless, from the beginning, it seems that fuzzy integrals have been essentially used as an aggregation operator in the field of multicriteria analysis, mainly subjective evaluation problems [8, 16] and pattern recognition [5].

In most of the cases, the reason researchers have used fuzzy integrals as an aggregation operator is based only on the intuitive meaning of fuzzy measure and integral, i.e. a kind of average value taking into account weights on criteria and subsets of criteria, without bothering on their exact properties (however, see an attempt of justification in the case of subjective evaluation in [4]). A question arises here:

*What is the situation of fuzzy integrals (in its most general form) in the field of all known fuzzy connectives, as t-norms, t-conorms, averaging operators, medians, compensatory operators, etc... ? In short, what is their expressive power?*

To our opinion, this is an important question, because if it is intuitively felt that fuzzy integrals have some expressive power, their practical use is not so easy, because for a $n$ criteria problem, $2^n - 2$ coefficients must be identified.

After some thought, the above question can be considered essentially under three viewpoints:

**set relations :** find all relations of intersection, inclusion between *sets* of connectives, i.e. what kind of fuzzy integrals are averaging operators, medians, etc... and reciprocally.

**equivalence relations :** define an equivalence relation between connectives, having some sense in applications, and find the equivalences classes of fuzzy integrals.

**approximation :** find the class of connectives which can be approximated in some sense by fuzzy integrals.

The first viewpoint is exactly what did Banon [2] with fuzzy measures, belief functions and possibility measures. In this respect, some results are already available, for example Kandel and Byatt [9] have shown that the Sugeno integral is a median, which is in fact the only associative averaging operator [3]. We intend to further clarify this kind of relations in a first part of this paper.

The problem of approximation (third viewpoint) of a given function seems to have become recently a preoccupation of many researchers in the field of fuzzy logic control (see for example interesting results of Wang [18], Kosko [10]), so it is tempting to see what kind of connectives can be approximated by fuzzy integrals. We will address this point in the last part of the paper.

However to our opinion, the preceding viewpoints are not of so great interest as far as practical applications in the field of multicriteria decision are concerned. In fact, in fuzzy logic control, it is of great importance that a given control function can be approximated with the highest precision, for any

0-7803-0614-7/93$03.00 ©1993IEEE

deviation may seriously damage the performance of the controller. On the other hand, the aim of any decision theory is to *rank* alternatives, so that if a connective $\mathcal{H}$ is used to aggregate partial decisions $a_1, \dots, a_n$ concerning a given alternative, it is not the *value* of $\mathcal{H}(a_1, \dots, a_n)$ itself which is important, but its *relative position* with respect to the value of the other alternatives. Hence, approximation problems are meaningless here, and we must find equivalence classes of operators, of which members will always lead to the same ranking of the alternatives. We will address this point in a second part of our paper.

For the sake of completeness, we begin by some basic definitions on fuzzy connectives and fuzzy integrals.

## 2 Basic definitions and properties

As we deal with fuzzy integrals considered as connectives, we will consider only discrete spaces $X = \{x_1, \dots, x_n\}$, and associated algebra $\mathcal{X}$ will be always the power set $\mathcal{P}(X)$. Moreover, a *connective-like* notation will be adopted, instead of the usual $\int$ form.

**Definition 1** *A* fuzzy measure *$\mu$ defined on the measurable space $(X, \mathcal{X})$ is a set function $\mu : \mathcal{X} \longrightarrow [0,1]$ verifying the following axioms :*

**(i)** $\mu(\emptyset) = 0, \mu(X) = 1.$

**(ii)** $A \subseteq B \Rightarrow \mu(A) \leq \mu(B)$

*$(X, \mathcal{X}, \mu)$ is said to be a fuzzy measure space.*

Fuzzy measures include as particular cases probability measures, possibility and necessity measures, etc....

**Definition 2** [19] *Let $\perp$ be a t-conorm and $\mu$ a fuzzy measure. $\mu$ is said to be $\perp$-*decomposable *if $\mu(A \cup B) = \mu(A) \perp \mu(B)$* whenever *$A \cap B = \emptyset$.*

A possibility measure is a $\vee$-decomposable measure, and a probability measure is a $\hat{+}$-decomposable measure, where $\hat{+}$ indicates the bounded sum, $a \hat{+} b = (a+b) \wedge 1$.

**Definition 3** *The* Sugeno integral *of a function $f : X \longrightarrow [0,1]$ is defined by :*

$$\mathcal{S}_\mu(f(x_1), \dots, f(x_n)) \triangleq \bigvee_{i=1}^{n} (f(x_i) \wedge \mu(A_i)) \quad (1)$$

*where it is assumed without loss of generality that $0 \leq f(x_1) \leq \dots \leq f(x_n) \leq 1$, and $A_i \triangleq \{x_i, \dots, x_n\}$.*

**Definition 4** *The* Choquet integral *of a function $f : X \longrightarrow [0,1]$ is defined by*

$$\mathcal{C}_\mu(f(x_1), \dots, f(x_n)) \triangleq \sum_{i=1}^{n} (f(x_i) - f(x_{i-1}))\mu(A_i) \quad (2)$$

*with the same notations as above, and $f(x_0) = 0$.*

We now proceed to define fuzzy t-conorm integrals. First we need some elementary facts and definitions about t-conorms:

- a t-conorm is said to be strict if it is continuous on $[0,1]^2$ and strictly increasing in each place.
- a t-conorm is said to be Archimedean if it verifies: $\forall x \in ]0,1[,\ x \perp x > x$
- for every continuous Archimedean t-conorm there exists a continuous and strictly increasing function, called the *generation function* $k : [0,1] \longrightarrow [0, \infty]$, with $k(0) = 0$ such that

  $$x \perp y = k^{-1}[k(1) \wedge (k(x) + k(y))]$$

  T-conorms such that $k(1)$ is finite are said nilpotent, otherwise they are strict.

In order to avoid unnecessary developments and to focus on the multicriteria analysis problem, we will slightly restrict the definition of fuzzy t-conorm integrals. Original definitions can be found in [13] and [6].

**Definition 5** *Let $\mathcal{F} = (\triangle, \perp)$ be a system of Archimedean t-conorms whose generation functions are respectively $h, g$, with $g(1) = 1$, i.e. a nilpotent t-conorm, and $(X, \mathcal{X}, \mu)$ a fuzzy measure space. The* fuzzy t-conorm integral *of a function $f$ based on $\mathcal{F}$ with respect to $\mu$ is defined by*

$$\mathcal{F}_\mu(f) \triangleq h^{-1}[\, \mathcal{C}_{g \circ \mu}(h \circ f)\, ] \quad (3)$$

Remark that the Choquet integral is recovered with $\triangle = \perp = \hat{+}$. Note that as $\vee$ is not Archimedean, the Sugeno integral is not recovered by this definition, but this can be done using the general definition (see [13]). Otherwise indicated, $\mathcal{F}_\mu$ will indicate the restricted definition.

We give now some properties of fuzzy integrals.

**Property 1** *For every $\mathcal{F}_\mu$, including the Sugeno integral, $\mathcal{F}_\mu(a, a, \dots, a) = a$.*

**Property 2** *For the particular measure $\mu_{\min}$ defined by $\forall B \in \mathcal{X}, B \neq X$, $\mu_{\min}(B) = 0$, and $\mu_{\min}(X) = 1$ (resp. $\mu_{\max}$ defined by $\forall B \in \mathcal{X}, B \neq \emptyset$, $\mu_{\max}(B) = 1$, and $\mu_{\max}(\emptyset) = 0$), $\mathcal{F}_{\mu_{\min}}$ (resp. $\mathcal{F}_{\mu_{\max}}$) reduces to the minimum operator (resp. maximum) (same property for Sugeno integral).*

**Property 3** *Let $f, f'$ be two functions on $X$ and $\mu$ a measure on $(X, \mathcal{X})$. Then for every $\mathcal{F}_\mu$, including the Sugeno integral, $f(x) \leq f'(x)$ for every $x \in X \Rightarrow \mathcal{F}_\mu(f) \leq \mathcal{F}_\mu(f')$*

**Property 4** *Let $\mu, \mu'$ be two measures on $(X, \mathcal{X})$. Then for every $\mathcal{F}_\mu$, including the Sugeno integral, $\mu(B) \leq \mu'(B)$ for every $B \in \mathcal{X} \Rightarrow \mathcal{F}_\mu(f) \leq \mathcal{F}_{\mu'}(f)$*

**Property 5** *Using properties 2 and 4, we can deduce $\bigwedge_{j=1}^n a_j \leq \mathcal{F}_\mu(a_1, \ldots, a_n) \leq \bigvee_{j=1}^n a_j$.*

**Property 6** *For every additive measure, the Choquet integral reduces to the usual Lebesgue integral, that is: $\mathcal{C}_\mu(a_1, \ldots, a_n) = \sum_{i=1}^n a_i \mu(\{x_i\})$*

**Property 7** *The Sugeno integral is a median (see definition 7 below):*

$$\mathcal{S}_\mu(a_1, \ldots, a_n) = \text{med}(a_1, \ldots, a_n, \mu(A_2), \ldots, \mu(A_n))$$

*where it is assumed that $a_1 \leq \cdots \leq a_n$ and $A_i = \{x_i, \ldots, x_n\}$.*

This result was established by Kandel and Byatt [9].

**Property 8** *For every fuzzy measure, fuzzy integrals are continuous functionals, i.e, for every sequence of functions $\{f_n\}_{n \in N}$ on $X$ we have:*

$$\lim_{n \to \infty} \mathcal{F}_\mu(f_n) = \mathcal{F}_\mu(\lim_{n \to \infty} f_n)$$

**Property 9** *[6] For every $\mathcal{F}_\mu$ such that $\triangle, \perp$ are both nilpotent,*

$$\mathcal{F}_\mu(f) = 1 -_\triangle \mathcal{F}_{\mu^\perp}(1 -_\triangle f)$$

*where $-_\triangle$ is a $\triangle$-difference operator defined by $a -_\triangle b \stackrel{\Delta}{=} h^{-1}(0 \vee (h(a) - h(b)))$, and $\mu^\perp$ is the $\perp$-dual of $\mu$, i.e. $\mu^\perp(A) = 1 -_\perp \mu(A^c)$. The Sugeno integral verifies this property for ordinary difference and dual measure only.*

We conclude this section by giving the definition of averaging operators and related concepts. See for example [3, 17] for details and proofs.

**Definition 6** *An* averaging operator *or* mean operator *$m : [0,1]^n \longrightarrow [0,1]$ verifies the following properties:*

1. *$\forall a, m(a, a, \ldots, a) = a$ (idempotency)*
2. *the order of arguments is unimportant (commutativity)*
3. *$m$ is non decreasing in each place*

Let us remark that these properties implies that averaging operators lie between *min* and *max*. Some authors request also the continuity, and the fact that *min* and *max* are excluded of the family [3].

**Definition 7** *Consider a sequence of an odd number of real numbers in $[0,1]$, $a_1, \ldots, a_{2q-1}$. Then the* median *of this sequence is defined by*

$$\text{med}(a_1, \ldots, a_{2q-1}) \stackrel{\Delta}{=} a_{\pi(q)}$$

*where $\pi$ is a permutation of the indices such that $a_{\pi(1)} \leq \cdots \leq a_{\pi(2q-1)}$, i.e. the median is the middle value of the ordered sequence.*

Dubois and Prade [3] have proved the following:

**Theorem 1** *Let $m$ be an associative averaging operator on $[0,1]^n$. Then necessarily, $m$ is a median of $2n-1$ terms defined by:*

$$\begin{aligned} m(a_1, \ldots, a_n) &= \text{med}(a_1, \ldots, a_n, \alpha, \ldots, \alpha) \\ &= \text{med}(\bigwedge_{i=1}^n a_i, \bigvee_{i=1}^n a_i, \alpha) \end{aligned}$$

*where $\alpha = m(0,1)$.*

**Definition 8** *$S : [0,1]^n \longrightarrow [0,1]$ is said to be a* symmetric summation *if and only if $S$ is continuous, non decreasing with respect to each argument, commutative, verifying $S(0,0) = 0$, $S(1,1) = 1$ and is auto-dual, that is:*

$$S(a_1, \ldots, a_n) = 1 - S(1 - a_1, \ldots, 1 - a_n)$$

This family of operators has been considered by Silvert.

# 3 Set relations between fuzzy integrals and other connectives

In this paragraph, we will seek only relations with averaging operators, medians and symmetric summations, as from property 5 it is known that the intersection between fuzzy integrals and t-norms (resp. t-conorms) is reduced to *min* (resp. *max*).

## 3.1 set relations with averaging operators and medians

It is easy to see from the definition of averaging operators and properties 1,3 and 8 of fuzzy integrals, that all commutative fuzzy integrals are averaging operators, except $\mathcal{F}_{\mu_{\min}}, \mathcal{F}_{\mu_{\max}}$. We can prove the following, using the definitions and property 7 for the Sugeno integral:

**Theorem 2** *Let $\mathcal{F}_\mu$ be either a Sugeno or a fuzzy t-conorm integral. Then $\mathcal{F}_\mu$ is an averaging operator if and only if $\mu$ verifies:*

$$\forall A, B \subseteq X \text{ such that } |A| = |B|, \mu(A) = \mu(B)$$

*where $|A|$ indicates the cardinal of $A$.*

The next result follows directly from Th. 1 and 2 and property 7:

**Theorem 3** *Every associative averaging operator $m$ is a commutative Sugeno integral $\mathcal{S}_\mu$, with $\mu(\{x_1\}) = \mu(\{x_1, x_2\}) = \cdots = \mu(\{x_1, \ldots, x_{n-1}\}) = m(0, 1)$.*

### 3.2 set relations with symmetric summations

From def. 8 and property 9, it is easy to see that all commutative fuzzy integrals with $\Delta = \widehat{+}$, including Sugeno integral are symmetric summations provided $\mu^\perp \equiv \mu$. So we have the following result:

**Theorem 4** *Every commutative fuzzy integral with $\mathcal{F} = (\widehat{+}, \perp)$ is a symmetric summation if and only if:*

$$\forall A \subseteq X, \quad \mu(A) \perp \mu(A^c) = 1$$

*The property also holds for the Sugeno integral, with $+$ replacing $\perp$ in the above condition.*

Note that in particular, every $\perp$-decomposable measure verifies the above condition.

## 4 Equivalence classes of fuzzy integrals

Let $\underline{a}$ denote elements of $[0, 1]^n$. We introduce the following equivalence relations on the set of all connectives on $[0, 1]^n$:

**Definition 9** *Two connectives $\mathcal{H}_1$ and $\mathcal{H}_2$ are* strongly equivalent *(denoted by $\mathcal{H}_1 \sim \mathcal{H}_2$) if and only if:*

$$\mathcal{H}_1(\underline{a}) > \mathcal{H}_1(\underline{a'}) \Leftrightarrow \mathcal{H}_2(\underline{a}) > \mathcal{H}_2(\underline{a'})$$

*Next, two connectives are* weakly equivalent *(denoted by $\mathcal{H}_1 \approx \mathcal{H}_2$) if and only if:*

$$\mathcal{H}_1(\underline{a}) > \mathcal{H}_1(\underline{a'}) \Rightarrow \mathcal{H}_2(\underline{a}) \geq \mathcal{H}_2(\underline{a'})$$

Note that the first one is indeed an equivalence relation, but the second one has not the transitivity property.

It is clear that all connectives in the equivalence class of a given $\mathcal{H}$, denoted by $\mathcal{H}_\sim$, lead to exactly the same ranking of alternatives. However, with $\approx$, two alternatives may not be distinguished by some connectives of $\mathcal{H}_\approx$ but properly ranked by others. In any case, two operators in $\mathcal{H}_\approx$ will not lead to contradictory rankings.

Now we try to characterize the equivalence classes of fuzzy integrals. This will be done through the concept of level surfaces (or indifference surfaces), that is, considering a connective $\mathcal{H}$ on $[0, 1]^n$, its level surfaces are defined by $\mathcal{H}^{-1}(z) \triangleq \{(a_1, \ldots, a_n) | \mathcal{H}(a_1, \ldots, a_n) = z\}$, which is generally a $(n-1)$-dim surface.

We need the following result:

**Theorem 5** *Let $\mathcal{H}_1$, $\mathcal{H}_2$ be two connectives increasing in each place on a domain $D \subseteq [0, 1]^n$. Then $\mathcal{H}_1 \approx \mathcal{H}_2$ if and only if their level surface families $\{\mathcal{H}_1^{-1}(z)\}_{z \in [0,1]}$, $\{\mathcal{H}_2^{-1}(z)\}_{z \in [0,1]}$ don't intersect, i.e.*

$$\forall z, z' \in [0, 1], \text{ either } \mathcal{H}_1^{-1}(z) = \mathcal{H}_2^{-1}(z')$$

$$\text{or } \mathcal{H}_1^{-1}(z) \cap \mathcal{H}_2^{-1}(z') = \emptyset$$

**Sketch of the proof** take $n = 2$ and consider the figure below with two level surfaces $\mathcal{H}_1^{-1}(z_1^1), \mathcal{H}_1^{-1}(z_2^1)$ and $\mathcal{H}_2^{-1}(z_1^2), \mathcal{H}_2^{-1}(z_2^2)$, assuming w.l.o.g. that $z_1^i < z_2^i$. The $\mathcal{H}_1^{-1}$ and $\mathcal{H}_2^{-1}$ intersect each other means that there exists points $\underline{a} \in \mathcal{H}_1^{-1}(z_1^1)$ and $\underline{a}' \in \mathcal{H}_1^{-1}(z_2^1)$ which belong to $\mathcal{H}_2^{-1}(z_1^2)$, similarly for $\mathcal{H}_2^{-1}(z_2^2)$. But in this case we have simultaneously $\mathcal{H}_1^{-1}(\underline{a}'') < \mathcal{H}_1^{-1}(\underline{a}')$ and $\mathcal{H}_2^{-1}(\underline{a}'') > \mathcal{H}_2^{-1}(\underline{a}')$, which is equivalent to say that $\mathcal{H}_1^{-1} \approx \mathcal{H}_2^{-1}$ is false.

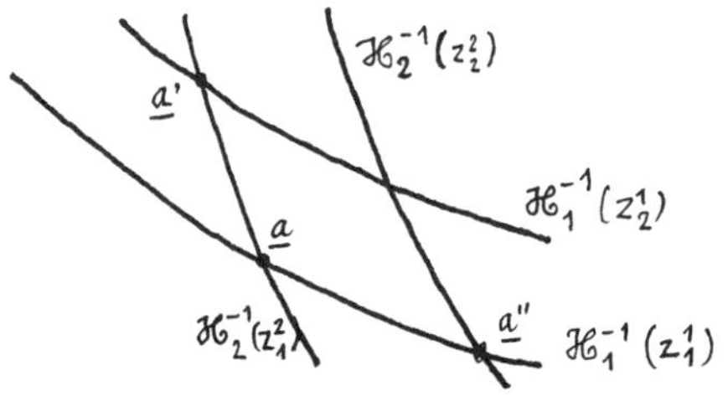


### 4.1 equivalence class of the Choquet integral

Let us now find the equivalence classes of the Choquet integral. We take $n = 2$ for simplicity, the generalization is straightforward. Consider first the region where $a_1 \geq a_2$. Then $\mathcal{C}_\mu(a_1, a_2) = a_2 + (a_1 - a_2)\mu(\{x_1\})$, so that the level curves are straight lines defined by:

$$\mathcal{C}_\mu^{-1}(z) = \{(a_1, a_2) | a_2 = (-a_1\mu(\{x_1\}) + z)/(1 - \mu(\{x_1\})), \ a_1 \geq a_2\}$$

and similarly for the region $a_1 \leq a_2$:

$$\mathcal{C}_\mu^{-1}(z) = \{(a_1, a_2) | a_2 = (-a_1(1 - \mu(\{x_2\})) + z)/\mu(\{x_2\}), \ a_1 \leq a_2\}$$

Note that the level curves are continuous on $[0,1]^2$ but have different slopes on the two regions, and that these slopes vary *independently* from 0 to $-\infty$ when $\mu(\{x_1\})$ and $\mu(\{x_2\})$ vary from 0 to 1. Equality of the two slopes implies $\mu(\{x_1\}) + \mu(\{x_2\}) = 1$, i.e. additivity, while symmetry with respect to the line $a_1 = a_2$ implies $\mu(\{x_1\}) = \mu(\{x_2\})$ (commutativity).

As any increasing connective has level curves with negative slope, we can state the following result:

**Theorem 6** *Any increasing continuous operator on $[0,1]^2$ of which level curves are straight lines of constant slope $k_1, k_2$ on each region $\{a_1 \geq a_2\}$, $\{a_1 \leq a_2\}$, is strongly equivalent to a Choquet integral $\mathcal{C}_\mu$ with $\mu(\{x_1\}) = k_1/(k_1-1)$, $\mu(\{x_2\}) = 1/(1-k_2)$.*

Weighted averages are trivial examples of such operators. More interestingly, it can be shown easily that the level curves of the OWA operators of Yager [17] have the requested properties so that theorem 6 applies. This means that, as far as the ranking of alternatives is concerned, Choquet integrals are more general than OWA operators.

## 4.2 equivalence classes of fuzzy t-conorm integrals

As above, we take $n = 2$ for simplicity. With the same notations as above, it follows that the level curves are defined by:

$$\mathcal{F}_\mu^{-1}(z) = \{(a_1, a_2) | a_2 = h^{-1}\left[\frac{h(z) - h(a_1) g \circ \mu(\{x_1\})}{1 - g \circ \mu(\{x_1\})}\right], a_1 \geq a_2\}$$

$$\mathcal{F}_\mu^{-1}(z) = \{(a_1, a_2) | a_2 = h^{-1}\left[\frac{h(z) - h(a_1)(1 - g \circ \mu(\{x_2\}))}{g \circ \mu(\{x_2\})}\right], a_1 \leq a_2\}$$

In fact, we can show that:

**Theorem 7** *Let $\Delta$ be a strict t-conorm. Then there exists a unique fuzzy measure $\mu$ such that $\Delta$ and $\mathcal{F}_\mu$ are strongly equivalent, with $\mathcal{F} = (\Delta, \perp)$, and $\perp$ is any nilpotent t-conorm. $\mu$ is defined by $\mu(\{x_1\}) = \mu(\{x_2\}) = g^{-1}(1/2)$.*

**Proof** first we characterize the level curves of $\Delta$. Using the generation function $h$, we see that $a_1 \Delta a_2 = z$ implies $h(a_1) + h(a_2) = h(z)$, because $\Delta$ is strict, so that the level curves are simply

$$\mathcal{H}_\Delta(z) = \{(a_1, a_2) | a_2 = h^{-1}(h(z) - h(a_1))\}$$

The problem consists in finding if we can equate $\mathcal{H}_{\mathcal{F}_\mu}(z)$ and $\mathcal{H}_\Delta(z')$ for some $(z, z')$. Consider $a_2 \geq a_1$. We have for a given $z$:

$$a_2 = h^{-1}\left[\frac{g \circ \mu(\{x_1\})}{1 - g \circ \mu(\{x_1\})}(h(K) - h(a_1))\right]$$

with $h(K) = h(z)/g \circ \mu(\{x_1\})$ (such a $K$ always exists since $h$ is strictly increasing). By identification we get $g \circ \mu(\{x_1\})/(1 - g \circ \mu(\{x_1\})) = 1$, so that $g \circ \mu(\{x_1\}) = 1/2$, and $K = z'$, i.e. $z' = h^{-1}(2h(z))$. Now since t-conorms are commutative, we must have $\mu(\{x_2\}) = \mu(\{x_1\})$. □

The theorem can be extended to nilpotent t-conorms, but only a weak equivalence is obtained (see beginning of §4).

Let us further consider the case of nilpotent t-conorms. For each nilpotent t-conorm $\Delta$, we can define an associated t-norm $\nabla$, called its $\Delta$-dual by the relation $a \nabla b = 1 -_\Delta [(1 -_\Delta a) \Delta (1 -_\Delta b)]$. Using the generation function $h$, we can show that:

$$a \nabla b = h^{-1}[0 \vee (h(x) + h(y) - h(1))]$$

and the reader can verify that it is indeed a t-norm. Then we can prove the following:

**Theorem 8** *Let $(\Delta, \perp)$ be two nilpotent t-conorms defining a fuzzy t-conorm integral $\mathcal{F}_\mu$. Then for the fuzzy measure $\mu$ defined by $\mu(\{x_1\}) = \mu(\{x_2\}) = g^{-1}(1/2)$, $\mathcal{F}_\mu$ is weakly equivalent to $\nabla$, the $\Delta$-dual t-norm of $\Delta$.*

**Proof** first remark that if two operators $\mathcal{H}_1$ and $\mathcal{H}_2$ increasing in each place are weakly equivalent, then for any nilpotent t-conorm $\Delta$, their $\Delta$-duals $1 -_\Delta \mathcal{H}_i(1 -_\Delta \cdot, 1 -_\Delta \cdot)$ are also weakly equivalent. This comes from the fact that if $a < b$ then $1 -_\Delta a > 1 -_\Delta b$.

Then applying this result to $\Delta$ and $\mathcal{F}_\mu$, and using property 9 and theorem 7, the theorem is proved. □

# 5 Approximation of connectives by the Choquet integral

We now turn to the approximation problem. Consider the unit hypercube $[0,1]^n$. Conditions such $a_1 \leq a_2 \leq \cdots \leq a_n$ and all its permutations lead to divide the hypercube into $n!$ regions, in which the Choquet integral is linear. Using the definition, remark that all the coefficients involved in the linear equations are given by the value of the Choquet integral on the vertices of the hypercube:

$$\begin{aligned}
\mathcal{C}_\mu(0,1,1,\ldots,1) &= \mu(\{a_2,\ldots,a_n\}) \\
\mathcal{C}_\mu(0,0,1,\ldots,1) &= \mu(\{a_3,\ldots,a_n\}) \\
&\cdots \\
\mathcal{C}_\mu(0,0,\ldots,0,1) &= \mu(\{a_n\})
\end{aligned}$$

and similarly for all the $2^n$ vertices.

Now consider a connective $\mathcal{H}$ defined on the unit hypercube, and increasing in each place. Then substituting in the above system $\mathcal{C}_\mu$ by $\mathcal{H}$, we define a unique Choquet integral which coincide with $\mathcal{H}$ on all the vertices of the hypercube (the fuzzy measure is well defined since $\mathcal{H}$ is increasing in each place). Thus:

**Theorem 9** *For any connective $\mathcal{H}$ on the unit hypercube $[0,1]^n$ being increasing in each place, there exists a unique fuzzy measure $\mu$ such that the Choquet integral with respect to $\mu$ coincide with $\mathcal{H}$ on the vertices of the hypercube.*

This means that the Choquet integral provides a "regionwise" linear approximation to any increasing operator. Further investigation must be done in order to characterize the precision of this approximation. In particular, when $n \to \infty$, it is known that for the weak topology, the vertices become dense in the hypercube [1]: this could be a useful result for our purpose, but needs careful consideration.

# 6 Conclusion

We have shown some results which clarify the use of fuzzy integral as a fuzzy connective. Particularly, we have exhibited the class of fuzzy integrals which are averaging operators, medians and symmetric summations.

In a second part, we have shown that the equivalence class of the Choquet integral contains all continuous operators of which level surfaces are hyperplanes in each of the regions $\{a_1 \geq a_2 \geq \cdots \geq a_n\}$ and all permutations.

Similarly, it was found that every t-conorm has its equivalent fuzzy integral. The result holds also for a more restricted class of t-norms. This means that the class of fuzzy integrals is much more general than t-norms and t-conorms.

These results, together with an approximation theorem shown in the last part of this paper, bring a new justification of the use of fuzzy integral in multicriteria decision making.

# References


[1] J.P. Aubin, *Mathematical methods of economics and game theory*, North Holland.

[2] G. Banon, Distinction between several subsets of fuzzy measures, *Fuzzy Sets and Systems* 5 (1981), 291-305.

[3] D. Dubois, H. Prade, Criteria aggregation and ranking of alternatives in the framework of fuzzy set theory, in *Fuzzy Sets and Decision Analysis*, Zimmermann, Zadeh, Gaines (eds), Elsevier Sc. Publ., 1984, 209-240.

[4] M. Grabisch, M. Yoneda, S. Fukami, Subjective evaluation by fuzzy integral: the crisp and possibilistic case, *Int. Fuzzy Engineering Symposium*, Yokohama, Japan, November 13-15th, 1991.

[5] M. Grabisch, M. Sugeno, Multi-attribute classification using fuzzy integral, *IEEE Int. Conf. Fuzzy Systems*, San Diego, CA, march 8-12, 1992, 47-54.

[6] M. Grabisch, T. Murofushi, M. Sugeno, Fuzzy Measure of Fuzzy Events Defined by Fuzzy Integrals, to appear in *Fuzzy Sets & Systems*.

[7] H. Ichihashi, H. Tanaka, K. Asai, Fuzzy integrals based on pseudo-addition and multiplication, *J. Math. Anal. Appl.* 130 (1988), 354-364.

[8] K. Ishii, M. Sugeno, A Model of Human Evaluation Process Using Fuzzy Measure, *Int. J. Man-Machine Studies* 22 (1985), 19-38.

[9] A. Kandel, W.J. Byatt, Fuzzy sets, fuzzy algebra, and fuzzy statistics, *Proc. of the IEEE* 66 (1978), 1619-1639.

[10] B. Kosko, Fuzzy systems as universal approximators, *IEEE Int. Conf. Fuzzy Systems*, San Diego, CA, march 8-12, 1992, 1153-1162.

[11] R. Kruse, Fuzzy integrals and conditional fuzzy measures,*Fuzzy Sets & Systems* 10 (1983), 309-313.

[12] T. Murofushi, M. Sugeno, An interpretation of fuzzy measure and the Choquet integral as an integral with respect to a fuzzy measure, *Fuzzy Sets & Systems* 29 (1989), 201-227.

[13] T. Murofushi, M. Sugeno, Fuzzy t-conorm integrals with respect to fuzzy measures: generalization of Sugeno integral and Choquet integral, *Fuzzy Sets & Systems* 42 (1991), 57-71.

[14] M. Sugeno, *Theory of fuzzy integrals and its applications*, Doct. Thesis, Tokyo Institute of Technology (1974).

[15] M. Sugeno, Fuzzy measures and fuzzy integrals — A survey, in *Fuzzy Automata and Decision Processes*, Gupta, Saridis, Gaines (eds), (1977), 89-102

[16] K. Tanaka, M. Sugeno, A Study on Subjective Evaluation of Color Printing Images, *Int. J. of Approximate Reasoning* 5 (1991), 213-222.

[17] R.R. Yager, Connectives and quantifiers in fuzzy sets, *Fuzzy Sets & Systems*, 40 (1991), 39-75.

[18] L.X. Wang, Fuzzy systems are universal approximators, *IEEE Int. Conf. Fuzzy Systems*, San Diego, CA, march 8-12, 1992, 1163-1169.

[19] S. Weber, $\perp$-decomposable measures and integrals for Archimedean t-conorms $\perp$, *J. Math. Anal. Appl.* 101 (1984), 114-138.

# Fuzzy Implication and Compatibility Modification

Valerie Cross and Thomas Sudkamp
Department of Computer Science
Wright State University
Dayton, Ohio 45435

***Abstract*— Compatibility modification (CM) fuzzy inference separates the evaluation of the antecedent of a rule from the generation of the output. The determination of the degree to which the input matches the antecedent is determined by a compatibility measure and an aggregation operator. The order in which these operations occur changes the set of applicable rules. Separability conditions are introduced to define circumstances in which rule evaluation is independent of the input evaluation strategy. It is shown that compatibility modification inference using fuzzy partial matching and Minkowski dissimilarity satisfy several separability conditions.**

## I. Introduction

Inference in fuzzy reasoning systems generally employs one of two predominant techniques: the compositional rule of inference or compatibility modification inference. Composition originated as a generalization of binary logical deduction to fuzzy logic. The compositional rule of inference analyzes the input and creates the output in a single step. Compatibility modification (CM) has been employed in fuzzy rule based systems to separate the evaluation of the antecedent of a rule from the generation of the output.

In CM inference, a compatibility measure and an aggregation operator are used to determine the degree to which the input matches the antecedent of a rule. The order in which these operations are performed may produce different degrees of agreement. When the compatibility measures and the aggregation operator satisfy a property called *aggregation separability*, the same result is obtained regardless of the order of the operations involved in the measurement of compatibility. This paper examines the separability of several common families of CM inference techniques.

The examination of fuzzy inference begins with a brief review of compatibility modification inference. This is followed by the introduction and analysis of aggregration separability. Finally, it is shown that CM inference based on compatibility and that based on dissimilarity have distinct separability properties. Throughout this paper, $A$, $B$, and $C$ will represent fuzzy sets over $U = \{u_1, \ldots, u_n\}$, $V = \{v_1, \ldots, v_m\}$, and $W = \{w_1, \ldots, w_p\}$, respectively.

## II. Compatibility-Modification Inference

Compatibility-modification (CM) inference first determines the degree to which the input matches the antecedent of a rule and then modifies the consequent of the rule accordingly. Dubois and Prade [2,6] have distinguished CM from compositional inference by labeling CM plausible reasoning. Plausible reasoning may be considered to follow the pattern

$X$ is $A'$
If $X$ is $A$ then $Y$ is $C$
$A'$ is $\gamma$ compatible with (similar to) $A$
---
$C'$ is $\gamma$ compatible with $C$

There are two distinct operations employed in CM inference, a compatibility measure and a consequent modification operation. The modification operation constructs the fuzzy set $C'$ that is $\gamma$ compatible with $C$. The term CM inference, which has also been called approximate analogical reasoning [8], is used to encompass all inference techniques that separate measurement of the compatibility of the antecedent from the modification of the consequent fuzzy set.

Zadeh's fuzzy interpolation [11] provides an example of CM inference. The compatibility of the input $A'$ and the antecedent $A$ is determined by a sup-min comparison of the two fuzzy sets

$$\gamma = \sup_{u \in U} \min(\mu_{A'}(u), \mu_A(u)).$$

The resulting compatibility measure $\gamma$ is applied to the consequent fuzzy set $C$ to produce the output fuzzy set

0-7803-0614-7/93$03.00 ©1993IEEE

$C'$ defined by the membership function

$$\mu_{C'}(w) = \min(\gamma, \mu_C(w)).$$

Thus Zadeh's interpolation is a CM inference technique that uses sup-min as the compatibility measure and the T-norm min modifies the output. The preceding inference system can be generalized by replacing min in the compatibility measurement with an arbitrary T-norm. This type of inference will be referred to as sup-T CM.

Yager [9] formalized CM inference based on a dissimilarity measure $J$ and an implication operator $I$. Following Yager's approach, the dissimilarity of $A$ and $A'$ is used to modify $C'$ by $\mu'_C(w) = I(J(A, A'), \mu_C(w))$. Soula [7] proposed the Hamming distance between $A$ and $A'$ and Dubois and Prade [2] suggested the Hausdorff metric for measuring the dissimilarity of fuzzy sets for CM inference.

Magrez and Smets [5] introduced a CM technique that measures the compatibility of $A$ and $A'$ as the necessity of $A$ knowing that $A'$ is true [3],

$$\begin{aligned} \delta &= \sup \min(1 - \mu_A(u), \mu_{A'}(u)) \\ &= \inf \max(\mu_A(u), 1 - \mu_{A'}(u)). \qquad (1) \end{aligned}$$

The compatibility measure is subtracted from 1 to produce a dissimilarity meaure $1 - \delta$, which is then added (using bounded sum addition) to the membership value of each element in the consequent $C$,

$$\mu_{C'}(w) = \min(1, \mu_C(w) + 1 - \delta). \qquad (2)$$

Thus this inference technique falls within the class of CM inference described by Yager. Adding the dissimilarity measure to the fuzzy consequent is equivalent to applying the Lukaciewicz implication operator between the compatibility measure $\delta$ and the fuzzy set $C$, i.e., $I_L(\delta, \mu_C(w)) = \min(1, \mu_C(w) + 1 - \delta)$.

Dissimilarity and compatibility measures are often considered to be interchangable since a compatibility measure may be constructed from dissimilarity and vice versa. Given a dissimilarity measure $J$ whose maximal value is 1, then $1 - J(A, A')$ is a measure of the similarity (compatibility) of $A$ and $A'$. It is shown in Section 4 that in CM inference analyzing the degree of match of the antecedent and the input via compatibility or dissimilarity affects the rules that will be processed.

## III. Properties of CM Inference

CM inference permits flexibility in processing rules in which the antecedent has multiple clauses. In this section, the two distinct approaches to CM inference, referred to as *aggregation-compatibility* and *compatibility-aggregation*, are examined. The name assigned to each technique describes the order of the occurrence of the operations that are used to determine the compatibility of the antecedent with the input data. There is no distinction between these techniques when the antecedent of a rule consists of a single linguistic term. Throughout this analysis, we will consider inference using a rule of the form 'If $X$ is $A$ and $Y$ is $B$ then $Z$ is $C$' with input $A'$ and $B'$. A property of the order of the determination of compatibility, *aggregation separability*, is defined based on the relationship between these two techniques.

Upon the receipt of input in a fuzzy inference system using compositional inference, an output $C'$ is constructed for every rule in the rule-base since the evaluation of the input and the construction of the output are combined in a single step. In CM inference, however, these two processes are decoupled. Thus, it is possible to limit the generation of the output fuzzy sets to rules whose antecedents match the input to a high degree. Following a strategy often employed in rule-base expert systems a threshold, $\tau$, $0 \leq \tau \leq 1$, is assigned to each rule. This value represents the degree of agreement required between the antecedent of the rule and the input in order to process the consequent (fire the rule). The possibility of not firing a rule adds significance to the manner in which the compatibility of the input is measured.

### A. Aggregation-Compatibility Evaluation

A fuzzy relation $R_{A \times B}$ may be used in the definition of the antecedent of a rule such as 'If $(X, Y)$ is $R_{A \times B}$ then $Z$ is $C$'. The fuzzy relation $R_{A \times B}$ is a fuzzy set whose members are tuples of the form $(u_i, v_j)$ where $u_i \in U$ and $v_i \in V$. For example, let $R_{A \times B}$ be the fuzzy set LARGE that describes the size of a carpet [10]. The fuzzy relation LARGE consists of tuples whose first components are widths and whose second components are lengths. The membership degree of each tuple represents the degree to which these values are perceived to describe a large carpet. The antecedent of a rule describing the size of a carpet may have the form "If CARPET-SIZE is LARGE".

One approach to the construction of a fuzzy relation $R_{A \times B}$ uses fuzzy sets $A$ and $B$ over the constituent domains. The membership value for $(u_i, v_j)$ is obtained using a T-norm, $R_{A \times B}(u_i, v_j) = T(u_i, v_j)$ (generally the constituent clauses are assumed to be non-interactive and the T-norm min is used). Observations $A'$ and $B'$ are combined in the same manner to produce the relation $R_{A' \times B'}(u_i, v_j) = T(\mu_{A'}(u_i), \mu_{B'}(v_j))$. In some applications, however, the entire relation might be constructed directly without employing separate fuzzy sets.

Regardless of how the rule antecedent and input fuzzy relations are created, a compatibility measure $C$ is used to determine the compatibility $C(R_{A \times B}, R_{A' \times B'})$ of the input with the relation. As before, the compatibility is then used to modify the consequent. Aggregation-compatibility CM inference first combines the clauses in the antecedent to produce a relation over the Cartesian product of the domains and then compares the input with this relation. Letting $\cap$ be the aggregation operation and $C$ the compatibility measure, aggregation-compatibility can be considered to have the form

$$C(A \cap B, A' \cap B')$$

where $A'$ and $B'$ are the input fuzzy sets.

*B. Compatibility-Aggregation Evaluation*

When the antecedent of a rule is specified as a relation $R_{A\times B}$ based on fuzzy sets $A$ and $B$, it is often a translation of the rule antecedent from 'If $X$ is $A$ and $Y$ is $B$' to 'If $(X, Y)$ is $R_{A\times B}$'. Continuing with the preceding example, the antecedent 'If CARPET-SIZE is LARGE' might also be written as 'If CARPET-WIDTH is WIDE and CARPET-LENGTH is LONG'. The historical motivation for the translation of the fuzzy sets in the antecedent into a fuzzy relation is the suitability of the latter format for compositional inference.

In CM inference, there is no specific requirement for the construction of the relation $R_{A\times B}$. When the observations '$X$ is $A'$' and '$Y$ is $B'$' are input, the compatibility between $A$ and $A'$, $C(A, A')$, and the compatibility between $B$ and $B'$, $C(B, B')$, are determined. The two compatibilities are combined using an aggregation operator $\oplus$ to define the overall compatibility between the rule's antecedent and the fuzzy input. Thus the general form of compatibility-aggregation CM inference may be written as $\oplus(C(A, A'), C(B, B'))$.

*C. Aggregation Separability*

The order of the operations has a significant effect on the efficiency of evaluating the antecedent of a rule. In the compatibility-aggregation approach, measuring compatibility requires $n + m$ operations (recall that $n$ and $m$ are the cardinalities of $U$ and $V$, respectively). A single application of the aggregation operator combines the two compatibility measures. In the aggregation-compatibility method, if the fuzzy input is not already provided as a fuzzy relation, producing the input fuzzy relation requires $nm$ operations and measuring the compatibility also requires $nm$ operations. Thus, it is advantageous to compute the compatibility of the antecedents independently. This section examines the consequences of the order of compatibility evaluation on the processing of rules.

It has been suggested [1] that a desirable property of a compatibility measure is

$$C(A \cap B, A' \cap B') \geq \oplus(C(A, A'), C(B, B')) \qquad (3)$$

where $\oplus$ is the aggregation operator. The satisfaction of (3) guarantees that whenever the compatibility-aggregation evaluation of the input exceeds the threshold, the aggregation-compatibility would also. However, there may be input $A'$ and $B'$ for which $C(A \cap B, A' \cap B') \geq \tau > \oplus(C(A, A'), C(B, B'))$, creating a situation where the aggregation-compatibility method would cause the rule's consequent to be processed but not the compatibility-aggregation method. When (3) is satisfied, Bilgic and Turksen [1] propose a two step procedure for firing rules: if $\oplus(C(A, A'), C(B, B')) \geq \tau$, the consequent is automatically processed; otherwise, $C(A \cap B, A' \cap B')$ is computed to decide whether to process the consequent.

The opposite inequality,

$$C(A \cap B, A' \cap B') \leq \oplus(C(A, A'), C(B, B')), \qquad (4)$$

is briefly mentioned in [1], but no complete description of its usefulness is provided. This property might be considered desirable since it would guarantee that whenever the compatibility-aggregation method does not fire a rule, the aggregation-compatibility method would not either. The satisfaction of (4), however, does not eliminate situations where the threshold $\tau$ falls between $C(A \cap B, A' \cap B')$ and $\oplus(C(A, A'), C(B, B'))$. In this case, the compatibility-aggregation evaluation would fire the rule but aggregation-compatibility would not. If a compatibility measure that satisfies (4) is selected, the algorithm for processing rules could test if $\oplus(C(A, A'), C(B, B')) < \tau$, then the consequent is automatically not processed; otherwise, $C(A \cap B, A' \cap B')$ must be computed to decide if the consequent is processed.

The property of compatibility measures specified by (3) is termed *greater than* aggregation separability and that specified by (4) is termed *less than* aggregation separability. A family of compatibility measures based on fuzzy partial matching is shown to satisfy both properties. Another family based on the Minkowski $r$-metrics is shown to satisfy a variant of greater than separability for dissimilarity measures.

First we consider one of the most popular family of compatibility measures, the consistency or partial matching indices [3]

$$PM(A, A') = \sup_i T(a_i, a'_i) \qquad (5)$$

where $T$ is any T-norm. $PM$ compatibility is that used in sup-$T$ CM. Proposition 1 establishes conditions under which PM compatibility has both greater than and less than separability.

**Proposition 1.** Let PM be a partial matching index defined by T-norm $T$. Then

$$PM(A \cap B, A' \cap B') = T(PM(A, A'), PM(B, B')) \qquad (6)$$

when the aggregation operator $\cap$ is $T$.

**Proof**: For simplicity, we let $a_i = \mu_A(u_i)$. Rewriting the left side of (6) by substituting the definition of PM yields

$$\begin{aligned} PM(A \cap B, A' \cap B') &= \sup_{ij} T(T(a_i, b_j), T(a_i', b_j')) \\ &= \sup_{ij} T(T(a_i, a_i'), T(b_j, b_j')) \end{aligned}$$

The final step follows from the associativity of T-norms.

Now let $k$ be the subscript for which $\sup_i T(a_i, a_i') = T(a_k, a_k\prime)$, that is, $T(a_k, a_k') \geq T(a_i, a_i')$ for $1 \leq i \leq n$. Similarly, let $\sup_j T(b_j, b_j') = T(b_l, b_l')$. By the monotonicity of T-norms,

$$T(T(a_k, a_k'), T(b_l, b_l')) \geq T(T(a_i, a_i'), T(b_j, b_j')),.$$

for $1 \leq i \leq n$ and $1 \leq j \leq m$.
The right side of (6) may be rewritten as $T(PM(A, A'), PM(B, B'))$

$$\begin{aligned} &= T(\sup_i T(a_i, a_i'), \sup_j T(b_j, b_j')) \\ &= T(T(a_k, a_k'), T(b_l, b_l')) \\ &= \sup_{ij} T(T(a_i, a_i'), T(b_j, b_j')). \end{aligned}$$

Both the left side and the right side of (6) are equal to $\sup_{ij} T(T(a_i, a_i'), T(b_j, b_j'))$. Thus, (6) holds when the same T-norm is used consistently throughout and, under these conditions, *PM* compatibility measure possesses both greater than and less than aggregation separability. □

It is reasonable to assume that the same T-norm should be used for both the $T$ and the $\cap$ in (6) since this T-norm represents a logical aggregation of the antecedent clauses, the only difference is when the aggregation is performed. On the left side of (6), the aggregation is performed before the compatibility measure is taken; the compatibility measure is between the complete antecedents. On the right side, the aggregation is done after the compatibility measure is taken for each clause in the antecedent; the aggregation is over the compatibility measures for each clause in the antecedent.

The assumption of the same T-norm for $T$ and $\cap$ in (6), however, does not mandate that the T-norm in the definition of a *PM* compatibility measure must be the same T-norm as used for the aggregation. This flexibility permits the definition of the compatibility measure to be independent of the definition for the logical aggregation represented in the rule's antecedent. For example, it might be useful to employ the same compatibility measure for all rules but vary the aggregation operator used on the clauses in the rule's antecedent, or vice versa.

| $i$ | $A$ | $B$ | $A'$ | $B'$ |
|---|---|---|---|---|
| 1 | 0.7 | 0.0 | 0.0 | 0.6 |
| 2 | 1.0 | 0.5 | 0.3 | 1.0 |
| 3 | 0.8 | 1.0 | 1.0 | 0.7 |
| 4 | 0.0 | 0.5 | 0.4 | 0.0 |

Table 1:

| $i$ | $T_3(a_i, a_i')$ | $T_3(b_j, b_j')$ | $T_2(a_i, a_i')$ | $T_2(b_j, b_j')$ |
|---|---|---|---|---|
| 1 | 0.0 | 0.0 | 0.0 | 0.0 |
| 2 | 0.3 | 0.5 | 0.3 | 0.5 |
| 3 | 0.8 | 0.7 | 0.8 | 0.7 |
| 4 | 0.0 | 0.0 | 0.0 | 0.0 |
| sup | 0.8 | 0.7 | 0.8 | 0.7 |

Table 2:

Pursuing this latter possibility, we show that (6) is not guaranteed when the T-norm in the *PM* compatibility measure (5) differs from the one used for both the $\cap$ and $T$ in (6). Choosing $T_2$ to be the aggregation operator and sup-$T_3$ as the compatibility measure produces

$$\sup_{ij} T_3[T_2(a_i', b_j'), T_2(a_i, b_j)] \tag{7}$$

for the left side of (6) and

$$T_2[\sup_i T_3(a_i', a_i), \sup_j T_3(b_j', b_j)] \tag{8}$$

for the right side of (6).

The associativity of a fixed T-norm was essential in the proof of Proposition 1. Since the T-norms in (7) and (8) are different, the preceding argument does not carry over to this case. A simple counterexample is constructed to show that (7) and (8) are not equivalent.

Tables 1 and 2 gives the values of fuzzy sets $A$, $B$, $A'$, and $B'$ and the pairwise $T_2$ and $T_3$ combinations of $A$ and $A'$ and $B$ and $B'$. The value of (8) is $T_2(0.8, 0.7) = 0.56$. For this assignment input $A'$ and $B'$, the value of (7) is $\sup_{ij} T_3[T_2(a_i', b_j'), T_2(a_i, b_j)] = 0.7$.

Interchanging the roles of $T_3$ and $T_2$ in (7) and (8) produces

$$\sup_{ij} T_2[T_3(a_i', b_j'), T_3(a_i, b_j)] = 0.56 \tag{9}$$

and

$$T_3[\sup_i T_2(a_i', a_i), \sup_j T_2(b_j', b_j)] = 0.7, \tag{10}$$

respectively. In this case (6) still is not satisfied. This example shows one pair of selections, $T_3$ ($T_2$) for the $T$ in

| Element | $A$ | $B$ | $A'$ | $B'$ |
|---|---|---|---|---|
| 1 | 1.0 | 1.0 | 1.0 | 0.5 |
| 2 | 1.0 | 1.0 | 1.0 | 0.5 |

Table 3:

$PM$ and $T_2$ ($T_3$) for both the $T$ and $\cap$ for which (6) does not hold.

Though not presented in this manner, the property of aggregation separability is addressed in [4]. Lee is concerned with determining whether performing sup-$T$ CM inference with the aggregation of the clauses produces the same result as performing a separate sup-$T$ CM for each clause and then aggregating the results. In this case, the analysis is not restricted to the antecedent but carries through to the generation of output. The relationship between the two techniques can be illustrated by considering a rule 'If $X$ is $A$ and $Y$ is $B$ then $Z$ is $C$' where $C$ is precise, i.e., $k = 1$ and $c_1 = 1.0$. Under these conditions, sup-$T$ CM is equivalent to determining the PM compatibility (5). Lemmas 2 and 2′ in [4] suggest that (6) should hold for T-norms $T_2$ and $T_3$. A counterexample to this assertion has been provided above. The results stated in [4] hold only when the same T-norm is used throughout the expressions.

## IV. Compatibility vs Dissimilarity

The class of CM inference techniques defined by Yager (see Section 2) uses dissimilarity rather than compatibility to measure the degree of matching between the antecedent and the input. The dissimilarity formulation of greater than aggregation separability is

$$diss(A, A') \oplus diss(B, B') \geq diss(A \cap B, A' \cap B') \quad (11)$$

where $diss$ is a dissimilarity measure and $\oplus$ is an aggregator. In a domain using dissimilarity measures, the firing of the rule might be defined to occur when a dissimilarity measure is less than a threshold $\tau$. The satisfaction of 11 guarantees than whenever the dissimilarity-aggregation evaluation of the input does not exceed the threshold, the aggregation-dissimilarity would not either.

Metric functions provide a dissimilarity measure of fuzzy sets. Fuzzy sets $A$ and $A'$ over a universe of cardinality $n$ are considered as points $[\mu_A(u_1), \ldots, \mu_A(u_n)]$ and $[\mu_{A'}(u_1), \ldots, \mu_{A'}(u_n)]$ in $\mathbf{R}^n$. The normalized Minkowski $r$-metric is defined by

$$d_r/n^{\frac{1}{r}}(A, A') = \frac{(\sum_{i=1}^{n} |\mu_A(u_i) - \mu_{A'}(u_i)|^r)^{1/r}}{n^{\frac{1}{r}}}, \quad (12)$$

with $r \geq 1$. It can be shown that inference using a dissimilarity measure derived from a normalized Minkowski distance and addition for $\oplus$ satisfies (11).

The compatibility measure associated with a Minkowski dissimilarity measure is $1 - d_r/n^{\frac{1}{r}}$. Substituting this compatibility measure and addition for $\oplus$ into the definition of greater than aggregation separability(3) produces

$$1 - \frac{d_r(A \cap B, A' \cap B')}{(nm)^{\frac{1}{r}}} \quad (13)$$

for the left side of (3) and

$$1 - \frac{d_r(A, A')}{n^{\frac{1}{r}}} + 1 - \frac{d_r(B, B')}{m^{\frac{1}{r}}} \quad (14)$$

for the right side of (3). The fuzzy sets given in Table 3 show that the compatibility measure obtained from $d_1$ does not satisfy greater than aggregation separability. The expression for the left side evaluates to 0.5 and the right side to 1.5.

The preceding shows that the standard conversion between compatibility measures and dissimilarity measures does not preserve separability. When using a threshold to determine the applicability of rules, the method used to determine the degree of matching may alter the set of rules to be fired.

## V. Conclusion

Compatibility modification inference separates the analysis of the antecedent of a rule from the generation of the output. This separation permits the introduction of a threshold in the determination of the set of rules to be fired upon the acquisition of input. It has been shown, however, that the manner in which the compatibility is measured may affect the selection of the rules. When a compatibility measure satisfies both greater than and less than separability, the order of the evaluation is irrelevant. Satisfying either condition provides information that may reduce the computational requirements needed to determine the applicable rules.

## References


[1] Taner Bilgic and I. B. Turksen, "Effective search methods for pattern matching inferencing using specific similarity measures," in *Proceedings of the IEEE International Conference on Fuzzy Systems*, San Diego, CA, pp. 161–168, 1992.

[2] D. Dubois and H. Prade, "The generalized modus ponens under sup-min composition a theoretical study," in *Approximate Reasoning in Expert Systems*, M. M. Gupta, A. Kandel, W. Bandler, and J. Kiszka, Eds. Amsterdam:North-Holland, 1985, pp. 217–232.

[3] D. Dubois and H. Prade, "A unifying view of comparison indices in a fuzzy set-theoretic framework," in *Fuzzy Set and Possibility Theory Recent Developments*, Ronald R. Yager, Ed. New York:Pergamon Press, 1982, pp. 3–13.

[4] C. C. Lee, "Fuzzy logic in control systems: fuzzy logic controller (Part I and II)," *IEEE Trans. on Systems, Man, and Cybernetics*, vol. 20, no. 2, pp. 404–432, 1989.

[5] P. Magrez and P. Smets, "Fuzzy modus ponens: A new model suitable for applications in knowledge-based systems," *Int. J. Intelligent Systems*, vol. 4, pp. 181–200, 1989.

[6] H. Prade, "Approximate and plausible reasoning," in *Fuzzy Information, Knowledge Representation, and Decision Analysis*, E. Sanchez, Ed. Oxford:Pergamon Press, 1984.

[7] G. Soula, *Aide a la Decision en Logique Floue: Application en Medecine*. 1981.

[8] I. B. Turksen, "An analogical approximate reasoning based on similarity measures," *IEEE Trans. on Systems, Man and Cybernetics*, vol. 18, pp. 1049–1056, 1988.

[9] R. R. Yager, "Aspects of Possibilistic Uncertainty", *Intern. J. Man-Machine Stud.*, vol. 12, pp. 283–298, 1980.

[10] L. A. Zadeh, "Fuzzy sets as a basis for a theory of possibility," *Fuzzy Sets and Systems*, vol. 1, pp. 3–28, 1978.

[11] L. A. Zadeh, "The role of fuzzy logic in the management of uncertainty in expert systems," *Fuzzy Sets and Systems*, vol. 11, pp. 199–227, 1983.

# Conjunction and disjunction with synergistic effect

Yasunori Kato, Thierry Arnould, Tsutomu Miyoshi and Shun'ichi Tano
Laboratory for International Fuzzy Engineering Research
Siber Hegner Bldg. 4Fl., 89-1 Yamashita-Cho, Naka-Ku,
Yokohama 231 Japan

***Abstract*** - **This paper presents a method for forming combination operators that can account for synergism between the items that make up the conditions of a rule. When the condition part of a rule contains several conditions, the global truth value of the rule is given by the combination of the truth values of all the conditions. Usual combination operators, however, cannot be used to express synergistic relations that may exist between the different items in the conditions. This paper therefore proposes a method for defining new combination operators that take those relations into account.**

***Key-words*** - **fuzzy inference, fuzzy rule, conjunction, disjunction, synergy, combination operator**

## I. Introduction

At the Laboratory for International Fuzzy Engineering Research, we have developed an expert system called FOREX (FOReign exchange trade support EXpert system), that predicts trends of the exchange rate between the yen and the dollar [1] [2]. Several expert systems for the purpose of predicting such things as trends in the foreign exchange market and stock prices have already been developed. All of these systems base their predictions on numerical input only, and provide nothing more than predictions based on expert knowledge concerned with technical analysis. FOREX predicts are based not only on numerical data representing economic indicators but also on news items that concern the activities of relevant individuals (and that are input into the system in the form of natural language expressions).

When we developed this system, we found that economic knowledge representing relations between economic items can be expressed by linguistic expressions such as : "IF FF-rate becomes high, then short-term-interest will be high." This expression of economic knowledge contains vague natural language, so to express this kind of knowledge we used fuzzy production rules. But because of the gradual nature of the relation between the condition part of the rule and the conclusion part of the rule, and because of the possible synergism between different items of the condition part, the usual fuzzy inference methods were not suitable.

0-7803-0614-7/93$03.00 ©1993IEEE

For example, the economic knowledge represented in FOREX by the expression: "IF FF-rate becomes high, then Short-term-interest becomes high" actually means "The higher FF-rate becomes, the higher Short-term-interest becomes." That is, the higher the degree to which items in the condition part satisfy the predicates in the condition part, the higher is the degree to which the result in the conclusion part satisfies the predicate in the conclusion part. This property is called the gradual property. The gradual property can be expressed using the Gradual Equivalence [3] [4]. We can shift the result distributions on the universe of discourse, according to the truth value of the condition part, avoiding an increase of fuzziness.

Another kind of expressions used in FOREX is "the higher FF-rate is and the lower the US-official-rate is, the higher the US-short-term-interest is." When expressions contain several items in the condition part and when each item of the condition part satisfies the condition's predicate to a high degree, then the result in the conclusion part should satisfy the conclusion's predicate to a degree higher than that to which the predicates of the condition part are satisfied. We call this the synergistic effect.

For example, if "FF-rate" and "US-official-rate" have synergistic effect, when the state of "FF-rate" is "almost high" and the state of "US-official-rate" is "almost low", as each item satisfies ones predicate with a rather high grade, each item strengthens the other item's truth value and the value of "US-short-term-interest", that is the conclusion item, will be "very high" instead of "almost high."

When there are several items in the condition part of the rule, the total truth value of the condition part is usually obtained by combination using t-norm operators [5]. But these operators are not suitable for representing the synergistic effect we used in FOREX. This paper proposes a method for expressing the synergistic effect. We propose a new kind of 'and' (resp. 'or') operator denoted by '*and*' (resp. '*or*').

## II. The *and* Operator

The '*and*' operator is a truth value operator used to combine the truth values of the conditions of the rule. So, it can be defined as follows :

$$*and*: [0,1] \times [0,1] \rightarrow [0,1].$$

When all the input truth values are "rather high", (that is, when each item in the condition part satisfies the predicate with a "rather high" degree), the output value of the '*and*' operator must be higher than that of the t-norm operator. For other input values, the '*and*' operator is a usual t-norm operator. Fig. 1 shows the area where the synergistic effect is required.

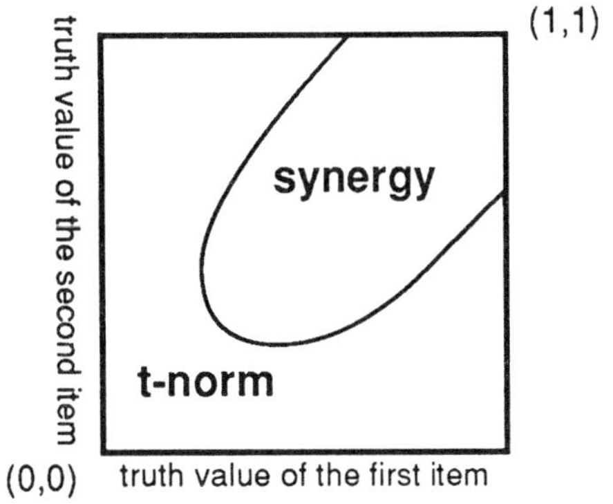


Fig. 1. Area where the synergistic effect is required.

## III. The Construction Algorithm

This section mentions about the algorithm used to define the '*and*' operator. We use three functions *basic function*, *synergy function* and *area definition function*, as follows :

**Basic function**:
basic combination operator (t-norm)
**Synergy function**:
expresses the synergistic property
**Area Definition function**:
defines the area where the synergistic effect is required

Using these functions, the '*and*' operator is defined as the weighted average of the *basic function* and *synergy function* using the *area definition function.* Fig. 2 shows the basic algorithm used to construct the '*and*' operator.

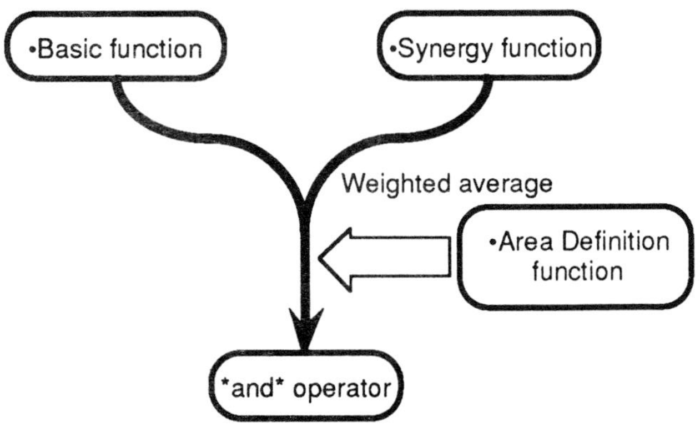


Fig. 2. Structure of the '*and*' operator.

This section describes the construction algorithm and the definition of each function. We first consider the case when there are only two input values. We will extend the operator to the n-variable case later.

*a) The basic function*

When no synergistic effect is required, the '*and*' operator must be equivalent to the usual 'and' operator, which is represented by a t-norm. So, we define the *basic function* as a normal t-norm operator.

$$basic(x,y) = T(x,y) \tag{1}$$

where T(x,y) is a t-norm operator.

*b) The synergy function*

The output value of this function must be greater than the output of the t-norm operator. Whatever the input values, however, the greatest possible output value is equal to 1. We therefore define the *synergy function* as

$$synergy(x,y) = 1. \tag{2}$$

*c) The area definition function*

This function defines this area where the synergistic effect is required. To define this area, we consider the two following properties :

*(i) almost equal*
*(ii) rather high*

*(i) almost equal*

To define the area where the two input truth values are considered to be *almost equal*, we first define, given a real number *a*, a fuzzy number "*almost a*" on the interval of truth values [0,1]. The membership function can be given by one of the following two functions:

$$almost(a,x,\alpha) = \begin{cases} \max\left(\frac{1}{\alpha}(x-a)+1, 0\right) & (if \quad x < a) \\ \max\left(\frac{-1}{\alpha}(x-a-\alpha), 0\right) & (if \quad x \geq a) \end{cases} \tag{3}$$

or

$$almost(a,x,\alpha) = e^{\ln 0.5 \bullet (x-a)^2 / \alpha^2}. \tag{4}$$

Equation (3) leads to a triangular membership function, whose width can be expressed using the parameter $\alpha$. Equation (4) leads to a bell-shaped membership function, and the parameter $\alpha$ indicates the width of the membership function, when the membership grade is equal to 0.5.

The shape of these membership functions are shown in Fig. 3 and Fig. 4.

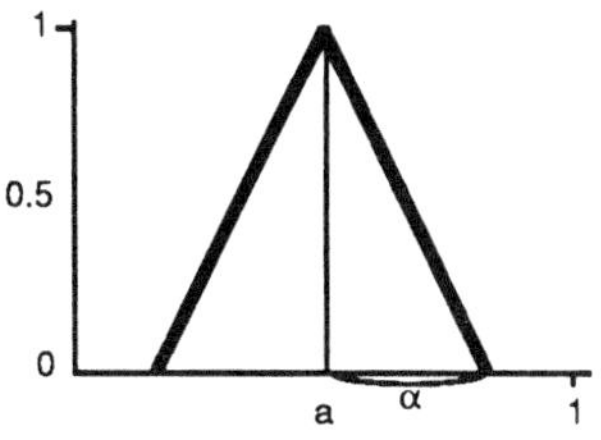


Fig. 3. Triangle membership function.

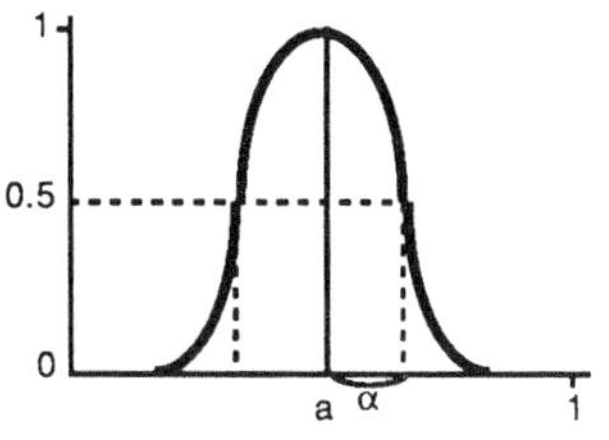


Fig. 4. Bell-shaped membership function.

Using either (3) or (4), we define the *equal* function, that indicates the degree to which two numbers *"almost a"* and *"almost b "* are *almost equal* :

$$equal(a,b) = \sup_x \left( almost(a,x,\alpha) \wedge almost(b,x,\alpha) \right).(5)$$

Fig. 5. shows the graph of this function, using triangular membership functions, when the parameter α is equal to 0.3.

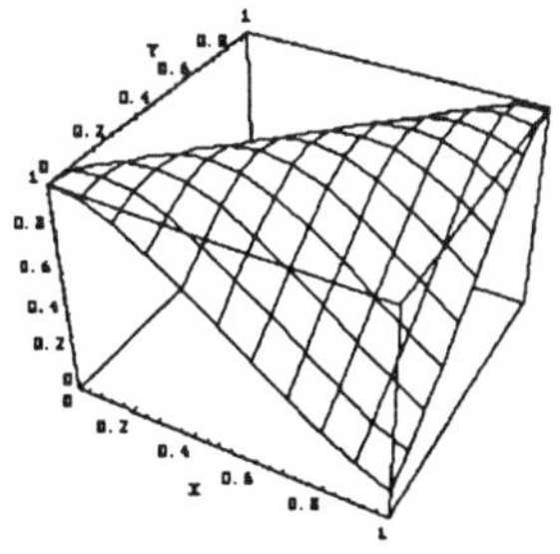

Fig. 5. Graph of equal(x,y).

*(ii) rather high*

To define the area where both the input data are considered to be *rather high*, we define the *high* function as

$$high(x,y) = \frac{\max(\min(x,y) - \beta, 0)}{1-\beta} \qquad (6)$$

where the parameter β is a threshold under which the input data can not be considered as *rather high*. Fig. 6 shows the graph of this function, when the parameter β is equal to 0.3.

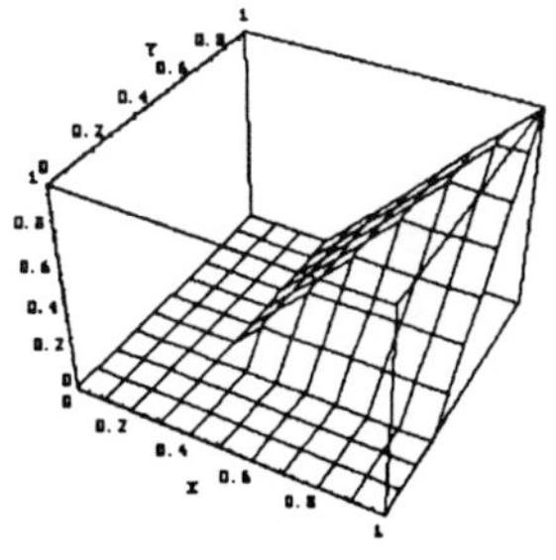

Fig. 6. Graph of high(x,y).

Using these two qualitative functions, we define the area where the synergistic effect will take place by :

$$area(x,y) = equal(x,y) \times high(x,y). \qquad (7)$$

Fig. 7 shows the graph of this function, when α is equal to 0.3 and β is equal to 0.3. So, we can define the *area definition function.*

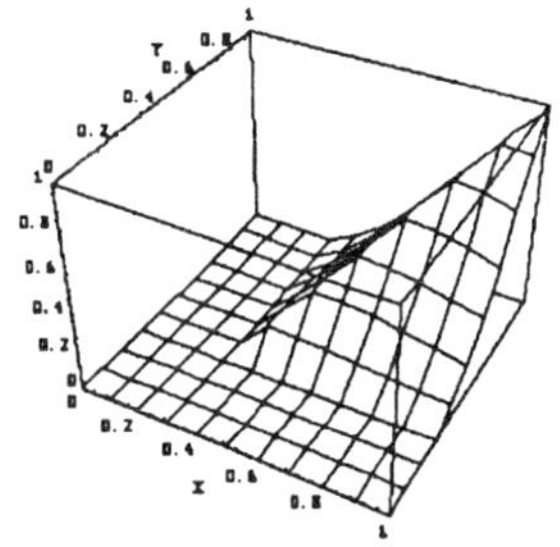

Fig. 7. Graph of area(x,y).

*d) Construction of the '*and*' operator.*

We defined three fundamental functions. Using these functions, the '*and*' operator is defined as the following weighted average of the *basic function* and *synergy function* :

$$*and*(x,y) = w \times synergy(x,y) + (1-w) \times basic(x,y)$$
$$\text{where } w = area(x,y). \qquad (8)$$

For example, when the *basic function* is the algebraic product (Fig. 8),

$$basic(x,y) = T(x,y) = x.y \qquad (9)$$

and when we assign the value 0.3 to the parameters α and β, the resulting '*and*' operator is shown in Fig. 9 and TABLE I.

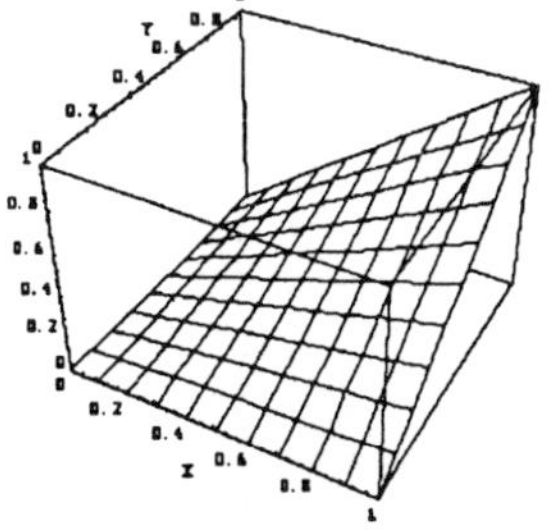

Fig. 8. t-norm operator . (algebraic product)

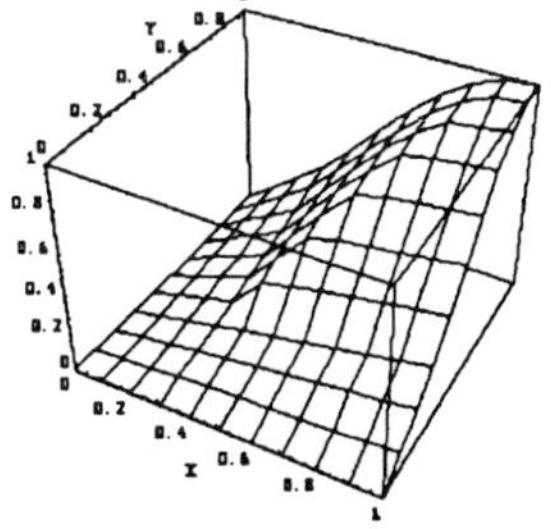

Fig. 9. '*and*' operator.

TABLE I

*and* OPERATOR OUTPUT

| | 0.0 | 0.1 | 0.2 | 0.3 | 0.4 | 0.5 | 0.6 | 0.7 | 0.8 | 0.9 | 1.0 |
|---|---|---|---|---|---|---|---|---|---|---|---|
| 0.0 | 0.000 | 0.000 | 0.000 | 0.000 | 0.000 | 0.000 | 0.000 | 0.000 | 0.000 | 0.000 | 0.000 |
| 0.1 | 0.000 | 0.010 | 0.020 | 0.030 | 0.040 | 0.050 | 0.060 | 0.070 | 0.080 | 0.090 | 0.100 |
| 0.2 | 0.000 | 0.020 | 0.040 | 0.060 | 0.080 | 0.100 | 0.120 | 0.140 | 0.160 | 0.180 | 0.200 |
| 0.3 | 0.000 | 0.030 | 0.060 | 0.090 | 0.120 | 0.150 | 0.180 | 0.210 | 0.240 | 0.270 | 0.300 |
| 0.4 | 0.000 | 0.040 | 0.080 | 0.120 | 0.280 | 0.312 | 0.341 | 0.367 | 0.391 | 0.417 | 0.443 |
| 0.5 | 0.000 | 0.050 | 0.100 | 0.150 | 0.312 | 0.464 | 0.496 | 0.522 | 0.544 | 0.566 | 0.588 |
| 0.6 | 0.000 | 0.060 | 0.120 | 0.180 | 0.341 | 0.496 | 0.634 | 0.664 | 0.686 | 0.706 | 0.726 |
| 0.7 | 0.000 | 0.070 | 0.140 | 0.210 | 0.367 | 0.522 | 0.664 | 0.781 | 0.807 | 0.826 | 0.844 |
| 0.8 | 0.000 | 0.080 | 0.160 | 0.240 | 0.391 | 0.544 | 0.686 | 0.807 | 0.897 | 0.916 | 0.932 |
| 0.9 | 0.000 | 0.090 | 0.180 | 0.270 | 0.417 | 0.566 | 0.706 | 0.826 | 0.916 | 0.973 | 0.984 |
| 1.0 | 0.000 | 0.100 | 0.200 | 0.300 | 0.443 | 0.588 | 0.726 | 0.844 | 0.932 | 0.984 | 1.000 |

## IV. NATURE OF THE PARAMETERS α AND β

We used two parameters α and β to define the *area definition function*. This section mentions about the nature of these parameters.

### A. The Parameter α

The parameter α is used to define the fuzziness of the fuzzy number "*almost a.*" An increase of this parameter increases the fuzziness of the fuzzy number. Thus, if α has a high value, the *area definition function* can no longer be used to express the concept *almost equal* because almost *a* and almost *b* are always *almost equal.*

### B. The Parameter β

The parameter β is the threshold value under which the input data cannot be considered to be *rather high.* An increase of this parameter decreases the area where the synergistic effect is applied. When β=1, the '*and*' operator has no synergistic effect (it is then equivalent to a usual 'and' operator).

Fig. 10 illustrates the property of the parameters α and β : the parameter α expresses the width of the synergistic area, whereas the expresses β indicates the length of the area.

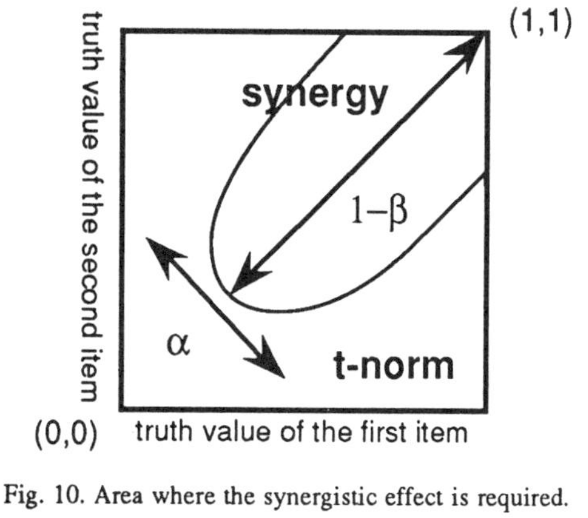


Fig. 10. Area where the synergistic effect is required.

## V. EXTENSIONS

We defined the '*and*' operator in the case of two variables. In this section, we extend the '*and*' operator to the case of n-variables and to the definition of an '*or*' operator.

### A. Extension to n Variables

The functions we used to define the '*and*' operator have to be redefined and extended to n-variable functions. As usual 'and' operators satisfy the associative law, it is easy to extend them as n-variable functions, as follows :

$$and(x,y,z) = T(T(x,y),z) = T(x,T(y,z)) \qquad (10)$$

where T(x,y) is a t-norm operator.

However, the '*and*' operator does not satisfy the associative law, so that we extended each fundamental function to a 3-variable function, as follows :

- Basic function

$$basic(x,y) = T(x,y)$$
$$\rightarrow basic(x,y,z) = T(T(x,y),z) = T(x,T(y,z))$$

- Synergy function

$$synergy(x,y) = 1$$
$$\rightarrow synergy(x,y,z) = 1$$

- almost function

$almost(a,x,\alpha)$ is not changed

- equal function

$$equal(a,b) = \sup_x(almost(a,x,\alpha) \wedge almost(b,x,\alpha))$$
$$\rightarrow equal(a,b,c) = \sup_x(almost(a,x,\alpha) \wedge almost(b,x,\alpha) \wedge almost(c,x,\alpha)$$

- high function

$$high(x,y) = \frac{\max(\min(x,y) - \beta, 0)}{1-\beta}$$
$$\rightarrow high(x,y,z) = \frac{\max(\min(x,y,z) - \beta, 0)}{1-\beta}$$

- area definition function

$$area(x,y) = equal(x,y) \times high(x,y)$$
$$\rightarrow area(x,y,z) = equal(x,y,z) \times high(x,y,z)$$

• *and* operator

$$*and*(x,y) = w \times synergy(x,y) + (1-w) \times basic(x,y)$$

where $w = area(x,y)$

$$\rightarrow *and*(x,y,z) = w \times synergy(x,y,z) + (1-w) \times basic(x,y,z)$$

where $w = area(x,y,z)$

As mentioned before, each operator can be extended to a n-variable operator. The construction algorithm is the same as for the two-variable '*and*' operator.

### *B. Extension to an '*or*' Operator*

We discussed only about 'and' operators. An other kind of operators are the 'or' operators. When the *basic function* is a t-conorm operator instead of a t-norm operator, we denote the corresponding operator by '*or*'. The basic algorithm is the same as for the '*and*' operator.

For example, when the *basic function* is the algebraic sum (Fig. 10) ,

$$basic(x,y) = x + y - x.y \tag{11}$$

and when we assign the value 0.3 to $\alpha$ and $\beta$, the resulting '*or*' operator is shown in Fig. 11 and TABLE II.

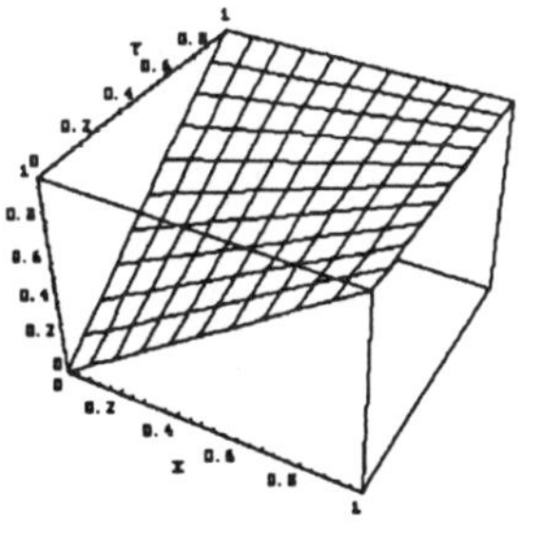

Fig. 10. t-conorm operator (algebraic sum).

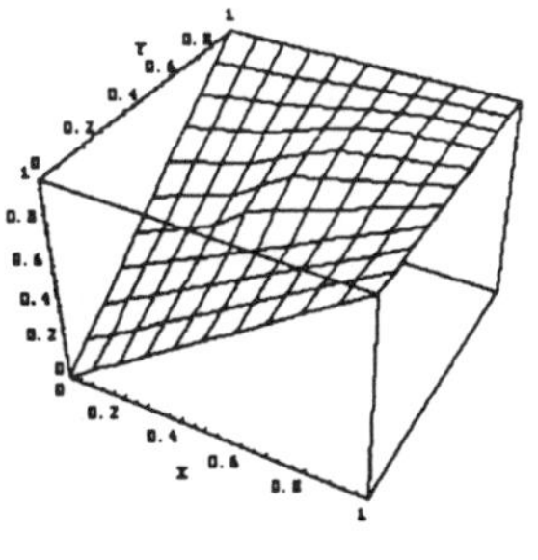

Fig. 11. '*or*' operator.

TABLE II

'*or*' Operator Output.

| | 0.0 | 0.1 | 0.2 | 0.3 | 0.4 | 0.5 | 0.6 | 0.7 | 0.8 | 0.9 | 1.0 |
|---|---|---|---|---|---|---|---|---|---|---|---|
| 0.0 | 0.000 | 0.100 | 0.200 | 0.300 | 0.400 | 0.500 | 0.600 | 0.700 | 0.800 | 0.900 | 1.000 |
| 0.1 | 0.100 | 0.190 | 0.280 | 0.370 | 0.460 | 0.550 | 0.640 | 0.730 | 0.820 | 0.910 | 1.000 |
| 0.2 | 0.200 | 0.280 | 0.360 | 0.440 | 0.520 | 0.600 | 0.680 | 0.760 | 0.840 | 0.920 | 1.000 |
| 0.3 | 0.300 | 0.370 | 0.440 | 0.510 | 0.580 | 0.650 | 0.720 | 0.790 | 0.860 | 0.930 | 1.000 |
| 0.4 | 0.400 | 0.460 | 0.520 | 0.580 | 0.691 | 0.742 | 0.792 | 0.841 | 0.893 | 0.945 | 1.000 |
| 0.5 | 0.500 | 0.550 | 0.600 | 0.650 | 0.742 | 0.821 | 0.856 | 0.890 | 0.924 | 0.960 | 1.000 |
| 0.6 | 0.600 | 0.640 | 0.680 | 0.720 | 0.792 | 0.856 | 0.909 | 0.930 | 0.952 | 0.974 | 1.000 |
| 0.7 | 0.700 | 0.730 | 0.760 | 0.790 | 0.842 | 0.890 | 0.930 | 0.961 | 0.974 | 0.986 | 1.000 |
| 0.8 | 0.800 | 0.820 | 0.840 | 0.860 | 0.893 | 0.924 | 0.952 | 0.974 | 0.989 | 0.994 | 1.000 |
| 0.9 | 0.900 | 0.910 | 0.920 | 0.930 | 0.945 | 0.960 | 0.974 | 0.986 | 0.994 | 0.999 | 1.000 |
| 1.0 | 1.000 | 1.000 | 1.000 | 1.000 | 1.000 | 1.000 | 1.000 | 1.000 | 1.000 | 1.000 | 1.000 |

## VI. Simulation With Real Knowledge

In this section, we use real economic knowledge and simulate inference using the '*and*' and '*or*' operators. The rule we use is :

"the higher FF-rate is *and* the lower US-official-rate is, the higher US-short-term-interest is"

that is, a rule written gradual property [3] [4]. The predicate of each condition is defined as shown in Fig. 12.

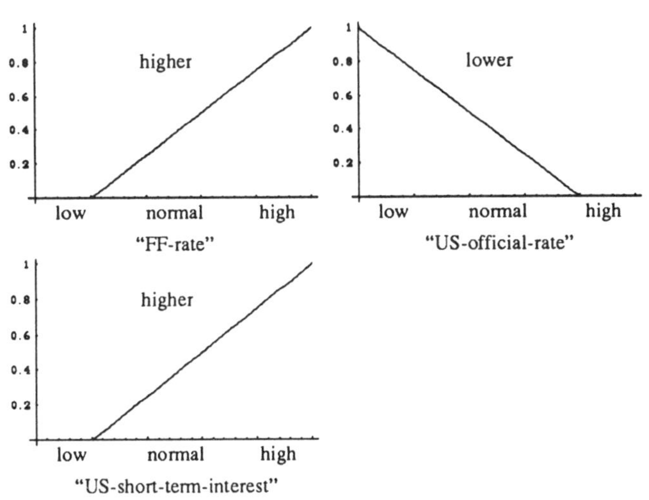


Fig. 12. Predicates of the rule.

The input data are given by the possibility distributions shown in Fig. 13. The input distributions denote "almost high" and "almost low" respectively.

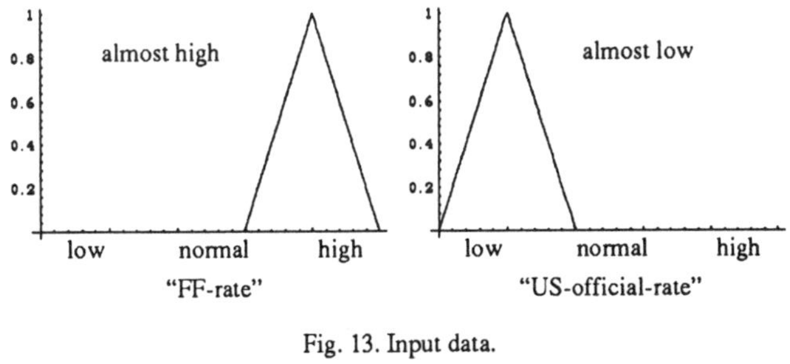


Fig. 13. Input data.

Fig. 14 shows the output distribution obtained by using the 'and' and '*and*' operators to combine the two conditions of the rule, and the Gradual Equivalence [3] as an inference method. To compare the results given by the two operators, we use the same t-norm as the 'and' operator and as the *basic function* of the '*and*' operator, in our example the algebraic product (see equation (9) and Fig. 8). In this example, the two input data satisfy each predicate with a high degree, and the inference result satisfies the predicate more than with a normal 'and' operator. This result of the simulation is quite consistent with the expert's use of the rule. It can be interpreted as the following rule :

"the higher FF-rate is and the lower US-official-rate is,
the much higher US-short-term-interest is."

Next, we use the same rule, where we replace the '*and*' operator by the '*or*' operator. It becomes :

"the higher FF-rate is *or* the lower US-official-rate is,
the higher US-short-term-interest is."

The predicates and input data are the same as in the previous case (Fig. 12 and Fig. 13). As the 'or' operator, and as the *basic function* of the '*or*' operator, we choose the algebraic sum (see equation (11) and Fig. 10). The chosen inference method is the Gradual Equivalence.

Figure 15 shows the output distribution obtained by using the 'or' and '*or*' operators to combine the two conditions of the rule. In this example the inference result satisfies the predicate more than with a normal 'or' operator. When we developed FOREX, we had interviews with economic analysts and found that the knowledge expressed just before had several meanings : if one of the two condition items, FF-rate or US-official-rate, satisfies the condition's predicate, the conclusion item US-short-term-interest becomes high ; if both condition items satisfy the condition predicates, the conclusion item US-short-term-interest becomes very high. To express this knowledge with usual operators, the following three rules are necessary :

1) "the higher FF-rate is,
the higher US-short-term-interest is"
2) "the lower US-official-rate is,
the higher US-short-term-interest is"
3) "the higher FF-rate is and the lower US-official-rate is,
the much higher US-short-term-interest is"

By using the '*or*' operator, this knowledge can be expressed with one rule only , used in this simulation. This feature is desirable not only from the viewpoint of simplicity of the rule coding but also from the viewpoint of consistency with expert's knowledge representation.

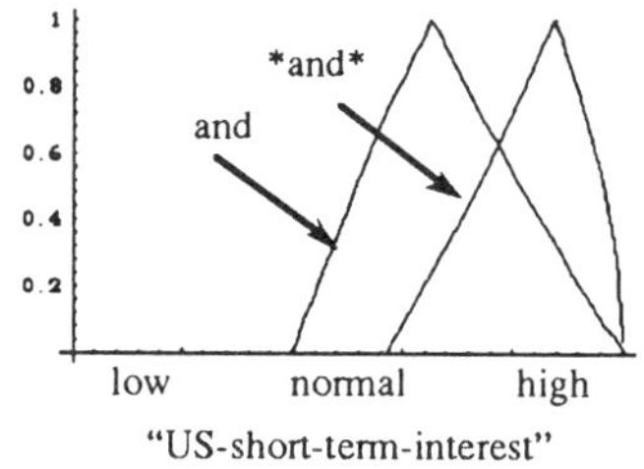


Fig. 14. Output using '*and*'.

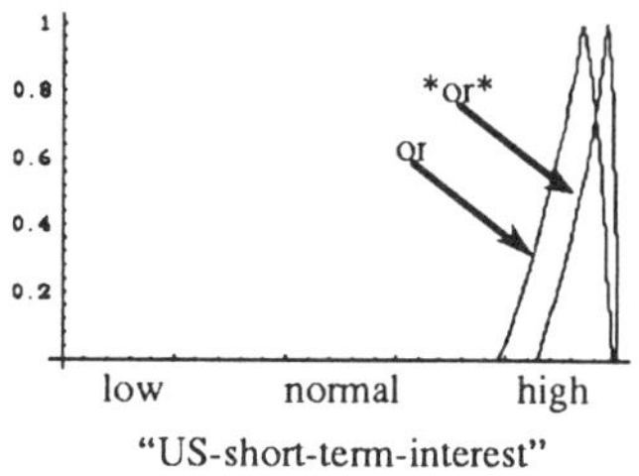


Fig. 15. Output using '*or*'.

## VII. CONCLUSION

This paper proposed a method for defining new combination operators that can take into account synergism between the conditions of a rule. This operator is constructed using weighted average and simulation showed that it could be used to express complex knowledge in a simple way. This constructing method should also be useful for constructing operators that take into account other properties, such as cancellation (by cancellation, we mean the opposite property of synergy). In further work, we will study operators for rules that have several conditions when these conditions do not all have the same synergistic effect.

## REFERENCES


[1] T. Yagyu, H. Yuize, M. Yoneda, M. Grabisch, and S. Fukami : "Foreign Exchange Trade Support Expert System," *Proceedings of IFSA'91 Brussels*, Artificial Intelligence, pp. 214-217 (1991).

[2] H. Yuize, T. Yagyu, M. Yoneda, Y. Katoh, S. Tano, M. Grabisch, and S. Fukami : "Decision Support System for Foreign Exchange Trading - Practical Implementation -," *Proceedings of IFES'91 Japan*, pp. 971-982 (1991).

[3] Y. Katoh, H. Yuize, M. Yoneda, K. Takahashi, S. Tano, T. Yagyu, M. Grabisch, and S. Fukami : "Gradual Rules in a Decision Support System for Foreign Exchange Trading," *Proceedings of IIZUKA'92 Japan*, pp. 625-628 (1992).

[4] D. Dubois and H. Prade : "Gradual Inference Rules in Approximate Reasoning", *Information Sciences*, Vol.61, pp. 103-122 (1992).

[5] Godo L., Jacas J., and Valverde L. : "Fuzzy values in fuzzy logic," *Proceedings of IFSA USA*, pp.829-832 (1989).

# Completeness of Fuzzy Controller Carrying a Mapping $f : R^1 \rightarrow R^1$

Jihong Lee and Seog Chae
Dept. of Control Engineering
Kum-oh National Institute of Technology
188 Shinpyoung-dong, Kumi, Kyoungbuk 730-701, Korea

*Abstract-* **Even though the representations and the processings of data and informations in the fuzzy controller are quitely different from other control algorithms, the information processing operation that it carries out is basically a function $f : A \subset R^n \rightarrow R^m$, from a bounded subset $A$ of $n$-dimensional Euclidean space to a bounded subset $f[A]$ of $m$-dimensional Euclidean space, where $n$ and $m$ are the number of measured states and the number of control inputs of controlled system, respectively. Under the assumptions of Mamdani's direct reasoning method, and C.O.A.(center of area) defuzzification method, the fuzzy controllers are proven to perform the mappings of any given functions with physically meaningful fuzzy sets. The mapping capabilities of the fuzzy controllers are analyzed in detail for the case; $f : R^1 \rightarrow R^1$.**

## I. INTRODUCTION

Application of fuzzy logic to control dates back to the works of Zadeh[1] and Chang and Zadeh[2]. Thereafter, many theoretical advances have been made related to the "fuzzy control"[3,4,5], and numerous successful applications have been reported[6,7]. Typical structure of fuzzy feedback controller is shown in Fig. 1.

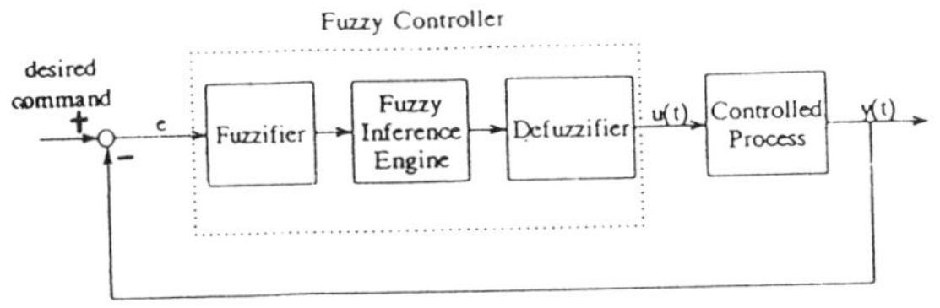


Fig. 1. A fuzzy feedback controller

In this case, the fuzzy controller generates control input from measured output of a controlled process and a reference command through a series of fuzzy data handling techniques.

One of the general concerns of many researchers related to the fuzzy controllers has been "how to determine many design factors to organize a successful fuzzy logic controller, such as fuzzy sets, inference rules, defuzzification methods, knowledge bases, and so on".

Looking into the structures of the fuzzy feedback controllers shown in Fig. 1, however, one can observe the fact that information processing of the fuzzy controller, in spite of the fuzzy characteristics of internal processing, is represented as a crisp function $f : E \subset R^1 \rightarrow R^1$, where $R^1$ is one dimensional Euclidean spaces. Here, a natural question arises : can the fuzzy controllers generate any functions defined on some prescribed domains? or equivalently, are the function spaces generated by the fuzzy controllers compact? The concern of the paper is to answer the question with the typical fuzzy controllers of Fig. 1. In conclusion, it is shown that a class of fuzzy systems can duplicate any arbitrary function with finite number of physically meaningful fuzzy sets as long as the input domain is partitioned into finite number of regions with boundaries defining local maxima or minima, while Kosko[11] and Wang[12] showed that a class of fuzzy systems are capable of approximating any real function to arbitrary accuracy, provided that sufficiently many rules are available.

In proving that there exists a structure of fuzzy controller performing the mapping of an arbitrarily given function, first, the domain of the function is partitioned according to the behavior of the function, second, it is shown that there exist a group of fuzzy sets making the fuzzy controller perform the given function on each partitioned region, and next, the result is extended to overall domain.

0-7803-0614-7/93$03.00 ©1993IEEE

## II. THE FUNCTIONS CARRIED OUT BY FUZZY CONTROLLERS

In this section, the mapping capability of the fuzzy controller shown in Fig. 1 is investigated. The internal model of the fuzzy controller is defined in Section II-A, and an illustrative example is given in Section II-B.

### A. The model

As shown in Fig. 1, the fuzzy controllers performs a mapping described by a real valued function $f$ :

$$f : E \rightarrow U, \quad E \subset R^1, \; U \subset R^1 \tag{1}$$

where $R^1$ is one dimensional Euclidean space, and $E$ and $U$ are universes whose generic elements are $e$ and $u$, respectively. The fuzzy set $\tilde{A}$ in $E$ and the fuzzy set $\tilde{B}$ in $U$ are assumed to be described in the following forms.

$$\tilde{A} = \int_E \mu_{\tilde{A}}(e)/e \tag{2}$$

$$\tilde{B} = \sum_{i=1}^{n} \mu_{\tilde{B}}(u_i)/u_i \tag{3}$$

Also, Mamdani's direct method, C.O.A.(Center Of Area) defuzzification method, and control rules described by the followings are assumed for the fuzzy controller.

$$If\ e\ is\ \tilde{A},\ then\ u\ is\ \tilde{B}. \tag{4}$$

Then, an important property of the fuzzy controller is proposed.

*Lemma 1*: Let a bounded function $f$ described by (1) be given, and a fuzzy controller described above be given. Then, there exists a finete number of continuous convex fuzzy set[1] of (2) in $E$ and a group of discrete fuzzy set of (3) in $U$ with which the fuzzy controller can perform the mapping exactly the same with $f$, if $f$ is either monotonic increasing or monotonic decreasing function on $E$.

*Proof* : First, let's consider the case where $f$ is monotonic decreasing function in $E$, and denote $E$ as an interval $[e_r, e_v]$. Then, $f(e_r)$ and $f(e_v)$ is the supremum and infimum of $f$ in $E$, respectively. Here, we select two control rules:

$$If\ e\ is\ \tilde{E}_r,\ then\ u\ is\ \tilde{U}_r. \tag{5}$$

$$If\ e\ is\ \tilde{E}_v,\ then\ u\ is\ \tilde{U}_v. \tag{6}$$

The fuzzy sets in (5)-(6) is defined as

$$\tilde{E}_r = \int_E \mu_{\tilde{E}_r}(e)/e \tag{7}$$

$$\tilde{E}_v = \int_E \mu_{\tilde{E}_v}(e)/e \tag{8}$$

$$\tilde{U}_r = \mu_{\tilde{U}_r}(u_r)/u_r = 1.0/u_r \tag{9}$$

$$\tilde{U}_v = \mu_{\tilde{U}_v}(u_v)/u_v = 1.0/u_v \tag{10}$$

With Mamdani's direct inference method and C.O.A. defuzzification, the control input $u_0$ for a measured $e_0 \in E$ is calculated by the fuzzy controller as the following:

$$u(e_0) = \frac{\mu_{\tilde{E}_r}(e_0)u_r + \mu_{\tilde{E}_v}(e_0)u_v}{\mu_{\tilde{E}_r}(e_0) + \mu_{\tilde{E}_v}(e_0)}, \quad e_0 \in E \tag{11}$$

Defining $\mu_{\tilde{E}_r}$ and $\mu_{\tilde{E}_v}$ as

$$\mu_{\tilde{E}_r}(e_0) = \frac{f(e_0) - f(e_v)}{f(e_r) - f(e_v)} \tag{12}$$

$$\mu_{\tilde{E}_v}(e_0) = \frac{f(e_r) - f(e_0)}{f(e_r) - f(e_v)} \tag{13}$$

and noting that $u_r = f(e_r)$ and $u_v = f(e_v)$, we are given from (11)

$$u(e) = f(e), \quad e \in E \tag{14}$$

Also, it easily checked that $\tilde{E}_r$ and $\tilde{E}_v$ are convex fuzzy sets, and have maximum values of one. For the case of monotonically increasing case, the procedure is basically the same with the one described above, hence omitted here.-Q.E.D.

The calculations of (11) is graphically represented in Fig. 2.

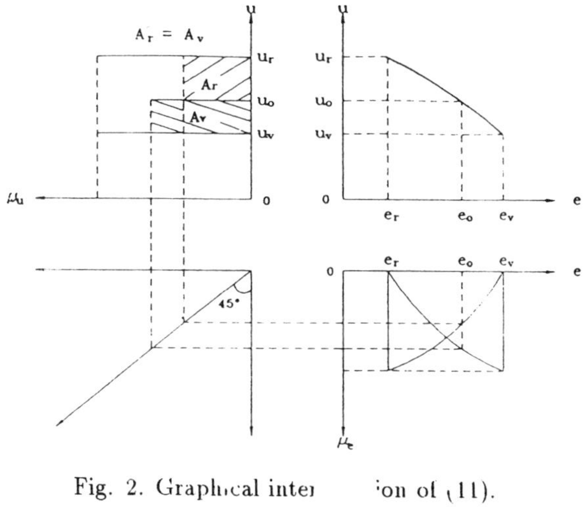


Fig. 2. Graphical inte[illegible]on of (11).

We extend the result of the *Lemma* 1 to more general case.

*Theorem* 1 : Let a bounded function $f$ described by (1) be given, and a fuzzy controller described above be given. Then, there exists a group of continuous convex fuzzy set[1] of (2) in $E$ and a group of discrete fuzzy set of (3) in $U$ with which the fuzzy controller can perform the mapping exactly the same with $f$.

*Proof* : Let's partition the domain $E$ to a set of subset $E^k$ , $k = 1, 2, ..., N$ satisfying the followings: on each $E^k$ $f$ is monotonically varying(either monotonic increasing or monotonic decreasing function), and satisfying (15) and (16)

$$E = \bigcup_k E^k \qquad (15)$$

$$E^i \bigcap E^j = \emptyset , \quad if\ i \neq j \qquad (16)$$

Applying *Lemma* 1 to every $E^k$, we have proven the theorem.-Q.E.D.

*B. Illustrative example*

The details of the procedures described in Section II-A are given in this section with an arbitrarily chosen function. Let's assume that function(in reality, input-output relation for a controlled process) shown in Fig. 3 is given, and denote it as $u = f(e)$. How to get the function is not our concern at this time. The function may be a control law obtained from an expert or a solution derived by some optimization methods.

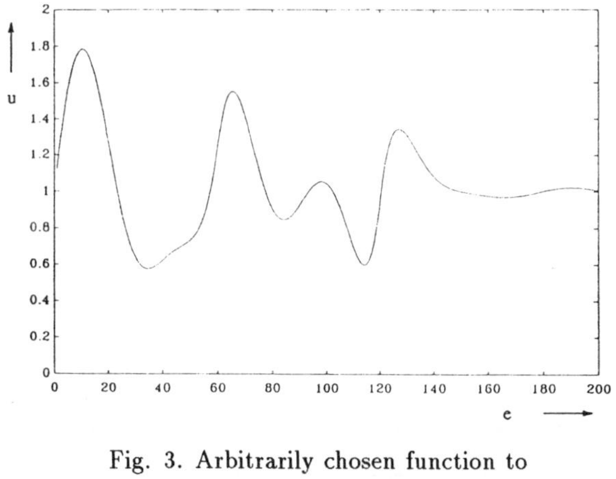


Fig. 3. Arbitrarily chosen function to be reconstructed by fuzzy controller.

The concerns of the work is to show that there exists a fuzzy controller which can perform the mapping of the given function. The overall domain of $f$ is partitioned to subsets $E^k$, $k = 1, 2, \cdots,$ [illegible] in which $f$ is either monotonically increasing or monotonically decreasing(Fig. 4). The intervals in which $f$ increases monotonically are denoted as

$$E^{2i-1} = [e_v^i,\ e_r^i) \qquad i = 1, 2, \cdots, 5 \qquad (17)$$

and the intervals in which $f$ decreases monotonically are described as

$$E^{2i} = [e_r^i,\ e_v^i + 1) \qquad i = 1, 2, \cdots, 5 \qquad (18)$$

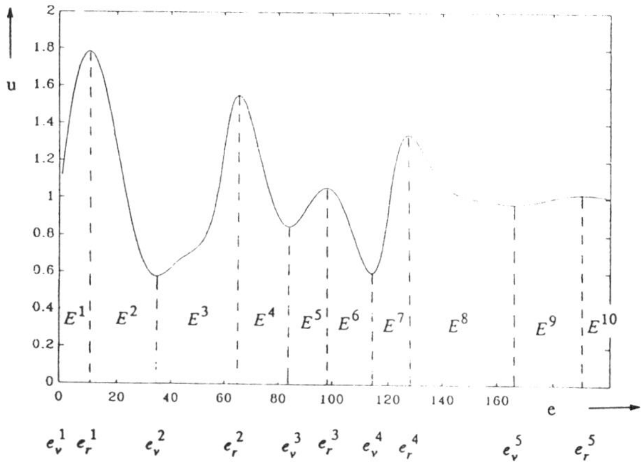


Fig. 4. Fartitioned domains for the example function of Fig. 3.

For the convenience of explaining, we are using the initials of the words "ridge" and "valley" in describing special points on the curve of $f$, such as $e_r$, $e_v$, $u_r$, or $u_v$. Then, for a measured value of input $e$, the degree of the membership for a fuzzy set $\tilde{E}_r^k$ is determined by

$$\tilde{E}_r^k = \int_E \mu_{\tilde{E}_r^k}(e)/e \qquad (19)$$

$$\mu_{\tilde{E}_r^k}(e) = \begin{cases} \frac{f(e)-f(e_v^k)}{f(e_r^k)-f(e_v^k)} & e \in E^{2k-1} \\ \frac{f(e)-f(e_v^{k+1})}{f(e_r^k)-f(e_v^{k+1})} & e \in E^{2k} \\ 0 & otherwise \end{cases} \qquad (20)$$

And, a fuzzy set $\tilde{E}_v^k$ is determined by

$$\tilde{E}_v^k = \int_E \mu_{\tilde{E}_v^k}(e)/e \qquad (21)$$

$$\mu_{\tilde{E}_v^k}(e) \begin{cases} \frac{f(e)-f(e_r^k)}{f(e_v^k)-f(e_r^k)} & e \in E^{2k-1} \\ \frac{f(e)-f(e_r^{k+1})}{f(e_v^k)-f(e_r^{k+1})} & e \in E^{2k} \\ 0 & otherwise \end{cases} \qquad (22)$$

Note that the membership degr[illegible] r al[illegible] uzzy sets outside of $E$ is given zero. The [illegible] tan[illegible] zzy sets $\tilde{E}_r^i$ , $i = 1, 2, \cdots, 5$ and $\tilde{E}_v^j, j = 1, 2$ [illegible] are shown Fig.

5. From the Fig. 5, one can easily know that each fuzzy set satisfies convexity and has only one maximum point. With the fuzzy sets, the fuzzy controller performed the function $f$ exactly the same manner.

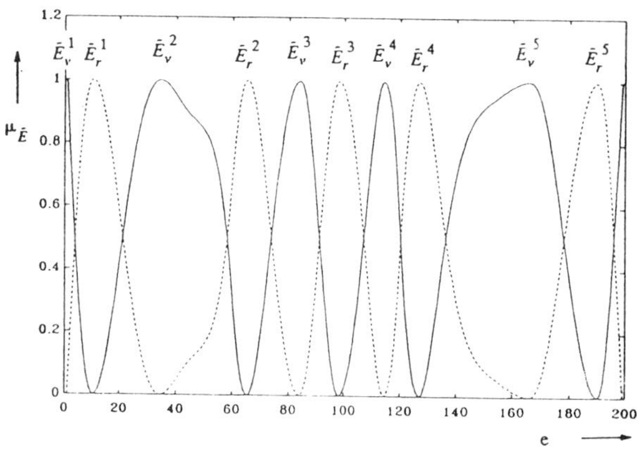


Fig. 5. Resultant fuzzy sets for the example function of Fig. 3.

## III. DISCUSSIONS AND CONCLUDING REMARKS

Even though the controlled systems considered in Section II only one control input, the results obtained in the sections can be applied to a system having control inputs more than one. For a system having $N$ control inputs, assume that a fuzzy control rule be given as

$$If\ error\ is\ \tilde{A},\ then\ u_1\ is\ \tilde{B}_1\ , \tag{23}$$

$$u_2\ is\ \tilde{B}_2\ ,\cdots,\ and\ u_N\ is\ \tilde{B}_N. \tag{24}$$

This rule is, then, equivalently expanded to $N$ rules as the followings.

$$If\ error\ is\ \tilde{B},\ then\ u_1\ is\ \tilde{B}_1\ . \tag{25}$$

$$If\ error\ is\ \tilde{B},\ then\ u_2\ is\ \tilde{B}_2\ . \tag{26}$$

$$\cdots \tag{27}$$

$$If\ error\ is\ \tilde{B},\ then\ u_N\ is\ \tilde{B}_N\ . \tag{28}$$

Each rule of (23) maps from the space of *error* to $R^1$, and hence the analysis made in Section II can be applicable to the cases of multi-dimensional control input.

As shown in the Section II, the type of the fuzzy sets on the $u$ space is not so heavily related to the mapping capability of the fuzzy controllers. Moreover, the fuzzy partition of the input space is closely related with the mapping capability, which is intuitively acceptable.

In summary, the operation that a fuzzy controller carries out was looked upon as a mapping between two crisp sets, and the mapping capability of the fuzzy controller was analyzed in detail, while general concerns related to fuzzy controllers have been to design efficient controllers by combining rules of experts and mathematical tools. The case, $f\ :\ R^1\ \rightarrow\ R^1$ was handled in detail, and it was also shown that the results can be extended to the case of multi-input cases.

Under the assumptions of direct reasoning method, and center of gravity defuzzification methods, the fuzzy controllers were proven to perform mappings of given functions with appropriately defined fuzzy sets.

In the analysis, the domain was partitioned a primary region in which the given function behaves in *monotonic* manner. This approach is based on the observation that an expert divides the possible cases into several groups and assign a rule or rules to each group. With the fuzzy sets, it was shown that the operation carried out in the fuzzy controller can be represented by the mapping of the given function.

The studies on the limit of the performance of a given fuzzy controller should run parallel with the studies on the design of efficient fuzzy controllers such as rule identification, fuzzy set identification and so on.


## ACKNOWLEDGMENT

The authors would like to thank Mr. Byoung-Hyoun Joung for his assistence in simulation and printing results.


## REFERENCES


[1] L. A. Zadeh, "Fuzzy sets,", *Imformt. Control,* vol. 8, pp. 338-353, 1965.

[2] S. S. L. Chang and L. A. Zadeh, "On Fuzzy Mapping and Control," *IEEE Trans. Syst. Man Cybern.*, vol. 15, no. 1, pp. 175-189, 1985.

[3] C. C. Lee, "Fuzzy Logic in Control Systems : Fuzzy Logic Controller - Part I, II," *IEEE Trans. Syst. Man Cybern.,* vol. 20, no. 2, pp. 404-435, 1990.

[4] R. M. Tong, "A Control Engineering Review of Fuzzy Systems," *Automatica,* vol. 13, pp. 559-569, 1977.

[5] R. M. Tong, "Some Properties of Fuzzy Feedback Systems," *IEEE Trans. Syst. [illegible] Cy[illegible]*, vol SMC-10, no. 6, pp. 327-330, [illegible].

[6] E. H. Mamdani and S. Assilia[illegible] [illegible] Ex[illegible] [illegible] in Linguistic Synthesis with [illegible] [illegible]y Lo[illegible] [illegible]

troller," *Int. J. Man-Machine Studies,* vol. 7, no. 12, pp. 1-13, 1975.

[7] M. Sugeno and K. Murakami, "Fuzzy Parking Control of Model Car," *Proc. 23rd IEEE Conf. on Decision and Control,* Las Vegas, NV, 1984.

[8] M. Stinchcombe and H. White, "Approximating and Learning Unknown Mappings Using Multilayer Feedforward Networks," *Proc. Int. Joint Conf. on Neural Networks,* vol III, pp. 7-16, 1990.

[9] R. H.-Nielsen, "Theory of the Backpropagation Neural Network," *Proc. Int. Joint Conf. on Neural Networks,* vol I, pp. 593-605, 1989.

[10] D. G. Luenberger, *Optimization by Vector Space Methods.,* John Wiley & Sons, 1969. pp. 11-45,

[11] B. Kosko, "Fuzzy Systems as Universal Approximators," *proc. Int. Conf. on Fuzzy Systerms,* pp. 1153-1162, San Diego, USA, 1992.

[12] L.-X. Wang, "Fuzzy systems are universal Approximators.," *proc. Int. Conf. on Fuzzy Systerms,* pp. 1163-1170, San Diego, USA, 1992.

# FUZZY PREDICTIVE CONTROL FOR THE TIME-DELAY SYSTEM


Chieh-Li Chen and Ming-Jyi Jong

Institute of Aeronautics and Astronautics
National Cheng-Kung University
Tainan, Taiwan, Republic of China

Department of Mechanical Engineering
Far East Junior College
Tainan, Taiwan, Republic of China



**Abstract-- In this paper, an alternative procedure for constructing a fuzzy model of the given system is proposed associated with a digital fuzzy filter. Based on the structure of the Smith predictor, a hybrid model predictive control approach is also presented. An experimental result has also been illustrated. It shows that the proposed control structure is a feasible approach to control a system with time-delay.**


## I. Introduction

The application of the fuzzy set theory to control system designs is a feasible approach for complex systems. The main advantage of using fuzzy control approach is that no mathematical model is needed. However, in some control problems (e.g., a time-delay system) a model is required for good predictive control. If the mathematical model is difficult to obtain then a fuzzy model which consists of a set of linguistic fuzzy process rules can be employed. In general, the fuzzy dynamic system modelling is based on the idea of determining a set of input-output relations corresponding to the plant.

A time-delay system is a high-order non-minimum phase system. Therefore, there exist some design limitations and the performance of the resulting control system is difficult to achieve. To this problem, the conventional architecture of the Smith predictor [1] can be employed to solve the difficulties due to the time-delay. However, this will not be the case, if a large modelling error occurs [2,3,4].

In this paper, an enhanced Smith control structure which is composed of a fuzzy model and a digital fuzzy filter is proposed. Based on this structure, two functional mechanisms ,i.e., the fuzzy model updating and the fuzzy controller tuning mechanisms, are applied to reduced the model mismatch and to improve system performance. An experimental result using the analog simulator has also been illustrated. It shows that the proposed control structure can control a perturbed time-delay system well with an identified fuzzy model and a fuzzy controller tuning mechanism.


0-7803-0614-7/93$03.00 ©1993IEEE


## II. Fuzzy Dynamic modelling

### A. Fuzzy model

The fuzzy model is based on fuzzy implications, which describes the input-output relation of the plant.There are two types of implications with regard to the fuzzy modelling:

1) Construct the fuzzy relation matrix according to the input-output relation of the plant and the membership function of the fuzzy controller. The fuzzy relation matrix can be considered as a fuzzy model which predicts the system output using fuzzy set operations [5].

2) Construct a locally linear multi-model using fuzzy implications and the fuzzy reasoning method suggested by Takagi and Sugeno [6]. Statistic methods and a successive identification algorithm [7,8] are used to identify the structure of a fuzzy model. This modelling method is very useful for complex or nonlinear systems.

### B.Fuzzy dynamic model

In some cases, it may be difficult to identify a dynamic system using the methods described above. In this paper, an alternative modelling method, which is proposed in the spirit of the linear multi-model approach, is presented as follows.

Consider the regulating problem, the time response of the control system can, somehow, be shown as Fig. 1. It can be divided into three stages (I, II and III). Each stage is corresponding to certain system behaviour. The response in stage I represents the order of the system. The response in stage II corresponds to the maximum changing rate of the system output, if the maximum control input is applied. The third stage is the settling stage which can be described using a linear model. The system modelling for each stage is proposed as follows.

1) Obtain the time responses corresponding to stage I and II using the maximum control input. The models of these two stages are used to predict system output, which will not be used in an on-line tuning process. For this reason, they are used in a feedforward control sense and is adjust using

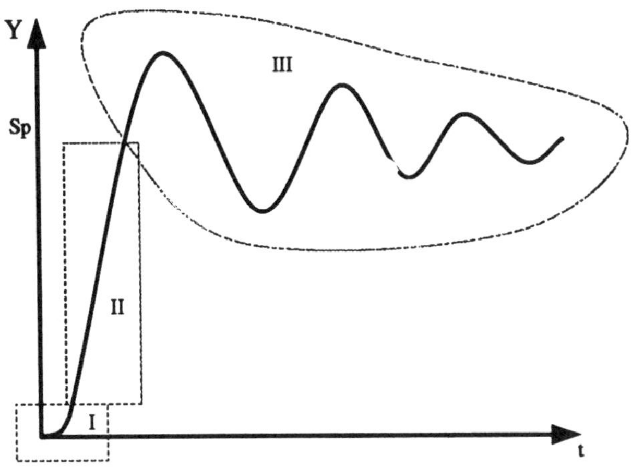


Fig. 1. System time response of a regulating system

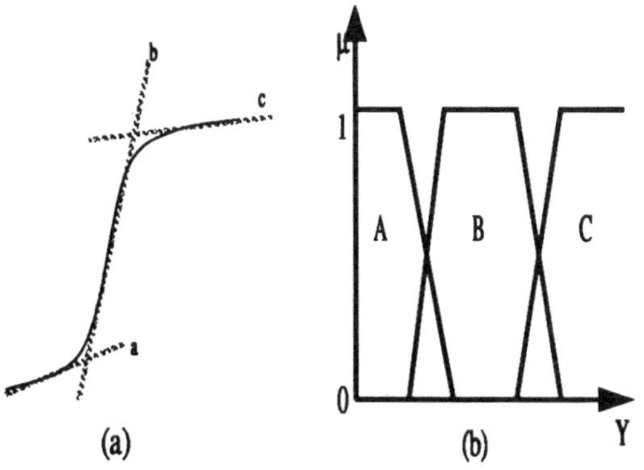


Fig. 2. Fuzzy partitation for the response of stage II.

an off-line procedure. The fuzzy model corresponding to stage II can be described as follows:

If y(t) is A then Δy is a
If y(t) is B then Δy is b
If y(t) is C then Δy is c

where A , B and C are fuzzy variables which can be defined using a set of suitable membership functions shown in Fig.2(b). The gradients a, b, and c, as shown in Fig.2(a), are tuned using an off-line procedure. In this way, the predictive system output will be

$$y(t+1) = y(t) + \Delta y \tag{1}$$

This can be used in the circumstance where the error is large.

2). The fuzzy model for stage III will very much like a linear model. It can be tuned on-line and can be described as

$$\Delta y(t+1) = \alpha(t) \bullet \Delta u(t) \tag{2}$$

where $\alpha(t)$ is a mapping factor corresponding to the input-output response of the system in certain circumstance. A two-step on-line adjustment procedure is presented as follows.

Step 1: determining of the system mapping factor $\alpha$. Let $y(t)$ and $y_e(t)$ be the output and the predictive output at time t, respectively. Then, the model mismatch is $e(t) = y(t) - y_e(t)$, and

$$\Delta y(t) = \Delta y_e(t) + \Delta e(t) \tag{3}$$

If the model mismatch is to be minimised at time t+1,i.e., the estimated output increment $\Delta y(t+1)$ and the system output increment $\Delta y(t+1)$ are equal

$$\alpha(t+1) \bullet \Delta u(t+1) = \alpha(t) \bullet \Delta u(t+1) + e(t)$$

Then,

$$\alpha(t+1) = \alpha(t) + \Delta e(t)/\Delta u(t+1) \tag{4}$$

With this tuning algorithm, a will converged in a noise free case . In a noisy environment , (4) is modified as

$$\alpha(t+1) = \alpha(t) + \Delta e(t)/(\Delta u(t+1) \bullet \beta_1) \tag{5}$$

where $\beta_1$ is a factor, which is increased with respect to time and should be carefully selected such that a is converged to the neighbourhood of the correct value. In the worst case, $\alpha$ will still be converged with certain modelling error.

step 2: on-line adaptation of the fuzzy model. As $\alpha$ has been roughly determined (converged to certain value), an on-line adaptation algorithm is further used to reduced model mismatch as the following.

$$\Delta y_e(t+1) = \alpha \bullet \Delta u(t+1) + \beta_2 \tag{6}$$

where $\beta_2$ is an adaptation term determined by a linguistic tuning rules.

Using the proposed procedure, a linear model can described well. However, in the case of the occurrence of large disturbances, an extra parameter can be introduced to minimise the model mismatch and (6) is modified as

$$\Delta y_e(t+1) = \alpha \bullet \Delta u(t+1) + \beta_2 + \gamma \bullet (y(t) - S_p) \tag{7}$$

where $S_p$ is the set point, and $\gamma$ can be tuned like $\alpha$. Therefore, the mechanism of the proposed fuzzy dynamic modelling can be shown in Fig.3. In general, this procedure can successfully be applied to the open-loop stable system. It will be shown in the latter section.

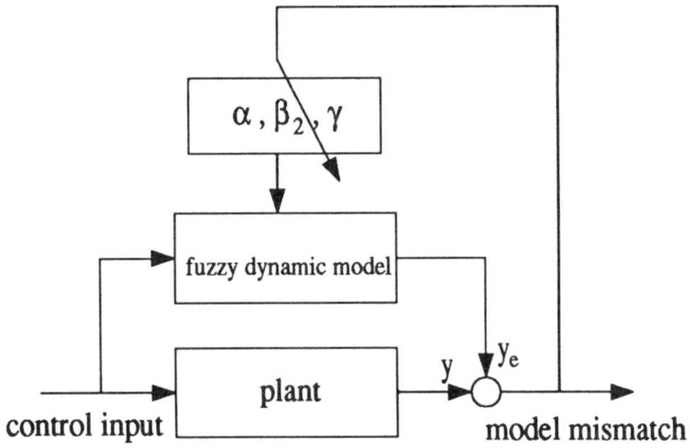


Fig. 3. Schematic diagram of the fuzzy dynamic modelling procedure.

## III. An Enhanced Smith Control structure

The time-delay (dead-time) compensation technique known as the Smith predictor,which has reduced the problem due to the time-delay. The major defect to this approach is its sensitivity to modelling error. Consider a system with transfer function G(s) and a pure time- delay exp(-sT) represented as G and T, respectively. An enhanced Smith control structure can be proposed using the fuzzy dynamic modelling procedure along with a fuzzy filter(FF)[9] and a self-tuning fuzzy controller (FC), as shown in Fig.4, where G and T are the predict model of G and T ,respectively. Two main functional blocks: (1) the fuzzy model updating mechanism and (2) the fuzzy controller tuning mechanism, are described as follows.

### A. The fuzzy model updating mechanism

The procedure for the fuzzy model updating can be presented in the spirit of the fuzzy dynamic modelling process proposed in section II as follows.

1) The knowledge of the system time response in stage I, as shown in Fig.1,does not have strong effect on the resulting performance, if a fast response is always required. The lag is due to the order of the system. Therefore, e knowledge of this stage, which includes the pure time-delay term T and the initial lag, comes from the last experience. Any mismatch can be compensated in the next response.

2) The system behaviour of stage II can be described as that in section II . The updating of the gradient (m) corresponding to each fuzzy variable (A, B, and C) is used to minimised the model mismatch and is processed as

$$m(n+1) = m(n) + f_2 \quad (8)$$

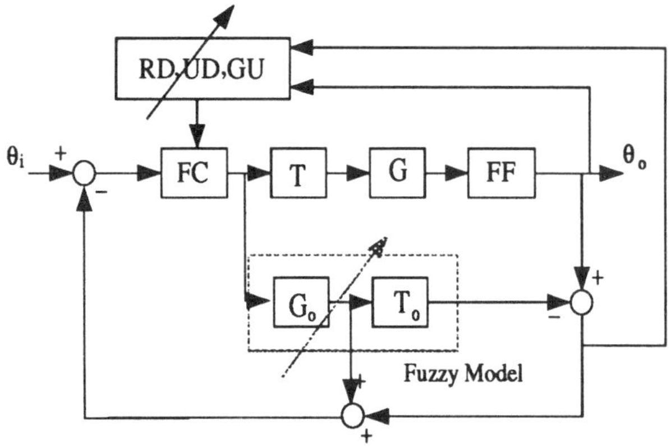


Fig. 4. The enhanced Smith control structure.

where $f_2$ is a tuning factor determined by a set of tuning rules using the mean value of model mismatch (MV) and its derivative(ΔMV) corresponding to stage A , B and C of the last experience (the time response with respect to the nth stimulation, a step input in general ). The rule table is established by a linguistic tuning rules.

3) The behaviour of the settling stage III is described by a mapping factor α and adaptation parameter $\beta_2$ and γ shown in section II. However , the success of the on-line tuning process is strongly dependant on the success of the previous stages.

### B.The fuzzy controller tuning mechanism

In the regulating control system design, it is usually required that the resulting system time response should has short rise time with small overshoot , if is possible to achieved. For this reason , the controller should provide maximum control energy (UM) when a significant error has suddenly ocurred.Then,switch to certain control energy level (UD) with respect to certain index(RD) and hopefully the zero steady state is achieved using the feedback loop, as shown in Fig.5. This strategy may lead to a good control from the sense of the minimum time control process.

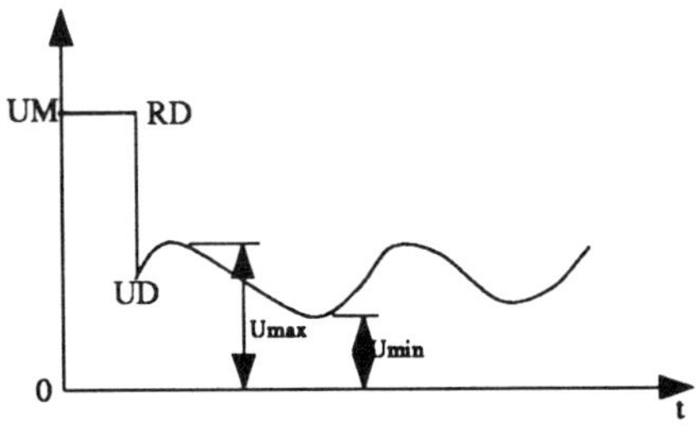


Fig. 5. Control input of the regulating control system.

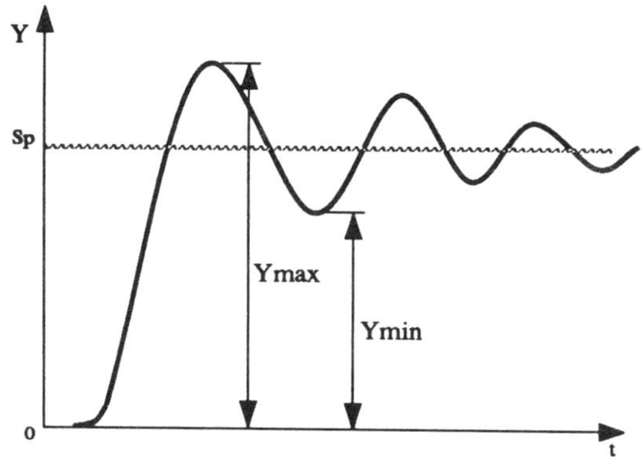


Fig. 6. System time response of the resulting control system.

Using the idea above, the fuzzy controller output can be divided into two stages, one is for the occurrence of a significant error and the other is for small error or settling stage. The former stage uses a feedforward control strategy and the latter stage uses traditional fuzzy feedback compensations. The key parameters of the fuzzy controller , RD , UD and GU (the output scaling factor) are tuned in the time history. The tuning procedure is stated as follows.

1) RD

RD indicates the circumstance for the controller output switching from UM to UD. It can be defined as (i) the time after a small error is occurred, or (ii) a certain degree of error ,e.g., the controller output switches from UM to UD as the error is decreasing and |error| < RD is satisfied. In the latter case,the tuning of RD to achieve design requirement is performed

$$RD(n+1) = RD(n) + rd \tag{9}$$

where rd is a tuning factor which is proportional to the maximum overshoot ($OV=Y_{max} - S_p$ ).

2) UD

UD can be considered as the predictive steady-state control energy.It can be updated using the following interpolation-like algorithm.

$$UD(n+1) = UD(n) - (U_{max}(n) - U_{min}(n)) \times \frac{(Y_{max}(n) - S_p)}{(Y_{max}(n) - Y_{min}(n))} \tag{10}$$

where $U_{max}$ ,$U_{min}$ and $Y_{max}$ ,$Y_{min}$ are shown in Fig. 5 and 6 ,respectively.

3) GU

GU is the output scaling factor of the fuzzy controller output. It can be considered as a system loop gain from the classical control point of view. A system with time-delay will result in certain performance limitation because high gain is impossible to achieve. Therefore, the scaling factor GU should be small when model mismatch is large, and GU can be increased as the accuracy of the fuzzy model is increased (in this case the time-delay effect is reduced). The tuning

algorithm used in this paper is

$$GU(n+1) = GU(n) + \Delta GUf_3 \tag{11}$$

where $\Delta GU=MV(n-1)-MV(n)$ and $f_3$ is a scaling factor which dependent on the system time constant.

Using the modified Smith control structure with a self-tuning fuzzy controller, a time-delay system can be controlled well. The resulting system time response will have short rise time and small overshoot as required. In the following section, one experimental result is used to illustrate the effectiveness of the proposed control scheme.

## IV. Analog simulations

In this study, the adaptation capability of the proposed control scheme is demonstrated using an analog simulator which is used to construct a perturbed plant. The controller is implemented using a personal computer with sampling interval being 0.05 sec , and the maximum control input voltage is selected as 5 volts.

Case 1. (overdamped system with time-delay)

An initial model

$$G(s) = \frac{1.4 \bullet \exp(-0.5s)}{(s+1)(s+1.4)} \tag{12}$$

is first identified using the off-line fuzzy dynamic modelling method. As G(s) been perturbed to be

$$G(s) = \frac{1.96 \bullet \exp(-0.85s)}{(s+1)(s+1.4)(s+1.4)} \tag{13}$$

the fuzzy model of this perturbed plant and the fuzzy controller are then updated using the proposed tuning procedure in a few step input stimulations. The resulting control system output and model mismatch for the first four steps are shown in Fig.7-8, respectively.

Case 2. (underdamped system with time-delay)

An initial model

$$G(s) = \frac{0.9 \bullet \exp(-1.5s)}{(s^2+0.7s+1)} \tag{14}$$

is first identified using the off-line fuzzy dynamic modelling method. In a similar manner, as the time-delay term of G(s) been perturbed to be 0.85 sec, the use of the proposed control scheme can still control the perturbed system well. The resulting control system output corresponding to step stimulation 1, 3, 5, and 7 are shown in Fig.9. It shows that even in a strongly perturbed case, the proposed control

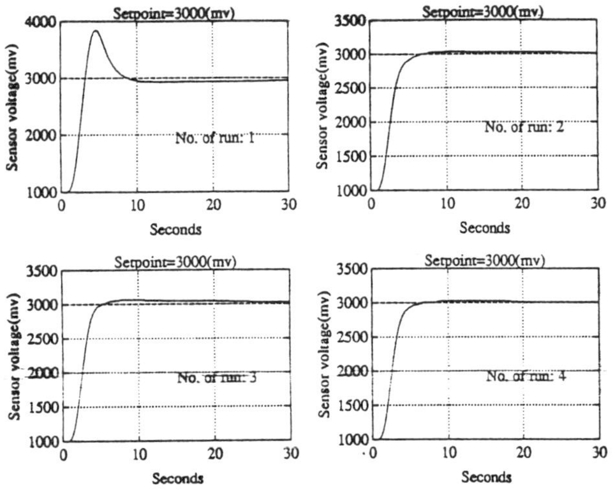


Fig. 7. System time response corresponding to each step stimulation in case 1 (overdamped system).

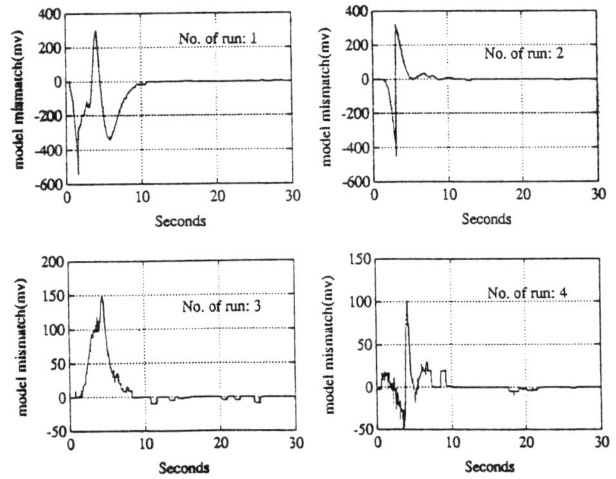


Fig. 8. Model mismatch corresponding to each step stimulation in case 1 (overdamped system).

structure is capable of achieving required regulating performance , i.e., responses with short rise time and small overshoot.

## V. Conclusions

In this paper,a fuzzy dynamic modelling procedure is proposed. The robustness of the Smith control structure is improved using a model updating and a self-tuning fuzzy controller. The resulting control scheme is particularly useful in the system with time-delay and large time constant. Simulation results were also illustrated to shown the effectiveness of the proposed control structure.

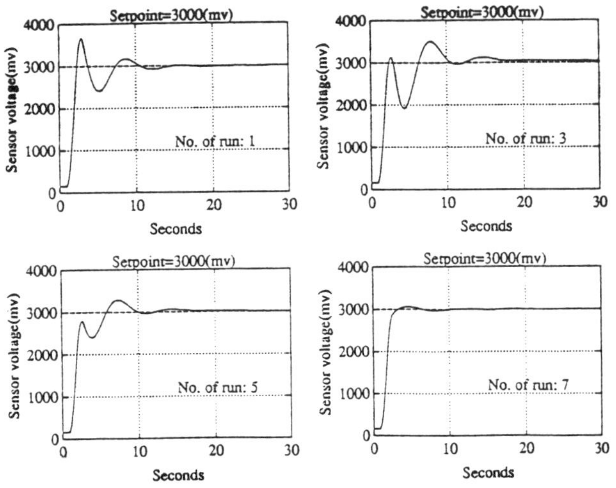


Fig. 9. System time response corresponding to each stimulation in case 2 (underdapmed system).

## References

[1] O. J. M. Smith, "Closer Control of Loops with Dead Time",Chemical Engineering Progress, vol.53 ,no.5, pp.217-219 , 1957

[2] Z.Palmor , " Stability Properties of Smith Dead- Time Compensator",INT.J.Control, vol.32 ,no.6 , pp.937-949 ,198?

[3] I. Horowitz, "Some Properties of Delayed controls (Smith Regulator) ", INT. J. Control, vol.38 ,no.5, pp.977-990 ,1983

[4] J. E. Marshall , Control of Time-Delay Systems , PhD thesis ,University of Bath , England , 1978

[5] M. M. Gupta , J. B. Kiszka and G. M. Trojan , "Multivariable Structure of Fuzzy Control Systems ",IEEE Transactions on Systems, Man, Cybernetics ,vol.SMC-16 , no.5 , pp.638 -656 , 1986

[6] T. Tagagi and M. Sugeno, " Fuzzy Identification of Systems and Its application to Modelling and Control " , IEEE Transactions on Systems , Man , Cybernetics , vol.SMC-15 ,no.1,pp.116-132,1984

[7] M.Sugeno and G.T.Kang ,"Structure Identification of Fuzzy Model ", Fuzzy Sets and Systems,vol.28 ,pp.15-33,1988

[8] M.Sugeno and K.Tanaka, " Successive Identi- fication of a Fuzzy Model and Its Applications to Prediction of a Complex System " , Fuzzy Sets and Systems, vol.42. pp.315-334 ,1991

[9] K.Arakawa and Y.Arakawa,"Digital Signal Processing Using Fuzzy Clustering for Nonstationary Signals " , Fuzzy Engineering Toward Human Friendly Systems IFES'91, pp.877-888,1991

# Learning Fuzzy Control with Hybrid Symbolic, Connectionist Networks

Steve G. Romaniuk
Department of Information Systems and Computer Science
National University of Singapore
10 Kent Ridge
Singapore, 0511

***Abstract*— The problem of deriving membership functions as a means for describing linguistic variables (for some control process) and the choice of fuzzy inference operators and connectives is at the heart of developing fuzzy control systems. To this date much of the selection process is under the control of the system engineer, and dependent on his or her ability to make the right choice for the right application. In this paper, it will be shown by means of a real world example for controlling a steam engine, how hybrid learning systems can be employed for automating the design of fuzzy controllers. Deriving the necessary linguistic variables and accompanying membership functions from raw data by use of machine learning is addressed. Finally, the viability of such a system is emphasized to act not only as a fuzzy controller, but more importantly independent of human intervention, automatically derive acceptable control strategies.**

## I. Introduction

### A. Fuzzy Control

In the classical case of fuzzy control a process operator may formulate a verbal description of some process behavior using fuzzy sets to formalize the verbal description. This results in building a fuzzy model of the real process versus constructing a mathematical model of the system, which may be cumbersome to realize due to its increased complexity. Models derived from fuzzy set theory may be equivalent to simulated models with respect to their performance capabilities, but easier to construct. Nonetheless, they require human intervention at two levels: First, a technique for formulating the verbal description of process behavior that can yield an adequate description of the linguistic variables, needs to be developed (Rules and Membership functions). Second, a reliable mathematical apparatus that is used by the system designer for formalizing accurate verbal description for the fuzzy model, needs to be created (Selection of fuzzy inference operators) [3]. The purpose of this paper is to point out how learning systems - specifically hybrid symbolic/connectionist - can be utilized to support the fuzzy control process. Emphasis will be placed on practical results showing how membership functions and linguistic variables can be derived by means of an autonomous learning system. The problem to be studied is the control of a steam engine.

### B. Overview of SC-net

SC-net has been primarily developed as a tool to support the knowledge engineer in the difficult task of knowledge acquisition [4] [6] [8] for expert systems [10]. Machine learning and knowledge representation in a hybrid/symbolic connectionist environment, together with uncertainty management through means of fuzzy logic, form the corner stones of the system. Learning is analogous to instance-based learners in that only a single pass through the training data is required and examples are encoded through Recruitment of Cells Algorithm (RCA). During RCA pass an instance is either identified (Difference in actual and expected output of network within $\epsilon$), the bias of one or more cells - representing encoded instances - is modified (Difference within $5\epsilon$), or a new cell is recruited to encode the presented instance. For more detail on the RCA algorithm see [7] [8].

The choice of a connectionist architecture was motivated by the following three desirable features:

1. A highly parallel and uniform representation of knowledge (the SC-net network).
2. Fault tolerance and noise resistance.
3. A built-in ability to deal with non-crisp inputs and outputs.

On the other hand it proved beneficial to incorporate some of the strong points of symbolic machine learning. From the symbolic side we can identify:

0-7803-0614-7/93$03.00 ©1993IEEE

1. The ability to encode rules to support knowledge refinement.

2. Allow for rule extraction as a direct means to elicit learned knowledge and support the implementation of expert system standards such as, consulation and explanation facilities.

3. Provide a means to represent symbolic constructs such as variables, comparators and quantifiers. This leads to a more powerful language for describing knowledge and augmenting the learning process through use of domain specific meta knowledge.

The quintessence of SC-net is to combine the virtues of both symbolic and connectionist representations, that is to yield a more powerful environment for automating the knowledge acquisition process.

## II. Fuzzy Variables and their Adaption through Learning

SC-net supports the representation of fuzzy variables, which allow either the user or the system itself to divide the numerical range of a variable into its fuzzy equivalent. In general fuzzy variables are described by a set of membership functions, where each function is associated with a linguistic hedge (variable) such as *high*, *small*, etc [11]. These membership functions correlate a given numerical value with a degree of membership indicating the strength (membership) of the numerical value being a member of the predefined fuzzy sets. In SC-net only pi-shaped membership functions are supported. An extension to more complicated membership functions is forthcoming [9]. A linguistic hedge in SC-net is defined by the 4 quantities:

$$< HedgeId >: Bound_{Lower} \ldots Bound_{Upper} \quad (Plateau_{Lower}, Plateau_{Upper}) \tag{1}$$

Whenever the value of a given fuzzy variable lies between $Bound_{Lower}$ and $Bound_{upper}$ $< HedgeId >$ takes on a membership value of 1. If the value falls outside the $Plateau_{upper}$ and $Plateau_{lower}$ range a membership value of 0 is assigned. In every other case a graded membership response is returned which in turn is described by a linear function (arms of pi-shaped membership function). Figure 1 shows the general network structure used by SC-net to represent a fuzzy variable (labeled attribute) and a linguistic hedge (labeled attribute[value]). Cells labeled with the numerical value of -1 return the minimum of the incoming activations, whereas cells with a 0 label return the strong negation $(1 - Activation)$ of the incoming activation.

The weights are calculated as follows:

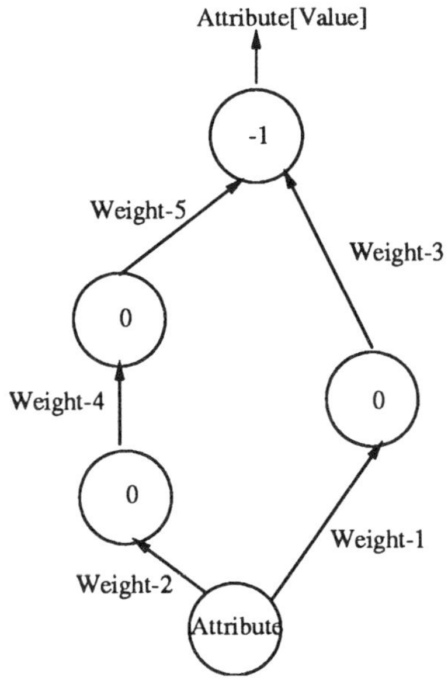


Figure 1: SC-net Network Representing Fuzzy Attribute

$$\begin{aligned} Weight-1 &= \frac{1}{Plateau_{Upper}} \\ Weight-2 &= 1 \\ Weight-3 &= \frac{Plateau_{Upper}}{Plateau_{Upper} - Bound_{Upper}} \\ Weight-4 &= \frac{1}{1 - Plateau_{Lower}} \\ Weight-5 &= \frac{1 - Plateau_{Lower}}{Bound_{Lower} - Plateau_{Lower}} \end{aligned} \tag{2}$$

Finally, SC-net allows the arms of fuzzy membership functions to be dynamically adapted through use of the Dynamic Plateau Modification Algorithm (dubbed DPM) [7]. Initially $Plateau_{Lower}$ and $Plateau_{Upper}$ are set to the smallest and the largest variable range value, respectively. By presenting encoded instances of examples learned by RCA and represented into a network structure (the SC-net network) the membership arms of the pi-shaped functions are modified. The central idea of the algorithm is to place constraints on the degree of generalization provided by each of the arms. If the degree of membership calculated by either arm is too high (for a given encoded example), it is lowered by appropriately moving either the $Plateau_{Lower}$ value closer to the $Bound_{Lower}$ or the $Plateau_{Upper}$ value closer to $Bound_{Upper}$. The amount of adjustment is determined by comparing the actual and the expected output response of a cell. For further detail on fuzzy variables, activation functions used by SC-net and the Dynamic Plateau Modification algorithm refer to [7, 8].

## III. Experiments

### *A. Domain of Study*

The control problem to be studied is concerned with the heat-pressure loop of a steam engine. The pressure error (PE) and the change in the pressure error (CPE) are two system inputs, whereas the heat input change (HC) is the only system output. Table 1 lists the abreviations of other linguistic variables used throughout the remainder of the paper.

Table 1. Abreviations for linguistic variables

| | |
|---|---|
| Positive Big | PB |
| Positive Medium | PM |
| Positive Small | PS |
| Positive Zero | PO |
| Zero | ZO |
| Negative Zero | NO |
| Negative Small | NS |
| Negative Medium | NM |
| Negative big | NB |

In Table 2 a set of sample rules used for the steam engine control problem is displayed. By decomposing the given rules into conjuncts (Ex. PE[x] and CPE[y]) a total of 42 conjuncts were obtained, and used during the first experiment testing SC-net's ability to act as a fuzzy controller.

Table 2. Sample of rules used for steam engine control

Rule 1:
if and(or(fuzzy(PE[NB])=1.0, fuzzy(PE[NM])=1.0), fuzzy(CPE[NM])=1.0) then HC_PM (1.0).

Rule 2:
if and(or(fuzzy(PE[NS])=1.0, fuzzy(PE[PS])=1.0), fuzzy(CPE[NS])=1.0) then HC_ZO (1.0).

The membership functions for both pressure error and change in pressure error are shown in Tables 3 and 4. Since SC-net to this point only supports pi-shaped membership functions some change from the original membership functions used in [5] resulted. This change is only minor in that the drop-off of the original membership arms is not linear, but experiences a slight decrease.

Table 3. Membership definitions for pressure error PE

| Label | LB | UB | LP | UP |
|---|---|---|---|---|
| PB | 6.0 | 6.0 | 3.0 | 7.0 |
| PM | 4.0 | 4.0 | 1.0 | 7.0 |
| PS | 2.0 | 2.0 | -0.5 | 5.0 |
| PO | 0.0 | 0.5 | -0.5 | 3.0 |
| NO | -0.5 | 0.0 | -3.0 | 0.5 |
| NS | -2.0 | -2.0 | -5.0 | 0.5 |
| NM | -4.0 | -4.0 | -7.0 | -1.0 |
| NB | -6.0 | -6.0 | -7.0 | -3.0 |

Table 4. Membership definitions for change in pressure error CPE

| Label | LB | UB | LP | UP |
|---|---|---|---|---|
| PB | 6.0 | 6.0 | 3.0 | 7.0 |
| PM | 4.0 | 4.0 | 1.0 | 7.0 |
| PS | 2.0 | 2.0 | -1.0 | 5.0 |
| ZO | 0.0 | 0.0 | -3.0 | 3.0 |
| NS | -2.0 | -2.0 | -5.0 | 1.0 |
| NM | -4.0 | -4.0 | -7.0 | -1.0 |
| NB | -6.0 | -6.0 | -7.0 | -3.0 |

Table 5 shows the original membership definitions for linguistic variables PB and PM which are representative for the remaining variable definitions.

Table 5. Sample of original fuzzy sets

| | -6-1 | 2 | 3 | 4 | 5 | 6 |
|---|---|---|---|---|---|---|
| PB | 0.0 | 0.0 | 0.0 | 0.3 | 0.7 | 1.0 |
| PM | 0.0 | 0.3 | 0.7 | 1.0 | 0.7 | 0.3 |

### *B. Description of Experiments*

The first experiment attempts to determine SC-net's applicability as a fuzzy controller based on a connectionist architecture. The question that needs to be answered is by what degree do SC-net's control responses for any given input of PE and CPE differ from those of a classical fuzzy control system? Secondly, what impact does a priori knowledge have on the outcome of these results? Table 6 displays the original fuzzy decision matrix for the control problem, which will form the basis for all further comparisons. SC-nets performance (using rules) is shown in Table 7. Here, the breakdown of differences in the original and the SC-net generated fuzzy decision table is given in terms of responses, which differ by 0, 1, or 2+ (2 or more) linguistic variables. The first row of the table presents the results for using the membership functions defined earlier in Tables 3 and 4. As can be seen, over 87% of the time SC-nets response is identical to that of the fuzzy controller. Only in little over 3% of the cases the difference is greater than or equal to 2 in the predicted responses. This clearly shows that SC-net can perform adequately as a controller for the steam engine problem. For the results shown in the second row of Table 7, the same partitions (linguistic variables) were used as in the first experiment, but the Lower-Plateau and Upper-Plateau values were all reset to the minimum and maximum range values, respectively. This time DPM was applied to automatically determine the best set of membership arms for this problem. As the results indicate the performance of SC-net clearly improved. Over 91% of all responses were identical to those made by the original controller. More importantly, the number of false responses of 2 or more linguistic variables difference decreased by 33%. Finally, in the last experiment, using control rules SC-net is forced to derive its own linguistic variables using the APG algorithm [7] and then applying DPM to derive the final set of membership functions. Though no increase in the performance of the number of correctly given responses is achieved, the number of errors made in the third category is again decreased by 33% (moved into second category).

Table 6. Fuzzy logic decision table for controlling steam engine

| | -6 | -5 | -4 | -3 | -2 | -1 | 0 | 1 | 2 | 3 | 4 | 5 | 6 |
|---|---|---|---|---|---|---|---|---|---|---|---|---|---|
| -6 | 0 | 0 | 4 | 4 | 4 | 5 | 6 | 6 | 6 | 6 | 5 | 6 | 6 |
| -5 | 0 | 0 | 4 | 4 | 4 | 5 | 6 | 6 | 6 | 6 | 5 | 6 | 6 |
| -4 | -2 | -2 | 1 | 4 | 4 | 5 | 6 | 6 | 6 | 6 | 4 | 6 | 6 |
| -3 | -2 | -2 | -2 | 1 | 2 | 3 | 5 | 5 | 5 | 5 | 6 | 6 | 6 |
| -2 | -2 | -2 | -2 | -1 | 0 | 2 | 4 | 4 | 4 | 5 | 6 | 6 | 6 |
| -1 | -3 | -3 | -3 | -2 | -1 | 1 | 2 | 2 | 4 | 5 | 5 | 5 | 5 |
| -0 | -4 | -4 | -4 | -3 | -2 | -1 | 0 | 0 | 2 | 4 | 4 | 4 | 4 |
| +0 | 4 | 4 | 4 | 3 | 2 | 1 | 0 | 0 | -2 | -4 | -4 | -4 | -4 |
| 1 | 3 | 3 | 3 | 2 | 1 | -1 | -2 | -2 | -4 | -5 | -5 | -5 | -5 |
| 2 | 2 | 2 | 2 | 1 | 0 | -2 | -4 | -4 | -4 | -5 | -6 | -6 | -6 |
| 3 | 2 | 2 | 2 | -1 | -2 | -3 | -5 | -5 | -5 | -5 | -6 | -6 | -6 |
| 4 | 2 | 2 | -1 | -4 | -4 | -5 | -6 | -6 | -6 | -6 | -4 | -6 | -6 |
| 5 | 0 | 0 | -4 | -4 | -4 | -5 | -6 | -6 | -6 | -6 | -5 | -6 | -6 |
| 6 | 0 | 0 | -4 | -4 | -4 | -5 | -6 | -6 | -6 | -6 | -5 | -6 | -6 |

Table 7. Steam Engine control results with using rules

| | 0-LVD | 1-LVD | 2+-LVD |
|---|---|---|---|
| PP and PMF | 87.4% | 9.3% | 3.3% |
| PP | 91.2% | 6.6% | 2.2% |
| Rules only | 91.2% | 7.7% | 1.1% |

The generated membership functions for pressure error and change in pressure error are displayed in Tables 8 and 9. It is interesting to note, that the number of partitions derived for PE and CPE is identical to the number original defined by human experts [5]. In the final experiment SC-net's learning ability is tested. For this test random partitions of the original decision matrix were generated (10) for various partitions sizes (10%-90%). The selected points of the partitions were then used for training. No information with regard to the fuzzy membership functions was provided. The results of this experiment are shown in Table 10. Figure 2 demonstrates graphically SC-net's generalization ability as the size of the training sets is increased in 10% increments.

Table 8. Membership functions derived by APG and DPM for pressure error

| Partitions | Lower-Bound | Upper-Bound |
|---|---|---|
| p1 | -6.0 | -5.0 |
| p2 | -5.0 | -3.0 |
| p3 | -3.0 | -1.03 |
| p4 | -1.03 | -0.0 |
| p5 | -0.0 | 1.02 |
| p6 | 1.02 | 3.0 |
| p7 | 3.0 | 5.0 |
| p8 | 5.0 | 6.0 |

Table 9. Membership functions derived by APG and DPM for change in pressure error

| Partitions | Lower-Bound | Upper-Bound |
|---|---|---|
| p1 | -6.0 | -5.0 |
| p2 | -5.0 | -3.0 |
| p3 | -3.0 | -1.0 |
| p4 | -1.0 | 1.0 |
| p5 | 1.0 | 3.0 |
| p6 | 3.0 | 5.0 |
| p7 | 5.0 | 6.0 |

Table 10. Steam Engine control results for various training sizes anD different train set partitions

| | 0-LVD | 1-LVD | 2+-LVD |
|---|---|---|---|
| 10% | 63.0% | 21.4% | 15.6% |
| 20% | 75.8% | 13.7% | 10.5% |
| 30% | 80.5% | 13.1% | 6.4% |
| 40% | 83.9% | 10.4% | 5.7% |
| 50% | 88.1% | 7.3% | 4.6% |
| 60% | 90.6% | 5.4% | 4.0% |
| 70% | 93.3% | 3.8% | 2.9% |
| 80% | 95.7% | 2.4% | 2.0% |
| 90% | 97.8% | 1.2% | 1.0% |
| 100% | 100.0% | 0.0% | 0.0% |

As expected SC-net shows a high degree of performance early on. At about 60% SC-net achieves over 90% exact responses in its choice of linguistic control variables. The other two categories of misses are about equal (difference around 1%). Importantly, throughout all tests the number in differences of 2 or more always remains lower than

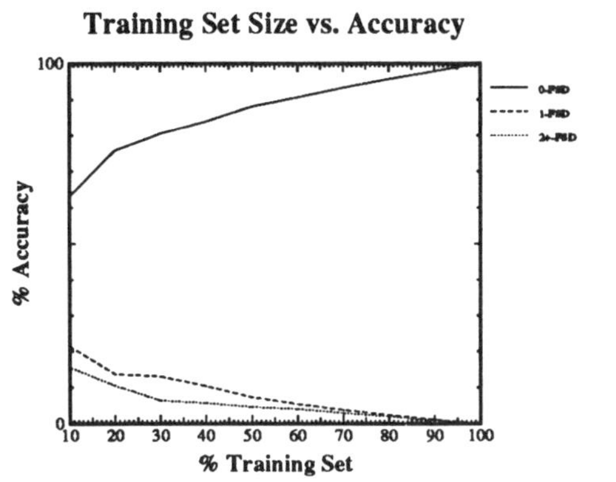


Figure 2: Performance Results for Steam Engine Control

that of 1 difference. For control purposes, a difference of 0 or 1 in the choice of linguistic variable, will only have minimal affect on the controllers performance. It is therefore crucial that the third category remains low. This is exactly what can be observed in Figure 2. From the last experiment we can safely conclude that SC-net is capable of not only acting as a fuzzy controller, but can also be trained to perform the same task. Additionaly, no a priori knowledge in form of user defined fuzzy membership functions or fuzzy rules is required. The system can act as an autonomous entity and derive the best decisions without any human intervention. Lastly, Table 11 displays the fuzzy responses provided by SC-net for steam engine control.

Table 11. SC-net derived decision table using rules

| | -6 | -5 | -4 | -3 | -2 | -1 | 0 | 1 | 2 | 3 | 4 | 5 | 6 |
|---|---|---|---|---|---|---|---|---|---|---|---|---|---|
| -6 | 1 | 5 | 4 | 4 | 4 | 5 | 6 | 6 | 6 | 6 | 6 | 6 | 6 |
| -5 | -2 | 2 | 4 | 4 | 4 | 5 | 6 | 6 | 6 | 6 | 6 | 6 | 6 |
| -4 | -2 | -2 | 1 | 4 | 4 | 5 | 6 | 6 | 6 | 6 | 6 | 6 | 6 |
| -3 | -2 | -2 | -2 | 1 | 2 | 3 | 5 | 5 | 5 | 5 | 6 | 6 | 6 |
| -2 | -2 | -2 | -2 | -1 | 0 | 2 | 4 | 4 | 4 | 5 | 6 | 6 | 6 |
| -1 | -4 | -4 | -4 | -3 | -2 | -2 | 4 | 4 | 4 | 4 | 4 | 4 | 4 |
| -0 | -4 | -4 | -4 | -3 | -2 | -2 | 0 | 0 | 0 | 4 | 4 | 4 | 4 |
| +0 | 4 | 4 | 4 | 3 | 2 | 1 | 0 | 0 | 0 | -4 | -4 | -4 | -4 |
| 1 | 4 | 4 | 4 | 3 | 2 | 1 | 0 | 0 | -4 | -4 | -4 | -4 | -4 |
| 2 | 2 | 2 | 2 | 1 | 0 | -2 | -4 | -4 | -4 | -5 | -6 | -6 | -6 |
| 3 | 2 | 2 | 2 | -1 | -2 | -3 | -5 | -5 | -5 | -5 | -6 | -6 | -6 |
| 4 | 2 | 2 | -1 | -4 | -4 | -5 | -6 | -6 | -6 | -6 | -6 | -6 | -6 |
| 5 | 2 | -2 | -4 | -4 | -4 | -5 | -6 | -6 | -6 | -6 | -6 | -6 | -6 |
| 6 | -1 | -5 | -4 | -4 | -5 | -6 | -6 | -6 | -6 | -6 | -6 | -6 | -6 |

## IV. Summary

This paper provided a series of experiments targeted to investigate the salience of the prototypical hybrid symbolic/connectionist expert system development tool SC-net to not only act as a fuzzy controller, but more importantly to independently derive the necessary fuzzy control strategies required to adequately control a steam engine. Deriving linguistic variables and their associated membership functions was stressed, as well as SC-nets ability to construct control strategies from raw training data, without intervention from any human party. In light of the positive results obtained, it would seem justified to conclude the viability of knowledge acquisition for fuzzy control problems - at least in the domain studied - by means of machine learning, and warrant to continue investigating SC-net's applicability in this area.

## References

[1] Hall, L.O. and Romaniuk, S.G. (1990), A Hybrid Connectionist, Symbolic Learning System, AAAI-90, Boston, Ma.

[2] Kibler, D. and Aha, D.W. (1990), Learning representative Exemplars of Concepts: An Initial Case Study, in Readings in Machine Learning (ed. Shavlik and Dietterich), Morgan Kaufman, Los Gatos, Ca.

[3] Kiszka, J.B., Kochanska, M.E., Sliwinska, D. S. (1985) The Influence of some Fuzzy Implication Operators on the Accuracy of a Fuzzy Model - Part II, Fuzzy Sets and Systems 15, pp. 223-240.

[4] Lee, H., Romaniuk, S.G., Hall, L.O. (1991), A Study of Machine Learning Approaches for some Classification Domains, FLAIRS-91, Coco Beach, Florida, April.

[5] Mamudani, Ebraham H. (1977) "Applications of Fuzzy Logic to Approximate Reasoning Using Linguistic Synthesi s" IEEE Transactions on Computers, Vol. C-26, No. 12, December 1977 pp. 1182-1191

[6] Perez, R.A., Hall, L.O., Romaniuk, S.G., Lilkendey, J,T. (1992), Inductive Learning For Expert Systems In Manufacturing, HICCS-25, Hawaii International Conference on Systems Sciences, Koloa, Hawaii, January.

[7] Romaniuk, S.G. (1991) Extracting Knowledge from a Hybrid Symbolic, Connectionist Network, PhD Dissertation, University of South Florida.

[8] Romaniuk, S.G., Hall, L.O. (1992) SC-net: A Hybrid Connectionist, Symbolic Network, To appear: Journal of Information Sciences.

[9] Romaniuk, S.G., (1993) Representing Complex Fuzzy Membership Functions in a Connectionist network, International Fuzzy Systems and Intelligent Control Conference, Kentucky.

[10] Waterman, Donald A. (1986), *A Guide to Expert Systems.* Reading, Mass:Addison-Wesley.

[11] Zimmermann, H. J. (1991) "Fuzzy Set Theory - and Its Applications, Second, Revised Edition" Kluwer Academic Publishers

# CELL STATE SPACE ALGORITHM AND NEURAL NETWORK BASED FUZZY LOGIC CONTROLLER DESIGN

BaoSheng Hu(IEEE Senior Member) GenYa Ding
The system enginerring institute of
Xian Jiao University,Xian 710049
China

Abstract-This paper presents a new method to automatically design Fuzzy logic controller(FLC).The main problems of designing FLC are how to optimally and automatically select the control rules and the parameters of membership function(MF). Cell state space algorithms(CSS), differential competitive learning(DCL) and multilayer neural network are combined in this paper to solve the problems. When the dynamical model of a control process is known.CSS can be used to generate a group of optimal input output pairs(X,Y) used by a controller.The(X,Y) then can be used to determine the FLC rules by DCL and to determine the optimal parameters of MF by multilayer neural network trained by BP algorithm.

Keywords: Cell state space algorithm, Neural networks,Fuzzy logic controller design

## INTRODUCTION

Fuzzy logic has been widely used in control systems.Nonlinear,time-varying,ill-defined systems can be efficiently controlled by FLC. When the dynamic model of process is known, we can develop FLC for this process either by the expert experience to the process or by the known dynamic model[9].For very complex nonlinear system it is difficult to use traditional techniques to control it. Developing FLC to deal such complex system is an alternative.CSS[6][7][8] introduced by C.S.Hsu is an effective algorithm to globally analyse nonlinear systems.The optimal control of nonlinear system has been realized by CSS[8]. The optimal control tabel generated by CSS is used to realize optimal discrete control. Optimal continuous control can be realized by FLC after the input output data pairs are extracted from the optimal control table and are used to generate the FLC rules and to fine tune the parameters of MF.

When a group of data pairs of a control process are known,DCL can be used to generate the FLC rules[1][2].A multilayer neural network is used to realize the optimal seclection of the parameters of MF.

The paper in [3] proposes using CSS to generate the optimal control inputs-outputs,and uses the inputs-outputs to determine the consequent part of the fuzzy rules.The conseqent is not fuzzy set, but is a linear function of the fuzzy rules' input variable(the form is a0+a1*x1+a2*x2...).The a0,a1,a2... are determined by gradient decent algorithm.The paper in [4] pretends the inputs-outputs of a process are known .Then the paper trys to automatically generate FLC rules by a multilayer neural network.This paper integrates the ideas in [1][2][3][4][9] and forms a new method to generate FLC rules.

## CELL STATE SPACE ALGORITHM

CSS is an algorithm to generate the optimal control table. CSS is developed by cell to cell mapping which is originated from point to point mapping represented by x(n)=G(x(n-1),u(n),t(n)) ,x(n) is state vector, t(n) is the time interval during control u(n). For every control process only a finite region to which the state vectors belong is needed to be considered.So we can represent the finite region by a number of cells. Each N-dimensional vector (x1,x2,...,xn) correspond to N-dimensional cell(z1,z2,...,zn),z1,z2,...,zn are integers,(zi-1/2)*hi=<xi<=(zi+1/2)*hi ,hi is the interval size for state xi.After the cells(z1,z2,...,zn) are generated.we have a cell to cell mapping equation:z(n)=C(z(n-1),u(n),t(n)

0-7803-0614-7/93$03.00 ©1993IEEE

),u(n) belongs to U,t(n) belongs to J. There are Nu elements in U, and Nt elements in J. z(n)= C(z(n-1),u(n),t(n)) represents Nu*Nt possible mappings for each state z(n-1). When z(n-1) is mapped to z(n), there is a cost increment w(n)= Q(z(n-1),u(n),t(n)). If a dynamic process moves from z(0) through z(1),z(2),...,to arrive at z(ne),under a sequence of controls{u(j),t(j)},j= 1,2,...,ne,then the value of the cost function V for this dynamic process is given by V(p)=V(z(0),z(ne),u,t)=Q(z(0),u(1),t(1))+Q(z(0),u(2),t(2))+ ...+Q(z(ne-1),u(ne),t(ne)).To realize the CSS,we must first specify the possible cost increments w1<w2<w3<....From the possible increments we can form all the possible cost levels of the process v1<v2<v3<....,vi=k1*w1+k2*w2+...+kp*wp,ki>=0. If the target state is z(nt), the aim of the CSS is to find the optimal control table for each state z(j). The optimal control table includes:

U(j):identifies the control vector u to state z(j).

T(j):identifies the control time t during U(j)

Image(j):identifies the image state of z(j) under this optimal interval control.

W(j):cost increment for z(j).

V(j):total cost moving to target state z(nt) for z(j).

S(j): the number of steps moving to target state z(nt) for z(j).

The optimal control table is generated by dynamic programming approach.Starting with target state z(ne) and the smallest cost level v1,we find all the states that map to z(nt) at cost v1. We place the states and the z(nt) in set Y. Then starting with cost level v2 and each state in the Y,we find all the states that map to z(nt) at cost level v2. This process continues until every state is processed.

After the optimal control table is generated, the array U(j) denotes the control input for each state z(j),z(j) is the cell of state x(j). So we can use (x(j),U(j)), j=1,2,...,N. (N is the total number of cells that can reach to the target state) to design a FLC by DCL and multilayer neural network.

## DIFFERENTIAL COMPETITIVE LEARNING

For a controlller ,the input is x(j),the output of the controller is U(j). We pretend x(j) is two dimensional,represented by E(j) and CE(j).U(j) is one dimensional.We divide E(j),CE(j),U(j) into d1, d2 and d3 intervals,respectively.The intervals are not overlapped, each interval denotes a term of linguistic variable. So it is possible to have D d1*d2*d3 Fuzzy rules.We then use a two-layered network(Fig.1) to train (E(j),CE(j),U(j)),j=1,2,... N.

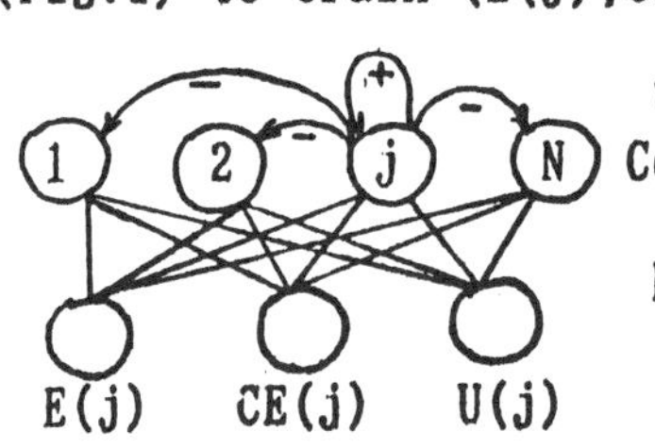


Fig.1. Topology of the laterally inhibitive DCL network

$$
M=\begin{vmatrix} m11 & m21 & \dots & mn1 \\ m12 & m22 & \dots & mn2 \\ m13 & m23 & \dots & mn3 \end{vmatrix}
$$

$$
W=\begin{vmatrix} 2 & -1 & -1 & \dots & -1 \\ -1 & 2 & -1 & \dots & -1 \\ & & \cdot & & \\ & & \cdot & & \\ -1 & -1 & -1 & \dots & 2 \end{vmatrix}
$$

mij is the weight between neuron i in input field Fx and neuron j in competition field Fy. The DCL algorithm that trains the network(Fig.1) is as follows: We represents (E(i),CE(i),U(i)) by x(i).

(1).initialize mi(0)=x(t).i=1,2,...,D.

(2).Find winning mj(t):||mj(t)-x(t)||=min||mi(t)-x(t)||.i=1,2,...,D.

(3) Update winning mj(t):

mj(t)=mj(t)+Ct*ISj(yj(t))[S(x(t))-mj(t)] if the jth neuron wins,mi(t+1)=mi(t) if the ith neuron loses. ISj(yj(t)) denotes the time change of the jth neuron's competition signal Sj(yj) in the competition layer Fy:

ISj(yj(t))=sgn[Sj(yj(t+1))-Sj(yj(t))].

sgn(x)=1 if x>0,sgn(x)=0 if x=0,sgn(x)=-1 if x<0.

yj(t+1)=yj(t)+ Sj(xi(t))*mij(t)+ Sk(yk(t))*Wkj.

After all the data pairs E(i),CE(i),U(i) i=1, 2,...,N, have been trained, the N columns of matrix M indicate the distribution of E,CE and U in the three dimensioanl space.Each column (mi1,mi2, mi3) represents a fuzzy rule whose interval is (It1,It2,It3),mi1,mi2,mi3 are in interval It1,It2 ,It3,respectively.The corresponding fuzzy rule is If E is L(it1) and CE is L(it2) then U is L(it3). L(iti) corresponds to the term whose interval is iti.

The intervals of MF in fuzzy rules produced by DCL is not overlapped,so the output of the FLC is not continuous and smooth.We can use multilayer neural network to fine tune the MF parameters.

## MULTILAYER NEURAL NETWORK

A neural network which realizes a FLC is illustrated at Fig.2. Two inputs,one output Fuzzy controller is represented by Fig.2.The linguistic variable E, CE and the output linguistic variable have three terms only.

The neural network is composed of three parts: the first part is fuzzify that turns linguistic variable into MF values. The second part is the inference part.Every link in this part represents a rule.So there are nine rules at Fig.2.The lower nodes in this part execute the precondition of a rule and the higher nodes execute the consequence of a rule. The third part is the defuzzify which is used to get the nonfuzzy control values.

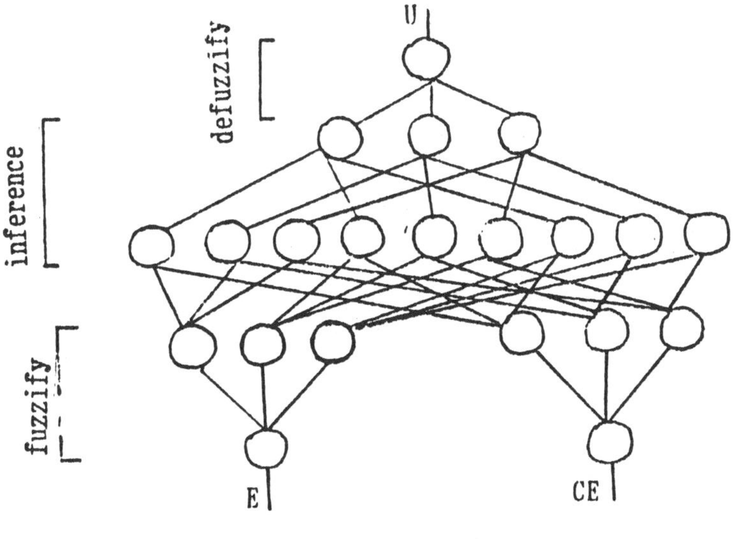


Fig.2 A multilayer neural network used to fine tune the parameters of the membership functions.

All MF are represented by bell-shaped function exp(-(ui-mij)/oij), ui is the input, mij is the center of the MF,oij is the width of the MF.The weights in the fuzzify part is mij.The nodes function in this part is exp(-(ui-mij)/oij). The weights between the fuzzify part and the inference part are 1.the weights in the inference part are 1.The lower nodes function in this part is min(u1,u2,..,up). The higher nodes function is min(1,u1+u2+...+up).The weights in the defuzzify part is mij*oij.The nodes function in this part is(mij*oij)*ui/ oij*ui.We have used DCL to determine the FLC rules. The intervals of MF have been fixed by hand. So we can map rules and the MF on this neural network.For a linguistic variable E,it has three terms corresponding to three intervals[I1,I2],[I2,I3] and [I3,I4].The first term width o11 is I2-I1,its center m11 is(I1+I2)/2.So o12=I3-I2,m12=(I2+I3)/2,o13=I4-I3,m13=(I3+I4)/2.

The parameters oij,mij can be fine tuned by this neural network using BP-algorithm and the training datas(x(j),U(j)) j=1,2,...,N,which generated by the CSS.The training time is substantially reduced for the oij and mij have been determined coarsely.

## CONCLUSION

Using the modern computer speed and storage to solve problems will get great benefits. The method proposed in this paper is computational-oriented. It is interesting to get a group of rules automatically from some complex linear or nonlinear differential equations.The CSS produces the optimal input U(j) for each state z(j).If all the optimal control data pairs (z(j),U(j)) are trained on the multilayer neural network successfully, then the parameters oij and mij is optimal. If the training does not converge,then it is necessary to redivide the state z(j) intervals needed by DCL and to use DCL to get FLC rules once more,and to map the FLC rules to the multilayer neural network, and to train this network again until the training finally converges.The final FLC produced in this way will be near optimal.

REFERENCES


[1] B.Kosko,"Neural Network and Fuzzy systems: A Dynamical systems approach to machine intelligence,"Englehood Cliffs,NJ: Prentice-Hall, 1992

[2] Seong-gon Kong and Bart Kosko,"Adaptive Fuzzy systems for backing up a truck-and-trailer," IEEE Tran. on Neural networks,vol.3,no.2, march 1992

[3] Samuel M.Smith and David J.Comor,"Autoamted Calibration of a fuzzy logic controller using a cell state space algorithm,"IEEE control systems,august 1991

[4] Chin-Teng Lin and C.S.George Lee, "Neural-network-based fuzzy logic control and decision system,"IEEE Tran.on Computer,vol.40,no .12,dec.1991

[5] C.S.Hsu, "A theory of cell-to-cell mapping dynamical systems,"J.Appl.Mechan.,vol.47,pp. 931-939,Dec.1980

[6] C.S.Hsu and R.S.Guttalu,"An unravelling algorithm for global analysis of dynamical systems:an application to cell to cell mapping," J.Appl.Mechan.vol.47,pp.940-948,dec.1980

[7] C.S.Hsu,"A discrete method of optimal control based upon the cell state space concept,"J. optimization theory and Appl.vol.46,no.4,pp. 547-569,aug,1985

[8] Peng Xian-Tu,"Generating rules for fuzzy logic controller by functions," Fuzzy sets and systems 36(1990) 83-89

# A PARALLEL FUZZY REASONING MODEL FOR FUZZY PRODUCTION SYSTEMS ON A MULTIPROCESSOR

Xiaozhe Zhao and Zhongtuo Wang

Institute of Systems Engineering
Dalian University of Technology
Dalian 116024, China

Abstract—This paper discusses a parallel fuzzy reasoning model for implementation of fuzzy production systems on a multiprocessor constructed by Transputers. Firstly, it introduces a new fuzzy reasoning model for sequential fuzzy production systems, which characterized by the meta-rules for the firing rule selection. Secondly, it is shown, by means of analysing of the interrelation among fuzzy rules, how sequential fuzzy production systems can be transformed into its parallel forms. Finally, three types of multiprocessor architecture used for implementing the new parallel fuzzy reasoning model are suggested.

## I. INTRODUCTION

A fuzzy production system (FPS) or a fuzzy-rule-based system is a production system in which knowledge representation and processing are all based on fuzzy logic. A FPS model is composed of 1) fuzzy database; 2) fuzzy rules; and 3) fuzzy inference mechanism. The objects and relations contained in the fuzzy database are represented in linguistic variables or fuzzy predicates. The fuzzy rule is represented as: IF LHS THEN RHS. LHS is a conjunction of condition elements which indicated whether or not the rule can be used in the sense of fuzziness, it takes the form of $C_1$ and $C_2$ and ... and $C_n$ where $C_i$ is a fuzzy proposition of the form: $X_i$ is $A_i$, and RHS is a set of actions of firing the rule, it takes the form of $Q_1$, $Q_2$, ..., $Q_m$, where $Q_j$ takes the form as: $Y_j$ is $B_j$ which expresses adding a proposition, $Y_j$ is $B_j$, to the database, or $-(Y_j$ is $B_j)$ which expresses eliminating a proposition, $Y_j$ is $B_j$, from the database. In the form shown above, $X_i$ and $Y_j$ are names of objects or relations, and $A_i$ and $B_j$ are fuzzy values of them respectively. The inference mechanism using fuzzy reasoning technique deduces a group of imprecise conclusions from the initial conditions. The process of a FPS to infer the conclusions is a process of it to change the state of fuzzy database by using fuzzy rules. These changes include adding (or eliminating) some fuzzy objects and relations into (or from) database and re-assigning the possibility distributions of any objects or relations. In other words, the solution of a special problem is not only the states of the database satisfying the goal condition, but also the pathes, the sequence of rules having been used, to the goal. The path from initial conditions to the goal is nondeterministic, it depends both on the initial condition and the current state of the rules.

The fuzzy inference mechanism repeatedly executes the Match-Select-Act cycle choosing and firing the rules to solve the given problem. In a sequential FPS only one cycle is executed each time, but, in a parallel FPS, instead of only one cycle being executed, multiple cycles are allowed to execute in different processors simultaneously. The simultaneous executing of the Match-Select-Act cycles include parallel matching, parallel selecting and parallel firing of the fuzzy rules.

This paper proposes a parallel fuzzy reasoning model for implementation of fuzzy production systems on a multiprocessor. The issues include:

1. A new sequential fuzzy-rule-based inference model characterized by the integration of the linguistic values, the meta-rules of firing rule selection, the compositional rules of fuzzy inference, and the syllogistic rules based on the fuzzy-quantifier-tables for inference using propositions with fuzzy quantifiers;

2. Parallelism in FPS and how to determine the rule sets in which the rules can be used parallelly;

3. The partitioning and mapping of FPS into multiprocssor;

4. The architectures of multiprocessor suggested to implement the parallel FPS (PFPS).

The rest of the paper is organized as follows. In section II we put forward and discuss the new sequential fuzzy-rule-based inference model. In section III parallelism in fuzzy-rule-based model is discussed. In section IV, a PFPS model is suggested, including the parallel fuzzy-rule-based model and several hardware architecture models used by it. Finally, section V is a summary of the paper.

## II. NEW FUZZY-RULE-BASED INFERENCE MODEL

The new fuzzy-rule-based inference model is composed of three parts, 1) a fuzzy knowledge representation model in which fuzzy knowledge is represented using possibility distribution associated with linguistic variable of synthetic language system, 2) a fuzzy inference model for categorical inference with some matching meta-rules for firing-rule selection and, 3) a fuzzy-quantifier-table model for syllogistic inference. Categorical inference rules and syllogistic inference rules were suggested by L.A.Zadeh and were summarized in [2].

### Knowledge Representation

Knowledge in the new inference model is represented with fuzzy propositions, fuzzy predicates and possibility distributions. Two representation forms, possibility distribution form and linguistic variable form, are equally used to represent the knowledge in the model. Using test-score semantics and translation rules, such as Rules Pertaining to Modification, Rules Pertaining to Quantification, Rules Pertaining to Composition and Rules Pertaining to Qualification, the meaning of each proposition or predicated can be translated into its possibility distribution form which is connected with the possibility distribution of its components. The universe of discourse of possibility distribution function may be either numerical sets or linguistic value sets. Propositions in the system may also be represented by linguistic variables and their values directly, take the canonical form [1] such as "X is F", "if X is F then Y is G" or "QA's are B's". All of the propositions contained in the system, facts and knowledge, are taken canonical form associated with their possibility distributions respectively.

0-7803-0614-7/93$03.00 ©1993IEEE

## Inference Method

Inference mechanism in the new fuzzy-rule-based inference model is nothing but a fuzzy inference model described by the collection of categorical rules associated with a collection of inference meta-rules called firing-rule-matching rules, or matching-rule for short.

We know from our early experiences that an important problem which arises in the operation of any fuzzy production system is the following. Suppose that the user supplies a fact which in its canonical form may be expressed as X is F, where F is a fuzzy or nonfuzzy predicated or a linguistic value. Furthermore, suppose that there is no rule in knowledge-base whose antecedent (LHS) matches F exactly. The question which arises is: Which rules should be fired and how should the results be combined?

The approach suggested by fuzzy logic involves the use of an interpolation technique which is based on the P-rule and is in the spirit of the generalized modus ponens. The results came from the use of this method are the combinations of all rules in KB whose antecedents are connected with the fact to a great or less extent. For example, there are n rules in KB:

$R_1$: if $X_{11}$ is $A_{11}$ and ... and $X_{1m}$ is $A_{1m}$ then $Y_1$ is $B_1$;

............

$R_n$: if $X_{n1}$ is $A_{n1}$ and ... and $X_{nm}$ is $A_{nm}$ then $Y_n$ is $B_n$;

The fact is stated by the form as:

F: $X_1$ is $A_1$ and ... and $X_l$ is $A_l$;

The possibility distributions induced by them are:

$R_1$: $H_1$; ... ; $R_n$: $H_n$; and F: A;

The possibility distribution of the inference result be represented as:

$A \cdot H_1 \cup \ldots \cup A \cdot H_n$;

From the view of surface, this method seems to make the consequences meaningful by taking all of the information about the fact in KB, but in the essential aspect, from our experiences of fuzzy inference, this method takes almost all of the noise (useless information) in KB into the consequences and at a higher computing-cost.

The new inference model solves the problem by means of using matching-rules. The matching-rules are a set of meta-rules for selecting the rules in KB whose antecedents match the fact in a sense of fuzzy match. The spirit of the matching-rule lies in the estimate of the matching degree between the rule and the fact. The estimation is focused the attention on three points:

1. The matching degree between the linguistic variables in the antecedent of the rule and the linguistic variables in the fact;

2. The matching degree between the linguistic variable values in the antecedent of the rule and the values of the same linguistic variables in the fact;

3. The compatibility between the meaning of the antecedent of the rule and the meaning of the fact.

Suppose that the rule in KB takes the form as:

R: if $X_1$ is $A_1$ and ... and $X_n$ is $A_n$ then Y is B;

and the fact takes the form as:

F: $Z_1$ is $C_1$ and ... and $Z_m$ is $C_m$;

the focus points mentioned above can be represented by following functions:

1. $$f(R,F)=\frac{\Sigma_N(\{X\}\cap\{Z\})}{\Sigma_N(\{X\})}$$

where $\{X\}=\{X_1, \ldots, X_n\}$ is the set of the linguistic variables in the rule, $\{Z\}=\{Z_1, \ldots, Z_m\}$ is the set of the linguistic variables in the fact, and, $\Sigma_N(\{X\})$ is the number of elements in the set of $\{X\}$ . The function f(R, F), as its definition, is the proportion of the number of linguistic variables in the antecedent of the rule which are in the fact simultaneously. The larger the f-value (the value of f(R, F)), the higher the matching degree of the rule and the fact in the discourse level.

2. $$g(R,F)=\frac{\Sigma\left(\frac{\Sigma_N(M(A_i)\cap M(C_j))}{\Sigma_C(M(A_i))}\right)}{\Sigma_N(\{X\})}$$

where $A_i$ and $C_j$ are the linguistic values of the same linguistic variables in the antecedent of the rule and the fact, $M(A_i)$ and $M(C_j)$ are the meanings of $A_i$ and $C_j$ which are represented by possibility distribution respectively, and, $\Sigma_C(M(A_i))$ is the sigma-count of $M(A_i)$. The large the g-value (the value of g(R,F)), the higher the matching degree of the rule and the fact in the characteristic level

3. $$h(R,F)=\frac{\Sigma(\delta(A_i,C_j))}{\Sigma_N(\{X\})}$$

where $\delta(A_i,C_j)=\frac{\Sigma(\mu_{Ai}(u_1)*\mu_{Cj}(u_1))}{\Sigma_C(M(A_i)\cup M(C_j))}$ is defined as the compatibility of the meaning of $A_i$ with the meaning of $C_j$, in which the meaning of $A_i$ and the meaning of $C_j$ are represented in the forms of $M(A_i)=\int\mu_{Ai}(u_1)/u_1$ and $M(C_j)=\int\mu_{Cj}(u_1)/u_1$ respectively. The function h(R,F) is the average compatibility of the meaning of the antecedence of the rule with the meaning of the fact. The larger the h-value (the value of h(R,F)), the higher the matching degree of the rule and the fact in the meaning level.

The matching-rule can be described by means of the functions f(R,F), H(R,F), and g(R,F) as following: For any given fact, the f-value, g-value and h-value of every rule in the KB can be calculated, the rules to be executed are selected according to their f-value, g-value and h-value, the selecting sequence is f-, g-, h-, and from large to small value.

When a rule is selected using the matching-rules, the conclusion is conbined from the rule and the fact by means of categorical rules.

## Quantifier-table Model for Syllogistic Inference

In the new fuzzy reasoning model, the syllogistic rules may be described as:

$Q_1$A's are B's
$Q_2$B's are D's
---
Q E's are F's

From the syllogistic rules, we know that the quantifier in conclusion, Q, can be calculated only by $Q_1$ and $Q_2$, and E, F are merely the combination of A, B, C, D. To utilize this feature of syllogistic rules, we can establish a quantifier-table for each of the syllogistic rule to store the quantifiers of calculating results. A quantifier-table is a matrix of quantifiers. The following table shows the quantifier-table of Intersection/Product syllogism in a synthetic language system in which a, b, c, d, e are the quantifiers involved only:

| | a | b | c | d | e |
|---|---|---|---|---|---|
| a | d | b | e | c | a |
| b | e | c | ... | | |
| c | | ... | | | |
| d | | ... | | | |
| e | | ... | | | |

For example, if $Q_1$=a, $Q_2$=c then Q=e. The fuzzy variables in the results, E and F, can be calculated from A, B, C, D directly according to the syllogistic rule which defined the calculating form. In Intersection / Product syllogizm, we have E=A and F=C $\wedge$D.

## III. PARALLELISM IN THE FUZZY-RULE-BASED INFERENCE MODEL

Although the advantages of the new inference model mentioned above have provided substantial performance improvements for FPS, speed improvements are still required for FPS to solve the real-world complex problems. Parallel processing technique can be used to increase the processing speed of a FPS.

### Parallelism in FPS

There are four levels of parallelism in FPS:

1) parallelism in the representation of fuzzy rules;
2) parallel rule matching;
3) parallel rule firing; and
4) Parallelism in the process of firing a fuzzy rule (Parallelism within the fuzzy rule).

This paper focus on the discussing of the parallelism in the representation of fuzzy rules and parallel fuzzy rule firing, because that parallel rule matching does not bring out any trouble to the working process of FPS, but parallel rule firing does. In FPS, only fuzzy-compatible rules can be fired concurrently. The result of parallel execution of a set of fuzzy-compatible rules is same as, under a special fuzzy judgment, the result of sequential execution in any order of these rules.

### Fuzzy-Compatiblism Analysis

The analysis of fuzzy rule interrelation to find the fuzzy-compatible rule sets is fundamental to implement PFPS. For any pair of two fuzzy rules, several interrelations can be defined to determine whether or not they are compatible in a fuzzy sense. As stated before, a fuzzy rule can be represented as:

R: IF LHS THEN RHS.

The interrelations are defined as following:

DEFINITION 1.1: Rule $R_1$ is input interrelated to rule $R_2$ in the sense of $\alpha$-cut, if there is, at least, one $Q_j$ in $RHS_2$ taking the form as -($Y_j$ is $B_j$), and Sup($LHS_1 \cap$ ($Y_j$ is $B_j$))$\geqq \alpha$. Where Sup(P) is the maximum of the possibility distribution values of P. This indicates that, in the sense of $\alpha$-cut, firing $R_2$ will destroy the conditions to fire $R_1$.

DEFINITION 1.2: Rule $R_1$ is adding/eliminating interrelated to rule $R_2$ in the sense of $\alpha$-cut, if there is, at least, one $Q_j$ in $RHS_2$ taking the form as -($Y_j$ is $B_j$), and Sup($THS_1 \cap$($Y_j$ is $B_j$))$\geqq \alpha$. Where Sup(P) is the maximum of the possibility distribution values of P. This indicates that, in the sense of $\alpha$-cut, firing $R_2$ will eliminate the same element(s) as the element(s) added by firing $R_1$.

DEFINITION 1.3: Rule $R_1$ is output-input interrelated to rule $R_2$ in the sense of $\alpha$-cut, if there is, at least, one $Q_j$ in $RHS_1$ taking the form as -($Y_j$ is $B_j$) or ($Y_j$ is $B_j$), and Sup(($Y_j$ is $B_j$)$\cap LHS_2$)$\geqq \alpha$. Where Sup(P) is the maximum of the possibility distribution values of P. This indicates that, in the sense of $\alpha$-cut, firing $R_1$ will create or destroy the conditions to fire $R_2$.

DEFINITION 2: Two rules $R_1$ and $R_2$ are fuzzy-compatible (that is compatible in the sense of $\alpha$-cut) if they are neither input interrelated nor adding/eliminating interrelated in the sense of $\alpha$-cut in both directions.

THEOREM 1: If two rules R1 and $R_2$ are fuzzy-compatible in the sense of $\alpha$-cut, then they can be fired concurrently without any problem in the sense of $\alpha$-cut.

According to the THEOREM 1, the fuzzy rules of a FPS can be partitioned into several sets, in which the rules can be fired concurrently in the sense of $\alpha$-cut. It is useful for mapping a FPS into a multiprocessor and is useful for parallel rules firing. DEFINITION 1.3 gives important information about how the firing of a fuzzy rule affects the firing of other rules, it is useful for controlling the communication in the multiprocessor.

## IV. PARALLEL FUZZY PRODUCTION SYSTEM MODEL

### Hardware Architecture Model of PFPS

The hardware model used for implementing PFPS is a message-passing multiprocessor and it can be constructed using a host computer and transputers. The simple one is shown in Fig. 1. Its derivates are shown in Fig.2 and Fig.3. Each architecture is suitable for some problems according to the structure of the problems.

In the simple architecture shown in Fig.1, Tran1 is the controller called control-Transputer (CT) of the system which contains a global fuzzy database and a scheduler of the system. Parallel reasoning is done by Tran21, Tran22 and Tran23, which are called reasoning-Transputer (RT), each of them contains a distributed fuzzy database, a distributed fuzzy rulebase stored the rules assigned to this processor and a control memory used to store the parallelism and communication information of the rules. The links connected to the processors in the system are the links provided by Transputer,

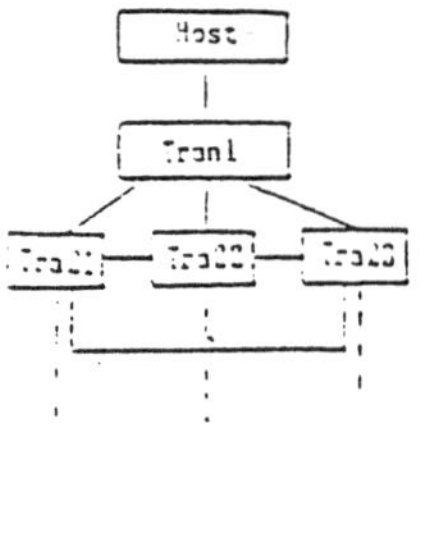


Fig.1

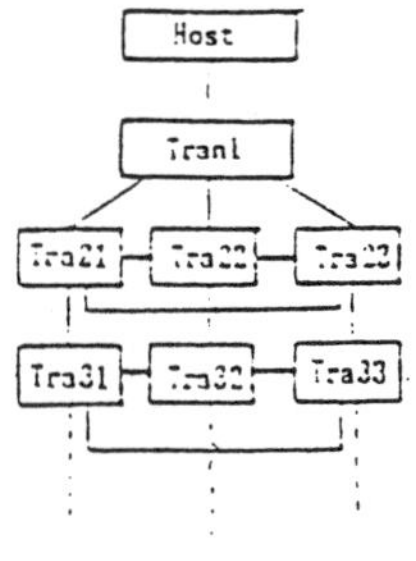


Fig.2

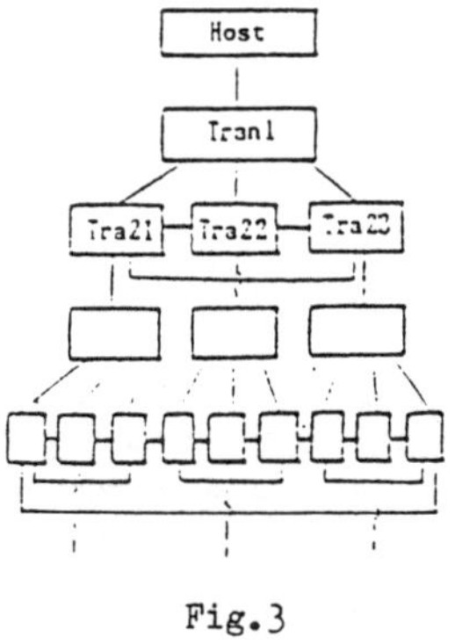


Fig.3

they play the part of the message-passing channels of the system, in which the control information and the consequences of rules firing are transmitted to coordinate the working process of the system. The architecture shown in Fig.2 is suitable for the problems in which there is a vast amount of calculation for rule application. In this architecture, Tran31 through Tran 33 are used for calculation, each of them can be used to deal with different rules concurrently or to deal with a single rule together. The last one is designed for the problems with hierarcheal structure. The host computer in each system takes the part of user interface. There are three links left in every system. These links can be used to connect other multiprocessors, therefore, a large amount of varieties can be constructed to meet the needs of any kind tasks by this way.

Software Model of PFPS

1. Parallelism and communication representation

In order to express the parallelism and the communication connection in the PFPS, two matrices are constructed: 1) parallelism matrix P; 2) communication matrix C. Matrices P, C are all the binary square matrices with the dimensionality equal to the number of rules. each $p_{ij}$ is defined as:

$$p_{ij}=\begin{cases}1 & \text{if } R_i \text{ and } R_j \text{ can be used in parallel in the sense of } \alpha\text{-cut} \\ 0 & \text{otherwise}\end{cases}$$

Matrix P carries information about the parallel firing of rules, that is, if $p_{ij}=1$ then $R_i$ and $R_j$ can be fired concurrently.

Matrix C is defined as following. each $c_{ij}$ is defined as:

$$c_{ij}=\begin{cases}1 & \text{if } R_i \text{ and } R_j \text{ is output-input interrelation in } \alpha\text{-cut} \\ 0 & \text{otherwise}\end{cases}$$

Matrix C carries information about the communication relation of the rules, that is, if $c_{ij}=1$ then the consequences of firing $R_i$ must be transmitted to the processor contained $R_j$.

2. Rules and initial data assignment

An important thing is the proper assignment of rules and initial data to each processor. Matrices P and C are used for assignment. The objective is to decrease the communication cost while maintaining the inherent parallelism. The principles of assignment are:

1) Using P to identify the sets of parallel rules;
2) Assigning parallel rules to different RT as much as possible;
3) Using C to identify the sets of rules which must communicate each other;
4) Assigning the rules in the same set identified from 3) to same RT as much as possible;
5) The initial data assigning to a RT would be consistent with the LHSs of the rules assiged to the RT in a fuzzy sense as much as possible. The data left, after initial data assignment, can be stored in CT.

3. The Scheduler

There are two ways to control the working process of a PFPS:

1) Forced synchronization: The CT synchronizes the beginning of Match-Select-Act cycle in each RT. When it sends a start-signal to all of the RTs, every RT begins its own Match-Select-Act cycle to do the rule matching, parallelism checking, updating and communicating each other, and then wait for the next start-signal.

2) Competing: There is no start-signal in this controlling method. The RT match rule respectively. When any RT finds a rule matched the current database state, it checks whether or not this rule can be used with the rules being used by other RTs concurrently. If the rule can be used now, the RT sends a message to CT to inform the system which rule is being used, then it fires the rule and transmits the results to other RTs if necessary. After these, it receives the information coming from CT and other RTs and modifies its database state to maintain the consistency of whole system, and then it begins the next rule using cycle.

Implementation

System Fig.1 has been implemented on a system with PC 386 and 4 T800 Transputers. The language is OCCAM. A simple example has been run successfully on it.

## V. CONCLUSION

In this paper, a new model for implementation of PFPS on a multiprocessor is introduced. The model is a new fuzzy-rule-based inference model. Parallelism is expoited at both rule matching level and rule firing level. Three hardware architectures are introduced to implement the PFPS, there are all the message-passing multiprocessors. Further work about this kind of PFPS will focus on the improvement of the inference methods of the model, analysing algorithm of run-time parallelism of fuzzy rules and the architecture of the systems. Another important work is performance evaluation on several FPS applications.

## REFERENCES


[1] L.A.Zadeh, "The role of fuzzy logic in management of uncertainty in expert systems, "Fuzzy Sets and Systems, 11(1983), 199-227

[2] L.A.Zadeh, "Knowledge representation in fuzzy logic," IEEE Trans. on Knowledge and Data Engineering, Vol.1, No.1 (1989), 89-100

[3] N.S.Freeling, "Fuzzy sets and decision analysis," IEEE Trans. on Syst. Man, and Cybern., Vol.10, No.7 (1980), 341-354

[4] H.J.Zimmermann, "Fuzzy sets, decision making, and expert system," Kluwer Academic Publishers, USA,(1987)

[5] D.Moldovan, "RUBIC: A multiprocessor for rule-based systems," IEEE Trans. on Sys. Man, and Cybern., Vol.19, No.4 (1989), 699-706

[6] T.Ishida, "Parallel rule firing in production systems," IEEE Trans. on Knowledge and Data engineering, Vol.3, No.1 (1991), 1-17

[7] A.Carling, "Parallel processing-Occam and the transputer," Sigma Press, Wilmslow, Cheshire, U.K. (1988)

[8] X.Zhao & Z.Wang, "A multiprocessor with fuzzy reasoning for decision support systems." in.S.Hu & Z. Jiang Eds. "Information and systems," International Academic Publishers, China (1991)

# Real-time Fuzzy Control of Mean Arterial Pressure in Postsurgical Patients in an Intensive Care Unit

Hao Ying[1,2,3], Michael McEachern[4], Donald W. Eddleman[4] and Louis C. Sheppard[1,2]

1) Department of Physiology and Biophysics
2) Biomedical Engineering Center
3) Office of Academic Computing
University of Texas Medical Branch, Galveston, TX 77555, USA
4) Carraway Methodist Medical Center, Birmingham, AL 35234, USA

**Abstract — We have developed a fuzzy control system to provide closed-loop control of mean arterial pressure (MAP) in postsurgical patients in a cardiac surgical intensive care unit setting, by regulating sodium nitroprusside infusion. The core of the control algorithms was a nonlinear proportional-integral (PI) controller, which precisely represents a simplest fuzzy controller. The proportional-gain and integral-gain adjusted continuously according to error and rate change of error of MAP. Twelve postoperative patients who exhibited elevated MAP following coronary artery bypass grafting procedures took part in the study. The length of time that 12 patients were on the fuzzy control system ranged from 1 hour 45 minutes to 18 hours 7 minutes. The total fuzzy-controller-run time was 95 hours 13 minutes. Clinical results showed that the average percentage of time in which MAP stayed between 90% and 110% of the MAP setpoint was 89.31%, with a standard deviation of 4.96%.**

## I. INTRODUCTION

The fast-acting vasodilator drug sodium nitroprusside (SNP) is used to treat patients who demonstrate elevated systemic arterial blood pressure after open-heart surgery. The rapid and powerful action of SNP imposes upon nursing personnel the task of frequent monitoring of mean arterial pressure (MAP) followed by adjustment of SNP infusion rate. Because nurses have many other duties, inappropriate or infrequent control actions on SNP adjustment may occur, which may lead to poor system performance.

To improve the quality of patient care, automatic closed-loop control SNP delivery systems have been developed. A nonlinear proportional-integral-derivative (PID) control system was first built and used clinically in the mid-1970s [7]. Various control algorithms including nonlinear adaptive control, multiple-model adaptive control and adaptive multivariable control were developed and tested [1][2][3][4][5][6][8][9]. The success of developing the above-mentioned drug delivery control systems, especially the adaptive control systems, heavily depended on mathematical models of patients. However, accurately identifying a mathematical model of patients is a very difficult task, due to the complexity of the human body. Alternatively, fuzzy control may be used.

We previously developed a generalized expert-system-shell-based fuzzy controller [10], which we then utilized to control MAP by regulating SNP infusion, in both digital computer simulation and real-time in pigs [11][12]. In this paper, we report the results of fuzzy control of MAP in postsurgical patients clinically (see also [14]).

## II. NONLINEAR FUZZY CONTROL ALGORITHMS

The generalized expert-system-shell-based fuzzy controller had two input fuzzy sets, three output fuzzy sets, four fuzzy control rules, fuzzy logic AND and OR and a center of gravity defuzzification algorithm. To greatly reduce execution time and reveal the structure of the fuzzy controller, we analytically converted the expert-system-shell-based fuzzy controller into nonfuzzy controller which turned out to be a nonlinear PI controller [13]. The inputs of the fuzzy controller at sampling time nT are the scaled error and rate change of error of MAP:

$$GE \cdot e(nT) = GE[\ \text{setpoint} - MAP(nT)\ ], \qquad (2.1)$$

$$GR \cdot r(nT) = GR[\ MAP(nT-T) - MAP(nT)\ ] / T \qquad (2.2)$$

where GE and GR are the input scalars, T is sampling period and the setpoint is the desired MAP level. The nonlinear PI controller is described as:

0-7803-0614-7/93$03.00 ©1993IEEE

$$GI \cdot \delta SNP(nT) = K_i \cdot e(nT) + K_p \cdot r(nT) \quad (2.3)$$

where the nonlinear proportional-gain and integral-gain are

$$K_p = \frac{L \cdot GI \cdot GR}{3L - input} \quad (2.4)$$

$$K_i = \frac{L \cdot GI \cdot GE}{3L - input} \quad (2.5)$$

and

$$input = \begin{cases} GE|e(nT)|, & GR \cdot |r(nT)| \le GE \cdot |e(nT)| \le L \\ GR|r(nT)|, & GE \cdot |e(nT)| \le GR \cdot |r(nT)| \le L \end{cases}. \quad (2.6)$$

$\delta SNP(nT)$ is the incremental SNP infusion rate. GI is the output scalar for $\delta SNP(nT)$. It is obvious that $K_p$ and $K_i$ change with e(nT) and r(nT). The larger the absolute value of e(nT) or r(nT), the larger $K_p$ and $K_i$, which help reduce e(nT) and r(nT) quickly. On the other hand, the smaller the absolute value of e(nT) and r(nT), the smaller $K_p$ and $K_i$, which drive SNP gradually to the setpoint in a stable manner. Hence, the fuzzy controller is a nonlinear adaptive PI controller.

A new SNP infusion rate, SNP(nT), is computed as

$$SNP(nT) = SNP(nT\text{-}T) + GI \cdot \delta SNP(nT) \cdot T. \quad (2.7)$$

It should be noted that above nonlinear PI control algorithms are the analytical description of the expert-system-shell-based fuzzy controller, precisely representing the fuzzy controller.

## III. CLINICAL SETTING

The fuzzy controller was used to maintain desired MAP in patients in the Cardiac Surgical Intensive Care Unit (CICU) of the Carraway Methodist Medical Center. Fig. 1 is a block diagram of the implemented fuzzy control SNP delivery system. A Hewlett-Packard 78534 Monitor/Terminal was used to collect, process and display MAP, systolic pressure, diastolic pressure, left atrial pressure, right atrial pressure, heart rate and the electrocardiogram. A Puritan-Bennett 7200a Microprocessor Ventilator was connected to the patients to maintain respiration. MAP values were fed from the Hewlett-Packard Monitor into an IBM PS/2 Model 70 computer, ran the fuzzy controller in the form of the nonlinear control algorithms encoded in C programming language. SNP infusion rate calculated by the fuzzy controller was sent to an Abbott/Shaw LifeCare™ Pump Model 4. The pump infused SNP to patients.

Twelve postoperative patients who exhibited elevated MAP following coronary artery bypass grafting procedures took part in the study. Typically the trials began within one to two hours after the patients arrived in CICU. The typical MAP setpoint, determined by the attending medical doctors or the nurses, was 80 mm Hg. The fuzzy control system was started by technical personnel when the attending nurses thought SNP was needed for a patient. The fuzzy control system was always initiated at a SNP infusion rate of zero.

## IV. CLINICAL PERFORMANCE OF THE FUZZY CONTROL SNP DELIVERY SYSTEM

During the trials, all normal patient care duties were performed by the nurses. The duties included sampling patient blood, suctioning the patient to clear his/her airway, bathing the patient, changing bed linen, injecting drugs other than SNP, infusing blood, and so on. MAP in the patient frequently fluctuated considerably when the above-mentioned duties were being carried out. Besides these situations, other factors also affected MAP. Substantial fluctuation of MAP took place as body temperature of the patients was changing or if the patients were in pain. Spontaneous fluctuation of MAP occurred as well. In addition to these, sensitivity of the patients to SNP changed with time. The response delay to SNP varied among the patients.

Fig. 2 shows a typical trend plot of both MAP and the corresponding SNP infusion rate obtained from a fuzzy controller controlled patient. For this specific patient, blood was sampled at 12:57, 13:42, 15:56 and 17:50. Suctioning the patient began at 13:04, 17:00 and 19:17. The patient was bathed between 15:36 to 15:50. Changing bed linen started at 19:45 and lasted for several minutes. Injection of Vallium took place at 13:09, 14:41 and 17:57. The drugs Pavulon and Morphine were injected into the patient at 14:50 and 17:10, respectively. As the result shows, the fuzzy control SNP delivery system regulated MAP satisfactorily even with the fluctuation of MAP caused by the various factors stated above. For this patient, the percentage of time in which MAP stayed within the band between 90% and 110% of the MAP setpoint was 86.5%. The trial lasted 7 hours and 49 minutes.

The length of time that 12 patients were on the fuzzy control system ranged from 1 hour 45 minutes to 18 hours 7 minutes. The total fuzzy-controller-run time was 95 hours 13 minutes. For the sampling period T=10 seconds, 34,278 MAP samples were collected from the patients. The overall performance of the fuzzy control SNP delivery system in 12 patient trials is summarized in Table 1. The table exhibits that MAP is tightly controlled around the desired MAP level.

## V. DISCUSSION

A wide variation of patient sensitivity to SNP was experienced during the clinical trials. The fuzzy controller could cope with different sensitivity by continuously adjusting its nonlinear proportional-gain $K_d$ and integral-gain $K_i$. Fig. 3 shows simulated results using a patient model [8], which indicate that the fuzzy control SNP system could adapt to a wide range of patient sensitivity, from the sensitive patients (K=-2.88) to the insensitive patients (K=-0.18), a ratio of 16:1. This range of sensitivity covers that of most patients.

## VI. CONCLUSION

Results of the clinical trials on 12 patients revealed that the performance of the fuzzy control SNP delivery system was clinically acceptable. Based on the clinical results and the simulated results, we expect the fuzzy control SNP delivery system to perform well for most patients.

## REFERENCES


[1] R. A. de Asla, A. M. Benis, R. A. Jurado, and R. S. Litwak, "Management of postcardiotomy hypertension by microcomputer-controlled administration of sodium nitroprusside," *J Thorac Cardiovasc Surg*; 89: 115-120, 1985.

[2] J. J. Hammond, W. M. Kirkendall, and R. V. Calfee, "Hypertensive crisis managed by computer controlled infusion of sodium nitroprusside: a model for the closed loop administration of short acting vasoactive agents," *Comput Biomed Res*; 12: 97-108, 1979.

[3] J. F. Martin, A. M. Schneider, and N. T. Smith, "Multiple-model adaptive control of blood pressure using sodium nitroprusside," *IEEE Trans Biomed Engineering*; 34: 603-611, 1987.

[4] L. J. Meline, D. R. Westenskow, N. L. Pace, and M. N. Bodily, "Computer-controlled regulation of sodium nitroprusside infusion," *Anesth Analg*; 64: 38-42, 1985.

[5] J. S. Packer, D. G. Mason, J. F. Cade, and S. M. Mckinley, "An adaptive controller for closed-loop management of blood pressure in seriously ill patients," *IEEE Trans Biomed Engineering*; 34: 612-616, 1987.

[6] J. H. Petre, D. M. Cosgrove, and F. G. Estafanous, "Closed loop computerized control of sodium nitroprusside," *Trans Am Soc Artif Intern Organs*; XXXIX: 501-505, 1983.

[7] L. C. Sheppard, "Computer control of the infusion of vasoactive drugs," *Ann of Biomed Engineering*; 8: 431-444, 1980.

[8] J. Slate, and L. C. Sheppard, "Automatic control of blood pressure by drug infusion," *IEE Proc*; 129, Pt. A, No. 9, 1982.

[9] G. I. Voss, P. G. Katona, and H. J. Chizeck, "Adaptive multivariable drug delivery: control of arterial pressure and cardiac output in anesthetized dogs," *IEEE Trans Biomed Engineering*; 34: 617-623, 1987.

[10] H. Ying, W. M. Siler, and D. M. Tucker, "A new type of fuzzy controller based on fuzzy expert system shell FLOPS," Proceedings of IEEE International Workshop on Expert Systems and Their Application in Industry, Japan, May, 1988.

[11] H. Ying, L. C. Sheppard, and D. M. Tucker, "Expert-system-based fuzzy control of arterial pressure by drug infusion," *Med Prog thr Technol*; 13: 202-215, 1988.

[12] H. Ying, and L. C. Sheppard, "Real-time expert-system-based fuzzy control of mean arterial pressure in pigs with sodium nitroprusside infusion," *Med Prog thr Technol*; 16: 69-76, 1989.

[13] H. Ying, W. M. Siler, and J. J. Buckley, "Fuzzy control theory: a nonlinear case," *Automatica*; 26: 513-520, 1990.

[14] H. Ying, M. McEachern, D. W. Eddleman, and L. C. Sheppard, "Fuzzy control of mean arterial pressure in postsurgical patients with sodium nitroprusside infusion," *IEEE Trans Biomed Engineering*; 39(10): 1060-1070, 1992.

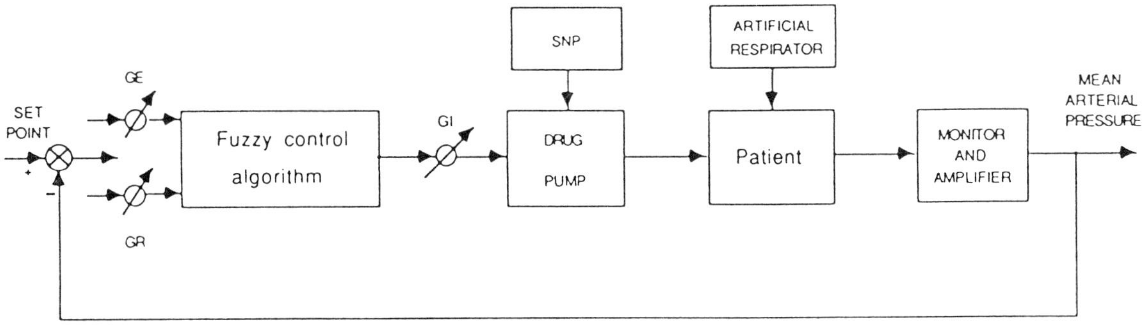


Fig. 1. Block diagram of the fuzzy control SNP delivery system.

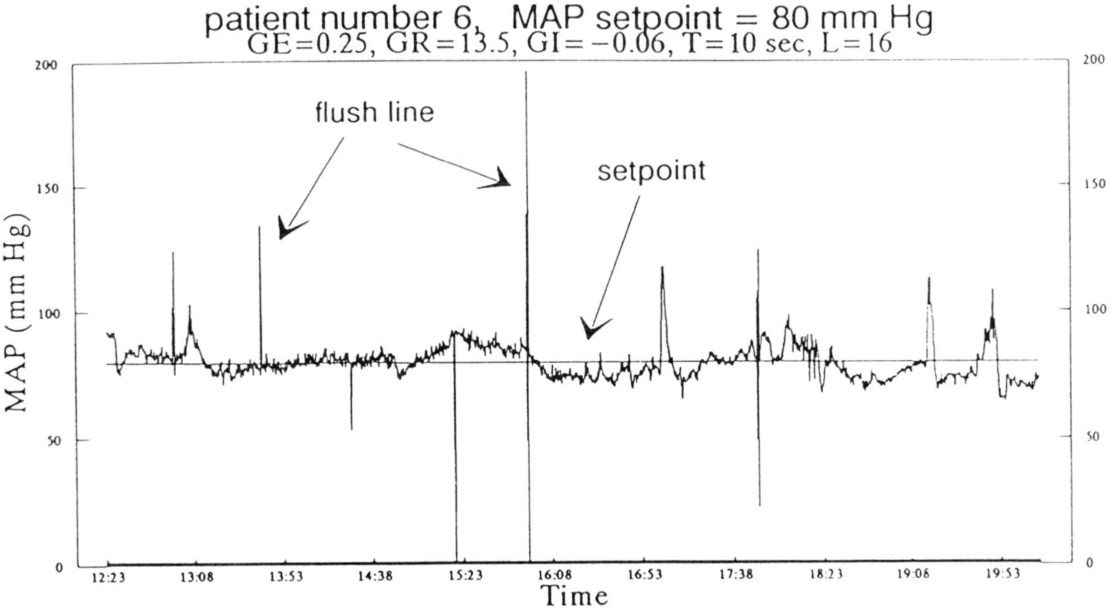


Fig. 2(a). MAP response for a single patient obtained by using the fuzzy control SNP delivery system clinically.

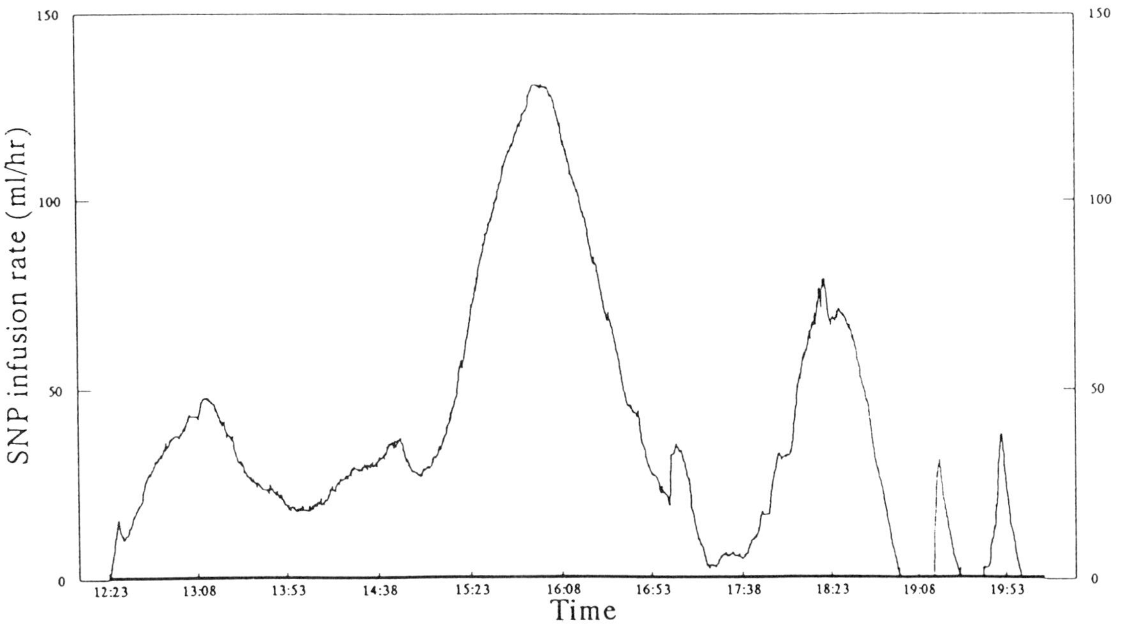


Fig. 2(b). The corresponding SNP infusion rate.

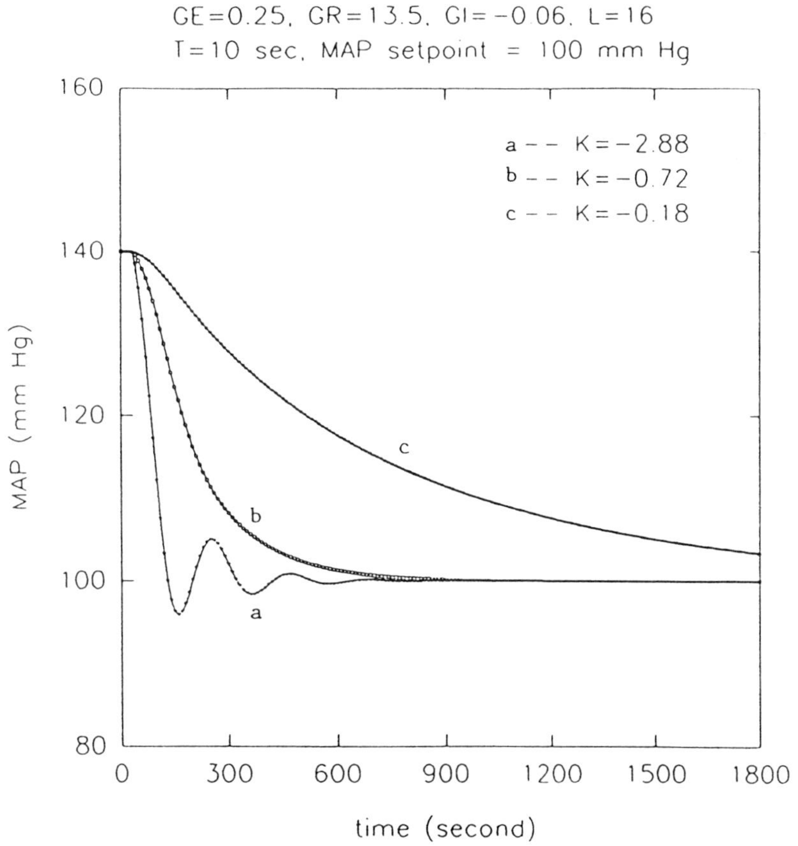


Fig. 3. Simulated MAP for the sensitive patients (K = -2.88), the normal patients (K = -0.72) and the insensitive patients (K = -0.18).

Table 1. Mean ($\mu$) percent of total fuzzy-controller-run time (and standard deviation $\sigma$) for different MAP intervals. The calculation is based on 12 patient trials. $MAP_d$ is the desired MAP.

| | $< 0.8MAP_d$ | $(0.8\text{-}0.9)MAP_d$ | $(0.9\text{-}1.1)MAP_d$ | $(1.1\text{-}1.2)MAP_d$ | $> 1.2MAP_d$ |
|---|---|---|---|---|---|
| $\mu$ | 1.00 | 3.92 | 89.31 | 3.85 | 1.92 |
| $\sigma$ | 1.09 | 2.72 | 4.96 | 1.84 | 1.14 |

# AN AUTOMATED RULE DESIGN OF FUZZY LOGIC CONTROLLERS FOR UNCERTAIN DYNAMIC SYSTEMS

Hoon Kang

Dept. of Control & Instrumentation Engineering, Chung-Ang University
221 Huksuk-dong, Dongjak-gu, Seoul 156-756, Korea
(hkang@krcaucc1.bitnet)

## ABSTRACT

*We investigate a systematic design procedure of* ***automated rule generation*** *of fuzzy logic based controllers for uncertain dynamic systems such as an engine dynamic model. By "automated rule generation" we mean autonomous clustering or collection of such* ***meaningful transitional relations*** *from one conditional subspace to another. During the design procedure, we also consider* ***optimal control strategies*** *such as minimum squared error, near minimum time, minimum energy or combined performance criteria. Fuzzy feedback control systems designed by the* ***cell-state transition method*** *have the properties of* ***closed-loop stability, robustness*** *under parameter variabtions, and a certain degree of* ***optimality****. Most of all, the main advantage of the proposed approach is that* ***reliability*** *can be potentially increased even if a large grain of uncertainty is involved within the control system under consideration. A numerical example is shown in which we apply our strategic fuzzy controller design to a highly nonlinear model of engine idle speed control.*

## 1. Introduction

Fuzzy logic/linguistic control can be catagorized as a knowledge-based system or an expert control paradigm, the reason for which is that every control action derived by the fuzzy inference engine is based on some a priori knowledge source (learning) [1]. However, the difficulties in constructing a rule base have prevented the fuzzy engineers from approaching to a generalized methodology for fuzzy rule based control systems [2] in Figure 1.

We propose a systematic design procedure of automated rule generation[1)] for uncertain dynamic processes. Fuzzy logic based feedback control is suitable for our physical target of automated rule generation. Membership functions stored in a fuzzy logic controller can be easily modified and updated without repetitive tedious re-evaluation of different dynamic models. This procedure of generating the rules required in a fuzzy logic controller should guarantee stability of the closed-loop system and robustness under parameter variations. We utilize the cell-to-cell mapping theory originally introduced by Hsu[3] and later applied to fuzzy dynamic systems by Chen et al.[4] The key point in making a stable control rule base is that every stabilizable feedback system has a chain of state transitions from one cell-state to another. Consequential elements of such transitions are anticipated according to applied control action of each rule. Data required for this transitional set of rules can be collected via [5]

* *a priori* information such as experimental results
* numerical simulation runs based on dynamic models
* expertise and heuristics.

Specifications, accuracy and precision, of the system tolerances can be arbitrarily adjusted and are a function of resolution of design parameter[6]. The next section deals with the step-by-step procedure as to the synthesis of a fuzzy logic control rule base based on the given input-state data pairs of a particular nonlinear dynamic model. These are the training data for approximate learning.

## 2. Automated Rule Generation for Fuzzy Logic Controllers

### 2.1 Fuzzy Logic Control Based On Cell-State Transitions

General fuzzy controllers have four components: fuzzifier, rule base, fuzzy inference engine, and defuzzifier. The control rules can be determined by using the cell-to-cell mapping theory[3] and the cell-state transitions that are introduced here. Comparing with the point-to-point mapping theory, this concept makes use of the intervals and a finite number of cells in the cell-state. The dynamical characteristics are preserved as accurately as the resolution allows.

DEFINITION 1: A "Cell" z is defined as an n-tuple of integers in the cell-state space Z such that

$$z = \{z_1, z_2, \ldots, z_n\} = \sum_{i=1}^{n} z_i\, e_i \qquad (1)$$

where $e_i$ is the unit vector in the direction of $z_i$ and the corresponding $x_i$ is represented by an interval $X_i$.

$$(z_i - 0.5)\, h_i < x_i < (z_i + 0.5)\, h_i \qquad (2)$$

$h_i$ ... interval size, $z_i$ ... integer representing $x_i$ ■

DEFINITION 2: A "Cell-to-Cell Mapping" F is a relation between cells in the cell-state space, F: Z —> Z, and the function values have one-to-one correspondence with the point-to-point mapping f such that

$$x_{k+1} = f(x_k) \in z^* \Longleftrightarrow z^* = F(z_k) \qquad (3)$$ ■

DEFINITION 3: An "Equilibrium Cell" $z^*$ (Invariant Cell) of F satisfies

1) This research is partially supported by the KOSEF under grant no. 92-22-00-08.

0-7803-0614-7/93$03.00 ©1993IEEE

$$z^* = F(z^*) \quad (4)$$ ■

DEFINITION 4: A k-Period Motion Cell is the distinct k cells $z(1),\ldots,z(k)$ that satisfy

$$z(1) = F^k(z(1)) \text{ and } z(m+1) = F^m(z(1)).$$ ■

As an example, four 4-period motion cells are represented in Figure 2.

## 2.2 Systematic Procedure for Automated Rule Generation

STEP 1: Consider a two-state dynamic model given by

$$dx_1/dt = f_1(x_1, x_2, \delta, \theta) \quad (5a)$$
$$dx_2/dt = f_2(x_1, x_2, \delta, \theta) \quad (5b)$$

where $x_1$, $x_2$ are the states (or the errors); $f_1$, $f_2$ are the nonlinear functions; and $\delta$, $\theta$ are the control inputs. From the admissible controls, we select finite representative constant values of $\delta$'s and $\theta$'s and we call them $\delta_i$'s and $\theta_j$'s. These crisp numbers will be fuzzified later after the performance is satisfied. Moreover, we choose finite representative points in the state space to anticipate the trajectories from one subspace to another. These countless trajectories are called a 'manifold' and one set of data is collected by setting $\delta_i$ and $\theta_j$ constant. Repeated collections of such information are used next in order to obtain a rule base for feedback regulation of $x_1$ and $x_2$, i.e., $x_1$, $x_2$ —> 0 as t increases. We denote $X_{1m}$, $X_{2n}$ the m-th and the n-th interval sets of $x_1$ and $x_2$, respectively. Linguistically, we define

$$L_{mn} = X_{1m} \times X_{2n} \quad (6)$$

and then $L_{mn}$ is a finite region in the state space. By applying fixed controls, $\delta_i$ and $\theta_j$, one set of transitional relations is obtained for example,

**Rules of Dynamical Behavior**
**(Cell-State Transitions by $(\delta_1, \theta_1)$):**

$$\begin{aligned} (\delta_1,\theta_1) &: z\{L_{12}\} \longrightarrow z\{L_{34}\} \\ (\delta_1,\theta_1) &: z\{L_{23}\} \longrightarrow z\{L_{33}\} \\ &\ldots \qquad \ldots \\ (\delta_1,\theta_1) &: z\{L_{55}\} \longrightarrow z\{L_{73}\} \end{aligned} \quad (7)$$

The above transitions are valid if the average dynamic behavior of the system shows the above rules. We continue to change $(\delta_i, \theta_j)$ to get other sets of transitional rules and exhaustively gather finite number of transitional relations. When we store the above information, we include time required during transitions, and other optimal performance indices such as energy, squared errors, etc. These transition relations are stored in the table which we call a "cell-state transition table". An example of cell-state transition table in the 3 dimensional cell-state space is shown in Table 1. Changing $(\delta_i, \theta_j)$, we may have the same $L_{mn}$ for the source and the destination and this is called an 'invariant cell' or an 'invariant manifold'. For an invariant cell, there should be a design limit in the transition time since it is an indefinite stay in that $L_{mn}$. It is emphasized that the target $L_{mn}$ (the specified goal) must have an invariant manifold for some fixed controls $(\delta_i, \theta_j)$ for convergence and asymptotic stability. This is equivalent to the 'reachability condition' in the classical control theory. The target is denoted as $L^*$.

STEP 2: From the collected data, we generate an N-ary tree that connects from one node $L_{mn}$ to another. The root of the tree is $L^*$ and we avoid any looping structures. We proceed with a backward tracking search technique in artificial intelligence (AI) and the search procedure is initiated by finding all possible dynamic transitions to the target region $L^*$. As shown in Figure 3(a), there may be multiple paths from one region to another and we eliminate multiplicity and extract only one transition by considering the following concepts:

Priority #1: Minimum Euclidean Distance
Priority #2: Optimal Strategies --
Minimum Energy,
Minimum Time,
Minimum Squared Errors,
or Combinations
Priority#3: Redundancy in Controls, Transitions

The elements of the finalized tree constitute a set of control rule base and an example is represented in Figure 3(b). These rules are automatically generated on the basis of optimal performance criteria such as minimum time or minimum energy concepts. In brief, the transitional relations that force the state trajectories to move from any points in the state space to the desired goal within the prescribed tolerances are themselves the control rules for feedback regulation. We store the membership functions for the transition relations in matrices $P_i$ and $C_j$, in which rows the numerical values in [0,1] are the chosen membership functions.

STEP 3: The fuzzification procedure undertakes the rule generation so that the crisp transitions between the regions in the state space can be smoothed out. To each $L_{mn}$, assigned are as many elements as accuracy and precision can allow. In a practical sense, five to seven elements are suitable for fuzzification of $L_{mn}$ in the state space. Membership functions may be triangular or of simple functional type. The trade-off's between the number of quantization in Lmn and the transition smoothness, the total numbers of $X_{1m}$, $X_{2n}$ and the performance are important and these issues are related to heuristics. For each rule, numbers between 0 and 1 are stored for each vector array of one membership function. In our example, a two-input two-output fuzzy controller has two vector arrays for the conditional parts and two vector arrays for the action parts for each rule. The fuzzy sets for control inputs $\delta_i$ and $\theta_j$ are denoted as $\Delta_i$ and $\Theta_j$, respectively while $\chi_1$ and $\chi_2$ are the fuzzy sets for $x_1$ and $x_2$, respectively.

## 2.3 Fuzzy Inferencing Using Decomposition of Fuzzy Hypercubes

For the practical purpose, a discrete version of fuzzy controllers is needed and it is convenient if we utilize a decomposed fuzzy hypercube [7] which is suitable for implementing a fuzzy logic controller with the cell-state mapping concept. Each rule numerically stored in a fuzzy hypercube corresponds with each cell in the cell-state. For each rule, one membership function in $X_{1m}$ is stored in the premise matrix no.1, $P_1$, and so is another in the premise matrix no.2, $P_2$, and so on. Each row in $P_1$ or $P_2$ is the membership function obtained in the stepwise procedure stated earlier. The same is true for the consequence matrices, $C_1$ and $C_2$, representing $\Delta_i$ and $\Theta_j$ for each rule. Let the max-min product be denoted as "$\circ$", then for given fuzzy sets $\chi_1$ in $X_{1m}$ and $\chi_2$ in $X_{2n}$, the control input fuzzy sets, $\Delta$ and $\Theta$, are obtained as

$$\Delta = C_1^T \circ \{ (P_1 \circ \chi_1) \otimes (P_2 \circ \chi_2) \} \quad (8a)$$
$$\Theta = C_2^T \circ \{ (P_1 \circ \chi_1) \otimes (P_2 \circ \chi_2) \} \quad (8b)$$

where $C_i^T$ is the transpose of the matrix $C_i$ and "$\otimes$" is the element-wise minimum operator. The crisp results of $\Delta$ and $\Theta$ are

$$\delta = \text{DEFUZZIFIER}(\Delta) \tag{9a}$$
$$\theta = \text{DEFUZZIFIER}(\Theta) \tag{9b}$$

where DEFUZZIFIER(.) is a defuzzification operator chosen among the maximum criterion method, the mean of maxima procedure, and the centroid algorithm.

## 3. Application: Design of An Engine Idling Speed Fuzzy Controller

Simulation Model: The well-known model for engine idling speed control has been rigorously studied in [8,9]. For simulation input-state training pairs, we collect the data exhaustively for the given fixed control values from the following nonlinear model with uncertainty:

$x_1$ = N (Engine Rotor Speed [rpm]),
$x_2$ = P (Manifold Pressure [kPascal])

$$\text{Rotating Dynamics: } dx_1/dt = K_n (T_i - T_L) \tag{10a}$$
$$\text{Manifold Dynamics: } dx_2/dt = K_p (m'_{ai} - m'_{ao}) \tag{10b}$$

where $m'_{ai} = (1 + 0.907\theta + 0.0998\theta^2)\, g(x_2)$

$$g(x_2) = \begin{cases} 1 & x_2 < 50.66 \\ 0.0197\,(101.325\, x_2 - x_2^2)^{1/2} & x_2 \geq 50.66 \end{cases}$$

$$m'_{ao} = -0.0005968\, x_1 - 0.1336\, x_2 + 0.0005341\, x_1 x_2 + 0.000001757\, x_1 x_2^2$$

$$T_i = -39.22 + 325024.0\, m_{ao} - 0.0112\,\delta^2 + 0.000675\,\delta x_1 (2\pi/60) + 0.635\,\delta + 0.0216\, x_1 (2\pi/60) - 0.000102\, x_1^2 (2\pi/60)^2$$

$$T_L = (x_1/263.17)^2 + T_d$$

($T_L$:Load Torque, $T_d$:Disturbance Torque)

$m_{ao} = m'_{ao}(t-\tau)/120x_1$,

$\tau = 120/4x_1$ (Induction Power Delay), $K_p, K_n$: constants

Equations (10) are a highly nonlinear two-state engine model for idle speed control. We will obtain cell-state transitions from the above model in order to derive fuzzy logic control rules that stabilize and regulate the state trajectories toward the goal state, and in our case, the goal is N = $x_1$ = 750.0 (rpm), P = $x_2$ = 34.0 (kPascal). It is noted that $L^* = L_{43}$ in the state trajectories in Figures 4-6 where the interval of $X_{14}$ is $750.0 \pm 166.67/2$, and that of $X_{23}$ is $34.0 \pm 15.0/2$.

Cell- State Transitions: As in Step 1, we gathered 9 kinds of the complete state trajectories by using 9 fixed controls from $(\delta_1, \theta_1)$ to $(\delta_3, \theta_3)$. Among them, the state trajectories of every initial conditions for $(\delta_2, \theta_1)$, $(\delta_2, \theta_2)$, and $(\delta_2, \theta_3)$, are shown in Figures 4-6. Initial states start from the representative positions in the cell-state space. In Figure 5, we can find an equilibrium cell with fixed $(\delta_2, \theta_2) = U_{22}$ in $L* = L_{43}$.

The next step (Step 2) is to find a chain of connections among the cells with the assigned controls according to the chosen optimal strategy. This procedure is the most important one in the design of fuzzy logic controllers. The finalized control rules are determined by using the backward tracking algorithm as shown in Figure 7. $L_{43}$ is the root of N-ary tree in the backward tracking search algorithm. In our example, 5 elements are assigned to each cell, and the support of each membership function has 7 elements, thus making 2 overlapping elements. (Step 3)

Figure 7 represents the results of the performance criterion for the minimum squared error control and the simulation results are shown in Figure 8. We have used the centroid algorithm for defuzzification.

In Figure 9, a different optimal strategy has been chosen to compare the results and this case is the minimum energy/control effort criterion. Moreover, Figure 10 represents the corresponding simulation runs on different initial conditions.

In Figure 11, the responses of the minimum squared error (MSE) and the minimum control effort (MCE) control results are shown with the inferred control actions together. In Figure 11, (a) and (b) are the idle speed control results for MSE and MCE, respectively, while (c) and (d) represents the throttle angles for MSE and MCE. The membership functions for each fuzzy subsets in the minimum energy based strategic control rule base are shown in Figure 12 where the premise part and the consequence part of fuzzy implications are represented.

## 4. Conclusions

With a two-input two-output multivariable fuzzy logic control scheme for the automated design of a fuzzy controller rule base, we can easily generalize the systematic procedure for a m-input n-output multivariable fuzzy control system. The automated production design of fuzzy logic control rule bases for different optimal control strategies and the associated simulation results ensure **versatility** and **flexibility** of the proposed **cell-state transition method**. Emphasis is placed upon the fact that, for given arbitrary systems, we can make fuzzy logic based control rule bases that **stabilize** the closed-loop feedback control systems, and that the design procedure is totally **automated**. Furthermore, the rules are determined according to the chosen **optimal strategy**. Numerical simulation results strongly suggest that the automated rule design of fuzzy logic controllers for uncertain dynamic systems guarantees a promising controller design paradigm for intelligent control.

## References


[1] H.-J. Zimmermann, Fuzzy Set Theory and Its Applications, 2nd ed., Kluwer-Nijhoff, Boston, 1985

[2] Togai InfraLogic, Inc., TIL Shell & Fuzzy C-Development System User's Manuals, 1988

[3] C. S. Hsu, "A Theory of Cell-to-Cell Mapping Dynamical Systems", Trans. of ASME, Journal of Applied Mechanics, vol.47, pp.931-939, Dec 1980

[4] Y. Y. Chen and T. C. Tsao, "A Description of the Dynamical Behavior of Fuzzy Systems", IEEE Trans. on Systems, Man, and Cybernetics, vol.SMC-19, no.4, pp.745-755, Jul/Aug 1989

[5] H. Kang and G. Vachtsevanos, "Nonlinear Fuzzy Control Based On The Vector Fields of The Phase Portrait Assignment Algorithm", Proc. American Control Conference, San Diego CA, pp.1479-1484, May 1990

[6] M. Braae and D. A. Rutherford, "Selection of Parameters for a Fuzzy Logic Controllers", Fuzzy Sets and Systems, vol.2, pp.185-199, 1979

[7] H. Kang and G. Vachtsevanos, "Fuzzy Hypercubes: A Possibilistic Inferencing Paradigm", Proc. IEEE Conf. on Fuzzy Systems, San Diego CA, pp.553-560, Mar 1992

[8] A. W. Olbrot and B. K. Powell, "Robust Design and Analysis of Third and Fourth Order Time Delay Systems with Application to Automotive Idle Speed Control", Proc. American Control Conference, vol.2, pp.1029-1039, May 1989

[9] B. K. Powell and A. W. Olbrot, "Robust Analysis of Automotive Idle Speed Control System", FORD, Technical Report No. SR-90-09, Jan 1990

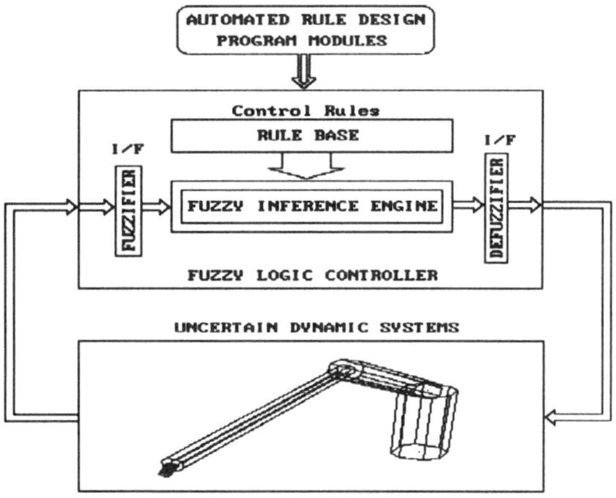


Fig. 1 Block Diagram of Fuzzy Logic Control Systems

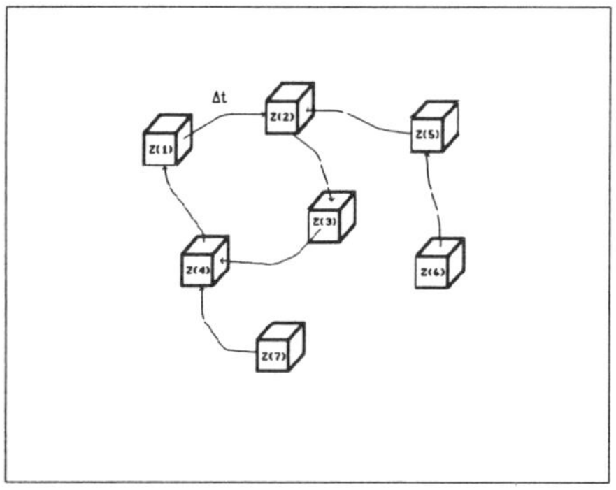


Fig. 2 4-Period Motion Cells z(1), z(2), z(3), z(4)

| Controls | Prev.State | Next State | Time t | Perf.Index |
|---|---|---|---|---|
| | {1,2,2} | {2,1,3} | 0.5 sec | 5.2 |
| u = + 1 | {1,1,1} | {3,2,1} | 0.3 sec | 7.7 |
| | {2,1,3}* | {2,1,3}* | ∞ sec | 0.0 |
| | {1,3,1} | {1,2,2} | 0.4 sec | 4.3 |
| u = - 1 | {2,1,2} | {3,2,1} | 0.3 sec | 11.2 |
| | ... | ... | ... | ... |

Table 1. Cell-State Transition Table in the 3-dim. Cell-State Space

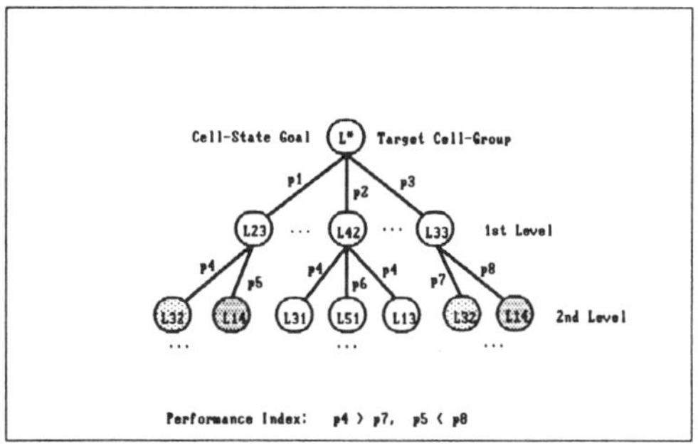


Fig. 3 (a) Exemplary N-ary Tree from Cell-State Transition Table

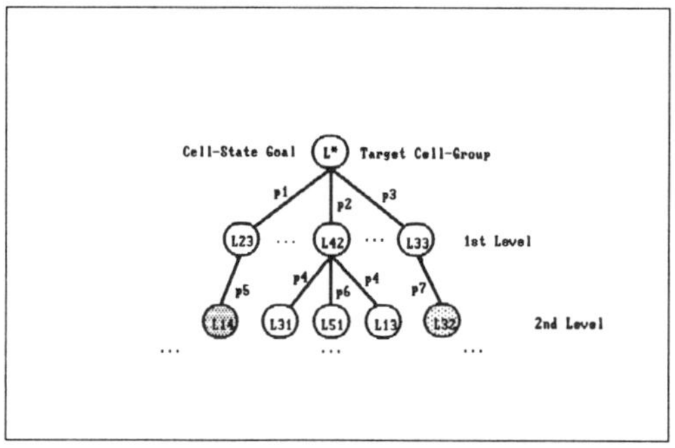


Fig. 3 (b) Result of Tree by Optimal Search Technique w/ Redundancy Removed

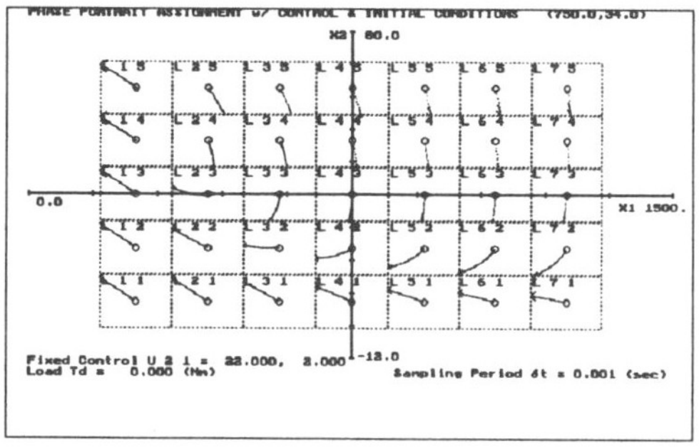

Fig. 4 State Trajectories for ($\delta_2$=22.0, $\theta_1$=2.0) = $U_{21}$

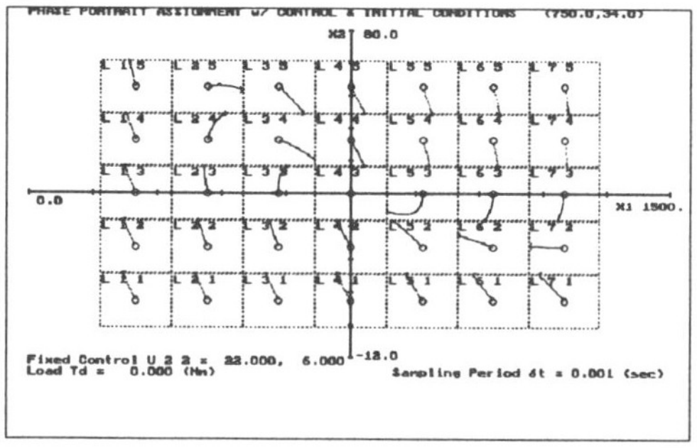

Fig. 5 State Trajectories for ($\delta_2$=22.0, $\theta_2$=6.0) = $U_{22}$

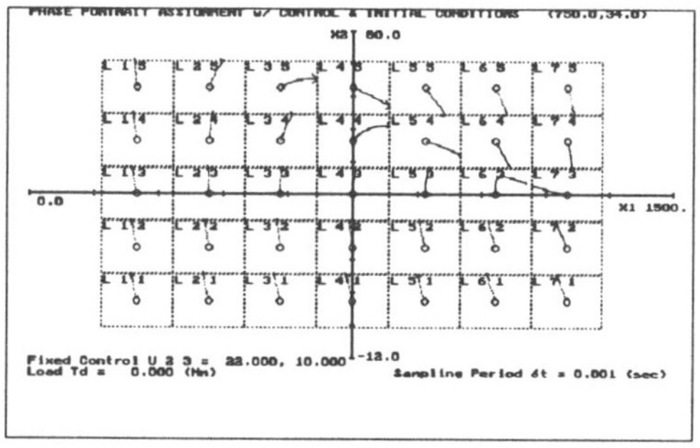

Fig. 6 State Trajectories for ($\delta_2$=22.0, $\theta_3$=10.0) = $U_{23}$

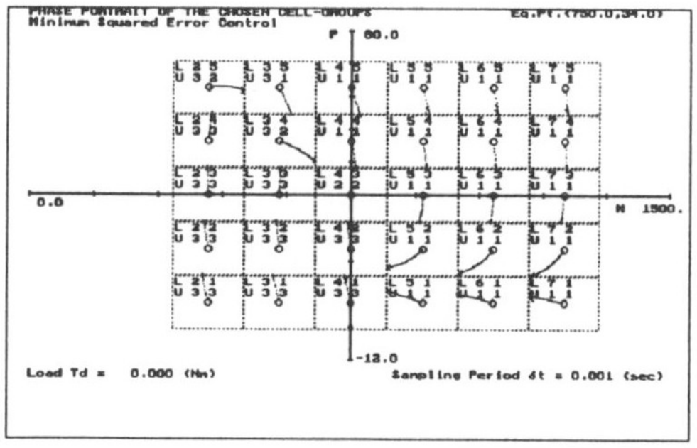

Fig. 7 Control Rules Generated by The Minimum Squared Error Strategy (MSE)

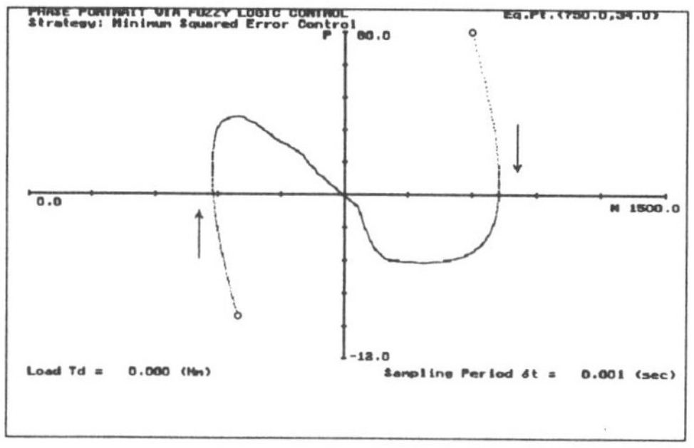

Fig. 8 Simulation Results for The Minimum Squared Error Strategy (MSE)

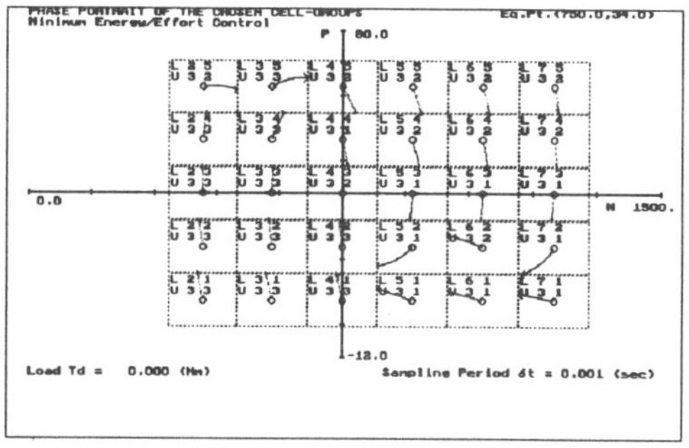

Fig. 9 Control Rules Generated by The Minimum Energy/Control Effort Strategy (MCE)

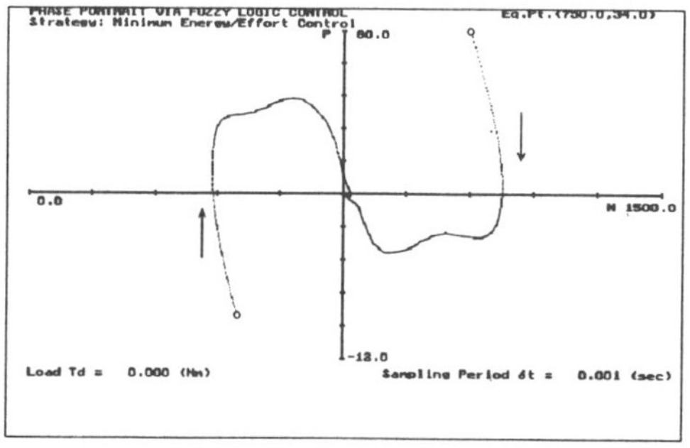

Fig. 10 Simulation Results for The Minimum Energy/Control Effort Strategy (MCE)

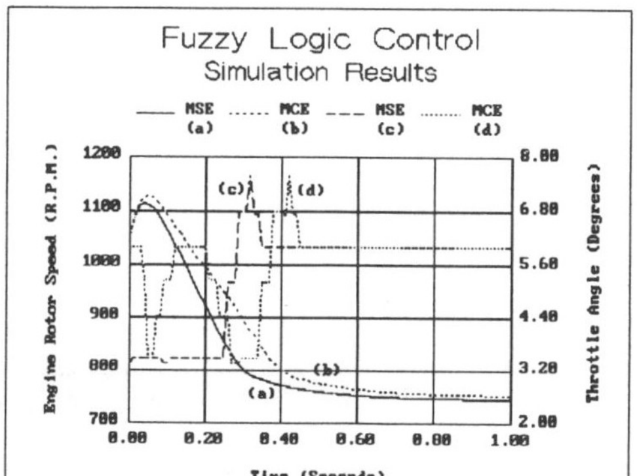


Fig. 11 Comparison of Simulation Results between MSE and MCE

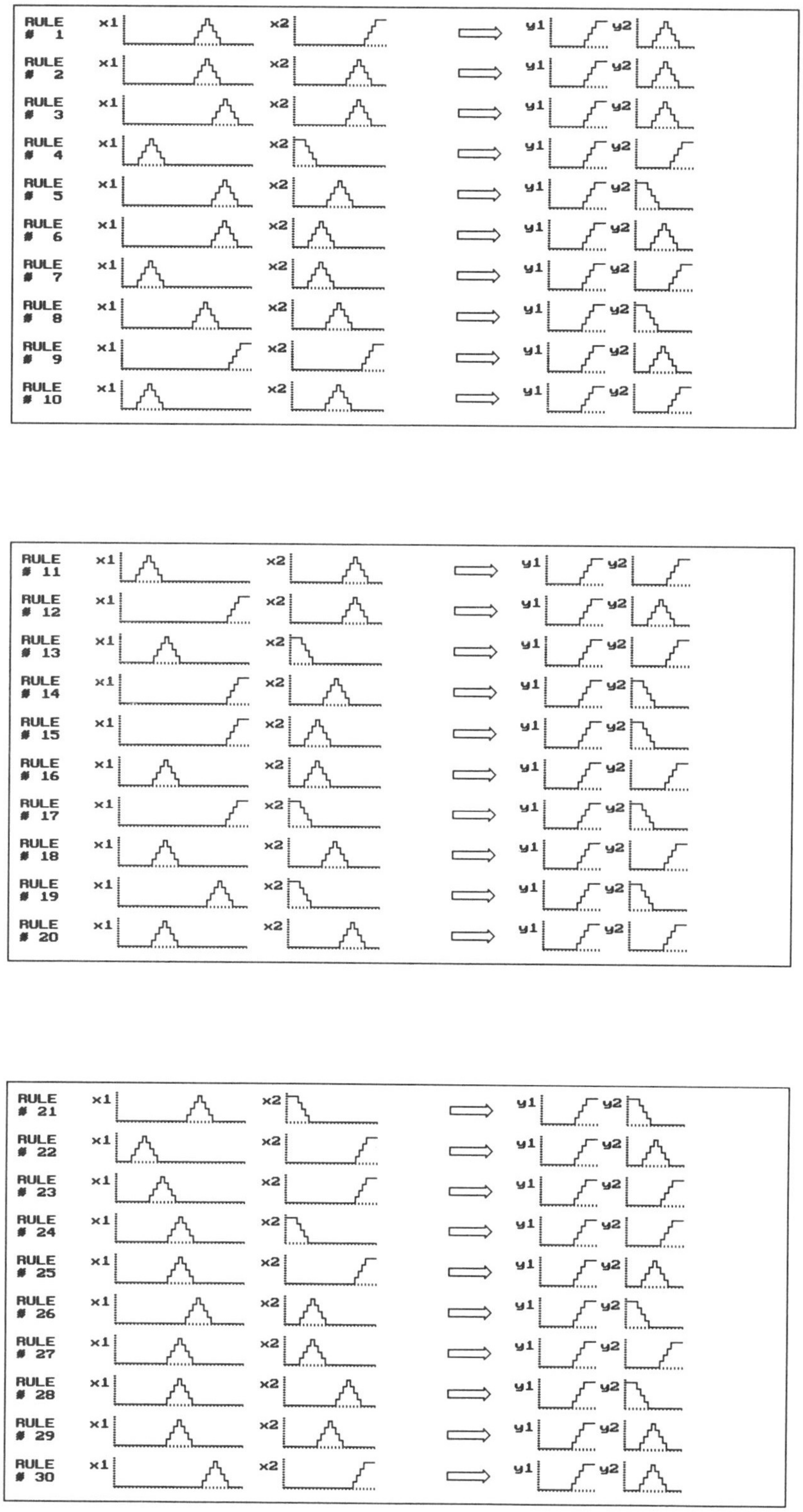


Fig. 12 Fuzzy Logic Control Rule Base (30 Rules) of The Minimum Energy/Control Effort Strategy (MCE)

NOTE: The control output y1 is chosen to be always 'POSITIVE LARGE' in this case which is correct in the engine spark advance ($\delta$=22.0). Also, the total number of cell-groups was 35 among which 30 discrete rules were selected by this computer-aided design.

# A FUZZY COORDINATION APPROACH FOR MULTI-OBJECTIVE VOLTAGE AND REACTIVE POWER SCHEDULING OF AN ELECTRIC POWER SYSTEM

**Takahide Niimura**
**Fuji Electric Corporate R&D**
**1 Fuji-machi, Hino**
**Tokyo, 191 Japan**

**Ryuichi Yokoyama**
**Tokyo Metropolitan University**
**2-1-1 Minami-Osawa,**
**Hachioji, 192-02 Japan**

**Brian J. Cory**
**Imperial College**
**Exhibition Road**
**London SW7 2BT, U.K.**

*Abstract* - This paper presents a multi-objective optimal scheduling procedure of voltage and reactive power in power systems based on fuzzy set theory. The optimization procedure includes a transmission loss and bus voltage error from a target voltage as conflicting objectives to be coordinated. We apply fuzzy sets to measure adaptability of objectives to the operational goals, and solve the multi-objective optimization by maximizing a composite decision-making function. The definition of fuzzy sets can provide system operators an opportunity to decide on different preferences according to system operating conditions, thus resulting in a more flexible operation. Model analysis is employed to demonstrate the effectiveness of the proposed approach.

## I. INTRODUCTION

In the power system operation field, many studies report optimization procedures on voltage-reactive power control [1][2]. There are two major objectives of the optimization [3]. One is to improve economy of power system operation by minimizing transmission losses, and the other is to maintain security by improving voltage profiles. Conventional approaches often translate the latter criterion into certain constraints and optimize voltage-reactive power allocation for the minimum losses. However, such procedures often result in a solution at one of the limits on voltage constraints. Therefore, we find it prone to cause voltage violations by slight disturbances. This is mainly because the original two-fold goals of the optimization are in a trade-off relationship and non-commensurable, having different units of measurement.

One systematic way to solve such a conflict is to apply multi-objective programming [4], making both system losses and voltage errors objective functions. In multi-objective programming (MOP), however, solutions are given by a set of Pareto optimal solutions, i.e., feasible solutions which may improve one objective at the expense of the other. Therefore, we have to provide a means to choose a suitable solution which satisfies the decision-maker. Man-machine interface is also desired to reflect the operators', sometimes indecisive, concept by the interactive feature in decision-making.

In this paper we observe voltage-reactive power scheduling as multi-objective optimization, and apply a coordination technique based on fuzzy set theory [5] for loss-voltage compromises and improved man-machine interface. First, we introduce fuzzy sets to measure the adaptability of real power loss of transmission lines and of the errors of bus voltage from operational goal. Then the multi-objective optimization is solved by maximizing the composite fuzzy decision-making function, which combines the objective functions by a numerical sum of grades in the fuzzy sets.

The proposed approach offers the following advantages:

(1) we can make decisions by compromise between minimum loss and desirable system voltages with the intention of making both values as close to the optimum as possible.
(2) solutions are evaluated by adaptability indices which are given by the grade of membership to the fuzzy sets representing the operational goals. We can readily perceive the degree of compromise and attainment of different objectives.
(3) adjustment of parameters which define the fuzzy sets offers the decision-maker a means to reflect preference for one of the objectives. This feature enables flexible operation of power system voltages.

This paper first formulates the multi-objective problem, and introduces a decision-making algorithm based on fuzzy set theory. Numerical examples are then demonstrated to show the effectiveness and unique features of this approach.

## II. FORMULATION OF MULTI-OBJECTIVE VOLTAGE-REACTIVE POWER SCHEDULING

Transmission losses and system voltages are controlled by generator terminal voltages and bus reactive power injections. Using the bus voltage vector $\boldsymbol{x} = [x_1, x_2, ..., x_N]^T$ as variables of the optimization problem, we obtain the solutions for generator/synchronous compensator AVR setting at a voltage-controlled bus, and reactive power injection by switchable capacitors and shunt reactances, approximated by continuous value, at a reactive power-controlled bus. An on-load tap changer, another common voltage control device, is not included as a control variable because transformer taps may affect voltage stability and, if local voltage control devices such as voltage relays are applied, the control signal is given by the voltage settings, i.e., voltage-controlled. Throughout this paper, we designate the N-th bus as slack in an N-bus system.

### *2.1 Objective Functions*

In this study we have chosen transmission losses and

0-7803-0614-7/93$03.00 ©1993IEEE

voltage errors as two objectives to minimize:

(1) Loss index: We aim at real power loss reduction by minimizing the following objective function:

$$L(x) = \sum_{i,j \in I} | P_{ij}(x) + P_{ji}(x) | \qquad (1)$$

where

| | | |
|---|---|---|
| $L(x)$ | : | total real power loss in the system [p.u.] |
| $\mathbf{x}$ | : | bus voltage vector |
| $I$ | : | a set of busses connected by transmission lines |
| $P_{ij}(x)$ | : | real power flow from i-th bus to j-th bus |

(2) Voltage index: We have used the sum of voltage errors as the second index to minimize.

$$E(x) = \sum_{i \in B} | V_i(x) - V_i^G |^2 \qquad (2)$$

where

| | | |
|---|---|---|
| $E(x)$ | : | total voltage error from specified bus voltage [p.u.] |
| $B$ | : | a set of load busses |
| $V_i(x)$ | : | voltage magnitude at i-th bus [p.u.] |
| $V_i^G$ | : | the goal of voltage magnitude at i-th bus [p.u.] |

*2.2 EQUALITY AND INEQUALITY CONSTRAINTS*

Constraints considered in this problem include bus real and reactive power, bus voltage magnitude, and slack generator output as follows:

(1) Specified bus real power: We need to supply real power to meet the demand at load busses. The real power output of generators is assumed previously determined by economic dispatch, except for a slack generator on N-th bus. The load is also assumed previously known.

$$P_i(x) - P_i^s = 0 \quad \text{for} \quad i = 1, 2, ..., N-1 \qquad (3)$$

where

| | | |
|---|---|---|
| $P_i(x)$ | : | i-th bus real power [p.u.] |
| $P_i^s$ | : | specified real power at i-th bus [p.u.] |
| $N$ | : | number of busses |

(2) Upper and lower limits of bus reactive power: For generator buses reactive power limits are given by generators' capability to supply reactive power, whereas for load busses compensator limits must be observed such that:

$$Q_i^{min} \leq Q_i(x) \leq Q_i^{max} \quad \text{for} \quad i = 1, 2, ..., N \qquad (4)$$

where

| | | |
|---|---|---|
| $Q_i(\mathbf{x})$ | : | i-th bus reactive power [p.u.] |
| $Q_i^{max}$ | : | maximum reactive power at i-th bus [p.u.] |
| $Q_i^{min}$ | : | minimum reactive power at i-th bus [p.u.] |

Bus reactive power requirements at load busses without compensators can be also expressed by replacing $Q_i^{max}$ and $Q_i^{min}$ with $Q_i^s$ in the above condition (4) such that:

$$Q_i(x) - Q_i^s = 0 \quad \text{for} \quad i \in D \qquad (5)$$

where

| | | |
|---|---|---|
| $Q_i^s$ | : | reactive power demand at i-th bus [p.u.] |
| $D$ | : | a set of load busses without voltage control equipment |

(3) Upper and lower limits of bus voltage: Each bus voltage is constrained by its upper and lower limits of operational voltage to ensure security condition.

$$V_i^{min} \leq V_i(x) \leq V_i^{max} \quad \text{for} \quad i = 1, 2, ..., N \qquad (6)$$

where $V_i^{min}$ and $V_i^{max}$ are the lower and the upper limits of voltage magnitude at i-th bus in p.u., respectively.

(4) Slack generator output: The real power balance between generation and load must be maintained by adjustment of a slack generator output $P_N(x)$ such that:

$$P_N(x) = \sum_{i=1}^{N-1} P_i^s + L(x) \qquad (7)$$

## III. ALGORITHM OF FUZZY COORDINATION FOR VOLTAGE-REACTIVE POWER SCHEDULING

An optimization problem with two or more objective functions under a given set of constraints is called multi-objective programming (MOP)[6]. Optimal solutions for the multi-objective programming is generally given by a set of non-inferior, or Pareto optimal, solutions. A Pareto optimal solution is the solution where any improvement of one objective function can be achieved only at the expense of others. Several multi-objective optimization techniques have been developed to obtain a set of non-inferior solutions. However, few techniques provide solutions which are Pareto optimal as well as satisfactory in a decision-maker's view point.

In addition, as a practical matter, we need to transform the original multi-objective optimization problem into a single objective form by some means, because commercial optimization software packages are usually applicable to a single objective problem. The transformed problem should maintain the Pareto optimality as well as flexible user interface for interactive decision-making.

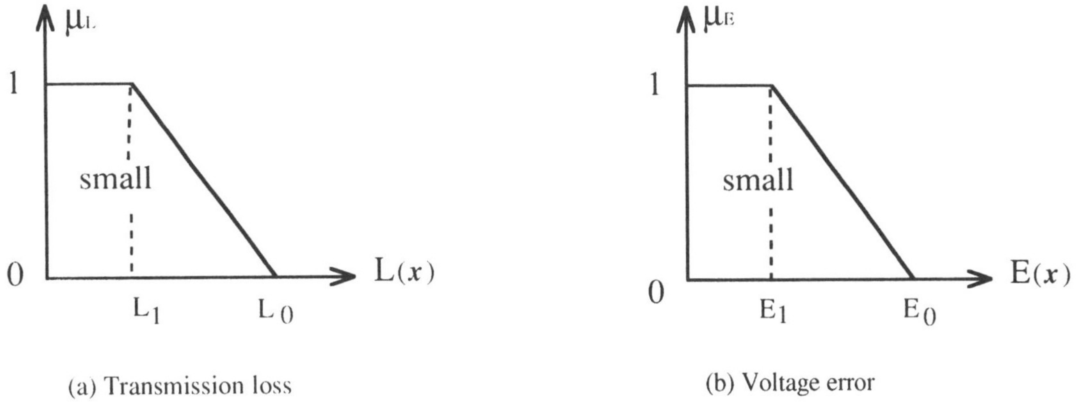


Fig. 1 Fuzzy sets for voltage-reactive power scheduling

In this paper, we represent the operational goals for decision-making in the multi-objective problem by fuzzy sets [7]. The two goals are given in linguistic variables such as:

(a) small loss

(b) small errors from voltage goal

and defined by fuzzy sets shown in Fig. 1. In the definitions $L_1$ and $E_1$ are the objective value when they are fully satisfactory for the decision-maker; $L_0$ and $E_0$ are the least permissible value of the transmission loss and the bus voltage error, respectively.

The fuzzy sets define numerically how the decision-maker is satisfied by the value of attribute [8], in this case, system loss L and voltage deviation E. Therefore, the grade of membership in the fuzzy sets indicates the adaptability of the voltage index.

The optimal solution of the MOP is then given by maximizing the following composite objective function subject to constraints (3)-(7) :

$$DM(\boldsymbol{x}) = \mu_L(\boldsymbol{x}) + \mu_E(\boldsymbol{x}) \tag{8}$$

Fig. 2 shows the algorithm of the proposed optimization procedure. To determine the fuzzy sets we first minimize objectives (1) and (2) separately subject to constraints (3) - (7) . If the approximate maxima of objectives are empirically given, we can apply the overestimate of the maxima to define the fuzzy sets. Otherwise, we can also maximize each objective in order to obtain the range of objective value to ensure the Pareto optimality. It is shown by Sakawa and Yano [5] that the solution of (8) is Pareto-optimal, when the solution satisfies the condition (9) and (10) at the same time:

$$0 < \mu_L(\boldsymbol{x}) < 1 \tag{9}$$

$$0 < \mu_E(\boldsymbol{x}) < 1 \tag{10}$$

Therefore, it is secure to define fuzzy sets by parameters indulging the following conditions:

$$L1 \leq L(\boldsymbol{x})^{min} \quad \text{and} \quad L(\boldsymbol{x})^{max} \leq L0 \tag{11}$$

$$E1 \leq E(\boldsymbol{x})^{min} \quad \text{and} \quad E(\boldsymbol{x})^{max} \leq E0 \tag{12}$$

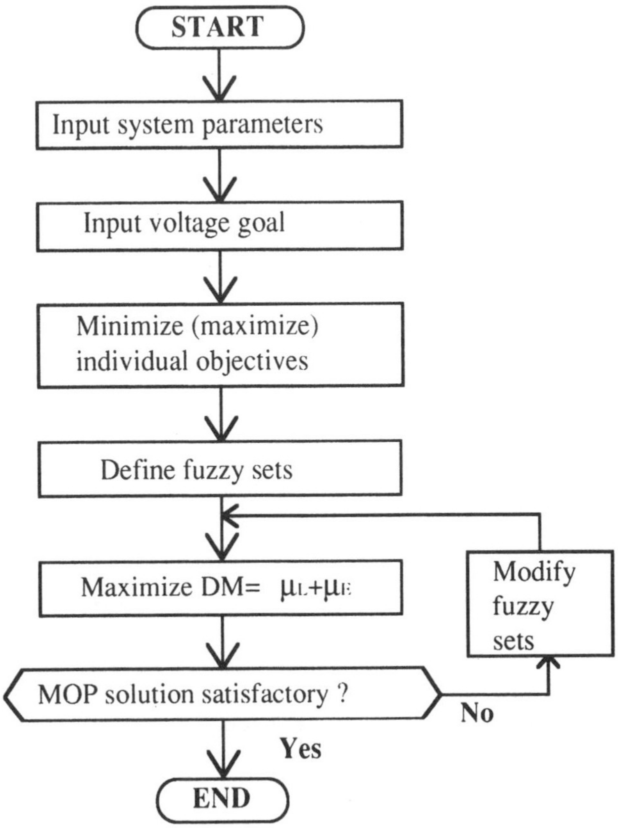


Fig. 2 The procedure to determine voltage-reactive power scheduling based on fuzzy coordination technique.

IV. MODEL ANALYSES

For a model analysis we have employed a modified IEEE-14 bus system shown in Fig.3. In the figure, bus #14 is assumed as slack. Buses #1-9 are load busses and #10-13 are assumed voltage controlled. The operational goal of bus voltage, including the limits of bus voltage (Vi) and reactive powers (Qi) are listed in Table 1. Other system parameters are shown in [7]. All load busses except bus #7 are assumed to be reactive-power controlled. We have solved the nonlinear programming problems by MINOS 5.1 developed by Murtagh and Saunders [9].

First, we have solved a single objective optimization with respect to real power loss in order to determine the fuzzy sets. Solutions are listed in Minimum Loss column of Table 3. The maximum voltage errors are obtained by the comparison of the voltage goal with the upper or lower limits. Naturally, the minimum value of voltage error is zero with definitions of fuzzy sets given in Table 2. Then, the multi-objective optimization is solved without priority of objectives. The solution is listed in Case 1 column of Table 3. The solution is acquired as close as possible to the optimal solutions of the original single objective problem for the minimum loss of transmission.

Next, we consider putting greater weights to the voltage error. Case 2 column of Table 3 lists the result. This is done by relaxing the L() value so that the definition of fuzzy set for the voltage error becomes relatively more strict compared to Case 1. As we expect, the voltage solution is closer to the goal than in Case 1.

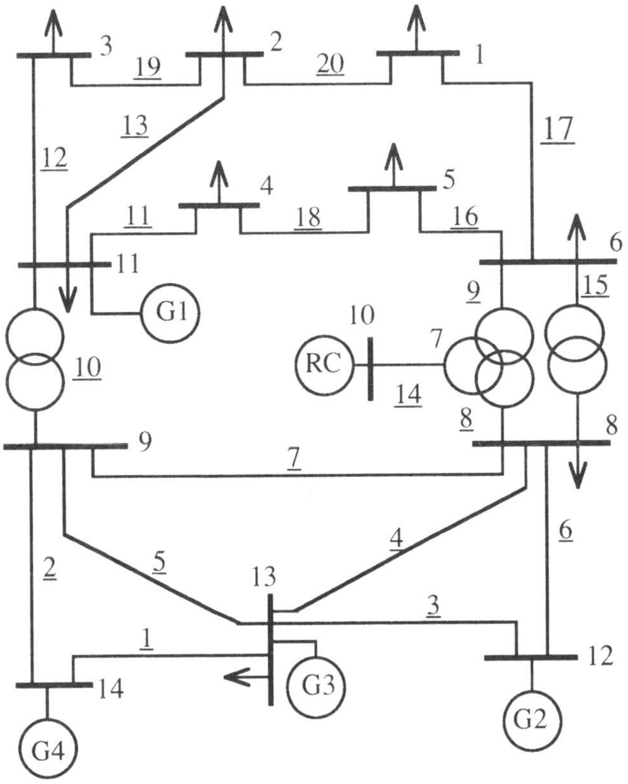


Fig. 3 Model system

TABLE I
Bus V-Q Specifications

| Bus No. | Type | $V_i^G$ | $V_i^{min}$ | $V_i^{max}$ | $Q_i^{min}$ | $Q_i^{max}$ |
|---|---|---|---|---|---|---|
| 1 | P-Q | 0.995 | 0.900 | 1.100 | -0.083 | 0.017 |
| 2 | P-Q | 1.017 | 0.900 | 1.100 | -0.058 | 0.032 |
| 3 | P-Q | 1.021 | 0.900 | 1.100 | -0.031 | 0.010 |
| 4 | P-Q | 1.000 | 0.900 | 1.100 | -0.023 | -0.003 |
| 5 | P-Q | 0.999 | 0.900 | 1.100 | -0.085 | -0.025 |
| 6 | P-Q | 1.006 | 0.900 | 1.100 | -0.190 | 0.006 |
| 7 | P-Q | 1.020 | 0.900 | 1.100 | 0.000 | 0.000 |
| 8 | P-Q | 0.998 | 0.900 | 1.100 | -0.120 | 0.198 |
| 9 | P-Q | 1.004 | 0.900 | 1.100 | -0.420 | 0.009 |
| 10 | P-V | 1.044 | 0.940 | 1.148 | -0.300 | 0.400 |
| 11 | P-V | 1.034 | 0.931 | 1.138 | -0.200 | 0.200 |
| 12 | P-V | 1.005 | 0.904 | 1.106 | 0.000 | 0.600 |
| 13 | P-V | 1.022 | 0.920 | 1.124 | -0.700 | 0.300 |
| 14 | Slack | 1.060 | 1.060 | 1.060 | ----- | ----- |
| Unit | ----- | p.u. | p.u. | p.u. | p.u. | p.u. |

TABLE II
Membership Functions

| Index | Min | Max |
|---|---|---|
| L(*x*) | 0.1339 | 0.1798 |
| E(*x*) | 0 | 0.1486 |

TABLE III
Coordinated Solutions

| Bus | Minimum loss | | Case 1 | | Case 2 | |
|---|---|---|---|---|---|---|
| No. | Vi | Qi | Vi | Qi | Vi | Qi |
| 1 | 1.044 | 0.005 | 1.037 | 0.017 | 1.025 | 0.017 |
| 2 | 1.052 | 0.016 | 1.034 | 0.032 | 1.020 | 0.015 |
| 3 | 1.054 | 0.008 | 1.043 | 0.097 | 1.022 | 0.010 |
| 4 | 1.040 | -0.013 | 1.037 | -0.014 | 1.028 | -0.014 |
| 5 | 1.047 | -0.025 | 1.045 | -0.025 | 1.035 | -0.025 |
| 6 | 1.055 | 0.006 | 1.053 | -0.006 | 1.043 | -0.006 |
| 7 | 1.069 | 0.000 | 1.065 | 0.000 | 1.055 | 0.000 |
| 8 | 1.028 | 0.120 | 1.016 | 0.028 | 1.006 | 0.012 |
| 9 | 1.033 | 0.009 | 1.019 | 0.009 | 1.010 | 0.009 |
| 10 | 1.128 | 0.379 | 1.107 | 0.266 | 1.098 | 0.269 |
| 11 | 1.061 | 0.200 | 1.034 | -0.412 | 1.026 | -0.303 |
| 12 | 1.018 | 0.083 | 1.012 | 0.143 | 1.007 | 0.223 |
| 13 | 1.047 | 0.236 | 1.037 | 0.079 | 1.023 | -0.263 |
| 14 | 1.060 | -0.103 | 1.060 | 0.006 | 1.060 | 0.293 |
| Unit | p.u. | p.u. | p.u. | p.u. | p.u. | p.u. |

TABLE IV
Multi-Objective Optima

| Objectives / Approach | L(x) | E(x) |
|---|---|---|
| Minimum loss | 0.1339 | 0.0178 |
| Minimum error | 0.1429 | 0.0000 |
| Coordination (Case 1) | 0.1356 | 0.0071 |
| Coordination (Case 2) | 0.1380 | 0.0019 |

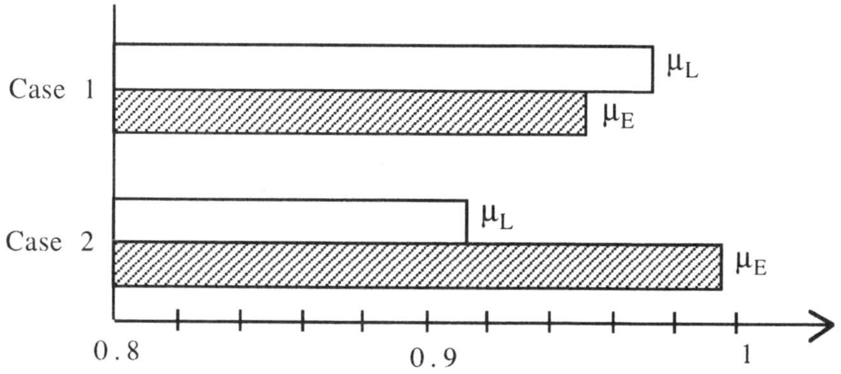


Fig. 4 Adaptability indices

We compare the objective value obtained in the above cases in Table 4. With the priority (Case 2) the voltage index is apparently improved, and the trade-off between each objective is easily seen in the difference of adaptability according to changes in preferences. Fig.4 shows the adaptability indices by the grade of membership for each objective in Cases 1 and 2.

## V. CONCLUSIONS

This paper has presented a multi-objective fuzzy coordination approach for voltage-reactive power scheduling. Simulation results clarify the following four advantages of the new approach for the multi-objective optimization:

(1) we can make compromise decisions between minimum loss and desirable voltages with the intention of making both values as close to the optimum as possible.

(2) solutions are evaluated by adaptability indices which are given by grades of membership in fuzzy sets representing operational goals. Using the adaptability indices we can readily perceive the degree of compromise and the attainment of goals.

(3) adjustment of parameters which define fuzzy sets provide the decision-maker with a means to reflect preference for one of the objectives. This feature is particularly effective in flexible scheduling of voltage-reactive power allocation.

(4) the proposed algorithm transforms the original multi-objective optimization problem into a single objective maximization of the decision making function, and therefore, we can readily apply a conventional nonlinear programming software package.

Further work should be encouraged in the following fields:

(a) a systematic approach is needed to define the fuzzy sets which can truly represent the priority order of the performance indices.

(b) inclusion of dynamic effects of on-load tap changers and load voltage characteristics.

(c) development of voltage instability countermeasures.

## REFERENCES


[1] I. Hano, Y. Tamura, S. Narita, and K. Matsumoto, "Real time control of system voltage and reactive power," *IEEE Trans. PAS*, Vol. 88, No.5, pp.1544-1558, Nov/Dec 1969.

[2] C. J. Rehn, J. A. Bubenko, and D. Sjelvgren, "Voltage optimization using augmented Lagrangian functions and quasi-Newton techniques," *IEEE PES Winter Meeting*, 89WM 186-8 PWRS, February, 1989.

[3] K. R. C. Mamandur and R. D. Chenoweth, "Optimal control of reactive power flow for improvements in voltage problems and for real power loss minimization," *IEEE Trans. PAS*, Vol. 100, No.7, pp.3185-3194, July, 1981.

[4] R. Yokohama, S. H. Bae, T. Morita, and H. Sasaki, "Multi-objective optimal generation dispatch based on probability security criteria," *IEEE Trans. PWRS*, Vol. 3, No.1, pp.317-324, February, 1988.

[5] M. Sakawa and H. Yano, "An interactive fuzzy satisfycing method for multi-objective nonlinear programming problems with fuzzy parameters," *Fuzzy Sets and Systems*, Vol.30, 1989, pp.221- 238, North-Holland.

[6] T. Sawaragi, H. Nakayama, and T. Tanino, "*Theory of Multi-Objective Optimization*," Academic Press, 1985.

[7] T. Niimura, "*Intelligent Operation and Control of Electrical Power Systems Based on Operators' Heuristics*," Ph. D. Dissertation, pp.66-85, Tokyo Metropolitan University, July, 1992.

[8] H. J. Zimmermann, "*Fuzzy Set Theory and its Applications*," Kluwer Academic Publishers, Dordrecht, 1985.

[9] B. A. Murtagh and M. A. Saunders, "*MINOS 5.1 User's Guide*", SOL83-20R, Stanford University, 1987.

# Architecture for Event-Driven Intelligent Fuzzy Controller

Janos Grantner
Department of Electrical Engineering
University of Minnesota, Minneapolis, MN 55455

Marek J. Patyra and Marian S. Stachowicz
Department of Computer Engineering
University of Minnesota, Duluth, MN 55812

***Abstract*--Most linguistic models known are essentially static, that is, time is not a parameter in describing the behavior of the object's model. In this paper we show a model for synchronous finite state machines based on fuzzy logic. Such finite state machines can be used to build both event-driven time-varying rule-based systems and also the control unit section of a fuzzy logic computer. The architecture of a pipelined intelligent fuzzy controller is presented, and the linguistic model is represented by an overall fuzzy relation stored in a single rule memory. A VLSI integrated circuit implementation of the fuzzy controller is suggested. At a clock rate of 30 MHz, the controller can perform 3M FLIPS on multi-dimensional fuzzy data.**

## I. INTRODUCTION

The general model of a finite state machine (FSM) is illustrated in Fig. 1. Formally, a sequential circuit is specified by two sets of Boolean logic functions:

$$f_z(X, y) \to Z$$
$$f_y(X, y) \to Y \qquad (1)$$

where X, Z, y, and Y stand for a finite set of input, output, the present and next states of the state variables, respectively.

Function $f_z$ maps the input and the present state of the state variables to the output, and function $f_y$ maps the input and the present state of the state variables to the next state of the state variables.

The current states of the memory elements hold information on the past history of the circuit. The behavior of a synchronous sequential circuit can be defined from the knowledge of its signals at discrete instants of time. Those time instants are determined by a periodic train of clock pulses. The memory elements hold their output until the next clock pulse arrives.

We extend this model by introducing membership functions and fuzzy relations to map the changes which take place in fuzzy input data to both the fuzzy output and next state of the state variables.

With the model presented in this paper, the definition of states will remain crisp- that is, the state of the system can be represented in one of the usual ways (i.e. by isolated flip-flops, registers or a microprogrammed control 1 unit). The fuzzy outputs will be devised from a dynamically changing linguistic model, since the response to a specific change in fuzzy input will vary with different states of the FSM. We will refer to this model as Crisp-State-Fuzzy-Output FSM or CSFO FSM. A block diagram of the CSFO FSM is shown in Fig. 2. X and Z stand for a finite set of fuzzy input and output, respectively.

R stands for the object's model which is now function of the y present state of the state variables, and o is the operator of composition. The zc crisp values of the fuzzy output are obtained by computing the DF defuzzification strategy. B stands for the transformation which maps the linguistic values of the X linguistic (fuzzy) variables to the $X_B$ Boolean (two-valued) logic variables. Function $f_y$ maps both the $X_b$ Boolean logic variables and the y present state variables to the Y next state of the state variables. To accelerate the mapping of the fuzzy input data X to both the new set of fuzzy

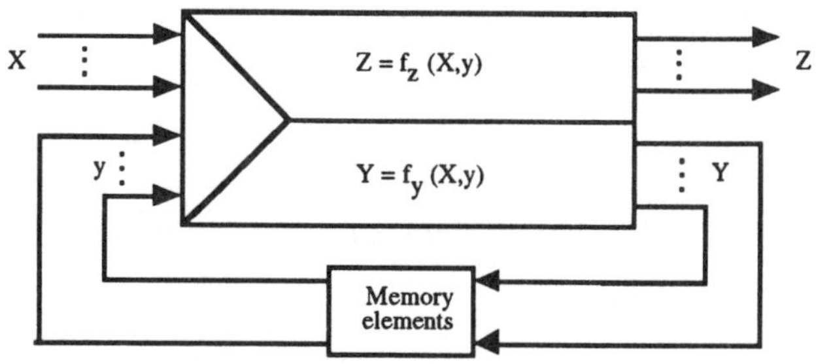


Fig. 1. General model of a Finite State Machine (FSM).

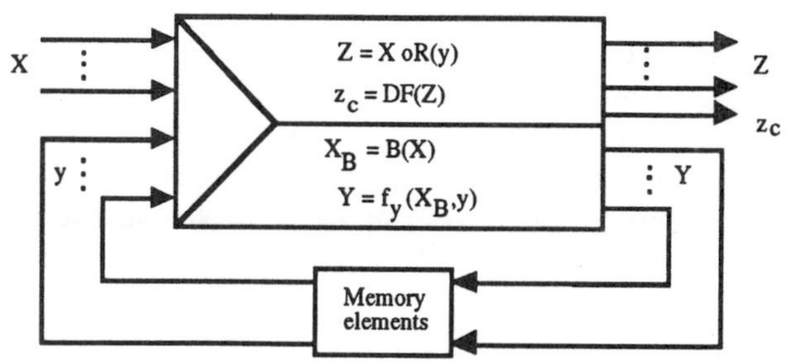


Fig. 2. General model of a Crisp-State-Fuzzy-Output Finite State Machine (FSM).

output Z and crisp output $z_c$ (i.e. to compute fuzzy inference and the DF defuzzification strategy), our pipelined fuzzy logic hardware accelerator model [5] will also be employed with the CSFO FSM. The next state of the state variables will be devised from both the present state variables and the $X_B$ Boolean logic variables. For instance, a Boolean variable X1LOW is true if the position of the maximum in the membership function for linguistic variable X1 falls in the range 1 to 5. X1LOW is otherwise false.

The state transients will be completed simultaneously with the fuzzy pre-processing pipeline step.The new SK state of the CSFO FSM will then determine the overall fuzzy relation RK, which will in turn be used as the linguistic model in the fuzzy inference pipeline step while the system is in state SK. With this model, the state variables will take their new values at the rate at which the pipeline steps proceed. The fuzzy outputs will be defuzzified in the last pipeline step. In the course of the learning process (eq. 3), an overall relation RI is created for each state SI (I = 1,..., N) of the CSFO FSM.

## II. Algorithm for Creating a Multiple-Input Fuzzy Model

A linguistic model of a process can be built using a specialized software; fuzzy inference and defuzzification strategies can also be computed without using any dedicated hardware. However, in the case of real-time control applications, the pure software approach may not offer sufficient performance. We suggest using a hardware accelerator for a multiple-input fuzzy logic controller based upon the mathematical model as follows. The process operation control strategy is created by analysis of input and output values, in which not only measurable quantities are taken into account but also parameters which cannot be measured, only observed [1]. On the basis of the verbal description, which is called a linguistic model, a fuzzy relation R is created:

$$R = *_{I=1}^{N}(XI \rightarrow YI) \qquad (2)$$

In formula (2) $\rightarrow$ denotes the fuzzy implication operator, and the symbol * represents the sentence connective ALSO.

We shall present the algorithm not only intended for creating a fuzzy model when a verbal description is given, but also for determining the model's answer to a given input [2].

The verbal description of the process performance contains N relations, and fuzzy sets describe the particular states which occur in the verbal description of inputs $X^{(1)}$ and $X^{(2)}$ and output Y be given in Formula 2. The graphic interpretations [4] of fuzzy sets $X^{(1)}$, $X^{(2)}$, and Y are illustrated in Fig. 3.

R1: IF $X^{(1)}$ is very small $((X^{(1)}1)$ AND $X^{(2)}$ is medium $((X^{(2)}1)$ THEN Y is medium (Y1)

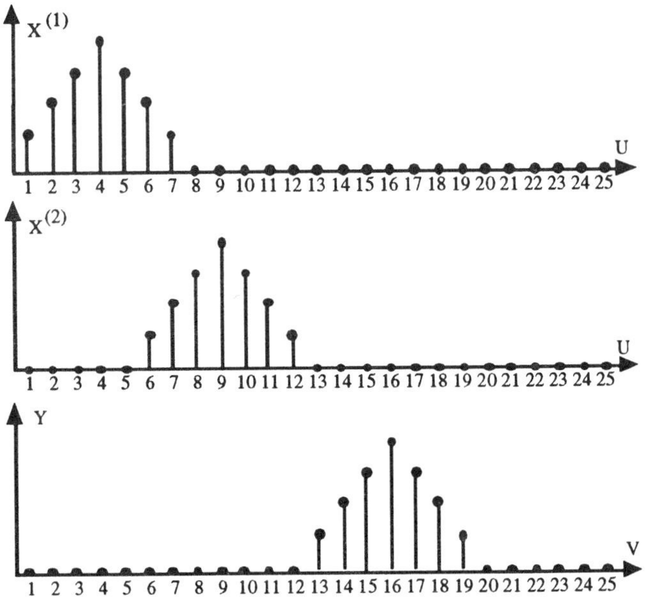


Fig. 3. Graphical interpretation of fuzzy sets $X^{(1)}$, $X^{(2)}$, and Y.

ALSO

:

:

RN: IF $X^{(1)}$ is very big $((X^{(1)}N)$ AND $X^{(2)}$ is medium $((X^{(2)}N)$ THEN Y is medium (YN)

The paragraphs below illustrate in turn:

### A. Fuzzy Learning

A method of creating fuzzy relation R1 which represents the first fuzzy implication in the verbal description is interpreted as intersection. The remaining relations *R2, R3,..., RN* are created analogously by application of the same definition of fuzzy implication.

$$R = X1 \times Y1 \qquad (3)$$

$$\forall(our) \in U \times W \; R1(u, w) = \min(X1(u), Y1(w)) \qquad (4)$$

$$\forall u \in U \; X1(u) = \min(X^{(1)}(u), X^{(2)}(u)) \qquad (5)$$

The final relation R (being the object's model) is obtained as the union of *R1, R2,..., RN*, since the sentence connective ALSO is defined as union.

$$R = R1 \cup R2 \cup \ldots \cup RN \qquad (6)$$

$$\forall(u, w) \in U \times W \; R(u, w) = \max(R1(u, w), R2(u, w), \ldots, RN(u, w)) \qquad (7)$$

### B. Fuzzy Inference

The method of finding fuzzy output Y given the fuzzy input X and the relation R applies max-min composition.

$$\forall w \in W \quad Y(w) = \max_{u \in U} (\min(X(u), R(u,w))) \qquad (8)$$

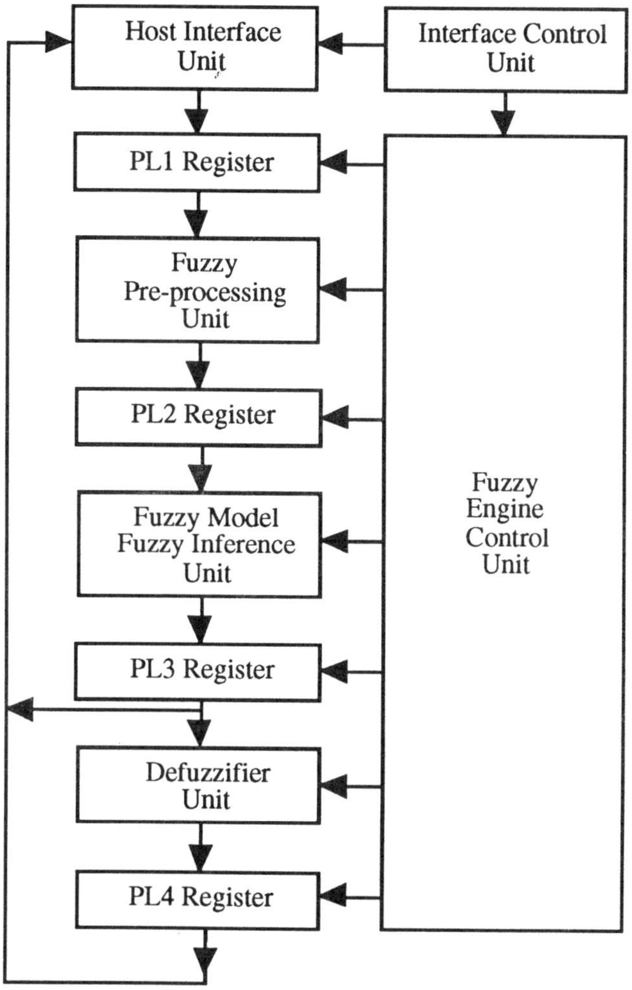


Fig. 4. Pipeline architecture of the Hardware Accelerator. (Reproduced by permission of Larry L. Kinney.)

### C. Defuzzification Strategy

The deterministic value of the answer (crisp value) is determined using the formula

$$y_c = \frac{1}{L} \sum_{J=1}^{L} w_J \quad (9)$$

where L is the number of points $w_J \in W$ for which output set Y reaches the maximum.

## III. HARDWARE ACCELERATOR

The hardware accelerator which performs fuzzy learning, fuzzy inference, and defuzzification computation - that is, which maps the fuzzy inputs to fuzzy and/or crisp outputs - is summarized in this section. In our research, the membership function is a discrete valued function with a 5-element domain set. With two-valued logic, three bits are used to represent each element of the set. The number of levels, of the membership function, can be extended up to eight. The universe of discourse of a fuzzy subset is limited to a finite set with 25 elements ($u_{max} = w_{max} = 25$). Seventy five bits are used for digitization of the membership function. The accelerator consists of four basic units: the host interface, the fuzzy pre-processing unit, the combined fuzzy model/fuzzy inference unit, and the defuzzifier unit. The last two are referred to as the fuzzy engine [3]. The functional block diagram of the accelerator is shown in Fig. 4. To achieve a high-processing rate for real time applications, the units are connected in a four-level pipeline.

The core of the hardware accelerator is a fuzzy engine which implements the formulae (6) to (8). It is split into the fuzzy model/fuzzy inference unit and the defuzzifier unit. The functional block diagram of the fuzzy model/fuzzy inference unit (without increased parallelism) is shown in Fig. 5. After the XI and YI registers have been loaded, learning a multi-dimensional rule RK takes umax clock periods. The MUX2 multiplexer at the input of the minimum unit selects the YI register. During the first clock step, uI is paired with all w elements of YI and these pairs are fed to the inputs of the minimum unit. If the current rule is the first in a learning sequence, throughout the learning cycle, 0-valued (nonmembership) elements will be paired with the outputs of the minimum unit and fed to the inputs of the maximum unit. The whole word of maximum values is stored at the first location of the R rule memory. During the ACTH clock step, up is compared to all

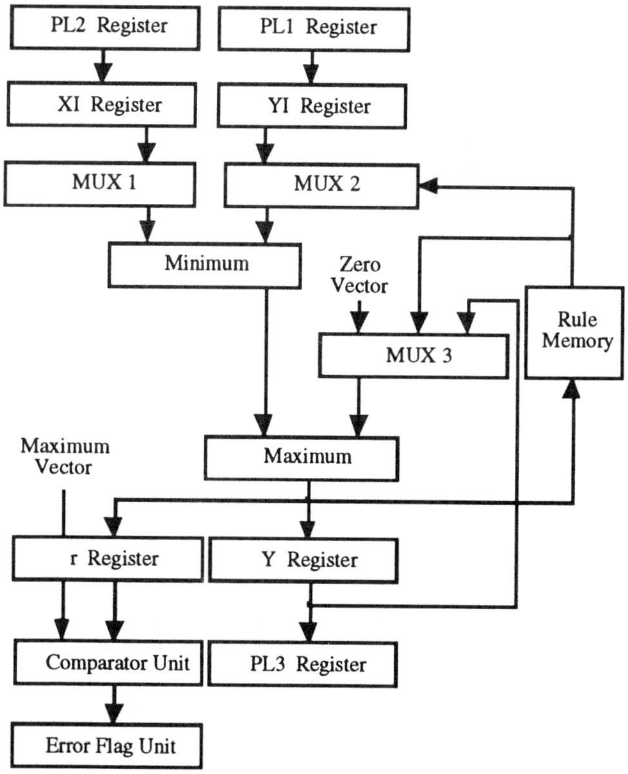


Fig. 5. Functional block diagram of the fuzzy model/fuzzy inference unit.

w elements of YI simultaneously and the vector of the max elements is stored in the jth location of R. If the current rule is not the first one in the learning process, the MUX3 multiplexer at the input of the maximum unit selects the ith row of $R(1 \leq i \leq umax)$ during the ith clock step, and the contents of this row in R are updated from the outputs of the maximum unit.

Therefore, the learning process of N rules takes N×umax clock periods with the architecture shown in Fig. 5. The clock steps needed to load registers XI and YI are ignored at this point. Computing the fuzzy inference (max-min composition) also takes umax clock periods. This time the MUX2 multiplexer at the input of the minimum unit pairs the ui element of XI with all r elements of the ith row in R. If i = 1 (first clock step), then the MUX3 multiplexer at the input of the maximum unit selects 0 as the other operand for each element at the output of the minimum unit. The outputs of the maximum unit are fed to the inputs of the Y register. From the second to the last clock steps, outputs of the Y register are fed back to the inputs of the maximum unit through the MUX3 multiplexer.

Contents of the R rule memory remain unchanged during the fuzzy inference process. After the last clock step, register Y holds the result of the XoR operation in the digitized fuzzy data format. To detect whether the condition: $\forall(u, w) \in U \times W$, $R(u,w) = 1$ is met, an error flag was added to the fuzzy engine. If the error flag is activated at the completion of the learning of a new rule, then all elements in the R rule memory become equal to 1 (full membership). This flag can be used to generate an interrupt request to the host machine. The system can then recover from this erroneous state by either downloading a "safe" model to the R memory or starting over the learning process with a modified model. Due to the linear property of the max-min composition, by quadrupling the functional units of the basic architecture, the time required to complete the pipeline steps for either the fuzzy learning or the fuzzy inference process can be reduced to [umax+ 4]+ 2 clock periods. Since the precedence relation of the subtasks (I/O data transfer ($T_1$), the pre-processing of the multiple fuzzy

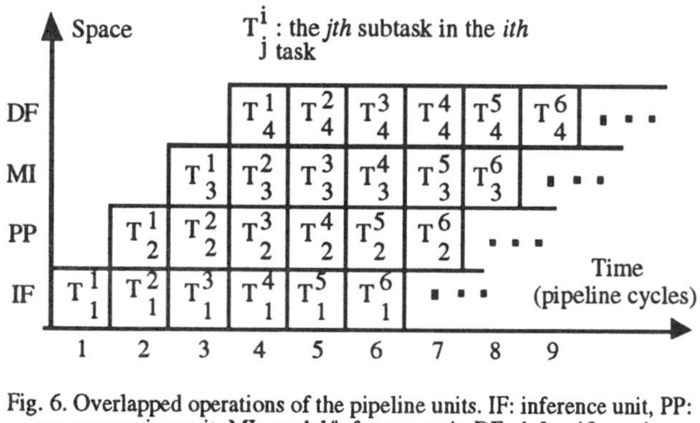


Fig. 6. Overlapped operations of the pipeline units. IF: inference unit, PP: pre-processing unit, MI: model/inference unit, DF: defuzzifier unit.

inputs ($T_2$), the learning of a new rule or the performing a fuzzy inference operation ($T_3$), and the defuzzification ($T_4$)) are all linear operations, the four basic units of the hardware accelerator form a linear pipeline. The pipeline architecture allows the simultaneous operation of the four units. The space-time diagram in Fig. 6 illustrates the overlapped operations of the pipeline units. Assuming that the downloading of the fuzzy data from the host system to the accelerator and the reading of the resultant fuzzy and/or crisp output data (subtasks $T_1$) does not exceed [umax+4]+2 (9) clock periods, the accelerator produces new fuzzy and/or crisp output data every [umax+4]+2 clock periods once the pipeline is filled. Thus, at a clock rate of 30 MHz, the fuzzy engine can perform over 3,000,000 fuzzy logical inferences per second with the current fuzzy data format. The defuzzifier unit performs the defuzzification operation. Finding both the maximum and its position takes at most 4 clock periods. Two parallel adder networks are used to sum up the position codes of the maximum and obtain the total number of maximum simultaneously. Then, the crisp value is taken from the look-up table. If maximum value is zero then the crisp value is flagged.

## IV. VLSI IMPLEMENTATION

One of the most difficult issues coming from practical realization is associated with the VLSI implementation. Therefore the information provided in this section is based on our estimates and previous experience with projects of a similar nature. Due to our objective constraints, i.e. the MOSIS service is available for chip fabrication at this time, the full design version of the proposed controller will be designed, along with a scaled-down version, which will pass the constraints and will finally be fabricated. There are two different versions of the fuzzy logic controller that could be useful in most practical implementations: a controller working stand alone (SA) or with an appropriate host computer (HC). These options will be taken into consideration. Let us discuss the VLSI implementation issues in more detail starting with the full scale design. According to our preliminary assumption, we formulated the descriptions of design signals summarized in Table I and Table II. We assume that the proposed fuzzy controller will have three basic cycles of operation: fuzzy learning, fuzzy inference and stand-by. In case of the fuzzy learning and fuzzy inference operations, the HC version will be supplied with fuzzy data through the host computer which performs the fuzzification of the analog inputs. It is obvious that HC version will be able to process only digital representation of the fuzzy data prepared by the host computer. In our first approach this version will not be cascadable. The SA version of the chip will input the analog data and perform the fuzzification operation by itself. The stand-by mode will be common for both versions. One can also see that the HC version will require a very detailed design of the interface to the bus system used by the host computer for data transmission

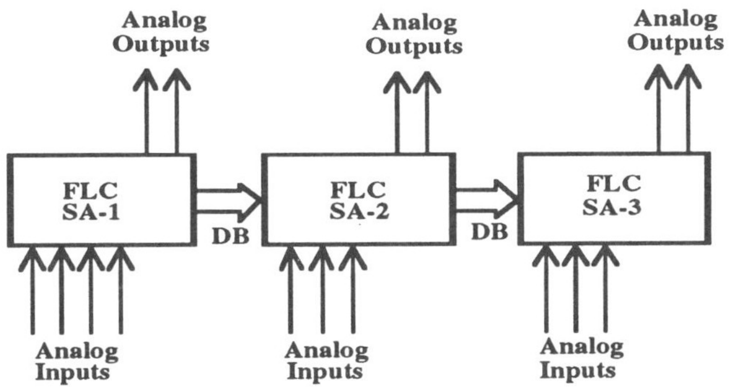


Fig. 7. Serial mode of operation of cascaded SA version of fuzzy logic controller.

while the SA version will need an A/D converter and a few D/A converters, which will be included in the chip design. It is assumed that such a version will be communicating with the host computer through FUTUREBUS. Each version has its own advantages and disadvantages basically due to communications issues, the number of pins and the design effort. It is important to point out that one can expect some instant differences in the performance of the four versions which will further be investigated in detail. Let us now focus on the scaled down implementations of the SA version of the proposed fuzzy logic controller. Two basic modes for chip operation will be designed: normal and programmed. The normal mode will include cascaded (parallel or serial) and noncascaded operations. The block diagrams illustrating these modes are shown in Fig. 7 and Fig. 8. In the programming mode, we assume that it will be possible to preprogram the fuzzifier, defuzzifier or inference engine, or any combination of these, in order to preserve a flexible operation tuned to the actual system under control. In order to achieve programmability, an EPROM/EEPROM type of memory block will be built-into the chip and will be controlled by the external source through both a memory I/O port and control signals. Our decision to fabricate this version is based on both the

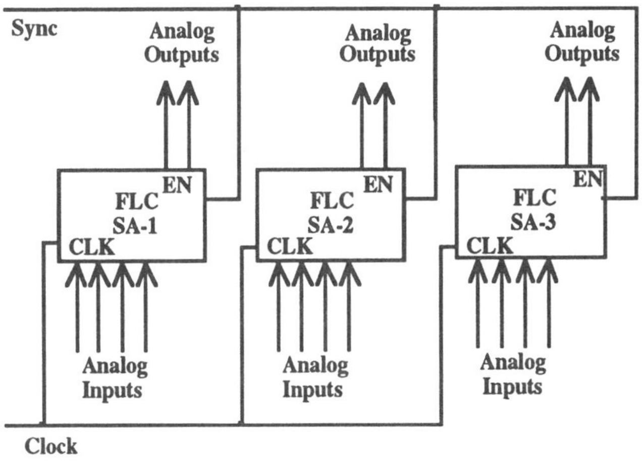


Fig. 8.Parallel mode of operation of cascaded SA version of fuzzy logic controller.

number of pins and also the number of signals needed to implement this version (no data interface is needed). The scaled down implementation of our design matching the objective constraints is presented as follows with respect to major design issues.

### *A. Chip Size and Package*

A reasonable MOSIS package has 132 pins and contains a chip occupying 7.8mm*9.2mm$^2$ of silicon area (max). The choice of CMOS technology leads us to the variety of available processes starting from lambda=1μm to lambda=0.6μm. Keep in mind that the maximum signal frequency for chip operation, which was originally set between 25MHz and 30MHz, as well as the maximum chip size, (the n-well, double-metal CMOS technology with lambda=0.6μm) will be adequate to achieve the design goals.

### *B. Chip Area and Number of Transistors*

The maximum chip area of 72.68mm$^2$ (132 pins package) can contain about 600000 transistors for highly regular struc-

TABLE I
PRELIMINARY DEFINITIONS OF SIGNALS FOR FULL SCALE VERSION OF HC FUZZY LOGIC CONTROLLER

| DESCRIPTION OF THE SIGNAL | Number |
|---|---|
| Scalable address/data bus | 64 |
| Data bus control | 55 |
| Fuzzification control | 3 |
| Inference control | 3 |
| Defuzzification control | 3 |
| Bus/address control | 10 |
| Standard chip control (RES, CLK, CS) | 3 |
| Power supply | 8 |

TABLE II
PRELIMINARY DEFINITIONS OF SIGNALS FOR FULL SCALE VERSION OF SA FUZZY LOGIC CONTROLLER

| DESCRIPTION OF THE SIGNAL | Number |
|---|---|
| Scalable address/data bus | 64 |
| XPROM interface/control signals | 32/4 |
| Fuzzification control | 3 |
| Inference control | 3 |
| Defuzzification control | 2 |
| Analog inputs/outputs | 4/2 |
| Standard chip control (RES, CLK, CS, EN, STB) | 5 |
| Power supply | 8 |

ture with the standard CMOS technology (lambda=0.6μm). According to our estimations, we will be able to put at most four parallel fuzzy data processing paths into the chip. The single data processing path including programming options (memoryless) is estimated to have about 50,000 transistors.

### *C. Clock Strategy and Clock Distribution*

We decided to use a single external clock signal (CLK) to generate on-chip, two-phase non-overlapping internal clock signals (Φ1 and Φ2), with the two phase-clock system having the advantage of making hazard problems within the pipeline paths more identifiable. These phases will be distributed over the whole chip using a second metallization layer. Because the longest possible metal line is about 10mm, we chose a tree-like structure for phase distributions driven by high gain clock drivers. These drivers will be designated to drive an appropriate capacitive load of the whole clock line tree. According to the results of our previous research, the single processing path will have the ability to operate at 8 clock cycles/pipeline step. Setting the external clock rate at around 30MHz will enable us to operate the processor at a high processing rate. A future investigation will help us to determine the highest possible clock rate.

### *D. Rule Memory*

The major problem with the limited capacity of the internal SRAM (for HC version) of EPROM/EEPROM (for SA version) memory for the storage of rules of inferences in previous works [6-8] does not exist in our approach due to the strategy of building a global rule for the whole linguistic model described in Section II. In our case only a 1/4 Kilobyte SRAM or EPROM/EEPROM is needed to store the global rule. Such an approach creates a luxury of increasing the parallelism of the internal structure by a factor of four, which is discussed in the next section.

It also should be noted that the idea of CSFO FSM is intended to be implemented in the SA version. Furthermore, the required extension of the rule memory (every FSM state will have assigned rule memory) will be evaluated. It is, however, unlikely that overall number of transistors for a single path will reach 100,000 transistors.

### *E. Pipeline Architecture*

The estimated number of transistors for a single fuzzy data processing path is around 50,000. This means that the chip capacity under consideration can contain at least four separate fuzzy data processing paths plus rule memory, which gives total estimation of about 300,000 transistors (look-up table used for defuzzification is included). The estimated area occupied by transistors is about 45 mm2. The rest of the chip area will be used to provide high speed communication between processing units and the built-in memory (EPROM/EEPROM). It is expected that four parallel data paths designed in the chip, could possibly increase the actual speed of operation twice. In the proposed design, 3M FLIPS performance is expected to feature a clock rate of 30MHz.

## V. CONCLUSIONS

This paper describes the general model for fuzzy state machine (FSM) which is used to formulate the fuzzy controller for event-driven real-time systems. As a result the improved architecture for fuzzy logic controller has been defined. The improvement over already published architectures [5-9] lies in the novel strategy for fuzzy model building, which enables fuzzy inferences to be performed in a single stage of a hardware accelerator. As it has been estimated, the proposed architecture, appropriately pipelined for the hardware accelerator, will profit in reaching at least 3M fuzzy logical inferences per second (FLIPS). The presented approach can be utilized for fuzzy controller hardware accelerators intended to work in the real time environment.

## REFERENCES


[1] L. A. Zadeh, "A Fuzzy Algorithmic Approach to the Definition of Complex or Imprecise Concepts", International Journal Man Machine Studies, 8, pp. 249-291, 1976.

[2] M. S. Stachowicz, "The Application of Fuzzy Modeling in Real-Time Expert Systems for Control", Proc. 49th Ironmaking Conference, Detroit, pp. 503-512, March 25-28, 1990.

[3] M. S. Stachowicz, J. Grantner, L. L. Kinney, "Two-Valued Logic for Linguistic Data Acquisition", NAFIPS Workshop '91, University of Missouri-Columbia, Proc. pp. 168-172, May 14-17, 1991.

[4] M. S. Stachowicz, M. E. Kochanska, "Graphic Interpretation of Fuzzy Sets and Fuzzy Relations", Mathematics at the Service of Man, (eds. A. Ballester, D. Cardins, E. Trillas), Springer-Verlag, West Berlin, pp. 620-629, 1982.

[5] M. S. Stachowicz, J. Grantner, L. L. Kinney, "Pipeline Architecture Boosts Performance of Fuzzy Logic Controller", IFS-ICC'92 International Fuzzy Systems and Intelligent Control Conference, Louisville, Kentucky, Proc. pp. 190-198, March 15-18, 1992.

[6] M. Togai, H. Watanabe, "Expert System on a Chip: An Engine for Real-Time Approximate Reasoning", IEEE Expert, pp. 55-62, Fall 1986.

[7] H. Watanabe, W. D. Dettloff, K. E. Yount, "A VLSI Fuzzy Logic Controller with Reconfigure, Cascadable Architecture", IEEE Journal of Solid-State Circuits, pp. 376-381, Vol. 25, No. 2, April 1990.

[8] FC 110 Digital Fuzzy Processor DFPTM. Togai InfraLogic, Inc. 10/1991.

[9] M. J. Patyra, "VLSI Implementation of Fuzzy-Logic Circuits", International Fuzzy Systems Association World Congress, Brussels, Belgium, June, 1991.

# Fuzzy Logic Based Banknote Transfer Control

Masayasu Sato[†], Tohru Kitagawa[†], Takehito Sekiguchi[†], Keisuke Watanabe[†] and Masao Goto[††]
[†]Human Interface Laboratory, Oki Electric Industry Co., Ltd.
550-5, Higashiasakawa, Hachioji, Tokyo 193, Japan
[††]Human Interface Technology R&D Center, Oki Electric Industry Co., Ltd.
3-1, Futaba, Takasaki, Gunma Prefecture 370, Japan

*Abstract*-**A fuzzy logic based banknote transfer control mechanism is proposed. In this mechanism, linear stepping motors are used for adjusting the gap between a feed roller and a gate roller. The gap is automatically adjusted to an optimal value through fuzzy reasoning, based on the statistical mean value of the feeding pitch and on the statistical mean value of the skew of the banknotes that have been fed.**

**Experimental results show that the mean value of the feeding pitch and the mean value of the skew are successfully adjusted automatically to error of ±0.28 msec and ±0.25° or less, respectively, by performing control within one to four times. We have confirmed that fuzzy logic is applied effectively to a banknote separation and feeding mechanism.**

## I. Introduction

A large number of automatic teller machines (ATMs) and cash dispensers (CDs) have been put into use to streamline and improve the efficiency of the teller window business at banks. In recent years, substantial liberalization of financial markets has occurred in Japan, and ATMs and CDs have come to be installed not only in bank premises but also in train stations, department stores, offices, convenience stores and the like, to be used much more widely. In addition, their operating hours have been extended and they are now available for certain time of periods at night and on holidays. These factors - installation in more varied places, longer operating hours and automated operation on holidays - have brought about an even greater demand for highly reliable machines that are free from troubles or failure. The separation and feeding technology is the most important for obtaining high reliability in ATMs and CDs.

Friction separation methods are commonly used at present for the separation and feeding of the banknotes, as they enable high speed separation and feeding as well as downsizing of machines. In these methods, the banknotes are separated and fed one by one from the gap between the feed roller and the gate roller. This separation and feeding capability varies by assembly error and changes in the operating environment and the passage of time. For example, the interval between one banknote and another may become abnormally small or a large skew may occur, thus affecting reliability of ATMs and CDs. Therefore, initial adjustment of the gap and subsequent adjustment must be performed frequently. And yet it is delicate to adjust the gap properly, as it requires skill.

In this paper, a fuzzy logic based banknote transfer control mechanism is proposed that automatically adjusts the gap to an optimal value through fuzzy reasoning based on the statistical mean value of the feeding pitch and the statistical mean value of the skew obtained from the banknotes that have been fed, and the results of experiments are described.

## II. Friction Separation Mechanism

Friction separation mechanisms commonly used at present consist of two pickup rollers, two feed rollers and two gate rollers (see Fig. 1, but ignore linear stepping motors and the control system). The surfaces of these rollers are covered with a rubber material. Banknotes are pressed by spring force onto the pickup rollers. As the pickup rollers rotate, the top banknote is separated from the other banknotes and moved in the direction of rotation of the pickup rollers. The banknote then passes through the left and right gaps ($X_L$ and $X_R$) between the feed rollers and the gate rollers (which are installed face to face against these feed rollers to prevent the feeding of more than one banknote at a time). The gaps ($X_L$ and $X_R$) can be varied independently by the gap adjustment mechanism. In the friction separation methods, the conditions of feeding banknotes such as the feeding pitch, the skew, and the duplicate feeding are substantially affected by the coefficient of friction of the rubber on each roller and the gaps. The gaps have an especially substantial effect on the characteristics of the separation and feeding, depending on how they are adjusted. Thus adjusting the gaps demands skill.

## III. Principle of Fuzzy Logic Based Control Mechanism

Fig. 1 shows the banknote separation and feeding mechanism based on fuzzy logic proposed in this paper.

For the following reasons, we have analyzed a system for automatically adjusting the gaps of a banknote separation and feeding mechanism using fuzzy logic.

(1) It is difficult to know the quantitative characteristics of the controlled object because of a large number of parameters.

(2) It is quite difficult to physically model the controlled object and the control system.

(3) The manual adjustments performed by an expert through understanding the qualitative characteristics of the controlled object and the control system can be partly expressed linguistically and turned into rules.

0-7803-0614-7/93$03.00 ©1993IEEE

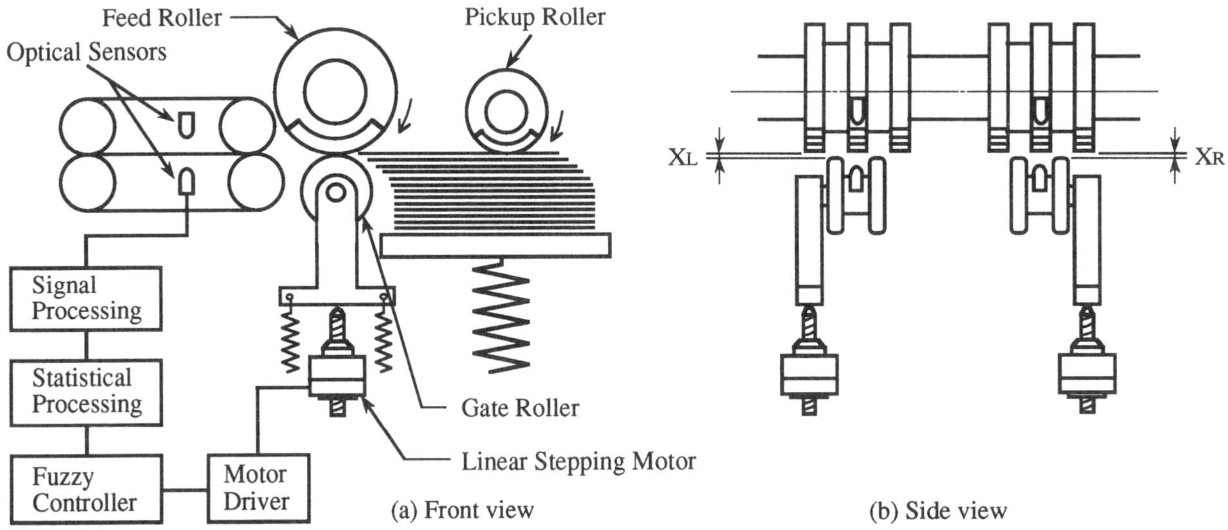


Fig. 1. Schematic diagram of fuzzy logic based banknote transfer control mechanism.

### *A. Control System*

In the mechanism proposed, conventional manual adjustment mechanism is replaced by two linear stepping motors (25 μm/step) . Two pairs of optical sensors in the banknote feed path detect the feeding pitch and the skew of the banknotes, and a feedback loop is structured so that the gaps ($X_L$ and $X_R$) are controlled through fuzzy reasoning based on these data.

This is a two-input, two-output control system. The input fuzzy variables are the statistical mean value of the feeding pitch, and the statistical mean value of the skew of the banknotes that have been fed. The output fuzzy variables are the manipulating variable for the left motor and the manipulating variable for the right motor.

It is difficult to know the correlation between a banknote that has just passed through the optical sensors and the banknote to be separated and fed next. Therefore it would make the control system unstable if the gaps were controlled for each banknote that had been fed. This could cause repetitive fluctuations on the feeding pitch and the skew. In our system, therefore, the mean value of the feeding pitch and the mean value of the skew are used for control. This reduces the effect of the differences in the characteristics of the banknotes.

The feeding pitch and the skew of a banknote fed into the feed path through the gaps are detected by the optical sensors. As each banknote is fed into the feed path, these data detected are accumulated successively. Then each mean value is obtained. The difference between the mean value (pm) of the feeding pitch (pi) and the target value (pr) is represented by $\Delta p$ (= pm- pr), while the difference between the mean value ($\theta m$) of the skew ($\theta i$) and the target value ($\theta r$) is represented by $\Delta\theta$(= $\theta m$ - $\theta r$). Through fuzzy reasoning based on $\Delta p$ and $\Delta\theta$, drive signals to the left and right motors are calculated independently. In response to these signals, the left and right gaps are adjusted independently by the motors so that the left and right gaps become optimal.

### *B. Fuzzy Control*

Fuzzy control is studied on the basis of the specifications of fuzzy chip set developed by our company. Its specifications are shown in Table I. The Oki fuzzy chip set has two chips, a fuzzification chip (MSM91U044) and a defuzzification chip (MSM91U045). The max- min-gravity method proposed by Mamdani [1] is used. Up to 16 input variables and up to 2 output variables are available. For each input variable, up to 14 membership functions (up to 56 membership functions for all of the input variables) can be set. For the output variables, up to 15 membership functions can be set for each variable. The universe of discourse of the membership functions in the if-part is fixed at 64 elements. The universe of discourse of the membership functions in the then-part is 32 elements with a single output variable and 16 elements with double output variables. There are 16 grades in both the if-part and the then-part.

TABLE I

BASIC SPECIFICATIONS OF OKI FUZZY INFERENCE CHIP

| Parameter | Condition | Standard Value | | | Unit |
|---|---|---|---|---|---|
| | | MIN | TYP | MAX | |
| Inference Speed | | | | 7.5 | MFLIPS |
| Number of Input Valuables | | 1 | | 16 | |
| Number of Output Valuables | | 1 | | 2 | |
| Number of Rules | | 1 | | 960 | Rule |
| Number of Membership Function | IF-part | | | 56 | Kind |
| | THEN-part | | | 30 | |
| Universe of Discourse | IF-part | — | 64 | — | Element |
| | THEN-part | 16 | | 32 | |
| Grade | | — | 16 | — | Step |
| Clock Frequency | | | 20 | 30 | MHz |

An example of the basic characteristics of the gap and the feeding pitch is shown in Fig. 2. The horizontal axis represents the relative value of the gap, while the vertical axis represents $\Delta p$ and $\Delta\theta$. As the gap is made larger, the feeding pitch becomes shorter than the target value. Conversely, as the gaps are made smaller, the feeding pitch becomes longer. The linear area is small, exhibiting nonlinear characteristics.

Table II shows the rules that are prepared for our system in consideration of nonlinear characteristics. The meanings of the labels given to $\Delta p$ in Table II are as follows. ZR means that the mean value of the feeding pitch is almost equal to the target value, with a negligible difference ($\Delta p=0$); N means that the mean value of the feeding pitch is shorter than the target value ($\Delta p<0$), that is, the interval between one banknote and the next one is shorter than the target value, and P means that the mean value of the feeding pitch is longer than the target value ($\Delta p>0$), that is, the interval between one banknote and the next one is longer than the target value. As for the meanings of the labels given to $\Delta\theta$, ZR means that there is little skewing of banknotes ($\Delta\theta=0$); N means that the left side of the banknotes is ahead of the right side (left ahead) ($\Delta\theta<0$) and P means that the right side of the banknotes is ahead of the left side (right ahead) ($\Delta\theta>0$). For both $\Delta p$ and $\Delta\theta$, L, M and S respectively mean "the difference from the target value is LARGE," "the difference is MEDIUM" and "the difference is SMALL". With respect to the meanings of the labels given to the output for the motor, ZR means that the output is made so as to hardly drive the motor; N means that the output is made so as to drive the motor in the direction of reducing the gap, and P means that the output is made so as to drive the motor in the direction of widening the gap. In addition, L, M, S and VS respectively indicate "drive to a LARGE degree," "drive to a MEDIUM degree" "drive to a SMALL degree" and "drive to a VERY SMALL degree". There are nine labels for the output to the motors.

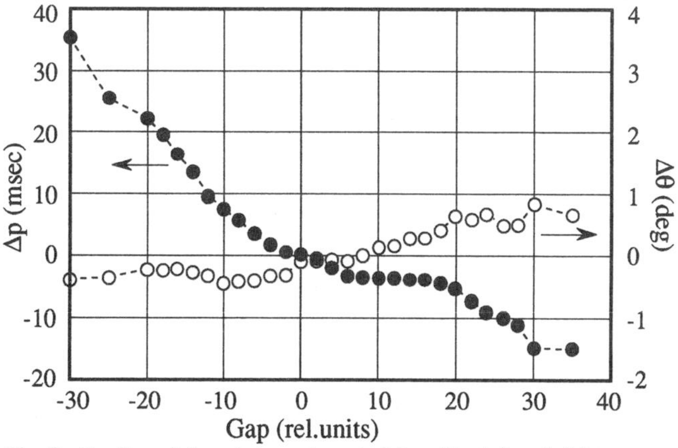


Fig. 2. Feeding pitch and skew characteristics. The left and right gaps are varied by a certain quantity at the same time.

Fig. 3 shows the membership functions. In Fig. 3, (a) is the membership function for the feeding pitch; (b) is the membership function for the skew, and (c) is the membership function for the output to the motor. The shapes of the membership functions for the skew are steep, to ensure more sensitive control . Moreover, for the membership functions for output to the motor, two functions, NVS and PVS, have been added to the seven general functions, for a total of nine functions, so that fine control can be achieved in the neighborhood of $\Delta p=0$ and $\Delta\theta=0$.

In Fig. 4, the results of calculation of the outputs to the left and right motors by the max-min-gravity method with $\Delta p$ and $\Delta\theta$ as parameters are shown three-dimensionally using the rules and the membership functions described above. The (a) and (b) graphs are the results of calculation of the outputs to the left and right motors respectively. The X axis represents the skew. The center is where $\Delta\theta=0°$. Moving right from the center, the right-ahead skew becomes larger, while when moving left from the center, the left-ahead skew becomes larger. The Y axis represents the feeding pitch. The center of the Y axis is where $\Delta p=0$ msec. Moving forward, the feeding pitch becomes shorter than the target value, while when moving backward, the feeding pitch becomes longer than the target value. The Z axis represents the output to the motor. The center of the Z axis is where the output to the motor is zero. Moving up from the center, the output is such that the motor is driven in the direction of widening the gap. Moving down from the center, the output is such that the motor is driven in the direction of reducing the gap.

TABLE II

RULE TABLES FOR THE MOTOR CONTROL

(a) Rules for controlling the left motor

| | | Left | | | $\Delta\theta$ | | | Right |
|---|---|---|---|---|---|---|---|---|
| | | NL | NM | NS | ZR | PS | PM | PL |
| Short | NL | NL | NL | NL | NL | NM | NS | ZR |
| | NM | NM | NM | NM | NL | NS | ZR | PS |
| | NS | NS | NS | NS | NS | ZR | PS | PM |
| $\Delta p$ | ZR | NM | NS | NVS | ZR | PVS | PS | PM |
| | PS | NM | NS | ZR | PS | PS | PS | PS |
| Long | PM | NS | ZR | PS | PM | PM | PM | PM |
| | PL | ZR | PS | PM | PL | PL | PL | PL |

(b) Rules for controlling the rigth motor

| | | Left | | | $\Delta\theta$ | | | Right |
|---|---|---|---|---|---|---|---|---|
| | | NL | NM | NS | ZR | PS | PM | PL |
| Short | NL | ZR | NS | NM | NL | NL | NL | NL |
| | NM | PS | ZR | NS | NL | NM | NM | NM |
| | NS | PM | PS | ZR | NS | NS | NS | NS |
| $\Delta p$ | ZR | PM | PS | PVS | ZR | NVS | NS | NM |
| | PS | PS | PS | PS | PS | ZR | NS | NM |
| Long | PM | PM | PM | PM | PM | PS | ZR | NS |
| | PL | PL | PL | PL | PL | PM | PS | ZR |

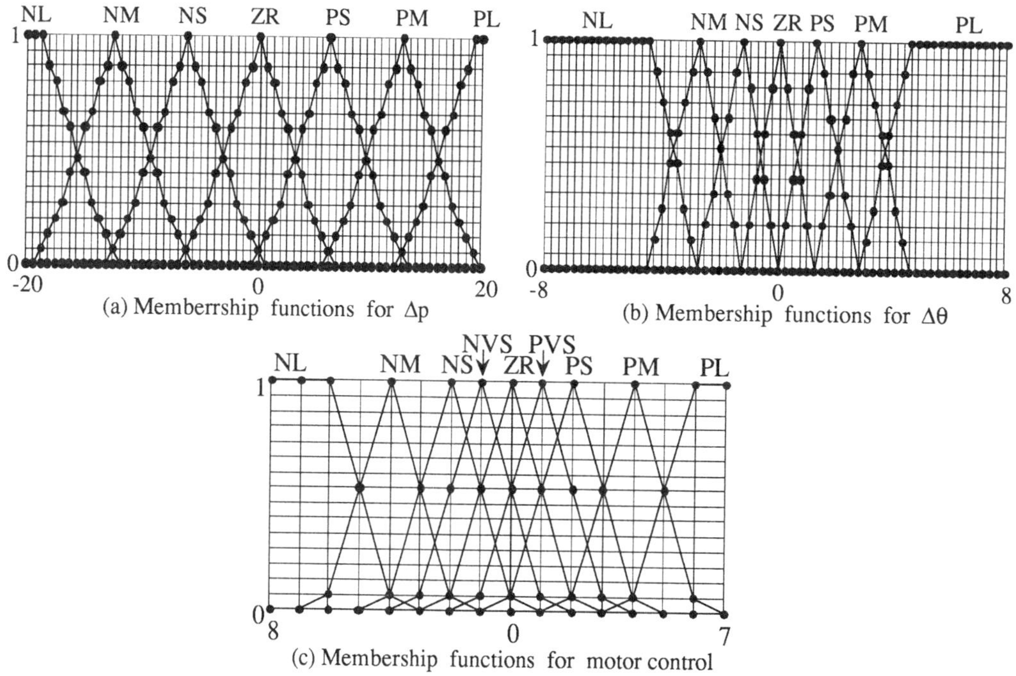


(a) Memberrship functions for Δp

(b) Membership functions for Δθ

(c) Membership functions for motor control

Fig. 3. Membership functions for the motor control

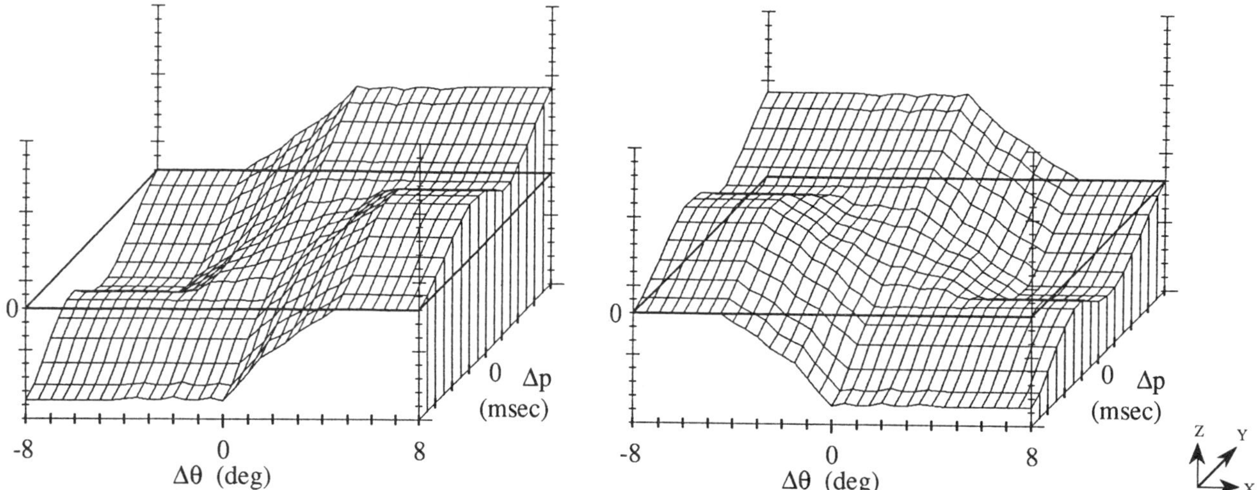


(a) Output profile for the left motor

(b) Output profile for the right motor

Fig. 4. Three dimensional fuzzy control profile.

## IV. Experimental Results

Experiments are performed using 500 in-circulation banknotes. They vary widely in paper quality, extent of damage, and other conditions. They are not arranged uniformly in terms of either direction or faces. The pickup roller, the feed roller, and other parts are not new, which means that they have been already used for some time. Therefore, they have some abrasion.

The experiments are performed according to the following steps:

1) The 500 in-circulation banknotes are used. The mean value of the feeding pitch (pm) and the mean value of the skew (θm) of the 500 banknotes that have passed through are obtained.

2) The differences from the respective target values (Δp and Δθ respectively) are obtained. Fuzzy control is performed based on those values.

3) Operations 1) and 2) are repeated until the outputs to both the left and the right motors become zero.

The results of the experiment are shown in Fig. 5, Fig. 6 and Fig. 7.

Fig. 5 shows the results in histogram form. The upper graphs are histograms of the feeding pitch, while the lower graphs are histograms of the skew. In the upper graphs, the horizontal axis represents the difference between the feeding pitch of the banknote that have been fed and the target value. In the lower graphs, the horizontal axis represents the difference between the skew and the target value. In both the upper and lower graphs, the vertical axis represents frequency. The (a) graphs are histograms of the feeding pitch and the skew for the 500 banknotes passed through without the fuzzy control. The (b) graphs indicate the results of the fuzzy control performed on (a) (first control). And the (c) graphs indicate the results of the fuzzy control performed on (b) (second control). In the initial state, or in (a), $\Delta p$=23.6 msec and $\Delta\theta$=-2.4°. After the first control, $\Delta p$=2.1 msec and $\Delta\theta$=0.05°. That is, the skew had been already controlled to within the target value. After the second control, or in (b), $\Delta p$=0.23 msec and $\Delta\theta$=0.09°. That is, the feeding pitch was also controlled to within the target value.

Fig. 6 shows the process of the feeding pitch and the skew converging on the respective target values as a result of the fuzzy control. These are data for a set of 100 banknotes (1

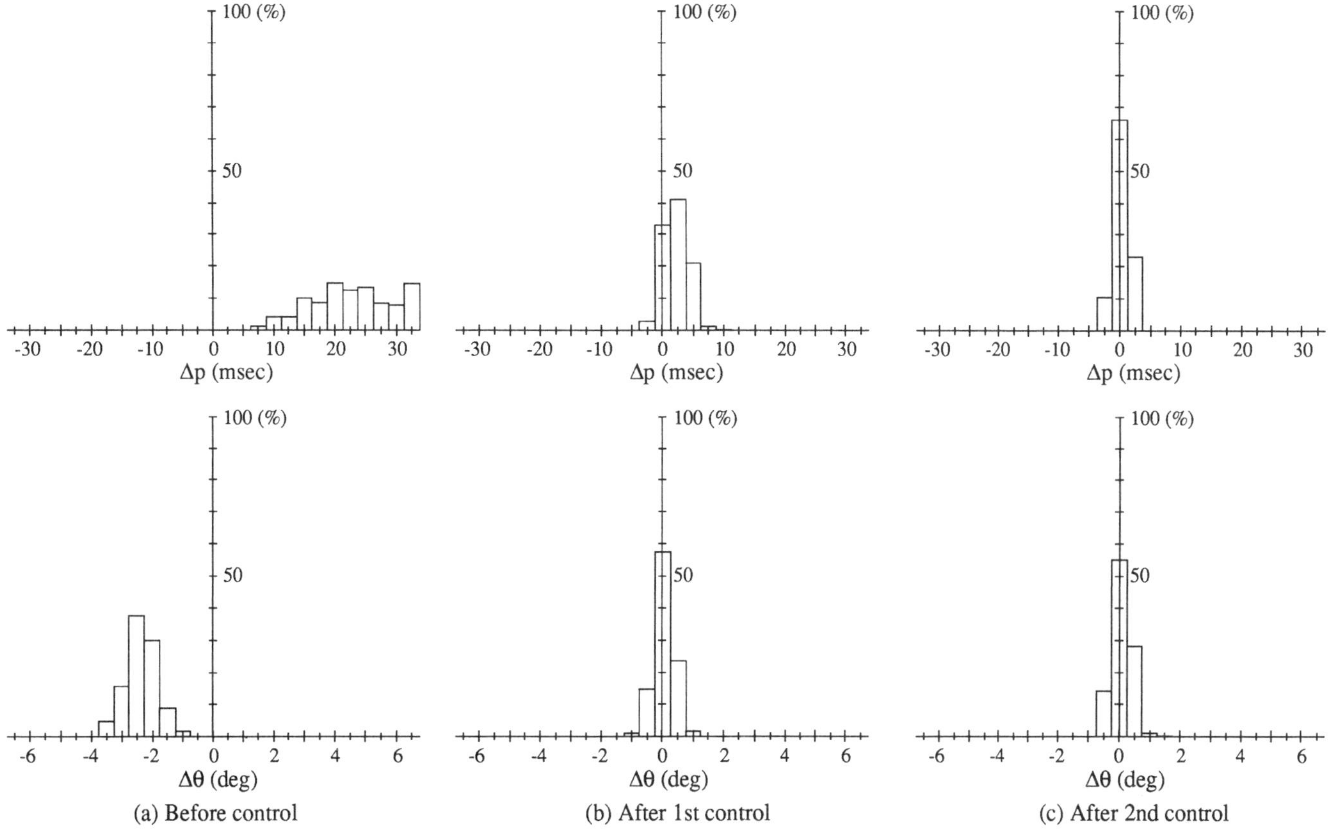

Fig. 5. Histograms of the feeding pitch and the skew.

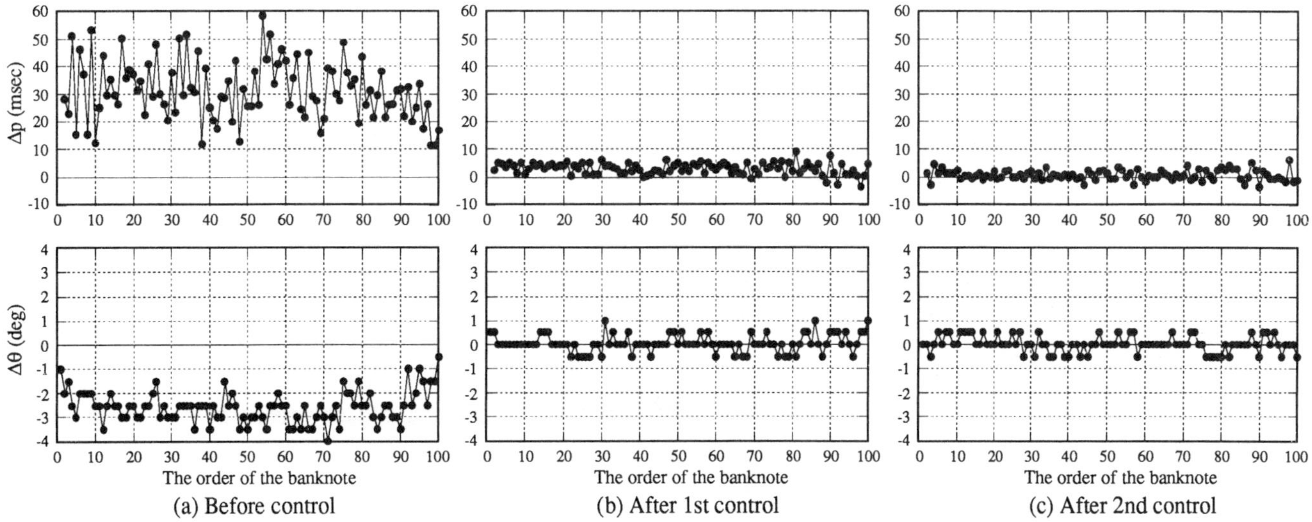

Fig. 6. Variations of the feeding pitch and the skew measurement when a series of banknote feeding by fuzzy logic control.

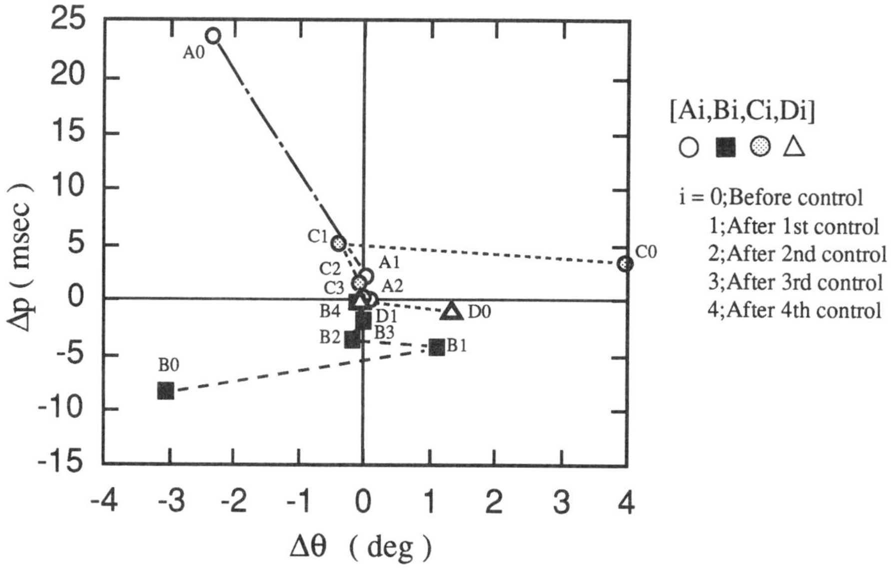


Fig. 7. Feeding pitch and skew trajectory by fuzzy logic control

unit) among the data on the feeding of the banknotes shown in Fig. 5. The (a), (b) and (c) graphs in Fig. 6 indicate "(a) before control," "(b) first control" and "(c) second control," respectively as in Fig. 5. The horizontal axis represents the order of banknote separated and fed. The vertical axis in the upper graphs indicates the difference between the feeding pitch that have been fed and the target value, while the vertical axis in the lower graphs indicates the difference of the skew and the target value. These graphs show the process by which the uneven feeding pitch and skew in the initial state converge on the respective target values through the repetition of control. In the initial state, the standard deviation of the feeding pitch is 10.7 and that of the skew is 0.66. After the second control, these values are 1.81 and 0.32 respectively.

Fig. 7 shows the results of examining the effect of fuzzy control by changing the initial state in various ways. The vertical axis represents the feeding pitch and the horizontal axis represents the skew. The point of intersection of the vertical axis and the horizontal axis is a target point ($\Delta p=0$ and $\Delta\theta=0$). The each marked points ( ○, ■, ◎ and △ ) represent the mean values of the feeding pitch and the skew obtained from the experiment on the 500 in-circulation banknotes. As a result of the experiment, the mean values of the feeding pitch and the skew are successfully controlled within ±0.28 msec and ±0.25°, respectively, by performing control once or twice, at most four times. Even when control had to be repeated four times in order to reduce the differences to ±0.28 msec and ±0.25° or less, differences of ±2.8 msec and ±0.5° or less were achieved in the third control. Moreover, when the initial state is in the neighborhood of the target values, the convergence is achieved by performing control only once.

## V. Conclusion

The adjustment of a banknote separation and feeding mechanism, which is manually performed by experts, has been successfully automated by applying fuzzy control. Through the performance of control from one to four times, the mean value of the feeding pitch and the mean value for skew are automatically adjusted to differences of less than ±0.28 msec and ±0.25°, respectively, from the target values. When the differences are ±2.8 msec or less and ±0.5° or less, control are required only once. In short, a highly precise adjustment mechanism with a simple structure is realized by grasping the basic characteristics of a friction separation mechanism. Thus, we have confirmed that fuzzy control can be applied effectively to a friction separation mechanism.

## Acknowledgments

We express our deep gratitude to Mr. Tuji, Mr. Koshida, Mr. Ishidate and Mr. Ozawa, who helped us prepare rules and offer the experimental mechanism, and Mr. Kuroe who helped us design control system and further gave us precious advice in our execution of the experiment.

## Reference

[1] E. H. Mamdani : "Application of Fuzzy Algorithms for Control of Simple Dynamic Plant ", Proc. IEE, vol. 121, no. 12, pp. 1585-1588, 1974

# Lexicographic Tuning of a Fuzzy Controller Using Box's "Complex" Algorithm

Thomas Whalen and Brian Schott
Decision Science Department
Georgia State University
Atlanta, Georgia 30303-3083

Abstract: A fuzzy control system typically requires "tuning," or adjustment of the parameters defining its linguistic variables. Automating this process amounts to applying a second "metacontrol" layer to drive the controller and plant to desired performance levels. Current methods of automated tuning rely on a single crisp numeric functional to evaluate control system performance. A generalization of Box's complex algorithm allows more realistic tuning based on lexicographic aggregation of multiple ordinal scales of performance, such as effectiveness and efficiency. The method is presented and illustrated using a simple inverted pendulum control system.

## *1. Control and Metacontrol*

Figure 1 presents the basic idea of a control system. The controller compares the output of some physical process (the "plant") against a desired value (the "control objective" or "set point") and applies a control signal to the plant. The goal is to drive its output toward the control objective. In general the plant's behavior is also subject to an uncontrolled disturbance. The disturbance may be an initial difference between the plant output and the control objective, it may consist of later fluctuations that tend to push the plant's output away from the control objective, or both. (More detailed treatments of control dynamics make a distinction between the internal state of a plant and its external output. The present discussion suppresses this distinction without undue loss of generality since observable states can be mapped into the output vector while unobservable states can be approximated as disturbances.)

Figure 1: Basic Control System

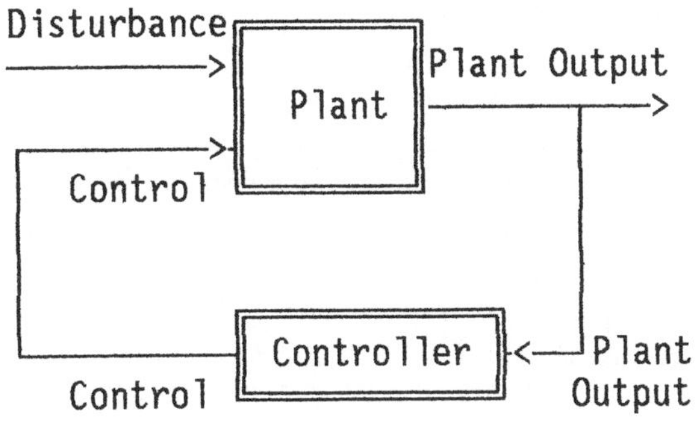


Figure 2 adds another element to the picture. The "metacontroller" observes the inputs and outputs of the controller. These inputs and outputs are accumulated and compared against metacontrol objectives; on the basis of this comparison the metacontroller adjusts the parameters of the controller to bring the behavior of the system as a whole closer to the metacontrol objective.

The most common application of metacontrol is during the design and implementation of a control system. The basic design of the controller under investigation may be a traditional Proportional Integral Differential (PID) controller, a neural net, a fuzzy logic control system, or some other design. In any case, the controller design will typically include several parameters whose values collectively specify a particular member of a general family of controllers. These parameters define a mathematical space that must be searched to find a satisfactory or optimal controller for the system in question.

The "metacontroller" in the early stages of implementation is generally the members of the design team themselves, aided by general purpose hardware and software. The design team specifies various prototype versions of the controller, each of which defines a point in parameter space. They then test each prototype against a real or simulated plant. The evaluation criteria for judging prototype controllers can be conceptualized in terms of two broad categories. The first category, effectiveness, concerns how well the controlled system approximated the control objective.

0-7803-0614-7/93$03.00 ©1993IEEE

Figure 2: MetaControl System

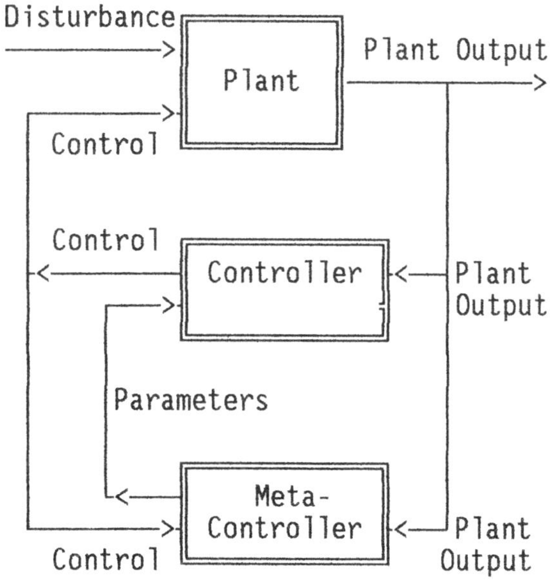


The second category, efficiency, concerns how well the controller itself performed in terms of resources consumed and undesirable side effects produced. The design team uses information about the effectiveness and efficiency of the controller prototypes investigated so far to pick new points in parameter space to be investigated. The process continues iteratively until the system is good enough to be released.

The process described above relies very heavily on human judgment and expertise; we may call it the "manual metacontrol" paradigm. There are several good reasons to try to automate parts of the metacontrol process. One strong impetus comes from the area of adaptive control, in which the metacontroller is integrated with the controller. An adaptive system continually evaluates the effectiveness and efficiency of the control process and continually or intermittently modifies the values of the control parameters to improve them. It has become customary to refer to this updating of control parameters as "learning" by analogy with the way a human operator improves his or her control of a system with increasing experience.

Another reason to automate the metacontrol process is to be able to deliver more control systems to the market in less time. Advances in hardware and software have opened up many opportunities to market "smart" devices of all sorts, as soon as control systems exist to implement them. In this context, automated metacontrol systems become an important component of computer aided design, greatly increasing the productivity of control engineers. A final reason for automated metacontrol is documentation. A controller that arises from a standardized search algorithm with a well defined stopping criterion may be easier to recognize as "good enough" than one which is simply the best one seen so far in an undocumented process of trial and error search.

## 2. *Fuzzy Logic Control*

Fuzzy controllers have received considerable attention, both practical and theoretical, because domain experts without special training in control engineering can qualitatively frame fuzzy rules for narrowly defined systems. [Sugeno and Yasukawa, 1991] Although the general structure of such rules can be acquired rather directly because of their linguistic flavor, tuning or calibration of the fuzzy variables can nevertheless be very challenging.

We consider standard fuzzy controllers which encode their knowledge as rules comprised of combinations of subrules. A typical subrule i has the form "If the value of $x_i$ is $\mathcal{X}_i$ and the value of $y_i$ is $\mathcal{Y}_i$, then the value of $z_i$ should be $\mathcal{Z}_i$." Lower case letters x and y signify the names of antecedent variables such as position and velocity while $\mathcal{X}_i$ and $\mathcal{Y}_i$ are fuzzy linguistic values describing these variables. Similarly, z and $\mathcal{Z}_i$ are a consequent variable and its fuzzy value. The rule contains subrules $i = 1,\ldots,\boldsymbol{i}$ which are fused into the overall rule by the fuzzy operator minimum or maximum, depending on the multivalued logic employed in the system. The term set for the fuzzy values $\mathcal{X}$, $\mathcal{Y}$, and $\mathcal{Z}$ commonly includes: Large negative; Negative; Small negative; Zero; Small positive; Positive; and Large positive. A typical subrule is "If the error angle is Small negative and the angular velocity is Small negative, then the force of the push should be Small positive."

In operation, the fuzzy controller observes the actual data values for the antecedent variables x and y, denoted $\mathbf{x}$ and $\mathbf{y}$. In a practical fuzzy control system, the actual values are observed in the form of crisp numeric singletons. Also the operational controller defuzzifies the rule's detached consequent value $\mathbf{z}$ into a crisp numeric singleton that is the actual control signal to the plant. The current study uses a system that contains one rule with eleven subrules.

Common performance variables for mobile systems are safety, fuel economy, smoothness of ride, and speed of recovery. Performance factors of the controller itself include speed, robustness, memory needs, physical dimensions, and cost. We are concerned in this study with performance factors that result from tuning decisions. We attempt to optimize system performance in relation to these criteria, or at least to satisfy the more important ones. The methodology employed does not assume that the controllable factors and the performance variable have continuous numeric values.

## 3. *Metacontrol for Fuzzy Logic Controllers*

Researchers have proposed and tried many approaches to searching parameter spaces of controllers in general and fuzzy logic controllers in particular. Nearly all the automated approaches involve defining a single numeric objective functional[1] or "figure of merit" to express both the effectiveness and the efficiency of the control process. Given such an objective functional, various workers have optimized it analytically [Kirk, 1970], using Response Surface Methodology [Schott & Whalen, 1992], using neural networks [Hayashi et al, 1992; Kosko, 1992; Berenji, 1992; Keller & Tahani, 1992], as well as other approaches. However, the use of a single numeric objective functional seems to work directly contrary to the principal advantages of fuzzy control systems over their non-fuzzy counterparts.

In a fuzzy control system, the control objectives themselves may be ordinal; they need not be restricted to an interval or ratio scale. Fuzzy control objectives can reflect and exploit the fact that many real situations are more or less tolerant of imprecision. For example, a fuzzy controller may strive to maximize a global assessment of "comfort" in a transportation system while a nonfuzzy system can only optimize some mathematical function combining acceleration, vibration, and noise. As a result, it seems questionable to tune a fuzzy controller using optimization procedures that attempt to estimate first and second derivatives of the degree to which the system meets its fuzzy control objectives.

Ordinal scales also facilitate lexicographic and other nonlinear approaches for dealing with multiple objectives. The use of a single numeric objective functional to capture all aspects of effectiveness and efficiency trade-offs can be problematic even in nonfuzzy control environments. And it is doubly questionable to represent the trade-offs between fuzzy effectiveness and fuzzy efficiency with a single crisp functional.

A classic algorithm, coincidently published in the same year as Zadeh's original article on fuzzy sets, provides a solution to both these problems. The algorithm is due to M.J. Box [1965; Himmelblau 1972 p. 177-178]; it is called the "complex" algorithm not because it is especially intricate but because it involves a set of points in parameter space consisting of more than the minimum number of points necessary to span the space. Box called such a set of points a *complex* to distinguish it from a simplex which contains only the minimum number of points, which is one more than the number of dimensions.

The great advantage of Box's complex algorithm for tuning a fuzzy controller is that it uses only ordinal evaluations of the quality of points in parameter space. In fact, the complex algorithm can seek an optimum even when the quality of the points is incompletely ordered. (The algorithm also has some faults; when the objective is single and differentiable, derivative based approaches require fewer evaluations of the objective function [Himmelblau, 1972]. Also, the algorithm can fail when the minimum lies along a sharply curving valley.)

## 4. *Tuning a Fuzzy Controller By Box's Complex Algorithm*

The following discussion presents a generalization of Box's original algorithm as applied to tuning a fuzzy control system with some free parameters. The approach generalizes that of Box by explicitly considering the possibility that the quality of points might not be completely ordered. It is possible that two control systems may perform equally well within the limits of our ability to judge them. It is also possible that two control systems may be clearly different in their performance, but we are still unwilling or unable to say one is better and the other worse. This can happen when one is clearly more effective, but the other satisfies minimal effectiveness requirements and is much more efficient.

***Step 1:*** To begin tuning a fuzzy controller using the complex algorithm, select an initial complex of points in the mathematical space defined by the parameters of the control system to be implemented. (Box suggests that the number of points be three times the number of parameters.)

------------------------------

*1. The term "functional" is used rather than "function" to emphasize the fact that the argument of the objective functional is itself a function of time, which tracks system performance throughout the test period.*

Each point defines a control system; run each control system with a real or simulated plant for a standard trial period. It is important that the points span the space of parameters; this can be accomplished either by using a design matrix as in [Schott & Whalen, 1992] or by random perturbation from an initial value as suggested by Box.

***Step 2:*** Rank the points from best to worst with respect to the performance of the corresponding controllers. (Ties and incomparables are allowed.) For example, an engineer might rank the performance of control systems for a vehicle in terms of safety while a human factors expert ranked them in terms of comfort and an accountant ranked them in terms of cost. The final ranking might depend on safety, with ties on safety broken by trading off comfort and cost.

***Step 3:*** Select the worst point in the current set. If there is no unique worst point, randomly select a point from among those that are not ranked better than any other point. Construct a line in parameter space from the worst point to the centroid of the other points. Multiply the distance from the worst point to the centroid by an "overexpansion factor" (Box suggests 1.3), and extend the line beyond the centroid by a distance equal to the result. This defines the new candidate point in parameter space.

***Step 4:*** Run the corresponding control system with the real or simulated plant for a standard trial period, and compare its performance with the worst point in the complex. If the new point is better than the old worst point, replace the latter with the new point and return to step 2. If the new point is worse than the old worst point, if the two points are tied, or if the two points are not comparable, then create a new candidate point half way between the old candidate point and the centroid. Make this the new candidate point and repeat Step 4.

Continue these steps until the points in the complex are all within a predetermined radius of one another or until some other criterion is met. If the algorithm seems to be stuck in Step 4, check the performance of the control system defined by the centroid of the complex. If the centroid control system performans worse than any of the points in the complex, Box's algorithm cannot proceed. To get further improvement in this case, re-start the algorithm with a new complex in the vicinity of the best points seen so far.

## 5. Example: Tuning an Inverted Pendulum Controller

Control of an inverted pendulum has become a common benchmark problem among fuzzy researchers. A cart on a straight track is pushed with varying degrees of force according to the controller's instructions. A sensor detects the angle $\theta$ in radians that the pole makes with the vertical. The angular velocity of the pole angle, $\theta'$, is computed approximately based on the change in $\theta$. Another sensor measures the cart's position $\phi$ relative to its starting position. A pushing force $\Gamma$ is applied to the cart. $\theta$, $\theta'$, $\phi$, and $\Gamma$ can take on positive and negative values depending on leftward or rightward orientation.

The fuzzy controller uses eleven sub-rules containing $\theta$ and $\theta'$ as antecedent variables, and with $\Gamma$ as the consequent variable. Five terms were defined for each variable: Negative; Small negative; Zero; Small positive; and Positive. All fuzzy (linguistic) variables were represented as symmetrical trapezoids. The scales of all the trapezoids on each universe of discourse were uniform relative to one another, but the scales on different universes were independent.

The controller was tuned (metacontrolled) by calibrating the scale[2] of the axes of the three universes: $\Theta$, $\Theta'$, and $\Gamma$. Two criteria were used for optimization. The most important goal of the system, its "effectiveness," was simply to balance the pole. The system was run for a simulated period of 5 seconds, divided into 250 "ticks" of the simulation clock. If the pole angle $\theta$ passed out of the controllable range during this time, the number of ticks remaining in the test period was reported. If the pole was still standing at the end of the simulation, this figure was equal to zero. Effectiveness scores are presented in Figures 1 and 2 in the column headed "time left;" the smaller this number, the more effective the control.

A second goal, "efficiency," was to achieve a smooth, steady balancing of the pole rather than a jittery or runaway one. (Cart-pole systems are subject to a "runaway" condition, in which the pole remains balanced at an angle while the cart accelerates continuously until it runs into the end of the track. The second goal was represented by the product of two quantities: the integral of the absolute angle $|\theta|$ and the integral of the absolute cart position $|\phi|$. This number is

-------------------------------

*2. Each continuous variable's axis was discretized at 17 equidistant values. The "scale" value is the distance between adjacent points, with the median (eighth) point always anchored at zero.*

very large under runaway conditions and moderately large for an inefficient, jittery control system. Effectiveness scores are presented in Figures 1 and 2 in the column headed "instability;" the smaller this number, the more efficient the control.

The two goals were combined lexicographically. In other words, any control system that balanced the pole for a longer period ranked better than any control system that balanced the pole for a shorter period, regardless of their relative efficiency ratings. If the control systems both balanced the pole for the entire experimental period, or if they both balanced the pole for equal periods of time before losing control, then the more efficient control system ranked higher. (The actual scores used were time remaining when control was lost and a measure of inefficiency, so optimization was by minimization.)

Following Box's suggestion, we used nine original sample points to tune the three parameters of the system. The triads ($\theta$, $\theta'$, and $\Gamma$ scales) for each of the 9 original sample points in Table 1 were set judgmentally to include a broad range of reasonable designs. Each point specifies the scale values of the 3 variables: pole angle scale in radians, pole angular velocity scale in radians per second, and pushing force scale in newtons. The initial nine points consisted of the controller for which all three scales equalled 2.0 plus the eight controllers formed by all combinations of angular ($\theta$) scale = .03 or 1.0 radians; angular velocity ($\theta'$) scale = .02 or 2.0 radians per second; and push ($\Gamma$) scale = 1 or 10 newtons. The smaller the scale for $\theta$ and $\theta'$, the more sensitive the controller is with respect to changes in angular position and velocity. The larger the scale for $\Gamma$, the stronger the output of the controller. At each stage of tuning, the nine points are presented in sorted order, so the ninth row is always the worst of the nine points currently under consideration.

Every experiment was run with a starting angle $\theta$ = 0.05, and all other transient variables set to 0. Time was incremented every 0.02 seconds, cart mass was 1.0 Kg, pole mass was 0.1 Kg, pole length was 0.5 m, and acceleration due to gravity was 9.8 $m/s^2$. The simulation was based on differential equations provided by Hamid Berenji [Berenji, 1992]. The simulation assumed a frictionless plant and certain other simplifications, and is not intended to precisely represent a real inverted pendulum. Box's complex algorithm was implemented as a Lotus 123 spreadsheet.

Table 1 shows the original set of nine points in parameter space along with the two criterion variables for each one. At the bottom of the table the coordinates of the centroid point and the vector from the worst point to the centroid appear. In the box at the top of the table appear the coordinates of the next candidate point. The box also contains the locations where the user will enter the values of the criterion variables from the simulation using the controller defined by the candidate point.

Table 2 shows the spreadsheet after approximately 65 iterations. Note that all nine points are much closer together than in the original spreadsheet. The first of the nine points would define the system to be implemented and marketed if this were an actual design project.

## References

Berenji, H. R. "A reinforcement learning-based architecture for fuzzy logic control," **Int. J Approx Reasoning 6**:2, pp. 267-292, 1992.

Box, M.J. "A New Method of Constrained Optimization and a Comparison with Other Methods," **Computer J. 8**, p.42-52,1965.

Hayashi, I.; Nomura, H.; Yamasaki, H. & Wakami, N. "Construction of fuzzy inferences rules by neural network driven fuzzy reasoning and neural network driven fuzzy reasoning with learning functions," **Int. J Approx Reasoning, 6**:2, pp. 241-266, 1992.

Himmelblau, D. **APPLIED NONLINEAR PROGRAMMING**, McGraw-Hill, New York, 1972.

Keller, J. M. & Tahani, H. "Implementation of conjunctive and disjunctive fuzzy logic rules with neural networks," **Int J Approx Reasoning 6**:2, pp. 221-240, 1992.

Kirk, D. **OPTIMAL CONTROL THEORY: AN INTRODUCTION**, Prentice-Hall, Englewood Cliffs, 1970.

Kosko, B. **Neural Networks and Fuzzy Systems**, Prentice-Hall, Englewood Cliffs, 1992.

Schott, B. & Whalen, T. "Tuning a Fuzzy Controller Using Quadratic Response Surfaces," **Proc. N Amer. Fuzzy Information Processing Society Conf**, 1992.

Sugeno, M. & Yasukawa, T. "Linguistic modeling based on numerical data," **Proceedings of the 4th International Fuzzy Systems Association Congress**, 1991, pp. 264-267.

### Table 1: Initial Spreadsheet Tableau

| | INPUT TO SIMULATOR | | | OUTPUT FROM SIMULATOR | | |
|---|---|---|---|---|---|---|
| | 0.582 | 0.954 | 3.606 | | | Better |
| | newΘ | newΘ′ | newΓ | newT | newI | |
| | Θ | Θ′ | Γ | time left | insta-bility | |
| 1 | 0.03 | 0.2 | 1 | 0 | 83.2758 | |
| 2 | 0.03 | 0.2 | 10 | 0 | 100.012 | |
| 3 | 0.03 | 2 | 1 | 30 | 569 | |
| 4 | 1 | 0.2 | 10 | 154 | 123000 | |
| 5 | 1 | 0.2 | 1 | 213 | 2570000 | |
| 6 | 0.03 | 2 | 10 | 219 | 723000 | |
| 7 | 1 | 2 | 1 | 223 | 792000 | |
| 8 | 2 | 2 | 2 | 223 | 4600000 | |
| 9 | 1 | 2 | 10 | 223 | 5090000 | |
| | badΘ | badΘ′ | badΓ | badT | badI | |
| | 0.64 | 1.1 | 4.5 | centroid | | |
| | -0.36 | -0.9 | -5.5 | vector | | 0.1625 coefficient |

### Table 2: Final Spreadsheet Tableau

| | INPUT TO SIMULATOR | | | OUTPUT FROM SIMULATOR | | |
|---|---|---|---|---|---|---|
| | 0.028 | 0.485 | 6.601 | | | Better |
| | newΘ | newΘ′ | newΓ | newT | newI | |
| | Θ | Θ′ | Γ | time left | insta-bility | |
| 1 | 0.023 | 0.347 | 7.243 | 0 | 12.2303 | |
| 2 | 0.027 | 0.435 | 6.034 | 0 | 22.6592 | |
| 3 | 0.031 | 0.62 | 6.522 | 0 | 31.7899 | |
| 4 | 0.031 | 0.515 | 4.781 | 0 | 35.0634 | |
| 5 | 0.027 | 0.444 | 6.598 | 0 | 39.342 | |
| 6 | 0.032 | 0.59 | 7.159 | 0 | 39.9405 | |
| 7 | 0.028 | 0.578 | 9.925 | 0 | 46.3209 | |
| 8 | 0.022 | 0.333 | 4.546 | 0 | 46.7025 | |
| 9 | 0.021 | 0.116 | 6.561 | 0 | 49.9 | |
| | badΘ | badΘ′ | badΓ | badT | badI | |
| | 0.027 | 0.485 | 6.601 | centroid | | |
| | 0.007 | 0.369 | 0.04 | vector | | 0.0013 coefficient |

# COMPUTER AIDED TUNING TOOL FOR FUZZY CONTROLLERS

**A . Boscolo**
ZELTRON S.p.A. - Via Principe di Udine 114, 33030 Campoformido Italy
DEEI Department of Electrical , Electronic and Computer Science Trieste University

**F. Drius**
ZELTRON S.p.A. - Via Principe di Udine 114, 33030 Campoformido Italy
LASA ZELTRON-DEEI Trieste University Via Valerio 10 34100 Trieste

*Abstract* - **The paper describes a procedure concerning the fuzzy controller tuning. The target is reached through an interactive procedure based on largely confirmed linearization and optimization methods, aimed to maintain the process physical image.**
**A real application example is given regarding a non well defined non-linear mechanical process.**

## I. INTRODUCTION

The synthesis of a control system generally requires a deep knowledge about the process that has to be controlled. When the process is too complex to be given a good physical description, the controller synthesis must be based only on intuitions and on heuristic knowledge. In this context the fuzzy approach gives a good support in translating into numerical algorithms the heuristic rules given, in linguistic form, by a skilled process operator able to manage the process manually. The fuzzy techniques are successful in a wide class of applications that overcome the pure commercial and fashion aspects, as the fuzzy techniques are able to transfer the problem of the process knowledge collection into the emulation of a skilled operator able to manage it.
In spite of the previous large capabilities, there are some troubles in identifying the fuzzy system able to reproduce a given behaviour with an adequate global accuracy.
From on other point of view, the tuning of fuzzy systems in order to fit the expert behaviour is the more complex step in the system design procedure.
About this specific item, lots of different approaches exist. Among these methods, the trivial "trial and error" is a very popular one: it is very deeply based on the intuition and deduction capability of the system designer.
Some researchers give an answer to the tuning problems through the neural networks learning. Other reasearchers have proposed an assisted procedure in order to tune and validate fuzzy systems based on the global errors minimization and on the entropy evaluation.

## II. SCENARIO AND ACTIVITY MOTIVATION

The previous methods give the best results when the system specifications cover, with adequate accuracy, all the system state space and when the fuzzy system is asked to fit only some given expert behaviour examples.
When the fuzzy system is designed to work in closed loop with a not well defined process the previous methods show some performance degradation so that more attention is required.
Unfortunately, when we speak about controllers, very often the expert is not able to give significant examples that allow to perform the tuning of a rough controller.
In a closed loop control the operator has an unconscious image of the process state: he is not actually looking for the reason of his intervention: he is rather interested in the process state evolution. In other words, the manual control action is performed by the operator who is rather engaged in sensing the process behaviour instead of thinking about what his hands are doing.
In this scenario, where the expert is not able to describe his control procedure and to give significant control action examples, it is possible to design only "rough control rules" that still need a tuning procedure.
In many real processes the controller on-line tuning is very expensive both from the economics and time consumption point of view, or physically impossible.
In these situations some researchers have proposed a controller design procedure based on a process mapping and on a rule searching through the learning of a devoted neural network .
During this activity very often the poor physical knowledge available about the process is masked in the searching procedure. In this way any support coming from the intuition of the designer is lost.

0-7803-0614-7/93$03.00 ©1993IEEE

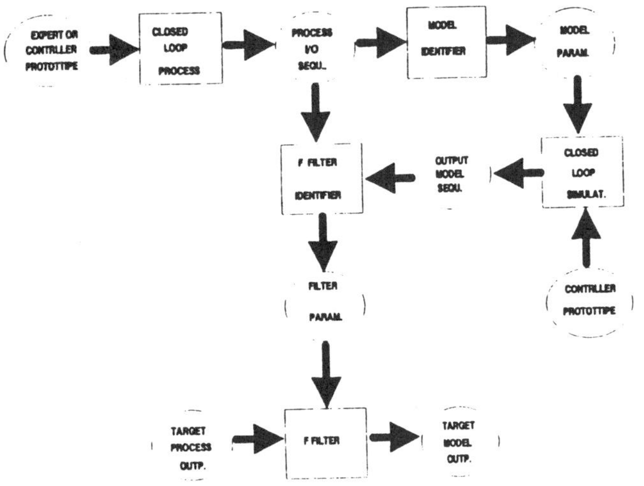


Fig.1) Proposed procedure

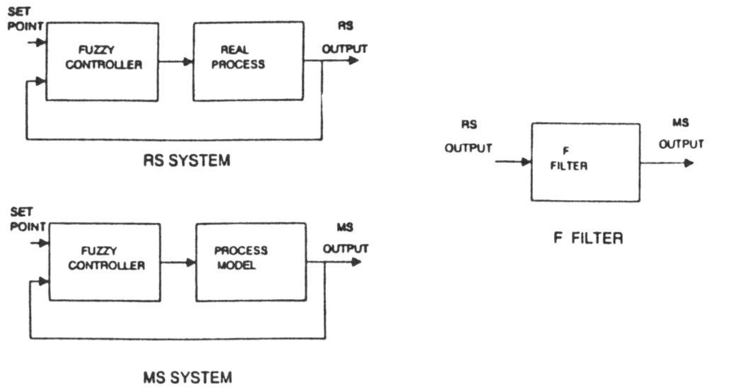


Fig.3) RS, MS structure, F filter

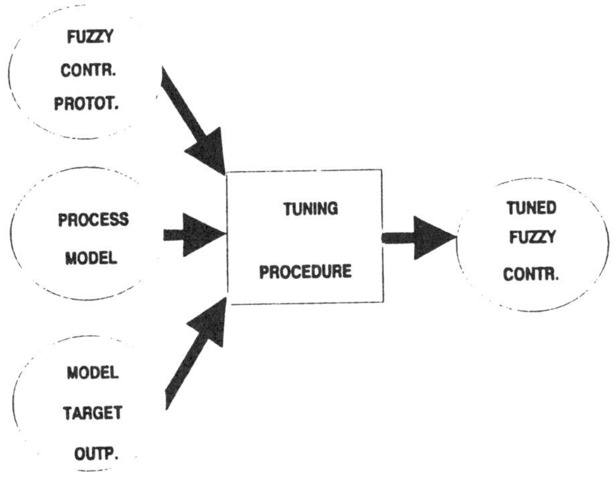


Fig.2) Tuning procedure

In the wide spectrum of applications an emerging set exists which can be identified with the embedded controllers. They are engaged in social relevant industrial activity like automotive and consumer appliance industries.
This kind of system is typically no more complex than a high performances speed controller for electric motors, and they are conceptually and physically "merged" with the process and the machine which it belongs to Moreover the embedded controller must be robust, full reliable, simple and able to adapt its behaviour to the large spectrum of operative situations. In order to achieve these performances the controller design phase requires a very deep optimization and the fuzzy approach becomes mandatory.
From the industrial point of view, a very important aspect related with the embedded controllers design procedure is the design tools performances. If do not exist design tools which are able to give a real support to the designer in evaluating the quality and the distance of the controller prototype performances from the optimal one, a large amount of investments may be lost because of a solution proposal more closed to the optimal one can be done.

## III. PROPOSED TOOL

In a previous work the authors have presented a tool able to perform the optimal tuning of any fuzzy system starting from a set of rules and a set of examples to emulate (B1).
Following the style, proposed in the previous paper, and based on the synthesis of computer aided tool on well known and validated procedures this paper suggests a tool evolution. This evolution makes the previous tool able to support the tuning of controllers and, expecially, of embedded controllers, devoted to not well defined processes closed loop control. The improvement is based on the substitution of the real process with a mathematical model able to fit, even crudely, the process behaviour in a narrow working field. This model allows the tuning of the controller out of loop by an optimization procedure that reduces the behaviour errors of the process output. The innovative contribute of the proposal is obviously more focalized on the tuning methodology then on the involved tools.

The synthesis of a fuzzy system proceeds typically through three phases. In the first phase the knowledge base is builded up through a very deeply interview with the expert and through the recording of his control actions. In the second phase a searching for possible and exhaustive rules is performed. The third one is the tuning of the prototype in order to fit the collected set of examples.
The resulting controller is not yet able to fit optimally the closed loop target but it will guide the process on a trajectory roughly closed to the closed loop specifications.
After the rough controller synthesis, it is possible to choose two distinct but not mutually exclusive ways.
In the first we may put in the real control loop the controller prototype and adjust it.
In the second we may adjust the controller as best as possible before putting it in the control loop.
In order to follow the second way the procedure that is illustrated in the figures 1 and 2 is proposed.
In order to improve, with a numerical algorithm, the quality of the closed loop system it is necessary to work at least with a model of the physical process. After the previous considerations on the distance between the obtained and the asked state trajectory, the adoption of a linear model of the unknown physical process makes sense. The order of the selected model influences strongly the capability to emulate the system behaviour in different working conditions. Lots of model selection criteria are described in literature and the best choice is related to the particular process to be modelled [B7], [B8].
The model identification procedure is the most important step in achieving good results. In our experience the Generalized Recursive Least Squares (GRLS) algorithm is a good choice. This algorithm works well even with the correlated noise that typically pollutes the input-output recorded data sequence of real complex processes [B6].
The behaviour of the closed loop system formed by fuzzy controller prototype and process model may be distant both from the specifications and the real system behaviour.
In order to reduce this distance, the proposed method is the following. With the same GRLS algorithms, used in the system model generation, it is possible to identify a linear filter (F) so that when its input is the output sequence of the real closed loop system (RS), its output fits the output of the modelled closed loop system (MS) (fig. 3) .
If the process model fits well the real process behaviour even in a limited space around the given trajectory, the tuning of the controller makes sense in order to fit the system target after filtering (F), instead of the system target itself. If this kind of tuning is performed, the controller is able to drive the real process giving an output more closed to the target then the previous controller realization, since the tuning procedure forces the closed loop simulated output of MS to fit the (F) filtered goal trajectory.
In this way, if the behaviour of the initial RS is near to the target trajectory, the tuned fuzzy controller with its new membership functions is able to drive the process actually a trajectory more closed to the target.
From an other point of view the suggested method performs an "implicit tuning like" procedure.
Namely:

A(z) : transfer function of the linearized RS
B(z) : transfer function of the linearized MS
Ra(z) : target transfer function for A(z)
Rb(z) : target transfer function for B(z)
F(z) : transfer function of the F filter
Bopt(z) : transfer function of the linearized tuned MS
Aopt(z) : transfer function of the linearized tuned RS.

We can write:

$$B(z) = A(z) * F(z)$$

After the tuning:

$$Bopt(z) = Rb(z)$$
$$Rb(z) = Ra(z) * F(z)$$
$$Bopt(z) = Aopt(z) * F(z).$$

Then:

$$Aopt(z) = Ra(z).$$

The previous considerations are correct only when all the systems are linear: nevertheless, also in non-linear systems it is possible to achieve good results when the MS output trajectory is near enough to the RS.
In order to achieve the optimal controller realization, the proposed tuning method is suitable to a recursive approach. The new iteration selection criteria are obviously based on global errors or on the evaluation of the distance between the RS and MS behaviour and the target.
In some real applications, the first iteration require more attention because the distance between MS and RS behaviour may be too wide.
Typically, the behaviour of the process controlled by the human operator is often closed to the behaviour of the process controlled by the rough controller. This assumption is justified because the rough controller is already optimized to emulate the operator actions. In this case it is possible to identify the first process model directly on the input-output data collected during the human control operation.
If the behaviour of the MS should not be near enough to the real system one, it would be necessary to connect the rough controller to the real process in order to collect more realistic input-output data and to achieve a more realistic starting model. This short time loop connection is, in any case, less traumatic then any complete on-line tuning procedure.

After that it is possible to achieve the optimal controller realization through a small number of iterations where the model and the filter are updated by putting the current tuned controller in the loop.

## IV. OBTAINED RESULTS

In order to give a feeling about the proposed tool capability the following example shows the tuning of an adaptive PI fuzzy controller applied to an AC universal motor working in series connection. The load of the motor contains non-linear elastic transmission elements, non-linear variable geometry rotating masses and non-linear friction. The motor is driven by a TRIAC phase partialization system.

The controller is a fuzzy adaptive gain scheduling like PI. Its set of rules reflects a popular method of improving the PI performances.

The proportional and integral gains of the PI controller are initially fixed through a standard approximated method (B2).

The preliminary fuzzy controller is tuned in order to fit a required behaviour of its proportional and integral gains. This rough controller is already able to pilot the process but the motor speed trajectory (fig. 4) is far from the system specification (fig. 5). The laboratory conditions allow to connect the rough controller to the process: then the input-output behaviour of the process is recorded during the rough fuzzy controller operation (fig.6). Thanks to this input-output sequence recorded in the real working conditions, the

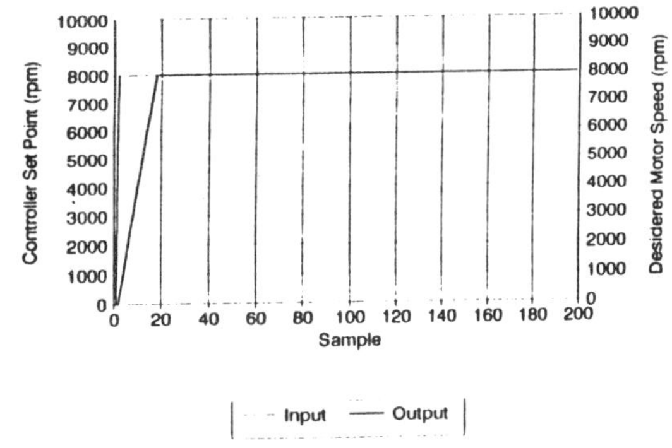

Fig. 5) Process specification

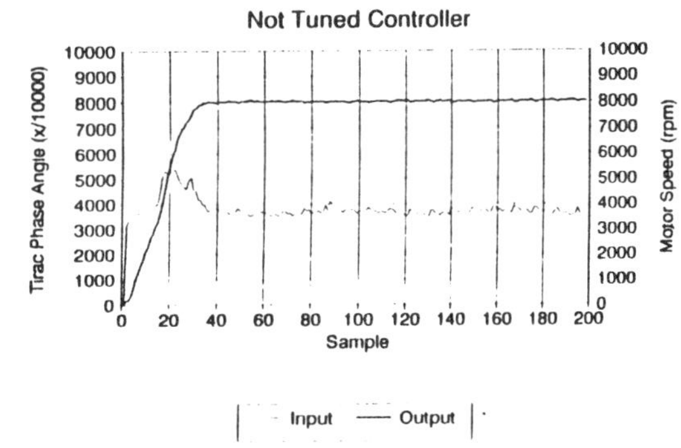

Fig. 6) Input-output process response

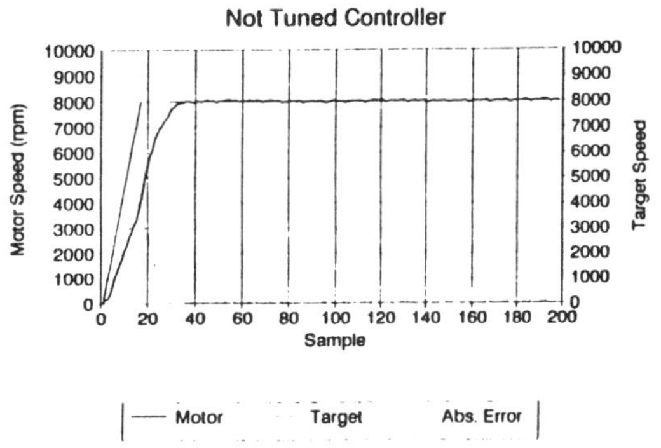

Fig. 4) Process response

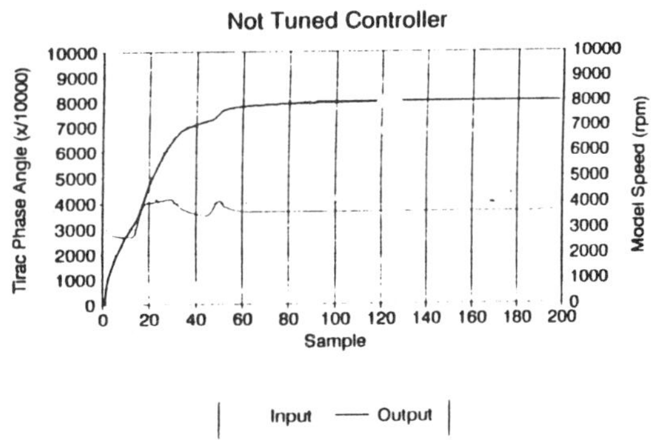

Fig 7) Input-output model response

identification of a more approximated linear model of the process is possible.

The identified process model is a traditional ARMAX model and its structure, without the dynamic noise generators is

$$Y(k+1) = a_1*Y(k-1) \ldots a_n*Y(k-n)+b_1*U(k-1)+\ldots+B_r*U(k-r)$$

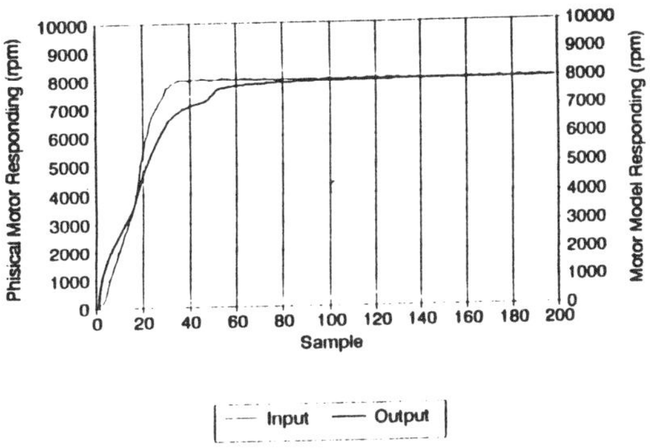


Fig. 8) Input-output sequence of the F filter

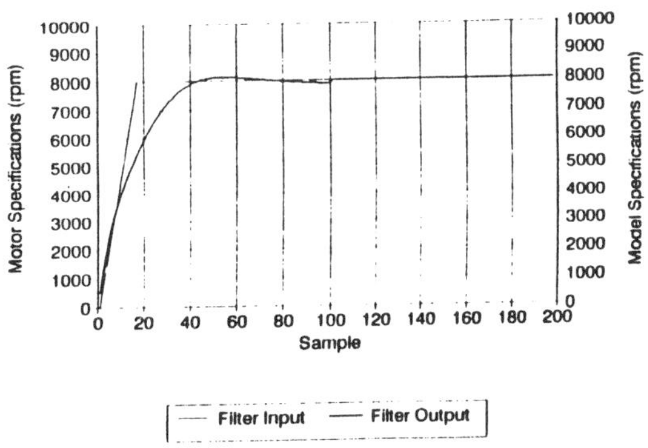


Fig. 9) F filtered specification

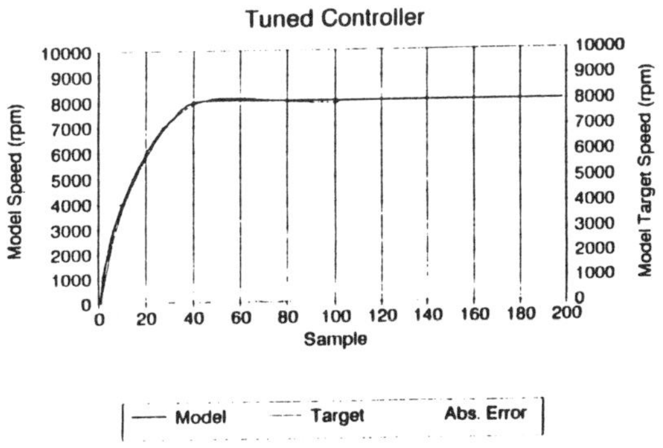


Fig. 10) Tuned MS response

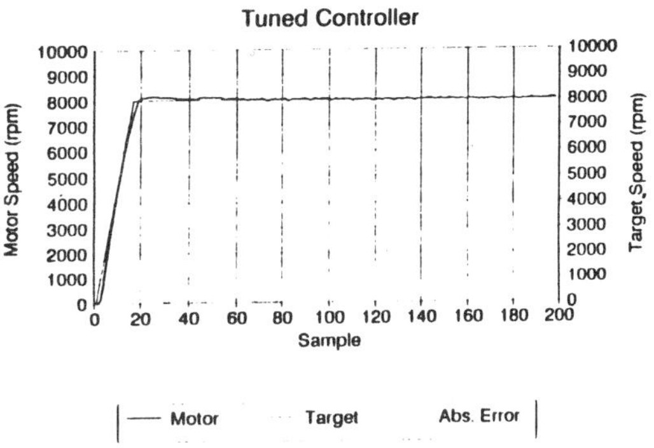


Fig.11) Tuned real system reponse

In this example the optimal model has n=3 and r=2.
The process model, connected in closed loop with the rough fuzzy controller, responds in similar qualitative manner to the real system (fig. 7). Nevertheless the time constants of the MS are larger than the corresponding time constants in the real system.
The 'F' filter is identified on its input-output target sequence, where its input is the real process response and its output is the MS response (fig. 8).
The filtered specification to be applied to the process simulation is shown in (fig. 9).
Now it is possible to tune the parameters of the membership function of the fuzzy rules in order to force the MS output to fit the filtered specification (fig. 10).
The RS with the tuned controller applied to the process has an output very closed to the required specification (fig. 11).

## V. CONCLUSIONS

This paper shows how by linearization and optimization techniques it is possible to design and tune fuzzy controllers that are devoted to closed loop control of not well defined non-linear processes. We have shown also how it is possible to avoid trial and error procedures that are not able to give a good validation of the result and other techniques that do not allow the process designer to get in touch with the physical feeling.
The proposed method is suitable for all the applications that need a development time compression without degradation in terms of quality.

## ACKNOWLEDGEMENTS

**The research activities have been supported by ZELTRON S.p.A. and by the Italian Ministry for the University and the Scientific and Technological Research (MURST).**

## BIBLIOGRAPHY

(B1) A. Boscolo, F. Drius, "Computer aided tuning and validation of fuzzy systems", IEEE International Conference on Fuzzy Systems 1992.

(B2) A. Boscolo, C. Mangiavacchi, F. Drius, M. Goiak "Fuzzy controller for generally loaded dc electric motor", IFAC Symposium on Intelligent Components and Instruments for Control Applications SICICA 92.

(B3) D.G. Burkhardt, P.P. Bonissone, "Automatic Fuzzy Knowledge Base Generation and Tuning", IEEE Int. Conf. on Fuzzy Systems 1992.

(B4) T. Yamakawa, M. Furukawa, "A Design Algorithm of Membership Functions for a Fuzzy Neuron Using Example-Based Learning", IEEE Int. Conf. on Fuzzy Systems 1992.

(B5) S.Z He, S.H. Tan, C.C. Hang, P.Z. Wang, "Design of an On-Line Rules Adaprive Fuzzy Controller System", IEEE Int. Conf. on Fuzzy Systems 1992.

(B6) S.Sagara, Z.J. Yang, K. Wada, "Recursive Identification Algorithms for Continuous Systems in Adaptive Procedure", Int. J. Contr., 1991, vol. 53, N. 2.

(B7) S. Bittanti, "Parametric Identification", ed. CLUP Milano, 1982.

(B8) H. Akaike, "On Model Structure Testing in System Identification", Int. J. Contr. 1978, vol. 27.

# Fuzzification of Control Timing in Driving Control of a Model Car

Katsumi Nishimori, Susumu Hirakawa, Heizo Tokutaka, Satoru Kishida and Naganori Ishihara

Department of Electrical and Electronic Engineering, Tottori University, Koyama, Tottori 680, JAPAN.

Abstract- This paper shows that a new fuzzy control method can be realized by the fuzzification of control timing in the left (or right) turning of a model car. The fuzzy set was introduced to construct the timing of the steering operation for the left turning. The steering angle was determined as a function of the timing. In the simulation, the car trajectory of the fuzzy model curved very smoothly. For any turning angles of the corner, the trajectory was easily improved by changing the shape of the timing membership functions.

## I. INTRODUCTION

Recently, fuzzy reasoning methods have been widely used for the control of systems[1]. Many studies of fuzzy control are associated with their application to the feedback system which are used in the conventional controls such as PID control and optimal control etc..

Sugeno and Nishida [2] reported that the fuzzy reasoning method well worked for the driving control of a model car in the turning at the crank course. However, they used many different membership functions and 20 fuzzy control rules because of the employment of 'If...Then...' production rules. Since the optimization of many membership functions and control rules needs a lot of trials for the driving operation of the car, it takes long time and much cost.

In this report, we propose the new fuzzy reasoning method to solve this problem. First we consider the crisp model in which, when the turning angle at the corner is given, the target trajectory of the car can be geometrically determined from the shape of the corner. Second, the fuzzy theory is used to obtain the timings at the beginning and the end of the steering operation for the turning. This procedure is very close to the human operation of the car at the corner.

## II. CRISP MODEL FOR THE LEFT TURNING OF THE CAR

Let us consider the crisp model to control the left turning of the car for simplicity. The model car runs at the constant speed V[cm/s] on the straight road. When we push the switch button of the controller in front of the corner, the steering angle is set at the constant steering angle $\phi_0$[rad]. That is, the car begins to turn to the left with a constant radius of curvature. When we release the switch button, the car goes straight ahead again.

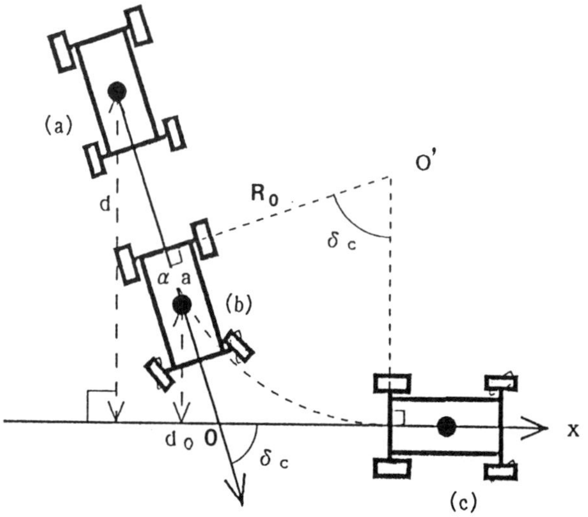


Fig. 1. Crisp model for left turning of a car.

As shown in Fig. 1, the car trajectory is determined by the corner angle $\delta_c$[rad] and the curvature radius $R_0$[cm] at the center of the rear wheels. First, the car goes straight on toward the X axis and the driving direction makes an angle of $-\delta_c$ together with X axis. Next, the car curves and rides on the X axis by the switch operation of the left turning. The wheel base, which is the distance between the front and the rear wheels, is 'a'[cm] and the wheel tread, which is the distance between the left and the right wheels, is 'b'[cm]. The center of gravity (marked by the solid circle) of the car is located at the point on the center line of the car by $\alpha$ a[cm] ($0 \leq \alpha \leq 1$) apart from the center of the rear wheels. The car position and the speed are measured at this point. The control switch is pushed at the very moment when the car just reaches the position of Fig. 1(b). The distance $d_0$ given by the following equation (eq.);

$$d_0 = R_0(1 - \cos\delta_c) - \alpha a \sin\delta_c \quad (1).$$

The switch button is released, when the car just reaches the position (c) of Fig. 1. At this time, the center line of the car is on the X axis, where the angle $\delta$ between the car direction and the X axis equals zero.

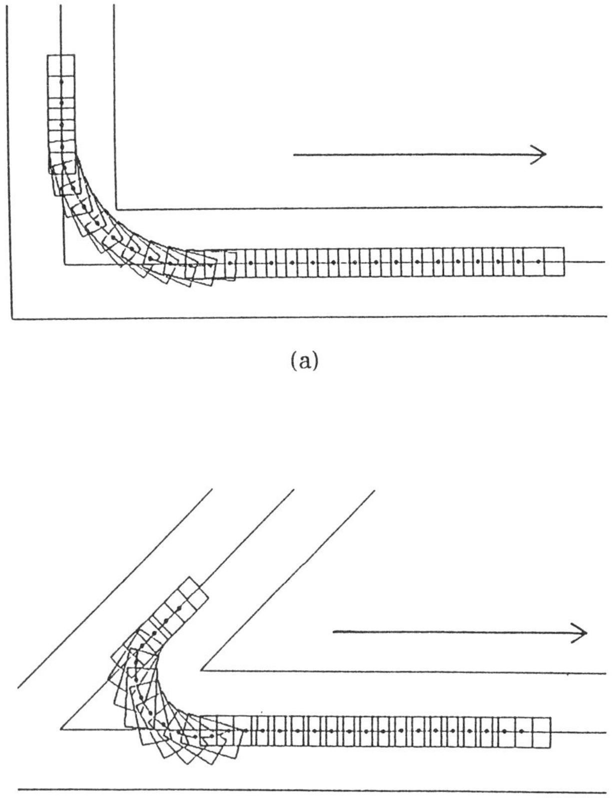

(a)

(b)

a=20[cm], b=10[cm], $\alpha$=0.5, L=50[cm]
(a) $\delta_c = 1/2 \cdot \pi$ [rad], $R_0$=50[cm]
(b) $\delta_c = 3/4 \cdot \pi$ [rad], $R_0$=20[cm]

Fig. 2. Car trajectories taken from crisp model simulation.

The simulation is carried out when the car is controlled by this method with the best timings to push and to release the switch. The results are shown in Fig. 2. The left and right side lines of the road are indicated by L[cm] apart from the center line. The car trajectories are very similar to those drawn with a linear ruler and compasses.

We assume in the above model that the steering controller of the front wheels ideally responds to the output signal of the on/off switching without time delay. However, if we need a constant response time for the model, the timing of the on/off switching should be set early by the response time. In the next section, we consider the case that the steering angle $\phi$ is given by a value between zero and $\phi_0$. For this case we should introduce the fuzzy theory to the model.

## III. FUZZY MODEL FOR THE LEFT TURNING OF THE CAR

Let us apply a fuzzy theory to the above crisp model. We consider the replacement of the switch button in the crisp model with a variable electric resister (dial) which has a continuous resistance degree of zero to 1. The timing of the switching operation from the straight running to the left turning or the reverse are continuously altered with the dial. The steering angle $\phi$ can be adjusted by the dial operation. When the dial is turned clockwise, $\phi$ increases, and when returned counterclockwise, $\phi$ decreases. The relation between the angle $\phi$ and the fuzzy grade for the left turning is defined by the membership function $\mu_{Left}(\phi)$, as shown in Fig. 3. The membership function $\mu_{Left}(\phi)$ is expressed by the following eq.;

$$\mu_{Left}(\phi) = \phi / \phi_0, \quad (2).$$

where $\phi$ is the value of $0 \leq \phi \leq \phi_0$.

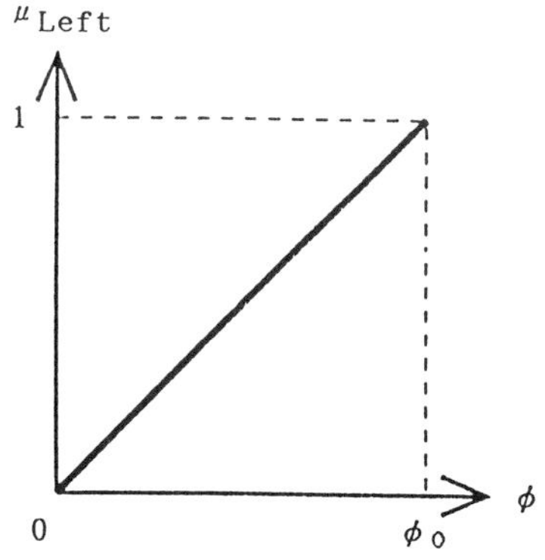


Fig. 3. Membership function for left turning as a function of steering angle $\phi$.

When the control signal $\mu_V$ from the dial turning is equivalent to the membership function $\mu_{Left}$, the steering angle $\phi$ is immediately determined from the relation of Fig. 3 and the steering controller works with no time delay. The timing obtained from the dial turning is given by the fuzzy set of Fig. 4(a) as a function of the distance d. This 'timing fuzzy set' determines the dial fuzzy set of the left turning in Fig. 4(b). The grade of the dial fuzzy set is proportional to $\phi$ as described in eq. (2). In a similar way, the end timing of the left turning is shown in Fig. 5 as a function of the angle $\delta$ defined in Fig. 1.

Let us calculate the dial fuzzy set of left turning from the start and end timing fuzzy sets. The timing sets are expressed by the fuzzy number defined on the support set which consists of the distance d from the car center to X axis. We assume that the dial fuzzy set is defined by the set including the real number less than the above fuzzy number. The start timing set of the left turning is given by the triangular membership function with maximum grade 1.0 at the distance $d_0$ of eq. (1), as shown in Fig. 4(a). Since $d_0$ is the function of $\delta_c$ and $R_0$ as described in eq. (1), the start timing set $\mu_{Ts}(d)$ of left turning also has the fuzzy relation with $\delta_c$ and $R_0$. This fuzzy relation $R_{Ts}$ can be given by using the fuzzy relation determination method[3]. That is,

$$\mu_{R_{Ts}}(d, \delta_c, R_0) = \Lambda_{Ts}\left(\frac{d - d_0(\delta_c, R_0)}{\varepsilon_{Ts}(\delta_c, R_0)}\right) \qquad (3).$$

$\Lambda_{Ts}(u)$ is the $\Lambda$-shaped function defined on the u value region of [0,1]. The start timing fuzzy set is expressed by the following eq.

$$\Lambda_{Ts}(u) = \begin{cases} 1-|u| & (|u| < 1) \\ 0 & (|u| \geqq 1) \end{cases} \qquad (4).$$

$\varepsilon_{Ts}(\delta_c, R_0)$ is the parameter relating to the width of the membership function $\Lambda_{Ts}$ and here assumed to be 20 % of $d_0$;

$$\varepsilon_{Ts}(\delta_c, R_0) = d_0(\delta_c, R_0) \times \frac{20}{100} \qquad (5).$$

The membership function $\mu_{Vs}(d)$ of the dial fuzzy set for the start of the left turning is calculated from the start timing membership function $\mu_{Ts}(d)$;

$$\mu_{Vs}(d) = Z_{Ts}\left(\frac{d - d_0(\delta_c, R_0)}{\varepsilon_{Ts}(\delta_c, R_0)}\right) \qquad (6).$$

$Z_{Ts}$ is the Z-shaped function as shown in Fig. 4(b), and expresses the fuzzy set with the fuzzy number less than that of $\Lambda_{Ts}$ function. $Z_{Ts}$ is defined by the following eq.;

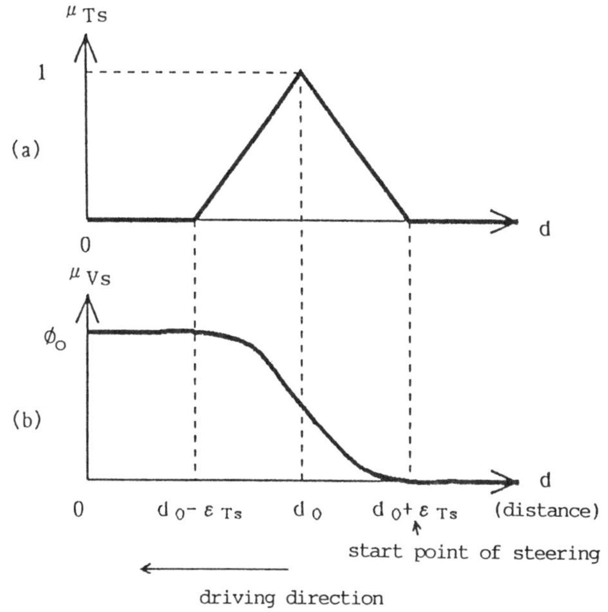


Fig. 4. Relation between (a) start timing fuzzy set and (b) dial fuzzy set at the beginning of left turning.

$$Z_{Ts}(u) = \int_u^{\infty} \Lambda_{Ts}(\upsilon)\, d\upsilon \qquad (7).$$

Next, as shown in Fig. 5, the end timing fuzzy set of the left turning is described by the membership function which is a function of the angle $\delta$ between the car direction and the X

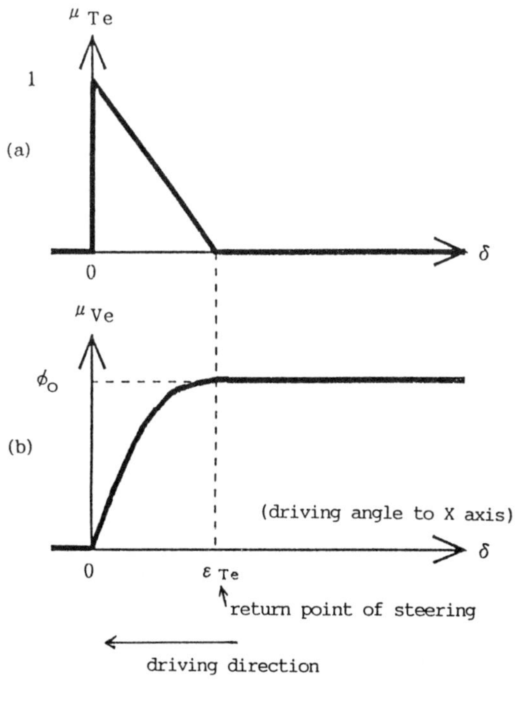


Fig. 5. Relation between (a) end timing fuzzy set and (b) dial fuzzy set at the end of left turning.

axis. The end timing membership function $\mu_{Te}(\delta)$ is expressed by the following eq.;

$$\mu_{Te}(\delta) = \Lambda_{Te}\left(\frac{\delta}{\varepsilon_{Te}(R_O)}\right) \tag{8}$$

Here, $\varepsilon_{Te}(R_0)$ is;

$$\varepsilon_{Te}(R_O) = \phi_0(R_O)\times\frac{20}{100} \tag{9}$$

and $\phi_0$ is;

$$\phi_0(R_O) = \tan^{-1}\left(\frac{a}{R_O}\right) \tag{10}$$

$\Lambda_{Te}(u)$ is defined as follows;

$$\Lambda_{Te}(u) = \begin{cases} 1-u & (0 \leqq u < 1) \\ 0 & (u < 0 \text{ or } 1 \leqq u) \end{cases} \tag{11}$$

Using these eqs. (8) - (11), the membership function $\mu_{Ve}(\delta)$ of the dial fuzzy set for the end stage of left turning is calculated in the same way as the start timing membership function $\mu_{Vs}(d)$. That is;

$$S_{Te}(u) = 2\int_{-\infty}^{u} \Lambda_{Te}(\upsilon)\,d\upsilon \tag{12}$$

and

$$\mu_{Ve}(\delta) = S_{Te}\left(\frac{\delta}{\varepsilon_{Te}(R_O)}\right) \tag{13}$$

Now, we can obtain the start and the end dial fuzzy sets on d-$\delta$ plane by the cylindrical extension. The total dial fuzzy set for left turning is given by the AND (product) operation between the start and the end dial fuzzy sets. Then, the dial operation of the left turning is carried out by using the membership function $\mu_V(d, \delta)$ of the total dial fuzzy set. Namely,

$$\mu_V(d, \delta) = \mu_{Vs}(d) * \mu_{Ve}(\delta) \tag{14}$$

where the operator * shows the T-norm operation for which the algebraic product (·) are used here.

The simulation of the left turning of the car is carried out by using this fuzzy method. The result is shown in Fig. 6. In each car trajectory of these figs. (a) and (b), the curvature radius of the fuzzy model varies more smoothly at the start and the end points of the left turning than that of the crisp model of Fig. 2. However, when the car goes over the end point of the left turning, the every car trajectory in (a) and (b) is slightly apart from the X axis. This comes from the fact that the car has to take the direction parallel to the X axis until the car steering operation of the left turning

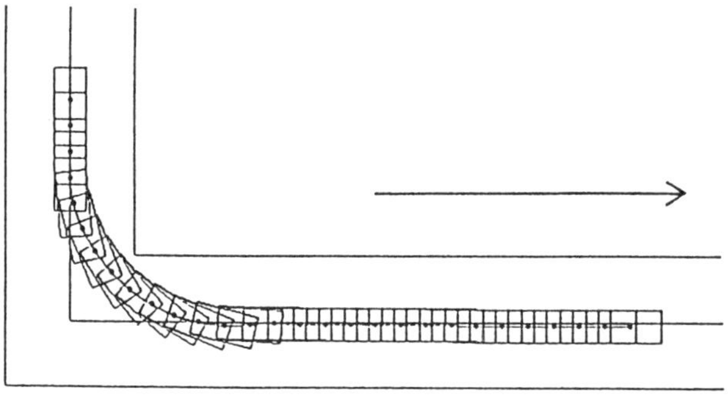

(a)

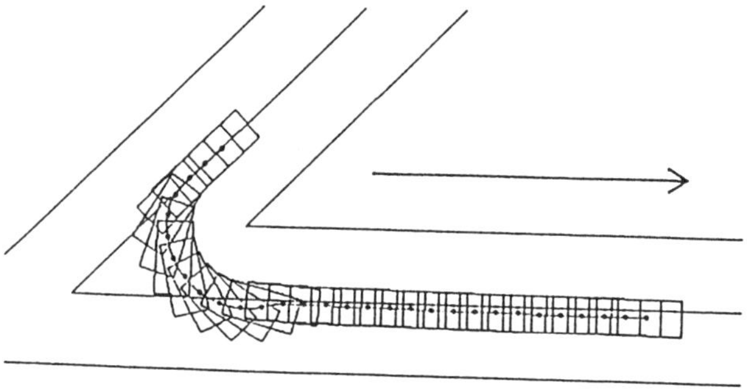

(b)

a=20[cm], b=10[cm], $\alpha$=0.5, L=50[cm]
(a) $\delta_c=1/2\cdot\pi$ [rad], $R_0$=50[cm]
(b) $\delta_c=3/4\cdot\pi$ [rad], $R_0$=20[cm]

Fig. 6. Car trajectories taken from fuzzy model simulation.

finishes. That is, the angle $\delta$ between the car and the X axis is the negative value (the angle direction is clockwise) at the beginning of the left turning. The angle does not become zero even over the point crossing the X axis and remains to have a small negative value at the end stage of the left turning until the car direction becomes parallel to the X axis ($\delta$=0). This situation is clearly shown in Fig. 6(b).

As a result, since the car direction is returned to be parallel to the X axis ($\delta$=0), the car trajectory crosses the X axis and goes round, as shown in Fig. 6(b). The steering operation of the left turning is designed with respect to the motion of rear wheels as shown in Fig. 1, the steering angle $\phi$ of the front wheels is returned to make the car direction parallel to the X axis.

## IV. IMPROVEMENT OF THE FUZZY MODEL

The timing to return the steering angle has to be corrected before the car reaches the end point of the left turning as described in the last part of section 3. Let us consider the case that the car steering angle is gradually returned to be close to the X axis until the car goes over the X axis. The return timing of dial operation is chosen at the time when the angle $\delta$ has the value of $\varepsilon_{Te}'$ as in figs. 7 and 8. At this time, the following eq. holds;

$$\varepsilon_{Te}' = \phi_0 = \tan^{-1}\left(\frac{a}{R_0}\right) \qquad (15).$$

Since the timing to finish the dial returning is when the angle $\delta$=0, $\Lambda_{Te}(u)$ in eq. (8) is replaced by the square-shaped function of $\Pi_{Te}(u)$. That is,

$$\Pi_{Te}(u) = \begin{cases} 1 & (0 \leqq u \leqq 1) \\ 0 & (u < 0 \text{ or } 1 < u) \end{cases} \qquad (16).$$

Then, the front wheels can keep the direction parallel to the X axis during the return steering operation as shown in Fig. 7, but the car goes over the X axis. Therefore, the return timing of the left turning has to be set earlier than the timing of $\delta$=0. The correction value $\Delta d_0$ of the distance is added to the $d_0$ of eq. (1). That is,

$$d_0' = d_0 + \Delta d_0 \qquad (17).$$

where

$$\Delta d_0 = R_0\left(\sqrt{1+\frac{a^2}{R_0^2}} - 1\right) = R_0\left(\frac{1}{\cos\phi_0} - 1\right) \qquad (18).$$

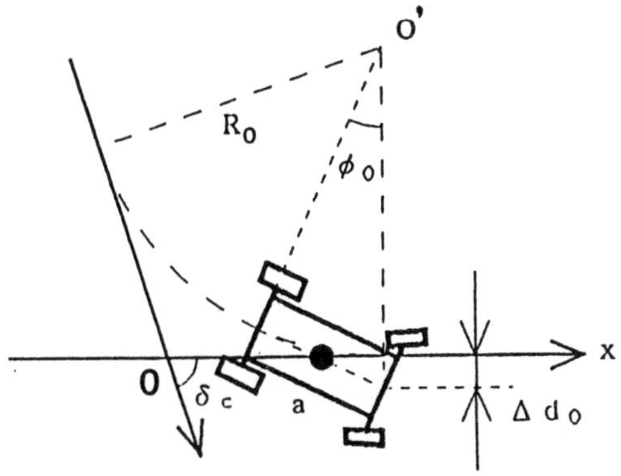


Fig. 7. Timing to return steering angle, and correction $\Delta d_0$ of the distance $d_0$ for improved fuzzy model.

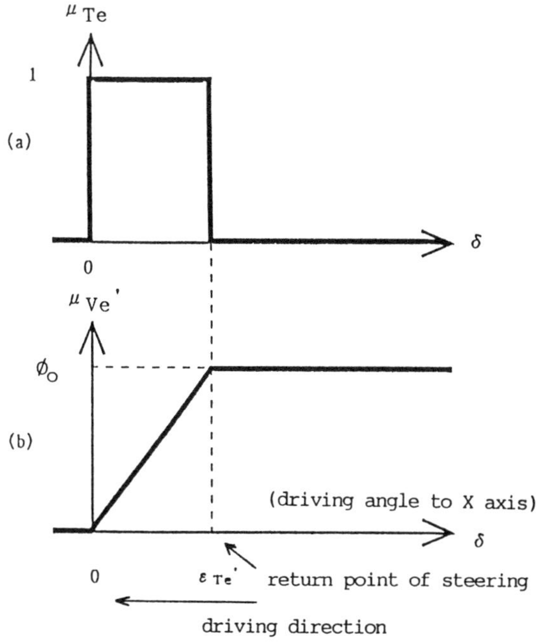


Fig. 8. Relation between (a) improved end timing fuzzy set and (b) improved dial fuzzy set at the end of left turning.

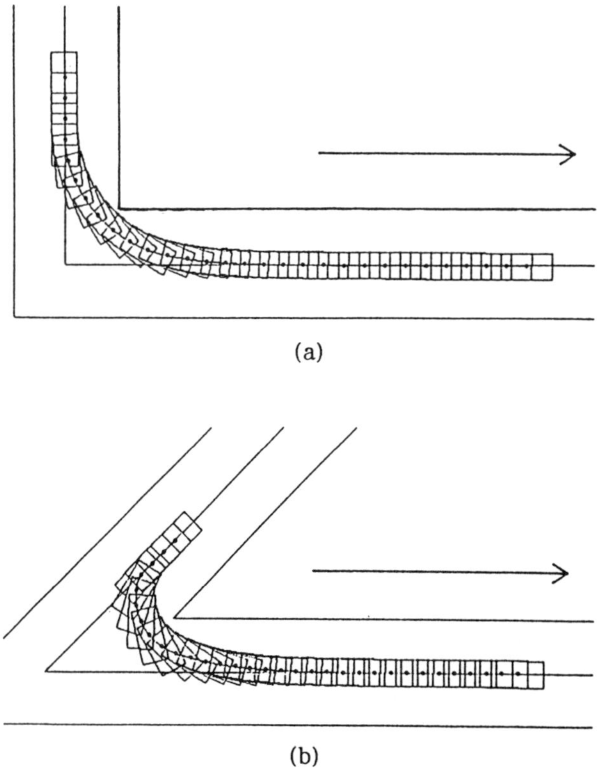


a=20[cm], b=10[cm], $\alpha$=0.5, L=50[cm]
(a) $\delta_c$=1/2·$\pi$ [rad], $R_0$=50[cm]
(b) $\delta_c$=3/4·$\pi$ [rad], $R_0$=20[cm]

Fig. 9. Car trajectories taken from improved fuzzy model simulation.

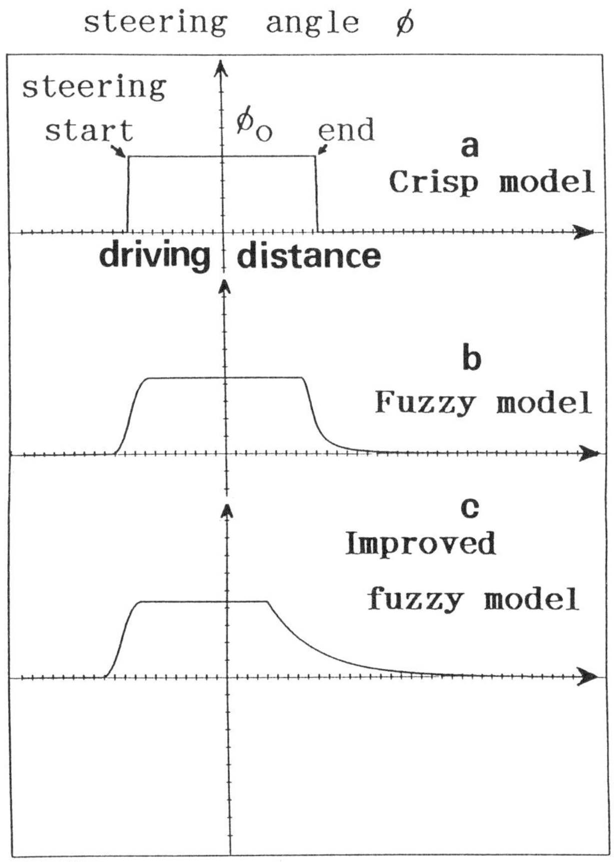


Fig. 10. Steering operations by the three kinds of fuzzy control methods for the left turning at $\delta_C = 1/2 \cdot \pi$ [rad].

The simulation of the left turning operation was carried out by this method. The result is shown in Fig. 9. In each car trajectory of the figs. 9(a) and (b), the center of the front wheels does not go across the X axis.

As a conclusion, Fig. 10 shows the steering operations at the corner angle $\delta_C = 1/2 \cdot \pi$ [rad] based on the fuzzy control timings for three kinds of simulation methods described above as a function of the driving distance. The curves (b) and (c) successfully simulate human operations for the driving control in the turning of the car.

## V. CONCLUSION

We have applied the idea of the fuzzy theory to the left turning for a car. For the left turning, the start and the end timings of the steering operation were treated with the fuzzy theory. The steering angle was determined by the fuzzy sets of the start and the end timings. The simulations show that the fuzzy method is useful for the control of the car because of only a few fuzzy rules and the simple control procedures.

## REFERENCES

[1] E.H.Mamdani, "Applications of Fuzzy Algorithm for Control of a Simple Dynamic Plant", Proc. of IEEE, Vol.121, pp.1585-1588, 1974.

[2] M.Sugeno and M.Nishida, "Fuzzy Control of Model Car", Fuzzy Sets and Systems, Vol.16, pp.103-113, 1985.

[3] Y.Akiyama, T.Abe, Y.Mitsunaga and H.Koga: "A Conceptual Study of the Max-* Composition on the Total Spaces Correspondence, and its Application in Determining Fuzzy Relations" (in Japanese), J. Jpn. Soc. Fuzzy Theory and Systems, Vol.3, No.2, pp.303-317, 1991.

# Self Organizing Fuzzy Logic Control of a Level Control Rig

Nader Vijeh[1]
Department of Engineering Sciences
University of Exeter
Exeter, Devon, UK

*Abstract* - Self Organizing Fuzzy Controllers are able to generate and modify their Control Policy based on a given Performance Criteria. This ability allows these controllers to be used in applications were the knowledge to control the process does not exist or the process is subject to changes in its dynamic characteristics. The Self Organizing Controller (SOC) described here is applied to a Level Control Rig and the results are presented.

**Introduction**

The simple fuzzy controllers, with a fixed set of rules, are used successfully when the control protocol is available in a textual format and the process under control does not exhibit dynamic changes due to aging or variation in the set-point. The problem of rule acquisition manifests itself particularly when the rules have to be obtained from a human operator whose knowledge is based on experience and he applies them in an intuitive manner.

A possible solution to these problems is a Self-Organizing Controller (SOC) which is capable of generating and modifying the Control Protocol by a learning process based on measuring performance. The controller described here is based on that presented by Procyk and Mamdani [3].

The learning process of the SOC is comprised of algorithms that allow the controller to assess its own performance on the basis of a set of predetermined performance rules. At every sampling point the controller uses a Performance Decision Table to determine the performance in terms of Error and Change in Error variables. If the performance is unsatisfactory, then either an existing rule, leading to the present poor performance, is modified or a new one generated and the redundant rules are deleted. The block diagram below represents the conceptual structure of the SOC.

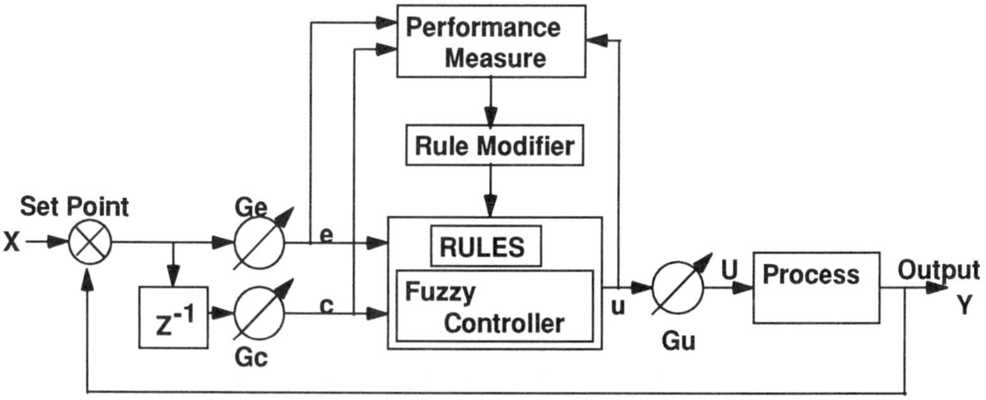


**Fig 1. Conceptual Block Diagram of SOC**

The controller can be viewed as a hierarchical rule based system, where a set of fixed performance rules are used to modify a second set of control rules. As can be seen from the block diagram, the lower half of the structure is essentially identical to that of a simple fuzzy controller.The performance rules represent the acceptable performance of the controller. In the case of a single-input single-output (SISO) process an assumption is made that the process is monotonic and an increase in the input will eventually cause an increase in the output.

---

[1]Author can be reached at: Advanced Micro Devices Inc., 901 Thompson Place, M/S 45, Sunnyvale, CA 94088.
Email: nader.vijeh@amd.com.

0-7803-0614-7/93/$03.00 ©1993 IEEE

## Protocol modification Using a Performance index

**Table 1. Performance Decision Table**

| ERROR | -6 | -5 | -4 | -3 | -2 | -1 | 0 | 1 | 2 | 3 | 4 | 5 | 6 |
|---|---|---|---|---|---|---|---|---|---|---|---|---|---|
| -6 | 6 | 6 | 6 | 6 | 6 | 6 | 6 | 0 | 0 | 0 | 0 | 0 | 0 |
| -5 | 6 | 6 | 6 | 6 | 6 | 6 | 6 | 3 | 2 | 2 | 0 | 0 | 0 |
| -4 | 6 | 6 | 6 | 6 | 6 | 6 | 6 | 5 | 4 | 2 | 0 | 0 | 0 |
| -3 | 6 | 5 | 5 | 4 | 4 | 4 | 4 | 3 | 2 | 2 | 0 | 0 | 0 |
| -2 | 6 | 5 | 4 | 3 | 2 | 2 | 2 | 0 | 0 | 0 | 0 | 0 | 0 |
| -1 | 5 | 4 | 3 | 2 | 1 | 1 | 1 | 0 | 0 | 0 | 0 | 0 | 0 |
| -0 | 4 | 3 | 2 | 1 | 0 | 0 | 0 | 0 | 0 | 0 | 0 | 0 | 0 |
| +0 | 0 | 0 | 0 | 0 | 0 | 0 | 0 | 0 | 0 | -1 | -2 | -3 | -4 |
| +1 | 0 | 0 | 0 | 0 | 0 | 0 | -1 | -1 | -1 | -2 | -3 | -4 | -5 |
| +2 | 0 | 0 | 0 | 0 | 0 | 0 | -2 | -2 | -2 | -3 | -4 | -5 | -6 |
| +3 | 0 | 0 | 0 | -2 | -2 | -3 | -4 | -4 | -4 | -4 | -5 | -5 | -6 |
| +4 | 0 | 0 | 0 | -2 | -4 | -5 | -6 | -6 | -6 | -6 | -6 | -6 | -6 |
| +5 | 0 | 0 | 0 | -2 | -2 | -3 | -6 | -6 | -6 | -6 | -6 | -6 | -6 |
| +6 | 0 | 0 | 0 | 0 | 0 | 0 | -6 | -6 | -6 | -6 | -6 | -6 | -6 |

**CHANGE IN ERROR**

The SOC modifies the control protocol according to the performance decision table. The desired performance is described in a per sample basis and is therefore a local criterion as opposed to a global performance one such as Integral Square of Error (ISE). Although this table can be tuned to a specific process, minor changes in this table were found to cause insignificant changes in the controller performance.

The non-zero entries in this table represent cases where the performance is not satisfactory and a modification to the rules is necessary. The upper left hand and lower right hand corners of the table present situations where there is a large error and the output is moving away from the set- point. The lower left hand and upper right hand corners contain zeros as these are the situations where there is a large error but the output is moving towards the set-point rapidly. The middle region of the table represents zero error and stationary output, therefor no modification to the rules is necessary.

The region of table containing zeros, stretching from the lower left hand corner to the upper right hand corner, represents the desired trajectory of the process output.

The rule modification algorithm operates on previously stored states of the controller. The computer program stores up to 40 previous Error (e), Change in Error (c) and the Controller output (u) instances.

If the normalized error and change in error in a given sample point (**n**) are given by $\mathbf{e_n}$ and $\mathbf{c_n}$, then the performance obtained from the above table is given by:

$\mathbf{p_n = P [ e_n, c_n ]}$.

Where, P stands for the Performance decision table. If $\mathbf{p_n}$ is not zero, then a delay factor (d) is used to retrieve a previous state of the controller from the delay buffer. The three values extracted are: $\mathbf{e_{n-d}}$, $\mathbf{c_{n-d}}$ and $\mathbf{u_{n-d}}$. The delay in reward factor (**d**) depends on the delays in the process and is generally equal to one for a first order system.

A new rule is then formed by generating *antecedents* that correspond to $\mathbf{e_{n-d}}$ and $\mathbf{c_{n-d}}$ and a *consequence* that corresponds to $\mathbf{u_{n-d} + p_n}$. This rule is then added to the control protocol and redundant rules with similar antecedents are deleted. Note that if no rule had existed for the **n-d** sample, then the corresponding change in control action ($\mathbf{u_{n-d}}$) would be zero and the new rule would simply contain the consequence corresponding to $\mathbf{p_n}$.

### Self-Organizing Control Algorithm

Before describing details of the SOC algorithm it is necessary to clarify the manner in which rules are stored internally. Rules are made-up of fuzzy subsets with a fixed shape. In general these are vectors containing a single maximum value of 10 with two other values of 7 and 3 surrounding the maximum. The exceptions to this are boundary values and error subsets NZ and PZ. It is therefore possible to store each rule as a set of three numbers ($e_m$, $c_m$ and $u_m$), corresponding to the position of the maximum value in the vector. Tables 2 and 3 describe the shape of Fuzzy Variables referenced here. For example the rule: *If* E is PM *and* C is NB *then* U is ZE can be stored as three integers: 12, 1 and 7.

The fuzzy subsets for each rule is then reconstructed using a ***Fuzzification*** procedure which simply recreates each subset in the shape of a vector, as specified in the following tables.

**Table 2. Normalized ERROR (e)**

| Fuzzy Sets | -6 | -5 | -4 | -3 | -2 | -1 | -0 | +0 | +1 | +2 | +3 | +4 | +5 | +6 |
|---|---|---|---|---|---|---|---|---|---|---|---|---|---|---|
| NB | 10 | 7 | 3 | 0 | 0 | 0 | 0 | 0 | 0 | 0 | 0 | 0 | 0 | 0 |
| NM | 3 | 7 | 10 | 7 | 3 | 0 | 0 | 0 | 0 | 0 | 0 | 0 | 0 | 0 |
| NS | 0 | 0 | 3 | 7 | 10 | 7 | 3 | 0 | 0 | 0 | 0 | 0 | 0 | 0 |
| ZE- | 0 | 0 | 0 | 0 | 3 | 7 | 10 | 0 | 0 | 0 | 0 | 0 | 0 | 0 |
| ZE+ | 0 | 0 | 0 | 0 | 0 | 0 | 0 | 10 | 7 | 3 | 0 | 0 | 0 | 0 |
| PS | 0 | 0 | 0 | 0 | 0 | 0 | 0 | 3 | 7 | 10 | 7 | 3 | 0 | 0 |
| PM | 0 | 0 | 0 | 0 | 0 | 0 | 0 | 0 | 0 | 3 | 7 | 10 | 7 | 3 |
| PB | 0 | 0 | 0 | 0 | 0 | 0 | 0 | 0 | 0 | 0 | 0 | 3 | 7 | 10 |
| **Index** | 1 | 2 | 3 | 4 | 5 | 6 | 7 | 8 | 9 | 10 | 11 | 12 | 13 | 14 |

**Table 3. Normalized Change in Error (c) and Change in Control Action (u)**

| Fuzzy Sets | -6 | -5 | -4 | -3 | -2 | -1 | 0 | +1 | +2 | +3 | +4 | +5 | +6 |
|---|---|---|---|---|---|---|---|---|---|---|---|---|---|
| NB | 10 | 7 | 3 | 0 | 0 | 0 | 0 | 0 | 0 | 0 | 0 | 0 | 0 |
| NM | 3 | 7 | 10 | 7 | 3 | 0 | 0 | 0 | 0 | 0 | 0 | 0 | 0 |
| NS | 0 | 0 | 3 | 7 | 10 | 7 | 3 | 0 | 0 | 0 | 0 | 0 | 0 |
| ZE | 0 | 0 | 0 | 0 | 3 | 7 | 10 | 7 | 3 | 0 | 0 | 0 | 0 |
| PS | 0 | 0 | 0 | 0 | 0 | 0 | 3 | 7 | 10 | 7 | 3 | 0 | 0 |
| PM | 0 | 0 | 0 | 0 | 0 | 0 | 0 | 0 | 3 | 7 | 10 | 7 | 3 |
| PB | 0 | 0 | 0 | 0 | 0 | 0 | 0 | 0 | 0 | 0 | 3 | 7 | 10 |
| **Index** | 1 | 2 | 3 | 4 | 5 | 6 | 7 | 8 | 9 | 10 | 11 | 12 | 13 |

A procedure known as ***Quantification*** is used to transform the inferred consequence fuzzy subset into a single value. The procedure used here is the 'Mean of Maxima', which is described in [3] as well as other related literature.

The overall algorithm for the SOC can be described as follows; At every sampling point **n**:

**(a)** The Error is computed by subtracting the process output (y) from the desired set-point (**x**):

$$e_n = y_n - x_n$$

**(b)** The Change in Error (**c**) is computed as:

$$c_n = e_n - e_{n-1}$$

**(c)** These values are then multiplied by their corresponding Gain factors and are scaled into values in the range 1 to 14 for error and 1 to 13 for change in error.

$$e^o_n = f(e_n \cdot G_e)$$
$$c^o_n = f(c_n \cdot G_c)$$

**(d)** These values are used to find the performance index, $p^o$, from performance decision table **P**:

$$p^o_n = P\,[\,e^o_n, c^o_n\,]$$

if $p^o_n$ is not zero then create a rule with the following elements:

$$e' = e^o_{n-d}$$
$$c' = c^o_{n-d}$$
$$u' = u^o_{n-d} + p^o_n$$

**(e)** A consequence vector, U, is initialized with all members being zero. Next steps (f) through (h) are repeated for all the rules in the protocol.

**(f)** For rule, i, in the control protocol, the three numbers ($e^i_m$, $c^i_m$ and $u^i_m$) are used to generate fuzzy subsets, $E^i$, $C^i$ and $U^i$, using the Fuzzification procedures, $\phi$ for error and $\phi'$ for change in error and change in control action:

$$E^i = \phi\ (e^i_m)$$
$$C^i = \phi'\ (c^i_m)$$
$$U^i = \phi'\ (u^i_m)$$

**(g)** If $p^o_n$ is not zero then move to step (h). if Rule **i** is dissimilar to the rule generated by step (d) then continue to step (h), else delete rule **i** and repeat this procedure for the next rule in the protocol.

**(h)** Then Membership Grades ($\mu$) of the consequence vector, **U**, are computed from:

$$\mu_U(u) = \mathrm{Max}_u\{\mu_U(u), \mathrm{Min}(\alpha_i, \mu_{U_i}(u)\},$$

where: $\alpha_i = \mathrm{Min}\{\mu_{E_i}(e_0), \mu_{C_i}(c_0)\}$.

**(i)** If $p^o_n$ is not zero then add the new rule to the protocol and repeat step (h) for this rule.

**(j)** Quantify the consequence fuzzy subset, **U**, thus generated into value $u^o_n$ and store the values $e^o_n$, $c^o_n$ and $u^o_n$ in the First-in First-out (FIFO) delay buffer for future use by the rule modification algorithm.

**(k)** Compute the controller output from:

$$u_n = u_{n-1} + G_u \cdot u^o_n$$

## SOC Applied to a Level Control Rig

The controller was applied to a Level Control Rig in order to examine the controller behavior in dealing with a real plant and demonstrate the learning capability of the controller in dealing with changes in the characteristics of the process under control. The rig used for these experiments was located at the University of Exeter Chemical Engineering Laboratories and consisted of a liquid flow pipe network and a system of pneumatic valve control lines. This rig was used to conduct experiments with a variety of control algorithms and exhibits inherent non-linear characteristics. Experiments with the conventional PID controllers were found to have difficulty performing satisfactorily at different operating points, once these are tuned for a specific set-point [6].

The block diagram below shows the general set-up for these experiments. The liquid pipe network consisted of an open tank (T) of approximately one cubic meter capacity. The drain from the bottom of the tank leads to an electric pump (P) as shown in the diagram. A pneumatically operated control valve (V) adjusts the flow of water downstream from the valve. If the valve is fully closed, then the water is returned to the tank. If the valve is open, then a proportion of the flow is passed through an orifice plate and is then forced up a distance of over two meters.

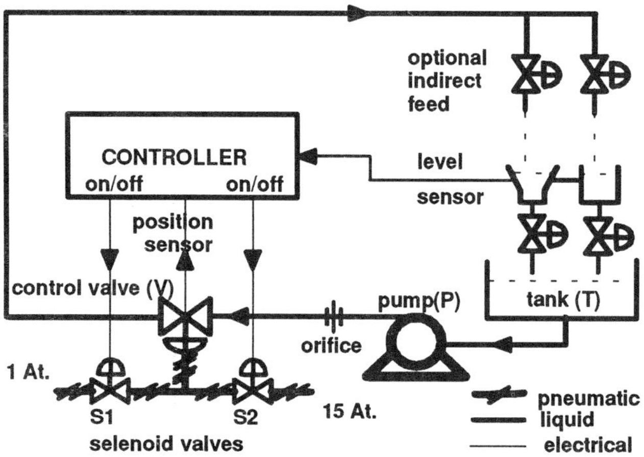


**Fig 2. Simplified block diagram of level control rig**

The flow can be directed towards either a narrow cylindrical vessel of approximately 50 cm height or a vessel of similar height, but a variable cross section, ranging from about 2 cm at the bottom to 20 cm at the top. The outflow from these vessels can be altered by means of a manually operated valve.

The position of the control valve can be adjusted by the sequential control of two electrically driven solenoid-pneumatic valves (S1 and S2). These valves are linked in series arrangement as shown and are supplied with pressurized air at approximately 15 psig at one side and atmospheric pressure at the other side.

The position of the control valve is sensed through a transducer and is fed to one of the analog inputs of the controller. Two of the digital outputs from the controller are used to turn on or off the solenoid valves, responsible for the lowering or raising the control valve.

A second transducer is responsible for measuring the level of the water in vessels and its output is also fed to one of the analog inputs of the controller. A third analog input is connected to a potentiometer and can be used to adjust the desired set-point for the water level.

The overall characteristics of the rig correspond to a real application and as such exhibit typical inherent non-linearities and delays.

**Controller modification**

The basic SOC algorithm described earlier is modified in order to set the position of the pneumatically operated control valve.

This modification consisted of a simple on-off control loop that would take the SOC output as the desired set-point for the control valve. The control valve would then be raised or lowered and its position monitored until the desired set-point is reached. To raise the valve, the solenoid-pneumatic valve S1 is closed and S2 is opened. To lower the control valve the opposite procedure is performed. Once the desired valve position is reached both solenoid valves are closed.

**Experimental Results**

**Cylindrical vessel**

A number of experiments with the cylindrical vessel were performed in order to demonstrate the learning behavior and overall response of the SOC. Figures below show the results of one of the experiments performed with the cylindrical vessel and a direct feed. The Sampling interval for this experiment was set to 2 seconds. Delay in reward parameter was set to 2, in order to account for long delays inherent in the system. The first graph shows the water level in the vessel versus the desired set-point and the second graph shows the controller output (Control valve position).

The controller is started with no rules and several changes to the set-point are made. The graph of the output (next page) shows a good performance by the controller, after the initial learning period. The control rules generated by the SOC are also shown on the next page.

Other experiments [6] with this setup showed the SOC to be relatively insensitive to minor changes to the performance rules. Also, as noted, the controller is started with no rules and is able to adapt to the process relatively quickly, without causing the process to become unstable.

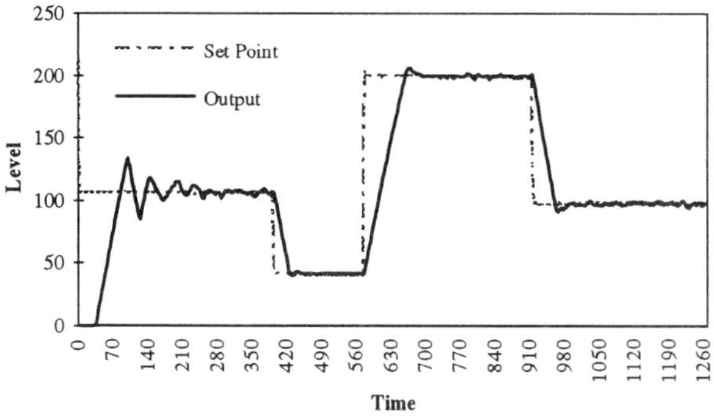


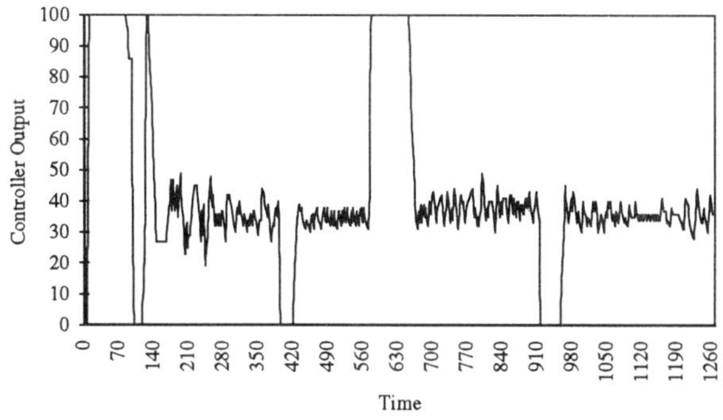


| Rule: | If E | and C | then U |
|---|---|---|---|
| 1 | NB+ | PB- | PB |
| 2 | NB | ZE | PB |
| 3 | NB | PS | PB |
| 4 | NS+ | PB- | NB |
| 5 | PS | PB- | NB |
| 6 | PB- | PB | NB |
| 7 | PB- | PM- | NB+ |
| 8 | ZE+ | NB+ | PB |
| 9 | NS | NB | PB |
| 10 | NM | NB | PB |
| 11 | ZE+ | PB | NB |
| 12 | ZE- | NM+ | PB- |
| 13 | ZE- | ZE | PS- |
| 14 | ZE+ | ZE | ZE |
| 15 | ZE+ | PS | NM+ |
| 16 | ZE+ | NS | NS+ |
| 17 | NS+ | PS | NS |
| 18 | ZE- | NB | PB |

**Variable Cross Section Vessel**

Experiments with the variable cross section vessel demonstrate the ability of the controller to deal with a process with variable characteristics. The results of one experiment is shown in the following graphs. The process characteristics in this case depended on the set-point. It was observed that the controller had to adjust the control policy for a change in the set-point and as the result the initial response to a change in the set-point was poorer than the cylindrical vessel. The total number of rules generated in this case was 20.

For this experiment the sampling interval was set to 2 seconds and the delay in reward parameter was set to 2.

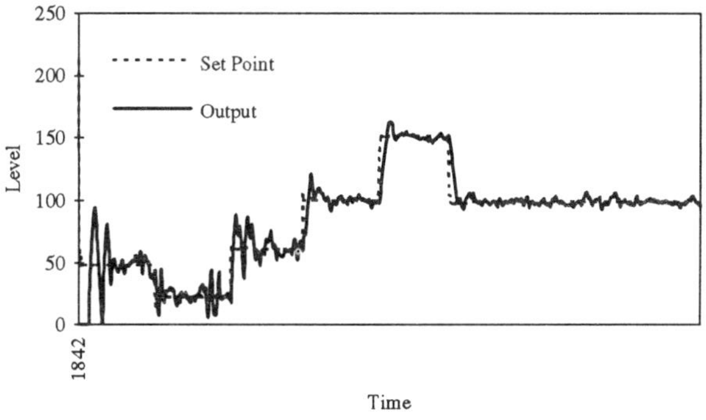


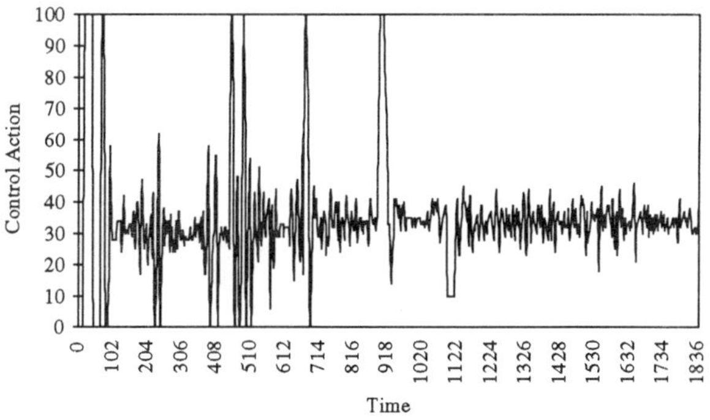


**Conclusions and Recommendations for future work**

The learning behavior of the SOC is based upon using a set of performance rules that describe the desired performance of the controller in terms of the deviation of the process output from a desired set-point. This learning method generates rules that are produced with a reward mechanism based on a local performance measure. There are no mechanisms for producing rules that result in an optimal form of control.

The controller can be viewed as a hierarchical decision making system, where the decision rules are generated by a second set of fixed performance rules. Further work is needed to explore the possibility of extending this hierarchical structure to include a higher level set of rules that would be used to generate the local performance measure on the basis of a more global performance criterion. These criteria could be based on describing the optimal performance of the controller based on the controller output and the process output.

This implies rules with multiple antecedents, referring to a number of previous states of the system and the consequence would be the changes made to the performance index.

The controller described here was implemented in the form of a simple microprocessor based system with 64K byte of memory and an 8 bit Analog to Digital converter. This system can be implemented in a very cost effective manner and applied to a variety of applications.

**Acknowledgments**

I would like to thank Professor John O. Flower for the opportunity to carry out this research at the University of Exeter and for his invaluable advice.

**References**

[1] King, P. J., Mamdani,E.H. (1977). "The Application of Fuzzy Control Systems to Industrial Processes", Automatica, Vol.13, 235-242.
[2] Mamdani,E.H. (1974). "Application of Fuzzy Algorithms for the control of a Dynamic Plant", Proc. IEE, Vol.121.
[3] Procyk,T.J., Mamdani,E.H. (1979). "A linguistic Self- Organising Process Controller", Automatica Vol.15, 15- 30.
[4] Tong,R.M. (1977). "A Control Engineering Review of Fuzzy Systems", Automatica, Vol.13, 559-569.
[5] Tong,R.M., Beck,M.B., Latten,A. (1980). "Fuzzy Control of Activated Sludge Wastewater Treatment Process", Automatica Vol. 16, 659-701.
[6] Vijeh, N. (1988). "Microprocessor Engineering Aspects of a Fuzzy Logic Self-Organizing Controller", Ph.D. Thesis, Department of Engineering Science, University of Exeter, Exeter, Devon, UK.
[7] Zadeh,L.A. (1965). "Fuzzy Sets", Information and Control, Vol.8, 338-353.
[8] Zadeh,L.A. (1973). "Outline of a New Approach to the Analysis of Complex Systems and Decision Processes", IEEE Transaction on Systems, Man and Cybernetics, SMC- 1, 28-44.
[9] Zadeh,L.A. (1976). "A Fuzzy Algorithmic Approach to the definition of Complex or Imprecise Concepts", Int. Journal of Man-Machine Studies, Vol.8, 249-291.

# A NEW SELF TUNING FUZZY CONTROLLER DESIGN AND EXPERIMENTS

C-Y Shieh, Graduate Student
Satish S. Nair, Assistant Professor
Department of Mechanical and Aerospace Engineering
University of Missouri - Columbia
Columbia, MO 65211

**Abstract**
A new fuzzy logic controller design with self tuning features for both the data base and the rule base is proposed. Real-time implementation of the controller using an experimental setup shows that the design performs very well for a complex nonlinear system.

## 1.0 INTRODUCTION

Fuzzy Logic Controllers (FLCs) [1,2] are popular for control applications, due in part to the fact that they can incorporate operator knowledge about a complex nonlinear system in an easy way. FLCs are also relatively easy to construct and special purpose hardware is increasingly being available for real time implementation. Most commercial products currently employ a constant look-up table type approach for implementing fuzzy control. Several researchers have been studying approaches to incorporate learning into the fuzzy control architecture. Most of these algorithms are, however, heuristic and subjective and there is no systematic procedure to design and analyze self-tuning fuzzy controllers. Along these lines, a self-tuning strategy with experimental validation was presented by Wu et al. [3,4] to tune the data base for a nonlinear time varying system. The technique essentially tuned the data base using a gradient descent technique. These and other studies reported in the literature illustrate the capability of fuzzy control to incorporate operator control experience in a natural way. Minimal 'prior knowledge' about the system is required as compared to modern control techniques. For instance, adaptive controllers for such systems are complex in structure requiring detailed knowledge of the system parameters [4]. Recently, preliminary simulation studies using a composite controller were reported by Shieh and Nair [5]. However, no real time experimental studies or comparisons had been performed. This paper reports a modified version of the design which performs well in real-time experiments.

<u>Description of the Experimental Example Case System</u>

Figure 1 shows the example case considered, a DC motor-drive four-bar linkage system, which is a representative transmission used in several machines. The parameter values for the linkage load and the DC motor are listed in Table 1. The governing equations for the load is given by Eq.1,

$$M(\theta)\ddot{\theta} + V(\theta)\dot{\theta}^2 + G(\theta) = T(t) \qquad (1)$$

where θ is angular position of link 2, M and V are complex nonlinear functions of θ representing the reflected inertia and the centrifugal and coriolis force terms respectively, and T is the torque applied by the motor [3]. The system thus has geometric load nonlinearities and coulomb friction in the motor. Fig. 2 illustrates the effect of the nonlinearities on the control of the speed of link 2. For a constant motor voltage of 5 volts, an error of approximately 39% is seen. The control objective is to reduce this error.

A 486-CPU based 50 MHz computer with a 50 KHz A/D board is used for data acquisition and control. The voltage output from the computer is limited to ± 10 volts (20 mA). This is amplified, with a gain setting of 7.2, using a high power high frequency switching amplifier. The output of the preamplifier drive is a PWM current source with a bridge type output. An optical encoder, with a resolution of 1024 pulses per revolution, is used for angular position measurement. An HCTL 2020 16-bit chip is used to decode the encoder output. Angular velocity is calculated from this position measurement by differentiation as well as measured using a tachometer that is integral with the motor. The tachometer has a gain of 31.5 volts per 1000 rpm. The A/D board features include 16 single ended analog inputs with 16-bit resolution, two analog outputs with 12-bit resolution, a timer, and a frequency output. Real-time control is implemented using C language, which is the language used for the simulation studies also.

## 2.0 COMPOSITE CONTROLLER

The composite controller proposed is of a 'velocity' type (Figure 3), and generates the control output Δu by adding two distinct components. One of the components, $\Delta u_d$, is from a data-base-tuning controller $FLC_d$, while the other component, $\Delta u_r$, is from a rule-base-tuning controller $FLC_r$. as follows :

$$\Delta u_d = f_1(e, \Delta e)$$
$$\Delta u_r = f_2(\theta, e) \qquad (2)$$

In the composite design proposed, $FLC_d$ is tuned first and then $FLC_r$.

0-7803-0614-7/93$03.00 

## 2.1 Linguistic Term Sets

The error (e), change of error ($\Delta e$), and the angular position of link 2 ($\theta$), are used as the control variables, as follows :

$e(k) = s(k) - y(k)$

$\Delta e(k) = e(k) - e(k-1)$ (3)

where k: sampling instant
k - 1: previous sampling instant
s(k): setpoint at instant k
y(k): output of plant at instant k

Triangular and trapezoidal membership functions are used to interpret the term sets of the linguistic variables. Based on this interpretation, the term sets in the data base can be represented by functions of the position of the fuzzy sets heights as

$E = E(0, s_e, m_e, b_e)$,

$\Delta E = \Delta E(0, s_{\Delta e}, m_{\Delta e}, b_{\Delta e})$,

$\Delta U = \Delta U(0, s_{\Delta u}, m_{\Delta u}, b_{\Delta u})$ (4)

The maximum overlap of membership functions of two adjacent fuzzy sets is 0.5 and three fuzzy sets do not overlap. The rule base controller inputs are e(k), $\theta(k)$. The term sets for e(k) are the same while $\theta(k)$ is divided into 5 degree intervals resulting in 73 term sets.

## 2.2 Self Tuning Strategy

Starting with an 'approximate' rule base and data base definitions for both $FLC_d$ and $FLC_r$, the data base is tuned for $FLC_d$ keeping the same rule base while the reverse is true for $FLC_r$. Details of the self tuning techniques are as follows :

Self Tuning $FLC_d$

A tuning factor $\alpha$ is introduced to modify the support of every fuzzy set of the term set simultaneously, keeping the same completeness, as

$F' = F [0, \alpha s, \alpha m, \alpha b]$

where F can be any fuzzy term set of E, $\Delta E$ and $\Delta U$, and F' is the modified fuzzy term set (Figure 4). Note that the rule base does not change in this case. IAE, in the following, represents the integral squared error calculated from $4{*}t_r$ to $t_f$, where $t_r$ is the system rise time and $t_f$ is the total observation time. This algorithm can be briefly stated as follows :

1. set all factors $\alpha_i = 1$
2. select linguistic variable $F_i$ to be tuned
3. run the control program and obtain $IAE_o$
4. modify $\alpha$ to $\alpha$-0.1 and get the new membership functions
5. run the control program to get $IAE_i$
6. if $0 \leq \alpha$ goto 4
7. get the minimum $IAE_i$ and select the corresponding value of $\alpha_i$ as optimal
8. go to 2 for the next linguistic variable $F_i$ until all are complete
9. repeat (2) through (8) until $\alpha_i(new) = \alpha_i(old)$, for all i.

Self Tuning $FLC_r$

This part of the algorithm is implemented on-line after time > $4{*}t_r$, where $t_r$ is the system rise time. The algorithm is structured as follows :

if $\Delta IAE(t+\Delta t) > \Delta IAE(t)$ then
  if $\omega_2 > \omega_d$, reduce the consequent term set of the predominant rule by one level
  if $\omega_2 < \omega_d$, increase the consequent term set of the predominant rule by one level.

If $\Delta IAE(t+\Delta t) \leq \Delta IAE(t)$, no changes are made. The term 'predominant' refers to the antecedent sets of the rules that are fired, which have membership function values $\mu(x) > 0.2$. In our case, they are $\mu_e(x) > 0.2$ and $\mu_\theta(x) > 0.2$, where 0.2 has been selected without rigorous optimization. The contribution from $FLC_r$ has been scaled at present to provide small correction inputs based on $\theta$, the position of link 2, and error e.

## 3.0 EXPERIMENTAL RESULTS AND DISCUSSION

The original rule base and the data base of the proposed fuzzy controller are based on the designer's 'experience' which may be heuristic and subjective. Control implementation is performed using C language and the hardware configuration described in Section 1. Two sampling rates are implemented experimentally : 150 Hz. and 500 Hz. The membership functions are used directly rather than by lookup tables. Training is continued for 30 seconds in all cases with a speed set point of 100 rpm (10.47 rad/s). Figures 5a,b and Figures 6a,b depict the variation of output speed seen experimentally for the two control configurations. The data base tuning was accomplished in three passes through the loop, i.e. steps 2 to 8 in the data base tuning algorithm above. For the $FLC_d$ only case, the integral absolute error reduces by about 13.85 for a 150 Hz. control update rate and by about 12.27 for the 500 Hz. update rate. The maximum error in speed for this controller with the two update rates are 34.7% and 26.8% respectively. Considerable improvement in control is found with the composite controller (Table 2) with the maximum errors being 25.1% and 9.3% for 150Hz. and 500 Hz. update rates. The control program has an update capable capability of approximately 700 Hz. A computed torque scheme gave a maximum error of 6% for the same speed setpoint of 100 rpm. The approach was complex comparatively requiring considerable information about the system [4].

Starting with a rudimentary controller, it is shown that the proposed composite fuzzy controller is capable of learning the system characteristics and compensating for the nonlinearities. Qualitatively, the two components of the controller can be viewed as performing different tasks. The data base part, FLCd, provides the coarse control, while the rule base part FLCr provides the fine control. In the example case system considered, the data base tuning by itself does not suffice since it does not capture the spatial variation effects. Appropriate rule base modification based primarily on the input angle θ is found to be effective. It should be noted that the fuzzy logic controller allows for the inclusion of this information in a simple way as compared to the classical and adaptive ones [5]. Another 'intelligent' technique that has been found to be effective for such systems is neural network control [6].

## 4.0 ACKNOWLEDGEMENTS

This work was partially supported by the National Center for Supercomputing Applications under grant number CEE920013N and utilized the computer system CRAY Y-MP4/464 at the National Center for Supercomputing Applications, University of Illinois at Urbana-Champaign.

## 5.0 REFERENCES

[1] T. J. Procyk and E. H. Mamdani, 1979, "A Linguistic Self-Organizing Process Controller," *Automatica*, vol. 15, pp. 15-30.

[2] C. Batur and V. Kasparian, 1989, "A Real Time Fuzzy Self-Tuning Control," *American Control Conference*, pp. 1810-1815.

[3] Wu, K, Outangoun, S., and Nair, S. S., March 1992, "Modeling and Experiments in Fuzzy Control", *First IEEE International Conference on Fuzzy Systems*, San Diego, CA , pp.725-731.

[4] Outangoun, S., Wu, K, and Nair, S. S., 1991, "Experimental Comparison of Fuzzy Logic and Adaptive Control Techniques," *North American Fuzzy Information Processing Society Annual Meeting*, Columbia, Missouri, pp 388-392.

[5] Shieh, C-Y., and Nair, S. S., 1992, "A Composite Self Tuning Strategy for Fuzzy Control of Dynamic Systems," *North American Fuzzy Information Processing Society Annual Meeting (NAFIPS-92)*, Puerto Vallarta, Mexico, pp 118-124.

[6] Outangoun, S., and Nair, S. S., "Neural Network Controller Designs and Comparisons for Dynamic Systems," *Modelling and Scientific Computing*, International Journal, Pergamon Press (accepted for publication.)

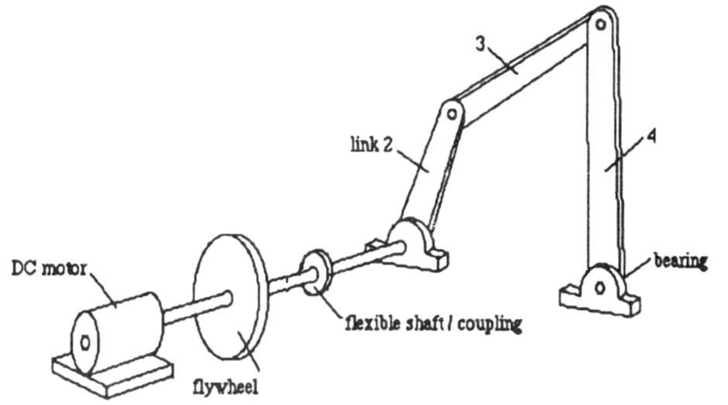


Figure 1. Schematic of the experimental example case system

**Table 1. System Parameters for the Example Case**
**(a) DC motor**

| Parameter | Symbol | Si Units |
|---|---|---|
| Torque constant | $K_t$ | 0.41 (N-m/ampere) |
| Voltage constant | $K_b$ | 0.41 (Volt-sec/rad) |
| Armature resistance | $R_m$ | 1.9 (Ohm) |
| Armature inductance | $L_m$ | 4.18E-3 (Henry) |
| Viscous damping | $B_m$ | 2.08E-4 (N-m/rad/sec) |
| Armature moment of inertia | $J_m$ | 1.102E-3 ($Kg\text{-}m^2$) |

**(b). Four bar linkage**

| | Link 1 | Link 2 | Link 3 | Link 4 |
|---|---|---|---|---|
| Mass (kg) | - | 0.147 | 0.277 | 0.277 |
| Mass moment of inertia c.g. (kg.mm ) | - | 278.82 | 1686.97 | 1686.97 |
| Length (mm) | 190.5 | 61 | 215 | 215 |

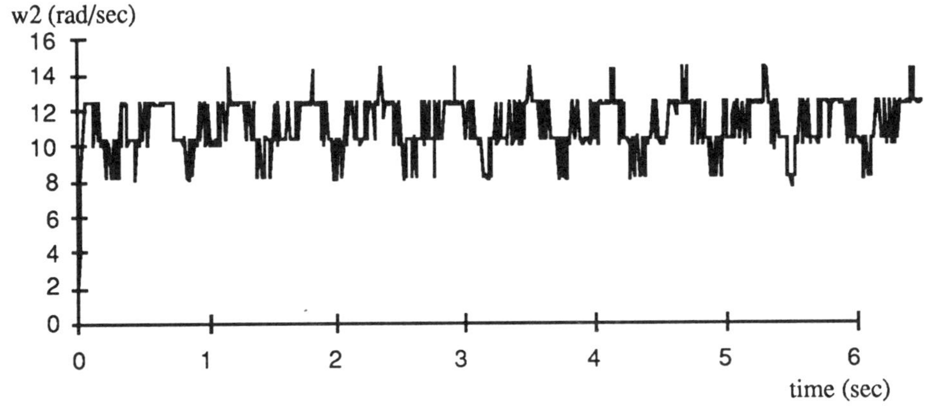


Figure 2. Variation of $\omega_2$ when $e_a$ = 5.0 volts

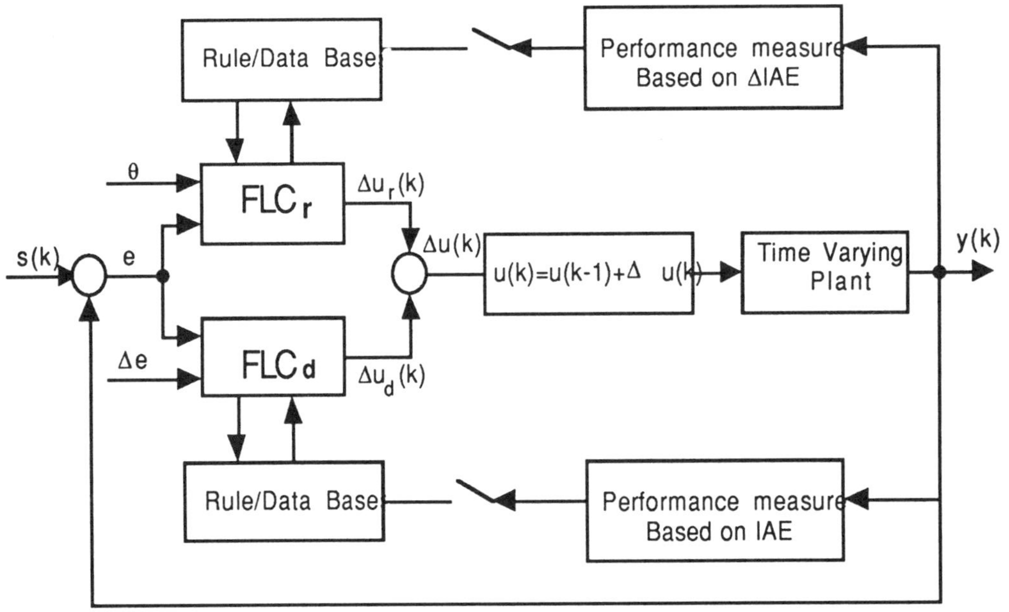


Figure 3. Composite fuzzy logic controller

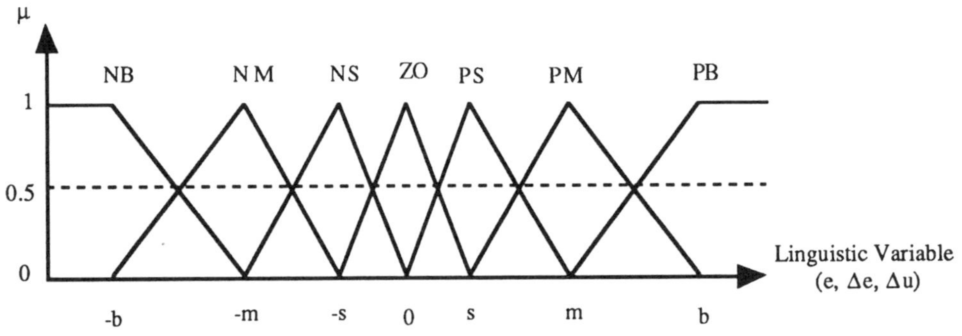


Figure 4. Membership functions for the fuzzy term sets

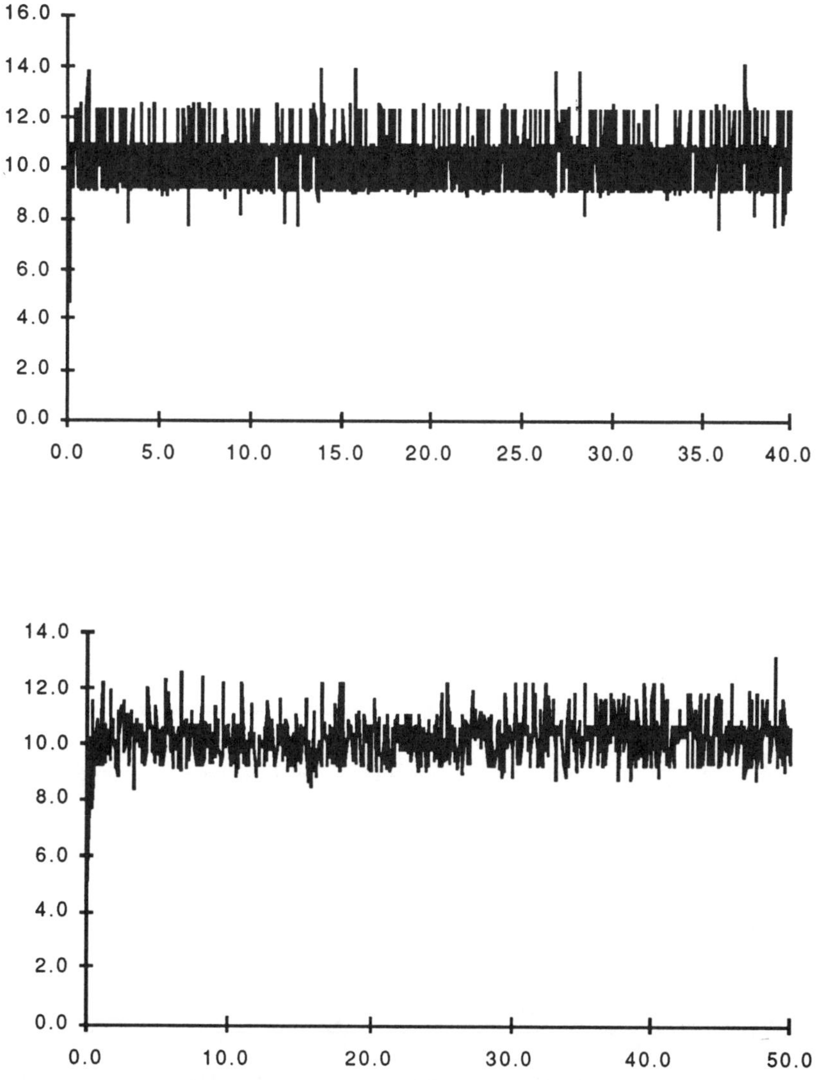


Figure 5. Variation of speed at 150 Hz. control update frequency: a) only $FLC_d$ b) $FLC_d$ and $FLC_r$

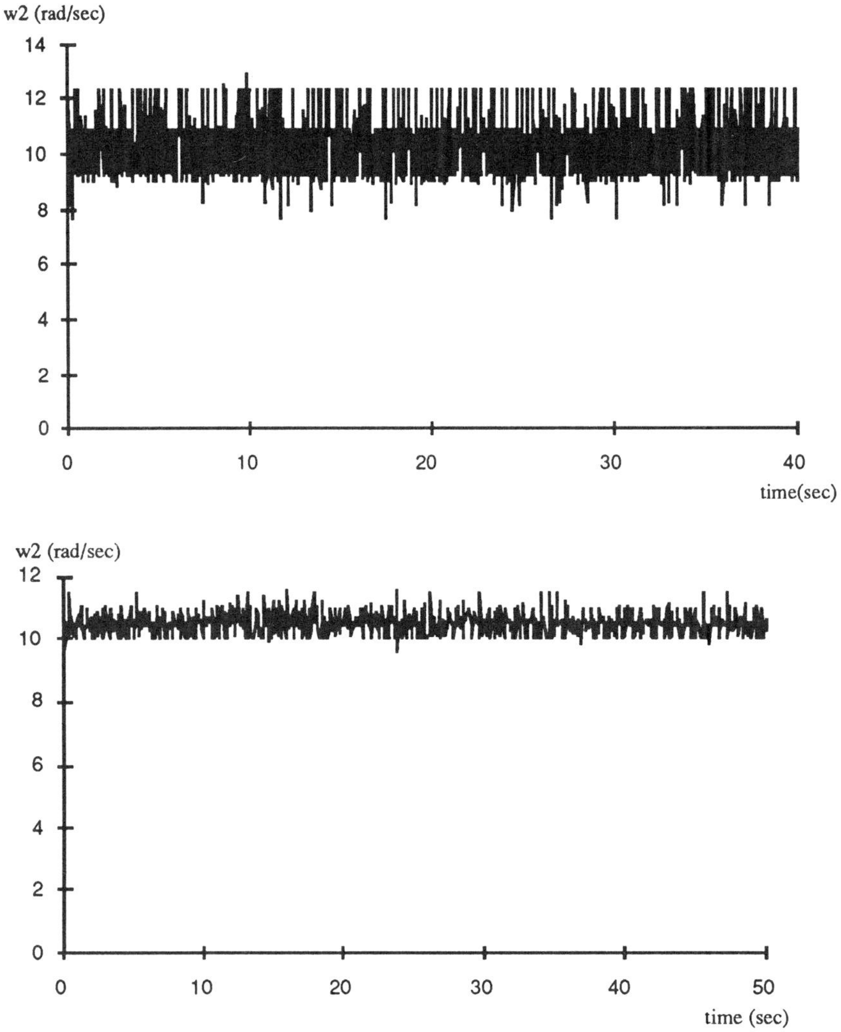


Figure 6. Variation of speed at 500 Hz. control update frequency: a) only $FLC_d$ b) $FLC_d$ and $FLC_r$

**Table 2. Experimental Comparison of Control Designs**

| Strategy | Update Frequency | ΔIAE | Max. Error |
|---|---|---|---|
| **Data Base Tuning** | 150Hz | 13.85 | 34.67% |
| | 500Hz | 12.27 | 26.75% |
| **Combined self-tuning** | 150Hz | 12.33 | 25.11% |
| | 500Hz | 4.48 | 9.33% |

# Real-Time Operation of an Autonomous Fuzzy Controller *

David Gardner, Chun-Hsin Chen, Kaveh Ashenayi, and Sujeet Shenoi
Center for Intelligent Systems
Keplinger Hall, University of Tulsa
Tulsa, Oklahoma 74104, USA

***Abstract*—This paper describes our efforts at developing a practical autonomous fuzzy logic controller. The autonomous controller is provided with basic knowledge which enables it to continuously observe plant performance while performing its control actions and to use the outcomes of its actions in shaping its control policy in real time. The controller remains dormant when the plant is operating satisfactorily but it rapidly and autonomously initiates adaptation in real time when adverse performance is observed. The autonomous controller can learn to control certain systems even when provided with a vacuous rule base. It can effectively deal with individual variations in plant characteristics, compensate for component wear and tear, and handle dynamic interactions with the environment.**

**The hardware implementation employs an Intel MCS-51 series micro-controller board incorporating an inexpensive 8-bit Intel 8031 processor. The design, which is geared for extensibility and the option of adding analog I/O, allows the 11.06 MHz micro-controller to operate at a rate exceeding 200 look-up table reinforcements per second. This is important when developing practical on-line adaptive controllers for relatively fast systems.**

## I. INTRODUCTION

The ultimate goal of on-going research at the Center for Intelligent Systems at the University of Tulsa is to implement an entirely autonomous controller in hardware. The autonomous controller would be endowed with basic control meta-knowledge which would enable it to learn to control plants in real time without prior knowledge of their characteristics, and possibly even without an initial crude control policy. It would continuously observe system performance while implementing its control actions and would use the outcomes of its actions in adjusting its control policy [1-3]. The autonomous controller would be designed to lie dormant when system behavior is satisfactory but would rapidly and autonomously initiate adaptation in real time when adverse performance is observed. Such an autonomous controller could be directly introduced into a given application where it would quickly learn to control the system in a reasonable manner. The advantages of a truly autonomous controller are numerous – it could deal with individual variations in system characteristics, compensate for degrading characteristics caused by system wear and tear, and handle dynamic interactions with the environment. It could also effectively handle black box systems and novel control scenarios.

Our efforts at implementing autonomy have focused on look-up-table-based fuzzy logic controllers [3,4]. A look-up table configuration is appealing because the controller demands minimal computational overhead during real-time operation. The controller employs a look-up table generated off-line from common sense rules using a fuzzy inference algorithm (see, e.g., [5,6]). The initial common sense rules represent "source code" [7]. The look-up table corresponds to "compiled object code" geared specifically for real-time control. The look-up table for a two-term fuzzy controller is a discrete function mapping error and error-change inputs to the appropriate controller output. It generates a 3-dimensional "control surface."

The main challenge in developing practical on-line adaptive look-up table controllers is to reduce the on-line computational overhead involved in adaptation [2,3]. This can be significant because each modification to a linguistic rule or membership function requires a recomputation of the entire look-up table. In terms of the compiler analogy, adaptation involves the production of new object code by the repeated recompilation of modified source code.

Our approach to adapting an initial control policy embodied in a look-up table is similar to the strategy employed in optimizing compilers. The idea is that repeated

*Research supported by NSF Grant IRI-9110709, OCAST Grant AR9-010, and by grants from the Oklahoma Center for Integrated Design and Manufacturing and from Sun Microsystems, Inc.

0-7803-0614-7/93/$03.00 ©1993 IEEE

on-line modification and recompilation of source code can be bypassed by modifying or "optimizing" the object code itself. Controller adaptation can therefore be implemented by "hammering" and/or "stretching" the control surface itself. However, as in the case of optimizing compilers, the changing 3-dimensional control surface must faithfully reflect the desired modifications in the source code.

The autonomous control algorithm presented in this work employs a simple reinforcement learning technique which adjusts clusters of table entries at a time. This strategy, designed specifically for hardware implementation, significantly reduces the computational effort and gives rise to rapid on-line adaptation. The controller also autonomously initiates its own adaptation based on its evaluation of system performance. Adaptation and performance monitoring/evaluation engage system-independent meta-knowledge which greatly increases the applicability of the autonomous controller.

The hardware implementation employs an Intel MCS-51 series micro-controller board incorporating an inexpensive 8-bit Intel 8031 processor. The design, which is geared for extensibility and the option of adding analog I/O, allows the 11.06 MHz micro-controller to operate at a rate exceeding 200 look-up table reinforcements per second. The resulting autonomous controller is able to learn to control a plant even while it is actually controlling it.

## II. AUTONOMOUS CONTROL ALGORITHM

In this section we highlight the main features of the autonomous control algorithm. We describe the basic controller configuration, the notions of sleep-awake and awake-sleep transitions, the performance measures used for guiding adaptation, and the strategy employed for updating look-up table entries.

### A. Look-Up Table Controller Configuration

The basic fuzzy control system considered in this work has two inputs, error and change in error, and a single controller output. The look-up table has 169 ($13 \times 13$) entries as the linguistic values for error and error-change (for the non-adaptive controller) are defined using a 13-value discrete scale $L_{13} = \{-6, -5, ..., -1, 0, 1, ..., 5, 6\}$. Actual analog error and error-change values are converted to digital values and are scaled to values in $L_{13}$. The final control action is obtained by adding the look-up table value corresponding to the digital error and error-change values to the previous controller output. This is done regardless of whether or not the controller is learning. We assume that the controller initially has a "vacuous" rule-base, i.e., the entire look-up table has zero values. The goal of adaptation is to hammer and stretch the control surface to obtain satisfactory performance subject to the restriction that the surface is always smooth, i.e., neighboring look-up table entries have values which are relatively close to each other.

### B. Sleep/Awake Transitions

At any instant the autonomous controller is either in a "sleep" or in an "awake" state. No adaptation occurs in the sleep state, but the autonomous controller continuously monitors system performance. When adverse performance is observed for a specified number of sampling instants, the controller moves to the awake state. While in the awake state the controller undergoes adaptation as detailed below. When acceptable performance is observed for a specified number of sampling instants, the controller moves to the sleep state where adaptation is terminated. It is important that the transition from the sleep state to the awake state be made as soon as possible, preferably within a single sampling instant, so that the controller can quickly respond to adverse performance.

### C. Performance Measures

A performance table provides the autonomous controller with a mechanism for assessing its performance. The performance table, like the look-up-table, has 169 entries. The table is divided into seven zones: $(-3, -2, -1, 0, 1, 2, 3)$ as shown in Fig. 1. Zone 0 entries correspond to system states requiring no corrective actions. A system in Zone 0 is either at the set-point, or moving towards the set-point in a reasonable manner. Zones -3 and 3 correspond to system states requiring maximal corrective actions. The controller must make large output adjustments in order to ensure that the system moves rapidly toward the set-point. On the other hand, states in Zones -1 and 1 require only minimal corrections as the system is already close to the set-point.

The performance table employed is symmetric in absolute values along the secondary diagonal (see Fig. 1). It conveys the notion of a "minimum tolerable response." The idea is that the further away the system states are from the desired trajectories the greater must be the corrections in the controller output. The zones take into account the distance from the set-point as well as the rate of approach to it. A key advantage of the performance table is that it embodies system-independent knowledge and can therefore be applied with minor adjustments to a variety of process systems or plants.

### D. Look-Up Table Modification

Adjustments to the control surface are performed using a special "weighted access count table" which records the number of times a given entry in the look-up table is "accessed" since its last modification. This weighted access count table has 169 entries, the same number as in the look-up table. When the number of accesses of given en-

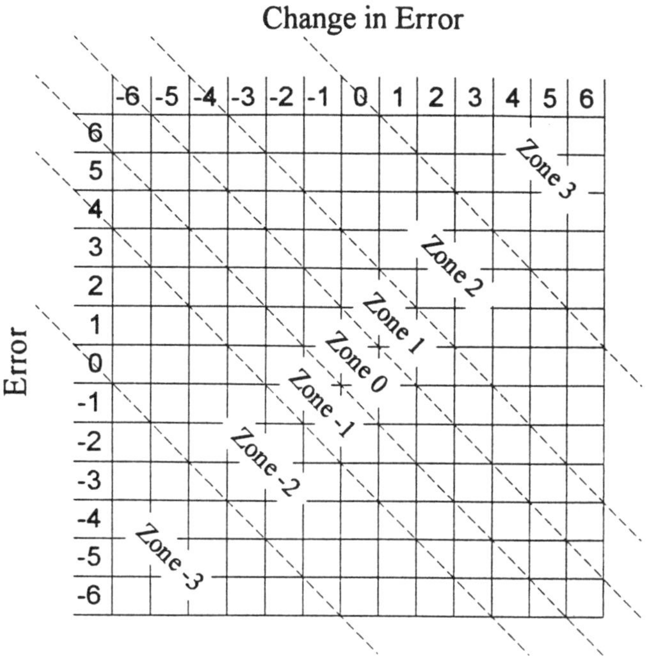


Fig. 1. Performance measure table.

try exceeds a pre-specified threshold, i.e., the entry repeatedly gives rise to inadequate performance, its value in the look-up table is appropriately modified.

The weighted access count table is maintained in the following manner. When an element in the look-up table is accessed, the corresponding element in the weighted access count table is incremented. The amount of the increment depends on the performance table zone in which the element belongs (see Fig. 1). The reasoning is that when the system is operating in Zone -3 or 3 the controller should be able to adapt more quickly than when the system is in Zone -1 or 1. This is accomplished by using larger weighted access count table increments for elements in the outer performance table zones than those for elements lying in the inner zones.

In order to make the controller learn in a neighborhood or "context" [3], increments to the look-up table entries propagate within a 9-element cluster comprising the look-up entry actually referenced and its 8 immediate neighbors. States in Zones -3 and 3 cause the weighted access counts for all the elements in the cluster to be incremented. Learning for entries in Zones -1 and 1 is more localized. Only the referenced element is incremented; its neighbors are not modified.

When an entry in the weighted access count table exceeds a preset threshold, look-up table values for all the elements in its cluster are adjusted. The modifications to the look-up table entries are scaled according to the weighted access count and the performance zone in which the entry is located.

System states in Zones -3 and 3 are relatively distant from the stable state, where the error and change in error both equal zero. Larger weighted access count increments for entries in these zones imply that fewer accesses are required before look-up table values are adjusted. The controller thus exhibits less "patience" in these zones. Furthermore, as the system output is usually in the rise time phase in Zones -3 and 3, the look-up table values must push the system to its stable state as rapidly as possible. This is achieved by maximally changing the look-up table values for the entries in these zones.

When the system is close to the stable state, for example in Zone -1 or 1, frequent look-up table modifications often cause undesirable oscillations in the system response. Consequently, smaller weighted access count increments are used in these zones. The result is that the threshold is not reached as frequently. Furthermore, as the system is typically in the response time phase in Zones -1 and 1, the corresponding look-up table values are updated with less vigor even when the threshold is ultimately exceeded.

Whenever a look-up table entry is adjusted all the elements in its cluster in the weighted access count table are reset to zero. Thus, the weighted access count table models a "short term memory." A separate "cumulative weighted access count table" is maintained so that the total number of accesses for each look-up table entry can be recorded over the "long term." The cumulative count is maintained over a "cycle" which is defined as extending from a sleep-awake transition to the following awake-sleep transition. The idea is as follows: The autonomous controller should learn rapidly during its initial cycles, but it should develop more patience and learn more slowly in later cycles. This ensures that the controller does not over-learn or unlearn the knowledge it has already picked up in preceding cycles.

When the controller awakes at the start of a cycle the cumulative access count table has zero values for all its entries. Then, whenever an access count value in the weighted access count table is zeroed, i.e., when the look-up table value of a cluster entry is adjusted, the corresponding entry in the cumulative access count table is incremented by its weighted access count. Entries in the cumulative access count table are incremented until the end of a sleep-awake-sleep cycle, at which time the entire cumulative access count table is reset to zero. Just before this is done, the values are compared with and are stored in a separate "maximum weighted access count table."

The maximum weighted access count table is designed

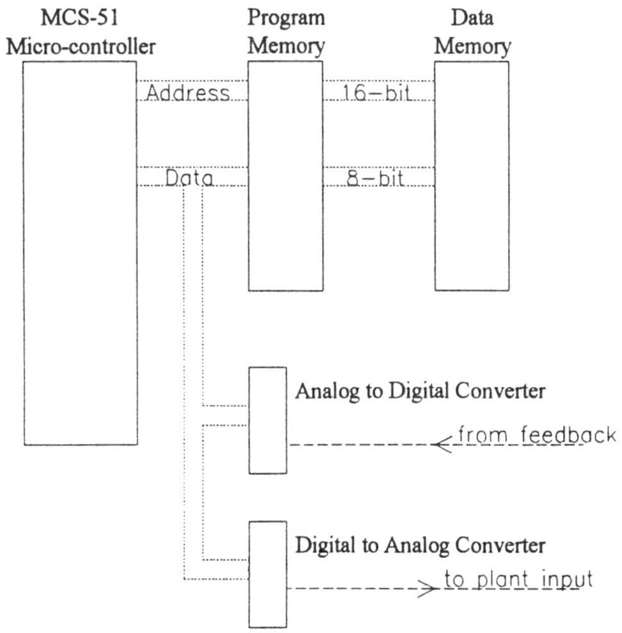


Fig. 2. Block diagram of autonomous controller.

to record the maximum number of accesses for each look-up table entry over the "lifetime" of the autonomous controller. We view a cycle as corresponding to a "day in the life" of the controller. Values in the maximum weighted access count table are first set at the end of the first cycle. From the second cycle onwards a look-up table entry is not updated – even though its threshold might be exceeded – unless the value of the entry in the cumulative access count table exceeds the corresponding value in the maximum access count table. This implies that the autonomous controller becomes more patient as time goes by. More table accesses are required in a given cycle that in all previous cycles before a change is made to a look-up table entry.

## III. HARDWARE IMPLEMENTATION

The hardware implementation of the algorithm is based on an Intel MCS-51 series micro-controller board (see Fig. 2). The micro-controller system offers up to 64K bytes of program space and 64K bytes of data space although only a very small portion of each of these is required for the implementation of the autonomous controller. Control and feedback I/O can be accomplished either by using digital/analog converters or by using the serial port of the micro-controller.

The entire autonomous control system is implemented on a printed circuit board measuring just 3.25in.×3.75in. The hardware version of the autonomous control algorithm is somewhat simplified in that it uses 8-bit integers for all its computations. The simplified algorithm allows the 11.06 MHz micro-controller to operate at a rate exceeding 200 look-up table reinforcements per second. This is important when developing practical on-line adaptive controllers for fast systems. It is also significant because an initial controller look-up table could be incorrect or vacuous. The high learning rate enables the controller to learn to control a system even while it is controlling it.

### *A. Simplifications for Hardware Implementation*

The algorithm was originally coded in the C programming language for simulation on a Sun SPARCstation. The original code, which used 32-bit floating point values for all computations, had to be modified for efficient micro-controller implementation. In an effort to maintain high sampling rates and low RAM requirements all of the 32-bit floating point values were converted to 8-bit integers. Also, many of the routines in the original C source code were optimized to keep the size of the micro-controller code to a minimum.

Assuming a symmetric performance table eliminated the need to use signed values. In Fig. 1 it can be seen that all the values in Zones -3, -2, and -1 would cause the controller output to be increased while those in Zones 1, 2, and 3 would cause the controller output to be decreased. Also, if the error and error-change indices for a system state are added, the sign of the result determines whether the state is in a positive zone or in a negative zone. The sum of the indices is positive when the system is in Zone 1, 2, or 3 and is negative when the system is in Zone -1, -2, or -3.

Another convenient simplification involved scaling values to obtain reasonable ranges using the available 8-bits. Elements in Zones -1 and 1 give rise to much smaller changes in output than elements in Zones -3 and 3. Therefore, to maintain a reasonable range with 8-bit integers the values in the look-up table are scaled according to how close they are to Zone 0. For example, if an element in Zone 1 is never greater than 20, the elements in the zone can be scaled up, say by a factor of 12, while they are stored in memory and then scaled down by the same factor before they are added to the output. This does not give the controller better resolution; however, the truncated portion of the integers can be summed over several samples and this sum can be added to the controller output when it becomes large. Using scaled values also makes it easier to make minor adjustments to the look-up table.

*B. Controller Input/Output*

As stated previously, there are two inputs to the control algorithm, error and error-change, and a single controller output. The hardware implementation is designed so that the change in error is calculated rather than measured. Thus, the autonomous control system has a single input and a single output. The I/O can be implemented in a variety of ways such as analog/digital converters or timers, depending on the application. In the majority of our tests we have made use of the serial communication features which are built into the micro-controller by using a PC to link the controller to the process being controlled. This approach was chosen for two reasons. First, inexpensive data acquisition devices for PCs are readily obtained for almost any application. Second, and more importantly, a serial link helps monitor learning in an autonomous controller when it is applied to a new process system or plant. For example, in one experiment the PC was set up to send the controller an error at each sampling instant. The autonomous controller would then return a packet containing information such as the sleep/awake state, quantized error and error-change, and the new controller output. Such monitoring is particularly helpful when setting up an autonomous controller in a new application. The serial link also allows downloading the look-up table to the PC for storage and further analysis.

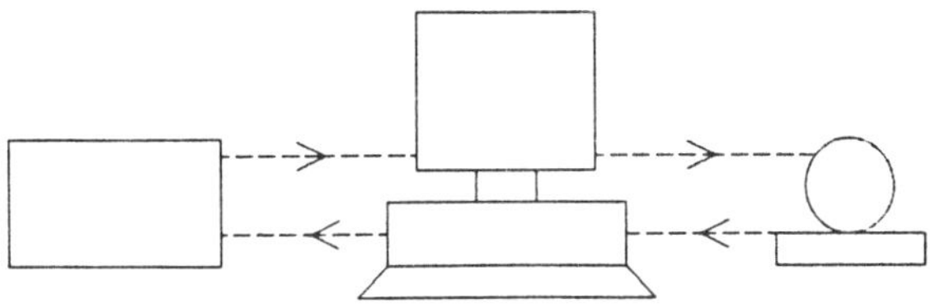


Fig. 3. Test setup.

## IV. SIMULATION RESULTS

The autonomous control algorithm has been thoroughly tested on a variety of simulated and real systems. The simulated systems include both linear as well as certain non-linear systems with time delays. The simulation results demonstrate that the autonomous control algorithm generally gives rise to acceptable performance even when the look-up table of the controller is initially vacuous, i.e., all of its entries are zero.

The hardware version of the algorithm was first tested by interfacing it to a computer which simulated different systems. The results obtained were nearly identical to those obtained with the original software version.

The autonomous control hardware was also tested in a thermal control application (see Fig. 3). In this experiment the temperature near an incandescent lamp was maintained constant by adjusting the power delivered to it. Starting with a vacuous look-up table, the autonomous controller was able to build an adequate control surface and learn to maintain the temperature at the desired level within two or three set-point changes. Subsequent set-point changes gave rise to relatively minor adjustments in the learned control policy and small improvements in performance. The results obtained for the first two runs in the incandescent lamp control experiment are presented in Fig. 4. The experimental results obtained with the autonomous control hardware are very encouraging. We are currently testing it in more complex control scenarios, including flying a 4 ft. model helicopter.

Extensive simulation experiments indicate that the autonomous fuzzy controller works best with stable systems. However, an autonomous controller with a vacuous control policy can be trained to control an unstable system by first applying it to a stable system similar to the original system. Thereafter, the partially learned controller is progressively applied to less stable systems even more similar to the original unstable system before being applied to the original unstable system.

## V. CONCLUSIONS

The adaptive fuzzy controller described in this paper displays all of the characteristics required in a truly autonomous controller. It incorporates basic system-independent learning principles and additional constructs for implementing autonomous learning. The constructs include the notion of a minimum tolerable response, and short-term and long-term control memories to ensure that the controller learns appropriately and does not unlearn what it has already learned. The controller is designed to lie dormant when system behavior is satisfactory and autonomously initiates adaptation in real time when adverse performance is observed.

The hardware implementation incorporating a relatively inexpensive 8-bit Intel 8031 processor allows the autonomous controller to operate at a rate exceeding 200 look-up table reinforcements per second. The high learning rate enables the autonomous controller to learn to control relatively fast systems even while it is controlling them. The controller is provided with battery-backed static RAM so that knowledge gained with time is retained when power is lost. The autonomous controller has demonstrated that can effectively control a variety of simulated and real systems even when provided with a vacuous rule base. It is currently being tested in more complex control scenarios, including flying a 4 ft. model helicopter, and the results are very promising.

## References


[1] T.J. Procyk and E.H. Mamdani, "A self-organizing fuzzy logic controller," *Automatica*, vol. 15, pp. 15-30, 1979.

[2] S.Z. He, S.H. Tan, C.C. Hang and P.Z. Wang, "Design of an on-line rule-adaptive fuzzy control system," *Proceedings of the 1992 IEEE International Conference on Fuzzy Systems*, San Diego, California, pp. 83-90, 1992.

[3] D. Mallampati and S. Shenoi, "On-line adaptive fuzzy logic controllers," *Proceedings of the 1992 International Fuzzy Systems and Intelligent Control Conference*, Louisville, Kentucky, pp. 68-80, 1992.

[4] S. Shenoi, C.H. Chen and A. Ramer, "Towards autonomous fuzzy control," *Proceedings of the Third International Workshop on Neural Networks and Fuzzy Logic*, NASA Johnson Space Center, Houston, Texas, pp. 53-55, 1992.

[5] M. Braae and D.A. Rutherford, "Theoretical and linguistic aspects of the fuzzy logic controller," *Automatica*, vol. 15, pp. 553-557, 1979.

[6] D.A. Rutherford and J.C. Bloore, "The implementation of fuzzy algorithms for control," *Proceedings of the IEEE*, vol. 64, pp. 572-573, 1976.

[7] P. Bonissone, "Fuzzy logic controllers: A knowledge-based system perspective," *Proceedings of the Third International Workshop on Neural Networks and Fuzzy Logic*, NASA Johnson Space Center, Houston, Texas, pp. 8-9, 1992.


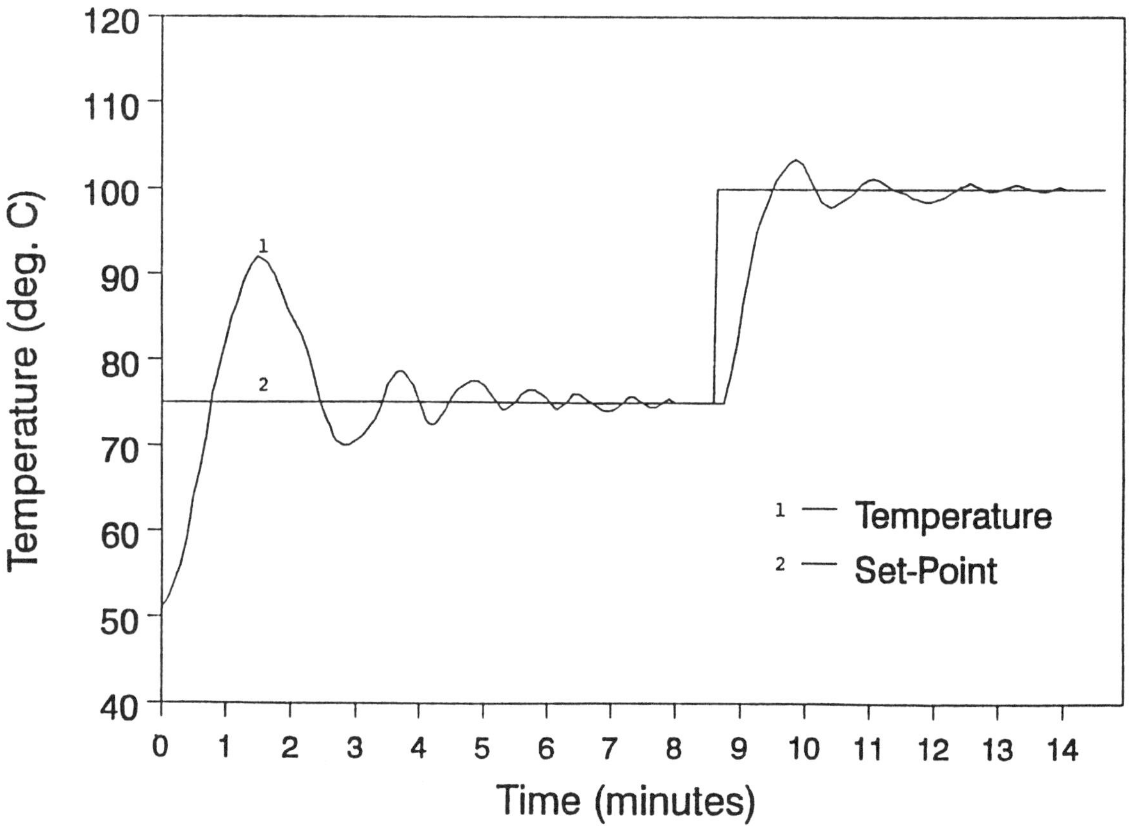


Fig. 4. Performance of autonomous controller with vacuous rule-base.

# Combining Fuzzy Logic and Neural Networks to Control an Autonomous Vehicle

Bernd Freisleben and Thomas Kunkelmann
Department of Computer Science (FB 20), University of Darmstadt
Alexanderstr. 10, D–6100 Darmstadt, Germany

***Abstract*—In this paper we present an approach to design a controller that enables a simulated car to drive autonomously around a race track. The input to the controller is the current speed of the car and several sensor signals indicating the properties of the race track, and as its output the controller is supposed to determine the car's change of direction and its change of speed in response to the information received. The basic idea of our proposal is to let a fuzzy controller supply the training data for a backpropagation neural network and use the trained network to drive the car on an unknown race track. The implementation of the proposal is described and the driving performance of the car is evaluated. The results indicate that the combined neural/fuzzy approach is superior to solutions where either the fuzzy controller or the neural network alone are used to drive the car.**

## I. Introduction

Designing a vehicle for totally autonomous operation is a difficult task which encompasses several aspects common to complex control applications, such as path planning, sensory–motor control and obstacle detection/avoidance. The dynamical properties and real–time requirements of vehicle control, together with the limited success of conventional computational methods, have fostered the development of neural network and fuzzy logic techniques as promising approaches to the problem. For example, neural networks have been used to back up a simulated truck [7], and fuzzy controllers have been developed to control a model race car [1, 5, 9].

In this paper we present an approach to the problem of controlling a car to drive autonomously around an unknown race track. The car is simulated in software; it is assumed to be equipped with a number of sensors pointing at different directions in order to view the road conditions. The information delivered by the sensors and the current speed of the car is used to determine its change of direction and speed. The basic idea of our solution is to train a neural network on several, differently structured race tracks and then evaluate its driving behaviour on an unknown race track. The most obvious way to train the network is to use a supervised learning algorithm, where the network tries to learn the mapping between the input data (the sensor signals and the current speed) and the desired actions it should perform (changes of direction and speed). The training set of input/desired output pairs must be delivered to the network, and the question arises where the information should come from. If a human expert is chosen to create the training set, he or she will be busy for quite some time due to the large number of different cases which must be considered in order to prepare the car for all potentially possible driving situations occuring in an unknown terrain. Furthermore, the human expert will probably not be able to always determine the optimal actions for a given input vector, and in most cases he or she will base the decisions for the desired actions on rules of thumb.

A better approach for creating the training data is to develop a set of basic rules from which the desired driving behaviour can be infered automatically. There are several ways to realize this approach, and the one most promising for our application is the use of a *fuzzy* rule base, together with suitable *fuzzy* operators and inference strategies [10]. Thus, the neural network designed for driving the car is trained with a set of input/output examples which are supplied by a *fuzzy controller* after having driven the car on the track. In other words, the fuzzy controller acts as a training preprocessor to the neural network, in contrast to other combined neural/fuzzy proposals [4], where the neural units consider the incoming signals as fuzzy sets and process them according to the mechanisms employed in fuzzy theory.

The proposal of combining a fuzzy controller with a neural network as described above is compared to an approach where a fuzzy controller alone is used to drive the car and also to an approach where a neural network, trained with hand–coded data, is responsible for performing the task. It will be shown that the combined approach is superior to the other two, both in terms of driving quality and efficiency. This, however, does not imply that a properly designed fuzzy controller cannot achieve the same

0-7803-0614-7/93$03.00 ©1993IEEE

performance as the combined approach. Our intention is to demonstrate that the large effort required for developing a high-quality fuzzy controller can be avoided by designing a fuzzy controller in a quick-and-dirty manner and combining it with a simple to implement neural network as described above.

The paper is organized as follows. Section 2 describes the general idea behind our proposal in more detail. In section 3 the design of the fuzzy controller is presented, whereas section 4 is devoted to the neural network model used. The implementation of our proposal and its performance are discussed in section 5. Section 6 concludes the paper and outlines directions for future research.

## II. General Approach

The general architecture of our system is shown in Fig. 1.

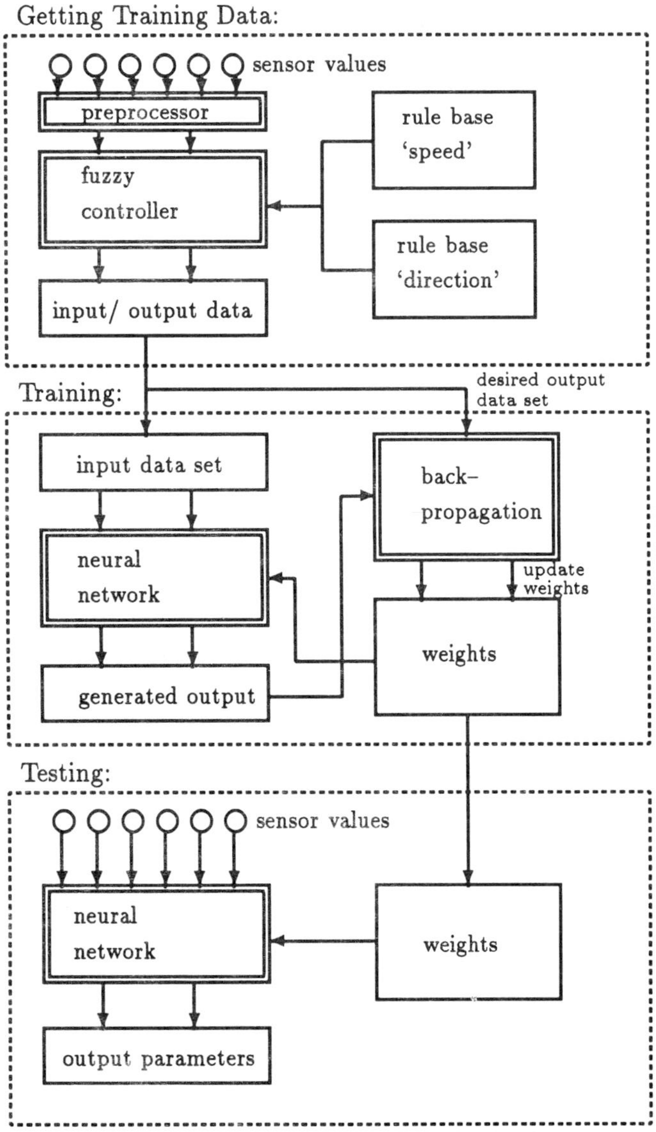


Fig. 1. General architecture of the proposed system

The training data for the neural network is obtained by letting the fuzzy controller drive the car around several sample tracks. The fuzzy controller successively receives the input vectors, suitably preprocessed and transformed into fuzzy sets (see section 3), and processes each of them individually by applying its fuzzy rules, operators and inference strategy to determine what the car should do in response to the input. Since the two possible actions are a change of the direction of movement and a change of the speed (either accelerating the car or slowing it down), the fuzzy controller is internally divided into two nearly identical parts which operate on different rule bases, one for the steering and the other one for the acceleration.

The input/output pairs obtained on the various race tracks are stored and the neural network is then put in charge of driving the car. It is trained with the back-propagation algorithm [8] on the data set created by the fuzzy controller until it has learned the output actions determined by the fuzzy controller, i.e. it basically becomes a "clone" of the fuzzy controller in the sense that its driving behaviour on the training tracks mimics the one of the fuzzy controller. The trained network is then used to drive the car around an unknown race track, the implicit assumption being that the neural network's generalization ability will be superior to the driving capabilities of the fuzzy controller exposed to the same unknown race track.

## III. The Fuzzy Controller

In this section we present the functionality of the fuzzy controller developed for our application (see Fig. 2). The design of the fuzzy controller is based on the standard procedure of developing a rule base containing *fuzzy if-then rules*, defining *linguistic variables* and using *defuzzification* heuristics [10]. A linguistic variable can adopt different values; each of these is a fuzzy set which is used to represent a particular interval within the range of possible values for the linguistic variable considered. Usually, all input and output parameters of the fuzzy controller are treated as linguistic variables, and it is therefore required to define appropriate fuzzy sets for each of them.

The raw input data is preprocessed to obtain the appropriate fuzzy sets. The fuzzy sets for the input parameter *speed (ISP)* are relatively straightforward to determine. The idea is to divide the range of possible speed values (in our case integer values between 0 and 30, normalized to [0,1]) into suitably sized intervals; we have decided to use the 6 intervals depicted in Table I.

In order to determine the linguistic variables and their values for the sensors, it is necessary to explain how they are installed on the simulated car and what information they deliver. We assume that sets of 5 or 6 different sensors are grouped together to approximately point to one

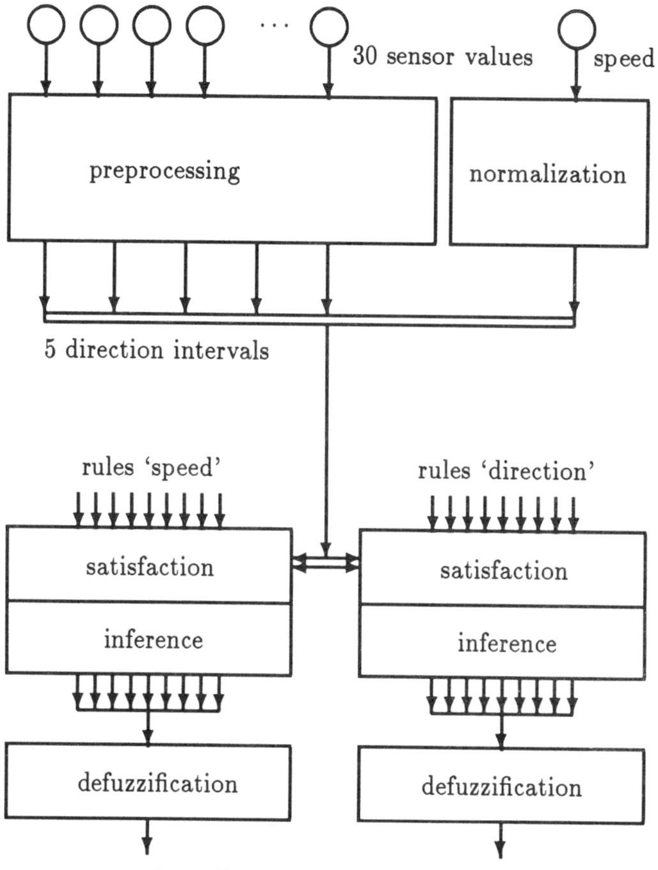


Fig. 2. Architecture of the fuzzy controller

TABLE I
Fuzzy Sets for Input Speed

| ISP | input speed | intervals |
|---|---|---|
| HA | halt | 0 – 5 % |
| SD | slide | 5 – 20 % |
| SL | slow | 20 – 40 % |
| NO | normal | 40 – 65 % |
| FS | fast | 65 – 85 % |
| TS | topspeed | 85 – 100 % |

out of 5 different directions, as shown in Fig. 3. The values on the two axis are distances (in meters).

The 5 directions are denoted as: *Direction Front Left (DFL)*, *Direction Front Middle (DFM)*, *Direction Front Right (DFR)*, *Direction Side Left (DSL)* and *Direction Side Right (DSR)*. The sensors pointing to a particular direction return integer values between 0 and 15 to indicate what they see on the track (pavement, border, obstacle etc.), each of them being responsible for a particular distance in that direction. This distance is the second linguistic variable used in our model, and since there are

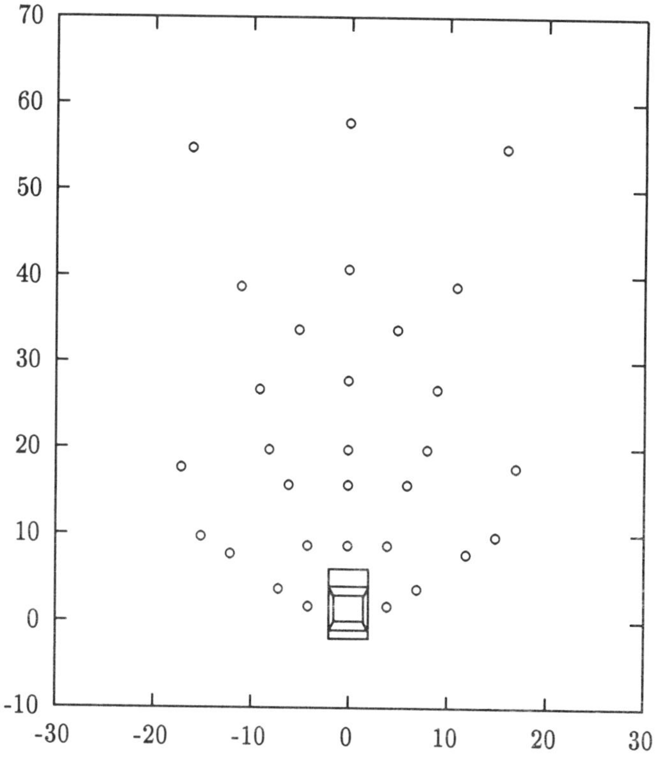


Fig. 3. Sensor positioning

directions emanating from the front and the side of the car, it seems reasonable to distinguish between the two linguistic variables *Frontal Distance (FDI)* and *Side Distance (SDI)*. The fuzzy sets for these linguistic variables have been determined as *Touch (FT/ST)*, *Accidental (FA/SA)*, *Critical (FC/SC)*, *Normal (FN/SN)*, *Far (FF/SF)* and *Infinite (FI/SI)*, where the letter *F* in the abbreviations indicates *Frontal* and *S* indicates *Side*. The fuzzy numbers for the linguistic variables associated with the sensors are summarized in Table II.

The two outputs of the fuzzy controller are also represented as linguistic variables. The range of the parameter *Change of Speed (COS)* (integer values between −4 and 2) is divided into 6 intervals, and the range of the parameter *Change of Direction (COD)* (values between −30 and +30 degrees, normalized to [-1,1]) is divided into 9 distinct intervals, as shown in Table III. For both of them, the interval ranges defined exceed the total range possible, in order to ensure that the maximal values can be definitely reached.

The first step the fuzzy controller has to perform is to calculate for each rule contained in the rule base the degree to which the rule in discussion is satisfied. In order to do so, the fuzzy controller matches the membership function of the inputs with the linguistic values present in the rule. This is achieved by applying a *fuzzy-AND* operator to the *AND-terms* of the rule, which in our design is the standard

TABLE II

Fuzzy Sets for the Sensor Signals

| FDI | frontal distance (*DFL, DFM, DFR*) | values |
|---|---|---|
| FT | frontal touch | 0 m |
| FA | frontal accidental | 3 m |
| FC | frontal critical | 6 m |
| FN | frontal near | 9 m |
| FF | frontal far | 15 m |
| FI | frontal infinite | > 15 m |

| SDI | side distance (*DSL, DSR*) | values |
|---|---|---|
| ST | side touch | 0 m |
| SA | side accidental | 1 m |
| SC | side critical | 2.5 m |
| SN | side normal | 4.5 m |
| SF | side far | 7 m |
| SI | side infinite | > 7 m |

TABLE III

Fuzzy Sets for the Output Parameters

| COS | change of speed | intervals |
|---|---|---|
| EMB | emergency brake | -4.5 – -3 |
| STB | strong brake | -3 – -2 |
| SOB | soft brake | -2 – -0.5 |
| CON | continue | -0.5 – 0.5 |
| ACC | accelerate | 0.5 – 1.5 |
| FSP | full speed | 1.5 – 2.5 |

| COD | change of direction | intervals |
|---|---|---|
| LL | left always | ≪ -100 % |
| LB | left big | -100 – -75 % |
| LM | left medium | -75 – -45 % |
| LS | left small | -45 – -15 % |
| ZE | zero | -15 – 15 % |
| RS | right small | 15 – 45 % |
| RM | right medium | 45 – 75 % |
| RB | right big | 75 – 100 % |
| RR | right always | ≫ 100 % |

*minimum* operator.

The resulting match value is used to compute the inference result. The method used in our design is to determine the minimum between the match value and the result of the rule, i.e. the fuzzy set. The fuzzy sets originating from the inference computation are then used to determine the final result. A *fuzzy-OR* operator, realized by the *maximum* operator, is applied to all fuzzy sets to obtain the final result. The result is again a fuzzy set which is transformed into a particular crisp value (*defuzzification*) by computing the center of the area below the membership function.

The two rule bases developed consist of about 100 *if–then* rules each, which were defined on the grounds of plausibility and were successively extended or refined to yield the desired behaviour of the car. An example of such an intuitive rule for the change of speed is:

```
ISP = NO & DFM = FI & DFL = FI : FSP
```

which indicates that if the car drives at normal speed (`ISP = NO`) *and* if no obstacle has been recognized by the sensors positioned at the middle front (`DFM = FI`) *and* the left front (`DFL = FI`) of the car, then it should accelerate as much as possible (`FSP: Full Speed`). Once a basic set of such rules has been set up to model the car's fundamental capabilities, the car will make its way through the race track. In our application, about 50 basic rules were required in each of the rule bases, but these rules are not sufficient to achieve a satisfactory driving performance, particularly in somewhat extreme situations (U–turns etc.). Refining the control strategy requires modifying, deleting or adding rules in a trial–and–error fashion, which often is a time–consuming process. Several proposals have been made to remedy this problem [1, 2, 6].

## IV. The Neural Network

The neural network used in our proposal is a standard three–layer feedforward architecture, where the input layer consist of 31 neural units (one for each of the 30 sensors and one for the current speed of the car), and the output layer has two units to determine the change of direction and the change of speed, respectively (Fig. 4).

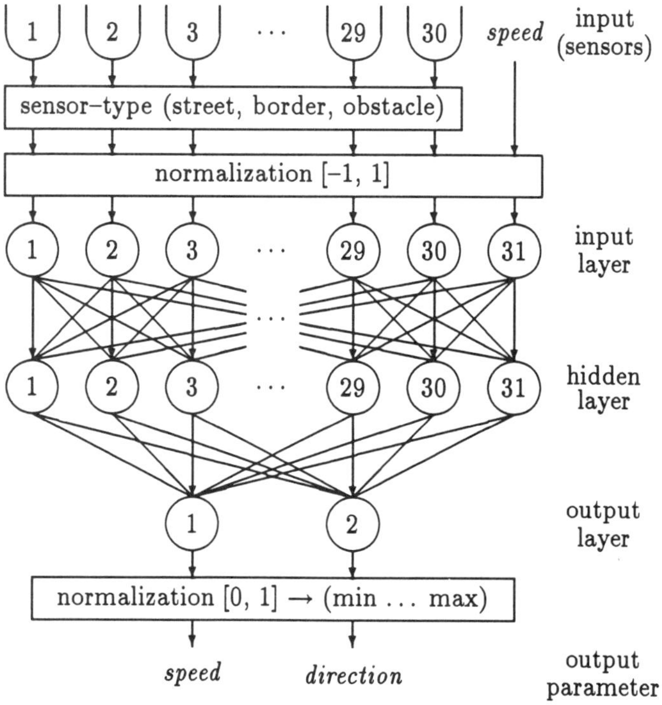


Fig. 4. Neural network architecture

The number of units in the hidden layer has been determined empirically, and the best results were obtained with 31 hidden units. The backpropagation algorithm [8] with momentum term [3] is used for training the network. Since the sigmoid activation functions of the output units produce values between 0 and 1, the network outputs are converted to the ranges adopted for the change of speed and the change of direction.

Since the car is not only supposed to drive safely through the race track and avoid any fixed obstacles, but also should compete against another car simultaneously on the track, the opponent car, which effectively constitutes a moving obstacle, must be appropriately encoded. This is achieved by determining the sensor which has detected the opponent car and check whether the nearby sensors signal the left or right border of the lane; the opponent's car is then treated like a border.

In the training mode, each input vector, stemming from one out of three rounds on three different race tracks, has been presented to the network about 100 times until the network had reduced the error below a predefined threshold. The training times on a SUN Sparcstation were about 60 minutes for processing the whole training set consisting of about 2500 input vectors.

## V. Implementation and Performance

Both the fuzzy controller and the neural network have been implemented in *C*. The neural network was trained on a SUN Sparcstation, but the trained network used to control the car was run on an IBM–PC, since the race track was simulated by using the graphics features of the PC. Several artificial race tracks were generated to train the network. An example of such a race track is shown in Fig. 5.

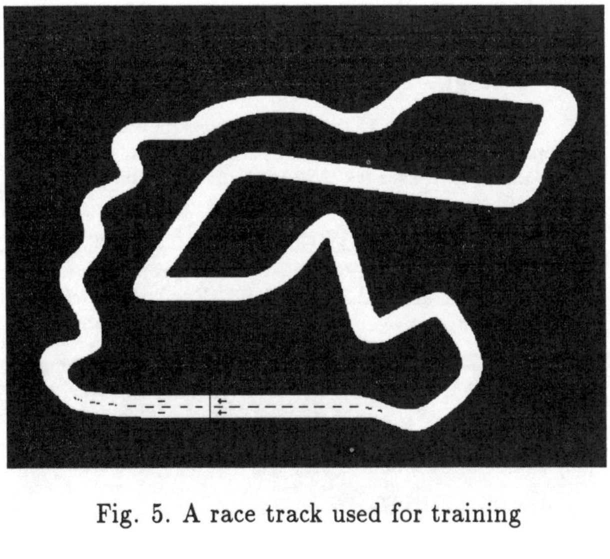

Fig. 5. A race track used for training

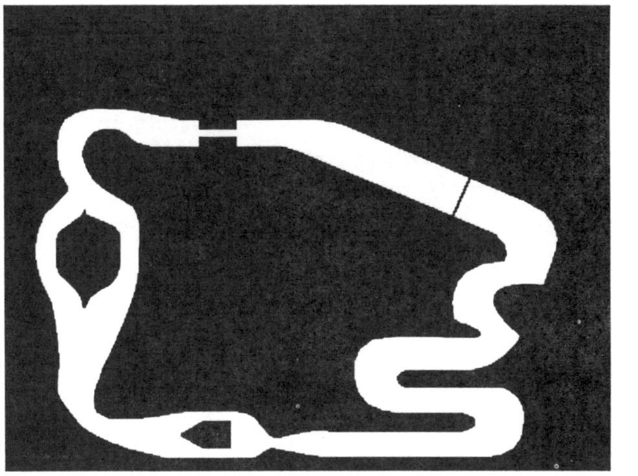

Fig. 6. A race track used for testing

The performance of the three controllers developed for driving the car, the fuzzy controller, the neural network trained with hand–coded training data and the combined fuzzy/neural approach, was measured in several rounds on several unknown test tracks. One of these is graphically shown in Fig. 6.

The first set of experiments conducted was to evaluate the driving capabilities of the three different controllers on a training track. In each experiment, two cars, driven by two different controllers, were competing against each other by permanently switching between the two after each of them has performed one simulation step to update the car's position. Whenever a car ran over the border of the track or touched the other car or an obstacle, a handicap, represented by some simulation steps of forced inactivity, was burdened on the car. All possible combinations of competing cars (fuzzy controller vs. hand–trained neural network, fuzzy controller vs. fuzzy/neural network, hand–trained neural network vs. fuzzy/neural network) were investigated two times. The second set of experiments was conducted analogously on an unknown test track. The performance obtained in the experiments is summarized in Table IV.

The values shown for the total number of simulation steps used and the number of steps considered as handicaps are average values of all races performed. On the training track, the driving performance of the controllers is not significantly different. All three controllers won and lost two races each; the neural network drives the car faster than the other two, but produces more accidents. The best driving quality is achieved by the fuzzy/neural network, since it never caused a crash.

On the test track, the situation is different. Although the relationships between the number of active steps are similar to those obtained on the training track, the driving quality of the fuzzy/neural network outperforms the other

TABLE IV

DRIVING PERFORMANCE

| car driver | training track #_steps + | #_handicap = | #_total | won : lost | test track #_steps + | #_handicap = | #_total | won : lost |
|---|---|---|---|---|---|---|---|---|
| fuzzy controller | 87 + | 6 = | 93 | 2 : 2 | 176 + | 50 = | 226 | 3 : 1 |
| neural net | 77 + | 23 = | 100 | 2 : 2 | 153 + | 93 = | 246 | 0 : 4 |
| fuzzy/neural net | 96 + | 0 = | 96 | 2 : 2 | 186 + | 14 = | 200 | 3 : 1 |

controllers, due to the fact that it produces far less accidents and therefore achieves the best result for the total number of simulation steps needed. It did, however, not succeed in winning all races, but lost in one race against the fuzzy controller.

Considering that in the combined fuzzy/neural approach the neural network was supposed to learn from the fuzzy controller how to drive, it seems somewhat surprising that the neural network is better than its teacher. One possible explanation for this is that the network is probably able to generalize better than the fuzzy controller. It should be noted that this does not imply that a fuzzy controller or a neural network cannot be designed in a way such that their individual driving performance outperforms the combined approach. However, supposing that the results obtained in our experiments do also hold in other driving environments, it might be reasonable to assume that a well designed fuzzy controller employed as the teacher of a neural network will lead to further performance improvements in a combined fuzzy/neural approach.

## VI. CONCLUSIONS

In this paper we have presented a combined neural/fuzzy approach to drive a simulated car around a race track. Assuming that the car is equipped with a set of sensors recognizing the properties of the track and the current speed is known, the task of the controller is to determine the car's change of direction and change of speed in response to the information received. The basic idea of our proposal was to develop a fuzzy controller for driving the car and use its outputs for creating the set of training patterns required to let a standard backpropagation neural network learn how to drive. We have implemented both the fuzzy controller and the neural network and have evaluated the driving performance obtained on several unknown race tracks simulated on an IBM-PC. The results have shown that the combined approach is superior to solutions where either the fuzzy controller or the network alone were used to drive the car. There are several issues for future research, such as using more sophisticated fuzzy design techniques than the simple *if–then* rules and *MAX/MIN* operators [1], testing other neural architectures and learning algorithms, and investigating if a combination of a fuzzy controller and a neural network in the manner proposed will be beneficial for other control applications.

## REFERENCES


[1] C. von Altrock, B. Krause, and H.J. Zimmermann, "Advanced fuzzy logic control of a model car in extreme situations," *Fuzzy Sets and Systems*, 48(1):41–52, 1992.

[2] J.C. Fodor, "On fuzzy implication operators," *Fuzzy Sets and Systems*, 42:293–300, 1991.

[3] J.A. Hertz, A. Krogh, and R. Palmer, *Introduction to the Theory of Neural Computation*, Addison–Wesley, Reading, Massachusetts, 1991.

[4] C. Lin and C.S. Lee, "Neural–network–based fuzzy logic control and decision system," *IEEE Transactions on Computers*, 40(12):1320–1336, 1991.

[5] M. Maeda, Y. Maeda, and S. Murakami, "Fuzzy drive control of an autonomous mobile robot," *Fuzzy Sets and Systems*, 39:195–204, 1991.

[6] M. Mitsumoto and H.-J. Zimmermann, "Comparison of fuzzy reasoning methods," *Fuzzy Sets and Systems*, 8:253–285, 1992.

[7] D. Nguyen and B. Widrow, "The truck backer–upper: An example of self–learning in neural networks," *Proc. of the Int. Joint Conference on Neural Networks*, vol. 2, pp. 357–364, 1989.

[8] D.E. Rumelhart, G.E. Hinton, and R.J. Williams, "Learning internal representations by error propagation," In: D.E. Rumelhart and J.L. McClelland (Eds.), *Parallel Distributed Processing*, vol. 1, pp. 318–362, MIT Press, Cambridge, 1986.

[9] M. Sugeno, T. Murofushi, T. Mori, T. Tatematsu, and J. Tanaka, "Fuzzy algorithmic control of a model car by oral instructions," *Fuzzy Sets and Systems*, 32:207–219, 1989.

[10] H.-J. Zimmermann, *Fuzzy Set Theory – and its Applications*, Kluwer Academic, Boston, 1991.

# Travelling Experiment of an Autonomous Mobile Robot for a Flush Parking

Masaaki OHKITA*, Hitoshi MIYATA**, Masatugu MIURA* and Hidemasa KOUNO*

*Department of Electrical and Electronic Engineering
Faculty of Engineering, Tottori University
4-101, Koyama-Minami, Tottori, Tottori, Japan 680
**Department of Electrical Engineering
Yonago National College of Technology
4448 Hikona, Yonago, Tottori, Japan 683

**Abstract—To control an autonomous mobile robot for the flush parking, the fuzzy rules can be derived by modelling driving actions of a car. Locus of the robot for the parking were drawn through the computer simulation. In such a simulation, the forward and reverse movements of the robot can be controlled smoothly by using simple membership functions.**

**For travelling control of a mobile robot, we have designed and constructed an autonomous mobile robot with four wheels 「DREAM-1」 driven by the fuzzy control theory. In this robot, the position and attitude of the robot can be recognized with an aid of the distance information obtained by six supersonic transducers installed in the upper part of the mobile robot.**

**Finally, 「DREAM-1」 was well controlled for the flush parking based on the computer simulation.**

## I. INTRODUCTION

The fuzzy theory is considered as a new tool for expressing the subjective uncertainties in our human mind. Hence, the theory has potentialities applicable to them such that it has been used in the fields of control, artificial intelligence, expert systems and so on [1]. The fuzzy logic consists of three basic ideas i.e., linguistic variable, canonical form and interpolative reasoning [2]. In the fields of control, these ideas have been applied to the controls of trains, home appliances and so on [1], [3].

We have made an autonomous mobile robot with four wheels 「DREAM-1」 on an experimental basis that is controlled by the fuzzy theory [4]. In case of driving a car, a flush parking necessitates a trained driving action for the car and is a little difficult especially for the beginners.

In this paper, we consider, by computer simulation, the fuzzy control of our robot for the flush parking. Next, based on these simulation results, travelling control of our robot 「DREAM-1」 for the flush parking is described here below.

## II. FUZZY REASONING METHOD

In a multi-input and single-output controller we use the simplified fuzzy control reasoning whose fuzziness has been introduced into the multi-valued logic [5]. It is assumed that, for its control, the membership functions used here increase or decrease monotonously. Let the i-th fuzzy variable for the j-th rule be $G_{ij}(X)$ for positive or negative value of X. Then, as an example, the membership functions are expressed as

$$G_{ij}(X) = a_{ij}X + b_{ij}, \quad -1 \leqq X \leqq 1. \tag{1}$$

where $0 \leqq a_{ij}X + b_{ij} \leqq 1,$
$i = 1, 2, \ldots, m, \quad j = 1, 2, \ldots, n.$

Coefficients $a_{ij}$ and $b_{ij}$ can be determined according to the shape of a course and the configuration and location of the robot.

Consider the j-th control rule :

$$R^j : \text{if } X_1 \text{ is } G_{1j}(X_1),\ X_2 \text{ is } G_{2j}(X_2),\ \ldots,\ \text{and } X_m \text{ is } G_{mj}(X_m), \text{ then } y \text{ is } Y_j(y), \tag{2}$$

where $X_1, X_2, \ldots, X_3$ designate the information

0-7803-0614-7/93$03.00 ©1993IEEE

on the state of a plant. A variable y designates an input to the plant. $Y_j(y)$ shows the fuzzy variables for y; $Y_j(y) = c_j y + d_j$, where $c_j$ and $d_j$ can be determined according to the properties of the steering mechanism and the dimensions of the robot.

For inputs to a fuzzy controller,

$$X_1 = X_1^0,\ X_2 = X_2^0,\ \ldots\ X_m = X_m^0 . \tag{3}$$

Suitableness for the j-th rule $R^j$ is

$$w_j = G_{1j}(X_1^0) \cap G_{2j}(X_2^0) \ \ldots \cap G_{mj}(X_m^0) = \bigcap_{i=1}^{m} G_{ij}(X_i^0) . \tag{4}$$

Next, in the consequence of the control rules, we obtain $y_j$ so that

$$w_j = Y_j(y_j). \tag{5}$$

$y_j (j=1,2,\ldots,n)$ can be written as

$$y_j = Y_j^{-1}(w_j),\quad j = 1,2,\ldots,n. \tag{6}$$

Finally, the input to the controller, $y^0$, can be determined as

$$y^0 = \sum_{j=1}^{n} w_j y_j \Big/ \sum_{j=1}^{n} w_j . \tag{7}$$

## III. FUZZY CONTROL RULES FOR THE FLUSH PARKING

For realizing travelling control of the autonomous mobile robot 「DREAM-1」, consider a robot moving reversely into a parking lot in a course as seen in Fig. 1. In the control for the flush parking, to evaluate the locations of the robot, four variables $X_1$, $X_2$, $X_3$ and $X_4$ are chosen as indicated in Fig. 1, where

- $X_1$: distance from the center of the left side of the robot to the right sided wall,
- $X_2$: distance from the center of the right side of the robot to the left sided wall,
- $X_3$: distance measurable by the supersonic sensor from the center of the rear and
- $X_4$: distance measurable by the supersonic sensor from the center of the front.

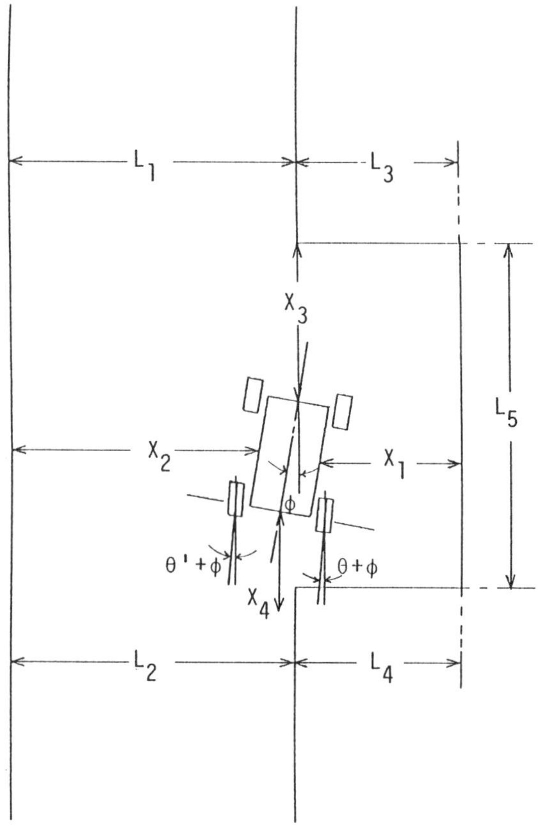


Fig. 1. Definition of the input variables for a flush parking.

Besides $\theta$ indicates the steering angle of the mobile robot, and five variables $L_1, L_2, \ldots,$ and $L_5$ indicate dimensions of the course.

To derive the fuzzy control rules, we considered three sorts of situation which are caused in the case of flush parking.

Situation 1
A stuation where the robot is moving outside the parking lot.

Situation 2
A stuation where the robot is moving forward in the parking lot.

Situation 3
A situation where the robot is moving reversely in the parking lot.

Fuzzy control rules can be easily derived dividing the situations into these three ones.

Figs. 2 (a), (b) and (c) illustrate these three situations respectively.

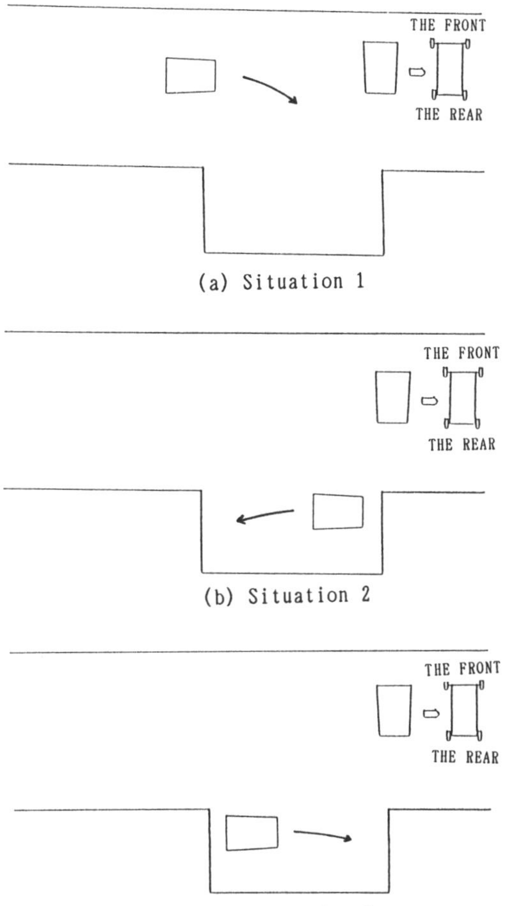


(a) Situation 1

(b) Situation 2

(c) Situation 3

Fig. 2. Three situations on controlling 「DREAM-1」 for a flush parking

The fuzzy reasoning rules for the flush parking can be expressed in linguistic form as follows:

Rules 1 (for the robot moving reversely along the right side wall on this side of the parking lot);

$R^{11}$: If $X_1$ is BIG and $X_3$ is BIG
then $\theta$ is to be LEFT,
$R^{12}$: If $X_1$ is MEDIUM and $X_3$ is SMALL
then $\theta$ is to be RIGHT,
$R^{13}$: If $X_1$ is SMALL and $X_3$ is SMALL
then $\theta$ is to be RIGHT,
$R^{14}$: If $X_3$ is BIG then $\theta$ is to be LEFT.

Rules 2 (for the robot moving forwardly in the parking lot);

$R^{21}$: If $X_1$ is SMALL and $X_3$ is BIG
then $\theta$ is to be LEFT,
$R^{22}$: If $X_1$ is SMALL and $X_3$ is SMALL
then $\theta$ is to be RIGHT.

Rules 3 (for the robot moving reversely in the parking lot);

$R^{31}$: If $X_1$ is SMALL and $X_3$ is BIG
then $\theta$ is to be LEFT,
$R^{32}$: If $X_1$ is SMALL and $X_3$ is SMALL
then $\theta$ is to be RIGHT.

The robot starts backward from the starting position (see Situation 1 in Fig. 2 (a)). The robot moving reversely can be controlled by the Rules 1 ($R^{11}$, $R^{12}$, $R^{13}$ and $R^{14}$). After the whole and a part of the body of the robot enter the parking lot, the robot moving forward can be controlled by Rules 2 ($R^{21}$ and $R^{22}$)(see Situation 2 in Fig. 2 (b)), and the one moving reversely can be controlled by Rules 3($R^{31}$ and $R^{32}$)(see Situation 3 in Fig. 2 (c)) [6] . One of these three rules are chosen properly in each situation.

The fuzzy inference is executed by using these control rules. Through the process of the defuzzification, steering angles of the robot can be computed as the weighted mean value discribed in section II [5] .

For tunning the membership functions to the movements of the robot, we simulated the travelling loci of the mobile robot. Results of the computer simulation for the flush parking are shown in Fig. 3. The robot is assumed to start from various positions in the course. The movements of the robot especialy for its reverse movements in the flush parking are well controlled by the fuzzy control rules.

## IV. TRAVELLING EXPERIMENT

### A. Autonomous Mobile Robot

We have designed and constructed an autonomous mobile robot with four wheels 「DREAM-1」 on an experimental basis, that was controlled by the fuzzy control theory. The outside view of 「DREAM-1」 is shown in Photo I.

The robot consists of three microcomputer systems for performing the fuzzy reasoning and determining its travelling direction, six supersonic distance meters for recognizing the location and position of the robot, a stepping motor for determining the steering angle of the front wheels and a direct-current motor for its rear wheel driving. Specification of 「DREAM-1」 is given in Table I.

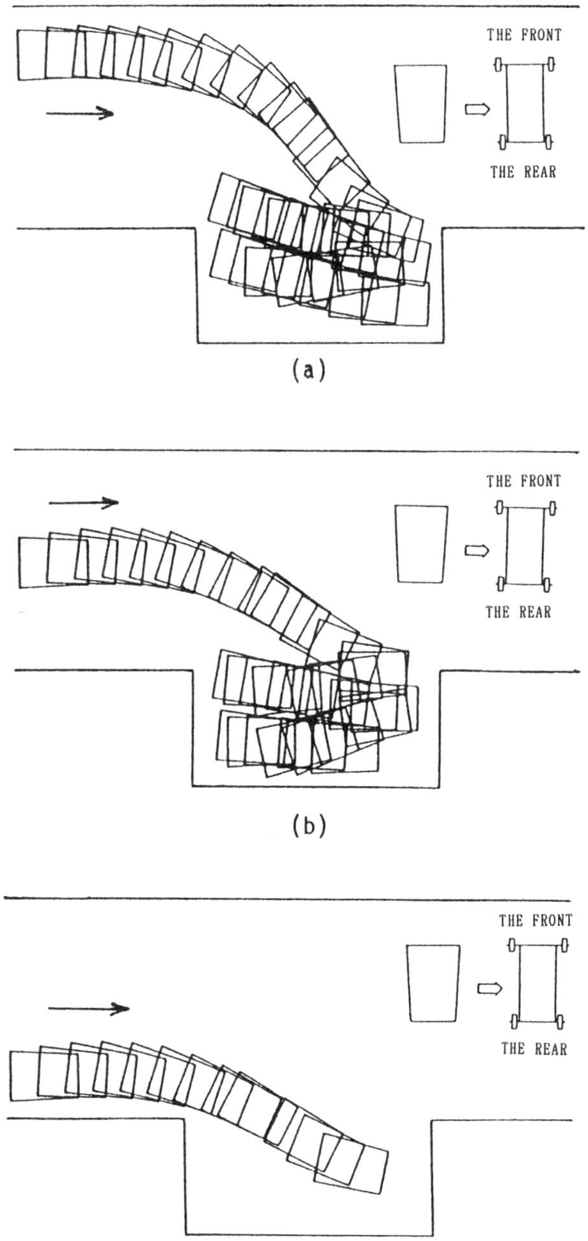


Fig. 3. Results of the computer simulation of a mobile robot for a flush parking. A robot starts (a) on the opposite side of a parking lot, (b) near the center of course, and (c) on this side of a parking lot.

Photo I. The outside view of 「DREAM-1」.

Table I. Specification of 「DREAM-1」.

| | |
|---|---|
| Size of the body | 700×470×470<br>Tread of front wheels 340 mm<br>Tread of rear wheels 300 mm<br>Wheelbase 355 mm |
| Weight of the body | 16 kg |
| Computation and Control | CPU···Z80(clock 2MHz)×3<br>ROM···32 KB ×3<br>RAM···32 KB ×3 |
| Sensors | Sensors with 40KHz supersonic transducers for measuring distance and direction<br>Infrared sensors for sending travelling instructions to the robot |
| Driving force | Front wheel control : Stepping motor Two-phases and 1.8 deg./step<br>Rear wheel driving : DC motor with a reduction gear and a differential gear |
| Power supply | Batteries : 12V4AH×2 |

### B. Procedure of the Experiment

We assume that, for the travelling of 「DREAM-1」, the dimensions and shape of a parking lot are determined as shown in Fig. 4.

Upon travelling in the course, shapes of the membership functions are adjusted until the robot can travel smoothly without colliding with the walls. The starting location is limited to the shaded area in Fig. 4.

For convenience of measuring distances, we use veneers with a height of 75 centimeters as the walls of the course.

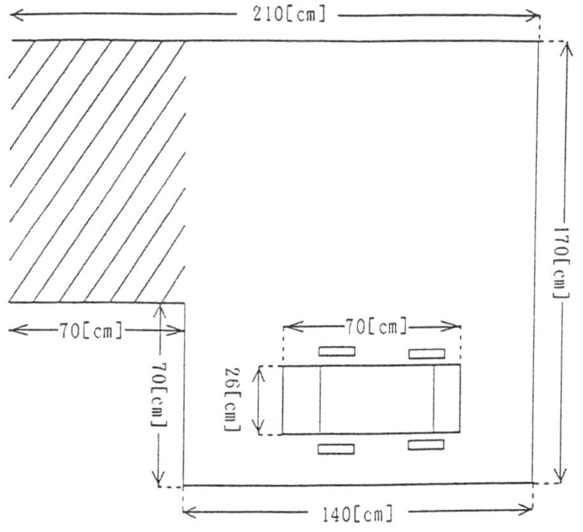


Fig. 4. Dimensions of the course and 「DREAM-1」

### C. Process of the Travelling Control

For the flush parking, the robot starts backward outside the parking lot and measures the distances between the robot itself and walls with the aid of the supersonic transducers.

Based on the distance information, the robot judges whether the robot collides with the walls or not. After the robot recognizes that it does not collide with the walls, the robot begins to recognize the environment in the course: i.e., the robot computes the premises in the corresponding rules and recognizes the position and the attitude of the robot. The steering angle will be decided through the process of the defuzzification of the inference and the stepping motor is driven and controlled according to the magnitude of the steering angle. When the robot travels in the parking lot and gets close to the wall up to about 25 centimeters, the robot halts and goes backward, reversing the steering angle. These process is detailed in a flowchart of Fig. 5.

As a result, the robot could enter the parking lot and halted at the desired destination smoothly though the robot repeated the forward and reverse movements ( approximately four times ) within the parking lot.

## V. DISCUSSION

When the robot enters the parking lot backward and gets close to the wall, the reasoning rules are turned to $R^{21}$ and $R^{22}$ from $R^{11}$-$R^{14}$ following the forward movement of the robot. After its movement, the robot gets close to the opposite wall and then the rules are changed from $R^{21}$ and $R^{22}$ to $R^{31}$ and $R^{32}$ for the reverse movement. Such movements of reverse and forward travelling are repeated until the robot reaches the desired destination.

In this experiment, travelling characteristics of the mobile robot are limited to a low speed. For speedy travelling, the centrifugal force working on the robot itself and slip of the wheels have to be considered.

When the robot enters the parking lot backward and leans largely toward the wall, the supersonic transducers cannot measure distances between the robot itself and the walls, because of the charateristics of the transducers. In other words, there is a dead space so as to be seen in the case of driving a car.

In these situations, the robot cannot recognize the environment so that the robot cannot travel safely in the course.

In our present travelling experiment, we adjust shapes of membership functions so that the angles of the robot with the walls do not become too big, i.e., less than approximately 20 degrees. In the next stage of our development, the dead space caused by a directivity of the supersonic transducers which are installed in our mobile robot is required to be decreased.

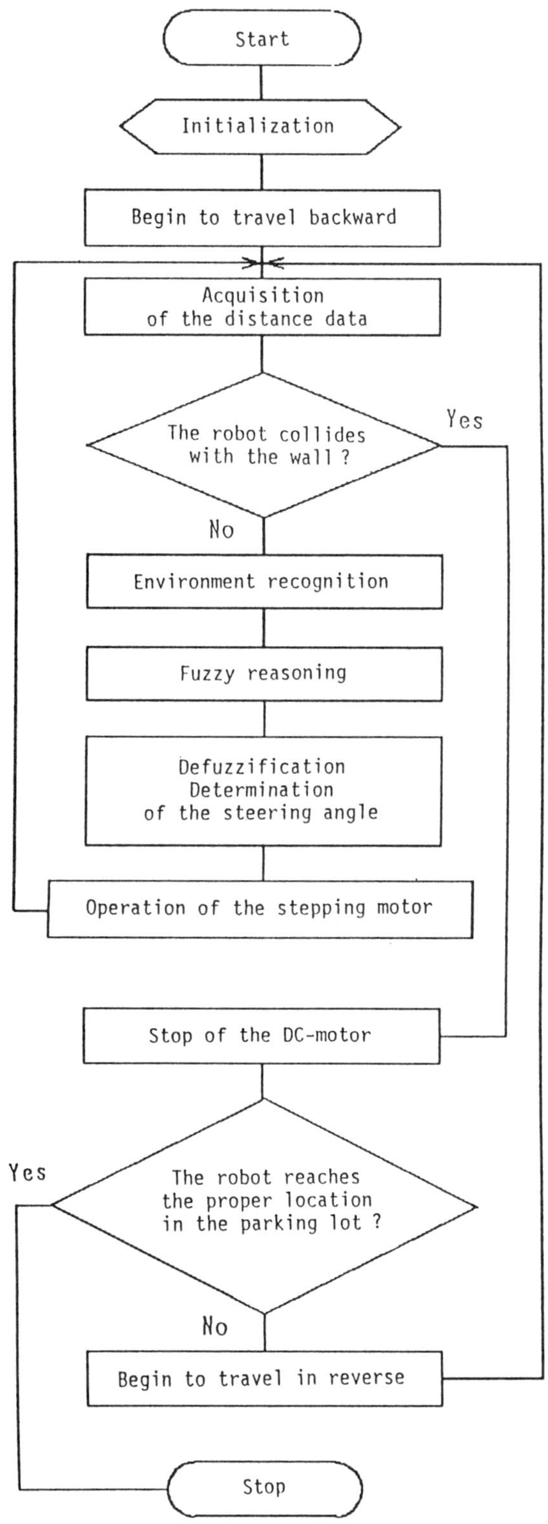


Fig. 5. Process of travelling control of 「DREAM-1」 for the flush parking.

## VI. CONCLUSIONS

Upon making an autonomous mobile robot to implement a flush parking, we computed the travelling locus of the mobile robot by the fuzzy control theory. The flush parking of the mobile robot can be controlled by combining simple rules of the fuzzy reasoning. In such a case, the reverse movement of the robot can be controlled smoothly by using simple fuzzy variables.

Based on the fuzzy control rules obtained in the above computer simulation, our mobile robot 「DREAM-1」 was well controlled for its flush parking.

## REFERENCES

[1] Nikkei Electronics, Fuzzy theory has been practized, Nikkei Electronics(Japan), No.426, pp.129-152, July 27, 1987.
[2] L.A.Zadeh, " Birth and evolution of fuzzy logic - Expectations of Japan's role- " J. of Japan Soc. for Fuzzy and Systems, Vol.2 No.2, 1990, pp.182-195.
[3] M.Sugeno and M.Nishida, " Fuzzy control of model car, " Fuzzy Set and Systems, Vol.16 1985, pp.103-113.
[4] M.Ohkita, H.Miyata, S.Maeda, T.Kamii and Y.Kobayashi, "A prototype of travelling robot controlled by the fuzzy theory, " Reports of Faculty of Engng., Tottori Univ. Vo.21, No.1, 1990, pp.75-90.
[5] M.Sugeno, Fuzzy Control, Nikkei-Kogyo Shinbun(Japan), 1988, p.84-87.
[6] M.Ohkita, H.Miyata, T.Ohishi, M.Kanezaki and Y.Takeda, "Control of a mobile robot by the fuzzy theory, " Proc.of the 13th IMACS World Congress on Computation and Applied Mathematics ( July 22-26, Trinity College Dublin, Ireland), vol.3, 1991, pp.1217-1219.

# Resolved Motion Rate Control of Redundant Robots using Fuzzy Logic

Sung-Woo Kim and Ju-Jang Lee
Department of Electrical Engineering
Korea Advanced Institute of Science and Technology
373–1 Kusong-Dong, Yusong-Gu, Taejon 305–701, Korea

*Abstract*— **The resolved motion rate control (RMRC) is used to map the Cartesian space trajectory to the Joint space trajectory. The RMRC for the redundant robot requires the pseudo-inverse of Jacobian matrix. However the pseudo-inverse is not easy to be implemented on a digital computer in real time as well as mathematically complex. In this paper, a simple fuzzy resolved motion rate control (FRMRC) that can replace the RMRC using pseudo-inverse of Jacobian is proposed. The proposed FRMRC with appropriate fuzzy rules, membership functions and reasoning method can solve the mapping problem between the spaces without complexity. Moreover we will show that this FRMRC is robust even when passing the singularity points. The mapped Joint space trajectory can be directly used to control redundant manipulators. The simulation results verify the efficiency of the proposed method.**

## I. Introduction

Conventional robot controllers are designed in the Joint space. However the desired trajectory is usually given in the Cartesian space, so it is necessary to convert between two spaces. This is generally known as kinematics when the Joint space is converted to the Cartesian space and inverse kinematics when the Cartesian space to the Joint space. Because of the high nonlinearity of a robot mechanism, the closed form solution of inverse kinematics is coupled and very complex. Instead of solving kinematics and inverse kinematics, conversion between two spaces can be done by using the differential relationship, so-called Jacobian. The Jacobian matrix determines the relationship between the change of the linear position and the change of the angular position. Using this property, one can find the current position by integrating the change of the position. These can be represented mathematically as follows.

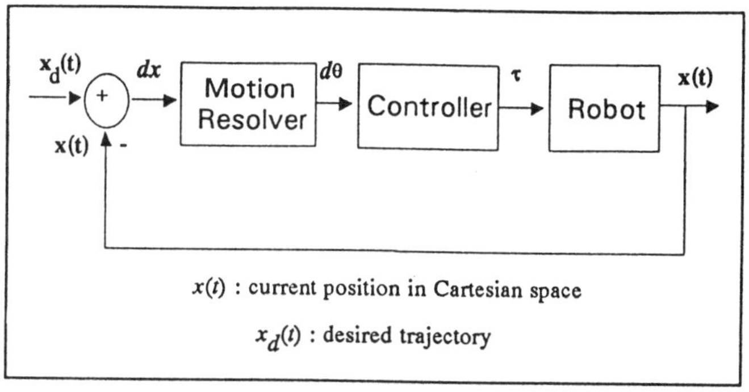


Fig. 1: Robot controller containing the motion resolver

Generally, the kinematic relationship between the Cartesian space and the Joint space is given by

$$x = f(\theta). \tag{1}$$

Their differential relationships are

$$dx = J(\theta)d\theta \tag{2}$$

where $J = \partial f/\partial x \in \Re^{m\times n}$ is the Jacobian matrix. Computing $d\theta$ by solving the linear equation of (2) for the given $\theta$ and $dx$ is proposed by Whitney as the *resolved motion rate control* [1]. Fig. 1 shows the block diagram of the robot control containing the motion resolver.

The general solution of (2) using the pseudo-inverse is obtained as follows.

$$d\theta = J^{\#}(\theta)dx + [I_n - J^{\#}(\theta)J(\theta)]\cdot y \tag{3}$$

where $J^{\#}(\theta) \in \Re^{n\times m}$ is the pseudo inverse of $J(\theta)$, $y \in \Re^n$ is an arbitrary vector and $I_n$ is an identity matrix. If the exact solution does not exist, (3) covers all the least-squares solutions that minimize $\| dx - J(\theta)d\theta \|$.

Even though the general solution of (2) is given by (3), finding out the pseudo-inverse of the Jacobian matrix is not easy. It needs much more calculation than the conventional matrix inverse. Thus the resolved motion rate



0-7803-0614-7/93$03.00 

control using equation (3) has difficulties to implement on a digital computer in real time. To overcome the above drawbacks, a simple algorithm is proposed to replace the pseudo-inverse of the Jacobian matrix. In other words, by solving equation (2) using the fuzzy logic, one can build a simple and fast calculating resolved motion rate controller.

In this paper, a fuzzy resolved motion rate controller (FRMRC) which converts the Joint space to the Cartesian space is proposed. It will be shown that conversion between two spaces can be done without solving the inverse kinematics nor the pseudo-inverse of Jacobian matrix. Moreover, by choosing the fuzzy variables and rules as small as possible, its structure is simplified. We extend this FRMRC for the n-DOF redundant manipulator by introducing the hierarchical structure. Next, to track the generated trajectory from FRMRC, we use the control law, *computed torque method*. Finally the simulation results will verify that the proposed scheme is sufficiently accurate for the control and it is singularity-robust.

## II. Fuzzy Resolved Motion Rate Controller

The basic structure of the fuzzy resolved motion rate controller (FRMRC) for redundant manipulators is like that the inverse Jacobian is replaced by the fuzzy reasoning. In other words, instead of calculating the pseudo-inverse of the Jacobian matrix, we define fuzzy rules, membership functions, fuzzy reasoning method, and universe of discourse (Fig. 2).

Consider a 3-DOF kinematically redundant manipulator (Fig. 3). Let $C$ and $S$ denote $\cos\theta$ and $\sin\theta$, respectively. We get the end-position from kinematics,

$$\begin{bmatrix} x \\ y \end{bmatrix} = \begin{bmatrix} \ell_1 C_1 + \ell_2 C_{12} + \ell_3 C_{123} \\ \ell_1 S_1 + \ell_2 S_{12} + \ell_3 S_{123} \end{bmatrix}. \quad (4)$$

The differential relationship is

$$\begin{bmatrix} dx \\ dy \end{bmatrix} = \begin{bmatrix} J_{11} & J_{12} & J_{13} \\ J_{21} & J_{22} & J_{23} \end{bmatrix} \begin{bmatrix} d\theta_1 \\ d\theta_2 \\ d\theta_3 \end{bmatrix} \quad (5)$$

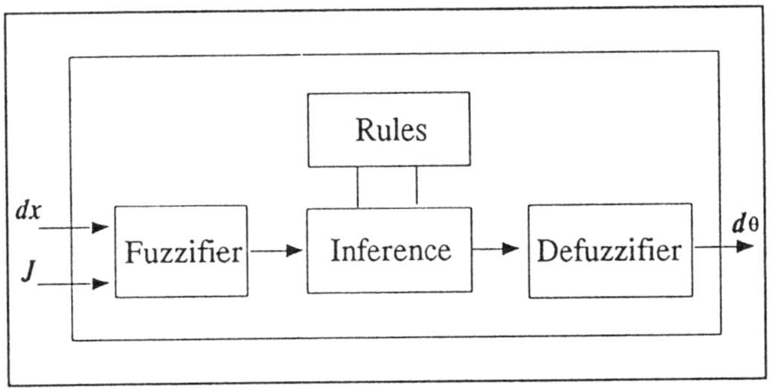


Fig. 2: Basic structure of FRMRC

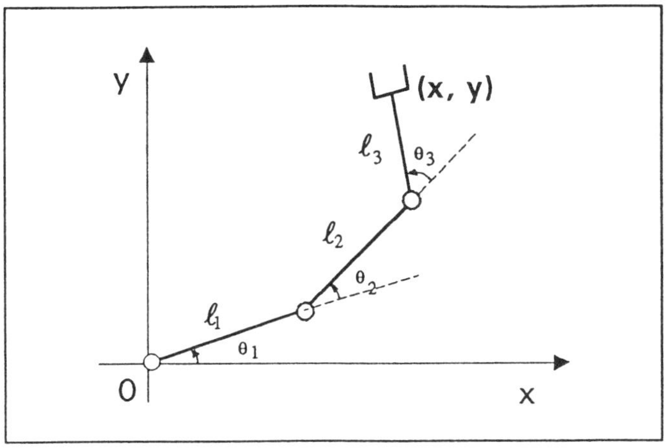


Fig. 3: 3-DOF redundant manipulator

$$= \begin{bmatrix} J_{11}d\theta_1 + J_{12}d\theta_2 + J_{13}d\theta_3 \\ J_{21}d\theta_1 + J_{22}d\theta_2 + J_{23}d\theta_3 \end{bmatrix}$$

where

$$\begin{aligned} J_{11} &= -\ell_1 S_1 - \ell_2 S_{12} - \ell_3 S_{123} \quad (6) \\ J_{12} &= -\ell_2 S_{12} - \ell_3 S_{123} \\ J_{13} &= -\ell_3 S_{123} \\ J_{21} &= \ell_1 C_1 + \ell_2 C_{12} + \ell_3 C_{123} \\ J_{22} &= \ell_2 C_{12} + \ell_3 C_{123} \\ J_{23} &= \ell_3 C_{123} \end{aligned}$$

From equation (5), $dx$ and $dy$ can be considered as superposition of $J_{1i}d\theta_i$ and $J_{2i}d\theta_i$, respectively, for $i = 1, 2, 3$. We can determine $d\theta_i$ by splitting equations (5) into three sub-equations of (7), (11) and (13).

First, $d\theta_1$ is calculated from the following equation

$$\begin{bmatrix} dx \\ dy \end{bmatrix} = \begin{bmatrix} J_{11}d\theta_1 \\ J_{21}d\theta_1 \end{bmatrix}. \quad (7)$$

We can see such facts that from (7) that if $dx$ is positive and $J_{11}$ is positive then $d\theta_1$ should be positive, and if $dx$ is negative and $J_{11}$ is positive then $d\theta_1$ should be negative. These facts are investigated more formally by introducing the linguistic variables and the fuzzy logic. Define $dx, J_{11}, d\theta_1$ as following linguistic variables

$$\begin{aligned} dx, d\theta_1 &= \{N, Z, P\} \quad (8) \\ J_{11} &= \{N, P\} \end{aligned}$$

where $N, P$ and $Z$ denote *Negative*, *Positive* and *Zero*, respectively. The membership functions corresponding the fuzzy variables are shown in Fig. 4. Based on the

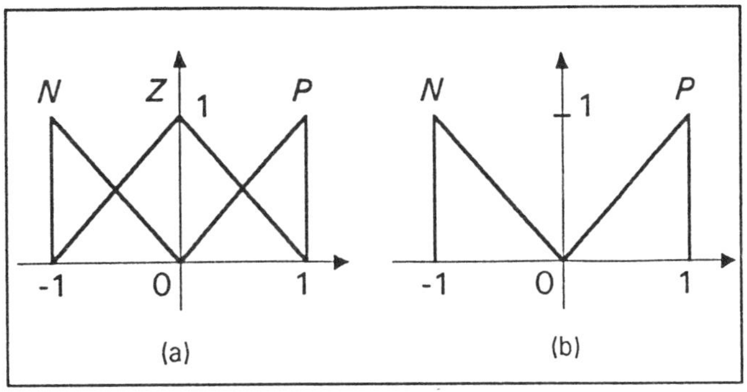


(a) Membership functions of $dx$ and $d\theta$ ,

(b) Membership functions of $J$

Fig. 4: Membership functions

fuzzy variables and membership functions of (7), we define five fuzzy rules,

| | |
|---|---|
| **Rule 1** | If $J_{11}$ is *Negative* and $dx$ is *Negative*, then $d\theta_1$ is *Positive*. |
| **Rule 2** | If $J_{11}$ is *Positive* and $dx$ is *Negative*, then $d\theta_1$ is *Negative*. |
| **Rule 3** | If $J_{11}$ is *Negative* and $dx$ is *Positive*, then $d\theta_1$ is *Negative*. |
| **Rule 4** | If $J_{11}$ is *Positive* and $dx$ is *Positive*, then $d\theta_1$ is *Positive*. |
| **Rule 5** | If $J_{11}$ is *Zero*, then $d\theta_1$ is *Zero*. |

For the case $dy = J_{21}d\theta_1$ of (7), the rules are same except that $dy$ and $J_{21}$ replace $dx$ and $J_{11}$, respectively. In order to determine one variable $d\theta_1$ from two equations $dx = J_{11}d\theta_1$ and $dy = J_{21}d\theta_2$, we introduce fuzzy $OR$ logic. There are a lot of candidates of fuzzy $OR(x, y)$, for example,

$$\begin{aligned} \text{Zadhe's } OR &= MAX(x,y) \qquad (9) \\ \text{Lukasiewicz's } OR &= MIN(x+y,1). \end{aligned}$$

In this paper we prefer Zadhe's. Thus equation (7) can be merged into one equation as follows

$$(dx, J_{11})\ OR\ (dy, J_{21}) \rightarrow d\theta_1. \tag{10}$$

Next, $d\theta_2$ is calculated from (11) that is the second sub-equation of (5).

$$\begin{bmatrix} dx' \\ dy' \end{bmatrix} = \begin{bmatrix} dx - J_{11}d\theta_1 \\ dy - J_{21}d\theta_1 \end{bmatrix} = \begin{bmatrix} J_{12}d\theta_2 \\ J_{22}d\theta_2 \end{bmatrix}. \tag{11}$$

In analogy with the previous case of $d\theta_1$ , the linguistic variables, membership functions and fuzzy rules are defined for (11), and $d\theta_2$ is inferred as follows

$$(dx', J_{12})\ OR\ (dy', J_{22}) \rightarrow d\theta_2. \tag{12}$$

Finally, $d\theta_3$ is calculated from (13), the third sub-equation of (5).

$$\begin{bmatrix} dx'' \\ dy'' \end{bmatrix} = \begin{bmatrix} dx' - J_{12}d\theta_2 \\ dy' - J_{22}d\theta_2 \end{bmatrix} = \begin{bmatrix} J_{13}d\theta_3 \\ J_{23}d\theta_3 \end{bmatrix}. \tag{13}$$

and

$$(dx'', J_{13})\ OR\ (dy'', J_{23}) \rightarrow d\theta_3. \tag{14}$$

By splitting (5) into (7), (11) and (13), we use the fuzzy rules, membership functions and reasoning method derived for equation (7) without modification. Generally, we can use the same rules and membership functions for the n-DOF manipulator by considering the sub-equations from the first link to the end link hierarchically.

| **Hierarchical FRMRC algorithm** |
|---|
| $dx = x_d(t) - x(t)$ |
| $dy = y_d(t) - y(t)$ |
| Do while $1 \le k \le n$ |
| $\quad (dx, J_{1k})\ OR\ (dy, J_{2k}) \rightarrow d\theta_k$ |
| $\quad dx = dx - J_{1k} \cdot d\theta_k$ |
| $\quad dy = dy - J_{2k} \cdot d\theta_k$ |
| $\quad k = k + 1$ |
| EndDo |

## III. Controller Design

The dynamic equations of the 3-DOF redundant robot with the joint torque, $\tau = [\tau_1\ \tau_2\ \tau_3]^T$ is

$$\begin{aligned} &\begin{bmatrix} M_{11} & M_{12} & M_{13} \\ M_{21} & M_{22} & M_{23} \\ M_{31} & M_{32} & M_{33} \end{bmatrix} \begin{bmatrix} \ddot{\theta}_1 \\ \ddot{\theta}_2 \\ \ddot{\theta}_3 \end{bmatrix} \\ + &\begin{bmatrix} V_{11} & V_{12} & V_{13} \\ V_{21} & V_{22} & V_{23} \\ V_{31} & V_{32} & 0 \end{bmatrix} \begin{bmatrix} (\dot{\theta}_1)^2 \\ (\dot{\theta}_1 + \dot{\theta}_2)^2 \\ (\dot{\theta}_1 + \dot{\theta}_2 + \dot{\theta}_3)^2 \end{bmatrix} \\ + &\begin{bmatrix} G_1 \\ G_2 \\ G_3 \end{bmatrix} = \begin{bmatrix} \tau_1 \\ \tau_2 \\ \tau_3 \end{bmatrix} \end{aligned} \tag{15}$$

where

$$\begin{aligned} M_{11} &= (\frac{1}{3}m_1 + m_2 + m_3)\ell_1^2 + (\frac{1}{3}m_2 + m_3)\ell_2^2 + \frac{1}{3}m_3\ell_3^2 \\ &+ (m_2 + 2m_3)\ell_1\ell_2C_2 + m_3\ell_1\ell_3C_{23} + m_3\ell_2\ell_3C_3 \\ M_{12} &= (\frac{1}{3}m_2 + m_3)\ell_2^2 + \frac{1}{3}m_3\ell_3^2 \\ &+ (\frac{1}{2}m_2 + m_3)\ell_1\ell_2C_2 + \frac{1}{2}m_3\ell_1\ell_3C_{23} + m_3\ell_2\ell_3C_3 \end{aligned}$$

$$
\begin{aligned}
M_{13} &= \frac{1}{3}m_3\ell_3^2 + \frac{1}{2}m_3\ell_1\ell_3C_{23} + \frac{1}{2}m_3\ell_2\ell_3C_3 \\
M_{22} &= (\frac{1}{3}m_2 + m_3)\ell_2^2 + \frac{1}{3}m_3\ell_3^2 + m_3\ell_2\ell_3C_3 \\
M_{23} &= \frac{1}{3}m_3\ell_3^2 + \frac{1}{2}m_3\ell_2\ell_3C_3 \\
M_{33} &= \frac{1}{3}m_3\ell_3^2 \\
V_{11} &= (\frac{1}{2}m_2 + m_3)\ell_1\ell_2S_2 + \frac{1}{2}m_3\ell_1\ell_3S_{23} \\
V_{12} &= -(\frac{1}{2}m_2 + m_3)\ell_1\ell_2S_2 + \frac{1}{2}m_3\ell_1\ell_3S_3 \\
V_{13} &= -\frac{1}{2}m_3\ell_1\ell_3S_{23} - \frac{1}{2}m_3\ell_2\ell_3S_3 \\
V_{21} &= V_{11} \\
V_{22} &= \frac{1}{2}m_3\ell_2\ell_3S_3 = -V_{23} \\
V_{31} &= \frac{1}{2}m_3\ell_1\ell_3S_{23} \\
V_{32} &= \frac{1}{2}m_3\ell_2\ell_3S_{23} \\
G_1 &= \frac{1}{2}\ell_1C_1m_1g + (\ell_1C_1 + \frac{1}{2}\ell_2C_{12})m_2g \\
&+ (\ell_1C_1 + \ell_2C_{12}\frac{1}{2}\ell_3C_{123})m_3g \\
G_2 &= \frac{1}{2}\ell_2C_{12}m_2g + (\ell_2C_{12}\frac{1}{2}\ell_3C_{123})m_3g \\
G_3 &= (\frac{1}{2}\ell_3C_{123})m_3g
\end{aligned}
$$

The above dynamics can be compactly expressed by $M(\theta)\ddot{\theta} + C(\theta,\dot{\theta}) + G(\theta) = \tau$. To achieve tracking control tasks, we use the following control law, so-called computed torque method,

$$\tau = M(\theta)u + C(\theta,\dot{\theta}) + G(\theta) \tag{16}$$

where

$$u = \ddot{\theta}_d - 2\lambda\dot{\tilde{\theta}} - \lambda^2\tilde{\theta} \tag{17}$$

with $u = [u_1\ u_2\ u_3]^T$ being the equivalent input, $\tilde{\theta} = \theta - \theta_d$ being the position tracking error and $\lambda$ being a positive constant. The desired $\theta_d$ is given by the fuzzy resolved motion rate control. Then the computed torque control makes the tracking error go to zero, since $\ddot{\tilde{\theta}} + 2\lambda\dot{\tilde{\theta}} + \lambda^2\tilde{\theta} = 0$ converges to zero exponentially.

## IV. Simulation and Results

As derived previous Section, simulate the proposed algorithm for a 3-DOF manipulator. Its link lengths are 0.5, 0.5 and 0.4 meters, respectively and the masses of each link are 1.0, 1.0 and 0.8 kilograms, respectively. The gravity force is applied in y-direction.

The concerning motion trajectories in the Cartesian space are circular, linear, and passing a singularity point that is located at the workspace boundary. For linear motion, its trajectory is generated by a cubic polynomial [7].

Fig. 5 shows the circular motion using FRMRC. The manipulator travels the circle in 10 sec. We see that the errors are sufficiently small (order of milli-meter) and the torques for controlling each link are moderate.

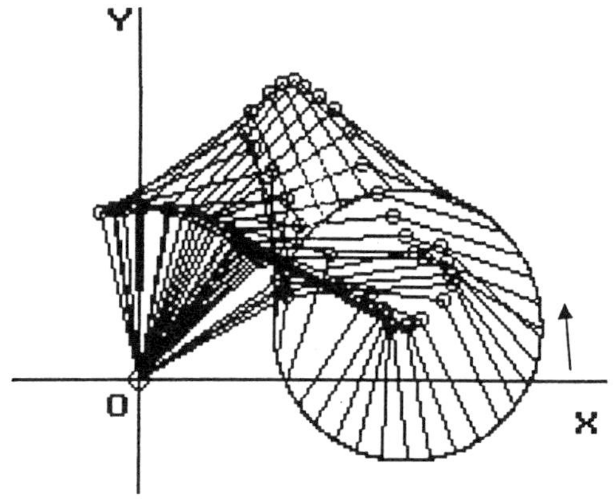


(a) Snapshots of circular motion

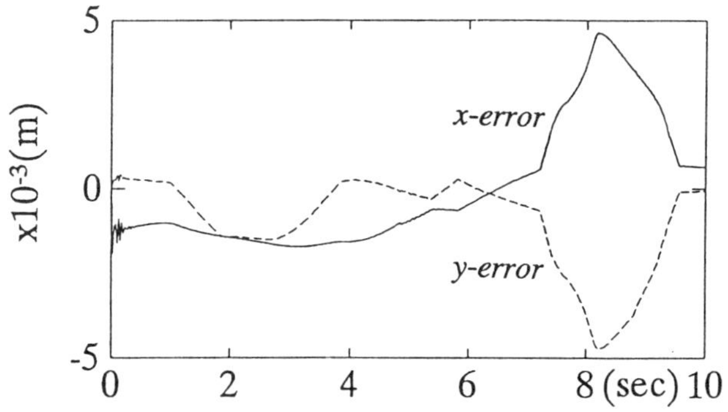


(b) Tracking errors

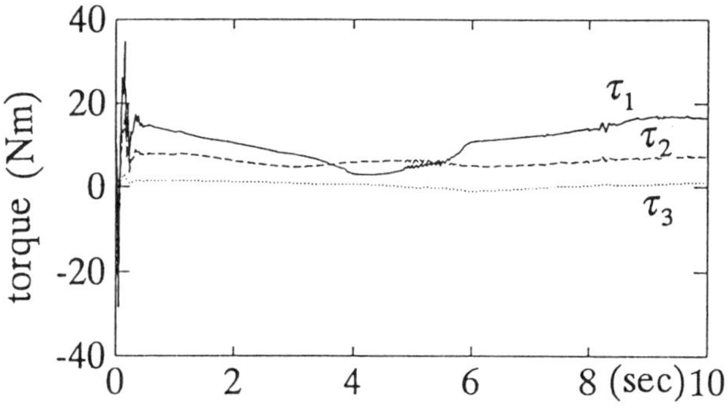


(c) Joint torques

Fig. 5: Tracking of circular trajectory using FRMRC

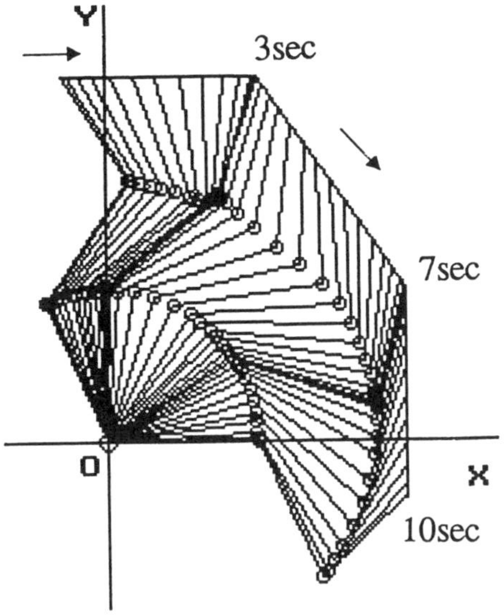


(a) Snapshots of linear motion

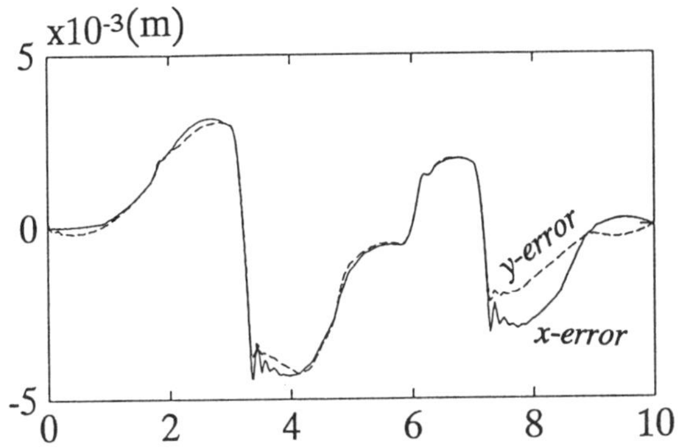


(b) Tracking errors

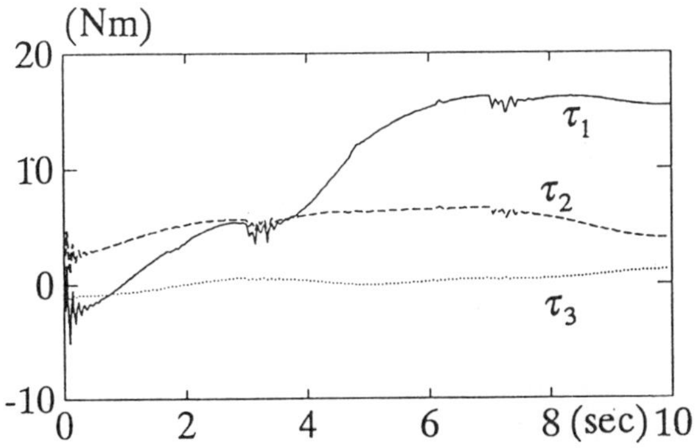


(c) Joint torques

Fig. 6: Tracking of linear trajectory using FRMRC

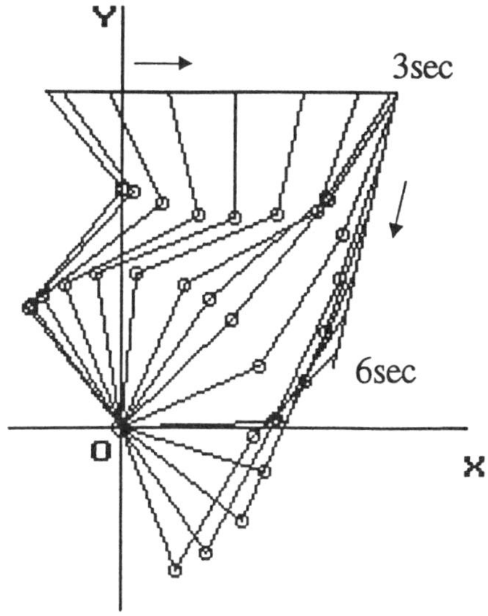


(a) Snapshots of motion

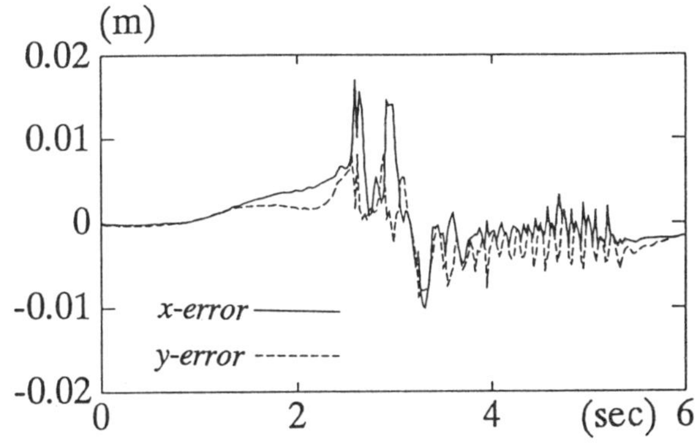


(b) Tracking errors using FRMRC

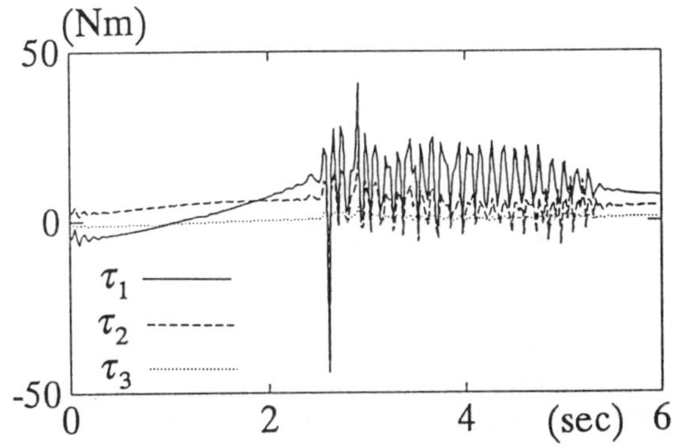


(c) Joint torques using FRMRC

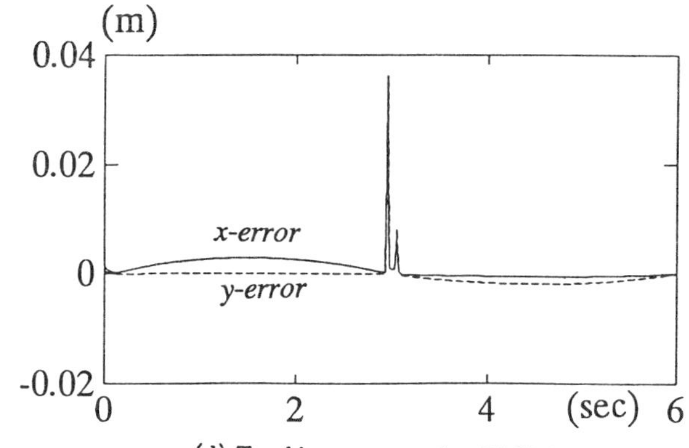

(d) Tracking errors using RMRC

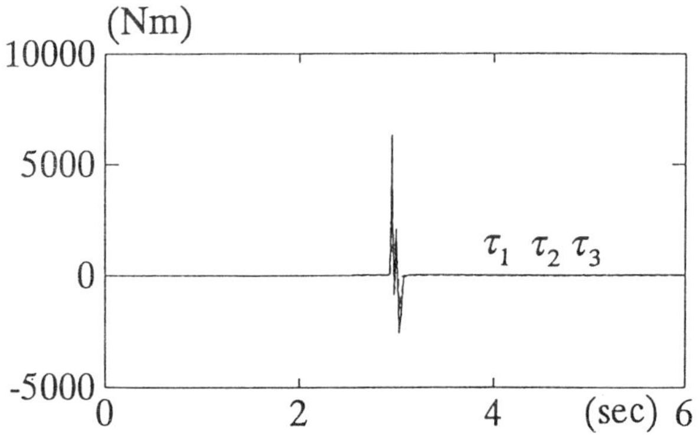

(e) Joint torques using RMRC

Fig. 7: Comparison of FRMRC and RMRC when passing a singularity point

Fig. 6 shows the linear motion. the direction of the trajectory changes at 3 and 7 second. There are small errors while the manipulator is moving, but the error goes to zeros at the final position.

Fig. 7 shows a case of passing a singularity point. The errors and the torques of RMRC using pseudo-inverse scheme and FRMRC are compared. We use Greville's algorithm [2] for calculating the pseudo-inverse. For RMRC using pseudo-inverse, the torques blow up to infinity at the singularity point, but for FRMRC, the torques are within moderate ranges with some oscillation. We see that FRMRC has smaller errors than RMRC at the singularity point.

## V. Conclusion

Since the redundant robot manipulator has extra degree-of-freedoms, its Jacobian matrix is not square. Therefore the resolved motion rate control of the robot that solves the mapping problem of the Cartesian space and the Joint space, uses so called a pseudo-inverse. The pseudo-inverse is an extension of conventional square matrix inverse and the calculation of the pseudo-inverse is much more complex than the inverse of a square one.

In this paper, we have proposed a simple algorithm using the fuzzy logic for the resolved motion rate control, and have showed that this can replace the RMRC using the pseudo-inverse of the Jacobian matrix. This FRMRC has advantages as follow. First, complexity and computation burdens are excluded, since it is sufficient to use only 5 rules and 2 (or 3) membership functions for each fuzzy variable in stead of calculating the complex pseudo-inverse. This algorithm can be applied to not only a 3-DOF manipulator but also multi-DOF robots using the hierarchical FRMRC. If the fuzzy reasoning method is realized by a look-up table, then the FRMRC can be implemented in real time by a digital computer. These advantages are summarized as simplicity and rapidness. For further studies, we try to extend the FRMRC to obstacle avoidance.

## References

[1] D. E. Whitney, "Resolved Motion Rate Control of Manipulators and Human Prostheses," *IEEE Trans. Man, Machine Sys.* MMS-10 (2), pp. 47-53

[2] Y. Nakamura, *Advanced Robotics – Redundancy and Optimization*, Addison-Wesley, 1991

[3] A. Nedungadi, "A Fuzzy Robot Controller – Hardware Implementation," *IEEE International Conf. on Fuzzy Sys.*, 1992, pp. 1325-1331

[4] S.-W. Kim and J.-J. Lee, "Fuzzy Logic Control for a Redundant Manipulator – Resolved Motion Rate Control," *Proceedings of 1992 Korean Automatic Control Conf. (KACC) - international*, pp. 479-484, 1992

[5] M. Sugeno, *Industrial Application of Fuzzy Control*, Elesevier Science Publishers, 1985

[6] C. C. Lee, "Fuzzy Logic in Control Systems: Fuzzy Logic Controller – Part I and II," *IEEE Trans. on Sys. Man and Cyb.* Vol. 20, No. 2, pp.404-435, March and April, 1990.

[7] J. J. Craig, *Introduction to Robotics - Mechanics and control*, second ed., Addison-Wesley, 1989.

[8] A. J. Koivo, *Fundamentals for Control of Robotic Manipulators*, John-Wiley & Sons, 1989.

# Twenty years of Fuzzy Control: Experiences Gained and Lessons Learnt

E. H. Mamdani
Dept. of Electronic Engineering
Queen Mary & Westfield College
Mile End Road
London E1 4NS
UK

***Abstract*—This paper is based on 20 years of involvement with Fuzzy control since its first inception. The main lesson one learns from this involvement is that process control can now be considered a multi-paradigm discipline, involving not just the dynamic system theory but also decision-making approaches gleaned from the field of artificial intelligence. Experience has shown that such approaches may provide better ways of implementing controllers than using conventional approaches based on dynamic system theory.**

## I. Introduction

This paper examines fuzzy control from its inception to its recent wide-spread acceptance in industrial use in order to reflect upon matters such as: its earlier dismissal, the reasons for its appeal, the nature of scientific research and engineering innovations. This paper only deals with fuzzy control and not other applications of fuzzy sets. Nor is the intention here to provide a retrospective look at fuzzy control. The fuzzy control area – for it is a whole area judging by the explosion of recent research into fuzzy control and neural networks, self-organising control, genetic algorithms and so on – is still continuing to expand with new results appearing regularly. (Such is the advance of this new work that any list of references to it runs the risk of errors caused by omitting important works of major significance.) Nevertheless, it is worth learning some lessons after some 20 years since the idea first appeared.

The dominant position of analytic control theory prevented fuzzy control from being taken seriously until its increasing application in Japan. Many of the arguments against fuzzy control were framed in the language of that dominant theory and also countered in that language. This paper argues that fuzzy control and allied other techniques such as self-organising fuzzy control, neural networks, genetic algorithms and so on, provide an alternate paradigm to the analytic control theory. This paradigm consists of non-analytic approaches to control and is based on decision-making approaches from artificial intelligence.

## II. A Brief Historical Look

Fuzzy control [1] came out of a doctoral research work that was concerned with the application of learning techniques to process control. At that time learning and adaptive control systems were well established research topics. There were already many results in these areas within the established control systems theory. Our work, however, was drawing on the emerging learning techniques from the field of artificial intelligence and pattern recognition: rule–based systems, protocol analysis, neural networks and statistical pattern recognition. In a sense this was one of the first studies to investigate the application of AI to control. This is nowadays sometimes referred to as intelligent control but this paper proposes the alternative reference to it as non-analytic control for reasons that will be given in the sequel.

The main part of the study concentrated upon the use of a supervised Bayesian learning system for learning the state–action statistics to control a laboratory prototype steam engine from the control behaviour of a human operator. The state of the steam engine (the boiler pressure and the engine speed) was periodically displayed to the human operator on a teletype. The human responded to this by typing the changes in the inputs (the boiler heat and the engine throttle) he wanted made and this action was implemented by the computer. At the same time the displayed state and the human action were noted by the Bayesian learning system algorithm which proceeded to learn the state–action statistics. After the learning control algorithm had converged, it was used on its own to control the steam engine. We noted that its behaviour was consistently inferior to that of the human. As

0-7803-0614-7/93$03.00 ©1993IEEE

experimenters our feeling was that we could tell a machine better in more abstract terms how to behave than just showing how as operators we behaved at each instance on observing the state of the engine.

The above should not be read to mean that learning controllers can never perform well. There is no doubt that we could have persevered and improved the behaviour of the learning controller. However, it was felt that a rule–based approach such as had been demonstrated in studies like DENDRAL [2] would be more effective. Indeed this was also tried, but the LISP systems of that period were primitive and slow. It was Zadeh's paper [3] published at that time which persuaded us to use a fuzzy rule–based approach. Between reading and understanding Zadeh's paper and having a working controller took a mere week and it was "surprising" how easy it was to design a rule–based controller. Other workers also have commented on the "surprising" ease with which a fuzzy controller can be implemented. This surprise reflects that results are different from what conventional wisdom would lead one to expect. That in turn begs the question as to what that conventional wisdom is. The answer seems to be that conventional control theory and the way it is taught suggests that controller design requires a good analytical treatment and any other technique such as one borrowed from AI would need to be sufficiently complex and involved to match the apparent sophistication involved in that analytical theory. Therefore, the use of a few simple fuzzy control rules is counter to that intuition.

On reflection it becomes clear that our approach to control was based on a different view of control (that based on the AI paradigm) than the conventional view of control (that based on the analytical control theory). The latter was at that time a very successful discipline and thus all powerful. So much so that any discussion on the problem of control could only be expressed in the terminology of that established paradigm. This culture clash was at the very heart of the early criticisms of fuzzy control. With the hindsight of twenty years and the industrial success of fuzzy control with us we can examine the arguments that took place then.

## III. THE CULTURE CLASH OF TWO PARADIGMS [4] [5]

It was the insight by research workers during the '40s and the '50s that many dynamical systems can be mathematically modelled using differential equations which led to the foundations of control theory. With the addition of transform theory, this provided an extremely powerful means of studying (i.e. both analysing and designing) control systems. First as the theory of servo-mechanisms and later as control systems theory, the discipline blossomed and continued to do so until the '70s. These were such heady days that often the area was called simply systems theory to indicate its definitiveness. The theory provided many powerful insights about control systems and although not all of these needed to be expressed using the body of mathematics of the area, it became mandatory to do so in keeping with definitive high ground that the paradigm soon staked for itself. Unfortunately, in too many instances and for a variety of reasons this approach could not be sustained because many systems are simply not amenable to the full force of mathematical analysis as dictated by the control theory.

### *A. The limitations of control systems theory*[1]

There were attempts to apply control theory to other systems, that is systems whose behaviour cannot be derived from first principles of physics and described as differential equations: systems such as management systems, economic systems even telecommunications systems. Such systems have many of the properties of industrial processes such as inertia, momentum, damping, time lags, oscillations and other forms of instability and so on; however, there is no intrinsic and immutable physical law of dynamics that governs their behaviour. In economic systems for example, various parameters of the economy are measured such as inflation rate, unemployment, money growth and so on and the goal of control is to manipulate fiscal and monetary variables in order to maintain the measured variables at some desired values. Similarly in telecommunications networks, the purpose of network management is to measure the network performance in terms of quality of service metrics and to maintain these at some desired values (sometimes decreed by regulations and / or contracts) by taking control actions such

---

[1] In passing let it be said that in no other discipline is it more difficult to distinguish an engineer from a scientist. For many, science is a precursor to engineering. This is a misguided view of the relationship between science and engineering. Engineering is related to technology and in human civilisation, technology came long before science. It may be permissible to say that engineering is concerned with creating technology by making use of the best available science, but this is a far cry from saying that technology is the application of science.

In control engineering journals, many of the papers had nothing to do with technology or engineering whatsoever. They were increasingly mathematical and often only about mathematics. Zadeh has commented many times that it was this trend that motivated him to come up with Fuzzy sets theory. The year of his first paper was 1965 so the origins of the trend were fully established in the mid-sixties.

as reconfiguration of the network. The behaviour of these type of systems is governed by humans who interact with them to make them work. That is to say, it is the human behaviour with respect to these systems that produces their dynamic behaviour. While it may be tempting to describe such systems by discovering a mathematical law that fits their input / output behaviour, the long term validity of any such law cannot be guaranteed because it is ultimately dependent on humans who may change their behaviour suddenly and in an unpredictable way.

There are yet other systems – process control systems no less – for which the mathematical law of system behaviour is difficult to derive from basic physical principles because the physical processes taking place are too complex. Cement kilns and Steel making are the best examples of such systems. Again it may be possible to discover the mathematical law governing their behaviour by an input output analysis. However, as with human systems mentioned above, the validity of that law would be limited. Indeed such systems have more in common with human systems than with classical linear dynamic systems.

When faced with such difficult systems, control engineers have always invoked the ultimate catch phrase: non-linearity. This is used in such an all encompassing way that it has never made much sense; for it is unclear where non-linearity ends and non-mathematical (non-analytic to be more accurate) begins. Non-linearity ought only to refer to those systems which have to be described by non-linear mathematical equations resulting in a degradation of analyticity; however, it is frequently applied to any non-analytic system. This discussion has been introduced here to point to an important attribute that has come about within control theory: the ***cult of analyticity***. This cult had much to do with early criticisms of fuzzy control resulting in a slow acceptance of the approach outside Japan. The criticism centred around the lack of stability analysis of fuzzy logic controllers.

*B. Argument concerning system stability*

Some of the insight on systems in control theory concerned the "stability" of the systems. It is worth noting that those well trained in the discipline will conjure up poles and zeroes in the s-plane and certain properties of matrices in order to properly discuss stability. To put it another way, it is often difficult for a control systems practitioner to define stability to a non-practitioner without using the mathematical theory. It is in this context that the main criticism of fuzzy control has to be understood - that it does not provide stability analysis. The answer provided at the time against this criticism was that in order to carry out such an analysis it was first necessary to have a reliable mathematical model of the system, and complex systems such as cement kilns (the first industrial application which used fuzzy control [6]) cannot be accurately modelled mathematically. This argument continued but the matter rested there with the implied suggestion that fuzzy control was best suited only for those complex systems such as cement kilns that could not be modelled mathematically. To-day, given the large number of other industrial applications of fuzzy control in Japan, one needs to modify that view considerably. To do so we need to examine two issues: firstly, the requirement for stability analysis from an industrial perspective; and secondly, the problem of control as viewed from within the AI paradigm.

*C. Industrial requirement for stability analysis*

Industry has never put forward a view that mathematical stability analysis is a necessary and sufficient requirement for the acceptance of a well designed control system. That is merely the view that control system scientists wished to put forward, but it has never gained currency outside academic circles. It is noteworthy that control system area never developed a comprehensive and prescriptive design methodology like we have for Information System Design such as SSADM. Control system methodologies are at best, recommendations on how to use a set of computer based analytic tools. They do not provide an industry approved standard for a structured step by step approach for the analysis and design of a system. An industrially accepted methodology that included stability analysis as an essential validation step will have a very limited use as so few systems would be amenable to that approach.

Prototype testing is more important than stability analysis; stability analysis by itself can never be considered a ***sufficient*** test. Moreover, in any practically useful methodology, a stability analysis step would need to be made a desirable but an optional step; it cannot be a ***necessary*** step. Control system scientists have continued to stress stability analysis with great vigour, but the argument is in fact disingenuous and should not have been treated by fuzzy logic control workers with the seriousness with which it was. Stability as understood by a control system scientist is not the property of the process being controlled but an operational definition expressed in the mathematical language of analytic control theory. It turns out on reflection that fuzzy control is one of the techniques that forms part of an alternative non-analytic control theory. Stability is still an important issue but a different way has to be found to study it. In the final analysis all one may be able to do is to build prototypes for the purpose of approval certification. That is a well tried and tested approach used in industry and there is no reason why it may not suffice with control systems as well.

## IV. Non-analytic control systems - an alternate system design paradigm

The problem of control is also one of decision making, viz.: given the observation of the state of the process to decide from encoded knowledge what action to take. Knowledge based systems and, in particular rule-based approaches, are ideally suited for such a decision making task. On the one hand one may use deep knowledge, perhaps in the form of qualitative models along with qualitative reasoning for such a decision making. On the other, a simple rule–base derived from the experience of a process operator can also be used. The only requirement is that decision making is automated so that the KBS system does not simply give advice on what action is best suited but actually goes ahead and implements the decision without checking with an operator. Control systems based on knowledge based systems, neural networks, genetic algorithms are all viable alternatives to those derived from conventional control theory.

The fuzzy controller is one such simple rule-based control system. The knowledge used for this does not derive only from an expert operator. It can also come from the designer of the system. Some of the knowledge can be based on the understanding of the behaviour of the class of dynamic systems in general to which the particular system belongs, and this can be refined further to suit the particular system if need be. Indeed if one looks at the rules in a fuzzy controller, one finds three sets of rules: a set dealing with providing a rapid response to large errors (rise time rules); a set for preventing overshoots (damping rules); and a set maintaining errors near to zero (steady state rules).

Whatever knowledge (intuition) that is available about dynamic systems, but which does not need to be expressed only mathematically can be encoded as rules. There is much knowledge on dynamic systems that can only be expressed using the mathematical language of control system theory; but there is equally much knowledge about particular dynamic systems that definitely cannot be expressed in any mathematical way. Such knowledge can be used in rule–bases but not in the conventional control paradigm. For some kind of knowledge a mathematical expression would be much more cumbersome even if it existed than a rule–based expression. Thus often a rule–based non-analytic control paradigm may in fact produce better controllers than the analytic control theory paradigm. It must now be stated clearly and forcefully that the goal of pursuing research on any kind of non-analytic controller is first and foremost to prove that the technique is applicable, and not just to find ways of designing controllers for only those systems for which the conventional theory cannot be used. Even if a controllers can be designed using the conventional analytic theory, a designer may still prefer to use the non-analytic approach because the resulting controller may be easy to design; it may be easy to implement in hardware; and / or, it may result in a more robust controller. (A robust controller is less susceptible to system parameter changes or to noise.) Recent industrial experience has shown that all these three criteria apply to fuzzy logic based controllers [7].

### *A. The advantages of using fuzzy control*

One of the main advantages of using a fuzzy approach is that fuzzy logic provides the best technique for knowledge representation that could possibly be devised for encoding knowledge about continuous (analogue[2]) variables. This can be briefly explained with reference to the figure below. Assume the system to have two measured variables and one control action variable. The controller will be composed of a set of control rules with two antecedents and one consequent.

The antecedents of each fuzzy rule describes a fuzzy region in the state-space (see figure below). By so doing one effectively quantizes an otherwise continuous state–space so that it is covered by a finite number of such regions (and consequently fuzzy rules). There is a specific fuzzy action associated with each such fuzzy rule (fuzzy region). During the decision making process (control process) each point in the state–space is differently affected by the actions associated with all the fuzzy regions in whose footprint the point falls. The fuzzy aggregation rule and the defuzzification method then yield a specific action for that point. As the point moves, the action also changes smoothly. This means that an effectively quantized representaion of the state–space nevertheless yields a smooth action surface over the state–space.

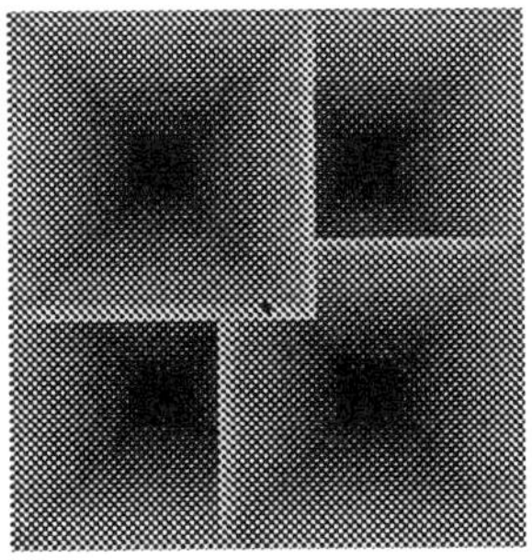

[2] "analogue" is not a synonym of "continuous" in English. Such usage in electrical engineering owes much to the primacy of control system theory and involves a convoluted reasoning: "analogue" is contrasted with "digital"; digital quantities are discrete; the opposite of discrete is continuous; hence etc.

In effect given only the actions associated with the centre point of overlapping rectangular regions of the state–space, fuzzy logic algorithm provides the action for any individual point in the state–space. Obviously all the other values are interpolated from the available finite sets of values by a method which depends upon the choice of the mathematical expression used for the various fuzzy logic connectives (AND, OR, ELSE etc.) as well as the choice of the defuzzification method. It is agreed by the workers in fuzzy control that one should limit the choice of mathematical expressions to those that would result in a smooth action surface. The result is that the control can be specified compactly with as few rules as possible.

This form of knowledge representation is applicable in any rule–based application dealing with continuous variables, not just for feed–back control. Many of the consumer products in Japan using fuzzy logic use it as a feed–forward decision maker. The washing machine is a case in point. Having introduced a new sensor for measuring the "turbidity" of the water, the manufacturers needed to introduce the associated intelligence to respond to this new information. It would be pointless to display the turbidity value to the user of a washing machine for it would not make much sense on how to interpret the information so conveyed and what to do with this new information. Because the engineers discovered what useful information this new sensor conveyed (how the articles being washed were soiled and how dirty they were), and thus what to do with it (adjust the washing time), this information could be encoded as rules. Furthermore as the information was itself "analogue", fuzzy control approach was the best knowledge representation method.

It is instructive to note that in most of its industrial applications (both in consumer products and in industrial systems) the fuzzy controller comprises only a few rules (around 7 to 15). It represents a tiny amount of intelligence being embedded into the overall system. In this respect it is analogous to the use of tiny electric motors in applications such as automobile rear view mirrors and wind-up mechanisms of even cheap cameras. Often there is more technical detail in the sensors and actuators of the overall system than in the controller itself. This shows that one of the problems with fuzzy control is that it is percieved by some (both control system and AI scientists) as too simple. While this factor provides technological strength it is also seen as a scientific weakness [3]. Outside Japan, where intellectual sophistication is revered, fuzzy control has been seen as scientifically weak and this is another strong reason why its is only belatedly being commercially exploited there. In Japan its technical strengths have been most apparent and it is not at all surprising that large scale exploitation has first occurred in Japan.

## V. Conclusions

This paper has concentrated mainly on the humble rule–based fuzzy controller. The experience gained with it over the past several years has shown that fuzzy control may often be a preferred method of designing controllers for dynamic systems even if traditional methods can be used. Experience has shown that a fuzzy controller is often more robust than a PID controller in the sense that it is less susceptible to noise and system parameter changes. Furthermore, a fuzzy controller may also be easier to design and, thanks to the fuzzy chips and so on, easier to implement. Its application is not limited to dynamic system control. Its main merit is that it gives the most efficient knowledge representation method that can be devised for rule–based systems that deal with continuous variables. It can be used wherever such knowledge exists. For example, in mathematical programming there is often more specific domain related problem knowledge available than can be expressed in the framework of the mathematical model.

Its delayed exploitation outside Japan teaches several lessons. This paper has argued that there are two main reasons for this: the tradition of intellectualism in engineering research in general and the cult of analiticity within control engineering research in particular. The former is responsible for continuing to produce a number of unscientific distortions in engineering research, albeit in the name of science. For example, journal editors and conference organis-

[3] To continue with the theme of the first footnote, the aims of scientific enquiery and technological implementation are not the same. Science delights in complexity as it discovers ever more subtle phenomena. Technology abhors complexity as it tends to render the product in question more difficult to use and maintain. In order to make use of advanced scientific concepts and techniques into products, it is necessary to encapsulate the complexity of a subsystem within a black box both functionally and physically. Functionally, the modules and subsystems that contain complexity must not be allowed to impose their subtleties on other modules that make up the system. Physically the complex parts must be tightly sealed so that when they fail they can be replaced as a whole and not require in situ maintenance. The purpose of the black box is to hide the scientific complexities in order to make the product usable. Designers of many products have not been able to resist showing the beauties of the complex science to the user.

ers have been heard to assert that they would like to see more application papers *even if this means lowering the quality*. Quality according to what criteria? Should one not question these criteria themselves? Evidence suggests that the quality of engineering as seen in all manners of applied work including the technology in the end products is of a higher a standard than anything seen in run-of-the-mill journal papers. Secondly, the authors of technical papers feel obliged to introduce a large amount of symbolism (again in the name of good science) even to the point of obscuring the main message[4].

The main lesson learnt is that the subject of automatic control can no longer be the exclusive preserve of conventional analytic control theory. An alternative paradigm of non-analytic control systems now exists. The advantage of the analytic approach is that it gives an abstract view of system dynamics which applies to a whole variety of linear systems and many non-linear systems as well. The problem is that often more is known about a particular system than can be expressed analytically and this knowledge may thus not be captured effectively. The alternative paradigm is extremely well suited to express special characteristics and knowledge as rule–bases. However, at the current state of research, the more abstract knowledge about system dynamics is not well captured.

New research should be aimed at strengthening the concepts and techniques about system dynamics available to this new paradigm. To achieve this it would prove useful to study the powerful control system concepts that already exist within the conventional theory and then see if they can be redefined from the mathematical language of that theory to be meaningful within the new paradigm (for example, by introducing qualitative modelling techniques). Some workers have suggested that a synthesis of the two paradigms is what should be sought. Aims of synthesising theories are fine as ideals, but in practice they lead to artificial shoe-horning of one set of concepts within the framework of an alien theory. It is far better to accept that control engineering has now become a multi–paradigm discipline.

[4]It is possible to think of a paper on, let us say, fuzzy control that can either be written as plainly as possible (a really good scientific practice) or with as much symbolism as possible (the more involved Greek symbols that can be used the better the scientific impact!!). A good heuristic for potential referees to use is that if an author has anything of substance to say in a paper than he would not wish to obfuscate it with excessive symbolism.

## References


1. Mamdani, E. H. and Assilian S. "An Experiment with in Linguistic Synthesis with a Fuzzy Logic Controller", *Int. J. Man-Machine Studies*, vol. 7 pp 1-13, 1975
2. R. K. Lindsay, B.G. Buchanan, E. A. Feigenbaum, J. Lederber *Applications of Art. Int. for Organic Chemistry: The DENDRAL PROJECT* New York, McGraw-Hill, 1980
3. Zadeh, L. A. "Outline of a new approach to the analysis of a complex system and decision processes", *IEEE Trans. SMC*, vol. 3, pp 28, 1973
4. T. S. Kuhn, *The Structure of Scientific Revolutions* Chicago, 1962
5. Brian Easlea, *Liberation and the Aims of Science*, Chatto & Windus (Sussex University) 1973.
6. Holmblad, L.P. and Ostergaard, J. J. , "Control of a Cement Kiln by Fuzzy Logic", *Fuzzy Information and Decision Processes* (Ed M.M. Gupta et al) North Holland 1982
7. Sugeno, M. (Ed) *Industrial Application of Fuzzy Control*, North-Holland, 1985

# A Simple Direct Adaptive Fuzzy Controller Derived from its Neural Equivalent

Hugues Bersini*, Jean-Pierre Nordvik** and Andrea Bonarini***

* IRIDIA- CP 194/6 - Universite Libre de Bruxelles - 50, av. Franklin Roosevelt - 1050 Bruxelles - Belgium
** Commission of the European Communities - Joint Research Centre, Institute for Systems Engineering and Informatics - 21020 Ispra (Va) - Italy
*** Dipartimento di Eletronica e Informazione - Politecnico di Milano - AI & Robotics Project - Piazza Leonardo da Vinci, 32 - 20133 Milano - Italy

***Abstract* - The main point argued and illustrated in this paper is the ease to import methods and ideas emerging in the connectionist community for control applications as soon as the fuzzy controller is supplied with a gradient-based automatic tuning of its parameters. Such a gradient method is applied for the simplest member of Sugeno's fuzzy systems. Then an example of direct adaptive fuzzy controller derived from its neural equivalent and comparing favorably with it is exposed and discussed.**

## I. INTRODUCTION

It is commonly accepted today that control methods grounded into either a neural or a fuzzy system offer several advantages with respect to more classical approaches:

a) they are simple to implement and allow to bypass constraining mathematical analysis by a pure algorithmic process

b) it has been established that multilayer neural networks [23] as well as fuzzy systems [6], provided an adequate adjustment of their intrinsic parameters, are capable to approximate in a satisfactory way any continuous function. This essential property justifies their study and use in the framework of nonlinear system theory and nonlinear control where mathematical analysis appear very laborious unless forcing to resort to rough linear approximations.

c) a straight consequence of b) and excessively repeated in the literature both neural and fuzzy is that to some extent control methods based on such systems can become "emancipated" from a precise knowledge of the process to control or from a rough linearisation. A "black box" control is then favored departing from classical approaches in which a prior description of the process is required.

d) a welcome robustness with respect first to undesirable perturbations occurring in the process itself, due to the neural and fuzzy capacity to generalize and extrapolate from referential situations, but also with respect to damages affecting the control systems, due to the redundancy induced by the parallel architecture of both fuzzy and neural systems.

0-7803-0614-7/93$03.00 ©1993IEEE

e) an adaptive capacity deriving from automatic tuning methods efficient and easy to implement. If studies of neural nets have always been indissociable of such tuning technics (the learning phase), the enrichment of fuzzy systems with similar methods has become a more recent concern which, as the next chapters will show, make the comparison of these two approaches more delicate.

f) finally, fuzzy and neural systems lend themselves to a material realization ("fuzzy chip", "neural hardware" ...) which allows a greater flexibility and an important acceleration as compared with the use of conventional binary and serial computer architecture.

For a couple of years when experts are asked to risk a comparative analysis between neural and fuzzy systems, both considered as alternative to classical methods of control, the opinions quite often are hesitant, evasive and cautious mainly because of the important continuous developments which characterize both approaches. A general opinion is that (see Kong and Kosko [12] and Werbos [22]) fuzzy control, on account of its intrinsic linguistic nature, turns to be more appropriate when the control system owes a lot to the interfacing with human operators. A linguistic-based control facilitates the operator elicitation, understanding and debugging, in short, intensifies his involvement during the conception of the system. The fuzzy controller can even be exploited for the training phase of novice operators. On the other hand, neural nets, which need to be coupled with a learning method, should be more appropriate for applications where only numerical data are available as clues to the control system to set up. Based on these data, the neural net has its parameters (the synaptic weights) adjusted in order to reproduce the desired control (during the learning phase).

However, the rapid progress within both fields tend to reinforce the "fuzziness" of the frontiers separating them. Indeed, on the neural side, more and more transparency is pursued and obtained either by a pre-structuring of the neural net to improve its performances, or by a possible interpretation of the synaptic matrix following the learning stage. On the fuzzy side, the development of methods allowing an automatic tuning of the parameters which characterize the fuzzy system largely draws its inspiration from similar methods used for long in the connectionist community. In brief, neural nets can improve their transparency which make them closer to fuzzy systems while fuzzy systems can self-adapt which makes them closer to neural nets. The main objective of this paper will be to illustrate how fuzzy systems, provided a gradient method for the automatic tuning of their parameters akin to the well-known backpropagation for multilayer neural nets, can benefit from ideas and methods appeared in the field of neurocontrol during these last years.

Fuzzy systems, when supplied with an automatic tuning of their parameters such as the membership functions, become easy to generate and can exhibit adaptive capacities. The project of automatic generation of fuzzy controllers is easily justifiable given the difficulties and reliability problems posed by the verbal elicitation of human operator expertise. As a matter of fact, it has already received large attention since the very first work of Procyk and Mamdani [16]. Their method, labeled self-organizing fuzzy controller, and subject to successive improvements [5] [19], demands a second level of rules (meta-rules) to adjust on-line a first level of fuzzy rules really in charge of the control. On the other hand approaches coming from the connectionist community aim to decrease more and more the part played by the human programmer in the generation of the control system and the conception of its adaptive capacities, in substance to guarantee more autonomy to the resulting controller.

On the whole, fuzzy controllers require to types of tuning which Lee [13] designates as structural and parametric tuning. The first one concerns the structure of the rules: the variables to account for, for each variable the partition of the universe of discourse, the number of rules and the conjunctions which constitute them ... A complementary paper [4] describes a possible use of genetic algorithms to contribute to this structural tuning by searching in the space of all possible fuzzy decision table. Once available a satisfactory structure, a fine adjustment of the membership functions remains to be done. Since it is at that level that a simple and easy to derive gradient method can be added, the paper is restricted to that type of tuning. The second chapter will briefly review some works (at the cross-roads of connectionism and fuzzy logic) which aim at automating the fuzzy systems parametric tuning. In the third chapter, we will propose and present our own gradient-based method for the automatic research of membership function. This method can only be applied to Sugeno's type of fuzzy system and improve a method developed by Nomura et al. and described in [15]. Once in possession of this gradient method for the adjustment of membership functions, the fourth chapter will discuss what lessons fuzzy control can learn from neurocontrol. The fifth chapter will present a simple DAFC (Direct Adaptive Fuzzy Controller) derived from a previous work involving one of the authors but based on neural nets: a simple DANC (Direct Adaptive Neural Controller [17] [18]). Finally we will present results obtained when applying the two methods (DAFC and DANC) for the control of the well known cart-pole and will discuss why DAFC seriously challenges DANC.

## II. Automatic Parametric Tuning of Fuzzy Systems: A Survey

It is likely that the recent attention paid to the automatic parametric tuning of fuzzy systems be the consequence of the explosion of interest in neural nets and their associate learning algorithms. Indeed, expressions like "fuzzy neural nets", "fuzzy neural control", "neural fuzzy systems" invade the literature on fuzzy systems and fuzzy control. Clearly the connectionist appeal in the fuzzy community can be ascribed more to the learning and self-tuning capacities than to anything else. When searching to merge the plasticity of neural nets with the inferential dynamics of fuzzy systems, two types of attitudes are possible (this is far to be an exhaustive survey but rather a rapid glance through the fuzzy-neural literature). A first attitude intends to respect the basic properties and architecture of neural nets and simply "fuzzify" some of their elements. A crisp neuron can become fuzzy and the response of a neuron to its lower layer activation signal be more of a fuzzy relation type (T-norm and T-conorm) than of a sigmoid type [8] [9]. Since the neural architecture is conserved, what is varying (provided the new learning rule has been derived) is still some kind of synaptic weights connecting low level to high level neurons. To some extent, these approaches sacrifice their "fuzzy authenticity" namely their linguistic transparency lost somewhere into the neural architecture (nothing is really analogous to a synaptic matrix in a genuine fuzzy system) and one remains to be convinced on the advantages of such merging with respect to non-fuzzy neural nets (a more universal non-linear mapping ?? learning faster ?? more robust ??)

A second attitude more respectful of the "fuzzy quality" and in so eventually more satisfactory aims at providing fuzzy systems with the kind of automatic tuning methods typical of neural nets, but to be applied to fuzzy systems as such without altering their classical presentation (a bunch of fuzzy production rules). Since this presentation is left unchanged, what is classically adjusted by means of these methods are the characteristic membership functions, which up to now were obtained via manual trial-and-error procedures. Neural nets can participate in this automatic adjustment in three ways. First, one can use neural learning capacity as such, separated from the fuzzy system, for an automatic determination and adjustment of fuzzy rules (the design of membership functions). Takagi and Hayashi [21] designate this method as NN-driven fuzzy reasoning i.e. the use of fuzzy reasoning supplied with the learning facilities of NN. In that case the fuzzy system is not really affected by the neural architecture. In a second attempt, the fuzzy system can be installed in an architecture isomorphic to neural nets i.e. a multilayer network, where each node performs a function such as to make the entire network perfectly equivalent to the fuzzy system [10]. The advantage if one offered by this "neural reading" of the fuzzy system is to facilitate the computation of the gradient method (akin to the backpropagation).

The third way that we believe to be both the most immediate and the most respectful of fuzzy logic (because the less affected by the connectionist flavor) is simply to make the adjustment of the membership functions to rely on a gradient method. This gradient method will play a similar role as the one played in neural nets but also, and in general, in any other kind of parametric systems seeking to optimize their parameters. A clear example is the algorithm developed by Nomura et al . [15] relying on a gradient-based method for the optimization of the parameters of a Sugeno's type of fuzzy system [20]. The method we have implemented and which is the key of the adaptive capacities of our DAFC is very close, just slightly improved, to this latter and will be the subject of the next chapter.

## III. A Gradient Method for Sugeno's Fuzzy Systems

In this chapter we will concentrate on the simplest member of Sugeno's fuzzy systems given by the following type of mapping of $R^n$ in R, $y = f(x_j)$:

IF $x_1$ IS $A_1^a$ AND $x_2$ IS $A_2^b$ AND $x_3$ ... THEN $y = F^d$

where 1,2 .. index the variable, α the rule (we suppose nr rules) and a,b,c,d ... the type of linguistic term. For reasons of homogeneity, the output y although crisp can only has a fixed number of discrete values corresponding to large, small, very large, very small,... The linguistic terms $A_j^a$ are fuzzy sets characterized by isosceles triangular shape given by:

$$A_j^a(x_j) = \max\left(1 - \frac{2\left|x_j - a_j^a\right|}{b_j^a}, 0\right)$$

then each linguistic term for each variable $A_j^a$ is described by two parameters $a_j^a$ and $b_j^a$ that a gradient method will try to optimize. The crisp value $F^d$ for the output will also be subject to an optimization process.

$$\text{if } \mu^\alpha = \prod_{j=1}^{4} A_j^a \quad (\text{for the } A_j^a \text{ that appear in rule } \alpha)$$

The final output y is given by :

$$y = \begin{cases} \dfrac{\sum_{\alpha=1}^{nr} \mu^\alpha F^\alpha}{\sum_{a=1}^{nr} \mu^\alpha} & \text{if } \sum_{\alpha=1}^{nr} \mu^\alpha \neq 0 \\ 0 & \text{if } \sum_{\alpha=1}^{nr} \mu^\alpha = 0 \end{cases}$$

The central difference (and we believe improvement) of the algorithm we propose with respect to Nomura et al. work is that we don't really try to optimize the rules but the linguistic terms instead. Briefly, we try to adjust $a_j^a$ and not at all $a_j^{\alpha,a}$. At the end of the optimization process, the linguistic term "small" for the variable $x_j$ will always mean the same independently on the rule in which it appears. This alternative method presents various benefic consequences: mainly it accelerates the optimization process and it keeps the final fuzzy system more coherent, easier to understand and to interface with. If $K_a$, $K_b$ and $K_f$ are the learning parameters respectively associated to the position of the triangle a, its base b and the output of the rule F, three expressions can easily be derived:

$$\Delta F^d = -K_f \sum_{\substack{\text{all rules } \alpha \\ \text{in which} \\ F^d \text{ appears}}} \mu^\alpha \Big/ \sum_{\alpha=1}^{nr} \mu^\alpha$$

$$\Delta a_j^a = -K_a \frac{1}{\sum_{\alpha=1}^{nr} \mu^\alpha} \left( \sum_{\substack{\text{all rules } \alpha \\ \text{in which} \\ A_j^a \text{ appears}}} \mu^\alpha F^\alpha - y \sum_{\substack{\text{all rules } \alpha \\ \text{in which} \\ A_j^a \text{ appears}}} \mu^\alpha \right) sgn\left(x_j - a_j^a\right) \frac{2}{b_j^a A_j^a(x_j)}$$

$$\Delta b_j^a = -K_b \frac{1}{\sum_{\alpha=1}^{nr} \mu^\alpha} \left( \sum_{\substack{\text{all rules } \alpha \\ \text{in which} \\ A_j^a \text{ appears}}} \mu^\alpha F^\alpha - y \sum_{\substack{\text{all rules } \alpha \\ \text{in which} \\ A_j^a \text{ appears}}} \mu^\alpha \right) \frac{1 - A_j^a(x_j)}{A_j^a(x_j)} \frac{1}{b_j^a}$$

The difference with [15] is clear since the sum is computed only on the rules in which a specific linguistic term associated to a specific variable, the one to be adjusted, appears.

## IV. What Fuzzy Control can Learn from Neural Control

Suppose in general a nonlinear mapping from input I to output O (we will call it a FUNNY system (a mixture of FUzzY and NN)). This mapping is characterized by a set of parameters symbolically designated by W hence $O = F_W(I)$.

Fig. 1 A FUNNY system

The partial derivatives ∂O/∂W will be assumed to be known. We computed them for the Sugeno's mapping tackled in the previous chapter. Once furnished with a gradient method for the parametric tuning of whatever fuzzy system ($\Delta W = -n\partial O/\partial W$), every use of neural nets in a control problem can be similarly contemplated using this fuzzy system. As Werbos [22] points out, it is quite immediate to substitute the neural net box by the fuzzy system box in any neural control application. Suppose the process to control indicated in fig.2, the objective being to tune the control parameter U to an unknown value Ud such as to drive the process output Y to a desired value Yd.

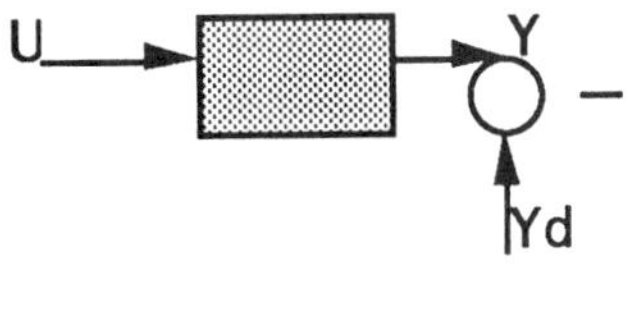


Fig. 2 The Process to control

A first control approach amounts to teach FUNNY to reproduce the human operator successful actions. One observes this operator and collects a set of efficient actions he executed on the process. These actions are expressed as pairs Y-->Ud. Then FUNNY must encode through the gradient method this set of pairs: $\Delta W = -n\partial(O-Ud)^2/\partial W$. A further approach called "general learning" and represented fig.3 presents a prior training stage which consists in using the process to produce a set of input-output pairs U -->Y. These pairs are then used as patterns for training FUNNY. During the training stage, the control parameters U are chosen randomly to scan over a certain working range. Those parameters are then injected into the process which supplies output values Y. The associated gradient method is used to train FUNNY to relate its output U to its input Y. After this learning stage, the FUNNY controller is able to provide the correct control parameter to reach any target Yd. This method has several drawbacks: first it raises enormous problems in case of non-invertible process; second, learning must be performed off-line and it is hard to call it an adaptive control, finally FUNNY cannot limit its working range to the Y that are actually relevant.

Another possible approach is called the "iterative inversion technique" in which first FUNNY learns the forward model of the process in terms of pairs: U-->Y. Then the control results from the inversion algorithm which backpropagates the error signals $(Y - Yd)^2$ down to the input layer to update the activation values of input (and no more W) so that the output error is decreased. In the indirect learning architecture (fig.4), the desired target Yd propagates through FUNNY producing the process input U1. This same input is tried on the process which responds by an output Y distinct, at the beginning, from the desired target. Then Y is injected as input of FUNNY and gives U2 as output. The learning phase immediately follows aiming at minimizing the error $(U1 - U2)^2$ obtained when presenting Y as input of FUNNY. However simulations show that FUNNY can settle to a solution that maps all the desired targets to a single plan input, which gives zero training error, but obviously a non-zero total error. Learning is indirect because the difference between the output of the process and the desired output are not directly reduced.

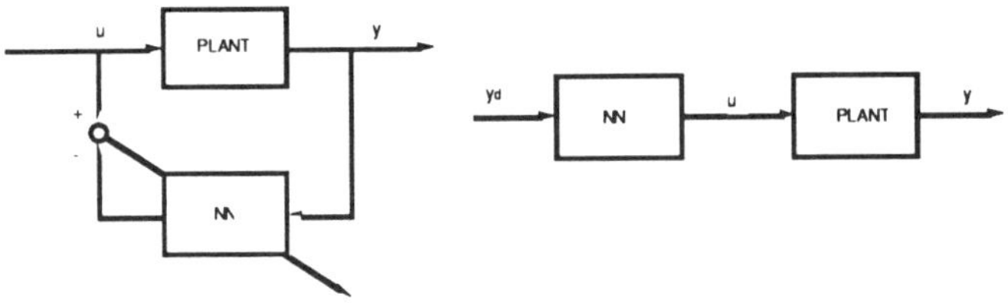


Fig. 3 General Learning Architecture

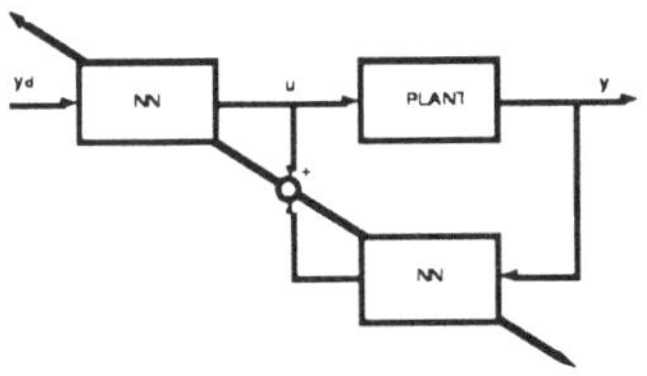


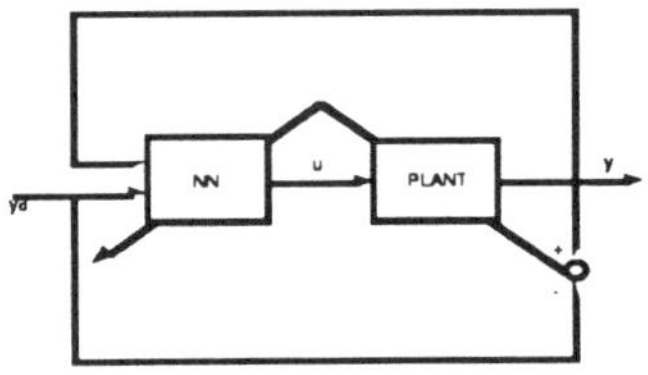


Fig. 4 Indirect Learning Fig.5 Specialized Learning

The final approach, the one that we have adopted, has been called specialized learning in the NN literature (fig.5). Specialized learning differs from general and indirect learning by the fact that the FUNNY controller learns no longer from input-output pairs but from a direct evaluation of FUNNY accuracy with respect to the output of the process. FUNNY uses the difference between the actual output of the plant Y and the desired output Yd to adapt its parameters W. Specialized learning avoids several drawbacks of general learning: there is no longer a specific training stage during which the controller is not operational, and FUNNY learns directly on the domain of relevant Y. It learns continually and is therefore adaptive. In fact, the specialized learning approach fits in perfectly with the classical adaptive control attitude. FUNNY controls and learns to control simultaneously. However, to make specialized learning possible, we need some prior knowledge on the way the plant reacts to slight control modifications i.e. the Jacobian of the plant. The reason for such requirement is obvious when computing the gradient no more on basis of the error obtained at the output of FUNNY: $(O-Od)^2$ but at the output of the process instead: $(Y-Yd)^2$. Od is all the less unknown since it is what is looking for.

$$\Delta W = -n\partial(Y-Yd)^2/\partial W = -2n(Y-Yd)\partial Y/\partial W$$
$$= -2n(Y-Yd)\partial Y/\partial O.\partial O/\partial W$$

$\partial O/\partial W$ corresponds to the gradient method akin to the backpropagation for multilayer neural nets and derived in the previous chapter in case of Sugeno's fuzzy system. What is unknown in the absence of an analytical model of the process is the Jacobian $\partial Y/\partial O$.

One possible strategy consists in approximating the partial derivatives by plotting the process reactions to slight control modifications at the operating points. Another recent and sophisticated technique has been proposed by Jordan [11] which incorporates the default prior knowledge in another FUNNY system and links the FUNNY controller to this FUNNY emulator of the plant - a fuzzy or neural system that identifies the process. It is easily shown that the Jacobian of the process can be derived by a mechanism of backpropagation applied no more on the weights but on the input of the FUNNY emulator (like for the iterative inversion technique mentioned before). The most salient application of such double FUNNY techniques has been, in the case of neural nets, the control and the backing up of a truck-and-trailer [14]. A similar double fuzzy systems is easy to conceive. Interestingly enough, Kosko [12] has turned to be one of the lead advocates of the use of fuzzy logic for control while generating a fuzzy controller which compares favorably with the neural controller in terms of black-box development effort, black-box computational load, smoothness of truck trajectories, and robustness. The resulting control neither requires any automatic parametric nor a fuzzy emulator. The rules were based on common-sense and gradually improved through manual trial-and-error. Nothing is really surprising in this double absence since the emulator is a direct consequence of the automatic tuning method and the non-necessity of one entails the non-necessity of the second.

The double FUNNY method corresponds to the classical indirect control approach where the parameters of the process are estimated at any instant and the parameters of the controller are adjusted assuming that the estimated values of the process represent the true values. This technique makes the control quite flexible for indeterminate learning problems, allowing generic constraints to be expressed separately at the control and at the behavioral level. However it demands, in compensation, either a preceding learning stage (the identification of the process) or a "neurally or fuzzily" expressed prior knowledge of the dynamic of the process. The specialized learning technique that we will present in the next chapter makes the bet that in many circumstances a satisfactory control can be attained even on basis of the least knowledge of the process, namely the sign of its Jacobian. Indeed the gradient method indicates that the adjustment of W follows the sign and amplitude given by $\partial Y/\partial O.\partial O/\partial W$. Since substituting $\partial Y/\partial O$ by its sign $\pm$ 1 just alters the amplitude not the direction of the variation, the gradient method can still give a adequate value of the control parameters. Reducing to such minimum the knowledge of the process necessary to its control makes the resulting control "direct". The process can remain a black-box with just few holes allowing to observe qualitatively the way it reacts to increases and decreases of its input. This type of knowledge is quite common and certainly the first shallow type of knowledge a user or observer of the process is aware of and able to communicate.

In order to conclude this chapter about what fuzzy control can learn from neurocontrol, it is appropriate to mention reinforcement learning strategy dating back to the works of Sutton and Barto [1] [2] and whose most convincing achievements address problems which present a delayed reinforcement occurring at the end of a sequence of actions. A technique to optimize this multi-steps choice, taking into account that indications on the goodness of the choice can be very delayed, called "temporal difference" [2] has been developed and has appeared, slightly adapted, within several fuzzy controllers [3] [7].

## V. A SIMPLE DAFC DERIVED FROM A SIMPLE DANC

In a previous work [17] [18] a simple Direct Adaptive Neural Controller (DANC) has been achieved and tested among other toy problems on the cart-pole. One and only one multilayer neural net was necessary, the sign of the cart-pole Jacobian was known, and this suffices to achieve a very satisfactory control with respect to the numerous other neurocontrols we were aware of and with respect to more conventional, like MRAC, approaches [18]. In this chapter we will describe in more details the equivalent simple Direct Adaptive Fuzzy Control (DAFC) similarly applied on the cart-pole problem (a problem frequently met in the NN and fuzzy control literature, see [1]). In our application the objective of the control is to bring the cart-pole, whatever its initial conditions, to a stable equilibrium position with the four variables describing the system: $x,v,\theta,\omega$, driven to 0.

The fuzzy controller is of the Sugeno's basic type with four variables in the IF part and only one variable (given by a "fuzzy set" restricted to a single value), the force to exert on the cart, in the THEN part. An example of rule is:

IF x is large and v is very small THEN f = 10 N (large)

The parameters to adjust are the memberships functions coding the linguistic terms appearing in the premise of the rules and the different possible values of the force appearing in the consequent part. The gradient method is the one described in chapter 3. With regard to the Jacobian, the signs of the four partial derivatives: $\partial x/\partial F$, $\partial v/\partial F$, $\partial \theta/\partial F$, $\partial \omega/\partial F$ are all negative. The error is given by :

$$Err = 0.5(\alpha_1 x^2(t-1) + \alpha_2 v^2(t-1) + \alpha_3 \theta^2(t) + \alpha_4 \omega^2(t))$$

where $\alpha_i$ weights the relative contribution of each variable.

The reason for both t and t-1 is linked to the slower reaction (a larger delay) of x and v in comparison with $\theta$ and $\omega$. The global gradient method immediately follows: $\Delta W = -n\partial Err/\partial W = -\partial Err/\partial F.\partial F/\partial W$ with $\partial F/\partial W$ given by the method presented in chapter 3 and $\partial Err/\partial F$ easily obtained by substituting $\partial x/\partial F$, $\partial v/\partial F$, $\partial \theta/\partial F$, $\partial \omega/\partial F$ all by -1.

The results were very impressive. For a large range of initial conditions even drastic, the equilibrium position was reached after an average of 5 falls and an average of 2500 learning steps. For these experiments and in order to make possible the comparison with the results of our neural controller for the same problem, the rules were generated, like a synaptic matrix at random. More precisely, any initial set of rules contains 30 rules. In each rule, the probability of appearance on one of the four variables is 0.3 (and no rule can be generated without variables), each variable is associated to a linguistic terms taken randomly among seven possibilities. On the whole the learning was faster than for the neural equivalent and, confirming results exposed by Kosko [12] for another application, the final controller tolerated more initial conditions and resulted more robust than the neural controller.

Furthermore and here is a very crucial issue and determining advantage of fuzzy controller with respect to neural controller, the equilibrium position could even be reached without a fall provided the initial set of rules was taken no more randomly but on basis on some simple common sense (a same a priori is quite difficult to install in the synaptic matrix of a NN). Finally, the idea of approximating the Jacobian by a rough very qualitative knowledge like its sign fits in perfectly with the philosophy of fuzzy logic, where the world is represented in a qualitative, linguistic way, avoiding very precise numerical expression. We are currently working on a technique to derive this sign directly from a linguistic, fuzzy model of the process. It is easily conceivable that a model like: "IF the input increase THEN the output decrease", could have its Jacobian reduced to a negative sign and this would be so just because "increase" is the contrary of "decrease". The substitution of decrease by "very decrease" would just amplify the negative value. In a nutshell, the sign of the Jacobian could be obtained on basis of a semantic analysis of the fuzzy rules and no more of the numerical mapping concealed behind the rules.

## VI. Conclusion

The main point developed in this paper is the ease to import methods and ideas appeared in the connectionist community for control problems as soon as the fuzzy controller is supplied with a gradient-based automatic tuning of its parameters. An example of direct adaptive fuzzy controller derived from a neural equivalent and comparing favorably with it has been exposed. Indeed, even in the random case, the worst that can be done with a fuzzy system (since so much a priori can be easily installed in the fuzzy rules), the learning attains performances similar to neural ones. The imbrication of these two fields: fuzzy logic and neural net for control applications makes very delicate a definitive comparison between them. It would be prudent right now to avoid any premature conclusion. But even adopting the greatest prudence, something will certainly remain at the credit of fuzzy logic to the detriment of NN: the linguistic nature of fuzzy logic which enormously facilitates any interface with the human teacher, programmer, user, debugger and even learner.

## Acknowledgment

This paper would not have been possible without the precious collaboration and software developments of Vittorio Gorrini.

## REFERENCES

[1] Barto, A., Sutton, R. and C. Anderson. 1983: Neuronlike adaptive elements that can solve difficult learning control problems. *IEEE Transactions on Systems, Man and Cybernetics*, 13(5).

[2] Barto, A. 1990: Connectionist Learning for Control. In *Neural Networks for Control* (Thomas Miller III, Richard Sutton and Paul Werbos eds.) MIT Press.

[3] Berenji, H.R. 1992: A Reinforcement Learning-Based Architecture for Fuzzy Logic Control. In International Journal of Approximate Reasoning; 6; 267-292.

[4] Bersini, H. 1992: Generation Automatique de Systèmes de Commande Floue par les Méthodes de Gradient et les Algorithmes Génétiques. In Proceedings of the Deuxièmes Journées Nationales sur les Applications des Ensembles Flous.

[5] Burkhardt, D.G. and P.P. Bonissone 1992: Automated Fuzzy Knowledge Base Generation and Tuning. In *Proceedings of the IEEE International Conference on Fuzzy Systems* pp 179 - 188

[6] Cao, Z. 1990: Mathematical principle of fuzzy reasoning, in Proceedings of NAFIPS'90 Toronto.

[7] Chen, Y. , Lin, K. and S. Hsu 1992: A Self-Learning Fuzzy Controller. In *Proceedings of the IEEE International Conference on Fuzzy Systems* pp 189-196.

[8] Gupta, M.M. and G.F. Knopf 1992: Fuzzy neural network approach to control systems. In *Analysis and MAnagement of Uncertainty. Theory and Applications*. B. Ayyub, M.M. Gupta and L.N. Kanal (Eds.) Elsevier Science Publishers

[9] Hayashi, Y., Czogola, E. and J.J. Buckley 1992: Fuzzy Neural Controller. In *Proceedings of the IEEE International COnference on Fuzzy Systems*.

[10] Jang R. J.S. 1992: Fuzzy Controller Design without Domain Experts. In *Proceedings of the IEEE International Conference on Fuzzy Systems* pp 289-296.

[11] Jordan, M.I. and D. Rumelhart 1991: Internal world models and supervised learning. In *Proceedings of the eight International Workshop on Machine Learning*, pp 70-74.

[12] Kong S.G. and B. Kosko 1992: Adaptive Fuzzy Systems for Backing up a Truck-and-Trailer. In *IEEE Transactions on Neural Networks*, Vol. 3, No 2, March 1992, pp 211 - 223

[13] Lee, C.C. 1990: Fuzzy logic in control systems: Fuzzy logic controller - Parts I, II. *IEEE Trans. Syst. Man Cybern.*, vol 20, no 2, pp 404-435

[14] NGuyen D. and B. Widrow 1989: The truck backer-upper: an example of self-learning in neural networks. In Proceedings of International Joint Conference on Neural Networks, Washington, Vol. II, pp 347 -353.

[15] Nomura, H. , Hayashi, I. and N. Wakami 1992: A Learning Method of Fuzzy Inference Rules by Descent method. In *Proceedings of the IEEE International Conference on Fuzzy Systems* pp 203 - 210.

[16] Procyk, T. J. and E.H. Mandani 1979: A linguistic self-organizing process controller. In *Automatica*, vol. 15, no 1, pp 15-30.

[17] Saerens, M. and A. Soquet 1991: Neural Controller Based on Back-Propagation Algorithm. *IEE Proceedings-F*, 138 (1), pp. 55-62.

[18] Saerens, M., A. Soquet, J.M. renders and H. Bersini 1992: A Preliminary Comparison between a Neural Adaptive Controller and a Model Reference Adaptive Controller. In *Proceedings of the Workshop on Neural Networks in Robotics*.

[19] Shao, S. 1988: Fuzzy self-organizing controller and its application for dynamic processes. Fuzzy Sets and Systems, vol. 26, pp 151-164.

[20] Tagaki, T. and M. Sugeno 1985: Fuzzy identification of systems and its application to modelling and control. *IEEE Trans. Syst. Man Cybern.*, vol SMC-15, no. 1, pp 116-132.

[21] Tagaki,T. and I. Hayashi 1991: NN-Driven Fuzzy Reasoning. In *International Journal of Approximate Reasoning* 5.

[22] Werbos, P.J. 1992: Neurocontrol and Fuzzy Logic: Connections and Designs. In *International Journal of Approximate Reasoning*; 6; pp 185-219

[23] White, H. 1989 Learning in artificial neural networks: A Statistical perspective - Neural Computataion, 1.

# INFERENCE, INQUIRY AND EXPLANATION IN EXPERT SYSTEMS BY MEANS OF FUZZY NEURAL NETWORKS

Ricardo José Machado
IBM Rio Scientific Center
Av. Presidente Vargas, 824
20071-001 Rio de Janeiro, Brasil

Armando Freitas da Rocha
Instituto de Biologia - UNICAMP
13081 Campinas, Brasil

*Abstract* — **A new field of research focus on integrating the paradigms of expert systems and neural networks. In this paper we show how basic functions of expert systems, such as inference, inquiry and explanation can be implemented by means of fuzzy neural networks.**

## I. INTRODUCTION

Hybrid architectures for intelligent systems is a new field of Artificial Intelligence research concerned with the development of the next generation of intelligent systems. Current research interests in this field focus on integrating the computational paradigms of expert systems and neural networks in a manner that exploits the strengths of both systems expanding the applications to which either system could be applied individually. Such systems are called *Connectionist Expert Systems* (CES) [4,18].

The ability to learn in uncertain or unknown environments is an essential component of any intelligent system and is particularly crucial to its performance. This ability, which is lacking in traditional expert systems, can be achieved by incorporating neural network learning mechanisms into expert systems. These learning techniques enable expert systems to modify and/or enrich their knowledge structures autonomously. As pointed out by Kandel and Langholz [9], intelligent hybrid systems offer the means to overcome some of the major drawbacks of conventional expert systems, such as: 1) their total reliance on consultation with human experts for knowledge acquisition (the knowledge acquisition bottleneck), 2) their inability to synthesize new knowledge, 3) their inability to allow for dynamic environments by modifying knowledge whenever it becomes necessary.

However, in combining such methods we should take care of not violating basic characteristics of expert systems [8], such as:

- the ability of receiving and representing knowledge elicited from human experts,
- the capability of dealing with data absence, and inquiring the user when additional data is necessary,
- the intelligibility of the knowledge base,
- the capability of explaining and justifying solutions or recommendations to convince the user that its reasoning is in fact correct.

Unfortunately this list of requirements imposes severe restrictions on the use of classical artificial neural networks for constructing CESs. Neural networks cannot directly encode structured knowledge. This makes difficult for them to receive knowledge elicited from an expert, and also to justify their conclusions. Neural nets superimpose several input-output samples on a black-box web of synapses. It is quite hard to know what a neural net has learned or forgotten. Many models (e.g., Backpropagation) are unable to perform incremental learning (that is imperative for a neural net being able to receive an initial load of expert knowledge, and to refine it with the experience [17]). It is also difficult (or impossible) for most neural net models to represent and deal with data absence, which is crucial for the inquiry process.

*Fuzzy Neural Networks* (FNN) appear as a solution able to meet all the above mentioned requirements [5,6,7,10,12,14,20,21]. Such networks, based on the fuzzy logic theory, encode directly structured knowledge, but in a numerical framework. They have been studied recently by many researchers. In this paper we will describe the FNN model introduced by Machado and Rocha for building CES, called *Combinatorial Neural Model* (CNM) [16,17], aimed at solving classification tasks.

As an additional advantage, FNNs provide to CESs powerful uncertainty management procedures for approximate reasoning. The fuzzy logic framework allows CESs to deal systematically with the vagueness, inaccuracy, incompleteness, and inconsistency frequently associated to the human reasoning [10]. This would enable CESs to better emulate human decision-making processes as well as allow for imprecise information and/or uncertain environment.

The purpose of present paper is to introduce some measures of uncertainty to be used by the connectionist inference machine of a classification expert system when performing inference, inquiry and explanation. Here, we

0-7803-0614-7/93$03.00 ©1993IEEE

do not describe the neural network structure in detail, nor the techniques for building/training them, because it was done before in [14,16,17], but we will focus our attention upon the uncertainty processing by the different elements of the network.

To accomplish this task, the paper is divided in the following sections: Fuzzy Neural Networks, which gives a summarized description of the CNM connectionist model, and introduces the uncertainty measures employed in this model; Consultation, which describes the control strategy used during consultation; Inference, which describes how classification is performed by FNNs; Inquiry, where the algorithm for the inquiring process is discussed in detail; Explanation, where the algorithms allowing the system to justify how a conclusion was achieved, or why an evidence was asked are introduced.

## II. FUZZY NEURAL NETWORKS

In CESs, the knowledge base is replaced by a fuzzy neural network called the *Connectionist Knowledge Base* (CKB). The concepts of the problem domain are represented by neurons whose activations can be interpreted as their degrees of possibility. The neurons are connected by synapses whose weights represent the degree of adhesion between the corresponding neuron concepts. The resulting network constitutes the CKB [15,18].

The fuzzy neural network associated to a classification task assumes a feedforward topology with three or more layers: the *Input layer* for evidences, *Hidden layers* for intermediate abstractions and the *Output layer* for hypotheses. It uses several types of neurons: *Fuzzy-Number* at the input layer, *Fuzzy-AND* at hidden layers, *Fuzzy-OR* at the output layer.

The Fuzzy-Number cells located at the input layer may receive input data in a symbolic or numeric form. In the second case, such neurons will perform the fuzzification of the input numeric data into possibility degrees, using the membership functions corresponding to their associated concepts [15].

The fuzzy-AND cells of the hidden layers implement a fundamental characteristic in the reasoning of humans: to chunk input evidences into clusters of information for representing regular patterns of the environment. These clusters are intermediate abstractions, used to reduce the computational complexity of performing the classification task.

The fuzzy-OR cells of the output layer compute the degree of possibility of each hypothesis, i.e., the degree of membership of the object under analysis to each class of a classification task. Note that every pathway reaching a fuzzy-OR neuron can be seen as an independent module competing with other modules for establishing the decision of the case.

Each class of the classification task has an independent network called *Hypothesis Network*. For example, Figure 2-I presents one hypothesis network in a FNN. The FNN architecture is strongly inspired in the *knowledge graphs* elicited from experts by the application of the knowledge acquisition technique of Rocha et al. [11,13]. In this technique, experts express their knowledge about each hypothesis of the problem domain by selecting a set of appropriate evidences and building an acyclic weighted AND-OR graph (called *Knowledge Graph*) to describe how these evidences must be combine to support decision making. The similitude between knowledge graphs and fuzzy neural nets allows the direct and easy translation of expert knowledge into the FNNs of the CKB [17,18].

### A. Uncertainty Management in FNNs

A FNN is capable of reasoning with three types of uncertainty: Fuzziness, Imprecision and Incompleteness [15]. For coping with these uncertainties, the FNN computes for every neuron (and consequently for every problem domain concept) a *Possibility Value Interval* (PVI), which contains its unknown possibility value. The PVI's lower bound represents the minimal degree of confirmation for the possibility value assignment. The PVI's upper bound represents the degree to which the evidence failed to refute the possibility value assignment. The interval's width represents the amount of ignorance associated with the possibility value assignment. Consummate concepts have PVI of width equal to zero. The absence of information is represented by the PVI [0, 1]. The fuzzy negation of an evidence presenting a PVI: [a, b] is given by the interval [1-b, 1-a]. A similar approach was used in the system PRIMO [1].

The above uncertainty management scheme can be easily implemented in fuzzy neural networks by defining two different activations for neurons: *Current Activation* (CA) representing the PVI's lower bound, and *Potential Activation* (PA) representing the PVI's upper bound, which is the maximum degree of activation a neuron will reach. We define the *Ignorance* (IG) of a neuron: IG = PA - CA. At the input layer, IG expresses the imprecision of input data. Neurons corresponding to unmeasured input concepts present IG equal to 1. At the hidden and output layers, IG expresses the maximum gain that neurons can obtain in their CA's if ignorance is solved at all neurons of the input level.

## III. THE CONSULTATION

The consultation process in a CES follows the hypothetico-deductive approach [19] mimicking human information processing characteristics. The consultation is organized in two different phases: the *passive* and the *active phase.* During the passive phase the user enters a set of triggering data into the system, which will be used to trigger hypotheses, forming the *Consultation Focus* (also called *Context* or *Differential*). Hypotheses that present, at the end of the passive phase, CA larger than a pre-defined *Triggering Threshold* are included into the consultation focus.

During the active phase the system tries to prove or refute the hypotheses belonging to the focus by actively inquiring the user. Only hypotheses with possibility degree larger than a predefined *Acceptance Threshold* ($T_{acc}$) are presented to the user as the problem solution. The focus may be revised periodically to account for new data.

The processes of *Inference, Inquiry*, and *Explanation* are required during the consultation. They are described in the following sections.

A. Inference

The goal of the inference process is to compute the degree of possibility of each hypothesis, and to point those classes having possibilities greater than the Acceptance Threshold as the problem solution. This is done by propagating the available input evidences forward in the network through two information flows. We define the *Current Evidential Flow* ($CEF_{ij}$) and the *Potential Evidential Flow* ($PEF_{ij}$) in a synapsis (i,j) as

$$CEF_{ij} = CA_i \cdot w_{ij} \qquad PEF_{ij} = PA_i \cdot w_{ij}$$

where $CA_i$ = current activation of neuron i, $PA_i$ = potential activation of neuron i, $w_{ij}$ = weight of the synapsis (i,j). If the synapsis (i,j) is inhibitory (fuzzy negation) then

$$CEF_{ij} = (1 - PA_i) \cdot w_{ij} \qquad PEF_{ij} = (1 - CA_i) \cdot w_{ij}$$

The inference process performs the forward propagation of the current and potential activations of the neurons of the input layer, according to the fuzzy evidence aggregation rules of the neurons in the network. The Fuzzy-AND and Fuzzy-OR neurons aggregate the incoming evidential flows according to the classical rules (minimum and maximum) of fuzzy logic, shown below:

For Fuzzy-AND neurons:

$$CA_X = \min_{i \in R_X} (CEF_i) \qquad PA_X = \min_{i \in R_X} (PEF_i)$$

For Fuzzy-OR neurons:

$$CA_X = \max_{i \in R_X} (CEF_i) \qquad PA_X = \max_{i \in R_X} (PEF_i)$$

where $R_X$ is the set of neurons that can send directly messages to the neuron X. Other norm and co-norms can be used for implementing Fuzzy-AND and Fuzzy-OR [3].

In FNNs based on CNM, if the PVIs of the input neurons are not divergent intervals across time, then the PVIs of the output neurons also will not be divergent intervals with time (See proof at appendix). Figure 1 shows the evolution of CA and PA of a hypothesis neuron H with time as evidences are measured. At time t, CA represents the support achieved for concept H, MA-PA is the concept refutation (MA being the potential activation at the beginning of the consultation), and PA-CA is the ignorance of H. When the curves of PA and CA meet, it means that there is no more ignorance to be solved in relation to the neuron H.

Only hypotheses exhibiting $CA \geq T_{acc}$ are accepted by the system as the problem solution. Hypotheses having $PA < T_{acc}$ are rejected by the system. Hypotheses showing $CA < T_{acc}$ and $PA \geq T_{acc}$ are considered undecided, requiring additional information for decision making. Figure 2 shows for example the inference calculation in two different networks. (Synaptic weights are denoted at the side of arcs. In both nets the hypotheses remained undecided for $T_{acc}$).

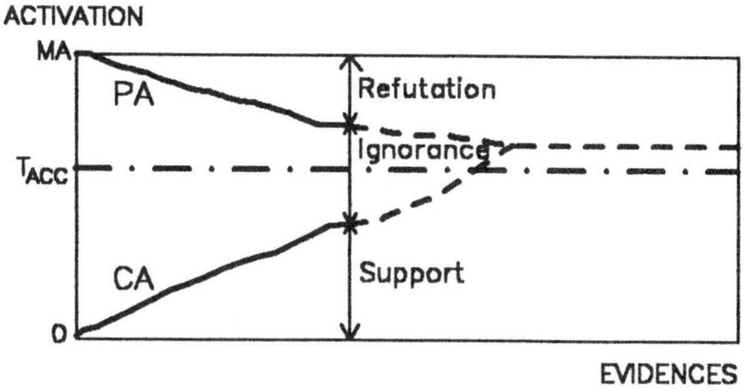


Fig. 1 - Potential and current activations of a hypothesis as a function of the evidences measured during a consultation.

B. Inquiry

The inquiry process has as goal to find the best question to be asked next to the user, among the input nodes having ignorance. If a neuron located at any level presents IG > 0, it implies that there is ignorance at the input layer, that if solved can potentially improve the current activation of this neuron. This makes possible to determine the best question through local decisions, producing a very efficient inquiry algorithm (no backtracking required), which is shown next:

***Algorithm INQUIRY***

1- Select the hypothesis H of the focus with PA > $T_{acc}$ and the largest IG
2- BACKWARD (QUERY, H)
3- If QUERY is not empty then ask to the user the question QUERY.
4- Stop

***Algorithm BACKWARD( QUERY, N)***

0- Comment. N: neuron from which we want to determine the best question QUERY: input node selected as the best question
1- Determine R = {X | ($PEF_{XN}$ - $CEF_{XN}$ > 0) and ($PEF_{XN}$ > $CA_N$)}
2- Select X ∈ R that presents minimum PA if N is a Fuzzy-AND neuron, or the maximum PA if N is a Fuzzy-OR neuron
3- If X is an input node
   then QUERY ← X
      Stop
   else BACKWARD(QUERY,X)
      Stop

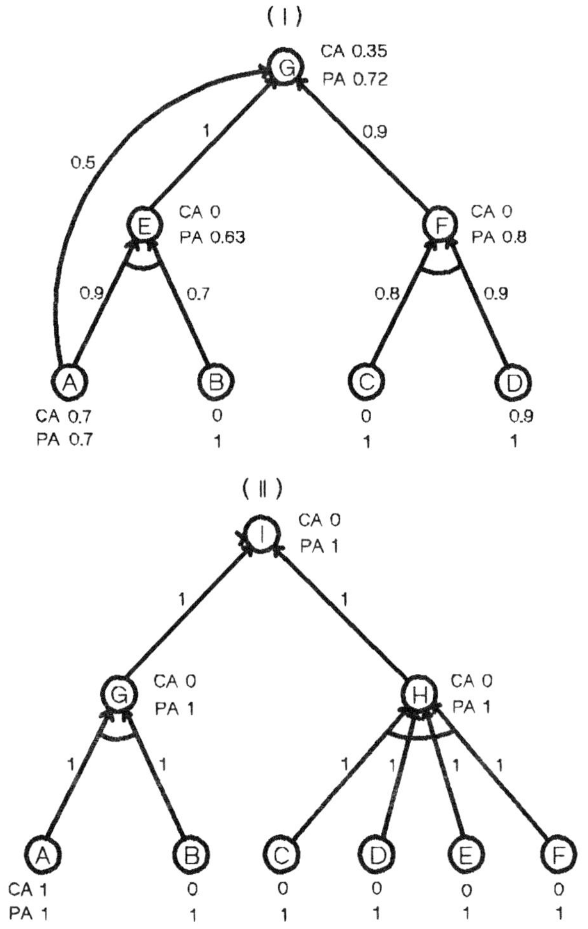


Fig. 2 - Inference process. In the network I the evidences B and C are unknown and the evidence D is imprecise. In the network II only the evidence A is known.

This algorithm can be applied to general networks using Fuzzy-AND and Fuzzy-OR nodes at any layers. The heuristic rules employed in this algorithm provide a powerful pruning mechanism, avoiding to make questions and to perform tests that cannot contribute to solve the problem. Hypotheses of the focus whose PA's become smaller than $T_{acc}$ are not more investigated during a consultation (rejected hypotheses). For instance, in Figure 2-I if the neuron C receives CA=PA=0.9 then the evidence associated to the neuron B will not be more asked to the user (assuming $T_{acc}=.7$). The algorithm pursues all hypotheses of the focus, being able to deal with multiple fault problems. Sometimes the algorithm may ask the user to repeat a previous measurement, which was early informed with too much imprecision. If we are interested in stopping the investigation of hypotheses as soon as they are accepted, the step 1 of the algorithm Inquiry should be replaced by:

*1- Select the undecided hypothesis of the Focus with the largest IG.*

This inquiry algorithm takes into consideration only the informational power of the neurons. This can produce some troubles. For instance, at the node I in the network II of the figure 2, the system does not have any information to select between B or C. However, the choice B is clearly more adequate, because it suffices to make one question to the user to eventually solve the problem. In the network I of the figure 2, if B represents a X-Ray and C a biopsy, the natural choice should be the X-Ray, a procedure of lower cost and risk. However the presented algorithm would choose the biopsy. To avoid this problem, it is fundamental to make the system sensitive to the costs (and or risks) associated to the measurement procedures. With this intent, we will define for every neuron a variable called *Neuron Utility* (NU), that expresses the utility associated to to solving the ignorance associated to the neuron. To be able to make local decisions we need to propagate the utility information of input evidences in the network. This is done defining a third flow in the synapses, called the *Utility Flow* (UF):

$$UF_{ij} = NU_i \,.\, s(PA_i - CA_i) \quad = \text{utility flow at synapsis } (i,j)$$

where $s(x) = s(PA_i\text{-}CA_i)$ is the unitary step function: $s(x)=1$ if $x>0$, $s(x)=0$ if $x \leq 0$. Each neuron computes its utility aggregating the utility flows received from other neurons. Fuzzy-AND neurons aggregate cost utilities according to

$$NU_j = \sum_{i \in R_{AND}} UF_{ij} \qquad \text{where } R_{AND} = \{ i \mid CEF_{ij} < PA_j \}$$

Fuzzy-OR neurons aggregate cost utilities according to

$$NU_j = \min_{i \in R_{OR}} \{ UF_{ij} \}$$

where $R_{OR} = \{ i \mid (PEF_{ij} > T_{acc}) \text{ and } (PEF_{ij} > CA_j) \text{ and } (PEF_{ij} - CEF_{ij} > 0)\}$

These aggregation functions should be modified if we intend to work with risks rather than costs.

To be able to describe the benefit/cost relation of a pathway, we define the variable *Merit* ($M_{ij}$) of an arc (i,j)

$$M_{ij} = (PEF_{ij} - CEF_{ij}) \,/\, UF_{ij}.$$

This division is defined equal to 0 if the numerator and the denominator become simultaneously equal to 0.

Now a *Utility Conscious Inquiry Algorithm* can be easily obtained by replacing the line 1 of the algorithm INQUIRY by:

*1- Select the hypothesis H of the focus with PA > $T_{acc}$ and the largest ratio IG/NU* and by replacing the lines 1 and 2 of the algorithm BACKWARD by:

*1- Determine $R = R_{AND}$, or $R_{OR}$ depending on the type of the neuron N.*
*2- Select $X \in R$ that presents the largest merit $M_{XN}$.*

For example, assuming NU=1 for every input neuron in the network II of Figure 2, the algorithm INQUIRY will select the node B to be inquired first.

## C. Explanation

We provide explanations for the following two types of user inquiries during a consultation: those that ask HOW a particular conclusion was reached, and those that ask WHY a particular question was formulated [8].

The structure of FNNs is highly modular making easy to justify their conclusions and questions. Each hypothesis network is formed by a set of pathways that compete for sending the largest evidential flow to the hypothesis neuron. Since each neuron in a pathway can be seen as a heuristic fuzzy rule, it is quite simple to provide a trace style explanation of the system response just by showing the structure of the winning pathway as a chain of pseudo-production rules.

A pseudo-production rule is a practical way of describing neurons in the winning pathway. For a hidden neuron it has the following format:

If $Y_1(w_1,t_1)$ $\Psi$ ... $\Psi$ $Y_n(w_n,t_n)$ then N

where

- X = hidden neuron that is being described
- $Y_i$ = code (or name of the associated concept, if available) of the $i^{th}$ neuron of the fan-in of X
- $(w_i,t_i)$ = weight and synaptic type (excitatory or inhibitory) of the synapsis connecting $Y_i$ to X
- $\Psi$ = logical connective associated to neuron X (fuzzy-AND or fuzzy-OR)
- N = Neuron to which X is connected in the pathway under consideration

For an input neuron belonging to a pathway linking it directly to the hypothesis neuron, the pseudo-production rule is expressed as:

If $X(w_X, t_X)$ then H

where $(w_X,t_X)$ = weight and synaptic type of the synapsis from X to N.

An algorithm for HOW-inquiries, based on this approach, is presented next:

***Algorithm HOW(H)***
0- Comment: Justification of the acceptance of hypothesis H
1- Present $CA_H$ and show that $CA_H \geq T_{acc}$
2- Select the neuron X sending the largest CEF to H
3- If X is an input neuron
then Present its pseudo-production rule and show $CA_X$
else Trace(X, H)
4- Stop

***Algorithm TRACE(X,N)***
0- Comment: Display of the synapsis X→N and its antecedents
1- Present the pseudo-production rule describing neuron X
2- Present the CA of neurons belonging to the fan-in R of X
3- For each neuron $Y \in R$ do
- If Y does not belong to the input layer
then TRACE(X, N)
Endfor
4- Stop

The algorithm for answering WHY an evidence E is being asked by the system in the context of a hypothesis H is presented next:

***Algorithm WHY(E, H)***
0- Comment: Justification why E is needed in the context of H
1- Justify why H is being investigated: it belongs to the focus, it was not yet rejected, and it presents the best ratio IG/NC
2- Justify why in this context E was selected to be inquired, showing the sequence of selections of synapses with largest merit, that were made departing from H and going backwards in direction of evidence E.
3- Show the PVI that H will have if we receive a PVI equal to [1, 1] for E.
4- Stop

The algorithm HOW may be invoked by the user at the end of the consultation, whereas the algorithm WHY may be invoked after every question formulated by the system.

# IV. CONCLUSION

Expert systems and neural networks represent complementary approaches to knowledge representation: the logical, cognitive, and mechanical nature of expert systems versus the numeric, associative, and self-organizing nature of the neural network. The use of fuzzy neural networks for building connectionist expert systems has several advantages. Besides the capabilities of easy construction of the FNN from expert knowledge, incremental learning, and knowledge base readability, it provides an easy implementation of basic functions of expert systems, such as inference, inquiry, and explanation, as well as the ability to deal with the vague, uncertain and partial data, commonly found in practical applications.

The inference, inquiry, and explanation processes in CESs present low computational cost because they are performed using local decisions on an acyclic network. No backtracking is required. Local decisions in the inquiry process were made possible by propagating simultaneously in the network three different informations: the current evidential flow, the potential evidential flow and the utility flow. The complexities for the above mentioned processes are:
$O(n+s)$ for inference, where n = number of neurons and s = number of synapses;
$O(f.l+h)$ for inquiry, where l = number of layers, f = maximum size of the fan-in of a hidden neuron (usually limited to the number 7 of Miller) and h = number of hypotheses;

O(l.f+h+n) for HOW explanation;
O(l.f+h) for WHY explanation.

In the sense of the Dreyfus' definition [2], we can tell that the system seems to have intuition: "the ability of effortlessly and rapidly to associate with one's present situation an action or decision which experience has shown to be appropriate". The utility conscious inquiry process introduced in this paper allows to lower significantly the consultation cost/risk and can give to the system the common sense property presented by experts when selecting tests to be performed.

The CES scheme described in this paper may be seen as a building block for construction of more complex expert systems, whose cognitive task can be decomposed in different interconnected elementary classification tasks. For instance, in the medical domain the physician usually performs a diagnostic task, before proceeding the treatment prescription and the prognostic tasks. Each of these tasks should be implemented by a different FNN.

The model and the methods described in this paper were implemented successfully in the system NEXTOOL [18], a shell for building classification expert systems able to learn from experience.

## REFERENCES


[1] J.K. Aragones, P.J. Bonissone, and J. Stillman, "PRIMO: a tool for reasoning with incomplete and uncertain information," *Proceedings of the Third International Conference IPMU*, Paris, 1990, pp. 325-327.

[2] R. Davis, (1989). "Expert systems: how far they can go?," *AI Magazine*, pp. 65-84, Spring 1989.

[3] D. Dubois, and I.I. Prade, "A class of fuzzy measures based on triangular norms," *Int. J. General Systems*, vol.8, pp. 43-61, 1992.

[4] S.I. Gallant, "Connectionist expert systems," *Communications of the ACM*, vol. 31, pp. 152-169, 1992.

[5] T. Hashiyama, T. Furuhashi, Y. Uchikawa, "An interval fuzzy model using a fuzzy neural network," *International Joint Conference on Neural Networks*, Baltimore, 1992.

[6] Y. Hayashi, "A neural expert system using fuzzy teaching input," *IEEE International Conference on Fuzzy Systems*, San Diego, 1992.

[7] L.S. Hsu, H.H. Teh, P.Z. Wang, and K.F. Loe, "Fuzzy neural-logic system," *International Joint Conference on Neural Networks*, Baltimore, 1992.

[8] P. Jackson, *Introduction to Expert Systems*, Workingham: Addison Wesley, 1990.

[9] A. Kandel, and G. Langholz, *Hybrid Architectures for Intelligent Systems*, Boca Raton, Fl: CRC Press, 1992.

[10] B. Kosko, *Neural Networks and Fuzzy Systems*, Englewood Cliffs, NJ: Prentice Hall Inc., 1992.

[11] B.F. Leão, "Proposed methodology for knowledge acquisition - a study on congenital heart diseases diagnosis," *Methods of Inf. in Medicine*, vol. 29, pp. 30-40, 1990.

[12] C.T. Lin, and C.S.G. Lee, "Real-time supervised structure / parameter learning for fuzzy neural network," *IEEE International Conference on Fuzzy Systems*, San Diego, 1992.

[13] R.J. Machado, A.F. Rocha, and B.F. Leão, "Calculating the mean knowledge representation from multiple experts," in *Multi-person Decision Making Models Using Fuzzy Sets and Possibility Theory*, J. Kacprzyk and M. Fedrizzi, Eds. Dordrecht: Kluwer Academic Publishers, 1990, pp. 113-127.

[14] R.J. Machado, and A.F Rocha, "The combinatorial neural network: a connectionist model for knowledge based systems," in *Uncertainty in Knowledge Bases*, B. Bouchon, L. Zadeh and R. Yager, Eds. Berlin: Springer Verlag, 1991, pp. 578-587.

[15] R.J. Machado, and A.F. Rocha, "A hybrid architecture for fuzzy connectionist expert systems," in *Hybrid Architectures for Intelligent Systems*, A. Kandel, G. Langholz, Eds. Boca Raton, Fl: CRC Press Inc, 1992, pp. 136-152.

[16] R.J. Machado, and A.F. Rocha, "Evolutive fuzzy neural networks," *IEEE International Conference on Fuzzy Systems*, San Diego, 1992.

[17] R.J. Machado, C. Ferlin, A.F. Rocha , and G.J. Erthal, "Incremental learning in fuzzy neural networks," *International Conference on Information Processing and Management of Uncertainty in Knowledge-Based Systems, IPMU*, Palma de Mallorca, 1992.

[18] R.J. Machado, A.F. Rocha, and C. Ferlin, "NEXTOOL - an environment for connectionist expert systems," *Third Annual Symposium of the International Association of Knowledge Engineers*, Washington, 1992.

[19] R.S. Patil, "Artificial intelligence techniques for diagnostic reasoning on medicine," in *Exploring Artificial Intelligence: Surveys Talks from the National Conference on AI*, H.E. Schrobe and AAAI, Eds. San Mateo: Morgan Kauffmann, 1988, pp. 347-379.

[20] S.G. Romaniuk, and L.O. Hall, "Learning fuzzy information in a hybrid connectionist, symbolic model," *IEEE International Conference on Fuzzy Systems*, San Diego, 1992.

[21] F. Yuan, A. Feldkamp, L.I. Davis, and G.V. Puskorios, (1992). "Training a hybrid neural-fuzzy system," *International Joint Conference on Neural Networks*, Baltimore, 1992.


## APPENDIX

**Theorem 1**: The partial derivatives of activations of output neurons with respect to the activations of input neurons in a FNN based on the CNM model are positive or null.

Proof: Let H be an output neuron, and X an input neuron of a FNN with activations $p_H$ and $p_X$ respectively. Let i be anyone of the pathways towards H departing from the neuron X.

$$CEF_i = w_i \cdot \min(w_1 \cdot p_1, \ldots, w_X \cdot p_X, \ldots, w_n \cdot p_n)$$

$$p_H = \max(CEF_1, \ldots CEF_i, \ldots)$$

Let $\Delta p_X$ be a positive variation on $p_X$. It will produce

$\Delta CEF_i \geq 0$ and $\Delta p_H \geq 0$

Hence we have the ratio $\Delta p_H / \Delta p_X \geq 0$

For $\Delta p_X \to 0$ the theorem is proved.

**Theorem 2**: If the PVIs of input neurons of a FNN based on CNM are not divergent across time, then the same will happen with the PVIs of output neurons.

Proof: Let X be an input neuron presenting at time t a PVI = [p'(t), p''(t)], and H be an output neuron presenting at this time PVI = [$\pi'(t)$, $\pi''(t)$].

At time $t+\Delta t$ we will have $p'(t+\Delta t) \geq p'(t)$ and $p''(t+\Delta t) \leq p''(t)$ (non-divergent PVI),

and by theorem 1:

$\pi'(t+\Delta t) \geq \pi'(t)$ and $\pi''(t+\Delta t) \leq \pi''(t)$

Hence the PVI of H is not divergent.

# Measurement of Fuzzy Values Using Artificial Neural Networks and Fuzzy Arithmetic

Andreas Ikonomopoulos, Lefteri H. Tsoukalas, Robert E. Uhrig
Center for Neural Engineering Applications
Department of Nuclear Engineering
The University of Tennessee, Knoxville
Knoxville, TN, 37996-2300

***Abstract***-**A methodology for monitoring complex systems utilizing artificial neural networks and fuzzy arithmetic is presented. It employs the notion of a virtual measuring device, i.e., a software based instrument for the "measurement" of user-specified dynamic variables with operational significance. Neural networks are utilized for mapping a set of complex, temporal, input patterns to a simplified set of membership functions. The resultant membership functions are supplied to a rule-based system which draws a conclusion about the validity of the network responses based on the statistical characteristics of the original training signals. The notion of time is explicitly incorporated into the decision-making procedure, rendering the measuring process a dynamic one. The rule-based system is supplemented by fuzzy algebraic techniques composing an optimization algorithm capable of identifying the most appropriate shape and position of the resultant membership functions in the universe of discourse. Data obtained from an experimental nuclear reactor during a start-up period are utilized to demonstrate the excellent tolerance of the methodology to noisy and faulty signals.**

## I. Introduction

It may be argued, rather persuasively, that much of the fame surrounding the theory of fuzzy sets is due to the efficient and successful implementation of various fuzzy reasoning schemes in industrial applications. The approximate reasoning method introduced by Mamdani, as early as 1974, has been the milestone for further development of fuzzy controllers in industrial applications [1]. The calculus of linguistic control rules utilized in fuzzy inferencing, is simple and almost intuitive. The practice in the design of systems employing fuzzy rules is based on a cut-and-trial procedure [2]. The designer of the basic tools for fuzzy reasoning assumes initially a set of prototype membership functions and then examines if they satisfy the needs of the application environment. If they are found to be unfit, a tuning procedure is applied which is based on the experience of the modeller. This means that there is no formal procedure for designing membership functions. In addition, once the membership functions have been posed in the universe of discourse, they remain unchanged throughout the operational life of the fuzzy algorithm. As a consequence, the fuzzy system does not follow possible changes in the application environment. This approach renders the system static rather than dynamic. Takagi [3] and Zadeh [2] have proposed the utilization of artificial neural networks (ANNs) as means for a systematic approach to the issues of inference rule development and calibration. Significant work on this field has been produced by Takagi and Hayashi [4], Hayashi et al. [5], as well as, Keller and Tahani [6].

The present work is concerned with the development of a formal approach for designing membership functions utilized in a general purpose approximate reasoning scheme and not with the inference scheme itself. The methodology employs ANNs as membership function generators based on a library of prototype membership functions. A set of neural networks receives a number of spatiotemporal patterns reflecting the present state of the system being monitored. These patterns are composed of directly measurable system parameters. The set of neural networks is trained to compute the state of another system parameter not in the form of a time series, but rather in the form of membership functions describing the state of a non-directly measurable system parameter. The membership functions computed from independently firing neural networks are used as input to a rule-based system that decides upon the validity of the network responses. Fuzzy arithmetic operations are used to enhance the decision-making abilities of the rule-based diagnostician, constituting an optimization algorithm which draws a conclusion concerning the shape and position of the most appropriate membership functions in the universe of discourse.

The rule-based system along with the set of neural networks establish a monitoring device that evaluates the state of non-physically measurable parameters. This type of monitoring device is a software-based instrument, suitable for measuring user-specified dynamic variables with operational significance [7]. Usually, these variables can not be measured directly. Their measurement may contain significant time lags and in general the failure of a particular sensor requires that this variable should be computed from other system parameters [8]. A major advantage of virtual instruments is that they are software, rather than hardware based. Thus, their function can be modified by software re-programming, not hardware alteration. Furthermore, the output of the measuring device in the form of a membership function can be used as input to a fuzzy controller thus, simplifying the inference scheme currently employed in this type of controllers since there is no need for fuzzification of crisp input values.

## II. Model Description

The outline of the proposed methodology is demonstrated utilizing actual data obtained during a start-up period of the High Flux Isotope Reactor (HFIR) at the Oak Ridge National Laboratory. The monitored parameter is the position of the secondary flow control valve which controls the water flow in the secondary side of the system. Five parameters have been chosen for describing the secondary flow control valve position: neutron flux, primary flow

0-7803-0614-7/93$03.00 ©1993IEEE

pressure variation (DP), core inlet temperature, core outlet temperature and secondary flow. The time series of these five parameters are used to train a set of five neural networks, where each one them is responsible for recognizing a specific position of the valve-disc. The output of every neural network is a membership function uniquely labeling a particular position of the secondary flow control valve. The behavior of the valve is represented in the universe of discourse of human linguistics with the fuzzy variable VALVE_POSITION that takes five fuzzy values: *closed, partially_closed, medium, partially_open* and *open*. A membership function, namely: $\mu_{closed}$, $\mu_{partially\_closed}$, $\mu_{medium}$, $\mu_{partially\_open}$ and $\mu_{open}$ is defined for every fuzzy value, respectively. These five membership functions compose a library of prototypes and are considered sufficient for describing the position of the valve at every instant during a start-up period. A schematic representation of the universe of discourse of the fuzzy variable VALVE_POSITION, is given in Fig. 1.

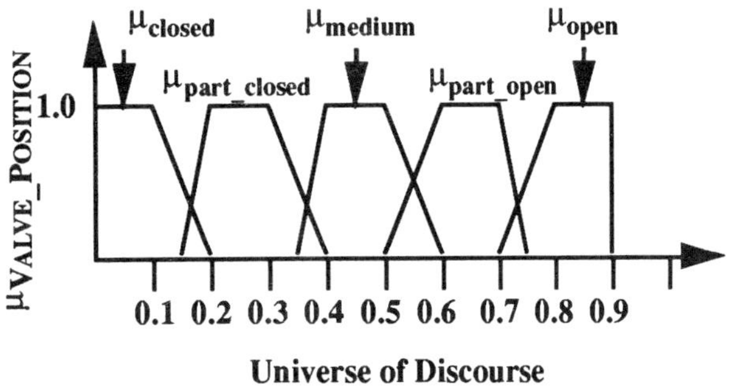


Fig. 1. Membership function labeling of possible valve positions.

The area occupied by every membership function in the universe of discourse depicts the uncertainty associated with this particular class [9]. The fuzzy subset (or class) *open* in the universe of discourse is characterized by the membership function $\mu_{open}$ (Fig. 1). The membership function assigns a degree of membership (or level of presumption) to a value of the base variable in the class *open*. This framework provides a natural way of dealing with problems where the source of imprecision is the absence of sharply defined boundaries of class membership, rather than the presence of random variables [10]. The membership functions developed for this particular study have trapezoidal shape, which is very useful for performing computations in the area of fuzzy control [11]. The shape of the membership function assigned to each state of the valve is unique, and therefore there is a sharp distinction between different states. Furthermore, an ANN trained to recognize a specific complicated time pattern will lose much of its ability to sustain noisy input signals. This stems from the network's tendency when presented with distorted inputs to produce averaged forms of the desired output, missing therefore vital pieces of information [8]. This handicap is overcome by the proposed technique which has an output that is a simple membership function.

### A. *Artificial Neural Networks*

Artificial neural networks learn from experience through a "learning algorithm," generalize from previous examples to new ones, and abstract essential characteristics from inputs containing irrelevant data [12]. The most successful learning algorithm devised for feed-forward architectures is known as backpropagation. It was invented independently several times, though the scheme most widely used is the one proposed by Rumelhart, Hinton, and Williams [13]. Backpropagation is a learning algorithm for neural networks that seeks to find weights, such that given an input pattern from the training pairs of input/output patterns, the network will produce the output of the training set. Having learned the underlying relation between inputs and outputs from the training set, a neural network provides an output when supplied with a previously "unseen" input. The network response is based on its ability to learn through the "learning algorithm" the fundamental relationship between input and output patterns from the training sets.

Cybenko [14] has shown that sufficiently complex multilayer feed-forward networks are capable of approximating an unknown mapping, $f$: $\mathbf{R}' \rightarrow \mathbf{R}$ arbitrarily well. One of the major advantages of artificial neural networks is their unique ability to be trained in a *user-defined environment*. Funahashi [15] has proved that any continuous mapping can be approximately realized by Rumelhart-Hinton-Williams' neural networks with at least one hidden layer having a sigmoid transfer function. Lapedes and Farber [16] have shown experimentally, that a neural network is capable of making predictions of arbitrarily complex mappings with very good accuracy. This property is utilized in the present work where the ANNs predict the possible states of the VALVE_POSITION in the near future. Thus, the neural networks receive information concerning the state of the system at time $t$ and estimate what the position of the valve may be at time $t+\Delta t$. Since the present work is based on sampled data provided from HFIR, the time interval $\Delta t$ is set equal to one sampling interval (i.e., 16 seconds). A higher value for $\Delta t$ would significantly affect the accuracy of the prediction [17]. Furthermore, the time interval of 16 seconds is considered sufficient for decision-making procedures, and offers adequate time margin for performing computational tasks without using special hardware to speed up the computation.

The neural networks contemplated in this research are three-layer feed-forward networks. The algorithm for training is backpropagation using generalized delta learning and a momentum term [12]. The input layer of the network consists of five neurons (processing elements), each one receiving input from a particular time series, i.e., the network receives simultaneous values of five variables. The output layer consists of ten neurons corresponding to the coordinates of ten points on the trapezoidal membership function describing the position of the valve in the universe of discourse. Typically, 7,000 training cycles will produce a sum square error of 0.02 when 30 nodes are used in the hidden layer.

### B. *Dissemblance Index*

Generally, the outputs of the neural networks are somewhat different than the membership functions for which the networks are trained. Moreover, one or two membership functions (since overlap has been allowed) represent correct values, while the rest need to be ignored. In order to identify the correct output and eliminate the information surplus created by network abundance, the membership functions calculated by the neural networks are treated as *fuzzy numbers*. The *dissemblance index* [18] is an estimate of the distance between the prototype membership function and the membership function predicted from the network. Suppose, for example, that the network that recognizes the VALVE_POSITION fuzzy value *closed* has been trained on the output membership function $\mu_{closed}$, and after training it produces an actual output membership function, $\mu^*_{closed}$. The two membership functions are treated as fuzzy numbers, call them **C** and **C***, each with a trapezoidal shape and support on the universe of discourse [0,1]. The support of each function is an interval, i.e., $C=[c_1,c_2]$ and $C^*=[c_1^*,c_2^*]$. We can compute a real number, $\delta(C,C^*) \in [0,1]$, (called the *dissemblance index*) which is the distance between **C** and **C*** in terms of the support of each fuzzy value and the α-cuts, $\alpha \in [0,1]$:

$$\delta(C,C^*) = \frac{1}{2}\int_{\alpha=0}^{\alpha=1}\left(\left|c_1^{(\alpha)} - c_1^{*(\alpha)}\right| + \left|c_2^{(\alpha)} - c_2^{*(\alpha)}\right|\right)d\alpha \qquad (1)$$

The integral presented in (1) is multiplied by 1/2 so that the result of the integration will be normalized in the interval [0,1]. If $\delta(C,C^*) \approx 0$, then **C** and **C*** are almost identical. On the other hand, if $\delta(C,C^*) \approx 1$ then **C** and **C*** are totally different. The computed membership functions are compared with the prototype membership functions, and the *dissemblance index*, $\delta_k(j)$, is calculated for every input vector, **j**, and network **k**. The maximum value, **max**_$\delta_k$, of all $\delta_k$'s is computed from the responses of every network **k** and is set as the upper limit of the dissemblance index values attained by every network-predicted membership function. Hence, when the set of networks is tested on a new (previously "unseen") set of input vectors **j***'s, the membership functions estimated from every network will be accepted only if the corresponding dissemblance index values are less or equal to the **max**_$\delta_k$ value for network **k** (*dissemblance index criterion*). Since all five neural networks will fire when presented with an input vector, it is the *dissemblance index criterion* that provides a basis for accepting or discarding a network response.

### C. *Information Fusion*

It is in the nature of neural memory to associate new patterns with existing ones in a manner that can implicitly construct the variable *time*. In other words, neural network learning algorithms lack a mechanism for associating time labels with the corresponding patterns. Therefore, every backpropagation neural system is by nature *static*, rather than *dynamic*. On the other hand, the notion of measurement is associated with the variation of system parameters in time. The notion of time is incorporated into the measurement process through a *rule-based* system that takes a decision based on the outputs of all firing networks satisfying the *dissemblance index criterion*. Its operation is dynamic in the sense it takes into consideration the previous and current responses, as well as predictions about the future from every firing neural network. The acceptance, or rejection, of a network response depends only on the responses of this particular network in the past, the present, and its predicted values about the near future. In this approach the rule-based system "fuses" information pertaining to a single network rather than combining information supplied from different networks. The decision made is a dynamic one, since the variable time is explicitly incorporated into the decision process. Therefore, although static tools (ANNs) are used for mapping purposes, it is the information fusion tool (rule-based system) which makes the virtual measuring device a dynamic one.

The neural networks used for computing the membership functions are well known for their ability to preserve, as well as analyze all statistical properties of the incoming signals. Apparently, these statistical characteristics are inherent properties of the system generating the input signals (i.e., the HFIR). Thus, models containing as independent variables the training signals offer a very good estimation of the statistical properties implied in the membership functions calculated from the neural networks. A number of multi-regression models has been developed containing the position of the valve at the next time step as dependent variable and lagged values of the input signals as independent parameters. The model selection criteria are the coefficient of multiple determination $R^2$, the Mallow conditional mean-square error criterion (MC), the prediction criterion (PC), and the Akaike Information Criterion (AIC) [19]. The results obtained suggest that the optimum number of time lags required for a global system representation is four.

The rule-based system takes into consideration the responses of each network during the previous four time steps, the present time step, as well as the prediction about the next time step. Since six consecutive network responses are included in the decision process, there exist 64 possible dyadic combinations of network estimations. The dyadic representation of the network responses stems from the satisfaction, or not, of the *dissemblance index criterion* at every time step. The rule-based procedure is designed to abide with the following general criterion, called the *dynamic behavior criterion*:

**IF**
*FOR EVERY NETWORK **k***
*THE DISSEMBLANCE INDEX CRITERION IS SATISFIED*
*(1) AT THE PRESENT TIME STEP,*
*(2) AT THE FUTURE TIME STEP,*
*(3) DURING AT LEAST THREE CONSECUTIVE TIME STEPS*
**THEN**
*THE NETWORK **k** ESTIMATION FOR THE PRESENT TIME STEP MUST BE ACCEPTED*
**OTHERWISE**
*THE NETWORK **k** ESTIMATION FOR THE PRESENT TIME STEP MUST BE DISCARDED*

The rule-based procedure operates in a dynamic fashion in a sense that it automatically accepts or discards present responses, and then considers the new values when it draws conclusions during future evaluations. As an example, consider the case where the present network evaluation satisfies the *dissemblance index criterion*, but the rule-based procedure decides to discard this value according to the *dynamic behavior criterion*. At the next time step when it will consider this value as a previous network response, it will conceive it as not having satisfied the *dissemblance index criterion*. The same applies in reverse. In order to model the *dynamic behavior criterion* stated above, a total of 11 propositions, in the form of IF-THEN rules, is adequate to satisfy all possible response scenarios from an independent network.

The overlap of adjacent membership functions creates a decision-making environment where the state of the valve can be simultaneously classified in two different categories (i.e., *medium* and *partially_open*). Apparently, this type of representation contradicts the deterministic behavior of the physical system since it can not be at the same time in two mutually exclusive or even neighboring states. This issue is resolved by introducing the grade of satisfaction of the *dissemblance index criterion*. The characteristic equation of the set of *satisfactory_responses (SR)* is considered to be continuous, therefore rendering the set *SR* fuzzy. The grade of membership to the set of *satisfactory_responses* depends on how close to $\mathbf{max\_\delta_k}$ is every network $\mathbf{k}$ prediction, $\delta_k(j)$, for the input vector $\mathbf{j}$. The membership function assigning a certain degree of belief to the network $\mathbf{k}$ prediction which has satisfied the *dissemblance index criterion*, is given by the following equation

$$\mu_{SR_k}(\delta_k(j)) = \frac{1 - \dfrac{\delta_k(j)}{\max\_\delta_k}}{\sum_k \left(1 - \dfrac{\delta_k(j)}{\max\_\delta_k}\right)} \tag{2}$$

where, the summation in the denominator holds for all possible firing networks $\mathbf{k}$, and the valuation set for the elements of *SR* is the real interval [0,1]. The membership function $\mu_{SR_k}$ attains its minimum value when $\delta_k(j) = \mathbf{max\_\delta_k}$ and its maximum value when the denominator is equal to the nominator (i.e., one firing network only). The satisfaction of the *dissemblance index criterion* defines the domain of $\mu_{SR_k}$ for every network $\mathbf{k}$, in the interval $[0, \mathbf{max\_\delta_k}]$. Equation (2) defines the level of presumption of satisfaction of the *dissemblance index criterion* from network $\mathbf{k}$. The network predicted membership functions are treated as fuzzy numbers, and their product with the level of presumption of their own satisfactory responses is calculated. The multiplication of a fuzzy number $A$ having an interval of confidence $A_\alpha = [a_1^{(\alpha)}, a_2^{(\alpha)}]$ for the level of presumption $\alpha \in [0,1]$, with a crisp number $K \in \mathbf{R}$ is defined as [18]

$$K * A_\alpha = [K,K](*)[a_1^{(\alpha)}, a_2^{(\alpha)}] = [Ka_1^{(\alpha)}, Ka_2^{(\alpha)}] \tag{3}$$

where the result of the multiplication is a new fuzzy number. The fuzzy numbers resulting from the multiplication of the prototype membership functions with the degree of satisfaction of the *dissemblance index criterion* $\mu_{SR_k}(\delta_k(j))$, are added in the universe of discourse creating a new fuzzy number. The addition of two fuzzy numbers $A$ and $B$, with $A_\alpha = [a_1^{(\alpha)}, a_2^{(\alpha)}]$ and $B_\alpha = [b_1^{(\alpha)}, b_2^{(\alpha)}]$ their intervals of confidence for the level of presumption $\alpha \in [0,1]$, is defined as [18]

$$A_\alpha(+)B_\alpha = [a_1^{(\alpha)} + b_1^{(\alpha)}, a_2^{(\alpha)} + b_2^{(\alpha)}] \tag{4}$$

The resultant membership function computed from the summation of independent membership functions satisfying the *dissemblance index* and *dynamic behavior criteria* describes uniquely and unambiguously the state of the valve.

## III. Model Testing and Results

The proposed methodology has been tested by substituting each one of the five time series with a signal corresponding to 100% random noise. This random signal along with the remaining four original signals compose the test vectors supplied to the measuring device during the recall process. The results presented in Fig. 2 depict the substitution of the secondary flow signal by a totally random one. In Fig. 2, the responses of the virtual measuring device are plotted with a frequency of appearance equal to one every 100 time steps for the total start-up duration. The measuring device has successfully recorded the abrupt opening of the valve at around 100 time steps after the start-up initiation, where the VALVE_POSITION has been identified as *medium*.

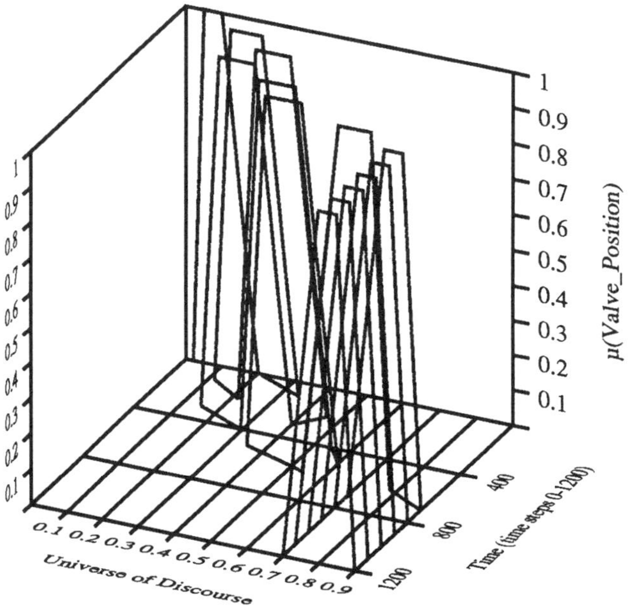


Fig. 2. Virtual measurements for time steps 0-1200 when the secondary flow signal has been substituted with 100% noise.

Considering that the crisp measurements of the position of the valve attain a maximum value in the vicinity of 0.5, one may deduce from Fig. 1 that the characterization is correct. Around 700 time steps after the transient initiation the valve-position is characterized as *open*, and the same characterization is repeated thereafter. By taking into account that at the same moment the crisp value of the valve-position becomes 0.8 and later increases to 0.9, it may be concluded that the identification of the state of the valve is accurate. Furthermore, during the time interval between time steps 200 and 700 the position of the valve-disc is initially defined as *medium* and later as *partially_open*, following the transition of the valve from one state to another.

The accuracy of the responses obtained from the virtual measuring device is checked by calculating the centroids of the resultant membership functions at every time step. The results are compared to those obtained when all five input signals are supplied to the virtual measuring device (training period), and the Euclidean distance is computed between the centroids of the prototype and actual membership functions. The results of the comparison are shown in Fig. 3, where more than half of the membership functions have been predicted with 100% accuracy. In addition, a quarter of them has been calculated with a norm equal to or less than 0.05. For the remaining time steps, the membership functions have been estimated with an Euclidean norm from the expected values, equal to or less than 0.35.

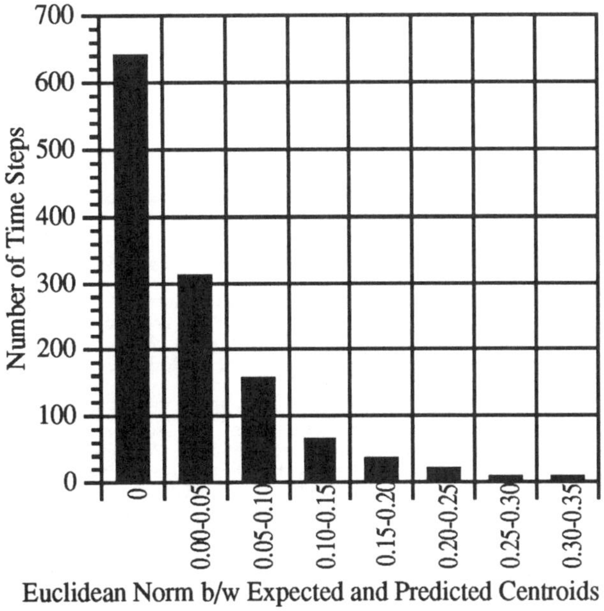


Fig. 3. Distribution of the Euclidean norms b/w expected and predicted centroids when the secondary flow signal has been eliminated.

Similar results are obtained when any one of the remaining four input time series is replaced by a random signal. The average Euclidean norm is computed for every substitution of the original signals with a random one and the results are plotted in Fig. 4. A look at Fig. 4 reveals a striking difference between the case where the inlet-temperature signal has been eliminated and the other four cases. Although, the average Euclidean norms computed for the other four tests are around 0.035, the replacement of the inlet-temperature signal by random noise increases the average Euclidean norm to 0.064. Therefore, the elimination of this signal has a more significant impact on the measurement process than the substitution with noise of any of the four remaining time series. In other words, the information about HFIR encoded in the average inlet-temperature signal can not be sufficiently recovered from the remaining four signals. This behavior implies a low correlation between the average inlet-temperature and the rest of the signals and appears to be characteristic of the methodology developed.

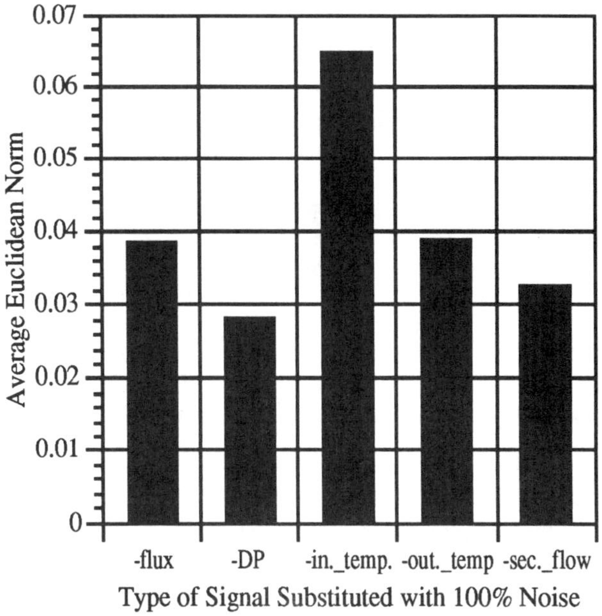


Fig. 4. Values of the average Euclidean norm for different signal eliminations.

In order to examine whether the evaluations of the measuring device are consistent with the behavior of the system under investigation, the degree of dependence among the five signals used as input to the virtual measuring device has been computed. The correlation between two different signals $x$ and $y$ is quantified by the *correlation coefficient*, $r_{xy}$ [19]. In this case there exist five signals and the *correlation matrix*, $C$, describing the correlation among the five time series, is calculated. The results of the computations are presented in the matrix shown in Table I.

TABLE I

CORRELATION MATRIX, $C$, OF THE FIVE INPUT SIGNALS

| | *flux* | *DP* | *in.-temp.* | *out.-temp.* | *sec-flow* |
|---|---|---|---|---|---|
| *flux* | 1.0 | 0.9361 | 0.3374 | 0.9878 | 0.9955 |
| *DP* | | 1.0 | 0.5568 | 0.9615 | 0.9341 |
| *in.-temp.* | | | 1.0 | 0.4952 | 0.3575 |
| *out.-temp.* | | | | 1.0 | 0.9857 |
| *sec-flow* | | | | | 1.0 |

The numbers given in the $C$ matrix are in full agreement with the qualitative behavior of the virtual measuring device. Four out of the five signals have a correlation coefficient close to one, which actually means that the signals are highly dependent. The exception to this pattern is the average inlet-temperature signal which exhibits a relatively low correlation with all four signals. This means that the degree of dependence between this time series and the rest is very low. Thus, the elimination of this signal from the input patterns presented to the measuring device, results in loosing some information which *can not* be obtained from the remaining information supplied from the other signals. This does not happen when one of the other four signals is replaced by random noise, since the high correlation existing among them suggests that the information carried by each signal is very similar (if not identical) to the information existing in the other three time series. Therefore, the behavior exhibited by the virtual measuring device is supported by the results obtained from the correlation matrix. This is sufficient to reassure that the methodology developed embodies all statistical properties existing in the original training signals.

## IV. Conclusions

A new approach has been presented for measuring linguistic (fuzzy) variables. The developed methodology (overall schematic is shown in Fig. 5) is based on the utilization of artificial neural networks as generators of membership functions. This novel implementation offers the unique advantage of describing the state of a system in a condensed form suitable for control applications. The membership functions computed from the *virtual measuring device* can be used as input to a fuzzy controller without the prerequisite of fuzzifying the crisp input values obtained from measurements.

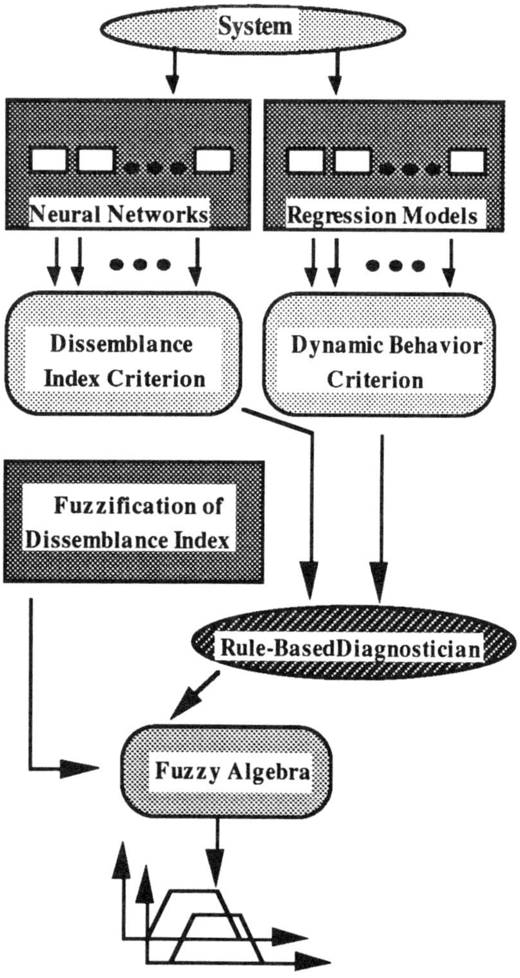


Fig. 5. Schematic representation of the software-based virtual measuring device.

The methodology has been subjected to numerous tests exhibiting robustness to noisy inputs and faulty signals. The virtual measuring device was capable of identifying the correct system state even in the case where one of the five input signals was replaced by 100% random noise. The optimization algorithm developed considers the responses of the independently firing neural networks and concludes the membership function which most appropriately describes the state of the system in the universe of discourse. The decision-making procedure computes the shape and position of the resultant membership function in response to changes in the operating environment. Furthermore, the utilization of artificial neural networks and information criteria enabled the analysis and "learning" of the statistical properties embedded into the original signals. These statistical properties are preserved during the mapping of measurable system parameters - in the form of time series - to a set of fuzzy values accurately representing the state of the system on which a measurement is performed.

## References

[1] E. H. Mamdani, "Application of fuzzy algorithms for control of a simple dynamic plant," *Proc. of IEE, Control and Science*, 121, pp. 1585-1588, 1974.

[2] L. A. Zadeh, "The calculus of fuzzy if/then rules," *AI Expert*, pp. 23-27, March 1992.

[3] H. Takagi, "Fusion technology of fuzzy theory and neural networks: survey and future directions," *Proc. Int. Conf. on Fuzzy Logic and Neural Networks*, Iizuka, Japan, 1, pp. 13-26, 1990.

[4] H. Takagi and I. Hayashi, "NN-driven fuzzy reasoning," *Int. J. Approx. Reasoning*, 5, pp. 191-212, 1991.

[5] I. Hayashi, et al., "Construction of fuzzy inference rules by NDF and NDFL," *Int. J. Approx. Reasoning*, 6, pp. 241-266, 1992.

[6] J. Keller and H. Tahani, "Implementation of conjunctive and disjunctive fuzzy logic rules with neural networks," *Int. J. Approx. Reasoning*, 6, pp. 221-240, 1992.

[7] L. H. Tsoukalas and A. Ikonomopoulos, "Modeling complex systems with artificial neural networks," in *Intelligent Engineering Systems Through Artificial Neural Networks*, C. Dagli, S. Kumara and Y. Shin, Eds. New York: ASME Press, 1991, pp. 581-587.

[8] A. Ikonomopoulos, L. H. Tsoukalas, J. A. Mullens and R. E. Uhrig, "Monitoring nuclear reactor systems using neural networks and fuzzy logic," *Proc. of the 1992 ANS Topical Meeting on Advances in Reactor Physics*, Vol. 2, Charleston, S.C., pp. 140-151, 1992.

[9] L. A. Zadeh, "Fuzzy sets," *Information and Control*, 8, pp. 338-353, 1965.

[10] L. A. Zadeh, "A fuzzy-algorithmic approach to the definition of complex or imprecise concepts," Intl. *Journal Man-Machine Studies*, 8, pp. 249-291, 1976.

[11] L. A. Zadeh, "Fuzzy logic," *IEEE Computer*, pp. 83-93, April 1988.

[12] J. Hertz, A. Krogh and R. Palmer, *Introduction to the Theory of Neural Computation*, Redwood City, MA: Addison-Wesley, 1991.

[13] D. E. Rumelhart, G. E. Hinton and R. J. Williams, "Learning representations by back-propagating errors," *Nature*, 323, pp. 533-536, 1986.

[14] G. Cybenko, "Approximation by superposition of a sigmoidal function," *Mathematics of control, signals and systems*, 2, pp. 303-314, 1989.

[15] K. Funahashi, "On the approximate realization of continuous mappings by neural networks," *Neural Networks*, 2, pp. 183-192, 1989.

[16] A. Lapedes and R. Farber, "Nonlinear signal processing using neural networks: predictions and system modeling," *Los Alamos Nat'l. Lab. report LA-UR-87-2662* , 1987.

[17] A. G. Ivakhnenko and V. G. Lapa, *Cybernetics and Forecasting Techniques*, New York: American Elsevier Publishing Company, 1967.

[18] A. Kaufmann and M. M. Gupta, *Introduction to Fuzzy Arithmetic*, New York: Van Nostrand Reinhold, 1991.

[19] G. G. Judge, W. E. Griffiths, H. Lutkepohl and T. C. Lee, *The Theory and Practice of Econometrics*, New York: John Wiley and Sons, 1985.

# An Analysis Of Fuzzy Control Systems Using Vector Space

Yang-Ming Pok
Ngee Ann Polytechnic
Department of Electrical Engineering
535, Clementi Road, Singapore 2159

Jian-Xin Xu
National University Of Singapore
Department of Electrical Engineering
10, Kent Ridge Crescent, Singapore 0511

**Abstract- This paper describes the use of vector spaces to analyze fuzzy control systems. Using Zadeh's fuzzy-set theory, the outputs of a fuzzy controller are expressed as parametric equations in terms of its inputs and plotted as a contour map in a modified state-space known as a vector space, thereby displaying visually its dynamic behavior and offering new insights in designing fuzzy control systems.**

Keywords: Fuzzy Control, Fuzzy PID, Vector Control, Fuzzy Vector Space

## I. INTRODUCTION

The input-output relationship of a rule-based fuzzy controller is obtained by Zadeh's rules of inference [1] and visually displayed as a contour map in a vector space. Section 2 defines the fuzzy vector space and fuzzy vector [2]. Section 3 shows how controller outputs are obtained through conventional fuzzy-set mathematics and displayed in a contour map. Section 4 describes the linearization of controller output by means of input quantization and derives the control laws from the vector spaces. Section 5 analyses fuzzy control modes. The steady-state error analysis is covered in Section 6. Section 7 discusses a new design method for fuzzy control systems, comparing the performance of two classes of fuzzy controllers with computer simulation results. The last section concludes the paper.

## II. FUZZY VECTOR SPACE

A fuzzy controller has 2 or more input variables. For tuning purposes, each input variable, denoted by $x_i$, where $i = 1, 2, .., n$, is scaled by a weight $w_i$ during the process of fuzzification. Also, the weighted input variables are normalized to a common universe of discourse in the interval $[-L, L]$, where $L$ is a convenient number. A fuzzy vector space is defined by a Cartesian system of coordinates such that each of its axes represents a weighted fuzzy input variable. As an example, a 3D vector space is simply a 3D Cartesian space bounded between +L and -L in each axis, with $E = w_1.x_1$, $R = w_2.x_2$ and $A = w_3.x_3$ as its coordinate axis, where $x_1$ = error $e$, $x_2$ = rate (of change of error) $r$ and $x_3$ = acceleration $a$.

Each input variable is defined by means of a finite number of linguistic variables, such as "Positive-Big", "Positive", "Zero", "Negative" and "Negative-Big". Each linguistic variable is transformed into a fuzzy variable, known as a fuzzy member, by means of a unique central value and a membership function. A fuzzy member becomes a fuzzy subset of the input variable and is represented in the vector space by a fuzzy vector. This fuzzy vector is constructed along the axis of the input variable such that its magnitude is equal to the central value of the fuzzy member.

The AND operation of the input variables in an antecedent of a control rule amounts to specifying the input states and therefore corresponds to a point in the vector space, known as a grid point [2]. The line joining the origin of the coordinate system to a grid point is a vector-sum. Since a grid point or its corresponding vector-sum represents graphically the antecedent of a control rule, a vector space that encompasses all the grid points therefore represents the complete knowledge base of the fuzzy system.

0-7803-0614-7/93$03.00 ©1993IEEE

III. FUZZY CONTROLLER OUTPUT

## A. *Single-Cell Fuzzy Controller*

Fig. 1 shows a typical 2D fuzzy vector diagram (or a cross-sectional view of a 3D fuzzy vector space with $w_3x_3$=0) with triangular membership functions for each axis. It corresponds to a fuzzy knowledge base with 4 control rules, and 2 fuzzy members per axis, linguistically termed as Positive P and Negative N. Here, a square, with 4 grid points as its vertices, constitutes the most elemental geometrical form which is known as a fuzzy cell [2]. The fuzzy outputs corresponding to an arbitrary operating point $A_1$ in Fig. 1 calculated from it's 4 control rules using Zadeh fuzzy-set theory are given as follows:.

Rule/Grid Point:

1. (P, P) = min[$g(L+E)$, $g(L+R)$]= $g(L+R)$,
2. (P, N) = min[$g(L+E)$, $g(L-R)$] = $g(L-R)$,
3. (N, P) = min[$g(L-E)$, $g(L+R)$] = $g(L-E)$,
4. (N, N) = min[$g(L-E)$, $g(L-R)$] = $g(L-E)$,

where $g = 1/(2L)$ is the absolute gradient of the triangular membership function curves.

The antecedent or grid point (P, P) corresponds to a controller output of $-L$. The antecedents (P, N) and (N, P) both correspond to an output of zero, while the antecedent (N, N) corresponds to an output of $L$. Using the center of gravity defuzzication algorithm, the crisp output becomes:

$$U \text{ (or } \Delta U) = \sum_{j=1}^{k} c_j . u_j \Big/ \sum_{j=1}^{k} u_j \,, \qquad (1)$$

where $c_j$ is the central value of the output fuzzy member $j$, $u_j$ is the output truth value from rule $j$ and $k$ is the total number of rules. (The terms in the left-hand-side of (1) are explained in Section V.)

The crisp output at the point A1, (or A2, A5, A6) is given by:

$$U(A_1) \text{ or } (\Delta U(A_1)) = -\frac{L}{(2L-E)} \cdot \frac{(E+R)}{2} \,. \qquad (2)$$

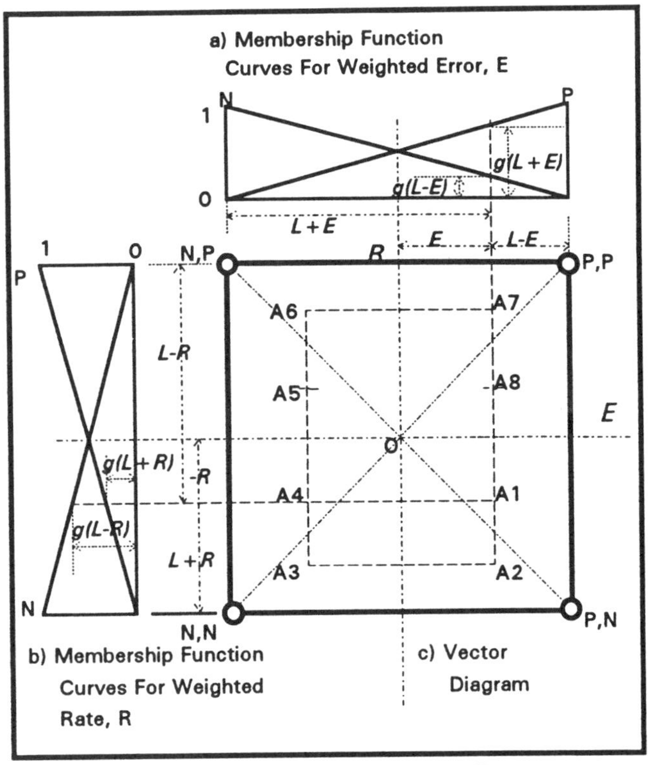


*Fig. 1: A Single-Cell Fuzzy Vector Diagram*

Similarly, the crisp output for point $A_3$, (or $A_4$, $A_7$ $A_8$) is given by:

$$U(A_3) \text{ (or } \Delta U(A_3)) = -\frac{L}{(2L-R)} \cdot \frac{(E+R)}{2} \,. \qquad (3)$$

Hence, we can generalize the control law from (2) and (3) as follows:

$$U \text{ (or } \Delta U) = -\, w_0 \,.N(E,R)\, M(E,R) \,, \qquad (4)$$

where $w_0$ is the output weight,

$$N(E,R) = L/\min[(2L-|E|), (2L-|R|)],$$
$$M(E,R) = (E+R)/2. \qquad (5)$$

Equation (4) is called an elemental-cell or single-cell control law. The term $N(E, R)$ contributes to the dynamic-gain behavior of the fuzzy controller [3]. The term $M(E, R)$ is simply the mean value of the two weighted inputs. The contour lines for $N(E, R)$ or $N$-Contours are shown in Fig. 2 as concentric squares. Similarly, the $M$-Contours and output $O$-Contours are shown respectively as straight lines and curves generally parallel to the global diagonal.

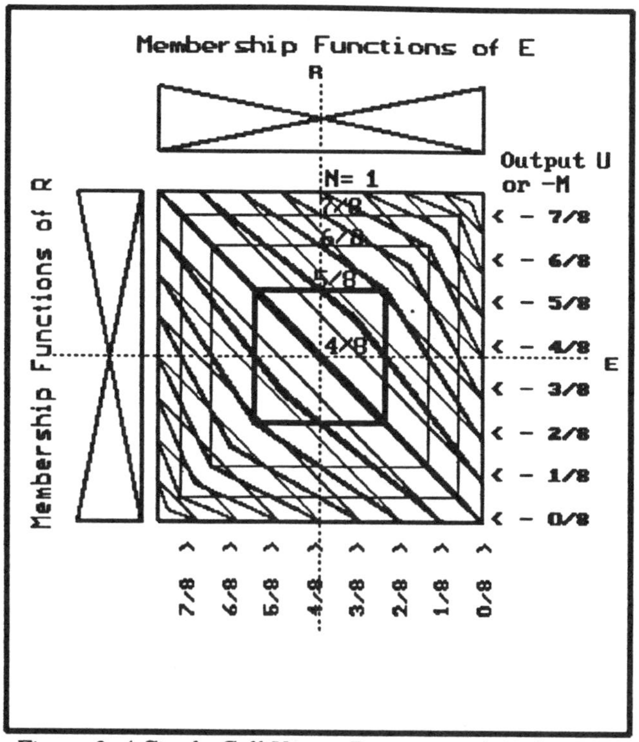


*Figure 2: A Single-Cell Vector Diagram with N-Contours M-Contours and O-Contours*

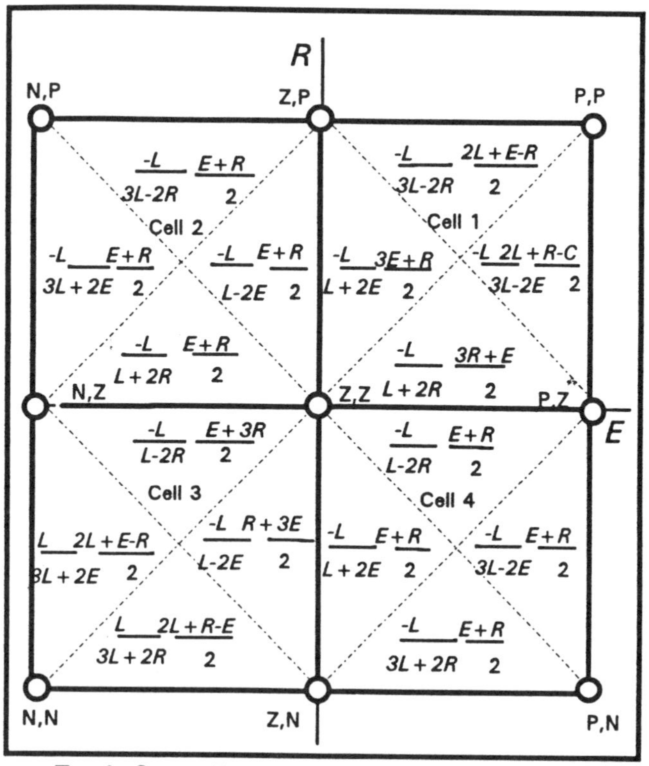


*Fig. 3: Output Functions of 4-Cell Vector Diagram*

For a 3-input single-cell fuzzy controller, the control law becomes

$$U \text{ (or } \Delta U) = -w_0 . N(E, R, A).(E + R + A), \qquad (6)$$

where $N(E, R, A)$ is a nonlinear function of the three input variables $E$, $R$ and $A$ and $w_0$ is a constant which incorporates the divider 3 accompanying $(E+R+A)$.

## *B. Multi-Cell Fuzzy Controller*

Similar approach can be applied to a multi-cell vector diagram. Fig. 3 shows a 4-cell vector diagram that corresponds to 9 control rules. Each input variable or axis has 3 fuzzy members, namely Positive P, Zero Z and Negative N, with triangular membership functions. This gives rise to a total of 16 output functions as shown. The resultant crisp controller outputs are shown in Fig. 4 as O-contour lines.

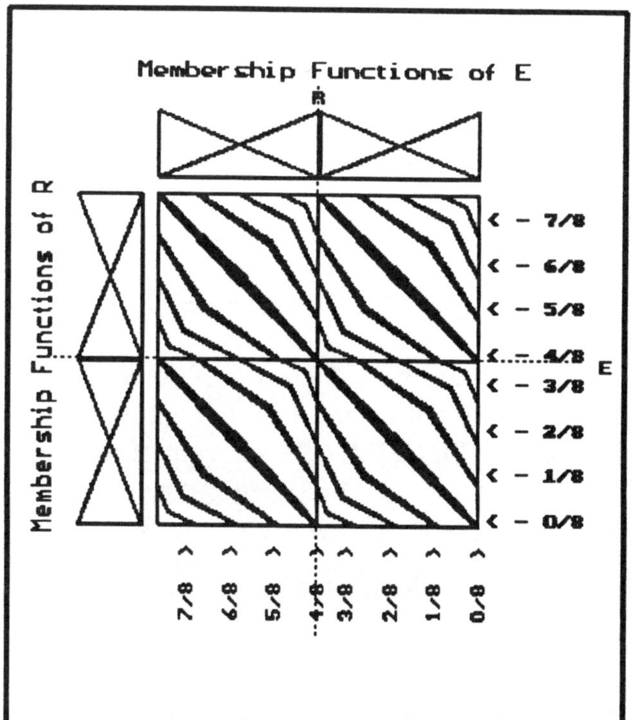


*Fig. 4: O-Contour Map of a 4-Cell Fuzzy Controller*

## IV. QUANTIZATION

Input quantization for a multi-cell fuzzy controller is defined as a process of rounding of the weighted input variable $w_i.x_i$ to its nearest central value in the universe of discourse [2]. This process effectively forces the operating point to coincide with one of the grid points nearest to it, thereby holding the dynamic gain $N(E, R)$ at constant unity value and eliminates nonlinearity completely.

Fig. 5 shows the output contour of a 64-cell fuzzy controller with input quantizations. It is seen that the contour lines are all straight lines parallel to the global diagonal. Also, the output functions can be obtained directly from the vector space by means of point to point mappings. This is illustrated by an operating point at a grid position A. The crisp output at point A is measured from the length $OQ = OP + PQ = E \cos 45 + R \sin 45 = (E + R)/2$.

Therefore, taking into account the output weight $w_0$ and the correct sign, we have

$$U \text{ (or } \Delta U) = -w_0.(E + R)/2. \quad (7)$$

This equation that is derived from the vector space is known as a vector control law and is applicable to a multi-cell fuzzy controller with input quantization.

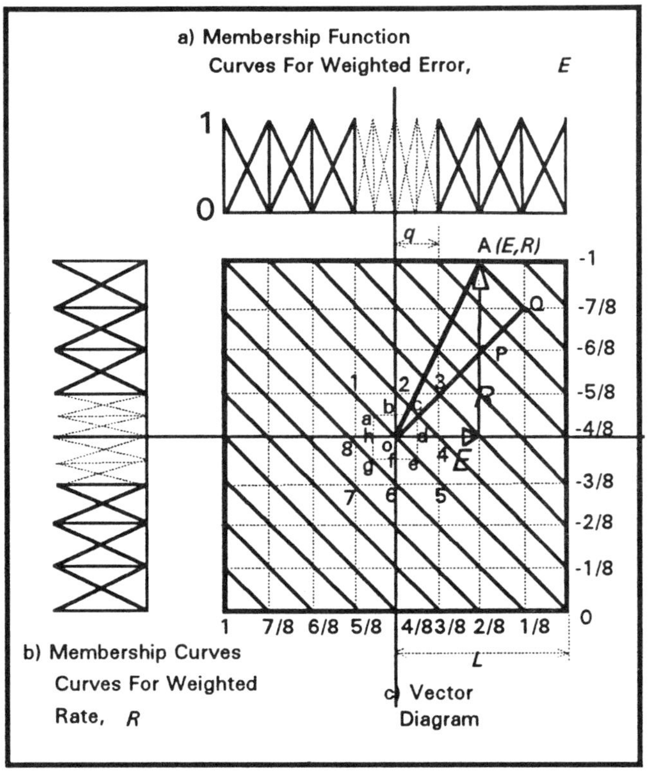


*Fig. 5: 64-Cell With Input Quantization and Intermediate Grid Points*

## V. FUZZY CONTROL MODES

This section describes 2 classes of control rules, namely Class A and B, and 4 fuzzy control modes, abbreviated as PIDA [2]. Class A control rules are of the form: "If $X$ and $Y$ ... then output $U = M$," while Class B control rules are of the form: "If $X$ and $Y$ ... then change of output $\Delta U = M$."

### *A. PID+A (acceleration) Controller*

Consider an analog controller with a control law given by:

$$U(t) = K_p.e + K_i\int e.dt + K_d\frac{de}{dt} + K_a\frac{d^2e}{dt^2}. \quad (8)$$

In discrete form, we have

$$\Delta U(nT)/T = -K_p.r(nT) + K_i.e(nT) + K_d.a(nT) + K_a.b(nT), \quad (9)$$

where $r(nT) = [e(nT)-e(nT-T)]/T$,
$a(nT) = [r(nT)-r(nT-T)]/T$, and
$b(nT) = [a(nT)-a(nT-T]/T$,
$T$ = sampling time.

### *B. Class A Fuzzy Controller*

Restating (6) for Class A control rules gives

$$\Delta U(nT)/T \cong -w_0.N(nT).[w_1 r(nT) + w_2 a(nT) + w_3 b(nT)], \quad (10)$$

provided that $N(nT) \cong N(nT-T)$. Comparing (10) with (9), it is seen that a Class A three-input fuzzy controller is approximately a PDA controller with dynamic gains, $K_P = w_0 w_1.N(nT)$; $K_d = w_0 w_2.N(nT)$; $K_a = w_0 w_3.N(nT)$. Also, Class A two-input controller is approximately a PD controller. This can be deduced by simply equating the acceleration term to zero.

### C. *Class B Fuzzy Controller*

Similarly, restating (6) for Class B control rules gives

$$\Delta U(nT)/T = -(w_0.N(nT)/T).(w_1\, e(nT) + w_2.r(nT) + w_3.a(nT)) \tag{11}$$

Comparing (11) with (9), we see that Class B three-input fuzzy controller is equivalent to a PID controller with dynamic gains of $K_P = w_0 w_2.N(nT)/T$; $K_i = w_0 w_1.N(nT)/T$; $K_i = w_0 w_3.N(nT)/T$. Similarly, a Class B two-input fuzzy controller is equivalent to a PI controller with dynamic gains, $K_P$ and $K_i$ as given above.

## VI. STEADY-STATE ERROR ANALYSIS

The absence of an integral term in a Class A controller results in steady-state errors $e_{ss}$ in Type 0 processes. For a linear process $G_P(s)$ with a unit step input $R(s) = 1/s$, the final value theorem gives

$$e_{ss} = \lim_{s\to 0} \frac{sR(s)}{H(1 + G_P.G_{css}.H)}$$

$$= \frac{1}{(1 + w_0 w_1.\lim_{s\to 0} G_P)}, \tag{12}$$

where $G_{css}(s) = w_0 w_1$ is the approximate controller transfer function at steady-state or near steady-state conditions and the feedback transfer function is $H(s)=1$.

An additional error is introduced by input quantization. This can be visualized in Fig. 5. Under near steady-state condition, the operating point hunts around the origin, jumping from one grid point to another around the origin, marked as 1 to 8 in the vector diagram. The incremental controller output that causes a maximum quantization error at the grid point 3 or 7 is obtained from (4) to give

$$\delta U(nT) = \pm\, w_0.1.(q + q)/2 = \pm w_0\, q, \tag{13}$$

where $q = L/N$ is the quantization interval, noting that $(2N+1)$ is defined as the number of fuzzy members per input variable. As an approximation, this incremental output can be viewed as a step disturbance signal, $D(s) = \pm w_0.q/s$, superimposed onto an idealized fuzzy vector control system as shown in Fig. 6.

$Y_d(s)$ can be obtained by letting $R(s) = 0$ and solving the loop equation, so
$(-Y_d(s).G_{css} + D(s)).G_P(s) = Y_d(s)$, therefore

$$Y_d(s) = \frac{D(s).G_P(s)}{(1 + G_{css})}. \tag{14}$$

Using final value theorem, its time-domain value is

$$y_d(t) = \pm \frac{w_0.q.\lim_{s\to 0} G_P}{1 + w_0 w_1}. \tag{15}$$

This leads to the conclusion that quantization error can be reduced by adding intermediate grid points around the origin, such as points a to g in

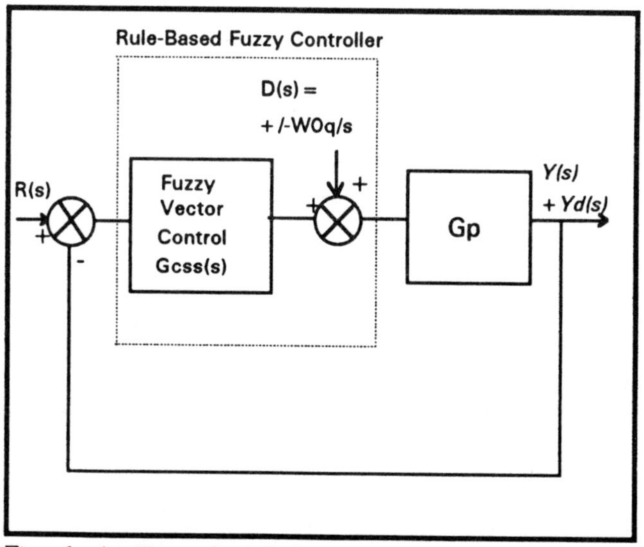


*Fig. 6: An Equivalent Rule-Based Fuzzy Controller near steady-state conditions.*

Fig. 5. This amounts to additional rules and also narrower membership function curves. On the other hand, for a Class B fuzzy controller, there is no steady-state error in Type 0 processes, although quantization error still exists, which can be evaluated from (14) to give

$$Y_d(s) = \pm \frac{q.T.\lim_{s \to 0} G_P}{w_2}. \qquad (16)$$

VII. DESIGN OF FUZZY CONTROL SYSTEMS

In processes using fuzzy controllers, proper control can be achieved through careful selection of the linguistic statements in the consequent part of the control rules. For example, Class B control rules should be specified in Type 0 processes to eliminate steady-state error and Class A control rules should be specified for Type 1 and 2 processes. For first-order processes, a two-term fuzzy controller is adequate. Second-order processes may have two or preferably three control modes. Higher-order processes require three-term control. As an example, we consider a 2nd-order process with a transfer function given by: $G_P = \frac{2e^{-0.5S}}{(s+1)(s+2)}$. Fig. 7 shows simulation results of 2 linearized, 16-cell fuzzy controllers with

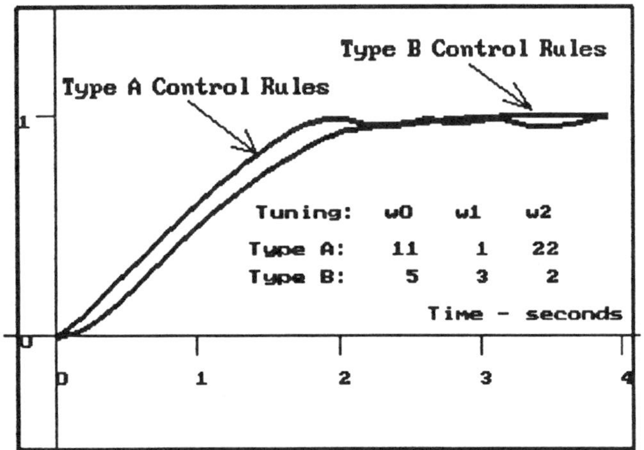


*Figure 7: Time Responses of Class A and B Linearized Fuzzy Controllers in a 1st Order LTI Process*

Class A and B fuzzy control rules respectively. Quantization error in Class B fuzzy control from (16) is ±0.8% for L=1 and N= 2 and there is no steady-state error. On the other hand, the quantization error in Class A fuzzy control from (15) is ±0.4% and its steady-state error from (12) is 4.5%, giving a maximum estimated root-mean-square error of 4.52% Clearly, Class B controller is more accurate. Also, its non-quantized version is preferred.

VIII. CONCLUSIONS

A fuzzy vector space is a bounded modified state-space representing the input structure of a fuzzy controller. This paper describes only 2D and 3D vector spaces. However, visualization techniques for high-dimensional spaces or hyperobjects using multi-dimensional graphics are available [3]. Insights obtained from the vector spaces enable one to select suitable class and modes of fuzzy control, taking into account the process characteristics, thereby ensuring fast response and zero steady-state error.

REFERENCES


[1] L.A. Zadeh, "Outline of A New Approach to the Analysis of Complex Systems and Decision Process", IEEE Trans., Systems, Man, and Cybernetics, Vol. SMC-3, No. 1, Jan 1973.

[2] Yang-Ming Pok, "Analysis, Synthesis and Development of Fuzzy Control Systems", PhD thesis, National University of Singapore, Singapore, unpublished.

[3] Hao Ying, W.Silver & J.J. Buckley, "Fuzzy Control Theory, A Nonlinear Case", Automatica 90,Vol. 26, No. 3, pp. 513-520.

[4] Yang-Ming Pok and Yeong-Kong Huen, "Visualisation of Hyperobjects in Hgram-Space by Computers", International Workshop on "Modern Geometric Computing for Visualization", Japan, 1992, Springer-Verlag, pp. 141-159.

# An Application of Fuzzy Neural Networks to a Stability Analysis of Fuzzy Control Systems

Takeshi Furuhashi, Shin-ichi Horikawa, Yoshiki Uchikawa
Department of Electronic-Mechanical Engineering, Nagoya University
Furo-cho, Chikusa-ku, Nagoya, 464-01, Japan
Tel.+81-52-781-5111 ext.2792 Fax. +81-52-781-9263
E-mail: furu@uchikawa.nuem.nagoya-u.ac.jp

***Abstract-This paper presents an analysis method on stability of fuzzy control systems using fuzzy neural networks(FNNs). The FNNs are capable of acquiring fuzzy rules and tuning the membership functions automatically with the Back Propagation(BP) learning algorithm. One FNN is used for obtaining a fuzzy model of the controlled object. The other FNN is trained to acquire a fuzzy model of the controller. A new definition of the stability of fuzzy control systems is also presented in this paper. Using the fuzzy controller and the obtained fuzzy model of the controlled object, the stability analysis based on the proposed definition is done linguistically and is very easy to understand.***

## I.INTRODUCTION

Fuzzy control has a distinguishing feature in that it can incorporate expert's control rules using linguistic expressions. One of the main problems of the fuzzy control is the difficulty of acquiring the fuzzy rules and tuning the membership functions. There have been many researches on applications of neural networks to fuzzy reasonings[1]~[5]. The authors have also proposed three types of fuzzy neural networks and have studied an automatic identification method of fuzzy models of controlled objects/controllers[6]~[9].

Another main problem of the fuzzy controller is that the stability is hard to guarantee. Many research works have been done on the stability analysis[12]~[18]. These works have been done describing the control system in numerical equations. These method do not make use of the distinguishing feature of the fuzzy controls, i.e. easily understandable linguistic expressions.

This paper presents a new stability analysis method using FNNs. One FNN is used for obtaining a fuzzy model of the controlled object. The other FNN is trained to acquire a fuzzy model of the controller. Since both the controlled object and the controller are described with linguistic rules, i.e. fuzzy *if - then* rules, it is possible that the stability analysis is done linguistically. A new definition of stability of fuzzy control systems is also presented in this paper. Although the new definition of stability is not mathematically rigorous, the process of the analysis based on the the definition is very simple and easy to understand even to those who have little knowledge of control theory.

For the verification of the proposed method, simulations are done using a simple nonlinear plant.

## II.FUZZY NEURAL NETWORK

The FNNs presented by the authors are the multi-layered back-propagation(BP) models of which the structures are designed to realize the processes of fuzzy reasoning and to make the connection weights of the networks correspond to the parameters of the fuzzy reasoning. Through the learning with the BP algorithm, the FNNs can identify the fuzzy rules and tune the membership functions of fuzzy reasoning automatically. The fuzzy model of the controlled object as well as the fuzzy controller are to be obtained with the fuzzy neural

0-7803-0614-7/93$03.00 ©1993IEEE

networks(FNNs). This paper uses "Type I" of the FNNs in [8].

Figure 1 shows an example of the configuration of the FNN. The FNN realizes a simplified fuzzy inference of which the consequences are described with singletons. The BP algorithm can be applied for adjusting the weights in the neural networks. The inputs are non-fuzzy numbers. The simplified fuzzy inference with two inputs $x_1, x_2$ and one output $y$ is written as follows:

$$R^i : \text{If } x_1 \text{ is } A_{i,1} \text{ and } x_2 \text{ is } A_{i,2} \text{ then } y = f_i \quad (i = 1,2,\cdots,n) \qquad (1)$$

$$\mu_i = A_{i,1}(x_1)A_{i,2}(x_2) \qquad (2)$$

$$y^* = \frac{\sum_{i=1}^{n} \mu_i f_i}{\sum_{i=1}^{n} \mu_i} = \sum_{i=1}^{n} \hat{\mu}_i f_i \qquad \hat{\mu}_i = \frac{\mu_i}{\sum_k \mu_k} \qquad (3)$$

where $R^i$ is the $i$-th fuzzy rule. $A_{i,1}, A_{i,2}$ are fuzzy variables. $f_i$ is a constant. $n$ is the number of fuzzy rules. $\mu_i$ is the truth value of $R^i$. $\hat{\mu}_i$ is the normalized truth value so that the sum of $\hat{\mu}_i$ is unity. $y^*$ is the inferred value.

The FNN in Fig.1 has two inputs $x_1, x_2$ and one out-put y* and three membership functions in each premise. The circles and squares in the figure mean units of the neural network and the denotation $w_c, w_g, w_f$ and 1, -1 are the connection weights. The FNN realizes the inference in (1)- (3) in the neural network structure. The connection weights of the network $w_c, w_g, w_f$ corresponding to the parameters of fuzzy inference are updated with the BP learning algorithm. Figure 2 shows the membership functions in the premise $A_{1j}(x_j)$, $A_{2j}(x_j)$, $A_{3j}(x_j)$ realized in (A)-(D)-layers. The connection weights $w_c$, $w_g$ determine the positions and gradients of the sigmoid functions in the units in (C)-layer, respectively. Each membership function consists of one or two sigmoid functions. The outputs of the units in (D)-layer are the grades of the membership functions. The products of the grades are fed to the units in (E)-layer and the

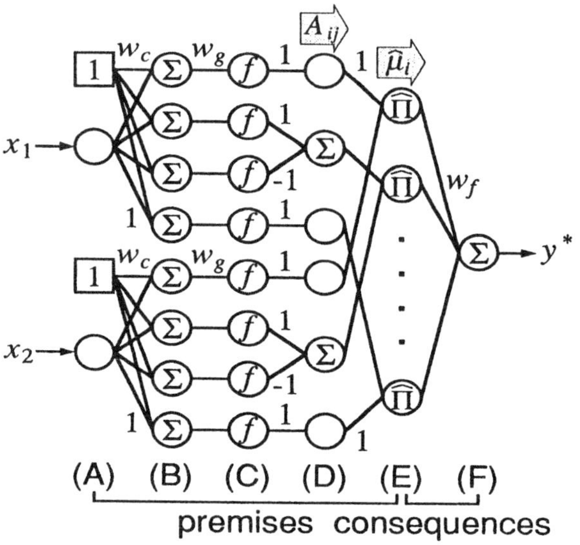


Fig.1 Fuzzy neural network

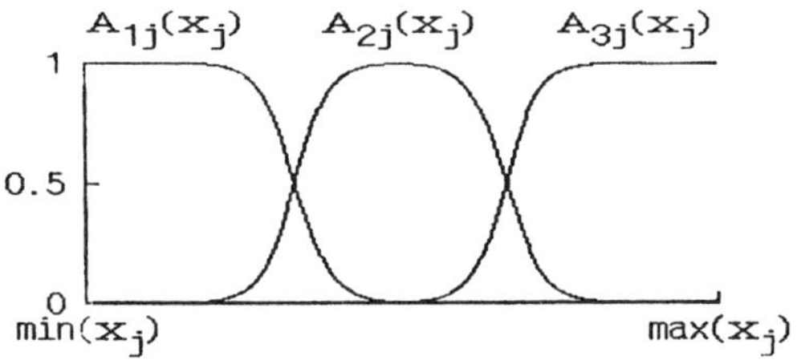


Fig.2 Membership function in premise

output of the units are normalized truth value in the premises $\hat{\mu}_i$. The output of the unit in (F)-layer is the sum of the products of the connection weights $w_f$ and $\hat{\mu}_i$. The connection weights $w_f$ correspond to the singletons in the consequence $f_i$. The output in (F)-layer is, therefore, the inferred value $y^*$.

The FNN tunes the membership functions in the premises and identifies the fuzzy rules by adjusting the connection weights $w_c, w_g$ and $w_f$, respectively. $w_f$ are initialized to be zero. The FNN has no rules at the beginning of the learning.

Since the center-of-gravity method is used in (E)-layer, the updating method of connection weights, i.e. BP algorithm, needs some modifications. The learning algorithm for the FNN is well described in [8].

## III.STABILITY OF FUZZY CONTROLLER

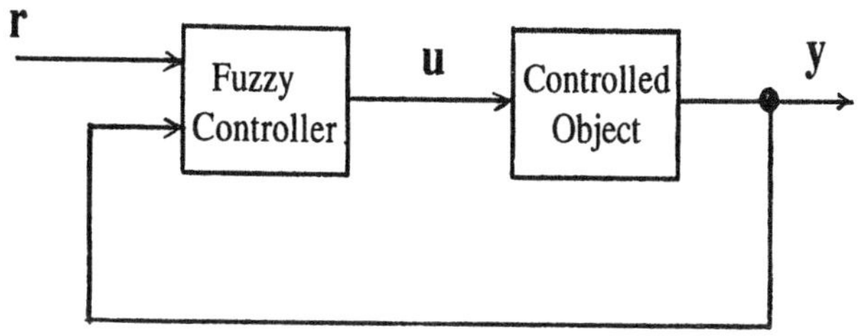


Fig.3 Control system

Figure 3 shows a fuzzy control system. The controlled object $P$ is a nonlinear system. A feedback control is done by the fuzzy controller $C$. Both the controlled object and the controller will be identified by the FNNs. In this section, a definition of the stability of the fuzzy control system is given. The control system to be discussed in this section is more general than that to be identified by the FNNs. The plant $P$ is the system with $m_u$-inputs and $m_y$-outputs. The manipulating variables at the $t$-th time are denoted by $\mathbf{u}_t = (u_{1t}, u_{2t}, \cdots, u_{m_u t})^T$. The outputs of the system are denoted by $\mathbf{y}_t = (y_{1t}, y_{2t}, \cdots, y_{m_y t})^T$. The commands of the system are $\mathbf{r}_t = (r_{1t}, r_{2t}, \cdots, r_{m_r t})^T$. For simplicity, the elements of the variables $\mathbf{u}_t$, $\mathbf{y}_t$, $\mathbf{r}_t$ are considered to be non-fuzzy numbers. The nonlinear plant $P$ can be described with a fuzzy relation $R_P$ [10][11] as

$$
\begin{aligned}
\mathrm{R}_P^i : \text{IF } &\mathbf{u}_t \text{ is } \mathbf{A}_{i,t}, \mathbf{u}_{t-1} \text{ is } \mathbf{A}_{i,t-1}, \\
&\cdots, \mathbf{u}_{t-p} \text{ is } \mathbf{A}_{i,t-p}, \\
&\mathbf{y}_t \text{ is } \mathbf{B}_{i,t}, \mathbf{y}_{t-1} \text{ is } \mathbf{B}_{i,t-1}, \\
&\cdots, \mathbf{y}_{t-q} \text{ is } \mathbf{B}_{i,t-q} \\
\text{THEN } & \\
&\mathbf{y}_{i,t+1} \text{ is } \mathbf{C}_{i,t+1}
\end{aligned} \tag{4}
$$

where $\mathrm{R}_P^i (i = 1, 2, \cdots, np)$ is the $i$-th fuzzy rule, $np$ is the number of rules, $\mathbf{A}i,j$ $(j = t, \cdots, t - p)$, $\mathbf{B}i,h$ $(h = t, \cdots, t - q)$, and $\mathbf{C}i,t+1$ are $mu$-, $my$-, $my$-dimensional vectors with elements of fuzzy numbers, respectively. The integers $p$, $q$ are to be determined with the characteristics of the plant. $\mathbf{y}_{i,k+1}$ is the inferred value of the $i$-th fuzzy rule at the time of $t$+1.

The crisp inferred value of the output of the plant $P$ at the $t$+1-th time $y^*t+1$ is obtained by defuzzifying the total inferred value $\mathrm{y}t+1$ given by

$$
\mathbf{y}_{t+1} = \bigcup_i \mu_{Pi} \wedge \mathbf{C}_{i,t+1} \tag{5}
$$

where $\mu_{Pi}$ is the truth value of the $i$-th fuzzy rule and $\wedge$ means *min*-operation. The center-of-gravity method is one of the main defuzzification methods.

The fuzzy controller $C$ is described in the same way as

$$
\begin{aligned}
\mathrm{R}_C^i : \text{IF } &\mathbf{r}_t \text{ is } \mathbf{D}_{i,t}, \mathbf{r}_{t-1} \text{ is } \mathbf{D}_{i,t-1}, \\
&\cdots, \mathbf{r}_{t-s} \text{ is } \mathbf{D}_{i,t-s}, \\
&\mathbf{y}_t \text{ is } \mathbf{E}_{i,t}, \mathbf{y}_{t-1} \text{ is } \mathbf{E}_{i,t-1}, \\
&\cdots, \mathbf{y}_{t-p'} \text{ is } \mathbf{E}_{i,t-p'} \\
\text{THEN } & \\
&\mathbf{u}_{i,t} \text{ is } \mathbf{F}_{i,t}
\end{aligned} \tag{6}
$$

The defuzzification of the total inferred value of the fuzzy rules makes it difficult to keep the correspondence between the linguistic descriptions of the fuzzy rules and the total inferred value.

This paper defines the fired fuzzy rules as

**Def.1**: (Fired Fuzzy Rule)

The rule $\mathrm{R}^i$ is said to be the fired fuzzy rule at the $t$-th time if the truth value of the rule is the highest at $t$, i.e.

$$
(\mu_i = \max_j \mu_j, \quad j = 1, \cdots, n). \tag{7}
$$

The fired fuzzy rule is denoted by $\mathrm{R}^{i_t}$.

A sequence of the fired fuzzy rules is described using the fired fuzzy rule of the fuzzy controller at the $t$-th time $\mathrm{R}_C^{i_t}$ and the fired fuzzy rule of the fuzzy model of the plant $P$ at the $t$-th time $\mathrm{R}_P^{i_t}$ as

$$
\mathrm{R}_C^{i_t} \to \mathrm{R}_P^{i_t} \to \mathrm{R}_C^{i_t+1} \to \mathrm{R}_P^{i_t+1}. \tag{8}
$$

The fired fuzzy rules at the $t$-th time can be described as the following:

$$
\mathrm{R}^{i_t} = \mathrm{R}_C^{i_t} * \mathrm{R}_P^{i_t} \tag{9}
$$

where $*$ is a composition operation with which $x*y$ means that the rule $y$ will be fired after the rule $x$.

**Def.2:** (Stability)
The fuzzy control system is stable if there exists a positive integer $\tau$ such that:

$$\mathrm{R}^{i_{t+\tau}} = \mathrm{R}^{i_t} \qquad (10)$$

where $t$ is an integer and $t \geq T_0 \geq 0$. $T_0$ is also an integer and is determined depending on the initial condition of the control system.

**Def.3:** (Asymptotic Stability)
The fuzzy control system is asymptotically stable if there exists $\mathrm{R}^{i_\infty}$ such that:

$$\lim_{t \to \infty} \mathrm{R}^{i_t} = \mathrm{R}^{i_\infty} \qquad (11)$$

## IV.FUZZY MODELING

The controlled object used in this paper is simple and is expressed as

$$T \cdot \frac{dy}{dt} + y = u^{3/2} \qquad (12)$$

where $u$ is the input and $y$ is the output. $T$ is the time constant. The system has nonlinearity in its input. The training data of the FNNs are made by manually controlling the controlled object by a human operator. The fuzzy model of the controlled object is to be identified from the input-output pairs of data of the controlled object[8][9]. Figure 4 shows the fuzzy model of the system. Since the system is a first order system, the system can be described by the model with inputs $u_t$, $y_t$ and output $y_{t+1}$. The model estimates the output of the system one sampling ahead $y_{t+1}$ from the current manipulating variable $u_t$ and the current output $y_t$. The obtained fuzzy rules of the controlled object

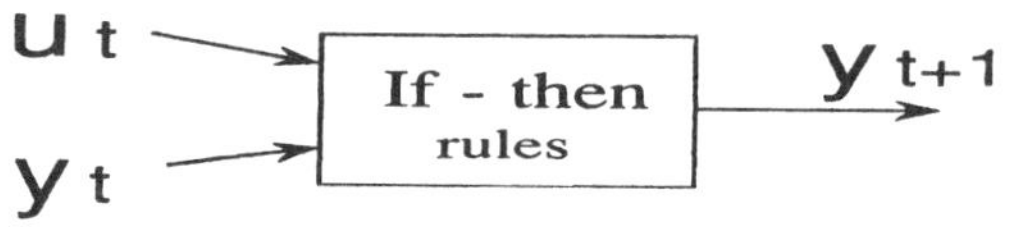


Fig.4 Fuzzy model of controlled object

Table I Fuzzy model of plant

T =50

| $y_{t+1}$ | | yt | | | | | | |
|---|---|---|---|---|---|---|---|---|
| | | NB | NM | NS | ZO | PS | PM | PB |
| ut | NB | -0.96 | -0.55 | -0.30 | -0.02 | 0.27 | 0.52 | 0.92 |
| | NM | -0.95 | -0.54 | -0.29 | -0.01 | 0.27 | 0.53 | 0.93 |
| | NS | -0.94 | -0.54 | -0.29 | 0 | 0.28 | 0.53 | 0.94 |
| | ZO | -0.94 | -0.53 | -0.28 | 0 | 0.28 | 0.53 | 0.94 |
| | PS | -0.94 | -0.53 | -0.28 | 0 | 0.29 | 0.54 | 0.94 |
| | PM | -0.93 | -0.53 | -0.27 | 0.01 | 0.29 | 0.54 | 0.95 |
| | PB | -0.92 | -0.52 | -0.27 | 0.02 | 0.30 | 0.55 | 0.96 |

with the FNN are shown on Table I. The labels on the table *NB, NM, NS, ZO, PS, PM, PB* are *Negative Big, Negative Medium, Negative Small, Zero, Positive Small, Positive Medium, Positive Big,* respectively. The time constant of the object was 50. The unit of time is the sampling time.

Figure 5 shows the fuzzy controller, i.e. the fuzzy model of the human operator. Since the controlled object has the nonlinearity in the input, the controller uses the current command $r_t$, the current output $y_t$ and the output one sample before $y_{t-1}$ as the input variables instead of the errors between the command and the output. The acquired fuzzy control rules with another FNN are shown on Table II. The data used for the learning of the FNN were based on the successful control by the human operator. The obtained control rules do not cover all the input subspaces. Many of the rules of $u_t = 0$ mean that there were no data in th corresponding subspaces.

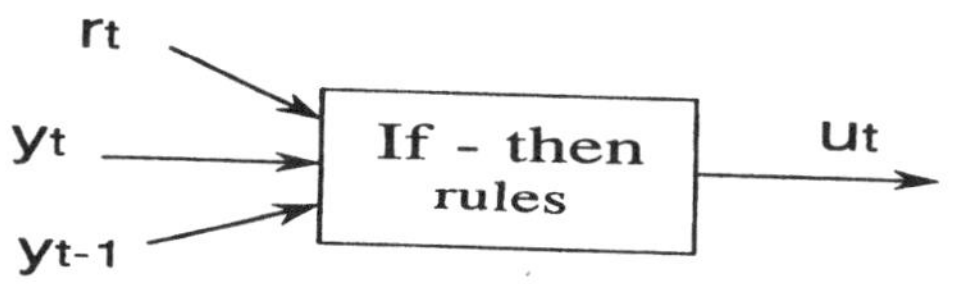


Fig.5 Fuzzy controller

Table II Fuzzy control rules

rt is NB

| ut | | yt NB | NM | NS | ZO |
|---|---|---|---|---|---|
| yt-1 | NB | -0.89 | -0.79 | 0 | 0 |
| | NM | -1.25 | -0.99 | -0.70 | 0 |
| | NS | 0 | -1.11 | -1.00 | -0.47 |
| | ZO | 0 | 0 | -1.03 | -1.00 |

rt is NM

| ut | | yt NB | NM | NS | ZO |
|---|---|---|---|---|---|
| yt-1 | NB | 0 | 0.17 | 0 | 0 |
| | NM | 0.28 | -1.05 | -0.50 | 0 |
| | NS | 0 | -1.07 | -1.03 | -0.56 |
| | ZO | 0 | 0 | -1.03 | -0.96 |

rt is NS

| ut | | yt NM | NS | ZO |
|---|---|---|---|---|
| yt-1 | NM | 0.28 | 0.26 | 0 |
| | NS | 0.24 | -0.82 | -0.41 |
| | ZO | 0 | -1.03 | -1.27 |

rt is ZO

| ut | | yt NS | ZO | PS |
|---|---|---|---|---|
| yt-1 | NS | 0.86 | -0.01 | 0 |
| | ZO | -0.12 | 0 | 0.12 |
| | PS | 0 | 0.01 | -0.86 |

rt is PS

| ut | | yt ZO | PS | PM |
|---|---|---|---|---|
| yt-1 | ZO | 1.27 | 1.03 | 0 |
| | PS | 0.41 | 0.82 | -0.24 |
| | PM | 0 | -0.26 | -0.28 |

rt is PM

| ut | | yt ZO | PS | PM | PB |
|---|---|---|---|---|---|
| yt-1 | ZO | 0.96 | 1.04 | 0 | 0 |
| | PS | 0.56 | 1.03 | 1.07 | 0 |
| | PM | 0 | 0.50 | 1.05 | -0.28 |
| | PB | 0 | 0 | -0.17 | 0 |

rt is PB

| ut | | yt ZO | PS | PM | PB |
|---|---|---|---|---|---|
| yt-1 | ZO | 1.00 | 1.03 | 0 | 0 |
| | PS | 0.47 | 1.00 | 1.11 | 0 |
| | PM | 0 | 0.70 | 0.99 | 1.25 |
| | PB | 0 | 0 | 0.79 | 0.89 |

## V.STABILITY ANALYSIS

The stability analysis of the control system is done in the way shown in Fig. 6. The figure shows the case where the command is "Positive Big" and the initial states of the controlled object $u_{-1}$, $y_{-1}$ and $y_0$ are all "Zero". From the table of fuzzy control rules, the manipulating variable $u_0$ is known to be "Positive Big (1.0)". Then from the table of the fuzzy model of the controlled object is very large (50) compared with the sampling time (unity), the outputs $y_t$ and $y_{t-1}$ are nearly equal. The operation of the controller is found to shift along the diagonal of the table indicated as 1→2→3→4. The state of the controlled object moves along the row of "$u_t$ is Positive Big" also indicated as 1→2→3→4. The control system settles at "4" on the tables. This case shows that the system is asymptotically stable. The corresponding step response of the system controlled by the acquired fuzzy controller is shown in Fig. 7.

Then the time constant of the controlled object is changed from 50 to 2. The fuzzy model of the controlled object is updated with new input-output pairs of data from the new object. Figure 8 shows the stability analysis with the new object and the old fuzzy controller. In this case, the time constant is comparable to the sampling time. The output $y_t$ can be different from $y_{t-1}$. The operation of the controller shifts along $1 \to 2 \to 3 \to 4 \to 5 \to 6 \to 7 \to 3 \to \cdots$. The state of the controlled object also moves along $1 \to 2 \to 3 \to 4 \to \cdots$. The control system has a limit cycle. The corresponding step response of the new system controlled by the old fuzzy controller is shown in Fig. 9.

Fuzzy controller

| ut | | yt ZO 0.0 | PS 0.3 | PM 0.6 | PB 0.9 |
|---|---|---|---|---|---|
| yt-1 | ZO 0.0 | 1.00 ① | 1.03 | 0 | 0 |
| | PS 0.3 | 0.47 | 1.00 ② | 1.11 | 0 |
| | PM 0.6 | 0 | 0.70 | 0.99 ③ | 1.25 |
| | PB 0.9 | 0 | 0 | 0.79 | 0.89 ④ |

rt is PB (0.9)

Controlled object

| yt+1 | | yt ZO 0.0 | PS 0.3 | PM 0.6 | PB 0.9 |
|---|---|---|---|---|---|
| ut | ZO 0.0 | 0 | 0.28 | 0.53 | 0.94 |
| | PS 0.3 | 0 | 0.29 | 0.54 | 0.94 |
| | PM 0.6 | 0.01 | 0.29 | 0.54 | 0.95 |
| | PB 0.9 | 0.02 ① → | 0.30 ② → | 0.55 ③ → | 0.96 ④ |

Fig.6 Stability analysis (T=50)

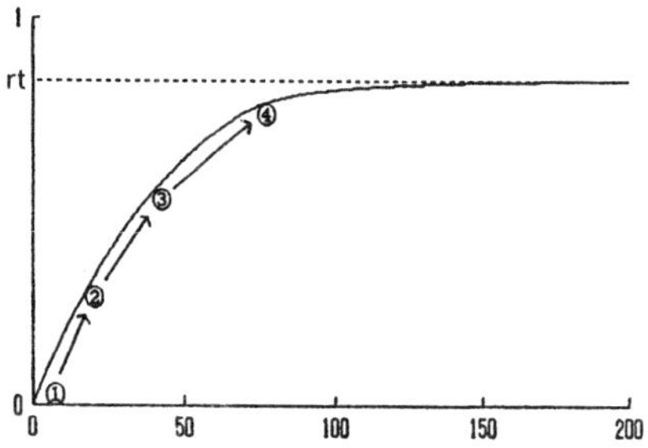


Fig.7 Step response (T=50)

Fuzzy controller

| $u_t$ | | $y_t$ ZO 0.0 | $y_t$ PS 0.3 | $y_t$ PM 0.6 | $y_t$ PB 0.9 |
|---|---|---|---|---|---|
| $y_{t-1}$ | ZO 0.0 | 0.96 ① | 1.04 ② | 0 | 0 |
| $y_{t-1}$ | PS 0.3 | 0.56 | 1.03 ⑦ | 1.07 ③⑧ | 0 |
| $y_{t-1}$ | PM 0.6 | 0 | 0.50 ⑥ | 1.05 | 0.28 ④ |
| $y_{t-1}$ | PB 0.9 | 0 | 0 | 0.17 ⑤ | 0 |

rt is PM (0.6)

Controlled object

| $y_{t+1}$ | | $y_t$ ZO 0.0 | $y_t$ PS 0.3 | $y_t$ PM 0.6 | $y_t$ PB 0.9 |
|---|---|---|---|---|---|
| $u_t$ | NS -0.3 | -0.06 | 0.12 | 0.28 ⑤ | 0.50 ④ |
| $u_t$ | ZO 0.0 | 0 | 0.17 | 0.34 | 0.56 |
| $u_t$ | PS 0.3 | 0.06 | 0.23 | 0.39 | 0.61 |
| $u_t$ | PM 0.6 | 0.16 | 0.34 ⑥ | 0.50 | 0.72 |
| $u_t$ | PB 0.9 | 0.34 ① | 0.52 ②⑦ | 0.68 ③⑧ | 0.90 |

Fig.8 Stability analysis (T=2)

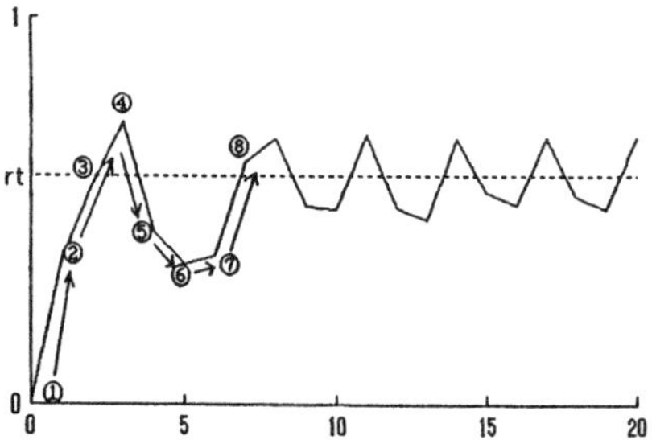


Fig.9 Step response (T=2)

## VI.CONCLUSIONS

This paper presented a stability analysis method using fuzzy neural networks. A new definition of the stability of the fuzzy control systems was also presented in this paper. Although the new definition was not rigorous, the stability analysis based on the definition was very simple and easy to comprehend linguistically.

## References


[1]M. M. Gupta and J.Qi, IJCNN-91-SEATTLE, Vol.II, pp. 431-436 (1991)

[2]M. M. Gupta and M. B. Gorzalczany, IEEE Int'l. Conf. on Fuzzy Systems(FUZZ-IEEE'92) pp. 1271-1274 (1992)

[3]C. T. Liu and C. S. G. Lee, IEEE Int'l. Conf. on Fuzzy Systems(FUZZ-IEEE'92) pp. 1283-1291 (1992)

[4]F. Wong and P. Z. Wang, Proc. of the Int'l. Fuzzy Eng. Symp.(IFES'91), pp. 535-545 (1991)

[5]Y. Wang, Proc. of the Int'l. Fuzzy Eng. Symp.(IFES'91), pp. 546-549 (1991)

[6]S. Horikawa, T. Furuhashi, S. Ohkuma, Y. Uchikawa, Proc. of Int. Conf. on Fuzzy Logic & Neural Networks, pp.103-106, 1990.

[7]S. Horikawa, T. Furuhashi, S. Ohkuma, Y. Uchikawa, Conf. Record of IEEE/IECON'90, pp.1253-1258.

[8]S. Horikawa, T. Furuhashi, Y. Uchikawa and T. Tagawa, Proc. of the Int'l. Fuzzy Eng. Symp.(IFES'91), pp. 562-573 (1991)

[9] S.Horikawa, T.Furuhashi, Y.Uchikawa, IEEE Trans. on Neural Networks, vol.3, no.5, 1992 (to appear).

[10] V.Novak, Adam Hilger, 1986.

[11] T.Takagi and M.Sugeno, IEEE Trans. Syst., Man, Cybern., vol.SMC-15, no.1, pp.116-132, 1985.

[12] K.Tanaka and M.Sugeno, Fuzzy Sets and Systems, vol.45, 1992, pp.135-156.

[13] S.Kawase and N.Yanagihara, The 3rd IFSA Congress, 1989, pp.67-70.

[14] S.Kitamura, Trans. of SICE, vol.27, no.5, 1991, pp.532-537 (in Japanese).

[15] T.Hojo, T.Terano, S.Masui, 6th Fuzzy Systems Symposium, 1990, pp.357-360 (in Japanese).

[16] M.Maeda, S.Murakami, K.Inage, 5th Fuzzy System Symposium, 1989, pp.493-498 (in Japanese).

[17] S.Singh, Proc. of IEEE Int. Conf. of Fuzzy Systems 1992 (FUZZ-IEEE'92), pp.527-534.

[18] S.Kawamoto, K.Tada, A.Ishigame and T.Taniguchi, Proc. of IEEE Int. Conf. of Fuzzy Systems 1992 (FUZZ-IEEE'92), pp.1427-1434.

# Application of Fuzzy Control in Chemical Distillation Processes

Gerhard Klett
BASF AG, ZXT/S Informatik - Sicherheit
Kaiser-Wilhelm-Straße 52
6700 Ludwigshafen
FRG

**Abstract—Fuzzy control, combined with common control techniques, offers new possibilties to implement descriptive models for complex control tasks in a brief and distinct manner. This paper describes the application of fuzzy control in a non-continuous chemical distillation process.**

## I. INTRODUCTION

Distillation is used widely in the chemical industry to separate components with different boiling temperatures of a liquid. The distillation apparatus generally consists of the distillation flask, a packed distillation column, a condensor and valves to adjust the reflux ratio and for the collection of the separate components. The distillation flask is placed in a heating bath with adjustable temperature. Sensors for pressure and temperature are attached in the apparatus. Fig. 1 is a schematic of the distillation apparatus.

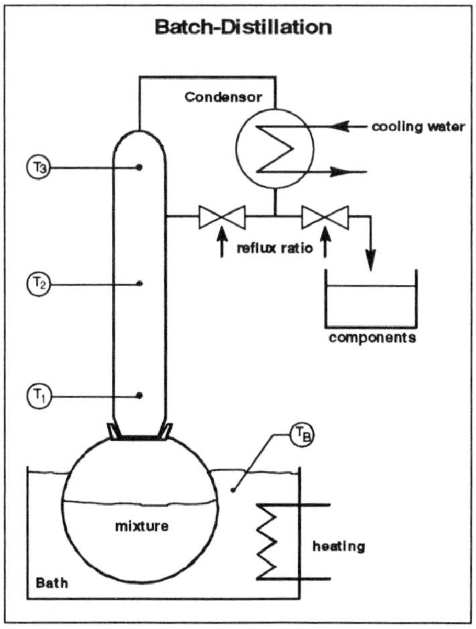


Fig. 1 Batch Distillation, note the location of the temperature sensors

The goal of the distillation is the separation of the mixture into its components with high purity. A manual control process is performed in the following way: depending on the temperature and pressure gradients at several locations inside the apparatus, the temperature of the bath $T_B$, the pressure p and the reflux ratio is varied inside the apparatus. Additionally, the human expert bases his control strategy on some nonquantifyable phenomena such as shape of the drops at the column head or the occurence of "streaks" in the column.

## II. FUZZY LOGIC BASED MODEL

For the configuration of a fuzzy control we must first extract linguistic variables (LV), their values and if/then rules from the description of a human expert. For controls, which form a closed loop within the process, it is a basic presumption to have linguistic variables with related technical values. Each value of a linguistic variable represent a fuzzy set. The degree of membership of a technical value x to a fuzzy set is determined by a so called membership function $\mu(x) \in [0,1]$ (see [1] and [2] for more details about fuzzy set theory). Fig 2. gives an overview of a fuzzy inference scheme.

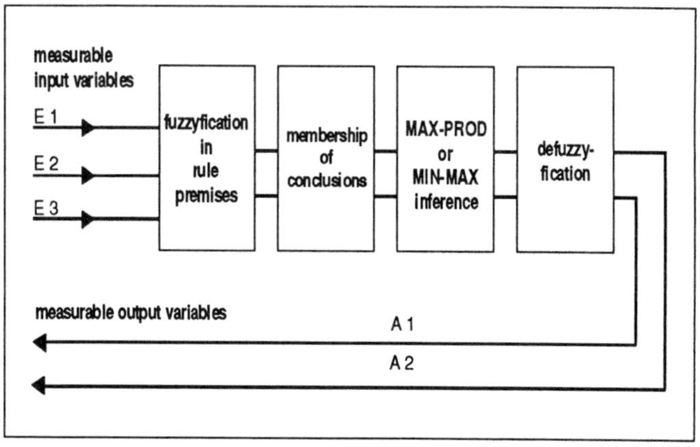


Fig. 2 fuzzy inference scheme

For the design of the control strategy two things are important: the separated components of the mixture should be as pure as possible and the stress of the apparatus, expressed as pressure in the apparatus, should be constant. From physics it is known, that a constant temperature at the column head $T_3$ (see Fig.1) is equivalent to a constant pressure during the separation of a component. Increase of this temperature identifies the separation of the next component. Fig. 3 shows the graph of $T_3$ over time for the separation of two components.

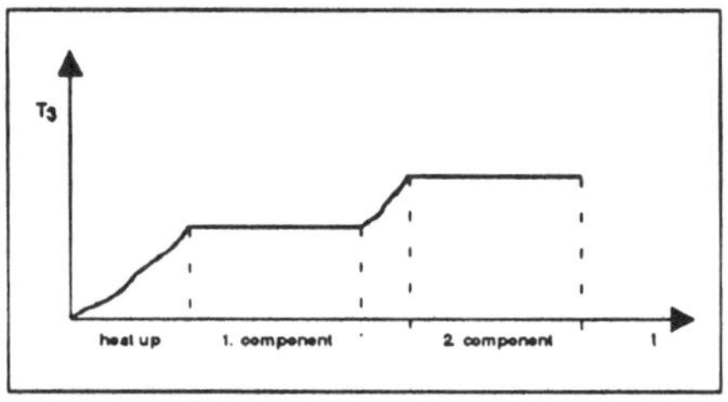


Fig. 3 Temperature $T_3$ in the head of the column during the separation of two components

After the knowledge acquisition, by interviewing the experts, we could isolate eight linguistic variables and their values: five for input and three for output (Table I).

TABLE I
LINGUISTIC VARIABLES FOR INPUT AND OUTPUT

| LV | Values | Symbol |
|---|---|---|
| | Input variables | |
| pressure differnce | low, middle, high, too high | $\Delta p$ |
| temperature difference | low, middle, high | $T_2$-$T_3$<br>$T_1$-$T_2$ |
| temperature gradients | low, middle, high | $\partial T_2/\partial t$<br>$\partial T_3/\partial t$ |
| | Output variables | |
| reflux ratio | very low, low, high, very high, too high | r |
| bath temperature | low, middle, high | $T_B$ |
| set value of pressure | low, middle, high | p |

We use linear membership functions in our model, giving the technical values linear weights [3]. The boundaries of the intervals were heuristically found. Fig. 4 shows the membership functions of bath temperature and pressure.

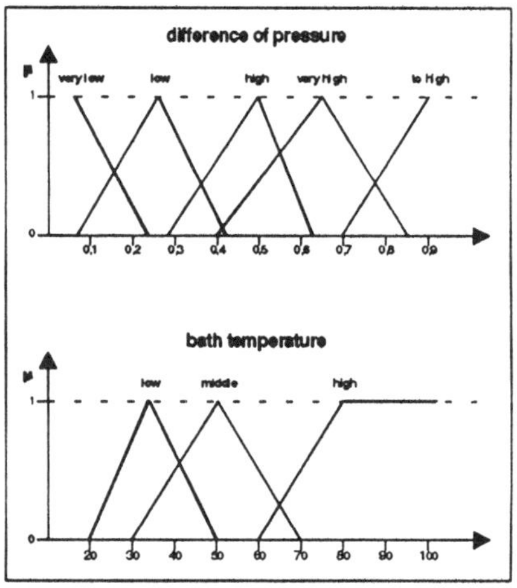


Fig. 4 Membership functions for bath temperature and pressure

As mentioned above, the expert uses two nonquantifyable values, streaks and dropshape, for his control strategies. A mathematical model was developed to substitute measurable signals for these subjective phenomena. Fig. 5 shows the structure of the fuzzy control with input and output variables.

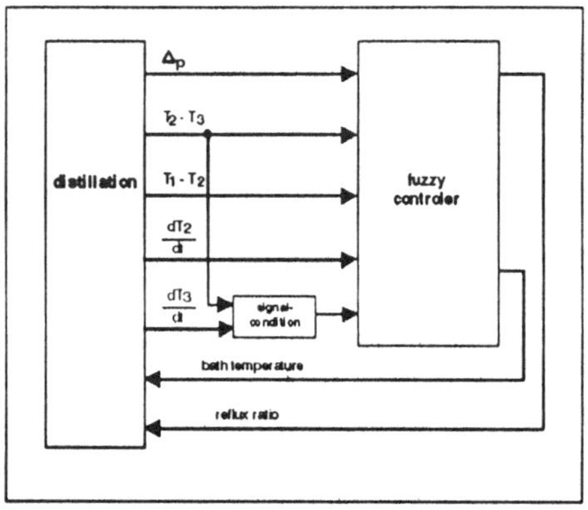


Fig. 5 Structure of the fuzzy control with input and output variables

III. HARD- AND SOFTWARE DEVELOPMENT ENVIRONMENT

One of the constraints of the design was the implementation of the fuzzy control on a inexpensive standard hardware platform. We used a 386 DOS PC with 20 Mhz clock rate, VGA graphics, 4 MB of RAM and 40 MB fixed disk for our development and run-time system. The software design was supported by the TIL (Togai InfraLogic) Shell [4] and MS-

Windows 3.1 as graphical interface. The TIL Shell contains powerful tools for the editing of rules, structuring of the knowledge base, linguistic variables and input/output variables. A graphical editor makes the creation and modification of membership function an easy task. After designing the rule base and defining input/output variables with their fuzzy sets and membership functions the TIL shell produces C source code which can be easily included in the program system.

## IV. ON - LINE CONNECTION TO THE PROCESS

The required time intervals for signal acquisitation is rather long and it is sufficient to have sample intervals of 30 seconds. We use an inexpensive and robust programmable data concentrator. This device is connected via RS232 to the serial port of the PC. The ADC's and DAC's are connected to the sensors and regulators of the distillation apparatus. The data concentrator also provide simple signal conditioning and sends digitized data to the PC on a specific software protocol. Fig. 6 shows the PC - process connection.

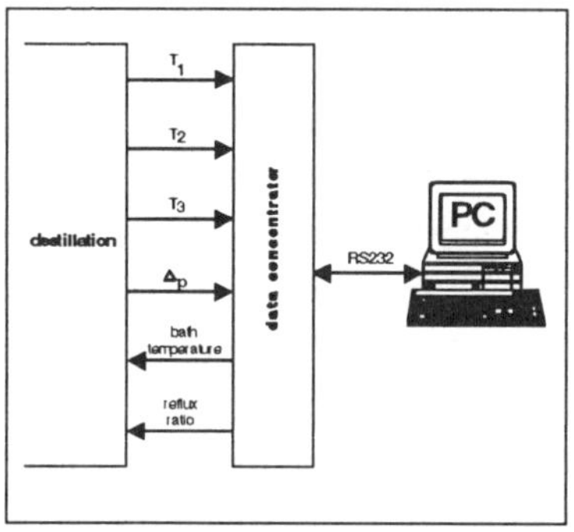


Fig. 6 Communication between the PC and the instrumentation of the distillation apparatus via data concentrator and RS232

## V. SOFTWARE STRUCTURE OF THE FUZZY DISTILLATION

When we talk about the application of fuzzy control for chemical distillations we have to mention that the fuzzy inference is a small, but important, part of the whole system. We combined the fuzzy inference with modules for signal conditioning, task monitoring, graphical user interface, data acquisition and communication with external controllers for bath temperature and pressure, written as "ordinary" programs in C++. Fig. 7 gives an overview of the structure of the control system.

Fig. 7 Structure of the control system

## V. RESULTS

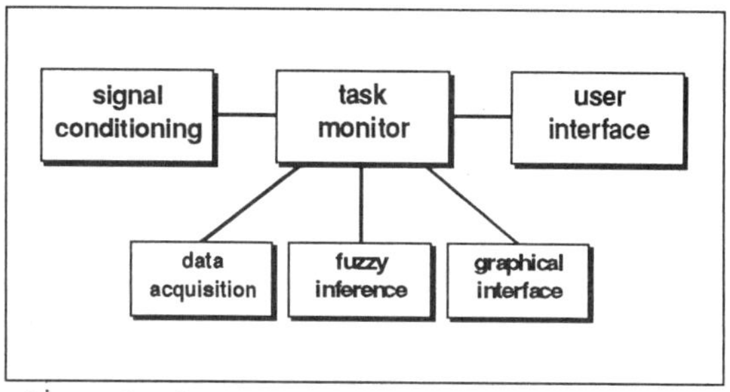


After 3 month of prototyping and testing we distilled several different mixtures of known composition with the fuzzy control device and via manual control by the expert. The fractions were analyzed after the separation by gas chromatography. The comparison shows the following results:

- there is no significant difference in purity of the fractions between the distillation with fuzzy control and under manual control.
- the stress of the distillation apparatus is constant with fuzzy control and varies in a 40 % range with manual control.
- the time required for separating an unknown mixture is, on the average, with fuzzy control, half that of the expert using manual control.

## VI. CONCLUSIONS

Application of fuzzy logic gives us possibilities to design multi-value control for the full automatization of a chemical distillation process. Some attempts in the past to design a controller with traditional techniques failed. The fuzzy controller system with data acquisition and a comfortable graphical user interface was developed in 5 months. It runs on inexpensive standard 386 PC's. It shortens the time for distillation by a remarkable factor of 2.

## REFERENCES

[1] H.J. Zimmermann, *Fuzzy Sets Theory and its Application*, 2nd Edition, Kluwer-Nijhoff Publishing, 1990

[2] H.J. Zimmermann, *Fuzzy Sets, Decision Making and Expert Systems*, Kluwer-Nijhoff Publishing, Boston, 1987

[3] C. v. Altrock, "Über den Daumen gepeilt", *c't*, vol 3, pp. 188-204, 1991

[4] Togai InfraLogic, *Fuzzy Control Development System*, User Manual, Irvine, 1991

# Multiple Attribute Decision Making based on Fuzzy Relationships between Objectives and its Application in Metal Forming

Rudolf Felix, Stefan Reddig
Fuzzy Demonstrations-Zentrum Dortmund
Martin-Schmeißer-Weg 18, D-4600 Dortmund 50, Germany,
Tel. +49 231 9 744 751, Fax. +49 231 9 744 777

Alois Adelhof
Universität Dortmund Lehrstuhl für Umformende Fertigungsverfahren
Postfach 50 05 00, D-4600 Dortmund 50, Germany,
Tel. +49 231 755-2680, Telex +49 231 822 465 unido d

**Abstract** – *A novel theory for multiple attribute decision making based on fuzzy relationships between objectives is presented. In contrast to other approaches the interactive structure of objectives for each decision problem is inferred and represented explicitely. This interactive structure reflects existing fuzzy relationships between objectives and provides for a better understanding of the decision problem. The relationships between objectives are evaluated for the determination of the so called* **focus of attention** *— a fuzzy set of alternatives which are consistent with the multiple objectives, their relationships and priorities. The practical relevance of the theory is shown on a complex multiple attribute decision problem in the field of beam bending, a part of modern metal forming.*

## 1 INTRODUCTION

Since the fuzzy set theory has been suggested as a suitable conceptual framework of decision making [2], two directions in the field of fuzzy decision making can be observed. The first direction reflects the fuzzification of conventional approaches like linear programming [7] or dynamic programming. The second direction is based on the assumption that the process of decision making can be modelled by axiomatically specified aggregation operators [3]. Although these approaches exist for a relatively long period of time and a lot of research work has been done, none of them sufficiently refers to the in our opinion most important aspect of decision making, namely to the aspect of an explicit modeling of relationships between objectives.

For instance, if there is a decision maker who intends to earn money (objective 1) and to have fun (objective 2) simultaneously, and the only way to earn money is to work, then at least two situations are possible:

• Situation 1: The decision maker does not like to work. Therefore, while working the decision maker will not have fun. The alternative *working* supports objective 1 but distracts objective 2.

• Situation 2: The decision maker likes to work. Therefore, while working the decision maker will have fun. The alternative *working* supports both objective 1 and objective 2.

In the first situation the objective "earn money" *hinders* the objective "have fun". In contrast to that, in the second situation the objective "earn money" *assists* the objective "have fun".

Already this simple example shows that there is a substantial need for a detailed automated reasoning about relationships between objectives when dealing with decision making. The related approaches like [7] and [3] lack in this important aspect and either require a very restricted way of the description of objectives or postulate that decision making shall be performed by aggregation operators based on a few, very general axioms like comutativity or associativity. In contrast to that, approaches, especially in the context of complex technical decision problems, human decision makers usually do not describe their objectives by formulas and renounce the use of axiomatically specified operators in order to work out their decisions. Instead of this, they concentrate on the estimation which objectives are supported by which alternatives and which objectives are distracted by which alternatives. Furthermore, they evaluate this information[1]

[1] which is in most cases uncertain.

in order to infer how the objectives do depend on each other and ask for the actual preferences of the objectives.

If we subsume for each objective $g$ the alternatives supporting $g$ in the set $S_g$ and the alternatives distracting the objective $g$ in the set $D_g$, we observe the following characteristics of the two situations of the small example drafted above:

- Situation 1: $S_{\text{earn money}} = \{\text{working}\} = D_{\text{have fun}}$
- Situation 2: $S_{\text{earn money}} = \{\text{working}\} = S_{\text{have fun}}$

Motivated by this observation, relationships between two objectives are defined based on a fuzzy inclusion and a fuzzy non-inclusion between the support and distraction sets of the corresponding objectives.

## 2 BASIC DEFINITIONS

Before we define relationships between objectives, we introduce the notion of support and distraction sets of objectives.

**Def. 1)** Let $A$ be a nonempty and finite set of potential alternatives, $G$ a nonempty and finite set of objectives, $A \cap G = \emptyset$, $a \in A$, $g \in G$, $\delta \in (0,1]$. For each objective $g$ we define the functions $s_g$, $d_g$ and $i_g$ each from $A$ into [0, 1] as follows:

1. *support function of the objective g*

$$s_g(a) := \begin{cases} \delta, & a \text{ supports } g \text{ with degree } \delta \\ 0, & \text{else} \end{cases}$$

2. *distraction function of the objective g*

$$d_g(a) := \begin{cases} \delta, & a \text{ distracts } g \text{ with degree } \delta \\ 0, & \text{else} \end{cases}$$

**Def. 2)** Let $s_g$, $d_g$, and $i_g$ be as defined in Def. 1. For each objective $g$ we define the two fuzzy sets $S_g$ and $D_g$ by the functions $s_g$ and $d_g$ as membership functions, respectively. $S_g$ is called the *support set of g*, $D_g$ the *distraction set of g*.

The set $S_g$ describes which alternatives support the objective $g$ and the certainty degree of the support. The set $D_g$ describes which alternatives distract the objective $g$ and the certainty degree of the distraction.

**Def. 3)** Let $A$ be a finite nonempty crisp set of alternatives. Let $P(A)$ be the set of all fuzzy subsets of $A$. Let $X, Y \in P(A)$, $x$ and $y$ the membership functions of $X$ and $Y$ respectively. The fuzzy inclusion $\mu_{\subset}$ is defined as follows:

$$\mu_{\subset} : P(A) \times P(A) \rightarrow [0,1]$$

$$\mu_{\subset}(X,Y) := \begin{cases} \dfrac{\sum_{a \in A} \min(x(a), y(a))}{\sum_{a \in A} x(a)}, & \text{für } X \neq \emptyset \\ 1, & \text{für } X = \emptyset \end{cases}$$

The fuzzy non-inclusion $\mu_{\not\subset}$ is defined as:

$$\mu_{\not\subset} : P(A) \times P(A) \rightarrow [0,1]$$

$$\mu_{\not\subset}(X,Y) := 1 - \mu_{\subset}(X,Y).$$

For simplicity we write $\subset$ instead of $\mu_{\subset}$ and $\not\subset$ instead of $\mu_{\not\subset}$

The inclusions indicate the existence of relationships between objectives. The higher the degree of inclusion between the support sets of two objectives, the more cooperative the relationship between them. The higher the degree of inclusion between the support set of an objective and the distraction set of the second, the more competitive the relationship. The non-inclusions are evaluated in a similar way. The higher the degree of non-inclusion between the support sets of two objectives, the less cooperative the relationship between them. The higher the degree of non-inclusion between the support set of an objective and the distraction set of the second, the less competitive the relationship.

## 3 RELATIONSHIPS BETWEEN OBJECTIVES

Based on the inclusion and non-inclusion defined above, 16 relationships between objectives are defined [5]. In the following we show the definition of the most important of them. The relationships cover the whole spectrum from the analogy between objectives to a trade-off. The independence of objectives and the case of an unspecied dependence are also considered.

**Def. 4)** Let $S_{g_1}, D_{g_1}, S_{g_2}, D_{g_2}$ be fuzzy sets given by the corresponding membership functions as defined in Def. 2. For simplicity we write $S_1$ instead of $S_{g_1}$ etc.. Let $g_1, g_2 \in G$ where $G$ is a set of objectives. The relationships between two objectives are defined as relations which are fuzzy subsets of $G \times G$ as follows:

1. $g_1$ is independent of $g_2$: $\Leftrightarrow$

$$\Phi_0 := \bigcap_{i,j=1,\ldots,n,i\neq j} F_j^i$$

it is checked, if there are decision alternatives common for all local focuses of attention. If not, those local focuses $F_j^i$, which refer to the least important objective and are characterized by a relatively small cardinality, are excluded from the successive intersecting. The process iterates until either a satisfying, nonempty intersection is found or there are no alternatives satisfying the current constellation of objectives, relationships and priorities. In order to determine which local focus should be excluded in the next iteration step, the ordering relation *before* is defined. For each two local focuses $F_j^i$ *before* $F_l^k$ means that $F_j^i$ will be excluded in a later step than $F_l^k$.

Let n>1 be the number of objectives, $m=n^2-n$, $I$ a sorted set of indices {1,...,n\}, $P_i$ the priority of the objective $g_i$ for $i=1,...,n$. Let $i, j, k, l$ be indices of objectives and $i \neq j$, $k \neq l$.

$F_j^i$ before $F_l^k$ : $\Leftrightarrow$

$$\begin{cases} \frac{P_i+P_j}{2} > \frac{P_k+P_l}{2}, & P_i+P_j \neq P_k+P_l; \\ \text{card}(F_j^i) > \text{card}(F_l^k), & P_i+P_j = P_k+P_l; \\ & \text{card}(F_j^i) \neq \text{card}(F_l^k); \\ i<k, & P_i+P_j = P_k+P_l; \\ & \text{card}(F_j^i) = \text{card}(F_l^k); \\ j<l, & \text{otherwise} \end{cases}$$

The relation *before* induces the following sequence of local focusses:

$F_0 := \left(F_j^i\right)_1, \ldots, \left(F_l^k\right)_m$ Using this sequence, each step $k, k \in \{0,\ldots,m-1\}$ of the process of succesive intersecting provides the corresponding set $\Phi_k$, which is a possible candidate for the global focus of attention $F$:

$$\Phi_k := \begin{cases} \bigcap_{p=1}^{m-k}\left(F_j^i\right)_p, & n>1 \\ \emptyset, & n=1 \end{cases}$$

Now the global focus of attention is defined as the first set in the process of successive intersecting which is large enough. For which $k$ the set $\Phi_k$ is large enough is determinated using the threshold value $t_F$ specified by the user[2].

As a measure of largeness of $\Phi_k$ the relative cardinality $\gamma_k$ defined as:

$$\gamma_k := \begin{cases} \frac{\text{card}(\Phi_k)}{\text{card}\left(\bigcup_{p=1}^{m-k}\left(F_j^i\right)_p\right)}, & \Phi_k \neq \emptyset \\ 0, & \Phi_k \neq \emptyset \end{cases}$$

is applied. Finally, the global focus of attention, which is the final decision set, is determined as follows:

$$F := \begin{cases} \Phi_f & f := \min\{k \mid k=0,\ldots,m-1, \gamma_k > t_F\} \\ S_b & \textit{otherwise} \end{cases}$$

In the "otherwise" case, that means in the case that no $\Phi_k$ exists which is large enough, the support set $S_b$ of a most important objective $g_b$ is chosen as the global focus $F$ This makes sense because in such a case no simultaneous persecution of more than one objective is possible.

## 5 APPLICATION IN MODERN BEAM BENDING TECHNOLOGY

Beam bending is a part of metal forming and includes bending methods which form bar profiles to different shapes of the workpiece. The decisions to be made when planning the bending process are strongly influenced by both the properties of the desired bending product and the capabilities of the bending methods.

The desired properties of the bending product define quality objectives to be obtained. For example, too big a crowning of the profile bottom and profile limbs swinging open too wide have to be avoided (**Fig 1**). On the other hand, a constant radius as well as a profile surface as intact as possible are desired. On the other hand, the different bending set-up and adjustment actions pursue the objectives with different degrees of efficiency and are more or less adequate for obtaining the entire objectives (including a priority setting).

---

[2] In most applications of the decision theory the value $t_F$=0,5 appeared to be appropriate.

$S_1 \not\subset S_2$, $S_1 \not\subset D_2$, $S_2 \not\subset D_1$, $D_1 \not\subset D_2$ and

$\mu_{\text{is independent of}}(g_1, g_2) :=$

$\min(\mu_{\not\subset}(S_1, S_2), \mu_{\not\subset}(S_1, D_2), \mu_{\not\subset}(S_2, D_1), \mu_{\not\subset}(D_1, D_2))$

2. $g_1$ assists $g_2$: $\Leftrightarrow$

$S_1 \subset S_2$, $S_1 \not\subset D_2$ and

$\mu_{\text{assists}}(g_1, g_2) := \min(\mu_{\subset}(S_1, S_2), \mu_{\not\subset}(S_1, D_2))$

3. $g_1$ cooperates with $g_2$: $\Leftrightarrow$

$S_1 \subset S_2$, $S_1 \not\subset D_2$, $S_2 \not\subset D_1$ and

$\mu_{\text{cooperates with}}(g_1, g_2) :=$

$\min(\mu_{\subset}(S_1, S_2), \mu_{\not\subset}(S_1, D_2), \mu_{\not\subset}(S_2, D_1))$

4. $g_1$ is analogue to $g_2$: $\Leftrightarrow$

$S_1 \subset S_2$, $S_1 \not\subset D_2$,

$S_2 \not\subset D_1$, $D_1 \subset D_2$ and

$\mu_{\text{is analogue of}}(g_1, g_2) :=$

$\min(\mu_{\subset}(S_1, S_2), \mu_{\not\subset}(S_1, D_2), \mu_{\not\subset}(S_2, D_1), \mu_{\subset}(D_1, D_2))$

5. $g_1$ hinders $g_2$: $\Leftrightarrow$

$S_1 \not\subset S_2$, $S_1 \subset D_2$ and

$\mu_{\text{hinders}}(g1, g2) :=$

$\min(\mu_{\not\subset}(S_1, S_2), \mu_{\subset}(S_1, D_2))$

6. $g_1$ competes with $g_2$: $\Leftrightarrow$

$S_1 \not\subset S_2$, $S_1 \subset D_2$, $S_2 \subset D_1$ and

$\mu_{\text{competes with}}(g_1, g_2) :=$

$\min(\mu_{\not\subset}(S_1, S_2), \mu_{\subset}(S_1, D_2), \mu_{\subset}(S_2, D_1))$

7. $g_1$ is in trade-off to $g_2$: $\Leftrightarrow$

$S_1 \not\subset S_2$, $S_1 \subset D_2$,

$S_2 \subset D_1$ ,$D_1 \not\subset D_2$ and

$\mu_{\text{is in trade-off to}}(g_1, g_2) :=$

$\min(\mu_{\not\subset}(S_1, S_2), \mu_{\subset}(S_1, D_2), \mu_{\subset}(S_2, D_1), \mu_{\not\subset}(D_1, D_2))$

8. $g_1$ is unspecified dependent from $g_2$: $\Leftrightarrow$

$S_1 \subset S_2$, $S_1 \subset D_2$, $S_2 \subset D_1$, $D_1 \subset D_2$ and

$\mu_{\text{is unspecified dependent from}}(g_1, g_2) :=$

$\min(\mu_{\subset}(S_1, S_2), \mu_{\subset}(S_1, D_2), \mu_{\subset}(S_2, D_1), \mu_{\subset}(D_1, D_2))$

The relationships between objectives are substantial for an adequate treatment of decision making because they reflect the structure of interaction between the objectives and describe the pros and cons of the decision problem. Together with information about actual preferences, the relationships between objectives are the basic aggregation guidelines for the decision maker. For example, for cooperative objectives a conjunctive aggregation is appropriate. If the objectives are independent or rather competitive, then a disjunctive aggregation can be recommended.

## 4 DETERMINING THE FOCUS OF ATTENTION

The *focus of attention F* is the final *decision set*. It contains alternatives which have to meet the following conditions:

1. The choice of each of the alternatives is consistent with the inferred relationships between objectives and the strategies associated to the relationships.

2. The choice of each of the alternatives is consistent with the user specified priorities of the objectives.

The focus of attention $F$ is determined in two steps. In the *first step* for each pair of objectives $g_i$ and $g_j$ their relationship is inferred and the *local focus of attention*, the fuzzy set $F_j^i$ is determined using an appropriate *local decision strategy*. For example, for the realtionship $g_1$ competes with $g_2$, the strategy *sacrifice less important objective* is used. This strategy recommends alternatives supporting the more important objective regardless of their impact on the other one. $F_j^i$ is the focus of attention from the point of view of solely $g_i$ and $g_j$. The strategies are specified as fuzzy if-then rules. The premise of each rule is conjunctively composed of fuzzy predicates referring to the degree of the relationship between the objectives, to the relation between the priorities of the objectives and to wether or not the actual degree of the relationship and the priorities are considered to be rather high or rather small.

In the *second step* the focus of attention $F$ is determined by considering all fuzzy sets $F_j^i$ and the priorities of all objectives simultaneously by a process called *successive intersecting*. Successive intersecting considers the sets $F_j^i$ for each pair of objectives $g_i$ and $g_j$ and the user specified priorities of objectives as well as the fuzziness of the $F_j^i$. The basic idea is to build a sequence of intersections among the different fuzzy sets $F_j^i$ (*i,j=1,...,n, where n is the number of objectives*) [4]. Starting with the set

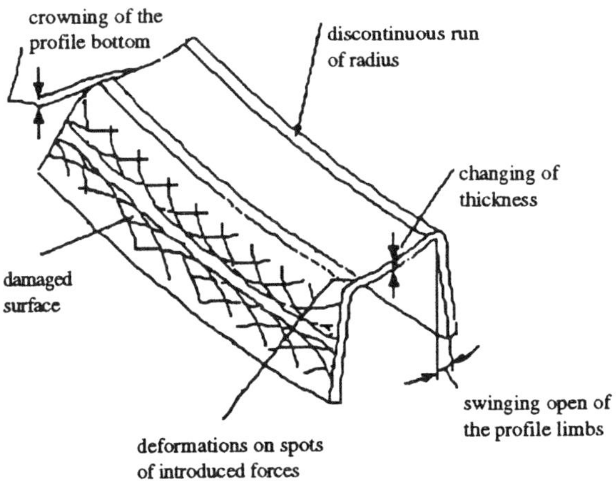


Figure 1: Examples for properties of the bending product.

Since the impact of the set-up and adjustment actions on the objectives represents the fundamental element of bending know-how and the informative structure of this know-how corresponds to the structure of the decision theory presented in previous sections, a novel CAD system based on the bending know-how and the decision theory has been succesfully implemented [1].

The system provides for adequate decisions on the bending method in order to correspond best to

- the quality objectives of the product,
- the technical requirements and the
- adequat bending know-how.

The formalization of the bending know-how is performed by defining the adjustment and set-up actions or actions in the sense of the decision theory. The bending know-how is formalized by support and distraction sets as shown in **Tab 1.**

| cross-section: shape of the workpiece: bending method: roll bending (3 rollers) — support = positive value, distraction = negative value | GOALS | | | | |
|---|---|---|---|---|---|
| ACTIONS | swinging open of the profile limbs must be avoided | small deformations at the spots of forces must be obtained | a discontinuous run of radius must be obtained | intact surface must be obtained | crowning of the profile bottom must be avoided |
| assign guiding elements | 0,9 | -0,3 | | -0,3 | |
| distribute the lateral power over the breadth of the workpiece | 0,4 | 0,8 | | -0,2 | 0,2 |
| superpose compressive strain in the profile bottom | -0,3 | -0,4 | -0,4 | | 0,7 |
| bend the ends of the profile | | -0,2 | 0,9 | | |

Table 1: A small part of the bending know how formalized in a impact set.

The quality objectives of the bending product are defined as objectives is the sense of the decision theory. By using the fuzzy relationships between objectives and the decision strategies of the decision theory, the system determines both the actions necessary to achieve the objectives and, if required, the extent to which these actions are recommended. Since the importance of objectives depends on the actual bending task, the user of the system is able to specify priorities which may differ from case to case exactly as proposed by the decision theory.

The decisions of the system have been experimentally proved on the two basic bending methods [9]

- rounding on a bending machine and
- roll bending on a 3-roller bending machine.

The experiments evaluated by human beam bending

experts demonstrate that the system integrates the bending knowledge appropriately and provides for adequate bending decisions.

## 6 CONCLUSIONS

A novel decision making theory based on fuzzy relationships between objectives is presented. In contrast to other approaches the interactive structure of objectives for each decision problem is inferred and represented explicitely. This interactive structure reflects existing relationships between objectives and provides for a better understanding of the decision problem. The relationships between objectives are evaluated for the determination of the so called *focus of attention* — a fuzzy set of alternatives which are consistent with the multiple objectives, their relationships and priorities.

Compared with former approaches like fuzzy linear programming or fuzzy dynamic programming, the presented decision theory is less restrictive with respect to the way of description of the decision problem. Compared with the approach of axiomatic aggregation operators, the decision theory represents better relationships between objectives and uses preference information adequately.

The application of the decision theory in the domain of modern metal forming demonstrates the high practical relevance of the approach and is an important contribution to the automation of engineering decisions. The universal character of the decision theory and application in other domains, for example VLSI design [6] indicate that using the theory may significantly help in modelling decision situations in which former approaches are either to restrictive or do not reflect properly the interactive structure of the decision problem [5].

## 7 REFERENCES


1. ADELHOF, A. FELIX, R., KLEINER, M., 1991
Modellierung von zielorientierten Entscheidungen am Beispiel eines Biegeverfahrens,
VDI-Z. Zeitschrift für integrierte Produktionstechnik, 9/91 VDI Verlag, Düsseldorf

2. BELLMANN, R. E., ZADEH, L. A., 1970
Decision Making in a Fuzzy Environment, Management Sciences 17

3. DUBOIS, D., PRADE, H., 1984
Criteria Aggregation and Ranking of Alternatives in the Framework of Fuzzy Set Theory,
Studies in Management Sciences, Volume 20

4. FELIX, R., 1990
Goal-oriented Selection of Alternatives Based on Fuzzy Sets,
IPMU Information Processing and Management of Uncertainty in Knowledge-based Systems}, Paris, France

5. FELIX, R., 1991
Entscheidungen bei qualitativen Zielen,
Universität Dortmund, Fachbereich Informatik, PhD Dissertation, (in German) Dortmund, Germany

6. FELIX, R., 1992
Multiple Attribute Decision Making based on Fuzzy Relationships between Objectives,
Proceedings of the 2nd International Conferenca on Fuzzy Logic and Neural Networks, Iizuka, Japan, 1992

7. FELIX, R., POSWIG, J. PROJECT-ALBERT GROUP, 1991
Goal-oriented high-level synthesis based on a fuzzy decision theory,
IEEE Clear Applications of Fuzzy Logic, Delft, The Netherlands

8. ZIMMERMANN, H.-J., ZYSNO, P., 1980
Latent Connectives in Human Decision Making,
Fuzzy Sets and Systems 4, 1980, North Holland, Amsterdam, The Netherlands

9. ADELHOF, A., REDDIG, S., 1991
Einsatz und Weiterentwicklung eines Expertensystems auf der Basis eines zielorientierten Entscheidungsmodells zur Auswahl von Profilbiegeverfahren,
Universität Dortmund, Fachbereich Maschinenbau, Diplomarbeit, Dortmund, Germany

# Fitting Fuzzy Switching Functions to Uncertain Knowledge

Hiroaki Kikuchi
Fujitsu Laboratories Ltd.
10-1 Morinosato-Wakamiya, Atsugi 243-01, Japan

***Abstract*— A new uncertain reasoning algorithm based on fuzzy switching functions is proposed. The algorithm takes a piece of uncertain knowledge which is a mapping with restricted domain and outputs the logic formula with the shortest distance to the given mapping. The execution is done in three steps; first, the given mapping is divided into some *Q-equivalent classes*; second, the distances between the mapping and each local fuzzy switching function are calculated by a simplified logic formula; and last, the shortest distance is obtained by a modified graph-theoretic algorithm.**

## I. Introduction

Humans are able to infer meaningful conclusions from uncertain information. This information can be written down in the form of if-then rules. Given a knowledge base of several if-then rules, fuzzy inference methods can be used to model uncertain reasoning. The manual derivation of rule bases is a current issue in this field, and several automated approaches have been attempted, including fuzzy neural expert systems, self-learning methods, and adaptive networks.

Instead of If-then rule base, we use a logic formula. A fuzzy switching function is a mapping that can be represented with a logic formula. We have already proposed two uncertain reasoning methods based on some properties of fuzzy switching functions[1, 2], which can treat any given response of a human expert with *consistency* and *uniqueness*. In this inference scheme, we suppose that uncertain knowledge is a mapping, say $h$, and find the fuzzy switching function $f$ that satisfies the exact value of $h$. However, some noise and incompleteness involved by human response could spoil the consistency of fuzzy switching functions. Here, we suppose a human expert's response based on a logic formula but with some noise, and then attempt to fit fuzzy switching functions to the underlying knowledge. Let us formalize the fitting problem.

**Fitting Problem** Find a 2-variable fuzzy switching function $f^*$ with the shortest distance to the following mapping $h$:

$$h(0.8,0.7) = 0.9,\ h(0.3,0.1) = 0.6,$$
$$h(0.4,0.8) = 0.3,\ h(0.7,0.6) = 0.8,$$
$$h(0.6,0.9) = 0.6.$$

Where a mapping $h$ is defined on $D = \{(0.8,0.7), (0.3,0.1), (0.4,0.8), (0.7,0.6), (0.6,0.9)\}$. We should also define a *distance* between two functions. Let $f$ and $g$ be mappings $f : A \to V$ and $g : B \to V$. The *distance* $E(f,g)$ between $f$ and $g$ is defined as

$$E(f,g) = \sum_{a \in A \cap B} (f(\boldsymbol{a}) - g(\boldsymbol{a}))^2.$$

The first thing we can do with this problem is to examine each fuzzy switching function one by one, which is called the brute force algorithm. This works well with a low number of variables up to 3; however, it is impossible for a large $n$. In this paper, we will divide the problem into some meaningful small problems, what we call *Q-equivalent classes*, and estimate each distance to the given $h$, and then combine them so as to represent a fuzzy switching function. This will allow us to find the fuzzy switching function $f^*$ with the shortest distance for all possible combinations of the classes. We use a graph-theoretic algorithm to find $f^*$, which is a variation of Dijkstra's shortest path algorithm.

## II. Fuzzy Switching Functions

### A. Basic Definitions

**Definition 1** Let $V = [0,1]$, $V_2 = \{0,1\}$ and $V_3 = \{0, 0.5, 1\}$ be the sets of truth values. A *logic formula* consists of constants 0 and 1, variables $x_i$ ($i = 1,\ldots,n$), and logic connectives *and*($\wedge$), *or*($\vee$) and *not*(-), that are defined by $x_i \wedge x_j = x_i x_j = \min(x_i, x_j)$, $x_i \vee x_j = \max(x_i, x_j)$, and $\overline{x_i} = 1 - x_i$.

A *fuzzy switching function* is a mapping from an $n$-dimensional Cartesian product $V^n$ to $V$ which is represented by a logic formula.

0-7803-0614-7/93$03.00 ©1993IEEE

**Definition 2** Let $a$ and $b$ be elements of $V$. Then, $a \succeq b$ if and only if either $0.5 \geq a \geq b$ or $b \geq a \geq 0.5$. The relation $\succeq$ can be extended to $V^n$ by letting $\boldsymbol{a} = (a_1, \ldots, a_n)$ and $\boldsymbol{b} = (b_1, \ldots, b_n)$ be elements of $V^n$, $\boldsymbol{a} \succeq \boldsymbol{b}$ if and only if $a_i \succeq b_i$ for each $i$ $(i = 1, \ldots, n)$. Any two elements $a$ in $[0, 0.5)$ and $b$ in $(0.5, 1]$ are not comparable with respect to $\succeq$. We denote this by $a \not\succeq b$. We write $a \succ b$ to mean that $a \succeq b$ and $a \neq b$.

**Definition 3** Let $x$ and $\lambda$ be elements of $V$. A *quantization* $\overline{x}^{\lambda}$ of $x$ by $\lambda$ is an element of $V_3$ defined by:

$$\overline{x}^{\lambda} = \begin{cases} 0 & \text{if } 0 \leq x \leq \min(\lambda, 1-\lambda) \leq 0.5,\ x \neq 0.5, \\ 1 & \text{if } 1 \geq x \geq \max(\lambda, 1-\lambda) \geq 0.5,\ x \neq 0.5, \\ 0.5 & \text{otherwise.} \end{cases}$$

Let $\boldsymbol{x} = (x_1, \ldots, x_n)$ be an element of $V^n$. A quantization $\overline{\boldsymbol{x}}^{\lambda}$ of $\boldsymbol{x}$ by $\lambda$ is defined by $\overline{\boldsymbol{x}}^{\lambda} = (\overline{x_1}^{\lambda}, \ldots, \overline{x_n}^{\lambda}) \in V_3^n$.

A fuzzy switching function is representable by a disjunctive form which is a disjunction (*or*) of some conjunctions (*and*). There are two types of phrases. One is called a *complementary phrase* which contains a literal and its negation such as $x_i \wedge (\overline{x_i})$. The other is called a *simple phrase*.

**Definition 4** Let $\boldsymbol{a}$ and $\boldsymbol{b}$ be elements of $V_3^n$. A *simple phrase* $\alpha^a$ corresponding to $\boldsymbol{a}$ and a complementary phrase $\beta^b$ corresponding to $\boldsymbol{b}$ are defined by $\alpha^a = x_1^{a_1} \wedge \cdots \wedge x_n^{a_n}$ and $\beta^b = x_1^{b_1} \wedge \cdots \wedge x_n^{b_n}$ where

$$x_i^{a_i} = \begin{cases} x_i & \text{if } a_i = 1 \\ \overline{x_i} & \text{if } a_i = 0 \\ 1 & \text{if } a_i = 0.5, \end{cases} \qquad x_i^{b_i} = \begin{cases} x_i & \text{if } b_i = 1 \\ \overline{x_i} & \text{if } b_i = 0 \\ x_i \overline{x_i} & \text{if } b_i = 0.5. \end{cases}$$

for every $i$ $(i = 1, \ldots, n)$.

*B. Fundamental Properties of Fuzzy Switching Functions*

**Theorem 1 (Normality)** [4] Let $f$ be a fuzzy switching function.

$$\boldsymbol{a} \in V_2^n \Rightarrow f(\boldsymbol{a}) \in V_2$$

**Theorem 2 (Monotonicity)** [4] Let $\boldsymbol{a}$ and $\boldsymbol{b}$ be elements of $V^n$.

$$\boldsymbol{a} \succeq \boldsymbol{b} \Rightarrow f(\boldsymbol{a}) \succeq f(\boldsymbol{b})$$

**Theorem 3 (Quantization)** [4] Let $f$ be a fuzzy switching function, and $\boldsymbol{a}$ be an element of $V^n$. Then

$$\overline{f(\boldsymbol{a})}^{\lambda} = f(\overline{\boldsymbol{a}}^{\lambda}) \quad \text{for all } \lambda \text{ in } V.$$

**Theorem 4** [4] Let $f$ and $g$ be fuzzy switching functions. Then, $f(\boldsymbol{a}) = g(\boldsymbol{a})$ for every element $\boldsymbol{a}$ of $V_3^n$, if and only if $f(\boldsymbol{a}) = g(\boldsymbol{a})$ for every element $\boldsymbol{a}$ of $V^n$.

**Theorem 5 (Representation)** Any fuzzy switching function $f$ is represented by

$$f(\boldsymbol{x}) = \bigvee_{a \in V_3^n} \left( \bigwedge_{a \succ b} f(\boldsymbol{b})\alpha^a(\boldsymbol{x}) \vee f(\boldsymbol{a})\beta^a(\boldsymbol{x}) \right)$$

where $\alpha^a$ is a simple phrase corresponding to $\boldsymbol{a}$ and $\beta^a$ is a complementary phrase corresponding to $\boldsymbol{a}$.

**Proposition 1** Let $\alpha^a$ be a simple phrase corresponding to $\boldsymbol{a} \in V_3^n$.

$$\boldsymbol{a} \succeq \boldsymbol{b} \Leftrightarrow \alpha^a(\boldsymbol{b}) = 1$$
$$\boldsymbol{b} \succ \boldsymbol{a} \Rightarrow \alpha^a(\boldsymbol{b}) = 0.5$$

**Proposition 2** Let $\boldsymbol{a}$ be in $V^n$ and $\lambda, \tau$ be in $V$.

$$\lambda \succ \tau \quad \Rightarrow \quad \overline{\boldsymbol{a}}^{\tau} \succ \overline{\boldsymbol{a}}^{\lambda}$$

## III. Fitting Fuzzy Switching Functions

*A. Dividing into Q-equivalent Classes*

*A.1. Quantized Sets and Q-equivalent Classes*

**Definition 5** Let $\boldsymbol{a} = (a_1, \ldots, a_n)$ in $V^n$. A quantized set of $\boldsymbol{a}$ is a subset of $V_3^n$ defined by

$$C(\boldsymbol{a}) = \{\overline{\boldsymbol{a}}^{\lambda} \in V_3^n \mid \lambda \in V\}.$$

(Example 1)

$$\begin{aligned} C((0.4, 0.8)) &= \{\overline{(0.4, 0.8)}^{0.4}, \overline{(0.4, 0.8)}^{0.8}, \overline{(0.4, 0.8)}^{1}\} \\ &= \{(0, 1), (0.5, 1), (0.5, 0.5)\} \end{aligned}$$

**Definition 6** Elements $\boldsymbol{a}$ and $\boldsymbol{b}$ of $V^n$ are *Q-equivalent* if and only if the quantized sets satisfy $C(\boldsymbol{a}) = C(\boldsymbol{b})$. We write $\boldsymbol{a} \approx \boldsymbol{b}$ in this case.

(Example 2)
$(0.8, 0.7) \approx (0.7, 0.6)$, $(0.3, 0.8) \approx (0.2, 0.9) \approx (0.4, 0.9)$

**Definition 7** Let $A$ be a subset of $V^n$ and $\boldsymbol{a} \in A$. A *Q-equivalent class* containing $\boldsymbol{a}$ is a subset of $A$ defined by

$$[\boldsymbol{a}] = \{\boldsymbol{b} \in A \mid \boldsymbol{a} \approx \boldsymbol{b}\}.$$

The set of all equivalence classes of $A$ is denoted by $[A]$, i.e., $[A] = \{[\boldsymbol{a}] \subset A | \ \boldsymbol{a} \in A\}$.[1]

[1] In general, a domain of fuzzy switching functions can be divided into a few subsets in which variables are in order[3]. This is called *cell space* and corresponds to a Q-equivalent class. It should be noted that their boundary conditions are slightly different. For example, both $(0.5, 0.2)$ and $(0.6, 0.2)$ are in the same cell space, while $C(0.5, 0.2) = \{(0.5, 0), (0.5, 0.5)\} \neq C(0.6, 0.2) = \{(1, 0), (0.5, 0), (0.5, 0.5)\}$.

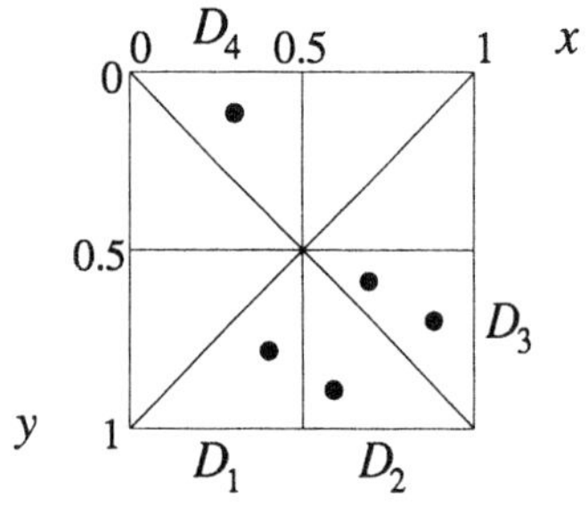


Figure 1: Partition $[D]$

Since $\approx$ is an equivalence relation, we have a partition $[A] = A_1, \ldots, A_m$, that is,

$$A_i \cap A_j = \phi \quad \text{for every } i \neq j,$$
$$A = A_1 \cup \cdots \cup A_m.$$

(Example 3) Recall $D$ in the introduction. We have the partition

$$\begin{aligned} D &= \{(0.4,0.8),(0.6,0.9),(0.7,0.6),(0.8,0.7),(0.3,0.1)\} \\ &= \{(0.4,0.8)\} \cup \{(0.6,0.9)\} \\ &\quad \cup \{(0.7,0.6),(0.8,0.7)\} \cup \{(0.3,0.1)\} \\ &= D_1 \cup D_2 \cup D_3 \cup D_4. \end{aligned}$$

Figure 1 shows how we divide $D$ into Q-equivalence classes.

*A.2. Some Properties of Q-equivalent Classes*

**Proposition 3** Let $C(\boldsymbol{a}) = \{\boldsymbol{a}_1, \boldsymbol{a}_2, \ldots, \boldsymbol{a}_m\}$ be a Q-equivalent class in $V_3^n$. Then

$$\boldsymbol{a}_i \succeq \boldsymbol{a}_j \text{ or } \boldsymbol{a}_j \succeq \boldsymbol{a}_i \quad \text{for every } i,j \in \{1,\ldots,m\}.$$

This means that there are subscripts $i_1 < i_2 < \cdots < i_m$ so that $\boldsymbol{a}_{i_1} \succeq \boldsymbol{a}_{i_2} \succeq \cdots \succeq \boldsymbol{a}_{i_m}$.

**Proposition 4** Let $\boldsymbol{a} = (a_1, \ldots, a_n)$ and $\boldsymbol{b} = (b_1, \ldots, b_n)$ be elements of $V^n$. Then $\boldsymbol{a} \approx \boldsymbol{b}$ if and only if $\overline{a_i}^{a_j} = \overline{b_i}^{b_j}$ for every $i,j$ in $\{1,\ldots,n\}$.

**Proof.** Suppose $\overline{a_i}^{a_j} \succeq \overline{b_i}^{b_j}$ for some $i,j$ when $\boldsymbol{a} \approx \boldsymbol{b}$. Then, $\overline{\boldsymbol{a}}^{j} \succeq \overline{\boldsymbol{b}}^{b_j}$. Consider $\lambda \in V$ such that $\overline{b_i}^{\lambda} = \overline{a_i}^{a_j} = 0.5$. Since $b_j \succeq \lambda$, $\overline{b_j}^{\lambda} = 0.5 \neq \overline{a_j}^{a_j} \in \{0,1\}$. Thus, there is no $\lambda$ that satisfies $\overline{\boldsymbol{a}}^{a_j} = \overline{\boldsymbol{b}}^{\lambda}$, that is, $\overline{\boldsymbol{a}}^{a_j} \notin C(\boldsymbol{b})$. When $\overline{b_i}^{b_j} \succeq \overline{a_i}^{a_j}$, similarly $\overline{\boldsymbol{a}}^{a_j} \notin C(\boldsymbol{b})$. When $\overline{a_i}^{a_j} = 0$ and $\overline{b_i}^{b_j} = 1$, apparently there is no $\lambda$ such as $\overline{b_i}^{\lambda} = \overline{a_i}^{\lambda}$. Therefore, in any case, $\overline{a_i}^{a_j} = \overline{b_i}^{b_j}$.

Conversely, suppose $\overline{a_i}^{a_j} = \overline{b_i}^{b_j}$ for every $i,j$. For any $\lambda \in \{0,1\}$, $\overline{\boldsymbol{a}}^{\lambda} = \overline{\boldsymbol{b}}^{\lambda} = (0.5,\ldots,0.5)$. For any $\lambda \in V - \{0,1\}$, there is a $k$ that holds for $\overline{\boldsymbol{a}}^{\lambda} = \overline{\boldsymbol{a}}^{a_k}$ and $\overline{\boldsymbol{b}}^{\lambda} = \overline{\boldsymbol{b}}^{b_k}$. Thus, we have $\boldsymbol{a} \succeq \boldsymbol{b}$. (Q.E.D.)

**Theorem 6** Let $f$ and $g$ be fuzzy switching functions, and $\boldsymbol{d} \in V^n$. Then, $f(\boldsymbol{a}) = g(\boldsymbol{a})$ for all $\boldsymbol{a} \in C(\boldsymbol{d})$ if and only if $f(\boldsymbol{b}) = g(\boldsymbol{b})$ for all $\boldsymbol{b} \in V^n$ such that $\boldsymbol{b} \approx \boldsymbol{d}$.

**Proof.** Suppose $f(\boldsymbol{b}) \neq g(\boldsymbol{b})$ for some $\boldsymbol{b} \in V^n$ when $f(\boldsymbol{a}) = g(\boldsymbol{a})$ for all $\boldsymbol{a} \in C(\boldsymbol{d})$. By Theorem 3, $\overline{f(\boldsymbol{b})}^{\lambda} = f(\overline{\boldsymbol{b}}^{\lambda}) \neq \overline{g(\boldsymbol{b})}^{\lambda} = g(\overline{\boldsymbol{b}}^{\lambda})$ for some $\lambda \in V$. This is contradictory to the hypothesis.

Conversely, suppose $f(\boldsymbol{a}) \neq g(\boldsymbol{a})$ for some $\boldsymbol{a} \approx \boldsymbol{d} \in C(\boldsymbol{d})$ when $f(\boldsymbol{b}) = g(\boldsymbol{b})$ for every $\boldsymbol{b} \approx \boldsymbol{d}$. Then, there is a $\boldsymbol{b} \in V^n$ such as $\overline{\boldsymbol{b}}^{\lambda} = \boldsymbol{a}$, and for $\boldsymbol{b}$, $f(\boldsymbol{a}) = f(\overline{\boldsymbol{b}}^{\lambda}) = \overline{f(\boldsymbol{b})}^{\lambda} \neq g(\boldsymbol{a}) = g(\overline{\boldsymbol{b}}^{\lambda}) = \overline{g(\boldsymbol{b})}^{\lambda}$. Since $\boldsymbol{b} \approx \boldsymbol{d}$, this contradicts the hypothesis. (Q.E.D.)

(Example 4) Given $f(0,1)$, $f(0.5,1)$ and $f(0.5,0.5)$ in $V$, $f(0.3,0.8)$ is determined uniquely by them.

*B. Estimation of Local Distances*

*B.1. Restrictions of Fuzzy Switching Function*

**Definition 8** Let $\boldsymbol{a} \in V_3^n$, and $b \in \{0,1\}$. Then $F|_{F(a)=b}$ is a fuzzy switching function such that $F|_{F(a)=b}(\boldsymbol{a}) = b$ and $F|_{F(a)=b}(\boldsymbol{b}) = 0.5$ for every $\boldsymbol{b} \succeq \boldsymbol{a} \in V_3^n$. We say $F|_{F(a)=b}$ is a *restriction* of a fuzzy switching function with $F(\boldsymbol{a}) = b$.

Note that a restriction is not unique. For example, here are two restrictions $f|_{f(0)=1}$:

$$f_1(x) = \overline{x}, \qquad f_2(x) = x \vee \overline{x},$$

both of which satisfy $f(0) = 1$ and $f(0.5) = 0.5$.

**Lemma 1** Let $f$ be a fuzzy switching function and $\boldsymbol{d} \in V^n$. If $f(\boldsymbol{d}) \notin V_3$ then

$$f(\boldsymbol{a}) = 0.5 \quad \text{for all } \boldsymbol{a} \in C(\boldsymbol{d}) \text{ such as } \boldsymbol{a} \succ \boldsymbol{a}^*,$$
$$f(\boldsymbol{a}^*) \in \{0,1\},$$

where $\boldsymbol{a}^* = \overline{\boldsymbol{d}}^{f(\boldsymbol{d})}$.

**Proof.** Since $f(\boldsymbol{d}) \notin V_3$, obviously, $\overline{f(\boldsymbol{d})}^{f(\boldsymbol{d})} = f(\overline{\boldsymbol{d}}^{f(\boldsymbol{d})}) = f(\boldsymbol{a}^*) \in \{0,1\}$. For any $\boldsymbol{b} \succ \boldsymbol{a}^*$, there is a $\lambda \in V$ such that $\boldsymbol{b} = \overline{\boldsymbol{d}}^{\lambda} \succeq \overline{\boldsymbol{d}}^{f(\boldsymbol{d})}$ and $f(\boldsymbol{d}) \succ \lambda$. Thus $\overline{f(\boldsymbol{d})}^{\lambda} = 0.5$, and thus, $f(\overline{\boldsymbol{d}}^{\lambda}) = f(\boldsymbol{b}) = 0.5$. (Q.E.D.)

**Theorem 7** Let $\boldsymbol{d} \in V^n$ and $f$ be a fuzzy switching function. For every $\boldsymbol{c} \in V^n$ with $\boldsymbol{c} \approx \boldsymbol{d}$,

$$f(\boldsymbol{c}) = F|_{F(a^*)=f(a^*)}(\boldsymbol{c}),$$

where $\boldsymbol{a}^* = \overline{\boldsymbol{d}}^{f(\boldsymbol{d})}$.

**Proof.** For any $c = a^*$, obviously $F|_{F(a^*)=f(a^*)}(c) = f(c)$. For any $c \succeq a^*$, $F|_{F(a^*)=f(a^*)}(c) = 0.5 = f(c)$ [Lemma 1]. For any $a^* \succeq c$, $F|_{F(a^*)=f(a^*)}(c) = f(a^*) = f(c)$ [Theorem 2]. Thus, $F|_{F(a^*)=f(a^*)}(c) = f(a^*) = f(c)$ for all elements of $C(d)$, and by Theorem 6, $F|_{F(a^*)=f(a^*)}(c) = f(a^*) = f(c)$ for all elements of $[d]$. (Q.E.D.)

This theorem implies that a fuzzy switching function can be determined uniquely within the given Q-equivalence class, and that for any fuzzy switching function $F$, there must be $F|_{F(a)=b}$ in a given Q-equivalence class. Hence, combinations of $(a, b)$ are enough to represent all fuzzy switching functions.

**Corollary 1** Let $D = \{d_1, \ldots, d_m\}$ be a subset of $V^n$, $D_1, \ldots, D_s$ be Q-equivalent classes of $D$, and $f$ be a function $f : D \to V$. For any fuzzy switching functin $F$, there are $a_1, \ldots, a_s \in V_3^n$ and $b_1, \ldots, b_s \in \{0, 1\}$ that satisfy

$$E(F, f) = \sum_{i=1}^{m} (f(d_i) - F(d_i))^2 = \sum_{i=1}^{s} E(F|_{F(a_i)=b_i}, f|_{D_i})$$

where $f|_{D_i}$ is a restriction of $f$, that is, $f|_{D_i} : D_i \to V$.

(Example 5) Let $D_i$ be as in Example 3. For $F(x, y) = \overline{y} \vee \overline{x}y$,

$$\begin{aligned} E(F, f) &= E(F|_{F(0,1)=1}, f|_{D_1}) + E(F|_{F(1,0)=0}, f|_{D_2}) \\ &\quad + E(F|_{F(1,0)=0}, f|_{D_3}) + E(F|_{F(0.5,0)=1}, f|_{D_4}). \end{aligned}$$

*B.2. Representation of Restrictions*

**Theorem 8** Let $d \in V^n$, $c \in C(d)$, $b \in V_2$, and $f$ be a fuzzy switching function. For $a \in V^n$ such as $a \approx d$,

$$f|_{f(c)=b}(a) = \begin{cases} \alpha^c(a) & \text{if } b = 1 \\ \overline{\alpha^c}(a) & \text{if } b = 0. \end{cases}$$

**Proof.** For $c \succeq a$, $f|_{f(c)=1}(a) = 1 = \alpha^c(a) = 1$ and $f|_{f(c)=0}(a) = 0 = \overline{\alpha^c}(a) = 0$ [Proposition 1]. For $a \succ c$, $f|_{f(c)=b}(a) = 0.5 = \alpha^c(a) = \overline{\alpha^c}(a) = 0.5$ [Proposition 1]. Thus, the theorem holds for all elements of $C(d)$. Therefore, by Theorem 6, we have $f|_{f(c)=1}(a) = \alpha^c(a)$ and $f|_{f(c)=0}(a) = \overline{\alpha^c}(a)$ for every $a \approx d$. (Q.E.D.)

(Example 6) Let $d = (0.3, 0.8)$, and $c = (0, 1) \in C(d)$. For any $a \in V^n$ such as $a \approx (0.3, 0.8)$,

$$\begin{aligned} f|_{f(0,1)=1}(a) &= \alpha^{(0,1)}(a) = \overline{x}y(a), \\ f|_{f(0,1)=0}(a) &= \overline{\alpha^{(0,1)}}(a) = \overline{\overline{x}y}(a) = (x \vee \overline{y})(a). \end{aligned}$$

Since $0 < \overline{y} < x < 0.5 < \overline{x} < y < 1$ [Proposition 4], we have

$$f|_{f(0,1)=1}(x) = \overline{x}, \quad f|_{f(0,1)=0}(x) = x.$$

Table 1: Distances $E(f, h)$

$E(f|_{f(a)=b}, h|_{D_1})$

| $a \backslash b$ | 1 | 0 |
|---|---|---|
| (0,1) | 0.09 | 0.01 |
| (0.5,1) | 0.25 | 0.01 |
| (0.5,0.5) | 0.49 | 0.09 |

$E(f|_{f(a)=b}, h|_{D_2})$

| $a \backslash b$ | 1 | 0 |
|---|---|---|
| (1,1) | 0 | 0.04 |
| (0.5,1) | 0.09 | 0.04 |
| (0.5,0.5) | 0.16 | 0.36 |

$E(f|_{f(a)=b}, h|_{D_3})$

| $a \backslash b$ | 1 | 0 |
|---|---|---|
| (1,1) | 0.08 | 0.52 |
| (1,0.5) | 0.02 | 0.74 |
| (0.5,0.5) | 0.05 | 1.45 |

$E(f|_{f(a)=b}, h|_{D_4})$

| $a \backslash b$ | 1 | 0 |
|---|---|---|
| (0,0) | 0.01 | 0.09 |
| (0.5,0) | 0.09 | 0.25 |
| (0.5,0.5) | 0.14 | 0.36 |

**Corollary 2** Let $[d]$ be Q-equivalent class of $d \in V_n$, $a \in C(d)$ and $b \in V_2$.

$$E(f|_{f(a)=b}, h|_{[d]}) = \begin{cases} \sum_{d' \in [d]} (h(d') - \alpha^a(d'))^2 & \text{if } b = 1 \\ \sum_{d' \in [d]} (h(d') - \overline{\alpha^a}(d'))^2 & \text{if } b = 0 \end{cases}$$

where $h|_{[d]}$ be a restriction of $h$ with domain $[d]$, that is, $h|_{[d]} : [d] \to V$.

(Example 7) Let $h$ and $D$ be as for the fitting problem shown in the Introduction. We calculate every distance to $h$ for each Q-equivalent class in $[D]$. For $d = (0.8, 0.7)$ in $D$, we have the quantized set $C(d)$ and the Q-equivalent class $[d]$:

$$\begin{aligned} C(d) &= \{(1,1), (1,0.5), (0.5,0.5)\}, \\ [d] &= \{(0.8,0.7), (0.7,0.6)\}. \end{aligned}$$

By Corollary 1, the distances for every fuzzy switching function are given by examining each $a$ in $C(d)$, that is, $|C(d)| \cdot |V_2| = 3 \cdot 2 = 6$ restrictions in $[d]$. Theorem 8 is useful to simplify the calculation of distances. For example,

$$\begin{aligned} E(f|_{f(1,1)=1}, h|_{[d]}) &= \sum_{a \in [d]} (h(a) - \alpha^{(1,1)}(a))^2 \\ &= (h(0.8,0.7) - (y)(0.8,0.7))^2 \\ &\quad + (h(0.7,0.6) - (y)(0.7,0.6))^2 \\ &= (0.9 - 0.7)^2 + (0.8 - 0.6)^2 = 0.08 \\ E(f|_{f(1,1)=0}, h|_{[d]}) &= (0.9 - 0.3)^2 + (0.8 - 0.4)^2 = 0.52 \end{aligned}$$

Table 1 lists the distances to $h$ for each Q-equivalent class, $D_1, \ldots, D_4$.

*C. Combining Local Solutions*

In the previous section, we have obtained all possible local distances for each Q-equivalence class in the partitions

[D]. The next problem is how to combine them and find the global solution $f^*$ with the shortest distance. We use a graph-theoretic algorithm, which is a variation of Dijkstra's shortest path algorithm[5].

*C.1. Condition for Two Functions to Be Consistent*

**Theorem 9 (Consistency)**
[1] Let $A = \{\boldsymbol{a}_1, \boldsymbol{a}_2, \ldots, \boldsymbol{a}_m\}$ be a subset of $V^n$, and $f$ be a mapping $f : A \to V$. There is a fuzzy switching function $F$ such that $F(\boldsymbol{a}) = f(\boldsymbol{a})$ for all $\boldsymbol{a}$ in $A$ if and only if $f$ satisfies both of the following conditions:

$$(R)\quad C_i^*(f) \cap C_j^*(f) = \phi \qquad \text{for every } i \neq j,$$
$$(N)\quad C_U(f) \cap V_2^n = \phi,$$

where $C_i(f)$ and $C_i^*(f)$ are called *quantized sets* and *expansions* that are defined as follows:

$$C_i(f) = \{\overline{\boldsymbol{a}}^\lambda \mid \boldsymbol{a} \in A, \lambda \in V, \overline{f(\boldsymbol{a})}^\lambda = i\} \; (i = 0, 1, 0.5),$$
$$C_i^*(f) = C_i(f) \cup \{\boldsymbol{a} \in V_3^n \mid \boldsymbol{b} \in C_i(f), \boldsymbol{b} \succeq \boldsymbol{a}\} \; (i = 0, 1),$$
$$C_{0.5}^*(f) = C_{0.5}(f) \cup \left\{\boldsymbol{a} \in V_3^n \mid \begin{array}{l} \boldsymbol{b} \in C_1^*(f),\ \boldsymbol{a} \succeq \boldsymbol{b}, \\ \boldsymbol{c} \in C_0^*(f),\ \boldsymbol{a} \succeq \boldsymbol{c} \end{array}\right\}$$

**Corollary 3** Let $A$ and $B$ be distinct subsets of $V_3^n$, $f$ and $g$ be restrictions of a fuzzy switching function such that $f : A \to V$ and $g : B \to V$. There is a fuzzy switching function $F$ that satisfies $F(\boldsymbol{c}) = f(\boldsymbol{c})$ for all $\boldsymbol{c} \in A$ and $F(\boldsymbol{c}) = g(\boldsymbol{c})$ for all $\boldsymbol{c} \in B$ if and only if

$$\text{(R)}\ (C_i^*(f) \cup C_i^*(g)) \cap (C_j^*(f) \cup C_j^*(g)) = \phi \qquad \text{for every } i \neq j.$$

(Example 8) For the following restrictions of fuzzy switching function $f_1, f_2$ :

$$f_1(0,1) = 1, \quad f_1(0.5,1) = 1, \quad f_1(0.5,0.5) = 0.5,$$
$$f_2(1,1) = 0, \quad f_2(1,0.5) = 0, \quad f_2(0.5,0.5) = 0.5,$$

there is no fuzzy switching function that satisfies both. That is because element $(1,1) \in C_1^*(f_1) \cap C_0^*(f_2)$ violates the condition (R) in Corollary 3.

*C.2. Replacement with Undirected Graph G*

**Definition 9** Let $D = \{\boldsymbol{d}_1, \ldots, \boldsymbol{d}_m\}$ be a subset of $V^n$. A (undirected) *graph G* characterized by $D$ consists of two nonempty sets, the set $V(G)$ of vertices of $G$ and the set $Ed(G)$ of edges of $G$ as follows:

$$V(G) = \{f|_{[\boldsymbol{d}]} \mid f \in \mathcal{F}_n, \boldsymbol{d} \in D\} = Ve_1 \cup Ve_2 \cup \cdots \cup Ve_s$$

where $\mathcal{F}$ is a set of all $n$-variable fuzzy switching functions and $Ve_i$ $(i = 1, \ldots, s)$ is defined by

$$Ve_i = \{f|_{[d_i]} \mid f \in \mathcal{F}_n\} = \{v_{i1}, v_{i2}, \ldots, v_{i(2(n+1))}\}.$$

If vertices $u, v \in V(G)$ satisfy the following conditions:

(R) $(C_i^*(u) \cup C_i^*(v)) \cap (C_j^*(u) \cup C_j^*(v)) = \phi$ for every $i \neq j$,
(A) there is no vertex $w$ in $V(G)$ with edges $\{u, w\}$ and $\{w, v\}$,

then edge $\{u, v\}$ is in $Ed(G)$.

A *weight* $W(v)$ of a vertex $v$ in $V(G)$ is defined by

$$W(v) = E(v,h) = \sum_{\boldsymbol{d} \in C(v)} (h(\boldsymbol{d}) - v(\boldsymbol{d}))^2$$

where $E(v,h)$ is a distance between $v$ and $h$, and $C(v)$ is the quantized set of $v$, which corresponds to the Q-equivalent class.

(Example 9) Figure 2 illustrates the undirected graph $G$ given by $D$ in Example 3, where restrictions $v_{ij}$ are indicated on the truth tables for $V_3^2$.

*C.3. Shortest Path Algorithm(SPA)*

To find $f^*$ with the shortest distance $D(f^*, h)$ is to find sequence $v_1^*, \ldots, v_s^*$ in $V(G)$ such that $\sum_i^s E(v_i^*, h)$ is the lowest value for all others. We prefer a quicker way without checking all possible combinations.

Assume $Ve_i, Ve_{i+1} \in V(G)$ are in order so that edge $\{v_i, v_{i+1}\} \in Ed(G)$ for some $v_i \in Ve_i$ and $v_{i+1} \in Ve_{i+1}$. The following shortest path algorithm works with min-weight $W_i$.

```
Shortest Path Algorithm(SPA).
For each v in Ve_1
    Set W^1(v) = E(v,h)
    Set P^1(v) = {v}
End for
For i = 2 to s
    For each v in Ve_i
        For each u in Ve_{i-1} with {u,v} in Ed(G)
            If W^i(v) > W^{i-1}(u) + E(v,h)
                Set W^i(v) = W^{i-1}(u) + E(v,h)
                Set u* = u
        End for
        Set P^i(v) = P^{i-1}(u*) ∪ {v}
    End for
End for
```

(Example 10) We apply the shortest path algorithm to the graph $G$ in Example 9 (in Table 2).

Clearly the shortest path algorithm always stops, and the minimum weight in final $W^s$:

$$W^* = \min_{v \in Ve_s} W^s(v) = E(f^*, h)$$

is the intended distance to $f^*$ and $P^* = P^s(v^*) = \{v_1^*, v_2*, \ldots, v_s^*\}$ gives $f^*$, that is, for any $\boldsymbol{d} \in D$,

$$f^*(\boldsymbol{d}) = v^*(\boldsymbol{d}) \qquad \text{for some } v^* \in P^*.$$

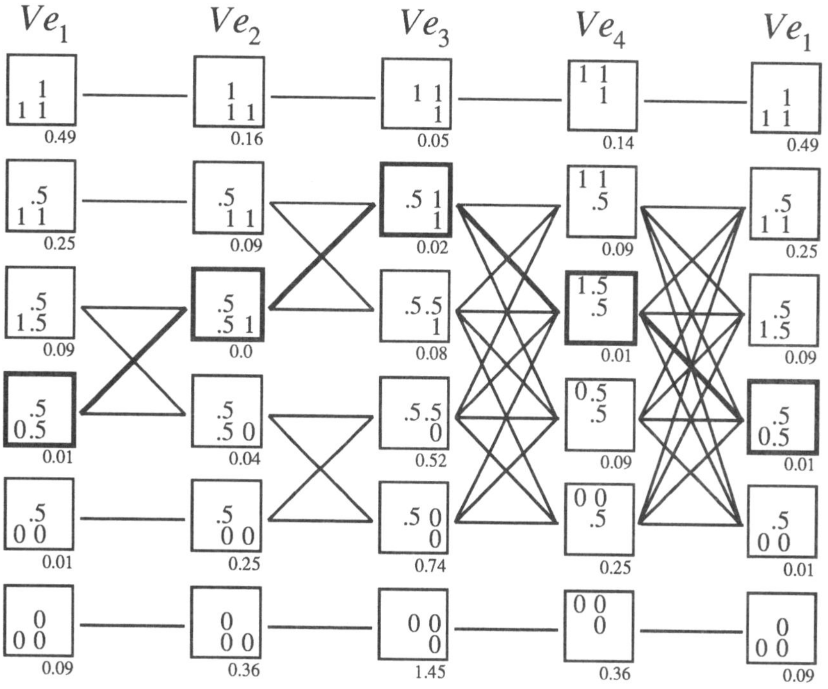


Figure 2: Graph $G$

(Example 11) Given $W^i$ and $P^i$ in Example 10, we have

$$\begin{aligned} W^* &= \min_{v \in Ve_4} W^4(v) = W^4(v_{43}) = 0.03, \\ P^* &= \{v_{14}, v_{23}, v_{32}, v_{43}\}. \end{aligned}$$

According to Theorem 5, we can reconstruct the logic formula representing $f^*$ as follows:

$$\begin{aligned} f^*(x,y) &= xy \vee x \vee x\overline{y} \vee \overline{x}\,\overline{y}x\overline{x}y \vee x\overline{x}\,\overline{y} \vee \overline{x}y\overline{y} \\ &= x \vee \overline{x}\,\overline{y}. \end{aligned}$$

This is the answer $f^*$ and we have solved the fitting problem. Note that $x \vee \overline{x}\,\overline{y} \neq x \vee \overline{y}$ in fuzzy logic.

*D. Estimation of SPA*

If we simply examine each fuzzy switching functions one by one, using a brute force algorithm (BFA), it takes as much time as the number of $n$-variable fuzzy switching functions $|\mathcal{F}_n|$, which is greater than the number of $n$-variable boolean switching functions $|\mathcal{B}_n| = 2^{2^n}$. The time to execute BFA is thus $O(2^{2^n})$.

For simplifying the estimation, we assume $D \in (V - V_3)^n$ as we have done in the examples. The number of Q-equivalent classes $|[D]|$ is at most $2^n n!$, because by Proposition 4, it is equal to the permutation of $n$ variables with two literals $x_i$ and $\overline{x_i}$. For each Q-equivalent class $[\boldsymbol{d}]$, there are $2(n+1)$ restrictions of fuzzy switching functions. The number of vertices $|V(G)|$ is thus less than or equal to $2(n+1) \cdot 2^n n! = 2^{n+1}(n+1)!$. For example, when $n = 2$, $|V(G)| = 2^{2+1} \cdot (2+1)! = 48$. If we examine all of the possible sequences $v_1, \ldots, v_s$ in $V(G)$ so that $v_i \in Ve_i$, then the total time is $O(2(n+1)^{2^n n!})$, which is worse than BFA.

The computation could be speeded up considerably by the shortest path algorithm (SPA). The comparison/replacement step inside the $u$-loop takes at most some fixed amount of time. The step is done at most $|Ve_{i-1}| = 2(n+1)$ times in the $u$-loop, which is done $|Ve_i| = 2(n+1)$ times for the $v$-loop, which is done $s - 1 = 2^n n! - 1$ times for the outside $i$-loop. The total time to execute the algorithm is $O(2(n+1) \cdot 2(n+1) \cdot 2^n n!) = O(2^{n+2} n!(n+1)^2)$. Table 3 is how long each algorithm takes for $n = 1, \ldots, 5$.

## IV. Conclusion

We have proposed an algorithm to find the fuzzy switching function with the shortest distance to a given partial function. We have proved that the division of the uncertain knowledge into some Q-equivalent classes is sufficient to calculate all possible local distances in Theorem 6, and have shown some useful properties with respect to restrictions of fuzzy switching functions. The shortest path algorithm can work with a large value of $n$ on weighted graph $G$ which corresponds to all possible $n$-variable fuzzy switching functions. We have also estimated the algorithm and shown that it can reduce computation in finding the

Table 3: Computational times

| algorithm \ $n$ | | 1 | 2 | 3 | 4 | 5 |
|---|---|---|---|---|---|---|
| BFA ($=\|\mathcal{F}_n\|$) | $O(?)$ | 6 | 84 | 43918 | 160297985276 | ? |
| BFA ($>\|\mathcal{B}_n\|$) | $O(2^{2^n})$ | 4 | 16 | 256 | 65536 | 4294967296 |
| SPA | $O(2^{n+1}n!(n+1)^2)$ | 16 | 144 | 1536 | 19200 | 276480 |

Table 2: Matrices $W(G)$

| $j\backslash i$ | 1 |
|---|---|
| 1 | .49 |
| 2 | .25 |
| 3 | .09 |
| 4 | .01 |
| 5 | .01 |
| 6 | .09 |

$W^1(v_{ij})$

| $j\backslash i$ | 1 | 2 |
|---|---|---|
| 1 | .49 | .65 |
| 2 | .25 | .34 |
| 3 | .09 | .01 |
| 4 | .01 | .05 |
| 5 | .01 | .26 |
| 6 | .09 | .45 |

$W^2(v_{ij})$

| $j\backslash i$ | 1 | 2 | 3 |
|---|---|---|---|
| 1 | .49 | .65 | .21 |
| 2 | .25 | .34 | .02 |
| 3 | .09 | .01 | .09 |
| 4 | .01 | .05 | 1.02 |
| 5 | .01 | .26 | 1.24 |
| 6 | .09 | .45 | 1.9 |

$W^3(v_{ij})$

| $j\backslash i$ | 1 | 2 | 3 | 4 |
|---|---|---|---|---|
| 1 | .49 | .65 | .21 | .35 |
| 2 | .25 | .34 | **.02** | .11 |
| 3 | .09 | **.01** | .09 | **.03** |
| 4 | **.01** | .05 | 1.02 | .11 |
| 5 | .01 | .26 | 1.24 | .34 |
| 6 | .09 | .45 | 1.9 | 2.26 |

$W^4(v_{ij})$

fuzzy switching function with the shortest distance for all others.


## Acknowledgments

The author thanks Mr. Tetsuji Morishita and Prof. Masao Mukaidono for their advice.


## References


[1] Kikuchi H., (1991) "Necessary and Sufficient Conditions for Restrictions to Be Fuzzy Switching Functions," *Proc. the International Fuzzy Engineering Symposium' 91*, vol.1, pp.91-102.

[2] Kikuchi H., Morishita T. and Mukaidono M., (1992) "Restrictions of P-Fuzzy Switching Functions – How we can get knowledge from fuzzy information," *Proc. IEEE International Conference on Fuzzy Systems 1992*, IEEE, pp.169-176

[3] Mukaidono M., (1979) "A Necessary and Sufficient Condition for Fuzzy Logic Functions," *Proc. the 9th International Symposium on Multiple-valued Logic*, IEEE, pp.159-166

[4] Mukaidono, M., "The Representation and Minimization of Fuzzy Switching Functions," *Analysis of Fuzzy Information*, pp.213-229 .

[5] K.A.Ross and C.R.B.Wright, (1988) *Discrete Mathematics*, Prentice-Hall

# FLIP-net: A Network Representation of Fuzzy Inference Procedure and its Application to Fuzzy Rule Structure Analysis

Akira Maeda, Toshihide Ichimori, and Motohisa Funabashi
Systems Development Laboratory, Hitachi, Ltd.
1099 Ohzenji, Asao-ku, Kawasaki 215, JAPAN

***Abstract*— A method for analyzing dynamic behavior of fuzzy rules using FLIP-net (Fuzzy Logic Inference Procedure NETwork), a network representation of fuzzy inference procedure, is proposed. Using FLIP-net representation, errors of fuzzy inference output can be backpropagated. Since backpropagated error indicates how each component of fuzzy logic contributes to the output error, analysis of the error distribution over FLIP-net gives useful information about dynamic behavior of fuzzy rules. It is shown that fuzzy rule deficiencies, such as redundant, missing, or contradictory fuzzy rules, can be detected by this method, and thus efficient addition and/or modification of fuzzy rules becomes possible.**

## I. Introduction

We have proposed a supervised learning algorithm for fuzzy membership functions using *FLIP-net* (Fuzzy Logic Inference Procedure NETwork)[1][2]. FLIP-net is a kind of computational flow network, and backpropagation-like learning algorithm can be applied for membership function tuning. However, the algorithm changes the shape of membership functions, not the structure of fuzzy rules because FLIP-net has one-to-one correspondence to the fuzzy knowledge structure.

It is very difficult, or almost impossible to give correct fuzzy rules from the first place. Therefore, in developing fuzzy knowledge base, it frequently happens that the designer of the fuzzy knowledge base should add or modify fuzzy rules to achieve the desired fuzzy inference output in a trial-and-error fashion. Even when membership functions can be automatically tuned by the learning method, the refinement of fuzzy rules still remains a hard task.

In this paper, we propose a method for analyzing behaviors of fuzzy rules and thus detecting deficiencies in the fuzzy rule structure, such as redundant, missing, or contradictory rules. Using a FLIP-net representation of fuzzy knowledge, this method analyzes the internal distribution of backpropagated errors that are obtained in membership function learning process. Since the errors will concentrate on structural defects in the FLIP-net which correspond to the deficiencies in the fuzzy rule structure, this method can easily identify rules and membership functions for efficient addition and/or modification of fuzzy rules.

## II. FLIP-net Representation

In this section, the FLIP-net representation and the supervised learning algorithm for fuzzy membership functions are briefly described [1][2].

FLIP-net is a computational flow network representation of fuzzy inference procedure. An example of FLIP-net is shown in Fig.1. In this paper, "MIN-MAX and center of gravity" method[3] is assumed as a fuzzy inference procedue. It is straightforward to give a slight modification of FLIP-net representation for other variations of fuzzy inference.

In this example, we have 2 fuzzy rules:

- Rule 1: if $x_1$ is small$_1$ and $x_2$ is mid-range$_2$ then $y$ is mid-range$_y$
- Rule 2: if $x_1$ is large$_1$ and $x_2$ is small$_2$ then $y$ is large$_y$

Rule grades are given by the minimum of the if-part grades:

$$\begin{aligned} W_1 &= \min(\text{small}_1(x_1), \text{mid-range}_2(x_2)) \\ W_2 &= \min(\text{large}_1(x_1), \text{small}_2(x_2)) \end{aligned} \quad (1)$$

Output membership function $B$ is given by:

$$B(z) = \max[(W_1(\text{mid-range}_y(z)), (W_2(\text{large}_y(z))] \quad (2)$$

where $z$ is a dummy variable for the defuzzification integral.

The output value of the fuzzy inference $y$ is:

$$\frac{\sum_k B(z_k)z_k}{\sum_k B(z_k)} \quad (3)$$

The $z_k$ variable represents the discretized value of $z$ for digital calculations.

0-7803-0614-7/93$03.00 

FLIP-net in Fig.1 gives a network representation of Eqs.(1)–(3). Using the FLIP-net, the output error $(y-y_s)$ ($y_s$: correct output) can be backpropagated through the network, and we can get the partial derivative coefficients $\partial E/\partial o_{n_j} = (y - y_s) \cdot \partial y/\partial o_{n_j}$ for all node output values $o_{n_j}$[4], where $E = \frac{1}{2}(y - y_s)^2$ is a target function to be minimized.

For membership function tuning, $\partial E/\partial \alpha_i$ ($\alpha_i$: membership function parameters) can be calculated by:

$$\frac{\partial E}{\partial \alpha_i} = \frac{\partial E}{\partial o_{n_j}} \frac{\partial o_{n_j}}{\partial \alpha_i} \tag{4}$$

where node $n_j$ includes the parameter $\alpha_i$. Using $\partial E/\partial \alpha_i$, a steepest descent method can be applied to minimize the target function $E$. Since the fuzzy inference procedure is represented by a FLIP-net, efficient calculation of the partial derivative coefficients becomes possible. This enables a high throughput of membership function tuning.

## III. Analysis of Fuzzy Rule Structure

In this section, we propose a method for analyzing behaviors of fuzzy rules and thus detecting deficiencies in the fuzzy rule structure. Using a FLIP-net representation of fuzzy knowledge, this method analyzes the internal distribution of backpropagated errors that are obtained in membership function learning process. Since the errors will concentrate on structural defects in the FLIP-net which correspond to the deficiencies in the fuzzy rule structure, this method can easily identify rules and membership functions for efficient addition and/or modification of fuzzy rules.

The purpose of the method is to identify fuzzy rules and membership functions which are responsible for the remaining error after membership function tuning. Large remaining error indicates the existence of deficiencies in fuzzy rule structure, because the remaining error should be sufficiently small if the fuzzy rules represent the correct structure of the problem.

The basic idea of the proposed method is to find nodes in the FLIP-net which mainly contribute to the remaining error. Since each node in a FLIP-net has a definite meaning in fuzzy logic, we can expect that the method can identify fuzzy rules and membership functions which should be modified to improve the fuzzy inference accuracy.

We introduce a quantity $\delta_{n_j}$ for each node $n_j$, which is defined by:

$$\delta_{n_j} = \frac{\partial E}{\partial o_{n_j}} \tag{5}$$

During the membership function tuning process described in the previous section, $\delta_{n_j}$ is calculated for every node in the FLIP-net. From the definition (5), $\delta_{n_j}$ means the amount of the node $n_j$'s responsibility to the output error $E$. Therefore, we can estimate in what amount each rule or membership function contributes to the output error, using the value of $\delta_{n_j}$.

We can regard the *squared average* $\langle \delta_{n_j}^2 \rangle$ as an indicator of how responsible the node is to the remaining output error $\langle E \rangle$, where $\langle \cdot \rangle$ denotes the average over teacher data set. Therefore, by calculating $\langle \delta_{n_j}^2 \rangle$ for all nodes $n_j$ in the FLIP-net, we can detect fuzzy rules and membership functions which mainly contribute to the remaining output error, so that the designer can efficiently improve the fuzzy knowledge base. In the same manner, we can also detect unnecessary fuzzy rules because $\delta_{n_j}$ becomes zero for unused fuzzy rules.

## IV. Simulation

In this section, a simple example is described to show how the proposed method works.

Fig.2 illustrates the problem taken as an example. We have 2 inputs $x_1$ and $x_2$, and one output $y$. The correct input-output relation is defined by fuzzy logic, such that "IF $x_1$ is negative AND $x_2$ is negative THEN $y$ is negative," "IF $x_1$ is positive AND $x_2$ is negative THEN $y$ is positive," and so on.

Assume that the input-output relation is described by the following 5 fuzzy rules:

**Rule#1:** IF $x_1$ is negative THEN $y$ is negative

**Rule#2:** IF $x_2$ is negative THEN $y$ is negative

**Rule#3:** IF $x_1$ is zero AND $x_2$ is zero THEN $y$ is zero

**Rule#4:** IF $x_1$ is positive AND $x_2$ is zero THEN $y$ is positive

**Rule#5:** IF $x_2$ is positive THEN $y$ is positive

These rules are also shown in Fig.2.

The structure of the given fuzzy rules is not correct, because for $x_1$ is positive and $x_2$ is negative, $y$ is positive instead of negative. Therefore, it is impossible to represent the input-output relation correctly by the above 5 fuzzy rules, whatever membership functions are assigned. In other words, the fuzzy rules have deficiencies. We will see how these deficiencies are detected by the proposed algorithm.

As teacher data set, 90 samples are generated, namely 10 samples for each box in Fig.2. Random Gaussian noise is added to input and output values of each sample.

Fig.3 illustrates the result of the proposed algorithm. First, the membership function learning algorithm is applied. Since the fuzzy rule structure is not correct, large remaining error $\langle E \rangle$ exists, after the algorithm converges.

Then, $\langle \delta^2_{n_j} \rangle$ for membership function nodes and rule nodes are calculated. In Fig.3, diameter of a node is shown to be proportional to $\sqrt{\langle \delta^2_{n_j} \rangle}$ for visual presentation.

For the rule nodes in Fig.3, the node corresponding to Rule#2 has the largest value. From Fig.2, it is easily understood that Rule#2 should be divided to correctly represent the input-output relation. The proposed method successfully detects the incorrect fuzzy rule in this example.

In Fig.3, the node corresponding to Rule.#4 has the next largest contribution to the output error. This can be interpreted as follows: In order to explain the correct output "positive" for "$x_1$ is positive and $x_2$ is negative", Rule#4 tried to cover the input variable region by changing the definition of the membership function "$x_2$ is zero" during the learning process. However, it is impossible to cover the region completely of course, because the membership function "$x_2$ is zero" is also used in Rule#3. During this process, Rule#4 shares a part of the responsibility of Rule#2 to the remaining error.

The situation becomes clearer when we look at the membership function nodes in Fig.3. Among output membership function nodes, the node corresponding to "$y$ is positive" has the largest contribution, just as expected. For input membership function nodes, the node "$x_2$ is negative" has the largest value, also as expected. However, the nodes "$x_1$ is positive" and "$x_2$ is zero" have also relatively large contribution. Both membership functions are used in Rule#5, and this result coincides with the above interpretation.

## V. Conclusion

A new method for the analysis of fuzzy rule structure is proposed. The method uses FLIP-net, which is a network representation of fuzzy inference procedure. This method provides a very useful information about the dynamic behavior of fuzzy rules. With this information, a designer of fuzzy knowledge base can easily identify fuzzy rules that mainly contribute to the remaining error. FLIP-net also provides a very efficient learning algorithm for fuzzy membership functions. Using these two methods, the efficiency of fuzzy knowledge base construction work can be greatly improved.

## References


[1] A. Maeda, R, Someya, and M. Funabashi, "A self-tuning algorithm for fuzzy membership functions using computational flow network," *Proceeding of IFSA'91, Brussels*, 1991.

[2] A. Maeda, R. Someya, S. Yasunobu, and M. Funabashi, "A fuzzy-based expert system building tool with self-tuning capability for membership functions," *Proceedings of the World Congress on Expert Systems*, pp.639–647, 1991.

[3] E. Mamdani, "Advances in the linguistic synthesis of fuzzy controllers," *Int. J. Man-Machine Studies*, vol. 8, no. 6, pp.669–679, 1976.

[4] M. Iri, "Simultaneous computation of functions, partial derivatives, and estimates of rounding errors," *Japan J. of Appl. Math.*, vol. 1, no. 2, pp.223–252, 1984.

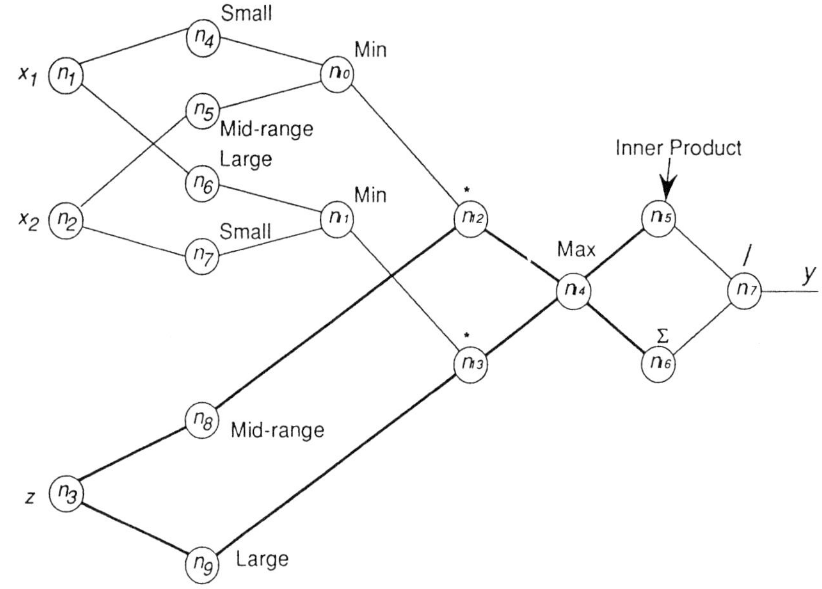


Figure 1: Example of FLIP-net representation

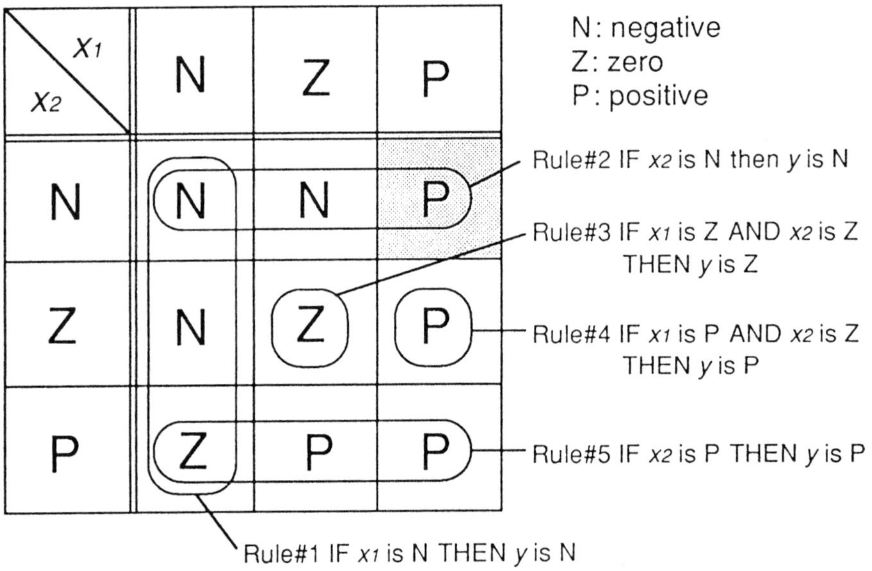


Figure 2: Input-output relation of the example.

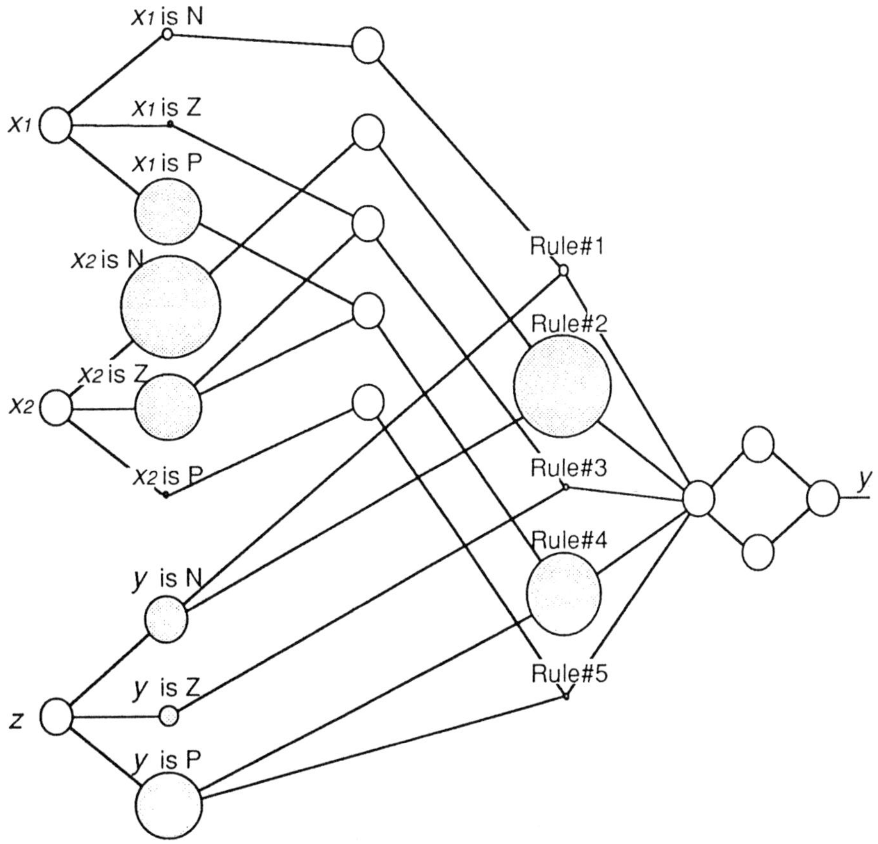


Figure 3: Result of the proposed algorithm.

# Approximate Reasoning with Fuzzy Petri Nets

Padmini Srinivasan and Denis Gračanin

***Abstract***— **A new approach to the fuzzification of the Petri Net is proposed. A token in the Fuzzy Petri Net (FPN) is described as a membership function of a linguistic term as opposed to a single number representing a membership value. A transition, specified as a production rule, can be fired if the "distance" of input tokens from the desired values is less than a given tolerance. This distance determines the difference between the output tokens produced and the output token values specified by the transition. The FPN is an excellent modeling and knowledge representation tool for all Petri Net modeling applications involving imprecise token values.**

## I. Introduction

A Petri net consists of a fixed number of places and transitions with tokens distributed over places. Each transition has an associated set of input places and a set of output places. When every input place of a transition has a token, the transition is enabled and may fire. When a transition is fired, a token is consumed (deleted) from every input place and produced (added) to every output place. A Petri net is an abstract, formal model of information flow, and functions as a very general knowledge representation scheme. Transition firing flexibly represents causality and inferencing relations. Several tools and techniques are available for the qualitative analysis of the PN representation of a system. PNs have been generalized by allowing multiple tokens, inhibitor arcs, place capacity, colored tokens, etc. [5].

The Petri net (PN) is an excellent tool for analysis of systems in diverse applications. However, PN based analysis is only as effective as the available PN description of the system. The use of the standard PN is inappropriate in situations where the system is difficult to describe precisely. Possible sources of imprecision in the system description are:

1. Input to the system:
   (a) System properties are not amenable to a well-defined measurement providing a mapping to numerical values, (e.g. cement production properties of the mixture, or sensor measurements in a noisy environment).
   (b) A well-defined measurement exists but is too costly, too slow or too redundant (example: washing machine evaluating dirtiness of clothes),
   (c) A well-defined and relatively inexpensive measurement exists, but fuzzy description is equally or more appropriate and effective, and less costly (e.g., traffic regulation in transportation systems).
2. Output from the system (e.g. qualitative description of required action such as grabbing, pressure control, etc.),
3. Fuzzy internal operations [7].

The FPN addresses two essential aspects of fuzzification of a given model: representation of fuzzy information, and control of fuzziness of the information. The latter aspect is important in modeling the acquisition of information to decrease imprecision, e.g to enable the production of crisp decisions following a fuzzy analysis.

## II. Literature Review

A straightforward, and rather naive fuzzification of the PN is presented in [2]. Every place has an associated proposition and every token in that place has an associated truth value. Depending on truth values of input tokens and a threshold value associated to a transition, the transition may or may not fire. This model has the representation capability of a rule-based system with fuzzy production rules. The PN structure explicitly represents antecedent consequence relationships between propositions. An example using this approach is presented in [1].

## III. Fuzzy Petri Net Model

The structure of a standard PN is [5]:

$$PN = (P, T, I, O)$$

where:

0-7803-0614-7/93$03.00 

$PN$ : Petri Net,

$P$ : Set of places $P = \{p_1, \ldots, p_{|P|}\}$,

$|P|$ : Number of places,

$T$ : Set of transitions $T = \{t_1, \ldots, t_{|T|}\}$,

$|T|$ : Number of transitions,

$I$ : Input function (determines the input places to a transition) $I : T \rightarrow P^\infty$,

$O$ : Output function (determines the output places from a transition) $O : T \rightarrow P^\infty$.

The Petri Net is a qualitative system analysis tool rather than a quantitative tool. An introduction of fuzziness involving explicit numerical manipulation of the membership function to be used in firing transition would constitute a contradiction. A more general approach is to assume that truth value is itself fuzzy so that token is not described by a number but a membership function. Therefore, the linguistic variable is used in the representation of a token's degree of membership in a property represented by a place. Based on this idea, a new model of FPN is derived and described in detail.

The following components of the standard Petri Net are redefined to incorporate fuzziness:

**fuzzy place** Each fuzzy place has a predicate or property associated with it. A token in the fuzzy place is characterized by that property and a level (membership function) to which it possesses that property or belongs in that place. In the standard PN, a token may have only two values:

**0** not in the place,

**1** in the place.

In the FPN, degree of membership in the property, property intensity, or degree of achievement [8] is a linguistic variable.

**fuzzy token** A fuzzy token is a generalization of the token in the standard PN which either does or does not belong to a place. A fuzzy token has a membership function which determines the degree of membership in a particular place, or the truth value of that proposition.

**fuzzy transition** A fuzzy transition corresponds to an if-then fuzzy production rule of the form given below, where $Lx_i$ corresponds to a value of a linguistic variable

**if** $x_1$ **is** $Lx_1$ **and** ... **and** $x_n$ **is** $Lx_n$
**then** $y_1$ **is** $Ly_1$ **and** ... **and** $y_m$ **is** $Ly_m$

**fuzzy arc** A fuzzy arc specifies the required linguistic value of corresponding input/output token, except when the fuzzy transition specifies a comparison not involving a predefined value, e.g. a comparisons between token values.

*A. Formal Description of FPN*

A formal description of FPN is derived from [2].

$$FPN = (P, T, D, L, M, I, O, f, \beta)$$

where:

$P = \{p_1, \ldots, p_{|P|}\}$ is a finite set of places,

$T = \{t_1, \ldots, t_{|T|}\}$ is a finite set of transitions,

$D = \{d_1, \ldots, d_{|D|}\}$ is a finite set of properties,

$L = \{l_1, \ldots, l_{|L|}\}$ is a finite set of linguistic values,

$M = \{m_1, \ldots, m_{|M|}\}$ is a finite set of linguistic modifiers,

$I : T \times M \times L \rightarrow P^\infty$ is the input function, a mapping from transitions and modified linguistic values to bags of places,

$O : T \times M \times L \rightarrow P^\infty$ is the output function, a mapping from transitions and modified linguistic values to bags of places,

$f : T \rightarrow [0, 1]$, is a function which assigns a real number to every transition. That number determines maximum allowed distance of input token from desired input linguistic value, and is a mechanism to control the degree of fuzziness in the FPN,

$\beta : P \rightarrow D$, is a mapping from places to properties, not necessary bijective.

*B. Linguistic values and modifiers*

Every place in the FPN has a property intensity described by the same set of possible linguistic values and modifiers. Arbitrary trapezoidal membership functions are used to represent the various linguistic terms' values (Figure 1) and the set of possible modifiers may be defined as in [8]:

**very** (concentration): $\mu_{very(A)}(x) = \mu_A^2(x)$,

**more or less** (dilatation): $\mu_{m.or\ l.(A)}(x) = \mu_A^{\frac{1}{2}}(x)$,

... etc.

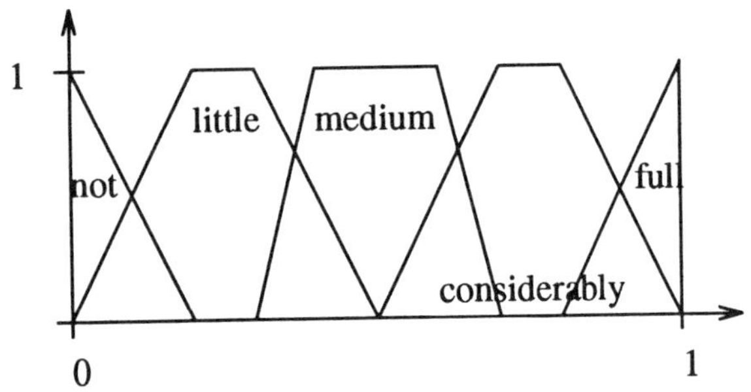


Figure 1: An example for linguistic values – a degree of achievement

Any combination of values and modifiers may be used within a rule. A token is described by its membership function. Whenever a token's membership function does not exactly match that required as input to a transition, it is necessary to calculate the closest approximation or a distance from the required input token value. The distance is calculated as follows [6]:

$$d^2 = \int_0^1 (\mu_A(x) - \mu_d(x))^2 dx$$

where:

$\mu_A(x)$ is a membership function of some token in the place,

$\mu_d(x)$ is a required membership function described in an input arc to some transition.

The integral goes over interval $[0, 1]$ i.e., over all possible values for a level of possessing a property (Figure 1). In general, different properties may have different intervals, especially when the property has discrete values or does not have straightforward numerical description, e.g. color. In that case, the distance may be calculated as follows:

$$d^2 = \frac{1}{n}\sum_{i=1}^{n}(\mu_A(x_i) - \mu_d(x_i))^2$$

where:

$n$ is a number of different values, $\{x_1, \ldots, x_n\}$,

Various other distance measures may be used.

### *C. Fuzzy transition*

For a given transition $t_i$, $I(t_i)$ determines a set (bag) of input places as well as desired linguistic values. If there are tokens in every input place of a transition, the token with the minimum distance from required value in each input place is selected. If the selected tokens in all the input places satisfy a condition that their distance is less than $f(t_i)$, then the transition is enabled and may fire. The maximum distance among input tokens is used to modify the transition's output linguistic values.

## IV. Example

### *A. Transition firing*

In order to illustrate concepts mentioned above, a following set of linguistic values (Figure 2) and modifier are defined:

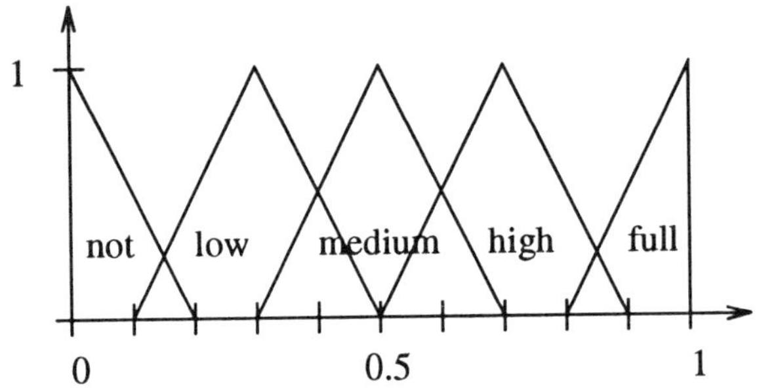


Figure 2: Linguistic values

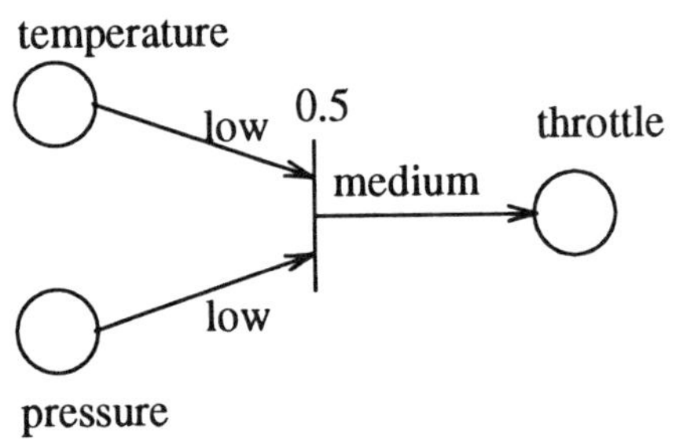


Figure 3: Set of linguistic values

**very** concentration: $\mu_{very(A)}(x) = \mu_A^2(x)$

Therefore, the set of all possible (modified) linguistic values is

$$\{not, very\ not, low, very\ low, medium,$$
$$very\ medium, high, very\ high, full, very\ full\}$$

Given the fuzzy rule from [8]

**if** temperature `is` low `or` pressure `is` low
**then** throttle `is` medium

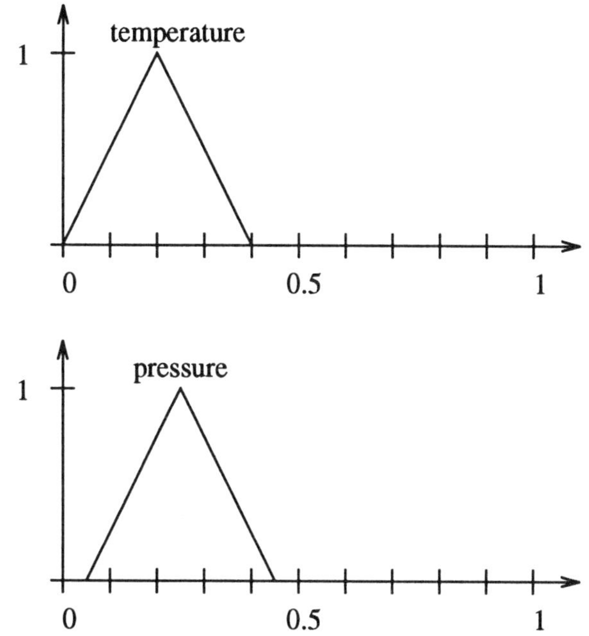


Figure 4: Values of temperature and pressure tokens

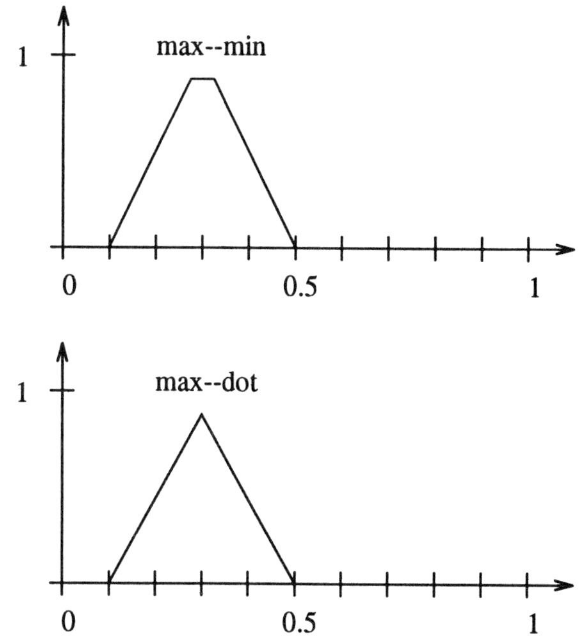


Figure 5: The value of output token for max–min and max–dot inference rule

a transition $t$ is shown in Figure 3, values of temperature and pressure tokens in Figure 4, and the value of the output token for max–min or max–dot inference rule [8] in Figure 5. The distance for temperature is 0.274 and for pressure even smaller and transition may fire since the value of $f(t)$ is 0.5.

The difference between this approach and one described in [1] is that here a token is described with a linguistic value while in [1] it is a single number.

*B. FPN Construction: A Fuzzy Logic Controller*

A Fuzzy Logic Controller (FLC) [3, 4] shown in Figure 6 may be modeled as a FPN. The dynamic behavior of a fuzzy system is characterized by a set of linguistic description rules or fuzzy conditional statements. These fuzzy control rules can be easily represented as a transition in the FPN. For example , for the rule:

$$\mathtt{R_1} : \textbf{if } x \textbf{ is } A_1 \textbf{ and } y \textbf{ is } B_1 \textbf{ then } z \textbf{ is } C_1$$

the corresponding transition before and after firing is shown in Figure 7. Each linguistic variable corresponding to a process state variable has a corresponding place in FPN. A linguistic value is represented as the membership function of the token in the place. Fuzzy control rules are represented by transitions with appropriate linguistic values associated to arcs.

Four parts of the FLC are modeled as follows:

**fuzzification interface :** For every input variable there is a transition which maps it to the corresponding linguistic value. A corresponding input place (on the side of the controlled system process) has a value which is a single point membership function ($\delta$ function) or some similar function describing measured value. A transition then produces fuzzy output token in the corresponding place (on the side of the FLC) which has the closest value to the input token as shown in Figure 8.

**knowledge base :** A knowledge base which consists of a data base and a fuzzy control rule base is represented as follows:

**data base** The set of possible membership functions (linguistic values) for tokens. This set is also used during the fuzzification and defuzzification.

**rule base** A set of transitions corresponding to the set of fuzzy rules. Different types of rules may be easily represented. According to [3] there are two types of fuzzy control rules currently in use, state evaluation fuzzy control rules and object evaluation fuzzy control rules. While state evaluation rules are easy to translate to fuzzy transitions, object evaluation rules re-

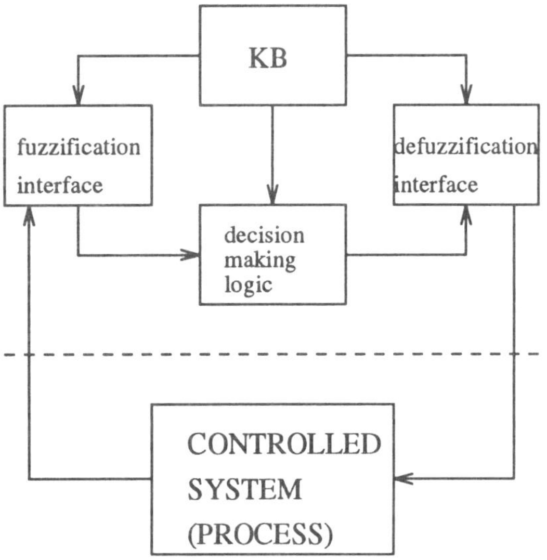


Figure 6: Fuzzy Logic Controller

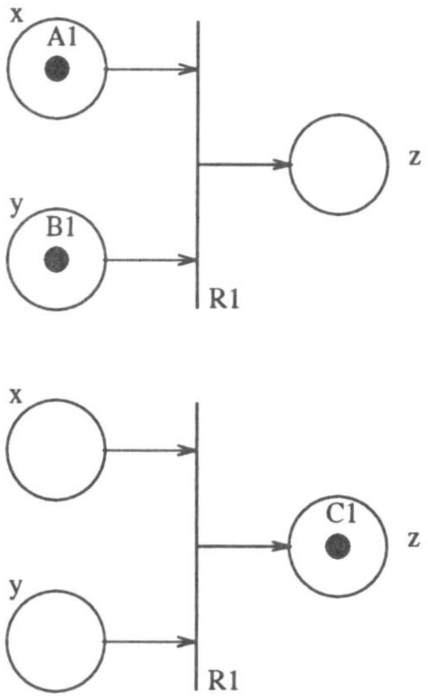


Figure 7: Transition

quire additional efforts in order to model predicting a result or to memorize previous results.

**decision making logic :** Simulating human decision making process or approximate reasoning is done by selecting how to apply fuzzy implication and the rules of inference. Usually, FLC is a one step inference process. Different types of approximate (fuzzy) reasoning may be applied [4], but that affects only the way how is a fuzzy transition fired while the FPN remains the same.

**defuzzification interface** As for the fuzzification, there are two places for each output state variable, one on the side of the FLC and the other one on the side of controlled system process. The corresponding transition produces an output token representing a control action which is the closest to the linguistic value of the input token (Figure 8).

This example shows how to map an existing fuzzy application (FLC) to the FPN model. FPN model enables easier representation and modeling. Petri Net (PN) model is not powerful enough to do that, in the same manner as the standard controller may not be directly constructed based on simulated human decision making.

## V. Conclusion

The FPN subsumes a standard fuzzy rule base as a knowledge representation scheme. It fundamentally

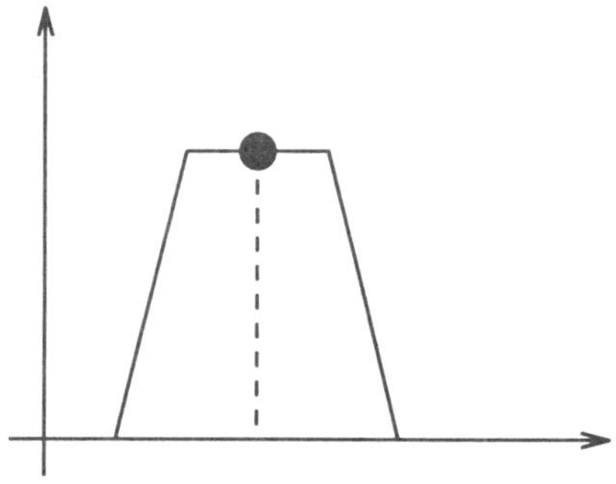

Figure 8: Fuzzification/defuzzification

extends the PN representation of a fuzzy rule–based system and includes an added dimension of imprecision of fuzziness. The PN inferencing mechanism is redefined to accommodate this fuzziness.

The token of the FPN may alternatively be described as a point of a type–II membership function. In defining a fuzzy set $A$ based on the set of all properties $D$, we describe an element of this set by a fuzzy set, or membership *function* for each property, instead of a numeric membership *value* for each property, i.e. $A$ becomes a type–II fuzzy set.

The membership values of tokens in the FPN in the type–II set of properties are represented as membership functions of linguistic terms of a standardized linguistic variable representing degree of possession or intensity of the specific property.

Introducing fuzziness in the Petri Net is useful in modeling problems where information is lacking or not precise enough. An important concern is to introduce fuzzy manipulation in a simple, appropriate, yet effective manner. Linguistic values describing a membership function of the token in a single place is reasonably simple yet quite descriptive. It is possible to define linguistic value and modifiers separately for every place depending on the property associated with the place. Future research includes the definition of properties of such a Fuzzy Petri Net.

## References


[1] Tiehua Cao and Arthur C. Sanderson. Sensor-based error recovery for robotic task sequences using fuzzy petri nets. In *Proceedings of the 1992 IEEE International Conference on Robotics and Automation*, volume 2, pages 1063–1069, Nice, May 1992.

[2] Shyi-Ming Chen, Jyh-Sheng Ke, and Jin-Fu Chang. Knowledge representation using fuzzy petri nets. *IEEE Transactions on Knowledge and Data Engineering*, 2(3):311–319, September 1990.

[3] Chuen Chien Lee. Fuzzy logic in control systems: Fuzzy logic controler–part I. *IEEE Transactions on Systems, Man and Cybernetics*, 20(2):404–418, March/April 1990.

[4] Chuen Chien Lee. Fuzzy logic in control systems: Fuzzy logic controler–part II. *IEEE Transactions on Systems, Man and Cybernetics*, 20(2):419–435, March/April 1990.

[5] James L. Peterson. *Petri Net Theory and the Modeling of Systems.* Prentice–Hall, Englewood Cliffs, New Jersey 07632, 1981.

[6] Kurt J. Schmucker. *Fuzzy Sets, Natural Language Computations, and Risk Analysis.* Computer Science Press, Rockville, Maryland, 1984.

[7] H.M. Stellakis and K.P. Valavanis. Fuzzy logic-based formulation of the organizer of intelligent robotic system. *Journal of Intelligent and Robotic Systems*, 4:1–24, 1991.

[8] Hans-Jürgen Zimmerman. *Fuzzy Set Theory and Its Applications, Second Edition.* Kluwer Academic Publishers, Boston, 1991.

# Interpolation in Structured Fuzzy Rule Bases

László T. Kóczy* and Kaoru Hirota+

*Dept. of Telecommunication and Telematics, Technical University of Budapest
Sztoczek u. 2, H-1111 Budapest, Hungary

+Dept. of Instrumentation and Control Engineering, College of Engineering, Hosei University
Kajino-cho, Koganei-shi, Tokyo 184, Japan

*Abstract*–**Fuzzy rule based systems are important in many control engineering applications. In real problems, often the number of input variables is very high. Time complexity of the fuzzy control algorithms based on either the CRI or Mamdani's technique is too high for real time control.**

**In the real situations, often a few or maximally a few dozen variables dominate the system in a certain area of the state space. This is why an experienced operator is able to control rather well, although overview of all the variables is virtually impossible. The subset of variables changes when the working point changes.**

**Hierarchically structured rule bases are introduced where "sub-rule bases" contain only subsets of the input variables. Fuzzy partition of the state space is applied for the structuring of the total input universe of discourse. While application of the corresponding sub-rule base in the "middle" of a fuzzy subset of $X$ is unambiguous, the "twilight zones", i.e. areas belonging to different subsets to a greater than 0 degree form a serious theoretical problem. The idea of interpolation based on cylindric extensions of the subspaces is introduced and the necessary formulas are presented. Some examples are shown.**

## I. Introduction

Fuzzy rule based control was originally proposed by Zadeh in his seminal paper [1] on modelling complex systems. The technique described there is the compositional rule of inference which preserves maximal amount of information contained in the rules and the observation, however, it can be algorithmized only in forms with high time complexity: exponential in terms of the number of input variables [2]). The simplified computational technique applied by Mamdani [3] has a convenient polynomial complexity (cf. [2]), but it loses essential information: the conclusion does not follow the observation with full sensitivity [4]. Although a compromise solution can be found ("compact rules algorithm" *ibidem*), a new problem is introduced: the rule base will often not cover the whole observation space, it will be sparse. There are also other considerations which point out the importance of dealing with sparse rule bases, as e.g. rules received by tuning the rule base [5] and the need of data compression e.g. for fast transmission [6].

In the case of sparse rule bases conclusion by the combination of the rule base and the observation can be obtained only by applying rule interpolation in the strict sense. A theoretical basis for rule base approximation is formed by the interpretation of rule bases as fuzzy points representing a fuzzy mapping ("function") from the cross product space of the input variables to the outputs (cf. the so named "graph view" in [7] and the considerations in [8]).

Applying the fuzzy mapping view combined with the resolution principle, traditional function approximation techniques can be applied for every level set. An overview of such techniques including extrapolation, regression and other methods can be found in [9].

## II. State Spaces with Extremely Large Number of Variables

Suppose that the rule base consists of $r$ rules, each of which contains $k$ input variables and the maximum resolution in all the dimensions is $N$(there are maximally $N$ discrete points in every $X_i$, i.e. evaluation of any membership function over $X_i$ requires maximally $N$ steps in uniform complexity). Then, the complexity of the rule base itself is

$$\mathcal{I} = \mathcal{O}(rkN)$$

0-7803-0614-7/93$03.00 ©1993IEEE

Using the CRI, time complexity of a single conclusion is

$$\mathcal{C}_1 = \mathcal{O}(rN^{k+1})$$

, using Mamdani's algorithm, this is reduced to

$$\mathcal{C}_2 = \mathcal{O}(rkN)$$

.

(For the above results, see [2], for uniform time complexity, etc. see e.g. [10].)

Clearly, the second algorithm is very well tractable, while the first one is applicable only in the case of low k. On the other hand, the simplified algorithm often completely or partially hides the exact nature of the observation, as e.g. when the core of the observation changes together with the degree of fuzziness in such a way that the height of the intersection of the observation with the only or dominant rule remains constant.

In some industrial systems, however, an even more serious difficulty arises. Suppose that there are several hundred or even several thousand input variables in the state space. An example for that can be a petroleum raffinery, etc. Here, the complexity of the very rule base (I) becomes already rather high for practical purposes. The hidden difficulty is especially in the value of r. If an acceptably complete rule base is constructed, the number of rules becomes very high: even if the number of linguistic terms for every input variable is bounded by a constant $L$, there are altogether $L^k$ possible input combinations. When using free membership shapes in the If parts, the situation is even worse. So we conclude that thru $r$, handling the rule base and calculation of the conclusion becomes less and less realistic with such systems.

It is surprising that in real life, still some well experienced operators are able to control even such very complicated systems with an acceptable success. How is that possible? Certain areas of the state space, i.e. "clusters" of states are dominated by certain subsets of the input variables. By recognizing the approximate situation of the system, it is possible to reduce the attention to a group of variables. The cardinality of these subsets is usually considerably lower than the cardinality of the total input variable set (k), and might be a few dozens even in the case of several thousand variables. Often, some input variables effect the behavior of the whole system only in an indirect way, i.e. by selecting the group of dominating variables. If a crisp partitioning of the state space is possible, the original problem can be reduced into several subproblems in the following way: The state space $X$ is partitioned into the finite family of subsets

$$X = \cup_{i=1}^{p} D_i, where D_i \cap D_j = \emptyset if i \neq j$$

For every $D_i$ another subset of $X_i$ is dominant ($X = \prod_{i=1}^{k} X_i$). Let us denote the subset for $D_i$ by $S_i = \{X_{i1}, \ldots, X_{\mathrm{in}_i}\}$ where $n_i << k$ and nothing is known about $S_i \cap S_j$(i.e., $S_i - s$ are usually not disjoint). In such a case, for every $S_i$, another rule base can be generated, where the number of necessary rules is considerably lower than in a space without partition. Compared to the orienting value $L^k$ for the unpartitioned space, the total number of necessary rules in all $D_i - s$ together is approximately

$r' = \sum_{i=1}^{p} L^{n_i}$, where $n_i << k$, as stated previously. The real total number of rules, however, includes also the rules concerning the partition, i.e. rules telling, which $D_i$ should be taken for a given state. This number can be approximated by $cp$, where $c$ is constant. The number of all rules in such a system is

$$r = cp + \sum_{i=1}^{p} L^{n_i} << L^k.$$

This kind of rule base is called a hierarchical, structured rule system. Succesful applications of this kind of rule system have been recently published by Sugeno *et al.* (see e.g. [11]).

## III. Fuzzy Partition of the State Space

The problem with the immediate application of the preceding style partition based hierarchical rule systems in many industrial processes is that no partition of $X$ exists in the crisp sense. There are areas of $X$ where a certain subset of $X_i - s$ dominates the process, and these areas are disjoint, as well, however the condition for the union of $D_i$ cannot be fulfilled as

$$\cup_{i=1}^{p} D_i \subset X$$

where $\subset$ denotes proper containment. What can be done in the case of observations lying in

$$T = X \setminus \cup_{i=1}^{p} D_i?$$

We shall call $T$ the "twilight zone" of the state space. In various parts of $T$, usually the effect of several subgroups of variables dominates jointly. In Figure 1, *a* simple state space ($X = X_1 \times X_2$) is shown where the space contains two domains ($D_1$ and $D_2$) dominated by different subsets (e.g. $X_1$ and $X_2$)$D_1 \cup D_2 \neq X$, so there is a twilight zone (T), the area where all subsets (in this case, both variables) influence the behavior. If the observation lies in $T$, none of the rule bases is applicable alone for computing the conclusion. In the Figure, $x_0$ represents a crisp observation, so that $x_0 \in$ T. Similarly, the fuzzy observation $A^*$ lies completely outside of both domains, supp($A^*$) $\subset$ T.

As a matter of course, in real problems, the crisp definition of $D_i$ is a rough estimation. The influence domain of a certain subset of variables can be more precisely given by a fuzzy set, so a fuzzy cover of $X$ is generated. This is a full cover if

$\cup_{i=1}^{p} supp(\tilde{D}_i) = X$, but $core(\tilde{D}_i) \cap core(\tilde{D}_j) = \emptyset$ is always necessary, otherwise the base is contradictory. Usually, $supp(\tilde{D}_i) \cap supp(\tilde{D}_j) \neq \emptyset$ , even if $i \neq$ j. Figure 2 depicts such a case for $X = X_1 \times X_2$, with two fuzzy domains ($\tilde{D}_1$ and $\tilde{D}_2$). Figure 2 is in accordance with Figure 1 in the sense that the cores of the fuzzy domains might be the crisp sets in the previous example. The twilight zone is also a fuzzy set ($\tilde{\mathrm{T}}$). Figure 3 shows an example, there are two overlapping fuzzy domains (denoted by $D_1$ and $D_2$, for simplicity) which form a complete cover of $X = X_1 \times X_2$, and the observation $A^*$= "$X_1$ is $A_1$ and $X_2$ is $A_2$" has a pyramidal membership function generated by $\min\{\mu_{A_1}, \mu_{A_2}\}$. Clearly, $A^*$ is completely in $\tilde{\mathrm{T}}$.

Observations in this new situation might be of two types: $A_1^*$ is completely contained in a single domain (in the sense that supp($A_1^*$) $\subseteq$ core($\tilde{D}_1$) = $D_1$), while $A_2^*$ lies in the twilight zone, so its support is contained in at least two domain supports. What to do in such a case?

It is clear that here both (or, in general, all) variable subsets and sub rule bases must be combined. Suppose that a given $A^*$ overlaps with two domains in the fuzzy cover of $X = X_1 \times \ldots \times X_6$ : $D_2$ and $D_3$(for simplicity, the tilde is omitted from now on), where the variable subsets $\{X_2, X_4, X_5\}$ and $\{X_2, X_3, X_6\}$ are occuring in the antecedent parts of the corresponding sub rule base. So, e.g. rules in $R_2$ have the form

"If $X_2$ is $A_{2i}$ and $X_4$ is $A_{4j}$ and $X_5$ is $A_{5k}$ then $Y$ is $B_l$"

where $A_{pq}$ and $B_l$ are fuzzy sets over the corresponding universe of discourse (e.g. linguistic terms). For $A^*$, all rules from $R_2$ and $R_3$ must be considered and so the input variable set $S = \{X_2, X_3, X_4, X_5, X_6\}$ must be used as the universe of discourse (reduced state space). In this example, $S$ contains only one less component than $X$ itself, but if $k$ is rather high, the subsets have much fewer elements, so their union $S$ will be also much less dimensional than X. All calculations can be done in $S$ and all the rules must be transformed into S. Here, the method of cylindric extension can be applied: If $R_i$ is described by a set of membership functions $\mu_{ij}(x_{i1}, \ldots, x_{ik_i})$ which correspond to the individual rules, and $R_j$ brings into $S$ the additional variables $x_{jl_1}, \ldots, x_{jl_m}$, so that $\{x_{i1}, \ldots, x_{ik_i}\} \cup \{x_{jl_1}, \ldots, x_{jl_m}\} = \{x_{S1}, \ldots, x_{Sk_S}\}$ and $\{x_{i1}, \ldots, x_{ik_i}\} \cap \{x_{jl_1}, \ldots, x_{jl_m}\} = \emptyset$ , the same rules (in $R_i$) are extended into

$$\mu_{ij}(x_{S1}, \ldots, x_{Sk_S}) =$$
$$\min\{\mu_{ij}(x_{i1}, \ldots, x_{ik_i}), 1(x_{jl_1}, \ldots, x_{jl_m})\} =$$
$$\mu_{ij}(x_{i1}, \ldots, x_{ik_i}).$$

This simple way of extension makes calculations with rules in different subspaces very simple, while the complexity increases only in the minimally necessary degree. Denoting the extended rule bases by $R_{ie}$ and $R_{je}$, the new rule base is obtained by

$$R_S = R_{ie} \cup R_{je}.$$

So the conclusion belonging to $A^*$ can be calculated by using theoretically any of the suitable algorithms (e.g. the CRI, or Mamdani's technique) with considering only $R_S$, and no other part of the total rule base. The idea of cylindric extension for two sub rule bases is illustrated in Figure 4. In the base $R_1$ there are 5 rules, in $R_2$ there are four rules, each of them is represented on level $\alpha$ by a single point (cf. [8]). There are approximations of the fuzzy mappings

$$\mathcal{F}_1 : X_1 \rightarrow Y and \mathcal{F}_2 : X_2 \rightarrow Y$$

in the subspaces $X_1 \times Y$ and $X_2 \times Y$(represented by the smoothly fitting curves), which are extended in the cylindric way (the two surfaces in the Figure). In the united (sub)space $X_1 \times X_2$, the two extended rule bases, or their approximations can be treated uniformly.

## IV. Interpolation and Other Approximations in the United Subspaces

In the case of very large number of variables even the possibility of handling complete covers is questionable. A hierarchical rule base might be sparse in two different senses: either the sub rule bases do not cover the whole input state space (i.e. the union of the supports of all fuzzy domains is a proper subset of the state space) or the individual sub rule bases might have "gaps" inside the given domain. As a matter of course, the combination of both can also occur. Another type of sparseness is if the sub rule bases themselves are sparse. This problem, however, has been discussed yet in previous literature [4,8] in the case of a single sparse sub rule base, i.e. a nonstructured rule base.

For an overview of the possible approximation techniques, especially the various types of interpolation, see e.g. [9]. The main statement of this paper is:

**Proposition** : Applying the method of cylindric extension to all sub rule bases flanking the observation (in the sense of the partial ordering introduced in [8]), including the case of partial overlapping, as well, the methods of rule interpolation are directly applicable for obtaining the conclusion, and this algorithm for the inference has a low computational complexity, maximally $\mathcal{O}(kL^{k_{\max}}) = \mathcal{O}(kL^c)$ , where $k_{\max}$ denotes $\max\{k_i\} << k$.

How is this technique applied in the practice? The location of the observation is determined first. Then, all those sub rule bases must be found of which the observation is contained in the support. These sub rule bases must be extended cylindrically. It is more difficult if the observation does not overlap with any sub rule bases. Then, all those sub rule bases must be found which are nearest to the observation measured along all the axes according to the variables of the state space, both in positive and negative direction. For nearness, the supports are considered. As an example, let us consider the situation on Fig. 1. The support of the observation does not overlap with any of the supports of the sub rule bases. The nearest sub rule bases (associated with the domains on the figure) are $D_2$ and $D_1$ along $X_1$ in positive and negative directions, resp., further on $D_1$ and nothing, along $X_2$. Uniting all these sub rule bases in the final set, only those must be considered in the extension, which are included (at least once) in this set. (As a matter of course, in the previously mentioned example, the final set will be $\{D_1, D_2\}$, but in general, a proper subset of all available sub rule bases (in well constructed cases, a small subset of them) will remain.

Cylindric extension is necessary, as the sub rule bases will refer to only a subset of all state variables, a different one in every case. When all variables occuring in one of these rule bases are united in the "actual reduced variable set", the interpolation problem must be solved in the subspace generated by the direct product of all variables in this "actual set". Rules referring to only a subset of the variables in the space of calculation must be extended cylindrically, so the $\alpha$-cuts of the rules can be considered and compared in the total (actual, reduced) space. The interpolated conclusion will be a function of all input variables in the "actual set", and will deliver results (action values) for all the output variables occuring in at least one sub rule base (as output variable). The number of both input and output variables in the actual reduced space might be considerably lower than the total number of variables, if the sub rule bases are constructed well.

In addition to the above technique, interpolatio may be applied also inside the rule bases.

The technique explained here opens the door to the construction of tractable hierarchically constructed rule bases even in the case of very large number of variables. Research is going on in this direction.

## References


[1] L. A. Zadeh: Outline of a new approach to the analysis of complex systems and decision processes. IEEE Trans. on Systems, Man and Cybernetics (1973). pp. 28-44.

[2] L. T. Kóczy: Computational complexity of various fuzzy inference algorithms. Annales Univ. Sci. Budapest 12 (1991). pp. 151-158.

[3] E. H. Mamdani and S. Assilian: An experiment in linguistic synthesis with a fuzzy logic controller. Int. J. of Man-Machine Studies 7 (1978). pp. 1-13.

[4] L. T. Kóczy and Á. Juhász: Fuzzy rule interpolation and the RULEINT program. Proc. Joint Hungarian-Japanese Symposium on Fuzzy Systems and Applications, Budapest, 1991. pp. 91-94.

[5] D. G. Burkhardt and P. P. Bonissone: Automated fuzzy knowledge base generation and tuning. Proc. IEEE Conference on Fuzzy Systems, San Diego, 1992. pp. 179-188.

[6] L. A. Zadeh: Interpolative reasoning as a common basis for inference in fuzzy logic, neural network theory and the calculus of fuzzy If-then rules. Opening talk, *2nd* Int. Conf. on fuzzy logic and neural networks, Iizuka, 1992.

[7] D. Dubois and H. Prade: Basic issues on fuzzy rules and their application to fuzzy control. Fuzzy Control Workshop, IJCAI-91, Sidney, 1991.

[8] L. T. Kóczy and K. Hirota: Interpolative reasoning with insufficient evidence in sparse fuzzy rule bases, to appear in Information Sciences, 1992.

[9] L. T. Kóczy: Techniques of inference in insufficient and inconsistent fuzzy rule bases. Proc. Non-Classical Logics and their Applications, Linz, 1992. pp. 46-50.

[10] A. V. Aho, J. E. Hopcroft and J. D. Ullman: The design and analysis of computer algorithms. Addison-Wesley, Reading, etc., 1974.

[11] M. Sugeno, T. Murofushi, J. Nishio and H. Miwa: Helicopter control based on fuzzy logic. Presented at the Second Fuzzy Symposium on Fuzzy Systems and Their Applications to Human and Natural Systems, Tokyo - Ochanomizu, 1991.

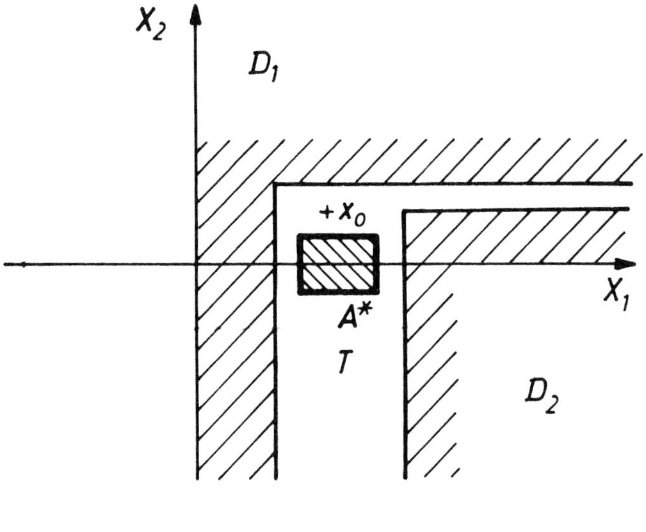


Figure 1.

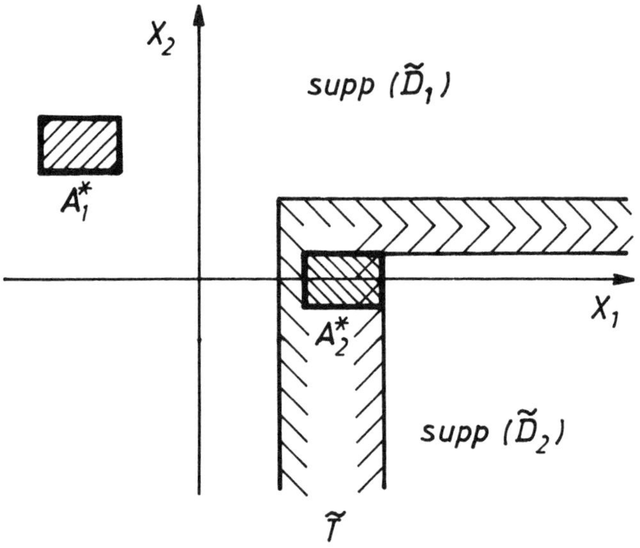


Figure 2.

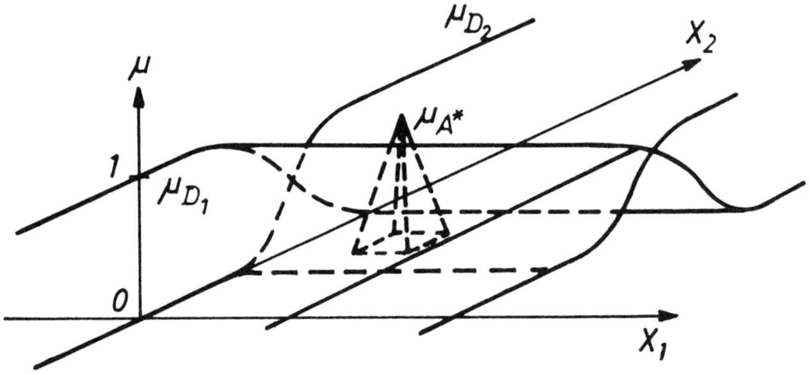


Figure 3.

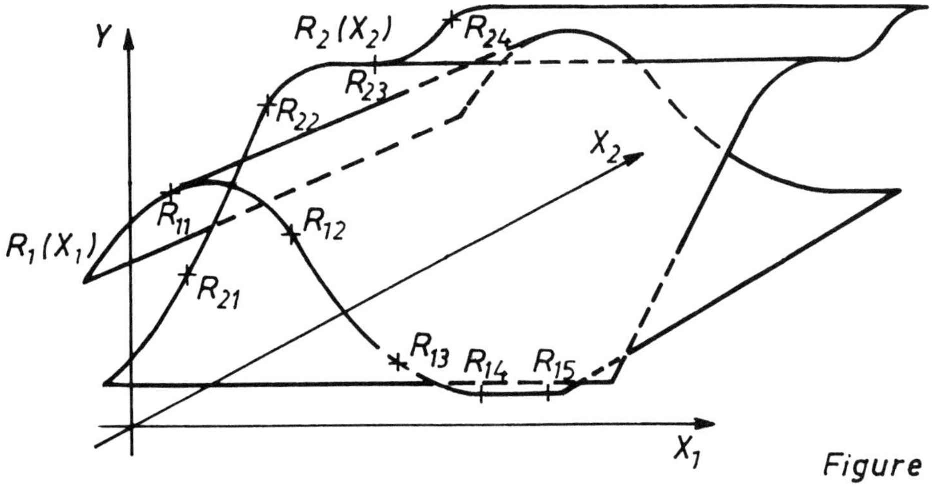


Figure 4.

# Self Generating Radial Basis Function as Neuro-Fuzzy Model and its Application to Nonlinear Prediction of Chaotic Time Series

Ryu Katayama, Yuji Kajitani, Kaihei Kuwata, and Yukiteru Nishida
Information & Communication Systems Research Center, Sanyo Electric Co., Ltd.
1-18-13, Hashiridani, Hirakata, Osaka, 573, Japan
E-mail: katayama@rd.sanyo.co.jp

## Abstract

In this paper, we propose a self generating algorithm for radial basis functions in order to automatically determine the minimal number of basis functions to achieve the specified model error. This model is also regarded as multi-layered neural network or fuzzy model of class $C^{\infty}$. The self generating algorithm consists of the following two processes such as the model parameter tuning process by gradient method for fixed number of rules, and basis function generation procedure by which new basis function is generated in such a way that whose center is located at the point where maximal inference error takes place in the input space, when the effect of parameter tuning is diminished.

Numerical example is shown that the proposed algorithm can achieve the specified model error with less number of basis functions than the other methods by which only coefficients of the basis functions are tuned. Furthemore, the proposed method is applied to the nonlinear prediction of optical chaotic time series.

## 1. Introduction

Multi-layered neural networks are widely used for a variety of applications such as pattern recognition, control, identification and prediction of nonlinear dynamical systems, as a universal function approximator. But slow convergence speed by the commonly used backpropagation algorithm[1] makes it difficult to put these applications into practice. On the other hand, radial basis function (RBF) proposed by Powell et al.[2] has very fast convergence property compared to the backpropagation type algorithm, since an arbitrary function can be approximated by the linear combination of locally tuned factorizable basis functions. This RBF model is also regarded as multi-layered neural network[3] or fuzzy model[4]. To identify the nonlinear systems by RBF, several algorithms such as k-means clustering method[3] or Gram-Schmidt orthogonalization procedure[4] have been proposed. But in these methods, relatively large number of the basis functions are required since the tuning parameters are limited to only the coefficients of RBF.

In this paper, we propose a new self generating algorithm[7] for RBF to satisfy the specified model error with relatively small number of basis functions compared to the

0-7803-0614-7/93$03.00 ©1993IEEE

conventinal methods, where not only the coefficients, but also the center values and widths of RBF are tuned. The proposed algorithm consists of the following two processes:

(1)model parameter tuning process by gradient method for fixed number of the basis functions,

(2)basis function generation procedure, which is invoked when the effect of parameter tuning is diminished, by which new basis function is generated in such a way that whose center is located at the point where maximal inference error occurs in the input space.

Numerical example is shown that the proposed algorithm can achieve the specified model error with less number of basis functions than the other methods by which only coefficients of the basis functions are tuned. Furthemore, the method is also applied to the nonlinear prediction of optical chaotic time series.

## 2. Nonlinear Identification Problem

Let $\underline{x} = (x_1, \cdots, x_i, \cdots, x_m) \in R^m$ be inputs and $\underline{y} = (y_1, \cdots, y_j, \cdots, y_s) \in R^s$ be outputs. The problem is to identify the nonlinear function $\underline{y} = \underline{f}(\underline{x}): R^m \rightarrow R^s$ with the given N input/output data

$$(\underline{x}^1, \underline{y}^1), \cdots, (\underline{x}^p, \underline{y}^p), \cdots, (\underline{x}^N, \underline{y}^N) \tag{1}$$

Also if $s \geqq 2$, output $y_j$, $j=1,\cdots,s$, are assumed to be normalized such that $y_j \in Y_j = \{y_j \mid -1 \leqq y_j \leqq 1\}$, $j=1,\cdots,s$, by dividing original $y_j$ by the $\max(|y_j|)$, $j=1,\cdots,s$, respectively.

In RBF model, output $y_j$ is given by (2):

$$y_j = \sum_{k=1}^{n} w_{jk} \cdot \mu_k(\underline{x}, \underline{a}_k, \underline{b}_k), \quad j=1,\cdots,s \tag{2}$$

where n is a number of RBF, $\underline{a}_k \in R^m$, $\underline{b}_k \in R^m$, $k=1,\cdots,n$, are the center value vector and the width value vector of RBF respectively, and $w_{jk}$, $k=1,\cdots,n$, $j=1,\cdots,s$, are coefficients of RBF. Moreover

$$\mu_k(\underline{x}, \underline{a}_k, \underline{b}_k) = \prod_{i=1}^{m} A_{ik}(x_i, a_{ik}, b_{ik}) \tag{3}$$

and $A_{ik}(x_i, a_{ik}, b_{ik})$ is a RBF for input $x_i$ where $a_{ik}, b_{ik}$ are the center value and the width of RBF, respectively. Several functions are used as $A_{ik}$, and in this paper we use Gaussian type RBF[3-7] which is given by (4):

$$A_{ik}(x_i, a_{ik}, b_{ik}) = \exp(-(x_i - a_{ik})^2 / b_{ik}) \tag{4}$$

In this case, the model is equivalent to GRBF[6] or fuzzy model of class $C^{\infty}$[4,7]. If we regard this model as a fuzzy model, the k-th fuzzy rule is expressed by (5):

Rule k :

If $x_1=A_{1k}$ and$\cdots$and $x_i=A_{ik}$ and$\cdots$and $x_m=A_{mk}$,

$$\text{Then } y_1=w_{1k},\ y_2=w_{2k},\cdots,\ y_s=w_{sk},\quad k=1,\cdots,n \tag{5}$$

Let us introduce the following parameter vectors $\underline{a}^n \in R^{mn}$, $\underline{b}^n \in R^{mn}$, $\underline{w}^n \in R^{sn}$ defined by (6) when the number of RBF is n:

$$\begin{aligned}
&\underline{a}_k=(a_{1k},a_{2k},\cdots,a_{ik},\cdots,a_{mk}) &&\in R^m,\ k=1,\cdots,n\\
&\underline{a}^n=(\underline{a}_1,\underline{a}_2,\cdots,\underline{a}_k,\cdots,\underline{a}_n) &&\in R^{mn}\\
&\underline{b}_k=(b_{1k},b_{2k},\cdots,b_{ik},\cdots,b_{mk}) &&\in R^m,\ k=1,\cdots,n\\
&\underline{b}^n=(\underline{b}_1,\underline{b}_2,\cdots,\underline{b}_k,\cdots,\underline{b}_n) &&\in R^{mn}\\
&\underline{w}_k=(w_{1k},w_{2k},\cdots,w_{jk},\cdots,w_{sk}) &&\in R^s,\ k=1,\cdots,n\\
&\underline{w}^n=(\underline{w}_1,\underline{w}_2,\cdots,\underline{w}_k,\cdots,\underline{w}_n) &&\in R^{sn}
\end{aligned} \tag{6}$$

Then, given the N input/output data (1), the identification problem P(n) with n basis functions is defined as (7).

Identification Problem P(n) :

$$\min_{(\underline{a}^n,\underline{b}^n,\underline{w}^n)} E_n(\underline{a}^n,\underline{b}^n,\underline{w}^n) = \frac{1}{2}\sum_{p=1}^{N}\sum_{j=1}^{s}(y_j^p - {}^*y_j^p)^2 \tag{7}$$

subj.to (2),(3),(4)

where $y_j^p$ and ${}^*y_j^p$ are the j-th output and the j-th inference output for p-th input $\underline{x}^p$, respectively. Let optimal solution for Problem P(n) be $({}^*\underline{a}^n, {}^*\underline{b}^n, {}^*\underline{w}^n)$ and the specified model error value be $\varepsilon > 0$, then the overall identification problem is formulated as follows.

Overall Identification Problem :

Given the N input/output data (1) and the specified model error $\varepsilon > 0$, obtain the minimal number n of RBF and optimal solution $({}^*\underline{a}^n,{}^*\underline{b}^n,{}^*\underline{w}^n)$ for Problem P(n) which satisfies the inequality (8):

$$E_n({}^*\underline{a}^n, {}^*\underline{b}^n, {}^*\underline{w}^n) < \varepsilon \tag{8}$$

## 3. Tuning Algorithm for P(n) by Gradient Method

In order to solve the identification problem P(n), we derive the gradients of $E_n$ with respect to $a_{ik}$, $b_{ik}$, $w_{jk}$ $i=1,\cdots,m$, $j=1,\cdots,s$, $k=1,\cdots,n$, under the equality constraints (2)~(4) as (9):

$$\frac{\partial E_n}{\partial a_{ik}} = \frac{-2}{b_{ik}} \times \sum_{p=1}^{N}\sum_{j=1}^{s}[w_{jk}\cdot \mu_k(\underline{x}^p,\underline{a}_k,\underline{b}_k)\cdot(y_j^p-{}^*y_j^p)\cdot(x_i^p-a_{ik})]$$

$$i=1,\cdots,m,\ k=1,\cdots,n \tag{9a}$$

$$\frac{\partial E_n}{\partial b_{ik}} = \frac{-1}{b_{ik}^2} \times \sum_{p=1}^{N} \sum_{j=1}^{s} [w_{jk} \cdot \mu_k(\underline{x}^p, \underline{a}_k, \underline{b}_k) \cdot (y_j^p - {}^*y_j^p) \cdot (x_i^p - a_{ik})^2]$$

$$i=1,\cdots,m, \quad k=1,\cdots,n \qquad (9b)$$

$$\frac{\partial E_n}{\partial w_{jk}} = -\sum_{p=1}^{N} [\mu_k(\underline{x}^p, \underline{a}_k, \underline{b}_k) \cdot (y_j^p - {}^*y_j^p)]$$

$$j=1,\cdots,s, \quad k=1,\cdots,n \qquad (9c)$$

By using the gradients (9), we can solve the identification problem P(n) by appropriate gradient methods such as steepest descent method, conjugate gradient method, and quasi-Newton method. The learning algorithm by steepest descent method is given by (10), where h is iteration number and $\alpha$ is an optimal step size obtained by solving a one-dimensional search problem.

$$a_{ik}(h+1) = a_{ik}(h) - \alpha \cdot \partial E_n / \partial a_{ik} \qquad (10a)$$

$$b_{ik}(h+1) = b_{ik}(h) - \alpha \cdot \partial E_n / \partial b_{ik} \qquad (10b)$$

$$w_{jk}(h+1) = w_{jk}(h) - \alpha \cdot \partial E_n / \partial w_{jk} \qquad (10c)$$

$$i=1,\cdots,m, \quad j=1,\cdots,s, \quad k=1,\cdots,n$$

## 4. Self Generating Algorithm for Overall Identification Problem

When the decreasing rate of $E_n$ becomes relatively small during the tuning process for fixed number of RBF, we generate new basis function. Althogh several generating strategies may be possible, we adopt such a strategy that a new basis function is generated in such a way that whose center is located at the point where maximum of absolute inference error occurs in the input space. The self generating algorithm based on this idea is described as follows.

【Step 1】

Set the model error $\varepsilon$ to be satisified and stop criteria constant $\varepsilon_1$ for the parameter tuning process for the fixed number of RBF.

【Step 2】

Let h be the iteration number for the fixed number of RBF, and st be the total iteration number for the entire procedure. Let h=0, st=0.

【Step 3】

For N input/output data (1), compute $(\underline{a}^n(h+1), \underline{b}^n(h+1), \underline{w}^n(h+1))$ from $(\underline{a}^n(h), \underline{b}^n(h), \underline{w}^n(h))$ by learning rule (10).

【Step 4】

Let $E_n(h) \equiv E_n(\underline{a}^n(h), \underline{b}^n(h), \underline{w}^n(h))$ for simplicity. Then, if inequality

$$E_n(h+1) < \varepsilon \qquad (11)$$

is satisfied, n is the least number of RBF which satisfies the inequality (8) for overall identification problem. In this case, let $(\underline{a}^n(h+1), \underline{b}^n(h+1), \underline{w}^n(h+1))$ be an optimal solution $(*\underline{a}^n, *\underline{b}^n, *\underline{w}^n)$ and terminate the procedure. If inequality (11) is not satisfied, go to step 5.

【Step 5】

Let us define the decreasing rate of $E_n$ by (12).

$$D_n(h+1) \equiv |(E_n(h+1) - E_n(h)) / E_n(h)| \tag{12}$$

If inequality (13)

$$D_n(h+1) < \varepsilon_1 \tag{13}$$

is satisfied, go to step 6 to generate a new basis function. If

$$D_n(h+1) \geqq \varepsilon_1 \tag{14}$$

holds, let h=h+1, st=st+1, and go back to step 3.

【Step 6】

Let $(\underline{x}^q, \underline{y}^q)$ be a input/output vector such that the absolute inference error takes maximum at this point among the N input/output data $(\underline{x}^p, \underline{y}^p)$, $p=1,\cdots,N$. That is to say, find $(\underline{x}^q, \underline{y}^q)$ satisfying (15):

$$|y_r^q - y_r^q| = \max_{1 \leqq p \leqq N} \max_{1 \leqq j \leqq s} |y_j^p - *y_j^p| \tag{15}$$

Then, generate a (n+1)-th new radial basis function accoriging to (16):

$$a_{i,n+1} = x_i^q, \qquad i=1,\cdots,m \tag{16a}$$

$$b_{i,n+1} = b_0, \qquad i=1,\cdots,m \tag{16b}$$

$$w_{j,n+1} = y_j^q - *y_j^q, \quad j=1,\cdots,s \tag{16c}$$

where $b_0$ is a given constant for the initial width of RBF. Let n = n+1 and h = 0, and go back to step 3. The initial values for other model parameters are set identical which are obtained by the just previous procedure. ■

## 5. Numerical Example for Function Approximation

In order to show the the effectiveness of the proposed algorithm, let us consider the following one dimensional function approximation problem defined by (17):

$$y = \begin{cases} (1/\pi^2) \cdot |(x+\pi) \cdot (x-\pi) \cdot \sin(2x)|, & (-\pi \leqq x < 0) \\ (1/\pi^2) \cdot |(x+\pi) \cdot (x-\pi) \cdot \sin(3x)|, & (0 \leqq x < \pi) \end{cases} \tag{17}$$

Total number of learning data contained in the interval $[-\pi, \pi]$ is N=101, specified model error $\varepsilon = 0.3$, and stop criteria $\varepsilon_1 = 0.0025$ for parameter tuning for fixed number of RBF. Initial number of RBF n =0, and we apply the proposed algorithm with such four tuning parameter cases as (1) $(\underline{w}^n)$, (2) $(\underline{a}^n, \underline{w}^n)$, (3) $(\underline{b}^n, \underline{w}^n)$, (4) $(\underline{a}^n, \underline{b}^n, \underline{w}^n)$.

Fig.1 shows the convergence curve for case (4), where "*" in Fig.1 represents the time when a new RBF is generated. From Fig.1, monotonous decrease of $E_n$ with respect to the total iteration number st is almost guaranteed for every st. This is because the learning results with n RBFs are effectively used as initial parameter values for the learning process of (n+1) RBFs, and this works well for factorizable RBF model. However, this initial value setting scheme may not work well in general for conventional neural network with sigmoid transfer functions. Also, an approximated result for this case is shown in Fig.2.

Fig. 3 represents the relationship between $E_n$ and the number of RBFs for above four cases. The result shows that the case (4) achieves the minimal value $E_n$ for every n, and it requires almost a half number of the basis functions compared to the case (1). Another interesting result is that the case (3) shows almost as good performance as the case (4). This suggests that if the initial center values are suitably selected, the center values may be fixed during the tuning process, and this reduces the computation steps at each parameter tuning process.

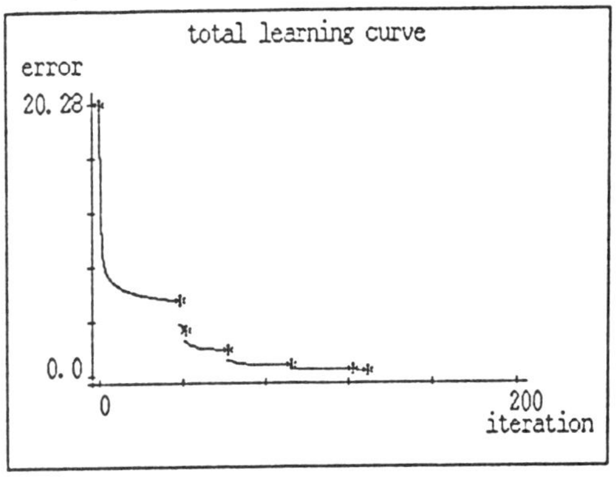


Fig. 1 Model Error vs Total Iteration Number (parameters: $(\underline{a}^n, \underline{b}^n, \underline{w}^n)$)

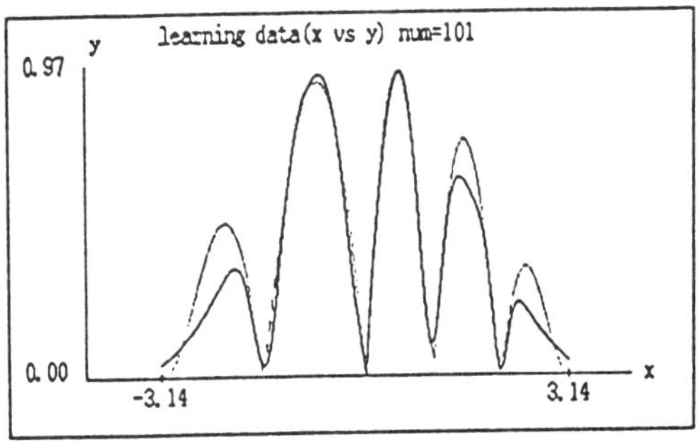


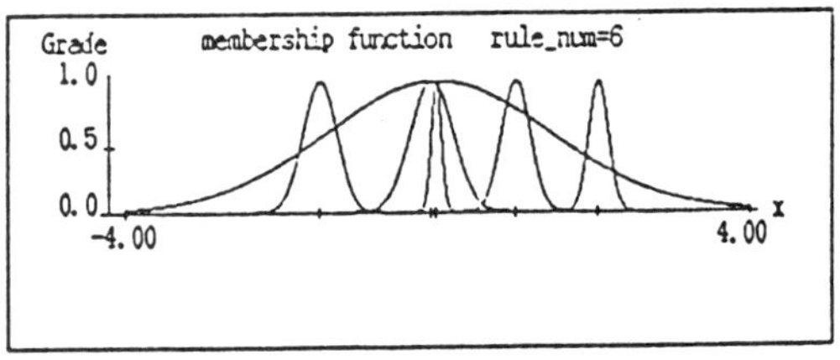


Fig.2 Approximated Result (parameters: $(\underline{a}^n, \underline{b}^n, \underline{w}^n)$)

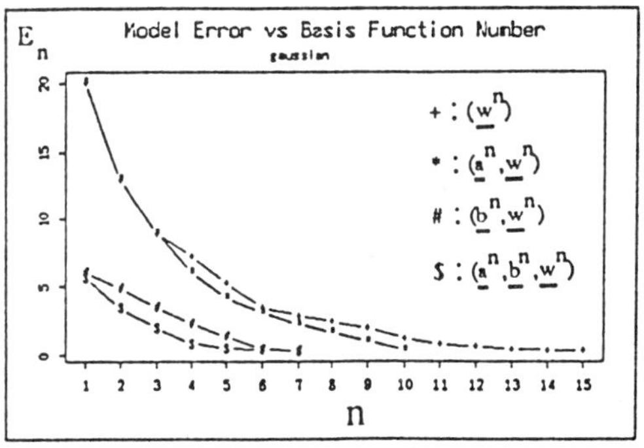


Fig. 3 Model Error vs Number of Basis Functions

## 6. Application to Nonlinear Prediction of Chaotic Time Series

In this section, we apply the self generating algorithm to the nonlinear prediction of optical chaotic time series. Let us consider the Ikeda map[8,9] defined by (18)

$$x(t+1) = a + b \cdot (x(t) \cdot \cos(z(t)) - y(t) \cdot \sin(z(t))) \tag{18a}$$

$$y(t+1) = b \cdot (x(t) \cdot \sin(z(t)) + y(t) \cdot \cos(z(t))) \tag{18b}$$

$$z(t) = 0.4 - 6.0/(1 + x(t)^2 + y(t)^2) \tag{18c}$$

where x, y are a real part and an imaginary part of electric field of a ring cavity[8], respectively. The model generates relatively simple strange attractor[9] with a=1.0, b=0.7. We apply 2 input / 2 output RBF model to identify the chaoic time series generated by (18). The learning data are obtained by solving (18) with initial condition x(0)=y(0)=0. We set N=100, $\varepsilon$ =0.05, $\varepsilon$ 1 =0.01, and we assume initial number of RBF n=0. By applying our method, the stop condition $E_n < \varepsilon$ is satisfyed when n=10. The self generating process of RBF in input space is shown in Fig.4. Thus, new basis functions are successively generated as if they are breeding cells in a living organism.

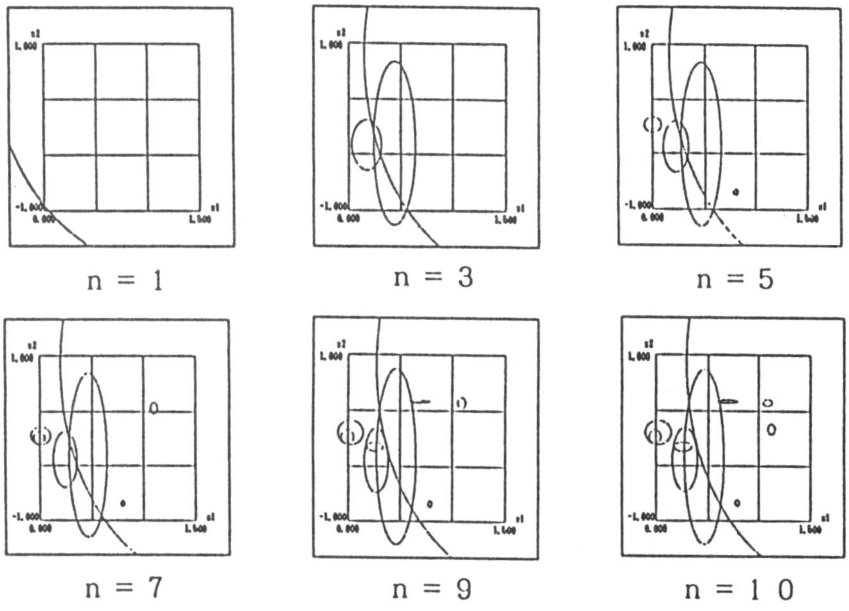


Fig.4 Self Generating Process of RBF for Problem (18)

In order to evaluate the prediction result, we computed correlation $\rho$ defined by (19) between original time series x(t) and predicted time series x'(t), t=1,···,M:

$$\rho = (1/M) \sum_t (x(t)-x_m)(x'(t)-x'_m) \Big/ \sqrt{(1/M) \sum_t (x(t)-x_m)^2} \sqrt{(1/M) \sum_t (x'(t)-x'_m)^2} \tag{19}$$

where $x_m$, $x'_m$ are mean value of x(t), x'(t), t=1,···,M, respectively. Now we show the relation between $\rho$ and prediction time $\tau$ for x(t) and $\tau$ step ahead predicted time series x'(t) in Fig.5, and the same relation for y(t) and y'(t) in Fig.6 when M=90. By these figures, $\rho$ gradually decreases with positive value for about 6 step ahead for both x(t) and y(t). This indicates that the prediction of chaotic trajectory defined by (18) is valid for about 6 steps by our proposed self generating RBF model.

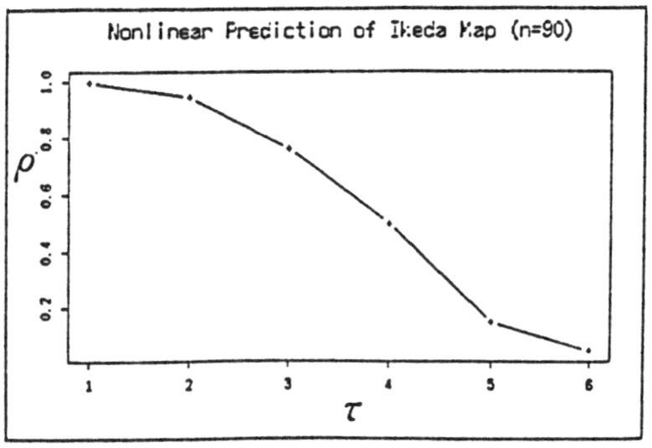


Fig.5 $\rho$ vs $\tau$ for x(t) and x'(t)

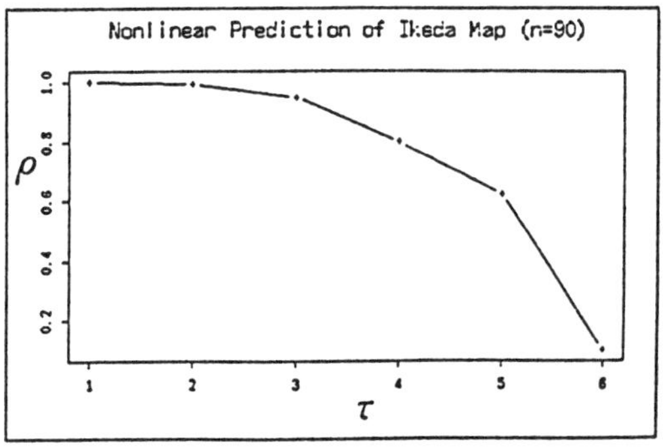


Fig.6 $\rho$ vs $\tau$ for y(t) and y'(t)

## 7. Conclusion

We have proposed a self generating RBF as Neuro-Fuzzy Model to automatically determmine the minimal number of RBF to satisify any desired model accuracy specified by the system designer. For the nonlinear function approximation problem we dealt with in this paper, almost only half number of RBF is required compared to the conventinal identification methods. Therefore, our method is usefull to design adaptive systems incorporating RBF with limited memory resources. We have also demonstrated that the proposed method can be successfully applied to the nonlinear deterministic prediction of optical chaotic time series.

## References

[1] D.E.Rumelhart, G.E.Hinton, R.J.Williams : " Learning internal representations by error propagation", Parallel Distributed Processing, pp.318-362, 1986.

[2] M.J.D.Powell : "Radial basis function approximations to polynomials", Proc. 12th Biennial Numerical Analysis Conference (Dundee), pp.223-241, 1987.

[3] J.Moody, C.J.Darken : "Fast Learning in Networks of Locally-Tuned Processing Units", Neural Computations, 1, pp.281-294, 1989.

[4] H.Ichihashi : "Iterative fuzzy modeling and a hierarchical network", Proc. of Fourth International Congress of the Internatinal Fuzzy System Association, Belgium, Vol.ENG., pp.49-52, 1991.

[5] S.Chen, S.A.Billings, C.F.N.Cowan, P.M.Grant : "Practical Identification of NARMAX Models with Radial Basis Functions", Int.J.Control, vol.52, no.6, pp.1327-1350, 1990.

[6] T.Poggio, F.Girosi : "A Theory of Networks for Approximation and Learning", A.I.Memo, No.1140, C.B.I.P.Paper, No.31, July, 1989.

[7] R.Katayama, K.Kuwata, Y.Kajitani, Y.Nishida : "A Self-Generating Fuzzy Model of Class $C^{\infty}$ and its Application to Identification of Chaotic Time Series" , Proc. of 8th Fuzzy System Symposium, pp.253-256, 1992. (In Japanese)

[8] K.Ikeda, H.Daido, O.Akimoto : "Optical Turbulence: Chaotic Behavior of Transmitted Light from a Ring Cavity", Physical Review Letters, vol.45, no.9, pp.709-712, 1980.

[9] M.Casdagli : "Nonlinear Prediction of Chaotic Time Series", Physica D, 35, pp.335-356, 1989.

# A Connectionist Production System with Approximate Matching Function

Katsuaki Sanou
Systems Laboratory, Kawasaki Steel Corporation,
2-2-3 Uchisaiwai-cho, Chiyoda-ku, Tokyo 100, Japan

Steve G. Romaniuk
Department of Information Systems and Computer Science
National University of Singapore, 10 Kent Ridge, Singapore 0511, Singapore

Lawrence O. Hall*
Department of Computer Science and Engineering
University of South Florida, Tampa, Florida 33620

Abstract — The task of implementing a simple rule-based production system in terms of a hybrid symbolic/connectionist architecture is discussed and it is shown that an approximate matching function can be incorporated into that architecture quite naturally. This paper shows how to convert sequential production rules into a parallel firing connectionist network consisting of simple cells and providing an approximate matching function. The architecture, described here, uses a local representation and builds upon prior work, which resulted in the symbolic/connectionist expert system development tool SC-net. The Hybrid Symbolic/Connectionist Production System (HSC-PS) supports two types of working memory elements, the attribute/value pair and the object/attribute/value triplet. An attribute may be either crisp or fuzzy. The degree of match is measured between a condition having a fuzzy attribute and the working memory element which exists for that attribute and is represented by a value between 0 and 1. The strength of firing for a rule is decided by the minimum match of all the conditions in the left hand side of the rule. Every rule whose strength of firing is greater than 0 performs the actions in the right hand side of the rule according to that strength of firing. If two or more actions are to be performed on the same fuzzy attribute, a new action is composed from them and executed. A classic problem for control is taken as an example and the results by HSC-PS are compared with those by a fuzzy logic controller.

## 1 Introduction

The Hybrid Symbolic/Connectionist Production System (HSC-PS) is based on the network structures of the connectionist expert system development tool SC-net [7, 8]. It has been pointed out that a connectionist-based production system provides advantages such as learnability, robustness, and parallelism over conventional schemes [10, 17]. It is approximate matching that is focused on in this paper. That is, it is shown here that an approximate matching function can be incorporated into HSC-PS quite naturally. In order to implement a connectionist-based production system, it is essential to be capable of representing explicit rules in terms of neuron-like computing units. The computing unit has to be primitive enough to be realized by means of one simple processor. The approach adopted is the implementation of a production system in the SC-net structure, which provides a clear "local" form of connectionist knowledge representation.

$CA_i$ ··· cell activation for cell $C_i$, $CA_i$ in [0..1]
$CW_{ij}$ ··· weight for connection between cell $C_i$ and $C_j$, $CW_{ij}$ in R
$CB_i$ ··· cell bias for cell $C_i$, $CB_i$ in [-1..+1]

$$CA'_i = \begin{cases} |CB_i| * \min_{j=0,\ldots,i-1,i+1,\ldots,n}\{CA_j * CW_{ij}\}, & C_i \text{ is a min cell} \\ |CB_i| * \max_{j=0,\ldots,i-1,i+1,\ldots,n}\{CA_j * CW_{ij}\}, & C_i \text{ is a max cell} \\ |CB_i| * \left|\sum_{j=0,j\neq i}^{n} CA_j * CW_{ij}\right| & C_i \text{ is a ltc cell} \\ 1-(CA_j * CW_{ij}), & C_i \text{ is a negate cell} \end{cases}$$

Figure 1 Cell activation formula

The SC-net network structure consists of simple cells modeling the fuzzy t-norm and co-norm operators min

*Research partially supported by grants from Florida High Technology and Industry Council, Computer Integrated Engineering and Manufacturing section and Software section.

0-7803-0614-7/93$03.00 ©1993IEEE

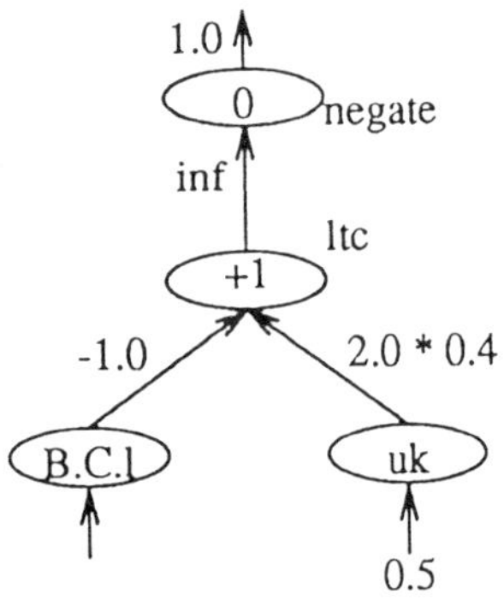


Figure 2 Network structure for (+ (B C 0.4))

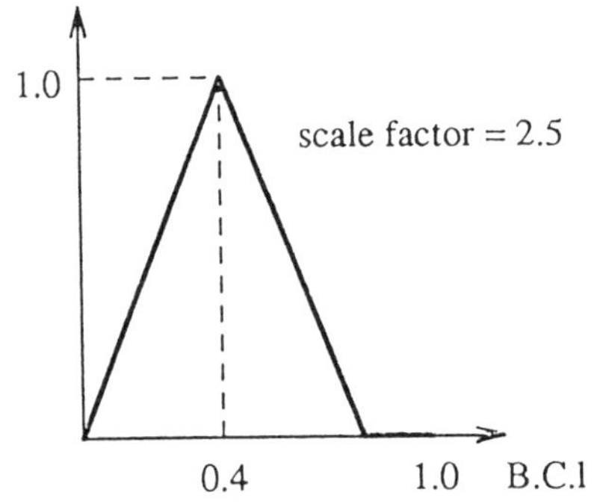


Figure 3 Output of the subnetwork in Figure 2

and max [18], linear threshold cells, or negate cells. Every cell in a network composed of n+1 cells accommodates n inputs, each of which is associated with a weight, and contains a bias value, which indicates what type of fuzzy function a cell models, and whose absolute value represents an upper threshold on the activation a cell can take on. Cell $C_i$ with a cell activation of $CA_i$ computes its new cell activation $CA'_i$ according to the formula in Figure 1. If cell $C_i$ and cell $C_j$ are connected then the weight of the connecting link is given as $CW_{ij}$, otherwise $CW_{ij}=0$. The output of cell $C_i$ is then calculated as $\mathbf{O_i=min\{1, max\{0, CA'_i\}\}}$. The SC-net network is a layered network, but the number of incoming links for one cell may be different from that for another cell and one cell can be connected to another cell in any different layer.

## 2 Architecture

HSC-PS consists of four separate but connected network structures, which feed into each other. The Left Hand Attribute Layer (LHAL) is a single layer of input cells which represent the different attributes that are known to the system. The inputs to the subset consisting of these cells are provided initially by the user or from some initial working memory configuration and in subsequent Recognize/Act cycles from the Right Hand Attribute Layer (RHAL). The production rules are represented in the Encoded Rule Network (ERN), which is composed of two network structures, the Left Hand Side Layer (LHSL) and the Right Hand Side Layer (RHSL), and finally leads into the Right Hand Attribute Layer (RHAL). The RHAL duplicates the LHAL and contains the new attribute values after one forward pass through the HSC-PS network. Finally, the Working Memory Recognition Network (WMRN) detects whether the activations stored in the LHAL are identical to the ones in the RHAL. The Recognize/Act cycle should be terminated when such is the case.

HSC-PS supports two types of working memory elements. One is the attribute/value pair and the other is the object/attribute/value triplet. The attribute of an attribute/value pair is represented by a single cell and its activation encodes the current value of the attribute. This cell and its duplication are placed in the LHAL and the RHAL respectively. For instance, the cells A.l and A.r are placed in the LHAL and RHAL respectively for the attribute A. The activation of a cell in the LHAL, for example A.l, encodes the current value of the attribute and the activation of a cell in the RHAL, for example A.r, encodes the new value of the attribute after one forward pass through the HSC-PS network. The same principle can be applied to the representation of the object/attribute/value triplet. The attribute of an object is represented by a single cell and its activation encodes the current value of the attribute. For instance, in order to handle the attribute C of object B, the cells B.C.l and B.C.r are placed in the LHAL and the RHAL respectively. The relation between the cell in the LHAL and its counterpart in the RHAL is the same as that in the case of the attribute/value pair. The attribute is grouped by means of another criterion. It is either "crisp" or "fuzzy" and every fuzzy attribute has to be declared. Because the value of the attribute is encoded by the activation of the cell which represents the attribute and a cell has only one output, the attribute can take only one value at any time. One object can have an arbitrary number of attributes at the same time **under the condition** that for every object all the attributes which may be used during the course of the inference are provided to HSC-PS in advance.

## 3 Left Hand Side Layer Network

In this section, the subnetworks to build the left hand side of a rule are developed.

(+ (B C 0.4)) is a typical example. This pattern is matched if the value of the attribute C of object B is 0.4

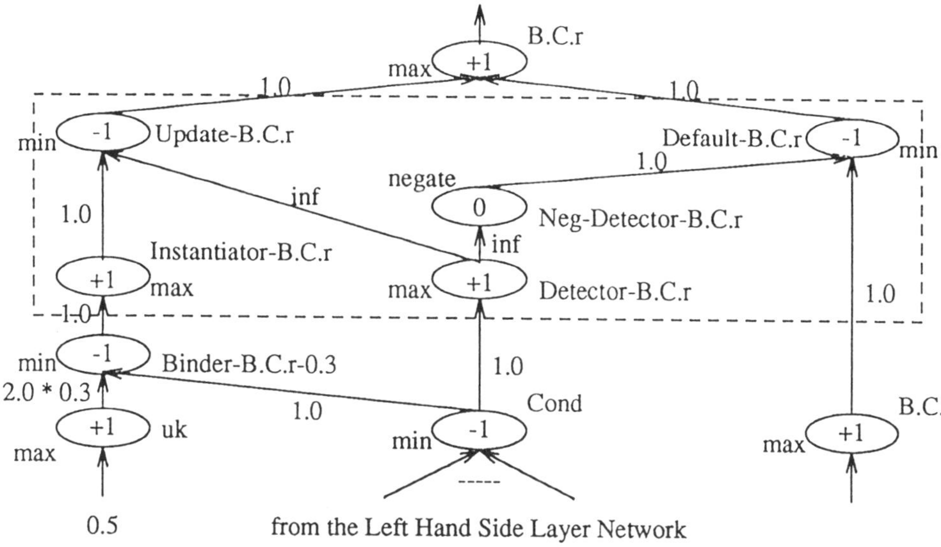


Figure 4 Network structure for (+ (B C 0.3))

and is not matched otherwise. In HSC-PS, an attribute is represented by a single cell, even if it belongs to some object, and its activation encodes the current value of the attribute. Therefore, the network structure for the pattern is shown in Figure 2. In Figure 2 and later figures, the weight "inf" means the output is multiplied with the largest representable positive number on a computer. The UK cell always takes on an activation of 0.5 and is used to provide a threshold. The input of the ltc cell from the UK cell is 0.4 because the activation of the UK cell is 0.5 and the weight of the link from the UK cell to the ltc cell is 0.8 ($= 2.0 * 0.4$). Hence, if the cell B.C.l receives an activation of 0.4 as its input, the ltc cell will take on an activation of 0 and the negate cell will be activated. Otherwise, the negate cell propagates an activation of 0 because the ltc cell propagates an activation greater than 0 and that activation is multiplied with an infinite number, which means the input to the negate cell is an infinite number. That is, when the weight associated with the link connecting the ltc cell and the negate cell is "inf", the subnetwork shown in Figure 2 outputs either 1 or 0, which expresses whether the condition is matched or not.

However, when the attribute C of object B is declared to be fuzzy, it is not enough to judge just whether the condition is satisfied or not. In order to express the situation of partial match, an approximate matching function has to be incorporated into that subnetwork. The idea is to measure the degree of match between the condition using a fuzzy attribute and the working memory element which exists for that attribute. This can be realized naturally by using that weight as a scale factor. When a value other than "inf" is set to the weight associated with the link between the ltc cell and the negate cell, the output of the subnetwork shown in Figure 2 takes the function depicted in Figure 3, where the weight is set to 2.5, and represents the closeness to the expected match by a value between 0 and 1.

In HSC-PS, the left hand side of a rule is a conjunction of conditions. Hence, the outputs of subnetworks for the conditions in a rule are fed into a min cell which is called Cond cell. When no fuzzy attribute is used in the left hand side of a rule, the Cond cell propagates an activation of either 1 or 0, which means the left hand side is satisfied or not. In the case that fuzzy attributes are used, it propagates as an activation a value between 0 and 1. That activation is regarded as the strength of firing for that rule. The situation of partial firing is represented in terms of some strength which is greater than 0 but less than 1.

## 4 Right Hand Side Layer Network

In this section, the subnetworks to build the right hand side of a rule are described.

(+ (B C 0.3)) is a typical example. This action requires the value of attribute C of object B be set to 0.3. Figure 4 shows the subnetwork for this action. The cell B.C.l comes from the LHAL and the cell B.C.r forms the corresponding RHAL. Whenever there exist in the rule-base one or more actions which are performed on an attribute $A_i$, one group of five cells are recruited. They are Detector-$A_i$.r, Neg-Detector-$A_i$.r, Instantiator-$A_i$.r, Update-$A_i$.r, and Default-$A_i$.r, which are surrounded by the dashed lines in Figure 4. The Update-$A_i$.r cell and the Default-$A_i$.r cell are connected to the attribute cell $A_i$.r in the

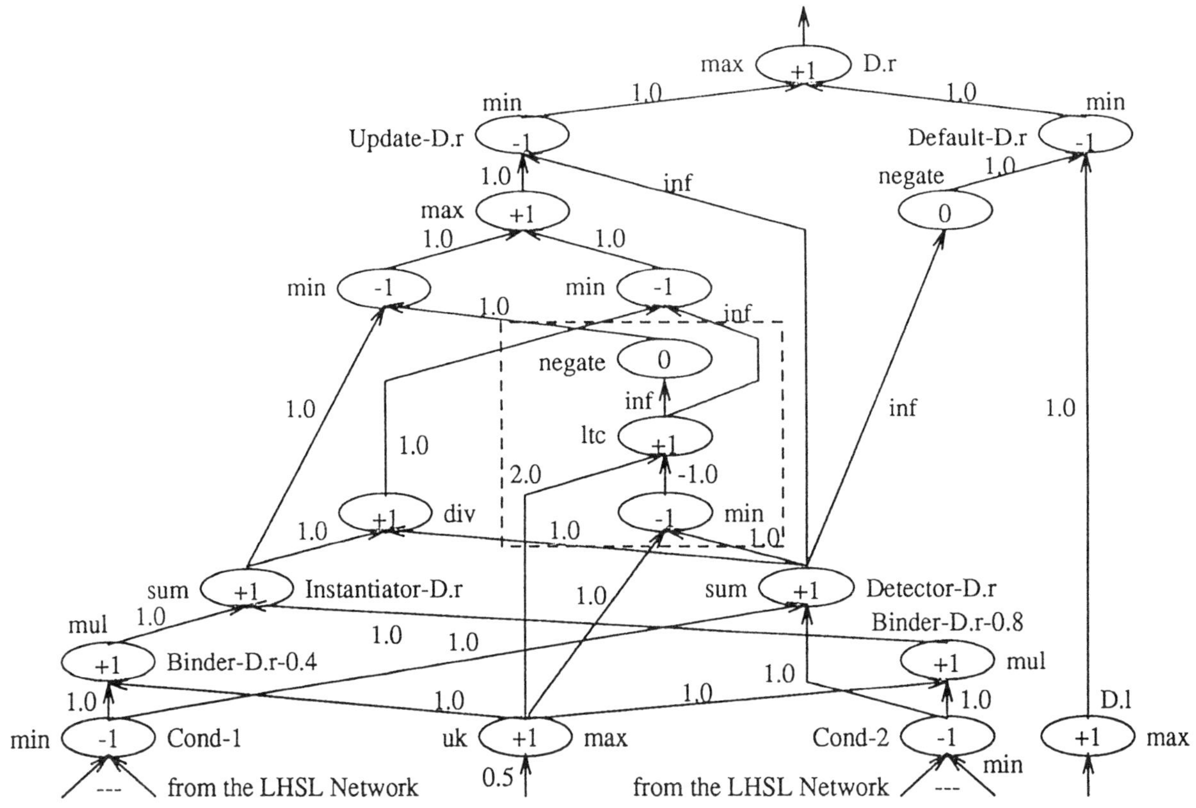


Figure 5 Network structure for composing a new action

RHAL. In case an attribute is not used in the right hand side of any rule, the attribute cell $A_i$.l in the LHAL is directly connected to $A_i$.r in the RHAL.

The Detector-$A_i$.r cell takes on an activation of 1.0 whenever some action to an attribute $A_i$ is to be performed and 0.0 otherwise. The Default-$A_i$.r cell propagates the activation value of the cell $A_i$.l when the Detector-$A_i$.r cell takes on an activation of 0.0, which means none of the rules whose actions are performed on the attribute $A_i$ fire. Otherwise, it propagates 0.0. The Update-$A_i$.r cell propagates either the activation value of the Instantiator-$A_i$.r cell when the Detector-$A_i$.r cell takes on an activation of 1.0, that is, when one or more rules whose actions are performed to the attribute $A_i$ fire, or 0.0 otherwise. The value the Instantiator-$A_i$.r cell propagates is the one to be set to the attribute $A_i$ when some action is performed. The Instantiator-$A_i$.r cell receives as inputs connections from all the Binder-$A_i$.r cells, each of which is recruited for the action of attribute $A_i$ and propagates the values to be set to the attribute $A_i$ when that action is done. One of the activations of the Binder-$A_i$.r cells appears as the output of the Instantiator-$A_i$.r cell by taking the maximum among them. In the above example, the value the Binder-B.C.r-0.3 cell propagates has to be 0.3 because it is the value for this assignment action. This can be represented as the input from the UK cell because the activation of the UK cell is 0.5 and the weight of the link from the UK cell to the Binder cell is 0.6 (= 2.0 * 0.3).

In the case that the attribute B of object C is a fuzzy one and the Cond cell propagates an activation between 0 and 1 as the strength of rule firing, the subnetwork in Figure 4 is modified slightly. In such a case the actual amount of action should be affected by the strength of rule firing. In the above example, it is going to be calculated as the product of 0.3 and the strength of rule firing. That is, *the actual amount of action = the amount of action specified in the rule * the strength of firing for the rule.* In order to realize this idea, the role of the Binder cell has to be changed. It would be most natural if the Binder cell propagates that product. Hence, the operation of multiplication has to be performed in the Binder cell instead of the minimum operation. This leads to the introduction of new kind of cell which is called "mul". Thus, to perform an action on a fuzzy attribute, one "mul" cell has to be recruited for the Binder cell. Other parts are the same as the subnetwork in Figure 4.

The next situation to be discussed is the one where two or more actions, each of which belongs to a different rule, are performed on the same attribute. Consider the follow-

Pressure Error (PE)

| | -6 | -5 | -4 | -3 | -2 | -1 | -0 | +0 | +1 | +2 | +3 | +4 | +5 | +6 |
|---|---|---|---|---|---|---|---|---|---|---|---|---|---|---|
| PB | 0.0 | 0.0 | 0.0 | 0.0 | 0.0 | 0.0 | 0.0 | 0.0 | 0.0 | 0.0 | 0.0 | 0.3 | 0.7 | 1.0 |
| PM | 0.0 | 0.0 | 0.0 | 0.0 | 0.0 | 0.0 | 0.0 | 0.0 | 0.0 | 0.3 | 0.7 | 1.0 | 0.7 | 0.3 |
| PS | 0.0 | 0.0 | 0.0 | 0.0 | 0.0 | 0.0 | 0.0 | 0.3 | 0.7 | 1.0 | 0.7 | 0.3 | 0.0 | 0.0 |
| PO | 0.0 | 0.0 | 0.0 | 0.0 | 0.0 | 0.0 | 0.0 | 1.0 | 0.7 | 0.3 | 0.0 | 0.0 | 0.0 | 0.0 |
| NO | 0.0 | 0.0 | 0.0 | 0.0 | 0.3 | 0.7 | 1.0 | 0.0 | 0.0 | 0.0 | 0.0 | 0.0 | 0.0 | 0.0 |
| NS | 0.0 | 0.0 | 0.3 | 0.7 | 1.0 | 0.7 | 0.3 | 0.0 | 0.0 | 0.0 | 0.0 | 0.0 | 0.0 | 0.0 |
| NM | 0.3 | 0.7 | 1.0 | 0.7 | 0.3 | 0.0 | 0.0 | 0.0 | 0.0 | 0.0 | 0.0 | 0.0 | 0.0 | 0.0 |
| NB | 1.0 | 0.7 | 0.3 | 0.0 | 0.0 | 0.0 | 0.0 | 0.0 | 0.0 | 0.0 | 0.0 | 0.0 | 0.0 | 0.0 |

Figure 6 Part of the fuzzy subsets

If PE = NM
    then If CPE = NS
        then HC = PM
or
If PE = PM
    then If CPE = NS
        then HC = NM
or
⋮

Figure 7 Part of fuzzy rules

ing two actions; one is (+ (D 0.4)) of rule-1 and the other is (+ (D 0.8)) of rule-2. In the crisp mode, if both rules fire, one of the two actions is selected by the mechanism of conflict resolution and executed. In the approximate mode, however, if both rules fire even partially, a new action is composed of the two actions because the action to be performed should be affected by both rules. That is, some proper amount of action on a fuzzy attribute has to be calculated, taking into account the strength of firing for each rule which has an action to be performed on that fuzzy attribute. A reasonable way to compute this would be to sum up the actual amount of action for each rule. In the example in question, the final amount of action would be the sum of *0.4 * the strength of firing for rule-1* and *0.8 * the strength of firing for rule-2.*

Some special consideration has to be given to the case where both rules fire relatively strong, strictly speaking, the case where the sum of activations of Cond cells for rule-1 and rule-2 exceeds 1. In this case the final amount of action calculated in the above way would be larger than allowed and some adjustment would be required. For another example, suppose the strength of rule-1 is 0.4 and that of rule-2 is 0.8. Then, the amount calculated in the above way is $0.80 = 0.4 * 0.4 + 0.8 * 0.8$, which is the same amount as that of rule-2. However, the preferable amount would be 0.67 which is obtained by the adjustment of $0.80 / (0.4 + 0.8)$. That is, when the sum of activations of Cond cells for the rules which have the actions to be performed on the same fuzzy attribute is greater than 1, the sum of actual amounts of actions has to be divided by the sum of strengths for normalization. To realize this idea, two new kinds of cells, "sum" and "div", have to be introduced. The sum cell performs the same calculation as that of the ltc cell except the summed result is not truncated to a value between 0 and 1. This type of cell is used to calculate the sum of activations of Cond cells and that of Binder cells. The div cell is used to perform the operation of division.

Figure 5 shows the subnetwork for the example. The cells in the box of dashed lines are recruited for checking whether the sum of the activation of Cond-1 cell and that of Cond-2 cell, which is propagated through the Detector-D.r cell, is less than or equal to 1 or not. If that sum is less than or equal to 1, then the activation of Instantiator-D.r cell, which is the sum of activations of two Binder cells, is propagated to the Update-D.r cell. Otherwise, the activation of Instantiator-D.r cell is divided by that of Detector-D.r, that is, normalized at the div cell and the quotient is propagated to the Update-D.r cell.

## 5 Experimental Results

In this section, the results obtained by a fuzzy logic controller are compared with the results derived from HSC-PS based on a classic problem for controlling a steam engine [2, 3]. In this system, the control is achieved by measuring the pressure error (PE) and the change in the pressure error (CPE) and by inferring the amount of the heat change (HC). Figure 6 shows part of the fuzzy subsets to define the linguistic values of various control variables and several rules to represent the control strategy are listed in Figure 7.

The two rules shown in Figure 7 can be expressed in HSC-PS by means of the two rules in Figure 8. There exists a subtle difference between the rule for the fuzzy logic controller and that for HSC-PS. In the rule for the fuzzy logic controller, the condition "If PE = NM" judges to what degree the value of PE is NM, where NM means

```
((rule-1)                    ;;; the name of rule
  ( (+ (PE  "-4"))
                             ;;; the left hand side
    (+ (CPE  "-2")))
  ( (+ (HC  "+4"))))         ;;; the right hand side

((rule-2)                    ;;; the name of rule
  ( (+ (PE  "+4"))
                             ;;; the left hand side
    (+ (CPE  "-2")))
  ( (+ (HC  "-4"))))         ;;; the right hand side
```

Figure 8 Part of HSC-PS rules

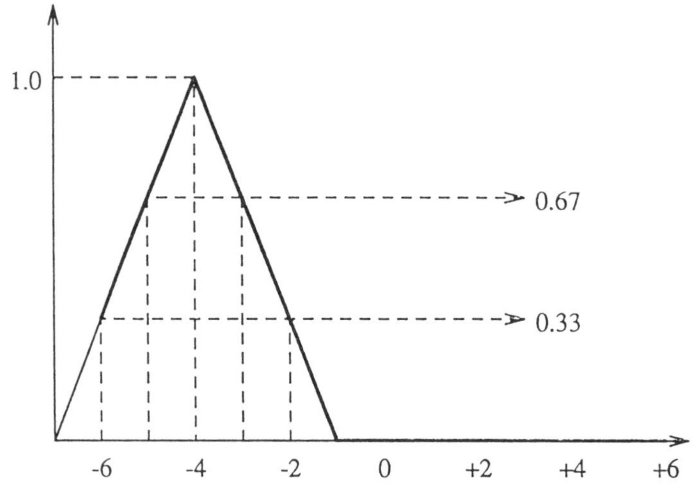


Figure 9 Degree of match to (+ (PE "-4"))

Negative Medium. On the contrary, the condition (+ (PE "-4")) measures how close the value of PE is to "-4" in the rule for HSC-PS. Because the membership value of NM takes 1.0 when -4 is given as the value of PE, that value is used to form the condition of HSC-PS rule corresponding to that of the fuzzy logic controller. 6.66 is used as the scale factor for the condition. Hence, the degree of match to the condition (+ (PE "-4")) is measured via the function shown in Figure 9. That function does not take exactly the same values as the membership values specified in the table in Figure 6. This shows the limitation of the mechanism HSC-PS provides in measuring the degree of match. It is restricted to being linear to the comparing value in HSC-PS but arbitrary values are available as membership values in fuzzy subsets and more flexible relations can be realized between the membership value and the value for the fuzzy variable. Thus, the fuzzy subsets, part of which are shown in Figure 6, can not be incorporated perfectly into the HSC-PS network but the scale factor of 6.66 used here is the best possible one to realize fuzzily similar behavior.

The decision table showing the amounts of HC for all possible combinations of PE and CPE is calculated as Table 1 to the fuzzy subsets and the set of rules. In the calculation, the mean of max-min criteria is used. Table 2 is the decision table obtained by running HSC-PS with the same set of rules. In more than 90% of all possible combinations, the difference between the amount of HC by the fuzzy logic controller and that by HSC-PS is less than or equal to 1. Hence, it can be regarded that the two tables express a similar control strategy, although they are not completely equivalent and it is impossible for us to decide which table is more suitable. In conclusion, it is shown that HSC-PS can realize an approximate matching function in a natural way and use the results from that function effectively in deciding the amount of action.

## 6 Summary

In this paper, a hybrid symbolic/connectionist implementation of a production system with an approximate matching function was presented. The model uses the previously developed SC-net architecture and expands upon its basic components. HSC-PS supports two types of working memory element; one is the attribute/value pair and the other is the object/attribute/value triplet. An attribute may be either crisp or fuzzy. By adjusting some weights in the HSC-PS network, the degree of match can be measured between the condition using a fuzzy attribute and the working memory element for that attribute, although the relation is limited to being linear between the degree of match and the value for that fuzzy attribute. The strength of firing for a rule is the minimum match of all the conditions in the left hand side of the rule. The strength of firing affects the actual amount of action to be performed on a fuzzy attribute. If two or more actions are to be performed on the same fuzzy attribute, the final amount of action is decided from the actual amounts of those actions by the idea of normalization. Further research remains on how to tune an HSC-PS controller and a network with more homogeneous cells may be pursued. From the results of the experiment using a classic problem for controlling a steam engine, it can be said that HSC-PS can use the outcome from an approximate matching function effectively in deciding the amount of action. In closing, HSC-PS will be able to provide good parallel fuzzy controllers because HSC-PS not only functions simply as some sort of fuzzy logic controller but also is expected to provide parallelism due to its connectionist architecture.

| | CPE | | | | | | | | | | | | |
|---|---|---|---|---|---|---|---|---|---|---|---|---|---|
| | -6 | -5 | -4 | -3 | -2 | -1 | 0 | +1 | +2 | +3 | +4 | +5 | +6 |
| -6 | 0 | 0 | 4 | 4 | 4 | 5 | 6 | 6 | 6 | 6 | 5 | 6 | 6 |
| -5 | 0 | 0 | 4 | 4 | 4 | 5 | 6 | 6 | 6 | 6 | 5 | 6 | 6 |
| -4 | -2 | -2 | 1 | 4 | 4 | 5 | 6 | 6 | 6 | 6 | 4 | 6 | 6 |
| -3 | -2 | -2 | -2 | 1 | 2 | 3 | 5 | 5 | 5 | 5 | 6 | 6 | 6 |
| -2 | -2 | -2 | -2 | -1 | 0 | 2 | 4 | 4 | 4 | 5 | 6 | 6 | 6 |
| -1 | -3 | -3 | -3 | -2 | -1 | 1 | 2 | 2 | 4 | 5 | 5 | 5 | 5 |
| P -0 | -4 | -4 | -4 | -3 | -2 | -1 | 0 | 0 | 2 | 4 | 4 | 4 | 4 |
| E +0 | 4 | 4 | 4 | 3 | 2 | 1 | 0 | 0 | -2 | -4 | -4 | -4 | -4 |
| +1 | 3 | 3 | 3 | 2 | 1 | -1 | -2 | -2 | -4 | -5 | -5 | -5 | -5 |
| +2 | 2 | 2 | 2 | 1 | 0 | -2 | -4 | -4 | -4 | -5 | -6 | -6 | -6 |
| +3 | 2 | 2 | 2 | -1 | -2 | -3 | -5 | -5 | -5 | -5 | -6 | -6 | -6 |
| +4 | 2 | 2 | -1 | -4 | -4 | -5 | -6 | -6 | -6 | -6 | -4 | -6 | -6 |
| +5 | 0 | 0 | -4 | -4 | -4 | -5 | -6 | -6 | -6 | -6 | -5 | -6 | -6 |
| +6 | 0 | 0 | -4 | -4 | -4 | -5 | -6 | -6 | -6 | -6 | -5 | -6 | -6 |

Table 1 Decision table by the fuzzy logic controller

| | CPE | | | | | | | | | | | | |
|---|---|---|---|---|---|---|---|---|---|---|---|---|---|
| | -6 | -5 | -4 | -3 | -2 | -1 | 0 | +1 | +2 | +3 | +4 | +5 | +6 |
| -6 | 0 | 0 | -1 | 4 | 5 | 5 | 5 | 6 | 6 | 6 | 6 | 6 | 6 |
| -5 | 0 | 0 | -1 | 4 | 5 | 5 | 5 | 6 | 6 | 6 | 6 | 6 | 6 |
| -4 | -5 | -5 | 1 | 2 | 3 | 4 | 5 | 5 | 6 | 6 | 6 | 6 | 6 |
| -3 | -2 | -2 | 0 | 1 | 2 | 3 | 4 | 5 | 5 | 5 | 6 | 6 | 6 |
| -2 | -3 | -3 | -2 | -1 | 1 | 2 | 3 | 4 | 4 | 5 | 5 | 5 | 5 |
| -1 | -3 | -3 | -2 | -2 | -1 | 1 | 1 | 3 | 4 | 5 | 5 | 5 | 5 |
| P -0 | -3 | -3 | -3 | -2 | -1 | 0 | 1 | 2 | 4 | 4 | 5 | 5 | 5 |
| E +0 | 3 | 3 | 3 | 2 | 1 | 0 | -1 | -2 | -4 | -4 | -5 | -5 | -5 |
| +1 | 3 | 3 | 2 | 2 | 1 | -1 | -1 | -3 | -4 | -5 | -5 | -5 | -5 |
| +2 | 3 | 3 | 2 | 1 | -1 | -2 | -3 | -4 | -4 | -5 | -5 | -5 | -5 |
| +3 | 2 | 2 | 0 | -1 | -2 | -3 | -4 | -5 | -5 | -5 | -6 | -6 | -6 |
| +4 | -2 | -2 | -1 | -2 | -3 | -4 | -5 | -5 | -6 | -6 | -6 | -6 | -6 |
| +5 | 0 | 0 | -6 | -4 | -5 | -5 | -5 | -6 | -6 | -6 | -6 | -6 | -6 |
| +6 | 0 | 0 | -6 | -4 | -5 | -5 | -5 | -6 | -6 | -6 | -6 | -6 | -6 |

Table 2 Decision table by HSC-PS

## References


[1] Kuo, Steve., Moldovan, Dan. (1991) "Performance Comparison of Models for Multiple Rule Firing" IJCAI-91, pp. 42-47

[2] Mamudani, E. H., Assilan S. (1974) "An Experiment in Linguistic Systhesis with a Fuzzy Logic Controller" Int. J. man-mac. Stud., Vol. 7, 1974, pp. 1-13

[3] Mamudani, Ebraham H. (1977) "Applications of Fuzzy Logic to Approximate Reasoning Using Linguistic Synthesis" IEEE Transactions on Computers, Vol. C-26, No. 12, December 1977, pp. 1182-1191

[4] Mettrey, William. (1991) "A Comparative Evaluation of Expert System Tools" Computer, Vol. 24, No. 2, February 1991, pp. 19-31

[5] Nilsson, Nils J. (1971) "Problem-Solving Methods in Artificial Intelligence" McGraw-Hill, Inc.

[6] Romaniuk, S. G., Hall, L. O. (1989) "Parallel Connectionist Expert Systems" IASTED, Applications and Theory Track, Zurich, June, pp. 241-244

[7] Romaniuk, S. G., Hall, L. O. (1989) "FUZZNET: Towards a Fuzzy Connectionist Expert System Development Tool" IJCNN-90, Washington D.C., January

[8] Romaniuk, S. G., Hall, L. O. (1990) "The Use of Fuzzy Variables in a Hybrid Connectionist Expert System" NAFIPS'90, Toronto, Canada, June

[9] Romaniuk, S. G., Hall, L. O. (1991) "Injecting Symbol Processing into a Connectionist Model" Neural and Intelligent Systems Integration, Branko Soucek, (Ed.), John Wiley, N.Y.

[10] Samad, Tariq. (1988) "Towards Connectionist Rule-Based Systems" In. Vol. II, Proceedings of the International Conference on Neural Networks

[11] Sanou, K., Romaniuk, S. G., Hall, L. O. (1992) "A Hybrid Symbolic/Connectionist Production System" 4th International Conference on Tools with Artificial Intelligence, Arlington, U.S.A., November

[12] Sanou, K., Romaniuk, S. G., Hall, L. O. (1993) "A Connectionist Implementation of a Production System on a Hypercube Multiprocessor" '93 Korea/Japan Joint Conference on Expert Systems Seoul, Korea, February

[13] Schmolze, James G., Goel, Suraj (1990) "A Parallel Asynchronous Distributed Production System" AAAI-90, pp. 65-71

[14] Shastri, Lokendra., Ajjanagade, Venkat. (1989) "A Connectionist System for Rule Based Reasoning with Multi-Place Predicates and Variables" Technical Report MS-CIS-89-06, University of Pennsylvania, Philadelphia, January 1989

[15] Shirai, Yoshiaki., Tsuji, Jun-ichi. (1982) "Artificial Intelligence - Concepts, Techniques and Applications" Iwanami Shoten, Publishers

[16] Simon, Herbert A., Newell, Allen. (1961) "Computer Simulation of Human Thinking and Problem Solving" Datamation Vol. 7, No. 6, pp. 18-20

[17] Touretzky, David S., Hinton, Geoffrey E. (1988) "A Distributed Connectionist Production System" Cognitive Science 12, pp. 423-466

[18] Zimmermann, H. J. (1991) "Fuzzy Set Theory - and Its Applications, Second, Revised Edition" Kluwer Academic Publishers

# Stability Analysis of Neural Networks Using Stability Conditions of Fuzzy Systems

Kazuo TANAKA and Manabu SANO

Department of Mechanical Systems Engineering
Kanazawa University
2-40-20 Kodatsuno Kanazawa 920 Japan
Tel. +81-762-61-2101 (ext. 405)
Fax. +81-762-63-3849

**Abstract** - This paper discusses stability of neural networks using stability conditions of fuzzy systems. The Parameter Region (PR) representation, which graphically shows location of fuzzy if-then rules in consequent parameters space, is proposed by introducing new concepts of *Edge Rule* (*Matrix*) and *Minimum Representation*. Stability criterion of neural networks is illustrated in terms of PR representation. Some properties for stability of neural networks are derived from the results of stability criterion.

## 1. Introduction

One of the most important concepts concerning the properties of dynamical system is stability. It is, however, difficult to analyze stability of nonlinear systems such as neural networks and fuzzy systems. Recently, Tanaka and Sugeno [1], [2] derived a stability theorem for fuzzy systems. This theorem gives a sufficient condition which guarantees stability of fuzzy systems in accordance with the definition of stability in the sense of Lyapunov.

The purpose of this paper is to discuss stability of neural networks using the Tanaka-Sugeno's Theorem. The stability analysis discussed in this paper can be applied not only to neural networks but also to other nonlinear systems if the nonlinear systems can be approximated by fuzzy systems. This paper is organized as follows. Section 2 shows a type of neural networks analyzed in this paper. Section 3 gives the Tanaka-Sugeno's stability theorem and proposes a representation method of parameter region (PR) which graphically shows location of fuzzy if-then rules in consequent parameters space. Section 4 illustrates stability criterion of neural networks using the PR method and derives some properties for stability of neural networks.

## 2. Neural network

In this paper, we analyze stability of neural networks shown in Fig.1, where x(k) ~ x(k-n+1) are state variables, and u(k) ~ u(k-m+1) are input variables.

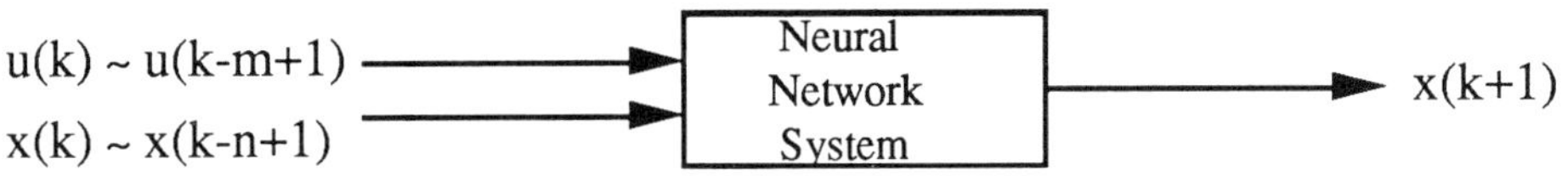


Fig.1 Neural network

Assume that all output functions f(v) of units in the neural network system are differentiable, f(0) = 0 and $f(v) \in [-1,1]$ for all v. Moreover, assume that u(k)=u(k-1)= $\cdots$ =u(k-m+1)=0 because we analyze stability of the equilibrium of neural networks. In this paper, f' (v) means df(v)/dv.

## 3. Tanaka-Sugeno's stability theorem

The fuzzy discrete free system, proposed by Takagi and Sugeno [3], can be described as follows.

0-7803-0614-7/93$03.00 ©1993IEEE

Rule i : IF $x(k)$ is $\mathcal{A}_{1i}$ and $\cdots$ and $x(k-n+1)$ is $\mathcal{A}_{ni}$

THEN $x_i(k+1) = a_{i1}\, x(k) + a_{i2}\, x(k-1) + \cdots + a_{in}\, x(k-n+1)$ , $i=1, 2, \cdots, r$ (1)

where r is the number of rules, $\mathcal{A}_{1i} \sim \mathcal{A}_{ni}$ are fuzzy sets, $a_{i1} \sim a_{in}$ are consequent parameters and $x(k) \sim x(k-n+1)$ are state variables. The final output is calculated as follows.

$$\mathbf{x}(k+1) = \sum_{i=1}^{r} w_i(k)\mathbf{A}_i\mathbf{x}(k) \Big/ \sum_{i=1}^{r} w_i(k) , \tag{2}$$

where

$$w_i(k) = \prod_{j=1}^{n} \mathcal{A}_{ij}\,(\, x(k-j+1)\,), \tag{3}$$

$$\mathbf{x}^T(k) = [x(k), x(k-1) , \cdots , x(k-n+1)],$$

$$\mathbf{A}_i = \begin{bmatrix} a_{i1} & \cdots & a_{i\,n-1} & a_{in} \\ 1 & 0 & \cdots & 0 \\ 0 & \ddots & \ddots & \vdots \\ \vdots & \ddots & \ddots & \vdots \\ 0 & \cdots 0 & 1 & 0 \end{bmatrix}.$$

$\mathbf{A}_i$ is called " consequent matrix" and each linear consequent equation represented by $\mathbf{A}_i\, x(k)$ is called "subsystem". Tanaka-Sugeno's stability theorem [2] for ensuring stability of Eq.(2) is given as follows.

**[Theorem 3.1]**

*The equilibrium of a fuzzy system described by Eq.(2) is asymptotically stable in the large if there exists a common positive definite matrix* **P** *such that*

$$\mathbf{A}_i^T \mathbf{P} \mathbf{A}_i - \mathbf{P} < \mathbf{0} \tag{4}$$

*for* $i = 1, 2, \cdots, r$.

This theorem is reduced to the Lyapunov stability theorem for linear discrete systems when r=1. Theorem 3.1 gives, of course, a sufficient condition for ensuring stability of Eq.(2). We may intuitively guess that a fuzzy system is globally stable if all subsystems are stable. However, this is not the case in general [2]. We must notice that a fuzzy system of Eq.(2) is not always asymptotically stable in the large even if all the $\mathbf{A}_i$ 's are stable matrices.

In order to check stability of fuzzy systems, we must find a common **P** such that

$$\mathbf{A}_i^T \mathbf{P} \mathbf{A}_i - \mathbf{P} < \mathbf{0}$$

for all r. We have some studies [3], [4] for finding a common positive definite matrix **P**. We should notice that the stability condition of Eq.(4) depends only on $\mathbf{A}_i$. In other words, the stability condition of Eq.(4) does not depend on $w_i(k)$. Next, we give a necessary condition [2] for ensuring existence of a common **P**.

**[Theorem 3.2]**

*Assume that* $\mathbf{A}_i$ *is a stable matrix for* $i=1, 2, \cdots, r$ . $\mathbf{A}_i \mathbf{A}_j$ *is a stable matrix for* $i, j = 1, 2, \cdots, r$ *if there exists a common positive definite matrix* **P** *such that*

$$\mathbf{A}_i^T \mathbf{P} \mathbf{A}_i - \mathbf{P} < \mathbf{0}$$

*for all* r.

The contraposition of this theorem means that if one of the $\mathbf{A}_i \mathbf{A}_j$ 's is at least an unstable matrix, then there dose not exist a common **P** such that Eq.(4) is satisfied.

Next, we propose a representation method of parameter region (PR) for Eq.(2). The PR representation graphically shows locations of fuzzy if-then rules in consequent parameters space. Let us give two simple examples of PR representation.

**[Example 3.1]**

Let us consider the following fuzzy system (Fuzzy System 1) .

Rule 1: IF x(k) is $\mathcal{A}_1$ THEN $x1(k+1) = 0.1x(k) + 0.1x(k-1)$
Rule 2: IF x(k) is $\mathcal{A}_2$ THEN $x2(k+1) = 0.3x(k) + 0.1x(k-1)$
Rule 3: IF x(k) is $\mathcal{A}_3$ THEN $x3(k+1) = 0.1x(k) + 0.3x(k-1)$

We can obtain

$$\mathbf{A}_1 = \begin{bmatrix} 0.1 & 0.1 \\ 1 & 0 \end{bmatrix} \quad \mathbf{A}_2 = \begin{bmatrix} 0.3 & 0.1 \\ 1 & 0 \end{bmatrix} \quad \mathbf{A}_3 = \begin{bmatrix} 0.1 & 0.3 \\ 1 & 0 \end{bmatrix}$$

from the consequent equations. Fig.2 shows PR of this fuzzy system.

**[Example 3.2]**

Let us consider the following fuzzy system (Fuzzy System 2) .

Rule 1: IF x(k) is $\mathcal{B}_1$ THEN $x1(k+1) = 0.1x(k) + 0.1x(k-1)$
Rule 2: IF x(k) is $\mathcal{B}_2$ THEN $x2(k+1) = 0.3x(k) + 0.1x(k-1)$
Rule 3: IF x(k) is $\mathcal{B}_3$ THEN $x3(k+1) = 0.1x(k) + 0.3x(k-1)$
Rule 4: IF x(k) is $\mathcal{B}_4$ THEN $x4(k+1) = 0.2x(k) + 0.2x(k-1)$
Rule 5: IF x(k) is $\mathcal{B}_5$ THEN $x5(k+1) = 0.15x(k) + 0.15x(k-1)$

We can obtain

$$\mathbf{A}_1 = \begin{bmatrix} 0.1 & 0.1 \\ 1 & 0 \end{bmatrix} \quad \mathbf{A}_2 = \begin{bmatrix} 0.3 & 0.1 \\ 1 & 0 \end{bmatrix} \quad \mathbf{A}_3 = \begin{bmatrix} 0.1 & 0.3 \\ 1 & 0 \end{bmatrix}$$

$$\mathbf{A}_4 = \begin{bmatrix} 0.2 & 0.2 \\ 1 & 0 \end{bmatrix} \quad \mathbf{A}_5 = \begin{bmatrix} 0.15 & 0.15 \\ 1 & 0 \end{bmatrix}$$

from the consequent equations. Fig.3 shows PR of this fuzzy system.

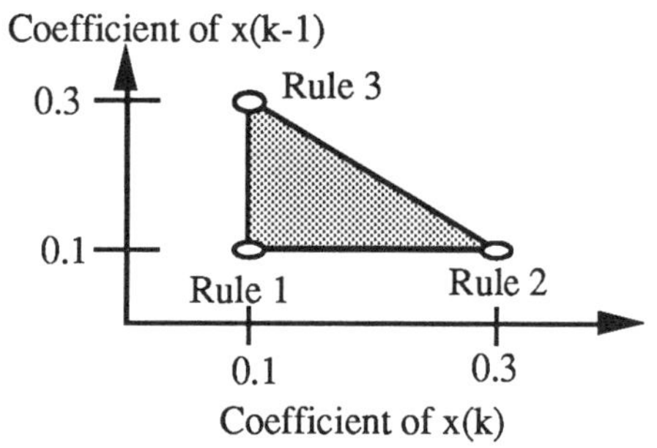


Fig.2 PR of Fuzzy System 1

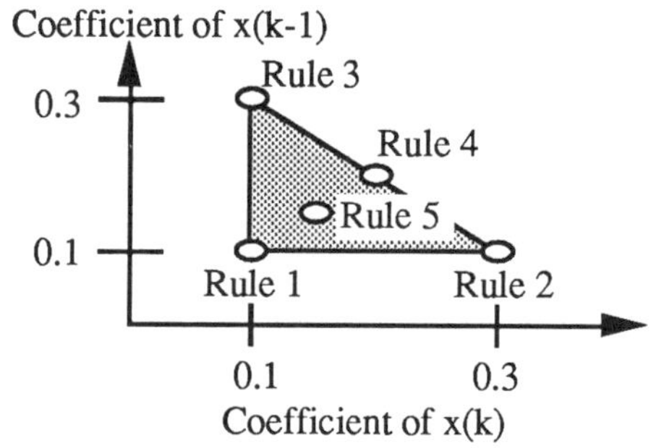


Fig.3 PR of Fuzzy System 2

We should notice a difference between PR of Fuzzy System 1 and that of Fuzzy System 2. In Fig.2, each plotted point corresponds to each edge of the parameter region. Conversely, in Fig.3, the parameter region constructed by the plotted points of Rule 1, Rule 2 and Rule 3 includes the plotted points of Rule 4 and Rule 5. Let us define ***Edge Rule*** (***Edge Matrix***) and ***Minimum Representation***.

[**Definition 3.1** : ***Edge Rule*** and ***Edge Matrix***]

Rule 1, Rule 2 and Rule 3 of Fuzzy System 1 or 2 in Examples 3.1 and 3.2, which correspond to edges of the PR respectively, are said to be ***Edge Rules***. The consequent matrices, $\mathbf{A}_1$, $\mathbf{A}_2$ and $\mathbf{A}_3$, in ***Edge Rules*** are said to be ***Edge Matrices***.

[**Definition 3.2** : ***Minimum Representation***]

A fuzzy system which consists only of ***Edge Rules*** is said to be a ***Minimum Representation***.

Of course, Fuzzy System 1 is a ***Minimum Representation***. Fuzzy System 2 is not a ***Minimum Representation***. Next, we derive an important theorem for checking stability in the case of non-***Minimum Representation***.

**[Theorem 3.3]**

*Assume that* **P** *is a positive definite matrix. If* $\mathbf{A}_i{}^T \mathbf{P} \mathbf{A}_i - \mathbf{P} < 0$ *for* i = 1, 2, ⋯, r, *then* $\mathbf{A}^{*T} \mathbf{P} \mathbf{A}^{*} - \mathbf{P} < 0$, *where*

$$\mathbf{A}^{*} = \sum_{i=1}^{r} s_i \mathbf{A}_i, \quad s_i \geq 0, \quad \sum_{i=1}^{r} s_i = 1.$$

*(proof) The proof is omitted due to lack of space.*

It is clear from Theorem 3.3 that $\mathbf{A}^{*}$ is not ***Edge Matrix***. Theorem 3.3 shows that stability of fuzzy systems can be checked by applying Tanaka-Sugeno's Theorem to a ***Minimum Representation*** of fuzzy systems. In Example 3.2, $\mathbf{A}_4 = 0.5\mathbf{A}_2 + 0.5\mathbf{A}_3$ and $\mathbf{A}_5 = 0.5\mathbf{A}_1 + 0.25\mathbf{A}_2 + 0.25\mathbf{A}_3$. Therefore, a ***Minimum Representation*** of Fuzzy System 2 is equivalent to Fuzzy System 1. It is found from Theorem 3.3 that Fuzzy System 2 is stable if Fuzzy System 1 is stable.

## 4. Stability analysis of neural networks

It is necessary to represent dynamics of a neural network by a fuzzy system, Eq.(2), in order to analyze stability of neural networks using Tanaka-Sugeno's Theorem. In this section, we present a procedure for representing dynamics of a neural network system by a fuzzy system. The basic idea is to represent each output function, f(v), in each unit by a fuzzy system. Let us consider a simple neural network which consists of a single layer, that is,

$$v = w_1\, x(k) + w_2\, x(k-1),$$
$$x(k+1) = f(v),$$

where $w1$ and $w2$ are connection weights. Assume that the output function f(v) is a sigmoid function,

$$f(v) = 2/(1 + \exp(-v/q)) - 1,$$

where q is a parameter of the function. We should notice that the output, f(v), satisfies

$$g_1 v \leq f(v) \leq g_2 v,$$

where $g_1$ and $g_2$ are the minimum value and the maximum value of f'(v), that is,

$$g_1 = \min_v f'(v) = 0, \qquad g_2 = \max_v f'(v) = 0.5/q.$$

Therefore, this neural network can be represented by the following fuzzy system.

$$x(k+1) = f(v) = (w_1(k)\, g_1 + w_2(k)\, g_2)\, v \,/\, (w_1(k) + w_2(k))$$
$$= \sum_{i=1}^{2} w_i(k)\, g_i\, (w_1\, x(k) + w_2\, x(k-1)) \,/\, \sum_{i=1}^{2} w_i(k),$$

where $w_1(k), w_2(k) \in [0,1]$ for all k. $w_1(k)$ and $w_2(k)$ are regarded as membership values. The ***Edge Matrices*** are obtained as follows.

$$\mathbf{A}_1 = \begin{bmatrix} g_1 w_1 & g_1 w_2 \\ 1 & 0 \end{bmatrix} \qquad \mathbf{A}_2 = \begin{bmatrix} g_2 w_1 & g_2 w_2 \\ 1 & 0 \end{bmatrix}$$

In this case, PR becomes a straight line. We can analyze stability of the simple neural network using the ***Edge Matrices*** and Theorem 3.3. If we can find a common positive definite matrix **P** such that

$$\mathbf{A}_i^T \mathbf{P} \mathbf{A}_i - \mathbf{P} < \mathbf{0}, \qquad i=1,2$$

this neural network is asymptotically stable in the large. We illustrate some examples of stability analysis based on this idea.

**[Example 4.1]**

Let us consider stability of a neural network shown in Fig.4. From Fig.4, we obtain

$$v_{11} = w_{111}\, x(k) + w_{121}\, x(k-1), \tag{5}$$
$$v_{12} = w_{112}\, x(k) + w_{122}\, x(k-1), \tag{6}$$
$$v_{21} = w_{211}\, f_{11}(v_{11}) + w_{212}\, f_{12}(v_{12}), \tag{7}$$
$$x(k+1) = f_{21}(v_{21}). \tag{8}$$

Next, we define

$$f_{11}(v11) = 2/(1 + \exp(-v_{11}/q_{11})) - 1 \quad (9)$$

$$f_{12}(v12) = 2/(1 + \exp(-v_{12}/q_{12})) - 1 \quad (10)$$

$$f_{21}(v21) = 2/(1 + \exp(-v_{21}/q_{21})) - 1 \quad (11)$$

as output functions of the units, where q's are parameters of the output functions. Then, Eqs.(9) ~ (11) can be represented by the following fuzzy systems, respectively.

$$f_{11}(v_{11}) = (w_{111}(k)\, g_{111} + w_{112}(k)\, g_{112})\, v_{11} / (w_{111}(k) + w_{112}(k)), \quad (12)$$

$$f_{12}(v_{12}) = (w_{121}(k)\, g_{121} + w_{122}(k)\, g_{122})\, v_{12} / (w_{121}(k) + w_{122}(k)), \quad (13)$$

$$f_{21}(v_{21}) = (w_{211}(k)\, g_{211} + w_{212}(k)\, g_{212})\, v_{21} / (w_{211}(k) + w_{212}(k)), \quad (14)$$

where

$$g_{111} = \min_v f'_{11}(v) = 0, \qquad g_{121} = \min_v f'_{12}(v) = 0, \qquad g_{212} = \min_v f'_{21}(v) = 0,$$

$$g_{112} = \max_v f'_{11}(v) = 0.5/q_{11}, \quad g_{122} = \max_v f'_{12}(v) = 0.5/q_{12}, \quad g_{212} = \max_v f'_{21}(v) = 0.5/q_{21}.$$

Next, assume that $w_{111} = 1$, $w_{112} = -1$, $w_{121} = -0.5$, $w_{111} = 1$, $w_{122} = -0.5$, $w_{211} = 1$, $w_{212} = 1$ and $q_{11} = q_{12} = q_{21} = 0.25$ ($g_{112} = g_{122} = g_{212} = 2.0 = g_2$, $g_{111} = g_{121} = g_{211} = 0 = g_1$). We can derive Eq.(15) from Eqs.(5) ~ (8) and Eqs.(12) ~ (14).

$$x(k+1) = \frac{\sum_{i,j,\kappa=1}^{2} w_{11i}(k) w_{12j}(k) w_{21\kappa}(k) \{ g_\kappa (g_i w_{211} w_{111} + g_j w_{212} w_{112}) x(k) + g_\kappa (g_i w_{211} w_{121} + g_j w_{212} w_{1222}) x(k-1) \}}{\sum_{i,j,\kappa=1}^{2} w_{11i}(k) w_{12j}(k) w_{21\kappa}(k)} \quad (15)$$

From Eq.(15), ***Edge Matrices*** are obtained as follows.

$$\mathbf{A}_1 = \begin{bmatrix} 0 & 0 \\ 1.0 & 0 \end{bmatrix} \quad \mathbf{A}_2 = \begin{bmatrix} -2.0 & -1.0 \\ 1.0 & 0 \end{bmatrix} \quad \mathbf{A}_3 = \begin{bmatrix} 2.0 & -1.0 \\ 1.0 & 0 \end{bmatrix} \quad \mathbf{A}_4 = \begin{bmatrix} 0 & -2.0 \\ 1.0 & 0 \end{bmatrix}$$

Fig.5 shows PR of this network. There does not exist a common positive definite matrix P since $\mathbf{A}_2$, $\mathbf{A}_3$ and $\mathbf{A}_4$ are not stable matrices. Fig.6 shows behavior of this neural network. As shown in Fig.6, this system is not asymptotically stable in the large. In the case of unstable neural systems, behavior of neural networks becomes limit cycle due to saturation of output of units, that is, $f(v) \in [-1,1]$ for all v.

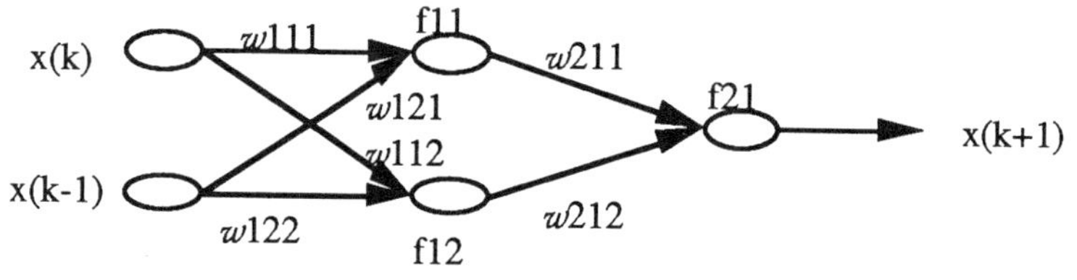


Fig.4 Three-layer neural network with two inputs

**[Example 4.2]**

In Example 4.1, assume that $q_{11} = q_{12} = q_{21} = 0.5$ ($g_{112} = g_{122} = g_{212} = 1.0$, $g_{111} = g_{121} = g_{211} = 0 = g_1$). Then, we can obtain the following ***Edge Matrices.***

$$\mathbf{A}_1 = \begin{bmatrix} 0 & 0 \\ 1.0 & 0 \end{bmatrix} \quad \mathbf{A}_2 = \begin{bmatrix} -1.0 & -0.5 \\ 1.0 & 0 \end{bmatrix} \quad \mathbf{A}_3 = \begin{bmatrix} 1.0 & -0.5 \\ 1.0 & 0 \end{bmatrix} \quad \mathbf{A}_4 = \begin{bmatrix} 0 & -1.0 \\ 1.0 & 0 \end{bmatrix}$$

Fig.5 shows PR of this network. It is found form Theorem 3.2 that there does not exist a common positive definite matrix **P** since $\mathbf{A}_1\mathbf{A}_2$ is not a stable matrix. Fig.6 shows behavior of this network.

**[Example 4.3]**

In Example 4.1, assume that $q_{11} = q_{12} = q_{21} = 0.75$ ($g_{112} = g_{122} = g_{212} = 0.67$, $g_{111} = g_{121} = g_{211} = 0 = g_1$). Then, we can obtain the following ***Edge Matrices***.

$$\mathbf{A}_1 = \begin{bmatrix} 0 & 0 \\ 1.0 & 0 \end{bmatrix} \quad \mathbf{A}_2 = \begin{bmatrix} -0.44 & -0.22 \\ 1.0 & 0 \end{bmatrix} \quad \mathbf{A}_3 = \begin{bmatrix} 0.44 & -0.22 \\ 1.0 & 0 \end{bmatrix} \quad \mathbf{A}_4 = \begin{bmatrix} 0 & -0.44 \\ 1.0 & 0 \end{bmatrix}$$

Fig.5 shows PR of this network. If we select

$$\mathbf{P} = \begin{bmatrix} 2.42 & -0.19 \\ -0.19 & 1.12 \end{bmatrix}$$

as a common positive definite matrix **P**, then $\mathbf{A}_i^T \mathbf{P} \mathbf{A}_i - \mathbf{P} < \mathbf{0}$ for i=1, ⋯ , 4. Therefore, this neural network system is stable. The **P** matrix was found by using the method proposed in the literature [3]. Fig.6 shows behavior of this network.

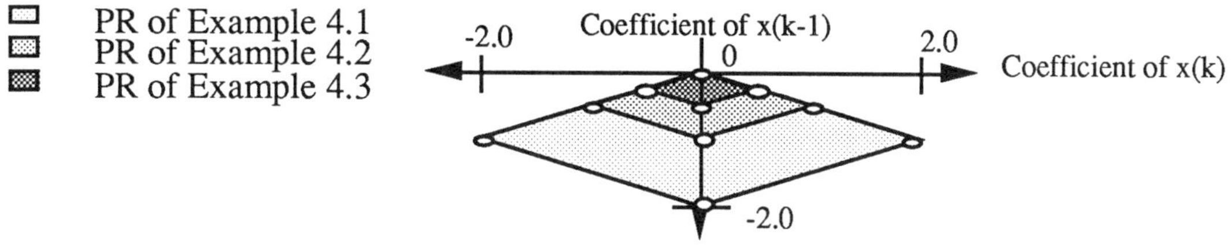


Fig.5 PR representation of Examples 4.1 , 4.2 and 4.3

We can obtain some properties for stability of neural networks from Examples 4.1, 4.2 and 4.3.

**[Property 1]**

Stability of neural networks is strongly related to the connection weights and the maximal gradient values $g_2$ of sigmoid functions. The eigenvalues of ***Edge Matrices*** are closer to origin of z-plane when the connection weights or $g_2$ is closer to 0. As shown in Fig.6, therefore, the behavior of neural networks becomes stable if $g_2$ is closer to 0.

**[Property 2]**

The situation where the connection weights or $g_2$ is closer to 0 means that the area of PR becomes smaller as shown in Fig.5. The area of PR is related to nonlinearity represented by neural networks. Of course, nonlinearity represented by neural networks is stronger when the area of PR becomes larger. Generally speaking, there is a contrary relation between guarantee of stability and degree of nonlinearity (area of PR).

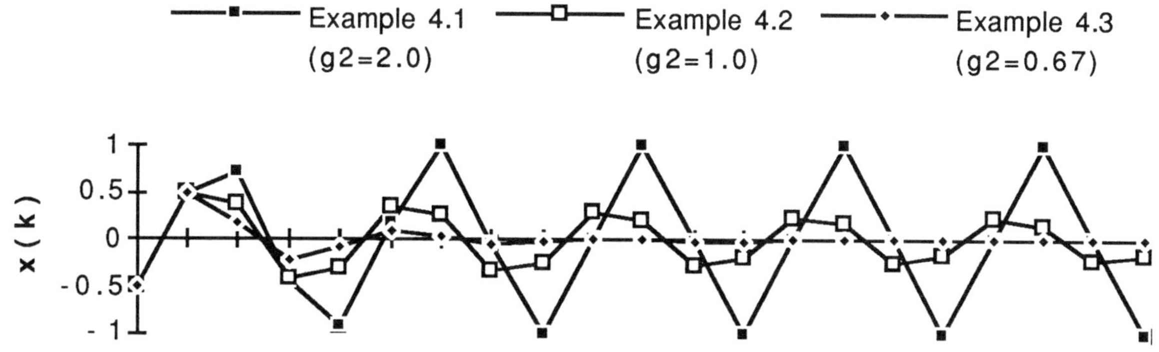


Fig.6 Behavior of neural networks of Examples 4.1, 4.2 and 4.3

**[Example 4.4]**

Let us consider stability of a neural network shown in Fig.7. Assume that $w_{111} = w_{112} = w_{121} = 0.3$, $w_{122} = -0.3$, $w_{131} = 0.2$, $w_{132} = -0.2$, $w_{211} = w_{212} = 1$, and $q_{11} = q_{12} = q_{21} = 0.5$ ($g_{112} = g_{122} = g_{212} = 1.0 = g_2$, $g_{111} = g_{121} = g_{211} = 0 = g_1$). From Fig.7, we obtain

$$x(k+1) = \sum_{i,j,\kappa=1}^{2} w_{11i}(k)w_{12j}(k)w_{21\kappa}(k) \Big\langle g_\kappa(g_i w_{211} w_{111} + g_j w_{212} w_{112})x(k) + g_\kappa(g_i w_{211} w_{121} + g_j w_{212} w_{122})x(k-1) + g_\kappa(g_i w_{211} w_{131} + g_j w_{212} w_{132})x(k-2) \Big\rangle \Big/ \sum_{i,j,\kappa=1}^{2} w_{11i}(k)w_{12j}(k)w_{21\kappa}(k) \quad (16)$$

in the same manner as Example 4.1. From Eq.(16), ***Edge Matrices*** are obtained as follows.

$$\mathbf{A}_1 = \begin{bmatrix} 0 & 0 & 0 \\ 1.0 & 0 & 0 \\ 0 & 1.0 & 0 \end{bmatrix} \quad \mathbf{A}_2 = \begin{bmatrix} 0.3 & -0.3 & -0.2 \\ 1.0 & 0 & 0 \\ 0 & 1.0 & 0 \end{bmatrix} \quad \mathbf{A}_3 = \begin{bmatrix} 0.3 & 0.3 & 0.2 \\ 1.0 & 0 & 0 \\ 0 & 1.0 & 0 \end{bmatrix} \quad \mathbf{A}_4 = \begin{bmatrix} 0.6 & 0 & 0 \\ 1.0 & 0 & 0 \\ 0 & 1.0 & 0 \end{bmatrix}$$

In this case, PR can be represented as a polyhedron since the number of state variables is 3. If we select

$$\mathbf{P} = \begin{bmatrix} 4.20 & -0.24 & -0.29 \\ -0.24 & 2.72 & 0.31 \\ -0.29 & 0.31 & 1.17 \end{bmatrix}$$

as a common **P**, then

$$\mathbf{A}_i^T \mathbf{P} \mathbf{A}_i - \mathbf{P} < \mathbf{0}$$

for $i=1, \cdots, 4$. Therefore, this neural network system is stable.

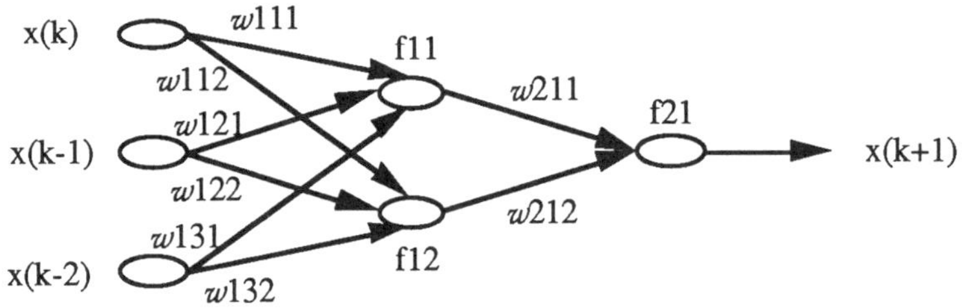


Fig.7 Three-layer neural network with three inputs

We can derive a property for stability by comparing Eq.(15) with Eq.(16).

**[Property 3]**

When the number of units in hidden layer or the number of layers does not change, the number of consequent matrices does not increase even if the number of state variables, that is, inputs of neural networks, increases. However, it exponentially increase if the number of units in hidden layer or the number of layers increases.

## 5. Conclusions

We have considered stability of neural networks using Tanaka-Sugeno's Theorem. The PR (Parameter Region) representation has been proposed by introducing new concepts of *Edge Rule (Matrix)* and *Minimum Representation*. Finally, we have illustrated stability criterion of some neural networks using the PR representation method and derived some properties for stability of neural networks. The stability analysis discussed in this paper can be applied not only to neural networks but also to other nonlinear systems if the nonlinear systems can be approximated by fuzzy systems.

### References


[1]K.Tanaka and M.Sugeno:Stability Analysis of Fuzzy Systems using Lyapunov's Direct Method, proceedings of NAFIPS'90, pp.133-136 (1990).

[2]K.Tanaka and M.Sugeno:Stability and Design of Fuzzy Control Systems, FUZZY SETS AND SYSTEMS 45, no.2, pp.135-156 (1992),

[3]K.Tanaka and M.Sugeno:Stability Analysis of Fuzzy Systems Using Lyapunov's Direct Method and Construction Procedure for Lyapunov Functions, 6th Fuzzy System Symposium pp.353-356 (1990). (in Japanese)

[4]S.Kawamoto et al.:An Approach to Stability Analysis of Second Order Fuzzy Systems, IEEE International Conference on Fuzzy Systems, pp.1427-1434 (1992).

# Processing of Incomplete Fuzzy Data Using Artificial Neural Networks

Marek J. Patyra and Taek Mu Kwon
Department of Computer Engineering
University of Minnesota, Duluth
271 MWAH, 10 University Dr.
Duluth, MN 55812

***Abstract*— In this paper, a degenerated fuzzy-number processing system based on artificial neural networks (ANN) is introduced. The digital representation of fuzzy numbers is assumed, where the universe of discourse is discretized into $n$ equally divided intervals. The representation of the membership function values is transformed into binary quantized values which have its maximum at $2^m - 1$ where $m$ is the number of data bits used in the system. It is proposed that fuzzy number processing be performed in two basic stages. The first stage performs the retrieval of fuzzy data consisting of degenerated fuzzy numbers, and the second stage performs the desired fuzzy operations on the retrieved data. The method of incomplete fuzzy-number retrieval is proposed based on an ANN structure which is trained to estimate the missing membership function values.**

Keywords: *fuzzy number processing, artificial neural network, approximation*

## I. Introduction

Signal processing using a fuzzy approach has become more attractive during the last few years when fuzzy sets and tools have been applied successfully to a variety of tasks [1-3]. These tasks cover different areas of applications from speech and image processing to various pattern classifications. Although the early stages of fuzzy signal processing mainly involving pattern recognition [2-3] have been successfully developed, fuzzy methods for data processing such as operations on noisy fuzzy numbers are yet to be developed. Recent theoretical developments on fuzzy arithmetics [4], such as fuzzy convolution and deconvolution, address several aspects of dealing with noisy fuzzy numbers, but practical applications of such operations have not been studied. In fuzzy-number processing systems, the major functions are performed by fuzzy processing elements like *min, max, bounded, absolute difference*, etc., which can be connected in different ways (for instance, *min-max-min*), depending on the desired structure. The fuzzy-number processing system may be implemented using standard or specialized software, but real time applications require implementation in hardware. Building a fuzzy-number processing system is not trivial in practice, because much of incoming data to the system (e.g. in control systems) is often incomplete (e.g., disturbed, noisy, or damaged). As a result, the outputs generated by the system can be incorrect or can contain unacceptable errors that may cause a series of problems. In this paper, we introduce a system that can effectively handle fuzzy operations on degenerated fuzzy numbers that are heavily damaged or have missing membership function values. This system significantly reduces the computational error of fuzzy operations resulting from degenerated fuzzy numbers.

The ANN (Artificial Neural Network) realization of the fuzzy operations related to addition, subtraction, multiplication, division, minimum, and maximum was developed in [5]. It was noted, however, that good results (in terms of average error) of fuzzy operations using ANN can be obtained only when the fuzzy operations are performed on nondegenerated fuzzy numbers. In contrast, the fuzzy operations on degenerated fuzzy-numbers using ANN result in large errors. In this paper, we propose a two-stage fuzzy-number processing system that drastically reduces such computational errors. In this new system, the first stage performs retrieval of the incomplete fuzzy-numbers, and the second stage processes the desired fuzzy operation using the retrieved fuzzy numbers.

This paper is organized in the following way. In Section II, the theoretical background for the retrieval of degenerated fuzzy-numbers is established. Section III discusses the ANN implementation of the fuzzy-number retrieval system. Section IV presents the ANN-based fuzzy-number processing system and its applications. Finally, simulation results of the proposed system are presented in Section V, followed by the conclusion.

## II. Fuzzy-Number Retrieval

We begin the discussion of fuzzy-number retrieval with the definition of the fuzzy number [4,6].

**Definition 1.A:** A fuzzy number $A$ is a normalized convex fuzzy set on the real line $R$ such that:

$$\max_{x_i \in X} \mu_A(x_i) = 1 \tag{1}$$

where $\mu_A(x_i)$ denotes the membership function of $x_i$ in $A$, $X$ is the support region of the fuzzy number $A$, and $x_0$ is referred to as the mean value of $A$ if $\mu_A(x_0) = 1$.

A fuzzy number can be also defined on discrete spaces.
**Definition 1.B:** Any fuzzy number $A$ can be described in a finite domain $\{x_i\}$, by

$$A = \sum_{i=1}^{n} \frac{\mu_A(x_i)}{x_i} \tag{2}$$

where $i = 1, \cdots, n$ and $n$ defines the number of equal intervals into which the fuzzy number $A$ is discretized and $\sum$ denotes the union operator.
Based on the Definition 1B, there exists a mean value for the ordinary fuzzy number, so that (2) can be rewritten separately for the left and right intervals around $x_0$ as follows:

$$A = \sum_{i=1}^{k} \frac{\mu_A(x_i)}{x_i} + \frac{1}{x_0} + \sum_{i=k+2}^{n} \frac{\mu_A(x_i)}{x_i}. \tag{3}$$

This representation is called the "discrete representation" of a fuzzy number. A special case of this discrete representation, which is referred to as the digital representation, is commonly used in current applications of fuzzy technology. In this representation, the universe of discourse is discretized into $n$ equally divided intervals, each one representing a "bit" in analogy with the digital representation of a crisp number.

**Definition 2:** The degenerated fuzzy number $Y'$ is the fuzzy number with missing membership-function values $\mu_{Y'}(x_i)$ on one or more bit positions [1] (see Fig. 1).

The degenerated fuzzy number $Y'$ and its retrieval system is defined in the following definition.

**Definition 3:** The fuzzy-number retrieval system is defined by a triplet: $(Y', Y'', \rho)$, where $Y'$ is the degenerated fuzzy number, $Y''$ is the retrieved fuzzy number, and $\rho$ is the retrieval function:

$$\rho : \mu_{Y'}(x_i) \longrightarrow \mu_{Y''}(x_i) \ \forall \ x_i \in X \tag{4}$$

where $\longrightarrow$ represents a mapping relation.

Hereafter we use the simplified notation: $\mu_Y(x_i) = \mu_i^Y$ representing the membership function value of $Y$ at the bit position $i$.

**Definition 4:** The fuzzy number retrieval function is defined by

$$\rho(\mu_i^{Y'}) = \mu_i^{Y''} \simeq \mu_i^Y \ \ \forall \ i \tag{5}$$

where $Y$ is the original fuzzy number which is free of missing bits (see Fig. 2).

The characteristics of the retrieval function depend on the particular application. The implementation of a linear approximation technique, which seems reasonable for most of fuzzy-data processing applications, is proposed.
Suppose that a membership function has $k$ consecutive missing values at the bit positions: $x_{s+1}, \cdots, x_{s+k}$. These consecutive missing values will be referred to as the missing segment of the membership function. In general, a degenerated fuzzy number may have multiple missing segments. We propose to interpolate each missing segment according to a linear approximation based on the nearest available bits surrounding this segment. If these bits are at $x_s$ and $x_{s+k+1}$, then the approximation is derived as:

$$\mu_{s+i}^{Y'} = \frac{\mu_{s+k+1}^{Y'} - \mu_s^{Y'}}{x_{s+k+1} - x_s} x_{s+i} - \frac{x_s \mu_{s+k+1}^{Y'} - x_{s+k+1}\mu_s^{Y'}}{x_{s+k+1} - x_s} \tag{6}$$

for $i = 1, \cdots, k$. If the missing segment includes the mean value, the mean value is first estimated using a simple linear extension of the available bits. The rest of missing bits are then linearly approximated as (6) by using the mean value as one of the surrounding bits. The linear approximation is further smoothed and quantized into the digital representation using the following relation:

$$\mu_k^{Y''} = int\left(\frac{\mu_{k-1}^{Y'} + 2\mu_k^{Y'} + \mu_{k+1}^{Y'}}{4}\right) \simeq \mu_k^Y \tag{7}$$

where the $int(\cdot)$ function denotes the quantization to the nearest digitized level. This retrieval relation is illustrated is Fig. 3.

## III. Neural Network Implementation of Fuzzy-Number Retrieval System

As it is proven in [9], any continuous function can be uniformly approximated by a feed-forward ANN structure

[1] A special case of the degenerated fuzzy number with missing membership function values for all bits is not discussed in this paper

with one hidden layer, given that the activation function of each node is continuous and nondecreasing. This characteristic can be utilized to approximate the missing membership-function values discussed in Section II.

Fig. 4. illustrates the structure of the ANN to be utilized for the fuzzy-number retrieval system. The ANN consists of an input layer (I), a hidden layer (H) and an output layer (O). Each layer is completely connected, meaning that each node from one layer is connected to all nodes of the next layer, and each connection is associated with a single weight.

The problem now can be reformulated into finding a set of weights that minimize the total error between the linear approximation and the actual output of the ANN network. The output of the ANN, which represents the retrieved membership-function value at the $i$th bit of the degenerated fuzzy number $Y'$, can be described as:

$$\mu_i^{Y''} = g\left(\sum_{j=1}^{m} V_{ij}\, g\left(\sum_{k=1}^{n} \mu_k^{Y'} w_{jk}\right)\right) \tag{8}$$

where $g(\cdot)$ denotes the sigmoid function [10], $V_{ij}$ represents the weight between the $i$th node in the output layer and the $j$th node in the hidden layer, while $w_{jk}$ represents the weight between the $j$th node in the hidden layer and the $k$th node in the input layer. In the actual implementation, we selected the number of nodes in each layer be *equal* to the number of bits in the fuzzy number (i.e. $n = m$) as shown in Fig. 4. A typical error function can then be formulated as [10]:

$$2\epsilon = \sum_{Y}\sum_{i=1}^{n}\left[\mu_i^{Y} - g\left(\sum_{j=1}^{m} V_{ij}\, g\left(\sum_{k=1}^{n} \mu_k^{Y'} w_{jk}\right)\right)\right]^2 \tag{9}$$

where $\mu_i^Y$ is the desired output of the membership function value at the $i$th bit of the degenerated fuzzy number $Y'$ and $\sum_Y$ denotes summation of the error over all the fuzzy numbers used for training.

The error function in (9) is absolutely continuous and differentiable function of weights, so the matrices $[V_{ij}]$ and $[w_{jk}]$ can be derived by minimizing the error using the common backpropagation training method [10]. Regarding the training set of the network, well represented fuzzy membership functions for the given application should be selected. In most cases, the desired form of each fuzzy number is known, since the membership functions are usually predetermined by a human expert. In this case, the input part of the training set is obtained from randomly degenerating the available fuzzy numbers. However, if we have only the degenerated fuzzy-number information and no information regarding the original membership functions, the desired output part of the training patterns must be obtained through an approximation method such as the linear estimation technique discussed in Section II.

## IV. Fuzzy-Number Processing System Based on ANN and its Applications

The proposed fuzzy-number processing system consists of two stages, the first stage performing fuzzy-number retrieval (described in Section III), and the second stage performing basic fuzzy operations. The architecture is shown in Fig. 5 where both stages are implemented using ANNs. Clearly, the main goal of the first stage is to obtain the retrieved fuzzy numbers and to make them available to the fuzzy data-bus for further processing. In the present system, the second-stage networks are designed to perform addition, subtraction, multiplication, division, maximum and minimum operations, which are required for basic fuzzy-data processing systems.

To illustrate the practical aspects of the above system, we discuss an application. Consider a control system of a factory environment that consists of several local machines and a central controller. Each local machine is connected to a local controller that monitors, measures, or controls the machine and reports the results to the central controller. The central controller distributes the necessary control signals to local controllers. We assume that the distance of the communication channel between the local machines and the central controller is sufficiently long to introduce noise. In traditional distributed-control systems, the local measurements are sampled through an A/D converter and transmitted to the central controller as a digital form of data. In this control method, missing a part of data in the communication channel can result in critical errors that may damage the whole system. If we consider the measurement and encoding errors of the local controller, the probability of the central-controller error can amount even further. In the proposed fuzzy system, the measurement itself is considered fuzzy (imprecise and noisy). Thus the local measurements are directly transformed into appropriate fuzzy numbers in contrast to the crisp numbers of the traditional systems. The fuzzy numbers are then transmitted over a noisy communication channel to the central controller. In the central controller, the proposed fuzzy-number processing system is implemented. Since the proposed system can recover the partially missing membership-function values through the retrieval function implemented in the first stage, the probability of critical control mistake induced by the communication channel can be tremendously reduced. In fact, our simulation study (Section V) shows that fuzzy numbers destroyed up to 50% still produce a very small error in the final fuzzy operations.

It should be mentioned that the proposed system is easy to implement either using a dedicated digital system or in combination with analog circuits. Moreover, since the system does not require training while in operation, the processing time is simply equal to the forward delay of the system. Thus the proposed fuzzy-number processing system is very attractive for real time distributed control applications.

## V. Computer Simulation Results

The testing data for the training of the retrieval ANN are generated based on triangular fuzzy numbers. Different types of testing fuzzy numbers are created (a) by changing the mean value, and (b) by varying the bit-width of the fuzzy number. After creation of various types of fuzzy numbers, the membership-function values are deleted up to 75% per each fuzzy number, resulting in a set of degenerated fuzzy numbers. The total number of fuzzy numbers used for training is 1024. Each degenerated fuzzy number becomes an input pattern of the ANN, and the desired output pattern is derived using the linear approximation discussed in Section II. In order to make the system be more practical, we do not use the error-free fuzzy numbers in the training of the ANN. The error-free fuzzy numbers are only used in the error measurement of the network performance. The PlaNet simulator [8] was used to train the ANN. The error is measured as the mean squared error between the target and actual output. The training is terminated when the mean squared error computed over all patterns used for training is less than 0.0001

Fig. 6. illustrates a fragment of the fuzzy-data processing system, extracted from the original design, including the retrieval stage. The values of membership function are coded in forms of sequences of squares. The area of a single square for a specific bit relates to the membership-function value in such a way that the largest square represents 1 and the smallest 0.1 (the empty place indicates 0). Two 32-bit long fuzzy numbers are set to Input1 and Input2. In this example, the fuzzy number on Input1 is degenerated, i.e. missing membership function values for two bits. The numbers are then processed in the retrieval ANN, i.e. Hidden1 and Hidden2, and the retrieved numbers are displayed in the Hidden3 layer. Notice that the missing membership-function values are completely recovered at the Hidden3 layer. Then, these two retrieved fuzzy numbers are processed in the subsequent layers (Hidden4 (64-bit), Hidden5 (64-bit)), producing the result of operation (in this case, addition) at the Output (64-bit).

The Table summarizes the results of mean squared errors of fuzzy-number arithmetics obtained from the error free fuzzy numbers, degenerated fuzzy numbers, and ANN trained fuzzy numbers. This error is computed over the training set consisting of 1024 fuzzy numbers. As one can see, the errors obtained from the fuzzy arithmetics using direct degenerated fuzzy numbers are eleven to thirty times greater than that of the arithmetics on error free fuzzy numbers. By including the ANN retrieval stage and training the network, the average error is significantly reduced (six to seventeen times less error after adding the retrieval stage) as shown in the table.

With respect to the percentage of missing membership-function values, the following results were obtained. When up to 30% of membership-function values are missing, the average error is slightly increased. When the missing values are increased from 30% to 50%, the average error is increased one order of magnitude higher. Finally, when the missing values are increased to the range of 50% to 75%, the error is sharply increased, i.e. ten to one hundred times greater than that of the error free data.

## VI. Conclusion

In this paper, the implementation of a fuzzy data processing system that is suitable for processing degenerated fuzzy-numbers is introduced. The system is implemented based on two cascading stages of ANNs. Upon completion of the ANN training, the first stage effectively recovers the membership functions from incomplete information. In the second stage, basic fuzzy operations i.e. addition, subtraction, multiplication, division, maximum, and minimum are implemented. These operations support the basic fuzzy modeling requirements. This architecture significantly improves the performance of fuzzy arithmetics on degenerated fuzzy numbers. More specifically, the average computation error of the proposed two-stage ANN structure is up to seventeen times less than that of the direct processing of the same degenerated-fuzzy numbers. It is obvious that allowing the degenerated fuzzy-numbers to enter a normal fuzzy processing system deteriorates the performance and produces output far away from the desired result. On the other hand, the proposed system effectively maintains its performance for the degenerated fuzzy numbers.

One should also notice that an additional advantage of the proposed ANN-based fuzzy-number processing system is in the processing time. Since the network is trained ahead of time, the only processing time required during the actual fuzzy arithmetics is the propagation delay of the feed-forward network. Hence, this data-processing system is fast and attractive in real-time applications with heavily damaged data or in applications where complete fuzzy numbers are not provided.

## References

[1] R. R. Yager and L. A. Zadeh eds., *An Introduction to Fuzzy Logic Applications in Intelligent Systems,* Norwell, MA: Kluwer Academic Publishers, 1992.

[2] S. K. Pal and D.K.D. Majumder, *Fuzzy Mathematical Approach to Pattern Recognition,* New York, NY: John Wiley and Sons, 1986.

[3] S. Miyamoto, *Fuzzy Sets in Information Retrieval and Cluster Analysis,* Norwell, MA: Kluwer Academic Press, 1990.

[4] A. Kaufmann and M. Gupta, *Introduction to Fuzzy Arithmetic: Theory and Applications,* New York, NY: Van Nostrand Reinhold Company, 1984.

[5] M. J. Patyra, "Implementation of fuzzy operations with neural network," *Proc. of the IEEE Conference on Fuzzy and Neural Systems and Vehicle Applications'91,* Tokyo, Japan, Nov. 1991.

[6] H. J. Zimmermann, *Fuzzy Set Theory and its Application,* 2nd Ed., Norwell, MA: Kluwer Academic Publishers, 1991.

[7] H. Takagi and I. Hayashi, "NN driven fuzzy reasoning," *International Journal of Approximate Reasoning,* pp. 191-212, 1991.

[8] Y. Miyata, "PlaNet User's Guide," Computer Science Department, University of Colorado, Boulder, CO, 1991.

[9] G. Cybenko, "Approximation by superposition of a sigmoid function," *Mathematics of Control, Signals and Systems,* pp. 303-314, 1989.

[10] D. E. Rumelhart, G. E. Hinton, and R. J. Williams, "Learning internal representations by error propagation," in *Parallel Distributed Processing,* vol. 1, D. E. Rumelhart, J. L. McClelland, and PDP research group, Eds., Cambridge, Mass: MIT Press, pp. 318-362, 1986.

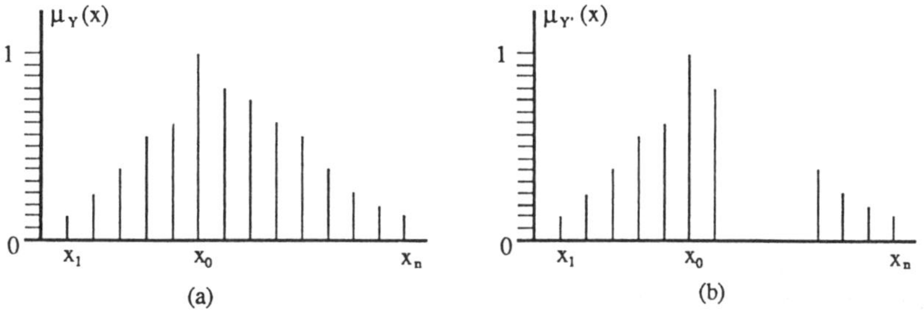


Fig. 1. (a) Digital representation of an ordinary fuzzy number; (b) an example of the corresponding degenerated fuzzy number.

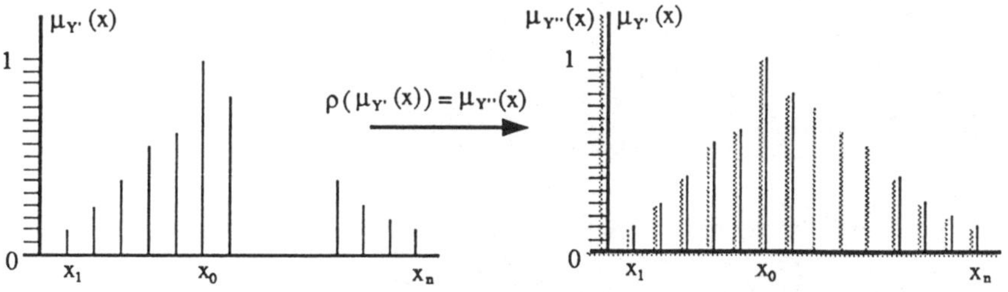


**Fig. 2. Interpretation of the definition of fuzzy retrieval system.**

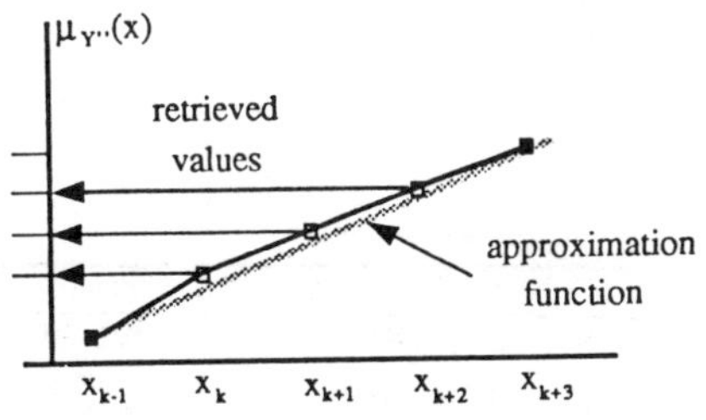


**Fig. 3. Example of membership function retrieval by linear approximation.**

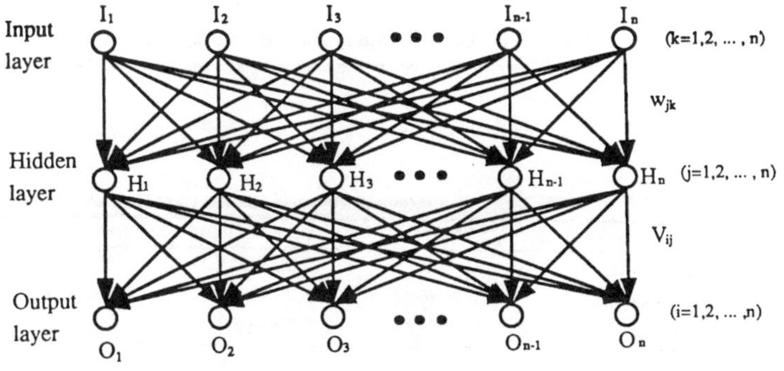


**Fig. 4. Structure of the ANN.**

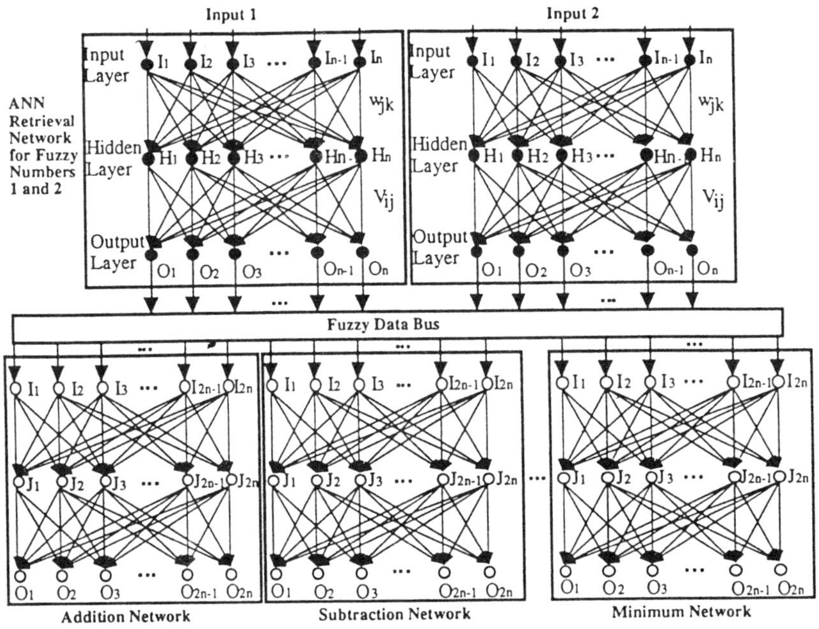


Fig. 5. General architecture of fuzzy data processing system using ANN.

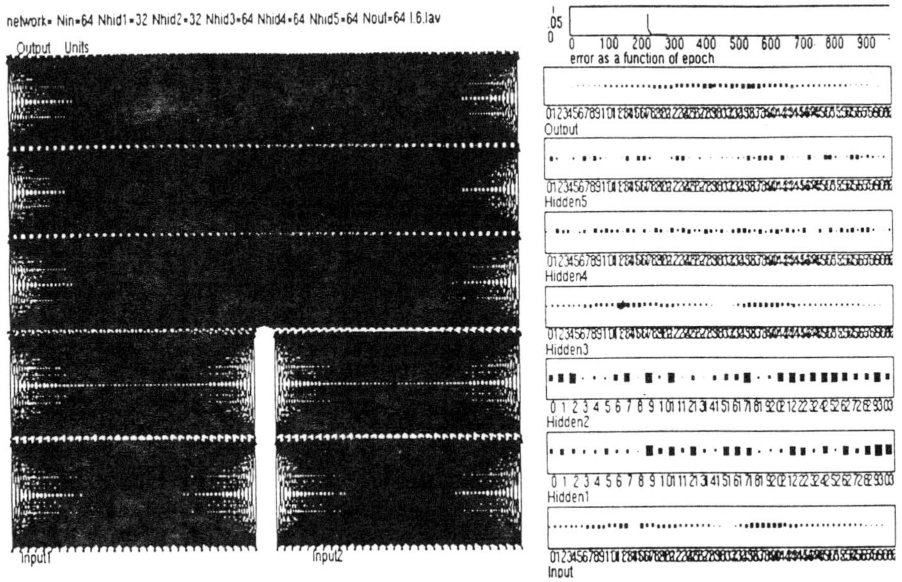


Fig. 6. Example of the fuzzy data processing system applying fuzzy data retrieval processor and the network performing fuzzy addition. At the top-right, the trace of average error is shown.

TABLE

COMPARISON OF AVERAGE ERRORS: ERROR FREE, DEGENERATED, AND TRAINED FUZZY NUMBER.

| FUZZY NUMBER | ADDITION | SUBTRACTION | MULTIPLICATION | DIVISION | MAXIMUM | MINIMUM |
|---|---|---|---|---|---|---|
| ERROR FREE | 0.000353 | 0.000335 | 0.000395 | 0.000219 | 0.000382 | 0.000331 |
| DEGENERATED | 0.004032 | 0.005630 | 0.009541 | 0.007680 | 0.006516 | 0.005766 |
| TRAINED | 0.000643 | 0.000761 | 0.000577 | 0.000867 | 0.000742 | 0.000522 |

# Design of Stable Fuzzy-logic-controlled Feedback Systems

*P.A. Ramamoorthy and Song Huang*
Department of Electrical & Computer Engineering,
University of Cincinnati, M.L. #30
Cincinnati, Ohio 45221-0030
FAX: (513) 556-7326; TEL: (513) 556-4757
Email: pramamoo@babbage.ece.uc.edu

## ABSTRACT

Recently, there is great deal of interest in the use of fuzzy expert systems in control applications. Controllers based on fuzzy logic belong to the class of static or memoryless nonlinear controllers and provide better control than is possible using linear control. The major strength of fuzzy controllers lies in the way a nonlinear output mapping of a number of inputs can be specified easily using fuzzy linguistic variables and fuzzy rules.

When we use fuzzy logic in feedback control, we in effect have a nonlinear system with feedback and the resulting system could potentially become unstable. In this work, we show how we can design feedback systems (with fuzzy control) that are guaranteed to be stable. This approach can also be used to design stand-alone feedback fuzzy systems (or recurrent fuzzy systems) that are guaranteed to be stable.

## INTRODUCTION

Linear control is a most commonly used method with a variety of techniques and has been successfully applied in a number of industrial applications. The reasons for their continued acceptance/use are that linear systems and linear controls are easier to characterize analytically and issues such as stability, performance and robustness can be analyzed and quantified accurately. However, linear controls have a number of limitations. First they rely on the linear model assumptions being valid. When this assumption is violated, the system performance can degrade and the system can even become unstable.

Linear control is the product of the technologies (or lack of it) of some two decades ago. Since then, the technology has changed rapidly and the complexity of systems to be controlled have increased dramatically. Thus, the need for nonlinear controllers have increased along with the feasibility for their cost-effective implementation.

Controllers based on fuzzy expert systems can be considered as a class of static (memoryless) nonlinear controllers [1,2] In the case of fuzzy controllers, the input(s) to output(s) mapping (in general nonlinear) is achieved through the following three steps:

a) fuzzification or input fuzzy set(s) activation from the crisp input sensor(s) values;

b) output fuzzy set(s) selection from the input fuzzy sets activated and the fuzzy rule base;

c) defuzzification or crisp output(s) values calculation based on the output fuzzy sets selected, the membership function values corresponding to the input(s) values and the membership functions corresponding to the output fuzzy sets selected.

The importance of fuzzy logic for control applications arise from the fact that a proper nonlinear mapping that would lead to a superior controller performance can be described easily using fuzzy linguistic variables and fuzzy rules.

Feedback architecture is the most preferred architecture (Figure 1) whether we use linear or nonlinear/fuzzy controllers due to reasons such as reduced sensitivity to plant coefficient variations and so on. The stability becomes an important issue in such feedback architectures. There is lot of research/literature on the stability of linear feedback systems. However, there are no known results that establish mathematically the stability of nonlinear fuzzy controller based feedback systems. Hence, there is hesitancy on the use of nonlinear/fuzzy controllers in applications where human safety etc. are involved. In this paper, we arrive at a new feedback architecture using fuzzy controllers. We will show that such an architecture will remain stable regardless of the fuzzy controller mapping used. This method can also be used to design stand-alone feedback fuzzy systems or recurrent fuzzy systems that are guaranteed to be stable.

## THEORY

An impressive number of tools for the analysis of nonlinear systems and techniques for controller design for nonlinear systems have been developed [3-6]. However, all of these techniques are analytically motivated (defining general form of nonlinear dynamical differential equations for representing systems and placing restrictions to satisfy stability etc.). However since nonlinear differential equations are highly complicated to be amenable for analytical approach, restrictions have been placed on the types of systems that can be analyzed and the types of controllers that can be employed etc.

Our approach for nonlinear system/controller design (that is equally applicable for design of feedback systems with fuzzy controllers and the thrust of this paper) is based on an entirely different paradigm. The proposed approach is based on *"Engineering"* and *"Reverse-Engineering"* philosophies as opposed to analytical/mathematical points of view adapted by earlier researchers. An engineering or physically motivated approach tries to take into consideration physical properties and constraints-based-on-physical-properties at every stage of design. Passivity formulation (to be defined shortly) is one such candidate that we use in our work. The term "Reverse-Engineering" is used here to imply learning from an existing system (not necessarily of the same kind) and plays an important role in our work. Thus, to design a nonlinear fuzzy controller, we force the closed-loop system dynamics to mimic[1] the dynamics of another system that is guaranteed to be stable, a process or technique similar to reverse engineering. Assuming such a system exists, making the closed-loop system mimic that system is indeed possible since we have the flexibility in the choice of the controller. We have used *passivity* concepts[2] to arrive at such a stable system, *a passive nonlinear electrical network.* The passivity concepts are used to first invent two new passive nonlinear electrical devices[3]: nonlinear transformers (2-port device) and multi-port gyrators. A proper interconnection of such elements with dynamical elements (capacitors and inductors) provides the target stable system. The controller dynamics are then chosen such that the closed-loop system dynamics mimic the dynamics of that stable electrical network.

Let us now illustrate the application of this approach to the design of fuzzy controllers using Figure 2. In Figure 2A, we show the classical architecture whose stability is open to question. In Figure 2B, we show the general architecture based on the passive nonlinear network approach and a nonlinear controller with a first-order dynamics[4]. Assuming that the plant can be represented by a 2nd-order transfer function with two state variables $y, \dot{y} = y_1$, the dynamics of the close-loop system can be written as

$$\begin{bmatrix} \dot{y} \\ \dot{y}_1 \\ \dot{k} \end{bmatrix} = \begin{bmatrix} 0 & 1 & 0 \\ -a & -b & F(e,\dot{e}) \\ 0 & \frac{-F(e,\dot{e})}{a} & 0 \end{bmatrix} \begin{bmatrix} y \\ y_1 \\ k \end{bmatrix} - \begin{bmatrix} 0 \\ 0 \\ f(k) \end{bmatrix} + \begin{bmatrix} 0 \\ 0 \\ u_1 \end{bmatrix}$$

where the dynamics of the controller are chosen so as to make the close-loop dynamics (third-order) mimic a nonlinear network with 3 dynamic elements, and the new device called nonlinear gyrator (a multi-port, lossless, and memoryless device) as shown in Figure 3. The mapping, f(k) has to be a passive (lossy) resistive mapping. That is, {k, f(k)} should be confined to the first and third quadrants. The input $u_1$ will be chosen so as to force k(t) to a particular constant as $e, \dot{e} \rightarrow 0$.

The above in a nut shell describes the methodology that would lead to the design of nonlinear fuzzy controllers that are guaranteed to be stable. Though the concept may sound simple, it is a very powerful methodology and can be applied to a number of applications other than fuzzy controller design and would silence critics who tend to raise questions such as "would you travel in an aircraft controlled by a fuzzy controller whose stability properties are unknown".

## SIMULATIONS

To illustrate this concept, we have taken a third order model example used in reference [7], retained only the two dominant poles and used the fuzzy look-up table given in that paper with some modifications to generate the fuzzy controller output $F(e,\dot{e})$. Denoting the transfer function of the plant as

$$H(s) = \frac{b}{s^2 + as + b} = \frac{Y(s)}{U(s)}$$

with u as input to the plant, and y the output of the plant, the dynamics of the complete system is given by

$$\dot{y} = y_1$$
$$\dot{y}_1 = -by - ay_1 + u$$
$$u = kF(e,\dot{e})$$
$$\dot{k} = -y_1F(e,\dot{e}) - k - \frac{4}{\pi}\tan^{-1}(k) + u_1$$

[1] The term "mimic" is used here to imply exactly the same performance. Thus, if the two systems were treated as two black boxes, we would not be able to identify them from external measurements.

[2] In this section, we provide only salient points of out approach. A complete description is provided in appendix A.

[3] We plan to obtain patent protection on these devices.

[4] The new architecture will be patented.

where $u_1$ is chosen to force k to a particular value as the plant output moves to the target value. The terms $k$ and $\frac{4}{\pi}\tan^{-1}(k)$ have been chosen based on the network requirements and correspond to the currents in two parallel resistors, one being nonlinear and the other linear. The form and the exact values were chosen rather arbitrarily for this example and the proper choice would lead to the optimal performance. The responses of the plant using the classical fuzzy control approach and the new network based approach for two values of $k(\infty)$ are shown in Figure 4. It can be noted that there is some improvement in the response. However, the key point here is that the system represented by the above set of equations will remain stable and robust for external disturbances.

## SUMMARY

We have proposed a new method that guarantees the stability of feedback systems with fuzzy controllers. Instead of using a fuzzy expert system as a simple static (no memory) nonlinear mapping device, the new approach arrives at a fuzzy controller with memory (Figure 2B). This approach is also suitable for stand-alone fuzzy expert systems with internal feedback.

## REFERENCES

[1] L. Zadeh. *Outline of a new approach to the analysis of complex systems and decision processes, IEEE Trans. Sys., Man, Cybern.*, vol. smc-3, pp. 28-44, 1973.

[2] S Chiu, S. Chand, D. Moore and A Chaudhary. *Fuzzy logic for control of roll and moment for a flexible wing aircraft, IEEE Control Systems Magazine*, pp. 42-48, June 1992.

[3] J.E. Slotine and W. Li. *Applied Nonlinear Control*, Prentice-Hall, 1991.

[4] S. Lefschetz. *Stability of Nonlinear Control Systems*, Academic Press, New York, pp. 114-188, 1965.

[5] V.M. Popov. *Hyper-Stability of Automatic Control Systems*, Springer-Verlag, New York, 1973.

[6] K.S. Narendra and J.H. Taylor. *Frequency Domain Criteria for Absolute Stability*, Academic Press, New York, 1973.

[7] S. Tzafestas and N.P. Papanikolopoulos. *Incremental Fuzzy Expert PID Control, IEEE Trans on Industrial Electronics*, pp. 365-371, vol. 37, No. 5, Oct. 1990.

## APPENDIX Stable Nonlinear System Design

In section II, we indicated that concepts such as passivity could be used effectively to design stable nonlinear systems. We explain here the fundamentals behind such an approach.

Passivity is a term commonly used in Electrical Network Theory to indicate consumption of energy. Thus, a passive electrical element (linear or nonlinear) is one which always consumes power/energy (lossy) or at the most, consumes no power/energy (lossless). They can be non-dynamic (no memory/ can't store energy) or dynamic (stores energy and gives it back at some other time). They can be two-terminal (one-port) elements or multi-terminal (multi-port) devices. A passive linear/nonlinear network is simply an electrical network formed by proper interconnection of various passive linear/nonlinear elements. The interconnections must be such that the basic circuit laws are obeyed. An important property of such networks is that they are stable and remain so as long as the values of individual elements remain in the permissive range for passivity. Thus, if we have proper nonlinear elements, we can form stable nonlinear networks, obtain dynamical equations describing such networks in terms of the element parameters and use them as target equations for the plant and controller system.

The above approach assumes that proper nonlinear elements exist. Only one element is available in the open literature, that of passive resistance. An example of a passive resistor is one with a current-voltage relationship given by

$$i_R(t) = G \tan^{-1} (v_R(t));\ G > 0$$

since the power p(t) consumed by this element given by

$$p(t) = i_R(t)\, v_R(t)$$

is always non-negative[1]

Other nonlinear elements do not exist in the literature. For this purpose, we have defined/invented a number of passive nonlinear devices. One such device is a nonlinear transformer, a two-port element and has the transfer characteristics given by

$$\begin{bmatrix} v_2(t) \\ i_2(t) \end{bmatrix} = \begin{bmatrix} N() & 0 \\ 0 & \frac{1}{N()} \end{bmatrix} \begin{bmatrix} v_1(t) \\ i_1(t) \end{bmatrix} \quad (3)$$

[1] The v-i characteristics of a general nonlinear passive resistor has to be confined to the first- and third-quadrants in the v-i plane and has to pass through the origin.

where N() is a nonlinear function of the currents(s) and voltage(s) in an electrical network in which the transformer is embedded. It should be noted that

$$v_1(t)\, i_1(t) + v_2(t)\, i_2(t) = 0 \tag{4}$$

regardless of what ever form N( ) takes. Thus, an ideal nonlinear transformer is a lossless, non-dynamic (or memoryless) two-port device[2].

Another device that we have invented and that is highly useful is that of a multi-port nonlinear gyrator. The nonlinear gyrator is described by the admittance matrix Y(=[$y_{ij}$( )], i,j=1 to N for an N-port device) where (T below denotes the transpose of a matrix)

$$Y + Y^T \equiv 0 \tag{5}$$

and the elements $y_{ij}()$ can be complex functions of the current(s) and voltage(s) in an electrical network. As an example, for a two-port nonlinear gyrator, we may have

$$\begin{bmatrix} i_1(t) \\ i_2(t) \end{bmatrix} = \begin{bmatrix} 0 & v_1^2 - v_2 \\ v_2 - v_1^2 & 0 \end{bmatrix} \begin{bmatrix} v_1(t) \\ v_2(t) \end{bmatrix} \tag{6}$$

Defining I and V to be the vectors of currents and voltages of a general N-port nonlinear gyrator as defined above, we can show that

$$[I]^T\,[V] \equiv 0 \tag{7}$$

for any $y_{ij}()$ function. That is, a nonlinear gyrator is a lossless and non-dynamic multi-port device.

We can also define nonlinear dynamical elements such as capacitors and inductors. However, such elements are more complex and are not needed for our application. Hence, we will restrict our attention to linear dynamical elements.

We can connect these devices to form a passive nonlinear electrical network.(an example of such a network will be as shown in Fig.3 of the Theory section) and use Kirchoff's current and voltage laws to write the dynamical equations for such a network. Such a nonlinear equation will represent all possible stable equations as we vary the coefficients of the various elements within their respective ranges.

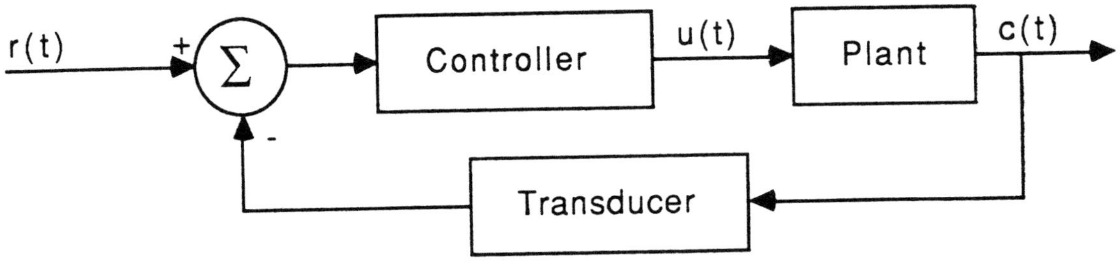


**Figure 1. General feedback control system architecture.**

2 Though an analog implementation of this device (and other devices invented) is feasible, we will be using digital implementation for the controller application. The digital implementation would allow us to realize such elements/devices with virtually no problem.

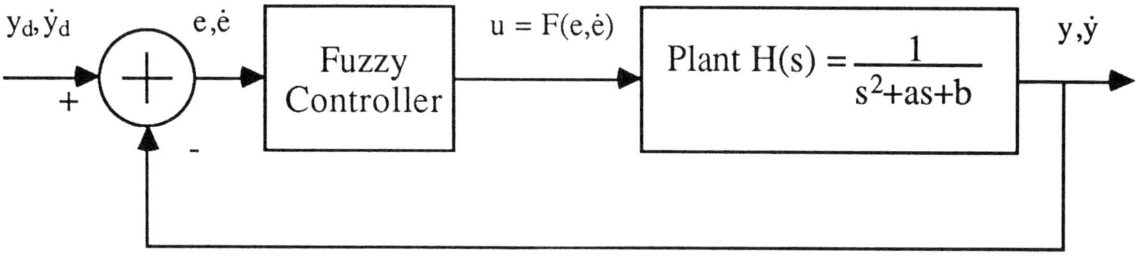


A. Classical Fuzzy Control Approach

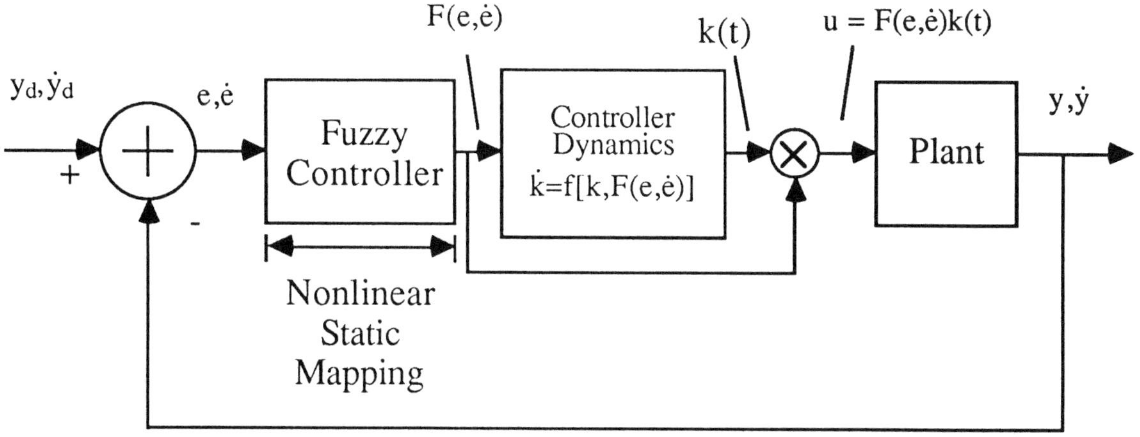


B. New Nonlinear Network Based Approach

**Figure 2. A. Classical approach. B. New Network based Architecture**

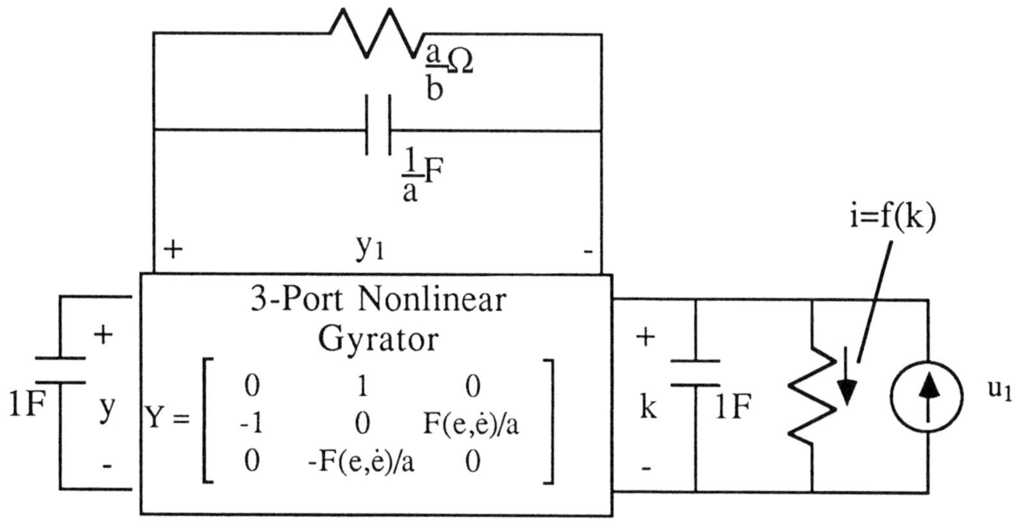


**Figure 3. Network Equivalent of Fuzzy Controller**

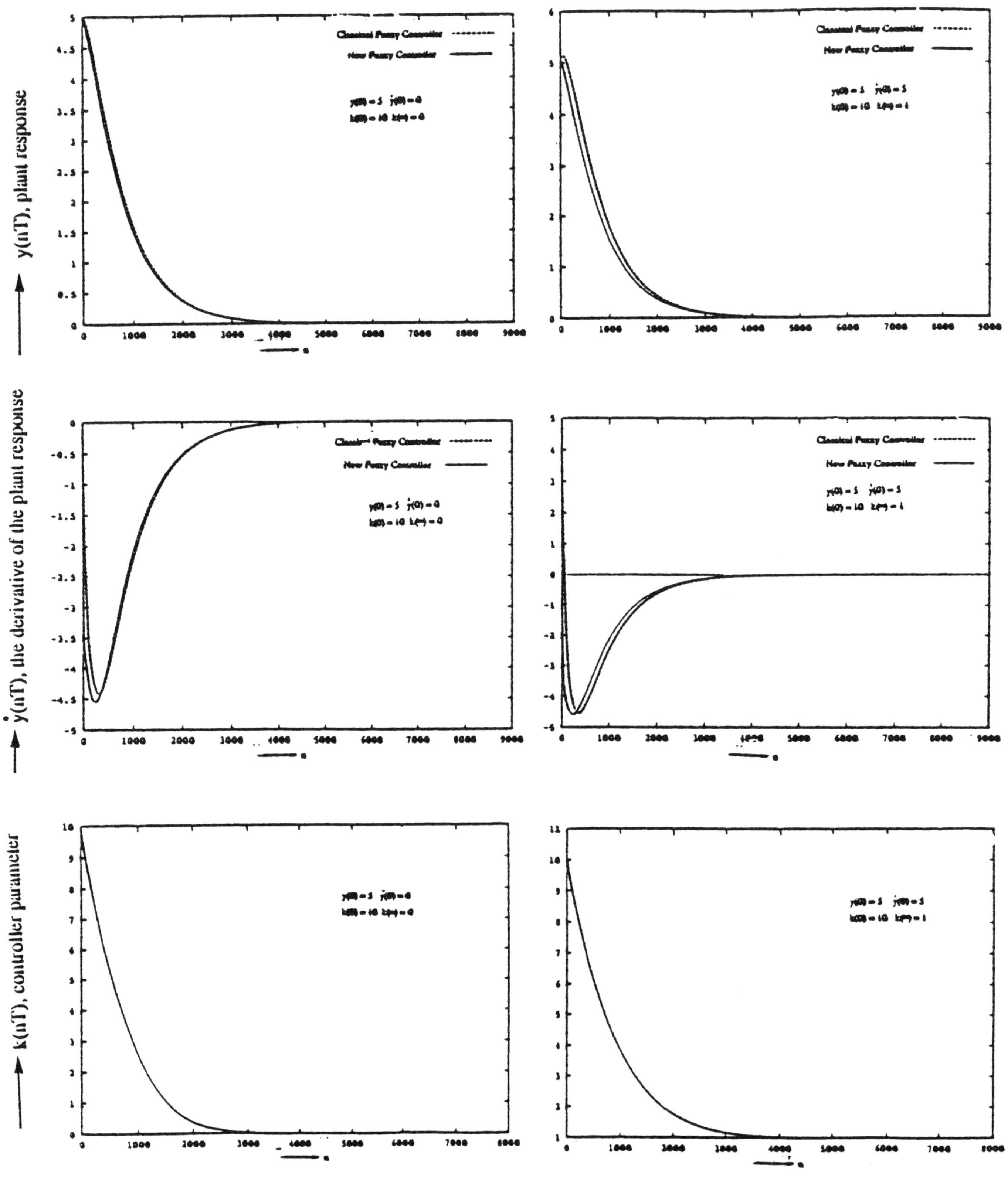


Figure 4. Responses of the plant using classical and network based Fuzzy Controller

# The Dimensions Effect of FAM Rule Table in Fuzzy PID Logic Control Systems

Chir-Ho Chang and John Y. Cheung

School of Electrical Engineering,
University of Oklahoma
Norman, Ok, 73019

*Abstract---To choose the proper number of generic elements in the discourse of universe in the design of a fuzzy logic controller (FLC) is always arbitrary and obscure. This paper examine the effects of the table dimensions when a domain expert doesn't exist. We test the PID FLC with three dimensional fuzzy associated memory (FAM) rule table whose entries are obtained from fuzzy addition on a dynamic second order tracking problem. Simulation results show that a fuzzy system which uses 'even' type FAM rule table (A 3D FAM rule table $T_{MxNxL}$ is called even-even-odd type and abbreviated even type if the first two dimensions M,N are even and the third dimension L is odd.) is more suitable for controlling a second order plant which demands slightly underdamped response.*

## 1. INTRODUCTIONS

When designing a PID FLC[3-6], we not only need to define the membership functions for each of the three input domains (error, Δerror, $\Delta^2$error), but also to decide the number of generic elements in those universe of discourses. The selection of proper number of generic elements in these universe of discourses is generally vague and arbitrary. Artificial neural networks are capable of doing the fuzzy clustering and form the membership functions automatically [2]. However, large crisp data samples are needed and are sometimes not possible.

We found two basic types of trapezoidal membership functions: even type and odd type. Fig. 1 and Fig. 2 show these two patterns. In Fig. 1, the universe of discourses for error and Δerror are subdivided into (even) four and (even) six generic elements respectively; while in Fig. 2 the universe of discourses for error and Δerror have (odd) five and (odd) seven generic elements respectively.

Most of the researchers assume the number of generic elements to be an odd number such that the middle element is always an fuzzy 'Zero' element[12-16]. However, if we assume the number of generic elements to be an even number, then this 'Zero' element will be replaced by two new generic elements. For example, FLC designer has to decide four generic elements (+Large, +Small, -Small, -Large) for error domain, six generic elements (+Large, +Medium, +Small, -Small, -Medium, -Large) for Δerror domain, and five generic elements (+Large, +Medium, **Zero**, -Medium, -Large) for the $\Delta^2$error domain. The numbers of fuzzy generic elements: four, six, and five in this example are according to the ad hoc application and largely depends on the expert's experience.

The objective of this paper is to examine and compare the effect of the types of tables through the transient response of a second order system.

We shall briefly review some of the definitions used in the literature[1,4,6] and summarize some of the basic terms used in this paper.

*Fuzzy subset*:
A set of ordered pairs whose first element is an element of the universe of discourse U, and whose second element is a real number between 0 and 1. i.e. fuzzy subset F $\triangleq$ { u, $\mu_F$ | u ⊂ U}.

*Membership functions (MBFs)*:
The mapping functions between the elements of the universe of discourse and their degrees of membership in fuzzy subset.

*Support*:
The support of a fuzzy set is the set of points in the universe of discourse at which the degree of the membership is greater than zero.

*Generalized Crossover Point*:
The crisp value at which $\mu_F(u_i)=\mu_F(u_{i+1})$. Where $\mu_F$ is the membership function, $u_i$ and $u_{i+1}$ are two adjacent fuzzy subsets of the universe of discourse. (In Zadeh's paper, the crossover point is the crisp value when the degree of membership equal to 0.5).

*Overlap*:
The distance between left zero MBF of $u_i$ and right zero

0-7803-0614-7/93$03.00 ©1993IEEE

MBF of $u_{i+1}$.

*Firm Support*:

The interval from two adjacent generalized crossover points.

*Mapping Factor*:

The ratio between two adjacent width of the firm support. Let $s_i$ and $s_{i+1}$ be the firm support of and $u_i$ and $u_{i+1}$ respectively, the mapping factor is defined as $s_{i+1}/s_i$.

## 2. PID FUZZY LOGIC CONTROLLER

We have developed a PID FLC[5]. A FLC usually constitutes the following fundamental building blocks:

*1) Membership functions*:

Fuzzy subsets are usually represented by membership functions. These functions define how big is big; how high is high in the universe of discourse. They play an important role in the process of fuzzy inference and approximate reasoning. To define membership functions by mapping factors and percentage of overlapping has been proposed by C. Chang[3,6]. By the use of these two new parameters, a special symmetrical type of membership functions could be uniquely determined.

*2) Three Dimensions FAM table (Inference engine)*:

The 3D FAM rule table contains the knowledge from an expert. A well tuned rule table is believed to be able to achieve almost any kind of control objectives[2]. However, there are two problems. First of all, the general the domain expert may give very little knowledge (few rules) as to the $\Delta^2$error input domain. And sometimes he can't give error free correct rules which benefit the overall performance. Secondly, the domain expert could be not even exist[9]. How to initialize the table without further tuning is not an easy job. To solve this, a fuzzy addition method is used under such situations. In previous work, the fuzzy addition and the concept of mapping factors are applied to odd number of generic elements only. As a matter of fact, fuzzy addition like mapping factors can be modified to handle even number of fuzzy generic elements. Given the number of generic elements M for error and N for $\Delta$error. The entries of the 2D odd FAM rule table are obtained from (1).

$$T_{ij} = -\{[i-\frac{M-1}{2}] + [j-\frac{N-1}{2}]\} \quad \text{If } M, N \text{ are odd number} \tag{1}$$

Where i=0..M-1, j=0..N-1 and Table 1 shows an odd-odd type 2D $FAM_{5x5}$ rule table.

Similarly, if we assume the number of fuzzy generic elements for $\Delta^2$error to be L, then the entries of an odd-odd-odd type 3D FAM rule table can be calculated from (2). Notice that i=0..M-1, j=0..N-1, k=0..L-1. And the number of fuzzy generic elements for the control output is assumed and is equal to M+N-1.

$$T_{ijk} = -\{[i-\frac{M-1}{2}] + [j-\frac{N-1}{2}] + [k-\frac{L-1}{2}]\} \quad \text{where } M,\ N,\ L \text{ are all odd number} \tag{2}$$

The entries of an even-even type 2D FAM rule table are computed from (3-5). Note that ceiling and floor functions are used. Table 2 shows an even-even type 2D $FAM_{4x6}$ rule table.

$$\begin{aligned} S_i &= \left\lceil i - \frac{M-1}{2} \right\rceil \quad \text{If } i > \frac{M-1}{2} \\ S_i &= \left\lfloor i - \frac{M-1}{2} \right\rfloor \quad \text{If } i < \frac{M-1}{2} \end{aligned} \tag{3}$$

$$\begin{aligned} S_j &= \left\lceil j - \frac{N-1}{2} \right\rceil \quad \text{If } j > \frac{N-1}{2} \\ S_j &= \left\lfloor j - \frac{N-1}{2} \right\rfloor \quad \text{If } j < \frac{N-1}{2} \end{aligned} \tag{4}$$

$$T_{ij} = -[S_i + S_j] \quad \text{Where } i = 0..M-1,\ j = 0..N-1 \tag{5}$$

Likewise, we also can extend even-even type 2D FAM rule entry formula to even-even-odd 3D FAM rule entry formula. (6) shows the modification. Notice that the number of fuzzy generic elements for control output is assumed and is equal to M+N+1.

$$T_{ijk} = -\{S_i + S_j + [k - \frac{L-1}{2}]\} \quad \text{Where } i=0..M-1,\ j=0..N-1,\ k=0..L-1 \tag{6}$$

*3) Fuzzifier and defuzzifier process*:

Fuzzification is the process to map from crisp values to fuzzy sets. On the other hand, defuzzification is the dequantizing process from fuzzy sets to crisp values. Three type of defuzzification methods are often used. Mean of height (MOH), center of area (COA) and maximum criterion method (MCM). Mean of maximum give the highest crisp value, while maximum criterion method gives lowest crisp value. Center of gravity defuzzification strategy is used in this paper.

Fig. 3 shows the schematic diagram of PID FLC which is used in our simulations.

## 3. SIMULATIONS

To compare the effect of odd-odd-odd type FAM rule table with even-even-odd type FAM rule table, we test them on a second order simplified automobile suspension system (parallel connection of dash pot and spring system). The simulation setups are:

| | |
|---|---|
| 1) Plant model: | $2/(s+1)(s+2)$ |
| 2) Sampling time: | .25 seconds |
| 3) Set points: | 2 for the first half cycle, 1 for the second half cycle |
| 4) Mapping factors: | $IMF_1=IMF_2=IMF_3=2$; $OMF=2$. |
| 5) Overlapping: | 25% |

Four experiments are tested using both type of tables: even-even-odd type $T1_{M1xN1xL1}$ and odd-odd-odd type $T2_{M2xN2xL2}$, and with all the other conditions remained the same. The subscripts of T1 and T2 represent the dimensions of the table. We assume the third dimension to be fixed (L1=L2=5) and the first two dimensions have this relation: M1xN1=M2xN2-1 for an attempt to limit the comparison domain.

Table 3 shows these four experiments. Notice the odd-odd-odd type table T1 has five more rules over the even-even-odd type table T2.

The simulation results are shown in Fig. (4-7). We know that a "hard riding" sports car may be overdamped while a car with weak shock absorbers is definitely underdamped. The best ride is achieved by designing the car to be slightly underdamped such that everybody feels comfortable. We found that a FLC with an even-even-odd type of FAM rule table gives quicker response with approximately the same amount or less than 2% higher overshoot (perfect case of slightly underdamped). Further research could be 1.) applying a self-tuning algorithm[6-8,10] to the FLC with even type table, 2.) test more complex plants.

## 4. SUMMARY

The transient response analysis in both mechanical and electrical control systems is very important. We have introduced a method for initializing a 3D FAM rule table in fuzzy system which don't have a domain expert. Since an odd-odd-odd type 3D FAM rule table which calculates its entries by previous fuzzy addition tends to overdamp a second order plant, especially when longer sampling time is used. We present a modified fuzzy addition formula for computing the even-even-odd type of 3D FAM rule table entries. Simulation results show that a fuzzy logic control system with even-even-odd type 3D FAM rule table has less damping control output, which means a shorter responding time and a little bit increase in the overshoot. A better design based on such table initialization scheme has been tested successfully on a simplified automobile suspension system.

## REFERENCE


[1] L.A. Zadeh, "Fuzzy Sets," Inform. and Control, vol.8, pp.338-353, 1965.

[2] C.T. Lin and C.S.G. Lee, "Real-Time Supervised Structure Parameter Learning for Fuzzy Neural Network," Proc. of FUZZ-IEEE, pp.1283-1291, March 1992.

[3] C.H. Chang, "Tuning PID Fuzzy Controller via I/O Mapping Factors", Master Thesis, Univ. of Oklahoma, 1989.

[4] C.C. Lee, "Fuzzy Logic in Control Systems: Fuzzy Logic Controller - Part I," IEEE Transactions on SMC, vol.20, no.2 pp.404-418, Mar./Apr. 1990.

[5] G.M. Abdelnour, C.H. Chang, F.H. Huang, and J.Y. Cheung, "Design of PID Fuzzy Controller" IEEE trans. on Systems, Man and Cybernetics Sept./Oct. pp. 955. 1991.

[6] C.H. Chang, J.Y. Cheung, "Tuning PID Fuzzy Controller By Membership Mapping Factors," to be published in the first international conference on Fuzzy Theory & Technology Control & Decision.

[7] S.Z. He, S.H.Tan, C.C. Hang, and P.Z. Wang, "Design of an On-line Rule-adaptive Fuzzy Control System," Proc. of FUZZ-IEEE, pp.83-91, 1992.

[8] David G. Burkhardt, Peiro P. Bonissone,"Automated Fuzzy Knowledge Base Generation and Tuning," Proc. of FUZZ-IEEE, pp.179-188, 1992.

[9] J.S.R. Jang, "Fuzzy Controller Design without Domain Experts," Proc. of FUZZ-IEEE, pp.289-296, 1992.

[10] W.C. Daugherity, B. Rathakrishnan, J. Yen, "Performance Evaluation of a Self-Tuning Fuzzy Controller," Proc. of FUZZ-IEEE, pp. 389-397, 1992.

[11] Bart Kosko, "Fuzzy Function Approximation," Proc. of IJCNN, vol.I, pp.209-213, 1992.

[12] H.C. Tseng, V. Hwang, and S.L. Lui, "Fuzzy Servocontroller; The Hierarchical Approach," Proc. of FUZZ-IEEE, pp.623-628, 1992.

[13] Li Zheng, "A Practical Guide to Tune of PI Like Fuzzy Controllers," Proc. of FUZZ-IEEE, pp.633-640. 1992.

[14] K. Wu and S. Outangoun, "Modeling and Experiments in Fuzzy Control," Proc. of FUZZ-IEEE, pp.725-731, 1992.
[15] J. Cleland and W. Turner, Paul Wang, T. Espy, P.J. Chappell, R.J. Spiegel, "Fuzzy Logic Control of AC Induction Motors," Proc. of FUZZ-IEEE, pp.843-850, 1992.
[16] J.Y. Yen, C.S. Lin, C.H Li, Y.Y. Chen, "Servo Controller Design For An Optical Disk Drive," Proc. of FUZZ-IEEE, pp.898-995, 1992.
[17] D.E. Schneider and Paul.P. Wang, "Design of a FLC for a Target Tracking System," Proc. of FUZZ-IEEE, pp.1131-1138.

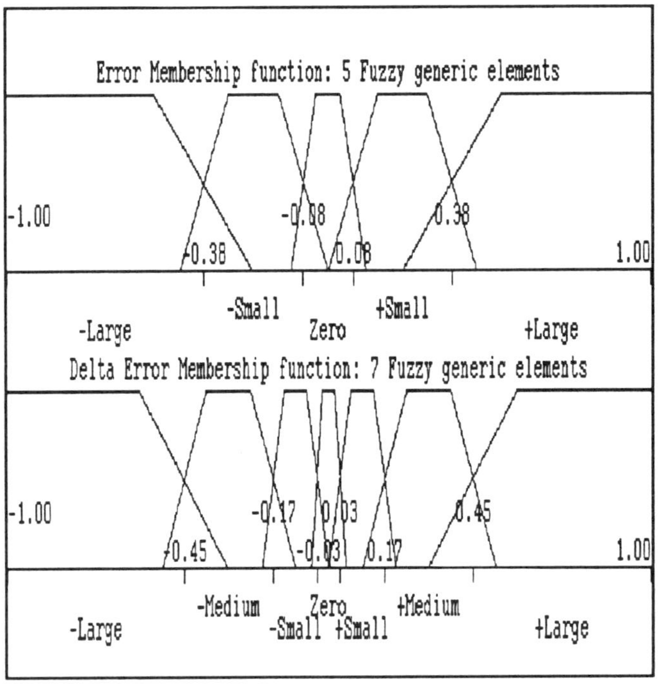

Fig. 2. Examples of odd fuzzy generic elements ($IMF_1 = IMF_2 = 2$.)

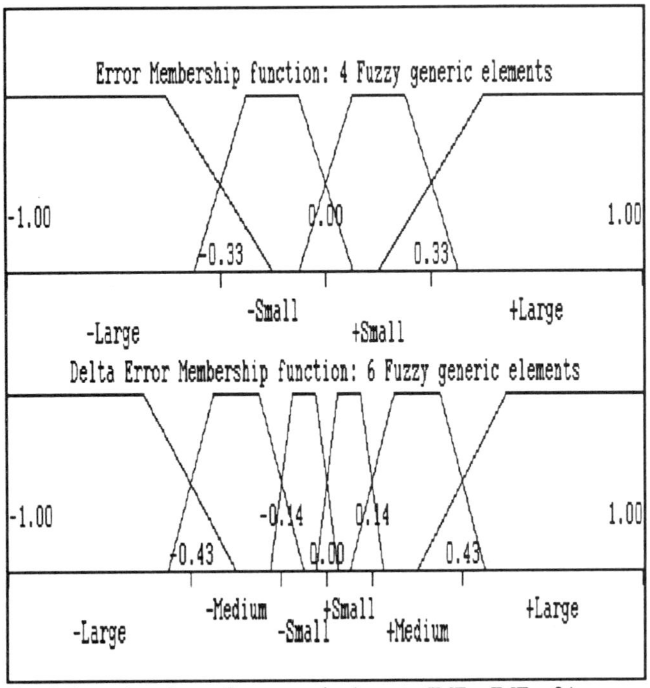

Fig. 1 Examples of even fuzzy generic elements ($IMF_1 = IMF_2 = 2$.)

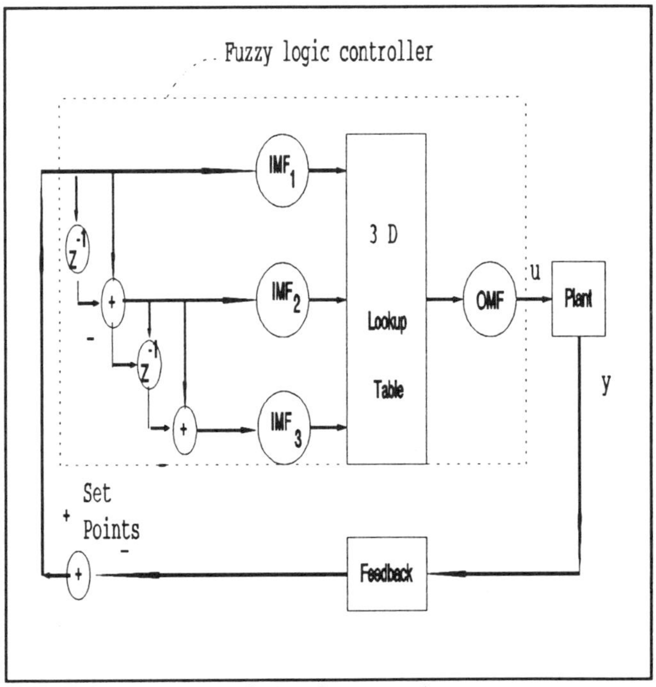

Fig. 3 Schematic diagram of a PID fuzzy control system

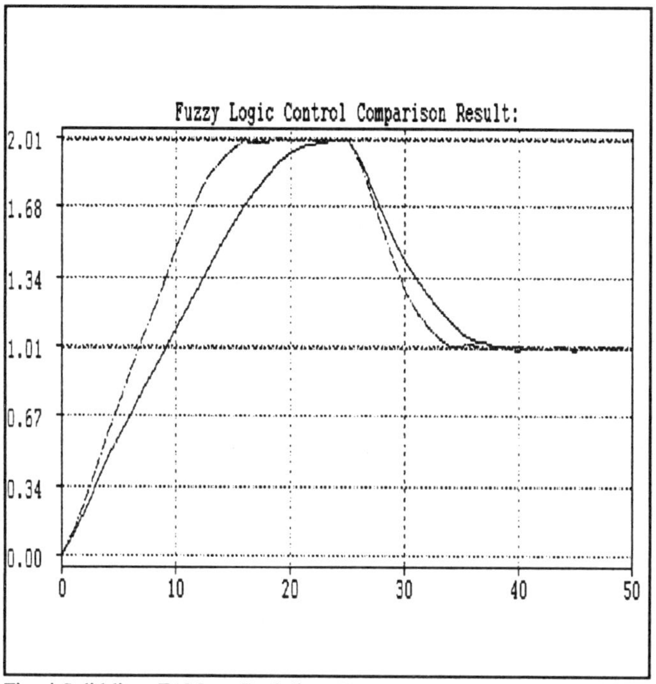


Fig. 4 Solid line: $FAM_{9x9x5}$ Dash line: $FAM_{10x8x5}$

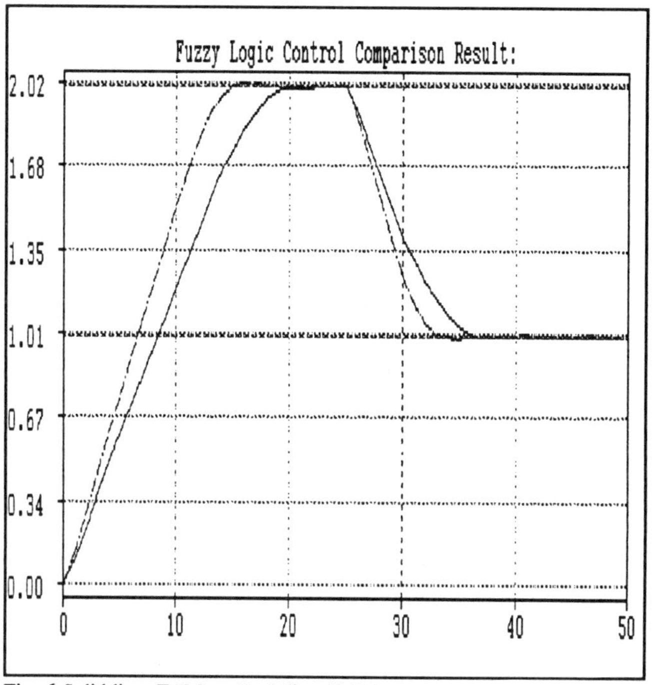


Fig. 6 Solid line: $FAM_{5x5x5}$ Dash line: $FAM_{6x4x5}$

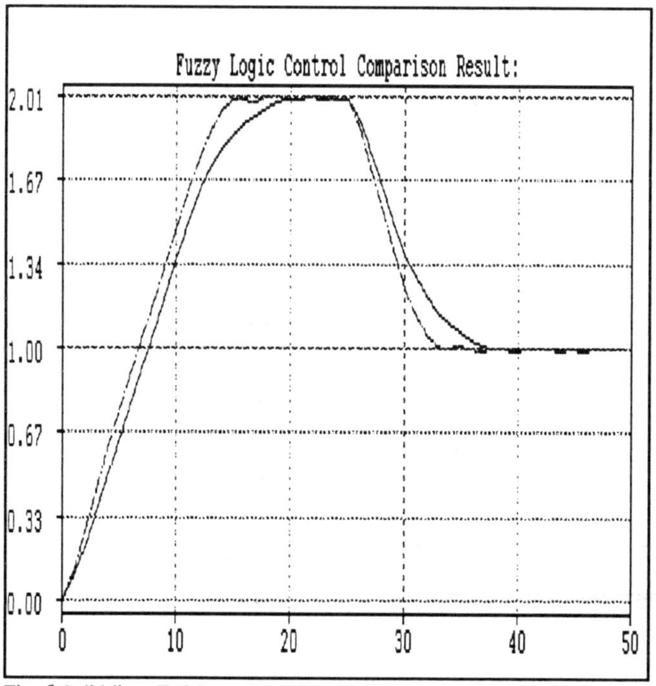


Fig. 5 Solid line: $FAM_{7x7x5}$, Dash line: $FAM_{8x6x5}$

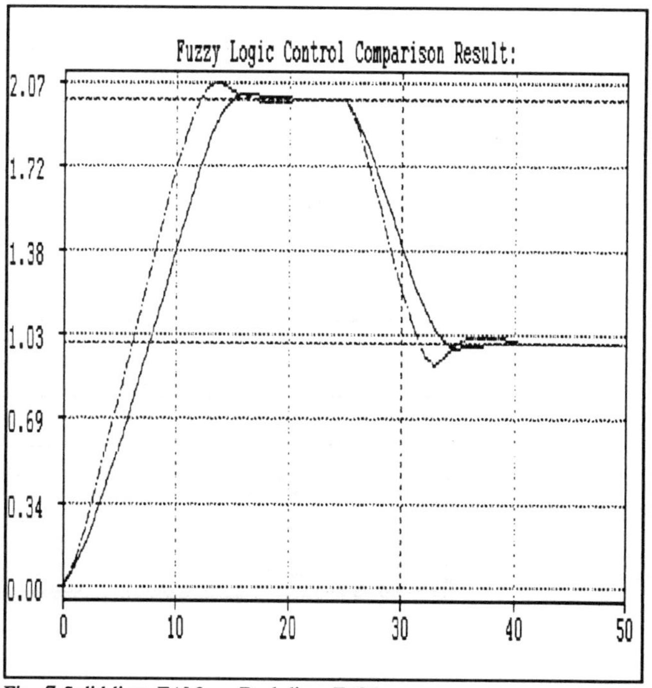


Fig. 7 Solid line: $FAM_{3x3x5}$ Dash line: $FAM_{4x2x5}$

Table 1 5x5 2D FAM rule table

| A 5x5 2D FAM rule table | | Five Generic Error Elements | | | | |
|---|---|---|---|---|---|---|
| | | -Large -2 | -Medium -1 | Zero 0 | +Medium +1 | +Large +2 |
| ΔError | -Large(-2) | +4 | +3 | +2 | +1 | 0 |
| | -Medium(-1) | +3 | +2 | +1 | 0 | -1 |
| | Zero(0) | +2 | +1 | 0 | -1 | -2 |
| | +Medium(+1) | +1 | 0 | -1 | -2 | -3 |
| | +Large(+2) | 0 | -1 | -2 | -3 | -4 |

Table 2 4x6 2D FAM rule table

| A 4x6 2D FAM rule table | | Four Generic Error Elements | | | |
|---|---|---|---|---|---|
| | | -Large -2 | -Medium -1 | +Medium +1 | +Large +2 |
| 6 ΔError GE | -Large(-3) | +5 | +4 | +2 | +1 |
| | -Medium(-2) | +4 | +3 | +1 | 0 |
| | -Small(-1) | +3 | +2 | 0 | -1 |
| | +Small(+1) | +1 | 0 | -2 | -3 |
| | +Medium(+2) | +0 | -1 | -3 | -4 |
| | +Large(+3) | -1 | -2 | -4 | -5 |

Table 3 Experiments

| | Even-Even-Odd FAM Rule Table | Odd-Odd-Odd FAM Rule Table |
|---|---|---|
| Experiment#1 | $T1_{10x8x5}$ | $T2_{9x9x5}$ |
| Experiment#2 | $T1_{8x6x5}$ | $T2_{7x7x5}$ |
| Experiment#3 | $T1_{6x4x5}$ | $T2_{5x5x5}$ |
| Experiment#4 | $T1_{4x2x5}$ | $T2_{3x3x5}$ |

# Architecture of a PDM VLSI Fuzzy Logic Controller with Pipelining and Optimized Chip area

Ansgar P. Ungering, Karsten Thuener*, Karl Goser

University of Dortmund, LS-BE, Emil-Figge-Str. 68, 4600 Dortmund, Germany

* ht Mikroelektronik GmbH, Pappenstrasse 36, 4100 Duisburg, Germany

*Abstract* — **We are describing the architecture of a fuzzy logic controller using pulse-width-modulation (PDM) technique and a pipeline structure. Features of this controller are: A new architecture for the inference unit, reduced chip area, variable resolution from 1, 2, 3...254, 255 (not in $2^x$-steps!) and less I/O-pins. Additionally the architecture has an optimized rule base and the operation time depends only on the resolution. A prototype with two inputs, one output and a resolution of 8-bit has been implemented on FPGAs and uses less then 10.000 gates including internal RAM. A prototype of the controller operates with 6 MHz and needs 170µs by 8-bit resolution or 22µs by 5-bit resolution for one control step, independent of the number of inputs, outputs and rules.**

## I. Introduction

In the recent years there has been an increasing interest in methods for realizing efficient digital fuzzy controller hardware suitable for integration [1,2,3]. First implementations [1] have a high flexibility but need a large amount of chip area. New architectures proposed in [2,3,4] restrict the degree of overlap of the membership functions (MFs) and lead to a reduced amount of chip area. This paper presents an extension of these methods.

A typical fuzzy hardware controller consists of the following building blocks:

1. A fuzzifier unit
2. A rule base
3. An inference unit
4. A defuzzification unit

In this communication, we will show that the first three of the building blocks mentioned above can be implemented with reduced hardware requirement but without reducing speed of operation. This is achieved if the input and output signals are pulse-width-modulated (PDM) and if the internal operation is also based on PDM-signals.

The authors would like to thank the VW-Stiftung Hannover for supporting this work.

0-7803-0614-7/93$03.00 ©1993IEEE

Chapter II explains the idea of PDM-signals for fuzzy logic. The following chapter discusses the MIN/MAX-algorithm under the aspects of minimizing chip area and operating time. Chapter IV shows some restrictions of the architecture. An overview of the general architecture is given in chapter V. Chapter VI will show the implementation of each block of the controller. The paper ends with a summary.

## II. Fuzzy operations with PDM-signals

The described controller uses only the minimum- (MIN) and maximum- (MAX) operators. Fig. 1 shows the simple realisation of MIN and MAX of three PDM-signals [5]. We only need one AND-gate for the MIN- and one OR-gate for the MAX-operation. Obviously, it is very easy to increase the numbers of inputs. Since it is not very difficult to build a voltage or current to PDM converter, a fuzzy controller using PDM signals usually requires less I/O-pins and less chip area than other methods.

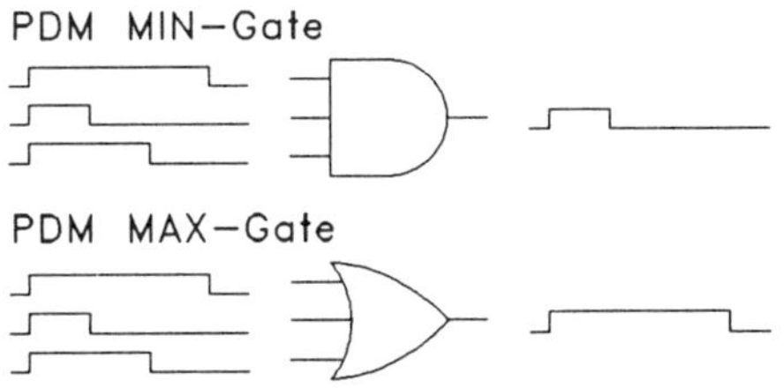


Fig. 1. MIN/MAX-operation for PDM-signals

## III. Analysis of the MIN/MAX-Algorithm

The algorithm is based on minimum- and maximum-operations. The "crisp" output value will be calculated using the center of gravity (COG) method. The basic principle of the well known algorithm is depicted in Fig. 2. The calculation of the "THEN-PART" (α-cut and combining the MFs Cx′ to one resulting output function C′) usually takes N time steps. N is equal to the x-resolution of the MFs. Therefore, if the x-resolution of the MFs is 256 (8-bit), N=256 time steps are required. This part of the controller limits the speed of operation.

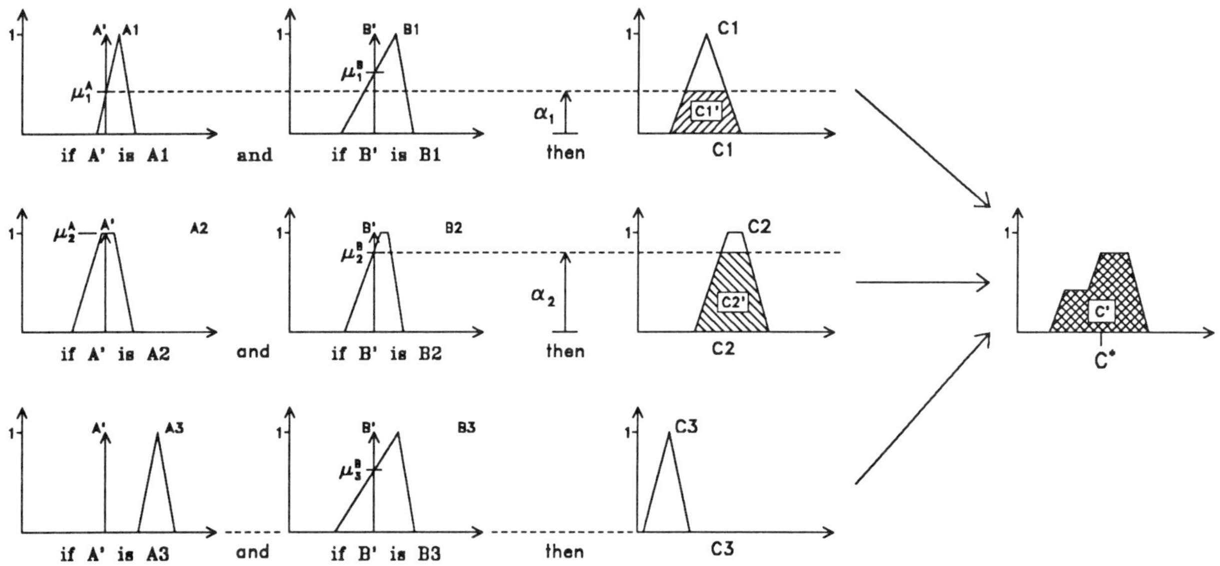


Fig. 2. Principle fuzzy controller algorithm

Clearly, using pipelining, all other parts of the controller do not reduce the speed of operation of the complete system, if their internal speed of operation does not exceed this limit. In the next chapters the architecture consisting of a 4-stage-pipeline will be described.

## IV. Necessary restrictions for saving chip area

In order to reduce the amount of chip area required by the controller, the following two conditions have to be obeyed:

1. The overlap degree of the MFs is two: Only the MFs in direct neighbourhood can overlap.

2. The rule base must be free of "inconsistency". This means that the resulting output function may not consist of more than 4 neighbouring MFs. Examples are given in Fig. 3.

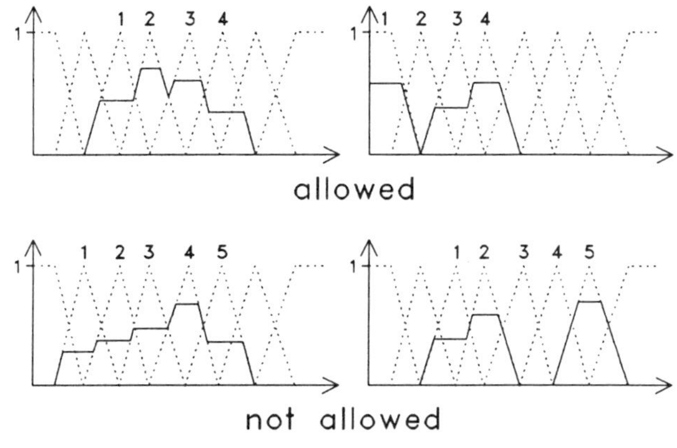


Fig. 3. The combined output MF must not consist of more than 4 neighbouring MFs

## V. Architecture

The following Fig. 4 depicts the basic architecture of the 8-bit fuzzy controller.

The system consist of the following blocks:

1. Fuzzifier
2. Rule base
3. Output MF-Generator
4. Defuzzifier
5. Control unit

The 4-stage-pipeline consists of the following stages: Fuzzifier, Rule base, Output-MF-Generator and Defuzzifier.

In the following paragraphs these blocks will be described. The resolution is 8-bit, so that one main clock cycle takes 256 system clock cycles. For a lower resolution the main clock cycle shows an equivalent reduction. Every stage of the pipeline requires one main clock cycle.

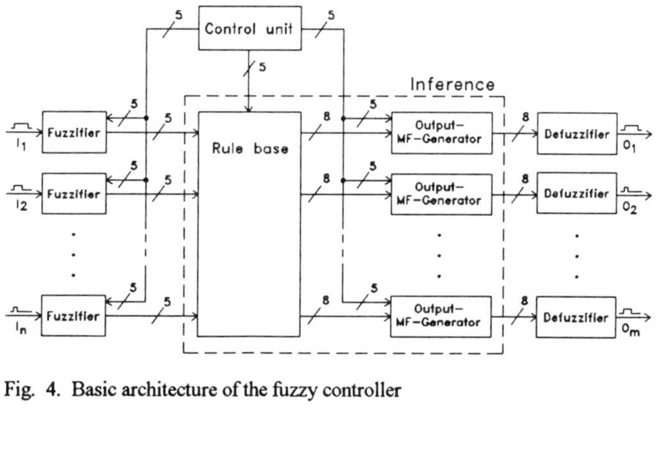


Fig. 4. Basic architecture of the fuzzy controller

## VI. Implementation of the Controller Blocks

### *A. Storing and reconstructing membership functions*

It is well known, that most fuzzy applications require only a limited number of MFs with an overlap degree of 2. Therefore, we restrict the number of MFs to 8 for each input and to 8 for each output. It should be noted that these restrictions are not disturbing. They lead, however, to simple circuits. The MFs (see Fig. 5.a) are stored by the memory blocks "even" and "odd" in the following manner [3,4]:

First, the MFs are numbered starting with the leftmost MF (MF0). The even MFs will be stored in the "even" memory block, all odd MFs will be stored in the "odd" memory block. Since the overlap degree is 2, none of the MFs overlap each other in each memory block.[1] In order to reduce the amount of the required RAM, the MFs are stored in compressed form (Fig. 5.b, 5.c): The first memory location k=0 contains the start value $m_0$. Each of the following memory locations $0 < k < k_{max}$ contains first the point on the x-axis $x_k$ where the next change of the slope occurs and second the slope $m_k$. Obviously triangular and trapezoidal MFs can be stored in a very compact form. The storage of MFs with arbitrary shapes is also possible.

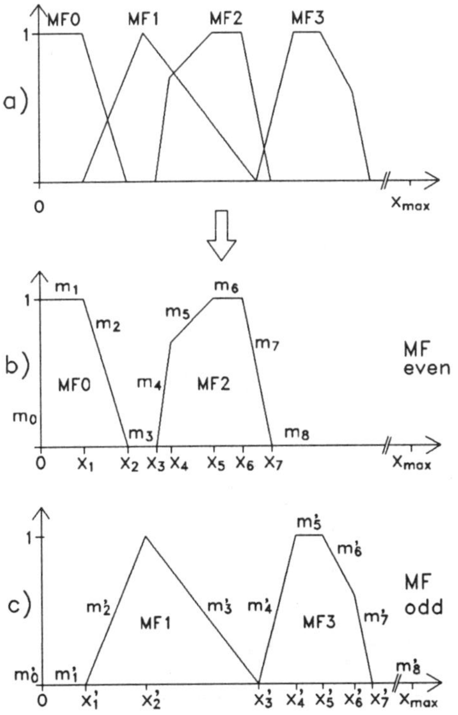


MF even

| k | x | m |
|---|---|---|
| 0. | — | $m_0$ |
| 1. | $x_1$ | $m_1$ |
| 2. | $x_2$ | $m_2$ |
| 3. | $x_3$ | $m_3$ |
| ⋮ | | |

MF odd

| k | x' | m' |
|---|---|---|
| 0. | — | $m'_0$ |
| 1. | $x'_1$ | $m'_1$ |
| 2. | $x'_2$ | $m'_2$ |
| 3. | $x'_3$ | $m'_3$ |
| ⋮ | | |

Fig. 5. Compressed storing of the MFs

[1] If the overlap degree q is greater than 2, we need q memory blocks.

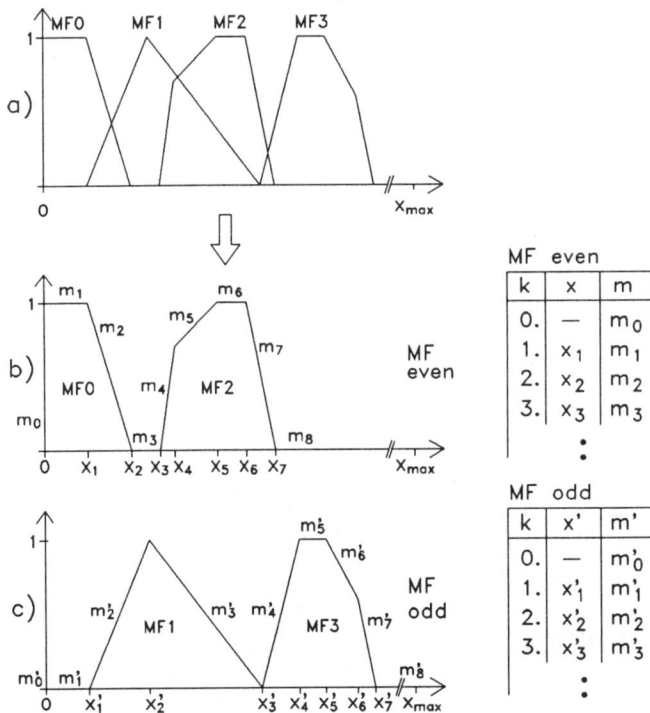

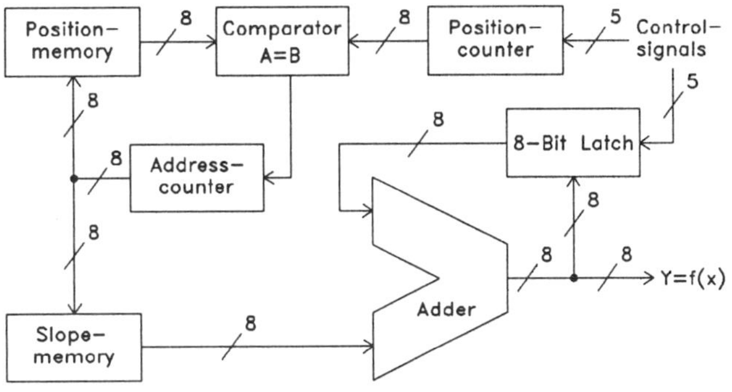


Fig. 6. MF-Generator

For 8 membership functions (see Fig. 7) we need no more than 56 bytes of memory (8-bit resolution).

The reconstruction of the original MFs from the stored information can be done by the following procedure (Fig. 6): At x=0, the start value $m_0$ will be read from address k=0 of the slope memory and stored into the adder. During the next system clock cycle (x=1), the contents of memory location k=1, $m_1$ (slope memory) and $x_1$ (position memory), will be read. During each of the subsequent system clock cycles, $m_1$ is added to the contents of the adder as long as $x \le x_1$. Now, the contents of the memory location k=2, $m_2$ and $x_2$, will be read. Again, $m_2$ is added at each system clock cycle to the contents of the adder as long as $x \le x_2$. This procedure will be repeated for $x \le x_{max}$.

### *B. Fuzzification of PDM-signals*

Fig. 7 depicts the architecture of the fuzzifier. In this and the following figures the latches which separate the pipeline stages are not included. With each system clock cycle new MF-values are produced by the MF-Generators. Therefore, a resolution of 8-bit results in 256 values for each main clock cycle (see chapter V). The MF-counter generates the numbers h and h+1 of the MFs which values are actually the outputs of the MF-Generators. In order to save wires, only the number h is delivers. After the input signal $I_n$ returns to zero, the corresponding MF-values and the MF-number h are stored for further processing. This completes the first stage of the pipeline.

The D/PDM converter transforms the stored MF-values into PDM-signals. This belongs to the second pipeline stage denoted prior by "rule base". Fig. 7 gives an example. The duration of the input signal $I_n$ is 183. The output of the even MF-Generator delivers a MF-value of 128 since MF 6 is active. In the odd MF-Generator, MF 5 is active and the corresponding MF-value is 204. Therefore h=5 is stored in the MF-counter.

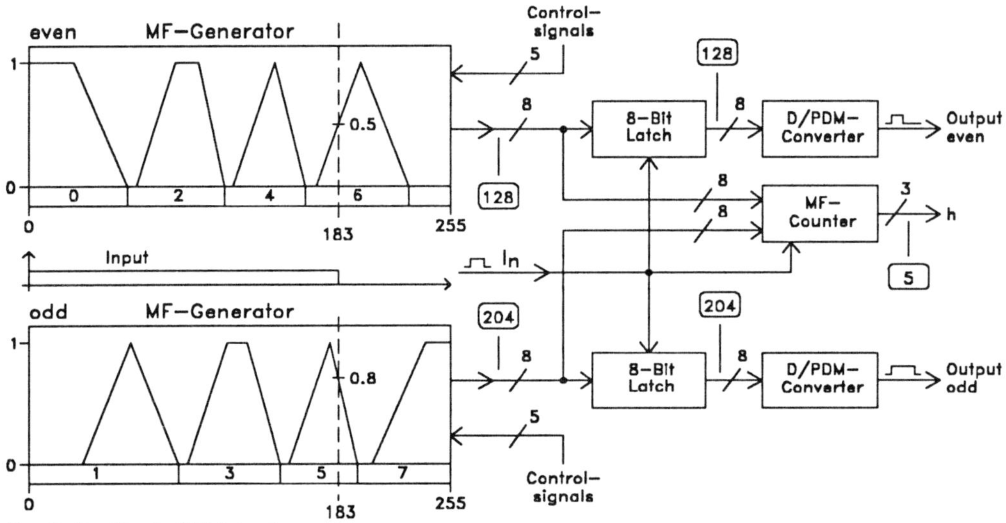


Fig. 7. Fuzzifier for PDM-signals

*C. Rule base*

In contrast to some traditional hardware realisations, this architecture allows the MIN- and MAX-operations on the fuzzified values without increasing the speed of operation and the amount of hardware substantially. It is also allowed to combine rules with the MIN- or MAX-operators. An example is given in Fig. 8.

Depending on the value of the MF-number h, the MF-DEMUX-unit switches the even and the odd signal to the wires MF0 to MF7. It should be noted that each rule contributes to one or more output blocks (MF′-MUX Output

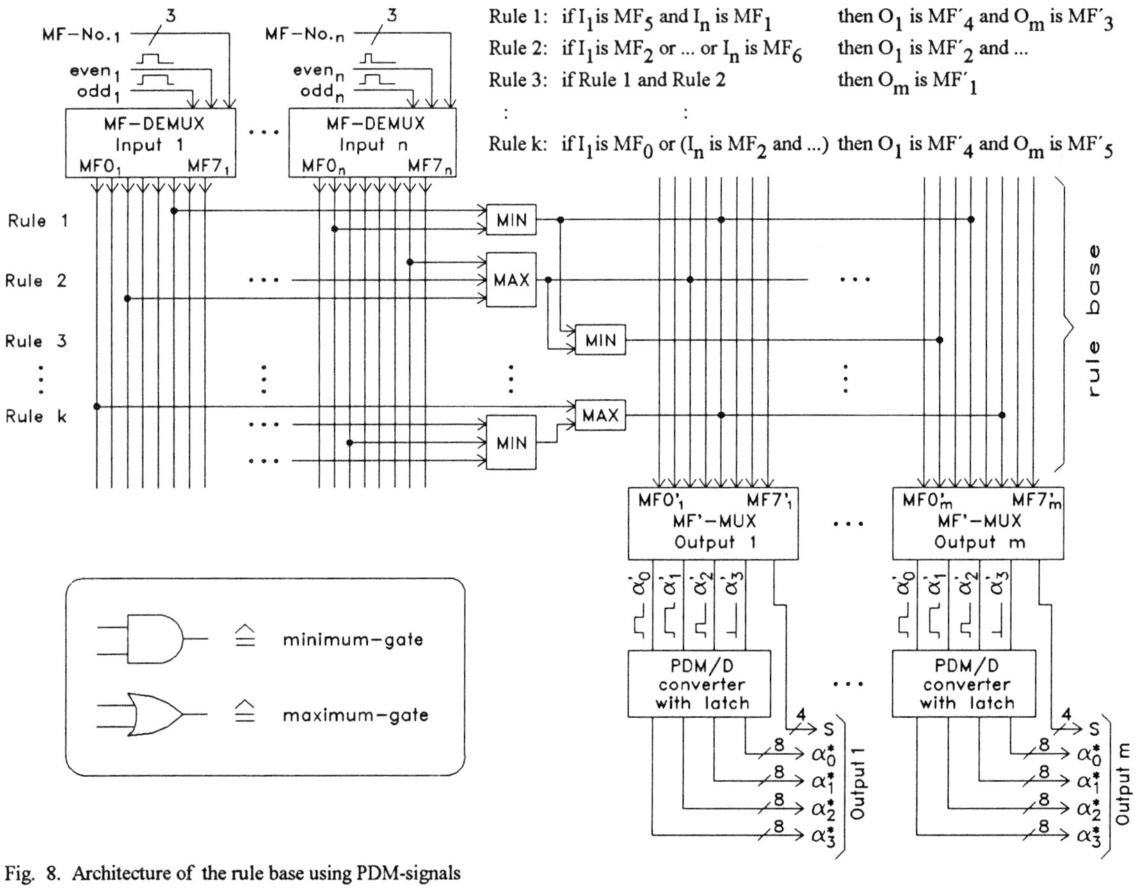


Fig. 8. Architecture of the rule base using PDM-signals

1 to m). If two or more rules using the same output MF′ (e.g. see Fig. 8: rule 1 and k), the rule with the highest α-value will be used. This is easily realized using a wired OR. In order to save chip area, (see chapter IV) only four of the eight MF′ input wires are switched to the output of the MF′-MUX. This is done by first evaluating the lowest MF′-number S of its active inputs and then switching this and the next three higher input signals to the four output wires $\alpha'_0$ to $\alpha'_3$. A MF′ is called "active" if the corresponding α-value is not zero. The MF′-MUX output signals $\alpha'_i$ are then converted by the PDM/D converter into digital numbers $\alpha^*_i$ and stored. This is the end of the second pipeline stage.

Obviously the rule base architecture offers flexibility, high speed operation and small chip area.

### *D. Inference unit*

The new architecture of the inference unit described in this chapter is based on the assumption that the degree of overlap of the membership functions is 2. Again two memory blocks denoted by odd and even are used for storing the output MFs. Fig. 9 depicts the architecture for one controller output with inference unit (third pipeline stage) and defuzzification. The operation of the MF-Generators are basically the same as described in chapter VI. A. The output values of the rule base $\alpha^*_i$ (see Fig. 8) are fed into two 3x8-bit multiplexers. The number S steers the select control unit. Depending on the value of S and the outputs of the MF-counters the outputs of both multiplexers contain either zero or one of the input values $\alpha^*_i$.

The inference for the even MFs will be shown in the following. On every system clock cycle the MF-Generator delivers a new value. If this value corresponds to an active MF′ the MUX switches the value $\alpha^*_i$ of this MF′ to the MIN-gate. Otherwise the MUX switches "zero" to the MIN-gate. The MIN-gate limits the values of the MF-Generator to the output of the MUX. Obviously the output of this MIN-gate corresponds to the "even" part of the resulting output function. The output of the "odd"-part is generated in the

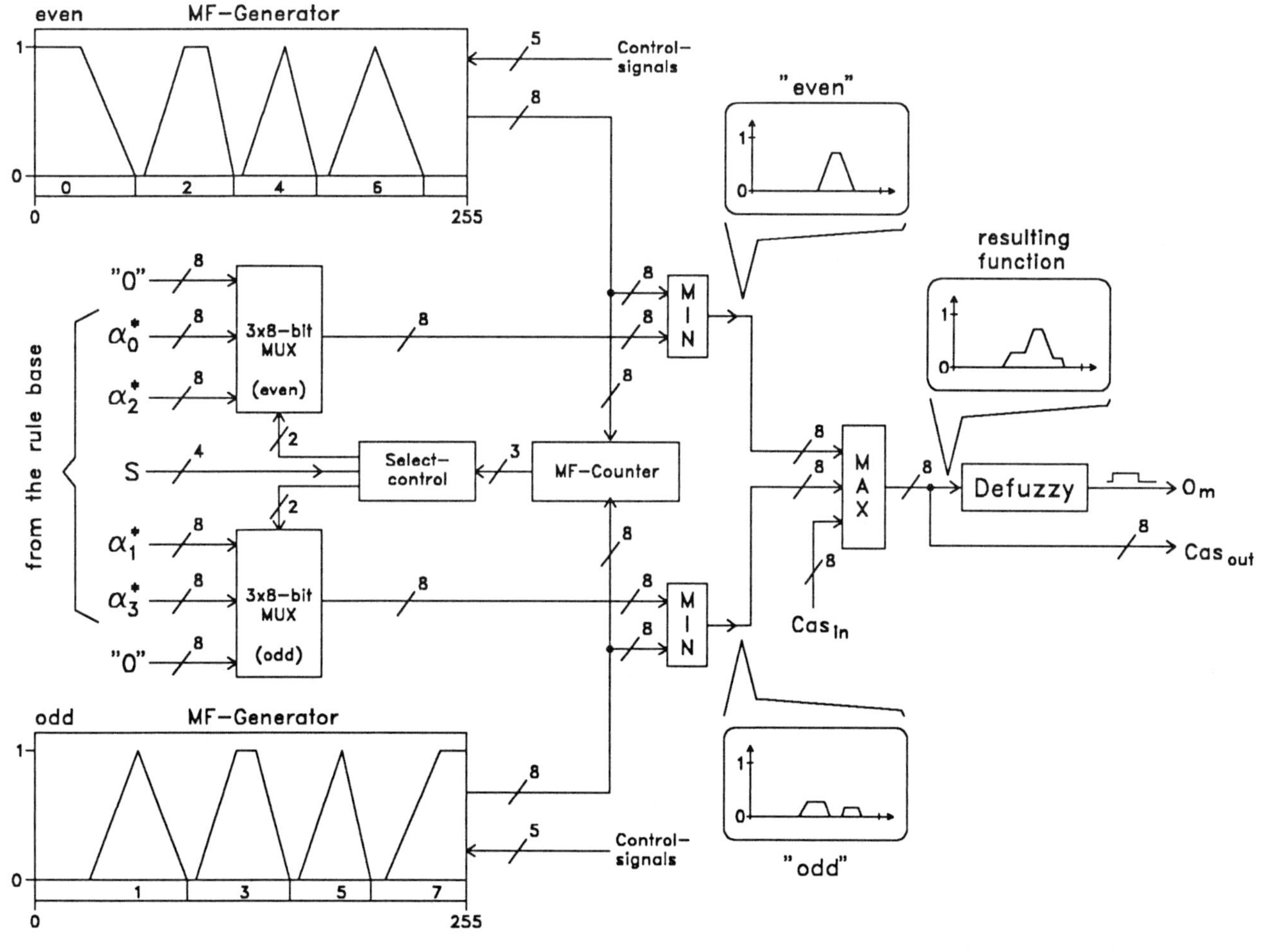


Fig. 9. Inference unit and defuzzification

same way. The MAX-gate combines the "even"- and the "odd"-part to one resulting output function. The controller is cascadable to build more complex systems by using the wires $CAS_{in}$ and $CAS_{out}$.

The advantages of the architecture are reduced memory (only two memory blocks) and a constant operation time not depending on the number of active rules.

### *E. Defuzzification*

The controller uses the COG method for defuzzification. For calculating the center of gravity we have used the algorithm shown in [1]. For an optimal pipeline architecture it was necessary to integrate the adders for calculation of the counter and of the denominator in the third pipeline stage. The division is executed in the last pipeline stage. Because every stage of the pipeline takes one main clock cycle, we implemented the devider by an adder to save chip area.

### *F. Control unit*

The control unit steers the pipeline and generates the following signals:

1. The system clock cycle (input clock dividing by four).
2. The main clock cycle (system clock cycle dividing by the x-resolution).
3. A power up reset.

## VII. SUMMARY

The use of PDM-signals allows to realize an architecture of a fuzzy controller with reduced chip area. A prototype of the controller with two inputs, one output and a resolution of 8-bit is implemented on FPGAs. We programmed the controller for an inverted pendulum with 19 rules and 7 MFs for each input and for the output. This takes less than 850 CLBs of the used four FPGAs (Xilinx: XC3090-100). This corresponds to about 10.000 gates. With an input clock of 6 MHz we achieved 6.000 controller operations per second.

Future work will speed up the controller with an ASIC-realisation. We expect to increase the operation time by at least 5 times.

## ACKNOWLEDGMENT

The authors would like to thank Prof. Frank (University of Dortmund) for supporting this work and Anke Goos for their assistence for preparing this paper.

## REFERENCES

[1] H. Watanabe, W.D. Dettloff and K.E. Yount, "VLSI Fuzzy Logic Controller with Reconfigurable, Cascadable Architecture," IEEE Journal of Solid-State Circuits, vol. 25, no. 2, pp. 376-382, April 1990

[2] H. Eichfeld, M. Löhner and M. Müller, "Architecture of a Fuzzy Logic Controller with optimized memory organisation and operator design," Int. Conf. on Fuzzy Systems, FUZZ-IEEE '92, San Diego, March 8-12, 1992, pp. 1317-1323

[3] Ansgar P. Ungering, Bashar Qubbaj und Karl Goser, "Geschwindigkeits- und speicheroptimierte VLSI-Architektur für Fuzzy-Controller," VDE-Fachtagung: "Technische Anwendungen von Fuzzy-Systemen", Dortmund, 12./13. November 1992, S. 317-325

[4] H.N. Teodorescu and T. Yamakawa, "Architectures for Rule-Chips Number Minimizing in Fuzzy Inference Systems," Proceedings of the 2nd International Conference on Fuzzy Logic & Neural Networks, Iizuka, Japan, July 17-22, 1992, pp. 547-550

[5] Omron Corp, "Programmable fuzzy logical circuit - converts input signals into pulse signal, with programmable logical circuit outputting pulse," patent, PN J03122720 A 910524 DW9127, 1989

[6] W.J.M. Kickert and E.H. Mamdani, "Analysis of a fuzzy logic controller", Fuzzy Sets and Systems 1, pp. 29-44, 1977

[7] L.A. Zadeh: "Fuzzy-Logik," IEEE Computer, pp. 83-92, April 1988

[8] H. Surmann, B. Moeller and K. Goser, "A distributing self-organizing fuzzy rule-based system," Neuro Nimes 92, november 2-6, 1992, pp. 187-194

[9] F. Deffontaines, A. Ungering, V. Tryba and K. Goser, "The concept of a RISC architecture for combining fuzzy logic and a Kohonen map on an integrated circuit," Neuro Nimes 92, november 2-6, 1992, pp. 555-564

# Efficient analog CMOS implementation of fuzzy rules by direct synthesis of multidimensional fuzzy subspaces

Oliver Landolt
Centre Suisse d'Electronique et de Microtechnique SA
Maladière 71
CH - 2007 Neuchâtel
Switzerland

***Abstract*** **- A novel way of evaluating the condition part of fuzzy rules is introduced in this paper. It requires only operations that can be efficiently performed with analog circuits. The degree of membership of an input vector to some fuzzy subspace is defined using a measure of distance between the input vector and some central point in the subspace. Since this approach avoids implementing operators like MIN or MAX, very dense analog hardware may be realised. A class of analog building blocks is presented, which allow a modular synthesis of arbitrary types of fuzzy rules. A circuit based on these blocks was integrated and tested successfully.**

## I. INTRODUCTION

The popularity of methods derived from the fuzzy set theory [1] is strongly increasing, particularly to solve control [1] problems. Fuzzy control [2] has proved to allow relatively short design times and good performance in a wide variety of low complexity applications. Implementing fuzzy control for a real world application may be done either purely by software on conventional processors, or by using dedicated hardware.

Although some work was done in the field of analog implementations [3,4], the major commercial products available now are digital processors improved for accelerating typical fuzzy operations. This type of circuits fits well for a lot of applications, as long as the constraints on size and speed of the circuits are not too high. However, many problems requiring tiny, low cost, stand-alone chips might be solved by the fuzzy approach as well, provided the corresponding hardware becomes commercially available. Analog circuits may fulfil these requirements, if very compact and simple basic cells are designed. The major features fuzzy controllers based upon such cells might provide include high speed due to a full parallelism, low power consumption, direct sensor or actuator interfacing and tiny chip area. The approach presented in this paper to design such basic cells consists of identifying the purpose of each step in the usual evaluation of a fuzzy rule, and choosing an equivalent function an opportunistic way to permit an efficient analog CMOS circuit implementation.

## II. CONVENTIONAL FUZZY CONTROLLER ARCHITECTURE

A fuzzy controller is an implementation of a set of rules of the type

"IF condition, THEN conclusion"

The premise of a rule consists of several single linguistic statements [5] involving input variables of the system, combined by operators such as "and" or "or". The conclusion is basically a set of statements assigning values to output variables. A rule defines the causal relationship between the inputs and the outputs for a given state of the system. Thus, a whole system may be described by a set of such rules in a finite number of states. Intermediate states are handled by taking several neighbouring rules into account in a weighted fashion, such as to provide an interpolation between the various explicit relations.

A description (i.e. specification or modelling) of a system by such a set of rules does not involve complex algebraic expressions, and is a way of translating intuitive knowledge into a machine usable form. Hence, it allows automatic processing of data and decision making by a computer or dedicated hardware.

The now classical way of dealing with the condition part of a rule is evaluating a membership function for each of the single statements, and use fuzzy logic gates such as MIN or MAX to compute the degree of truth of a complex condition out of the degrees of truth of the single statements. The overall result of this is defining indirectly a membership function of multiple variables, i.e. a *fuzzy subspace*. The example in Fig. 1 illustrates the equivalent function obtained by evaluating a condition with two inputs.

The actual shape a designer chooses for a membership function of an input variable is justified only by an intuitively satisfactory matching between its value and the correctness of the corresponding statement, over the definition range of the considered variable.

In the conclusion of each rule, a value is assigned to each output variable. These values may be represented by fuzzy sets, or simply by real numbers ("crisp values"). In the latter case, the overall output of a fuzzy controller is computed by the sum of the output values of the rules, weighted by their relative degrees of certainty.

---

[1] Patent registration in progress at submission time.

0-7803-0614-7/93$03.00 ©1993IEEE

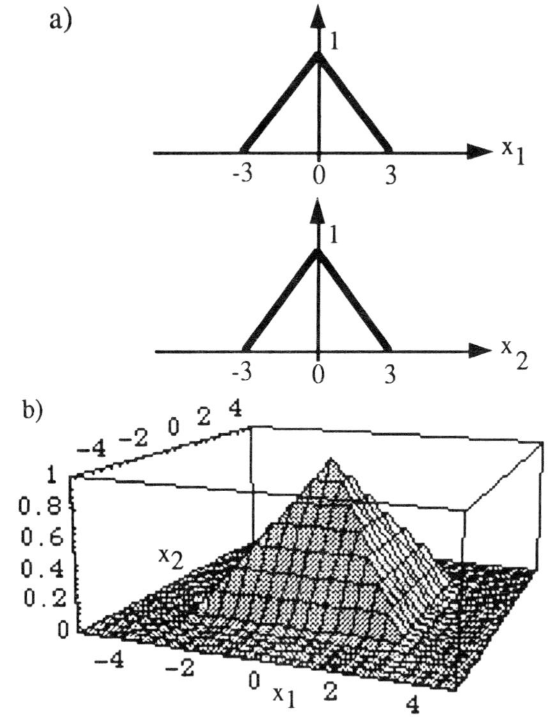


Fig. 1. Degree of truth of the expression "$x_1 \approx 0$ AND $x_2 \approx 0$"
a) membership function of each single statement
b) overall degree of truth for the combined expression

Realising an analog weighted sum circuit for the conclusion of a rule is simple. On the other hand however, the usual method for evaluating the premise of a rule doesn't suit well for a compact analog implementation. The approach presented in the following allows saving silicon by a one-stage evaluation of the premise.

## III. MULTIDIMENSIONAL MEMBERSHIP FUNCTIONS

The reason for defining first membership functions of one variable, and then combining the partial degrees of truth, is essentially formal. This procedure matches the structure of the linguistic expression that defines a rule. However, this fact is of limited importance, since the matching between MIN or MAX gates and the human's grasp of "and" and "or" is rather rough. Even though attempts have been made to define new operators to improve this matching [6], implementing those in dedicated hardware is unrealistic. Another reason for splitting an expression into its single components is the possibility offered, handling expressions of arbitrary complexity in a systematic way.

In many cases, specifying directly the membership function of an expression of several input variables, i.e. the fuzzy subspace in which a rule should be valid, may be preferable. The advantage of such an approach is not to be found on the theoretical level, but purely by practical considerations. In most applications of fuzzy control, the potential flexibility of defining conditions of arbitrary complexity *is not used.* Instead, the condition part is explicitly restricted to expressions using only ANDs, like the following:

IF $\text{condition}_1$ AND ... AND $\text{condition}_N$, THEN...

The subspace defined by such a condition is necessarily convex. A more direct description of such a subspace may be done by using the concept of *distance* between the input vector of a fuzzy controller and the center point of the subspace. For instance, the same condition as in Fig. 1 may be evaluated by an Euclidean measure of proximity, i.e. the complement to one of an Euclidean distance. The outline of the subarea obtained by this approach is depicted in Fig. 2. Since a membership function must take values in a bound range such as [0..1], the degree of membership in this example is defined as zero if the distance considered goes beyond a given limit.

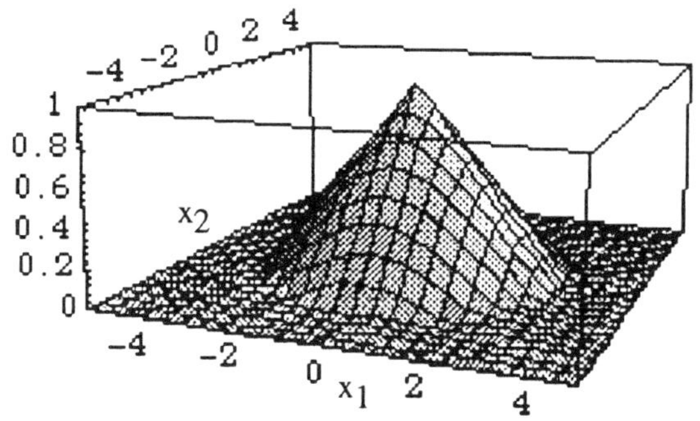


Fig. 2. Fuzzy subarea defined by a Euclidean measure of proximity

The major advantage of this approach resides in the possible efficiency of hardware implementations, especially for analog circuits. An opportunistic choice of the type of distance measure may lead to solutions with high performance as for circuit density, power consumption and accuracy. One possible realisation is presented further in this paper. The classical approach requires implementing operators, such as MIN or MAX, which are not well adapted to analog circuits. Although it is possible to realise such gates [7,8], the resulting circuits either offer poor performance, or are too space-consuming to fit high density integration requirements.

Note that using multidimensional membership functions has no incidence on the description of a system at the linguistic level. The intuitive matching between the expression of the first example, and the degrees of truth shown in Fig. 2, is as good as, or even better than, the pyramidal function displayed in Fig. 1.

## IV. ANALOG CMOS CURRENT-MODE IMPLEMENTATION

One possible implementation of this principle is being described here. We first consider a fuzzy controller with $n$ inputs $x_1..x_n$, and with rules of the type

IF $x_1 \approx A_1$ AND $x_2 \approx A_2$ ... AND $x_n \approx A_n$, THEN...

were the $A_i$'s are the central values of fuzzy subranges. The fuzzy subspace determined by the condition is chosen to be defined by the square of the Euclidean distance between the input vector ($x_1$ ... $x_n$) and the central position vector ($A_1$...$A_n$). The membership function writes

$$\mu(x_1,...,x_n) = 1 - \frac{(x_1 - A_1)^2}{K_1} - ... - \frac{(x_n - A_n)^2}{K_n} \qquad (1)$$

if this quantity is positive,

$$\mu(x_1,...,x_n) = 0 \qquad (2)$$

otherwise. This choice is made because squaring signals approximately may be done simply, by exploiting the characteristic of a saturated MOS transistor in strong inversion. In the above expression, the $K$'s are shape parameters that control the spread of the membership function. They might eventually be equal for each component.

The circuit shown in Fig. 3 evaluates a condition with one input only, for drawing convenience. This input is represented by a current $I_{in}$. The corresponding component of the position vector is $I_A$, the value of which may either be hard-wired in the circuit, or stored on some kind of memory to make it programmable. The main body of the circuit is based on Bult and Wallinga's current squarer [9]. The parameter $K$ in equation (1) is determined by the bias current $I_K$. Handling more inputs is done simply by duplicating this main body, and summing the output currents. The bias circuit may then be either common to the whole rule, or separate for each component depending on the desired flexibility.

The sum of the contributions of all components of a rule is simply subtracted from a unit current $I_{max}$. The difference is fed to the conclusion circuit of the rule through a diode to eliminate negative currents, and represents directly the degree of truth of the rule, i.e. the weight of this rule in the decision process.

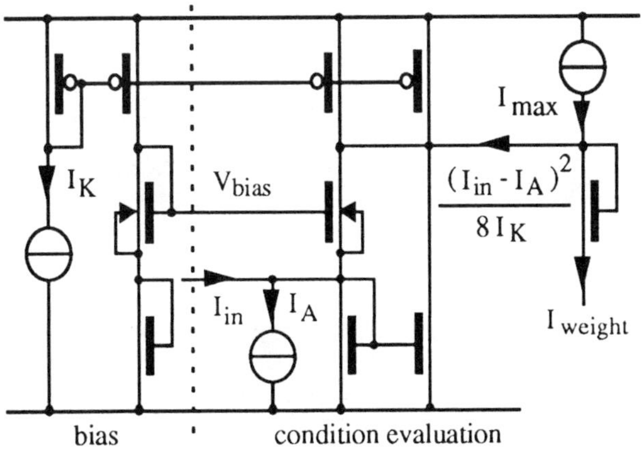


Fig. 3. Current-mode implementation of the condition evaluation

The circuit in Fig. 3 corresponds to the membership function in equation (1). It allows implementing bounded fuzzy subspaces only, i.e. conditions of the type "$x \approx A$". Variants of this circuit are shown in Fig. 4, which allow realising conditions of the type "$x < A$" and "$x > A$" as well. The three possible variants may be mixed arbitrarily for different components within a same rule.

The decision making consists of computing the weighted sum of the outputs, and may be implemented by a network of operational transconductance amplifiers (OTA) [10] as drawn in Fig. 5. One OTA connected in the follower configuration must be associated to each rule. The weight of the rule is represented by the transconductance of this OTA, which is set by its bias current. The output of the condition stage can thus simply be used as the bias current. If more than one output is required, this circuit may be duplicated, and the output of the condition stage must be replicated by a current mirror.

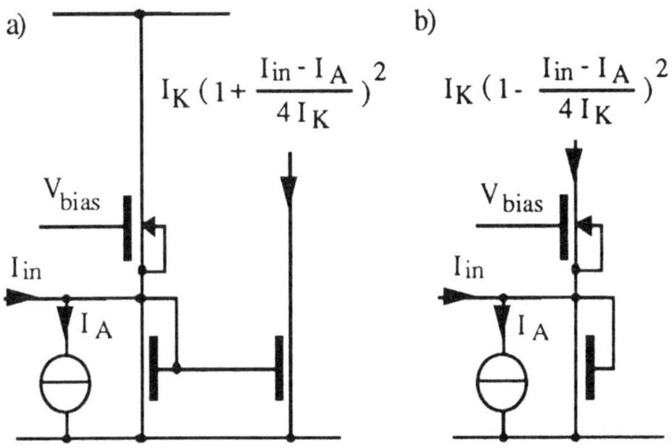


Fig. 4. Variants of the condition circuit
a) type "x < A"
b) type "x > A"

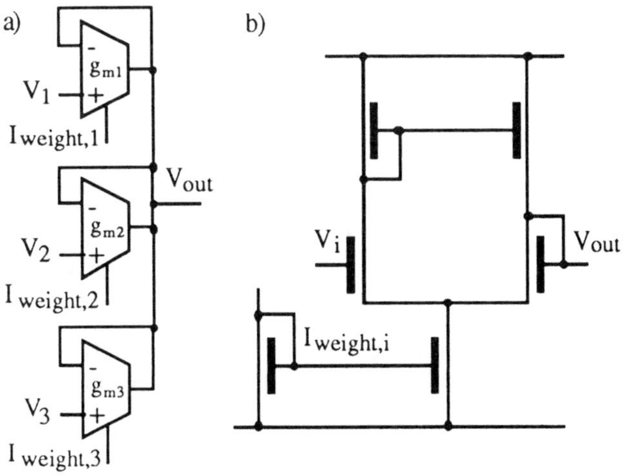


Fig. 5. Implementation of the conclusion part
a) principle of the weighted sum computation
b) circuit in each rule

## V. SIMULATION OF THE MEMBERSHIP FUNCTIONS

The shapes of the fuzzy subareas obtained as a translation of various linguistic expressions are shown in this section. The corresponding membership functions of several variables are displayed as a result of both the classical MIN approach and the circuit presented above. For the MIN approach, typical membership functions of one variable are chosen as shown in Fig. 6.

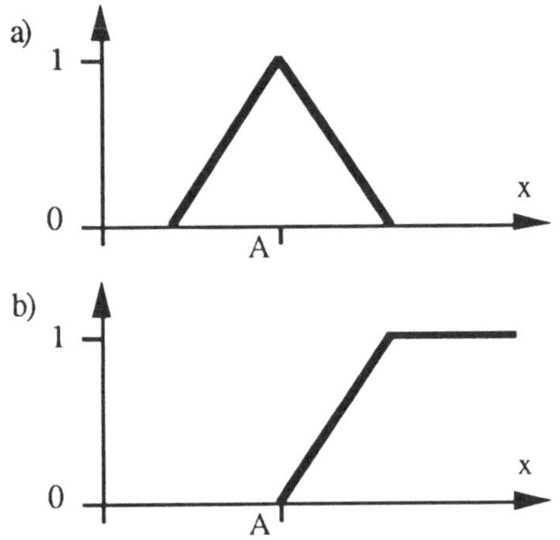


Fig. 6. Shapes of the membership functions used for the MIN approach
a) Expression of the type "$x \approx A$"
b) Expression of the type "$x > A$"

The following figures show membership functions of two variables for three different expressions. The value 1 represents the maximum influence of that rule. The case of a bound fuzzy subarea is illustrated in Fig. 7. Figures 8 and 9 represent expressions that remain valid in an unlimited region in one direction, respectively both directions.

With the particular shape chosen for the single membership functions, and the particular type of distance used here, the overall membership functions don't look very similar at first sight. However, they share the most important basic properties (continuity, monotony, convexity), which guarantees that the implementation of a set of rules by each of the two methods will yield control functions with equivalent properties. Moreover, the domain of the input space in which a rule is active, i.e. in which its membership function is higher than zero (or some other threshold), looks very similar in both cases.

In the particular case of the distance chosen here, one minor qualitative difference should be mentioned. The corners of the active region of a rule (as defined above) are smooth with the distance approach, whereas they are always sharp with the MIN approach. Covering the whole input space by a set of rules without leaving undefined regions will be slightly more difficult with rules with smooth corners. Overlapping adjacent rules sufficiently will overcome this minor drawback. Since the CMOS implementation of the distance approach yields so much simpler and more compact circuits, it is still highly justified using it.

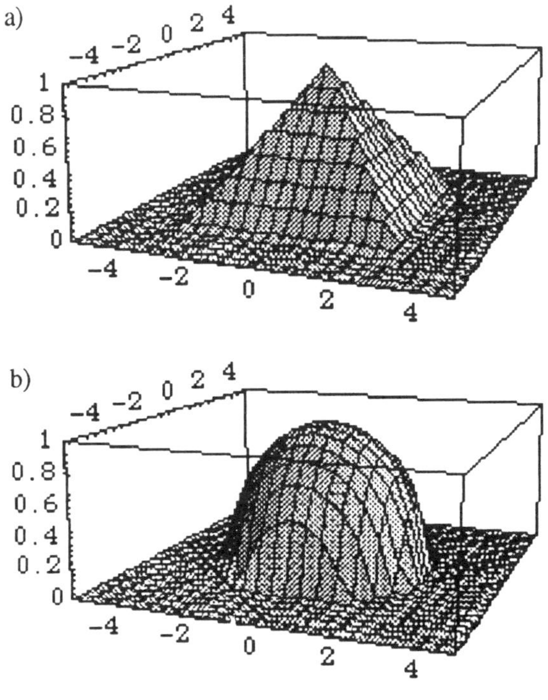


Fig. 7. Condition IF $x_1 \approx A_1$ and $x_2 \approx A_2$, THEN...
a) MIN approach
b) distance approach

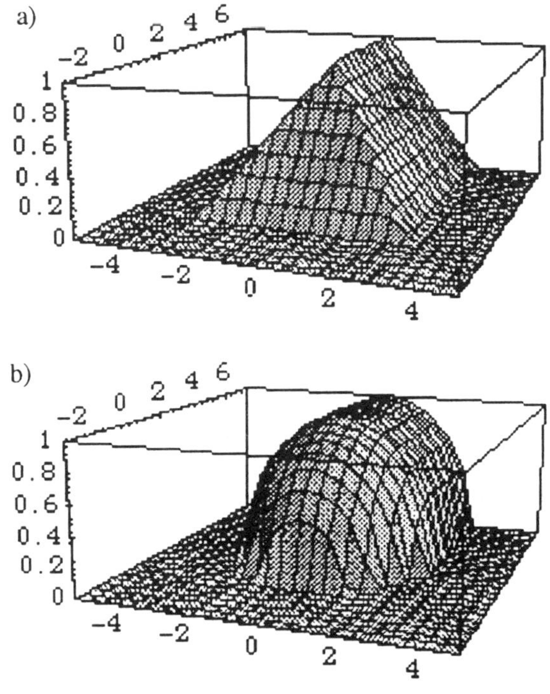


Fig. 8. Condition IF $x_1 > A_1$ and $x_2 \approx A_2$, THEN...
a) MIN approach
b) distance approach

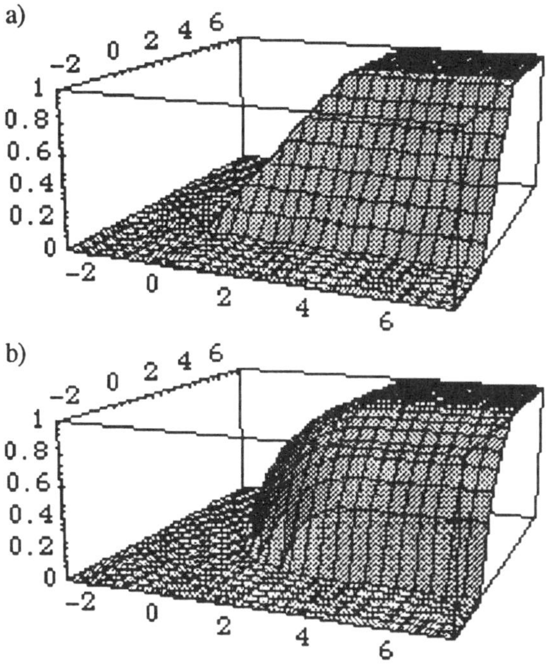


Fig. 9. Condition IF $x_1 > A_1$ and $x_2 > A_2$, THEN...
a) MIN approach
b) distance approach

## VI. TEST CIRCUIT

An experimental chip was designed using the building blocks described above. A group of two rules with four inputs and two outputs (group A) and a group of three rules with two inputs and two outputs (group B) are available. Each group of rules has its own set of inputs and outputs. Large rulebases may be created by using an arbitrary number of chips in parallel, and by connecting their outputs together. A proper current reference makes the resulting function independent of the spread of the process parameters from chip to chip.

The parameters describing each rule are stored on short-term capacitive memories, then converted to currents. The voltages on the capacitors may be applied from the outside of the chip. For this purpose, a common bus wide enough to describe entirely one rule runs through the whole chip. One of the rules may be selected by an addressing mechanism, and its storage capacitors may be refreshed to the current values on the bus. The content and structure of the chip are shown in Fig. 10.

The capacitive memories are able to keep their value within a few percent for at least 20 seconds, thus the refresh rate need not be very high. Nevertheless, some external refresh circuitry is necessary to use this chip. This is acceptable since the primary purpose of the circuit is testing and validating the new approach, rather than being used for real applications. Future work will include designing efficient long-term storage cells suitable for replacing the capacitors used here.

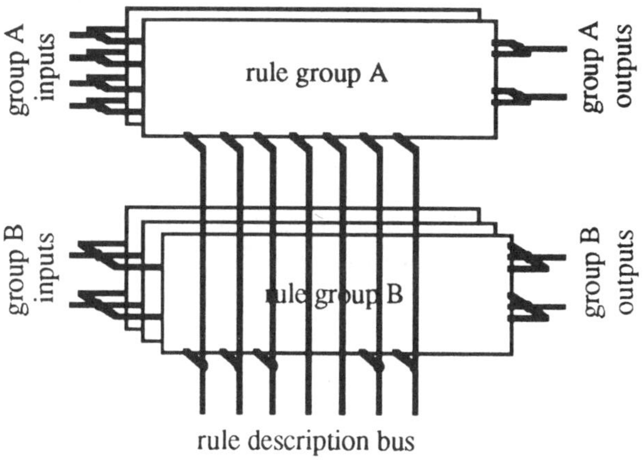


Fig. 10. Organisation of the test chip

The test chip was integrated using a 3µm, single-poly and single-metal process. A rule of type A requires less than $0.14mm^2$ of silicon, whilst a rule of type B needs about $0.1mm^2$. More than half of this surface is used by the storage capacitors and their associated V/I converters. More efficient storage and layout optimisation might lead to even much better density thank to the simple implementation of the main elements of the fuzzy rules. By using either hard-wired values in ASICs, or storage elements such as floating gates for programmable versions of a chip using these building blocks, it will be possible making the most of this approach.

## VII. CONCLUSION

The new way of evaluating fuzzy rules presented here offers practical advantages for hardware realisations. The proposed analog CMOS implementation, which is based on that principle, is believed to allow facing high constraints regarding circuit density, power consumption, speed and cost. Research is in progress designing an integrated circuit based on the same building blocks, and suitable for practical control or recognition applications.

## ACKNOWLEDGMENT

The author would like to thank Pierre Marchal for numerous discussions, and for doing the simulations for this paper.

## REFERENCES

[1] L. A. Zadeh, "Fuzzy sets", Information and Control 8, 1965, pp. 338-353

[2] L. A. Zadeh, "Outline of a new approach to the analysis of complex systems and decision processes", IEEE Transactions on System, Man and Cybernetics, Vol SMC-3, No 1, 1973, pp. 28-44

[3] T. Yamakawa, T. Miki, "The current mode fuzzy logic integrated circuits fabricated by the standard CMOS process", IEEE Transactions on Computers, vol. C-35, pp. 161-165, March 1986

[4] M. Sasaki, N. Ishikawa, F. Ueno and T. Inoue, "Current-mode analog fuzzy hardware with voltage input interface and normalisation locked loop", IEICE Trans. Fundamentals, Vol. E75-A, No 6, June 1992, pp. 650-654

[5] L. A. Zadeh, "The concept of a linguistic variable and its application to approximate reasoning", Information Sciences, Vol. 8, 1975, pp. 199-249

[6] H.J. Zimmermann, *Fuzzy Sets Theory - and Its Applications*, 2nd edition, Kluwer Academic Publishers 1991

[7] M. Sasaki, T. Inoue, Y. Shirai, F. Ueno, "Fuzzy multiple-input maximum and minimum circuits in current-mode and their analyses using bounded-difference equations", IEEE Transactions on Computers, vol. 39, No 6, June 1990, pp. 768-774

[8] K. Goser *et al*, "Hardware für Fuzzy-Controller", Proc. of 1st Dortmunder Fuzzy-Tag, June 1991

[9] K. Bult, H. Wallinga, "A class of analog CMOS circuits based on the square-law characteristic of an MOS transistor in saturation", IEEE Journal of Solid-State Circuits, vol. SC-22, No 3, June 1987, pp. 357-365

[10] C. Mead, *Analog VLSI and Neural Systems*, Addison Wesley, 1989

# FUZZY, PROBABILISTIC AND EVIDENTIAL REASONING IN FRIL

**J. F. Baldwin**
SERC Senior Research fellow
Engineering Mathematics Department
University of Bristol
Bristol BS8 1TR

**Abstract**
**The form, inference method and theory for the rules in the AI programming language Fril are discussed. The extended and evidential logic rules are additional to previous forms and allow for many new types of applications involving causal nets and case based reasoning. Semantic unification plays an important part in the inference process and this is also discussed.**

**Key Words** Fuzzy Sets, Logic Programming, Causal Nets, Expert System, Evidence

## I. INTRODUCTION

Two forms of uncertainty are important in the modeling of AI applications, namely, fuzzy, [Zadeh 1965], and probabilistic uncertainties. Fril, [Baldwin 1986, 87], [Baldwin et al 1988], is a programming language which allows both forms of uncertainty to be used for its basic knowledge representation. Prolog is a subset of Fril and PRUF, [Zadeh 1978], can be written in Fril.

Fril allows the following rules for knowledge representation involving uncertainty:

((h) ( (b1) (b2) ... (bn) ) ) :
( (u1 v1)(u2 v2) ... (un vn))

((h) (c1)(c2) ... (cn)) : ((x y)(s t))

((h) ( **f**(b1) n1 (b2) n2 ... (bn) nn) ) (u v)
where
"h" represent a sequence of words with the first being a predicate constant. "bi" represents (SEQ)...(SEQ) where SEQ is a sequence of words with the first being a predicate. "ci" represents SEQ.

For first rule
"((b1) ... (bn))" is the body of the rule and "(ui vi)" is an interval, called a support pair, containing Pr(h | bi).
It is also assumed that $\Sigma(\Pr(bi)) = 1$. This means that the conditions in the body of the rule are mutually exclusive and exhaustive.
For the second rule (c1) ... (cn) is the body of the rule, "(x y)" is an interval containing Pr(h|(c1)...(cn) ) and "(s t)" is an interval containing
Pr(h | ¬ {(c1)...(cn)} ). If the only the support pair (x y) is put at the end of the rule then it is assumed that (s t) = (0 1).
For the third rule
"((b1) n1 ... (bn) nn)" is the body of the rule, "bi" represent features and "ni" are importances of the features, (ni > 0). We choose $\Sigma$ ni = 1. **f** is a fuzzy set on [0, 1] and "(u v)" is a support pair where u is the support for (h | body) and 1 - v is the support for (¬ h | body).

## II. INFERENCE

### A INFERENCE USING POINT PROBABILITIES

Consider the rule
((h) ( (b1) (b2) ... (bn) ) ) : ( $(u_1)(u_2) \dots (u_n)$)
and facts
((bi)) : ($\alpha_i$) ; (all i)
then Fril concludes

$$((h)) : \sum_i u_i\alpha_i$$

This uses Jeffrey's rule, [Jeffrey 1965].
P'r(h) = Pr(h | b1)P'r(b1) + ... + Pr(h | bn)P'r(bn)
where P'r(bi) are the specific knowledge probabilities $\alpha_i$ and Pr(h | bi) are the general knowledge probabilities $u_i$. The use of Jeffrey's rule is consistent with the mass assignment updating rule given by [Baldwin 1991]. This is illustrated in the updating table 1 where xi = Pr(hbi) and yi = Pr(¬hbi).
Thus

$$P'r(h) = \sum_i Ki\ xi\ \alpha i = \sum_i \Pr(h \mid bi)\ \alpha i$$

since

$$\Pr(h \mid bi) = \frac{xi}{xi + yi} = Ki\ xi$$

The update rule used here results from minimizing the relative entropy of the update with the apriori, i.e.

$$\text{MIN} \sum_{H,B} P'r(HB) \operatorname{Ln} \frac{P'r(HB)}{\Pr(HB)}$$

If both the general knowlede and specific knowlege probabilities are derived from the same sample space then the rule is simply the total probabilities rule.

| Pr(HB) | $\alpha$1<br>b1 | ——— | $\alpha$n<br>bn | Update = P'r(HB) |
|---|---|---|---|---|
| x1 : hb1 | hb1 : K1x1$\alpha$1<br>Ø : 0 | ——— | Ø : 0 | hb1 : K1x1$\alpha$1 |
| xn : hbn | Ø : 0 | ——— | Ø : 0<br>hbn : Knxn$\alpha$n | hbn : Knxn$\alpha$n |
| y1 : ¬hb1 | ¬hb1 : K1y1$\alpha$1<br>Ø : 0 | | Ø : 0 | ¬hb1 : K1y1$\alpha$1 |
| yn : ¬hbn | Ø : 0 | ——— | Ø : 0<br>¬hb1 : Knyn$\alpha$n | ¬hbn : Knyn$\alpha$n |
| | K1=1 / (x1+y1) | ——— | Kn=1 / (xn+yn) | |

TABLE 1

### B INFERENCE USING SUPPORT PAIRS

Consider the inference for h when given
((h) ( (b1) (b2) ... (bn) ) ) :
( (u1 v1)(u2 v2) ... (un vn))
((bi)) : ($\alpha$i $\beta$i) ; (all i)

The inference is an interval version of that given for the point probability case. The inference is therefore

h : (z1 z2)
where

$$z1 = \text{MIN} \sum_i ui\ \theta i$$

$$\alpha i \le \theta i \le \beta i \quad \text{(all i)}$$

$$\sum_i \theta i = 1$$

$$z2 = \text{MAX} \sum_i vi\ \theta i$$

$$ai \le \theta i \le \beta i \quad \text{(all i)}$$

$$\sum_i \theta i = 1$$

These optimization problems are trivial requiring simple ordering of the {ui} and {vi} and an allocation algorithm for giving values to {θi} taking into account the sum equality constraint.

### III. CAUSAL NET EXAMPLE

Consider the example given by [Pearl 1988], Fig 1.
Pr(A) : Pr(a) = 0.2
Pr(B | A) : Pr(b | a) = 0.8 ; Pr(b | ¬a) = 0.2
Pr(C | A) : Pr(c | a) = 0.2 ; Pr(c | ¬ a) = 0.5
Pr(D | BC) :
Pr(d | bc) = 0.8 ; Pr(d | ¬bc) = 0.8
Pr(d | b ¬c) = 0.8 ; Pr(d | ¬b¬c) = 0.05

Pr(E | C) : Pr(e | c) = 0.8 ; Pr(e | ¬c) = 0.6

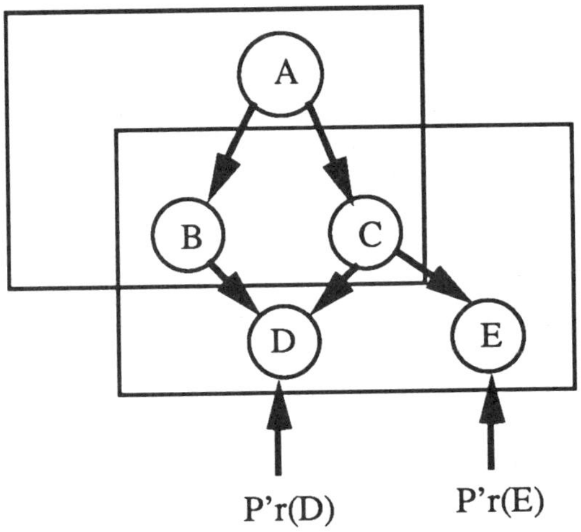


Fig 1

From these given values we can calculate
Pr(BC | DE), Pr(A | BC)
to provide the required conditional probabilities for the Fril rules.
The Fril rules are then given by
((a) ( ((bc)) ((b¬c)) ((¬bc)) ((¬b¬c)) ) :
((0.8) (0.4571) (0.2) (0.05))
((bc)
( ((d)(e)) ((d)(not e)) ((not d)(e)) ((not d)(not e)) ) ) :
((0.125) (0.0556) (0.0156) (0.006))
((b¬c)
( ((d)(e)) ((d)(not e)) ((not d)(e)) ((not d)(not e)) ) )
((0.6562) (0.7778) (0.0817) (0.0833))
((b¬c)
( ((d)(e)) ((d)(not e)) ((not d)(e)) ((not d)(not e)) ) )
((0.125) (0.0556) (0.0156) (0.006))
((b¬c)
( ((d)(e)) ((d)(not e)) ((not d)(e)) ((not d)(not e)) ) ) :
((0.0937) (0.1111) (0.8872) (0.9048))
((d)) : P'r(d)
((e)) : P'r(e)

These rules assume that the specific inputs for D and E are independent. If this assumption cannot be made then we replace all terms like
((d)( e)) by (conj (d) ( e)).
Fril will then calculate the conjunction as the support pair
$(0 \vee (P'r(d) + P'r(e) - 1) \quad P'r(d) \wedge P'r(e))$

The following results were obtained from Fril

| | | |
|---|---|---|
| For inputs | For inputs | For inputs |
| ((d)) : (0) | ((d)) : (0.1) | ((d)) : (0.2) |
| ((e)) : (1) | ((e)) : (0.9) | ((e)) : (0.9) |
| Fril gives | independent | independent |
| ((a)) : (0.0973) | ((a)) : (0.1297) | ((a)) : (0.1629) |

| | | |
|---|---|---|
| For inputs | For inputs | For inputs |
| ((d)) : (0 0.1) | ((d)) : (0.1) | ((d)) : (0 0.1) |
| ((e)) : (0.9 1) | ((e)) : (0.9 | ((e)) : (0.9 1) |
| independent | dependent | dependent |
| Fril gives | | |

((a)) : (0.1096 0.1337) ((a)) : (0.0925 0.1625) ((a)) : (0.0964 0.1625)

## IV. EVIDENTIAL LOGIC RULE

We introduce evidential support logic as an alternative to deductive logics for artificial intelligence applications which require a means of inductive and abductive reasoning. It is therefore applicable to such applications as case base reasoning, [Waltz 1990], in which concepts are defined in terms of examples illustrating the concept rather than by means of necessary and sufficient conditions. The logic allows generalization from examples to new situations. The body of a rule does not have to be fully supported in order that the conclusion of the rule be supported. If most of the terms in the body are supported then the head of the rule will have a high support. The terms in the body of the rule can be given different degrees of importances and these weights are taken into account when determining the support for the body. The support of the body is a function of the linear weighted sum of the individual term supports. This function can be chosen to give various interpretations for the body which can vary between the extremes of a conjunction of terms to a disjunction of terms. Intermediate interpretations can be thought of as softer forms of conjunction and softer forms of disjunction. Applications to handwritten character recognition have been shown to be successful.

The logic is derived for support pairs rather than single supports. A support pair consists of a lower support called the necessary support and an upper support called the possible support. One reason for deriving an evidential logic in terms of support pairs is to allow for partial matchings of terms. If we ask if a given object satisfies a certain concept which is defined in terms of a set of features then we can ask how many of these features are satisfied by the object. The answer might be that a certain percentage are satisfied, a certain percentage are not and no decision can be made concerning the others. This gives rise to a support for the match and a support against the match but these supports will not add up to 1. Thus a support pair is needed to represent the support for the match.

The form of representation allows the inclusion of fuzzy sets, [Zadeh 1965, 75, 78]. Fuzzy sets are also important for generalization in inductive reasoning. The fuzzification of precise values of features for a given example of a concept allows a new set of examples to represent the concept. This will be a fuzzy set so that there will be an ordering for this set of similar cases.
Partial matching in terms of fuzzy sets is derived using the notion of semantic unification. The theory is derived using a mass assignment theory given by the author, [Baldwin 1989, 1990, 1991, 1992]. The match of one fuzzy set given another is obtained as a mass assignment over the set {t, f}, i.e. a probability distribution over the power set of {t, f}. This is then converted to a support pair.
The latest version of the language Fril, [Baldwin 1988] incorporates this support logic.

The following are examples of valid sentences. Program in bold type.
**(good {a : 1, b : 0.7, c : 0.4})**
This defines the discrete fuzzy set good
**(recent [1970 : 0, 1982 : 1, 1992 : 1])**
This defines a continuous fuzzy set
**((book X is worth reading)**
**( ( (famous writer is author of X 0.2 )**
**(key words of text of X relevant 0.2)**
**(gives impression from scan through of X ok 0.3)**
**(review of X good 0.1)**
**(length of X suitable 0.05)**
**(publication date of X recent 0.1) ) )) : (0.9 1)**
(book ... reading) supported
IF
list of conditions,
famous writer ...
...
publication ....
'mostly' satisfied
**(not_too_big [150 : 0, 200 : 1, 250 : 1, 300 : 0])**
**(approx_250 [240 : 0, 250 : 1, 260 : 0])**
**(fairly_good {a : 0.3, b : 1, c : 0.5})**
More fuzzy set definitions
The following rules are used to define predicates

used in the evidential logic rule.
Support of head equal to support of body.

**((length of X suitable)**
**(conj (last page number of X is not_too_big)**
**(page size of X normal) ) ) : ((1 1)(0 0))**

Body is a conjunction of two conditions Normal Fril rule

**((famous writer is author of mind )) : (0.8 1)**

Support Logic fact

**((key words of text of mind relevant)) : (0.7 1)**
**((gives impression from scan through of mind ok )) : (0.7 0.9)**
**((review of mind fairly_good))**
**((publication date of mind 1980))**
**((last page number of mind is approx_250))**
**((page size of mind normal)) : 0.7**

The above statements make up a program. We could ask the query

**qs ((book mind is worth reading))**

support for the book is worth reading is ?

The evidential support logic system would return a support pair for the fact that the book "a theory of the mind" is worth reading. To determine this support pair the system determines a support for each of the terms of the body of the main rule and combines these in some way. The rule for this combining is part of the evidential support theory and takes account of the weights given in each term of the body and the function given as the first entry of the body. If this entry is missing then a default function is used. The support pairs for each of the terms are either found as facts or determined using other rules. These other rules may be evidential support rules involving or ordinary support rules. An example of an ordinary support rule is the one defining suitable length.

## V. Semantic Unification

Suppose a feature in the body of a rule has value **g** and the same feature associated with a given object has value **g'** where **g** and **g'** are fuzzy sets defined on G. We wish to find the mass assignment associated with the truth set of **g** given **g'**, denoted by $m_{(\mathbf{g} \mid \mathbf{g'})}$ defined over {t, f} where t represents true and f represents false.

Let $m_g$ = {Li : li} and $m_{g'}$ = {Mi : mi} be mass assignments associated with the fuzzy sets g and **g'**.

Form the matrix

$$M = \{T(Li \mid Mj) : li.mj\}$$

$$\text{where } T(Li \mid Mj) = \begin{cases} t \text{ if } Mj \subseteq Li \\ f \text{ if } Mj \cap Li = \emptyset \\ u \text{ otherwise} \end{cases}$$

u stands for uncertain and represents {t, f}.

The mass assignment $m_{(\mathbf{g} \mid \mathbf{g'})}$ is then given by

$$m_{(g \mid g')} = \begin{cases} t : \sum\limits_{\substack{i,j \\ T(Li \mid Mj) = t}} li.mj, \\ f : \sum\limits_{\substack{i,j \\ T(Li \mid Mj) = f}} li.mj, \\ u : \sum\limits_{\substack{i,j \\ T(Li \mid Mj) = u}} li.mj \end{cases}$$

This is called the multiplication model for semantic unification since the mass entries in the matrix M are obtained by multiplying the corresponding row and column masses. This model can be justified on the basis that the value of the feature for the object is not dependent on the value of the feature in the rule. Other semantic models can be defined.

We can rewrite the mass assignment for **g | g'** over {t, f} as a support pair for **g | g'**. The support pair for **g | g'** is given by

**Example**

Let **g** = a /1 + b / 0.7 + c / 0.2 ;
**g'** = a / 0.2 + b / 1 + c / 0.7 + d / 0.1

defined on G = {a, b, c, d, e} so that

$m_g$ = a : 0.3, {a, b} : 0.5, {a, b, c} : 0.2

$m_{g'}$ = b : 0.3, {b, c} : 0.5, {a, b, c} : 0.1, {a, b, c, d} : 0.1

giving matrix

| | 0.3 b | 0.5 {b, c} | 0.1 {a, b, c} | 0.1 {a, b, c, d} |
|---|---|---|---|---|
| 0.3 a | f 0.09 | f 0.15 | u 0.03 | u 0.03 |
| 0.5 {a, b} | t 0.15 | u 0.25 | u 0.05 | u 0.05 |
| 0.2 {a, b, c} | t 0.06 | t 0.1 | t 0.02 | u 0.02 |

t : 0.33
f : 0.24 giving **g | g'** = [0.33, 0.76]
u : 0.43

so that $m_{\mathbf{g} \mid \mathbf{g'}}$ = t : 0.33, f : 0.24, u : 0.43 giving **g | g'** = [0.33, 0.76].

**Theorem**

Let $g1, g'1 \subseteq_f G1, g2, g'2 \subseteq_f G2$

then

$$m_{(\mathbf{g1} \wedge . \mathbf{g2} \mid \mathbf{g'1} \wedge . \mathbf{g'2})} = m_{(\mathbf{g1} \mid \mathbf{g'1})} \wedge . \, m_{(\mathbf{g2} \mid \mathbf{g'2})}$$

where $\wedge .$ is the multiplication meet, [Baldwin 1992 ].

Furthermore if **g1 ∧. g2 | g'1 ∧. g'2** = [α, β], **g1 | g'1** = [α1, β1], **g2 | g'2** = [α2, β2] then
$\alpha = \alpha 1.\alpha 2$ ; $\beta = \beta 1.\beta 2$

## VI. EVIDENTIAL SUPPORT LOGIC

**Inference from single rule**

If the terms in the body of the rule of the form
(<head> (<func> ( <body>) ) : <supp_clause>
are replaced by the associated support pairs obtained directly from the program data or using semantic unification, then we obtain
(<head> (<func> $(\alpha_1\ \beta_1) : w_1 \ldots (\alpha_n\ \beta_n) : w_n)$) : <supp_clause>
where {wi} are the weights associated with the terms in the body and these sum to 1.

This is to be replaced by
(<head> body ) : <supp_clause>
body : $(\alpha\ \beta)$
from which we can infer
<head> : $(\gamma_1\ \gamma_2)$

In what follows we determine
(1) how to compute $(\alpha\ \beta)$ from the list of support pairs $\{(\alpha_i\ \beta_i)\}$, the function <func>, and the weights $\{w_i\}$
(2) how to determine $(\gamma 1\ \gamma 2)$ from $(\alpha\ \beta)$ and <supp_clause>.

Let <func> = S(x) where
$S(x) : [0, 1] \rightarrow [0, 1]$
and will be further discussed below.
The evidential support logic determines $(\alpha\ \beta)$ using

$$(\alpha\ \beta) = \left( S\left(\sum_{i}^{n} w_i\alpha_i\right)\ \ S\left(\sum_{i}^{n} w_i\beta_i\right) \right)$$

The choice of S(x) determines the nature of this combining operation.
Let <supp_clause> = ((x1 x2) (y1 y2))
The inference $(\gamma_1\ \gamma_2)$ is computed using the FRIL rule of inference, namely

$$\gamma_1 = \begin{pmatrix} x1.\beta + y1.(1 - \beta) \text{ if } x1 \leq y1 \\ x1.\alpha + y1.(1 - \alpha) \text{ if } x1 > y1 \end{pmatrix}$$

$$\gamma_2 = \begin{pmatrix} x2.\alpha + y2.(1 - \alpha) \text{ if } x2 \leq y2 \\ x2.\beta + y2.(1 - \beta) \text{ if } x2 > y2 \end{pmatrix}$$

This is in fact an interval version of Jeffrey's rule, [Jeffrey 1965], which is a special case of the mass assignment updating procedure given in [Baldwin 1992] and is the inference rule of support logic programming, [Baldwin 1986], used in FRIL.
A **special case** of this inference rule occurs when <supp_clause> = ((θ 1) (0 1)) which can be written as <supp_clause> = (θ 1)
The inference then gives the support pair $(\gamma_1\ 1)$ for the head where
$\gamma_1 = \theta\alpha$
This special case can be used for case based reasoning. A detailed treatment of case based reasoning is given in [Baldwin 1992].

The combining function S determines the interpretation of the body and hence the interpretation of the overall inference.

For example, if $w_i = 1/n$, (all i), and S(x) is defined as

$$S(x) = \begin{pmatrix} 1 \text{ if } x = 1 \\ 0 \text{ otherwise} \end{pmatrix}$$

then the body is equivalent to a logic conjunction of terms.
In this case $(\alpha\ \beta) = (1\ 1)$ or $(\alpha\ \beta) = (0\ 0)$ or $(\alpha\ \beta) = (0\ 1)$ depending on whether the support pairs $(\alpha_i\ \beta_i) = (1\ 1)$, (all i), or $(\alpha_i\ \beta_i) = (0\ 0)$ for some i, or $\alpha_i = 0$ for some i and $\beta_i = 1$, (all i), respectively.
If $w_i = 1/n$, (all i), and S(x) is defined as

$$S(x) = \begin{pmatrix} 1 \text{ if } x \geq 1/n \\ 0 \text{ if } x = 0 \end{pmatrix}$$

then the body, when the support pairs are (0 0) or (1 1), is equivalent to a logic disjunction of terms. If the support pairs are not restricted in this way, then the body is supported with support pair (1 1) if the lower supports of the terms of the body sum to at least 1. This is a generalization of the disjunction interpretation. interesting related work has been given by Yager (1992)

**Example**

We will consider the book example in the beginning of the paper concerning estimating whether to read a given book.

Body support pair calculation :-

| | Support | Weight |
|---|---|---|
| famous_writer is author of mind | [0.8 1] | 0.2 |
| key_words of text of mind relevant | [0.7 1] | 0.25 |
| scan_through ... impression | [0.7 0.9] | 0.3 |
| review of mind good | [0.55 0.79] | 0.1 |
| length of mind suitable | [0.63 0.7] | 0.05 |
| publication date of mind recent | [0.83 0.83] | 0.1 |

using

| Semantic Unification | Support |
|---|---|
| good \| fairly good | [0.55 0.79] |
| recent \| 1980 | [0.8333 0.8333] |

where support pair for length of mind suitable is calculated as

Semantic Unification | Support
---|---
not_too_big | approx_250 | [0.932 1]
normal | [0.7 0.7]
giving (conj ...) | [0.632 0.7]

Therefore the support pair for the body is given by
(α β) = (S(0.71493) S(0.91733)) = (0.7153 1)

Thus the head of the rule has support pair
(γ1 γ2) = ((0.9)(0.7153) 1) = (0.6438 1)

## VII COMBINING INFERENCES FROM VARIOUS RULES

Suppose we have several rules of the form
((variable Av is ai) ( (b1i) (b2i) ... (bni) ) ) :
((u_1i v_1i) ... (u_ni v_ni)) ; i= 1, ..., m
where {ai} are fuzzy sets on Sa.
Suppose
((bki)) : (α_ki β_ki) ; (all k, i)
then the inference methods above will give an inference of the form
((variable Av is ai)) : (xi yi)
which is converted to
((variable Av is a'i)) for i = 1, ..., m.
The final inference is
((variable Av is a'))
where

$$a' = \bigcap_i a'_i$$

## VIII. REFERENCES

**Baldwin J. F.**, 1986, "Support Logic Programming" in: Jones A.I et al., Eds., Fuzzy Sets Theory and Applications, Reidel, Dordrecht-Boston

**Baldwin J. F.**, 1987, "Evidential Support Logic Programming", Fuzzy Sets and Systems 24, pp 1-26.

**Baldwin J. F., Martin T. P. Pilsworth B. W.**, 1988, Fril Manual, Fril Systems Ltd.

**Baldwin J.F.**, (1989), A new approach to combining evidences for evidential reasoning,ITRC 151 Univ. of Bristol Report.

**Baldwin J.F.**, (1990), Computational models of uncertainty reasoning in expert systems, Computers Math. Applic., **19**, 105-119.

**Baldwin J.F.**, (1991), Combining evidences for evidential reasoning, Int. J. of Intelligent Systems, 6, 569-617

**Baldwin J, F**, (1992a), Evidential reasoning under probabilistic and fuzzy uncertainties, in An introduction to Fuzzy Logic Applications in Intelligent Systems, R. R. Yager and L. A. Zadeh, (Eds), Dordrecht: Kluwer

**Baldwin J.F.** (1992b), Inference for information systems containing probabilistic and fuzzy uncertainties, in Fuzzy Logic for the Management of Uncertainty, . L. A. Zadeh and J. Kacprzyk, (Eds), New York: Wiley, To Appear

**Baldwin J.F.** (1992c), Fuzzy and probabilistic uncertainties, in Encyclopedia of Artificial Intelligence, Second edition, S. C. Shapiro, (Ed), New York: John Wiley.

**Baldwin J. F.** (1992d), "A calculus for Mass Assignments in Evidential Reasoning", To Appear in "Advances in the Dempster-Shafer Theory of Evidence, (Ed Mario Fedrizza, Janusz Kacprzyk and Ronald Yager), Wiley & Sons, Inc. New York

**Baldwin J. F.** (1992e), "Mass Assignments and fuzzy sets for fuzzy databases", To Appear in "Advances in the Dempster-Shafer Theory of Evidence, (Ed Mario Fedrizza, Janusz Kacprzyk and Ronald Yager), Wiley & Sons, Inc. New York

**Baldwin J. F.** (1992f), "Evidential Support Logic, FRIL and Case Based Reasoning" To Appear in Int. J. Intelligent Systems.

**Baldwin J. F.** (1993), "Combining conclusions from support logic rules. Univ. of Bristol ITRC Report.

**Jeffrey R.**, (1965) 'The Logic of Decision", McGraw-Hill, New York

**Pearl J.**, (1988), "Probabilistic reasoning in Intelligent Systems", Morgan Kaufmann .

**Waltz D. L.**, (1990), Memory based reasoning, in Natural and Artificial Parallel Computing, (Eds M. A. Arbib, A. Robinson), MIT Press.

**Yager R. R.**, (1992), Families of OWA Operators, Tech. Report No. MII-1301

**Zadeh L**, (1965), Fuzzy sets, Information and Control, **8**, 338-353.

**Zadeh L**,. (1975), Fuzzy logic and approximate reasoning, Synthese, **30**, 407-428

**Zadeh L**, (1978), Fuzzy sets as a basis for a theory of possibility, Fuzzy Sets and Systems **1**, 3-28

**Zadeh L. A.**, 1978, "PRUF - A meaning representation language for natural languages", Int. J. Man-Machine Studies 10, 395-460.

**Zadeh L. A**, (1965), "Fuzzy sets", Information and Control, **8**,pp 338-353.

# A Mathematical Theory of Inference Channels

Sie-Keng Tan
Department of Mathematics, National University of Singapore
Singapore 0511, Republic of Singapore

Pei-Zhuang Wang
Institute of Systems Science, National University of Singapore
Singapore 0511, Republic of Singapore

***Abstract* — In this paper, we introduce the notion of inference channels and consider their truth values flowing among these channels. We shall study the mathematical structure of these inference channels based on a framework of knowledge representation and thus establish a theory of inference to gain an insight into the core of approximate reasoning.**

## I. INTRODUCTION

Since the emergence of approximate reasoning, several methodologies have been developed to manipulate imprecise and uncertain information. These techniques are primarily to assure an efficient derivation of reliable consequences from a knowledge base together with its rule base by means of generalized probabilistic reasoning, possibilistic reasoning or by the Dampster-Shafer calculus of evidence [4,5]. The main functions of these devices focus on the propagation schemes of fuzzy logic with truth values flowing from node to node through edges corresonding to some implication rules. In view of these varieties of different approaches of fuzzy inferences, we shall present a unified mathematical approach for the study of the theory of inference in this paper. The machinery of this model is the notion of inference channels with their truth value flow. We shall study the structure of these inference channels with respect to the framework of knowledge representation described in the theory of factor spaces [7], and then establish the foundation on which the present theory of approximate reasoning stands.

## II. INFERENCE CHANNELS

It is natural to view an inference as a process of determining the changes of truth values from a set of premises to another. Thus, if $P$ and $Q$ are extensions of some fuzzy concepts in the universes of discourse $X$ and $Y$ respectively, then an implication

$$(\forall(x,y)) \qquad \text{if P(x) then Q(y)}$$

where $(x,y) \in X \times Y$, can be considered as a device transferring the truth value of the antecedent $P(x)$ to that of the consequence $Q(y)$. Here, the truth value of $P(x)$ is considered as the membership degree of $x$ in $P$:

$$T(P(x)) = \mu_P(x).$$

In fact, more generally, the truth value need not be a numerical value between 0 and 1, it can also be a linguistic term such as rather true, very true and etc. We shall treat such an implication as an *inference channel*, or simply as a *channel*, denoted by an ordered pair $(P,Q)$ or by $(P \rightarrow Q)$, where truth values flow from one end to the other. We shall refer $P$ and $Q$ to as the *domain* and *codomain* of the channel $(P \rightarrow Q)$ respectively. The main difference between a channel and an implication is that a channel is an object consisting of a pair of extensions of concepts and it is not a proposition. The sole function of a channel is to convey truth values, it does not concern with the actual truth values involved.

In general, an implication $P \rightarrow Q$ may not be totally true and one may impose some degree of truth to the implication. Likewise, we can impose a *constraint coefficient* $\lambda$ to a channel $(P \rightarrow Q)$ in such a way that for any $x \in X$ and $y \in Y$,

$$\text{truth value of } Q(y) = \text{ truth value of } P(x) \wedge^* \lambda,$$

where the operation $\wedge^*$ may be interpreted as the min operation, the product operation or some appropriate operation. For convenience, we shall assume that $\wedge^*$ is the min operation in the sequel. In this case, we call $\lambda$ as the *capacity* of the channel $(P \rightarrow Q)$ and refer such a channel $(P \rightarrow Q)$ to as an $\lambda$ - *channel.* Thus each implication $P \rightarrow Q$ in the classical logic can be viewed as a 1-channel $(P \rightarrow Q)$ and vice versa.

Note that an implication $P \rightarrow Q$ is meaningful only when $P$ and $Q$ are related in a certain possible world.

Here we shall adopt the framework of knowledge representation described in the theory of factor spaces [7]. We also assume that the attributes of the objects in discourse can be unified and be represented in a factor space. The following is a brief account of the theory of factor spaces.

**Definition 1.** A *factor space* is a family of sets $\{X_f\}_{(f \in F)}$ where $F$ is a Boolean Algebra $F = (F, \vee, \wedge, ')$ satisfying the following two conditions:

(1) $X_0 = \emptyset$.

(2) If $T$ is an independent subset of $F$ (i.e. for any $t_1, t_2 \in T, t_1 \wedge t_2 = 0$ whenever $t_1 \neq t_2$.), then

$$X_{\vee\{t|t\in T\}} = \prod_{t \in T} X_t,$$

where the notations 0 and 1 are the least and greatest elements of $F$ respectively, and $\prod$ is the usual Cartesian product operation.

Each element $f$, $f \in F$, is called a *factor* and $X_f$ the corresponding *state space* of the factor $f$.

Now, let $C$ be a collection of objects in a universe of discourse. We say that $C$ can be represented by a factor space $\{X_f\}_{(f\in F)}$ provided that $F$ is composed by the factors describing the features of objects in $C$. That is, for each factor $f$ in $F$, there exists a (set-valued) mapping, again denoted by $f$, that maps each object $c$ in $C$ to the corresponding state space $X_f$ of $f$:

$$f: C \rightarrow X_f.$$

The image of $c$ under $f$ is said to be the *state* of $c$ in $X_f$. In terms of the Zadeh's possibility theory, we can interprete a factor as a linguistic variable, the state space of a factor $f$ as the domain of the base variable $f$, and the state of $c$ in $X_f$ as the linguistic value or possibility distribution of $c$ in $X_f$.

If $C$ can be represented by the factor space $\{X_f\}_{(f\in F)}$, then we say that the system $\Phi = (C, \{X_f\}_{(f\in F)})$ is a *knowledge framework* for $C$.

Now, for any set $X$, we shall denote by $\mathcal{P}(X), \mathcal{F}(X)$ the power set and the fuzzy power set of $X$ respectively:

$$\mathcal{P}(X) = \{0,1\}^X \quad \text{and} \quad \mathcal{F}(X) = [0,1]^X.$$

Given a knowledge framework $\Phi = (C, \{X_f\}_{(f\in F)})$ for $C$, let $\mathcal{D}_0 = \bigcup_{f\in F} \mathcal{P}(X_f)$ and $\mathcal{D} = \bigcup_{f\in F} \mathcal{F}(X_f)$. Then the set of all channels, denoted by $\tilde{\mathcal{L}}$, forms a fuzzy subset of $\mathcal{D}_0 \times \mathcal{D}_0$ or $\mathcal{D} \times \mathcal{D}$ by considering the capacity of the channel as the membership degree of that channel in $\tilde{\mathcal{L}}$. Let $\mathcal{L}_\lambda$ denote the $\lambda$ - cut of $\tilde{\mathcal{L}}$. Then we shall show that both $\tilde{\mathcal{L}}$ and $\mathcal{L}_\lambda$ have the struture of a lattice. To begin with, by observing some common nature of implications, we draw the following list of axioms for $\mathcal{L}_\lambda$.

**Axiom 1.** For any $P, Q \in \mathcal{D}$, if $P \subset Q$ then $(P \rightarrow Q)$ is a channel in $\mathcal{L}_\lambda$.

**Axiom 2.** If $(P \rightarrow Q)$ and $(Q \rightarrow R)$ are channels in $\mathcal{L}_\lambda$ then $(P \rightarrow R)$ is a channel in $\mathcal{L}_\lambda$.

From the above two axioms, we can easily derive the following properties:

**Proposition 1.** (i) If $(P \rightarrow Q)$ is a channel in $\mathcal{L}_\lambda$ and $P' \subset Q$ then $(P' \rightarrow Q)$ is a channel in $\mathcal{L}_\lambda$.

(ii) If $(P \rightarrow Q)$ is a channel in $\mathcal{L}_\lambda$ and $Q \subset Q'$ then $(P \rightarrow Q')$ is a channel in $\mathcal{L}_\lambda$.

Proposition 1 indicates that if we know that $(P \rightarrow Q)$ is a channel with capacity $\geq \lambda$, then we can obtain a collection of channels with capacity $\geq \lambda$ as long as the domain of the given channel becomes a smaller set or the codomain of the channel becomes a larger set. In other words, if $(P \rightarrow Q)$ is a $\lambda$ - channel, then $(P' \rightarrow Q')$ is also a $\lambda$- channel whenever $P' \subset P$ and $Q \subset Q'$. In view of this property, we introduce the following notion.

**Definition 2.** A channel $(P' \rightarrow Q')$ is said to be *deducible* from the channel $(P \rightarrow Q)$ provided that $P' \subset P$ and $Q \subset Q'$. In this case, we write

$$(P \rightarrow Q) \Longrightarrow (P' \rightarrow Q').$$

Definition 2 implies that if the implications in left hand side of '$\Longrightarrow$' attains a truth value of $\lambda$ then so is that in the right hand side. It also means that the statement represented in the right hand side of '$\Longrightarrow$' is more agreeable or acceptable than that in the left hand side. However, we would like to stress that this does not mean that the right hand side possesses more valuable information than the other. In fact, on the contrary, it possesses less valuable information by the principle of minimum specificity. Here we introduce the definition of valueness of information in the context of inference channels:

**Definition 3.** A channel $(P \rightarrow Q_1)$ is said to possess more valuable information than the channel $(P \rightarrow Q_2)$ provided that $Q_1 \subset Q_2$. A channel $(P \rightarrow Q)$ is *non - reducible* in $\mathcal{L}_\lambda$ if there is no channel $(P \rightarrow Q')$ in $\mathcal{L}_\lambda$ that possesses more valuable information than $(P \rightarrow Q)$.

**Axiom 3.** (i) If $(P \rightarrow Q_1)$ and $(P \rightarrow Q_2)$ are channels in $\mathcal{L}_\lambda$, then $(P \rightarrow Q_1 \cap Q_2)$ is a channel in $\mathcal{L}_\lambda$.

(ii) If $(P_1 \rightarrow Q)$ and $(P_2 \rightarrow Q)$ are channels in $\mathcal{L}_\lambda$, then $(P_1 \cup P_2 \rightarrow Q)$ is a channel in $\mathcal{L}_\lambda$.

**Proposition 2.** If $(P_1 \rightarrow Q_1)$ and $(P_2 \rightarrow Q_2)$ are channels in $\mathcal{L}_\lambda$, then

$(P_1 \cup P_2 \rightarrow Q_1 \cup Q_2)$ and $(P_1 \cap P_2 \rightarrow Q_1 \cap Q_2)$ are channels in $\mathcal{L}_\lambda$.

Proposition 2 can be verified easily, it indicates that if we have a set of inference rules in our knowledge base, then by taking the appropriate union and intersection of the domains and codomains of the corresponding channels, we can creat a larger collection of inference rules.

**Definition 4.** The operations of joint and meet in $\mathcal{L}_\lambda$ are defined as follows:

$$(P_1 \to Q_1) \vee (P_2 \to Q_2) \triangleq (P_1 \cup P_2 \to Q_1 \cup Q_2),$$

$$(P_1 \to Q_1) \wedge (P_2 \to Q_2) \triangleq (P_1 \cap P_2 \to Q_1 \cap Q_2).$$

It can be shown that for any $\lambda$, $(\mathcal{L}_\lambda, \vee, \wedge)$ is a distributive lattice. We shall call this lattice *a lattice of $\lambda$ - channels* based on the knowledge framework $\Phi = (C, \{X_f\}_{(f \in F)})$ for $C$. Similarly, it can be verified that $(\tilde{\mathcal{L}}, \vee, \wedge)$ also forms a lattice called the *the lattice of fuzzy channels* based on $\Phi$.

**Definition 5.** For any subset $\mathcal{D}$ of the lattice of $\lambda$ - channels $\mathcal{L}_\lambda$, let

$$[\mathcal{D}]_{\vee,\wedge,\Rightarrow}$$

$$\triangleq \bigcap \{\mathcal{E} | \mathcal{D} \subset \mathcal{E} \subset \mathcal{L}_\lambda, \mathcal{E} \text{ is closed under } \vee, \wedge \text{and} \Rightarrow\},$$

then we say that $[\mathcal{D}]_{\vee,\wedge,\Rightarrow}$ is the $\vee, \wedge, \Rightarrow$ - *closure* of $\mathcal{D}$ in $\mathcal{L}_\lambda$. If

$$[\mathcal{D}]_{\vee,\wedge,\Rightarrow} = \mathcal{L}_\lambda$$

then we say that $\mathcal{L}_\lambda$ is *generated* by $\mathcal{D}$ and that $\mathcal{D}$ is a *generating set* for $\mathcal{L}_\lambda$. If $\mathcal{D}$ is a finite set, then $\mathcal{L}_\lambda$ is said to be *finitely generated* by $\mathcal{D}$.

In practice, we usually need only to consider the knowledge framework of a class of objects on a smaller factor space. For simplicity, we shall only consider channels with capacity 1. Thus, assume that $\mathcal{L}$ is a lattice of channels on a knowledge framework $\Phi = (C, \{X_f\}_{(f \in F)})$ for $C$ and $F'$ a Boolean sublattice of the Boolean lattice $F$. Let $\mathcal{D}'_0 = \bigcup_{f \in F'} \mathcal{P}(X_f)$, and $\mathcal{L}'$ the restriction of $\mathcal{L}$ in $\mathcal{D}'_0 \times \mathcal{D}'_0$. That is,

$$\mathcal{L}' : \mathcal{D}'_0 \times \mathcal{D}'_0 \longrightarrow [0, 1],$$

where $\mathcal{L}' = \mathcal{L}$ when restricted on $\mathcal{D}'_0 \times \mathcal{D}'_0$. It can be readily shown that $\mathcal{L}'$ is a lattice of channels on the knowledge framework $\Phi' = (C, \{X_f\}_{(f \in F')})$. We shall call $\mathcal{L}'$ the *restriction* of $\mathcal{L}$ on $F'$. In particular, if $F' = \{0, f, g, 1\}$ is a Boolean lattice generated by two factors $f$ and $g$, and if $X = X_f$ and $Y = X_g$, then the restriction of $\mathcal{L}$ on

$$(\mathcal{P}(X) \times \mathcal{P}(X)) \cup (\mathcal{P}(X) \times \mathcal{P}(Y)) \cup (\mathcal{P}(Y) \times \mathcal{P}(Y)),$$

denoted by $\mathcal{L}^*$, is also a lattice of channels, called the *lattice of channels from $X$ to $Y$*.

## III. LATENT RELATIONS AND LATENT MATRICES

In this section, we shall investigate the basic core of approximate reasoning. For the ease of illustration, let us simply consider a knowledge framework $\Phi = (C, \{X_f\}_{(f \in F)})$ with $F = \{0, f, g, 1\}$. Let $X = X_f$ and $Y = X_g$ and assume that $\mathcal{D} = \{(P_i \to Q_i)\}_{(i=1,2,\ldots,n)}$ is a given set of channels with domain $X$ and codomain $Y$. Let $\mathcal{L}$ be the lattice generated by $\mathcal{D}$ and assume that $\mathcal{L}$ is *regular* in the sense that it satisfies the following conditions:

for any $x \in X$, there exists an $i$ such that $x \in P_i$ with $(P_i \to Q_i) \in \mathcal{D}$ and that

$$\bigcap \{Q_i | (P_i \to Q_i) \in \mathcal{D} \text{ and } x \in P_i\} \neq \emptyset.$$

Let us denote the set $\bigcap \{Q_i | (P_i \to Q_i) \in \mathcal{D} \text{ and } x \in P_i\}$ by $G(x)$, then with these notations, we define the notion of latent relation and latent matrix as follows:

**Definition 6.** Let $\mathcal{L}$ be a regular lattice generated by $\mathcal{D}$. Then the set $G$ defined by

$$G = \bigcup_{x \in X} \{(x, y) | y \in G(x)\}$$

is a relation from $X$ to $Y$ called the *latent relation* of $\mathcal{L}$. When both $X$ and $Y$ are finite sets, $G$ can be represented by a Boolean matrix called the *latent matrix* of the lattice $\mathcal{L}$.

We now state without proof the main theorem of this paper.

**Theorem 1.** ( Representation Theorem ) If $\mathcal{L}$ is a regular lattice finitely generated by a given set of channels $\mathcal{D}$, then the latent relation $G$ of $\mathcal{L}$ is uniquely determined by $\mathcal{L}$ and is independent of the choice of the generating set $\mathcal{D}$ of $\mathcal{L}$. Conversely, for any relation $G$ from $X$ to $Y$, a corresponding lattice $\mathcal{L}$ of $G$ can be constructed in the following way:

For any $P \in \mathcal{P}(X)$ and $Q \in \mathcal{P}(Y)$, $(P \to Q)$ is a channel in $\mathcal{L}$ if and only if the following inclusion relation holds:

$$(P \times Y) \cap G \subset (X \times Q) \cap G.$$

Moreover, the lattice $\mathcal{L}$ is a regular lattice such that the latent relation of $\mathcal{L}$ is the given relation $G$.

In general, the set $\mathcal{D}$ of channels is obtained from a rule base of knowledge acquired from experience, experiments or other means. It is usually imprecise or incomplete, and so the inferences derived from this knowledge base may not be reliable. Indeed, we have to have a certain criterion or point of reference to decide the reliability of the consequences. For this purpose, we have to consider the relation $R$ obtained from the set of images of all the elements $c$ in $C$ under the mapping

$$f \vee g : C \to X_{f \vee g}.$$

Since $X_{f \vee g} = X_f \times X_g = X \times Y$, R is in fact a relation from $X$ to $Y$. We shall call $R$ the *canonical relation* obtained from $C$. This canonical relation is actually the foundation on which our reasoning is based upon. An implication $P \rightarrow Q$, where $P \subset X$ and $Q \subset Y$, is valid or reliable only when the following property holds:

$$(P \times Y) \cap R \subset (X \times Q) \cap R.$$

Thus, the process of inference induced from the implication $P \rightarrow Q$ is as follows. Firstly we take the cylindrical extension of $P$ to $X \times Y$, then intersect with the canonical relation R, and then take the projection of the intersection to $Y$, the resulting subset $Q^*$ will then possess most valuable information that can be derived from the implication. In fact, the channel $(P \rightarrow Q^*)$ is a non-reducible channel in $\mathcal{L}$ with most reliable information. Thus the canonical relation $R$ can be considered as the point of reference for our approximate reasoning. The latent relation $G$ obtained from the lattice $\mathcal{L}$ generated by $\mathcal{D}$ is usually an approximation of $R$. Thus the accuracy of our knowledge of $C$ will depend on the relationship between $R$ and $G$. A necessary condition for our knowledge of $C$ to be accurate is that

$$G \supset R.$$

If $R - G \neq \emptyset$, then it indicates that our knowledge is not error-free and thus should be modified. Actually, the set $G-R$ represents an indicator of the accuracy of our knowledge; the smaller the set $G - R$ is, the better knowledge we possess. Thus it is an important topic to study how to minimise the difference set $G - R$.

In the applications of artificial intelligence, one of the main issues is on the integration of incomplete or unsubstantial knowledge to form a more reliable knowledge base. Here, based on the structure of inference channels, we suggest the following operation to integrate those information.

Assume that $\mathcal{L}_1$ and $\mathcal{L}_2$ are lattices of channels based on the same knowledge framework $\Phi$. Here, the assumption that $\mathcal{L}_1$ and $\mathcal{L}_2$ are based on the same framework indicates that the basic knowledge in discourse are of the same nature; whereas $\mathcal{L}_1$ and $\mathcal{L}_2$ are distinct implies that the corresponding knowledge base are different. It is our intention to form a new lattice of channels that represents a collection of more reliable and complete knowledge drawn from the given two knowledge bases. This new lattice will be constructed via the amalgamation of the latent relations and the latent matrices of $\mathcal{L}_1$ and $\mathcal{L}_2$ respectively.

**Definition 7.** Let $\mathcal{L}_1$ and $\mathcal{L}_2$ be lattices of channels from $X$ to $Y$. Assume that $G_1$ and $G_2$ are the latent relations of $\mathcal{L}_1$ and $\mathcal{L}_2$ respectively, then the *amalgamated product* of $G_1$ and $G_2$, denoted by $G_1 \wr G_2$, is a relation from $X$ to $Y$ defined in the following way:

For any $x \in X$, let $G_i(x) = \{y \in Y | (x, y) \in G_i\}, i = 1,2$ , and let

$$G(x) = \begin{cases} G_1(x) \cap G_2(x) & \text{if } G_1(x) \cap G_2(x) \neq \emptyset; \\ G_1(x) \cup G_2(x), & \text{otherwise.} \end{cases}$$

Then

$$G_1 \wr G_2 \triangleq \bigcup_{x \in X} \{(x, y) | y \in G(x)\}.$$

The corresponding lattice of $G_1 \wr G_2$ will be denoted by $\mathcal{L}_1 \wr \mathcal{L}_2$ and is called the *amalgamated product* of $\mathcal{L}_1$ and $\mathcal{L}_2$.

We can interpret the amalgamated product in the following way. In the case when $G_1(x)$ and $G_2(x)$, where $x \in X$, are not disjoint, then the inference channels obtained from the common information are certainly correct and so we have to select the intersection of $G_1(x)$ and $G_2(x)$. On the other hand, if $G_1(x)$ and $G_2(x)$ are disjoint, then it implies that there are some inconsistent or contradicting information in the knowledge base of $G_1$ and $G_2$, so we can only choose either $G_1(x)$ or $G_2(x)$ for our inference purpose, thus the union of $G_1(x)$ and $G_2(x)$ is selected.

We now give an example to illustrate this operation of amalgamated product. Assume that the latent matrices of $G_1$ and $G_2$ are given as follows:

$$G_1 = \begin{pmatrix} 1 & 1 & 0 & 0 & 1 \\ 1 & 0 & 1 & 0 & 0 \\ 1 & 1 & 1 & 0 & 0 \\ 0 & 0 & 1 & 1 & 0 \end{pmatrix} \qquad G_2 = \begin{pmatrix} 0 & 1 & 1 & 0 & 1 \\ 0 & 1 & 0 & 1 & 0 \\ 0 & 0 & 0 & 1 & 1 \\ 0 & 1 & 1 & 0 & 1 \end{pmatrix},$$

then the latent matrix of the amalgamated product of $G_1$ and $G_2$ is

$$G_1 \wr G_2 = \begin{pmatrix} 0 & 1 & 0 & 0 & 1 \\ 1 & 1 & 1 & 1 & 0 \\ 0 & 0 & 1 & 0 & 0 \\ 0 & 0 & 1 & 0 & 0 \end{pmatrix}$$

Finally, we would like to point out that the theory of inference channels is consistent with the existing theory of approximate reasoning. To quote a simple example, assume that an implication $P \rightarrow Q$ is given, where $P$ and $Q$ are crisp sets, then by choosing $\mathcal{D} = \{(P \rightarrow Q), (P^c \rightarrow Y)\}$, where $P^c = X - P$, to be the generating set for the lattice $\mathcal{L}$ of channels, we can show that the latent relation of $\mathcal{L}$ is precisely equal to $(P \times Q) \cup (P^c \times Y)$, which turns out to be the classical inference relation. When $P$ and $Q$ are fuzzy sets, then the theory of inference channels can be incorporated with that of falling shadows [9 ] to deduce those fundamental expressions of fuzzy inference relations [6,10].

## IV. CONCLUSION

It is our intention to establish a theory for the study of inference with the following features:

(1) Simplicity: for the theory to be implemented easily in the practical world.

(2) Competence: for the theory to be applied effectively and efficiently.

(3) Compatible: for the theory to be consistent and compatible with the existing theory of approximate reasoning.

As the theory of inference channels can also be presented in the form of graphs and networks, it can be integrated with the theory of neural networks and thus will be even more powerful in its applications.

## REFERENCES


[1] D. Dubois, H. Prade, *Necessity Measures and the Resolution Principle,* IEEE Transactions on Systems, Man, and Cybernetics, Vol. SMC - 17, Nov. 3, 1987, pp. 474 – 478.

[2] I.R. Goodman, H.T. Nguen, *Uncertainty Models for Knowledge-Based Systems,* Elsevier Science Publishers B.V. 1985.

[3] Ren Ping, *Generalized Fuzzy Sets and Representation of Incomplete Knowledge,* Fuzzy Sets and Systems, Vol 36, 1990, pp. 91 – 96.

[4] E.H. Ruspini, *Approximate Reasoning: Past, Present, Future,* Technical Note No. 492, SRI International, June 1990, pp. 1 – 22.

[5] G. Shafer, *A Mathematical Theory of Evidence,* Princeton, NJ: Princeton University, 1976.

[6] S.K. Tan, E. Stanley Lee, *Semantically Dependence Fuzzy Set Operations and Inference Relations,* Cybernetics and Systems Research '92. Vol. 1. 1992, pp. 463 –470.

[7] P.Z. Wang, *A Factor Space Approach to Knowledge Representation,* Fuzzy Sets and Systems, Vol. 36, 1990, pp. 113 – 124.

[8] P.Z. Wang, *Factor Space, in Approximate Reasoning Tools for Artificial Intelligence,* Verlag TUV Rheinland, 1990, pp. 62 –79.

[9] P.Z. Wang, *Fuzziness vs Randomness, Falling Shadow Theory,* BUSEFAL No. 48, 1991,

[10] P.Z. Wang, H.H. Teh, S.K. Tan, *Fuzzy Inference Relation Theory Based on the Shadow-Representation Approach,* Proceedings of the 8th International Conference of Cybernetics and Systems, Now York, 1990, pp. 30 – 31.

[11] P.Z. Wang, H.M. Zhang, X.T. Peng, W. Xu, *Truth-valued-flow Inference,* BUSEFAL No.38, 1989, pp. 130 – 139.

[12] L.A. Zadeh, *Fuzzy Sets as a Basis for a Theory of Possibility,* Fuzzy Sets and Systems, Vol. 1, 1978, pp. 3 – 28.

# A Calculation Method for Solving Fuzzy Arithmetic Equations with Triangular Norms

Mayuka F. KAWAGUCHI Tsutomu DA-TE
Department of Information Engineering, Faculty of Engineering, Hokkaido University,
Kita 13, Nishi 8, Kita-ku, Sapporo 060, JAPAN

***Abstract*— This paper deals with the equations involving fuzzy arithmetic based on the sup–(t–norm) convolution (i.e. fuzzy arithmetic equations). The authors apply the digital representation method, which has been proposed in our previous paper, to the procedure to solve fuzzy arithmetic equations (i.e. inf–$\varphi$ convolution). For this purpose, some new properties of $\varphi$–operators have been investigated. Also, we show some new results concerning (drastic product)–based operations on L–R fuzzy numbers.**

## I. Introduction

Fuzzy arithmetic have been studied actively since Zadeh[26] introduced the extension principle (i.e. sup–min convolution), which has been considered the most standard method of fuzzy arithmetic currently in use. Along this line, Dubois & Prade[7] derived fuzzy arithmetic based on a generalized extension principle (i.e. sup–(t–norm) convolution), and then many properties have been revealed through further studies [10,11,13,14,25]. A characteristic of this method is that the operations bring about less increase of fuzziness , in comparison with the sup–min convolution. Furthermore, it is possible to regulate the increase according to choice of types of t–norm and adjustment of its parameter value. Recently, the application to multi–objective linear programming is being tried out to make the best use of this feature[20].

The operations of fuzzy numbers, regardless of any t–norm, form a structure of monoid which does not have an inverse element generally, so that an algebraic equation related to fuzzy numbers (i.e. fuzzy arithmetic equation) can not be easily solved by similar methods to a usual algebraic equation. It is possible to consider that fuzzy arithmetic equation is a special case of fuzzy relation equation[21]. From this point of view, Sanchez[22], Gottwald[12], Di Nola et al.[3], Pedrycz[19] have shown through their studies the solution by the inf–$\varphi$ convolution. At the same time, Dubois & Prade[8,9] approached the problem by extending Minkowski operation to fuzzy sets and have arrived at the same conclusion. Moreover, Sanchez[23] has proposed an application involving fuzzy quantifiers in syllogisms as an example of fuzzy arithmetic equation.

Buckley et al.[2] and Zhao et al.[27] have shown the methods of calculation for solution of fuzzy arithmetic equation using $\alpha$–level sets. However, their methods are effective only in the case of the sup–min convolution, whereas a generalized method dealing with the case of the sup–(t–norm) convolution has not been investigated so far.

The authors have proposed a calculation method using digital representation relating to fuzzy arithmetic[13,14]. The studies have continued in this paper to obtain an approximate solution to fuzzy arithmetic equation based on the same method. In Section II, a t–norm and a $\varphi$–operator which is defined in connection with a given t–norm are summarized. Section III involves non–standard operations based on the inf–$\varphi$ convolution as solution for fuzzy arithmetic equation. Section IV leads to the formulae which involve the solution for the equations and the classifications of t–norms and $\varphi$–operators which are necessary for applying the formulae. Section V briefly illustrates some numerical examples.

## II. t–Norm and $\varphi$–operator

A t–norm $T$ is defined as a real function satisfying the four conditions [18,24], where $I = [0,1]$, $a, b, c \in I$:

(T1) $T(1,a) = a\ ,\ T(0,a) = 0;$

(T2) $a \leq b \longrightarrow T(a,c) \leq T(b,c);$

(T3) $T(a,b) = T(b,a);$

(T4) $T(a,T(b,c)) = T(T(a,b),c).$

A t–norm $T$ is called Archimedean iff $T$ satisfies the following two conditions:

(T5) $T$ is a continuous function;

(T6) $T(a,a) < a$ for any $a \in (0,1)$.

0-7803-0614-7/93$03.00 ©1993IEEE

Table 1 : $T(a,b)$ and $a\varphi b$. ($a\varphi b = 1$ for $a \le b$)

| triangular norm $T(a,b)$ | | continuity of $T$ | class of $T$ | $a\varphi b$ for $a > b$ | continuity of $a\varphi b$ | class of $\varphi$ |
|---|---|---|---|---|---|---|
| logical product | $\min(a,b)$ | cont. | A | $b$ | USC | A |
| drastic product | $\begin{cases} a & (b=1) \\ b & (a=1) \\ 0 & (\text{otherwise}) \end{cases}$ | USC | B | $\begin{cases} 1 & (a \ne 1) \\ b & (a=1) \end{cases}$ | $a$ : LSC<br>$b$ : cont. | B′ |
| Dubois & Prade ($0 \le p \le 1$) | $\dfrac{ab}{\max(a,b,p)}$ | cont. | A | $\dfrac{bp}{\min(a,p)}$ | USC | A |
| Bour et al. | $\begin{cases} 0 \\ (a+b-1 \le 0) \\ \min(a,b) \\ (a+b-1>0) \end{cases}$ | LSC | C | $\begin{cases} b \\ (1-a \le b < a) \\ 1-a \\ (b < \min(a, 1-a)) \end{cases}$ | USC | B′ |
| Archimedean t-norm: strict | $f^{[-1]}(f(a)+f(b))$ $f$:additive generator | cont. | A | $f^{[-1]}(f(b)-f(a))$ | cont. | A |
| Archimedean t-norm: non-strict | $f^{[-1]}(f(a)+f(b))$ $f$:additive generator | cont. | C | $f^{[-1]}(f(b)-f(a))$ | cont. | B′ |

An Archimedean t–norm $T$ is called strict iff $T$ satisfies

(T7) $T$ is a strictly monotone increasing function.

Every Archimedean t–norm $T$ is representable by a continuous and decreasing function $f$ such that

$$T(a,b) = f^{[-1]}(f(a)+f(b)), \tag{1}$$

and $f$ and $f^{[-1]}$ is called an additive generator of $T$ and a psudo inverse of $f$, respectively. Here, $f$ has the following properties. For strict $T$, $f:(0,1] \to [0,\infty)$ with $\lim_{x\to+0} f(x) = \infty$ and $f(1)=1$, $f^{[-1]} = f^{-1}$. For non-strict $T$, $f: I \to I$ with $f(0)=1$ and $f(1)=0$, $f^{[-1]}$ is defined by

$$f^{[-1]}(y) = \begin{cases} f^{-1}(y) & \text{for } y \in [0, f(0)] \\ 0 & \text{otherwise.} \end{cases} \tag{2}$$

A $\varphi$–operator $\varphi$: $I^2 \to I$ connected with a given t–norm $T$ is defined by [3,4,5,9,12,15,16]

$$a\varphi b = \sup\{x \mid T(a,x) \le b\}. \tag{3}$$

A $\varphi$–operator is decreasing in the first argument $a$ and increasing in the second argument $b$. Iff a t–norm $T$ is lower semicontinuous (LSC), the following holds[4]:

$$a\varphi b \in \{x \mid T(a,x) \le b\}, \tag{4}$$

for any $a, b \in I$ and the $\varphi$–operator connected with $T$.

And a $\varphi$–operator is representable by the additive generator of the connected $T$ such that

$$a\varphi b = \begin{cases} f^{[-1]}(f(b)-f(a)) & \text{for } a > b \\ 1 & \text{for } a \le b. \end{cases} \tag{5}$$

Bour et al.[1] have collected t–norms and $\varphi$–operators and tabulated their properties. We show in Table 1 their results with our new classifications (class of $T$, class of $\varphi$, continuity of $a\,\varphi\,b$), which will be explained in Section IV. The $\varphi$–operator connected with logical product is well-known as $\alpha$–operator by Sanchez[21], or Gödel's implication operation in multi-valued logic. Since drastic product is upper semicontinuous (USC), the connected $\varphi$ does not satisfy eq.(4). The t–norm by Bour et al. is an example of LSC t–norms. The t–norm by Dubois & Prade is non-Archimedean and continuous.

## III. Solution of fuzzy arithmetic equation and non-standard operation.

A fuzzy number $A$ in the wide sense is defined as a fuzzy subset on the real line $\boldsymbol{R}$ ( it is called a fuzzy quantity ). Usually, the membership function of $A$ i.e. $\mu_A : \boldsymbol{R} \to I$ is assumed to satisfy[6,17]

(FN1) normality: $\sup_x \mu_A(x) = 1$;

(FN2) convexity: $\mu_A(\lambda x_1 + (1-\lambda)x_2) \ge \min(\mu_A(x_1), \mu_A(x_2))$, for $^\forall\lambda \in I, ^\forall x_1 \in \boldsymbol{R}, ^\forall x_2 \in \boldsymbol{R}$;

(FN3) piecewise continuity.

Moreover, in practical use, $\mu_A$ is supposed to have[8]

(FN3') upper semicontinuity;

(FN4) boundedness of the support $\text{supp}\, A = \{x \mid \mu_A(x) > 0\}$.

A fuzzy number $A$ is called positive or negative if eq.(6) or eq.(7) holds respectively; otherwise, $A$ is called nearly equal to 0 ($A \simeq 0$).

$$A > 0 \quad \Leftrightarrow \quad \mu_A(x) = 0 \ ; \ {}^\forall x \leq 0 \tag{6}$$
$$A < 0 \quad \Leftrightarrow \quad \mu_A(x) = 0 \ ; \ {}^\forall x \geq 0 \tag{7}$$

Any binary operation $*$ on the real line $\boldsymbol{R}$ is extended to an operation of fuzzy numbers according to the extension principle, that is, sup–(t–norm) convolution [7,26]:

$$\mu_{A*B}(z) = \sup_{z=x*y} T(\mu_A(x), \mu_B(y)). \tag{8}$$

We will call it an operation based on the sup–(t–norm) convolution or a standard operation.

Only addition $+$ and multiplication $\times$ are considered as a fuzzy operation $*$ in the rest of this paper. Subtraction $A - B$ and division $A/B$ can be derived from addition and multiplication respectively, noting the properties $\mu_{-B}(y) = \mu_B(-y)$, $\mu_{1/B}(y) = \mu_B(1/y)$.

In contrast to standard addition and multiplication, non-standard subtraction and division are defined by inf–$\varphi$ convolution, respectively, as follows[3,12,22]:

$$\mu_{C\ominus A}(y) = \inf_{z=x+y} \mu_A(x) \,\varphi\, \mu_C(z), \tag{9}$$
$$\mu_{C\oslash A}(y) = \inf_{z=x\times y} \mu_A(x) \,\varphi\, \mu_C(z). \tag{10}$$

The rest of this paper will deal with the problem solving the equation $A + X = C$ or $A \times X = C$.

In regard to fuzzy numbers in the wide sense, the following theorem by Gottwald[12] holds.

**Theorem 1.**

Assume that a t–norm $T$ is LSC. The equation $A + X = C$ ( or $A \times X = C$ ) has a solution iff $A + (C \ominus A) = C$ ( or $A \times (C \oslash A) = C$ ). Moreover, when $A + X = C$ ( or $A \times X = C$ ) has a solution, $C \ominus A$ ( or $C \oslash A$ ) is the greatest one.

If $T$ is not LSC, for instance, $T$ is drastic product, Theorem 1 does not hold. In such a case, we have to represent $A$ and $C$ as $(m_A, \alpha_A, \beta_A)_{LR}$, $(m_C, \alpha_C, \beta_C)_{LR}$ according to L–R fuzzy numbers in [6]. Here, $m_A$,$\alpha_A$ and $\beta_A$ denote a mean value, left and right spreads of $A$, respectively; reference functions $L$ and $R$ are real functions satisfying $L(x) = L(-x)$, $L(0) = 1$, $L(x)$ is non-increasing on $[0, \infty)$. $\mu_A(x)$ is represented as

$$\mu_A(x) = \begin{cases} L((m_A - x)/\alpha_A) & \text{for } x \leq m_A \\ R((x - m_A)/\beta_A) & \text{for } x \geq m_A. \end{cases} \tag{11}$$

This L–R fuzzy number $A$ can be easily verified to have the above-mentioned properties (FN1), (FN2) and (FN3). Then we arrive at the following theorems in regard to the binary operations relating to drastic product.

**Theorem 2.**

If a t–norm $T$ is drastic product, the standard sum $A+B$ of $A = (m_A, \alpha_A, \beta_A)_{LR}$ and $B = (m_B, \alpha_B, \beta_B)_{LR}$ is a L–R fuzzy number with the same $L$ and $R$:

$$A + B = (m_A + m_B, \max(\alpha_A, \alpha_B), \max(\beta_A, \beta_B))_{LR}. \tag{12}$$

**Theorem 3.**

If a t–norm $T$ is drastic product, the non-standard difference $C \ominus A$ between $A = (m_A, \alpha_A, \beta_A)_{L'R'}$ and $C = (m_C, \alpha_C, \beta_C)_{LR}$ is a L–R fuzzy number with the same reference functions as $C$:

$$C \ominus A = (m_C - m_A, \alpha_C, \beta_C)_{LR}. \tag{13}$$

Moreover, if $\alpha_A \leq \alpha_C$, $\beta_A \leq \beta_C$, $L' = L$ and $R' = R$, then $A + (C \ominus A) = C$ holds, that is, $C \ominus A$ is a solution to $A + X = C$.

**Theorem 4.**

If a t–norm $T$ is drastic product, the standard product $A \times B$ is a L–R fuzzy number:

$$A \times B$$
$$= (m_A \times m_B, \max(m_A q, m_B p), \max(m_A s, m_B r))_{LR}. \tag{14}$$

The restriction of reference functions of $A$ and $B$ and the correspondence of meta-parameters $p, q, r$ and $s$ in eq. (14) are shown in Table 2 and Table 3 ,respectively.

Table 2 :
The restriction of reference functions in standard multiplication $A \times B$. ( sup–(drastic product) convolution )

| | | |
|---|---|---|
| $m_A$ | + | $B : LR$ |
| | − | $B : RL$ |
| | 0 | $B$ :free |
| $m_B$ | + | $A : LR$ |
| | − | $A : RL$ |
| | 0 | $A$ :free |

Table 3:
The correspondence of parameters in standard multiplication $A \times B$.

| | $p$ | $q$ | $r$ | $s$ |
|---|---|---|---|---|
| $A > 0, B > 0$ | $\alpha_A$ | $\alpha_B$ | $\beta_A$ | $\beta_B$ |
| $A < 0, B < 0$ | $-\beta_A$ | $-\beta_B$ | $-\alpha_A$ | $-\alpha_B$ |
| $A > 0, B < 0$ | $-\beta_A$ | $\alpha_B$ | $-\alpha_A$ | $\beta_B$ |
| $A < 0, B > 0$ | $\alpha_A$ | $-\beta_B$ | $\beta_A$ | $-\alpha_B$ |
| $A \simeq 0, B > 0$ | $\alpha_A$ | $-\beta_B$ | $\beta_A$ | $\beta_B$ |
| $A \simeq 0, B < 0$ | $-\beta_A$ | $\alpha_B$ | $-\alpha_A$ | $-\alpha_B$ |
| $A > 0, B \simeq 0$ | $-\beta_A$ | $\alpha_B$ | $\beta_A$ | $\beta_B$ |
| $A < 0, B \simeq 0$ | $\alpha_A$ | $-\beta_B$ | $-\alpha_A$ | $-\alpha_B$ |

**Theorem 5.**

If a t-norm $T$ is drastic product, the non-standard quotient $C \oslash A$ is a L-R fuzzy number:

$$C \oslash A = (m_C/m_A, q/m_A, s/m_A)_{LR} \quad (A > 0 \text{ or } A < 0). \quad (15)$$

Moreover, if $(m_C/m_A)p \leq m_A q$, $(m_C/m_A)r \leq m_A s$, and the reference functions of $A$ and $C$ satisfy the restriction shown in Table 4, then $A \times (C \oslash A) = C$ holds, that is, $C \oslash A$ is a solution to $A \times X = C$. Here, meta-parameters $q$ and $s$ correspond to the spread parameters of $A$ and $C$ as shown in Table 7.

It is clear that $A \times X = C$ has no solution if $A \simeq 0$ and $C > 0$ (or $C < 0$), so that we omit this case.

Table 4 :
The restriction of reference functions in non-standard division $C \oslash A$. ( sup–(drastic product) convolution )

| | | |
|---|---|---|
| $m_A$ | + | $C : LR$ |
| | − | $C : RL$ |
| $m_C$ | + | $A : LR$ |
| | − | $A : RL$ |
| | 0 | $A$ :free |

## IV. Approximate solution of equation by digital representation method.

### A. *Digital representation method*

As to methods for executing fuzzy number operations by computer, L–R fuzzy numbers [6] or $\alpha$–cut interval operations [17] are generally used. However, the former has strict restrictions in terms of membership function and the latter is only effective for the operations based on sup–min convolution.

In the digital representation method presented by the authors in [14], the membership functions of fuzzy numbers are digitalized on the real line, a finite number of sampling points are applied to the extension principle, and the result of an operation can be obtained as approximate discrete values of its membership function. This method is applicable to the extension principle with various kinds of t–norms, and there is no regulation against a unification of reference functions such as L–R representation. Hereafter, assume that fuzzy numbers $A, B$ and $C$ satisfy the properties (FN1),(FN2),(FN3) and (FN4) mentioned in Section II. We will use the notation $A = (m_A, \alpha_A, \beta_A)$, removing the indicator of reference functions from L–R representation, just because our method doesn't require the distinction by reference functions. It should be noted that $\alpha_A = m_A - \inf(\text{supp } A)$ and $\beta_A = \sup(\text{supp } A) - m_A$.

We consider at first, standard operation $A * B$ ( that is, $A + B$, $A \times B$ ). The membership function $\mu_A(x)$ may be sampled at the following $2n + 1$ points:

$$x_i = \begin{cases} m_A + i \times \alpha_A/n & (i = -n, \cdots, -2, -1) \\ m_A + i \times \beta_A/n & (i = 0, 1, \cdots, n). \end{cases} \quad (16)$$

Sampling points $y_i$ and $z_k$ of $B$ and $A * B$ can be obtained in a similar way. The mean value and spreads of $A * B$ should be obtained beforehand according to the formulae derived in [13,14]. Using the discrete fuzzy numbers $A = \sum_{i=-n}^{n} \mu_A(x_i)/x_i$ and $B = \sum_{j=-n}^{n} \mu_B(y_j)/y_j$, the binary operation $A * B = \sum_{k=-n}^{n} \mu_{A*B}(z_k)/z_k$ should be executed by following expression:

$$\mu_{A*B}(z_k) = \max_{z_k = x_i * y'} T(\mu_A(x_i), \mu_B(y')) \quad (i, k = 0, \pm 1, \cdots, \pm n). \quad (17)$$

Here, $y'$ is obtained from $x_i$ and $z_k$. In case $y'$ does not coincide with any $y_i$, $y'$ can be interpolated by two sampling values such that $y_i \leq y' \leq y_{i+1}$.

Next, we consider the calculations for non-standard operations $C \ominus A$ and $C \oslash A$. The sampling points $z_k, x_i$ and $y_i$ of $C, A, C \ominus A$ ( or $C \oslash A$) are obtained in the same way as in case of a standard operation. We can obtain the mean value and spreads of $C \ominus A$ and $C \oslash A$ by use of parameter formulae ( Table 5 – Table 7 ) which will be mentioned later on. The binary operation $C \ominus A$ and $C \oslash A$ are executed as follows:

$$\mu_{C \ominus A}(y_j) = \min_{z' = x_i + y_j} \mu_A(x_i) \, \varphi \, \mu_C(z'), \quad (18)$$

$$\mu_{C \oslash A}(y_j) = \min_{z' = x_i \times y_j} \mu_A(x_i) \, \varphi \, \mu_C(z') \quad (19)$$

$$(i, j = 0, \pm 1, \cdots, \pm n).$$

Here, $z'$ is obtained from $x_i$ and $y_i$. When $z'$ does not coincide with any $z_k$, the interpolation is done in the same way as a standard operation.

### B. *Parameter Formulae*

Before deriving parameter formulae related to the mean value and spreads of non-standard $C \ominus A$ and $C \oslash A$, we recall the classifications of t–norms which were introduced in [13] for standard operations.

Class A. the t–norms such that $T(a, b) \neq 0$ iff $a \neq 0$ and $b \neq 0$.

Class B. the t–norm in regard to drastic product: $T(a, b) \neq 0$ iff ($a = 1$ and $b \neq 0$) or ($a \neq 0$ and $b = 1$).

Class C. the t–norms other than drastic product, which satisfy $^{\exists}a, b \neq 0$ such that $T(a,b) = 0$.

When thinking of parameter formulae for non-standard operation, we can classify $\varphi$–operators into the following two classes:

Class A. the $\varphi$–operators such that $a \varphi b = 0$ iff $a \neq 0$ and $b = 0$.

Class B′. the $\varphi$–operators such that $a \varphi b = 0$ iff $a = 1$ and $b = 0$.

Here, it is easy to see that the classification is complete and Class A and Class B′ are mutually disjoint.

A $\varphi$–operator belonging to Class A is connected with a t–norm belonging to Class A; a $\varphi$–operator belonging to Class B′ is connected with a t–norm belonging to either Class B or Class C. ( See Table 1. )

Table 5 and Table 6 show the parameter formulae for non-standard subtraction and division, respectively. The meta-parameters $p, q, r$ and $s$ in Table 6 correspond to the spread parameters of $A$ and $C$ as shown in Table 7. It should be noted that $A \times X = C$ has no solution if $A \simeq 0$ and $C > 0$ (or $C < 0$).

Table 5 :
Parameter formulae for non-standard subtraction.

| | $m_{C\ominus A}$ | $\alpha_{C\ominus A}$ | $\beta_{C\ominus A}$ |
|---|---|---|---|
| Class A | $m_C - m_A$ | $\alpha_C - \alpha_A$ | $\beta_C - \beta_A$ |
| Class B′ | $m_C - m_A$ | $\alpha_C$ | $\beta_C$ |

Table 6 :
Parameter formulae for non-standard division.

| | $m_{C\oslash A}$ | $\alpha_{C\oslash A}$ | $\beta_{C\oslash A}$ |
|---|---|---|---|
| Class A | $\frac{m_C}{m_A}$ | $\frac{m_A q - m_C p}{m_A(m_A - p)}$ | $\frac{m_A s - m_C r}{m_A(m_A + r)}$ |
| Class B′ | $\frac{m_C}{m_A}$ | $\frac{q}{m_A}$ | $\frac{s}{m_A}$ |

Table 7 :
The correspondence of parameters in non-standard division $C \oslash A$.

| | $p$ | $q$ | $r$ | $s$ |
|---|---|---|---|---|
| $A > 0, C > 0$ | $\alpha_A$ | $\alpha_C$ | $\beta_A$ | $\beta_C$ |
| $A < 0, C < 0$ | $-\beta_A$ | $-\beta_C$ | $-\alpha_A$ | $-\alpha_C$ |
| $A > 0, C < 0$ | $-\beta_A$ | $\alpha_C$ | $-\alpha_A$ | $\beta_C$ |
| $A < 0, C > 0$ | $\alpha_A$ | $-\beta_C$ | $\beta_A$ | $-\alpha_C$ |
| $A > 0, C \simeq 0$ | $-\beta_A$ | $\alpha_C$ | $\beta_A$ | $\beta_C$ |
| $A < 0, C \simeq 0$ | $\alpha_A$ | $-\beta_C$ | $-\alpha_A$ | $-\alpha_C$ |

## V. Numerical examples.

In this section , we execute non-standard subtraction $C \ominus A$ corresponding to six kinds of $\varphi$–operators using the digital representation method, and illustrate their results. As mentioned in Section III, the results can be considered as the approximate greatest solutions for the equation $A + X = C$.

Numerical Examples.

$C \ominus A$ , $C = (5,2,2)_{\text{TFN}}, A = (2,1,1.5)_{\text{TFN}}$

Class A

(Ex.1) algebraic product:
$T(a,b) = ab$, $a \varphi b = b/a$ for $a > b$.
(Ex.2) logical product:
See Table 1.
(Ex.3) Hamacher product:
$T(a,b) = \frac{ab}{a+b-ab}, a \varphi b = \frac{ab}{ab+a-b}$ for $a > b$.

Class B′

(Ex.4) drastic product:
See Table 1.
(Ex.5) Bour's t–norm:
See Table 1.
(Ex.6) Yager's t–norm:
$T(a,b) = 1 - \min\{1, [(1-a)^p + (1-b)^p]^{1/p}\}$,
$a \varphi b = 1 - [(1-b)^p - (1-a)^p]^{1/p}$ for $a > b$. $(p = 2)$

Here, the subscript 'TFN' means a Triangular Fuzzy Number. Fig.1(a) shows the results of (Ex.1),(Ex.2) and (Ex.3), and Fig.1(b) shows those of (Ex.4),(Ex.5) and (Ex.6).

From Fig.1(a) and (b), we can see the followings.

- In (Ex.1) and (Ex.3), the membership functions $\mu_{C\ominus A}(y)$ are discontinuous at the bounds of their supports.
- In (Ex.2) and (Ex.4), $C \ominus A$ remain triangular.
- In (Ex.5), $\mu_{C\ominus A}$ is not so smooth.It is considered that this tendency is due to the calculation error of this method.
- In (Ex.6), $C \ominus A$ is not triangular in spite that the sum $A + B$ of two triangular fuzzy numbers remains triangular regardless of the parameter's value of Yager's t–norm[11,20].

## VI. Conclusions

This paper has described the approximate calculation method to solve a fuzzy arithmetic equation using the digital representation. We also derived the formulae for three kinds of parameters which are required in this method. The formulae have been obtained for each of two classes of $\varphi$–operators. On the contrary, as an interesting phenomenon, in the case of a standard operation by the same method, t–norms are classified into three classes. The adequate formulae have not yet been obtained for one of the classes[13].

Any type of approximation should require attention to deal with errors. In this method, in a similar way to other various approximate methods, increasing the number of sampling points (i.e. decreasing the sampling interval) improves accuracy and increases computational complexity. We previously considered the computational complexity in [14]. Thus, it is necessary to evaluate the error of the method for the next step. In further research, this problem will need to be considered from a broader viewpoint, that is, how errors are evaluated or treated in the method representing quantities with fuzziness.

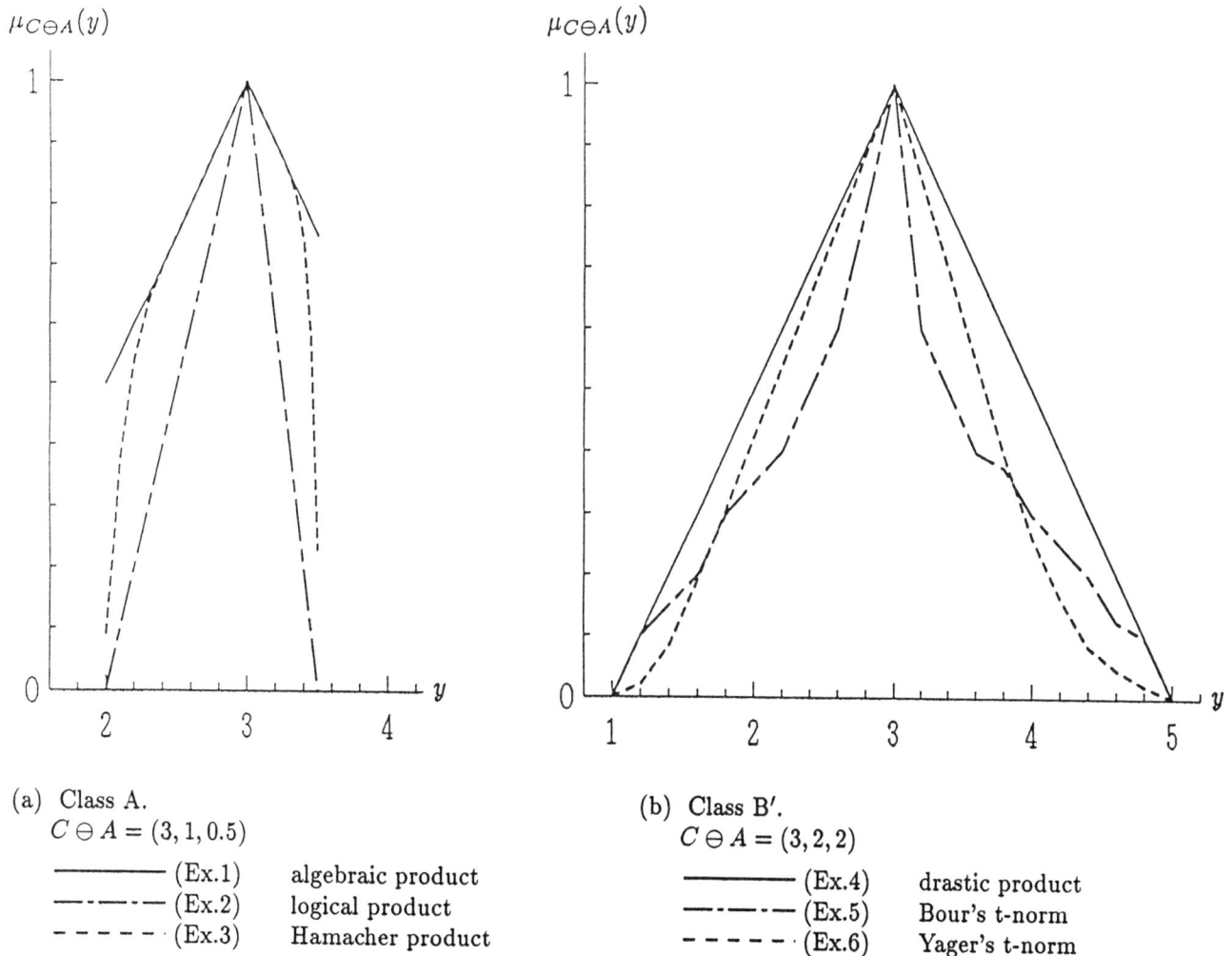

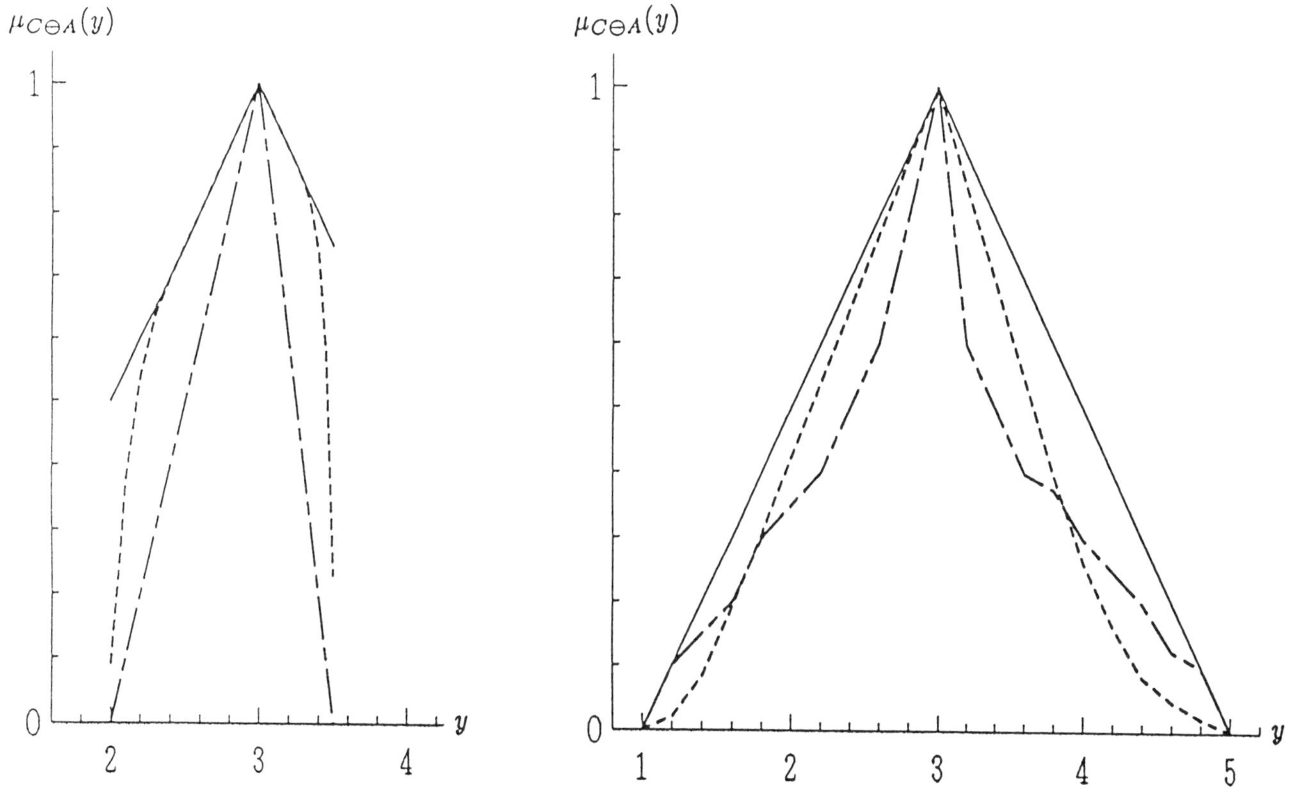


(a) Class A.
$C \ominus A = (3, 1, 0.5)$

| | | |
|---|---|---|
| ———— | (Ex.1) | algebraic product |
| —·—·— | (Ex.2) | logical product |
| - - - - - | (Ex.3) | Hamacher product |

(b) Class B′.
$C \ominus A = (3, 2, 2)$

| | | |
|---|---|---|
| ———— | (Ex.4) | drastic product |
| —·—·— | (Ex.5) | Bour's t-norm |
| - - - - - | (Ex.6) | Yager's t-norm |

Fig.1. Non-standard subtraction $C \ominus A$.
$C = (5, 2, 2)_{\mathrm{TFN}}, A = (2, 1, 1.5)_{\mathrm{TFN}}$
sampling points = 21 $(n = 10)$

## REFERENCES


[1] L.Bour, G.Hirsch & M.Lamotte: Détermination d'un opérateur de maximalisation pour la résolution d'équation de relation floue, BUSEFAL,25(1986)95-106.

[2] J.J.Buckley & Y.Qu: Solving linear and quadratic fuzzy equations, Fuzzy Sets and Systems, 38(1990)43-59.

[3] A.Di Nola, W.Pedrycz & S.Sessa: Processing of fuzzy numbers by fuzzy relation equations, Kybernetes, 15(1986)pp.43-47.

[4] A.Di Nola, W.Pedrycz & S.Sessa: Fuzzy relation equations under LSC and USC t-norms and their boolean solutions, Stochastica, XI-2,3(1987)151-183.

[5] A.Di Nola, S.Sessa, W.Pedrycz & E.Sanchez: Fuzzy relation equations and their applications to knowledge engineering, Kluwer Academic Publishers(1989).

[6] D.Dubois & H.Prade: Fuzzy sets and systems: Theory and applications, Academic Press,Inc.(1980).

[7] D.Dubois & H.Prade: Additions of interactive fuzzy numbers, IEEE Trans.Autom.Control,AC-26(1981)926-936.

[8] D.Dubois & H.Prade: Inverse operations for fuzzy numbers, Proc. of IFAC Fuzzy Information(1983)399-404.

[9] D.Dubois & H.Prade: Fuzzy-set-theoretic differences and inclusions and their use in the analysis of fuzzy equations, Control and Cybernetics, 13(1984)129-146.

[10] R.Fullér & T.Keresztfalvi: A note on t-norm based operations on LR fuzzy intervals, Supplement to Kybernetika, 28(1992)45-49.

[11] R.Fullér & T.Keresztfalvi: t-Norm based addition of fuzzy intervals, Fuzzy Sets and Systems, (to appear).

[12] S.Gottwald : Characterization of the solvability of fuzzy equations, Elektron.Inf.verarb.Kybern.EIK, 22(1986)67-91.

[13] M.F.Kawaguchi & T.Da-te: Parameter formulae for fundamental operations of weakly non-interactive fuzzy numbers, Proc. of Int. Conf. Fuzzy Sys.(FUZZ-IEEE'92), San Diego(1992)153-160.

[14] M.F.Kawaguchi & T.Da-te: On calculations of fundamental operations of weakly non-interactive fuzzy numbers using digital representation, Journal of Japan Society for Fuzzy Theory and Systems, 4(1992)491-503(in Japanese) , Japanese Journal of Fuzzy Theory and Systems (English version , in press).

[15] M.Miyakoshi & M.Shimbo: Composite fuzzy relations with t-norm and their applications to fuzzy inference, Trans.I.E.C.E. Japan, J67-D(1984)391-398(in Japanese).

[16] M.Miyakoshi & M.Shimbo: Solutions of composite fuzzy relational equations with triangular norms, Fuzzy Sets and Systems, 16(1985)53-63.

[17] M.Mizumoto & K.Tanaka: Some properties of fuzzy numbers, in Advances in Fuzzy Set Theory and Applications, M.M.Gupta(eds.), North-Holland (1979) 153-164.

[18] M.Mizumoto:Pictorial representations of fuzzy connectives, Part I: Cases of t-norms, t-conorms and averaging operators, Fuzzy Sets and Systems, 31(1989)217-242.

[19] W.Pedrycz: On solution of fuzzy functional equations, J.Math.Anal.Appl., 123(1987)589-604.

[20] H.Rommelfanger & T.Keresztfalvi: Multicriteria fuzzy optimization based on Yager's parametrized t-norm, Foundations of Computing and Decision Sciences, 16(1991)99- 110.

[21] E.Sanchez: Resolution of composite fuzzy relation equations, Information and Control, 30(1976)38-48.

[22] E.Sanchez: Solution of fuzzy equations with extended operations, Fuzzy Sets and Systems, 12(1984)237-248.

[23] E.Sanchez: Fuzzy quantifiers in syllogisms, direct versus inverse computations, Fuzzy Sets and Systems, 28(1988)305- 312.

[24] B.Schweizer & A.Sklar: Associative functions and statistical triangle inequalities, Publ.Math.,Debrecen,8(1961)pp.169-186.

[25] R.C.Williamson: The law of large numbers for fuzzy variables under a general norm extension principle, Fuzzy Sets and Systems, 41(1991)55-81.

[26] L.A.Zadeh: The concept of linguistic variable and its application to approximate reasoning: part I,Inf.Sci., 8(1975)199-249.

[27] R.Zhao & R.Govind: Solutions of algebraic equations involving generalized fuzzy numbers,Inf.Sci., 56(1991)199-243.

# Fuzzy Database Language and Library
## - Fuzzy Extension to SQL -

Hiroshi NAKAJIMA [*], Taiji SOGOH[†], Masaki ARAO[‡]
OMRON Corporation
Shimokaiinji, Nagaokakyo-City,
Kyoto, 617 JAPAN.
Phone: +81-75-951-5111
Fax: +81-75-955-2442

***Abstract*— In today's information-oriented society, the importance of accumulating and accessing information efficiently has become widely accepted. In response to this need, many database systems have been developed. A major limitation of these systems, however, is that many kinds of imprecise data exist which are not easily processed. To address this problem, the concept of a fuzzy database system has been developed.**

**Many studies have been done in the fuzzy database area and many useful proposals have been made. The following two points are generally recognized as critical to the design of a fuzzy database system:**

**(1) Using fuzzy database system, it should be easy to write fuzzy database applications.**

**(2) A fuzzy database system must be capable of processing information stored in conventional database systems.**

**This paper outlines the specification of a fuzzy database language called Fuzzy SQL and the FDL2 C library which is used to implement the Fuzzy SQL processor.**

## I. Introduction

A great number of database application systems have been developed using relational database systems and written in SQL language which is an international standard. These resources cannot be overlooked in developing fuzzy database applications. Accordingly, practical utilization requires relational model based fuzzy database systems and SQL based fuzzy database languages.

This paper presents the specification of the Fuzzy SQL language and the implementation aspects of *Fuzzy Database Library - FDL2*. The Fuzzy SQL language is extended from standard SQL using fuzzy theory. The specification includes both the Data Definition Language(DDL) and the Data Manipulation language(DML). FDL2 is designed to be built in a Fuzzy Database Management System(FDBMS) with a conventional Database Management System(DBMS). FDL2 is implemented using C language on an OMRON LUNA88K workstation.

## II. Fuzzy Relational Database

In the fuzzy relational database area, many useful proposals have been made [2] [3] [6] [7] [8]. In this section, the fuzzy relational database model and the fuzzy data representation method are described.

### A. Fuzzy Relational Database Model

A relational database consists of one or more relations. Each relations consists of one or more attributes. In a fuzzy relational database, a relation $R$ is characterized by following membership function.

$$\mu_R : D_1 \times D_2 \times \ldots \times D_n \rightarrow [0,1] \qquad (1)$$

where $D_i (i = 1, 2, \ldots, n)$ is the domain for each attribute $A_i$ of a relation $R$, and the symbol $\times$ signifies the Cartesian Product.

As defined above, the membership value named GRADE is added to a relation. The example of a relation *young people* is shown in Figure 1. This relation includes numerical fuzzy data like "*ABOUT 20*" or "*young*", string fuzzy data like "*like_red*", and reliability degree values like "(0.6)".

*Advanced Systems Research Center, Corporate Research and Development H.Q., E-Mail: nak@ari.ncl.omron.co.jp

†Fuzzy Technology Business Promotion Center, E-Mail: sgh@zoo.ncl.omron.co.jp

‡Fuzzy Technology Business Promotion Center, E-Mail: arao@zoo.ncl.omron.co.jp

0-7803-0614-7/93$03.00 ©1993IEEE

*young people*

| *Name* | *Age* | | *Hair Color* | *Hobby* | *GRADE* |
|---|---|---|---|---|---|
| Mary | 25 | (1.0) | red | Dancing | 0.5 |
| Smith | *ABOUT 20* | (0.6) | *like_red* | Sports | 1.0 |
| John | *young* | (0.7) | brown | Singing | 1.0 |
| Susan | *ABOUT 27* | (0.8) | *brown_red* | Sports | 0.2 |

Figure 1: An Example of Relation *young people*

### *B. Fuzzy Data in the Fuzzy Relational Database*

In the fuzzy relational database, various kinds of the fuzziness can be represented by the follow data types.

**Reliability Degree Value** represents reliability degree of each stored data for each attribute in a relation. This has [0,1] continuous values.

$$RV_i \rightarrow [0,1] \tag{2}$$

where $RV_i(i = 1,2,\ldots,n)$ is a Reliablity Degree Value on a attribute $A_i$ of a tuple.

**Fuzzy Number** represents fuzziness of numerical data like "*ABOUT 20*". This data is represented in a triangular shape fuzzy set which is normal and convex. A membership function $\mu_{FN}$ for a Fuzzy Number is defined as follows.

$$\mu_{FN}(x) = \begin{cases} 0 & (x_{N1} > x) \\ (x - x_{N1})/(x_{N2} - x_{n1}) & (x_{N1} \le x \le x_{N2}) \\ (x_{N3} - x)/(x_{N3} - x_{N2}) & (x_{N2} < x \le x_{N3}) \\ 0 & (x_{N3} < x) \end{cases} \tag{3}$$

where $x_{N1}$ is the coordinates of the left side of the base line of the triangular shape, $x_{N2}$ represents the vertex coordinates, and $x_{N3}$ is the coordinates of the right side of the base line.

These coordinates are defined by two methods. One method is the FIXED width method and the other is the RATED width method. These methods use given $x_{N2}$ coordinates to calculate the other coordinates, $x_{N1}$ and $x_{N3}$.

FIXED

$$\begin{cases} x_{N1} = x_{N2} - w \\ x_{N3} = x_{N3} + w \end{cases} \tag{4}$$

RATED

$$\begin{cases} x_{N1} = x_{N2} - x_{N2} \times p - w \\ x_{N3} = x_{N2} + x_{N2} \times p + w \end{cases} \tag{5}$$

where $w(> 0)$ denotes a parameter that specifies the width of the triangle and $p(> 0)$ denotes a parameter that specifies the ratio for the width of the triangle in only the RATED width method.

**Fuzzy Label** also represents fuzziness of numerical data like "*young*". This data is represented in a trapezoidal shape fuzzy set that is normal and convex. Additionally, Fuzzy Label has the name of a fuzzy set. A membership function $\mu_{FL}$ of Fuzzy Label is defined as follows.

$$\mu_{FL}(x) = \begin{cases} 0 & (x_{L1} > x) \\ (x - x_{L1})/(x_{L2} - x_{L1}) & (x_{L1} \le x \le x_{L2}) \\ 1 & (x_{L2} < x < x_{L3}) \\ (x_{L4} - x)/(x_{L4} - x_{L3}) & (x_{L3} \le x \le x_{L4}) \\ 0 & (x_{L4} < x) \end{cases} \tag{6}$$

where the $x_{L1}$ is the coordinates of the left side of the base line of a trapezoidal shape, $x_{L2}$ is the coordinates of the left side of the top line, $x_{L3}$ is the coordinates of right side of the top line, and $x_{L4}$ is the coordinates of the right side of the base line.

**String Fuzzy Set** represents fuzziness of strings data like "*like_red*". This fuzzy data is represented by pairs of strings and corresponding grade. The fuzzy set $\tilde{A}$ of Strings Fuzzy Set is defined as follows using the membership function $\mu_{\tilde{A}}$.

$$\tilde{A} = \sum_{i=1}^{n} \mu_{\tilde{A}}(x_i)/x_i \tag{7}$$

where the symbol / is a separator between an element $x_i$ and its membership value $\mu_{\tilde{A}}(x_i)$, and $\sum$ means a union operation.

## III. Language Specification

In this section, only extensions of Fuzzy SQL to the standard SQL specification are described. Additionally, the terms are unified with standard SQL. The standard SQL term "table" refers to "relation", "column" refers to "attributes", and "line" refers to "tuple".

### *A. Data Definition Language*

In Fuzzy SQL-DDL, Column constraint for fuzzy data and Fuzzy Data Definitions are added to the standard DDL.

```
CREATE TABLE People
( Name        CHAR(8)   NOT NULL,
  Age         DEC(3)    FUZZY RELDEG
              CHECK Age >= young
              WITH GRADE > 0,
  Hair_Color  CHAR(12) FUZZY,
  Hobby       CHAR(12),
  Height      DEC(3) FUZZY,
  Weight      DEC(3) FUZZY)
```

Figure 2: An Example of Table Definition

*A.1. Column Definition*

At the schema design, the columns that store fuzzy data should be explicitly specified because of design policy and for minimal memory usage. In Fuzzy SQL-DDL, this is implemented in Column constraint. An example of table definition is shown in Figure 2.

By using the keywords FUZZY and RELDEG, the columns that store fuzzy data and have Reliability Degree Value are specified. Additionally, search condition after the keyword CHECK and limitation condition after the keyword WITH have the following effect:

When data is specified in the INSERT or UPDATE statement to be stored into the column that is defined the CHECK option, the conditions are applied. Only data that has 0 grade value as the results of the conditions are stored.

*A.2. Fuzzy Data Definition*

Fuzzy data to be used in Fuzzy SQL-DML are defined in Fuzzy SQL-DDL. Fuzzy data is defined by specifying the table and column in which they are used. Fuzzy Number, Fuzzy Label, String Fuzzy Set, Linguistic Modifier, Fuzzy Relation, Operator Parameter are defined in the Fuzzy Data Definition. Fuzzy Data Definition examples are shown in Figure 3.

**Fuzzy Number** is defined by the CREATE FUZNUM statement. This statement includes three parameters: a triangle base line width decision method specified by the keywords FIXED and RATED, a width specification value, and a width constant value.

**Fuzzy Labels** are defined by CREATE FUZLAB. This includes fuzzy label names and the coordinates of trapezoidal shapes.

**String Fuzzy Sets** are defined by CREATE STRFSET. This includes the names of string fuzzy sets and the pairs of element and corresponding membership values.

**Linguistic Modifiers** are defined by CREATE HEDGE. This includes linguistic modifier names, shifting method types, and shifting volumes.

**Fuzzy Relation** defined by CREATE FUZREL and Operator Parameter defined by CREATE PARAM are for the fuzzy predicate. CREATE FUZREL defines a fuzzy matrix that represents a fuzzy relation between string type data. CREATE PARAM defines a parameter for follow functions [1]. The keywords APE, MGT, and MLT mean Approximately Equal, Much Greater Than, and Much Less Than respectively.

*Approximately Equal*

$$G_{X\sim=Y} = e^{-C_{ae}\times|X-Y|} \tag{8}$$

*Much Greater Than*

$$G_{X>>Y} = \begin{cases} 0 & (X \leq Y) \\ 1/(1+C_{mgt}/(X-Y)^2) & (X > Y) \end{cases} \tag{9}$$

*Much Less Than*

$$G_{X<<Y} = \begin{cases} 0 & (X \geq Y) \\ 1/(1+C_{mlt}/(X-Y)^2) & (X < Y) \end{cases} \tag{10}$$

*B. Data Manipulation Language*

In Fuzzy SQL-DML, fuzzy data can be used for any kind of operator such as predicates, connectives, linguistic modifier, aggregate functions, arithmetic operators, etc. Additionally, Limitation Condition which is useful to control the number of query results is added. Examples of Fuzzy SQL-DML are shown is Figure 4.

**Predicates** result in the grade values [0,1] which is calculated by the max-min composition. Available fuzzy predicates are =, <>, <, >, <=, >=,~=, <<, >>, BETWEEN, IN, LIKE, ANY, SOME, ALL, and EXISTS.

**Connectives** also result in the grade $[0,1]$ which is calculated by fuzzy logical operators and mean operators. Fuzzy Logical operators are logical product, algebraic product, bounded product, drastic product, logical sum, algebraic sum, bounded sum, and drastic sum. Mean operators are arithmetic mean, geometric mean, dual geometric mean, harmonic mean, and dual geometric mean.

```
CREATE FDD young_people.age
  CREATE FUZNUM
  ( RATED, 10, 2.0)
  CREATE FUZLAB
  ( young NMF(0, 0, 20, 30),
    old   NMF(40, 50, 150, 150))
  CREATE HEDGE
  ( very TIGHT 1, more_or_less WIDE 1 )
  CREATE PARAM
  ( APE 0.2, MGT 5, MLT 5)

CREATE FDD young_people.hair_color
  CREATE STRFSET
  ( like_red
      ( 'red' 0.9, 'pink' 0.7, 'orange' 0.3),
    brown_red ( 'brown' 0.3, 'red' 0.8 ))

CREATE FDD young_people.hobby
  CREATE FUZREL APE
  ( hobby_similarity
     ('Dancing' 'Sports' 0.8)
     ('Dancing' 'Singing' 0.7) ... )
```

Figure 3: Examples of Fuzzy Data Definitions

The last example in Figure 3 is a MEAN operator that calculates arithmetic mean. The values 0.6 and 1.0 are the Importance Degree Value of each argument of the mean operator. The keyword REL specifies Reliability Degree Value. Such as this example, all of the mean operators can calculate the grade with a weighting value for each argument.

**Linguistic Modifiers** are modifiers for numerical fuzzy data. They modify the shape of the membership function using a shifting method[9].

**Arithmetic Operators** can be applied to fuzzy data. Because the shapes of membership function are limited to trapezoidal, it is simple to calculate addition, subtraction, multiplication, and division. The results of multiplication and division are approximated to trapezoidal shape [5]. Fuzzy Numbers that have a triangular shaped membership function and Crisp Numbers that have a vertical line shaped membership functions are considered as a special case of trapezoidal.

**Aggregate Functions** are COUNT, SCOUNT, AVG, MAX, MIN, and SUM. The function COUNT counts the number of specified lines. The function SCOUNT calculates $\sum$counts [4]. The funcitons AVG and SUM denote average and sum that are calculated by using Arithmetic Operators. The functions MAX and MIN return the maximum and the minimum values that are calculated by ordering after defuzzification.

**Relational Operators** calculate the GRADE column value. In the case of UNION, the maximum value is calculated between the GRADE columns of the specified tables. Cartesian Product unifies GRADE columns of the producted line to the one GRADE column that has maximum value [2].

**Limitation Conditions** generate the derived table that is specified by the keyword WITH and quality limitation or quantity limitation. The searching conditions generate the derived table including lines that have the grade $[0, 1]$. Sometimes this derived table is not useful because the number of lines is too large for the user. The quality limitation generates the derived table including the specified grade of tuples. The quantity limitation generates a derived table including specified the number of tuples, ordered in ascending or descending sequence.

## IV. FDL2 and Implementation of Fuzzy Database

Fuzzy database library called FDL which can be applied to ordinary databases has already been developed [10] [11]. FDL2 is extended to be applied to Fuzzy SQL language based on FDL. In this section, implementation of fuzzy database using FDL2 is described.

### *A. Architecture of Fuzzy Database using FDL2*

The whole system of fuzzy database including the ordinary DBMS, the existing DB, and FDL2 is shown in Figure 5.

**Retrieval Code Buffer(RCB)** stores retrieval intermediate code that has a pre-ordered style. The codes are generated from search condition, SELECT statements by Fuzzy SQL processor and include the operations such as fuzzification, defuzzification, linguistic modifiers, arithmetic operation, predicate, connective, etc. RCB also contains the results of the operations that are calculated by Interpreter of FDL2.

**Database Buffer(DBB)** stores the one line of data read from the database. The line is of the derived table which is generated by statements like FROM, SELECT, etc.

**Fuzzy Data Dictionary Directory(FDD/D)** is a file including Fuzzy Data Definitions in Fuzzy SQL-DDL. Membership functions of Fuzzy Number,

```
''Find young people and
  the GRADE is greater then 0.5.''

  SELECT Name, Age,
    Hair_color, hobby FROM people
  WHERE age = young
  WITH GRADE > 0.5

''Find 20 people who are
  much  taller then  170cm
  and have like red hair  color.''

  SELECT name, height, weight
  FROM people
  WHERE (height >> 170)
    AND ( Hair_color = like_red)
  WITH LINES = 20

''Find the people who
  are about 10 cm taller
  more than high''

  SELECT name, height, weight
  FROM people
  WHERE height >= high + ABOUT 10

''Find the people who are age
  is young (0.6) and
  income is high (1.0).''

  SELECT name, age, income FROM people
  WHERE MEAN((age = young, REL(age)*0.6),
     (income >= high, REL(income)*1.0))
```

Figure 4: Examples of Fuzzy SQL-DML

Fuzzy Label, String Fuzzy Set, and Fuzzy Relation are generated by Fuzzy Data Definitions.

### *B. Library Functions*

FDL2 is mainly divided into three functions.

**FDD/D Handler** puts Fuzzy Data Definition of DDL into FDD/D, modifies, deletes, and refers them.

**Interpreter** interprets the RCB code to calculate the membership grade between the RCB code and data in DBB. One interpretation can calculate the grade of one line in the derived table.

**Basic Functions** provide arithmetic operations, linguistic modifiers operations, predicate operations, connectives operations, and relational operations.

FDL2 has two kinds of application interfaces, API1 and API2 as shown in Figure 5. API1 provides Interpreter level interface in which RCB, DBB, and Interpreter of FDL2 are used. API2 provides Basic Function level interface for basic fuzzy data operations.

## V. Conclusion

The Fuzzy SQL language specification and FDL2 that is used to develop the Fuzzy SQL processor are presented in this paper. By using FDL2, the development of the Fuzzy SQL processor is facilitated. Additionally, Fuzzy SQL language will make it easy to develop fuzzy database application programs.

In the future, other media such as image, text, etc will be increasingly used in database applications. On the database model side, object oriented database(OODB) will be come more common. The task of our project is to handle multimedia data and to extend the fuzzy database model under OODB. It will increase the practical utilization of fuzzy database systems.

## References


[1] L.A.Zadeh, *1971 Similarity Relations and Fuzzy Orderings*, Information Sciences 3, 177-200.

[2] M.Umano *1983 Retrieval from Fuzzy Database by Fuzzy Relational Algebra* Proceedings of the IFAC Symposium, Fuzzy Information, Knowledge Representation and Decision Analysis, 1-6.

[3] M.Zemankova L. and A. Kandel, *1984 Fuzzy Relational Data Bases a key to expert systems*, Verlag TÜV, Rheinland.

[4] A. Kandel, *1986 Fuzzy Mathematical Techniques with Application*, ADDISON-WESLEY PUBLISHING COMPANY.

[5] M.Mizumoto, *1988 Recent Studies on Fuzzy Theory*, Journal Information Processing Society Japan, Vol.29 No.1, 11-22.

[6] P.Bosc, M.Galibourg, and G.Hamon, *1988 Fuzzy Querying with SQL: Extension and Implementation Aspects*, Fuzzy Sets and Systems 28, 333-349.

[7] D.Li and D.Liu, *1990 A Fuzzy PROLOG Database System*, RESEARCH STUDIES PRESS LTD.

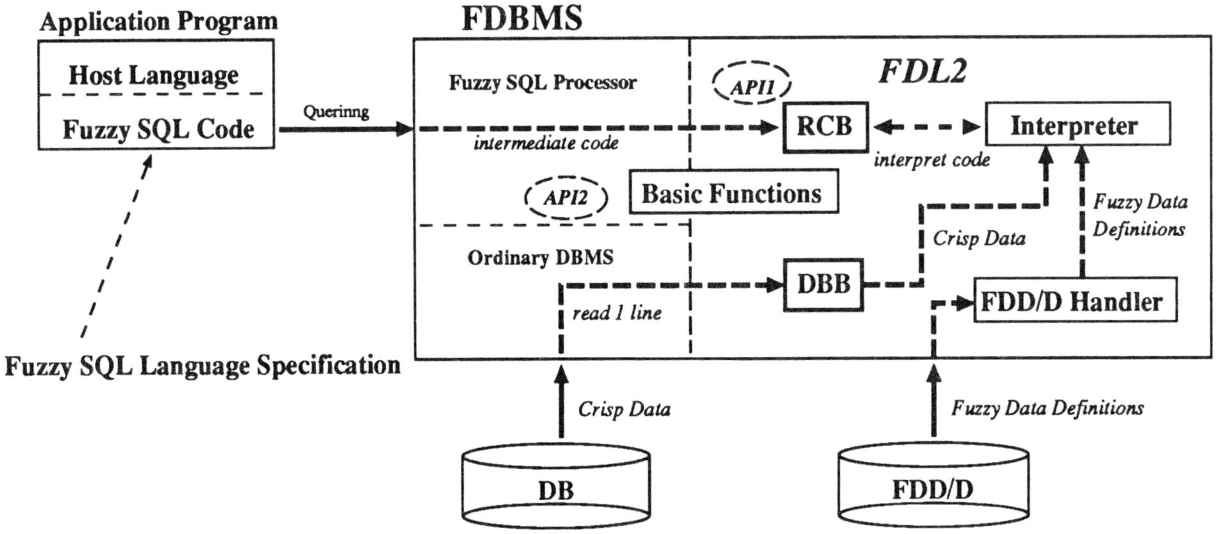


Figure 5: Architecture of Fuzzy Database using FDL2

[8] M.Umano and Y.Ezawa *1991 Implementation of SQL-type Data Manipulation Language for Fuzzy Relational Databases*, Proceedings of 7th Fuzzy System Symposium, 347-350

[9] Y.Ezawa and M. Umano *1991 Linguistic Approximation of Fuzzy Sets by Japanese Shift Hedges*, Proceedings of 7th Fuzzy System Symposium, 351-354.

[10] H.Kumamoto and H.Nakajima, *1992 Fuzzy Database Library*, OMRON TECHNICS Vol.32 No.1.

[11] H.Nakajima and T.Sogoh, *1992 Fuzzy Database Library*, Proceedings of 8th Fuzzy System Symposium, 333-336.

# Multilevel Database Security Using Information Clouding

Sujeet Shenoi *
Department of Mathematical and Computer Sciences
Keplinger Hall, University of Tulsa
Tulsa, Oklahoma 74104, USA

***Abstract*—Database security models typically employ information hiding together with polyinstantiation which helps create fictitious – and misleading – versions of confidential information. Although strict security-control is maintained, a high price is paid in terms of user convenience and data accuracy. Authorized users are often prevented from accessing necessary information which may not compromise security. Moreover, users accessing and unknowingly making decisions based on the misleading information used as cover stories can lead to disastrous consequences.**

**In this paper we employ fuzzy sets in a multilevel model for general-purpose database security. Sensitive information in database relations is meaningfully clouded by fuzzy sets; this is accomplished by broadening the possibility distributions constraining the values of sensitive attributes. The technique promotes the use of data and also maintains database security. Clouding with fuzzy sets is the middle ground between information release and information hiding/falsification. It nicely supplements the two security techniques and helps strike the right balance between user convenience and database security.**

## I. INTRODUCTION

The efficient implementation of strict database security is a problem of prime concern. A database must release information to support legitimate activities; at the same time, it must secure sensitive information from unauthorized users [1]. The simplest strategy is to hide information in views which present non-sensitive information to database users [1,3]. Although it is easily implemented, pure information hiding has several limitations. Authorized users can be prevented from accessing needed information which may not compromise database security. Moreover, in a multilevel database environment where information is tagged with different classification levels and is accessed by users with different clearance levels, information hiding can give rise to covert channels which permit the flow of high-level information to low-level users [3-7].

The SeaView multilevel database security model introduced polyinstantiation as a mechanism for eliminating covert channels [6,7]. It allows a database to hold different versions – or cover stories – for an entity in the real-world [4]. A cover story conveys "sanitized" information about an entity that can be securely released to users at a lower clearance level.

As an example, consider a relation containing the tuple:

| *Name* | *Occupation* |
|---|---|
| *Max Smart* (*c*) | *Spy* (*s*) |

The entire tuple is visible to users with clearance levels secret (*s*), or higher. A confidential (*c*) user would know of the existence of *Mr. Smart*; but his occupation at the confidential level would appear as a *null* value. The entire tuple is invisible to an unclassified (*u*) user. Now suppose a confidential user attempts to replace the *null* value listed as *Max Smart*'s occupation with the new value *Salesman*. Notifying the user about the conflict could lead to the disclosure of high-level information, or at the very least signal the existence of high-level information. Therefore, a "polyinstantiated element" or cover story is created for *Mr. Smart* as shown below.

| *Name* | *Occupation* |
|---|---|
| *Max Smart* (*c*) | *Spy* (*s*) |
| *Max Smart* (*c*) | *Salesman* (*c*) |

Users with secret clearance levels or higher see the real and the fictitious tuple. Confidential users are blissfully unaware of *Max Smart*'s real occupation. They are under the impression that he is a salesman.

*Research supported by NSF Grants IRI-9110709 and IRI-9244550 and by OCAST Grants AR2-002 and AR9-010.

Polyinstantiation is a powerful mechanism for securing database data. However, in attempting to hide the existence of sensitive information it releases misleading information to low-level users. Due to their very nature, military databases and certain corporate databases must necessarily release false information, e.g., to hide and/or cover the fact that an individual is an undercover agent or that an employee is involved in a secret company project. Nevertheless, implementing security-control by falsifying information is needlessly restrictive and is often unacceptable in general-purpose database environments.

Consider the case of a secretary in an investment firm who is asked to send a mailing to "middle-income" investors. In a polyinstantiation environment, since a cover story is not feasible, the secretary would have to be assigned a higher clearance level to access the correct salary information. Assigning the higher clearance level can compromise database security. It might permit the access of sensitive information that would otherwise not be available.

It is important to recognize that the availability of exact sensitive information is not essential to many database activities, e.g., the investment firm mailing. Rather, users can perform their tasks adequately using abstracted or clouded versions of sensitive information. Sensitive values with underlying unordered (i.e., scalar) or ordered (e.g., numeric) domains can be conveniently secured by information clouding. For example, the exact salaries of investors can be securely replaced by fuzzy sets. The "granularity" of the fuzzy sets should be large enough to ensure data security and small enough so as not to dilute the notion "middle-income". Such clouding with fuzzy sets captures natural data semantics and preserves confidentiality while supporting normal database activities.

In this paper we employ fuzzy sets in a model for database security. Although fuzzy sets have been extensively applied in databases geared for AI and expert system applications [2,9,10,13], security-control remains a novel and relatively unexplored area. Sensitive information in database relations can be meaningfully clouded using fuzzy sets. This technique promotes data usage and also maintains access- and inference-control. Clouding with fuzzy sets is the middle ground between information release and information hiding/falsification. More importantly, it can be used in combination with information hiding and polyinstantiation. It nicely supplements the two techniques and helps strike the right balance between user convenience and database security.

## II. FUZZY SETS AND CLOUDED INFORMATION

Sensitive information in a multilevel relation is clouded by simply replacing it with a meaningful fuzzy set. For example, *Mr. Smart*'s salary of $50K$ in the relation below can be clouded using the fuzzy set "moderate." The classification level ($s^*$) means that secret ($s$) information is clouded and is visible to users with the appropriate clearance level.

| *Name* | *Occupation* | *Salary* |
|---|---|---|
| *Max Smart* (*c*) | *Spy* (*s*) | $50K$ (*s*) |
| *Max Smart* (*c*) | *Spy* (*s*) | *moderate* ($s^*$) |

In terms of possibility theory [12], information clouding relaxes the possibility distribution constraining the value of a sensitive attribute. The precise fact, "*Salary* is $50K$," implies that

$$Poss\{Salary = x\} = \mu_{50K}(x), \quad x \epsilon D,$$

where $\mu_{50K}$ is the characteristic function of the classical set $50K$, and $Poss\{Salary = x\}$ is the possibility that *Salary* has the value $x$. On the other hand, the piece of clouded information "*Salary* is *moderate*," implies that

$$Poss\{Salary = x\} = \mu_{moderate}(x), \quad x \epsilon D.$$

Here *moderate* is a *normal* fuzzy set which properly contains the classical set $50K$ and which has several elements $x \epsilon X$ for which $\mu_{moderate}(x) = 1$.

Exact information is clouded using normal fuzzy sets whose $\alpha_{1.0}$-cuts are supersets of the original singleton sets. A fuzzy set with a larger $\alpha_{1.0}$-cut broadens the possibility distribution constraining the attribute value. This secures the exact information from unauthorized users who are presented with multiple elements with possibility values of 1.0. Note that normal "triangular" fuzzy sets cannot be used to cloud information. The cardinality of the $\alpha_{1.0}$-cut of a triangular fuzzy set is 1. ¿From the security point of view it is equivalent to exact information.

Since security considerations require only that the $\alpha_{1.0}$-cuts of clouded information have cardinalities greater than 1, it is possible to cloud information using classical sets and intervals. However, there are many advantages to using normal fuzzy sets in clouding sensitive information. Fuzzy sets are more general in that they include classical sets and intervals. More importantly, they better capture natural semantics. This helps release secure yet meaningful information to database users and promotes flexible (fuzzy) querying of the secured database.

A singleton classical set defined on a linearly-ordered domain and having a "spiked" characteristic function can be clouded by a "trapezoidal" fuzzy set. A trapezoidal fuzzy set is denoted by $(l_0, l_1, r_1, r_0)$. The support is the open interval $(l_0, r_0)$ and its $\alpha_{1.0}$-cut set is the closed interval $[l_1, r_1]$. Using this notation *Mr. Smart*'s *moderate* salary can be expressed as $(20K, 30K, 60K, 70K)$.

We now introduce the notion of a "context" [11] to formalize the treatment of clouded information and to enforce the compatibility of classical relations embodying exact information and fuzzy relations containing clouded information.

Definition:

A *context* $C$ is a partition defined by an equivalence relation $\rho$ on a set of elements $\hat{D}$. The set of elements $\hat{D}$ participating in $\rho$ is called its *restricted domain*; $\hat{D}$ is a subset of an underlying scalar database domain $D$. The set of all contexts capable of being constructed using restricted domains on $D$ is denoted by $\mathcal{C}_D$.

Contexts in $\mathcal{C}_D$ are ordered by the quantity of equivalences captured by the generating equivalence relations. Each equivalence class in a finer context is a subset of an equivalence class in a coarser context.

Definition:

Let $C$ and $C'$ be contexts in $\mathcal{C}_D$ induced by the equivalence relations $\rho$ and $\rho'$, respectively. Then, $C' \sqsubseteq_C C$, i.e., $C'$ is *coarser than* $C$, whenever $\rho \subseteq \rho'$.

A context acts as a *sieve* – with equivalence classes as its openings – for controlling the size of fuzzy sets released as tuple components.

Definition:

Let $t_i$ be normal fuzzy set. Then, $t_i$ is a *consistent* tuple component with respect to a context $C_i$ if and only if the $\alpha_{1.0}$-cut of $t_i$, denoted by $(t_i)_{1.0}$, is a non-empty subset of an equivalence class in $C_i$.

Since the $\alpha_{1.0}$-cut of a released fuzzy set can be no larger than an equivalence class, a context specifies the "high water mark" for clouded tuple components. "Precise" contexts with singleton equivalence classes release exact (classical) information, i.e., singleton classical sets with spiked characteristic functions. Contexts with larger equivalence classes permit the release of clouded information expressed as trapezoidal fuzzy sets. The coarsest context $\{D\}$ releases the most clouded information. Such clouding is equivalent to pure information hiding. Note that the null set is not a valid tuple component as the maximal chunk $D$ denotes an unknown, but defined value. Since contexts are derived from equivalences based on data semantics, the equivalence classes consist of closely related elements. This ensures that the clouded information chunks released as tuple components are always meaningful.

## III. MULTILEVEL CLOUDED RELATIONS

The first step in extending classical multilevel relations to embody clouded information is to define the access classes used for exact information and those used for clouded information. We consider the following set of linearly-ordered exact information access classes:

$$S = \{unclassified(u), confidential(c), secret(s), topsecret(ts)\}.$$

In addition to the access classes in $S$, multiple clouded access classes of varying degrees are employed for each $a \in S - \{u\}$. A clouded access class is denoted by $a^*$, e.g., $ts^*$ is a clouded class corresponding to top secret ($ts$). Note that clouded classes are not required for the unclassified ($u$) access class because unclassified information is non-sensitive by definition and is never clouded.

The set comprising an exact information access class $a$ and its clouded classes is denoted by $\mathcal{A}$. The elements in $\mathcal{A}$ are partially ordered by $\sqsubseteq_*$ ("more clouded than"). The exact access class $a$ dominates all the elements in $\mathcal{A}$ because it is associated with precise information. The expression $a^{**} \sqsubseteq_* a^*$ means that $a^{**}$ is associated with more clouded information than $a^*$.

One of the main advantages of the context formalism is that it provides a uniform mechanism for viewing exact and clouded information in database relations. A relational attribute $A_i$ with domain $D_i$ is assigned a separate "view context" $C_i$ (on $D_i$) for each distinct exact/clouded data classification level used in the extended relation. Within each access class set $\mathcal{A}_i$ used for attribute $A_i$, the assigned contexts must respect the natural ordering ($\sqsubseteq_*$) on the access classes. The dominant (top) element $a_i$ of each $\mathcal{A}_i$ (e.g., $s_i$) is assigned the finest context generated on the domain $D_i$. This context comprises singleton equivalence classes and only releases exact information. The more clouded the access class $a_i{}^*$ in $\mathcal{A}_i$ ($s_i{}^{**} \sqsubseteq_* s_i{}^*$), the coarser is the context assigned to it ($C_i{}^{**} \sqsubseteq_C C_i{}^*$), and the larger or more clouded are the information chunks released as tuple components.

Consider the following extended multilevel relation comprising exact and clouded information at various data classification levels.

| *Name* | *Occupation* | *Salary* |
|---|---|---|
| *Smart* ($c$) | *Spy* ($s$) | $50K$ ($s$) |
| *Smart* ($c$) | *Spy* ($s$) | ($20K$, $30K$, $60K$, $70K$) ($s^*$) |
| *Smart* ($c$) | *Spy* ($s$) | ($0K$, $0K$, $100K$, $100K$) ($s^{**}$) |
| *Smart* ($c$) | *Salesman* ($c$) | $24K$ ($c$) |
| *Smart* ($c$) | *Salesman* ($c$) | ($0K$, $20K$, $25K$, $45K$) ($c^*$) |

Since the secret ($s$) and confidential ($c$) classifications convey exact information, the *Salary* attribute is associated with the precise context on the *Salary* domain for each of these classifications. (For simplicity we denote exact information as an atomic value rather than as a singleton set). The coarser context { [$0K$, $29K$], [$30K$, $64K$], [$65K$, $100K$], [$101K$+] } is appropriate for the secret clouded classification $s^*$. ($[l_i, h_i]$ denotes an interval on the set of integers.) The context used for $s^{**}$, say { $[0K, 100K]$, $[101K+]$ }, is even coarser, as $s^{**}$ conveys more clouded information than $s^*$. The $s^{**}$ salary value is actually the classical set $[0K, 100K]$. Thus, exact information (singleton sets), classical sets, and more generally, fuzzy sets, can be uniformly stored/viewed in multilevel relations using the context mechanism.

The *Salary* context used for clouded confidential ($c^*$) information, say { $[0K, 50K]$, $[51K+]$ }, is independent of the contexts used for clouded secret ($s^*$ and $s^{**}$) information; however, it must be coarser than the precise context used for exact confidential ($c$) information.

Multilevel information clouding is implemented by providing a user with a clearance level $a \in S$ which permits access to exact information, and a clouded clearance level for each relational attribute $A_i$ which is to be clouded. To eliminate, or at least to reduce, covert channels, the clouded levels assigned to a user must correspond to his/her clearance level for exact information. Thus, a secret ($s$) user can only be provided with secret clouded clearance levels, say $s^*$ for one relational attribute and $s^{**}$ for another. The issue of clouded information flow through covert channels is discussed in a later section.

We are now in a position to define the actual information seen by users at various clearance levels. As in the classical multilevel model, users who have exact clearance levels would see information whose classifications do not dominate their clearance levels. Since such users see exact information conveyed at all (lower) levels, it is not necessary for them to see any clouded information. Secret ($s$) users thus see (exact) tuples 1 and 4 in their views of the multilevel relation above; confidential ($c$) users only see tuple 4. Likewise, users with clouded clearance levels for a certain attribute, say $s^*$, would only see $s^*$ information, and not $s^{**}$ information even if it exists. Thus, secret ($s$) users with $s^*$ clearance for *Salary* see tuples 2 and 4 (tuple 4 is visible because secret users are cleared for confidential information). Similarly, users with $s^{**}$ clearance for *Salary* see tuples 3 and 4. Those with $c^*$ clearance only see tuple 5.

The final issue to be disussed deals with updates to clouded databases. It is clear that update operations are crucial to achieving the desired levels of security in classical multilevel database systems; they are equally important in clouded multilevel database systems. Unfortunately, there is no clear consensus in the classical security community about the precise update semantics to be adopted in multilevel databases. Since the original SeaView model [6] several versions of an update semantics have been proposed for multilevel relations (see e.g., [5,7]. They differ in the number of core integrity properties employed and in the nature of these properties (e.g., whether they are state or transition properties). When discussing updates in the clouded model it is important to observe that clouding is designed to supplement rather than compete with the classical multilevel formulation. Two approaches to update handling are feasible as they support this view and would naturally integrate with any of the existing update semantics proposals. The simplest approach is to consider clouding purely as a mechanism for viewing database data. Thus, only users with exact clearance levels for relational attributes may update their values; users with clouded clearance levels are banned from performing updates. A better strategy is to allow updates by users with exact or clouded clearance levels as long as they input exact information which is subsequently tagged with the corresponding exact classification levels. For example, a user with an $s^*$ clearance level for *Salary* could enter an exact value, say $55K$ in place of $50K$ in the relation above, which would be tagged as secret ($s$). The authorization to make such updates would of course require a certain level of trust on the part of the user.

## IV. QUERYING CLOUDED RELATIONS

Contexts and their equivalences play an important role in query specification. The notion of equality used with exact information is generalized to a context-based notion of equivalence for uniformly querying exact and clouded/fuzzy information.

Definition:

The normal fuzzy sets, $t$ and $t'$, are formally *equivalent* with respect to a context $C$, denoted by $t \sim_C t'$, whenever their $\alpha_{1.0}$-cuts are subsets of a single equivalence class in $C$.

Equivalence classes comprise indistinguishable elements. Therefore, all non-empty subsets of an equivalence class (sieve opening) in a context are equivalent. Exact information (spiked fuzzy sets) and clouded information chunks (e.g., trapezoidal fuzzy sets, or normal fuzzy sets for unordered scalar domains) which pass through a given sieve opening are equivalent. Note that $\sim_C$ reduces to equality ($=_C$) for a precise context $C$.

A relation containing clouded information is queried by attaching contexts to values in query specifications. These "query contexts" act as sieves controlling the quality of the information recovered. A query specifies certain sieve openings (equivalence classes containing the specified values) and all information passing through these openings is retrieved. Classical query languages employ precise contexts. They formulate "precise queries" based on the equality of atomic values (equivalence in precise contexts). An imprecise query employs coarser contexts. The weaker equivalence permits the extraction of exact and clouded information which "approximately" matches the specifications.

Querying multilevel clouded relations actually involves the interaction of two types of contexts on the relational attributes. The first is the user-selected query context $C_q$ which determines the precision of the "submitted" query. The second is the "view context" $C_v$ assigned to a relational attribute for each exact/clouded data classification level in a multilevel relation. It determines the precision of the information released to the user. The two contexts are combined during query processing to produce an "effective context" $C_e$ used for information retrieval. The combination must ensure strict access-control: Users should not obtain less clouded versions of the information than they are cleared to receive by adjusting the precision of their query contexts or by carefully selecting the target values in their query specifications.

Definition:

An *effective context* $C_e$ is computed as $glb(C_v, C_q)$. If $\rho_v$ and $\rho_q$ are the equivalence relations generating $C_v$ and $C_q$, respectively, then $C_e$ is the context generated by $\rho_e = eq(\rho_v \cup \rho_q)$ ($eq$ is the equivalence relation closure operator).

$C_e$ is the finest context coarser than both $C_v$ and $C_q$. Therefore, regardless of how users phrase their queries, the effective precision is never finer than their view contexts. Since a multilevel view is already secured by information clouding, "creative querying" by snoopers only causes additional information to match their targets. The transformation coarsens precise queries to secure fuzzy queries and converts somewhat fuzzy queries into fuzzier and even more secure queries.

Proposition:

A query evaluated with effective contexts $C_e$ is *secure* with respect to access-control.

To understand the notion of an effective context consider a secret ($s$) user provided with an $s^*$ clearance for *Salary*. By definition, he/she also sees exact confidential ($c$) information. Assume that the coarse view context $C_v$ used for $s^*$ is { $[0K, 29K]$, $[30K, 64K]$, $[65K, 100K]$, $[101K+]$ }; a precise view context is used for exact confidential ($c$) information. Suppose the user issues the precise query ($C_q$ = { {$60K$} }):

$$Select(r) \ where \ Salary = 60K.$$

The effective context $C_e$ at the confidential classification is the precise *Salary* context. Thus, the user can query confidential ($c$) information precisely:

$$Select(r; c) \ where \ Salary = 60K.$$

On the other hand, since $C_e$ for $s^*$ information is { $[0K, 29K]$, $[30K, 64K]$, $[65K, 100K]$, $[101K+]$ }, the query is transformed and evaluated as:

$$Select(r; s^*) \ where \ Salary \sim_{C_e} 60K,$$

i.e., *Salary* = *moderate*. The view context for $s^*$ does not allow the user to distinguish between "moderate" salaries in $[30K, 64K]$. He/she can however differentiate between the fuzzy concepts "low," "medium," "high," and "very-high" salaries. Note that using coarser contexts $C_q$ for $s^*$ or $c$ information (e.g., those "spanning" two equivalence classes in $C_v$) only results in fuzzier queries.

Clouding eliminates inference attacks based on deduction [4,8] as users can only reason approximately with fuzzy information presented in their views. Consider the multilevel relation $R_1$(*Name*, *Occupation*, *Salary*) above, and the unclassified tax-table $R_2$(*Salary*, *Tax*) embodying the precise dependency between *Salary* and *Tax*. Since a secret user with $s^*$ clearance for *Salary* sees clouded salaries at the secret level, he/she can only obtain clouded tax assessments. A context-based *join* automatically produces clouded tax assessments based on $s^*$ information because it uses the *glb* of the *Salary* contexts of the two relations.

## V. CLOUDING AND COVER STORIES

Polyinstantiation plugs covert channels by eliminating the flow of high-level information to low-level users. E.g., updating *Mr. Smart*'s salary at the secret ($s$) level does not affect the value seen by confidential ($c$) users. Clouded multilevel relations are also "leakproof," e.g., $s$ to $c$, or $s$ to $c^*$. However, information flow is possible between an exact class and its clouded counterparts, e.g, $s$ to $s^*$ to $s^{**}$. This arises from the need to maintain the veracity of clouded information. A clouded cover story is useful because it does not lie; but it should reduce or perhaps even eliminate information flow.

Suppose the fuzzy set *moderate* is defined with respect to the $s^*$ context { $[0K, 29K]$, $[30K, 64K]$, $[65K, 100K]$, $[101K+]$ }. The veracity of the clouded cover story is maintained until *Mr. Smart*'s salary at the secret ($s$) level is updated to a value residing in a different equivalence class in the $s^*$ context. The $s^*$ salary must then be updated to "high" or "low" to preserve truth. Information flow to the $s^*$ level due to the update can be eliminated by using the coarsest possible context for $s^*$ salaries; but this is equivalent to providing no information at all.

| *Name* | *Occupation* | *Salary* |
|---|---|---|
| *Max Smart* ($c$) | *Spy* ($s$) | $50K$ ($s$) |
| *Max Smart* ($c$) | *Spy* ($s$) | *moderate* ($s^*$) |
| *Max Smart* ($c$) | *Salesman* ($c$) | $24K$ ($c$) |
| *Max Smart* ($c$) | *Salesman* ($c$) | *low* ($c^*$) |

Nevertheless, information flow can still be eliminated, or at least reduced, if the clouding context is selected based on likely future salaries. Appropriately selecting the context and the clouded salary can maintain the veracity of the clouded cover story without ever changing its value. However, its granularity should be sufficiently small to be useful to database users. It is interesting to note that the requirements for good clouded cover stories are similar to those placed on the depositions of well-coached defense witnesses!

The final point deals with the connection between clouded cover stories at different exact access classes: How should *Mr. Smart*'s clouded salary at the $c^*$ level relate to its value at the $s^*$ level? Although the clouded contexts and values are assigned independently, it is helpful if the clouded values are the same, e.g., *moderate* for both $s^*$ and $c^*$ above. This in turn implies that the exact values for different access classes should be semantically "close" to each other and the "fuzzy" dependencies between attributes should be preserved at the clouded levels. The present model based on fuzzy sets can help address this issue. A good cover story is sensible, but it must also be corroborated by other information sources. Otherwise it will soon be ridden with holes. Perhaps the best way to fabricate a cover story is to first ascertain that it is reconciled with the original information at the clouded level.

## VI. CONCLUSIONS

Multilevel database security is a novel and unexplored area for applying fuzzy set theory. Sensitive information in multilevel relations is meaningfully clouded using fuzzy sets. This is accomplished by broadening the possibility distributions constraining the values of sensitive attributes. Clouding is appealing because it maintains access-control and also eliminates inference attacks by enforcing fuzzy reasoning with sensitive information. It is the middle ground between information release and information hiding/cover stories. Moreover, unlike other multilevel security techniques, information clouding preserves truth. This promotes data usage and helps strike the right balance between user convenience and database security.

## REFERENCES


[1] N.R. Adam and J.C. Wortmann, "Security-control methods for statistical databases," *ACM Computing Surveys*, vol. 21, pp. 515-556, 1989.

[2] B.P. Buckles and F.E. Petry, "A fuzzy representation of data for relational databases," *Fuzzy Sets and Systems*, vol. 7, pp. 213-226, 1982.

[3] D.E. Denning, S.G. Akl, M. Heckman, T.F. Lunt, M. Morgenstern, P.G. Neumann and R.R. Schell, "Views for multilevel database security," *IEEE Transactions on Software Engineering*, vol. 13, pp. 129-140, 1987.

[4] T.D. Garvey and T.F. Lunt, "Cover stories for database security," *Proceedings of the Fifth IFIP WG11.3 Workshop on Database Security*, Shepherdstown, West Virginia, 1991.

[5] S. Jajodia and R. Sandhu, "Toward a multilevel secure relational data model," *Proceedings of ACM SIGMOD*, ACM Press, New York, pp. 50-59, 1991.

[6] T.F. Lunt, D.E. Denning, R.R. Schell, M. Heckman and W.R. Shockley, "The SeaView security model," *IEEE Transactions on Software Engineering*, vol. 16, pp. 593-607, 1991.

[7] T.F. Lunt and D. Hsieh, "Update semantics for a multilevel relational database system," in *Database Security IV: Status and Prospects*, S. Jajodia and C.E. Landwehr, Eds. New York: Elsevier Science, pp. 281-296, 1991.

[8] M. Morgenstern, "Security and inference in multilevel database and knowledge base systems," *Proceedings of ACM SIGMOD*, ACM Press, New York, pp. 357-373, 1987.

[9] K.V. Raju and A.K. Majumdar, "Fuzzy functional dependencies and lossless join decomposition of fuzzy relational database systems," *ACM Transactions on Database Systems*, vol. 13, pp. 129-166, 1988.

[10] S. Shenoi, A. Melton and L.T. Fan, "An equivalence classes model of fuzzy relational databases," *Fuzzy Sets and Systems*, vol. 38, pp. 153-170, 1990.

[11] S. Shenoi, K. Shenoi and A. Melton, "Contexts and abstract information processing," *Proceedings of the 4th International Conference on Industrial and Engineering Applications of AI and Expert Systems*, Kauai, Hawaii, pp. 44-50, 1991.

[12] L.A. Zadeh, "Fuzzy sets as a basis for a theory of possibility," *Fuzzy Sets and Systems*, vol. 1, pp. 3-28, 1978.

[13] M. Zemankova and A. Kandel, *Fuzzy Relational Databases: A Key to Expert Systems*. Cologne: Verlag TUV Rheinland, 1984.

# On The Use of Logical Definition of Fuzzy Relational Database

M.A .Vila, J.C. Cubero, J.M. Medina, and O. Pons
Departamento de Ciencias de la Computacion e I.A.
Facultad de Ciencias. Universidad de Granada
Campus de Fuentenueva
18071 Granada (Spain)

***Abstract*— Several approaches to fuzzy relational databases are analyzed by using a logical definition previously given by the authors. This work shows how in these previous approaches both the data structure and the query management can be considered particular cases of this definition.**

KEYWORDS: Fuzzy Relational Database, Logic Database

## I. Introduction

The Fuzzy Relational Database concept was introduced by several authors in order to deal with imprecise query and data. Two major approaches have appeared in the literature about this topic.

### *A. The homegeneous model approach*

(Buckles and Petry [3], Buckles, Petry and Sachar [3]) The basic features of this model are the following:

The attribute basic domain sets consist in scalar finite sets and the domain values themselves are ordinary sets whose cardinality increases with the degree of fuzziness. A fuzzy resemblance relation is also defined in every basic domain set.

The retrieval process may be carried out in two different ways. The first one consists in some extensions of "classical" relational algebra [2]. The second one is an appropriated modification of domain relational calculus [3], where wffs is considered to be accomplished with similarity level conditions. An equivalence theorem between these two ways is proved in [3].

### *B. The Possibility-based approach*

(Umano [9], Baldwin [1], Prade and Testemale [7], Zemankova-Leech and Kandel [11])

This approach is used, with some modifications, by the majority of authors and its main idea is to consider that the basic domains of attributes may be any scalar or numerical ones, but that the attribute values will be possibility distributions on the above mentioned basic domains. All the query evaluation processes are relational algebra based. The more developed models are presented in [7] and [11], they are very similar and both use upper and lower measures to manage imprecise query. In the following we will call them models P-T and Z-K respectively.

Although the possibilistic approach seems general enough to represent any fuzzy attribute, some problems exist in its practical applications. The main source of these problems is the following:

Whereas the possibility based approach assumes any attribute to be "totally" fuzzy, in a practical case we can find in the same relation scalar or numerical valued attributes, some of which may be totally crisp other "fuzzy" in the Buckles and Petry's sense and finally other attributes will take a possibility distribution, on their basic domains, as value in the tuples. We can also consider some crisp attributes could accept "fuzzy queries". So that, each attribute of a fuzzy database must have its specific way of fuzziness.

These considerations combined with the new perspectives appeared with the use of Logic in Relational Database Theory (Reiter [8], Gallaire et al [6]) led us to a new fuzzy database definition given by using logical concepts [10]. With this definition it is possible to consider several kinds of fuzziness for the database attributes. A Domain-Calculus based query language was also established for the model.

The present paper will deal with the relation between this definition and the above mentioned previous models. After a summary of the logical definition both homgeneous and possibilistic approaches are included in our model by means of the data structure and the query language identifications.

0-7803-0614-7/93$03.00 ©1993IEEE

## II. A Logic Definition of Fuzzy Database

### *A. Definition of Fuzzy Relational Language (FRL)*

We have defined fuzzy database as an interpretation of an special first order language. This be done by using previous concepts given by Reiter in [8]. So, first a "relational language", in Reiter's sense, $\mathcal{L} = (\mathcal{A}, \mathcal{W})$ is considered.

$\mathcal{A}$ is an alphabet with a finite number of constants and predicate symbols and without symbols to functions. $\mathcal{W}$ is the set of wffs obtained from $\mathcal{A}$. There is a non-empty set of distinguished unary predicates, called "types", standing for attribute basic domains. And a special binary predicate: $=$ is assumed to be also in $\mathcal{A}$.

To this relational language we introduce new special elements:

a) We assume $\mathcal{A}$ includes the **symbol MD**(Membership Degree) for a simple type which will represent the membership degree domain for all of $\mathcal{L}$ interpretations. A binary predicate symbol $\geq$ will be also included in $\mathcal{A}$ and it will appear associated with MD.

b) We also consider among the types of $\mathcal{A}$ that there are special types, called **fuzzy types**, which represent those attributes having fuzzy values and (or) fuzzy queries. Each fuzzy type $\tau$ is associated with three predicates symbols:

NOP$\tau$, simple type standing for the names of possibility distribution defined with $\tau$ as basic domain. POS$\tau$3-ary predicate, which will be used to represent the possibility distributions defined on $\tau$ . And $\cong \tau$, 3-ary predicate representing an "approximately equal relation" given between possibility distributions.

A relational language with these two additional elements is called **Fuzzy Relational Language.**

### *B. Definition of Fuzzy Interpretation (FI)*

Let $\mathcal{L}$ be a FRL and let us consider the interpretation $I = (D, F)$ where $D$ is the corresponding domain and $F$ the interpretation function. We will say I is a fuzzy interpretation (FI) of $\mathcal{L}$ if and only if it verifies:

i) $I$ is a relational interpretation of $\mathcal{L}$ in Reiter's sense

ii) A finite set $H \subset [0,1]$ exists such that:

- $0 \in H$ and $1 \in H$
- $H \subset D$
- $F(MD) = H$

iii) I is a model for the following set of integrity rules:

*B.1. Rules to establish the type MD semantic.*

- R1.-$\forall(x,y)(\geq(x,y) \rightarrow MD(x) \wedge MD(y))$
- R2.-$\forall(x)(MD(x)) \rightarrow \geq (x,x))$
- R3.-$\forall(x,y)(\geq(x,y) \wedge \geq(y,x) \rightarrow = (x,y))$
- R4.-$\forall(x,y,z)(\geq(x,y) \wedge \geq(y,z) \rightarrow \geq(x,z))$

Obviously, theses rules are to define $\geq$ as an order relation on MD.

*B.2. Rules to establish the fuzzy types semantic.*

For every fuzzy type $\tau$ being in $\mathcal{L}$

- R5.-$\forall(x,y,z)(\mathrm{POS}\tau(x,y,z) \rightarrow \mathrm{NOP}\tau(x) \wedge \tau(y) \wedge MD(z) \wedge \neg(=(z,0)))$
- R6.-$\forall(x)(\mathrm{NOP}\tau(x) \rightarrow \exists(y,z)(\mathrm{POS}\tau(x,y,z))$
- R7.-$\forall(x)(\tau(x) \rightarrow \mathrm{POS}\tau(x,x,1) \wedge \neg(\exists(y,z)$ $(\mathrm{POS}\tau(x,y,z) \wedge \neg = (x,y)))))$

R5 assures us POS$\tau$ is to decribe a possibility distribution on $\tau$. R6 imposes there is no possibility distribution name without its corresponding definition. Finally R7 includes the $\tau$ extension into the POS$\tau$ extension as crisp unary sets, whose names are the elements of $\tau$ themselves.

*B.3. Rules to establish the "approximately equal" semantic*

For every fuzzy type $\tau$ being in $\mathcal{L}$:

- R8.-$\forall(x,y,z)(\cong \tau(x,y,z) \rightarrow NOP\tau(x) \wedge NOP\tau(y) \wedge MD(z))$
- R9.-$\forall(x,y,z)(\cong \tau(x,y,z) \rightarrow \cong \tau(y,x,z))$

Theses last rules characterize the $\cong$ extension as a symmetric fuzzy relation on the possibility distributions domain, that is, as a ressemblance relation. Obviously with an additional rule, imposing some kind of transitivity, $\cong$ becomes a similarity fuzzy relation.

### *C. Definition of Fuzzy Relational Database (FRDB)*

We define Fuzzy Relational Database as the triple $(\mathcal{L}, I, IC)$ where $\mathcal{L}$ is a FRL , I is $aRI$ and $IC$ is a set of specific integrity rules, including $\forall Pn$-ary predicate of $\mathcal{L}$ different to $=$, the following one:

$\forall(x_1 \ldots x_n)(P(x_1 \ldots x_n) \rightarrow \tau_1(x_1) \wedge \ldots \tau_n(x_n))$

where $\tau_i$ are types.

### *D. Definition of a query language*

Since the FRDB definition given above characterizes it as a first order logical structure, it is possible to establish a query language by using the Predicate Calculus in the following way.

Let us consider a FRL $\mathcal{L} = (\mathcal{A}, \mathcal{W})$, we will call **query for $\mathcal{L}$** any expression of the form

$$< \bar{x}/\bar{\tau} \mid W(\bar{x}) > .$$

Where $\bar{x}/\bar{\tau}$ denotes the sequence $x_1/\tau_1 \ldots x_m/\tau_m$, being $x_j \mathcal{A}$ variables and $\tau_j$ simple types, and $W(\bar{x}) \in \mathcal{W}$ being among $x_1 \ldots x_m$ the only free variables of W.

Now, let us consider a FRDB , $B = (\mathcal{L}, I, IC)$ and a query for $\mathcal{L}$ : $Q = < \bar{x}/\bar{\tau} \mid W(\bar{x}) >$, we will say the constant tuple $\bar{c} = c_1 \ldots c_m$ is "an answer to $Q$ in $B$" if and only if:

- $\forall j \in \{1 \ldots m\} \tau_j(c_j)$ is true in $I$
- $W(\bar{c})$ is true in $I$

## III. Identification of the Homogeneous Model

Once we have established our logical approach as well as its associated query language, we are going to relate with to the previous models of fuzzy relational databases, i.e. the homogeneous and the posssibilistic ones. We will show how these models can be considered as particular cases of our approach, by adding a suitable integrity rule set.

We will start with the simplest model which is the homogeneous one. First we will identify the data structure which was presented in [2], and next the domains calculus presented in [3].

### A. Identification of the Data Structure

The data structure of the homogeneous model is that of a set of classical unnormalized relations, i.e. where the attribute values can be sets. It is also assumed that a resemblance fuzzy relation defined on each basic domain exists.

Now let us consider FRDB, $B = (\mathcal{L}, I, IC)$ defined according II.5 it will a fuzzy relational database in the homogeneous model's sense if it verify the following conditions:

#### A.1. Fuzziness condition

Any type $\tau \in \mathcal{L}$ representing an attribute domain and verifying $\tau \neq MD$ is a fuzzy type.

This condition asserts that any attribute in the database must be fuzzy. It is coherent with the homogeneous model definition. This model has no conditions about the existence in the database of some totally crisp attributes to define the tuples. It should be remarked that this condition plus the rules establishing the semantic of $\cong$ ($R8$ and $R9$), allows us to assure there exists a resemblance fuzzy relation associated with each attribute domain.

#### A.2. Condition of crisp values

For every fuzzy type $\tau$ :

- HM1.-$\forall(x, yz)(\text{POS}\tau(x, y, z) \rightarrow = (z, 1.0)$

This rule states that the only values of attributes must be crisp sets

#### A.3. Condition of resemblance relation definition

for every fuzzy type $\tau$:

- HM2.- $\forall(x, y, z)(\cong \tau(x, y, z) \wedge = (z, 1.0) \leftrightarrow = (x, y))$
- HM3.- $\forall(x, y, z)(\cong \tau(x, y, z) \leftrightarrow [\tau(y) \wedge (\forall w(\text{POS}\tau(x, w, 1.0) \rightarrow \exists v(\cong \tau(y, w, v) \wedge \geq (v, z)) \vee \tau(x) \wedge (\forall w(\text{POS}\tau(y, w, 1.0) \rightarrow \exists v(\cong \tau(x, w, v) \wedge \geq (v, z))])$

In this rule, we assume at least one of the values to be compared is a single value (i.e. an unitary set) and the resemblance level between a single value $x$ and a crisp set Y is given by:

$$z = \min_{w \in Y}(S(x, w))$$

where $S(.,.)$ is the resemblance relation associated with $\tau$ which is represented in our model by the predicate $\cong \tau$, as can be seen in [2] and this is the way used by the homogeneous model to compute resemblance levels

#### A.4. Identification of the query language

If we analyze the domain calculus language given by Buckles, Petry and Sachar in [3] for the homogeneous model ($B-P-S$ Language), we can see that the schema for any query is the same as which appear in the definition it II.D So we must only identify the formulae being on the right side of the query.

The formulae of $B-P-S$ language are sequences of the so called "atoms" connected by the usual logical operators and quantifiers. So, we only need to translate the atoms into our language to prove that any formula can be represented as an wff. There are two kind of atoms in the $B-P-S$ language:

A) $x\theta y$ WITH LEVEL $\alpha$

which represents a fuzzy comparison operation defined for crisp values belonging to any domain $\tau$.A pair of constants a and $b$ make this atom true when the accomplishment degree of the sentence "$a\theta b$" is greater than or equal to $\alpha$.

In our language, the fuzzy operator $\theta$ is represented by means of a 3-ary predicate symbol associated with the fuzzy type $\tau$ and verifying at least the rule:

$\forall(x, y, z)(\theta(x, y, z) \rightarrow (\tau(x) \wedge \tau(y) \wedge MD(z))$ and the atom is translated in the following wff

$\exists(w)(\theta(x, y, w) \wedge \geq (w, \alpha))$

B) $R(y_1 \ldots y_k)$ WITH LEVEL $(\alpha_1 \ldots \alpha_s)$

where $R$ is a k-ary predicate, $y_j, j \in \{1, 2 \ldots s\}$ are variables and $y_j, j \in \{s+1, s+2 \ldots k\}$ constants, with domains $\tau_j$ respectively.

According to the definition of acceptable instance given in [3], and taking into account that the $\cong$ operator has been defined coherently with the homogeneous model, it is easy to see that the translation of this atom will be:

$\exists(v_1 \ldots v_s)(R(w_1 \ldots w_s, y_{s+1} \cdots y_k)\wedge$
$(\cong \tau_1(w_1, y_1, v_1)\wedge \geq (v_1, \alpha_1)) \wedge \ldots \wedge (\cong \tau_s(w_s, y_s, v_s)\wedge$
$\geq (v_s, \alpha_s)))$

Therefore we can assure that any query written in the $B-P-S$ language can be translated into ours. Moreover, since these authors proved in [3] that the relational algebra associated with the homogeneous model is equivalent to their domains calculus, the above conclusion can be also applied to any query expressed in the relational algebra way.

## IV. Identification of Possibilisic Models

The main feature of possibilistic models that is any attribute can asumit any fuzzy value, which is represented by a possibility distribution. Therefore, it could be thought that the possibilistic models have a direct identification in our approach. However the problem is not so easy because of the way in which these models manage the queries. By analyzing this question in [7] or [11], we found the following key points which are common to both approaches, $P-T$ and $Z-K$ models:

- Any "atomic condition" $x\theta y$ defined between fuzzy values implies the existence of a previous fuzzy relation given between crisp values. This restriction is also applied to $\cong$.
- For every atomic condition there are two accomplishment degrees which are associated with the possibility and necessity concepts. These are computed by the above mentioned fuzzy relation and several operations: sum, product, max etc..., defined between membership degrees.

All these problems make the data structure identification quite complicated, because it is necessary to define several operations in the $MD$ domain and to establish two predicates for every fuzzy comparison operator.

### *A. Identification of the data structure*

We will make the identification in a general way by distinguishing between models $P-T$ and $Z-k$ only if it is necessary.

Let us consider again a FRDB $B = (\mathcal{L}, I, IC)$, it will be a fuzzy relational database in the possibilistic model's sense if it verifies the following conditions:

#### *A.1. Fuzziness condition*

Any type $\tau \in \mathcal{L}$ representing an attribute domain and verifying $\tau \neq MD$ is a fuzzy type.

We impose this conditions for the same reasons as in the above identification.

#### *A.2. Conditions to define some operations in $MD$ domain*

*a) Definition of minimum and maximum*

There are two predicates $MAX$ and $MIN$ associated to $MD$ and which verify

- E1.-$\forall(x,y)(MD(x) \wedge MD(y) \rightarrow \exists z(MD(z)\wedge MIN(x,y,z))$
- E2.-$\forall(x,y,z)(MIN(x,y,z) \leftrightarrow (MD(x) \wedge MD(y) \wedge MD(z) \wedge ((\geq (x,y)\wedge = (y,z)) \vee (\geq (y,x)\wedge = (x,z)))))$
- E3.-$\forall(x,y)(MD(x) \wedge MD(y) \rightarrow \exists z(MD(z)\wedge MAX(x,y,z))$
- E4.-$\forall(x,y,z)(MAX(x,y,z) \leftrightarrow (MD(x) \wedge MD(y) \wedge MD(z) \wedge ((\geq (x,y)\wedge = (x,z)) \vee (\geq (y,x)\wedge = (y,z)))))$

*b) Definition of addition and (1-...)*

There are two predicates $+(.,.,.)$ and $NEG(.,.)$ which verify:

- E5.-$\forall(x,y,z)(+(x,y,z) \rightarrow MD(x) \wedge MD(y)\wedge MD(z))$
- E6.-$\forall(x,y,z,v1,v2,w1,w2)(+(x,y,v1) \wedge +(v1,z,w1) \wedge +(y,z,v2) \wedge +(x,v2,w2) \rightarrow = (w1,w2))$
- E7.-$\forall(x,y,z)(+(x,y,z) \rightarrow +(y,x,z))$
- E8.-$\forall(x,y,z)(+(x,y,z) \rightarrow \geq (z,x)\wedge \geq (z,y))$
- E9.-$\forall(x)(MD(x) \rightarrow +(x,0,x))$
- E10.-$\forall(x)(MD(x) \rightarrow \exists y(MD(y) \wedge NEG(x,y))$
- E11.-$\forall(x,y)(NEG(x,y) \leftrightarrow +(x,y,1))$

As can be seen, we do not impose the addition to be defined for every pair of values of MD. In fact, this operation is introduced to obtain the operator NEG and the definition of the latter is totally guaranteed by E10 and E11.

*c) Definition of the product.*

A predicate $*(.,.,.)$ exists which verifies

- E12.-$\forall(x,y)(MD(x)\wedge MD(y) \rightarrow \exists z(MD(z) \wedge *(x,y,z))$
- E13.-$\forall(x,y,z,v1,v2,w1,w2)(*(x,y,v1)\wedge *(v1,z,w1)\wedge *(y,z,v2)\wedge *(x,v2,w2) \rightarrow = (w1,w2))$
- E14.-$\forall(x,y)(*(x,y,z) \rightarrow *(y,x,z))$

- E15.-$\forall(x)(MD(x) \rightarrow *(x,1,x))$
- E16.-$\forall(x)(MD(x) \rightarrow *(x,0,0))$
- E17.-$\forall(x,y,z)(*(x,y,z) \wedge \neg = (x,0) \wedge \neg = (y,0) \rightarrow \geq (x,z) \wedge \neg = (x,z) \wedge \geq (y,z) \wedge \neg = (y,z))$
- E18.-$\forall(x,y,z,v1,v2,v2,w1,w2)(+(x,y,v1) \wedge *(v1,z,w1) \wedge *(x,z,v2) \wedge *(y,z,v3) \wedge +(v2,v3,w2) \rightarrow = (w1,w2))$

These rules define * as an operation on [0,1] with the same properties as the ones of classical numerical product. It should be noted that the numerical product is not well defined in a finite subset of [0,1], but that a suitable rounding process solves this problem, and the usual product can be used as.

*A.3. Conditions for the definition of* $\cong$

In this paragraph we will give all the conditions to be verified by the predicate $\cong \tau$ associated with every fuzzy type $\tau$. These are formulated so that the accomplishment degree of the sentence "approximately equal" is the same one which we obtain if the computing ways ofthe models $P-T$ or $Z-K$ are used. As we said above, this implies the definition of three predicates instead of $\cong \tau$.

*a) Definition of the basic fuzzy relation.*

For every fuzzy type $\tau$a predicate $MU\tau(.,.,.)$ exists which verifies:

- A1.-$\forall(x,y)(\tau(x) \wedge \tau(y) \rightarrow \exists z(MD(z) \wedge MU\tau(x,y,z))$
- A2.-$\forall(x)(\tau(x) \rightarrow MU\tau(x,x,1))$
- A3.-$\forall(x,y,z)(MU\tau(x,y,z) \rightarrow MU\tau(y,x,z))$

As can be seen, we have imposed the minimal conditions to perform $MU\tau$ as a resemblance relation. $MU\tau$ represents the "approximately equal" relation defined in the domain of $/tau$.

*b) Definition of the predicate associated with possibility degree*

For every fuzzy type $\tau$a predicate $\cong \pi\tau(.,.,.)$ exists which verifies

- A4.-$\forall(x,y,z)(\cong \pi\tau(x,y,z) \rightarrow \text{NOM}\tau(x) \wedge \text{NOM}\tau(y) \wedge MD(z))$
- A5.-$\forall(x,y)(\text{NOM}\tau(x) \wedge \text{NOM}\tau(y) \rightarrow \exists z(\cong \pi\tau(x,y,z))$

The following rule refers to the concrete formula of possibility degree, so it must be model dependent:

$-P-T$ Model:

We define the following wff for the sake of simplicity as:

$F(x,y,zr,xr,yr) \equiv (\exists r1,r2,r3,r4)(\text{POS}\tau(x,xr,r1) \wedge \text{POS}\tau(y,yr,r2) \wedge MU\tau(xr,yr,r3) \wedge MIN(r1,r2,r4) \wedge MIN(r3,r4,zr))$

which is translation of:

$zr = \min(\text{pos}_x(xr), \text{pos}_y(yr), \mu(xr,yr)$ and then the rule which defines $\cong \pi\tau$ becomes

- A6.-$\forall(x,y,z)(\cong \pi\tau(x,y,z) \leftrightarrow (\exists x1,y1(F(x,y,z,x1,y1))) \wedge (\forall x2,y2,z2(F(x,y,z2,x2,y2) \rightarrow \geq (z,z2))$

The concordance with the corresponding definition given in [7] can be checked.

$-Z-K$ Model:

The wff $F$ is in this case:

$F(x,y,zr,xr,yr) \equiv (\exists r1,r2,r3,r4)$ $(POS\tau(x,xr,r1)POS\tau(y,yr,r2) \wedge$ $MU\tau(xr,yr,r3) \wedge MIN(r1,r3,r4) \wedge *(r2,r4,zr))$ which corresponds to:

$$zr = \min(\text{pos}_x(xr), \mu(xr,yr)), \text{pos}_y(yr))$$

The rule A6 remains the same.

It can be stated that $\cong \pi\tau$ verifies the rule $R8$ of definition B, and in the case of the $P-T$ model it also verifies $R9$.

*c) Definition of the predicate associated with necessity degrees.*

For every fuzzy type $\tau$ a predicate $\cong N\tau(.,.,.)$ exists which verifies

- A7.-$\forall(x,y,z)(\cong N\tau(x,y,z) \rightarrow \text{NOM}\tau(x) \wedge \text{NOM}\ \tau(y) \wedge MD(z))$
- A8.- $\forall(x,y)(\text{NOM}\tau(x) \wedge \text{NOM}\tau(y) \rightarrow \exists z(\cong N\tau(x,y,z))$

In this case, it should also distinguish between both models

$-P-T$ model:

now $F$ is given by

$F(x,y,zr,xr,yr) \equiv$ $(\exists r1,r2,r3,r4,r5,r6)(\text{POS}\tau(x,xr,r1) \wedge$ $\text{POS}\tau(y,yr,r2) \wedge MU\tau(xr,yr,r3) \wedge$ $NEG(r1,r4) \wedge NEG(r2,r5) \wedge$ $MAX(r4,r5,r6) \wedge MAX(r3,r6,zr))$

which corresponds to:

$$zr = \max(1-\text{pos}(xr), \mu(xr,yr), 1-\text{pos}(yr))$$

and the rule which defines the predicate is:

- A9.- $\forall(x,y,z)(\cong \pi N(x,y,z) \leftrightarrow$
  $(\exists x1, y1(F(x,y,z,x1,y1))\wedge$
  $(\forall x2, y2, z2(F(x,y,z2,x2,y2) \rightarrow \geq (z2,z))$

$-Z-K$ Model:
Both models are quite different in the case of lower measure (which is called "certainty" in [11]), so the wff $F$ and the rule itself will be different to . Let it be:
$F(x,yr,zr) \equiv \exists(r1,r2,r3)(\mathrm{POS}\tau(x,r1,r2)\wedge$
$MU\tau(r1,yr,r3) \wedge MIN(r2,r3,zr))\wedge$
$(\forall(r4,r5,r6,z1)(\mathrm{POS}\tau(x,r4,r5) \wedge MU\tau(r4,yr,r6)\wedge$
$MIN(r2,r3,z1) \rightarrow \geq (zr,z1))$
which represents

$$zr = \max_{x,r}\min(\mathrm{pos}_x(x,r), \mu(r,yr))$$

then the rule is:

- A10.- $\forall(x,y,z)(\cong N\tau(x,y,z) \leftrightarrow (\exists yr, zr, vr$
  $((F(x,yr,zr) \wedge \mathrm{POS}\tau(y,yr,vr) \wedge *(vr,zr,z))\wedge$
  $\forall(ys,zs,vs,z2((F(x,ys,zs) \wedge \mathrm{POS}\tau(y,ys,vs)\wedge$
  $*(vs,zs,z2) \rightarrow \geq (z2,z))$

This rule completes the definition of the fuzzy comparison "approximately equal", whose existence we have imposed for every fuzzy type. Obviously, it is possible to define suitable predicates for other imprecise comparisons, such as "quite greater than" etc... The rules to define them will be the same as in above case, only the basic fuzzy relation $MU\tau$ must be changed according to the semantic of the sentence. Therefore, in the following we will denote by $\theta\pi\tau$ and $\theta N\tau$ the respective upper and lower measures of the accomplishment degree of a fuzzy comparison $\theta$, defined for the fuzzy type $\tau$.

*B. Identification of the query language*

The identification is not immediate in this case, because no domain calculus is developed for the possibilistic models. Query management is done by using several operations based in the relational algebra. Three kinds of operations can be considered:

1. Those which are totally similar to the classical ones. In this class the operations are applied to classical (non-fuzzy) relations and the result is non-fuzzy too. It includes union, intersection, difference and cartesian product.
2. Another class includes those operations which are applied on classical relations given two fuzzy relations as a result. This is the case of the selection and joint.
3. Lastly, the projection must be separately considered. It can be applied to both classical and fuzzy relations, but in the second case the tuples which are identical with different membership levels, must be unified at maximum level.

*B.1. Identification of the union, intersection, difference and cartesian product.*

Let us consider two predicates symbols $R$ and $S$, corresponding to two $n$-ary relations with attributes in the same domains $\tau_1 \ldots \tau_n$. The following identifications are obvious:
$R \bigcup S \equiv < x_1 \ldots x_n / \tau_1(x_1) \ldots \tau_n(x_n) \mid$
$R(x_1 \ldots x_n) \vee S(x_1 \ldots x_n) >$
$R \bigcap S \equiv < x_1 \ldots x_n / \tau_1(x_1) \ldots \tau_n(x_n) \mid$
$R(x_1 \ldots x_n) \wedge S(x_1 \ldots x_n) >$
$R - S \equiv < x_1 \ldots x_n / \tau_1(x_1) \ldots \tau_n(x_n) \mid$
$R(x_1 \ldots x_n) \vee \neg S(x_1 \ldots x_n) >$
Let us suppose now the relation $R$ and $S$ have domains $\tau_1 \ldots \tau_n$ and $\nu_1 \ldots \nu_m$ respectively, we identify the cartesian product by the following:
$R \times S \equiv < x_1 \ldots x_n, y_1 \ldots y_m / \tau_1(x_1) \ldots \tau_n(x_n)$
$\nu_1(y_1) \ldots \nu_m(y_m) \mid R(x_1 \ldots x_m) \wedge S(y_1 \ldots y_m) >$

*B.2. Identification of Selection.*

Let a fuzzy comparison $\theta$ be associated with the fuzzy type $\tau, Y$ a fuzzy constant of $\tau$, i.e. $Y \in \mathrm{NOM}\tau$ and $R$ a predicate symbol of a n-ary relation whose i-th domain is also $\mathrm{NOM}\tau$. Then we have:
*a) Selection according possibility measure*
$R$ where $x_i\theta Y \equiv$
$< x_1 \ldots x_m, v/\tau_1(x_1) \ldots \tau_m(x_m), MD(v) \mid$
$R(x_1 \ldots x_m) \wedge \theta\pi\tau(x_i, Y, v) >$
*b) Selection according to necessity (certainty) measure*
$R$ where $x_i\theta Y \equiv$
$< x_1 \ldots x_m, v/\tau_1(x_1) \ldots \tau_m(x_m), MD(v) \mid$
$R(x_1 \ldots x_m) \wedge \theta N\tau(x_i, Y, v) >$
If we want to compare two attributes of the relation, $Y$ must be replaced by the corresponding $x_j$.

Obviously more complex expressions can be considered in the selection, all of them are a combination of atomic condition by means of "and", "or" and "not" operators. The identification in this case is done by using the corresponding predicates MAX, MIN and NEG defined in A.2. For instance, the case of operator "and" with possibility degree will be:
$R$ where $(x_i\theta_1 Y_1) \wedge (x_j\theta_2 Y_2) \equiv$
$< x_1 \ldots x_m, v/\tau_1(x_1) \ldots \tau_m(x_m), MD(v) \mid$
$R(x_1 \ldots x_m) \wedge (\exists v1, v2(\theta_1\pi\tau(x_i, Y_1, v1)\wedge$
$\theta_2\pi\tau(x_j, Y_2, v2) \wedge MIN(v1, v2, v) >$
The other cases are totally similar and therefore we haven't included them.

*B.3. Identification of the join*

According to its definition as an cartesian product plus a selection, the join is directly deduced from B.1and B.2. The only problem arises if we have an ≅ comparison and we want to avoid one of two involved attributes, as in the classicaL natural join. This is, in fact, an open problem which is closely related to the question of redundancy in fuzzy databases. It will be deal with in this context in a forthcoming work.

*B.4. Identification of the projection*

The following identification is given by considering the special characteristics of this operation, mentioned at the beginning of IV.2??.

Let us suppose an $n$-ary relation $R$ and we want to project it onto its $k$ first attributes.

$$R[A_1 \ldots A_k] \equiv < x_1 \ldots x_k / \tau_1(x_1) \ldots \tau_k(x_k) \mid$$
$$\exists(w_{k+1}, \ldots w_m)(R(x_1 \ldots x_k, w_{k+1}, \ldots w_k) \wedge$$
$$[(MD(x_1) \wedge$$
$$(\forall(x, v_{k+1}, \ldots v_m)(R(x, x_2, \ldots x_k, v_{k+1}, \ldots, v_m) \rightarrow$$
$$\geq (x_1, x))) \vee$$
$$\vdots$$
$$(MD(x_k) \wedge$$
$$(\forall(x, v_{k+1}, \ldots v_m)(R(x_1, \ldots, x_{k-1}, x, v_{k+1}, \ldots, v_m) \rightarrow$$
$$\geq (x_k, x)))]$$

This is the last identification of operations defined for the possibilistic models, and therefore the formulation of possibilistic models as an FRDB in our logical approach is completed.

## V. CONCLUDING REMARKS

We have identified the previous models of fuzzy relational database as particular cases of our logical definition, and this has been done in the two aspects of data structure and query management. This confirms the generality of the logical approach to the fuzzy relational databases.

But this result is not the final one which we can obtain by using this approach. In our opinion, the logical definition of fuzzy relational database will be useful to study the problem of imprecise data and knowledge in databases. Among the the possible research ways are the following:

- To study the problem of redundancy in fuzzy databases. That is, to analyze when it is possible to avoid any data (attribute value or tuple), because there is another one "approximately equal".
- To study Deductive Fuzzy Database models. In the same way as the logical approach to relational databases was the starting point for logic or deductive database models, we think the presented definitions could lead to new imprecise knowledge representation structures, where fuzzy databases and fuzzy rules will appear jointly represented in the same logical language

## VI. REFERENCES


[ 1] BALDWIN J.F. (1979) A new approach to approximate reasoning using a fuzzy logic Fuzzy Sets and Systems Vol. 2 pp.309-325

[ 2] BUCKLES W.P., PETRY F.E. (1982) A fuzzy representation of data for relational databases. Fuzzy Sets and Systems Vol. $7pp213 - 236$

[ 3] BUCKLES B.P., PETRY F.E., SACHAR H.S. (1989) A Domain Calculus for Fuzzy Relational Databases Fuzzy Sets and Systems Vol. 29 pp. 327-340

[ 4] CUBERO J.C., DIAZ J., MEDINA J.M.,PONS O.,PRADOS M. and VILA M.-A. (1991) A Knowledge representation model for fuzzy databases .Submitted to ACM Transc. on Database Systems.

[ 5] CUBERO J.C., VILA M.-A. (1992) A new definition of fuzzy functional dependency in fuzzy relational databases. Accepted for IJIS.

[ 6] GALLAIRE H., MINKER J., NICOLAS J.M. (1984) Logic and Databases a Deductive Approach ACM Computing Surveys Vol. 16 (2) 153-185

[ 7] PRADE H., TESTEMALE C. (1984) Generalizing Database Relational Algebra for the Treatment of Incomplete or Uncertain Information and Vague Queries Information Science Vol. 34 pp. 115-143

[ 8] REITER R. (1984) Towards a logical reconstruction of relational database theory. En Brodie M., Mylopoulos J. and Schmidt J.W. On Conceptual Modelingpp. 193-238 Springer- Verlag Berlin New York

[ 9] UMANO M. (1982) FREEDOM−0 : *A* Fuzzy Database Systems. En M.M. Gupta y E. Sanchez eds. Fuzzy Information and Decision Processes pp. 339-347 North Holland Pub. Co.

[10] VILA M.A., CUBERO J.C., MEDINA J.M., PONS O. (1992) A logic approach to Fuzzy relational Databases. To appear in International Journal of Intelligent Systems.

[11] ZEMANKOVA-LEECH M., KANDEL A. (1984) Fuzzy Relational Databases a key for Expert Systems Verlag TUV Rheiland

# An Integrity Constraint for a Fuzzy Relational Database

A. Yazıcı[1], E. Göçmen[1], B.P. Buckles[2], R. George[3] and F.E. Petry[2]

[1]Dep. of Computer Engineering
Middle East Technical University
06531, ANKARA/TÜRKİYE

[2]Dep. of Computer Science
Tulane University
N.O., LA 70118/USA

[3]Dep. of Comp. and Inf. Sc.
Clark-Atlanta University
Atlanta, GA 30314/USA

**Abstract-- Here in this paper, we formally define fuzzy multivalued dependencies on a chosen fuzzy database model. Since fuctional dependencies have already been extended for fuzzy database models [1,3], we only deal with an important integrity constraints, namely multivalued dependencies (MVDs), for the fuzzy relational database introduced by Buckles and Petry [7-8]. We show that the defined fuzzy MVDs (FMVDs) are consistent and reduces to the classic MVDs when they involve precise attribute values. We also give some examples to show how the integrity constraints imposed by the FMVDs are forced whenever a tuple is to be inserted or to be modified.**

Keywords: Fuzzy set theory, similarity relations, fuzzy relational database model and multivalued dependencies.

## I. INTRODUCTION

Real life applications involve a great deal of imprecise and inexact data, which stem from personal traits and judgments. A data representation which does not reflect the presence and degree of fuzziness of data would likely to fail in making precise decisions and inferences. The original database model originated by Codd [6] has no special feature to represent data and any integrity constraints (i.e. data dependencies) other than exact.

Fuzzy database models [2-3,7-10], in the light of fuzzy set theory introduced by Zadeh [10], have been introduced to handle inexact and imprecise data. Since integrity constraints play a critical role in a logical database design and among these constraints data dependencies are of more interest, functional dependencies (FDs) have been extended to represent fuzziness in the dependency. An example for a fuzzy functional dependency may be "The intelligence level of a person more or less determines the degree of success." There also exist some applications that may involve another very important integrity constraint, which is multivalued dependencies (MVDs) [4]. We can exemplify this by the following example.

Consider the instance of the fuzzy relational database consisting of the fuzzy relation WARDROBE shown in Figure 1.

The relation schema consists of three attributes, namely CLOTHES, PRICE and PLACE. For each clothe, there is a corresponding set of price levels and a set of places to go with that particular clothe. The attribute values of each set, Price and Place, have their own similarity relationships, depicted in Figure 2, which incorporate the fuzziness with the tuples.

The FMVDs CLOTHES-->->$_{F(\emptyset)}$PRICE/PLACE are involved with this relation instance. The FMVDs should be checked whenever a tuple is to be inserted or to be modified so that the integrity constraints imposed by the FMVDs are not violated. The dependency checking will force the user to add some other tuples together with the original tuple in order to satisfy the definition of the FMVD. Thus, incorrectly associated values of a set of attributes can be avoided. The detailed example representing various cases of forcing FMVDs whenever a tuple is to be inserted or to be modified is given in the Section 3. Let us first give the preliminary definitions next.

WARDROBE

| CLOTHES | PRICE | PLACE |
|---|---|---|
| jeans | {cheap,modest} | {home,picnic} |
| jeans | {cheap,modest} | {friends,university,shopping} |
| jeans | {affordable,acceptable} | {home,picnic} |
| jeans | {affordable,acceptable} | {friends,university,shopping} |
| jeans | {expensive} | {home,picnic} |
| jeans | {expensive} | {friends,university,shopping} |
| jacket | {expensive} | {friends,university,shopping} |
| jacket | {expensive} | {conference,coctail} |
| jacket | {very-expensive} | {friends,university,shopping} |
| jacket | {very-expensive} | {conference,coctail} |

Fig. 1: The Fuzzy Relation WARDROBE

| PRICE | very cheap | cheap | modest | affod. | accep. | exp. | very exp. |
|---|---|---|---|---|---|---|---|
| very cheap | 1 | 0.8 | 0.5 | 0.3 | 0.2 | 0 | 0 |
| cheap | 0.8 | 1 | 0.7 | 0.5 | 0.4 | 0.1 | 0 |
| modest | 0.5 | 0.7 | 1 | 0.8 | 0.6 | 0.1 | 0 |
| affordable | 0.3 | 0.5 | 0.8 | 1 | 0.9 | 0.4 | 0.1 |
| acceptable | 0.2 | 0.4 | 0.6 | 0.9 | 1 | 0.5 | 0.2 |
| expensive | 0 | 0.1 | 0.1 | 0.4 | 0.5 | 1 | 0.8 |
| very expensive | 0 | 0 | 0 | 0.1 | 0.2 | 0.8 | 1 |

| PLACE | home | pic. | friends | univ. | shop. | conf. | coct. |
|---|---|---|---|---|---|---|---|
| home | 1 | 0.7 | 0.5 | 0.4 | 0.4 | 0 | 0 |
| picnic | 0.7 | 1 | 0.7 | 0.6 | 0.5 | 0 | 0 |
| friends | 0.5 | 0.7 | 1 | 0.8 | 0.8 | 0.3 | 0.1 |
| university | 0.4 | 0.6 | 0.8 | 1 | 0.8 | 0.3 | 0.1 |
| shopping | 0.4 | 0.5 | 0.8 | 0.8 | 1 | 0.3 | 0.1 |
| conference | 0 | 0 | 0.3 | 0.3 | 0.3 | 1 | 0.9 |
| coctail | 0 | 0 | 0.1 | 0.1 | 0.1 | 0.9 | 1 |

Fig. 2: Similarity Relations for Domain PRICE and PLACE

## II. BACKGROUND

As background, we shall first describe the multivalued dependency concept [4], then we briefly discuss the fuzzy database model introduced in [7-8], since that model is our reference model. Finally, we give brief definitions of fuzzy functional dependencies showing how they have been transcribed to its fuzzy relational database context. We prefer the method presented in [1] for extending its fuzzy functional dependencies to fuzzy multivalued dependencies over the other approaches [2,3]. A detailed discussion on all of these existing approaches for fuzzy functional dependencies can be found in [1,2,3].

### 2.1. Multivalued Dependencies (MVDs)

Multivalued dependencies originated by Fagin [4] are one of the integrity constraints that are to be imposed on the tuples of a database. Informally, MVDs in databases relate a value of an attribute (or a set of attributes) to a set of values associated with a set of attributes, independent of the other attributes in the relation. Formally, MVDs can be defined as follows:

Definition: Let r be any relation on scheme $R(A_1,A_2,...,A_n)$, with X and Y subsets of $(A_1,A_2,...,A_n)$ and Z is equal to $(A_1,A_2,...,A_n)$-X-Y. Relation r satisfies the MVD X-->-->Y (reads as "X multi-determines Y") if, for every pair of tuples $t_1$ and $t_2$ in r, $t_1[X]=t_2[X]$, then there exists a tuple $t_3$ in r with $t_3[X]=t_1[X]$, $t_3[Y]=t_1[Y]$, and $t_3[Z]=t_2[Z]$. Whenever X-->-->Y holds in R, so does X-->->(R-X-Y) because of the symmetry in the definition. The value set of attribute set Y and those of Z are associated only with a value of X, and no association exists between Y and Z.

### 2.2. Fuzzy Relational Databases

A fuzzy relational database differs from the original relational model in two respects. Firstly, attribute values need not contain only atomic values in a fuzzy database, but the values are restricted to be drawn from a single set (the corresponding domain of the attribute). Secondly, the identity relationship among attribute values in the original relational model is replaced with a similarity relationship that reflects the closeness of attribute values. For example, an attribute may grade the success of a student, then the domain of the attribute can be the set D={poor, average,honor,high-honor}. Here the similarity between the attribute values poor and high-honor will certainly be less than the similarity between honor and high-honor.

A fuzzy relation in the fuzzy database model is defined as a subset of the cross product of the power sets ($2^{D1}$, $2^{D2}$, ..., $2^{Dn}$) of the domains of the attributes. Similarity relationships among the attribute values are to be decided purely on the semantics of the attribute values. A member of a fuzzy relation corresponding to a row of the table is called a tuple. More formally, a fuzzy tuple t is any member of both the fuzzy relation and $2^{D1}$ x $2^{D2}$ x...x $2^{Dn}$. An arbitrary tuple, $t_i$, is of the form

$t_i = (d_{i1},d_{i2},...,d_{in})$ where $D_j$ $d_{ij}$, where $1 \le i \le m$.

### 2.3. Fuzzy Functional Dependencies

Functional dependencies (FDs) have been the most popular integrity constraint. These are extended for various fuzzy relational database models to reflect fuzzy functional dependencies. An example of an FFD may be "The intelligence level of a person more or less determines the degree of success." Among all the extensions for FFDs, we concentrate on a specific one represented in [1]. Before we present the definition of the FFD represented in this study [1], we shall give some preliminary definitions.

Definition: The conformance of attribute $A_k$ defined on domain $D_k$ for any two tuples $t_1$ and $t_2$ present in relation R and denoted by $C(A_k[t_1,t_2])$ is given as

$C(A_k[t_1,t_2]) = \mathrm{Min}\{s(p,q)\}$,

where $p,q \in d_{1k} \cup d_{2k}$, and $d_{1k}$ is the value of attribute $A_k$ for tuple $t_1$ and $d_{2k}$ is the value of attribute $A_k$ for tuple $t_2$, and s(p,q) is a similarity relation for values p and q of a given attribute $A_k$. s(p,q) assigns a number in [0,1] to (p,q) for reflecting the user's view of the similarity between these two values.

Definition: Let relational schema $R(A_1,A_2,...,A_n)$ be defined over a finite set of domains $D=\{D_1,D_2,...,D_n\}$, where a set of discrete values or fuzzy numbers are represented, and let X and Y be the same as before. We say $X \rightarrow_F Y$ if

$C(Y[t_1,t_2]) \ge \phi * C(X[t_1,t_2])$

for all $t_1, t_2$ present in relation r. $C(X[t_1,t_2])$ is called the conformance for attribute set X for any two tuples $t_1$ and $t_2$ present in relation R and given as

$C(X[t_1,t_2]) = \text{Min}\ \{C(A_k[t_1,t_2])\}$ for all $A_k \in X$.

Here ø, which is an element of [0,1], is called linguistic strength term in functional dependency description and is optional. In the absence of linguistic term, the default value of ø is 1. In addition to this, if all attribute values are also singleton, the definition of fuzzy functional dependencies (FFDs) reduces to that of classical FDs for ordinary relational databases.

## III. EXTENDING FUZZY FDs TO FUZZY MVDs

The definition of MVDs for ordinary relational databases can not represent facts like a value of a set of attributes not fully determine a set of values. Therefore, a definition of MVDs for fuzzy relations should be given. The definitions for the conformance for a set of attributes and the conformance for an attribute were given above. In this study, we assume that if a value of an attribute fully multidetermines a set of values of an attribute, then that value multidetermines all values in that set with an equal strength. Although this assumption seems reasonable in real applications, in theory, there might still exist some cases in which some values of an attribute may be multidetermined with a different strength. This non-uniformity can be dealt with by reducing all the determination strengths of the values of an attribute to the same level of the lowest (or minimum) strength for the values of that attribute.

Definition: A fuzzy multivalued dependency (FMVD) $X\text{-->-->}_F Y$ specified on R, specifies the following constraints on any relation instance r of R; if two tuples $t_1$ and $t_2$ exist in r such that

$C(X[t_1,t_2]) \geq ø$,

then two tuples $t_3$ and $t_4$ should also exist in r with the following properties:

$C(X[t_3,t_1]) \geq ø$ and $C(X[t_4,t_2]) \geq ø$ and
$C(Y[t_3,t_1]) \geq ø$ and $C(Y[t_4,t_2]) \geq ø$,
$C(Z[t_3,t_2]) \geq ø$ and $C(Z[t_4,t_1]) \geq ø$.

Note that Z=(R-X-Y) and $X\text{-->-->}_F Y$ implies $X\text{-->-->}_F Z$. Therefore, it is written as $X\text{-->-->}_F Y/Z$. In the absence of the linguistic term, the default value of ø is 1, and in that case an FMVD becomes a crisp MVD.

Now let us show the consistency of this definition of FMVDs. In order to do that, we find a case where the definition of FMVDs for the fuzzy database reduces to that of the classic MVDs.

Lemma 1: The definition of the FMVDs is consistent.

Proof: If we take the case where ø=1 and apply it into the above definition of FMVDs, we can rewrite that as follows.

If $C(X[t_1,t_2])=1$, where ø=1, then
$C(X[t_3,t_1])=1$ and $C(X[t_4,t_2])=1$ and
$C(Y[t_3,t_1])=1$ and $C(Y[t_4,t_2])=1$,
$C(Z[t_3,t_2])=1$ and $C(Z[t_4,t_1])=1$.

Note that $C(X[t_1,t_2])=1$ is equivalent to $X[t_1]=X[t_2]$ by definition. By using the same reasoning,

$C(X[t_3,t_1])=1$ and $C(X[t_4,t_2])=1$ means
$(X[t_3] = X[t_1] = X[t_4] = X[t_2])$ and
$C(Y[t_3,t_1])=1$ and $C(Y[t_4,t_2])=1$ means
$(Y[t_3] = Y[t_1]$ and $Y[t_4] = Y[t_2])$ and
$C(Z[t_3,t_2])=1$ and $C(Z[t_4,t_1])=1$ means
$(Z[t_3] = Z[t_2]$ and $Z[t_4] = Z[t_1])$.

As can be seen from the definition of the classical MVDs, the resulting definition for the case where ø=1 is equivalent to the classical definition of the MVDs. So the definition of the FMVDs is consistent.

### 3.1. An Example

Let us consider again the fuzzy relation WARDROBE given in Figure 1, the similarity relations for domain PRICE and PLACE attributes given in Figure 2 and the following candidate tuple

$t_c$: <jacket,{affordable},{university}>.

The database designer may decide that when the statement of multivalued dependency is "CLOTHES more or less multidetermines PLACE/PRICE" then the possible linguistic strength term, ø, is equal to 0.6. In this case, every pair of tuples presented in the relation WARDROBE must satisfy the definition for the FMVDs given before. Therefore, insertion of $t_c$ into the relation WARDROBE necessitates to check the corresponding FMVDs to see whether the new tuple violates any of them. In general, in case of insertion of new tuples or updating an existing tuple, it must be ensured that the new tuple or updated tuple does not violate the constraints imposed by the given dependencies. Here, in this example, the consistency check will be performed on those tuples of the relation for which

$C(CLOTHES[t_c,t_i]]) \geq ø$

where $t_i \in$ WARDROBE.

Case-1:

$t_c$: <jacket,{affordable},{university}>
$t_7$: <jacket,{expensive},
{friends,university,shopping}>.

Since $C(PRICE[t_c,t_7]]) \leq ø$, insertion of $t_c$ will require two additional insertions. These are

$t_{c1}$:<jacket,{affordable},
{friends,university,shopping}>
$t_{c2}$:<jacket,{expensive},{university}>.

$t_{c2}$ will be merged with $t_7$ since $C(A_k[t_7,t_{c2}]])\geq ø$ for all $A_k \in$ WARDROBE and $t_{c1}$ will be merged with $t_c$ since $C(A_k[t_c,t_{c1}]])\geq ø$ for all $A_k \in$ WARDROBE. Hence, insertion of tc will not require any additional insertions.

Case 2:

$t_c$ : <jacket,{affordable},{university}>
$t_8$:<jacket,{expensive},
{conference,cocktail}>.

Since $C(PLACE[t_c,t_7]]) \leq ø$, two additional tuples must be inserted along with $t_c$ in order to preserve FMVDs of the relation,

$t'_c$:<jacket,{affordable},
{conference,cocktail}>
$t_{c2}$:<jacket,{expensive},{university}>.
$t_{c2}$ will be merged with $t_7$ since $C(A_k[t_7,t_{c2}])\geq\emptyset$ for all $A_k \in$ WARDROBE; hence, we are left with tuple $t'_c$.

Case-3:
$t_c$:<jacket,{affordable},{university}>
$t_9$:<jacket,{very expensive},
{friends,university,shopping}>.
Since $C(PLACE[t_c,t_9])\geq\emptyset$, but $C(PRICE[t_c,t_9])\leq\emptyset$, the additional insertions to satisfy the FMVDs are
$t_{c3}$:<jacket, {affordable},
{friends,university,shopping}>
$t_{c4}$:<jacket, {very expensive},
{university}>.
However, $t_{c3}$ and $t_{c4}$ will be merged with $t_c$ and $t_9$ respectively since $C(A_k[t_c,t_{c3}])\geq\emptyset$ for all $A_k \in$ WARDROBE and $C(A_k[t_9,t_{c4}])\geq\emptyset$ for all $A_k \in$ WARDROBE; hence, there will be no additional tuples to be inserted.

Case-4:
$t_c$ : <jacket,{affordable},{university}>
$t_{10}$:<jacket,{very expensive},
{conference,cocktail}>.
In this case the tuples to be inserted due to integrity constraints are:
$t'_c$:<jacket,{affordable},
{conference,cocktail}>
$t_{c4}$:<jacket,{very expensive},
{university}>
for all $A_k \in$ WARDROBE, $C(A_k[t_{c4},t_9]) \geq \emptyset$, as in the case-3 above. Therefore, from all of the above cases, $t'_c$ is the only tuple that must be inserted along with $t_c$ to satisfy the integrity constarint imposed by the FMVD given above.

Note that the values of attribute X in $X \rightarrow_F Y$ and $X \rightarrow\rightarrow_F Y$ dependencies does not have to be precise. The definition of the FMVDs given above can also handle the case where X representing CLOTHES atribute has fuzzy values such as {Jeans, Slacks}.

## IV. CONCLUSION

Multivalued dependencies stand as a generalization of functional dependencies and they are important data integrity constraints. In this paper, MVDs have been extended for the fuzzy database model [7-8]. As a base, the extension of a functional dependency to a fuzzy functional dependency is used and a particular extension has been chosen among existing ones. At the end, some examples were given to represent how the fuzzy multivalued dependencies are imposed on a fuzzy relation when a tuple is inserted or modified.

A further study involving the definitions of sound and complete inference rules for FMVDs and the definition of the normal form for the fuzzy relations have been going on.

## REFERENCES


[1] H. Sachar, "Theoretical Aspects of Design of and Rètrieval from Similarity-based Relational Database Systems," PhD Diss., 1986, Univ. of Texas at Arlington, TX, USA.

[2] M. Anvari and G.F.Rose, "Fuzzy Relational Databases," Analysis of Fuzzy Information, Vol II (J.C.Bezdek, editor), CRC Press, Boca Raton,FL, 1985, pp:204-212.

[3] K.V.S.V.N. Raju and A.K. Majumdar,"Fuzzy Functional Dependencies and Lossless Join Decomposition of Fuzzy Relational Database Systems," ACM Transactions on Database Systems, June 1988.

[4] R.Fagin, "Multivalued Dependencies and a New Normal Form for Relational Databases," ACM Transactions on Database Systems, Sept. 1977.

[5] R.Elmasri and S.B.Navathe, "Fundamentals of Database Systems," The Benjamin/Cumings Pub. Comp. Inc., Redwood City, CA, 1989.

[6] Codd, E.R. "A Relational Model of Data for Large Shared Databases", Communication of ACM, 13,6,1970.

[7] Buckles, B.P. and F. E. Petry, "A Fuzzy Representation of Data for Relational Databases", Fuzzy Sets and Systems, 7(1982), 213-226.

[8] Buckles, B.P. and F. E. Petry, "Uncertainty Models in Information and Database Systems", Journal of Information Science, 11, 1985.

[9] Petry, F.E., B.P.Buckles, A. Yazıcı and R. George "Fuzzy Information Systems," Proc. of IEEE International Conference on Fuzzy Systems, pp.1187-2000, March 8-12, 1992, San Diego, CA, USA.

[10] Zadeh, L.A. "Similarity relations and Fuzzy Orderings," Information Sciences, Vol. 3, No. 2, 1971, 177-200.

[11] Yazıcı, A., R.George, B.P. Buckles, Conceptual and Logical Data Models for Uncertainty Management," Fuzzy Logic for Management of Uncertainty, L.A. Zadeh, and J. Kacprzyk, eds.John Wiley and Sons, Inc., 1992, pp: 607-644, New York, USA.

# A Fuzzified CMAC Self-learning Controller

*Junhong Nie and D. A. Linkens*

Department of Automatic Control & Systems Engineering
University of Sheffield, Sheffield S1 3JD, U.K.

**Abstract: This paper presents a fuzzified CMAC network (FCMAC) acting as a multivariable adaptive controller with the feature of self-organizing association cells and the further ability of self-learning required teacher signals in real-time. In particular, the original CMAC has been reformulated within a framework of a simplified fuzzy control algorithm (SFCA) and the associated self-learning algorithms were developed as a result of incorporating the schemes of competitive learning and iterative learning control into the system. The approach described here can be thought of as either a completely unsupervised fuzzy-neural control strategy or equivalently an automatic real-time knowledge acquisition scheme. The proposed approach has been successfully applied to a problem of multivariable blood pressure control.**

## I. INTRODUCTION

Inspired by the neurophysiological theory of the cerebellum, Albus developed a mathematical model called Cerebellar Model Articulation Controller (CMAC) [1,2] in an attempt to derive an efficient computational algorithm for use in manipulator control. Functionally, a trained CMAC performs a multivariable function approximation in a generalized look-up table fashion. Structurally, it is equivalent to a network architecture with three layers. With increasing interest in neural networks, CMAC has gained more and more attention from, in particular, control engineering researchers, due to its unique characteristics such as fast training speed and localized generalization [6,11]. This paper presents a fuzzified CMAC network (FCMAC) acting as a multivariable adaptive controller with the feature of self-organizing association cells and the further ability of self-learning required teacher signals when applied to control problems. Two key ideas underlying the proposed approach are: a) the original CMAC is fuzzified based upon the observation of the close equivalence between the CMAC and a simplified fuzzy control algorithm (SFCA) developed by the authors [7]; b) the FCMAC is constructed automatically by introducing self-organizing and self-learning schemes operated in a real-time control environment. The approach described here is one of the results in our effort to functionally and structurally integrate fuzzy logic with neural networks [4,8-10]. As a demonstration, we have applied the proposed approach to a problem of multivariable control of blood pressure.

## II. SIMPLIFIED FUZZY CONTROL ALGORITHM

Assume that the controlled process is multivariable with $m$ inputs and $m$ outputs. The inputs to the fuzzy controller are various combinations of control error, change-in-error, and sum of error with respect to each controlled variable. In what follows, it is assumed that the controller input is composed of $n$ variables denoted by $u_i$. The output of the fuzzy controller consists of $m$ variables denoted by $v_k$. Furthermore, assume that there are $P$ rules in the rule-base, each of which has the form:

$$IF\ \bar{U}_1\ is\ A_1^j\ AND \cdots AND\ \bar{U}_n\ is\ A_n^j$$

$$THEN\ \bar{V}_1\ is\ B_1^j\ AND \cdots AND\ \bar{V}_m\ is\ B_m^j$$

where $\bar{U}_i$ and $\bar{V}_k$ are linguistic variables corresponding to the numerical variables $u_i$ and $v_k$, $A_i^j$ and $B_k^j$ are fuzzy subsets representing some linguistic terms and are defined on the corresponding universes of discourse $U_i$ and $V_k$ which are assumed to be compact on $R$.

A traditional fuzzy controller typically consists of three basic stages: fuzzification, fuzzy reasoning, and defuzzification. However by taking the nonfuzzy property regarding the numerical input/output of the fuzzy controller into account, we have derived a very simple but efficient MISO fuzzy control algorithm SFCA which consists of only two main steps, i.e pattern matching and weighted averaging, thereby eliminating the necessity for fuzzifying and defuzzifying procedures [7] and is described briefly below.

To begin with, we assume that $A_i^j$ and $B_k^j$ in the rules are normalized fuzzy subsets whose membership functions are defined uniquely by triangular forms on the corresponding universes, each of which is characterized only by two parameters, $M_{u,i}^j$ and $\delta_{u,i}^j$, or $M_{v,k}^j$ and $\delta_{v,k}^j$, where $M_{u,i}^j$ ($M_{v,k}^j$) is the center element of the support set of $A_i^j$ ($B_k^j$), and $\delta_{u,i}^j$ ($\delta_{v,k}^j$) is the half width of the support set. Hence the $j$th rule may be written as

$$IF\ \ (M_{u,1}^j, \delta_{u,1}^j)\ AND\ \ ......\ \ (M_{u,n}^j, \delta_{u,n}^j)$$

$$THEN\ \ (M_{v,1}^j, \delta_{v,1}^j)\ AND\ \ ......\ \ (M_{v,m}^j, \delta_{v,m}^j)$$

Let $M_u^j=(M_{u,1}^j, M_{u,2}^j, ......, M_{u,n}^j)$ and $\Delta_u^j=(\delta_{u,1}^j, \delta_{u,2}^j, ......, \delta_{u,n}^j)$ be two $n$-dimensional vectors. Then the IF part of the $j$th rule may be viewed as creating a subspace whose center and radius are $M_u^j$ and $\Delta_u^j$

0-7803-0614-7/93$03.00 ©1993IEEE

respectively or as defining a *rule pattern*. Similarly $n$ current inputs $u_{0i} \in U_i$ ( i=1, 2, ......, n), with $u_{0i}$ being a singleton, can also be represented as a $n$-dimensional vector $u_0$ and will be referred to as an *input pattern*. $P$ rule patterns partition the input space into $P$ subspaces which are typically overlapped to some degree along the boundaries due to the effect of fuzziness. On the contrary, a measured input is just a determined point situated in the same space as the rule pattern's.

The fuzzy control algorithm can be considered to be a process in which an appropriate control action is deduced from a current input and $P$ rules according to some prespecified reasoning algorithms. We split the whole reasoning procedure into two phases: pattern matching and weighted averaging. The first operation deals with the IF part for all rules, whereas the second one involves an operation on the THEN parts of the rules. From the pattern concept introduced above, we need to compute the matching degrees between the current input pattern and each rule pattern. Denote the current input by $u_0=(u_{01}, u_{02}, ...... ,u_{0n})$. Then the matching degree denoted by $s^j \in [0, 1]$ between $u_0$ and the $j$th rule pattern $M\Delta_u(j)$ can be measured by the relationship given by

$$s^j = 1 - D^j(u_0, M\Delta_u(j)) \tag{1}$$

where $D^j(u_0,M\Delta_u(j)) \in [0, 1]$ denotes *relative distance* from $u_o$ to $M\Delta_u(j)$. $D^j$ can be specified in many ways. With the assumption of an identical width $\delta$ being used for all fuzzy sets $A_i^j$, the computational definition of $D^j$ is given by

$$D^j = \begin{cases} \|M_u^j - u_0\|/\delta & \text{if } \|M_u^j - u_0\| \le \delta \\ 1 & \textit{otherwise} \end{cases} \tag{2}$$

where $\|.\|$ denotes distance metrics which can be Euclidean, Hamming, or Maximum.

Suppose that for a specific input $u_0$ and $P$ rules, after the matching process is completed, a matching degrees vector $s=(s^1,s^2,...,s^P)$ is obtained. Then the $k$th component of the deduced control action $v_k$ is given by

$$v_k = (\sum_{j=1}^{P} s^j \cdot M_{v,k}^j)/(\sum_{j=1}^{P} s^j) \tag{3}$$

It can be seen that eqn (3) gives a weighted averaging value with respect to the fired rule's THEN part. How large percentage a specific rule contributes to the global value is determined by the corresponding matching degree. Because only the centers of the THEN parts of the fired rules are utilized and they are the only elements having the maximum membership grade 1 in the corresponding support sets, the algorithm can be understood as a modified *maximum membership* decision scheme in which the global center is calculated by the *center of gravity* algorithm.

## III. FUZZIFIED CMAC

### A. *Description of the CMAC*

The CMAC is designed to represent approximately a multi-dimensional function by associating an input vector $\mu \in U \subset R^n$ with a corresponding function vector $v \in V \subset R^m$. As shown in Fig 1, the CMAC has a similar structure to a three layered network with association cells playing the role of hidden layer units. Mathematically, CMAC may be described as consisting of a series of mappings: $U \to A \to V$, where $A$ is a N-dimensional cell space.

A fixed mapping $U \to A$ transforms each $\mu \in U$ into an $N$-dimensional binary associate vector $a(\mu)$ in which only $N_L$ elements have the values of 1, where $N_L < N$ is referred to as the generalization width. In other words, each $\mu$ activates precisely $N_L$ association cells or geometrically each $\mu$ is associated with a neighbourhood in which $N_L$ association cells are included. An important property of the CMAC is local generalization derived from the fact that nearby input vectors $\mu_i$ and $\mu_j$ have some overlapping neighbourhood and therefore share some common association cells. According to the above principle, Albus developed a mapping algorithm consisting of two sequential mappings: $U \to M \to A$. $n$ components of $\mu$ are first mapped into $n$ $N_L$-dimensional vectors and these vectors are then concatenated into a binary associate vector $a$ with only $N_L$ elements being 1.

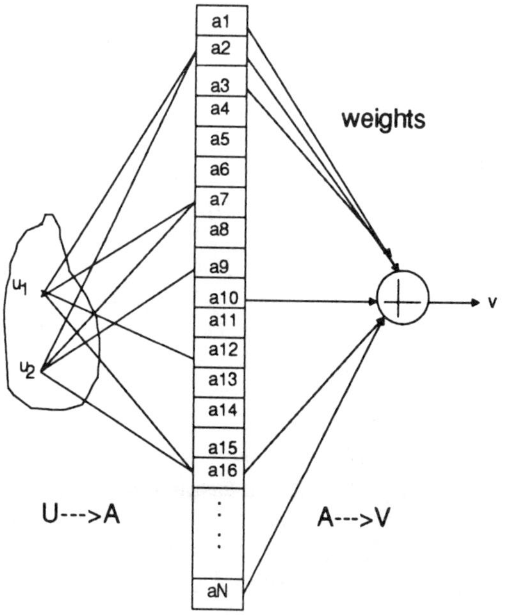


Fig 1 A structure of CMAC

The $A \rightarrow V$ mapping is simply a procedure of summing the weights of the association cells excited by the input vector $\mu$ to produce the output. More specifically, each component $v_k$ is given by

$$v_k = \sum_{j=1}^{N} a^j(\mu) \cdot \pi_k^j \quad (4)$$

where $\pi_k^j$ denotes the weight connecting the $j$th association cell to the $k$th output. Notice that only $N_L$ association cells contribute to the output.

B. *Fuzzified CMAC*

By carefully inspecting the SFCA and the CMAC, we have concluded that there exist some striking similarities between these two systems. Functionally, both of them perform a function approximation in an interpolative look-up table manner with an underlying principle of generalization and dichotomy: to produce similar outputs in response to similar input patterns and produce independent outputs to dissimilar input patterns. From the computational point of view, mapping $U \rightarrow A$ corresponds to the calculation of the matching degree in the SFCA, and mapping $A \rightarrow V$ to the weighted averaging procedure given by (3). While the later similarity is apparent by comparing (3) with (4), where $a^j$, $\pi_k^j$, and $N$ correspond to $s^j$, $M_{v,k}^j$, and $P$, the former equivalence can be made clearer as follows.

Instead of saying that each input $\mu$ is associated with a neighbourhood specified by $N_L$, we can equally consider that each association cell $\Psi^j$ is associated with a neighbourhood centered at, say $\omega^j \in U$ referred to as the reference vector, with the width controlled by $N_L$. If a current input $\mu$ is within the neighbourhood of $\Psi^j$, that cell is regarded as being active. In this view, the associate vector $a(\mu)$ can be derived by operating $N$ neighbourhood functions $\psi^j(\omega^j, \mu)$ with respect to $\mu$, where

$$\psi^j(\omega^j, \mu) = \begin{cases} 1 & \text{if } \mu \in \Psi^j \\ 0 & \textit{otherwise} \end{cases} \quad (5)$$

that is, $a(\mu) = (\psi^1, \psi^2, \ldots, \psi^N)$. By appropriately selecting the $\omega^j$, $a(\mu)$ can be made to contain only $N_L$ 1's. Now it becomes evident that the associate vector $a$ is similar to the matching degree vector $s$ except the former uses the crisp neighbourhood function, whereas the latter adopts the graded one. In fact, by letting $s^{*j}=1$ for $s^j>0$ and $s^{*j}=0$ for other cases, the vector $a$ will be precisely equal to the vector $s^*$, indicating that the former is a special case of the latter. We notice that a natural measurement of whether $\mu$ belongs to $\Psi^j$ in (5) is to use some distance metric relevant to the generalization width $N_L$.

Now we are in a position to implement a FCMAC by replacing eqn (4) with (3). Several advantages can be identified by this replacement. The concept of the graded matching degree not only provides a clear interpretation for $U \rightarrow A$ mapping, but also offers a much simpler and more systematic computing mechanism which is content-addressable than that proposed by Albus where some very complicated addressing techniques are utilized and further hashing code may be needed to reduce the storage. In addition, the graded neighbourhood functions overcome the problem of discontinuous responses over neighbourhood boundaries due to the crisp neighbourhood functions.

## IV. SELF-LEARNING IN REAL-TIME

Assuming that the control rule-base is unavailable, the SFCA cannot be applied directly. Unfortunately, the FCMAC is not feasible either. This is because the original CMAC is generally trained off-line by the supervised scheme, whereas in our case the FCMAC-based controller as shown in Fig 2 must be operated in real-time and more crucially there is no teacher signals supplied to the FCMAC. Therefore, the FCMAC must learn to construct itself including the association cell number $N$, reference vectors $\omega^j$, and weight vectors $\pi^j$ in a real-time control environment.

A. *Self-learning $N$ and $\omega$*

With a prespecified generalization width $N_L$, our concern is how a specific region of the input space can be partitioned into $N$ subregions without knowing $N$ in advance, each of which being represented by a reference vector $\omega^j$. We solve the problem using the Kohonen self-organizing scheme but with the following modifications. Instead of using absolute minimum distance as the winner selection criteria and a unique learning rate, we employ the matching degree given by (1) to determine the winner and many local learning rates. If one of the existing association cells is able to win the competition, the corresponding $\omega^j$ is modified; otherwise, a new cell is created. In any case, the graded associate vector $a$ is derived for computing the net output and modifying the corresponding weight vector $\pi^j$. In this manner, the required $N$ and $\omega$ are dynamically learned in response to the incoming controller input $u$. The algorithm is described as follows, where $t$ and $T$ denote the sampling instant and maximum sampling time respectively. In addition, $\alpha^j$ is the local gain controlling the speed of adaptive process of $\omega^j$ and is inversely proportional to the active frequency $n^j$ of the $j$th cell up to present time instant, $N(t)$ stands for the number of the cells at time $t$, and $0 \le a_0 < 1$ is a threshold controlling the cell number created.

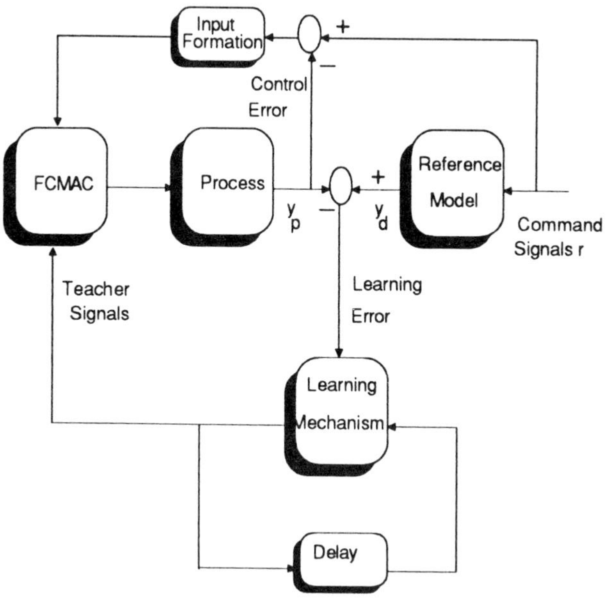


Fig 2 The FCMAC-based learning control system

a) Calculate the $a$ vector by (1), where

$$a=(a^1,a^2,\ldots,a^{N(t)})$$

b) Find the cell $J$ having the maximum matching degree:

$$a^J = \max_{j=1,N} a^j \tag{6}$$

c) Determine the winner using the following rule:

$$\begin{cases} \text{if } a^J \geq a_0 & \rightarrow J \text{ is winner} \\ \text{if } a^J < a_0 & \rightarrow \text{create a new cell} \end{cases} \tag{7}$$

d) Modify or intialize parameters:

If J is the winner:

$$\begin{aligned} n^J(t) &= n^J(t-1) + 1; \quad \alpha^J(t) = 1/n^J(t); \quad N(t) = N(t-1) \\ \omega^J(t) &= \omega^J(t-1) + \alpha^J \cdot [u(t) - \omega^J(t-1)] \end{aligned} \tag{8}$$

If a new cell is created:

$$N(t)=N(t-1) + 1; \quad \omega^{N(t)}=u(t) \quad n^{N(t)}=1; \, a^{N(t)}=1 \tag{9}$$

e) Output the $a$ vector.

B. *Self-learning* $\pi$

Suppose that the $k$ desired output $v_k^*$ corresponding to current CMAC input $u$ is known, Albus [1,2] developed a training algorithm spreading the output error evenly to a fixed $N_L$ association cell weights contributing to the present output $v_k(u)$. The algorithm is essentially a special case of gradient descent algorithm with the learning rate being $\beta/N_L$. More specifically, the algorithm is given by

$$\Delta\pi_k^j = \beta \cdot [(v_k^* - v_k(u)) \cdot a^j]/N_L \tag{10}$$

Notice that only $N_L$ weights are adjusted in the same amount due to $a^j$.

Following the standard gradient procedure, the update rule of FCMAC subject to the weighted average output scheme (3) can be easily derived and is given by

$$\Delta\pi_k^j = \beta \cdot [(v_k^* - v_k(u)) \cdot a^j)]/[\sum_{j=1}^{N} a^j] \tag{11}$$

It is evident that (10) is a special case of (11) when $a$ is a binary-valued vector with $N_L$ 1's.

As mentioned previously, the major difficulty in applying (10) or (11) lies in the unavailability of teacher signals $v_k^*$ guiding the supervised training. Here we propose a simple but efficient approach to carry out the task. First, the required teacher signals $v_k^*$ are explicitly constructed by an iterative learning approach [5]. Then the $v_k^*$ are supplied to the FCMAC so as to update the $\pi_k^j$ using (11).

As shown in Fig 2, the learning system consists of a reference model and a learning mechanism. The reference model is designed to specify what the process responses $y_p$ should be when both the model and the process are subject to the same command signal $r$. For the sake of simplicity, we adopt a non-interacting model with second-order linear transfer functions. Denoting the learning error by $e_L$, defined as the difference between the output $y_d$ of the model and the output $y_p$ of the process, the overall goal of the learning system is to force the learning error $e_L$(t) asymptotically to zero or to a predefined tolerant region $\varepsilon$ within a time interval of interest [0, T] by repeatedly operating the system.

By taking the process time delay into account, the learning law is given by

$$v^{*l+1}(t) = v^{*l}(t) + P_L \cdot e_{L,l}(t+\lambda) + Q_L \cdot c_{L,l}(t+\lambda) \tag{12}$$

where $v^{*l}, v^{*l+1} \in R^m$ are on-line learning teacher vector-valued functions at the $l$th and the $(l+1)$th iterations respectively, $e_{L,l}$, $c_{L,l} \in R^m$ are learning error and change of learning error defined by $c_{L,l}(t)=e_{L,l}(t+1)-e_{L,l}(t)$, $\lambda$ is an estimated time advance corresponding to the time delay of the process, and $P_L, Q_L \in R^{m \times m}$ are constant learning gain matrices.

## V. APPLICATION TO BLOOD PRESSURE CONTROL

It is required to regulate simultaneously the cardiac output (CO) and the mean arterial pressure (MAP) of a patient in hospital intensive care using various drugs. Two typical drugs used are dopamine (DOP), an inotropic drug, and sodium nitroprusside (SNP), a vasoactive drug. For the purpose of this simulation study, we adopt the same model as used before in [3] which is given by where $\Delta CO$ ( ml/s) and $\Delta MAP$ (mmHg) are the changes

$$\begin{bmatrix} \Delta CO \\ \Delta MAP \end{bmatrix} = \begin{bmatrix} 1.0 & -24.76 \\ 0.6636 & 76.38 \end{bmatrix} \begin{bmatrix} \dfrac{K_{11}e^{-\tau_1 s}}{sT_1+1} & \dfrac{K_{12}e^{-\tau_2 s}}{sT_1+1} \\ \dfrac{K_{21}e^{-\tau_2 s}}{sT_2+1} & \dfrac{K_{22}e^{-\tau_2 s}}{sT_2+1} \end{bmatrix} \begin{bmatrix} I_1 \\ I_2 \end{bmatrix}$$

in cardiac output and in mean arterial pressure due to $I_1$ and $I_2$ ; $I_1$ (μg /Kg/min ) and $I_2$ (ml/h) are the infusion rates of dopamine and nitroprusside; $K_{11}$, $K_{12}$, $K_{21}$ and $K_{22}$ are steady-state gains with typical values of 8.44, 5.275, – 0.09 and –0.15 respectively; $\tau_1$ and $\tau_2$ represent two time delays with typical values of $\tau_1$ = 60s and $\tau_2$ = 30s; and $T_1$ and $T_2$ are time constants typified by the values of 84.1s and 58.75s respectively.

In order to verify the applicability of the FCMAC-based controller and to investigate its self-learning behaviour, we carried out a set of simulations. Fig 3 shows an example of the output responses of the process after 10 iterations with the desired responses indicated by dashed lines, where the controller comprised 4 inputs (errors & change-in-errors) and 2 outputs using the following parameters: learning matrices $P_L$=diag{0.05,-0.05} and $Q_L$=0, learning rate $\beta$=0.1, threshold $a_0$=0.1, and the Euclidean distance with $\delta$=0.1. It can be seen the process followed the desired responses satisfactorily.

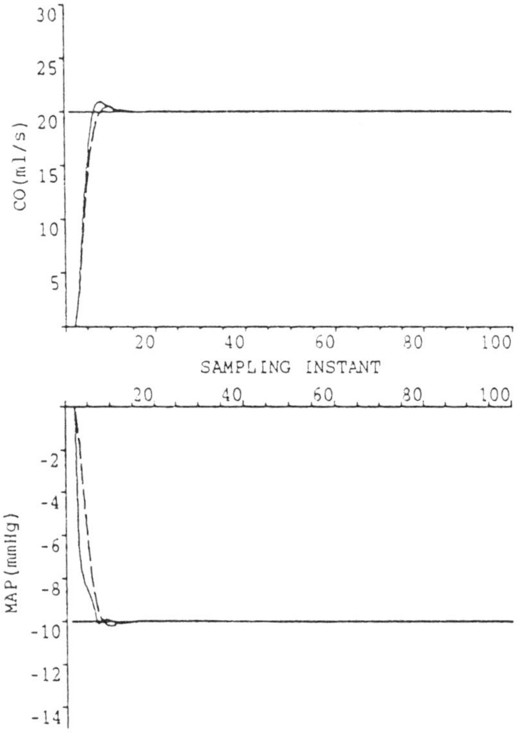


Fig 3 Output responses of the process

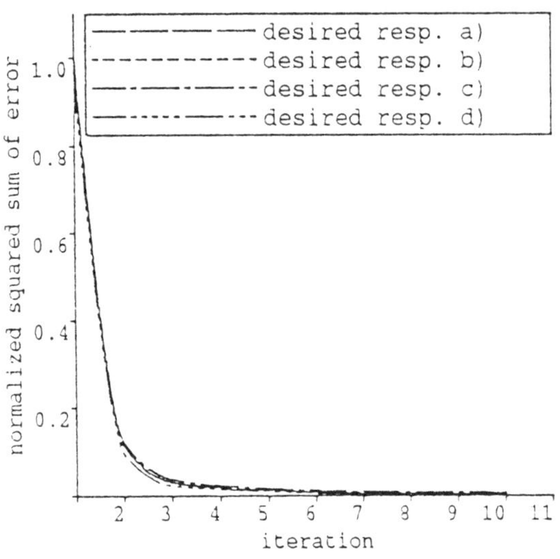


Fig 4 Adaptive ability to desired responses

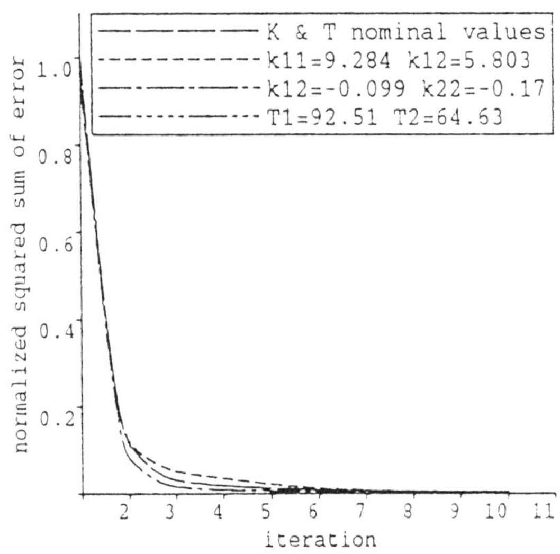


Fig 5 Adaptive ability to process parameters

By fixing the controller parameters as above, the adaptive ability of the controller was examined with respect to variations of the process parameters and to the desired performance requirements. Fig 4 shows the normalized squared sum of learning error $e_L$ (NSSE) against the iteration number when 4 different desired responses obtained by altering parameters in the reference model were required. Fig 5 gives the results when the process parameters were changed by 10% from their nominal values. It is evident that the controller is able to adapt to various per formance requirements and different process parameters in the sense that after a few iterations, the NSSE's tend to stable and small values.

The convergence property of the system with regard to the controller parameters was also examined. Fig 6 shows the results produced with different $\beta$ and

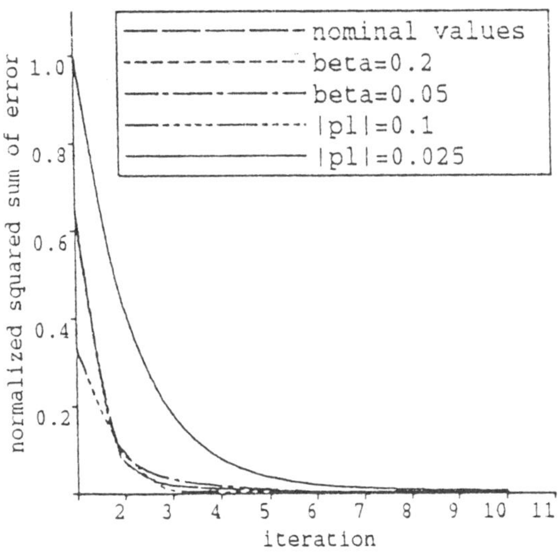


Fig 6 Convergence to learning rates

$P_L$. Compared with β, $P_L$ has a more significant effect on the learning convergence. As would be expected, however, the convergence property is not be very sensitive to the width δ and the threshold $a_0$. This is clearly indicated in Fig 7 where an almost identical curve was obtained for all cases. However, the number of created cells was strongly dependent on these parameters: the bigger δ, the fewer cells, and the bigger $a_0$, the more cells. Corresponding to Fig 7, the cells created after 10 iterations were 14, 11, 18, 16, and 13 respectively, which agrees with the above guideline.

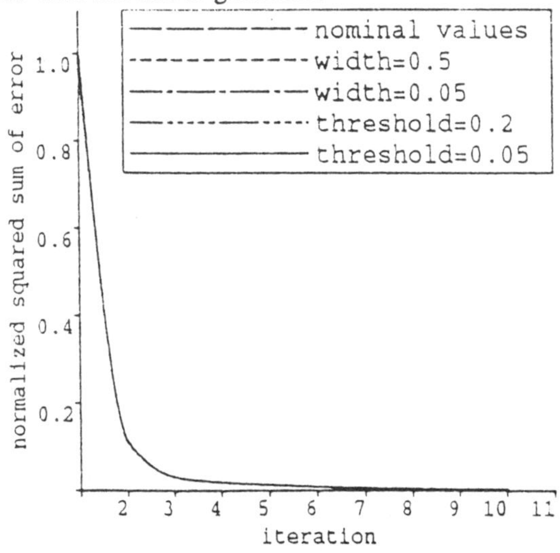


Fig 7 Convergence to width and threshold

## VI. CONCLUSIONS

We have presented an approach to fuzzifying the CMAC into FCMAC and extended this to self-constructing systematically a FCMAC-based multivariable controller. FCMAC possesses many advantages over CMAC. In particular, it provides a simple content-addressable scheme capable of dealing with arbitrary dimensional input space without involving complicated quantizing, addressing and hashing procedures whose complexity increases with the input space dimension so dramatically that the system may become intractable. In addition, the completely unsupervised learning characteristic of the system offers an efficient method for automatic construction of the fuzzy-neural controller directly from the process being controlled in a real-time manner and with arbitrary control performance requirements. The simulation results have demonstrated the feasibility of the proposed approach and revealed some useful properties relevant to its adaptive capability and learning convergence.

## REFERENCES


[1] J.S. Albus, A new approach to manipulator control: The Cerebellar Model Articulation Controller (CMAC), *J. Dynamic Syst. Meas. Contr.*, Vol 97, pp 220-227, 1975.

[2] J.S. Albus, Data storage in the Cerebellar Model Articulation Controller (CMAC), *J. Dynamic Syst. Meas. Contr.*, Vol 97, pp 228-233, 1975.

[3] D. A. Linkens and Junhong Nie, A unified real time approximate reasoning approach for use in intelligent control. Part 2: application to multivariable blood pressure control , *Int. J. Control*, Vol 56, pp 365-398,1992.

[4] D. A. Linkens and Junhong Nie, Rule extraction for BNN neural network-based fuzzy control systems by self-learning, in *Artificial Neural Networks 2* ,I. Aleksander and J. Taylor Eds., Noth-Holland, 1992, pp 459-462.

[5] D. A. Linkens and Junhong Nie, Constructing rule-bases for multivariable fuzzy control by self-learning. Part 1 : system structure and learning algorithms, *Int. J. System Science* , to appear.

[6] W.T. Miller, F.H. Glanz & L.G. Kraft, CMAC: an associative neural network alternative to backpropagation, *Proc. IEEE*, Vol 78, pp 1561-1567, 1990.

[7] Junhong Nie, A class of new fuzzy control algorithms, *Proc. IEEE Int. Conf. on Control and Applications*, Israel, 1989.

[8] Junhong Nie, *Fuzzy-neural control: principles, algorithms and applications*, Ph.D thesis, Department of Automatic Control and Systems Engineering, The University of Sheffield, 1992.

[9] Junhong Nie and D. A. Linkens, Neural network-based approximate reasoning: principles and implementation, *Int. J. Control*, Vol 56, pp 399-414, 1992.

[10] Junhong Nie and D. A. Linkens, Counterpropagation Network-based fuzzy controllers: explicit representation and self-construction of rule-bases, in *Artificial Neural Networks 2* ,I. Aleksander and J. Taylor Eds., Noth-Holland, 1992, pp 463-466.

[11] H.Tolle,P.C. Parks, E. Ersu, M.Hrmel, and J. Militzer, Learning control with interpolating memories, *Int. J. Control*, Vol 56, pp 291-317, 1992.

# Fuzzy Traffic Control: Adaptive Strategies

J. Favilla
IBM Brasil
C.P. 71
Campinas, SP - 13001 - Brazil

A. Machion, F. Gomide
UNICAMP/FEE/DCA
C.P. 6101
Campinas, SP - 13081 - Brazil

***Abstract* - This paper presents a Fuzzy Traffic Controller with adaptive strategies for fuzzy urban traffic control systems combined with two different defuzzification and decision-making criteria. The basic concept of adaptive strategies employed here is to adjust the membership functions according to the traffic conditions to optimize the controller's performance. The methods are: statistical-adaptive and fuzzy-adaptive, respectively.**

**The Fuzzy Traffic Controller (FTC), which is composed by a Fuzzy Logic Controller (FLC), a State Machine and an Adaptive Module, is described. A case study, concerning the application of the proposed FTC in the intersection of two major avenues of the city of Sao Paulo is also addressed. The results show that the FTC developed out perform conventional control strategies and alternative fuzzy control schemes.**

## INTRODUCTION

Traffic lights are usually used to control vehicle flows in cities' intersections. Since the control of the intersections may often be the critical factor to determine the overall capacity and performance of the urban network, adequate control of the traffic lights is an important problem to be addressed.

Often, control strategies consist in changing the green time (and consequently the cycle length) as a function of the incoming traffic such that the vehicles share the intersection area in an efficient way. This problem is complicated because each intersection has unique characteristics of physical layout, vehicle-flow rates, turning movements, pedestrian movements, and so forth [1]

The policies used in traffic control systems can be divided into two main categories [2]: **Fixed-Time Systems**, where traffic plans are generated off-line and applied on-line, and **On-Line Systems** where traffic plans are generated on-line and are directly used for traffic control. Both methods have advantages and disadvantages [3]. However, the common objective of all methods is to minimize the vehicles delay and average queue length caused by the intersection, as proposed by [4].

Previous works concerning the application of fuzzy systems theory in traffic control problems have been reported (see [5] and [6], for example). However their approaches are not adaptive in the sense considered in this paper. A first attempt to use adaptive strategies in traffic control has been described in [1], where it was shown that the adaptive schemes perform, in average, better than conventional and non adaptive fuzzy strategies.

This work presents a Fuzzy Traffic Controller which includes a Fuzzy Logic Controller, a State Machine and an Adaptive Module. The focus is the evaluation of two adaptive strategies: statistical-adaptive and fuzzy-adaptive and two different defuzzification and decision-making methods.

## THE FUZZY TRAFFIC CONTROLLER

Fig. 1 shows the block diagram of the FTC, which comprises the items described in the following :

1. **The sensing devices** are composed by a set of two inductive loops spaced by a distance *d* (one set per lane). This disposition allows the detection of the vehicles as well as the determination of their speed.
2. **The estimator** calculates the speed of each vehicle and estimates its time to cross the intersection, specially at the end of the green phase, and ends up with **Arrivals** and **Queue** of vehicles.
3. **The Fuzzy Logic Controller** is responsible for controlling the green time length according to the traffic condition.
4. **The State Machine** controls the sequence of states the FTC should go through.
5. **The Adaptive Module** changes the settings of the FLC in order to obtain a better performance of this controller.
6. **The Traffic Light Interface** has all the necessary circuits to turn on and off the lights, according to the controller's output.

0-7803-0614-7/93$03.00 ©1993IEEE

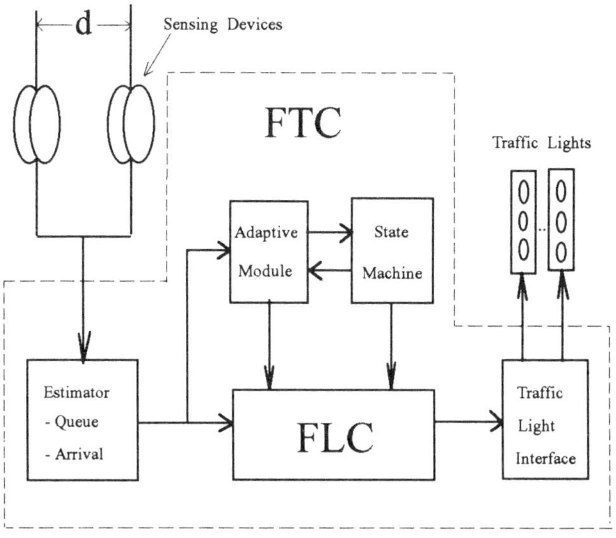


Fig. 1: Block diagram of the Fuzzy Traffic Controller

A description of the Fuzzy Logic Controller, the State Machine and the Adaptive Module is as follows. See [7] for a more detailed presentation.

## FUZZY LOGIC CONTROLLER

The FLC controls the traffic flow at an urban intersection. Basically it compares the incoming traffic of the approach that has the green phase with the vehicle queue formed in the other approaches. On the basis of this information, it decides to extend or not the current green time. If the decision is no extension of the current green time, the traffic light state will change to another state, allowing the traffic of another approach to flow. The proposed FLC has the following inputs: **Arrival** of vehicles in the approach that has the green phase, and **Queue** of vehicles in the approaches that has the red light. The controller output provides the **Extension** of the current green phase, which can be extended until a maximum previously defined value is reached. This condition also makes the state of the controller change.

The FLC has the following inputs (state variables) and their associated linguistic values: i) **Arrival** (almost_none, few, many, too_many), ii) **Queue** (very_small, small, medium, long). The output of the FLC (control variable) and its associated linguistic value is: **Extension** (very_short, short, medium, long). The graphical representation of the membership functions of these linguistic variables is presented in the Figs. 2a and 2b. The FLC rules are summarized in TABLE I.

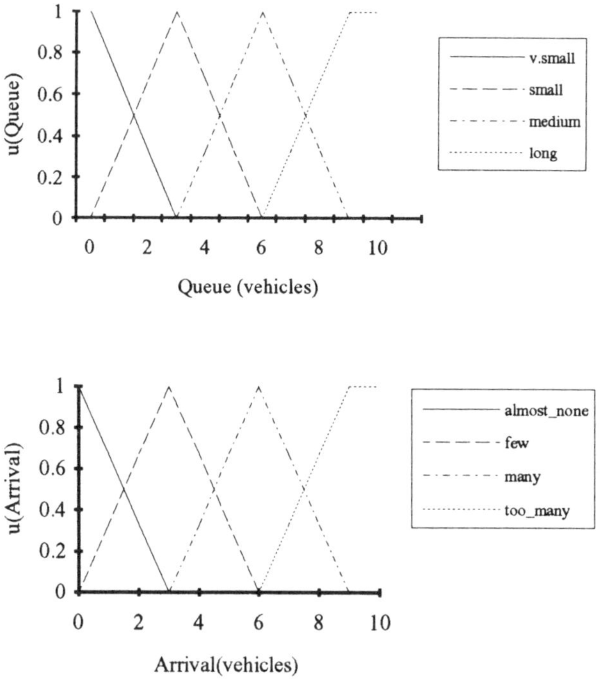


Fig. 2a: Graphical representation of antecedent membership functions

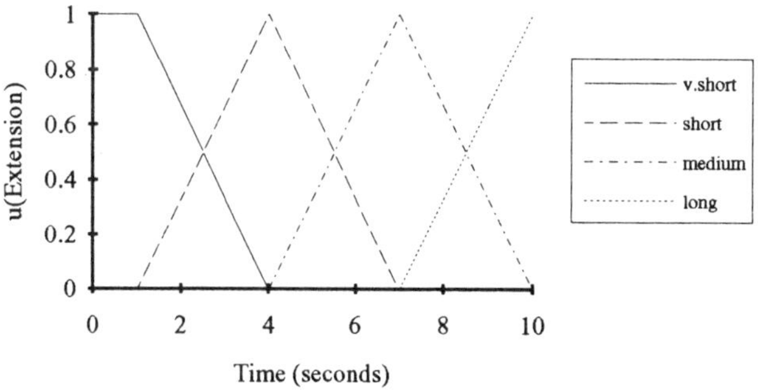


Fig.2b: Graphical representation of consequent membership function

TABLE I
FUZZY CONTROL RULES FOR THE FLC

| Arrival / Queue | almost_ none | few | many | too_many |
|---|---|---|---|---|
| v.small | v.short | short | medium | long |
| small | v.short | v.short | short | medium |
| medium | v.short | v.short | v.short | short |
| long | v.short | v.short | v.short | v.short |

The FLC has a total of eleven rules. The connectives *and* and *or* are interpreted as the operators *min* and *max* respectively. Based on the input information of the incoming vehicles (10 seconds away from the intersection) the *max-min* compositional rule of inference is applied. Two different defuzzification methods are a compared in this work: the *maximum* and the *center of area*. As the decision-making logic, the comparison considers the method proposed by [5] which evaluates the system state every second and then at the end of the current phase decides if an extension is necessary taking the worst case and, the one that evaluates the system state in every second too, but the decision is the weighted average.

## THE STATE MACHINE

The State Machine controls the sequence of states the FTC should cycle through. There is one state for each phase of the traffic light. There is also one default state which takes place when no incoming traffic is detected. This default state corresponds to the green time for a specific approach (usually to the main approach, where the probability of having incoming traffic is higher). Also, there is a pre-defined sequence of states. Three conditions are considered for state transition: the maximum time for the current state has expired or FLC decides that no extension is necessary or there is no incoming traffic being detected. In this sequence of states, a state can be skipped if there is no vehicle queues for the corresponding approach.

## THE ADAPTIVE STRATEGIES

Traffic flow conditions are dependent of the time of the day. During peak hours they are much higher traffic flow during than early morning hours. The main objective of the adaptive strategies proposed here is to optimize the controller's performance in a broader range of traffic conditions.

The FLC acts as a fine control mechanism while the adaptive module acts as a gross control mechanism, bringing the FLC to its best operational conditions. Two different adaptive strategies are compared. They are:

**1. Statistical-Adaptive**: This adaptive strategy consists in adjusting the membership functions of the variables in the antecedent of the fuzzy control rules (Arrival and Queue). The statistical calculation is the following: for every 10 seconds, the vehicle arrival is added up for each lane of each approach. Only the maximum value each lane of each approach is stored. This is done during 18 consecutive intervals of 10 seconds and all these measurements are stored. Then the average is calculated together with standard deviation ($\sigma$). The interval upper limits ($ul_{adaptive}$) for the membership function domains **Arrival** and **Queue** are updated according to the following:

$$ul_{adaptive} = \text{upper\_integer\_of}(\text{average} + 3\sigma)$$

if

$ul_{adaprive} \leq 10$ then

$ul_{new} = ul_{adaptive}$

else

$ul_{new} = ul_{old}$

The range of variation of the ***ul*** is from 4 to 10 vehicles. The maximum value 10 is due the physical constraint related to the positioning of the sensing devices (they are able to detect a Queue of up to 10 vehicles).

**2. Fuzzy-Adaptive**: this strategy consists in adjusting the membership function of the variable in the consequent of the fuzzy rules (extension). It employs another fuzzy logic monitor for this purpose. This fuzzy controller has as inputs (state variables) the **Residual-Queue** at the end of the green phase and the **Queue-Variation** during the green-phase. The output (control variable) ***ul* variation** is the variation to be applied to the upper limit of the membership function defined for **Extension** (refer to fig. 2b). The variation in the upper limit can range from 4 to 20 seconds. The graphical representation of the membership functions of these linguistic variables is presented in figs. 3a and 3b. The fuzzy adaptive rules are summarized in TABLE II.

TABLE II
FUZZY CONTROL RULES FOR THE ADAPTIVE MODULE

| Queue-Variation / Residual-Queue | small | medium | large |
|---|---|---|---|
| small | decrease | decrease | decrease |
| medium | increase | keep | decrease |
| large | increase | increase | keep |

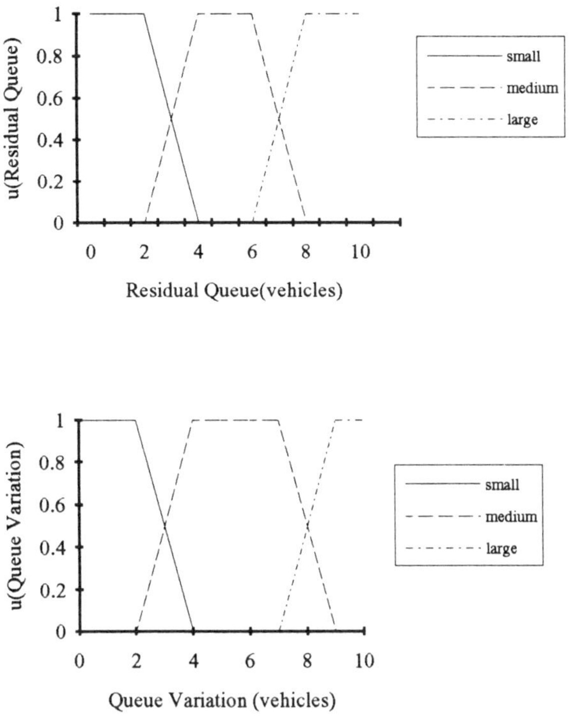


Fig. 3a: Graphical representation of antecedent membership functions

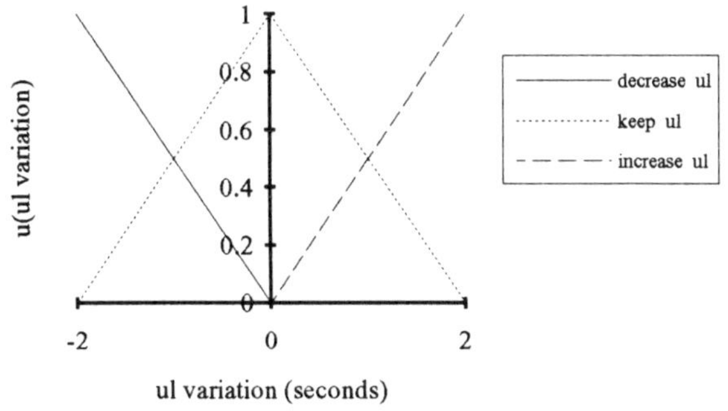


Fig. 3b: Graphical representation of consequent membership function

PERFORMANCE COMPARISON: A CASE STUDY

To analyze the effectiveness of the adaptive strategies, a comparison study has been carried out for the intersection of Brasil Avenue with Reboucas Avenue in the city of Sao Paulo. The layout of the intersection is shown in fig. 4.

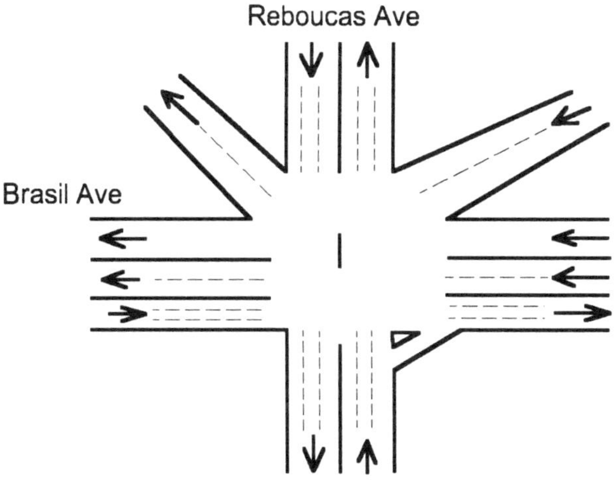


Fig. 4: Intersection of the Brasil avenue with Reboucas avenue

Based on the daily 15 minute average traffic flow volumes provided by CET - The Traffic Engineering Company of the City of Sao Paulo, a representative traffic flow volume profile was considered (fig. 5 shows this volume). The random traffic generated to simulate this profile is Poisson distributed as negative exponential, what assures that the average time between arrivals is constant.

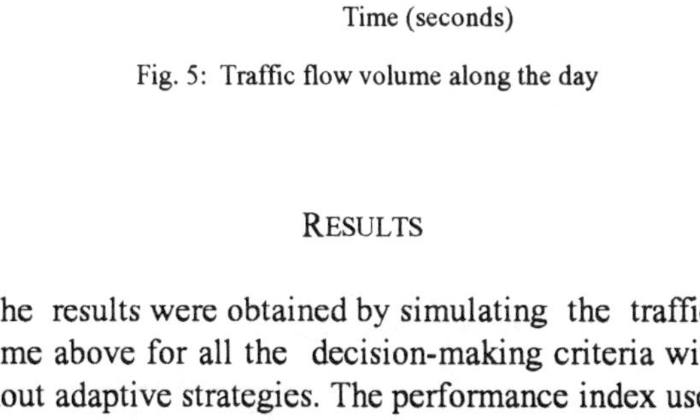

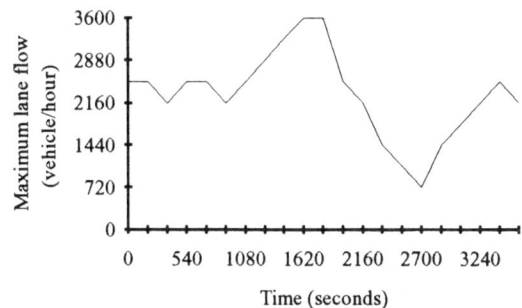

Fig. 5: Traffic flow volume along the day

RESULTS

The results were obtained by simulating the traffic flow volume above for all the decision-making criteria with and without adaptive strategies. The performance index used for comparison were the average queues. The profile of the extension membership function upper limit was also considered to show the adaptation performance. Fig. 6 shows the performance index for each decision-making criteria, and considering the different adaptation schemes.

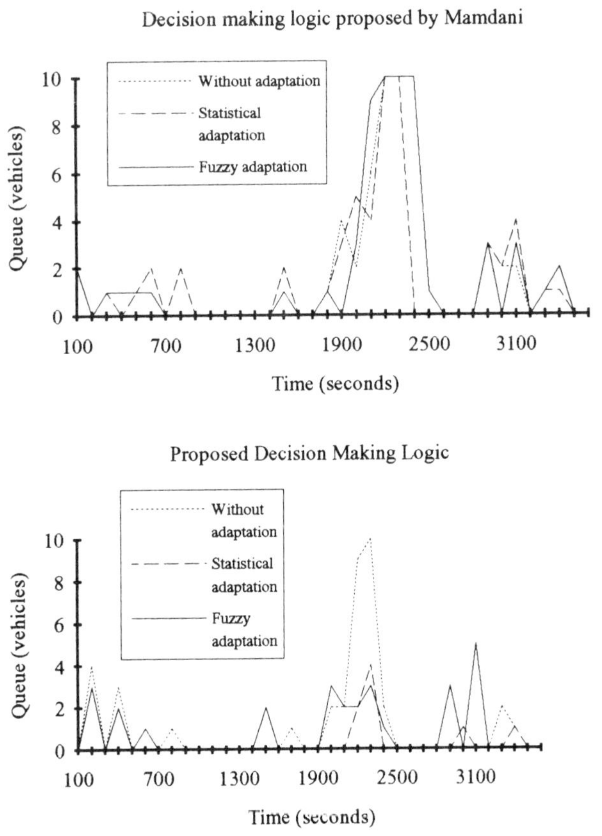


Fig. 6: Average queues versus time

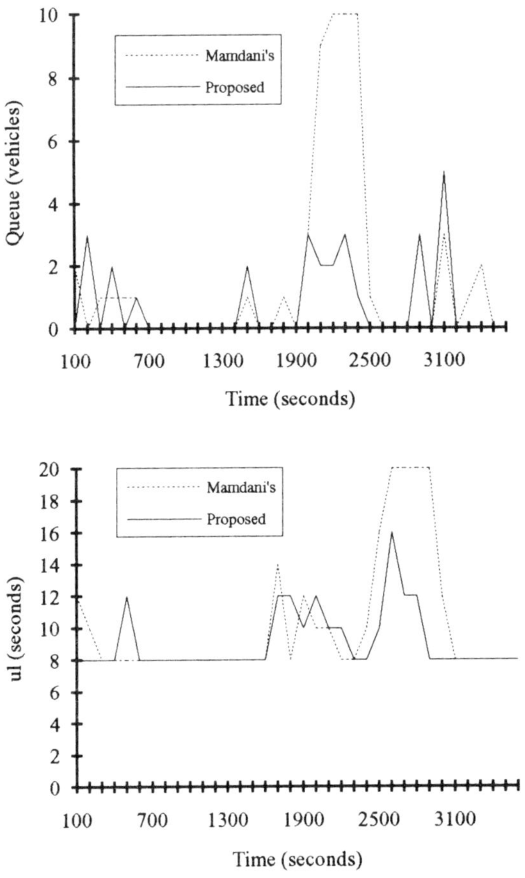


Fig. 7: Fuzzy adaptive scheme results for the different decision making criteria

Fig. 6 shows show that the fuzzy adaptive schemes perform better, in average, for all decision-making criteria.

Fig. 7 depicts the behavior of the best result of Mamdani's decision making scheme with the best of the decision making proposed in this paper. The changes in the upper limits, as defined by the fuzzy adaptive strategy is also shown. Note that, by using the proposed decision making and the fuzzy adaptive scheme, the FTC becomes more responsive to the traffic flow characteristics. Its better performance can also be observed when we compare the average delay provided by the two decisions making criteria with adaptation, as shown in table III below.

TABLE III
AVERAGE DELAY COMPARISON

| Decision Making Criteria<br>Adaptive Scheme | Mamdani | Proposed |
|---|---|---|
| Without Adaptation | 6.21 | 5.77 |
| Statistical Adaptation | 7.30 | 5.74 |
| Fuzzy Adaptation | 6.07 | 5.53 |

The worst performance provided by the statistical adaptive scheme in the case study considered is due to the time interval to compute the traffic statistics (180 seconds). A smaller period may be considered, but it increases the computational burden and storage requirements.

## CONCLUSIONS

A Fuzzy Traffic Controller with adaptive strategies for traffic control has been presented in this paper. Previous works have shown that the fuzzy control scheme provides a powerful approach to handle this class of problems. It has been shown here that adaptive strategies improve further the efficiency of fuzzy controllers in complex intersections. A case study was also presented, which confirms the usefulness of the approach proposed.

The performance of this controller under real traffic conditions (on the streets) is on-going and the results will be available soon.

Future work will address traffic pattern learning methods to be incorporated in advanced FTC. The symbioses of neural networks with fuzzy logic seems to be a promising approach to be explored.

## ACKNOWLEDGMENTS

The authors thank The Traffic Engineering Company - CET of the city of Sao Paulo, for their kind collaboration in providing data for the case study. The third author also acknowledges the CNPq for grant # 300729/86-3. The referees comments are also acknowledged.

## REFERENCES

[1] J.Favilla, A.Machion, F.Gomide and R.Gudwin, "Adaptive fuzzy logic based urban traffic control", Third Annual IAKE Symposium, November 16-19, Washington, 1992.

[2] A.J. Al-khalili, "Urban traffic control - a general approach", IEEE Trans. on Systems, Man, and Cybernetics, vol. SMC-15, no. 2, Mar. 1985, pp.260-271.

[3] D.J. Robertson, "Traffic models and optimum strategies of control: a review.", Proc. Int. Symposium Traffic Control Systems, University of California, vol. 1, 1979, pp. 262-288.

[4] F.V. Webster, "Traffic signal settings", Road Res. Tech. Paper 39, Road Research Laboratory, London, 1958.

[5] C.P. Pappis and E.H. Mamdani, "A fuzzy logic controller for a traffic junction", IEEE Trans. on Systems, Man, and Cybernetics, vol.SMC-7, no.10, Oct.1977, pp.707-717.

[6] M. Nabatsuyama, "Fuzzy logic controller for traffic junctions in one way arterial road", IFAC 9th World Congress, 2865 (1984)

[7] J. Favilla, A. Machion, F. Gomide, "Adaptive fuzzy logic control of urban traffic", Faculty of Electrical Engineering, Internal Report, UNICAMP-FEE-DCA, Campinas, 1992.

# Adaptive Fuzzy Controller Improves Comfort

L. Peters K. Beck R. Camposano
German National Research Center for Computer Science (GMD),
System Design Technology Institute (SET)
5205 St. Augustin, Germany

***Abstract*— We present a new control method based on fuzzy logic theory. It considerably reduces the number of rules in the input matrix, and adapts the rule set locally and temporarily to disturbances in the environment in which it is acting. Its main advantage is the capability to adapt the inference rules without the need of a learning phase.**

## I. Introduction

In thc last years considerable effort has been devoted to the development of suspension systems to improve performance, ride quality, and safety of automobiles. The control of such complex systems must respond properly under a wide variety of conditions such as temperature, changing road surfaces, vehicle speed and loading, tire performance etc. [5]. As a lot of suspension simulation models with classical control methods already exist, we chose a one-wheel suspension model to test the efficiency of our new fuzzy controlling method.

First we summarize the conflicting demands which must be satisfied to control a suspension system. Then we present the advantages of the new control method and its implementation for the chosen example. Finally the obtained results are compared with a standard suspension model.

## II. Adaptive Damping System

To improve the performance of suspension control systems, the following objectives have to be met: improvement of driving performance, functions, and safety; identification of driving environment (rough, bumpy, or smooth road); analysis and evaluation of driver's ride comfort feeling, preference, intentions, and operating capabilities [8]. Some requirements like safety and ride comfort have opposite demandings. To solve this conflict the requirements have to be analysed. The control system takes the final decision based on an intelligent merging of the conflicting demands.

In general safety rules require a rough damping, while the comfort feeling requests a soft one [6]. While the safety aspects depend mainly on the action of the driver and the car speed, the driver's perception of ride comfort depends on the environment. On one hand, when the driver is turning the steering wheel the safety requirements increase and require a high priority. On the other hand, when the car drives down a rough road at low speed, the ride comfort may get higher priority than safety. By means of this simple example it is easy to understand that safety and comfort requirements change steadily, and it is not an easy task to find an optimum for both.

The changing of the safety requirements can be correlated to the changing of the angle of the steering wheel, the vertical acceleration of the car body, and the vertical acceleration of the wheel. These values can easily be measured with available sensors mounted on the car.

One way of evaluating the ride comfort is by analysing the driver's environment. The quality of the road can be classified as rough, smooth, bumpy etc. dependent on its waviness $W$ and roughness $H$. The spectral analysis of the road $h(t)$ gives us a description of the ride comfort which is independent of the car speed. The frequency domain representation $H(f)$ has two interesting regions. The first one is near the resonance frequency of the car body (0.5...2 Hz), and the second one at higher frequencies (8...20 Hz) corresponds to the resonance frequency domain of the motor and car axle. The roughness $H$ represents the rms value of the Fourier transform calculated over the whole frequecy range $H$ (0.5...20 Hz) while the waviness $W$ is calculated from the rms value of the Fourier transform calculated over the two subranges mentioned above:

$$W = \frac{H(0.5\ldots2\,\mathrm{Hz})}{H(8\ldots20\,\mathrm{Hz})}\,.$$

For a very smooth road $W$ has its lowest value. The waviness $W$ reaches its peak values when the spectral lines in the low frequency region are predominant, which corresponds to a road with long waves. Although the general guideline for a smooth ride asks for soft damping in both

0-7803-0614-7/93$03.00 ©1993IEEE

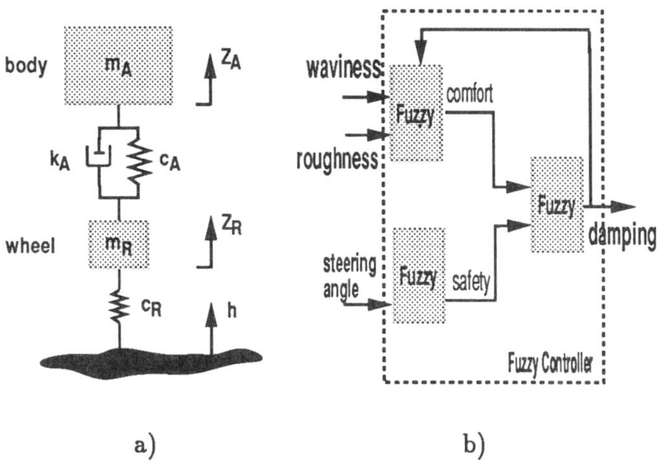


Figure 1: a) Body–wheel model of a quarter car. b) Fuzzy controlling system.

cases, test rides have shown that most people prefer more rigid damping on a wavy road, even though the car body acceleration increases [4]. This preference is well reflected by our fuzzy controller output in the simulation results.

As shown in this section the calculated values for $H$ and $W$ in the frequency domain can classify the road from smooth to rough and give a measurable value of the ride comfort.

If we model the car-wheel damping system according to Fig. 1a [7], neglecting the influence of wheel damping, the road profile $h(t)$ can be calculated from the vertical accelerations of the car body ($\ddot{z}_A$) and wheel ($\ddot{z}_R$) as follows:

$$m_A \cdot \ddot{z}_A + k_A(\dot{z}_A - \dot{z}_R) + c_A(z_A - z_R) = 0$$

$$m_R \cdot \ddot{z}_R - k_A(\dot{z}_A - \dot{z}_R) - c_A(z_A - z_R) + c_R \cdot z_R = c_R \cdot h$$

$$h = h(t) = z_R + \frac{m_A \cdot \ddot{z}_A + m_R \cdot \ddot{z}_R}{c_R} .$$

The variables — which are associated to the car body or the wheel by their indices $_A$ or $_R$, respectively — denote mass $m$, vertical displacement $z$, its derivatives $\dot{z}$ and $\ddot{z}$, spring constant $c$, and damping coefficient $k$.

In particular, the calculation of the road profile is independent of our targets, i. e. adjustable dampener and air spring. Thus, safety measures and ride comfort can be evaluated with a pair of acceleration sensors mounted on the chassis of a standard car.

## III. New Fuzzy Control Method

Fuzzy set theory has been accepted as an attractive tool for handling imprecision and uncertainty. It offers a good alternative to get an improved suspension control system where driver feelings like comfort and exact requirements like safety have to implemented. To overcome environmental variations the fuzzy rule-base needs to be modified during operation. In most cases, environmental conditions affect the entire rule-base. One way of solving this problem is to change between several rule sets [3] or to modify the definition of the linguistic variables [2]. In our method the number of rules remains the same, but the shape of the controlling surface is locally changed by small iterative steps.

To catch unforeseen changes in the environment, we introduced an internal feedback loop from the output to the inference block (Fig. 1b). Through this additional control input the weights of affected rules are changed so that their influence at the operating point on the controlling surface is increased or diminished. These changes are strictly local and do not influence the whole input space. When the disturbance disappeares the affected rules gradually reach their initial values by regressive iteration.

The second new feature which we introduced is the considerably reduced number of rules [1]. To ensure that a fuzzy controller always produces a meaningful output, at least one membership function must have a non-zero value for every input value. We are able to meet this requirement although we use a sparsely filled input matrix because of the properties of the chosen membership function. The chosen membership functions are of bell shape and therefore have non-zero values over all the input space. For a two dimensional input we used a set of 3 to 8 rules instead of the possible 49 (given 7 values for each dimension).

To control the suspension system we used a hierarchical fuzzy controller, where each fuzzy block has the features described above. One block gives the optimal damping necessary to ensure ride comfort. The inputs which reflect the road quality are the roughness $H$ and the waviness $W$ of the road surface. The second block gives the recommended damping based on safety rules and the action of the driver, which in our example is represented by the steering angle. The third block merges the two sometimes conflicting demands and produces the final value for the dampener. The adaptive tuning of the rules takes place in our simulations as the vehicle is driven down on the different road surfaces.

## IV. Simulations

The simulated system was limited to a quarter car (Fig. 1a). The simulations and fuzzy controller model are run on a SUN SPARCstation 2, which provides a control panel to access the simulator package. The user may vary parameters like road surface, tire pressure, dampener type and change the initial rule set. The software execution speed is controlled so that the simulation runs at a realistic speed.

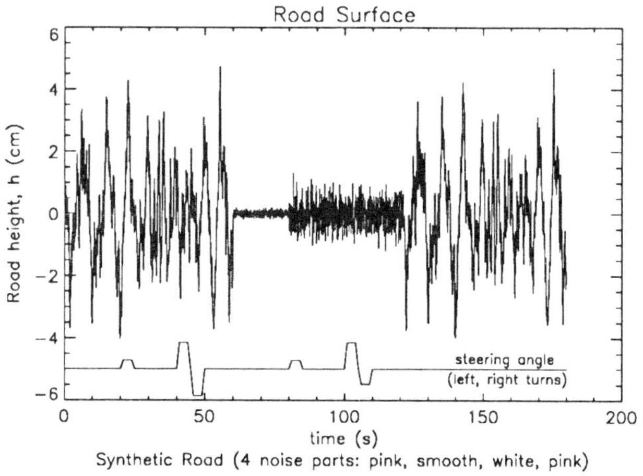


Figure 2: Road surface and steering wheel angles.

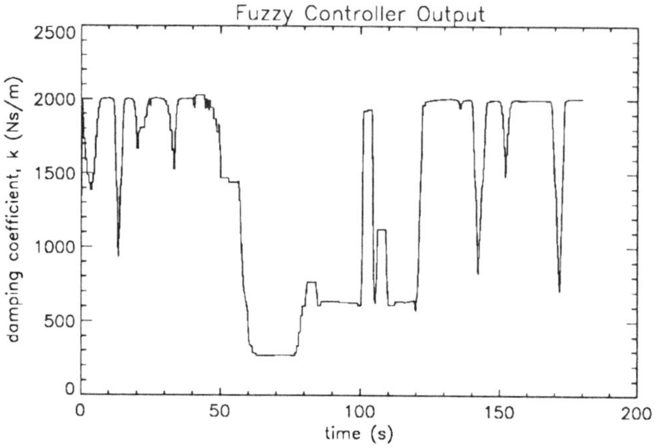


Figure 3: Fuzzy controller output corresponding to both road surface and steering angle (prior to smoothing for use with a four-level dampener).

Our fuzzy controlling system used a rule set fitted for a commercially available four-level dampener. (A slightly different rule set is necessary to drive a continuously variable dampener.) We compared our results with a passive damping system which is implemented in a standard car and is tuned up once in the factory at a fixed value either for a sportscar or a family car. Safety has priority for sportscars, and a rough damping is chosen. The damping for the family car takes comfort into account to a higher degree and the damping is softer (normal tuning).

We compared the two damping systems by using several realistic road surfaces created synthetically. A smooth road was simulated by a white noise signal, and a wavy road by a pink noise signal. Fig. 2 shows the road surface together with the steering angles used for the simulations. Fig. 3 shows the output of the fuzzy controller tuned for a four-level dampener. In most parts of the "pink road", between 0 s and 60 s, the changes between safety and comfort priority are not obvious. The output of the fuzzy controller sets the damping to high values (Fig. 3) because of the road's waviness (dominant low frequencies). These values are comparable with the passive dampener which is tuned for safety. However, the safety reaction on the steering angle (increased damping) is obvious in the white noise part between 60 s and 120 s where damping due to comfort requirements would otherwise be low. Here it is easy to see the difference between the passive and the fuzzy controlled damping system. This part of the simulation will be discussed further in detail.

Low car body acceleration is a measure for a comfortable ride, whereas low dynamic wheel load variation (i. e. deviation from the static load) gives a safer ride. In order to see the adaptability of the system to safety and comfort requirements we show details of the simulation, between 105 s and 115 s, in Figs. 4 and 5 (next page). In this cut-out the car is moving on a white road and the steering angle has changing values. Fig. 4 shows the vertical car body acceleration and the dynamical wheel load for the passive car dampener. The reaction of this damping system is merely dependent on the vertical accelerations of the car body and the wheel. In Fig. 5 the corresponding measurements are shown for the fuzzy controlled damping. Because of the steering angle below 110 s, the safety rules ask for a reduced wheel load variation, even if the car body acceleration is increased. As no steering angle is present after the 110 s time mark, comfort rules take over and reduce the car body acceleration.

## V. Conclusion

The presented fuzzy based control system delivers an improved ride comfort to drivers while guaranteeing high safety. The performance of this system shows the advantage over a conventional suspension system. It allows a continuous shift of priority between safety and comfort, using only standard sensor measurements. We have demonstrated that fuzzy controlling can be applied successfully to problems in control which are of current concern in the automobil industry.

## Acknowledgment

We wish to thank W. Kreft and H.-C. Wille from Volkswagen AG, Wolfsburg, for providing test data and for their useful and interesting discussions.

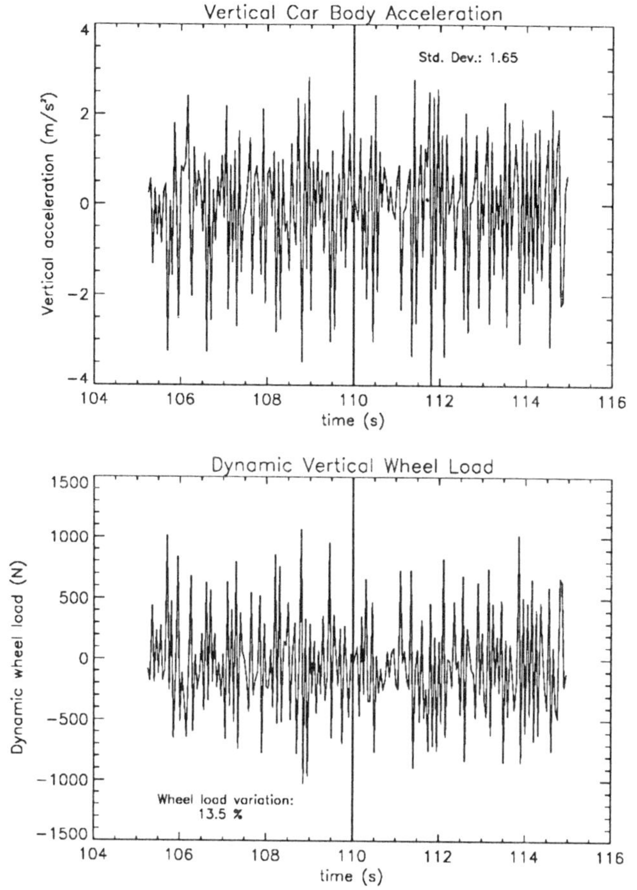


Figure 4: Body acceleration and dynamic wheel load using a passive dampener.

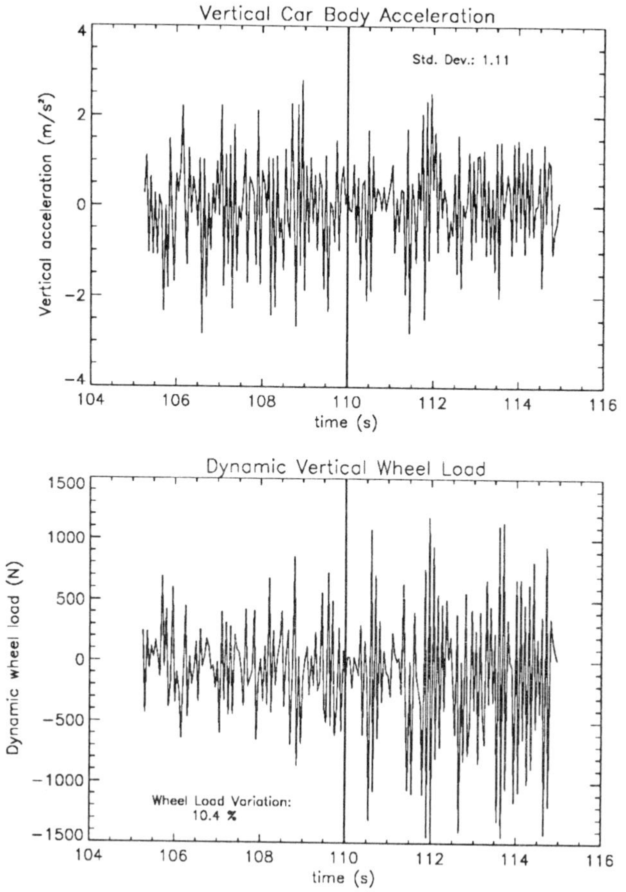


Figure 5: Body acceleration and dynamic wheel load using a *fuzzy controlled* dampener with safety priority (i.e. low wheel load variation, left part) and comfort priority (i.e. low body acceleration, right part).

## References


[1] K. Beck, L. Peters, *Ein Fuzzy Controller mit dynamischer Regelwichtung,* Technical Report GMD, no. 631, March 1992.

[2] J. Fei, C. Isik, *Adaptive Fuzzy Control Via Modification of Linguistic Variables,* IEEE International Conference on Fuzzy Systems, March 8-12, 1992.

[3] B. P. Graham, R. B. Newell, *Fuzzy Adaptive Control of a first-order Process,* Fuzzy Sets and Systems, no. 31, 1989, pp. 47-65.

[4] W. Klinkner, *Adaptives Dämpfungs-System (ADS) zur fahrbahn- und fahrzustandsabhängigen Steuerung von Dämpfern einer Fahrzeugfederung,* VDI Berichte, no. 778, 1989, pp. 323-356.

[5] K. A. Marko, B. D. Bryant, R. J. Hampo, J. V. James, *Vehicle applications of artificial neural systems: diagnostics and control,* Fuzzy and Neural Systems and Vehicle Applications, 8-9 November 1991, Tokyo.

[6] M. Mitschke, *Verbesserungsmöglichkeiten an Kraftfahrzeug-Schwingungssystemen,* ATZ Automobiltechnische Zeitschrift, vol. 89, no. 5, 1987, pp. 231-240.

[7] B. Richter, *Schwerpunkte der Fahrzeugdynamik,* Fahrzeugtechnische Schriftenreihe, TÜV Rheinland, Cologne, 1990.

[8] Y. Uyeda, *Improvement of Chassis Control by a Fuzzy Controller,* Fuzzy and Neural Systems and Vehicle Applications, 8-9 November 1991, Tokyo.

# A FUZZY CONTROLLER USING SWITCHED-CAPACITOR TECHNIQUES

**J. L. Huertas, S. Sánchez-Solano, A. Barriga, I. Baturone**

Dept. of Design of Analog Circuits.
Centro Nacional de Microelectrónica,
Edificio CICA, Avda. Reina Mercedes s/n, 41012-Sevilla (Spain)

**Key words:** fuzzy logic, discrete-time techniques, SC networks, CMOS integrated circuits.

*Abstract*

The use of Switched-Capacitor techniques to build a fuzzy controller is discussed in this contribution. Using a sequential architecture, the required building blocks are introduced and its realization is described. The proposed system can be considered as a starting point for exploring the future capabilities offered by SC networks to the hardware implementations of fuzzy systems.

## 1. INTRODUCTION

Fuzzy logic, although introduced many years ago [1], has not become until recently a practical alternative to conventional computers for performing inference operations. Many applications claiming the use of fuzzy concepts have recently appeared in the marketplace from japanese companies, but most of these products are based on a sort of software simulation of fuzziness using small wordlength conventional microcontrollers.

Besides these results obtained from purely digital approaches, it has been recognized [2, 3] the need of developing actual circuits performing the basic operations required for a fuzzy system to really take advantage of the full flavor of the fuzzy paradigm. Since a few years ago, attention is being paid to the development of circuit implementations for fuzzy logic, because actual fuzzy circuits might be the only way to extend the applicability of fuzzy logic to more demanding application areas. Then, hardware implementations have become a critical issue for adopting fuzzy solutions at a system level. The expected gains in terms of area reduction, operation speed (both at circuit and system levels), and functional flexibility turn out interesting to explore the possibilities offered by analog techniques, especially in those technologies fully compatible with digital circuits. Analog design techniques seem to be very appealing for the implementation of fuzzy circuits and systems. In particular, there exist a well-founded body of theoretical knowledge and practical experience related to linear (especially filters) and nonlinear switched-capacitor (SC) networks that can be applied to the realization of fuzzy controllers [4,5].

This communication addresses the design and implementation of a sequential microcontroller based on SC circuits. This is carried out at two levels: architecture, and cell design. At the first level, the bottlenecks of reported circuits are considered. In particular, we will focus on Yamakawa's [6] since, although his architecture is a valid solution in many practical cases, it seems interesting to look for modifications able to handle the design of systems with many rules in just one chip. A way to do that may be based on trading speed and interconnection complexity by resorting to the use of a sampled data approach. An additional advantage of this approach is the compatibility with sound analog techniques that can help in the design of the defuzzifier. The basic building blocks for the approach will be discussed as well as their use within the microcontroller. This system is conceived as a chip that can be operating embedded into a standard microprocessor environment.

## 2. SEQUENTIAL ARCHITECTURE DESCRIPTION

An architecture is proposed to deal with systems handling many rules. The new architecture employs essentially the same basic blocks proposed by Yamakawa, but the number of rules per chip can be significantly increased, and both the fuzzifier and the defuzzifier can be included in the same chip.

An overall view of the new architecture is shown in Figure 1, where its main blocks are detailed. This architecture is an adapted version of the one proposed in [7] for a current-mode fuzzy processor. Essential to this technique is the definition of an operation cycle (defined in terms of N cycles of a fundamental clock, Ck) whose duration will depend mainly on the precision we try to attain and the number of rules we consider. Such operation cycle will impose a limitation to the input signal bandwidth. Each Control Rule is implemented by an analog ROM, some Membership Function Circuits (MFC) and MAX/MIN gates. Started an operation cycle, the analog ROM will provide every clock cycle one value of voltage for truncating the values coming from the MFCs. Hence, the ROM performs as a serial Membership Function Generator (MFG) instead of working in parallel (as proposed by Yamakawa). This means that the M bus lines used by Yamakawa as a fuzzy word are replaced by a single wire that carries M successive samples representing such a word. The outputs from every Control Rule are processed by a MAX gate and fed the defuzzifier, which implements a center of gravity method. The first stage of this consists of two iterative summers preparing the numerator and the denominator of a discrete divider. After N clock cycles the divider will give the final output.

Basically, we divide an operation cycle into three phases. Phase 1 is devoted to sampling and holding the input variables as well as to pre-processing them trough the MFCs. Phase 2 is aimed to carry out the inference process by performing MIN-MAX operations on the input variables and the MFG outputs. Finally, in phase 3 the defuzzifying process is performed. For the sake of clarity we will call $N_j$ the number of fundamental clock cycles required for the j-th phase. In order to understand the overall structure, we will give in what follows a functional description of the blocks in Figure 1, detailing their circuit implementation and estimating the value for $N_j$ at every operation phase.

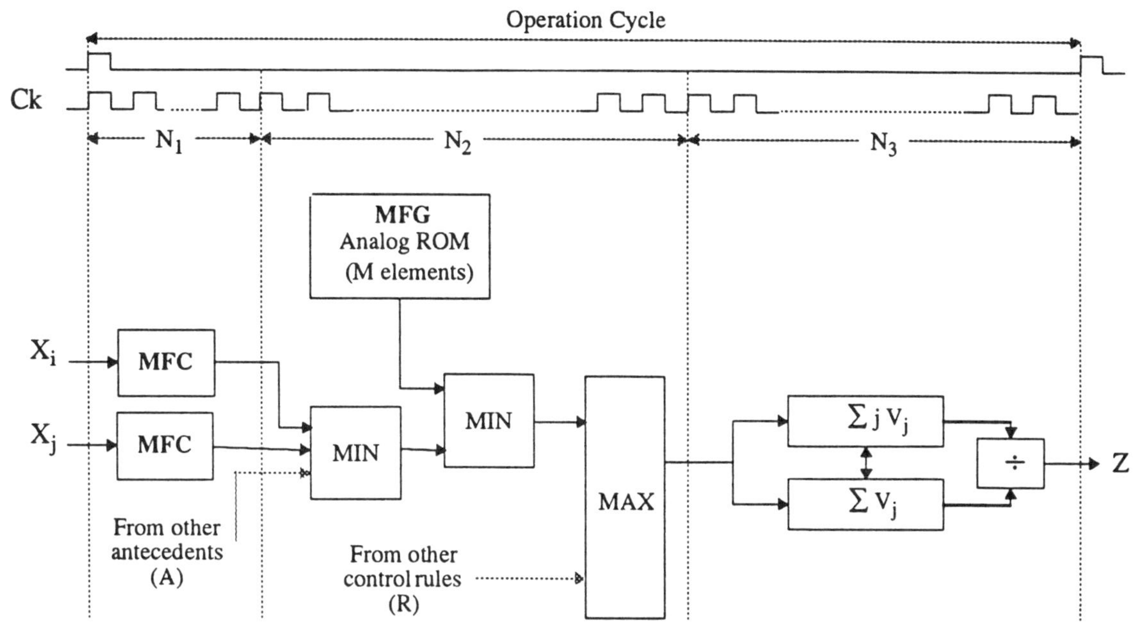


Fig. 1: Proposed architecture for a Switched-Capacitor fuzzy controller.

## 3. BASIC MFC

The basic element for fuzzy logic is a one-input operator performing a classification of its input variable according to a given membership function. Generally speaking, the functional transformation carried out by this operator is a nonlinear mapping (called a pertenence function), but in most common cases this transformation can be approximated by a symmetric piecewise-linear function of the shape shown in Figure 2-a, where the four parameters required to identify the breakpoints are depicted. To implement any transformation of this form, a possible solution is the circuit in Figure 2-b. The upper part of this circuit performs a piecewise-linear transformation under the control of the lower part. The input variable X is sampled and held sequentially to be compared with four voltage values defining the trapezoid breakpoints. The analog ROM1 in Figure 2-b provides such breakpoints. The results of the comparison are processed by a Finite Sequential Machine (FSM) that controls the switches of the upper circuit. The latter, depending on the result stored by the FSM, transforms the input in accordance with one of the five pieces of its piecewise-linear transformation. The Analog ROM2 supplies the required coefficients for this transformation.

As shown by its operation description, this circuit is active during phase 1 and must hold its output value whilst phases 2 and 3. The total time invested by this circuit to perform the described operation is 4 cycles for carrying out the comparison and 1 cycle for the nonlinear transformation, thus $N_1$=5.

A way to generate the voltages representing the breakpoints is shown in Figure 3-a [8]. Since current flowing out of the circuit to the comparator is neglectible, an almost ideal operation can be assumed. The form ratios of the different transistors will fix the comparison voltages. Since the output voltages are decreasingly ordered from top to bottom, a switching scheme successively addressing the different output voltages in increasing order is used. Typically, a claimed drawback for the circuit arrangement in Figure 3-a is its dependence on the power supply; however, since in our case the discourse universe is fixed by the bias voltages, this is not a problem any more.

Another solution to implement a membership function can be obtained from the former circuit just replacing the lower part

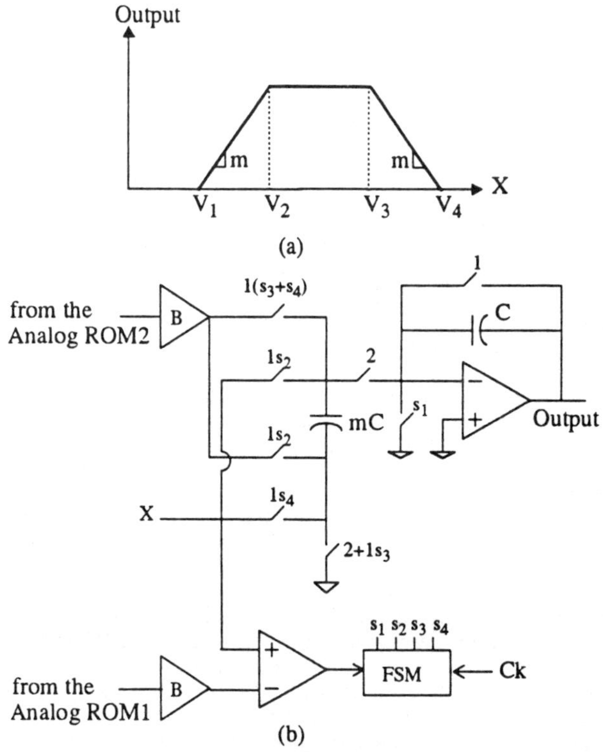


Fig. 2: (a) Symmetric trapezoid representing a membership function. (b) Circuit schematic of a MFC.

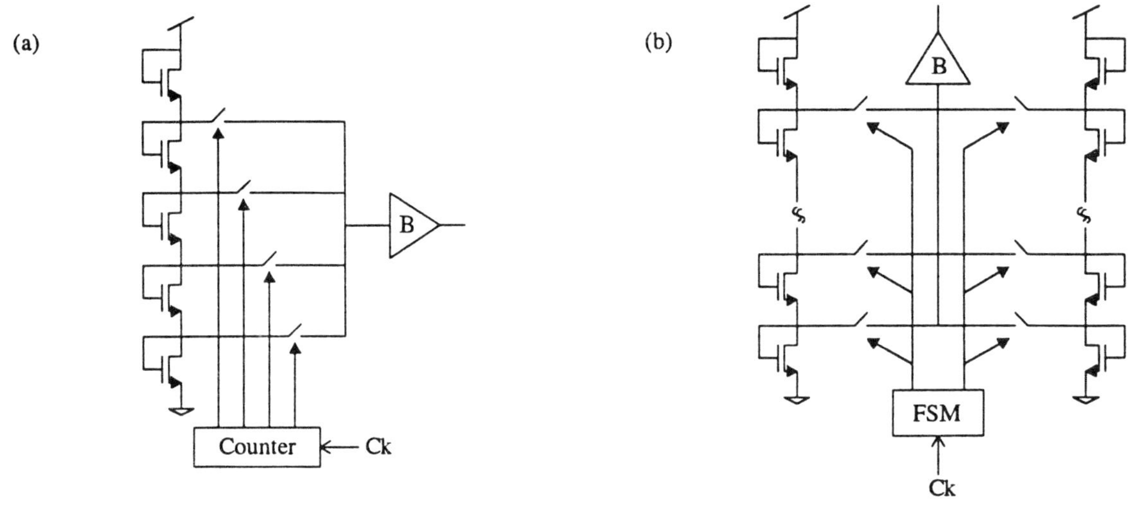


Fig. 3: (a) Analog ROM for the MFCs. (b) Analog ROM for the MFGs.

by four comparators. A simple combinatorial logic controls the switches so that the membership value is calculated in only one cycle and the ROM2 would not be needed. The four breakpoints are given in parallel by an analog ROM similar to that described, but the counter is not necessary now.

## 4. RULE EVALUATION CIRCUITRY

As was pointed out above, a good solution to avoid a large silicon area when many rules are considered, is achieved by the use of a serial MFG. Like MFCs, MFGs usually exhibit some symmetry, since they generate membership functions of fuzzy sets. Therefore, the number of membership values to be provided can be reduced so that the area occupation is still smaller.

The circuit in Figure 3-b has been used as the MFGs. The only difference with the circuit of Figure 3-a is due to the switching scheme. For the former a FSM must store the order in which the switches have to be closed. There is a practical limit to the number of transistors to be stacked up, roughly speaking this number is given by VDD/VT. In practice, it is preferable to derive the different voltage levels from several transistor poles. This is more flexible and avoids difficult trade-offs. Since the MFGs exhibit some symmetry, thus reducing the number of different voltage levels, the area occupation is not large.

Besides the MFG, MAX and MIN multi-input operators are required. A typical 4-input MIN gate is shown in Figure 4. In this figure, the four inputs are sequentially compared with the voltage previously stored by capacitor C. When the stored voltage is higher than an input voltage, the latter replaces the former since the corresponding switch is ON. Otherwise, the stored voltage does not change. Then, after four clock pulses we have the minimum of the four input signals. Either increasing the number of inputs or obtaining the MAX function is straightforward.

Taking into account its timing, the first level of MIN (in general MAX-MIN) operators will take a number of clock cycles that depends on the number of antecedents within every rule. Then, we must wait for a time equal to the slowest operation, which is equivalent to say a number of cycles equal to the highest number of antecedents in any rule. For the MFG operation, a two-input MIN is required, which means only a comparison and can be performed in just one cycle. Finally, for the last MAX stage, the time duration is equal to the number of rules times the clock cycle. Since these two-input MIN and multiple-input MAX operations have to be done for every element of the MFG:

$$N_2 = \max(A_j) + M\,(1 + R)$$

where $A_j$ applies for the number of antecedents within the j-th rule, M is the number of elements in the MFG, and R is the number of rules.

## 5. DEFUZZIFIER

The final stage of the fuzzy controller is implemented by two summers followed by a SC divider, as is illustrated in the block diagram of Figure 5-a. The circuit in Figure 5-b shows a circuit realization with a reduced number of elements; every time a

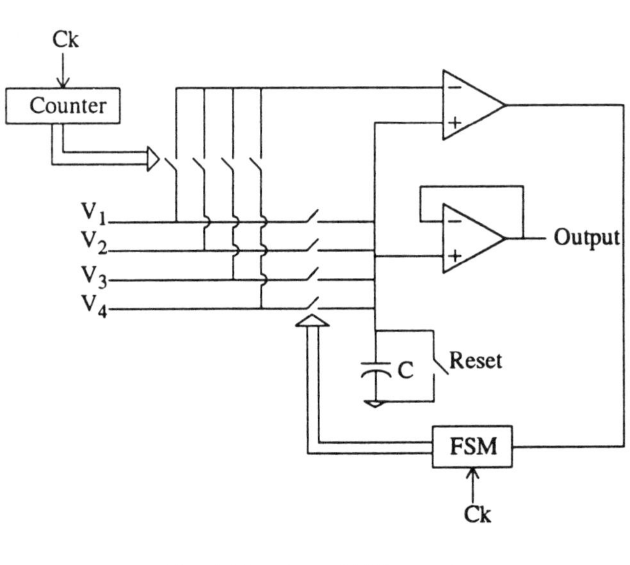


Fig. 4: Four-input MIN operator.

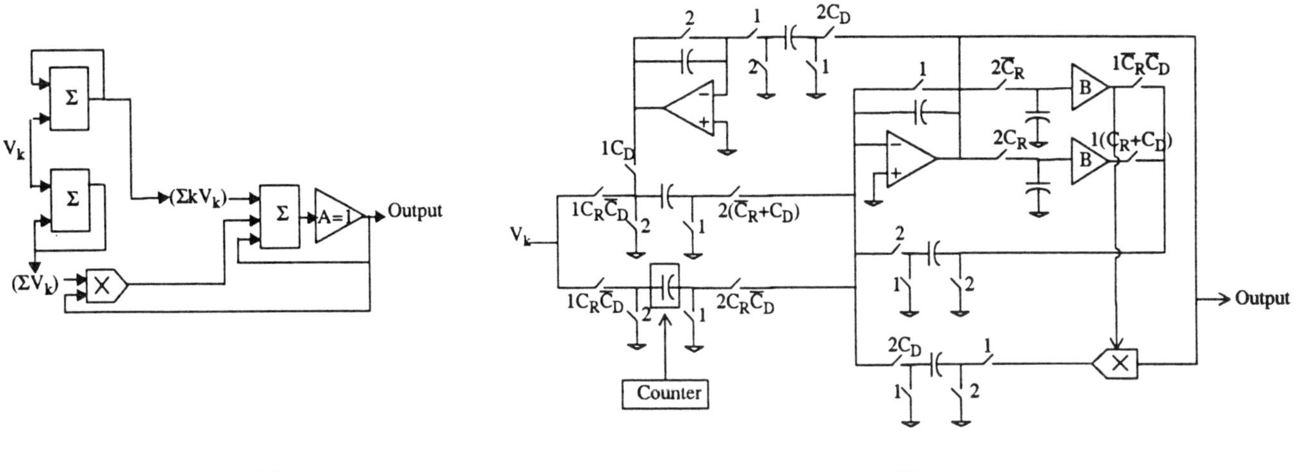


(a) (b)

Fig. 5: (a) Block diagram of a defuzzifier. (b) Circuit schematic.

MAX output ($V_k$) is validated, two partial sums ($\Sigma V_k$ and $\Sigma k V_k$) are stored in capacitors C1 and C2, respectively. In these capacitors, we are incrementally adding the voltages until the last value of a MFG is generated and processed. Then, a discrete-time divider (as can be seen in Figure 5) provides the final system output.

The counter in Figure 5-b is devoted to controlling a capacitor array which gives the weight k for the sum $\Sigma k V_k$. This type of arrays are frequently used in SC circuits but we can eliminate it and save on area by using two cascaded summers as is shown in the block diagram of Figure 6-a. Figure 6-b illustrates the resulting circuit.

In any case, the time required for the summer operation is 2 clock cycles for every MFG value, but since it can be done while the previous stage is processing, this value only accounts for the last step. The divider requires D pulses to operate, which gives for $N_3$ a value of $N_3 = D+2$.

## 6. PIPELINING

Because of the way we are implementing the different blocks, once the last value of a MFG cycle has been produced, we can start a new operation cycle (overlapping phase 3 with phase 1 and phase 2). It means a reduction on the overall operation cycle of D+2, the operation cycle can be expressed as:

$$N = b + 1 + \max(A_j) + M\,(R+1),$$

where b is the number of breakpoints in the MFC (normally 3 or 4).

The main limitation in this architecture is due to the sequential operation of the MAX circuit, which introduces the factor MR in the expression above. This is a consequence of having used only one MAX block in the microcontroller in order to minimize its area. However, if q MAX blocks are used the pipelining can be increased, dividing the factor MR by q. Hence, depending on the

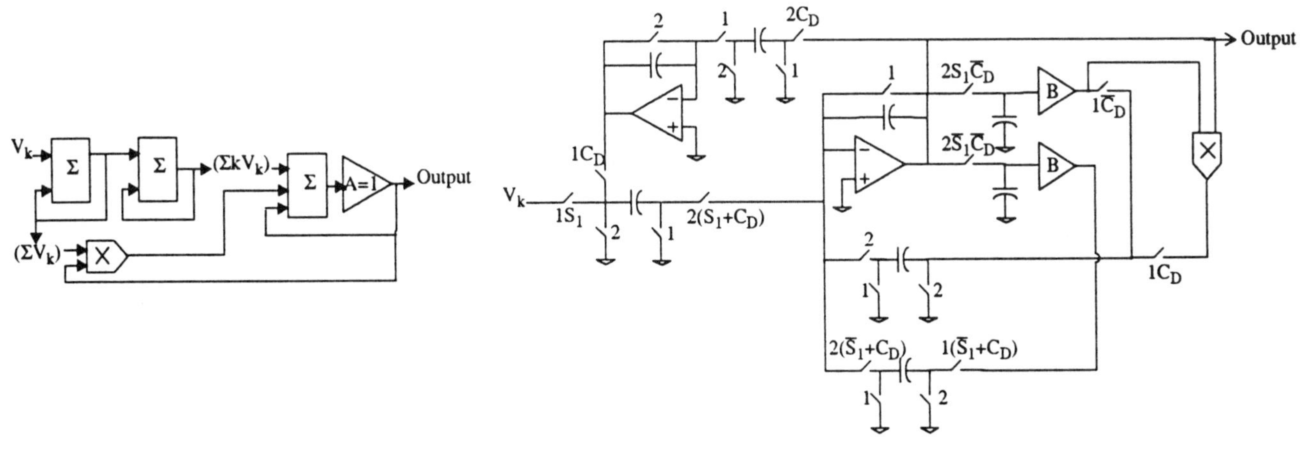


(a) (b)

Fig. 6: Alternative defuzzifier with cascaded summers: (a) Block diagram. (b) Circuit schematic.

applications, a compromise must be done between the number of MAX blocks (area) and the logic inference speed.

## 7. CONCLUDING REMARKS

In principle this architecture is slower than a parallel counterpart. However, since the area required for a parallel implementation is enormous (mainly because of the number of bus lines), solutions reported are based on connecting several (or even many) chips instead of a one-chip alternative. Then, the external interconnection delays bring forth a problem associated with higher delay time as compared with the ideal implementation in just one chip.

The aim of the proposed approach is to establish a trade-off between operational speed and silicon area occupation, but taking into account the value of actual delays when several chips must be connected. Thus, we can sacrifice a part of the internal speed (throughout a sequential operation) to be sure that many more rules can be implemented on-chip, this avoiding external connections other than I/O pins.

On the other hand, taking into account that SC circuits can operate at high frecuencies, the proposed microcontroller can be really competitive compared with digital implementations. In this sense, the main advantage of our system is its ability to deal with analog signals making possible the direct processing of membership values (what results in the elimination of A/D, D/A converters).

## REFERENCES


[1] L. A. Zadeh. *"Fuzzy Sets"*, Inform. and Control, 1965.

[2] T. Yamakawa, T. Miki and F. Ueno, *"The design and fabrication of the current mode fuzzy logic semi-custom IC in the standard CMOS IC technology"*, Proc. 15th IEEE Int. Symp. Multiple-Valued Logic, pp. 76-82, May 1985.

[3] T. Yamakawa and H. Kabuo, *"A Programable Fuzzifier Integrated Circuit: Synthesis, Design and Fabrication"*, Elsevier Science Pub. Comp. Inc., 1988.

[4] J. L. Huertas, A. Rodríguez-Vázquez, A. Rueda and L. O. Chua, *"Nonlinear Switched-Capacitors Networks: Basic Principles and Piecewise-Linear Design"*, IEEE Trans. Circuits and Systems, V. CAS-32, pp 305-319, 1985.

[5] A. Rodríguez-Vázquez, J. L. Huertas, A. Rueda, B. Pérez-Verdú and L. O. Chua, *"Chaos in Switched-Capacitors: Discrete Maps"* Proc. of the IEEE Vol. 75, pp 1090-1107, 1987.

[6] T. Yamakawa, *"High-Speed Fuzzy Controller Hardware System: The Mega FIPS Machine"*, Elsevier Science Pub. Comp. Inc., 1988.

[7] J. L. Huertas, S. Sánchez-Solano, A. Barriga, I. Baturone, *"Serial Architecture For Fuzzy Controllers: Hardware Implementation Using Analog/ Digital VLSI Techniques"*, 2nd. International Conference on Fuzzy Logic and Neural Networks, Iizuka, 1992.

[8] R. Gregorian and G. C. Temes, *"Analog MOS Integrated Circuits for Signal Processing"*, Wiley-Interscience Pub, 1986.

# Evaluation of Fuzzy Instructions in a RISC Processor

Hiroyuki Watanabe David Chen

Department of Computer Science
University of North Carolina
Chapel Hill, NC 27599-3175
TEL. (919) 962-1817
FAX. (919) 962-1799
Email: yuki@cs.unc.edu

**Abstract—We report the results of dynamic measurements which evaluate the effectiveness of an application specific microprocessor for fuzzy control and fuzzy information processing. We propose to specialize an architecture of microprocessor for fuzzy theoretic operations using quantitative techniques developed by designers of Reduced Instruction Set Computer (RISC). In particular, an introduction of specialized instructions is considered. Experimental results show that we can achieve as high as 2.5 speed-up of a program for fuzzy control by introducing two instructions, `min` and `max`†.**

## I. INTRODUCTION

In the past, we have designed fuzzy inference processors as application specific IC's (ASIC's) using full custom layout design [6, 7]. They are extremely fast but limited in generality. Many aspects of the design are limited or fixed. We have proposed to designing a fuzzy information processor as an application specific processor using a quantitative approach [9, 10]. The quantitative approach was developed by RISC designers [1].

As the first step, we consider very simple specialized instructions, `min` and `max`. These instruction implements *minimum (min)* and *maximum (max)* operations. They are proposed as intersection and union operations for fuzzy sets [11]. The `min` and `max` instructions are also useful in executing other fuzzy logical operations, *bounded product* and *bounded sum*. Then, we present the relative speeds of fuzzy inference among a MIPS R3000A microprocessor, a fictitious MIPS R3000A microprocessor with `min` and `max` instructions, and a full custom ASIC inference chip.

## II. MERIT OF FUZZY INSTRUCTION SET

One may question the advantage of designing a specially tuned microprocessor for fuzzy information processing when we can use a regular processor. At first, we show the advantage of a special purpose fuzzy instruction over using general purpose instructions. In fuzzy logic applications, one of the most common operations that we have to perform is the *min* operation, that is to take the minimum of two operands. The operation is expressed in the MIPS R3000A assembly language as in Figure 1. Here, the minimum of two numbers stored in registers `r2` and `r3` is moved to register `r1`. By introducing a new instruction `min` we can replace this sequence of 5 instructions by a single instruction:

```
min  r1,r2,r3  ;r1 gets the smaller of r2&r3.
```

Either 3 or 4 instructions are executed among the code in Figure 1 depending on whether register `r2` or register `r3` has the minimum value. That is, on average 3.5 instructions are replaced by a single `min` instruction. A reduction in executed instructions in this segment of code due to the new instruction is on average 71.4%.

The sequence shown in Figure 1 is intuitively clear. This is not, however, a sequence of codes actually generated by the MIPS optimizing C compiler. One of the main features of RISC microprocessors is their highly pipelined instruction execution [1]. Since the R3000A has five pipeline stages for instruction execution, its branch instructions are delayed by one cycle (*i.e.* delayed branch instruction) and its load instructions have one cycle delay before the loaded data can be used in following instructions. Its optimizing compiler attempts to fill such delay slot with useful instructions by moving instructions. If that is impossible, the compiler fills these slots with `nop` (no operation) instructions. Without code optimization, the code sequence in Figure 1 would actually look like that of Figure 2. Here, we added two `lbu` (load byte unsigned) instructions for loading data from memory. Two integers stored in memory are loaded into registers `r2` and `r3`. One `nop` is required in a load delay slot. Also, two `nop`'s are required to fill two branch delay slots.

† This material is based upon work supported by the National Science Foundation under Grant No. MIP-9103338. The Government has certain rights in this material.

```
        slt     r4,r2,r3        ; set r4 if [r2]<[r3] otherwise clear r4
        bne     r4,$49          ; branch to $49 if [r4]<> 0 ([r2]<[r3])
        move    r1,r3           ; move [r3] to r1
        j       $50             ; jump to $50
$49     move    r1,r2           ; move [r2] to r1
$50     <next instruction>
```

Figure 1: Most straightforward MIPS instructions for min operation.

```
        lbu     r2,0(r10)       ; load one data
        lbu     r3,0(r11)       ; load another data
        nop                     ; load-nop
        slt     r4,r2,r3        ; set r4 if [r2]<[r3] otherwise clear r4
        bne     r4,$49          ; branch to $49 if [r4]<>0 ([r2]<[r3])
        nop                     ; branch-nop
        move    r1,r3           ; move [r3] to r1
        j       $50             ; jump to $50
        nop                     ; branch-nop
$49     move    r1,r2           ; move [r2] to r1
$50     <next instruction>
```

Figure 2: Unoptimized MIPS instructions with `nop`'s

The sequence shown in Figure 1 is intuitively clear. This is not, however, a sequence of codes actually generated by the MIPS optimizing C compiler. One of the main features of RISC microprocessors is their highly pipelined instruction execution [1]. Since the R3000A has five pipeline stages for instruction execution, its branch instructions are delayed by one cycle (*i.e.* delayed branch instruction) and its load instructions have one cycle delay before the loaded data can be used in following instructions. Its optimizing compiler attempts to fill such delay slot with useful instructions by moving instructions. If that is impossible, the compiler fills these slots with `nop` (no operation) instructions. Without code optimization, the code sequence in Figure 1 would actually look like that of Figure 2. Here, we added two `lbu` (load byte unsigned) instructions for loading data from memory. Two integers stored in memory are loaded into registers `r2` and `r3`. One `nop` is required in a load delay slot. Also, two `nop`'s are required to fill two branch delay slots.

The actual code generated by the MIPS C compiler is shown in Figure 3. Two branch delay slots are filled with `move` instructions. These instructions that immediately follow the branch instructions are always executed before the actual execution of the branching operation. The compiler could not eliminate a `nop` instruction associated with the load delay. Note that the `j` (jump) instruction is unnecessary. The current version of the optimizing compiler failed to eliminate it, thus wasting one clock cycle.

## III. FUZZY INFERENCE SOFTWARE

We wrote a software simulator of a fuzzy inference board based on an ASIC fuzzy chip using the programming

```
        lbu     r2,0(r10)       ; load one data
        lbu     r3,0(r11)       ; load another data
        nop                     ; load-nop
        slt     r4,r2,r3        ; set r4 if [r2]<[r3] otherwise clear r4
        bne     r4,$50          ; if [r2]<[r3] branch to $50 (delayed branch)
        move    r1,r2           ;    after moving [r2] to r1
        j       $50             ; branch to $50 (delayed branch)
        move    r1,r3           ;    after moving [r3] to r1
$50     <next instruction>
```

Figure 3: Actual MIPS instructions for *min* operation.

language C. This software was delivered to Oak Ridge National Laboratory (ORNL). The software was used by ORNL researchers to develop control rules for autonomous robot navigation [4, 5] before we completed and delivered the inference board [8]. The format of a rule is as following:

If A and B and C and D
Then Do E, and Do F.

Each condition and consequent are represented by fuzzy sets. The universe of discourse of fuzzy set is represented by 64 elements. That is, the resolution in the x-direction is 64. The range between 0 and 1 for a membership function is represented by 16 levels. That is, each elements is represented by 4 bits. In the software simulator, data is not packed and a group of 4 bits occupies a byte of memory. Therefore, a single fuzzy set is represented by an array of size 64 of type character. This program performs the inference using the *max-min* compositional rule of inference performing *min* and *max* operations repeatedly. Incoming singleton input data are fuzzified. Both the if-part and the then-part of the computations are done by sweeping arrays of 64 elements. Due to the repeated usage of *min* and *max* operations, we included following macro definitions in the program header.

```
#define MIN(a,b)  ((a)<=(b)?(a):(b))
#define MAX(a,b)  ((a)>=(b)?(a):(b))
```

## IV. Method of Measurements

We used the above software to measure the run time characteristics of fuzzy inference on a MIPS R3000A (clock cycle 20 MHz) which is the central processor of the DEC workstation 5000/120. Then, we measured the performance of a modified program which predicts a run time if the MIPS R3000A had `min` and `max` instructions.

The proposed `min` and `max` instructions are three operand instructions. They are typical of the arithmetic and logical instructions in a RISC instruction set. In order to simulate these two instructions, we replaced the above macro definitions by the following.

```
#define MIN(a,b)  (a & b)
#define MAX(a,b)  (a | b)
```

Here, the *min* and *max* operations are substituted by bit-wise *and* and *or* operations. The result of the computation is not correct, however, what we are interested is the running time of the program execution. Here, an assumption is that the newly introduced instructions will run in the same cycle time with `and` and `or` instructions. This assumption is reasonable since the `min` and `max` instructions perform similar operations with comparison instructions such as the `slt` (set on less than) instruction. The only difference is that, in `min`/`max` instructions, the data in the source registers has

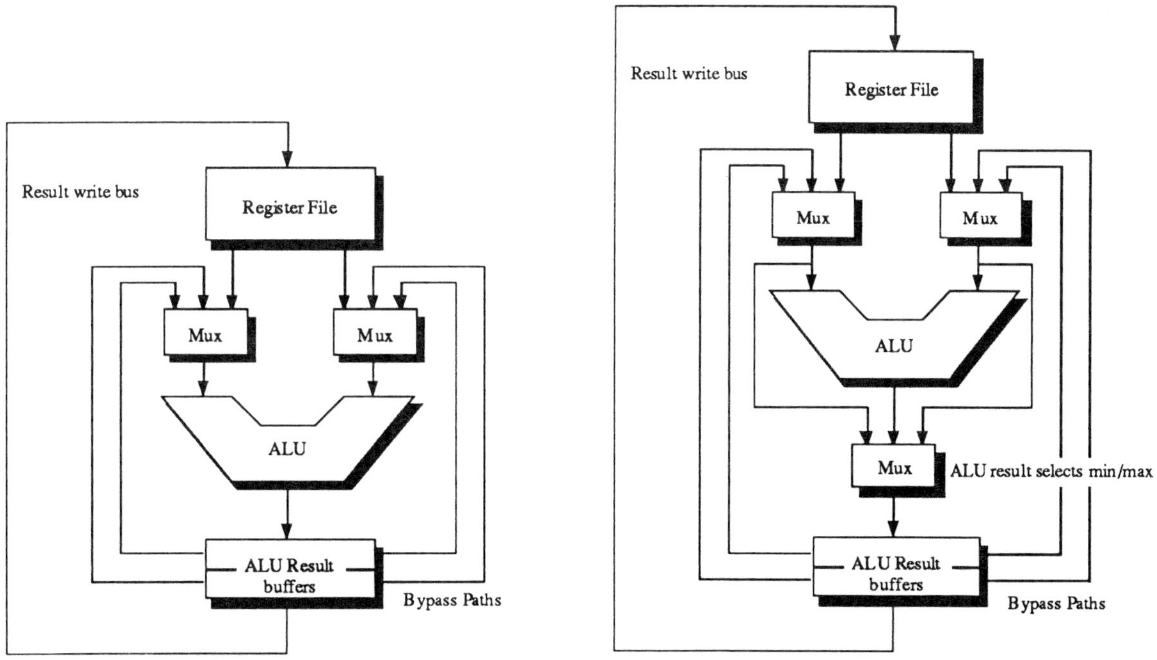


(a) Original datapath

(b) Addition of `min`/`max` instruction

Figure 4: RISC ALU datapath with bypass

TABLE I

RUN TIME STATISTICS FOR THE *MAX-MIN* INFERENCE

| | Regular | With min/max |
|---|---|---|
| No of instruction executed | $2{,}495 \times 10^6$ | $970 \times 10^6$ |
| No of nop's executed | $279 \times 10^6$ | $1.4 \times 10^6$ |
| Load nop's/load | 0.664 | 0.004 |
| Branch nop's/branch | 0.066 | 0.002 |
| Instruction/basic block | 3.9 | 18.0 |

to be moved to the destination register, whereas, in comparison instructions, the destination register has to be either set or cleared. A minor modification of a datapath in the ALU will accomplish this operation. Figure 4a shows a block diagram for a simplified RISC datapath with the capability of operand bypassing for efficient pipelining [1, p. 263]. We can implement the `min` and `max` instructions by introducing one more multiplexer as shown in Figure 4b.

We used 51 inference rules and executed the inference 6000 times using this inference software. Using profiling tools, *pixie*, *prof*, and *pixstats*, we collected run time statistics.

## V. RESULT OF MEASUREMENTS

Table 1 shows a dynamic instruction count and other key measurement figures. The number of instruction executed is reduced to 38.9% due to an introduction of the `min` and `max` instructions. This translates to a speed-up of 2.57. Note also the large reductions in the average numbers of `nop's` per load instruction and `nop's` per branch instruction. This reduction in load `nop's` is achieved by loop unrolling. As we have seen, data are loaded into registers before the *min* or *max* operations, and there is one `nop` instruction due to load delay. An optimizing compiler removed these load `nop` instructions by performing a loop unrolling. An example of the original code and the code resulting from loop unrolling is shown in Figure 5. The effect of loop unrolling is shown in the average number of instructions per basic block executed. This number increased from 3.9 to 18 due to the loop unrolling of inner loop blocks.

The dynamic code distribution of the executions are shown in Table 2. Here, the top 10 of the most frequently executed instructions are listed. In the actual R3000A codes, `lnop` (*i.e.* load nop) is the 6th frequently executed instruction. It is not in the list in `min/max` version due to effective filling of load delay slot. In the `min/max` version, entries for the `and` and `or` instructions represent the frequency of the `min` and `max` instructions. Though this program performs fuzzy logic inference repeatedly with the *max-min* method and nothing else, `min` and `max` instructions represent only 24.4% of the dynamic instruction count. The

```
lbu     r2,0(r10)   ; load one data
lbu     r3,0(r11)   ; load another data
nop                 ; load-nop
max     r4,r2,r3
sb      r4,0(r12)   ; store result
```

(a) Original code executed 64 times

```
lbu     r6,0(r10)   ; load 1st pair
lbu     r7,0(r11)
lbu     r9,1(r10)   ; load 2nd pair
lbu     r5,1(r11)
max     r8,r6,r7    ; get 1st max
max     r6,r9,r5    ; get 2nd max
sb      t8,0(r12)   ; store 1st max
sb      t6,1(r12)   ; store 2nd max
lbu     r6,2(r10)   ; load 3rd pair
lbu     r7,2(r11)
lbu     r9,3(r10)   ; load 4th pair
lbu     r5,3(r11)
max     r8,r6,r7    ; get max
max     r6,r9,r5    ; get max
sb      t8,2(r12)   ; store 3rd max
sb      t6,3(r12)   ; store 4th max
```

(b) Unrolled code executed 16 times

Figure 5: Loop unrolling

load/store instructions (*i.e.* `lbu`, `lw`, `sw` and `sb`) represents 53.6% of executed instructions. This is due to the fact that all fuzzy operations are performed on vectors, and they have to be loaded from and stored into the memory. Another frequent instruction, `addiu` (add immediate unsigned), is used for array address calculation.

TABLE II

DYNAMIC OPCODE DISTRIBUTION

| Regular | | With min/max | |
|---|---|---|---|
| addiu | 19.70% | lbu | 28.48% |
| lbu | 11.07% | addiu | 14.24% |
| addu | 9.95% | and(min) | 12.25% |
| bnez | 9.54% | or(max) | 12.12% |
| lnop | 9.52% | lw | 8.40% |
| j | 9.17% | sw | 8.40% |
| bne | 6.48% | sb | 8.32% |
| slt | 6.33% | bne | 4.47% |
| lw | 3.26% | addu | 1.58% |
| sw | 3.26% | lui | 0.52% |
| Total | 88.28% | Total | 98.78% |

TABLE III

RUN TIME STATISTICS FOR *MAX-BPD* INFERENCE

| | Regular | With max only | With max/bpd |
|---|---|---|---|
| No of instruction executed | $2{,}582 \times 10^6$ | $1{,}088 \times 10^6$ | $970 \times 10^6$ |
| No of nop's executed | $356 \times 10^6$ | $1.4 \times 10^6$ | $1.4 \times 10^6$ |
| Load nop's/load | 0.664 | 0.004 | 0.004 |
| Branch nop's/branch | 0.200 | 0.002 | 0.002 |
| Instruction/basic block | 4.0 | 20.3 | 18.0 |

## VI. BOUNDED SUM AND PRODUCT

Many operations other than *min* and *max* are proposed for fuzzy set intersection and union operations. We can perform other form of inference using some of these operators. For example, *bounded product (bdp)* and *bounded sum (bds)* are such a pair of operators. They are defined as :

$$\textit{bounded product}(x,y) = \max(0, x+y-1)$$
$$\textit{bounded sum}(x,y) = \min(1, x+y).$$

Using these operators, we can have alternative methods of inference [3]. One of such inference methods proposed by Mizumoto is the *max-bdp* method. In order to execute such inferences efficiently, we can support instructions, say `bdp` and `bds`, to perform these operations directly in hardware. This requires further modification of the ALU datapath. Usage of these operations must be frequent enough for such modification to be justifiable. An alternate method is to use the `min` and `max` instructions. In order to test the effectiveness of these two approaches, we performed experiments using the same inference software, but using a compositional rule of inference of *max-bdp* method. Table 3 shows the data measured by performing the inference 6000 times.

The first column shows the data for a regular MIPS R3000 processor. The *bounded product* operation requires more computation than the *min* operation. However, due to an efficiency of the optimizing compiler, an increase in a number of executed instructions is relatively small. The increase is 3.5% from the *max-min* method which is approximately $87 \times 10^6$ instructions. The second column shows the data for a R3000A with the `max` instruction. We can see that the `max` instruction is not only effective in implementing the *max* operation but also in implementing the *bounded product* operation. The resulting speed-up is 2.37, which is slightly lower than the speed-up of 2.57 for the *max-min* inference, but is still respectable. The final column shows the data for a R3000A with the `max` and `bpd` instructions. Here, the speed-up from the original processor is 2.66 but the incremental speed-up from the processor with only the `max` instruction is 1.12. The improvement by introducing the `bpd` instruction does not seem to warrant its cost of implementation.

We can conclude that `min` and `max` instructions are effective not only to implement the *min* and *max* operations but also to implement *bounded sum* and *bounded product* operations.

## VII. FURTHER IMPROVEMENTS

After the introduction of `min` and `max` instructions, there is a limited opportunity for further speed-up by additional improvements of the execution of fuzzy operations. For example, even if we were to design a super scalar architecture where one could execute `min`/`max` instructions in parallel with load/store instructions, the maximum possible speed-up would be 1.32 (=100/(100-24.4)). On the other hand, we have a good potential for speed-up by improving the load/store operations and memory cache performance. In particular, an introduction of the vector load/store operations can be effective.

## VIII. COMPARISON WITH ASIC CHIP

Table 4 shows the measured speed of the inference by a MIPS R3000A microprocessor and a fictitious MIPS R3000A microprocessor with `min` and `max` instructions. For comparison, the execution speed of a full-custom ASIC fuzzy chip, with a clock speed of 10MHz, is also shown [7]. This chip has a 3-staged pipeline and initiates a computation of new inference every 64 cycle. The first row is the

TABLE IV

INFERENCE TIME BY 51 RULES

| Time | MIPS R3000A | | ASIC |
|---|---|---|---|
| | Regular | With min/max | |
| 6000 inference | 125s | 49s | 0.0038s |
| 1 inference | 20.8ms | 8.2ms | 6.4μs |
| FLIPS | 48 | 122 | 156,250 |

computation time in seconds of 6000 inferences using 51 rules. The second row is the time required to perform a single inference. The last row is the fuzzy logical inferences per second (FLIPS) measured for each device.

There is a large gap in run time between the ASIC and software approaches even if we resort to a specialized fuzzy microprocessor. As for design time and cost, these two approaches represent two extremes. An ASIC approach is extremely expensive. It is, therefore, an important research topic to design a specialized computing architecture for fuzzy applications that falls between these two extremes both in run time and design time/cost. One such approach may be to use programmable devices such as an erasable programmable logic device (EPLD) or a field programmable gate array (FPGA) [2].

## IX. Conclusion

The experiments with a fuzzy inference program simulates the usage of an embedded processor. The embedded processor's main task is to control a device or a process. It usually runs a single or a limited number of programs from ROM (Read Only Memory) for this purpose.

Tuning a general purpose processor to create an embedded processor for fuzzy control is very effective. By introducing fuzzy logic specific instructions, `min` and `max`, some fuzzy control programs based on *max-min* inference run as much as 2.5 times faster than on a regular microprocessor. Also, these instructions can speed-up other fuzzy theoretic operations such as *bounded product* and *bounded sum*. Since a modification to a regular RISC core is modest, this approach is attractive.

## Reference


[1] Hennessy, J. L. and D. Patterson, *Computer Architecture: A Quantitative Approach*, Morgan–Kaufmann, 1990.

[2] Manzoul, M. A. and D. Jayabharathi, "Fuzzy Controller on FPGA Chip," *Proc. of IEEE International Conference on Fuzzy Systems*, pp. 1309–1316, March 1992.

[3] Mizumoto, M. "Comparison of Various Fuzzy Reasoning Methods," *Proc. of Second IFSA Congress*, pp. 2–7, July 1987.

[4] Pin, F. G., H. Watanabe, J. R. Symon, and R. S. Pattay, "Autonomous Navigation of a Mobile Robot Using Custom–Designed Qualitative Reasoning VLSI Chips and Boards," *Proc. of IEEE International Conference on Robotics and Automation*, pp. 123–128, May 1992.

[5] Pin, F. G., H. Watanabe, J. R. Symon, and R. S. Pattay, "Using Custom–Designed VLSI Fuzzy Inference Chips for the Autonomous Navigation of a Mobile Robot," *Proc. of IEEE/RSJ International Conference on Intelligent Robots and Systems*, pp. 790–795, July 1992.

[6] Togai, M. and H. Watanabe, "An Inference Engine for Real-time Approximate Reasoning: Toward an Expert on a Chip," *IEEE EXPERT*, Vol. 1, No. 3, pp. 55–62, August 1986.

[7] Watanabe, H., W. Dettloff and E. Yount "A VLSI Fuzzy Logic Inference Engine for Real–Time Process Control," *IEEE Journal of Solid–State Circuits*, Vol. 25, No. 2, pp. 376–382, April 1990.

[8] Watanabe, H., J. R. Symon, W. Dettloff and E. Yount ,"VLSI Fuzzy Chip and Inference Accelerator Board Systems," *Proc. of IEEE International Symposium on Multi–Valued Logic*, pp. 120–127, May 1991.

[9] Watanabe, H., "Some Consideration on Design of Fuzzy Information Processors –– From a Computer Architectural Point of View," *Proc. of International Fuzzy Engineering Symposium '91*, pp. 387–398, November 1991.

[10] Watanabe, H., "A RISC Approach to Design of Fuzzy Processor Architecture," *Proc. of IEEE International Conference on Fuzzy Systems*, pp. 431–440, March 1992.

[11] Zadeh, L. A., "Fuzzy set," Information and Control, Vol. 8 pp. 338–353, 1965.

# 7.5MFLIPS Fuzzy Microprocessor Using SIMD and Logic-in-Memory Structure

Mamoru SASAKI    Fumio UENO    Takahiro INOUE
Dept. of Electrical Engineering and Computer Science
Kumamoto University
Kumamoto, Kurokami 2-39-1, 860 Japan
(Tel.096-344-2111 Ext.3842)

## Abstract

*A fuzzy microprocessor, having SIMD and Logic-in-Memory structure, is developed using 1.2μm CMOS process. It can execute fuzzy inference for if-then fuzzy rules. The speed of inference including deffuzzification is 7.5MFLIPS, and it can process 960 rules and 16 input and output variables.*

## 1 Introduction

One of the most successful applications of the fuzzy theory[1],[2] is in the field of process control[3],[4]. In the applications, the models of the processes under control are often too complicated to be formulated mathematically. While, the fuzzy-logic rule-based system allow natural translations of domain knowledge into fuzzy rules. The inference procedure derives effective control actions using the fuzzy rules.

However, the computations required in the fuzzy inference procedure are usually too sophisticated to meet the real-time requirements of process controllers. The pioneering work of Togai and Watanabe marks the beginning of the hardware approach of implementing fuzzy-logic inference engine using digital circuits [5], [6]. Yamakawa adopts a different approach in that he used analog circuits [7].

In this paper, considering that the primary aim of hardware implementation is realization of high speed processing, we introduce effective control mechanism on parallel architecture, which can realize high speed fuzzy-logic inference. First, the fuzzy rules are reconstructed and the inference sequence for the reconstructed fuzzy rules is divided between if-part and then-part. Next, we introduce different two parallel architectures suitable for the if-part and then-part. In if-part, SIMD is introduced. In then-part, in order to effectively combine the results of all rules, bit serial max/min operations with multi-input are proposed and logic-in-memory suitable for implementing the bit serial max/min operations is introduced. Furthermore, in order to convert a fuzzy set being a final result to a certain value, a defuzzification operation ( Center of Area method ) is implemented with multi-operand parallel adders using Wallace tree.

## 2 Reconstruction of fuzzy rules

In the fuzzy inference sequence, the process of every fuzzy rule is independent each other. Considering parallel processing, we choose naturally one rule process as a parallel-processing unit. In fact, the previous fuzzy logic processor executes every rule in parallel [5], [6]. However, in case of many rules, it is impossible for restriction of hardware resources. So, we have to make the granularity of parallel process more large than one rule process.

The number of membership functions on one axis usually less than ten. Making use of this fact, we will reconstruct the fuzzy rules. The rules with the same membership function in then-part are gathered and a new rule is reconstructed by combining these rules. The process of one reconstructed rule is chosen as a parallel processing unit. In this case, the number of the processing elements can be equal to the number of the membership functions in then-part. Furthermore, the number of the operands of the max operation to combine the results of all rules can be also equal to the number. For example, the following rules are reconstructed as follows :

if $x_1$ is $A_{11}$ and $x_2$ is $A_{12}$ then $y$ is $B_1$
if $x_1$ is $A_{21}$ and $x_2$ is $A_{22}$ then $y$ is $B_2$
if $x_1$ is $A_{31}$ and $x_2$ is $A_{32}$ then $y$ is $B_2$
if $x_1$ is $A_{41}$ and $x_2$ is $A_{42}$ then $y$ is $B_1$

↓

if ($x_1$ is $A_{11}$ and $x_2$ is $A_{12}$) *or* ($x_1$ is $A_{41}$ and $x_2$ is $A_{42}$) then $y$ is $B_1$

if ($x_1$ is $A_{21}$ and $x_2$ is $A_{22}$) *or* ($x_1$ is $A_{31}$ and $x_2$ is $A_{32}$) then $y$ is $B_2$

Where $x_i (i = 1, 2)$ and $y$ are inputs and output, respectively. $A_{ji}(j = 1, \cdots, 4)$ and $B_k(k = 1, 2)$ are fuzzy subsets (membership functions).

Using max/min inference method, we can calculate firing grades of the rules, for example, in case that the inputs $x_1 = x_1'$ and $x_2 = x_2'$, a firing grade $w_1$ of the rule with $B_1$ is :

$$w_1 = \{A_{11}(x_1') \wedge A_{12}(x_2')\} \vee \{A_{41}(x_1')A_{42}(x_2')\} \quad (1)$$

Where $\wedge$ and $\vee$ are min and max operation, respectively. $A_{ji}(\cdot)$ is a grade of the membership function. Next, the inference result $B_k'$ of the rule with $B_k$ can be derived as follows.

$$B_k'(y) = w_k \wedge B_k(y) \quad (2)$$

Then, the final inference result $B'$ can be obtained by combining the result $B_k'$ of every rule.

$$B'(y) = \bigvee_k B_k'(y) = \bigvee_k \{w_k \wedge B_k(y)\} \quad (3)$$

When the certain value $y'$ is required, $y'$ can be calculated from the final result $B'(y)$ using the Center of Area method.

$$y' = \frac{\sum_y B'(y) \cdot y}{\sum_y B'(y)} \quad (4)$$

The inference sequence can be divided between if-part process (1), and then-part process (2), (3) and (4). Considering the if-part process, we understand that all firing grades can be calculated in parallel by using multiple processing elements. While, multi-input max operations are executed in then-part as mentioned (3) and it requires parallel processing on operation level. From the considerations, we introduce two parallel architectures suitable for if-part and then-part. Since same operations are executed to different data in if-part, SIMD is introduced for the if-part processor. While, logic-in-memory suitable for implementing the multi-input max operation is introduced for the then-part. The logic-in-memory can implement effectively multi-input operation by locating some logic circuits between memory elements.

# 3 An if-part processor

A structure of the if-part processor is shown in Fig.1. The function and feature of each block in Fig.1 are explained. A data memory can store the membership functions used in the if-part. The format of the membership function is that the number of elements is 64 ( 6 bits ) and the number of grade levels is 16 ( 4 bits ). Thus, one membership function can be represented with 256 bits. A structure of the data memory is shown in Fig.2. The data memory consist of four memory banks. Each memory bank can store 14 kinds of the membership functions. The labels in Fig.2 represent the kinds of the membership functions. The grade are read by 7-bit address, which consist of 1 bit being an identification of a bank and 6 bits being an identification of an element. All grade data of 14 membership functions are read on 56bits data bus. As shown in Fig.2, high-speed access time can be realized by using interlace method.

A rule memory is a local memory prepared for each processing element (refer Fig.1). The memory stores the labels representing the kinds of the membership functions. The labels correspond to $A_{ji}$ in the rule expressions and all labels in a rule are stored in a rule memory. In the processing element, the labels become select signals for obtaining the needed grade data. The labels are read sequentially by the instruction counter. A label consists of 4 bits and can specifis 14 membership functions and 2 NOPs for the max and min operations. Each rule memory is also divided between L-block and H-block, and high-speed access-time is realized by using the interlace method.

The instruction memory stores the instruction sequence for the processing elements, the input-interface block and the variable counter. The contents in the memories are read sequentially by the instruction counter. The instruction for the processing elements consists of min-end-flag and max-end-flag. The flags correspond to the timing signals to control the processing elements. The instruction for the input-interface block consists of an address of the input registers in INPUT I/F and a identification of the data memory bank storing the membership functions of the input. In a processing element (PE), first, the selector chooses the needed grade data from the 14 kinds grade data sent from the data memory. The grade data sets of every input is sequentially sent from the data memory and the selecting signal is sent from the rule memory. Next, the min circuit calculates sequentially each min-term in a reconstructed rule. Then, the max circuit executes sequentially max operation among all min-terms. A pipeline process is implemented by preparing the register between the min circuit and the max circuit. Finally, since bit serial operations are used in the then-part processor, the shift register converts the firing grades from parallel to serial and each single bit is outputted sequentially from MSB to LSB.

# 4 A then-part processor

The then-part processor gets the firing grades from the if-part processor and executes (2) and (3), and calculates the center of area shown in (4). The processor consists of a logic-in-memory executing (2) and (3), a defuzzifier circuit calculating the center of area, and a sequencer controlling the two blocks. First, let us consider parallel processing of (2) and (3).

The grade $B'_k(y)$ representing a result of a rule can be calculated by min operation between the grade $B_k(y)$ and the firing grade $w_k$. So, the process can be executed in parallel every element in the membership function. The grade $B(y)$ representing a final result can be calculated by max operation among all grade $B'_k(y)$. Here, a structure, in which min units are put in the form of matrix as shown in Fig.3, are introduced. In Fig.3, a row corresponds to one membership function in then-part. To broadcast the firing grade $w_k$, inputs of all min units on a row are connected with an input-bus. The firing grades of all rules are given to all input-buses at the same time and all min units execute (2) in parallel. Then, every result $B'_k(y)$ is outputted from every min unit. The outputs of the min units are connected to the multi-input max unit every column. Since there are the grades of same element on a column, the max unit can execute (3). Although the structure can realize massive scale parallel processing, many units are required. So, the units have to be constructed as simply as possible for restriction of hardware resources.

To overcome the problem, bit-serial operation is introduced. Bit serial max and min operations are explained. For example, $C = \max(A, B)$ is calculated, where $A, B, C$ are $n$-bit binary numbers, $a_i, b_i, c_i$ represent $i$th bits of $A, B, C$, respectively. $c_i$ can be calculated using the following iterative equations.

$$\begin{aligned} c_i &= (Qa_i \cdot a_i) + (Qb_i \cdot b_i) &(5)\\ Qa_{i-1} &= Qa_i \cdot (\overline{c_i} + a_i) &(6)\\ Qb_{i-1} &= Qb_i \cdot (\overline{c_i} + b_i) &(7) \end{aligned}$$

Where $\cdot$ and $+$ are AND and OR operators of binary logic, respectively. $\bar{\cdot}$ is NOT operation. $Qa_i$ and $Qb_i$ are status variables keeping comparison results. $C = \max(A, B)$ can be calculated by iterating the equations from $i = n-1$ to 0 with initial value $Qa_{n-1} = Qb_{n-1} = 1$. Iterative equations for the min operation are dual equations of the max operation as follows.

$$\begin{aligned} c_i &= (Qa_i + a_i) \cdot (Qb_i + b_i) &(8)\\ Qa_{i-1} &= Qa_i + (\overline{c_i} \cdot a_i) &(9)\\ Qb_{i-1} &= Qb_i + (\overline{c_i} \cdot b_i) &(10) \end{aligned}$$

Where initial value $Qa_{n-1} = Qb_{n-1} = 0$.
The iterative equations for the max/min operations can be easily expanded to multi-input case. For example, $C = \max(A1, A2, \cdots, Aj, \cdots, Am)$ is calculated, where $A1, \cdots, Aj, \cdots, Am$, and $C$ are the $n$-bit binary numbers, $aj_i$ and $c_i$ represent the $i$th bits of $Aj$ and $C$, respectively. $C = \max(A1, \cdots, Aj, \cdots, Am)$ can be calculated by iterating the following equations from $i = n-1$ to 0 with initial value $Qaj_{n-1} = 1$ $(j = 1, 2, \cdots, m)$.

$$\begin{aligned} c_i &= (Qa1_i \cdot a1_i) + \cdots + (Qam_i \cdot am_i)\\ &= c1_i + c2_i + \cdots + cm_i &(11)\\ cj_i &= Qaj_i \cdot aj_i &(12)\\ Qaj_{i-1} &= Qaj_i \cdot (\overline{c_i} + aj_i) &(13) \end{aligned}$$

A 2-input min circuit for the bit-serial operation can consists of some logic circuits and two flip-flops storing the status variables. Next, we consider a multiple inputs max circuit for bit-serial operation. The circuit can be implemented by referring the iterative equations (11), (12) and (13). Here, let us note that (12) and (13) can be locally calculated every element. Hence, the equation executed among multiple elements is only (11) and the operation is just multi-input OR operation of binary logic. The multi-input max circuit can be constructed with preprocess circuits operating locally (12) and (13), and one multi-input OR gate. The preprocess circuit consists of some logic circuits and one flip-flop storing the status variable.

A structure of the logic-in-memory is shown in Fig.4. A processing element (PE) consists of 4bit data Latch, selecter, 2-input min unit and the preprocess unit for multi-input max operation. The data Latch stores a grade data of a element. The grade data is read sequentially from MSB by the selector. In Fig.4, the outputs of the processing elements corresponding to $cj_i$ in (11) are connected with multi-input OR gates.

# 5 Defuzzification circuit

In this section, a defuzzifier circuit operating center of area method is explained. The feature of the center of area method is that there are additions with many operands in both the numerator and the denominator as shown in (4). Although multiplications between B'(y) and element order y are executed in the numerator, the multiplication can be implemented by adding the partial products generated by bit shifting. Because element order y is a constant, amounts of shifting and number of partial products for every y can be predetected and the shifting can be implemented by wiring. So, the calculation of both the numerator and the denominator can be implemented with multi-operand adders. In this case, however, the number of operands becomes large. One of the effective implementation of the adders with many operands is using Wallace-tree, which is applied to high-speed multiplier.

A numerator calculation unit and a denominator calculation unit are implemented using the Wallace-tree. Here, we exchange calculation order in numerator. To understand easily, a case that $0 \leq y \leq 7$ is considered. As shown in the following equations, operands are classified among some groups with same shifting amount.

$$\begin{aligned}\sum_y B'(y) \cdot y = \\ &\{B'(1) + B'(3) + B'(5) + B'(7)\} \cdot 1 + \\ &\{B'(2) + B'(3) + B'(6) + B'(7)\} \cdot 2 + \\ &\{B'(4) + B'(5) + B'(6) + B'(7)\} \cdot 4 \qquad (14)\end{aligned}$$

The operands are reduced every group using Wallace-tree and the reduced two operands are shifted. Then, the shifted operands of all groups are reduced to two operands using Wallace-tree again and the two operands are added by a typical adder. A defuzzification circuit consists of the numerator calculation unit, the denominator calculation unit and one divider.

# 6 VLSI fabrication and system configurations

The explained two processors have been fabricated as VLSI chips. $1.2\mu m$ n-well 2 metal CMOS process and standard cell design method were used. The performances are shown in Table 1.

A system configuration using the chips is shown in Fig.5. In Fig.5, a if-part chip works as master and some then-part chips work as slaves. The then-part chips get the trigger from the if-part chip and start the consequent inference sequence. Identification numbers are given to the then-part chips, and the then-part chips can detect their data and the trigger by comparing their identification numbers with a identification outputted from the if-part chip. The identification is the content of the output variable counter in the if-part chip.

Since a rule memory is constructed with 128 wards and the number of the rule memories is 15, 128 × 15 = 1920 antecedents (membership functions) can be specified by using all rule memories. Hence, the number of rules (typical rules ), which can be executed by the processor, is :

$$Number\ of\ rules = [1920/(i \cdot j)] \qquad (15)$$

where i and j are the number of inputs and outputs, respectively. $[\cdot]$ is gaussian operation.

Since the then-part chip can execute the consequent inference including defuzzification every 4 clock cycles and the clock frequency is 30MHz, the inference speed of the then-part chip is 7.5 MFLIPS(Fuzzy Logic Inference Per second). While, the inference speed of the if-part chip is dependent on the numbers of inputs, outputs and rules. First, let us consider the dependence on the numbers of inputs and outputs. When the number of inputs and outputs are $i$ and $j$ respectively, $i \cdot j$ clock cycles are needed at least for the process. So, the inference speed is $30/(i \cdot j)$ MFLIPS in case of 30 MHz clock frequency. Next, let us consider the dependence on the number of rules. Since each processing element executes one antecedent per one clock cycle, 30M (clock freacency) × 15 (the number of processing elements) = 450M antecedents can be executed per second. So, since $i \cdot j \cdot k$ antecedents are processed in case of $i$-input, $j$-output and $k$-rule, the inference speed is $450/(i \cdot j \cdot k)$ MFLIPS. Therefore, the inference speed of the system shown in Fig.5 can be expressed as follows:

$$Inference\ speed = \\ min\{7.5, 30/(i \cdot j), 450/(i \cdot j \cdot k)\} MFLIPS \qquad (16)$$

In case of one output, a relationship between the number of rules and the inference speed is described in Fig.6 with parameters being the number of inputs. A bold line in Fig.6 presents the performance of the previous fuzzy logic processor executing parallel processing every rule [6]. Under the number of rules which can be executed by the previous processor, 12.9 times higher performance can be realized in vertue of the high performance of the then-part chip and the flexiblity of the rule format in the if-part chip.

Next, an other system configuration, that is multiple if-part-chip configuration, is shown in Fig.7. The bit-serial max unit shown in Fig.7 can be easily implemented using PLA (Programable Logic Array). The inference speed can be independent of the number of rules using this configuration as shown dotted lines in Fig.6. The number of if-part chips and the inference speed is expressed as follows:

$$Number\ of\ if\ part\ processors = [k/15] + 1 \quad (17)$$

$$Inference\ speed = min\{7.5, 30/(i \cdot j)\}\ MFLIPS \quad (18)$$

where i, j and k are the numbers of inputs, outputs and rules, respectively.
A relationship between the number of if-part chips and the number of rules processed per second is described in Fig.8 with parameters being the number of inputs. As shown in Fig.8, the performances increase in proportion to the number of if-part chips without the overhead for the comunications.

# 7 Conclusions

Two fuzzy microprocessors have been developed as VLSI chips, one is the if-part processor with SIMD architecture and the other is the then-part processor with logic-in-memory. The features of the processors are ;

1. Maximum inference speed including defuzzification is 7.5MFLIPS (Mega Fuzzy Logic Inference Per Second). This inference speed is 12.9 times higher than the previous VLSI implementations.

2. Many fuzzy rules (960 rules), input variables (16 variables) and output variables (16 variables) can be processed. Furthermore, the rule format can be easily changed by rewriting the instructions stored in the memory.

3. The processors require no external memory since the knowledge-base (If-Then rules) can be stored in the internal memories.

## Acknowledgements

The authors wish to thank the electronic devices group of the Oki electric Industry corporation, especially T. Katashiro, for fabricating the circuits.

# References

[1] L.A.Zadeh, "Fuzzy sets," *Inform. Contr.*, vol.8, pp.338-358, 1965.

[2] L.A.Zadeh, "Outline of new approach to the analysis of complex systems and decision processes," *IEEE Trans. Syst., Man., Cybern.*, vol.SMC-3, pp.28-44, Jan. 1973.

[3] L.P.Holmbalad and J.J.Ostergaard, "Control of a cement kiln by fuzzy logic," in *Fuzzy Information and Decision Processes*. Amsterdam, The Netherlands: North-Holland, 1982, pp.389-399.

[4] S.Yasunobu, S.Miyamoto and H.Ihara,"A Predictive fuzzy control for automatic train operation," *Syst.Contr.Japan*, vol.28, pp.605-613, Oct. 1984

[5] M.Togai and S.Chiu, "A fuzzy chip and a fuzzy inference accelerator for real-time approximate reasoning," in *Proc. 17th IEEE Int. Symp. Multiple-Valued Logic*, May 1987, pp.25-29.

[6] H.Watanabe, W.D.Dettloff and K.E.Yount, "A VLSI fuzzy logic controller with reconfigurable, cascadable architecture," *IEEE J.Solid-State Circuits*, vol.SC-25, pp.376-382, Apr. 1990.

[7] T.Yamakawa, "Fuzzy microprocessors rule chip and defuzzifier chip," in *Proc. Int. Workshop on Fuzzy Syst. Appl.*, Aug. 1988, pp.51-52.

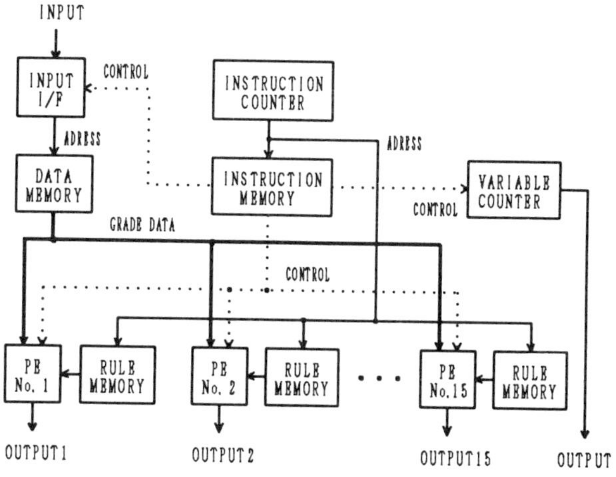


Fig.1. A structure of the if-part processor.

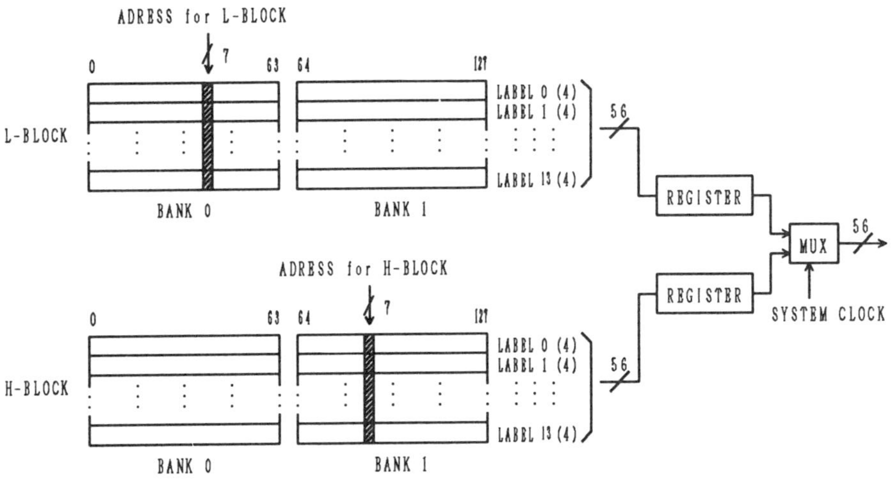


Fig.2. A structure of the data memory.

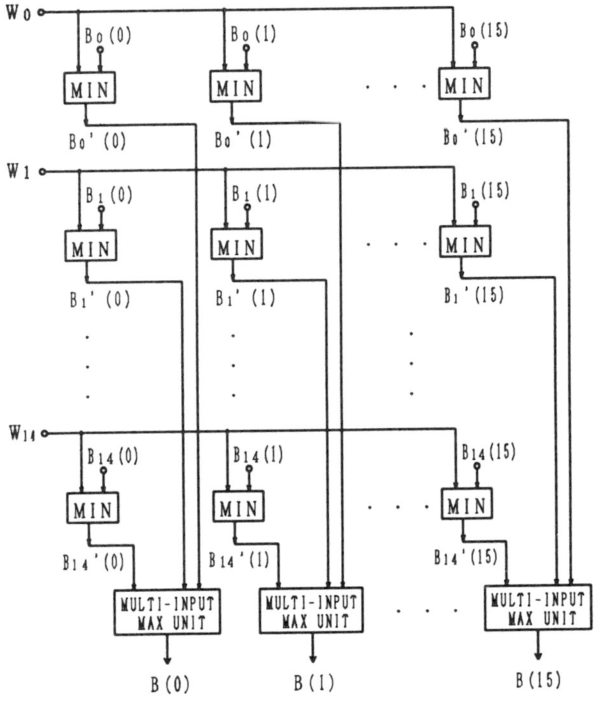


Fig.3. A matrix structure.

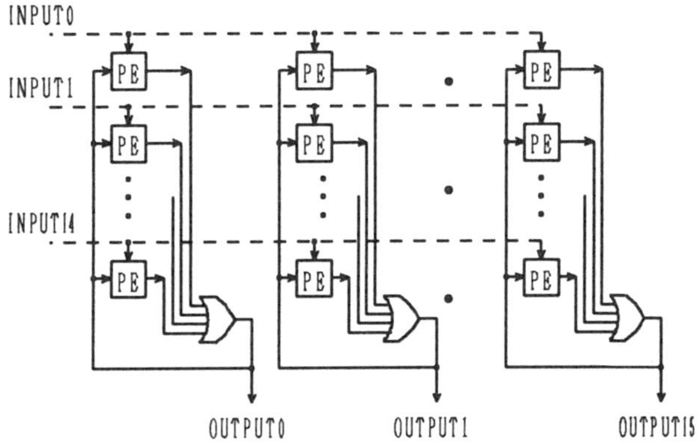


Fig.4. A structure of the logic-in-memory.

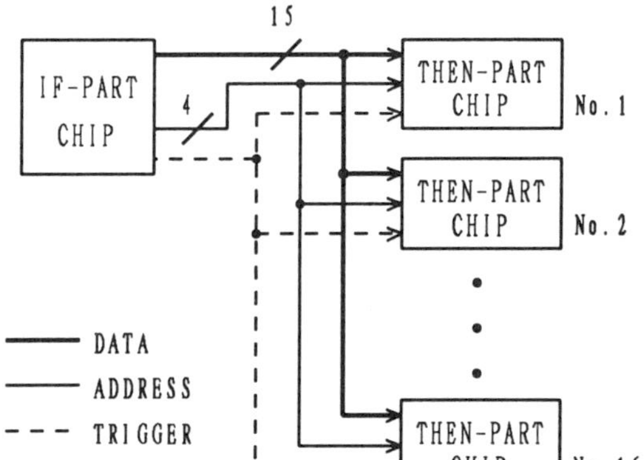


Fig.5. A system configuration (1).

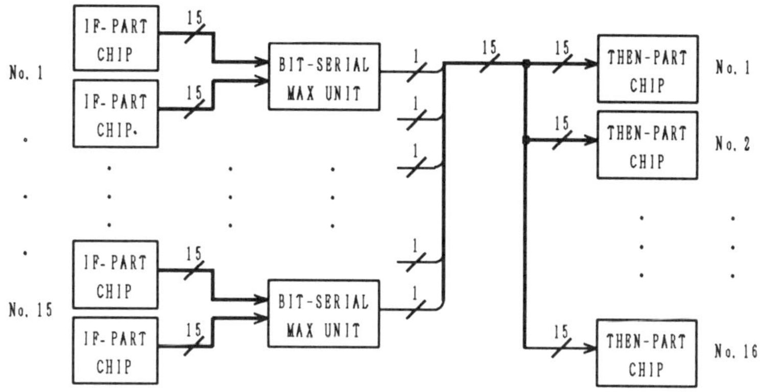


Fig.7. A system configuration (2).

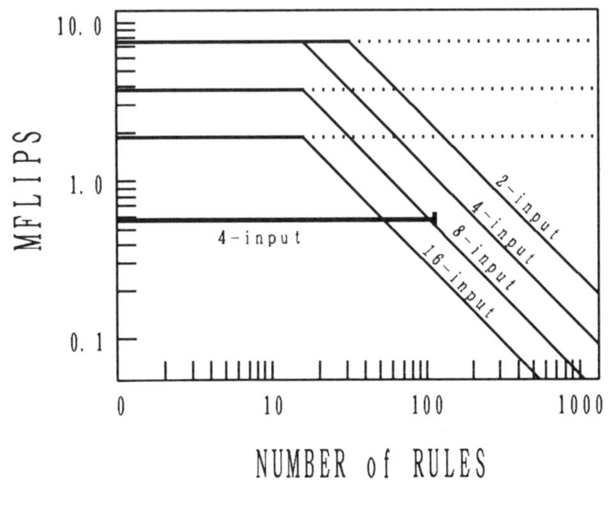


Fig.6. Number of rules vs inference speed.

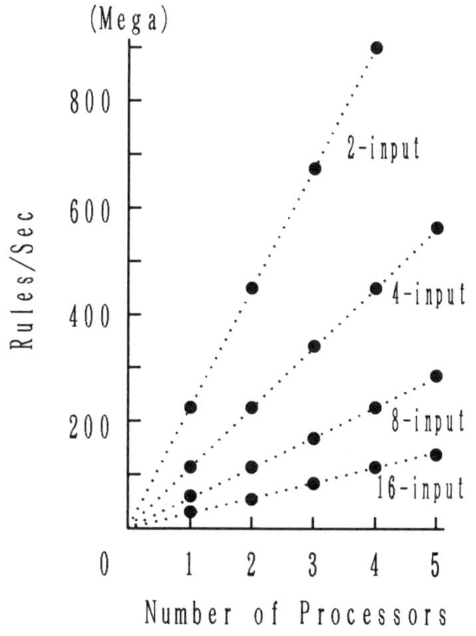


Fig.8. Number of processors vs rules/sec.

TABLE 1 Performances.

| | IF-PART Chip | THEN-PART Chip | |
|---|---|---|---|
| | | MODE 1 | MODE 2 |
| Clock Frequency | 30MHz | | |
| Inference Speed | 7.5M FLIPS (max) | | |
| Number of Rules | 960 (max) | | |
| Power Supply | 4.5 - 5.5 v | | |
| Power | 560 mW | | |
| Interface | TTL compatible | | |
| Package Type | 132 pins Ceramic PGA | | |
| Resolutions of Grade | 4 bits | | |
| Resolutions of Variables | 6 bits (input) | 8 bits (output) | |
| Number of Gates | 13 0k. gates | 29 2k. gates | |
| Number of RAMs | 18.7k bits | ——— | |
| Die Size (mm) | 13.79 x 13.72 | 13.65 x 13.65 | |
| Number of Pins | 93 | 95 | |
| Number of Variables | 16 (input) | 1 (output) | 2 (output) |
| Number of Labels | 14 x 4 | 15 | 15 x 2 |
| Number of Elements | 64 | 32 | 16 |

# Hardware Implementation of Fuzzy Filtering

**Takeshi Yamakawa*, Eiji Uchino*, Tsutomu Miki* and Shin Nakamura**
Kyushu Institute of Technology
Iizuka, Fukuoka 820, Japan
Phone : +81-948-29-7712 Fax : +81-948-29-7742
E-mail : yamakawa@ces.kyutech.ac.jp

**Key words :** fuzzy filtering, fuzzy inference, interpolation, smoothing, noise elimination

*Abstract*

This paper describes a simple interpolation algorithm for noisy signal data by using fuzzy inference and its hardware inplementation. Concretely, with the fact in mind that even if the signal data is disturbed by noise, the rough sketch of its true signal pattern can be generally made, a new interpolation algorithm based on fuzzy logic is proposed. The effectiveness of the proposed method is verified by computer simulations and the method is implemented by a hybrid electronic circuit on a breadborad.

## 1. INTRODUCTION

In an actual physical system, a fluctuation pattern of an objective signal is always disturbed partially, or sometimes almost completely, by an additive noise. It is essentially needed to reduce the effect of noise from the noisy data in order to evaluate the true fluctuation pattern of the objective signal.

Further, in addition to a non-stationary fluctuation of its noise, which is in many cases of arbitrary distribution type, the objective signal itself often fluctuates non-stationarily, and thus signal-to-noise ratio (SNR) changes momentarily. Kalman filter [1,2] is well-known in this field of noise reduction, however, in the application of this Kalman filter, the dynamics of its objective signal must be known or formulated somehow in advance, and linear Gaussian properties of the dynamics are strongly required. Therefore, in order to overcome these unrealistic idealized conditions, many works have been done. For example, in references [3,4], some new digital filters were proposed for a very wide general case and they showed indeed very good estimation performance, however, the algorithms are very complicated.

In this paper, we propose a simple algorithm of interpolation for noisy signal data by using fuzzy inference and it results eventually in the reduction of noise from the noisy signal data with no knowledge of its dynamics. This algorithm is realized in the electronic circuit in order to reduce the noise in the objective signal in real time.

## 2. PROBLEM STATEMENT

Consider the following observation mechanism:

$$y(t)=x(t)+v(t), \qquad (1)$$

where $x(t)$ is a time series signal of concern, $v(t)$ is an additive noise whose statistics are unknown, and $y(t)$ is a noisy observation disturbed by noise $v(t)$. The problem here is to estimate, based on fuzzy inference, the signal pattern of $x(t)$ by using the successive observation of $y(t)$.

## 3. INTERPOLATION BY FUZZY INFERENCE

The interpolation algorithm proposed here is fundamentally based on the fact that even if the signal pattern is disturbed by noise, one can somehow grasp the rough outline of its noiseless signal pattern by intuition. In that moment, in order to get an outline of this objective signal, we

* The authors are also with the Fuzzy Logic Systems Institute (FLSI), Yokota, Iizuka 820, Japan.

enclose the noisy signal pattern by the rectangles in which the objective signal pattern is involved. In the theoretical consideration, discussions are made for the sampled data for the simplicity.

The interpolation algorithm is briefly described as follows:

[i] Enclose the observed noisy signal pattern by the rectangles (which is refered to as "patch" here) so that the maximum and minimum values of its pattern are involved at top and bottom edges of this patch, respectively. Fig.1 shows an example of how to enclose the observed noisy signal pattern by rectangles. In this example, one rectangle is composed of 3 sampled points. Let $P_i$ be this patch and $G_i$ be the *center of gravity* of the patch $P_i$ (see Fig.2).

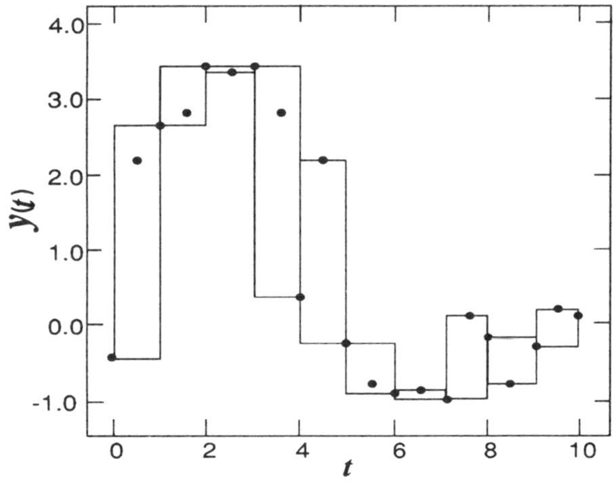


Fig.1. Sampled noisy signal and patches enclosing three sampled data for each. ●:Sample point of noisy observation; □: Patch.

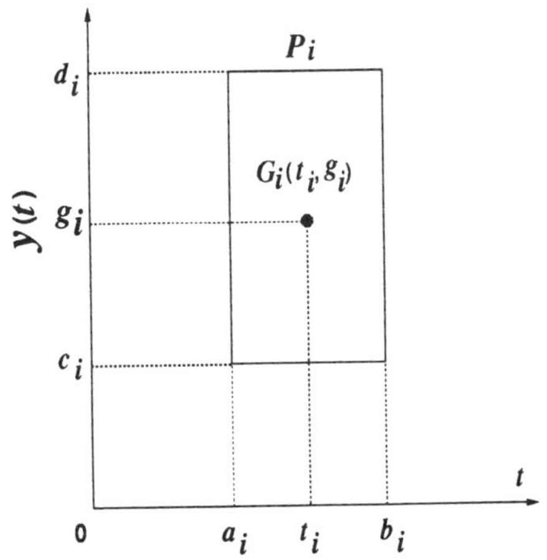


Fig.2. Parameters for a patch $P_i$.

[ii] Let $F_i(t)$ be the straight line which passes through the successive points of $G_i$ and $G_{i+1}$ (see Fig.3). This shows a slow and steady drift of signal at $t_i \leq t < t_{i+1}$, so that this line is refered to as a *main-line*.

[iii] $f_i(t)$, is determined in accordance with the positional relation among the successive three patches of $P_{i-1}$, $P_i$, and $P_{i+1}$ as follows:

(a) when $g_{i-1} < g_i < g_{i+1}$:

$$f_i(t) = \frac{d_i - c_i}{b_i - a_i}(t - t_i) + g_i, \tag{2}$$

(b) when $g_{i-1} > g_i > g_{i+1}$:

$$f_i(t) = -\frac{d_i - c_i}{b_i - a_i}(t - t_i) + g_i, \tag{3}$$

(c) otherwise:

$$f_i(t) = g_i, \tag{4}$$

where $a_i$, $b_i$, $c_i$, $d_i$, and $g_i$ are the parameters which characterize the patch $P_i$. These lines imply

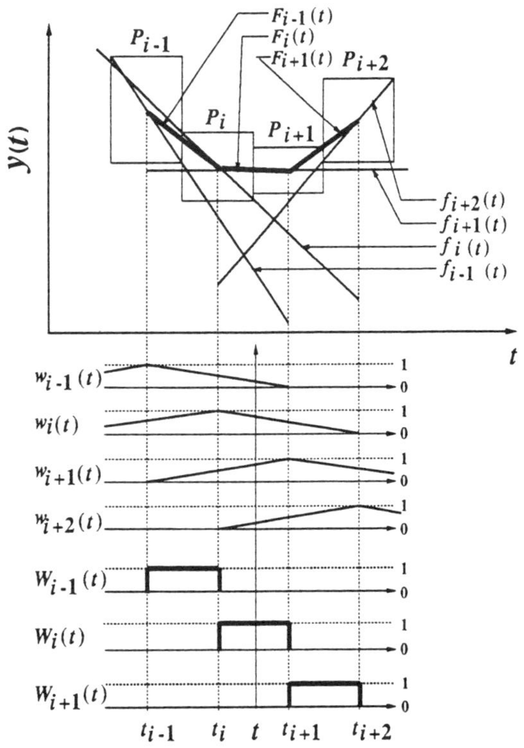


Fig.3. Membership functions assigned to main- and sub-lines.

the trend of fluctuation from the previous to the next stage and make a subsidary contribution to the smooth interpolation. So that thsese lines are refered to as *sub-lines*.

[iv] Assign a membership function to each of these main- and sub-lines. $W_i(t)$ and $w_i(t)$ are the membership functions assigned to $F_i(t)$ and $f_i(t)$ respectively (see Fig.3).

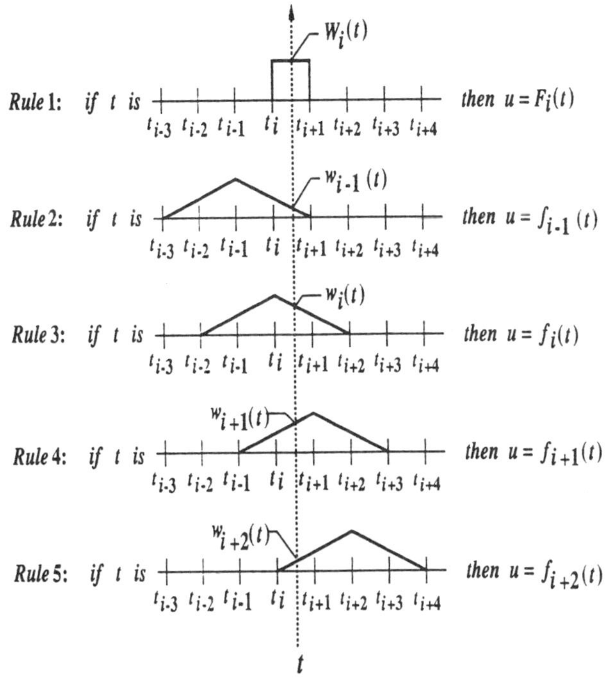


Fig.4. Rules for interpolation.

[v] Then, the interpolation of $x(t)$ in the interval $t_i \le t < t_{i+1}$ is accomplished according to the rules shown in Fig.4 and thus the interpolated value of $x(t)$ is given by [5, 6]:

$$\hat{x}(t) = \frac{w_{i-1}(t)f_{i-1}(t)+w_i(t)f_i(t)+2F_i(t)+w_{i+1}(t)f_{i+1}(t)+w_{i+2}(t)f_{i+2}(t)}{w_{i-1}(t)+w_i(t)+2+w_{i+1}(t)+w_{i+2}(t)}. \tag{5}$$

where

$$w_{i-1}(t) + w_i(t) + w_{i+1}(t) + w_{1+2}(t) = 2 \tag{6}$$

is always valid, so that the weight for the main-line $W_i(t)$ should also be 2. Thus the denominator of Eq.(6) is always 4 and the following equation is obtained.

$$\hat{x}(t) = \frac{1}{4}\{w_{i-1}(t)f_{i-1}(t) + w_i(t)f_i(t) + 2F_i(t) + w_{i+1}(t)f_{i+1}(t) + w_{i+2}(t)f_{i+2}(t)\} \tag{7}$$

## 4. SIMULATION EXPERIMENTS

In order to confirm the validity of the proposed method, computer simulation is made for the following model:

$$y(t)=x(t)+v(t), \tag{8}$$

$$x(t)=sin\omega_1 t + 2sin\omega_2 t + sin\omega_3 t \tag{9}$$

with $\omega_1$=0.32, $\omega_2$=0.64, and $\omega_3$=0.96. $v(t)$is a non-stationary noise which was generated artificially as:

$$v(t)=N(v(t);\, m_i,\sigma_i^2), \quad t_i \le t < t_{i+1}, \tag{10}$$

where $N(x;m,\sigma^2) \equiv (1/\sqrt{2\pi}\sigma)exp\{-(x-m)^2/2\sigma^2\}$ is a Gaussian distribution with mean $m$ and variance $\sigma^2$. The parameters are set as $m_1$=0, $\sigma_1^2$=1.0 at $0 \le t < 1$; $m_2$=0, $\sigma_2^2$=0.4 at $1 \le t < 3$; $m_3$=0, $\sigma_3^2$=1.0 at $3 \le t < 5$; and so on. In this simulation experiment, sampling interval of observation is $T$=0.5(sec) and the patch is composed of 3 sampled points. The simulation results are shown in Fig.5. The curve interpolated by the proposed method is catching well the true fluctuation curve of $x(t)$.

The interpolation results by using the conventional methods [7], e.g., Lagrange's method and a cubic spline method are also shown in Fig.6. This is only to show that the conventional interpolation methods can not be applied directly to this kind of noisy data.

## 5. APPLICATION TO ACTUAL DATA

To experimentally confirm the effectiveness of the proposed method, application is made to the actual acoustic data. Fig.7 shows the interpolated results for a road traffic noise disturbed by the

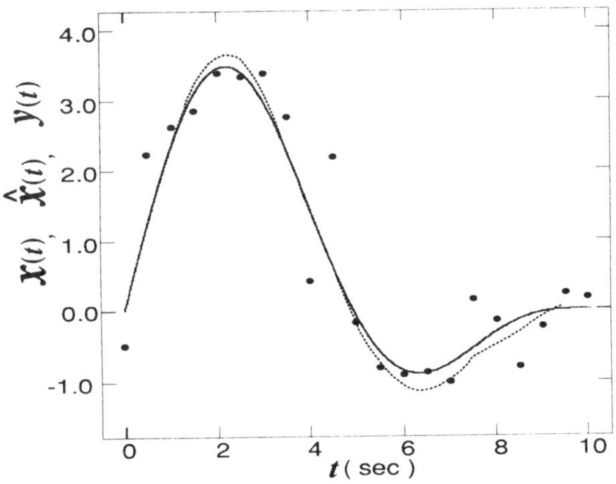


Fig.5. Interpolated results for the simulation model. ——— : True curve. ------- : Interpolated curve by the proposed method. ● : Sampled points of the noisy observation.

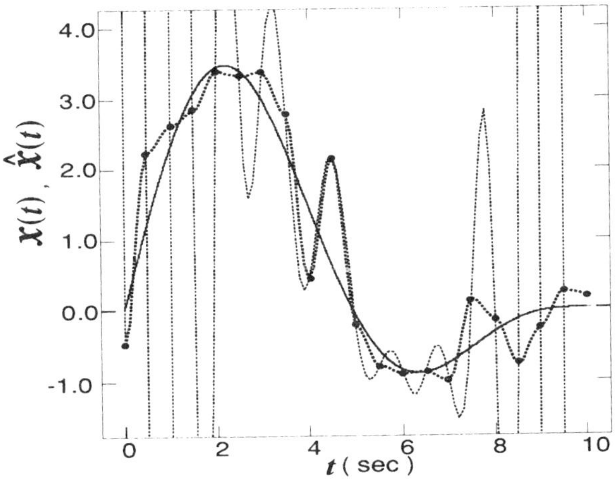


Fig.6. Interpolated results by using the conventional methods. ——— : True curve. ------- : Interpolated curve by the Lagrange's method. ▪▪▪▪▪▪▪▪▪▪▪▪: Interpolated curve by a cubic spline method; ● : Sample point of noisy observation.

non-stationary noise generated artificially. The sampling interval is $T$=0.2(sec). The rectangular region is composed of 3 sampled data.

Application is also made to a voice signal. Fig.8 shows a noisy signal pattern of voice /u/ sampled at 50KHz, which is disturbed by the non-stationary noise artificially generated. Fig.9 shows the interpolated results by the proposed method. The rectangular region is composed of 10 sampled data.

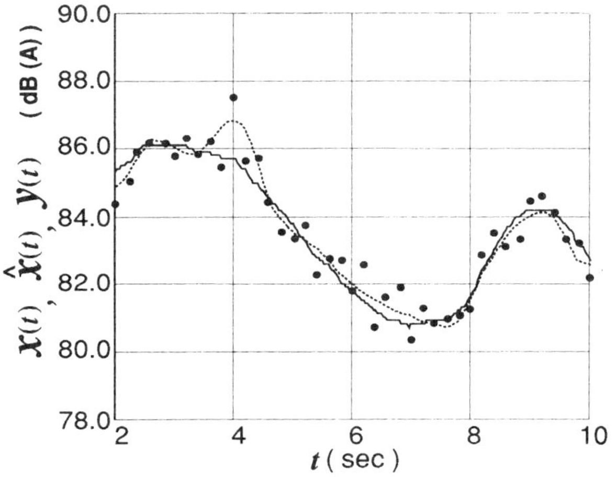


Fig.7. Interpolated results for a road traffic noise disturbed by artificial non-stationary noise. ——— : Road traffic noise. ------- : Interpolated curve by the proposed method. ● : Sampled points of the noisy observation.

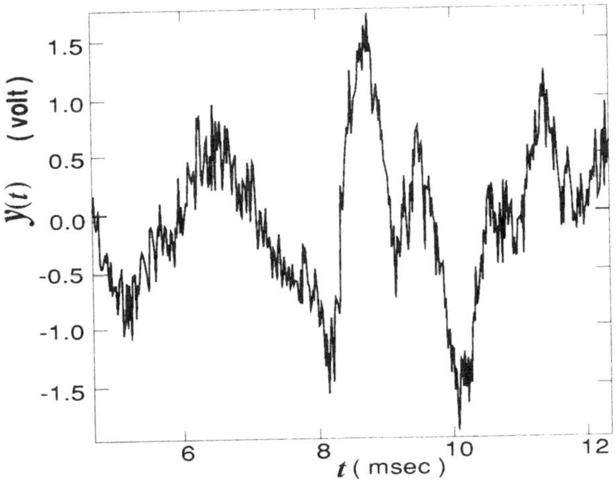


Fig.8. Noisy signal pattern of voice /u/.

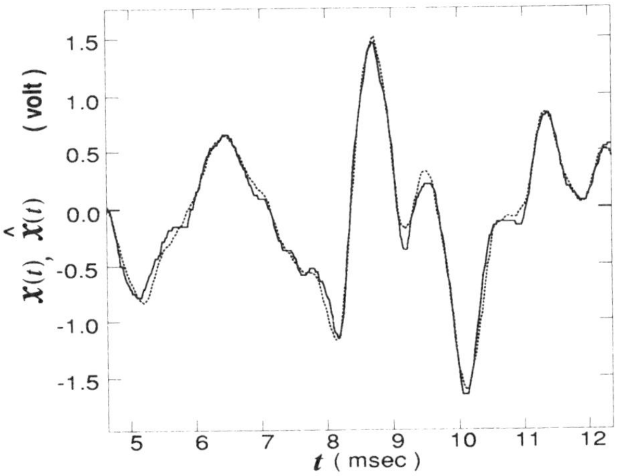


Fig.9. Interpolated results for the noisy voice signal. ——— : Voice signal of /u/.

------- : Interpolation by the proposed method.

Next, we apply this method to the restoration of the partially loosed signal data. Fig.10 shows the partially loosed voice signal of /u/. Fig.11 shows the results interpolated by the proposed method. The interpolated curve is catching well the true fluctuation pattern of voice signal.

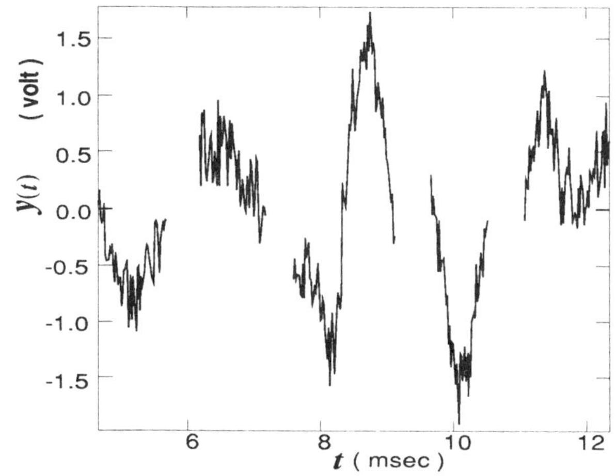


Fig.10. Partially lost voice signal.

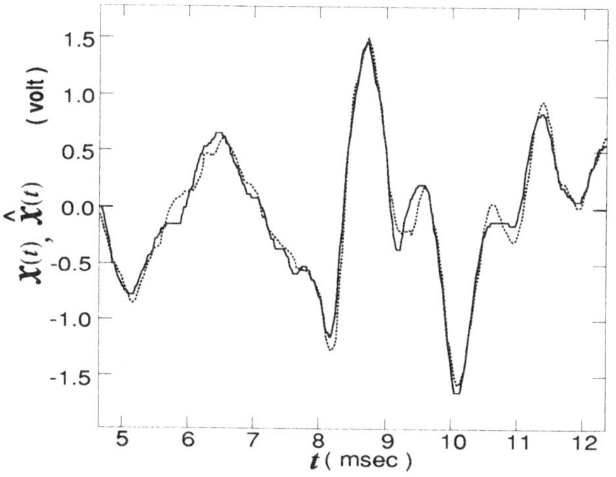


Fig.11. Interpolated results for the partially lost voice signal. ———— : Complete voice signal. ------- : Interpolation by the proposed method.

## 6. HARDWARE IMPLEMENTATION

An electronic circuit, which achieves the proposed interpolation in real time, is fabricated on a breadboard. It employs analog and digital circuits. This system is designed to accept contineous signals (not sampled data) and the patch width $T$ is assigned to be 0.4 msec, while it was assigned to include 3 sampled data in the previous simulation.

The block diagram is shown in Fig. 12. A noisy input signal is applied to the system and a max/hold and a min/hold circuit produce a maximum value $\vee_i$ and a minimum value $\wedge_i$ in the patch $P_i$ which are to be stored and shifted down through two rails of sample/hold circuits. In order to determine the four sub-lines around the main-line, a maximum value and a minimum value of each patch period should be stored for six periods.

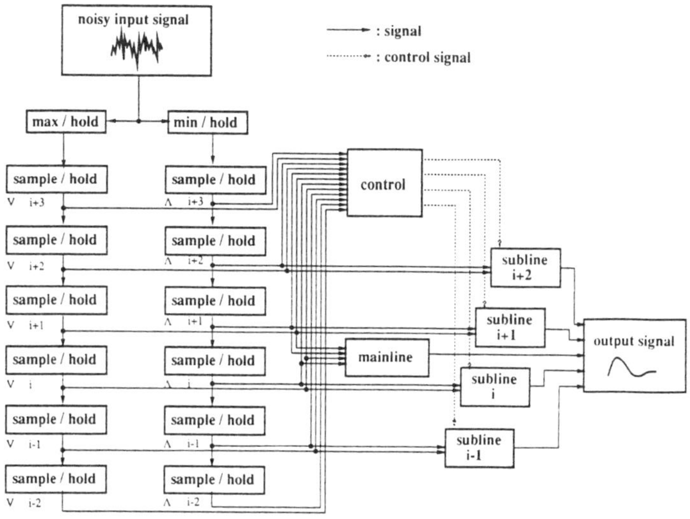


Fig.12. The Block diagram of the hardware system which facilitates smooth interpolation by fuzzy inference.

$2F_i(t)$ in Eq.(7) is represented by

$$2F_i(t) = \frac{1}{T}\{(\vee_{i+1} + \wedge_{i+1}) - (\vee_i + \wedge_i)\}t + (\vee_i + \wedge_i) \qquad (11)$$

where $\vee_{i+1}$ and $\wedge_{i+1}$ are a maximum value and a minimum value in the patch $P_{i+1}$, respectively. This signal is produced by the main-line circuit shown in Fig.13. Four input signals are applied from outputs of appropriate sample/hold circuits. The reset switch attached to the integrator is activated at each end of a patch period. A 4-bit digital switch changes the value of the resistor of the integrator to change the slope adequately, when a patch period $T$ is changed.

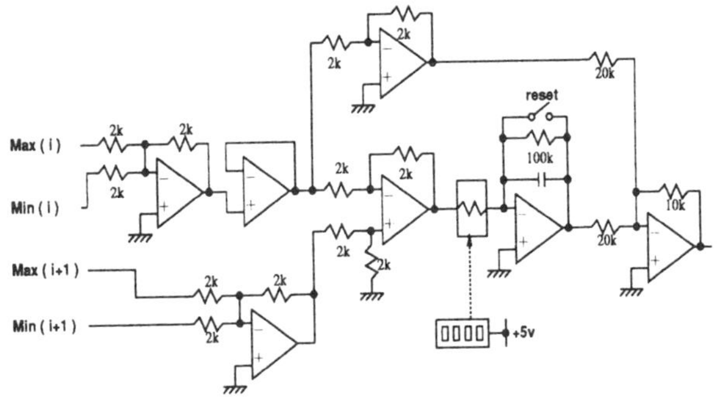


Fig.13. Main-line circuit.

A sub-line is represented by

$$f_{i+k}(t) = a\left\{\frac{1}{t}(\vee_{i+k} - \wedge_{i+k})t - k(\vee_{i+k} - \wedge_{i+k})\right\} + \frac{1}{2}(\vee_{i+k} + \wedge_{i+k}) \tag{12}$$

for $k = -1, 0, 1, 2$,
where coefficient $a$ is defined by

$$a = \begin{cases} -1 & (\vee_{i+k-1} + \wedge_{i+k-1} \geq \vee_{i+k} + \wedge_{i+k} \geq \vee_{i+k+1} + \wedge_{i+k+1}) \\ +1 & (\vee_{i+k-1} + \wedge_{i+k-1} \leq \vee_{i+k} + \wedge_{i+k} \leq \vee_{i+k+1} + \wedge_{i+k+1}) \\ 0 & otherwise \end{cases} \tag{13}$$

Weights for sub-lines are derived from Fig.3 as

$$\begin{aligned} w_{i-1}(t) &= -\frac{1}{2T}t + \frac{1}{2} \\ w_i(t) &= -\frac{1}{2T}t + 1 \\ w_{i+1}(t) &= \frac{1}{2T}t + \frac{1}{2} \\ w_{i+2}(t) &= \frac{1}{2T}t \end{aligned} \tag{14}$$

Therefore, weighted subline $w_{i-1}(t)f_{i-1}(t)$ , for example, can be represented by

$$w_{i-1}(t)f_{i-1}(t) = -\frac{a}{2T^2}(\vee_{i-1} - \wedge_{i-1})t^2 - \frac{1}{4T}(\vee_{i-1} + \wedge_{i-1})t + \frac{a}{2}(\vee_{i-1} - \wedge_{i-1}) + \frac{1}{4}(\vee_{i-1} + \wedge_{i-1}) \tag{15}$$

This is the first term of Eq.(7) and can be produced by the subline circut shown in Fig.14. Switches labeled "a", "b", "c", are turned on when the patch $P_{i-1}$ satisfies the following cases, respectively. Control signals a, b, c, $\bar{c}$ (logical complement of c), are delivered from the control circuit shown in Fig.15.

$(a)$ $\vee_{i-2} + \wedge_{i-2} \geq \vee_{i-1} + \wedge_{i-1} \geq \vee_i + \wedge_i)$

$(b)$ $\vee_{i-2} + \wedge_{i-2} \leq \vee_{i-1} + \wedge_{i-1} \leq \vee_i + \wedge_i)$

$(c)$ *otherwise*

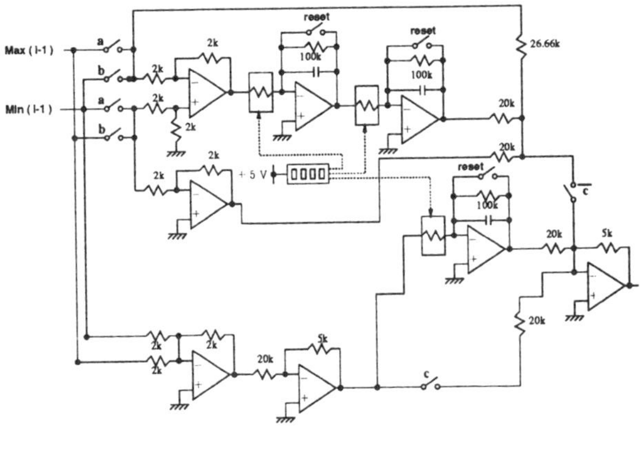


Fig.14. Sub-line circut.

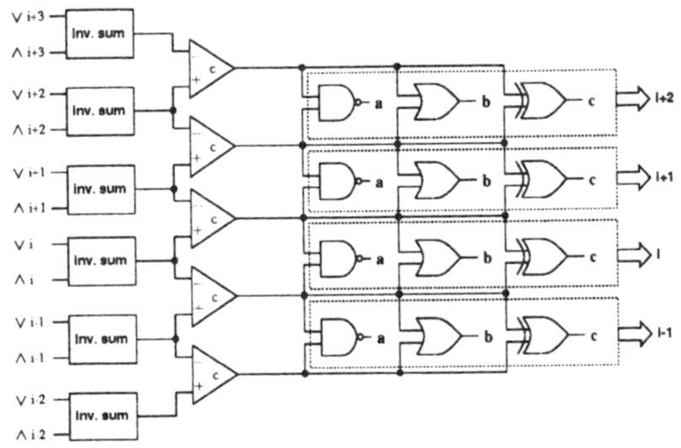


Fig.15. The control circuit.

In the similar manner, signals corresponding to other weighted sub-lines $w_i(t)f_i(t)$, $w_{i+1}(t)f_{i+1}(t)$, $w_{i+2}(t)f_{i+2}(t)$ are obtained. Consequently, the interpolated signal $\widehat{x}(t)$ is produced by a summing circuit weighted by 1/4.

Fig.16 shows the time chart of this system in case of patch period $T=0.4$ *msec*. $\phi_1$ transports the maximum and minimum values in the max/hold and min/hold circuits to the following sampl/hold cirucits, respectively. $\phi_1$ and $\phi_2$ are adopted to drive two rails of sample/hold circuits. The reset pulse initializes the max/hold and min/hold circuits and all the integrators in main-line and sub-line circuits at the end of each patch period.

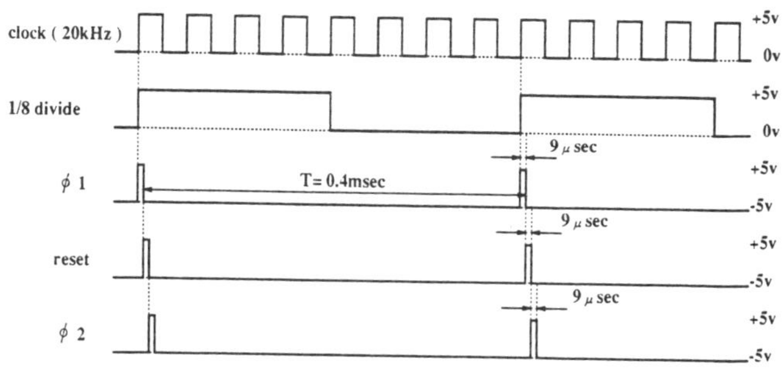


Fig.16. Time chart.

## 7. EXPERIMENTAL RESULTS

At first, in order to test the system, a sinusoidal wave (Fig.17(a)) is applied to this system, and the output of the main-line circuit (Fig.17(b)) and the outputs of four sub-line circuits (Fig.17(c), (d), (e), (f))were observed. All signals from the

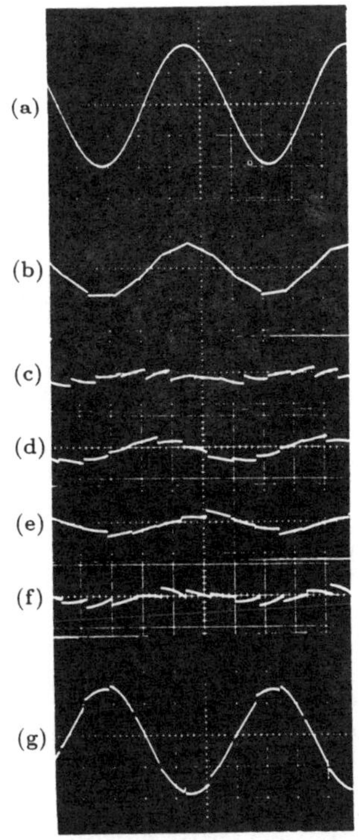


V: 0.5v/div, H: 0.5msec/div

Fig.17. Waveforms obtained from the test without noise. (a) A sinusoidal input signal, (b) an output of the main-line circuit, (c) an output of the sub-line circuit $w_{i-1}(t)f_{i-1}(t)$, (d) an output of the sub-line circuit $w_i(t)f_i(t)$, (e) an output of the sub-line circuit $w_{i+1}(t)f_{i+1}(t)$, (f) an output of the sub-line circuit $w_{i+2}(t)f_{i+2}(t)$, and (g) an output of the system. Patch period is $T$ = 0.4 *msec*.

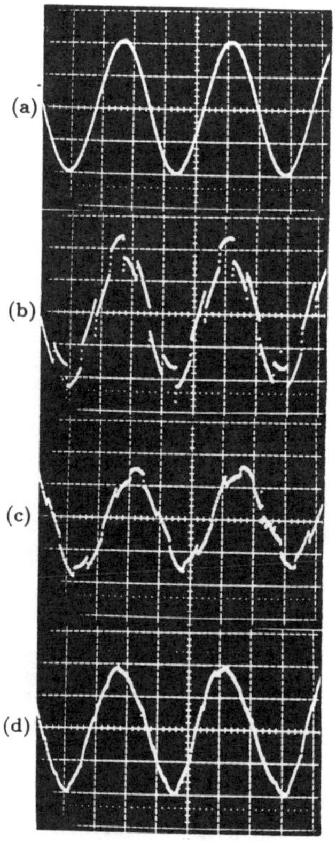


V: 0.5v/div, H: 1.0msec/div

Fig.18. Filtering of a sinusoidal wave superposed with a square wave. (a) a sinusoidal wave (290 Hz), (b) an input signal superposed with a square wave (1kHz), (c)an output obtained from a first order lowpass filter with a corner frequency of 400 Hz, and (d)an output from the system under test.

outputs of sub-line circuits are weighted ones by $w_{i+k}(t)$ (k= -1, 0, +1, +2). Therefore, they exhibit parabolic transients in each patch, while that of the main-line circuit exhibits linear transient in each patch.

Next, a sinusoidal signal superposed with a square wave, the frequency of which is closed to that of sinusoidal wave, is applied to the system input. Frequencies of the sinusoidal wave (Fig.18 (a)) and the square wave are 290 Hz and 1kHz, respectively.The noise-eliminated output of the system is shown in Fig. 18 (d). For the comparison, the same signal is applierd to the first order low pass filter, the corner frequency of which is 400 Hz, and the out put is observed as shown in Fig.18 (c). The square wave cannot be eliminated enough and the amplitude of the signal component is reduced. Two frequencies are so closed to each other that they cannot be filtered by the first order low pass filter.

A sinusoidal signal superposed with a white noise is applied to the system and the output is observed as shown in Fig.19.

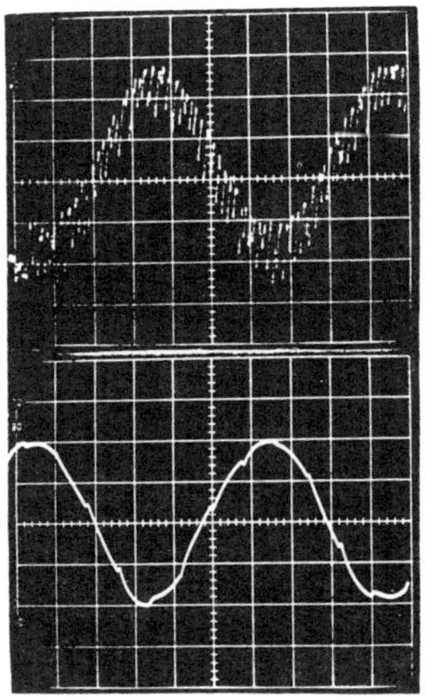


Fig.19. Noise elimination by the system proposed in this papwe. (a)An input signal with a white noise. (b) An output signal obtained from the system under test.

## 8. CONCLUSIONS

This paper described a simple method of interpolation for noisy signal data by using fuzzy inference. The results interpolated by the proposed method seem to be enough for a practical use in spite of its simplicity. This algorithm was implemented in the hardware employing digital and analog circuits and verified that it can eliminate the noise effectively.

## ACKNOWLEDGMENT

Many thanks are due to Prof. M.Ohta and his colleagues for providing the road traffic noise data.

## REFERENCES

[1] R.E. Kalman, "A New Approach to Linear Filtering and Prediction Problems," Trans. ASME, Ser. D. J. Basic Eng., Vol.82, pp.35-45, 1960.

[2] R.E. Kalman and R.S. Bucy, "New Results in Linear Filtering and Prediction Theory," Trans. ASME, Ser. D. J. Basic Eng., Vol.83, pp.95-108, 1961.

[3] M. Ohta, K. Hatakeyama and M. Nishimura, "Some Unified Methods in Multivariable Linear Filtering and Prediction Problems by Use of the Expansion Form of Bayes' Theorem," Proc. of 26th Jpn. Natl. Cong. for Appl. Mech., Vol.26, pp.383-398, 1976.

[4] M. Ohta and E. Uchino, "A Design for a General Digital Filter for State Estimation of an Arbitrary Stochastic Sound System," J. Acoust. Soc. Am., Vol.80, pp.804-812, 1986.

[5] M. Sugeno, "Fuzzy Control," Nikkan Kogyo, Tokyo, 1988.

[6] E. Uchino, T. Yamakawa and T. Yanaru, "How to Find out the Supplementary Rules Representing an Uncertain System," Proc. of 1990 Int. Conf. on Fuzzy Logic & Neural Networks, pp.533-536, 1990.

[7] M. Iri and Y. Fujino, "Common Sense in Numerical Calculation," Kyoritsu, Tokyo, 1991.

[8] A. Kandel, "Fuzzy Mathematical Techniques with Applications," Addison-Wesley, California, 1986.

[9] G.J. Klir and T.A. Folger, "Fuzzy Sets, Uncertainty, and Information," Prentice Hall, New Jersey, 1988.

# On Robustness of Fuzzy Logics

Hung T. Nguyen[1], Vladik Kreinovich[2], Dana Tolbert[2]
[1] Department of Systems Science, Tokyo Institute of Technology
4259 Nagatsuta-cho, Midori-ku, Yokohama 227 Japan
[2] Computer Science Department, University of Texas at El Paso, El Paso, TX 79968

***Abstract*—In this paper we propose to investigate the concept of robustness of combining operations in fuzzy systems. We use a concept similar to modulus of continuity in approximation theory to characterize robustness of fuzzy logic connectives. Specifically, it is shown that min and max are the most robust &− and ∨−operations.**

## I. Introduction

In the majority of expert systems and intelligent control systems uncertainty of expert's statements is represented by a number from the interval [0,1]: $t(A) = 1$ means that an expert is absolutely sure in $A$, $t(A) = 0$ means that the expert is absolutely sure that $A$ is false, and values from 0 to 1 represent different degrees of the expert's uncertainty. These numbers are called certainty values, degrees of belief, truth values, etc. If we know the degrees of belief $t(A)$ and $t(B)$ for two statements $A$ and $B$, and nothing else is known about $A$ and $B$, then what is the reasonable degree of belief in $A\&B$? in $A \vee B$? We must compute these estimates for $t(A\&B)$ and $t(A \vee B)$ from the only information that we have: from the values $t(A)$ and $t(B)$. The operations $f(a, b)$ that transform $t(A)$ and $t(B)$ into the estimates $f(t(A), t(B))$ for $t(A\&B)$ and $t(A\vee B)$ are called correspondingly an *&−operation* and an *∨−operation*.

The first paper by L. Zadeh [17] that introduced this approach to knowledge representation proposed $\min(a, b)$ and $ab$ as &−operations, and $\max(a, b)$ and $a + b - ab$ as ∨−operations. Zadeh himself stressed that these operations "are not the only operations in terms of which the union and intersection can be defined", and "which of these ... definitions are more appropriate depends on the context" ([18], pp. 225–226). Since then several dozens different &− and ∨−operations have been proposed and successfully used. Some operations have been discovered empirically while working on real expert systems (e.g., the famous MYCIN [3]) or while analyzing commonsense reasoning [15], [19]); some of them were proposed on a more theoretical basis (see, e.g., [6], [8]). A survey of such operations is given in [11].

One of the most successful applications of this type of uncertainty representation is in the area of intelligent control. This area was first outlined by L. Zadeh [4] and experimentally tested by E. Mamdani [14] in the framework of fuzzy set theory [17], therefore this area of research is also called *fuzzy control*. For the current state of fuzzy control the reader is referred to the surveys [16], [12], [1].

There is a necessity for robust operations. Indeed, there are many procedures to estimate degrees of belief ([6], [8]). Sometimes, an expert has a strong feeling about his degree of certainty; in such cases, this degree can be estimated more or less precisely. Sometimes, however, an expert is not sure about the strength of his own beliefs; in these cases the known procedures can lead to drastically different estimates $a$, $a'$ for the same degree of belief $t(A)$.

Some &− and ∨−operations are very sensitive to such uncertainty in the sense that small changes in $t(A)$ and $t(B)$ can lead to absolutely different estimates for $t(A\&B)$ and $t(A \vee B)$. We would like to restrict ourselves to the operations that are *robust* in the sense that they are the least sensitive to such changes.

## II. Definitions and the Main Results

Let us recall some tutorial facts.

**Definition 1.**
**(i).** By a *binary operation* (or *operation* for short) we mean a function $f(a, b)$ from $[0, 1] \times [0, 1]$ into [0,1].
**(ii).** A binary operation is called a *&−operation* if the following conditions are true:

- $f(0, 0) = f(0, 1) = f(1, 0) = 0, f(1, 1) = 1;$
- $f(a, b) = f(b, a)$ for all $a, b$;
- $f(a, b) \leq a$ for all $a$ and $b$.

**(iii).** A binary operation is called an *∨−operation* if the following conditions are true:

- $f(0, 0) = 0, f(0, 1) = f(1, 0) = f(1, 1) = 1;$
- $f(a, b) = f(b, a)$ for all $a, b$;
- $f(a, b) \geq a$ for all $a$ and $b$.

*Remark.* The above binary operations are slightly more general than the usual $t-$ norms and $t-$conorms in the

0-7803-0614-7/93$03.00 ©1993IEEE

literature [7].

**Definition 2.** Suppose that a binary operation $f(a,b)$ is given. We say that a $\delta$–input uncertainty leads to a $\leq \alpha$–output error, if for every $a, a', b, b'$, for which $|a-a'| \leq \delta$ and $|b-b'| \leq \delta$, we have $|f(a,b) - f(a',b')| \leq \alpha$.

**Definition 3.** Suppose that $f(a,b)$ is a binary operation, and $\delta > 0$ is a positive real number. By a *$\delta$–robustness* of an operation $f(a,b)$ we mean the smallest of real numbers $\alpha$, for which a $\delta$–input uncertainty leads to a $\leq \alpha$–output error. The $\delta$–robustness of an operation $f(a,b)$ will be denoted by $r_f(\delta)$.

*Remark.* It is easy to check that $r_f(\delta) = \sup\{|f(a,b) - f(a',b')| : |a-a'| \leq \delta, |b-b'| \leq \delta\}$. When $f$ is continuous, $r_f(\delta)$ is the well-known modulus of continuity of $f$ [13]. The above sup is in fact max: see Proposition 1 below; its proof, as well as all the proofs of the results in this section are in Section 3 for easy reading.

**PROPOSITION 1.** *For every operation $f(a,b)$, and for every $\delta > 0$, there exists a $\delta$–robustness (i.e., the smallest of real numbers $\alpha$, for which a $\delta$–input uncertainty leads to a $\leq \alpha$–output error).*

In order to compare different operations in terms of robustness, we will proceed as in standard decision theory, where $r_f(\delta)$ plays the role of the "risk" (see, e.g., [2]).

**Definition 4.** We say that operations $f(a,b)$ and $g(a,b)$ are *equally robust* if for every $\delta$, $r_f(\delta) = r_g(\delta)$. We say that an operation $f(a,b)$ is *more robust* than an operation $g(a,b)$, if for every $\delta$, $r_f(\delta) \leq r_g(\delta)$, and at least for one $\delta > 0$, $r_f(\delta) < r_g(\delta)$.

**Definition 5.** We say that an &–operation $f(a,b)$ is *the most robust &–operation*, if it is either more robust, or equally robust than any other &–operation. We say that an $\vee$–operation $g(a,b)$ is *the most robust $\vee$–operation*, if it is either more robust, or equally robust than any other $\vee$–operation.

**THEOREM 1.** *$f(a,b) = \min(a,b)$ is the most robust &–operation.*

**THEOREM 2.** *$f(a,b) = \max(a,b)$ is the most robust $\vee$–operation.*

*Remarks.*

1. In [9], [10], [11] general optimization problem are analyzed on the set of all possible &– and $\vee$– operations. As a result of this mathematical analysis, lists are given that include all &– and $\vee$– operations that can be optimal under reasonable optimality criteria. Our Theorems 1 and 2 are in good accordance with that general result, because both min and max are elements of those lists.

2. Similar questions of robustness in the context of neural networks are analyzed in [5].

3. It is interesting to know to what extent the robustness functions that correspond to min and max are smaller than those of the other binary operations: are they smaller in a few points only, or essentially smaller for all $\delta$? The answer is given by the following theorems:

**THEOREM 3.** *Suppose that $f(a,b)$ is an &–operation, and $f(a,b)$ is different from* min. *Then there exist a positive real number $\Delta > 0$ and a positive real number $C < 1$ such that for all $\delta < \Delta$, $r_{\min}(\delta) \leq C r_f(\delta)$.*

**THEOREM 4.** *Suppose that $f(a,b)$ is an $\vee$–operation, and $f(a,b)$ is different from* max. *Then there exist a positive real number $\Delta > 0$ and a positive real number $C < 1$* such that for all $\delta < \Delta$, $r_{\max}(\delta) \leq C r_f(\delta)$.

The following Theorem describes $\delta$–robustness for several other operations:

**THEOREM 5.**
*1) if $f(a,b) = ab$, then $r_f(\delta) = 2\delta - \delta^2$;*
*2) if $f(a,b) = a + b - ab$, then $r_f(\delta) = 2\delta - \delta^2$;*
*3) if $f(a,b) = \min(a+b, 1)$, then $r_f(\delta) = \min(2\delta, 1)$.*

*Remarks.*

1. In these three cases, $\lim_{\delta \to 0} r_f(\delta)/r_{\min}(\delta) = 2$, so in Theorems 3 and 4 we can take $C = 1/2 + \alpha$ for arbitrary small $\alpha > 0$.

2. From Theorem 5 we can conclude that the operations $ab$ and $a + b - ab$ are equally robust, and that $f(a,b) = \min(a+b, 1)$ is more robust than both of them.

3. The fact that $ab$ and $a + b - ab$ are equally robust stems from the fact that in general *dual* operations have the same modulus of continuity, where $g(a,b)$ is *dual* to an $f(a.b)$ if $g(a,b) = 1 - f(1-a, 1-b)$.

**PROPOSITION 2.** *Dual operations are equally robust.*

## III. Proofs

*Proof of Proposition 1*

The set $S$ of all real numbers $\alpha$, for which a $\delta$–input uncertainty leads to a $\leq \alpha$–output error, is bounded from below (by 0), and therefore, has an infimum (the greatest lower bound) $r$. $r$ is the value of $\delta$–robustness. Indeed, since $r$ is the greatest lower bound of the set $S$, for every positive integer $k$ there exists a number $r_k \in S$ such that $r_k < r + 1/k$. According to the definition of $S$, from $r_k \in S$ we conclude that if $|a-a'| \leq \delta$ and $|b-b'| \leq \delta$, then $|f(a,b) - f(a',b')| \leq r_k$. Letting $k \to \infty$, we conclude that $|f(a,b) - f(a',b')| \leq \lim_k r_k = r$. Q.E.D.

*Proof of Theorem 1*

$1^o$. Let us first prove that, as in the case of $t-$norm, we have $f(a,b) \le \min(a,b)$ for any $\&-$operation $f$.

Since $f(a,b) \le a$ and $f(a,b) = f(b,a) \le b$, it follows that $f(a,b) \le \min(a,b)$.

$2^o$. Next, $r_{\min}(\delta) = \delta$.

Indeed, for $|a-a'| \le \delta$, we have $a \le a' + \delta$ and likewise $b \le b' + \delta$. Hence,

$$\min(a,b) \le \min(a'+\delta, b'+\delta) = \min(a',b') + \delta,$$

therefore, $\min(a,b) \le \min(a',b') + \delta$. Likewise, $\min(a',b') \le \min(a,b) + \delta$, so

$$-\delta \le \min(a,b) - \min(a',b') \le \delta,$$

and $|\min(a,b) - \min(a',b')| \le \delta$. Take $a = b = \delta$, $a' = b' = 0$. Then $|\min(a,b) - \min(a',b')| = \delta$, and therefore, the output error is precisely $\delta$. So, we cannot take $\alpha < \delta$, and so the $\delta-$robustness of min is really equal to $\delta$.

$3^o$. For every $\&-$operation $f(a,b)$: $r_f(\delta) \ge r_{\min}(\delta) = \delta$.

Suppose that for some $\delta \in (0,1)$, $r_f(\delta) < \delta$. This means that if $|a-a'| \le \delta$ and $|b-b'| \le \delta$, then $|f(a,b)-f(a',b')| \le r_f(\delta) < \delta$. In particular, if we take $a = b = 1$ and $a' = b' = 1-\delta$, we conclude that $|f(1,1) - f(1-\delta,1-\delta)| < \delta$. But according to the definition of a $\&-$operation, $f(1,1) = 1$, therefore, this inequality turns into $|1-f(1-\delta,1-\delta)| < \delta$. Hence, $1-f(1-\delta,1-\delta) \le |1-f(1-\delta,1-\delta)| < \delta$, therefore, $f(1-\delta,1-\delta) > 1-\delta$. But we have already proved in $1^o$ that $f(a,b) \le \min(a,b)$, therefore, $f(1-\delta,1-\delta) \le 1-\delta$. These two inequalities contradict to each other. Therefore, our assumption that $r_f(\delta) < \delta$ is incorrect. Hence, $r_f(\delta) \ge \delta$.

$4^o$. $\min(a,b)$ is the only $\&-$operation, for which $r_f(\delta) = \delta$ for all $\delta$.

Indeed, suppose that $f$ is different from min. Then for some $a$ and $b$, $f(a,b) \ne \min(a,b)$, hence $f(a,b) < \min(a,b)$. Without loss of generality, assume $a \le b$, resulting in $f(a,b) < \min(a,b) = a$. For $a' = b' = 1$, we have $|a-a'| = 1-a$, $|b-b'| = 1-b \le 1-a$, but $|f(a,b)-f(a',b')| = 1-f(a,b) > 1-a$. So, for $\delta_0 = 1-a$, $r_f(\delta_0) > \delta_0$. Q.E.D.

*Proof of Theorem 2*

This proof is similar to the proof of Theorem 1.

*Proof of Theorem 3*

$1^o$. In part $4^o$ of the proof of Theorem 1, we showed that if an $\&-$operation $f(a,b)$ is different from $\min(a,b)$, then we can find the values $a$, $b$, $a' = 1$ and $b' = 1$, for which $a \le b$, $|a-a'| = \delta_0$, $|b-b'| \le \delta_0$, and $D > \delta_0$, where $D = f(a',b') - f(a,b)$ and $\delta_0 = 1-a$. From this we concluded that $r_f(\delta_0) \ge D > \delta_0$. Let us denote the ratio $\delta_0/D$ by $c$. From $D > \delta_0$ we conclude that $c < 1$.

$2^o$. Let us now prove that for every positive integer $n$, $r_f(\delta_0/n) \ge D/n$.

Indeed, let us divide the intervals $(a,a')$ and $(b,b')$ into $n$ equal parts, i.e., let us consider the values

$$a_i = a + i(1-a)/n$$

and $b_i = b + i(1-b)/n$, where $i = 0,1,2,...,n$. Then, for every $i$ from 0 to $n-1$, $|a_i - a_{i+1}| = (1-a)/n = \delta_0/n$, and $|b_i - b_{i+1}| = (1-b)/n \le \delta_0/n$. Therefore, according to the definition of $\delta_0-$robustness, $|f(a_i,b_i) - f(a_{i+1},b_{i+1})| \le r_f(\delta_0/n)$.

Let us now use the fact that $a_0 = a, b_0 = b, a_n = a', b_n = b'$, and $f(a_0,b_0) - f(a_n,b_n) = -D$. We have

$$f(a_0,b_0) - f(a_n,b_n) = (f(a_0,b_0) - f(a_1,b_1)) +$$

$$(f(a_1,b_1) - f(a_2,b_2)) + ... + (f(a_{n-1},b_{n-1}) - f(a_n,b_n)).$$

The absolute value of each of $n$ terms in the sum is bounded by $r_f(\delta_0/n)$, therefore, the absolute value of the entire sum is bounded by $nr_f(\delta_0/n)$. But this sum equals to $f(a_0,b_0) - f(a_n,b_n) = -D$, so its absolute value is $D$, and thus $D \le nr_f(\delta_0/n)$. Hence, $r_f(\delta_0/n) \ge D/n$.

$3^o$. In order to continue the proof, we need to prove the following simple fact: if $\delta < \delta'$, then $r_f(\delta) \le r_f(\delta')$ (it is well known for moduli of continuity, but we will reproduce a proof for completeness).

Indeed, from the definition of the function $r_f(\delta)$ we conclude that if $|a-a'| \le \delta'$ and $|b-b'| \le \delta'$, then $|f(a,b) - f(a',b')| \le r_f(\delta')$. But if $|a-a'| \le \delta$ and $|b-b'| \le \delta$, then (since $\delta < \delta'$) $|a-a'| \le \delta'$ and $|b-b'| \le \delta'$, and therefore, $|f(a,b) - f(a',b')| \le r_f(\delta')$. So, a $\delta-$input uncertainty leads to a $\le \alpha-$output error, where $\alpha = r_f(\delta')$. Since by definition $r_f(\delta)$ is the smallest of all such values $\alpha$, we conclude that $r_f(\delta) \le r_f(\delta')$.

$4^o$. Let us now take any real number $C$ between $c = \delta_0/D$ and 1 ($c < C < 1$), and prove that there exists a $\Delta > 0$ such that for all $\delta < \Delta$, we have $r_{\min}(\delta) \le Cr_f(\delta)$.

Since $r_{\min}(\delta) = \delta$ for all $\delta$, the inequality can be rewritten as $\delta \le Cr_f(\delta)$, or, equivalently, $r_f(\delta) \ge \delta/C$.

We already know how to estimate the values of $r_f(\delta)$ for $\delta = \delta_0/n$, where $n = 1,2,3,...$ So, to get the estimates for arbitrary $\delta$, we can use these known estimates. For every $\delta < \delta_0$, we want to find an $n$ such that

$$\delta_0/(n+1) \le \delta \le \delta_0/n.$$

This inequality is equivalent to $(n+1)/\delta_0 \ge 1/\delta \ge n\delta_0$, which, after multiplying both sides by $\delta_0$, turns out to be equivalent to the inequality $n \le \delta_0/\delta \le n+1$. Therefore, we can take for $n$ the integer part $\lfloor \delta_0/\delta \rfloor$ of the ratio $\delta_0/\delta$.

From $3^o$ we can conclude that $r_f(\delta) \geq r_f(\delta_0/(n+1))$. In $2^o$ we proved that $r_f(\delta_0/(n+1)) \geq D/(n+1)$. Therefore, $r_f(\delta) \geq D/(n+1)$. We defined $c$ as $c = \delta_0/D$; so, $D = \delta_0/c$, so $r_f(\delta) \geq \delta_0/(c(n+1))$.

We want to get an inequality $r_f(\delta) \geq \delta/C$. We will be able to deduce this inequality from the one that we have just proved if $\delta_0/(c(n+1)) \geq \delta/C$. Since $\delta \leq \delta_0/n$, this inequality is valid if $\delta_0/Cn \leq \delta_0/(c(n+1))$. Dividing both sides by $\delta_0$ and then inverting both sides, we get an equivalent inequality $Cn \geq c(n+1)$, which, in its turn, is equivalent to $(C-c)n \geq c$ and $n \geq c/(C-c)$. Therefore, if $n \geq c/(C-c)$, then for $\delta \leq \delta_0/n$ we get the desired inequality $r_f(\delta) \geq \delta/c$.

The inequality $n \geq c/(C-c)$ is valid for all $n$ starting from $N = \lfloor c/(C-c) \rfloor + 1$. Therefore, the desired inequality $r_f(\delta) \geq \delta/c$ is true for all $\delta < \Delta$, where $\Delta = \delta_0/N$. Q.E.D.

*Proof of Theorem 4*

This proof is similar to the proof of Theorem 3.

Before proving Theorem 5 let us prove Proposition 2.

*Proof of Proposition 2*

$1^o$. Let us first prove that if $f$ and $g$ are dual, i.e., $g(a,b) = 1 - f(1-a, 1-b)$, then for every $\delta$, $r_g(\delta) \geq r_f(\delta)$.

Indeed, suppose that $|a-a'| \leq \delta$ and $|b-b'| < \delta$, and let us prove that $|g(a,b) - g(a',b')| \leq r_f(\delta)$. Since $|a-a'| \leq \delta$ and $|b-b'| \leq \delta$, we have $|A-A'| = |a-a'| \leq \delta$ and

$$|B-B'| = |b-b'| \leq \delta,$$

where we denoted $A = 1-a$, $A' = 1-a'$, $B = 1-b$, and $B' = 1-b'$. Due to the definition of $r_f(\delta)$, we can conclude that $|f(A,B) - f(A',B')| \leq r_f(\delta)$. But $g(a,b) = 1 - f(A,B)$ and $g(a',b') = 1 - f(A',B')$, therefore

$$|g(a,b) - g(a',b')| = |f(A,B) - f(A',B')| \leq r_f(\delta).$$

So, for $\alpha = r_f(\delta)$, if $|a-a'| \leq \delta$ and $|b-b'| \leq \delta$, then $|g(a,b) - g(a',b')| \leq \alpha$. Since $r_g(\delta)$ is defined as the smallest of all $\alpha$ with this property, we conclude that $r_g(\delta) \leq r_f(\delta)$.

$2^o$. One can easily check that if $g$ is dual to $f$, then $f$ is dual to $g$. Therefore, we have both $r_g(\delta) \leq r_f(\delta)$ and $r_f(\delta) \leq r_g(\delta)$, hence $r_g(\delta) = r_g(\delta)$. Q.E.D.

*Proof of Theorem 5*

1) We must prove, first, that if $|a-a'| \leq \delta$ and $|b-b'| \leq \delta$, then $|ab - a'b'| \leq 2\delta - \delta^2$, and, second, that there exist such $a, b, a', b'$ for which $|a-a'| \leq \delta$, $|b-b'| \leq \delta$, and $|ab - a'b'| = 2\delta - \delta^2$.

The second statement is easy to prove: take $a = b = 1, a' = b' = 1-\delta$, then $|ab - a'b'| = 1 - (1-\delta)^2 = 2\delta - \delta^2$. Let us now prove the first one.

Let us denote $|a-a'|$ by $\Delta_a$, and $|b-b'|$ by $\Delta_b$. Then $\Delta_a \leq \delta$ and $\Delta_b \leq \delta$. Without losing any generality we can assume that $a \geq a'$. Then $a' = a - \Delta_a$. With respect to $b$ and $b'$, there are two possible cases: $b \geq b'$ and $b < b'$. Let us consider both of them.

If $b \geq b'$, then $b' = b - \Delta_b$, and $ab \geq a'b'$, so the desired absolute value $d = |ab - a'b'|$ can be computed as follows:

$$d = |ab - a'b'| = ab - a'b' = ab - (a - \Delta_a)(b - \Delta_b) = a\Delta_b + b\Delta_a - \Delta_a\Delta_b.$$

Since $a \leq 1$ and $b \leq 1$, we have $d \leq \Delta_a + \Delta_b - \Delta_a\Delta_b$. The right-hand side of this inequality can be expressed as $1 - (1-\Delta_a)(1-\Delta_b)$. Therefore, it is a monotonely increasing function of both $\Delta_a$ and $\Delta_b$. So, its maximal value is attained when both of these variables take their biggest possible values. Since $\Delta_a \leq \delta$ and $\Delta_b \leq \delta$, the maximal possible value is attained when $\Delta_a = \Delta_b = \delta$, and is equal to $2\delta - \delta^2$. Therefore, $d \leq \Delta_a + \Delta_b - \Delta_a\Delta_b \leq 2\delta - \delta^2$. So for this case the desired inequality is proved.

Let us now consider the case when $b < b'$. Then $b = b' - \Delta_b$, and $d = |ab - a'b'| = |a(b' - \Delta_b) - (a - \Delta_a)b'| = |a\Delta_b - b'\Delta_a|$. Let us consider two subcases: when the expression under the absolute value is positive or negative, i.e., when $a\Delta_b \geq b'\Delta_a$ and $a\Delta_b < b'\Delta_a$. In the first subcase, $d = a\Delta_b - b'\Delta_a$, therefore, $d \leq b'\Delta_a$. Since $\Delta_a \leq \delta$ and $b' \leq 1$, we get $d \leq \delta$.

In the second subcase similarly $d = b'\Delta_a - a\Delta_b \leq b'\Delta_a \leq \delta$. So, in both cases $d \leq \delta$.

So, to complete the proof, it is sufficient to show that $\delta \leq 2\delta - \delta^2$ for all $\delta$ from 0 to 1. Indeed, by dividing both sides by $\delta$ and moving all terms to the right-hand side, we conclude that this inequality is equivalent to $0 \leq 1 - \delta$, which is certainly true for $\delta \leq 1$.

2) follows from 1) and Proposition 2.

3) Let us first consider the case, when $\delta < 1/2$. Then $2\delta < 1$, and $\min(2\delta, 1) = 2\delta$. Let us prove that in this case, if $|a-a'| \leq \delta$ and $|b-b'| \leq \delta$, then $|f(a,b) - f(a',b')| \leq 2\delta$.

Indeed, if $|a-a'| \leq \delta$ and $|b-b'| \leq \delta$, then

$$|(a+b) - (a'+b')| = |(a-a') + (b-b')| \leq 2\delta.$$

In particular, this means that $a'+b' \leq a+b+2\delta$. Evidently, $a'+b' \leq 1$, therefore, $a'+b' \leq 1 < 1+2\delta$. So, $a'+b'$ is not bigger than the smallest of these two numbers: $a'+b' \leq \min(a+b+2\delta, 1+2\delta)$. But $\min(a+b+2\delta, 1+2\delta) = \min(a+b, 1) + 2\delta = f(a,b) + 2\delta$. So, $a'+b' \leq f(a,b) + 2\delta$. Since $f(a',b') = \min(a'+b', 1)$ and therefore, $f(a',b') \leq a'+b'$, we conclude that $f(a',b') \leq f(a,b) + 2\delta$. In a similar manner we can prove that $f(a,b) \leq f(a',b') + 2\delta$. Combining these two inequalities, we conclude that $|f(a,b) - f(a',b')| \leq 2\delta$. So, for $\delta < 1/2$, $r_f(\delta) \leq 2\delta$.

Let us now show that $\alpha = 2\delta$ is the smallest value, for which $\delta$–input uncertainty leads to a $\leq \alpha$–output error, and thus, $r_f(\delta) = 2\delta$. Indeed, if we take $a = b = 0, a' = b' = \delta$, then $|a-a'| \leq \delta$, $|b-b'| \leq 2\delta$, and

$$|f(a,b) - f(a',b')| = |0 - 2\delta| = 2\delta,$$

so the values $\alpha < 2\delta$ do not work in this case. So, for $\delta < 1/2$, we proved that $r_f(\delta) = 2\delta$.

Now let us consider the case when $\delta \geq 1/2$. In this case, $\min(2\delta, 1) = 1$. If we take $a = b = 0, a' = b' = \delta$, then $f(a, b) = 0$, $f(a', b') = 1$, $|a - a'| \leq \delta$, $|b - b'| \leq 2\delta$, and $|f(a, b) - f(a', b')| = |0 - 1| = 1$. Therefore, nothing smaller than 1 can serve as $\alpha$, hence $r_f(\delta) = 1$. Q.E.D.

## IV. Conclusions

As far as combining degrees of belief of experts is concerned, in situations where estimates can vary drastically, it is reasonable to use *robust* fuzzy logic connectives, which are the least sensitive to these variations. We have proved that in this situation, the dual pair $\min(a, b)$, $\max(a, b)$ are the most robust operations.

## Acknowledgment

This work was supported by a NSF Grant No. CDA-9015006, NASA Research Grant No. 9-482 and the Institute for Manufacturing and Materials Management grant. The work of the first named author was carried out at the Laboratory for International Fuzzy Engineering Research, chair of fuzzy theory, Japan.

## References

[1] H. R. Berenji, "Fuzzy logic controllers", *An Introduction to Fuzzy Logic Applications in Intelligent Systems* (R. R. Yager, L. A. Zadeh. eds.), Kluwer Academic Publ., 1991.

[2] D. Blackwell and M. A. Girshick, *Theory of Games and Statistical Decisions.* N.Y.: Dover, 1979.

[3] B. G. Buchanan and E. H. Shortliffe, *Rule-based Expert Systems. The MYCIN Experiments of the Stanford Heuristic Programming Project,* Reading, MA, Menlo Park, CA: Addison-Wesley, 1984.

[4] S. S. L. Chang and L. A. Zadeh, "On fuzzy mapping and control", *IEEE Transactions on Systems, Man and Cybernetics*, vol. SMC-2, pp. 30–34, 1972.

[5] P. Diamond and I. Fomenko, "Robustness and universal approximation in multilayer feedforward neural networks", *Proceedings of the ACNN92 (Conference on Neural Networks)*, Canberra, 1992.

[6] D. Dubois and H. Prade, *Fuzzy Sets and Systems: Theory and Applications.* N.Y., London: Academic Press, 1980.

[7] I. R. Goodman and H. T. Nguyen, *Uncertainty Models for Knowledge-based Systems,* Amsterdam: North Holland, 1985.

[8] G. J. Klir and T. A. Folger, *Fuzzy Sets, Uncertainty and Information*, Englewood Cliffs, NJ: Prentice Hall, 1988.

[9] V. Kreinovich and S. Kumar, "Optimal choice of &- and ∨-operations for expert values", *Proceedings of the 3rd University of New Brunswick Artificial Intelligence Workshop*, Fredericton, N.B., Canada, 1990, pp. 169–178.

[10] V. Kreinovich, C. Quintana, and R. Lea, "What procedure to choose while designing a fuzzy control? Towards mathematical foundations of fuzzy control", *Working Notes of the 1st International Workshop on Industrial Applications of Fuzzy Control and Intelligent Systems*, College Station, TX, 1991, pp. 123–130.

[11] V. Kreinovich, C. Quintana, R. Lea, O. Fuentes, A. Lokshin, S. Kumar, I. Boricheva, and L. Reznik, "What non-linearity to choose? Mathematical foundations of fuzzy control," *Proceedings of the 1992 International Fuzzy Systems and Intelligent Control Conference*, Louisville, KY, 1992, pp. 349–412.

[12] C. C. Lee, "Fuzzy logic in control systems: fuzzy logic controller.", *IEEE Transactions on Systems, Man and Cybernetics*, vol. 20, No. 2, pp. 404–435, 1990.

[13] G. G. Lorentz, *Approximation of Functions.* N.Y.: Holt, 1966.

[14] E. H. Mamdani, "Application of fuzzy algorithms for control of simple dynamic plant", *Proceedings of the IEE*, vol. 121, No. 12, pp. 1585–1588, 1974.

[15] G. C. Oden, "Integration of fuzzy logical information", *Journal of Experimental Psychology: Human Perception Perform.*, vol. 3, No. 4, pp. 565–575, 1977.

[16] M. Sugeno (editor), *Industrial Applications of Fuzzy Control*, Amsterdam: North Holland, 1985.

[17] L. Zadeh, "Fuzzy sets", *Information and Control*, vol. 8, pp. 338–353, 1965.

[18] L. A. Zadeh, "The concept of a linguistic variable and its application to approximate reasoning, Part 1", *Information Sciences*, vol. 8, pp. 199–249, 1975.

[19] H. J. Zimmermann. "Results of empirical studies in fuzzy set theory", *Applied General System Research* (G. J. Klir, ed.) N.Y.: Plenum, 1978, pp. 303–312.

# Backward-chaining with fuzzy "if... then..." rules

Thierry Arnould[1], Shun'ichi Tano, Yasunori Kato, Tsutomu Miyoshi
Laboratory for International Fuzzy Engineering Research
Siber Hegner Bldg. 4Fl., 89-1 Yamashita-Cho, Naka-Ku, Yokohama 231 Japan

***Abstract*** **- In this paper, we consider the issue of backward-chaining when using fuzzy "if...then..." rules. Usually, forward-chaining is used to induce new information given a set of rules and a set of data. On the other hand, backward-chaining is more efficient when one wants to check whether a given fact holds. Giving a mathematical formulation of the problem in the case of finite and continuous universes of discourse, we explain the method to solve the problem. We give an interpretation of the different possible solutions, and consider the problem of the choice of a particular solution. As the method is very sensitive to the given goal, in case no solution can be found, we see to which extent it is possible to modify the initial problem in order to find an approximate solution. Some results relative to finite sets are generalized to the case of continuous sets.**

***Key-words*** **- backward-chaining, fuzzy relational equations, inverse problem, principle of minimum specificity, approximate solutions**

## I. INTRODUCTION

Expert systems are usually composed of a rule base and a working memory. The rule base contains a set of rules, that is, the knowledge of an expert relative to a particular domain. The working memory contains all the information -the facts- that are known at a certain time. Most expert systems -except perhaps those designed for diagnosis- use a forward-chaining reasoning scheme only : considering the rules in the rule base, the facts in the working memory, the system looks for the rules and facts that can be combined to deduce new facts. The data in the working memory are matched to the conditions of the rules, and the rules having the highest matching degree are fired. FOREX, an decision support system for foreign exchange trading developped at LIFE, is such an expert system [5].

Another reasoning scheme, which is more efficient when goals are known, is backward-chaining : given a goal, the system looks for the sub-goals and rules that can be combined to deduce the given goal. To check if the initial goal holds, it is sufficient to check that one of the possible sub-goals holds. At present time, however, very few research has been carried out on fuzzy backward-chaining. In [4], Umano describes an implementation of backward chaining in a Fuzzy Production System. Contrary to [4], in this paper we focus our attention on backward-chaining when using fuzzy "if... then..." rules and the Generalized Modus Ponens as an inference scheme [6].

## II. THE FORMULATION OF BACKWARD-CHAINING

Backward-chaining in the crisp case can be expressed as follows :

*"Given a crisp rule "if x is A then y is B", it is sufficient to prove that the fact "x is A" holds in order to prove that the fact "y is B" holds."*

This can be expressed as follows

```
if x is A then y is B
               y is B ?
-----------------------------
        x is A ?
```

In the same way as the generalized Modus Ponens is a generalization to the fuzzy case of the classical Modus Ponens, it is possible to extend the previous reasoning scheme to the fuzzy case. The problem can then be written as follows :

*"Given a fuzzy rule "if x is A then y is B" and a goal "y is B'", where B' may be different from B, find a sub-goal "x is A'", such that when "x is A'" holds, "y is B'" too holds."*

Then, if we can find such a sub-goal "x is A'", to prove that the initial goal "y is B'" holds, it is sufficient to check that the fact "x is A'" holds.

We can express it as follows :

```
if x is A then y is B
               y is B' ?
------------------------------
        x is A' ?
```

The main problem in the fuzzy case is then, given the rule "if x is A then y is B" and the goal "y is B'", to "build" a suitable sub-goal "x is A'".

## III. MATHEMATICAL FORMULATION OF THE PROBLEM

X and Y are the respective universes of discourse on which the variables x and y take their values. A and A' are fuzzy sets on X, B and B' are fuzzy sets on Y. Each of the four fuzzy sets is characterized by its membership function. To model the existing link between x and y in the rule "if x is A then y is B", we use an implication operator I which, in fact, defines the conditional possibility distribution of y given x. More precisely, we have :

$$\forall x \in X, \forall y \in Y, \pi(y|x) = I(\mu_A(x), \mu_B(y)) \qquad (1)$$

[1] On leave from Rhône-Poulenc Industrialisation, Courbevoie, France

0-7803-0614-7/93$03.00 ©1993IEEE

As an inference scheme, we use the generalized Modus Ponens [6], so that the equation to be solved becomes :

$$\forall y \in Y, \mu_{B'}(y) = Sup_{x \in X}\left(\mu_{A'}(x) \wedge I(\mu_A(x), \mu_B(y))\right) \quad (2)$$

In forward-chaining reasoning, A, B, A' and I are given, and B' can be directly calculated using (2). In backward-chaining reasoning, on the other hand, A, B, B' and I are given and we must find A' such that (2) holds. Then, the backward-chaining problem is much more difficult to solve than the forward-chaining one.

As there are no general methods to solve (2) in the case of continuous universes of discourse, we first consider this equation in the case X and Y are discrete sets. We use the following notations :

$X = \{x_1, x_2, \ldots, x_i, \ldots, x_n\}$, $Y = \{y_1, y_2, \ldots, y_j, \ldots, y_m\}$,

$\forall i = 1,n,\ a'_i = \mu_{A'}(x_i)$, $\forall j = 1,m,\ b'_j = \mu_{B'}(y_j)$,

$\forall i = 1,n, \forall j = 1,m,\ I_{ij} = I(\mu_A(x_i), \mu_B(y_j))$

Equation (2) becomes :

$$\forall j = 1,m,\ b'_j = Max_{i=1,n}\left(a'_i \wedge I_{ij}\right) \quad (3)$$

Using matrix notations, we write $\mathbf{I} = (I_{ij})_{i=1,n,\ j=1,m}$, $\mathbf{A}' = (a'_1, a'_2, \ldots, a'_i, \ldots, a'_n)$ and $\mathbf{B}' = (b'_1, b'_2, \ldots, b'_j, \ldots, b'_m)$. Equation (3) is a usual fuzzy relational equation, written $\mathbf{A}' \circ \mathbf{I} = \mathbf{B}'$, where $\circ$ denotes the max-min composition operator. The problem of finding a sub-goal A' inducing the goal B' is then equivalent to solving the fuzzy relational equation $\mathbf{A}' \circ \mathbf{I} = \mathbf{B}'$. The problem of fuzzy backward-chaining in the discrete case can then be reformulated as follows :

*"Given a goal vector* **B'** *and an implication matrix* **I***, find a sub-goal vector* **A'** *such that* $\mathbf{A}' \circ \mathbf{I} = \mathbf{B}'$.*"*

## IV. Fuzzy Relational Equations

The so-called inverse problem of fuzzy relational equations can be expressed as follows :

*"Given a fuzzy relation $R \subset U \times V$ and a fuzzy subset $B \subset V$, find all the fuzzy subsets $A \subset U$ such that $A \circ R = B$."*

Many papers have already been written on the issue, and Sanchez [3] was the first author to provide a maximal solution to the equation. In another paper, Pappis and Sugeno [2] proposed an efficient method to calculate all the solutions of the equations. In the following, we give the main results of this method [2].

First, a partial ordering $\leq$ is defined on the set of all the fuzzy vectors of U by $\mathbf{A}^1 \leq \mathbf{A}^2 \Leftrightarrow \forall i = 1,n,\ a_i^1 \leq a_i^2$. Several composition operators are then defined as follows :

$\circ$-composition :
The $\circ$-composition of a row vector $\mathbf{A} = (a_i)_{i=1,n}$ with a column vector $\mathbf{B} = (b_i)_{i=1,n}$ is the scalar defined by $\mathbf{A} \circ \mathbf{B} = Max_{i=1,n}(a_i \wedge b_i)$. The $\circ$-composition of a row vector $\mathbf{A} = (a_i)_{i=1,n}$ and an $n \times m$ matrix $\mathbf{R} = (r_{ij})_{i=1,n,\ j=1,m}$ is the row vector $\mathbf{B} = (b_j)_{j=1,m}$ where $b_j = \mathbf{A} \circ \mathbf{R}_j$ and $\mathbf{R}_j$ is the j-th column vector of the matrix R. Then, $\forall j = 1,m,\ b_j = Max_{i=1,n}(a_i \wedge r_{ij})$.

α-composition :
The α-composition of two scalars x and y is given by

$$x\alpha y = \begin{cases} 1 & x \leq y \\ y & x > y \end{cases}$$

The α-composition of a column vector $\mathbf{A} = (a_i)_{i=1,n}$ with a scalar x is the column vector $\mathbf{B} = (b_i)_{i=1,n}$ given by $\forall i = 1,n,\ b_i = a_i \alpha x$. The α-composition of an $n \times m$ matrix $\mathbf{R} = (r_{ij})_{i=1,n,\ j=1,m}$ with a row vector $\mathbf{A} = (a_j)_{j=1,m}$ is an $n \times m$ matrix $\mathbf{S} = (s_{ij})_{i=1,n,\ j=1,m}$ where $\forall i = 1,n, \forall j = 1,m,\ s_{ij} = r_{ij} \alpha b_j$.

β-composition :
The β-composition of two scalars x and y is given by

$$x\beta y = \begin{cases} 0 & x < y \\ y & x \geq y \end{cases}$$

The β-composition of a column vector $\mathbf{A} = (a_i)_{i=1,n}$ with a scalar x is the column vector $\mathbf{B} = (b_i)_{i=1,n}$ given by $\forall i = 1,n,\ b_i = a_i \beta x$. The β-composition of an $n \times m$ matrix $\mathbf{R} = (r_{ij})_{i=1,n,\ j=1,m}$ with a row vector $\mathbf{A} = (a_j)_{j=1,m}$ is an $n \times m$ matrix $\mathbf{S} = (s_{ij})_{i=1,n,\ j=1,m}$ where $\forall i = 1,n, \forall j = 1,m,\ s_{ij} = r_{ij} \beta b_j$.

δ-composition :
The δ-composition of an $n \times m$ matrix $\mathbf{R} = (r_{ij})_{i=1,n,\ j=1,m}$ with a row vector $\mathbf{B} = (b_j)_{j=1,m}$ is an $n \times m$ matrix $\mathbf{S} = (s_{ij})_{i=1,n,\ j=1,m}$, where $\forall i = 1,n, \forall j = 1,m$, $s_{ij} = \left(Min_{k=1,m}(r_{ik} \alpha b_k)\right) \beta \left(r_{ij} \beta b_j\right)$. $\mathbf{R}\delta\mathbf{b}$ can be obtained from $\wedge(\mathbf{R}\alpha\mathbf{b})^T$ and $\mathbf{R}\beta\mathbf{b}$.

Φ-sets :
Given a column vector $\mathbf{A} = (a_i)_{i=1,n}$ such that $\forall i = 1,n, (a_i = 0 \text{ or } a_i = \hat{a})$ (two possible values only for all the $a_i$), the set $\Phi(\mathbf{A})$ of column vectors $\phi(\mathbf{A})$ is defined by

$\Phi(\mathbf{A}) = \{\phi(\mathbf{A})\}$ where $\phi(\mathbf{A}) = (\phi_i)_{i=1,n}$ and $(a_i = 0 \Rightarrow \phi_i = 0)$, $(a_i = \hat{a} \Rightarrow \phi_i = 0 \text{ or } \hat{a})$. Each vector $\phi(\mathbf{A})$ has only one component $\phi_i$ equal to $\hat{a}$, such that if there are k non-zero elements in A, there are k column vectors in $\Phi(\mathbf{A})$.

Given an $n \times m$ matrix $\mathbf{R} = (r_{ij})_{i=1,n,\ j=1,m}$, the set $\Phi(\mathbf{R})$ of matrices $\phi(\mathbf{R})$ is defined by $\Phi(\mathbf{R}) = \{\phi(\mathbf{R})\}$ where $\phi(\mathbf{R}) = [\phi(\mathbf{R}_1), \phi(\mathbf{R}_2), \ldots, \phi(\mathbf{R}_m)]$ and $\mathbf{R}_j$ is the j-th column vector of the matrix R.

The solutions of the inverse problem are given by the following theorems :

(i) Given an $n \times m$ matrix R and a row vector B, we have :

$\exists \mathbf{A} | \mathbf{A} \circ \mathbf{R} = \mathbf{B} \Leftrightarrow \wedge(\mathbf{R}\alpha\mathbf{B})^T \circ \mathbf{R} = \mathbf{B}$

$\exists \mathbf{A} | \mathbf{A} \circ \mathbf{R} = \mathbf{B} \Leftrightarrow \exists \phi(\mathbf{R}\beta\mathbf{B}) \in \Phi(\mathbf{R}\beta\mathbf{B}) | \vee(\phi(\mathbf{R}\beta\mathbf{B}))^T \circ \mathbf{R} = \mathbf{B}$

(ii) Given an $n \times m$ matrix R and a row vector B, assume that $\exists \mathbf{A} | \mathbf{A} \circ \mathbf{R} = \mathbf{B}$.

$\forall \mathbf{A} | \mathbf{A} \circ \mathbf{R} = \mathbf{B}$,

$\exists \phi(\mathbf{R}\delta\mathbf{B}) \in \Phi(\mathbf{R}\delta\mathbf{B}) | \vee(\phi(\mathbf{R}\delta\mathbf{B}))^T \leq \mathbf{A} \leq \wedge(\mathbf{R}\alpha\mathbf{B})^T$

$\forall \mathbf{A}, \forall \phi(\mathbf{R}\delta\mathbf{B}) \in \Phi(\mathbf{R}\delta\mathbf{B}) | \vee(\phi(\mathbf{R}\delta\mathbf{B}))^T \leq \mathbf{A} \leq \wedge(\mathbf{R}\alpha\mathbf{B})^T$,

$\mathbf{A} \circ \mathbf{R} = \mathbf{B}$

Algorithm :

• Step 1 : Calculate the row vector $\mathbf{A}^* = \wedge(\mathbf{R}\alpha\mathbf{B})^T$, i.e. $\forall i = 1,n,\ a_i^* = Min_{k=1,m}(r_{ik}\alpha b_k)$

• Step 2 : if $\mathbf{A}^*$ is such that $\mathbf{A}^* \circ \mathbf{R} = \mathbf{B}$, then the equation $\mathbf{A} \circ \mathbf{R} = \mathbf{B}$ has solutions and $\mathbf{A}^*$ is the least upper bound of the solution set, else the equation $\mathbf{A} \circ \mathbf{R} = \mathbf{B}$ has no solution.

• Step 3 : if the equation $\mathbf{A} \circ \mathbf{R} = \mathbf{B}$ has solutions, the minimal solutions $\mathbf{A}_{*i}$ of the equation are the minimal elements of the set $\{\vee(\phi(\mathbf{R}\delta\mathbf{B}))^T : \phi(\mathbf{R}\delta\mathbf{B}) \in \Phi(\mathbf{R}\delta\mathbf{B})\}$, and the solution set is given by the union of all the intervals that can be written using a minimal solution and the maximal one (see figure 1).

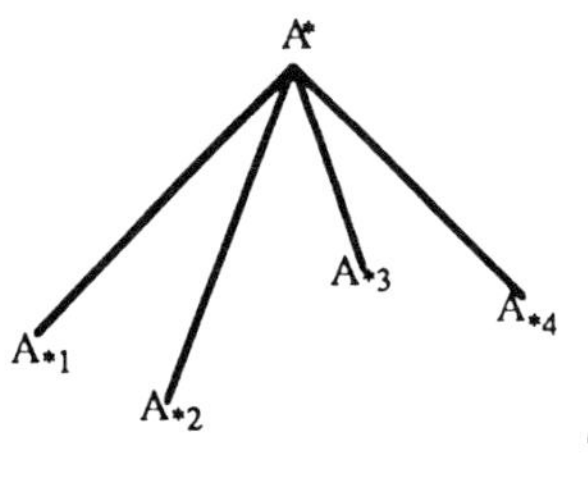


$$S = \{A | A \circ R = B\} = \bigcup_i \langle A_{*i}, A^* \rangle$$

Fig. 1. Structure Of The Solution Set

## V. Application To Backward-Chaining In The Discrete Case

The previous method can be used to look for the solutions of the backward-chaining problem. If the corresponding fuzzy relational equation $\mathbf{A}' \circ \mathbf{I} = \mathbf{B}'$ has no solution, it means that it is not possible to find a sub-goal A' which, combined with the fuzzy rule "if x is A then y is B", induces the initial goal B'. In that case, the rule "if x is A then y is B" can not be used to induce the goal B', and we must check whether other rules in the rule base can be used to induce the goal B' (backtracking technique).

On the other hand, if the fuzzy relational equation has solutions, it means that there is at least one sub-goal A' that can be combined with the rule "if x is A then y is B" to induce the goal B'. The use of the minimum and maximum operator in (3) leads to the non-uniqueness of the solution, and each solution of the equation $\mathbf{A}' \circ \mathbf{I} = \mathbf{B}'$ in fact corresponds to a possible sub-goal A'. In a one-step reasoning scheme, it is necessary and sufficient to check that at least one solution of the equation is present in the working memory. On the other hand, in a multiple-step reasoning scheme, it is necessary to choose one particular solution as a new goal, and to perform backward-chaining again with that new goal. A problem arises if we consider the choice of a particular solution as a new goal.

## VI. The Interpretation Of The Different Solutions Of The Equation

By definition, the maximal solution is greater than every other solution, and its membership function has the greatest possible value for all the elements of the universe of discourse. The corresponding possibility distribution defined for x is then the least specific one in accordance with the generalized modus ponens equation. The maximal solution can then be viewed as the solution giving the least specific (that is, least precise) information relative to the value of x.

On the other hand, each minimal solution is defined such that no smaller solution exist. A minimal solution is more restrictive than other greater solutions and then provide more precise information on a possible value of the variable x than greater solutions do. The existence of several minimal solutions can be interpreted by the existence of several possible values for x, the minimal solutions giving the most precise information on those possible values (see figure 2).

A general solution of the equation (that is, neither minimal nor maximal) is contained in an interval formed using a particular minimal solution and the maximal solution. Then, an "intermediate" solution can be considered as giving some information on one possible "value" of x (this particular value being given by the minimal solution), but less precise than the information given by the corresponding minimal solution (see figure 2).

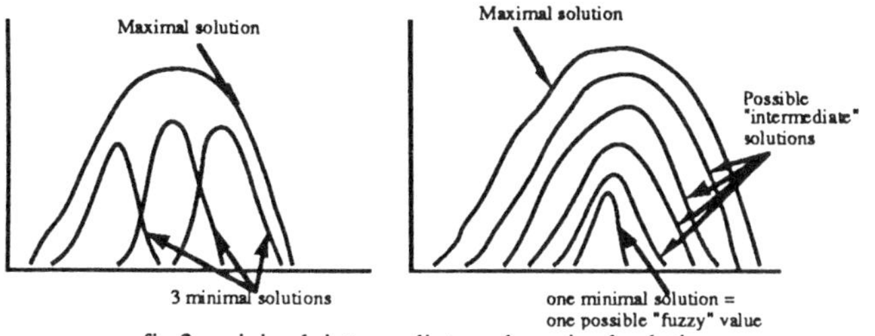

fig.2 : minimal, intermediate and maximal solutions

## VII. The Choice Of A Particular Solution As A New Sub-Goal

As the solution set is given by the union of several intervals, in order to choose a solution, it is first necessary to select a particular minimal solution, and then to choose the solution in the interval formed by the minimal solution and the maximal solution. According to the former interpretation, choosing a minimal solution is equivalent to assuming that we approximately know the value of the variable x, and choosing a solution between that minimal solution and the maximal one is equivalent to choosing the precision degree to which the value of the variable x is known. This shows that the choice of a particular solution is a very subjective choice.

The principle of minimum specificity [1] stipulates that when the available information stems from several sources (in our case, the different possible solutions), the possibility distribution that accounts is the least specific one satisfying the set of constraints induced by the pieces of information given by the different sources.

Then, the principle of minimum specificity on one hand, and the subjective aspect of the choice of a particular solution on the other hand, lead to choose, as a new goal, the least specific (that is, the least restrictive) solution, in our case the maximal solution of the equation.

## VIII. Sensitivity Of The Method To The Goal

To illustrate what we mean by sensitivity to the goal, let us consider the following example : X and Y are discrete sets, each one containing fifteen elements. The fuzzy sets A and B of the fuzzy rule are given by :

$$\mathbf{A} = \begin{pmatrix} 0 & 0 & 0 & 0 & 0.125 & 0.25 & 0.375 & 0.5 & 0.625 & 0.75 & 0.875 & 1 & 1 & 1 & 1 \end{pmatrix}$$

$$\mathbf{B} = \begin{pmatrix} 0 & 0 & 0 & 0 & 0.25 & 0.5 & 0.75 & 1 & 0.75 & 0.5 & 0.25 & 0 & 0 & 0 & 0 \end{pmatrix}$$

As an implication function, we choose the Gödel implication (the problem we want to mention here is a general problem and does not depend on the implication operator, the Gödel implication is just chosen here for the purpose of illustration.), the corresponding implication matrix **I** being given by :

$$\begin{pmatrix}
1 & 1 & 1 & 1 & 1 & 1 & 1 & 1 & 1 & 1 & 1 & 1 & 1 & 1 & 1 \\
1 & 1 & 1 & 1 & 1 & 1 & 1 & 1 & 1 & 1 & 1 & 1 & 1 & 1 & 1 \\
1 & 1 & 1 & 1 & 1 & 1 & 1 & 1 & 1 & 1 & 1 & 1 & 1 & 1 & 1 \\
1 & 1 & 1 & 1 & 1 & 1 & 1 & 1 & 1 & 1 & 1 & 1 & 1 & 1 & 1 \\
0 & 0 & 0 & 0 & 1 & 1 & 1 & 1 & 1 & 1 & 1 & 0 & 0 & 0 & 0 \\
0 & 0 & 0 & 0 & 1 & 1 & 1 & 1 & 1 & 1 & 1 & 0 & 0 & 0 & 0 \\
0 & 0 & 0 & 0 & 0.25 & 1 & 1 & 1 & 1 & 1 & 0.25 & 0 & 0 & 0 & 0 \\
0 & 0 & 0 & 0 & 0.25 & 1 & 1 & 1 & 1 & 1 & 0.25 & 0 & 0 & 0 & 0 \\
0 & 0 & 0 & 0 & 0.25 & 0.5 & 1 & 1 & 1 & 0.5 & 0.25 & 0 & 0 & 0 & 0 \\
0 & 0 & 0 & 0 & 0.25 & 0.5 & 1 & 1 & 1 & 0.5 & 0.25 & 0 & 0 & 0 & 0 \\
0 & 0 & 0 & 0 & 0.25 & 0.5 & 0.75 & 1 & 0.75 & 0.5 & 0.25 & 0 & 0 & 0 & 0 \\
0 & 0 & 0 & 0 & 0.25 & 0.5 & 0.75 & 1 & 0.75 & 0.5 & 0.25 & 0 & 0 & 0 & 0 \\
0 & 0 & 0 & 0 & 0.25 & 0.5 & 0.75 & 1 & 0.75 & 0.5 & 0.25 & 0 & 0 & 0 & 0 \\
0 & 0 & 0 & 0 & 0.25 & 0.5 & 0.75 & 1 & 0.75 & 0.5 & 0.25 & 0 & 0 & 0 & 0 \\
0 & 0 & 0 & 0 & 0.25 & 0.5 & 0.75 & 1 & 0.75 & 0.5 & 0.25 & 0 & 0 & 0 & 0
\end{pmatrix}$$

First, we take as a goal $\mathbf{B}'_1 = \mathbf{B} = \begin{pmatrix} 0 & 0 & 0 & 0 & 0.25 & 0.5 & 0.75 & 1 & 0.75 & 0.5 & 0.25 & 0 & 0 & 0 & 0 \end{pmatrix}$. The maximal solution of the equation is $\mathbf{A}'^{*}_1 = \begin{pmatrix} 0 & 0 & 0 & 0 & 0.25 & 0.25 & 0.5 & 0.5 & 0.75 & 0.75 & 1 & 1 & 1 & 1 & 1 \end{pmatrix}$ (it is worth noticing here that $\mathbf{B}'_1 = \mathbf{B}$ but $\mathbf{A}'^{*}_1 \neq \mathbf{A}$), and there are four minimal solutions.

In the two following examples, the goal $\mathbf{B}'$is a slightly modified version of **B** : the value of $b'_9$ is set to 0.749 and 0.751 respectively, instead of 0.75.

$\mathbf{B}'_2 = \begin{pmatrix} 0 & 0 & 0 & 0 & 0.25 & 0.5 & 0.75 & 1 & 0.749 & 0.5 & 0.25 & 0 & 0 & 0 & 0 \end{pmatrix}$ leads to a possible maximal solution $\mathbf{A}'^{*}_2 = \begin{pmatrix} 0 & 0 & 0 & 0 & 0.25 & 0.25 & 0.5 & 0.5 & 0.749 & 0.749 & 0.749 & 0.749 & 0.749 & 0.749 & 0.749 \end{pmatrix}$ given in step 1 of the former algorithm, but this is not a solution of the equation. Then, according to the criterion mentioned before for fuzzy relational equations, the equation has no solution.

$\mathbf{B}'_3 = \begin{pmatrix} 0 & 0 & 0 & 0 & 0.25 & 0.5 & 0.75 & 1 & 0.751 & 0.5 & 0.25 & 0 & 0 & 0 & 0 \end{pmatrix}$ leads to a possible maximal solution $\mathbf{A}'^{*}_3 = \begin{pmatrix} 0 & 0 & 0 & 0 & 0.25 & 0.25 & 0.5 & 0.5 & 0.75 & 0.75 & 1 & 1 & 1 & 1 & 1 \end{pmatrix}$ given in step 1 of the former algorithm, but this is not a solution of the equation. Then, the equation has no solution.

In fact, there is a very little difference between the three goals $\mathbf{B}'_1, \mathbf{B}'_2$ and $\mathbf{B}'_3$ but the fuzzy relational equation has solutions only in the first case. By nature, the information conveyed by a fuzzy set is not a precise information, and the solving method should not be so sensitive to the values of the goal $\mathbf{B}'$. To overcome this problem, in case the initial equation has no solution, we want to look for a (slightly) modified version of the initial goal $\mathbf{B}'$, such that the modified equation has solutions.

## IX. The Search For Approximate Solutions

As mentioned before, the necessary and sufficient condition for a fuzzy relational equation $\mathbf{A} \circ \mathbf{R} = \mathbf{B}$ to have solutions is $\wedge(\mathbf{R}\alpha\mathbf{B})^T \circ \mathbf{R} = \mathbf{B}$. Using the elements of the matrix **R** and the vector **B**, this condition can be written as follows :

$$S = \{\mathbf{A} | \mathbf{A} \circ \mathbf{R} = \mathbf{B}\} \neq \emptyset$$
$$\Leftrightarrow \forall j = 1,m, \exists i, \left( \left( b_j \leq r_{ij} \right) \text{ and } \left( \forall k : b_k < r_{ik}, b_j \leq b_k \right) \right) \quad (4)$$

In this section, we rewrite this criteria in a more simple way, in order to have indications on how the vector **B** could be modified in order to have solutions to the modified equation. To do so, we define the following matrices :

$$\mathbf{P} = (p_{ij})_{i=1,n,\, j=1,m} \text{ and } p_{ij} = \begin{cases} 1 & b_j \leq r_{ij} \\ 0 & b_j > r_{ij} \end{cases}$$

$$\mathbf{Q} = (q_{ij})_{i=1,n,\, j=1,m} \text{ and } q_{ij} = \begin{cases} 1 & b_j < r_{ij} \\ 0 & b_j \geq r_{ij} \end{cases}$$

$$\mathbf{C} = (c_i)_{i=1,n} \text{ and } c_i = \begin{cases} Min\{b_k | q_{ik} = 1\} & \{k | q_{ik} = 1\} \neq \emptyset \\ 1 & \{k | q_{ik} = 1\} = \emptyset \end{cases}$$

$$\mathbf{S} = (s_{ij})_{i=1,n,\, j=1,m} \text{ and } s_{ij} = \begin{cases} 1 & b_j \leq c_i \\ 0 & b_j > c_i \end{cases}$$

$$\mathbf{T} = (t_{ij})_{i=1,n,\, j=1,m} \text{ and } t_{ij} = Min\left(p_{ij}, s_{ij}\right)$$

The criterion to determine the non-emptiness of the solution set can now be expressed as follows :

$$\{\mathbf{A} : \mathbf{A} \circ \mathbf{R} = \mathbf{B}\} \neq \emptyset \Leftrightarrow \forall j = 1,m, \exists i, t_{ij} = 1 \quad (5)$$

Then, the equation has solutions if and only if each column of the matrix **T** has at least one non-zero element. On the contrary, if the matrix **T** is such that all the elements of one column are equal to 0, then the equation has no solution (on a calculation point of view, the criterion given in [2] and this one are of same computational complexity, but the criterion given in [2] requires to keep much more data into memory. Besides, using this criterion based on the matrix **T**, it is very easy to see if the equation has solutions).

Furthermore, we showed that the following result holds :

Theorem : In case the fuzzy relational equation $\mathbf{A} \circ \mathbf{R} = \mathbf{B}$ has solutions (that is, the matrix **T** verifies the above criterion), the maximal solution of the equation is given by the vector **C** defined above.

As a result of this, in case the matrix **T** verifies the above criterion, it is not necessary to calculate the maximal solution of the equation as it has already been calculated. This is one important advantage of this matrix-based method.

When the equation has no solution (one column j with all the elements $t_{ij}$ equal to 0), we want to modify the vector B such that each column of the matrix **T** has at least one non-zero element. We must consider four different cases :

- First case : $\begin{cases} p_{ij} = 1 \\ s_{ij} = 0 \end{cases}$, that is $c_i < b_j \leq r_{ij}$

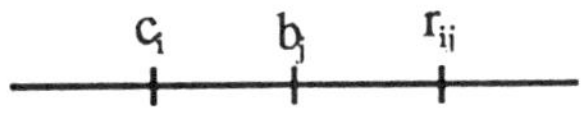


The value of $b_j$ and/or $c_i$ must be modified in order to have $b_j \leq c_i$. The smallest change will be obtained if we have $b_j = c_i$, and there are three possible methods to get this equality :

- modify only the value of $b_j$, that is, define the new vector B with the value $c_i$ instead of $b_j$,

- modify only the value of $c_i$, that is in fact, modify the value of one or several of the $b_k$ that are used to calculate $c_i$ so that $c_i$ becomes equal to $b_j$. We remind that $c_i = Min\{b_k : q_{ik} = 1\}$

- modify both $b_j$ and $c_i$ so that the two values become equal to a new value b.

- Second case : $\begin{cases} p_{ij} = 0 \\ s_{ij} = 1 \end{cases}$, that is $r_{ij} < b_j \leq c_i$

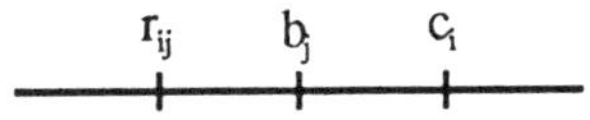


The only solution to have $p_{ij} = 1$ is to change the value of $b_j$, as the value of $r_{ij}$ can not be modified. The smallest change will be obtained if we have $b_j = r_{ij}$.

- Third case : $\begin{cases} p_{ij} = 0 \\ s_{ij} = 0 \end{cases}$ and $r_{ij} \leq c_i$, that is $r_{ij} \leq c_i < b_j$

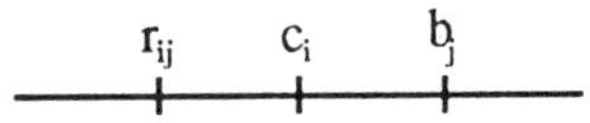


In order to have $p_{ij} = 1$ it is at least necessary to modify $b_j$ such that $b_j \leq r_{ij}$ but it will also be sufficient to have $b_j \leq c_i$ and $s_{ij} = 1$. The lowest modification will be obtained if we have $b_j = r_{ij}$.

- Fourth case : $\begin{cases} p_{ij} = 0 \\ s_{ij} = 0 \end{cases}$ and $c_i \leq r_{ij}$, that is $c_i \leq r_{ij} < b_j$

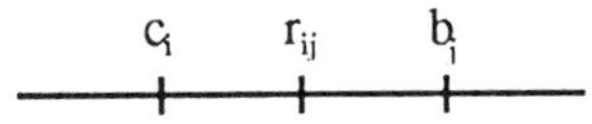


In order to have $p_{ij} = 1$it is at least necessary to modify $b_j$ such that $b_j \leq r_{ij}$ but, contrary to the previous case, it will not be sufficient to have $b_j \leq c_i$. A second modification will be necessary, either on the value of $b_j$, now equal to $r_{ij}$, or on the value of $c_i$ or on both values. The kind of modification to be done is the same as in the first case.

When all the possible modifications that can be done on the vector **B** in order to have at least one non-zero element in a column of the matrix **T** have been determined, the best modification to be done is that of lowest cost, that is the modification that requires the least global changes on the values of the vector **B**. One must be cautious however that the modifications that are done to have non-zero elements in the column j of the matrix **T** may have an influence on the value of the elements of other columns, such that some coefficients formerly equal to 1 may become equal to 0.

## X. Extension Of Some results Concerning The Maximal Solution Of The Equation To The Case Of Continuous Universes Of Discourse

As mentioned before, there is no general method to solve the equation (2) when the universes of discourse X and Y are continuous sets. However, some of the fundamental results established for discrete sets can be generalized to continuous sets. First, we need to extend some definitions to the continuous case.

$\circ$-composition

The $\circ$-composition of two fuzzy sets A and B on U is a scalar given by $A \circ B = Sup_{u \in U}(\mu_A(u) \wedge \mu_B(u))$. The $\circ$-composition of a fuzzy set A on U and a fuzzy relation R on $U \times V$ is a fuzzy set B on V given by $\forall v \in V$, $\mu_B(v) = A \circ R(v)$, where R(v) is the fuzzy set defined on U by $\mu_{R(v)}(u) = \mu_R(u,v)$.

Then, $\forall v \in V,\ \mu_B(v) = Sup_{u \in U}(\mu_A(u) \wedge \mu_R(u,v))$

$\alpha$-composition

The $\alpha$-composition of a fuzzy set A on U with a scalar x is a fuzzy set B on U given by $\forall u \in U, \mu_B(u) = \mu_A(u) \alpha x$. The $\alpha$-composition of a fuzzy relation R on $U \times V$ with a fuzzy set A on V is a fuzzy relation S on $U \times V$ given by $\forall u \in U, \forall v \in V, \mu_S(u,v) = \mu_R(u,v) \alpha \mu_A(v)$.

We showed that the following results hold :

Lemma 1 : Given a normalized fuzzy set B on U (that is, a fuzzy set B on U such that $\exists u \in U, \mu_B(u) = 1$) and a scalar x in [0,1], we have $\exists A: A \circ B = x \Leftrightarrow (B \alpha x) \circ B = x$

Lemma 2 : Given two fuzzy sets A and B on U, and a scalar x, $A \circ B = x \Rightarrow A \le (B \alpha x)$

Lemma 3 : Given a fuzzy relation R on $U \times V$ and two fuzzy sets A and B on U and V respectively, $A \circ R = B \Rightarrow A \le \wedge(R \alpha B)$, that is $\forall u \in U$, $\mu_A(u) \le Inf_{v \in V}(\mu_R(u,v) \alpha \mu_B(v))$.

Theorem : Given a fuzzy relation R on $U \times V$ and a normalized fuzzy set B on V, we have : $\exists A: A \circ R = B \Leftrightarrow \wedge(R \alpha B) \circ R = B$

This theorem gives a necessary and sufficient condition for a fuzzy relational equation to have a solution in the case of continuous sets, and provides the maximal solution when the solution set is not empty. The results relative to the minimal solutions of the equation can not be generalized to the continuous case.

## Conclusion

In this paper, we studied several issues related to backward-chaining when using fuzzy "if... then..." rules. We showed that in the case of discrete sets, the problem could be reduced to that of solving a fuzzy relational equation. Considering the interpretation of the different solutions, we decided to choose, as a new goal, the maximal solution of the corresponding equation, for which the results are also valid in the case of continuous sets. In the case no solution can be found to the fuzzy relational equation, we proposed a method to modify the initial goal so that the corresponding equation has solutions.

## References


[1] D. Dubois, H. Prade, Fuzzy sets in approximate reasoning - Part 1 : Inference with possibility distributions, *Fuzzy Sets and Systems* **40** (1991), 25th Anniversary Memorial Volume, pp. 143-202.

[2] C.P. Pappis, M. Sugeno, Fuzzy relational equations and the inverse problem, *Fuzzy Sets and Systems* **15** (1985), pp. 79-90.

[3] E. Sanchez, Resolution of composite relation equations, Inform. Control 30 (1) (1976), pp. 38-48.

[4] M. Umano, Y. Ezawa, Implementation of backward fuzzy reasoning in fuzzy production system, *Proceedings of IFES'91*, pp. 910-918.

[5] H. Yuize et al., Decision support system for foreign exchange trading ~ Practical implementation ~, *Proceedings of IFES'91*, pp. 971-982.

[6] L.A. Zadeh, A theory of approximate reasoning, *Memorandum M77/58, Electronic Research Laboratory*, University of California, Berkeley.

# Proposal of a Boolean Fuzzy Logic System Guaranteed a Logical Consistency

Hisashi Suzuki
Kyushu Institute of Technology
680-4 Kawazu
Iizuka, Fukuoka 820
Japan

Suguru Arimoto
The University of Tokyo
7-3-1 Hongo
Bunkyo-ku, Tokyo 113
Japan

***Abstract*— Based on a straightforward understanding of set operations, embedment of a fuzzy logic system into a Boolean algebra for guaranteeing a logical consistency contrarily to in traditional fuzzy logic systems is proposed, which is called a Boolean fuzzy logic system in this paper.**

## I. Introduction

One of regular formulations of fuzzy logic systems (cf. [1]-[5] for an overview of fuzzy theories) is an extension of Kleene's 3-valued logic system [6]. (Classic papers [6]-[8] cover all possibilities in formulating 3-valued logic systems.) Sentential operations, that is, a negation $\sim$, a disjunction $\vee$, a conjunction $\wedge$, and an implication $\rightarrow$ are characterized by a truth function $T_{\mathrm{K}}$ that maps sentences $p$ and $q$ (known their truth values $T_{\mathrm{K}}(p)$ and $T_{\mathrm{K}}(q)$) to a real in $[0,1]$ as follows: the truth value of $\sim p$ equals $1 - T_{\mathrm{K}}(p)$; that of $p \vee q$ equals the larger one of $T_{\mathrm{K}}(p)$ and $T_{\mathrm{K}}(q)$; that of $p \wedge q$ equals the smaller one of $T_{\mathrm{K}}(p)$ and $T_{\mathrm{K}}(q)$; and that of $p \rightarrow q$ equals the larger one of $1 - T_{\mathrm{K}}(p)$ and $T_{\mathrm{K}}(q)$.

Given a sentence "guess is right" with truth value 0.9, the value $T_{\mathrm{K}}$(guess is right $\rightarrow$ guess is right) equals the larger one of $1 - T_{\mathrm{K}}$(guess is right) and $T_{\mathrm{K}}$(guess is right), that is, $0.9 < 1$. This means that even though "guess is right with truth value 0.9," it is not always true that "guess is right with truth value 0.9." The substance of this semantic dilemma being inherent in fuzzy logic systems is detailed in [2], [9].

Axiomatically, the binary logic system corresponds to a special case of fuzzy logic system whose truth function $T_{\mathrm{B}}$ ranges over $\{0,1\}$. Since $T_{\mathrm{B}}(p)$ for any sentence $p$ can not assume values other than 0 and 1, the truth value of $p \rightarrow p$ given as the larger one of $1 - T_{\mathrm{B}}(p)$ and $T_{\mathrm{B}}(p)$ is identical with 1. However, in the above fuzzy logic system, the truth value of $p \rightarrow p$ given as the larger one of $1 - T_{\mathrm{K}}(p)$ and $T_{\mathrm{K}}(p)$ can not be 1 provided that $T_{\mathrm{K}}(p)$ is strictly between 0 and 1. In general, fuzzy logic systems do not guarantee the property $T_{\mathrm{K}}(p \rightarrow p) = 1$ called a "logical consistency."

In contrast, this paper suggests that there exists a method of embedding a fuzzy logic system into a Boolean algebra so that the logical consistency may be guaranteed. The suggested logic system is called a Boolean fuzzy logic system in this paper.

## II. Boolean Fuzzy Logic System

In this section, a Boolean fuzzy logic system is defined.

### *A. Membership, Grade, and Truth Functions*

We define a "membership function" $M$ as a mapping that associates a sentence $p$ with a set $M(p)$ comprising "members." We define a "grade function" $G$ as a mapping that associates a set of members with a nonnegative real, where let $G$ satisfy the additivity $G(A) + G(B) = G(A \cup B)$ for arbitrary exclusive sets $A$ and $B$ of members. Further, we define a "truth function" $T$ as a mapping that associates a sentence $p$ with a real in $[0,1]$ through a relation $T(p) = G(M(p)) \,/\, G(W)$, where $W$ denotes the whole set of members.

0-7803-0614-7/93$03.00 

When both a membership function and a grade function are clear with some physical images, computations of truth values can return to the Boolean operations on member subsets (cf. III). Even when the membership and the grade functions are not clear, the truth values are axiomatically computable without declaring these functions explicitly (cf. IV).

*B. Sentential Operations*

Considering any sentence $p$, we characterize its "negation" $\sim p$ as some sentence satisfying the equality $M(\sim p) = M^c(p)$, where $M^c(p)$ denotes $M(p)$'s complementary set $W - M(p)$. Considering any sentences $p$ and $q$, we characterize the "disjunction" $p \vee q$ of them as some sentence satisfying the equality $M(p \vee q) = M(p) \cup M(q)$. Similarly, we characterize the "conjunction" $p \wedge q$ as some sentence satisfying $M(p \wedge q) = M(p) \cap M(q)$, the "implication" $p \rightarrow q$ as some sentence satisfying $M(p \rightarrow q) = M^c(p) \cup M(q)$, and the "equivalence" $p \leftrightarrow q$ as the conjunction of $p \rightarrow q$ and $q \rightarrow p$.

*C. Truth Values*

The following properties are derived from the above characterizations of sentential operations.

*1) Negation:* The truth value of $\sim p$ is

$$T(\sim p) = \frac{G(M(\sim p))}{G(W)} = \frac{G(M^c(p))}{G(W)} \qquad (1)$$

that coincides with the ratio $[G(W) - G(M(p))] \; / \; G(W)$, that is, $1 - T(p)$ since the additivity on $G$.

*2) Disjunction:* The truth value of $p \vee q$ is

$$T(p \vee q) = \frac{G(M(p \vee q))}{G(W)} = \frac{G(M(p) \cup M(q))}{G(W)} \qquad (2)$$

that coincides with the ratio $G(M(p)) \; / \; G(W)$ if $M(q) \subseteq M(p)$ or with the ratio $G(M(q)) \; / \; G(W)$ if $M(p) \subseteq M(q)$, that is, with the larger one of $T(p)$ and $T(q)$ if $M(p) \subseteq M(q)$ or $M(q) \subseteq M(p)$.

*3) Conjunction:* The truth value of $p \wedge q$ is

$$T(p \wedge q) = \frac{G(M(p \wedge q))}{G(W)} = \frac{G(M(p) \cap M(q))}{G(W)} \qquad (3)$$

that coincides with the ratio $G(M(p)) \; / \; G(W)$ if $M(p) \subseteq M(q)$ or with the ratio $G(M(q)) \; / \; G(W)$ if $M(q) \subseteq M(p)$, that is, with the smaller one of $T(p)$ and $T(q)$ if $M(p) \subseteq M(q)$ or $M(q) \subseteq M(p)$.

*4) Implication:* The truth value of $p \rightarrow q$ is

$$T(p \rightarrow q) = \frac{G(M(p \rightarrow q))}{G(W)} = \frac{G(M^c(p) \cup M(q))}{G(W)} \qquad (4)$$

that coincides with the ratio $G(M^c(p)) \; / \; G(W)$ if $M(q) \subseteq M^c(p)$ or with the ratio $G(M(q)) \; / \; G(W)$ if $M^c(p) \subseteq M(q)$, that is, with the larger one of $1 - T(p)$ and $T(q)$ if $M^c(p) \subseteq M(q)$ or $M(q) \subseteq M^c(p)$. Especially, the truth value of the implication from $p$ to itself is

$$T(p \rightarrow p) = \frac{G(M^c(p) \cup M(p))}{G(W)} = \frac{G(W)}{G(W)} = 1, \qquad (5)$$

and the logical consistency is guaranteed.

*5) Equivalence:* The truth value of $p \leftrightarrow q$, that is, of $(p \rightarrow q) \wedge (q \rightarrow p)$ is computable by combining (4) and (3).

Thus, there is a certain relation between the truth values for sentential operations in the Boolean fuzzy logic system and those in the original fuzzy logic system.

## III. Explicit Computation

This section shows a method of computing explicitly the truth values for sentential operations when both the membership and grade functions are clear.

*A. Boolean Algebra*

We can ascertain the following properties for arbitrary member subsets $A \subseteq W$, $B \subseteq W$ and $C \subseteq W$. (i) $A$ is a subset of itself. $A$ equals $B$ if $A$ is a subset of $B$ and $B$ is a subset of $A$. Further, $A$ is a subset of $C$ if $A$ is a subset of $B$ and $B$ is a subset of $C$. Hence, the whole family $\mathcal{F}$ of $W$'s subsets with the relation $\subseteq$ forms a partially ordered set. (ii) $A \cup B$ is the least upper bound $A + B$ of $\{A, B\}$ under the partial order $\subseteq$ while $A \cap B$ is the greatest lower bound $A \cdot B$

of $\{A, B\}$, and the triple $(\mathcal{F}, +, \cdot)$ forms a lattice. (iii) The distributive law on $\cup$ and $\cap$ suggests the distributive law on $+$ and $\cdot$. Hence, $(\mathcal{F}, +, \cdot)$ forms a distributive lattice. (iv) There exist the smallest unit $\emptyset$ and the largest unit $W$ such that the equalities $A \cdot \emptyset = \emptyset$ and $A + W = W$ are valid whatever $A$ is. Hence, the quintuple $(\mathcal{F}, +, \cdot, \emptyset, W)$ forms a bounded lattice. (v) There exists an element $\bar{A} = A^c$ in $\mathcal{F}$ such that the equalities $A + A^c = W$ and $A \cdot A^c = \emptyset$ hold. Hence, the sextuple $(\mathcal{F}, +, \cdot, ^{-}, \emptyset, W)$ forms a complemented lattice. (Traditional fuzzy subsets do not conform to the complementary law, which causes lack of the logical consistency.) These properties (i)-(v) show that $\mathcal{F}$ forms a distributive, bounded and complemented lattice, that is, forms a Boolean algebra.

Now, by substituting the set operations $^c$, $\cap$ and $\cup$ in II.B respectively for the Boolean operations $^{-}$, $\cdot$ and $+$, we have the equality between $M(\sim p)$ and $\bar{M}(p)$, that between $M(p \vee q)$ and $M(p) + M(q)$, that between $M(p \wedge q)$ and $M(p) \cdot M(q)$, that between $M(p \rightarrow q)$ and $\bar{M}(p) + M(q)$, and that between $M(p \leftrightarrow q)$ and $(\bar{M}(p) + M(q)) \cdot (\bar{M}(q) + M(p))$. These equalities suggest the following.

A pair $(p, q)$ of sentences can be mapped by $M$ to a pair $(A, B)$ of member subsets. The least upper bound $C$ of $\{A, B\}$ is computable on the Boolean algebra. Further, $C$ can be mapped to a real $G(C)$ / $G(W)$ that is equal to $G(M(p \vee q))$ / $G(W) = T(p \vee q)$ since $C = M(p) + M(q) = M(p \vee q)$. This computation process is illustrated as follows.

$$\begin{array}{ccc} (p, q) & \xrightarrow{T(\vee)} & \frac{G(C)}{G(W)} \in [0, 1] \\ M \downarrow & & \uparrow G \\ (A, B) \in \mathcal{F} \times \mathcal{F} & \xrightarrow{+} & C \in \mathcal{F} \end{array} \tag{6}$$

Similarly, we have the equality between $T(\sim p)$ and $E(\bar{M}(p))$ / $E(W)$, that between $T(p \wedge q)$ and $G(M(p) \cdot M(q))$ / $G(W)$, that between $T(p \rightarrow q)$ and $G(\bar{M}(p) + M(q))$ / $G(W)$, and that between $T(p \leftrightarrow q)$ and $G((\bar{M}(p) + M(q)) \cdot (\bar{M}(q) + M(p)))$ / $G(W)$, each of which is illustrated like (6).

The truth values for sentential operations are computable in this way via the Boolean operations that are effectively manageable on high-level computer languages.

## IV. Implicit Computation

This section shows a method of computing the truth values for sentential operations without declaring the membership and grade functions explicitly.

### A. Disjunctive Normal Form

The formulation of sentential operations in II.B suggests the following properties.

*1) Negation:* Since

$$\begin{aligned} M^c(p) &= M^c(p) \cap (M(q) \cup M^c(q)) \\ &= (M^c(p) \cap M(q)) \cup (M^c(p) \cap M^c(q)), \end{aligned} \tag{7}$$

$T(\sim p)$ coincides with $T((\sim p \wedge q) \vee (\sim p \wedge \sim q))$ and, in this sense, the sentence $\sim p$ is equivalent to the disjunction of $\sim p \wedge q$ and $\sim p \wedge \sim q$.

*2) Disjunction:* Since

$$\begin{aligned} & M(p) \cup M(q) \\ &= M(p) \cup (M^c(p) \cap M(q)) \\ &= [M(p) \cap (M^c(q) \cup M(q))] \cup (M^c(p) \cap M(q)) \\ &= (M(p) \cap M^c(q)) \cup (M(p) \cap M(q)) \\ &\quad \cup (M^c(p) \cap M(q)), \end{aligned} \tag{8}$$

$T(p \vee q)$ coincides with $T((p \wedge \sim q) \vee (p \wedge q) \vee (\sim p \wedge q))$ and, in this sense, the sentence $p \vee q$ is equivalent to the disjunction of $p \wedge \sim q$, $p \wedge q$, and $\sim p \wedge q$.

*3) Implication:* Since

$$\begin{aligned} & M^c(p) \cup M(q) \\ &= (M^c(p) \cap M(q)) \cup (M^c(p) \cap M^c(q)) \\ &\quad \cup (M(p) \cap M(q)), \end{aligned} \tag{9}$$

$T(p \rightarrow q)$ coincides with $T((\sim p \wedge q) \vee (\sim p \wedge \sim q) \vee (p \wedge q))$ and, in this sense, the sentence $p \rightarrow q$ is equivalent to the disjunction of $\sim p \wedge q$, $\sim p \wedge \sim q$, and $p \wedge q$.

*4) Equivalence:* Since

$$\begin{aligned} & (M^c(p) \cup M(q)) \cap (M^c(q) \cup M(p)) \\ &= (M(p) \cap M(q)) \cup (M^c(p) \cap M^c(q)), \end{aligned} \tag{10}$$

the value $T(p \leftrightarrow q) = T((p \rightarrow q) \wedge (q \rightarrow p))$ coincides with $T((p \wedge q) \vee (\sim p \wedge \sim q))$ and, in this sense, the sentence $p \leftrightarrow q$ is equivalent to the disjunction of $p \wedge q$ and $\sim p \wedge \sim q$.

Any sentence composed with the sentential operations can be thus rewritten into some disjunctive normal form as well as in the traditional binary logic system.

*B. Synthesis*

Given a set $S = \{p_1, \cdots, p_m\}$ of sentences (where $m$ denotes the cardinality of $S$), let $x_i$ for every $i = 1, \cdots, m$ assume either $p_i$ or its negation $\sim p_i$. Further, suppose that the value of $T(x_1 \wedge \cdots \wedge x_m)$ for every $(x_1, \cdots, x_m)$ in $\{p_1, \sim p_1\} \times \cdots \times \{p_m, \sim p_m\}$ is known.

Now, given an arbitrary sentence $s$ composed of $p_1, \cdots, p_m$ with the sentential operations, we can rewrite $s$ into a disjunctive normal form (that is, the disjunction of some $x_1 \wedge \cdots \wedge x_m$ s) by referring to IV.A. The supposition of the additivity on $G$ suggests that $T(s)$ is then computable as the summation of the truth values for these $x_1 \wedge \cdots \wedge x_m$ s.

*C. Example*

Let a source generate 0 or 1 four times with an independent and uniform distribution, and regard the probability that "the $i$th symbol may be 1" $(i = 1, \cdots, 4)$ as the truth value of ONE($i$). When assuming $S$ as {ONE(1), ONE(2), ONE(3)}, the value of $T(x_1 \wedge x_2 \wedge x_3)$ is identical with $(1/2)^3$ for every $(x_1, x_2, x_3)$ in {ONE(1), ~ONE(1)} × {ONE(2), ~ONE (2)} × {ONE(3), ~ONE(3)}. Note that on this setting we declare neither membership function nor grade function explicitly.

By referring to IV.A, we can rewrite a sentence ~ONE(1) ∧ ~ONE(2) → ONE(3) into the disjunction of the following seven $x_1 \wedge x_2 \wedge x_3$s: ONE(1) ∧ ONE(2) ∧ ONE(3), ONE(1) ∧ ONE(2) ∧ ~ONE(3), ONE(1) ∧ ~ONE(2) ∧ ONE(3), ONE(1) ∧ ~ONE(2) ∧ ~ONE(3), ~ONE(1) ∧ ONE(2) ∧ ONE(3), ~ONE(1) ∧ ONE(2) ∧ ~ONE(3), and ~ONE(1) ∧ ~ONE(2) ∧ ONE(3). The truth value of ~ONE(1) ∧ ~ONE(2) → ONE(3) is computable as the summation of the truth values for these $x_1 \wedge x_2 \wedge x_3$s, that is, $(1/2)^3 \times 7 = 0.875$. This result means that the premises "the first symbol is 0" and "the second symbol is 0" conclude with truth value 0.875 "the third symbol is 1."

*Remark:* As far as understanding only in terms of probabilities, on condition that "the first symbol is 0" and "the second symbol is 0," the conditional probability that "the third symbol may be 1" is 0.5.

## V. Conclusion

A method of embedding a fuzzy logic system into a Boolean algebra to guarantee a logical consistency was suggested. By formulating the fuzzy logic system as in the way of this paper, there arises a possibility of inheriting a large store of axioms that are known in the traditional binary logic system. Developing effective applications of the Boolean fuzzy logic system is one of future problems.

## References


[1] L. A. Zadeh, "Fuzzy sets," *Information & Control*, vol. 8, pp. 338-353, 1965.

[2] L. A. Zadeh and R. E. Bellman, "Local and fuzzy logic," *Modern Uses of Multiple-Valued Logic*, J. H. Dunn and G. Epstein, Eds., D. Reidel, 1976.

[3] B. R. Gaines, "Foundations of fuzzy reasoning," *Int. J. Man-Machine Studies*, vol. 8, pp. 623-668, 1976.

[4] E. H. Mamdani, "Advances in the linguistic synthesis of fuzzy controller," *Int. J. Man-Machine Studies*, vol. 8, pp. 669-678, 1976.

[5] R. Turner, *Logics for Artificial Intelligence*, Ellis Horwood, 1984.

[6] S. Kleene, *Introduction to Metamathematics*, Van Nostrand, 1952.

[7] J. Lukasiewicz, "On 3-valued logic," *Polish Logic*, S. McCall, Ed., Oxford Univ. Press, 1967.

[8] D. Bochvar, "On three-valued calculus and its application to the analysis of contradictions," *Mathematiceskij Sbornik*, vol. 4, pp. 353-369, 1939.

[9] S. Haack, *The Philosophy of Logics*, Cambridge Univ. Press, 1981.

# Referential Modes of Reasoning

Kaoru Hirota* Witold Pedrycz**

*Dept. of Instrument and Control Engineering,College of Engineering
Hosei University, Koganei-city, Tokyo 184, Japan
**Dept. of Electrical and Computer Engineering
University of Manitoba, Winnipeg, Canada R3T 2N2
pedrycz@eeserv.ee.umanitoba.ca

**Abstract - We will propose and study models of referential structures and referential modes of reasoning for fuzzy data. The style of information processing considered there is aimed at reasoning about some global properties (such as, for instance similarity, dominance, inclusion, etc.)of the spaces in which the fuzzy data are situated. The distributed models designed in terms of logic-based neural networks realize the mapping of these properties between the spaces. The scheme of this type embraces reasoning about similarity, difference, dominance, and inclusion of the conclusions that is based on the corresponding relationships between the universes and their strength discerned for the antecedents. For instance, the conclusions issued within the scheme are of the form: b and b' are similar, b and b' are different, etc. It will be also clarified how, considering the available degrees of satisfaction of these properties, the corresponding fuzzy sets of conclusion can be determined.**

## I. INTRODUCTORY REMARKS AND PROBLEM STATEMENT

Let us consider pairs of fuzzy sets defined in the two discrete input and output universes of discourse[1]

$$(\mathbf{a}_1,\mathbf{b}_1)\ (\mathbf{a}_2,\mathbf{b}_2)\ ,\ldots,(\mathbf{a}_N,\mathbf{b}_N)$$

where $\mathbf{a}_k \in [0,1]^n$ and $\mathbf{b}_k \in [0,1]^m$ , k=1,2,...,N. The frequently accepted approach being used in describing these fuzzy data leads to identification of functional or relational relationships between the variables, say **b**=f(**a**) or $\mathbf{b}=\mathbf{a} \circ \mathbf{r}$ ,where **r** denotes relation existing between the data while "$\circ$" is used to denote a set-to-relation composition operator. Another alternative look at the fuzzy data can be derived by determining transformation of certain global properties defined in the input and output spaces. Formally speaking we will treat this transformation as a mapping

$$g: \mathcal{F}(\mathbf{a}',\mathbf{a}'') \rightarrow \mathcal{F}(\mathbf{b}',\mathbf{b}'')$$

where $\mathcal{F}(.,.)$ indicates a binary operation applied to any two elements in $[0,1]^n$ (and $[0,1]^m$, respectively) that describes a certain property of the space like nearness (similarity), difference, inclusion, dominance, etc. These properties constitute global rather than local (pointwise) characteristics of the space. They do not depend upon the single points of the space but take on a homogenous form that is valid across the entire universe. Depending on the nature of the mapping $\mathcal{F}$ one can envision several specific interpretations. Take, for instance, the property of similarity. This leads us directly to the scheme of reasoning by analogy [4]. In this form of reasoning one looks at the situation for which a single pair (**a**',**b**') is completely specified while the second one, say (**a**,**b**) is given partially, namely **b** has to determined while **a** is provided. In other words, **b** should adhere to the property of similarity (analogy), particularly the relationship $\mathcal{F}$(**b**,**b**') should be fulfilled to a certain degree depending on the level of similarity achieved for the first pair **a** and **a**', $\mathcal{F}$(**a**,**a**'). Once the degree of satisfaction of $\mathcal{F}$(**b**,**b**') has been computed then the fuzzy set **b** can be reconstructed as a solution to the corresponding inverse problem.

The paper is structured into sections. They will be dealing with the general architecture of the referential mode of reasoning involving their logic-based neural networks, knowledge representation aspects, learning issues and detailed schemes of reasoning.

## II. PRINCIPAL ELEMENTS OF THE ARCHITECTURE

The way of handling the global property of the spaces can be concisely summarized as below

[1]Support from the Natural Sciences and Engineering Research Council of Canada and MICRONET is gratefully acknowledged.

0-7803-0614-7/93$03.00 ©1993IEEE

$$\mathcal{F} \rightarrow g \rightarrow \mathcal{F}^{-1}$$

Each processing stage distinguished in this scheme takes on a specific interpretation:

(i) $\mathcal{F}$ operates on any two elements of $[0,1]^n$ ,say **a** and **a**' and returns a degree of satisfaction of the property characterized by $\mathcal{F}$,

(ii) the transformation "g" maps the level of satisfaction of $\mathcal{F}$ achieved in the input space into the level of satisfaction of this property in the output space,

(iii) finally, $\mathcal{F}^{-1}$stands for the operation inverse to that given by $\mathcal{F}$. The inverse problem arising now pertains to the determination of a specific object, say **b**', based on the level of satisfaction of the global property elicited in the output space and the characteristics of **b**.

Being acquainted how the scheme functions, we will now look at its functional components in more detail.

*A. Representing global properties of similarity, difference, inclusion and dominance.*

The introduced global properties are of a referential nature, namely they are determined with respect to any two points of the universe of discourse. The relationship $\mathcal{F}$(**a**',**a**) can be also expressed coordinatewise in the form,

$$\mathcal{F}(\mathbf{a}',\mathbf{a})= [\mathcal{F}(a_1',a_1)\ \mathcal{F}(a_2',a_2)\ ...\mathcal{F}(a_n',a_n)].$$

Several basic global properties are worth studying:

- similarity (equality). The degree of its satisfaction by x and a, x, a $\in$ [0,1] is embodied by considering the equality index defined as [6], EQ(x,a)=a $\equiv$ x

$$a \equiv x = \frac{1}{2}[(a\ \varphi\ x) \wedge (x\ \varphi\ a) + (\bar{a}\ \varphi\ \bar{x}) \wedge (\bar{x}\ \varphi\ \bar{a})]$$

where $\wedge$ stands for the minimum operation, overbar denotes complement and $\varphi$ is used to describe pseudocomplement (implication), cf. [6], $a\varphi x=\sup\{c \in [0,1] \mid atc \le x\}$.

- inclusion. The degree of inclusion of x in a, INCL(x,a), is expressed as

$$\text{INCL}(x,a)= x\ \varphi\ a$$

difference. The difference is viewed as a feature dual to the property of similarity, namely

$$\text{DIFF}(x,a)=1-\text{EQ}(x,a)$$

- dominance. The degree of dominance, DOM(x,a), is defined as

$$\text{DOM}(x,a) = a\varphi x$$

It represents a degree to which x dominates a.

The plots of the two first properties, EQ and INCL , are shown in Fig.1 (both of them are implemented with the use of the Lukasiewicz implication where the $\varphi$ -operator is given by

$$a\varphi x=\begin{cases} 1-a+x\ ,\text{if } a \ge x \\ 1,\text{otherwise} \end{cases}$$

This form of the implication used there implies a piecewise linear form of the derived indices.

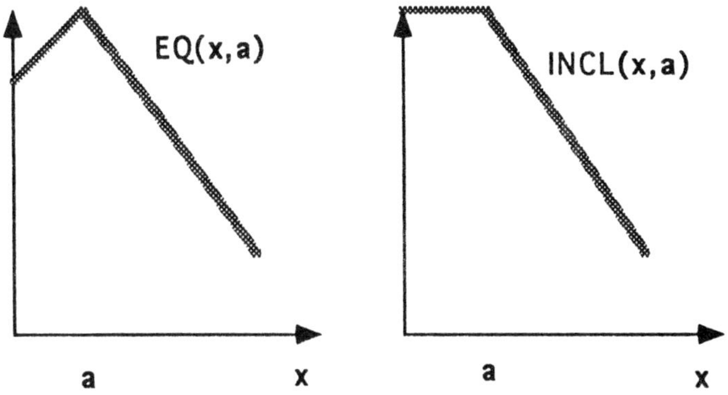


Fig.1. Plots of EQ(x,a) and INCL(x,a).

*B. Mapping the global properties between the spaces*

The global properties will be mapped from the input to the output space by considering logic-based neural networks. Denote by **x** and **y** the results of satisfaction of the global properties in the respective spaces, namely **x**=$\mathcal{F}$(**a**,**a**'), **y**=$\mathcal{F}$(**b**,**b**'). We will start from the two types of neurons that will be directly used in the design of the network.

OR neuron. The OR neuron is described as

$$y = \text{OR}\ (\mathbf{x};\ \mathbf{w})$$

or coordinatewise

$$y = \text{OR}\ [x_1 \text{ AND } w_1,\ x_2 \text{ AND } w_2,\ ...,\ x_n \text{ AND } w_n]$$

where $\mathbf{w} = [w_1, w_2, ..., w_n] \in [0,1]^n$ forms a vector of connections (weights) of the neuron. The standard implementation of the fuzzy set connectives standing in the formula of the neuron involves the usage of triangular norms,

$$y = \overset{n}{\underset{i=1}{S}} [x_i \, t \, w_i]$$

AND neuron. In comparison to the OR neuron, the AND neuron uses t-norm to aggregate partial results,

$$y = AND(\mathbf{x}; \mathbf{w})$$

Again in the notation of the triangular norms this translates into the expression

$$y = \overset{n}{\underset{i=1}{T}} [x_i \, s \, w_i].$$

The aim of the logic-based neural network consisting of these two type of the neurons is to approximate logical relationships between **x** and **y**, **y**=g(**x**). The basic architecture encompasses a single hidden layer composed of the AND nodes (neurons) followed by the OR nodes constituting the output layer of the network. Each AND neuron functions as a generalized minterm. The number of the neurons determines the size of the hidden layer and affects the representation capabilities of the network. The OR neurons placed in the output layer are used to build the generalized union of the minterms. Recall that any Boolean function can be represented as a sum of minterms; here the generalization includes standard forms of minterms and naxterms implemented via the AND and OR neurons. Thus the resulting neural network can be utilized to represent a fuzzy function and in this way approximate the available fuzzy data.

## *C. Inverse Problem*

The global property conveyed by $\mathcal{F}$ implies an associated inverse problem that is formulated as follows: Let us assume that the property $\mathcal{F}$ operating on **b** and **b**' satisfies a certain condition $\Gamma$. One of its arguments ,say **b**, is known. Determine all **b**'s satisfying the condition

$$\mathcal{F}(\mathbf{b}',\mathbf{b}) \in \Gamma$$

In particular, depending on the character of this property, we will be interested in more specific inequality problems where

$$EQ(\mathbf{b}',\mathbf{b}) \geq \gamma \qquad INCL(\mathbf{b}',\mathbf{b}) \geq \gamma$$
$$DOM(\mathbf{b}',\mathbf{b}) \geq \gamma \qquad DIFF(\mathbf{b}',\mathbf{b}) \leq \gamma.$$

The above conditions give rise to the families of scalar problems involving inequality constraints

$$\mathcal{F}(b'_j,b_j) \leq \gamma_j \qquad \text{or} \qquad \mathcal{F}(b'_j,b_j) \geq \gamma_j$$

j=1,2,...,m.

Depending on the character of $\mathcal{F}$ the following observations hold:

(i) similarity property: the solution set to $b'_j \equiv b_j \geq \gamma$ reduces to a single element $b'_j = b_j$ iff $\gamma_j = 1$. Otherwise the solution to this inverse problem forms a subinterval of [0,1], cf.[6 ]

(ii) the results for the difference property are dual to these summarized in (i), namely $b'_j = b_j$ iff $\gamma_j = 0$ In all the remaining situations higher values of $\gamma$ induce the membership intervals originating around $b_j$.

(iii) inclusion. This property induces the inverse problem in which $\gamma_j = 1$ produces the interval of membership values equal to $[0,b_j]$. Lower values of $\gamma$ generate broader ranges of admissible values of membership.

(iv) dominance. For the dominance property the condition $\gamma_j = 1$ produces subinterval $[b_j,1]$. Again, the lower the value of $\gamma$ ,the broader the range of the acceptable membership values.

It is worth mentioning that except $\gamma$ =0 (difference) and $\gamma$=1 (all the remaining problems), the solution to the inverse problem happens to constitute a certain interval of [0,1]. This means that even though **b** is a genuine fuzzy set, the result becomes an interval-valued fuzzy set [9].

We will conclude this section with the detailed architecture of the referential structure, see Fig. 2.

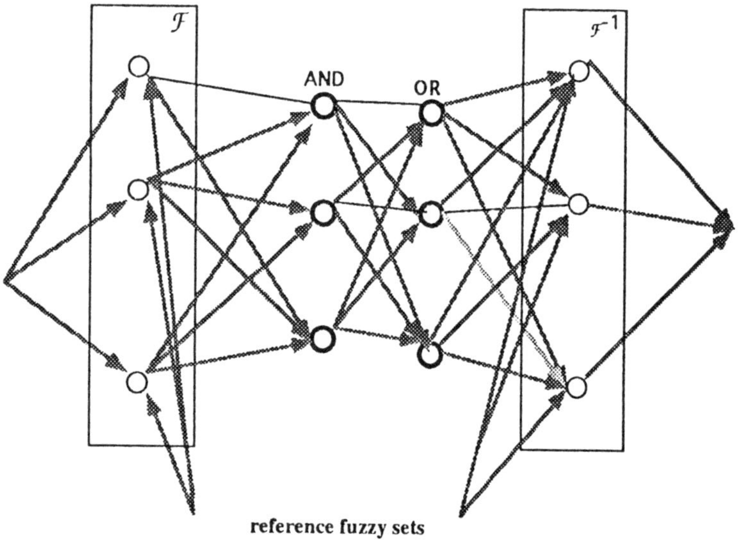


Fig. 2. Detailed architecture of the referential structure

## III. LEARNING IN THE REFERENTIAL STRUCTURE

The learning of the neural network is based on available cases and is carried out in a supervised mode. The original learning set consists of N cases (pairs) $r_k = (\mathbf{a}_k, \mathbf{b}_k)$. The process of learning proceeds in a referential format. This involves all different pairs of the cases. These cases used for training can be conveniently arranged in the matrix form,

$$\begin{bmatrix} (r_1, r_1) & (r_1, r_2) & (r_1, r_N) \\ (r_2, r_1) & (r_2, r_2) & (r_2, r_N) \\ (r_N, r_1) & (r_N, r_2) & (r_N, r_N) \end{bmatrix}$$

Each pair $(r_i, r_j)$ represents an individual training situation in which one of the cases, say $r_i$, is viewed as a point of reference. The matrix given above is symmetrical therefore only a single element out of each pair $(r_i, r_j)$ and $(r_j, r_i)$ will be utilized for training. This in total results in N(N + 1)/2 situations available in the training phase.

The learning set collected in that way is used to adjust the connections of the logic neurons of the network. Assuming a certain performance index Q we will be modifying the connections to achieve minimum of Q,

$$\min_{\text{connections}} Q$$

The standard gradient-like method yields the updates of the connections that are proportional to $-\dfrac{\partial Q}{\partial \text{ connections}}$. The resulting learning scheme is standard to a high extent; the reader is referred to [7] with respect to more specialized computational details as well as some improvements which could be particularly essential for some types of triangular norms being used in the network.

## IV. REASONING IN THE REFERENTIAL STRUCTURE

Using the structure given in Fig. 2 we will discuss in detail how the reasoning process is realized. The basic idea is to cycle through all the cases $r_k$ and for each of them summarize outcomes about the level of the global property satisfied in the output space. We will focus our analysis on a single element of the output space. Let us start from the first case $\mathbf{a}_1$ (input) and $b_1$ (output). The fuzzy set for which the reasoning will be realized is given as $\mathbf{a}$. The corresponding output of the network obtained in this situation is $\gamma_1$. Cycling through all the available cases $r_k$, k = 1,2,...,N, we obtain the pairs,

$$\begin{array}{c} b_1, \gamma_1 \\ b_2, \gamma_2 \\ \vdots \\ b_N, \gamma_N \end{array}$$

For all the discussed properties but difference we will select a maximal value in the above collection of $\gamma_i$'s as the one that produces the most specific reasoning results (viz. narrow intervals of membership grades constituting solutions to the inverse problem). Let us first assume that there exists only a single maximal element in the entire collection of $\gamma_i$'s,

$$\exists!_{i0}\ \gamma_{i0} = \max_{i=1,2,\ldots,N} \gamma_i.$$

Since $\gamma_{i0}$ and $b_{i0}$ are provided, the inverse problem can be solved.

For the difference property one is looking for the minimal value in the collection of $\gamma_i$'s. Assuming its uniqueness, namely

$$\exists!_{i0}\ \gamma_{i0} = \min_{i=1,2,\ldots,N} \gamma_i$$

the emerging inverse problem takes on the same format as in the previous situation.

The algorithm requires a slight modification if there are several cases for which this maximum (or minimum, respectively) holds.

Denote by $I_0$ the elements such that

$$I_0 = \{l=1,2,\ldots,p / \gamma_l = \max_{i=1,2,\ldots,N} \gamma_i \}$$

or, for the difference property

$$I_0 = \{l=1,2,\ldots,p / \gamma_l = \min_{i=1,2,\ldots,N} \gamma_i \}.$$

The inverse problem has to be tackled with respect to the set of the corresponding membership values. Let those associated values be equal to $b_{i1}, b_{i2}, \ldots, b_{ip}$, $i_1, i_2, \ldots, i_p \in I_0$. The inverse problem solved separately for each of them gives rise to the interval of the membership values,

$[b_{ij}^-, b_{ij}^+]$, j=1,2,...,p

The values are aggregated in a conservative way by forming the broadest possible interval based on the extremal values available in the set,

$$[\min_{j=1,2,\ldots,p} b_{ij}^{-}, \max_{j=1,2,\ldots,p} b_{ij}^{+}]$$

## V. ILLUSTRATIVE NUMERICAL STUDIES

This section will be devoted to the reasoning about similarities in the input and output spaces. This scheme pertains to reasoning by analogy being carried out for fuzzy data, cf. [1] [2] [3] [5] [8]. We will be using a simple numerical data set discussed in a two-dimensional space as shown in Fig. 3 Here n = 2 and m = 1. The property of similarity has been characterized through the use of the equality index implemented in terms of the Lukasiewicz implication, see Fig. 1. The learning has been performed in the neural network with the two nodes forming the hidden layer (the triangular norms used there were set as: t-norm: product, s-norm: probabilities sum). The optimized performance index was taken as a sum of squared errors between the target values and the corresponding results produced by the network. The obtained connections are equal to:
output-hidden layer:

$$\mathbf{w} = [0.022 \quad 0.982]$$

hidden-input layer:

$$\mathbf{v} = \begin{bmatrix} 0.553 & 0.000 \\ 0.000 & 0.545 \end{bmatrix}$$

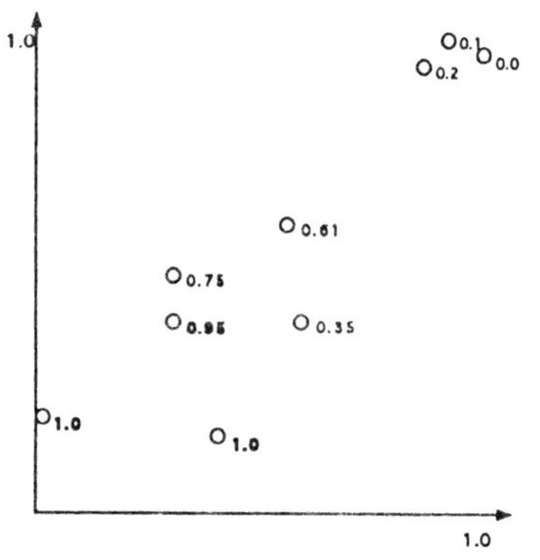


Fig. 3 Set of two-dimensional cases

The next plot, Fig. 4, summarizes the degrees of satisfaction of the similarity property in the output space for different entries of the neural network and selected reference points.

Fig. 5 illustrates a lower bound of the membership values derived from the values of the satisfaction of the similarity property produced by the network.

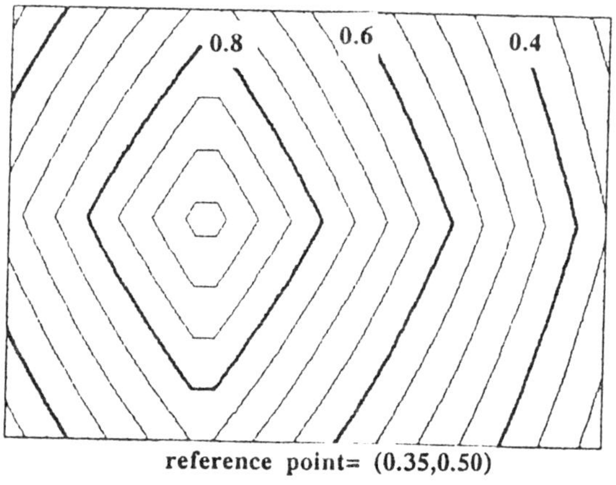


Fig. 4 .Values of $\gamma$ for different input values and selected reference points

## VI. REFERENTIAL SCHEMES OF REASONING

The general types of referential reasoning can be envisioned in the following formats,
(i)

$$\frac{\mathcal{F}(\mathbf{a}',\mathbf{a}) \qquad R=\{r_1,r_2,\ldots,r_N\}}{\mathcal{F}(\mathbf{b}',\mathbf{b})= ? \quad \mathbf{b}'=?}$$

where the case $(\mathbf{a},\mathbf{b})$ is fully specified, i.e., both $\mathbf{a}$ and $\mathbf{b}$ are available while R is used to express symbolically the cases encapsulated in the form of the neural network. Both the degree of the satisfaction of the property $\mathcal{F}$ and the values of the membership function of $\mathbf{b}'$ are to be inferred.

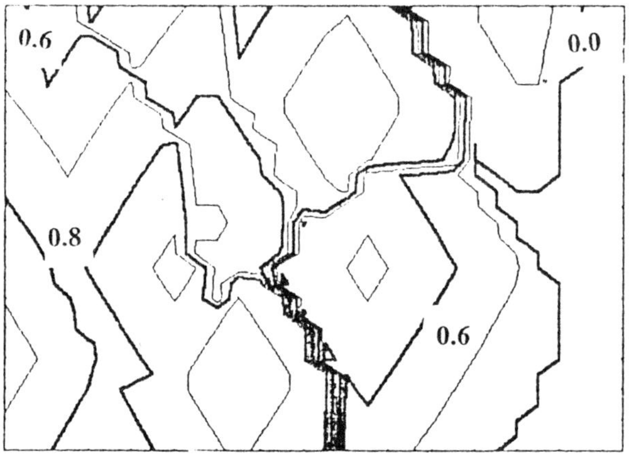


Fig. 5 Lower bound of the membership values for different values of the inputs

(ii). **a'** and **a** are given (viz. the case (**a**,**b**) is partially available). The inference is now focused on inferring the degree of satisfaction of $\mathcal{F}$ in the output space, $\mathcal{F}$(**b'**,**b**),

$$\frac{\begin{array}{l}\mathcal{F}(\mathbf{a}',\mathbf{a})\\ R=\{r_1,r_2,\ldots,r_N\}\end{array}}{\mathcal{F}(\mathbf{b}',\mathbf{b})=?}$$

Depending on the form of $\mathcal{F}$ one can study more specific examples of this class of inference schemes. For instance:

$$\frac{\begin{array}{l}\mathbf{a}' \text{ and } \mathbf{a} \text{ are } \textit{similar}\\ R=\{r_1,r_2,\ldots,r_N\}\end{array}}{\mathbf{b}' \text{ and } \mathbf{b} \text{ are } \textit{similar} = ?,\ \mathbf{b}'=?}$$

$$\frac{\begin{array}{l}\mathbf{a}' \text{ is } \textit{included} \text{ in } \mathbf{a}\\ R=\{r_1,r_2,\ldots,r_N\}\end{array}}{\mathbf{b}' \text{ is } \textit{included} \text{ in } \mathbf{b} = ?}$$

$$\frac{\begin{array}{l}\mathbf{a}' \text{ is } \textit{different} \text{ from } \mathbf{a}\\ R=\{r_1,r_2,\ldots,r_N\}\end{array}}{\mathbf{b}' \text{ is } \textit{different} \text{ from } \mathbf{b}= ?,\ \mathbf{b}'=?}$$

In general, the available cases can be used in constructing several specialized inference schemes that handle separately various global properties. The schemes are also constructed on a basis of different mutually exclusive families of cases utilized for training purposes. For instance, the inference to be realized for **a'** can invoke a series of schemes of reasoning,

$$\frac{\begin{array}{l}\mathcal{F}_1(\mathbf{a}',\mathbf{a})\\ R_1\end{array}}{\mathcal{F}_1(\mathbf{b}',\mathbf{b})= ?\ \ \mathbf{b}'=?}$$

$$\frac{\begin{array}{l}\mathcal{F}_2(\mathbf{a}',\mathbf{a})\\ R_2\end{array}}{\mathcal{F}_2(\mathbf{b}',\mathbf{b})= ?\ \ \mathbf{b}'=?}$$

...

$$\frac{\begin{array}{l}\mathcal{F}_c(\mathbf{a}',\mathbf{a})\\ R_c\end{array}}{\mathcal{F}_c(\mathbf{b}',\mathbf{b})= ?\ \ \mathbf{b}'=?}$$

The selection of the scheme that is the most suitable for **a'** can be accomplished by considering the maximum of the satisfaction levels obtained for the schemes. Denote the results produced by them by $\gamma_l = \mathcal{F}_l(\mathbf{b}',\mathbf{b})$, $l = 1,2,...,c$. The sum of the coordinates $\sum_{j=1}^{m} \gamma_{lj}$ computed for each of the schemes launches a linear order among them. Then the reasoning scheme with the highest value of this sum becomes selected.

## VII. CONCLUSIONS

We have developed a general scheme of referential reasoning. It enables us to reason about some general properties satisfied for the conclusion as well as determine this conclusion in the form of interval-valued set. The distributed model of the mapping of the global properties has been constructed with the aid of logic-based neural networks. The referential scheme of reasoning realized within the proposed framework embraces several types of the global properties one can reason about. The relevancy of the results of reasoning are quantitatively formulated by interval-valued fuzzy sets.

## REFERENCES


[1] Diamond, J., McLeod, R.D., Pedrycz,W., "A fuzzy cognitive structure: Foundations, applications and VLSI implementation," *Fuzzy Sets and Systems*, 47, 49-64, 1992.

[2] Di Nola, A., Pedrycz, W., Sessa, S., "Knowledge representation and processing in frame-based structures," In: *Fuzzy Engineering Toward Human Friendly Systems*, vol.1, Proc. Int. Fuzzy Engineering Symp. '91, Yokohama, pp. 461-470, 1991.

[3] Hirota K., Pedrycz,W, "Concepts formation: representation and processing issues," *Int. J. of Intelligent Systems*, 7, 3-13, 1992.

[4] Kling, R.E., "A paradigm for reasoning by analogy," *Artificial Intelligence*, 2, 147-178, 1971.

[5] Pedrycz, W., "A fuzzy cognitive structure for pattern recognition," *Pattern Recognition Letters*, 9, 305-313, 1989.

[6] Pedrycz, W., "Direct and inverse problem in comparison of fuzzy data," *Fuzzy Sets and Systems*, 34, 223-235, 1990.

[7] Pedrycz,W., "Neurocomputations in relational systems," *IEEE Trans. on Pattern Analysis and Machine Intelligence*, 13, 289-296, 1991.

[8] Pedrycz, W., Bortolan, G., Degani, R., "Classification of electrocardiographic signals:a fuzzy pattern matching approach," *Artificial Intelligence in Medicine*, 3, 331-46, 1991.

[9] Sambuc, R., "Functions ϕ-flous:Application de l'Aide a Diagnostique en Pathologie Thyroidienne," These Univ. de Marseille, Marseille, 1975.

# A Fuzzy Approach To Scene Understanding

Weijing Zhang
Aptronix, Inc., 2150 North First Street
San Jose, CA 95131, U.S.A.

Michio Sugeno
Tokyo Institute of Technology, 4259 Nagatsuta,
Midori-ku, Yokohama 227, Japan

*Abstract*: **In this paper, we suggest an approach to scene understanding by proposing a memory model that contains the necessary knowledge to understand a scene. Fuzzy sets theory is used for knowledge representation and reasoning. A system called SEE was developed based on the proposed approach and examples of natural color scenes are provided.**

## 1. Introduction

The purpose of computer vision is to enable a computer to understand its environment from visual information. Objects of computer vision may be outdoor scenes, indoor scenes, machine parts, characters and so on. Many image processing techniques [8,9,10,11] have been developed to extract features from input image data for recognition. However, aiming at natural scene understanding, computer vision has been proved to be one of the most difficult fields for scientists. Because the input image data is very sensitive to the environment, a scene in the real world may produce completely different data under different situations.

We can understand a scene with little difficulty no matter in what a situation we are. Why can we do that? Knowledge about the scene is a key point to this question [1,2,3,4,6,7]. We see things with our eyes, but we understand them with our brain. A baby may see a same scene of the real world, and do the same image processing with eyes as adults do, but certainly cannot understand what he/she is seeing because he/she knows very little about the real world.

If we want to build a system that is able to understand a scene in the same way as we human do, we must give the system the knowledge to understand the scene. In other words, in addition to conventional procedures to preprocess an input image and to extract a set of features, such as enhance an image or extract edges from the image, procedures of how to use these image processing results should be provided for to a vision system.

In the following sections, a memory model for scene understanding is introduced at first. This model contains knowledge we need to understand a scene. Fuzzy sets theory is employed for knowledge representation and processing. Recognitive processes using the memory model is provided. At last, system **SEE**, which has been developed based on the proposed approach, is introduced. Examples of natural color scenes are provided.

## 2. Knowledge Based Approach To Scene Understanding

### 2.1 *A Memory Model*

To understand a natural scene, the first thing we need is a plan for the process *see*. Then we need to know the objects in the scene, not only the features of the objects but also the relations among the objects. Also, we need to know how different a scene looks in different situations. From this observation, we proposed a memory model as a knowledge structure for scene understanding [3,4], which consists of following five sub-memories:

- Individual-Memory (IDM) is used to store particular knowledge about an individual object. In IDM we can find information like *my car is of red color.*

- Category-Memory (CM) contains general features of a class of objects. For instance, *a car has four wheels.* Information in CM is shared by a group of objects.

- Scene-Memory (SNM) provides information about relations among objects in scenes, such as *a car on a road.* We may define a scene as a set of relations among objects.

- Situational-Memory (STM) tells us how SNM and CM should be revised according to different situations. For instance, we know that *things look big when they are near* and *trees may turn yellow in Autumn.*

- Intentional-Memory (IM) plays a role of controlling the whole process of scene understanding. If we had no intention to find something, nothing would come to our mind even though we are seeing. When we are looking for something, we have a goal, and we try to create a plan to achieve the goal. For example, when we want to find a car, we try to find if there is a road; we do not look at sky to search for a car although this is what computers usually do. IM is a guidance when we are looking for something.

These five memories correspond to five different kinds of knowledge concerning a scene. All these memories are necessary when we want to understand a scene. Fig. 1 illustrates the basic idea of this memory model.

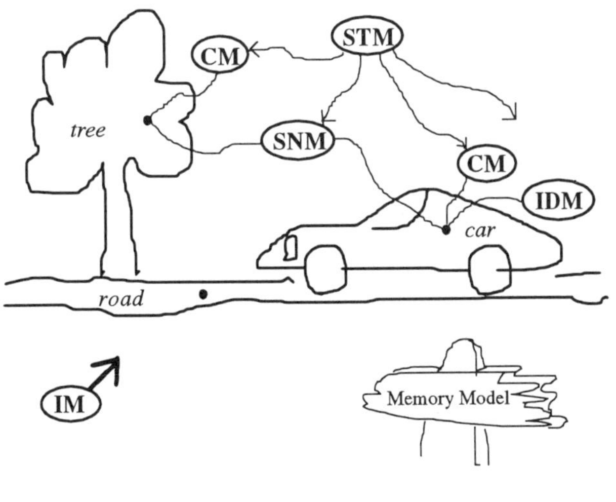


Fig. 1. Memory Model

2.2 *The Memory Model In System* ***SEE***

This section shows how to build a proposed memory model. The examples shown here are used in system **SEE**.

2.2.1 *Intentional Memory(IM)*

IM gives a procedure to achieve a goal. Table 1 shows an example of IM in system **SEE**. **vb, b, vps** are abbreviations of **very big, big,** and **vertical position** respectively. An IM record contains the items to be confirmed and the weights assigned to them. An item can be a feature stored in CM, such as color, vps, or a scene ID number in SNM, such as 10002, 10003, 10014, etc.

Table 1: IM-sky

| Object | Type | Weight | Item |
|---|---|---|---|
| sky | 1 | vb | color |
| sky | 1 | b | vps |
| sky | 2 | b | 10002 |
| sky | 2 | b | 10003 |
| sky | 2 | b | 10014 |
| sky | 2 | b | 10015 |
| sky | 2 | b | 10016 |

2.2.2 *Category Memory(CM)*

Table 2 gives an example of CM records in system **SEE**, where **cf** means certainty factor, **lblue**, **dblue**, **z** are abbreviations of **light blue**, **dark blue** and **zero**. Values of vps, **t**, **c**, **b** are abbreviations of **top**, **center** and **bottom** respectively; **tc** means the area between top and center, while **cb** means the area between center and bottom. **cf=zero** indicates the value is impossible for the feature of the object. For example, sky cannot have a **green** color as shown in Table 2.

Table 2: CM-sky

| Object | Feature | Type | Value | CF | STM-ID |
|---|---|---|---|---|---|
| sky | color | 2 | blue | vb | 00000 |
| sky | color | 2 | lblue | vb | 00000 |
| sky | color | 2 | dblue | b | 00000 |
| sky | color | 2 | white | vb | 00000 |
| sky | color | 2 | green | z | 00000 |
| sky | color | 2 | dgreen | z | 00000 |
| sky | color | 2 | lgreen | z | 00000 |
| sky | color | 2 | yellow | z | 00000 |
| sky | vps | 2 | t | vb | 00000 |
| sky | vps | 2 | tc | b | 00000 |
| sky | vps | 2 | c | m | 00000 |
| sky | vps | 2 | cb | s | 00000 |
| sky | vps | 2 | b | vs | 00000 |
| sky | size | 2 | vb | vb | 00000 |
| sky | size | 2 | b | b | 00000 |
| sky | size | 2 | m | m | 00000 |
| sky | size | 2 | s | s | 00000 |
| sky | size | 2 | vs | vs | 00000 |

Each color name like **blue, lblue** corresponds to a color distribution and represented by fuzzy sets[1].

2.2.3 *Situational Memory(STM)*

A lot of information can be stored as situation items in STM, such as weather, time, season, distance, angle, etc. Table 3 shows an example of records about weather. Although STM has not been used yet in current **SEE**, it is obvious that knowledge in STM is very useful. For instance, Table 4 gives some examples of how a feature value in CM varies in accordance with different situations. From the table we can read

In Situation No.00011 (Weather=**Clear**)
Possibility(is(*X*, sky) | color(*X*)=**blue**)=**very_big**

In Situation No.00012 (Weather=**Cloudy**)
Possibility(is(*X*, sky) | color(*X*)=**blue**)=**small**

Table 3: Examples of STM records

| STM-ID | Type | Situation | Value |
|---|---|---|---|
| 00000 | 0 | weather | unknown |
| 00011 | 1 | weather | clear |
| 00012 | 1 | weather | cloudy |
| 00013 | 1 | weather | rainy |
| 00014 | 1 | weather | snowing |

Table 4: CM-sky: STM effect

| Object | Feature | Type | Value | CF | STM-ID |
|---|---|---|---|---|---|
| sky | color | 2 | blue | vb | 00011 |
| sky | color | 2 | blue | s | 00012 |
| sky | color | 2 | blue | vs | 00013 |
| sky | color | 2 | blue | vs | 00014 |

2.2.4 *Scene Memory(SNM)*

Table 5 (see last page) shows SNM records about object sky. A SNM record gives a piece of information about the certainty of a region to be the object when a relation is satisfied. For example, SNM No.10011 says that

If there exists a region *Ax*, provided
color(*Ax*) = **darkblue**,
and there exists a region *Ay*, such that
**above**(*Ax*, *Ay*),
**connect**(*Ax*, *Ay*) and
**color**(*Ay*) = **blue**
then possibility(is(*Ax*, sky)) = **big**

2.3 *Recognitive Processes*

We can say a system understands a scene if the system can give reasonable solutions to following two types of problems. One is to find a particular object in a given scene, i.e. to answer whether the object is in the scene and give its location; another is to tell something about the whole scene, i.e., to give a linguistic description of the input image, such as *a red car on a road, trees around a park*, etc. The first one is a search problem while the second is an interpretation one. Recognitive process is different according to the problem type.

Procedure 1: Search for an object in a scene
1) Input scene and object name
2) Implement image segmentation
3) Set initial matching table
4) [Procedure-IM]
5) Output position of the object
if the answer is YES

Procedure 2: Interpretation of a scene
1) Input scene and scene type
2) Implement image segmentation
3) Set initial matching table
4) Read objects decided by scene type
4.1) Search for current object
4.2) [Procedure-IM]
4.3) Go to next object if any
5) Output scene description

[Procedure-IM]
a) Read IM of the object
b) Check current IM item
b1) CM type
b2) SNM type
c) Renew matching table
d) Go to next IM item if any

## 3. Fuzzy Knowledge

3.1 *Linguistic Values*

Knowledge in the memory model is represented by linguistic values[5], as we have seen in Section 2.2. Linguistic values, characterized by fuzzy sets, make it easy to denote vague and/or uncertain data such as **vb** (very big), **s** (small), for certainty factor; position **t** (top), color **dblue** (dark blue) for features of an object; **und** (under), **cnt** (connect) for relationship among objects in a scene. Definitions of this kind of values in **SEE** can be found in [1].

### 3.2 *Fuzzy Rules*

Knowledge for a vision system can be basically represented by following two types of implications,

(1) If $f(Ax) = v$ then $Ax = Obj$ : $CF$

(2) If $r(Ax, Ay)$ and $f(Ay) = v$ then $Ax = Obj$ : $CF$

where $f$ is a feature, such as color, $v$ is a value of $f$, such as **green**; $Ax$ and $Ay$ are regions in an image; $r$ is a relation defined among regions, such as **above**; $Obj$ is an object name in a natural scene of the real world, such as sky; $CF$ means certainty factor. $v$ is a linguistic value and $r$ is a fuzzy relation, both can be characterized by fuzzy sets. As we can see from the examples shown above, knowledge in CM is of type (1), and knowledge in SNM type (2). Each rule of either type is assigned to a certainty factor $CF$. When a value is impossible for a feature of an object, $CF$ is set to be **z(zero)**.

### 3.3 *Overall Evaluation of Matching Degrees*

Matching degrees of each region in an input image with the model data change whenever a new item in IM is matched. Suppose current matching degree of a region with the model data is $m$ and the matching degree of a new IM item is $m_i$ then the new value of $m$, denoted by $m_n$, is a function with two variables $m$ and $m_i$. Usually a product or minimum is taken in this case if the IM item is a CM type. If the IM item is an SNM, we cannot simply use product to calculate $m_n$. Because if there exists the relation then the whole matching degree will certainly get bigger, however if the relations can not be satisfied, we can say little about how should change. In **SEE** we use following to calculate new matching degree when an IM item is of SNM type.

$$m_n = \begin{cases} m + 4(1-m)(m-0.5)(m_i-0.5) & m > 0.5, m_i > 0.5 \\ m + 0.2(m_i - 0.5) & m = 0.5, m_i > 0.5 \\ m + 0.4m(m_i - 0.5) & m < 0.5, m_i > 0.5 \\ m & m_i = 0.5 \\ 2mm_i & m_i < 0.5 \end{cases}$$

## 4. Outline Of System SEE

**SEE** is a system we developed based on the approach proposed above, which runs on X windows at UNIX workstations (Fig.4, last page). Fig.2 shows the system diagram.

### 4.1 *Input Of SEE*

Inputs to **SEE** are color pictures of natural scene read from an image scanner. Each input scene can be written as a 256x256 matrix whose each element can take one of eight different values that represent eight different colors.

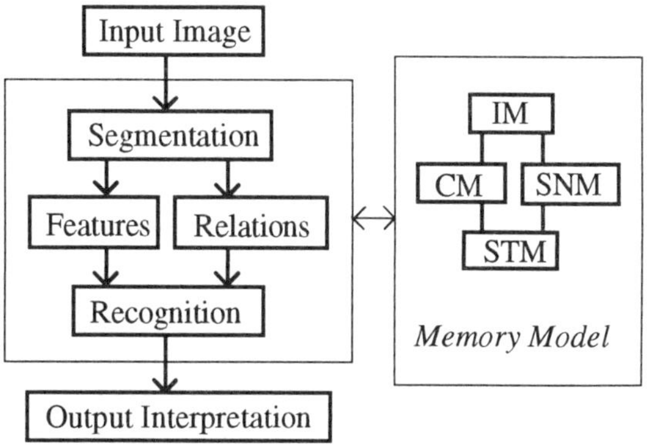


Fig.2. System Diagram of **SEE**

### 4.2 *Scene Type*

Usually we use a great deal of commonsense knowledge to understand a scene. It is almost impossible to input all commonsense knowledge into a computer so some assumptions are necessary. Suppose we know the picture is an outdoor scene, we would expect there is a blue or gray sky at the top of the scene, the areas in green color are trees, two parallel long lines make a road, small blocks on the road are cars, and so on.

Scenes can be divided into some groups, such as indoor scenes, outdoor scenes, etc. Scene type may be automatically decided by a computer, but in current **SEE**, we simply give system the scene type of an image as an input.

### 4.3 *Image Segmentation*

As we human see a scene, we grasp the rough impression of the scene from our first glance at it. If we have

Scene

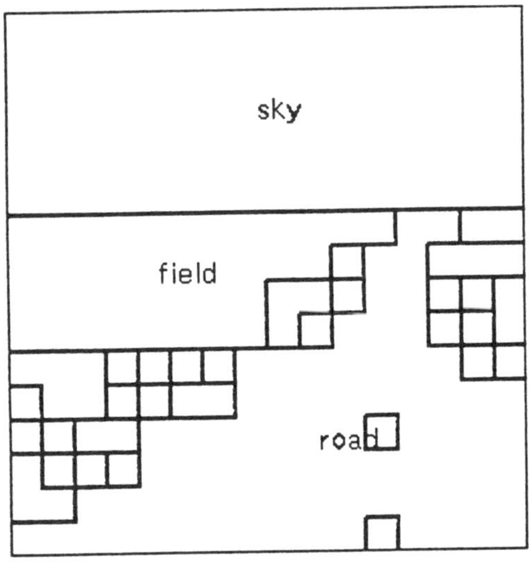


Interpretation

Fig. 3a. Example 1

Scene

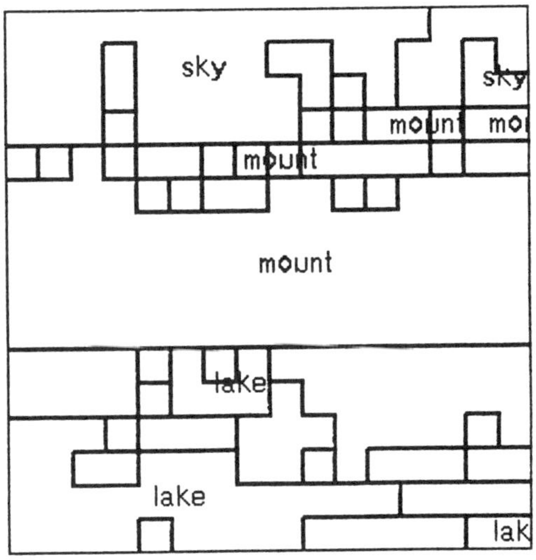


Interpretation

Fig.3b. Example 2

something interested in the image, we pay attention to that for details. This suggests to divide the whole image into some suitable sized regions at first. We think a 8x8 or 16x16 division of the original image is suitable. These correspond to the 3rd and 4th layers in a quadtree structure, which is often used to represent an image. Segmentation is carried out based the color distributions, denoted by fuzzy vectors, of the 8x8 and 16x16 partition of the original image[1].

4.4 *Examples*

Two types of natural scenes are tested in system **SEE**. One includes sky, field and road (Fig.3a) and another includes sky, mountain and lake (Fig.3b). With only little knowledge in the system, the outputs show satisfactory results.

5. CONCLUSIONS

Knowledge plays a critical role in an intelligent vision system. More attention should be paid to knowledge structure and knowledge processing. Conventional image processing techniques are necessary, but not enough. In the field related to knowledge processing, fuzzy sets theory can provide us with a powerful tool.

## REFERENCES


[1] W.Zhang and M.Sugeno, "SEE: a knowledge based scene understanding system using fuzzy reasoning," unpublished.

[2] W.Zhang and M.Sugeno, "Knowledge based scene understanding," *Proc. 7th Fuzzy System Symposium*, 1991.

[3] M.Sugeno and W.Zhang, "An approach to scene understanding using fuzzy case based reasoning," *Proc. Sino-Japan Joint Meeting on Fuzzy Sets and Systems*, 1990.

[4] W.Zhang and M.Sugeno, "Dynamic memory for scene understanding," *Proc. 6th Fuzzy System Symposium*, 1990.

[5] L.A.Zadeh, "The Concept of a Linguistic Variable and Its Application to Approximate Reasoning," I, II, III, *Information Sciences*, vol. 8, pp. 199-249, pp. 301-357, 1975, vol. 9, pp. 43-80, 1975.

[6] R.C.Schank, *Conceptual Information Processing*, North-Holland, 1975.

[7] R.C.Schank, *Dynamic Memory*, Cambridge University Press, 1982 (in Japanese translation).

[8] M. Ejiri, *Industrial Image Processing*, Shoukodo, 1988 (in Japanese).

[9] M. Nagao, *Image and Language Recognitions,* Corona Publishing Co. Ltd., 1989 (in Japanese).

[10] Y. Shirai, *Three-Dimensional Computer Vision*, Springer-Verlag, 1987.

[11] *Image Processing Handbook*, Shoukodo, 1987 (in Japanese).


Table 5 : SNM-sky(T:type abv:above und:under cnt:connect)

| ID | Obj | CF | STM | T | | | | | | | | | |
|---|---|---|---|---|---|---|---|---|---|---|---|---|---|
| 10001 | sky | b | 00000 | 1 | 0 | | | 1 | cnt | | 1 | color | white |
| 10002 | sky | vs | 00000 | 1 | 0 | | | 1 | und | | 1 | color | dgreen |
| 10003 | sky | s | 00000 | 1 | 0 | | | 1 | und | | 1 | color | dgrey |
| 10004 | sky | b | 00000 | 0 | 0 | | | 1 | cnt | | 1 | cloud | |
| 10011 | sky | b | 00000 | 6 | 1 | color | dblue | 2 | abv | cnt | 1 | color | blue |
| 10012 | sky | b | 00000 | 6 | 1 | color | dblue | 2 | abv | cnt | 1 | color | lblue |
| 10013 | sky | b | 00000 | 6 | 1 | color | blue | 2 | abv | cnt | 1 | color | lblue |
| 10014 | sky | s | 00000 | 6 | 1 | color | dblue | 2 | und | cnt | 1 | color | blue |
| 10015 | sky | s | 00000 | 6 | 1 | color | dblue | 2 | und | cnt | 1 | color | lblue |
| 10016 | sky | s | 00000 | 6 | 1 | color | blue | 2 | und | cnt | 1 | color | lblue |

Fig. 4. Scene Understanding System **SEE**

# A Fuzzy-Attributed Graph Approach to Handwritten Character Recognition

Gary, M.T. Man & Joe, C.H. Poon

Department of Electronic Engineering,
Hong Kong Polytechnic,
Hong Kong.

**Abstract - A handwritten character recognition system is described in this paper. It includes a simple but effective thinning algorithm to produce skeletons from raster scanned images which contain characters to be recognised. Based on this provision of skeletons, each character is decomposed into a number of connected strokes. Each stroke is classified as a primitive according to a fuzzy similarity criterion. When the structural information like the relations between contiguous primitives and the attribute of each primitive are extracted, a fuzzy-attributed graph can then be constructed to represent the character. In this paper, we propose a new method to measure the similarity between two fuzzy-attributed graphs of a known character class and an unknown character. In the recognition stage, when this similarity measure is applied, an input character can be correctly classified. Experiment shows that an accuracy rate around 93 percent has been achieved by our proposed approach.**

## 1. INTRODUCTION

The study of handwritten character recognition has been a considerable growth of scientific and practical interest in the past three decades. Numerous methods have been proposed and extensive literature in this area has been published [1,2]. Among various methods commonly used, statistical approaches [3] and structural approaches [4] have been sucessfully attempted and promising results have been produced. Nevertheless, a lot of problems still exist and need to be solved. The major problem is due to the large degree of variabilities of input data. Not only are there distortions of characters from one individual to another, but also there are even some variations produced from the same individual in different time. This situation is further worsen by the variation of shapes resulting from different writing habit, style, education, social environment, mood, health, as well as other factors such as the writing instrument, writing surface and scanning methods.

0-7803-0614-7/93$03.00 ©1993IEEE

Most researchers recognise that the weakness of the conventional structural approaches is partially due to the discreteness implicitly carried by the symbols and the production rules. Such discreteness implies that only discrete primitives can be analyzed. Continuous types of numerical features extracted from the primitives must be discretized first and then transformed into symbols for use in matching or parsing. In such a way, it leads to rounding or truncation errors and, consequently, reduces the recognition accuracy. On the other hand, because only exact matching or parsing is allowed in the conventional method, the structural variations due to pattern noise or distortion cannot be handled easily. In this paper, we intend to remove the above two weak points, discretness and exactness, of the conventional structural method by introducing numerical features called attributes into the symbolic representations for the primitives and their relations.

Attributed graph has been widely used as a hybrid approach to pattern representation and recognition. It is introduced by Fu and Tsai [5] . In this representation, a vertex with attributed values is used to represent a pattern primitive while the attributed arc is used to represent the relation between two primitives. When an unknown input pattern is represented by an attributed graph, the recognition procedure is to find an isomorphic reference pattern graph. The isomorphism between two attributed graphs means the corresponding matched pairs of vertices and arcs must have the same attribute values. However, such graph matching can not be applied in real world situations. Therefore, a number of algorithms using a similarity criterion or graph distance measure have been invented to solve this problem [5,6-9].

Fuzzy sets theory [10], introduced by L.A.Zadeh, can be used to cope with many natural properties and relations which are fuzzy in nature. The development of fuzzy sets has a strong impact on techniques of pattern recognition [11-13]. It is used as a reasonable tool for modeling and imitating cognitive processes of the human being, especially those concerning recognition aspects. Besides, it gives rise to a lot of novel algorithms which are useful for designing classification procedures [13]. Therefore, we try to extend the attributed graphs to fuzzy-attributed graphs (FAGs) by making the attributes fuzzy in order to increase the flexibility and power of the recognition and classification of real world images. In this paper, we will give a formal definition of FAGs in section 2 and describe a new method of similarity measure for FAGs in section 3. An example of applying the proposed method to the recognition of handwritten numerals is then demonstrated in section 4.

## 2. DEFINITIONS OF ATTRIBUTED GRAPHS AND FUZZY-ATTRIBUTED GRAPHS

The basic definitions of attributed graphs given below are adapted from [5].

### *a. Attributed Graphs*

Each vertex may take attributes from the set $Z=\{z_i| i=1,2,...,I\}$. For each attribute, $z_i$ it will have possible values $Si=\{s_{ij}|j=1,2,...,J\}$. The set of possible attribute value pairs of the vertices is denoted by the set $Lv=\{(z_i,s_{ij})| i=1,...,I;j=1,...,J_i\}$. A valid pattern primitive is just a subset of Lv in which each attribute appears only once and $\Pi$ denotes the set of all those valid pattern primitives. Thus, each node will be represented by an element of $\Pi$.

Similarly, for the arcs, we have the attribute set $F=\{f_i| i=1,...,I'\}$ in which each $f_i$ may take values $Ti=\{t_{ij}| j=1,...,J_i'\}$. $La=\{(f_i,t_{ij})|i=1,...,I';j=1,...,J_i'\}$ denotes the set of possible relational attribute value pairs. A valid relation is just a subset of La in which each attribute appears once. The set of all those valid relation is $\Theta$.

Definition 1 : An attributed graph G over $L=(L_v,L_a)$, with an underlying graph structure H=(N,E), is defined to be a pair (V,A) where $V=(N,\mu)$ is called an attributed vertex set and $A=(E,\delta)$ is called an attributed arc set. The mappings $\mu:N\rightarrow\Pi$ and $\delta:E\rightarrow\Theta$ are called vertex interpreter and arc interpreter, respectively.

Definition 2 : Two attributed graphs $G_1=(V_1,A_1)$ and $G_2=(V_2,A_2)$ are said to be structurally isomorphic if there exists an isomorphism $T:H_1\rightarrow H_2$ where $H_1=(N_1,E_1)$ and $H_2=(N_2,E_2)$ represent the structural aspects of $G_1$ and $G_2$, respectively. $G_1$ and $G_2$ are said to be completely isomorphic, written $G_1=G_2$, if there exists an attribute value preserving structural isomorphism T between $G_1$ and $G_2$.

### *b. Fuzzy-Attributed Graphs (FAGs)*

A fuzzy set (A) in a space of points $X=\{x\}$ is a class of events with a continuum of grades of membership and is characterised by a membership function $\mu_A(x)$ which associates with each point in X a real number in the interval [0,1] with the value of $\mu_A(x)$ at x representing the grade of membership of x in A. The membership function reflects the ambiguity in a set, and as it approaches unity, the grade of membership of an event in A becomes higher. For example, $\mu_A(x)=1$ indicates strictly the containment of the event x in A. If on the other hand x does not belong to A, $\mu_A(x)=0$. Any intermediate value would represent the degree to which x could be a member of A.

Because of this generalisation, we can modify our definition of attributed graph as follows [16]:

The set of all possible fuzzy attribute value pair is $L_v=\{(z_i,A_{Si})\,|\,i=1,2,...,I\}$ where $A_{Si}$ is a fuzzy set on the attribute-value set Si. A pattern primitive $\Pi$ is a subset of $L_v$ and let $\Omega$ be the set of all possible pattern primitives.

Similarly, we can assume the adjacency relations to be fuzzy, and then it can be generalized that the set of possible attribute-value pairs of the arcs denoted by $L_a=\{(f_i,B_{Ti})\,|\,i=1,2,...,I'\}$ where $B_{Ti}$ is a fuzzy set on the relational attribute-value set Ti. $\Phi$ denotes the set of all relations.

Definition 3 : A fuzzy-attribute graph G over $L=(L_v,L_a)$ with an underlying graph structure H=(N,E) is defined to be an ordered pair (V,A), where $V=(N,\sigma)$ is called a fuzzy vertex set and $A=(E,\delta)$ is called a fuzzy arc set. The mapping $\sigma:N\rightarrow\Omega$ and $\delta:E\rightarrow\Phi$ are called fuzzy vertex interpreter and fuzzy arc interpreter respectively.

## 3. A NEW METHOD OF SIMILARITY MEASUREMENT FOR FAGS

Formally, two graphs $G_1$ and $G_2$ are isomorphic if there exists a one-to-one and onto function f: $V(G_1)\rightarrow V(G_2)$ such that x and y belong to $V(G_1)$ if and only if f(x) and f(y) belong to $V(G_2)$. Similarly, two attributed graphs $G_1=(V_1,A_1)$ and $G_2=(V_2,A_2)$ are said to to isomorphic, denoted by $G_1=G_2$, if there exists an attributed value preserving isomorphism f between $G_1$ and $G_2$. However, in the case of fuzzy set, equality is too strict a condition for such a matching. Hence, we try to relax the requirement and define a similarity measure between FAGs.

Definition 4 : Let F(U) denotes all the fuzzy subsets in the universe U and A,B∈F(U). A fuzzy membership function α is defined as a relation between the Cartesian product, F(U)XF(U) and the real number interval [0,1], which is denoted by α: F(U) X F(U)→[0,1], (A,B)|→ α(A,B), if α satisfies

(i) α(A,B)=1 iff A=B;
(ii) α(A,B)=α(B,A);
(iii) A⊆B⊆C ⇒ α(A,C)≤α(A,B)∧α(B,C);

it denotes the similarity function of F(U). α(A,B) represents the degree of similarity between A and B. We proposed a similarity function as follows :

Let $F_{1xn} = \{(x_1,...,x_n) : 0 \le x_i \le 1,\ i=1,...,n\}$,

and $a,b \in F_{1xn}$, $\alpha(a,b) = \dfrac{\sum_{i=1}^{n}(a_i \wedge b_i)}{\sum_{i=1}^{n}(a_i \vee b_i)}$ satisfies the above conditions and α(a,b)=1 ⇒ a=b.

The above similarity function can be adopted as a similarity measure of two fuzzy attributed graphs and is defined as follows :

Definition 5 : Let $\alpha_1$ be the similarity function between vertices $v_1$ and $v_2$ of two FAGs $G_1$ and $G_2$. Let $A_{1Si}$ be the fuzzy subset that gives the attribute value for $z_i$ of $v_1$ and $A_{2Si}$ be that of $z_i$ of $v_2$.

$$\alpha_1(v_1,v_2) = \frac{\sum_{i=1}^{I}\sum_{j=1}^{J_i}(\mu_{A_{1Si}}(s_{ij}) \wedge \mu_{A_{2Si}}(s_{ij}))}{\sum_{i=1}^{I}\sum_{j=1}^{J_i}(\mu_{A_{1Si}}(s_{ij}) \vee \mu_{A_{2Si}}(s_{ij}))}$$

where $\mu(s_{ij})$ is the membership grade of $s_{ij}$ in the fuzzy subset $A_{ksi}$, k=1,2.

Definition 6 : Let $\alpha_2$ be the similarity function between arcs $a_1$ and $a_2$ of two FAGs $G_1$ and $G_2$. Let $B_{1Ti}$ be the fuzzy set that gives the attribute value of $t_i$ of $a_1$ and $B_{2Ti}$ be the attribute value of $t_i$ of $a_2$.

$$\alpha_2(a_1,a_2) = \frac{\sum_{i=1}^{I'}\sum_{j=1}^{J_i'}(\mu_{B_{1Ti}}(t_{ij}) \wedge \mu_{B_{1Ti}}(t_{ij}))}{\sum_{i=1}^{I'}\sum_{j=1}^{J_i'}(\mu_{B_{2Ti}}(t_{ij}) \vee \mu_{B_{2Ti}}(t_{ij}))}$$

where $\mu(t_{ij})$ is the membership value of $t_{ij}$ in the fuzzy set $B_{kti}$, k=1,2.

We combine the above two definitions to formulate a similarity function of two fuzzy-attributed graphs.

Definition 7 : Let α be the similarity function of two FAGs $G_1$ and $G_2$ such that the underlying graph $H_1$ of $G_1$ is isomorphic to the underlying graph $H_2$ of $G_2$.

$$\alpha(G_1,G_2) = \frac{\sum_{p} \alpha_1(p,h(p)) \underset{q \ge p}{\wedge} \alpha_2(a(p,q),a(h(p),h(q)))}{\sum_{p} \alpha_1(p,h(p)) \underset{q \ge p}{\vee} \alpha_2(a(p,q),a(h(p),h(q)))}$$

where p,q denote the vertices of $G_1$ and h(p),h(q) denote the corresponding vertices of $G_2$ under the isomorphism h. The value of $\alpha(G_1,G_2)$ represents the degree of similarity between $G_1$ and $G_2$.

In recognition, let B∈F(X) be the unknown input pattern to be classified and $A_i \in F(X)$, i=1,2,...,n be the reference patterns by the above similarity function. A selection threshold δ is defined as follows:

Definition 8 : Let $G_1$ and $G_2$ be two FAGs. $G_1$ is δ-isomorphic to $G_2$ if the degree of similarity $\alpha(G_1,G_2) \ge \delta$.

Then the unknown pattern B can be classified as reference pattern $A_i$ if

$$\alpha(B,A_i) = \max_{1 \le j \le n} \alpha(B,A_j) \ge \delta.$$

It is well-known that conventional graph isomorphism can be solved by tree-search methods [17]. The time complexity of these algorithms is a NP-complete problem. Various attempts have been tried to reduce the time requirement needed [14,17]. In this paper, we propose an algorithm for finding the optimal δ -isomorphism between two attributed graphs by adopting the branch-and-bound algorithm. The proposed similarity measure involves using an evaluation function in the branch-and-bound algorithm. Thus, the optimal isomorphism represents as the best matching case in which the degree of similarity is maximized. This search algorithm expands fewer number of nodes and so requires less CPU time. Its pseudo-code is given below :

Algorithm : Isomorphic Graph Matching

```
Begin
for all reference FAGs do
{
    initialize a linked list with all nodes of the
    unknown FAG;
    initialize the Upper Bound UB=0;
    while (the list is not empty) do
    {
        set the first node in the list be the current
        node,p;
        for each node q in the reference FAG do
        {
            compute α(p,h(p));
            UB= MAX(UB,α(p,h(p)));
        }
        record the node pair p and q in α_max(p,q);
        remove p from the list;
        add the upper bound to the total degree of
        similarity;
    }
    output the optimal solution with the total similarity
    score
}
assign the unknown FAG as the reference pattern FAG
with the greatest degree of similarity;
End.
```

During the search process, the degree of similarity of all combination of vertex pairs on the same level are computed. Only the most similar vertex pair is extended one level deeper and all the other possible paths are pruned. The procedure repeats until all the vertices of the input numeral are considered. Since the path with the maximum degree of similarity is chosen for extension, the path first reaches the last vertex of the input numeral is certain to be optimal.

## 4. APPLICATION TO HANDWRITTEN NUMERAL RECOGNITION

The overall block diagram of the handwritten numeral recognition system is shown in Fig.3. The system consists of five parts : 1)thresholding, 2)thinning, 3)extracting dominant points and segmenting the character into strokes, 4)classifying strokes as primitives based on their curvature and orientation parameters, 5)constructing FAG representation of the numeral and manipulating the similarity between different reference FAGs in order to classify the numeral correctly. In the training stage, a set of training samples from different writers are provided to build up a knowledge database which represents FAGs of all reference numerals.

The input numerals are scanned by a video camera. Therefore, we obtain a grey-level image but this may be converted to a binary pattern by thresholding. In the stage of thresholding,, a pixel is simply deemed as black or white according to a simple rule that whether it is larger or smaller than a threshold value. Afterwards, the binary image is thinned as skeletons by Hilditch's algorithm. This skeletonization algorithm [15] consists of many executing passes over the pattern, where point deletions are considered in each pass.:

Algorithm : Hilditch's Thinning Algorithm

```
Begin
Iteratively scan the pattern rowwise from left to right and
from top to bottom;
{
    For each point, test the following conditions :
    {
        i) it is an edge-point, that is along the edges of
           the image,
        ii) it is not an end-point, that is on the
            extremities of a stroke,
        iii) it is not a break-point, that is not a point of
             which the deletion would break the
             connectedness of the image,
    }the point will be deleted if all the above conditions
     are satisfied
}
the remaining points become the skeleton of the image;
End
```

After the process of thinning, some short spikes may be generated. A pruning algorithm is used to test the length of spikes and if it is smaller than some thresholds, it will be pruned.

When a skeleton is smoothed, its nodes will represent as tips, corners or junctions. We define tips are points that have one neighbor. Corners are points that have an abrupt change of line direction. Junctions are points that have three or more neighbors. Thus, tips and junctions can be easily detected by counting the number of its 8-neighbors wheras corner points can be determined by the slope of consecutive line segments. After identifying the nodes of the skeleton, the line segments between adjacent nodes are tested and

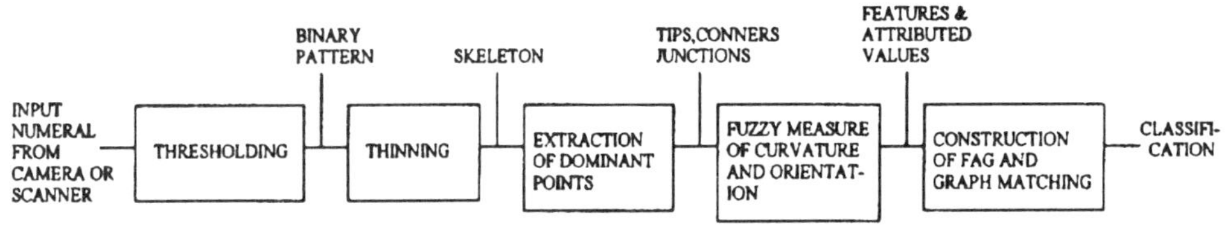

Fig.3 Block diagram of the numeral recognition system.

classified according to their straightness and orientation. The measure of straightness is defined as

$$\mu_{arc}(x) = (1 - \frac{a}{l})^{0.5}$$

where $a$ is the length of the straight line joining two extreme points of an arc and l is the arc length. It is obvious that the lower the ratio $(a/l)$ is, the less will be the degree of straightness. Futhermore, the measure of orientation is based on the slope of the line connecting the end points. If the line segment belongs to the straight line class, it will then be classified as one of the four fuzzy subsets which are labelled as "vertical", "horizontal", "positive slope" and "negative slope". Similarly, if the line segment represents a portion of circle, it may further be classified as one of the two fuzzy subsets which are labelled as "C-shaped" and "D-shaped". The fuzzy measure of these two fuzzy subsets is defined as follows :

$$\mu_C(x) = \begin{cases} \sqrt{\mu_{arc}(x)} & \text{if } x\_c < x\_d \\ 0 & \text{otherwise} \end{cases}$$

$$\mu_D(x) = \begin{cases} \sqrt{\mu_{arc}(x)} & \text{if } x\_c \geq x\_d \\ 0 & \text{otherwise} \end{cases}$$

where x_c denotes the x-coordinates of the cluster center of the line segment, and x_d denotes the x-coordinate of the midpoint of the line connecting the end points.

However, if the line segment represents a loop, it will further be classified as one of the five subsets: "loop at the left of the node", "loop at the right of the node", "loop above the node", "loop below the node" and "loop with no node attachment". The classification of these five types of loop is affected by two factors such as the ratio of the arc length and the perimeter of the loop. The above features are found sufficient for the structural description of any handwritten numeral. They are shown in Fig.4.

In the training stage, a set of 10 training samples for each numerals class is given for the construction of a knowledge database which contains all reference FAGs. Automatic supervised training is employed used to compute membership grades for each reference pattern class. The automatic supervised training algorithm is summarised as follows:

Algorithm : Supervised Training

```
Begin
for all numeral classes do
{
    for all training samples belonged to this numeral
    class do
    {
        extract primitives for each training samples;
        record the frequency of occurrence of each
        extracted primitives;
    }
    assign the membership grade of each primitive as
    (the frequency of occurrence of this primitive)/(no.
    of training samples of this class);
    store the membership grade of each primitive into
    the database;
}
End.
```

In the recognition stage, after the input numeral is represented by a FAG, it is matched with the reference FAGs of each numeral class with the proposed similarity measure algorithm and the branch-and-bound search algorithm. The matching process continues until a similrity score with one of the reference FAG is above a chosen threshold value. When all reference FAGs are compared but no successful matching is recorded, the input numeral is simply rejected. In our experiment, 500 numerals from different writers are tested. It is found that the percentage of correct classification is 93 percent. Some numerals are misclassified when they do share some similarity in appearance with other numeral classes. In general, to improve the recognition rate, a large training sample size from different writers is required. Besides, some poorly written numerals should also be trained so as to develop a more complete and flexible database. Some of the numeral samples are shown in Fig.5.

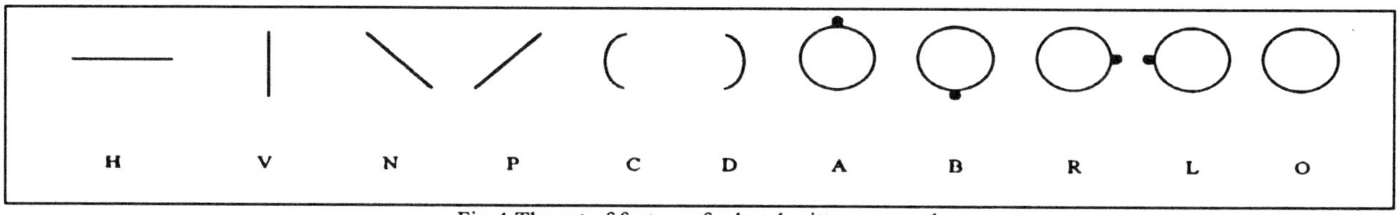


Fig.4 The set of features for handwritten numeral.

## 5. CONCLUSION

In this paper, the adoption of fuzzy-attributed graph is shown to be applicable in coping with real world images which are always corrupted by noise and distortion. The fuzzy set theory provides a well-founded mathematical base to encounter the uncertainty problems in pattern representation and classification. However, such extension will cause new problems in graph matching as we no longer easily check the equality of the fuzzy attributes. A new method of similarity measure for FAGs is proposed to solve the problem of graph matching. This algorithm reduces the time complexity for graph matching by using the branch-and-bound approach. Besides, it is a general approach applicable to the recognition of various types of patterns. Finally, the proposed algorithm is evaluated when it is applied to the recognition of handwritten numerals. The recognizer reduces as much as possible the written constraints while maintaining the overall recognition rate close to those existing methods.

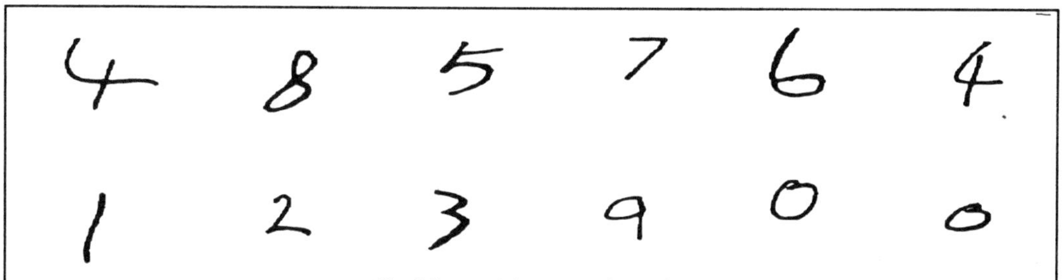

Fig.5 Some of the numeral samples

## REFERENCES


1. Charles C.Tappert, Ching Y.Suen and Toru Wakahara,"The State of the Art in On-Line Handwriting Recognition", IEEE Trans. on PAMI-12, No.8, Aug 1990, pp. 787-808.
2. J.Mantas,"An overview of character recognition methodologies", Pattern Recognition 1986, pp.425-430.
3. K.Fukunaga, "Introduction to Statistical Pattern Recognition", New York : Academic, 1972.
4. T.Pavlidis,"Structural Pattern Recognition.", New York : SpringerVerlag, 1977.
5. Wen-Hsiang Tsai and King-Sun Fu, "Error Correcting Isomorphisms of Attributed Relational Graphs for Pattern Analysis", IEEE Trans. on SMC-9, No.12, Dec. 1979.
6. A.Sanfeliu and K.S.Fu,"A Distance Measure Between Attributed Relational Graphs for Pattern Recognition." IEEE Trans. on SMC-13, 1983, pp. 353-362.
7. A.K.C.Wong and D.E.Ghahraman,"Random graphs :Structural Contextual Dichotomy." IEEE Trans. on PAMI-2, 1980, pp. 341-348.
8. A.K.C.Wong and Manlai You,"Entropy and Distance of Random Graphs with Application to Structural Pattern Recognition", IEEE Trans. on PAMI-7, No.5, Sept. 1985, pp. 599 - 609.
9. M.A.Eshera and K.S.Fu, "An Image Understanding System Using Attributed Symbolic Representation and Inexact Graph Matching", IEEE Trans. on PAMI-8, No.5, Sept 1986.
10. Sankar K.Pal and Dwijesh K.Dutta Majumder,"Fuzzy Mathematical Approach to Pattern Recognition", John Wiley & Sons, 1986.
11. Pepe Siy and C.S.Chen,"Fuzzy Logic for Handwritten Numeral Character Recognition", IEEE Trans. on SMC-4, Nov. 1974, pp.570-575.
12. Sankar K.Pal, Robert A.King and Abdullah A. Hashim, "Image Description and Primitive Extraction Using Fuzzy Sets", IEEE Trans. on SMC-13, No.1, 1983.
13. W.Pedrycz,"Fuzzy Sets in Pattern Recognition : Methodology and Methods", Pattern Recognition, Vol.23, No.1/2, 1990, pp. 121-146.
14. A.K.C.Wong, M.You and S.C.Chan,"An Algorithm for Graph Optimal Monomorphism", IEEE Trans. on SMC-20 No.3, May/June 1990, pp. 628-636.
15. Nabil Jean Naccache and Rajjan Shinghal, " An Investigation into the skeletonization approach of Hilditch", Pattern Recognition, Vol.17, No.3, 1984, pp. 279-284.
16. K.P.Chan and Y.S.Cheung, "Fuzzy-Attribute Graph and Its Application to Chinese Character Recognition".
17. J.R.Ullmann,"An Algorithm for Subgraph Isomorphism", J.ACM, Vol.23, No.1, 1976, pp. 31-42.

# Automatic Target Recognition Fuzzy System for Thermal Infrared Images

Christiaan Perneel*, Michel de Mathelin† and Marc Acheroy*

* Electrical Engineering Department, Royal Military Academy, 30 Renaissance avenue, 1040 Brussels, Belgium.
† Electrical & Computer Engineering Department, Carnegie Mellon University, Pittsburgh, PA 15213-3890.

**Abstract**

An expert system for automatic recognition of armored vehicles from short distance infrared images is presented in this paper. This expert system uses fuzzy logic to combine information coming from different heuristics and to guide the search of the solution to the recognition problem.

## 1 Introduction

The problem consists in identifying armored vehicles based on short distance 2-D infrared images. The major difficulty lies in the lack of knowledge of the position and orientation of the vehicle with respect to the camera. Indeed, it is completely unpractical to realize a close matching of the image of the vehicle with a template given all the possible positions and orientations of the vehicle, and also given all the possible vehicles. Therefore, the problem is divided in two subproblems where the first subproblem is the computation of the orientation and position of the vehicle based on a crude model of the vehicle and the second subproblem is the identification of the vehicle in a reference position and orientation.

A blackboard type expert system (*cf.* Nii [1, 2] or Hayes-Roth [3] for a definition and Murdock & Hayes-Roth [4] for an example of application) was designed to accomplish these two tasks. The first task, position and orientation detection, consists in putting a system of three axes, $\{X, Y, Z\}$, on the image of the vehicle according to predefined conventions. In our application, the conventions are the following:

- $X$ **axis**: the line on the side of the vehicle between the wheel train and the ground.
- $Y$ **axis**: the line on the front or on the rear of the vehicle between, respectively, the front or the rear wheels on each side of the vehicle and the ground.
- **origin**: the origin of the system of axes is, by convention, at the extremity of the $X$ axis on the engine side if the image gives a side view of the vehicle, on the left of the $Y$ axis if the image gives a front or rear view of the vehicle but the engine is on the front of the image, or on the right of the $Y$ axis if the image gives a front or rear view of the vehicle but the engine is on the rear of the image.
- $Z$ **axis**: the vertical direction of the vehicle, normalized on the wheel train height if the image gives a side view of the vehicle or on the distance between the floor of the vehicle body and the ground if the image gives a front or rear view of the vehicle.

An illustration of these conventions can be found in Fig. 1. The position and orientation detection is then

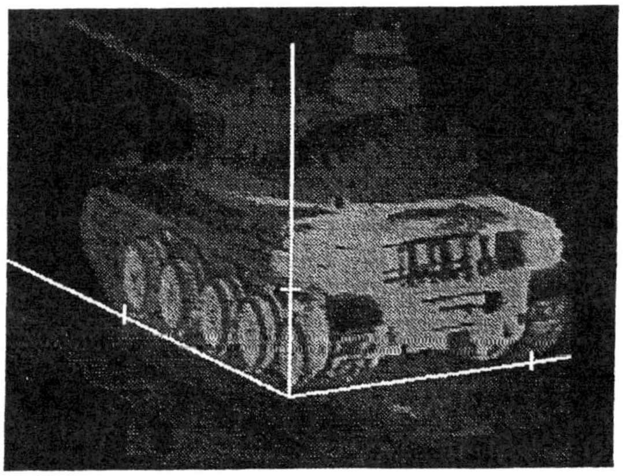

Figure 1: Axes conventions

divided into $N = 5$ subtasks. These 5 subtasks correspond to 5 different levels of increasing knowledge of the position and orientation of the vehicle.

- **Level 1**: Determination of one principal direction, either the direction of the $X$ axis if the image gives a side view, or the direction of the $Y$ axis if the image gives a front or rear view.
- **Level 2**: Determination of the line of the $X$ axis if it is a side view or the line of the $Y$ axis if it is a front or rear view.
- **Level 3**: Determination of the position of the origin.
- **Level 4**: Determination of the $Z$ axis.
- **Level 5**: Determination of the third axis, either the $Y$ axis if it is a side view, or the $X$ axis if it is a front or rear view.

The following assumptions are made about the expert system considered in this paper

0-7803-0614-7/93/$03.00 ©1993 IEEE

- Level 1 is the initial level of knowledge of the solution in the search process. When the expert system reaches the next level in the search process, some more information about a candidate solution is generated. A candidate solution consists in values for the $N$ predefined characteristics or sets of characteristics (one per level). A candidate partial solution up to level $i$ consists in values for $i$ predefined characteristics or sets of characteristics. Level $N$ is the last level of knowledge in the search process where a possible solution to the problem can be proposed.
- At each level $i$, at most $n_i$ new possible partial solutions or, equivalently, new branches in the search tree are proposed. The $n_i$ new candidate partial solutions consist in a candidate partial solution up to level $(i-1)$ combined with some incremental knowledge under the form of a possible value for one or more predefined characteristics of the solution. Consequently, the size of the candidate solution space $S \leq \prod_{i=1}^{N} n_i$. For our application, $\{n_1 = 8, n_2 = 30, n_3 = 2, n_4 = 8, n_5 = 8\}$. Therefore, the size of the solution space, $S = 30720$.
- At each level $i$, $m_i$ different heuristics give a rating to the $n_i$ new candidate partial solutions of level $i$. These ratings must be combined to give a global rating to the partial solutions. For our application, the number of heuristics is $\{m_1 = 3, m_2 = 12, m_3 = 2, m_4 = 3, m_5 = 6\}$.

The image processing algorithms involved in the computation of the different candidates and in the different heuristics are too time consuming to make possible an exhaustive exploration of the solution space. Therefore, a heuristic search strategy is implemented. Typically, the knowledge of an expert is required in order to guide the search toward the solution in a more efficient manner. This knowledge takes the form of heuristics and a search procedure using information of this sort is called heuristic search. At any given time in the search process all the possible paths are evaluated with the help of the heuristics and the most promising path according to the heuristics is selected. The heuristics are designed based on the knowledge and the experience of the expert. The main difficulty in this approach resides in finding an appropriate way to combine the knowledge coming from the heuristics into a unique evaluation function. Furthermore, as the search of the solution progresses, new paths must be compared with paths abandoned earlier. However, new paths are usually much more advanced in the building of a solution than old paths and it might become very difficult to find a common ground for comparison. In the section to follow, we will see how we can design a global rating function to evaluate the different candidates using fuzzy logic.

Once the first task of detecting the position and orientation of the vehicle is accomplished, it remains to identify the type of vehicle. This is done by defining characteristic details, such as, *e.g.*, number of wheels, engine position, tracks size, exhaust system, for each vehicle to be recognized. The location and the shape of these characteristic details is known *a priori*. Therefore, templates can be created with their location on the vehicle specified. Since the position and orientation of the vehicle is known, it is sufficient to verify that the templates match the corresponding areas on the image. This is done by using pattern recognition techniques as cross-correlation, or neural networks (*cf.* Duda & Hart [5] and Carpenter & Grossberg [6]).

# 2 Fuzzy reasoning

The heuristics rate the candidate solutions based on some measurements or observations. Very often, the range of the measured variable is more important than its exact value. The measurement can be often classified in simple categories describing how well it fits the hypothesis that the candidate is the solution to the problem. For example, **Good, Average**, and **Bad** could be the categories or linguistic terms describing how the observation fits the hypothesis. Furthermore, the transitions between these categories (*i.e.*, **Good** and **Average** or **Average** and **Bad**) are often blurred. Therefore, there must be also a smooth transition in the rating of the different candidate solutions.

## 2.1 Modelization

The following assumptions are made to model uncertainty and fuzziness (based on the terminology used in, *e.g.*, Bellman & Zadeh [7], or Mamdani [8])

- The values which define uniquely a solution (*cf.* Section 1) belong to nonfuzzy support sets of universes of discourse like, for example, the $k$-th characteristic or set of characteristics of the solution . Let $X_k$ be the set of possible values for the characteristics of the solution receiving a value at level $k$. Then, the $X_k$'s are the nonfuzzy support sets, having a finite number, $S_k$ where $S_k \leq \prod_{i=1}^{k} n_i$, of elements, of the universes of discourse "$k$-th characteristic or set of characteristics of the solution". Based on these definitions, the characteristics defining a candidate partial solution up to level $k$ will be represented by the nonfuzzy support set

$$Y_k = (\ X_1,\ \ \ldots,\ \ X_k\ )$$

Therefore, a partial solution up to level $i$ will be an element of $Y_i$ and a complete solution will be an element of $Y_N$.

- Suppose that the $n_i$ values $\{x_{ik}(y_{i-1})\}$, $k = 1,\ldots,n_i$, are all the elements depending on $y_{i-1}$ which belongs to the nonfuzzy support set $X_i$, where $y_{i-1}$ is an element selected in the previous levels nonfuzzy support set, $Y_{i-1}$. Suppose that there exist $M$ different categories or linguistic terms (*e.g.*, **Good, Average** and **Bad**, $M = 3$), for describing to what degree $x_{ik}$ fits the characteristics of the solution represented by $X_i$. These linguistic terms define fuzzy subsets of the universes of discourse "$i$-th characteristic or set of characteristics of the solution". Then, for each level $i$, there exists $M$ membership functions (one per linguistic term)

$$\begin{aligned} f_i^l(x_{ik}, y_{i-1}) &: X_i \times Y_{i-1} \rightarrow [0,1]\ k = 1,\ldots,n_i \\ x_{ik} &= x_{ik}(y_{i-1}) \end{aligned}$$

which define fuzzy subsets associated to the linguistic terms, $l$, describing to what degree the values $\{x_{ik}(y_{i-1})\}$ belongs or fits *a priori* the $i$-th characteristic or set of characteristics of the solution represented by $X_i$. Let's define the vector $f_i$ as the combination of these $M$ membership functions, $\{f_i^l\}$, in one vector. For example, if the linguistic terms are (**Good, Average, Bad**), the fuzzy subset $[x_1, x_2, x_3] = f_i(x_{ik})$ expresses that $x_{ik}$ belongs *a priori* to the **Good** fit category with a factor $x_1$, to the **Average** fit category with a factor $x_2$, and to the **Bad** fit category with a factor $x_3$, for the characteristics of the solution represented by $X_i$.
Without loss of generality, it can be assumed that the membership functions, $f_i^l$, are normalized, so that

$$\text{Normality rule:} \qquad \max_x f_i^l(x) = 1$$

Furthermore, to minimize the number of linguistic terms and avoid redundant categories, it is assumed that the $M$ linguistic terms describe different (but possibly overlapping) degrees of fit. Consequently, if the degree of fit of one category is 1, then the degree of fit of the others must be 0. This translates into the following rule

$$\text{Economy rule:} \qquad \sum_l f_i^l(x_{ik}) \leq 1 \quad \forall\, i,k$$

Furthermore, if it is assumed that the linguistic terms span all the possible degrees of fit, then the following rule applies

$$\text{Completeness rule:} \qquad \sum_l f_i^l(x_{ik}) \geq 1 \quad \forall\, i,k$$

Consequently, if the three rules apply

$$\sum_l f_i^l(x_{ik}) = 1 \quad \forall\, i,k$$

Finally, by default if the *a priori* degree of fit is unknown, a constant membership function will be assumed, *i.e.*,

$$\begin{aligned} f_i^{\textbf{Good}}(x_{ik}, y_{i-1}) &= 1 \\ f_i^{\textbf{Average}}(x_{ik}, y_{i-1}) &= 0 \\ f_i^{\textbf{Bad}}(x_{ik}, y_{i-1}) &= 0 \\ f_i(x_{ik}, y_{i-1}) &= [\ 1,\ \ 0,\ \ 0,\ ] \end{aligned}$$

- There are $m_i$ different rating functions at level $i$ (one per heuristic). Let $h_{ij}^l(p_{ij})$ be the $j$-th membership function of level $i$ associated to the linguistic term $l$, where $p_{ij}$ is an observation depending on the characteristics described by $Y_i$. Let $h_{ij}(p_{ij})$ be the vector of $M$ membership functions $h_{ij}^l(p_{ij})$. Given a candidate partial solution up to level $i$, $y_i \in Y_i$, an observation $p_{ij}(y_i)$ is made and the $j$-th heuristic returns a fuzzy subset, $h_{ij}(p_{ij})$, describing how the observation $p_{ij}(y_i)$ fits the hypothesis that $y_i$ is the solution up to level $i$.

  Without loss of generality, it can be assumed that the membership functions, $h_{ij}^l$, are normalized, so that $\max_p h_{ij}^l(p) = 1$. Furthermore, it is assumed that the Economy rule applies, *i.e.*, $\sum_l h_{ij}^l(p) \leq 1$, $\forall\, i,j$ and $\forall\, p$.

Given a measurement $p$ and assuming that $M = 3$, with linguistic terms (**Good, Average, Bad**), a typical heuristic membership function, $h(p)$, with an overlapping factor of 50%, may look like in Fig.2. The observation $p$ belongs to the **Good** fit category

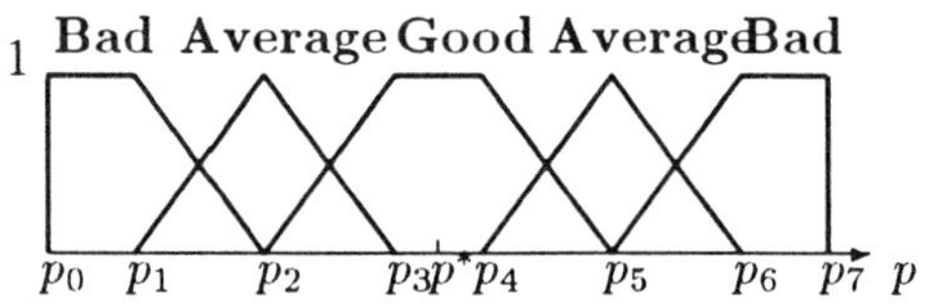


Figure 2: Heuristic fuzzy membership function, $h(p)$

with a factor 1 around a reference value, $p^*$, then slowly starts to belong to the **Average** category and finally to the **Bad** category outside this area. Finally, note that the overlapping factor between the different categories should be proportional to the fuzziness of the boundaries between categories. In his book, Kosko [9] recommends an overlapping factor of 25%. We use ourselves an overlapping factor of 50% in our application.

## 2.2 Rating function

A global rating function must be defined to follow an opportunistic search strategy. This global rating function should logically reinforce candidate partial solutions whose heuristics give mostly a **Good** rating and should disadvantage candidate partial solutions whose heuristics give mostly a **Bad** rating.

### 2.2.1 Heuristics rating functions

Each heuristics return a fuzzy vector of size $M$, the number of different linguistic terms. This fuzzy vector is made of the membership values for each of the $M$ different linguistic terms. Now, these $M$ different values must be combined in one unique value which is the rating given by the heuristic. This operation of transforming the fuzzy vector $h(p)$ in a unique nonfuzzy rating value is called défuzzification. This is usually done by assigning to each linguistic term, $l$, a rating membership function, $g^l(r)$, where $r$ is the nonfuzzy rating value, obeying to the following rules

- $g^l\colon \Re \longrightarrow [0, 1]$.
- $\forall\, r$, the rating must increase when the linguistic term expresses an improvement and must decrease when the linguistic term expresses a worsening. For example,

$$g^{\mathbf{Good}}(r) \geq g^{\mathbf{Average}}(r) \geq g^{\mathbf{Bad}}(r) \qquad \forall\, r$$

- Assuming that the rating values must belong to the interval $[K_{\text{min}}, K_{\text{max}}]$ then $g^l(K_{\text{min}}) = 1$ if $l$ is the "worst degree of fit" linguistic term and $g^l(K_{\text{max}}) = 1$ if $l$ is the "best degree of fit" linguistic term.

For example, suppose that $M = 3$ and the linguistics terms are (**Good, Average, Bad**) and assume that the minimum rating is $K_{\text{min}}$ and the maximum rating $K_{\text{max}}$, then typical rating membership functions $g^l$ are like in Fig.3. Once the rating member-

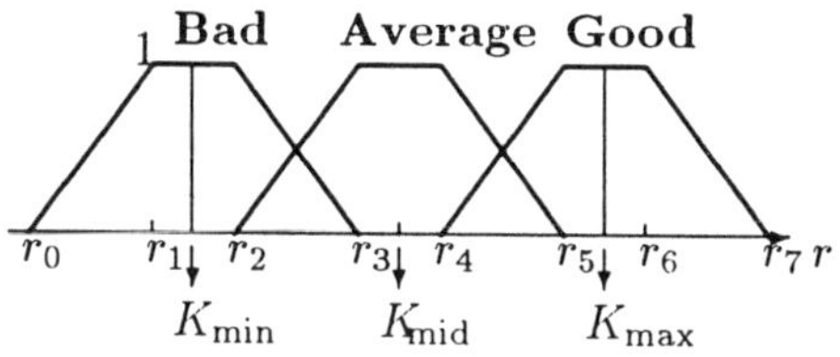


Figure 3: Rating membership functions

ship function have been defined, given an observation, $p$, given a heuristic with heuristic membership functions, $h^l(p)$, where $l$ spans all the possible linguistic terms, we can define the cumulated rating membership function, $g_c(r)$, as

$$g_c(r) = \sum_l \min(g^l(r), h^l(p)) \qquad \text{minimum inference}$$

For example, suppose that the the linguistics terms are (**Good, Average, Bad**), that $h(p)$ is defined as in Fig.2, that the observation $p$ is such that $h(p) = [4/5, 1/5, 0]$, and that the rating membership functions are defined as in Fig.3, then $g_c(r)$ will look like in Fig.4. Now, the simplest defuzzification scheme is

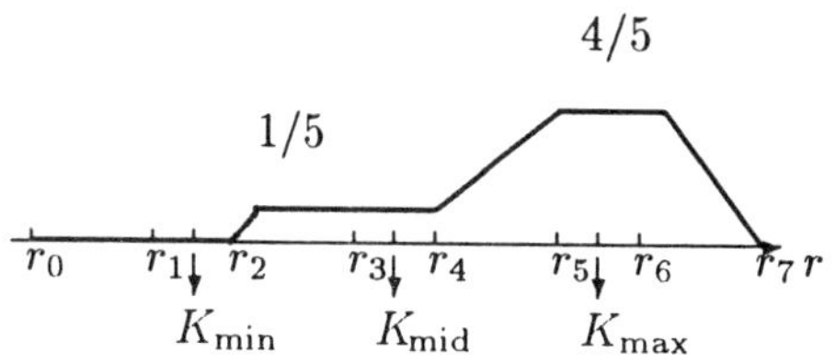


Figure 4: Cumulated rating membership functions

the *maximum-membership defuzzification* scheme defined as

$$R = \arg\max_r g_c(r) = \arg\max_r \left(\sum_l \min(g^l(r), h^l(p))\right)$$

However, this scheme has two important imperfections. First, $R$ is not always unique (see for example, Fig.4). Second, most of the information in the shape of the rating membership functions, $g^l$, is lost. A better alternative is the well known *fuzzy centroid defuzzification* scheme defined as

$$R = \frac{\int_{-\infty}^{+\infty} r g_c(r)\, dr}{\int_{-\infty}^{+\infty} g_c(r)\, dr} = \frac{\sum_l \int_{-\infty}^{+\infty} r \min(g^l(r), h^l(p))\, dr}{\sum_l \int_{-\infty}^{+\infty} \min(g^l(r), h^l(p))\, dr}$$

For the example in Fig. 4, the fuzzy centroid defuzzifier returns

$$R = \frac{\begin{array}{c} K_{\text{mid}} \int_{-\infty}^{+\infty} \min(g^{\mathbf{Average}}(r), 0.2)\, dr \\ + K_{\text{max}} \int_{-\infty}^{+\infty} \min(g^{\mathbf{Good}}(r), 0.8)\, dr \end{array}}{\begin{array}{c} \int_{-\infty}^{+\infty} \min(g^{\mathbf{Average}}(r), 0.2)\, dr \\ + \int_{-\infty}^{+\infty} \min(g^{\mathbf{Good}}(r), 0.8)\, dr \end{array}}$$

### 2.2.2 Level rating function

Now, the individual heuristic ratings must be combined for each levels. Suppose that $y_i$ is a candidate solution up to level $i$, the level rating function is the combination of the $m_i$ heuristic rating functions $h_{ij}$ and of the *a priori* rating function $f_i$ where all these rating functions represents fuzzy subsets. A classical approach in fuzzy logic (see, *e.g.*, Bellman & Zadeh [7] and Mamdani [8]) consists in taking the minimum of the different membership functions, *i.e.*, if the fuzzy centroid defuzzification scheme is adopted

$$\tilde{R}_i(y_i) = \frac{\sum_l \int_{-\infty}^{+\infty} r \min(\min_j(h_{ij}^l(p_{ij}(y_i))), f_i^l(y_i), g^l(r))\, dr}{\sum_l \int_{-\infty}^{+\infty} \min(\min_j(h_{ij}^l(p_{ij}(y_i))), f_i^l(y_i), g^l(r))\, dr}$$

Unfortunately, as it was pointed out by Kosko in [10], this type of combination rule is often too pessimistic, since the rating is always based on the heuristic giving the worst rating, even when all the other heuristics give a good rating. Much better, in our opinion, is the weighted sum approach where a weight, $w_{ij}$, is given to each heuristics according to its reliability and where a weighted sum of all the different membership functions is done

$$\tilde{R}_i(y_i) = \frac{\sum_l \int_{-\infty}^{+\infty} r(w_i \min(f_i^l(y_i), g^l(r)) + \sum_{j=1}^{m_i} w_{ij} \min(h_{ij}^l(p_{ij}(y_i)), g^l(r))) \, dr}{\sum_l \int_{-\infty}^{+\infty} (w_i \min(f_i^l(y_i), g^l(r)) + \sum_{j=1}^{m_i} w_{ij} \min(h_{ij}^l(p_{ij}(y_i)), g^l(r))) \, dr}$$

This rating function does not take into account the rating obtained at the previous levels ($< i$). Only the heuristics of the current level are combined in this manner. We propose different methods to take into account the ratings from the previous levels: either a weighted sum of the different level ratings, or a global centroid of all the different levels, or even the minimum of the different level ratings.

**1. Weighted sum level combination**

$$R_i(y_i) = \sum_{k=1}^{i} \tilde{R}_k(y_k)$$

$$= \frac{K_k \sum_l \int_{-\infty}^{+\infty} r(w_k \min(f_k^l(y_k), g^l(r)) + \sum_{j=1}^{m_k} w_{kj} \min(h_{kj}^l(p_{kj}(y_k)), g^l(r))) \, dr}{\sum_{k=1}^{i}(\sum_{k=1}^{i} K_k) \sum_l \int_{-\infty}^{+\infty} (w_k \min(f_k^l(y_k), g^l(r)) + \sum_{j=1}^{m_k} w_{kj} \min(h_{kj}^l(p_{kj}(y_k)), g^l(r))) \, dr}$$

**2. Centroid level combination**

$$R_i(y_i) = \frac{\sum_l \sum_{k=1}^{i} \int_{-\infty}^{+\infty} r(w_k \min(f_k^l(y_k), g^l(r)) + \sum_{j=1}^{m_k} w_{kj} \min(h_{kj}^l(p_{kj}(y_k)), g^l(r))) \, dr}{\sum_l \sum_{k=1}^{i} \int_{-\infty}^{+\infty} (w_k \min(f_k^l(y_k), g^l(r)) + \sum_{j=1}^{m_k} w_{kj} \min(h_{kj}^l(p_{kj}(y_k)), g^l(r))) \, dr}$$

**3. Minimum level combination**

$$R_i(y_i) = \min_{1 \le k \le i} (\tilde{R}_k(y_k))$$

$$= \min_{1 \le k \le i} \left( \frac{\sum_l \int_{-\infty}^{+\infty} r(w_k \min(f_k^l(y_k), g^l(r)) + \sum_{j=1}^{m_k} w_{kj} \min(h_{kj}^l(p_{kj}(y_k)), g^l(r))) \, dr}{\sum_l \int_{-\infty}^{+\infty} (w_k \min(f_k^l(y_k), g^l(r)) + \sum_{j=1}^{m_k} w_{kj} \min(h_{kj}^l(p_{kj}(y_k)), g^l(r))) \, dr} \right)$$

Which method should be selected depends on the application. For example, if it is critical to find the exact solution at each level and if the level have the same importance, the minimum combination is the most suitable method. However, if the different levels have different importance and if the different level characteristics $X_i$, $i = 1, \ldots, N$, are poorly related, it is much better to use the weighted sum method. Finally, if the different level characteristics $X_i$, $i = 1, \ldots, N$ are very interdependent, then the centroid combination method is better. In the case of our application the weighted sum level combination gives the best results. Finally, these level rating functions are unable to compare candidate partial solutions at different levels. To do so, a global rating function must be defined.

### 2.2.3 Global rating function

To compare two candidate partial solutions at different levels, the rating of the candidate at the lowest level should be extrapolated up to the highest level. In other words, an estimate of the rating that a candidate could possibly obtained at a higher level must be found. Therefore, to compare candidate solutions at every possible levels the rating of every candidates should be estimated up to the highest possible level, $N$. Assume that $y_i$ is a candidate solution up to level $i$, then an estimate of

$$h_{kj}(p_{kj}(y_k)) \quad \text{and} \quad f_k(y_k) \qquad k = i+1, \ldots, N$$

$$y_k = [\; y_i^T \quad (\text{unknown subvector})^T \;]^T$$

must be found. Let's define $E\{h_{kj}\}(y_i)$ and $E\{f_k\}(y_i)$, $k = i+1, \ldots, N$ as these estimates. Then, they could be computed different ways depending on the type of behavior which is desired for the expert system.

1. **Best case approach:** For each of the heuristics and *a priori* ratings to be estimated, the maximum value is always assumed. For example, if the linguistic terms are (**Good, Average, Bad**)

$$E\{h_{kj}\}(y_i) = [\; 1, \; 0, \; 0, \;]$$
$$E\{f_k\}(y_i) = [\; 1, \; 0, \; 0, \;]$$

   This is a cautious approach. Indeed, the candidate partial solutions at the lower levels might be advantaged with respect to the candidates at the higer levels.

2. **Arbitrary value approach:** If a less cautious approach is desired, expected values should be used. Estimates are selected arbitrarily and then later tuned for the application. For example, if the linguistic terms are (**Good, Average, Bad**)

$$E\{h_{kj}\}(y_i) = \left[\; \bar{h}_{kj}^{\mathbf{Good}}, \; \bar{h}_{kj}^{\mathbf{Average}}, \; 0, \;\right]$$
$$E\{f_k\}(y_i) = \left[\; \bar{f}_k^{\mathbf{Good}}, \; \bar{f}_k^{\mathbf{Average}}, \; 0, \;\right]$$

3. **Expected value approach:** Ofcourse, if the probability distributions of the observations $p_{kj}$, $k = i+1, \ldots, N$, are known for a partial solution equal to $y_i$, then the expected values could be directly computed. So, let $d_{kj}(p)_{|y_i}$, $k = i+1, \ldots, N$, be the corresponding density of probability functions, then

$$E\{h_{kj}\}(y_i) = \left[\; \int_{-\infty}^{+\infty} h_{kj}^{\mathbf{Good}}(p) d_{kj}(p)_{|y_i} dp, \; \int_{-\infty}^{+\infty} h_{kj}^{\mathbf{Average}}(p) d_{kj}(p)_{|y_i} dp, \; \int_{-\infty}^{+\infty} h_{kj}^{\mathbf{Bad}}(p) d_{kj}(p)_{|y_i} dp \;\right]$$

This is basically equivalent to the computation of the probability of the fuzzy event: "observation $p_{kj}$ is **Good**", as defined in Bellman & Zadeh [7].

Based on these estimates, the global rating function is defined the following way. Suppose that $y_i$ is a candidate solution up to level $i$

**Weighted sum level combination**

$$R(y_i) = \sum_{k=1}^{i} \frac{K_k}{\sum_{k=1}^{N} K_k} \frac{\sum_l \int_{-\infty}^{+\infty} r(w_k \min(f_k^l(y_k), g^l(r)) + \sum_{j=1}^{m_k} w_{kj} \min(h_{kj}^l(p_{kj}(y_k)), g^l(r))) \, dr}{\sum_l \int_{-\infty}^{+\infty} (w_k \min(f_k^l(y_k), g^l(r)) + \sum_{j=1}^{m_k} w_{kj} \min(h_{kj}^l(p_{kj}(y_k)), g^l(r))) \, dr} + \sum_{k=i+1}^{N} \frac{K_k}{\sum_{k=1}^{N} K_k} \frac{\sum_l \int_{-\infty}^{+\infty} r(w_k \min(E\{f_k^l\}(y_i), g^l(r)) + \sum_{j=1}^{m_k} w_{kj} \min(E\{h_{kj}^l\}(y_i), g^l(r))) \, dr}{\sum_l \int_{-\infty}^{+\infty} (w_k \min(E\{f_k^l\}(y_i), g^l(r)) + \sum_{j=1}^{m_k} w_{kj} \min(E\{h_{kj}^l\}(y_i), g^l(r))) \, dr}$$

## 3 Experimental results

Trapezoidal type rating functions, as in Fig.2, are used as rating functions. They are 3 different linguistic terms selected: **Good**, **Average**, and **Bad**. The overlapping factor is 50% for the heuristics and the rating membership functions. Finally, the parameters of the rating functions are tuned to maximize the quality of the results. The minimum inference is used, with the centroid defuzzifier for the level ratings, and the weighted sum level combination with the arbitrary value approach (for the estimation of the unknown ratings) for the global rating.

The results obtained for a database containing infrared images of 8 vehicles in 16 different positions are shown in Table 1. Perfect detection consists in finding the exact system of 3 axes. Almost perfect detection consists in finding the 3 axes directions, with small normalization or position errors. Sufficient detection consists in finding at least 2 axes (this is, in fact, sufficient for the vehicle identification). Partial detection consists in finding at least the direction of 2 axes. Even if this is not enough to identify directly the vehicle, the correct two axes are among the limited number, ($< 240$), of candidates having these same 2 directions. Once the first task of detecting the position and orientation of the vehicle is accomplished, it remains to identify the type of vehicle. This is done by defining characteristic details, such as, *e.g.*, number of wheels, engine position, tracks size, exhaust system, for each vehicle to be recognized. The location and the shape of these characteristic details is known *a priori*. Therefore, templates can be created with their location on the vehicle specified. Since the position and orientation of the vehicle is known, it is sufficient to verify that the templates match the corresponding areas on the image. Then, a final rating is given to the proposed solution based on the number of matching details and on the quality of this match. If the final rating is not judged sufficiently large, the expert system backtracks to the position and orientation task to propose another candidate solution.

| Quality of detection | Percentage |
|---|---|
| Perfect | 83.6 % |
| Almost perfect | 88.8 % |
| Sufficient | 94.0 % |
| Partial | 94.8 % |

Table 1: Position and orientation detection results

## References


[1] H. Penny Nii. Blackboard systems: the blackboard model of problem solving and the evolution of blackboard architectures. *The AI Magazine*, pages 38–53, Summer 1986.

[2] H. Penny Nii. Blackboard systems: blackboard application systems, blackboard systems from a knowledge engineering perspective. *The AI Magazine*, pages 82–91, August 1986.

[3] B. Hayes-Roth. A blackboard architecture for control. *Artificial Intelligence*, 26(3):251–321, 1985.

[4] J. L. Murdock and B. Hayes-Roth. Intelligent monitoring and control of semiconductor manufacturing equipment. *IEEE Expert*, 6(6):19–31, 1991.

[5] R. O. Duda and P. E. Hart. *Pattern classification and scene analysis*. John Wiley & Sons, New-York, NY, 1973.

[6] G. A. Carpenter and S. Grossberg. The art of adaptive pattern recognition by a self-organizing neural network. *Computer*, 21:77–88, 1988.

[7] R. E. Bellman and L. A. Zadeh. Decision making in a fuzzy environment. *Management Science*, 17:141–164, 1970.

[8] E. H. Mamdani. Application of fuzzy logic to approximate reasoning using linguistic synthesis. *IEEE Trans. on Computers*, C-26:1182–1191, 1977.

[9] B. Kosko. *Neural networks and fuzzy systems*. Prentice-Hall, Englewood Cliffs, NJ, 1992.

[10] B. Kosko. Fuzzy knowledge combination. *Int. J. of Intelligent Systems*, I:293–320, 1986.

# Fuzzy Techniques for Image Enhancement and Reconstruction

Osama K. AlShaykh, Srikrishna Ramaswamy, and Hsien-Sen Hung
Department of Electrical and Computer Engineering
Iowa State University
Ames, Iowa 50011

***Abstract*— A fuzzy-based enhancement algorithm that considers multiple-object images is presented. This algorithm is based on crisping the boundaries of the objects by minimizing their compactness. Further, this concept is incorporated in an iterative reconstruction-reprojection algorithm which compromises between the boundary crispness of the reconstructed image and the fidelity of the available projections in incomplete angular coverage situations. Simulation results are also provided to demonstrate the effectiveness of the proposed algorithm.**

## I. Introduction

Fuzzy sets were introduced in 1965 by Lotfi Zadeh as a new way to represent vagueness in everyday life [1]. They were introduced to capture the concept of imprecision in a way that would differentiate imprecision from uncertainity. Prewitt first noted that image subsets representing objects might be regarded as fuzzy. Later, several contributions were made. A. Rosenfeld extended the concepts of picture geometry to fuzzy geometry [3, 4]. S.K. Pal and R.A. King used fuzzy set theory for image enhancement [6]. Other researchers have also worked on colored image processing, edge detection, quality measures, and machine vision using fuzzy sets [5].

In this paper, a fuzzy-based image enhancement algorithm for multiple object images is first presented. The image is enhanced by minimizing the compactness of all objects in the image. The compactness measure is calculated for each gray level alone to avoid miscalculation of the compactness of boundaries when a sigmoidal function is used to represent the membership function of the gray levels. This enhancement method is a modification and generalization of the method proposed by S.K. Pal and A. Rosenfeld [7] for image thresholding by optimizing the compactness measure.

Next, an iterative fuzzy-based technique for tomographic reconstruction from incomplete angular coverage is presented. This reconstruction algorithm involves iterative reconstruction of the image using the available projections along with the enhanced projections in place of the missing ones to compromise between the fidelity of the available data and the boundary crispness of the reconstructed image.

This paper is organized as follows: Section two will briefly discuss image representation in terms of fuzzy sets and the concept of fuzzy compactness measure. Section three will propose the fuzzy-based image enhancement algorithm. Section four will present the fuzzy-based iterative image reconstruction-reprojection algorithm. Section five will discuss simulation results. Section six will draw some conclusions.

## II. Gray tone image in terms of fuzzy set theory

A gray tone image usually possesses fuzziness in each of its pixel values. This fuzziness results from the loss of data, lack of quantitative measurement of image quality and vagueness in the boundary of the objects as a result of the smoothness of natural images [8].

A gray tone image $X$ of dimension $M \times N$, and L levels, can be treated as an array of fuzzy singletons [5]:

$$X = \{(\mu_{mn}, x_{mn}); m = 1, \cdots, M; n = 1, \cdots, N\} \qquad (1)$$

where each pixel is characterized by the intensity value, $x_{mn}$, and its grade of possessing some membership, $\mu_{mn}(0 \leq \mu_{mn} \leq 1)$, relative to some brightness level $l, l = 0, \cdots, L-1$.

The fuzzy membership value, $\mu_{mn}$, may be defined in two ways, by the "S" function or by the "$\pi$" function, depending on the problem of interest. As an example consider a bimodal image $X$ of dimension $M \times N$, consisting of a dark background with gray level $l_{min}$ and a bright object region with gray level $l_{max}$. In this case it would be reasonable to assign membership values, $\mu_{mn}$, using the "S" function.

$$S(x;a,b,c) = \begin{cases} 0 & \text{if } x \leq a \\ S_1 & \text{if } a < x \leq b \\ S_2 & \text{if } b < x \leq c \\ 1 & \text{if } x > c \end{cases} \qquad (2)$$

where

$$S_1(x;a,b,c) = \frac{(x-a)^2}{(b-a)(c-a)} \qquad (3)$$

$$S_2(x;a,b,c) = 1 - \frac{(x-c)^2}{(c-b)(c-a)} \qquad (4)$$

0-7803-0614-7/93$03.00 ©1993IEEE

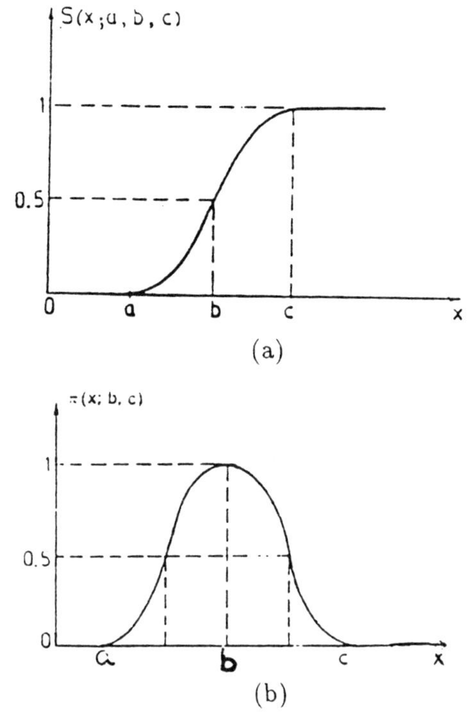


Figure 1: (a) The 'S' function. (b) The '$\pi$' function.

and

$$b = \frac{a+c}{2} \tag{5}$$

In this example, $a = l_{min}$, and $c = l_{max}$. The membership values are assigned such that pixels closer to $l_{max}$ are assigned values closer to 1, and those closer to $l_{min}$ are assigned values closer to 0. Pixels with gray levels closer to the cross-over point, b, are the fuzziest points and are assigned values closer to 0.5. The "S" function is illustrated in Figure 1.

The "$\pi$" function finds use in assigning membership values in multimodal images, for example, an image consisting of a dark background and several bright regions of different gray levels. The "$\pi$" function is defined as,

$$\pi(x;a,b,c) = \begin{cases} 0 & \text{if } x \le a \\ S(x;a,\frac{a+b}{2},b) & \text{if } a < x \le b \\ 1 - S(x;c,\frac{b+c}{2},b) & \text{if } b < x \le c \\ 0 & \text{if } x > c \end{cases} \tag{6}$$

Here, membership values are assigned in a way that pixels closer to the gray level corresponding to a chosen object gray level are assigned values closer to 1, whereas pixels with gray levels on either side of the peak are assigned decreasing values, with the pixels closest to the gray level of the background being assigned values close to 0. The "$\pi$" function is illustrated in Figure 1, and will be used in Section III.

A. Rosenfeld extended the concept of picture geometry to fuzzy sets [3, 4]. He defined several geometrical parameters in terms of fuzzy set theory. The compactness measure defined by him was used in image enhancement [7]. This measure is used to describe the fuzziness in the geometry of the image. Compactness of $\mu$ is defined by

$$comp(\mu) \triangleq \frac{a(\mu)}{p^2(\mu)} \tag{7}$$

where $a(\mu)$ is the area of the fuzzy set and is defined by

$$\begin{aligned} a(\mu) &\triangleq \int \mu \\ &\triangleq \sum_m \sum_n \mu_{mn} \end{aligned} \tag{8}$$

and $p(\mu)$ is the perimeter of the fuzzy set and is defined by

$$\begin{aligned} p(\mu) &\triangleq \int |\nabla \mu| \\ &\triangleq \sum_{m=0}^{M} \sum_{n=0}^{N-1} |\mu_{m,n} - \mu_{m,n+1}| \\ &\quad + \sum_{m=0}^{M-1} \sum_{n=0}^{N} |\mu_{m,n} - \mu_{m+1,n}| \end{aligned} \tag{9}$$

The definition of the area of a fuzzy set is a direct extension of the definition of the area of an ordinary set. The perimeter is defined as the derivative of the area in both horizontal and vertical directions, i.e. it represents the gradient of the object, which is just an extension of the definition of the perimeter of crisp sets where the perimeter is the derivative of the area. For crisp sets the compactness measure is largest for a disk and is equal to $\frac{1}{4\pi}$. So the compactness is smallest for a crisp version of the image. Hence, minimization of the compactness measure is expected to enhance the image [7].

## III. An Enhancement Algorithm for Multiple-Region Images

S. K. Pal and A. Rosenfeld [7] have proposed an algorithm to threshold images by optimizing the fuzzy compactness measure. First, they proposed to compute the compactness of each gray level by assigning to each pixel, a membership value according to the "S" function, a sigmoidal function [8]. The "S" function has that gray level value as its cross-over point, i.e. fuzziest point of membership value 0.5. The gray level corresponding to the smallest compactness measure is then used for thresholding.

However, it is anticipated that scanning the gray levels using the "S" function won't give good results for multiple object images. This is because the "S" function thresholds

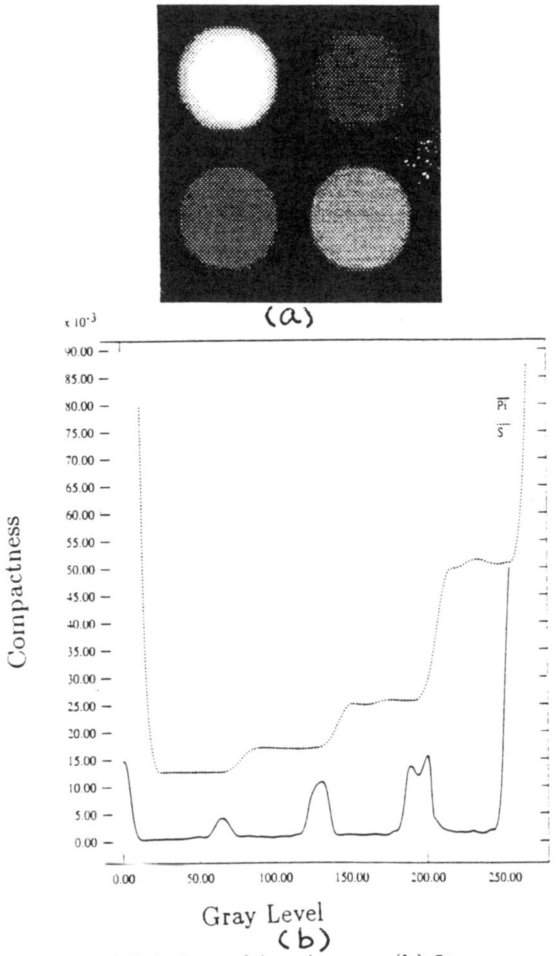


Figure 2: (a) A four object image. (b) Its compactness as measured using the "S" and "$\pi$" function.

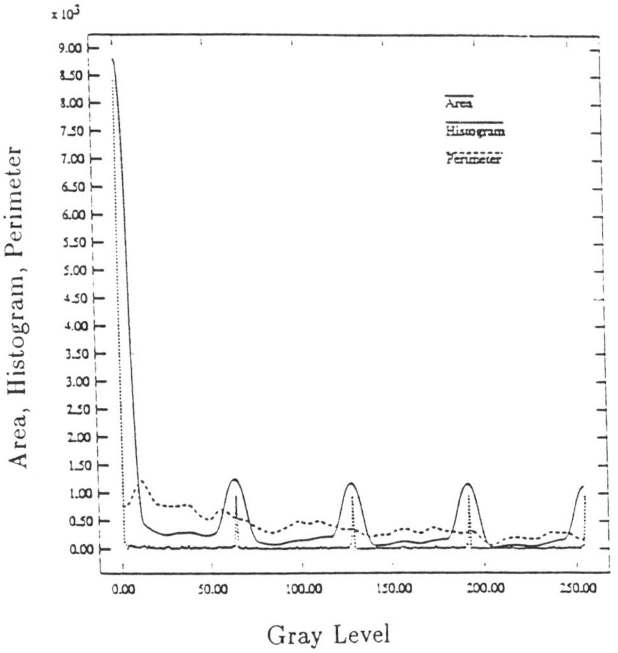


Figure 3: Area, perimeter, and histogram for the four object as a function of gray level.

the gray levels thus including the brighter regions while calculating the compactness. As a solution, we propose to use the "$\pi$" function which works as a window that includes only the gray level of interest while calculating the compactness.

Figure 2 illustrates the effectiveness of this proposed technique. The image consists of four blurred circles, each of a certain gray level. When compactness is calculated using the "S" function the fuzzy thresholding effect can be seen in the stair shape of the compactness measure. This shape is due to the fact that all the pixels that have a gray level larger than the cross over point will contribute to the compactness. It is evident that the compactness calculated using the "$\pi$" function overcomes this problem. This is because the "$\pi$" function will just include the gray level of interest when calculating the compactness.

When the gray levels are scanned using the "$\pi$" function the scanning is done by varying the "b" point of the "$\pi$" function. The compactness of fuzzy regions corresponding to each gray level is computed. Then, a new membership function based on the compactness values is constructed. The construction is done by alternate use of "$\pi$" and "$1-\pi$" functions. The fuzziest points of the new membership function will correspond to those levels having the local minima of the compactness measure and the crisp points to the local maxima. The new membership function is constructed so because the compactness of a gray level when representing an object is larger than that when representing a boundary. This is because the area of an object is much larger than that of its boundary while the relative change in the perimeter is minimal. This is illustrated in Figure 3.

Also observed is the effect of bandwidth on the scanning process. As can be seen, a false maximum has been detected close to gray level 190. This is because the chosen bandwidth value is quite small. The error is easily rectified by increasing the bandwidth value. But care should be taken while increasing the bandwidth value as larger values lead to loss of some minima and maxima. This is illustrated in Figure 4, where scanning at a larger bandwidth value, besides eliminating the false maximum around gray level 190, has also resulted in the loss of a maximum around gray level 65. Thus, the choice of bandwidth is critical.

After translation to fuzzy domain, the fuzziness of the image is reduced by crisping the boundaries using the contrast intensification function, INT operator [2]. Finally, The enhanced image is obtained by inverting the membership function.

The fuzzy based enhancement algorithm will be used as a functional block in the proposed algorithm, to be stated in Section IV, for image reconstruction.

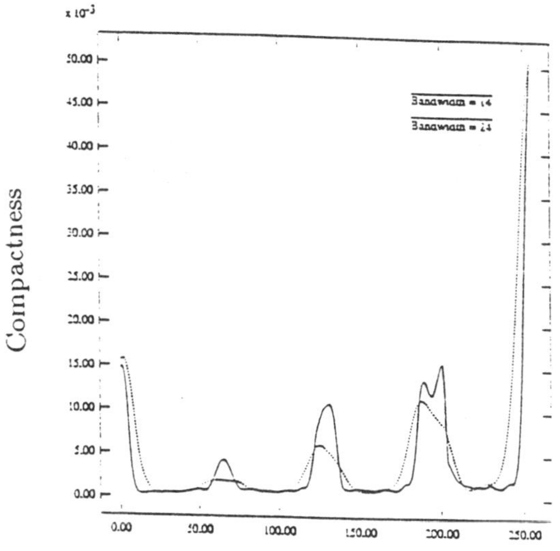


Figure 4: Effect of bandwidth on the scanning process.

The following summarizes the enhancement routine:

1. The "$\pi$" function with its "b" point varied over the entire range of gray levels is used as the membership function to find local minima and maxima of the fuzzy compactness measure.

2. A new membership function with alternating "$\pi$, $1-\pi$" functions is constructed based on the minima and the maxima obtained in step (1). The minima correspond to cross-over points associated with the boundaries of regions, whereas the maxima correspond to crisp points associated with background or object regions.

3. The contrast intensification function, INT operator, is used to reduce fuzziness.

4. The enhanced image is obtained by inverting the membership function.

## IV. Image reconstruction: Fuzzy-based iterative reconstruction-reprojection (FIRR)

Image reconstruction from limited projections has become increasingly important in computed tomography (CT). When Fourier-based reconstruction methods, for example filtered back projection (FBP) are used [16], lack of complete angular coverage in CT scanning usually results in considerable errors which introduce severe artifacts and also degrade accuracy in the reconstructed cross-sections. To improve reconstruction quality, various methods were attempted. Among these are, the method of moments, double orthogonal functions, exact integral formula equation, and approximate methods for solving interpolation integral equations [9]. Other methods that use prior knowledge about the object include Projection Onto the Convex Sets (POCS) [10], the Bayesian MAP method [11] and others [12, 13, 14].

Since most objects possess crisp (nonfuzzy) boundaries, this property can be effectively used to improve accuracy and reduce artifacts in the reconstructed image. To this end, a new approach, called Fuzzy-based Iterative Reconstruction-Reprojection (FIRR), is proposed to tackle the limited-view CT problem. This approach is novel in being able to extract missing projections by incorporating prior information on the non-fuzzy boundary only. Here, the filtered back projection algorithm is employed in an iterative fashion alternating with reprojection through the enhanced reconstructed image. The reprojection from image domain to projection (data) domain is required to estimate projection values in the angular sector of missing views. Before reprojection at each iteration, the fuzzy based enhancement routine stated in Section III is invoked to reduce fuzziness in the reconstructed image. This step incorporates information on the crisp boundary of the object which provides theoretical basis for the restoration of missing projections.

Missing projections introduce fuzziness in the reconstructed image. They affect mainly the boundaries and the shape of the object in locations perpendicular to the direction of the missing projections. Starting with the average value of available projections of the image instead of zero-valued projections in place of the missing projections is more appealing. Results so obtained show that the fuzzy characteristics of the reconstructed image are closer to those of the original image. After image reconstruction, fuzzy-based enhancement is applied to the image.

To compromise between the fidelity of the available projections and the crispness of the boundaries of the reconstructed image, the projections of the enhanced image in the missing regions are used along with the available projections to construct a new image. The aforementioned enhancement routine is applied to the new image. This procedure is iterated till an acceptable reconstruction is obtained.

Figure 5 shows the functional block diagram of FIRR.

## V. FIRR: Simulation results

To demonstrate the effectiveness of the proposed algorithm, the Shepp-Logan head phantom was used as a simulation example [16]. Parallel projection data were generated for 180 views with 127 rays per view. The 180 views were uniformly spaced over the full $180^o$ angular range. Missing views represent projections numbered 135-180 which cover an angular sector of $45^o$. Figure 6(a) shows the head phantom as reconstructed using FBP from 180 complete projections whereas Figure 6(b) is reconstructed using the first 135 projections with the average value of the available projections in place of the

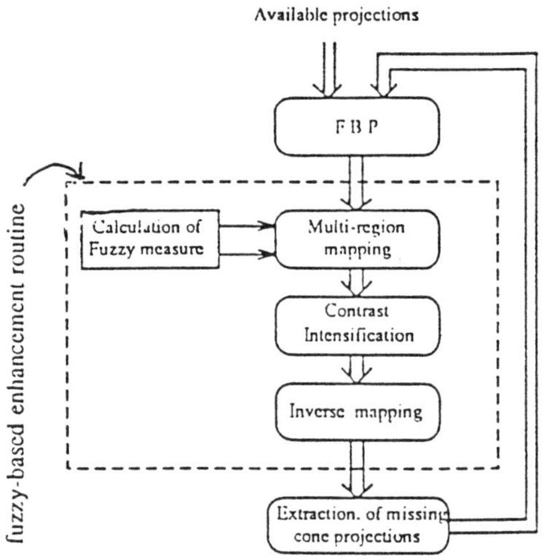


Figure 5: The functional block diagram of FIRR algorithm.

missing projections. Figure 6(b) can be regarded as the initial estimate for the subsequent processing by FIRR. Figures 6(c) through (f) show the enhanced and reconstructed images at the end of the first and second iterations. As is observed, the boundaries are crisper and the ellipses are converging to their original shape and size. It is also observed that FIRR improves on the reconstruction accuracy with each iteration, and converges to an image close to the original one. The effectiveness of restoration of missing projections based on the fuzzy-based enhancement routine is demonstrated in Figure 7. It is observed that the restored missing projection at 145° is converging to the profile of the actual projection. Figure 8 shows the normalized mean-square reconstruction error for two iterations. The corresponding errors after enhancement following reconstruction at each iteration are 14.6% and 13.7%.

When applying FIRR on the image, the image tends to converge to a binary one. So, the iterations should be terminated when an acceptable image is formed. It was observed that after some iterations the normalized mean-square error of the enhanced image started to increase. Hence, iterations should be terminated at this point.

## VI. Conclusions

A fuzzy-based image enhancement algorithm is proposed. This algorithm takes into account the presence of multiple objects with different or similar gray level. An

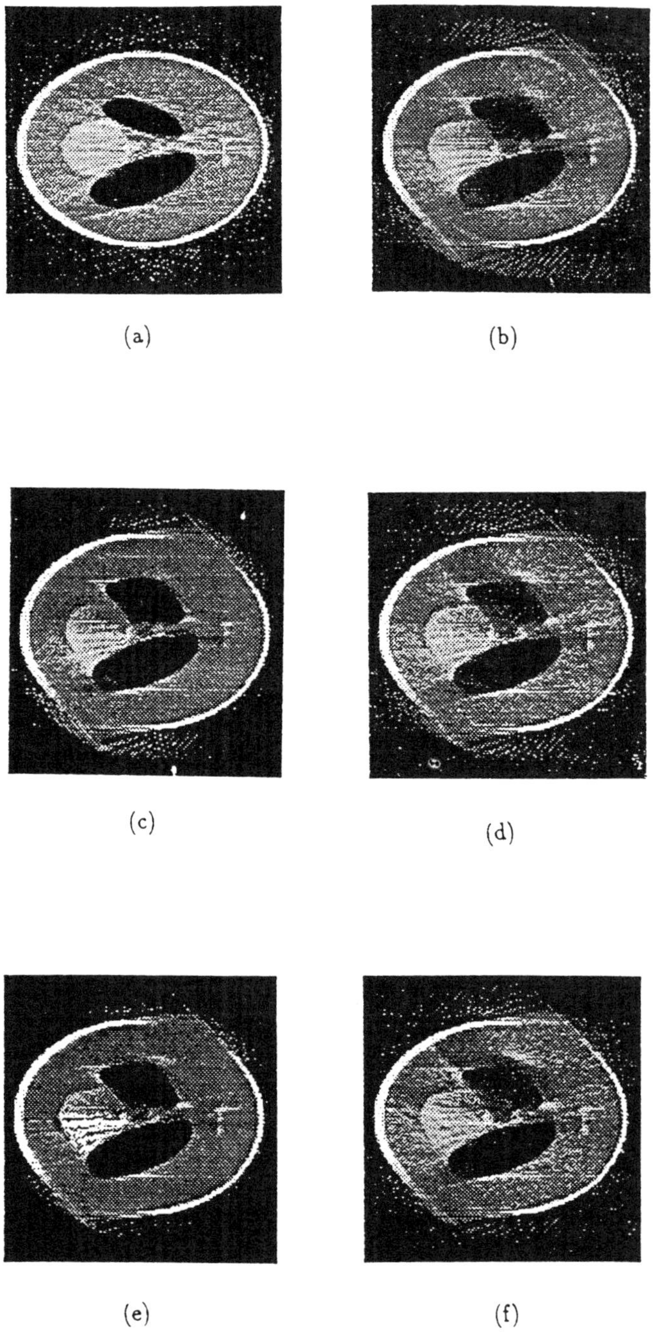


Figure 6: Reconstructed Shepp-Logan head phantom: (a)FBP based on complete 180 projections, (b)FBP based on the first 135 projections with their average as missing projections, (c)enhanced image at the first iteration of FIRR, (d)reconstructed image at the first iteration of FIRR, (e)enhanced image at the second iteration of FIRR, (f)reconstructed image at the second iteration of FIRR.

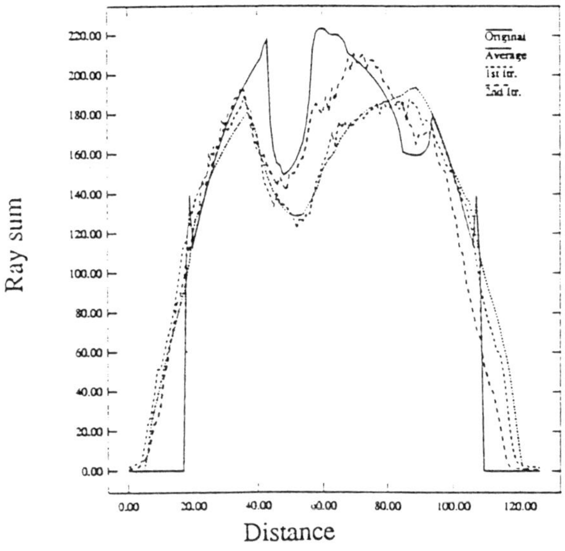


Figure 7: The estimated projection at 145° (The $10^{th}$ missing projection) while iterating.

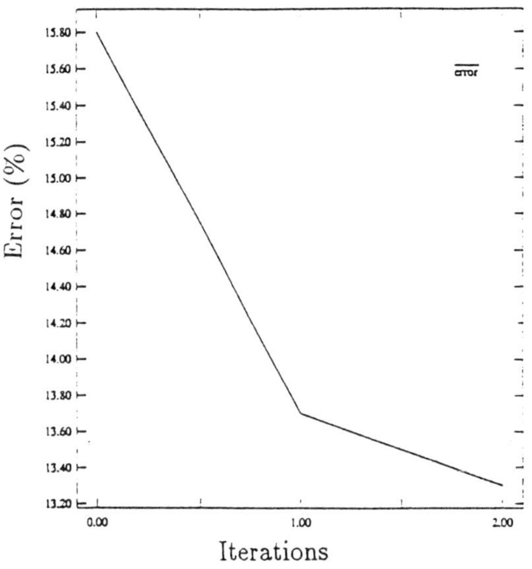


Figure 8: The normalized mean-square reconstruction error

illustration of the effectiveness of this algorithm in thresholding or enhancing multiple-object images is presented. The performance of the algorithm depends on the bandwidth of the function used for scanning the gray levels. Smaller bandwidths result in the detection of false minima and maxima, whereas larger bandwidths result in loss of some minima and maxima. A systematic way of choosing the bandwidth needs further investigation. Also in the need for further investigation, is an appropriate criterion to enable the termination of iterations at an acceptable image quality.

The proposed fuzzy image enhancement method is incorporated in an algorithm to reconstruct images from incomplete data sets. The reconstruction algorithm compromises between the fidelity of the available data and the crispness of the boundaries of the reconstructed image.

The example given illustrates the effectiveness of FIRR in utilizing the crispness of boundaries of an image as the only prior knowledge. FIRR can be modified to take into account any available prior knowledge such as the general shape of the missing profiles or some statistical parameters about them, due to its iterative nature.

## REFERENCES


[1] Lotfi A. Zadeh, "Fuzzy sets", *Inform, Control*, vol. 8, pp. 338-353, 1965.

[2] Lotfi A. Zadeh, "Outline of a new approach to the analysis of complex systems and decision processes", *IEEE Trans. Syst., Man, Cybern.*, vol. SMC-3, no. 1, pp. 28-44, January 1973.

[3] Azriel Rosenfeld, " Fuzzy digital topology", *Inform, Control*, vol. 40, no. 1, pp. 76-87, January 1979.

[4] Azriel Rosenfeld, "The fuzzy geometry of image subsets", *Pattern Recognition Letters*, vol. 2, pp. 311-317, September 1984.

[5] James C. Bezdek and Sankar K. Pal, "Fuzzy models for pattern recognition: methods that search for structures in data", *IEEE PRESS*, 1992.

[6] Sankar K. Pal and Robert A. King, "Image enhancement using smoothing with fuzzy sets", *IEEE Trans. Syst., Man, Cybern.*, vol. SMC-11, no. 75, pp. 494-501, July 1981.

[7] Sankar K. Pal and Azriel Rosenfeld, "Image enhancement and thresholding by optimization of fuzzy compactness", *Pattern Recognition Letters*, vol. 7, pp. 77-86, February 1988.

[8] Hua Li and Hyun S. Yang, "Fast and reliable image enhancement using fuzzy relaxation technique", *IEEE Trans. Syst., Man, Cybern.*, vol. SMC-19, no. 5, pp. 1276-1281, September/October 1989.

[9] V.P Palamodov, "Some singular problems in tomography", *Translations of mathematical monographs*, vol. 81, American mathematical society, 1990.

[10] M.I. Sezan and H. Stark. "Applications of convex projection theory to image recovery in tomography and related areas", *Image Recovery: Theory and Applications*, edited by H. Stark, Academic Press, 1987.

[11] K.M. Hanson and G.W. Wecksung, "Bayesian approach to limited angle reconstruction in computer tomography", *J. Opt. Soc. Amer.*, vol. 73, pp. 1501-1509, 1983.

[12] A. Safaeinili, J.P. Bassart, and H.S. Hung, "Reconstructing low contrast flows in limited-view-angle industrial tomograms using prototype flawless images", *Twenty-fourth Asilomar Conference on signals, Systems and Computers*, Pacific Grove, CA, 1990.

[13] M. Nassi, W.R. Brody, B.P. Medoff, and A. Macovski, "Iterative reconstruction-reprojection: An algorithm for limited data cardiac-computed tomography", *IEEE Trans. Biomed. Eng.*, vol. 29, pp. 331-341, May, 1982.

[14] R. Rangayan, A.P. Dhawan, and R. Gordon, "Algorithms for limited-view computed tomography; an annotated bibliography and a challenge", *Applied optics*, pp. 124, 1985.

[15] Sankar K. Pal and D. Dutta Majumder, *Fuzzy mathematical approach to pattern recognition*, Wiley (Halsted Press), New York, 1986.

[16] L.A. Shepp and B.F. Logan, "The Fourier Reconstruction of a Head Section", *IEEE Trans. Nucl. Sci. NS-21*, no. 3, pp. 21-43, 1974.

# Fuzzy Logic For Underwarter Target Noise Recognition

Jinlin Wang, Qihu Li and Wei Wei
Institute of Acoustics, Academia Sinica
P.O.Box 2712, Beijing 100080
People's Republic of China

Abstract-The underwater target noise recognition is one of the most important topic in modern sonar design. In this paper, the target noise are divided into five types, and their characteristics in frequency domain are described. The fuzzy logic algorithm for passive classi- fication is presented. An expert system based on this inference lo- gic for target noise recognition is introduced. Some test data for this system show that this expert system has a good performance and can be considered as an excellent tool of auxiliary decision -maker for sonar operator.

## I. Introduction

The underwater target noise classification is very important topic in the modern sonar system design. In fact, it is a branch of artificial intelligent(AI).

By theory, the target recognition is a typical problem of cluster analysis. There are two methods for solving this problem[7],namely:

(1) Statistical classifier
(2) Expert system approach

There are many successful examples by using method (1), such as: Bayes cluster analysis, nonparameter rank test, which has been applied in finger signature identification.

As mentioned by Horst Bendig[7], for underwater target noise recognition, there must be three basis requirements for statistical cluster, but it is difficult to satisfy in case of underwater target noise recognition. This is the reason why the expert system approach is favored.

By ES, we mean a new kind of software that simulates the problem-solving behavior of a human expert. By using inference mechanism, the expert system can simulate the problem-solving strategy of a human expert according to the architecture of knowledge base.

In real world, the expert knowledge contains many uncertain factors. These factors make big difficult in knowledge representation. For instance,the Doppler shift due to ship moving will result in the uncertainty of the frequency of line spectrum. Since 1965, the fuzzy set theory was introduced by Zadeh, L. A.[4], the theory has been used in many fields, such as: industry control, medical diagnosis. Fuzzy set theory provides a natural method for dealing with the linguistic terms by which an expert will describe a domain. An imprecise numeric term can be effectively described by a fuzzy number[1]. Hence, the use of fuzzy set theory in ES has caused an evolution of system design.

An expert system for ship radiated noise recognition named EXPLORE developed recently in Institute of Acoustics is a system based on fuzzy logic. The target noises are divided into five types: Submarine, Merchantship, Fishing vessel, Surface warship, and Oceanic noise, by using the membership functions

0-7803-0614-7/93$03.00 ©1993IEEE

which are represented by some numerical features of target noise. Using the criteria of maximum dependence principle of membership, EXPLORE will judge that which type mentioned above the input noise is belong to.

In section II, we will discuss the characteristics of warship briefly, then the membership functions will be constructed in section III, according to the analysis in section II. Finally, some test results will be shown in section IV.

## II. Characteristics Analysis

On the point of principle, the radiated noise can be analyzed in time domain or in frequency domain. Due to the Fourier transformation, there is one to one correspondence relationship between the two representations. For simplicity of algorithm, most results are discussed in frequency domain.

The EXPLORE discussed here is based on the frequency characteristics of radiated noise. It is necessary to introduce some related results briefly. For target noise recognition, the sole information source is the radiated noise from moving ship.

The source of radiated noise can be divided into three categories: machinery noise, propeller noise, and under water dynamic noise due to target moving. In general, the first two are main radiate noise source[5]. From the point on spectrum domain, the mixture of these sources will evolve strong line spectra and continuous spectrum. In higher frequency band, the spectrum level will decrease by 6-8dB per octave. In lower band, the spectrum level will increase with frequency increasing. This property results in a peak in radiate noise spectrum. Specially, for warship, this peak locates in the band between 100-1000Hz. Figure 1.

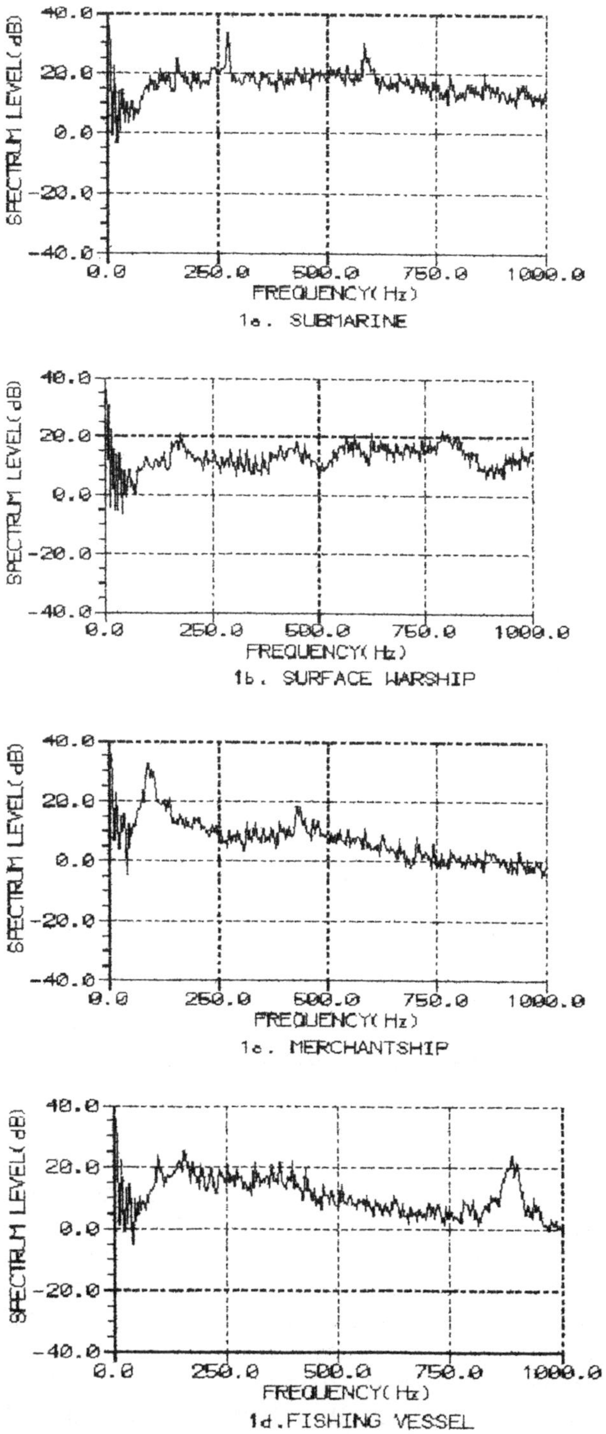


**Fig. 1 The average power spectrum in band 0-1KHz of the various ships.**

shows the spectra of the four types in band 0--1000Hz, respectively. From Figure 1a., we can see that there are strong line spectra with level higher than 10dB above average spectrum value in the band 200--700Hz for submarine. It is because that the magnetic field in electrical machinery of submarine generates period vibration between stator and rotor. For other types, because their power device is diesel, the frequency of line spectrum, if existed, is below 200Hz. So the line spectrum is selected as identification characteristics for submarine.

For merchantship and fishing vessel, we choose the position of the strongest frequency band as the characters, because there is a peak with stable position in band 0--1000Hz. This result can be seen in Figure 1c.,and 1d.

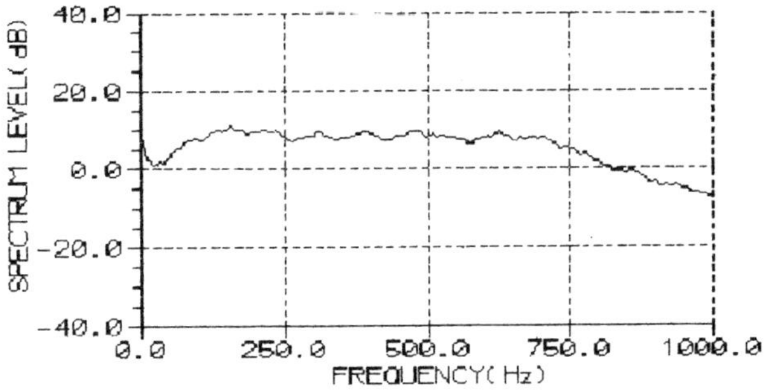


**Fig.2 The Average Power Spectrum On 0-1KHz of The Oceanic Noise.**

When the target is far away from the receiver, the signal received by sonar array is actually ambient noise due to tide, gush, and so on. In lower frequency band (<500Hz), the oceanic noise can be thought as Gaussian noise, and the curve of the spectrum in decibel scale is a horizontal line, approximately. In higher band(>500Hz),the spectrum level of oceanic noise is decreased by 6-8dB per octave, These feature mentioned above is selected as identification character of ambient noise. Figure 2. show the spectrum of oceanic noise.

## III. Decision Function Generation

Although there are many methods based on statistics for pattern recognition theory. In practice, these algorithms have not got success result. There will usually be a much large set of manifestation with vary degrees of equivocation[1]. To find the best explanation requires information search as well as mechanisms for reasoning with uncertain evidence. Uncertainty in EXPLORE arises from noise and distortion uncertainty in subjective inference rules obtained from human experts and the use of qualitative or linguistic terms often used by human experts. A formal way to copy with the uncertainty mentioned above is the theory of fuzzy sets[7].

There are two classic methods for fuzzy pattern recognition, one is based on the maximum dependence principle of membership, the other is based on the principle of the rule of nearest range. In EXPLORE we choose first method.

<u>Definition 3.1:</u>

Suppose the average power spectrum set in band 0-500Hz of a target noise is:

$$ASPL=\{X_0(0), X_0(1), \cdots, X_0(L_0-1)\} \quad (1)$$

where $L_0$ is the number of line spectrum. And

$$APSH=\{X_1(0), X_1(1), \cdots, X_1(L_1-1)\} \quad (2)$$

is the average power spectrum in band 500Hz-1000Hz.

Denote the $\overline{X_0}$ as the mean value of the set **ASPL**,

$$\overline{X_0} = \frac{1}{L_0}\sum_{i=0}^{L_0-1} X_0(i) \quad (3)$$

then the horizontal factor of the target noise is defined as:

$$HF = \frac{L_x}{L_0} \quad (4)$$

The slope factor of the target noise is defined as:

$$SF = \frac{L_h}{L_1} \quad (5)$$

where $L_x$ is the number of element in **ASPH**, which satisfies :

$$| X_0(i) - \overline{X_0} | \le T_0 \quad (6)$$

$T_0$ is the threshold of spectrum fluctuation. And $L_h$ is the number of element in ASPH, which satisfies:

$$X_1(i) - X_1(i+1) \ge T_1 \quad (7)$$

$T_1$ is the threshold of spectrum attenuation.

The symbol **U** denotes a universe of discourse, which is a collection of the spectrum characteristics of the ship radiated noise received by sonar array. That is:

$$U = LS \cup CS \quad (8)$$

**LS** is the character set of line spectrum which will be described lately. **CS** is the character set of continuous spectrum, which contains the position information of the strongest spectrum band **MAX_SP**, the horizontal factor **HF** and slope factor **SF**.

A fuzzy subset **A** of **U** is a set of order pair:

$$A = \{(u_i, \mu_A(u_i))\} \quad (9)$$

where the $\mu_A(u)$ represent the grades of membership which indicated the degree of membership.

For EXPLORE, there are five fuzzy subsets of **U** defined by five membership functions. The subsets are:

$A_1$ ={ spectrum characters of known submarine}
$A_2$ ={ spectrum characters of merchantship}
$A_3$ ={ spectrum characters of fishing vessel}
$A_4$ ={ spectrum characters of oceanic noise}
$A_5$ ={ the others}

The correspondence membership functions can be defined according to the analysis in section II.

$u_0$ is target noise to be recognized. $u_0 \in U$ .

Suppose there are $M_s$ submarines in knowledge base,specially,the character set of line spectrum is :

$$LS_i = \{(d_i(0),b_i(0)),(d_i(1),b_i(1)), \cdots,(d_i(L_i-1),b_i(L_i-1))\} \quad (10)$$

That is there are $L_i$ line spectrum in the spectrum set of i-th submarine, the frequency and level are denoted as : $(d_i(j),b_i(j))$ , $0<j<L_i-1$, $0<i<Ms$. The membership function $\mu_{A1}(u)$ is formulated as followed.

<u>Definition 3.2:</u>

The generalized intersection subset of $LS_i$ to $LS_j$ is denoted by $S_{ij}$,

$(x,y) \in LS_i$ if existing $(u,v) \in LS_j$ such as: $|x-u|<PW$

then $(x,y) \in S_{ij} \subset LS$ .

where **PW** is the threshold of the width of line spectrum.

<u>Definition 3.3:</u>

The order statistic set of the line spectrum set LS is defined by:

$$RLS = \{R(0),R(1),\cdots,R(L-1)\} \quad (11)$$

where **R(i)** is the order via level of ith line spectrum in overall line spectrum set.

The membership function of the fuzzy subset $A_1$ is defined as:

$$\mu_{A_1}(u) = \underset{0 \le i \le Ms}{Max}(f_{i0}(PW)) \quad (12)$$

where

$$f_{i0}(PW) = \frac{1}{L_{i0}} \sum_{k=0}^{L_{i0}-1} [1 - \frac{|RS_{0i}(k) - RS_{i0}(k)|}{L_{i0}} \quad (13)$$

and $RS_{0i}(k) \in RS_{0i}$ which is the order statistic set of generalized intersection subset of Generalization interior set of $RLS_0$ to $RLS_j$, $L_{i0}$ is the element number of the set $RS_{i0}$. And $RS_{i0}(k) \in RS_{i0}$ which is a generalized intersection subset of $RLS_j$ to $RLS_0$.

Suppose there are $L_m$ merchantships and $L_f$ fishing vessel in knowledge base and their strongest spectrum band position are: **Max_SP_M**$_i$ and **Max_SP_F**$_i$, respectively. The membership function of fuzzy subset $A_2$, $A_3$ are defined as followed, respectively.

$$\mu_{A_2}(u) = \underset{0 \le i \le L_m}{Max}(f_{M_i}(PW)) \quad (14)$$

where

$$f_{m_i}(PW) = [1 - \exp(-\alpha |Max\_SP\_M_i - Max\_SP|)] \quad (15)$$

and

$$\mu_{A_3}(u) = Max(f_{F_i}(PW)) \quad (16)$$

where

$$f_{F_i}(PW) = [1 - \exp(-\beta |Max\_SP\_F_i - Max\_SP|)] \quad (17)$$

where $0 \le \alpha \le 1$, $0 \le \beta \le 1$ are control factor.

For oceanic noise, the $\mu_{A_4}(u)$ is:

$$\mu_{A_4}(u) = WL \times HF + WH \times SF \quad (18)$$

where $WL + WH = 1$, $0 \le WL, WH \le 1$, are weighing factor.

For fuzzy subset $A_5$, according to its definition, we know that $A_5$ is a complemental subset of union set of $A_1, A_2, A_3, A_4$. That is

$$A_5 = \neg(\bigcup_{i=1}^{4} A_i) \quad (19)$$

where $\neg$ denotes the complemental operator. So:

$$\mu_{A_5}(u) = 1 - \underset{1 \le i \le 4}{Max}(\mu_{A_i}(u)) \quad (20)$$

So far, we define five subset on universe discourse **U** according to analysis in section **II**. It is obviously that the membership function correspondent to any subset can express the characteristics of identification of the type specialized.

## IV. The Experimental Result and Conclusion

A series test have been carried out for EXPLORE. The input signals originally are collect in some at sea experiment. It include a wide variety of warship and civil vessel class.

By adding the oceanic noise with different amplitude,we can get the input signals with different signal to noise ratio (SNR), and test the performance of the EXPLORE. Table 1.,2.,3. show the test results with different SNR.

It must point out that the SNR indicated in Table 1.,2.,3 is not real SNR, because the signal may contain the oceanic noise unavoid ably, which were recorded at the real sea environment. The real SNR is lower than that shown in Table

Table 1. The test results without noise

| TYPE | NUMBER OF SAMPLE | RATE OF RECOGNITION |
|---|---|---|
| SUBMARINE | 151 | 86% |
| MERCHANTSHIP | 51 | 88% |
| FISHING VESSEL | 19 | 86% |
| SURFACE WARSHIP | 130 | 90% |
| OCEANIC NOISE | 11 | 73% |

Table 2. The test results at 6dB

| TYPE | NUMBER OF SAMPLE | RATE OF RECOGNITION |
|---|---|---|
| SUBMARINE | 71 | 77% |
| MERCHANTSHIP | 17 | 100% |
| FISHING VESSEL | 6 | 100% |
| SURFACE WARSHIP | 69 | 90% |

Table 3. The test results at 3dB

| TYPE | NUMBER OF SAMPLE | RATE OF RECOGNITION |
|---|---|---|
| SUBMARINE | 62 | 52% |
| MERCHANTSHIP | 22 | 73% |
| FISHING VESSEL | 8 | 88% |
| SURFACE WARSHIP | 90 | 74% |

1.,2.,3.

From the Table 1.and Table 2., we can find that the performance is better at 6dB than that without noise. It is because that the line spectrum generated by some interference is strong enough to result in the fault judgement of submarine when there is not noise to added, which yield false alarm.

For fuzzy subset $A_5$, we divide it into two type, that is new submarine and surface warship according to the line spectrum level. In fact, the EXPLORE can judge six type of radiate noise.

**Reference:**


[1] Abraham Kandel, "Fuzzy expert systems", CRC Press, 1991, LONDON

[2] M.S. Fox, " AI and expert sys tem myths, legends and facts", IEEE expert Vol. 5, No.1(1990)

[3] Ed. by T. Yurban and P.R.Watl ins," Applied expert systems", North Holland, 1988

[4] Zadeh, L.A., "Fuzzy set", Inf. Control, 1965

[5] Wang Dezhao and Sang Erchang, " Underwater Acoustics",Academic Press, 1981,Beijing(In chinese)

[6] Ed. by C.H. Chen,"Pattern Rec ognition and Signal Process ing", SIJTHOFF & NOORDHOFF,1978

[7] Ed. by H.G. Urban,"Adaptive Methods in Underwater Acous tics", NATO ASI Series, 1985

[8] R.K.Blashfield et al.," Cluster analysis software", Ibid, 245 -264

[9] Dieter Nebendahl, "Expert Sys tems", John Wiley & Sons Limit ed, 1987.

# Recognition of Facial Expressions using Conceptual Fuzzy Sets

Hirohide USHIDA*, Tomohiro TAKAGI*, and Toru YAMAGUCHI**

*Laboratory for International Fuzzy Engineering Research,
SiberHegner Buil 3F, 89-1 Yamamashita-cho,
Naka-ku, Yokohama-shi, 231, JAPAN
(Tel.)+81-45-212-8227, (Fax.)+81-45-212-8255

**Systems & Software Engineering Laboratory, Research & Development Center, Toshiba Corp.
70 Yanagi-cho, Saiwai-ku, Kawasaki-shi, 230, JAPAN
(Tel.)+81-44-548-5637, (Fax.)+81-44-533-3593

*Abstract*

A facial expression is a vague concept that is difficult to explicitly describe. Conceptual Fuzzy Sets (CFS) have the ability to explicitly represent vague concepts. CFS are realized by using bi-directional associative memories, and a multi-layer structured CFS represents the meaning of a concept by various expressions in each layer. Multi-layered Reasoning in CFS has the following features: 1. Capability of simultaneous abstract and concrete representation 2. Capability of simultaneous top-down and bottom-up processing. In this paper, we apply CFS to the recognition of facial expressions and show that it can achieve context sensitive recognition.

**Key Words:** Facial Expression, Fuzzy Sets,
Bidirectional Associative Memory,
Context Sensitive, Multi-layered Reasoning

## 1. Introduction

Fuzzy sets theory provides an effective means to combine the real world (consisting of a very large number of instances of events and continuous numeric values) with human logical knowledge (consisting of abstracted concepts and symbols). It is difficult to delineate clear boundaries among human concepts because most of them are vague. Fuzzy sets theory is expected to provide knowledge processing in real-world applications, because it can support both symbolic processing and numeric processing, by connecting the logic-based world and the real world.

However, it is difficult to represent the meaning of a concept by using simple fuzzy sets because the meaning depends on the context. For example, a human facial expression consists of hazy and weak patterns on different parts of the face rather than one conspicuous pattern on one part of the face[1]. Although eyes look angry, for instance, the whole face may seem to be laughing depending on the expression of the mouth or other parts. Also it is difficult to explicitly represent the concepts of facial expressions by using ordinary fuzzy sets. For example, in the case of an angry face, it is easy to tell the degree of anger but not easy to explain why the face looks angry. Therefore, in the definition of a fuzzy set, a denotative description is generally used by means of the membership function, which associates each element with a grade of membership in the interval [0,1]. However the definition of the fuzzy set, described by instances "A, B, C", cannot determine the membership value of "D", which is a new instance. This means that a fuzzy set is not able to generalize knowledge from instances.

All these problems relate to the representation of the meaning of a concept. According to Wittgenstein[2], the meaning of a concept is represented by the totality of its uses. In this spirit we proposed the notion of Conceptual Fuzzy Sets (CFS)[3]. In the CFS the meaning of a concept is represented by the distribution of activations of labels that have concepts. Since the distribution changes, depending on the activated labels, to indicate a situation, CFS can represent context dependent meanings.

CFS also carry out Multi-Layered Reasoning (called MLR in this paper) based on association that is driven by propagation of activation of labels[4]. MLR by means of CFS has the following features:

1. Capability of simultaneous symbolic and quantitative processing (semantic guideline)
2. Capability of simultaneous top-down and bottom-up processing (context sensitive processing)

In this paper, we apply MLR by means of CFS to the recognition of human facial expressions. In section 2, we discuss the general characteristics of CFS. In section 3, we propose a CFS network for recognizing human facial expressions and discuss the results of recognition experiments.

0-7803-0614-7/93$03.00 

## 2. Conceptual Fuzzy Sets

### 2.1. Conceptual Fuzzy Sets for Concept Representation

A label of a fuzzy set represents the name of a concept and the fuzzy set represents the meaning of the concept. Therefore, the shape of a fuzzy set should be determined from the meaning of the label depending on various situations. According to the theory of meaning representation from use proposed by Wittgenstein, the various meanings of a label (word) may be represented by other labels (words). We can assign grades of activation showing compatibility degrees between different labels in CFS, so that the distribution of activation can represent the meaning of a label depending on context. In CFS, once a label is activated, the activation is propagated to other labels. The distribution determined by activations agrees with the region of thought corresponding to the word expressing its meaning and the grade of the activation expresses the degree of the relation among labels. Since situations are also indicated by activations, the meaning is expressed by overlapping the region of thought determined by these activations. Figure 2-1 illustrates the different meanings of the same label L1 in different situations S1 and S2.

CFS are realized using bi-directional associative memories, in which a node represents a concept and the strength of a link is determined by the strength of the relation between two connected concepts. Activations of nodes produce reverberation and system energy is stabilized to a local minimum. As a result, corresponding concepts are recollected. In this paper, the recollections are realized by means of Bidirectional Associative Memories (BAMs)[5]. During the association in BAMs reverberation is carried out according to:

$$Y(t) = \phi(M \cdot X(t)), \quad X(t+1) = \phi(M^T \cdot Y(t)) \quad (1)$$

where, $X(t) = [x1, x2, \dots, xm]^T$, $Y(t) = [y1, y2, \dots, yn]^T$ are activation vectors on x and y layers at the reverberation step t, and $\phi(\cdot)$ is a sigmoid function of each node. BAMs memorize corresponding pairs of elements at each layer in terms of a synaptic weight matrix M to memorize CFS. When the pairs of patterns $(X_1,Y_1)$, $(X_2,Y_2)$, ... , and $(X_p,Y_p)$ are given, M is calculated from these pairs with coefficient $\beta$:

$$M = \sum_{i=1}^{p} \beta Y_i X_i^T, \quad M^T = \sum_{i=1}^{p} \beta X_i Y_i^T \quad (2)$$

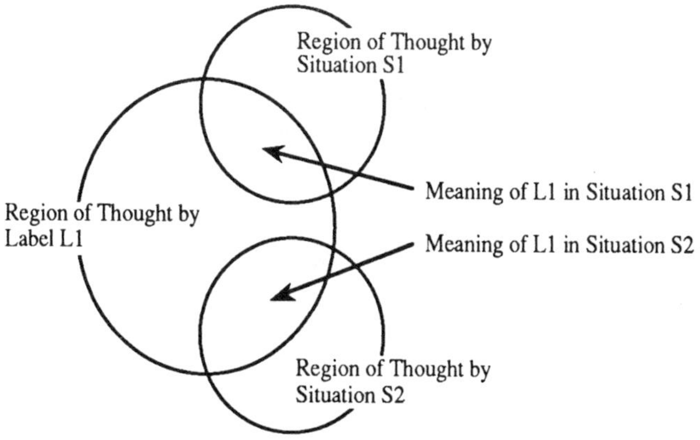


Fig. 2-1 Different meanings in different situations

### 2.2. Construction of CFS by Learning

CFS can be inductively constructed using neural network learning laws[6]. It means that the construction is carried out in terms of instances.

In this paper, a CFS is inductively constructed using the Hebbian learning law. The CFS is realized using associative memories in which a link represents the strength of the relation between two concepts. In the Hebbian learning law the strength $m_{ij}$ of a link is modified by the product of the activation of two nodes $x_i$ and $y_j$ according to:

$$\dot{m}_{ij} = -m_{ij} + x_i y_j \quad (3)$$

A complex CFS is realized by combining several pieces of associative memory structured individually. If $C_1, C_2, \dots, C_n$ denote individual CFSs and $M_1, M_2, \dots, M_n$ are their corresponding correlation matrices then we can combine them to obtain a CFS, whose correlation matrix M is given by:

$$M = M_1 + \dots + M_n \quad (4)$$

Such a combination of pieces of knowledge enables the CFS to realize context dependent representation.

### 2.3. Multi-Layered Reasoning by means of CFS [4]

Generally in image processing, recognition is carried out using characteristic values that are already obtained by low level image processing. However, when the model or context of an object is known, image recognition is more efficient. In fact, human mechanisms simultaneously realize both image processing and recognition, by means of the effective fusion of bottom-up and top-down processing supported by simultaneous information exchange and parallel processing. CFS can realize parallel processing to support the fusion of bottom-up and top-down processing by combining the semantic information processing in the upper layer and local processing in the lower layer. For example, in image recognition, the upper layer describes the knowledge about a context while the lower layer describes primitive concepts or instances. The concepts in the upper layer are explained by the instances in the lower layer. The characteristic values extracted from an original image activate the corresponding nodes in the lower layer. This results in the activation of the concept in the upper layer. At the same time the context described in the upper layer depresses the contradictory patterns of distribution of activation and promotes the meaningful patterns of activation in the lower layer. Thus the nodes denoting instances become active so as to satisfy both the characteristic values and the context. This context sensitive processing provides us with an accurate result. It uses the context to eliminate vagueness which may come from noisy and vague data and which would otherwise cause misunderstandings.

## 3. Recognition of Facial Expressions using CFS networks

Recently there has been much research in which human facial expression is regarded as a means of communicating human intention or emotion and recognition of these facial expressions is applied to human-machine interfaces[7]. However, it is difficult to explicitly describe a facial expression because of its vagueness. In this section, we apply MLR by means of CFS to the recognition of facial expressions.

### 3.1. Network Design

There are generally six types of basic facial expressions: surprise, fear, disgust, anger, happiness, and sadness. These expressions mostly consist of hazy and weak patterns on different parts of the face but do not very often strongly appear at one part of the face[1]. People judge other people's facial expressions by the whole face rather than one part of it. In psychology, various rules for facial expressions, AUs (Action Units), have been developed in the FACS (Facial Action Coding System) and most AUs are characteristics of eyebrows, eyes, or mouth[8]. Therefore characteristics of these three parts of the face are required for recognition of facial expressions. Also people usually judge by using knowledge obtained from experience. For example, if one has knowledge about angry faces and he look at an angry face, he can recognize that the face is angry by comparing the facial expression with his knowledge. Thus he can judge the expression of a new face by estimating similarities between it and his knowledge about several other facial expressions.

In this paper, we limit facial expressions to three categories: anger, happiness, and sadness, and propose a network model that recognizes these facial expressions. The network has links containing the knowledge that is constructed of characteristics extracted from instances (Fig. 3-1). The network consists of three stages and the lowest stage is composed of input layers. The network between the lowest stage and the middle stage is like an LVQ (Learning Vector Quantization) network. This network compares the knowledge in links with inputs by using the Kohonen algorithm[9] so that activations of nodes in the middle stage increase according to the degree of agreement. The network between the middle stage and the highest stage is composed of CFS networks and the distribution of activations of nodes in the network converges to satisfy contexts after the reverberation.

The details of the network design are as follows. The lowest stage of the network has three layers and each layer corresponds to an eyebrow, an eye, or a mouth. The characteristic vector extracted from each part of a face is input to the nodes of the corresponding layer. The middle stage has nine layers that consist of nodes denoting instances of people. The layers in the middle stage are divided to three parts: an eyebrow part, an eye part, and a mouth part; moreover, each part has three layers that

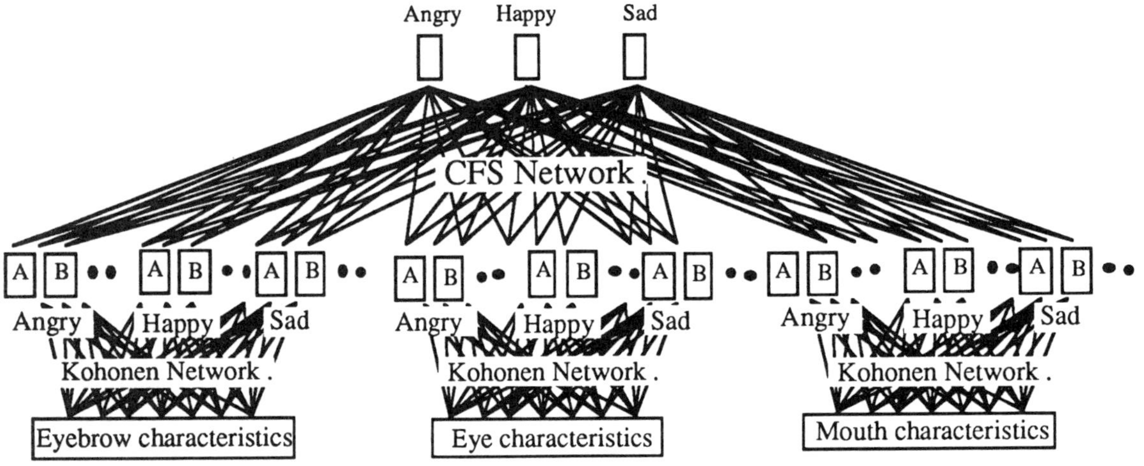


Fig. 3-1 CFS network for recognition of facial expressions

Table 1 Equations to obtain input values

| Eyebrow Data | Eye Data | Mouth Data |
|---|---|---|
| org_x = (x1 + x2) / 2<br>org_y = (y1 + y2) / 2<br>b1 = x9 - org_x<br>b2 = y9 - org_y<br>b3 = x10 - org_x<br>b4 = y10 - org_y<br>b5 = x11 - org_x<br>b6 = y11 - org_y<br>b7 = x12 - org_x<br>b8 = y12 - org_y<br>b9 = x13 - org_x<br>b10 = y13 - org_y<br>b11 = x14 - org_x<br>b12 = y14 - org_y | e1 = y3 - y1<br>e2 = y4 - y1<br>e3 = y5 - y1<br>e4 = y7 - y1<br>e5 = y8 - y1<br>e6 = y6 - y4<br>e7 = y3 - y9 | org_x = (x17 + x18) / 2<br>org_y = (y17 + y18) / 2<br>m1 = x15 - org_x<br>m2 = y15 - org_y<br>m3 = x16 - org_x<br>m4 = y16 - org_y<br>m5 = x17 - org_x<br>m6 = y17 - org_y<br>m7 = x18 - org_x<br>m8 = y18 - org_y<br><br>* ai = (xi, yi) |

correspond to three facial expressions. There are nodes of instances of people in each layer, and the node denoting the person who has the nearest characteristics is activated by the largest value when the values of the characteristics are inputted to the network. For Example, if the eyebrow characteristics of "Angry A" are the nearest to those of the input, the node denoting "Angry A" has the largest activation within the eyebrow layer. Also if the eye characters of "Sad B" are the nearest to those of the input, the node denoting "Sad B" has the largest activation within the mouth layer. Kohonen's network, i.e. LVQ network, is available in order to get activation that is proportional to the nearness between the characteristic vectors of people denoted by nodes and those from the input. Kohonen's network is composed of two layers: the input and output layers. The nodes in the output layer connect with all of the nodes in the input layer and the activation $\eta_j$ of the output node is proportional to the nearness between the node's vector (reference vector) $\boldsymbol{m}_j$ and input vector $\boldsymbol{x}$. Equation (5) shows this:

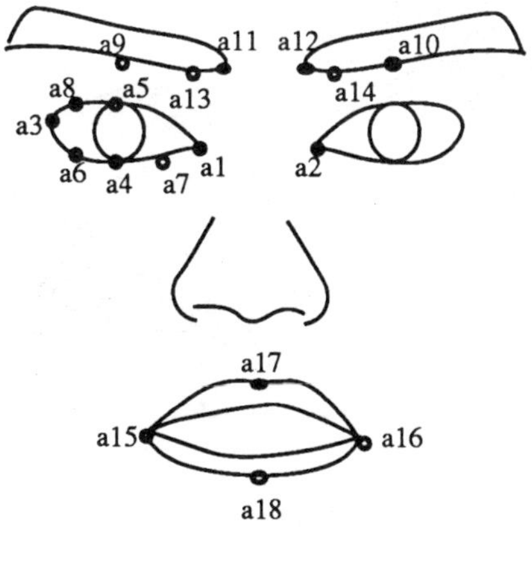


Fig. 3-2 Facial characteristic points

$$\eta_j = \Phi\left(\sum_{i=1}^{n} \mu_{ji} \cdot \xi_i\right)$$

$$\boldsymbol{m}_j = \{\mu_{j1}, \dots, \mu_{jn}\} \qquad (5)$$

$$\boldsymbol{x} = \{\xi_1, \dots, \xi_n\}$$

where $\phi(\cdot)$ is a monotone increasing function.

In this paper, the coordinates of each facial part illustrated in Fig. 3-2 are used as the characteristic values and transformed into relative coordinates for inputting to the network. The input nodes of each facial part connect with the nodes of the corresponding part in the middle stage as shown in Fig. 3-1. The weight vector of the links between the input nodes and the instance node in the middle stage is the vector whose components are the inputted relative coordinates. Thus if the input nodes are activated by a new pattern, the activations are propagated to the middle stage and the instance nodes have activations proportional to the nearness between the link vector and the input vector. Therefore these activations represent the instance of the middle stage that is the nearest to the input pattern in each layer. The distribution of the activations of the instance nodes can represent the facial expression nearest to the input pattern in each part but cannot do this for the whole face. For example, when the input vector of the eyebrow is near to "Angry A" but those of the mouth are near to "Sad B", we cannot explain the facial expression of the input. CFS networks between the middle stage and the highest stage are bi-directional associative memories in order to solve such problems. The activation in the middle stage is propagated to the upper stage. The highest stage has three nodes and each node denotes a facial expression: anger, happiness, or sadness. These nodes connect with all of the nodes in the middle stage. The Hebbian learning law is available to get the weight of the links in the CFS network. The network memorizes facial expressions as fuzzy sets. If a CFS network receives signals from the lower layers, reverberation occurs in the network and the distribution of activations converges in order to match the contexts. Thus the final distribution of activations in the highest stage represents the facial expression of the input. In this network, the contexts are standard patterns so a combination of characteristics similar to that of the standard patterns is promoted and the mismatch with the context is inhibited. Therefore the CFS network realizes context sensitive recognition. The result of recognition is represented by the activation of three nodes in the highest stage.

### 3.2. Facial Expression Recognition Experiment

In our experiment, 56 faces were used. These included three types of expressions: anger, happiness, and sadness. The data were the same as Kobayashi's[1]. Three facial expressions for each person, A, B, and C, were used to construct the knowledge of the network. The data were

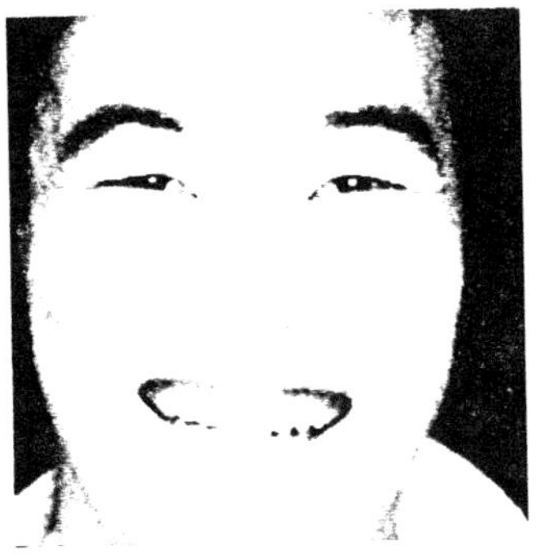

Fig. 3-3 Subject image

transformed into relative coordinates and normalized to make the length of the vector equal to 1. Table 1 shows equations to obtain the relative coordinates from the coordinates on the face. The Numbers of elements in the characteristic vectors of each facial part are 12 (eyebrows), 7(right eye), and 8(mouth). We used data only from the right eye, on the assumption that the face is symmetrical.

After training the network with 9 data, the network scored 100% correct faces about 9 training data. About 47 generalization testing data, the score was 78.7%. An example of a subject image is shown in Fig. 3-3 and the results of recognition are illustrated in Figs. 3-4 and 3-5. The face in Fig. 3-3 is a happy face and was not taught to the network. In Fig. 3-4, the activations of the nodes in the lowest stage denote values of characteristics of the face. The middle stage in the network shows the initial activation state (before reverberation). The state of the eyebrow part is near to "Angry" or "Sad" but the state of the eye is near to "Happy" or "Sad". The mouth part clearly shows nearness to "Happy". Thus we cannot judge the facial expression of the object from the output of the middle stage. Figure 3-5 illustrates the state of the network after reverberation. Fuzzy entropies have decreased during cycles of the reverberation and the distribution of activation of nodes converged to the state representing "Happy". The experimental results show that the distribution of activations converges to a state that is meaningful to the context after reverberation though the initial state of activation is too vague to judge the facial expression.

## 4. Conclusion

We described a network model in which Multi-Layered Reasoning using CFS was applied to recognition of vague facial expressions. MLR by means of CFS can simultaneously process both numeric values and abstract concepts, and achieve recognition depending on context by bi-directional parallel processing.

A facial expression is represented by vague and weak patterns at different parts of the face instead of one clear pattern at one part. We showed that the proposed network matches characteristics of each part with the knowledge constructed from instances and achieves context sensitive recognition by decreasing fuzzy entropies by means of bi-directional processing.


**Acknowledgements**

We express our deepest gratitude to Dr. Hara and Mr. Kobayashi, Science University of Tokyo, for useful advice and providing us with important data.


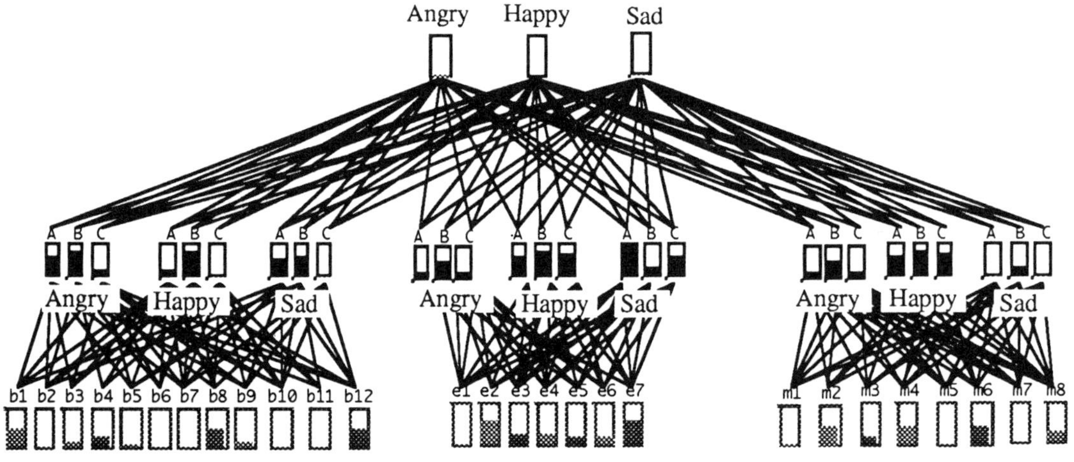


Fig. 3-4 Recognition of a facial expression (Initial State)

## References


[1] P.Ekman and W.V.Friesen: Unmasking The Face, Prentice-Hall, Inc., Englewood Cliffs, New Jersey (1975)
[2] Wittgenstein: Philosophical Investigations, Basil Blackwell, Oxford (1953)
[3] T.Takagi, T.Yamaguchi and M.Sugeno: Conceptual Fuzzy Sets, International Fuzzy Engineering Symposium '91, PART II, pp. 261-272 (1991)
[4] T.Takagi, A.Imura, H.Ushida and T.Yamaguchi: Multi-layered Reasoning by Means of Conceptual Fuzzy Sets, Third International Workshop on Neural Networks and Fuzzy Logic '92, NASA Johnson Space Center (1992)
[5] B.Kosko: Adaptive Bidirectional Associative Memories, Applied Optics, Vol.26, No.23, pp. 4947-4960 (1987)
[6] T.Takagi, A.Imura, H.Ushida and T.Yamaguchi: Inductive learning of Conceptual Fuzzy Sets, 2nd International Conference on Fuzzy Logic and Neural Networks IIZUKA '92 (1992)
[7] H.Kobayashi and F.Hara: The Recognition of Basic Facial Expressions by Neural Network, Proc. of IJCNN '91, pp. 460-466 (1991)
[8] P.Ekman and W.V.Friesen: The Facial Action Coding System, Consulting Psychologists Press, Inc., San Francisco, CA (1975)
[9] T.Kohonen: The neural phonetic typewriter, IEEE Computer, 21, 3, pp. 11-22 (1988)


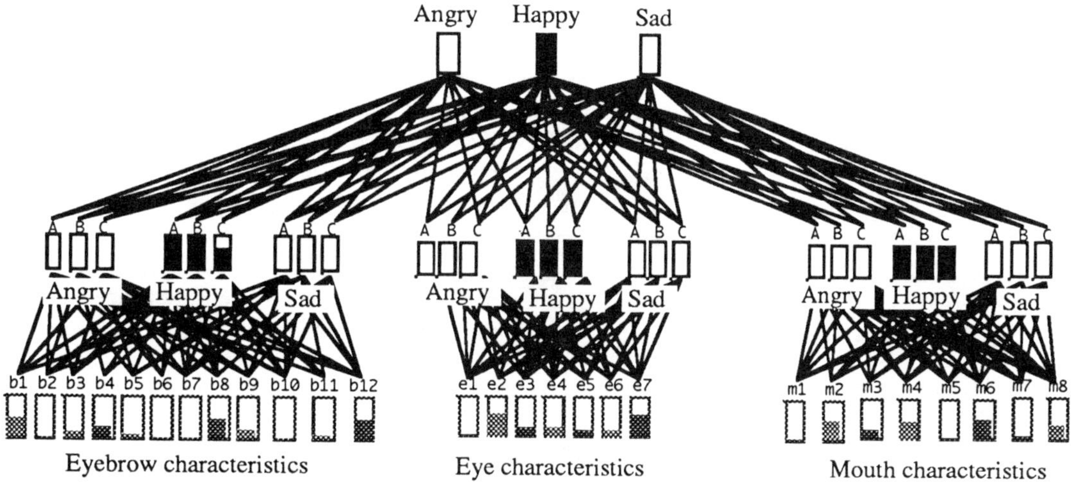


Fig. 3-5 Recognition of a facial expression (After reverberation)

# Image Processing - Enhancement, Filtering and Edge Detection Using the Fuzzy Logic Approach

*Ching-Yu Tyan and Paul P. Wang*
*Department of Electrical Engineering*
*Duke University, Box 90291*
*Durham, North Carolina 27708-0291*

## ABSTRACT

Novel algorithms for the image enhancement, filtering and edge detection using *the fuzzy logic* approach are proposed. An enhancement technique based on various combinations of *fuzzy logic* linguistic statements in the form of IF...., THEN.... rules, modifies the image contrast and dynamic range of gray level, and provides a linguistic approach to image enhancement. Fuzzy filtering technique based on the tuning of fuzzy membership functions in the *frequency domain* results in an improved restoration of an image which is degraded by additive random noise. An improved performance compared with traditional mask convolution filtering is also evident from the SNR (signal-to-noise ratio) improvement of 4.03 dB. The fuzzy edge detection algorithm provides a variety of edge information. A comparison of fuzzy edge detection algorithm with some existing edge detection algorithms by human observers is also shown to reveal the novelty.

## I. INTRODUCTION

In image enhancement, an image is enhanced by the way of modifying an image's dynamic range or contrast. In this method, the gray scale of an image is modified according to a specific transformation[1]. The dynamic shape and the range of the histogram are crucial factors for a suitable transformation which is usually hard to be identified. For this reason, we introduce the fuzzy linguistic rules as an alternative to enhance the image instead of mathematical mapping being used in all known conventional approaches.

In addition to the enhancement of images by contrast and dynamic modifications, it can also be enhanced by reducing degradations that may be present. In fuzzy image filtering, we discuss a technique using the *fuzzy logic* approach in *frequency domain* that attempts to reduce random noise.

Edge detection is very useful in a number of contexts. Edges characterize object boundaries and are, therefore, useful for segmentation, registration, and identification of objects in scenes. Edge detection algorithms, which are quite useful in a broad set of applications, have already been developed[2][3][4]. In this paper, we introduce a new algorithm which provides different edge information simply by using tuning techniques in the *fuzzy logic* approaches. These are proven to be very effective in control system designs in the *spatial domain*, and also increase the flexibility of edge information.

## II. IMAGE ENHANCEMENT

An effective way to apply *fuzzy logic* in image enhancement is to process each pixel of an image on the basis of rules applied to the pixels. Let G be the number of gray levels in a image (G = 256). Then the membership functions of fuzzy set suitable for image processing can be defined in the universe of discourse U = [0,....., G-1][5]. Let $f(n_1, n_2)$ be the fuzzy variable representing the intensity value of the pixel of coordinates $(n_1, n_2)$ in the source image and $p(n_1, n_2)$ be the corresponding fuzzy variable in the resulting image after being enhanced. A fuzzy set A of universe of discourse $U \in u$ is characterized by a membership function $u_A$, defined on U, taking values in the interval [0, 1][6].

$$A = \{(u,\ u_A) \mid u \in U \text{ and } u_A \in [0,\ 1]\} \qquad (1)$$

Then we define the fuzzy sets DARK, MED, and BRIGHT and their corresponding membership functions shown in Figure 1(b). A very simple example of aggregation rules which causes a "fuzzy reverse image" (Figure-1(m)-(o)) via the human linguistic language approach rather is shown as the following:

IF $f(n_1, n_2)$ is DARK, THEN $p(n_1, n_2)$ is BRIGHT.
IF $f(n_1, n_2)$ is MED, THEN $p(n_1, n_2)$ is MED.
IF $f(n_1, n_2)$ is BRIGHT, THEN $p(n_1, n_2)$ is DARK.

Figure-1(a) shows an original image of 320 ×200 pixels, and Figure-1(c) shows its corresponding histogram. Four possible sets of rules are given in Figure-1(d)(g)(j)(m) and their enhanced images are shown in Figure-1(e)(h)(k)(n). The differences can be seen from Figure-1(f)(i)(l)(o) as well as the original image Figure-1(c).

## III. IMAGE LOWPASS FILTERING

In order to reduce high-frequency noise while

0-7803-0614-7/93$03.00 ©1993IEEE

also preserving the low-frequency component, lowpass filtering can be used to reduce a large amount of noise at the expense of reducing a small amount of signal. A model of an image degraded by additive random noise is given by

$$fn(n_1, n_2) = f(n_1, n_2) + n(n_1, n_2) \tag{2}$$

where $n(n_1, n_2)$ represents the signal-independent additive random noise. In order to compare the performance among the original, degraded and processed images, some definitions are necessary[7]. We define the signal-to-noise ratio (SNR) as

$$\text{SNR in dB} = 10 \times \log_{10} \frac{\text{Var}[f(n_1, n_2)]}{\text{Var}[n(n_1, n_2)]}. \tag{3}$$

The normalized mean square error (NMSE) between the original image $f(n_1, n_2)$ and the processed image $p(n_1, n_2)$ is defined as

$$\text{NMSE}[f(n_1,n_2),p(n_1,n_2)]$$
$$= 100 \times \frac{\text{Var}[f(n_1, n_2) - p(n_1, n_2)]}{\text{Var}[f(n_1, n_2)]} \%. \tag{4}$$

where Var[.] is the variance. The SNR improvement due to processing is defined as

$$\text{SNR improvement}$$
$$= 10 \times \log_{10} \frac{\text{NMSE } [f(n_1, n_2), fn(n_1, n_2)]}{\text{NMSE } [f(n_1, n_2), p(n_1, n_2)]} \text{ dB}. \tag{5}$$

A $3 \times 3$ convolution mask $h(n_1, n_2)$ used for traditional lowpass filter is shown in Figure-2. Figure-3(a) shows an original noise-free image of $320 \times 200$ pixels, and Figure-3(b) shows an image degraded by wideband Gaussian random noise at an SNR of -1.26 dB with NMSE of 92.6%. The frequency response of a $3 \times 3$ convolution mask used for a traditional lowpass filter is shown in Figure-3(c). The frequency responses of a degraded image and a processed image by a traditional lowpass filter are shown in Figure-3(d)-(e).

Once we have the frequency information, we can apply the principle of *fuzzy logic* in designing a lowpass filter by designating frequency and gain components as fuzzy variables. The matrix in Figure-4 represents a set of IF...., THEN.... rules modeling the lowpass filter. As usual, Z, L, H, NH, and PH are fuzzy sets. Since the phase component often preserves the intelligibility of the image, we focus on the magnitude component in this nature. We define the fuzzy sets NH, L, and PH for the frequency variable *freq*, and their corresponding membership functions are shown in Figure-3(g). Similarly, we define the fuzzy sets Z, L, and H for the gain function *gain* for the original image and the gain function *new_gain* for the new image.

$h(n_1, n_2)$

| | | |
|---|---|---|
| 0 | $\frac{1}{6}$ | 0 |
| $\frac{1}{6}$ | $\frac{1}{3}$ | $\frac{1}{6}$ |
| 0 | $\frac{1}{6}$ | 0 |

Figure-2

*gain, new_gain*

| | Z | L | H |
|---|---|---|---|
| NH | Z | Z | Z |
| *freq* L | Z | L | H |
| PH | Z | Z | Z |

Figure-4

The SNR improvement shows the superiority of a fuzzy lowpass filter compared with a traditional mask convolution filter. Figure-3(i) shows that the traditional lowpass filter has an SNR improvement of 1.65 dB with NMSE of 66.2%, whereas the fuzzy lowpass filter has an SNR improvement of 4.03 dB with NMSE of 36.59% as shown in Figure-3(j).

## IV. IMAGE EDGE DETECTION

In this section, we shall be making the use of three of the well known and popular edge detection algorithms: *gradient* algorithm, *Laplacian* algorithm and *Sobel's* algorithm[8][9] and compare them against the fuzzy edge detection algorithm (Figure-5(a)-(h)).

An edge is the boundary between two regions with relatively distinct gray level properties, as shown in Figure-6, illustrates this concept. Figure-6(a) shows an image of a light stripe on a dark background, where Figure-6(b) shows the opposite situation.

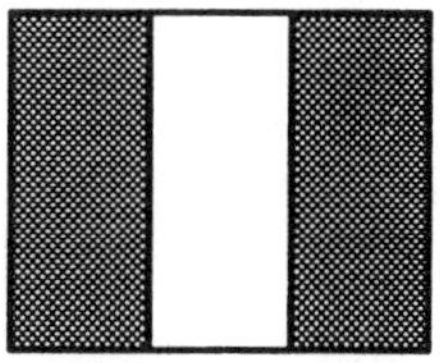

Figure-6(a)

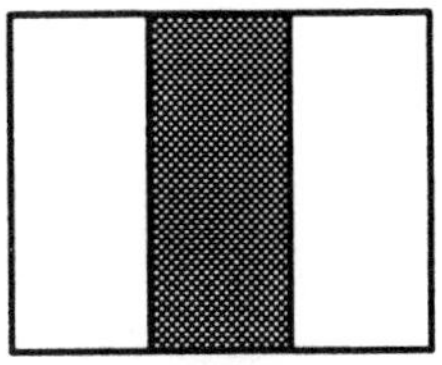

Figure-6(b)

The idea of fuzzy edge detection is to detect the transition associated with the dark and the bright sides of the edge. There are four cases of a $2 \times 2$ mask where edge occurs,

| | |
|---|---|
| *F11* | *F12* |
| *F21* | *F22* |

| | |
|---|---|
| D | B |
| D | B |

| | |
|---|---|
| B | D |
| B | D |

| | |
|---|---|
| D | D |
| B | B |

| | |
|---|---|
| B | B |
| D | D |

Where $F11 = f(n_1, n_2)$ $F12 = f(n_1, n_2 - 1)$
$F21 = f(n_1 - 1, n_2)$ $F22 = f(n_1 - 1, n_2 - 1)$ (6)

Then we define two fuzzy sets DARK(D) and BRIGHT(B) for four fuzzy variables *F11, F12, F21* and *F22* and let $p(n_1, n_2)$ be the fuzzy variable representing the image after the edges have been detected over the universe of discourse U = [0,....., G − 1] (G = 256). The corresponding membership functions and the processed images are shown in Figure-5(e)-(h). The fuzzy rules for this fuzzy edge detection algorithm are shown below:

Rule 1: IF *F11* is D AND *F12* is D AND *F21* is D AND *F22* is D, THEN $p(n_1, n_2)$ is D.
Rule 2: IF *F11* is D AND *F12* is D AND *F21* is D AND *F22* is B, THEN $p(n_1, n_2)$ is D.
Rule 3: IF *F11* is D AND *F12* is D AND *F21* is B AND *F22* is D, THEN $p(n_1, n_2)$ is D.
Rule 4: IF *F11* is D AND *F12* is D AND *F21* is B AND *F22* is B, THEN $p(n_1, n_2)$ is B.
Rule 5: IF *F11* is D AND *F12* is B AND *F21* is D AND *F22* is D, THEN $p(n_1, n_2)$ is D.
Rule 6: IF *F11* is D AND *F12* is B AND *F21* is D AND *F22* is B, THEN $p(n_1, n_2)$ is B.
Rule 7: IF *F11* is D AND *F12* is B AND *F21* is B AND *F22* is D, THEN $p(n_1, n_2)$ is D.
Rule 8: IF *F11* is D AND *F12* is B AND *F21* is B AND *F22* is B, THEN $p(n_1, n_2)$ is D.
Rule 9: IF *F11* is B AND *F12* is D AND *F21* is D AND *F22* is D, THEN $p(n_1, n_2)$ is D.
Rule 10: IF *F11* is B AND *F12* is D AND *F21* is D AND *F22* is B, THEN $p(n_1, n_2)$ is D.
Rule 11: IF *F11* is B AND *F12* is D AND *F21* is B AND *F22* is D, THEN $p(n_1, n_2)$ is B.
Rule 12: IF *F11* is B AND *F12* is D AND *F21* is B AND *F22* is B, THEN $p(n_1, n_2)$ is D.
Rule 13: IF *F11* is B AND *F12* is B AND *F21* is D AND *F22* is D, THEN $p(n_1, n_2)$ is B.
Rule 14: IF *F11* is B AND *F12* is B AND *F21* is D AND *F22* is B, THEN $p(n_1, n_2)$ is D.
Rule 15: IF *F11* is B AND *F12* is B AND *F21* is B AND *F22* is D, THEN $p(n_1, n_2)$ is D.
Rule 16: IF *F11* is B AND *F12* is B AND *F21* is B AND *F22* is B, THEN $p(n_1, n_2)$ is D.

Obviously, rule4, rule6, rule11 and rule13 are used to detect bright lines. From Figure-5(e)-(h), we can easily observe that a variety of edge images with different thicknesses can be obtained by simply changing the shapes of the fuzzy membership functions. This also proves that fuzzy algorithm serves better edge performance and provides more flexibility to edge information depending upon the need.

## V. CONCLUSION

In this paper, novel algorithms designed specifically for image processing using the *fuzzy logic* approach have been explored. All image information in *spatial domain* or *frequency domain* in image enhancement, filtering and edge detection can be represented by linguistic statements in natural language and the processing is performed by very practical and powerful *fuzzy approximate reasoning.* The processing performance relies on the fuzzy inference engine which has been demonstrated to be able to give better results than the traditional approaches. Experimentation using *fuzzy logic* algorithm leads to the conclusion that the adoption of fuzzy reasoning offers some significant advantages. This also proves that *fuzzy logic* theory is useful for the mutual exchange of information between a human and an image. This paper shows that it is possible to manipulate or to improve images via natural language making use of *fuzzy logic* methodology, hence the tasks can be accomplished with ease by an operator without demanding a sophisticated education background. Full scale research is currently underway, and the research report is expected to come out from our research laboratory in the future.

## REFERENCES


[1] R. C. Gonzales and B. A. Fittes, "Gray Level Transformation for Interactive Image Enhancement", Mech. Mach. Theory, Vol. 12, 1977, pp111-122.

[2] T. Peli and D. Malah, "A Study of Edge Detection Algorithms", Computer Graphics and Image Processing, Vol. 20, 1982, pp1-20.

[3] G. S. Robinson, "Edge Detection by Compress Gradient Mask", Computer Graphics Image Processing. 6, 1977, pp492-501.

[4] R. M. Haralick, "Zero Crossing of Second Directional Derivative Edge Detector", Robot Vision, SPIE 336, 1982, pp91-96.

[5] H. J. Zimmermann, "Fuzzy Set Theory and Its Applications", Kluwer Academic Inc., 1991, pp11-67.

[6] L. A. Zadeh, "Fuzzy Sets", Information and Control, 1965, pp338-353.

[7] Jae S. Lim, "Two-Dimensional Signal and Image Processing", Prentice Hall Inc., 1990, pp529-530.

[8] Anil K. Jain, "Fundamentals of Digital Image Processing", Prentice Hall Inc., 1989, pp342-353.

[9] R. C. Gonzalez and R. E. Woods, "Digital Image Proc.", Addison-Wesley Company Inc., 1992, pp413-429.

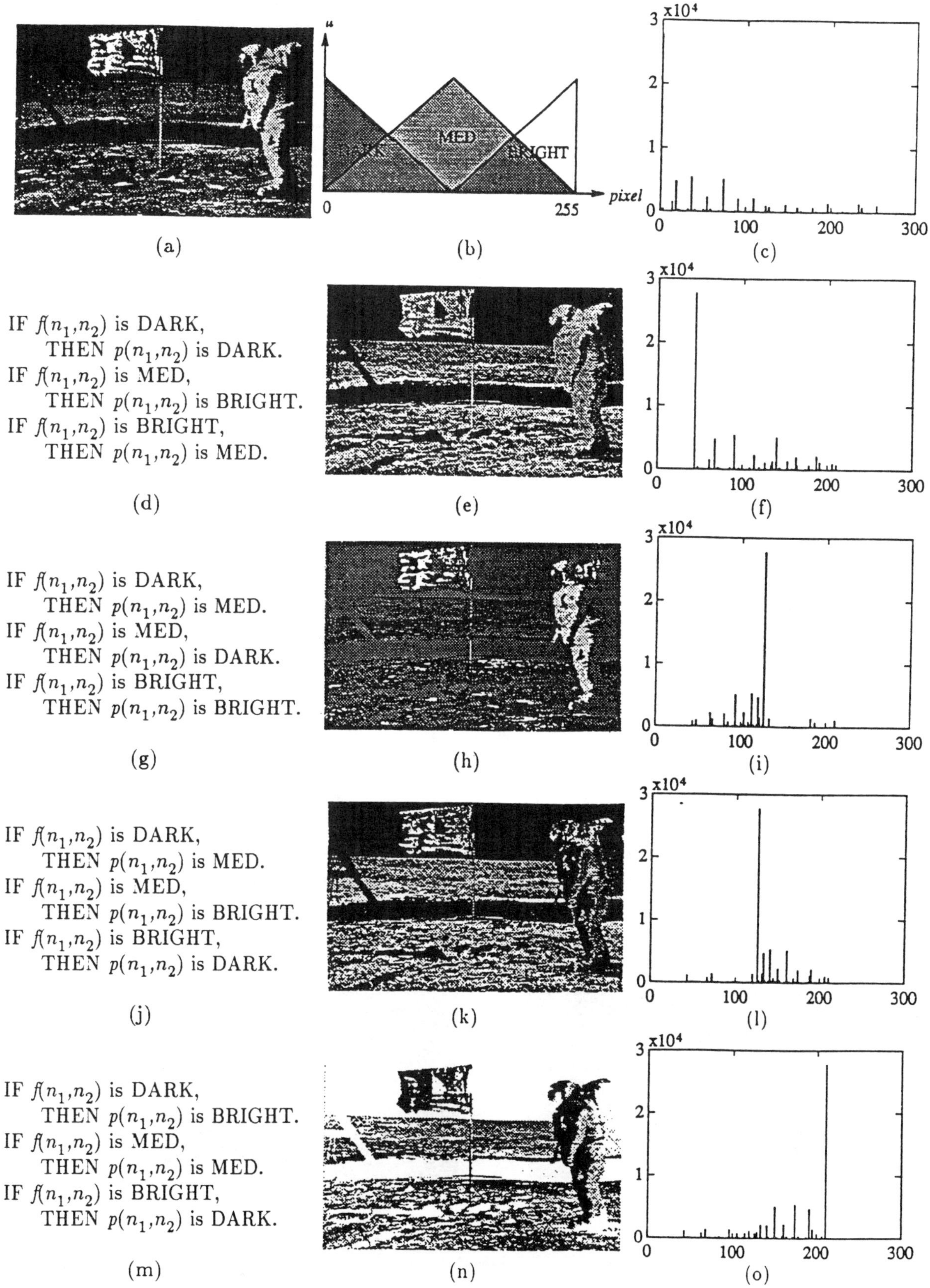

u
DARK
MED
BRIGHT
pixel
0
255
x10⁴
(a)
(b)
(c)
IF f(n1,n2) is DARK,
THEN p(n1,n2) is DARK.
IF f(n1,n2) is MED,
THEN p(n1,n2) is BRIGHT.
IF f(n1,n2) is BRIGHT,
THEN p(n1,n2) is MED.
(d)
(e)
(f)
IF f(n1,n2) is DARK,
THEN p(n1,n2) is MED.
IF f(n1,n2) is MED,
THEN p(n1,n2) is DARK.
IF f(n1,n2) is BRIGHT,
THEN p(n1,n2) is BRIGHT.
(g)
(h)
(i)
IF f(n1,n2) is DARK,
THEN p(n1,n2) is MED.
IF f(n1,n2) is MED,
THEN p(n1,n2) is BRIGHT.
IF f(n1,n2) is BRIGHT,
THEN p(n1,n2) is DARK.
(j)
(k)
(l)
IF f(n1,n2) is DARK,
THEN p(n1,n2) is BRIGHT.
IF f(n1,n2) is MED,
THEN p(n1,n2) is MED.
IF f(n1,n2) is BRIGHT,
THEN p(n1,n2) is DARK.
(m)
(n)
(o)

Figure-1. Image Enhancement: (a) Original image; (b) Membership functions for fuzzy image enhancement; (c) Histogram of original image; (d)-(o) Fuzzy rules, enhanced images and corresponding histograms.

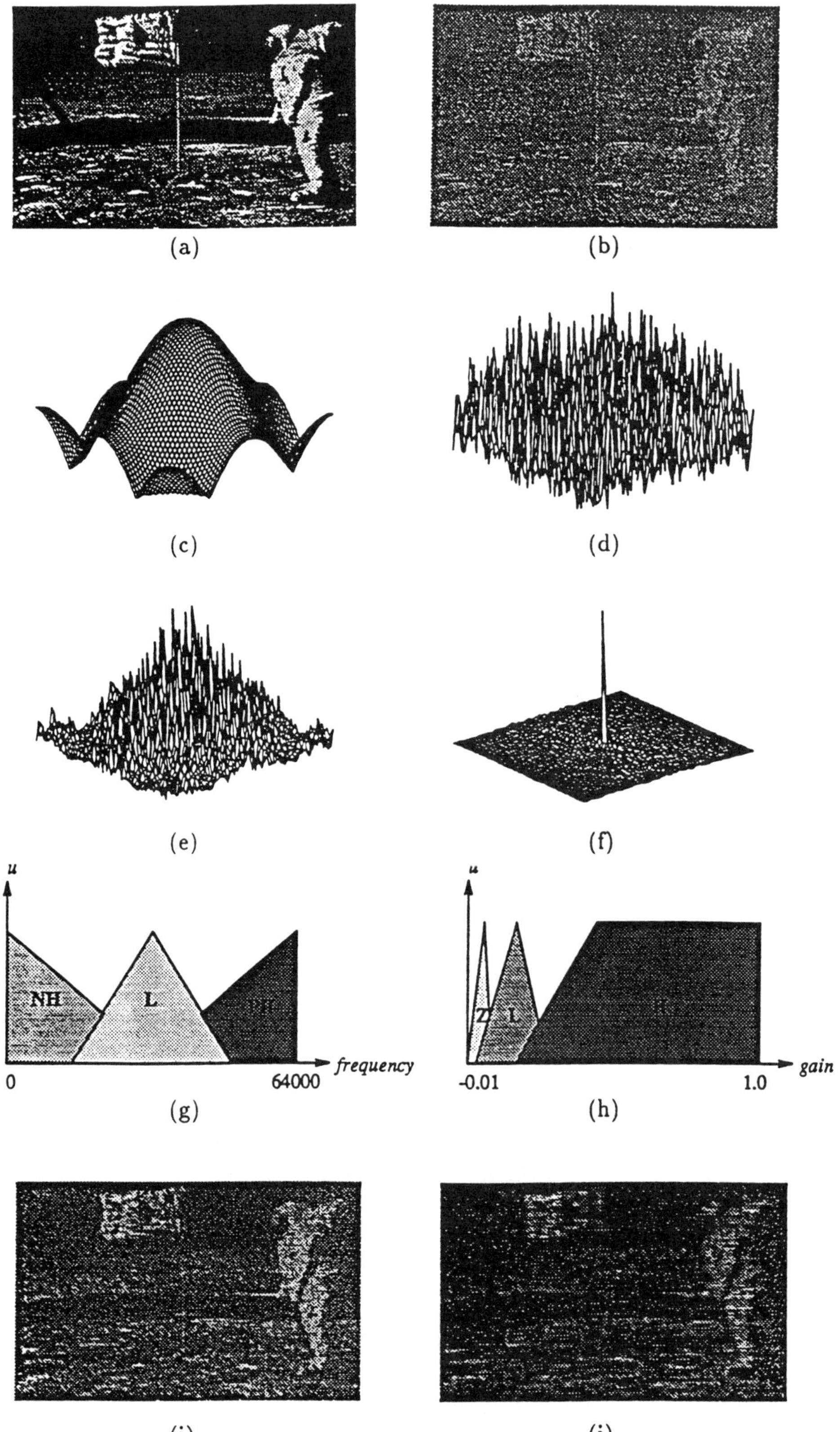


Figure-3. Lowpass Filtering Experiments: (a) Original image; (b) Image with additive wideband Gaussian random noise at SNR of -1.26 dB with NMSE of 92.6%; (c) Frequency response of a convolution-type lowpass filter; (d) Frequency response of degraded image; (e)-(f) Frequency responses after convolution lowpass filter and fuzzy lowpass filter; (g)-(h) Membership functions for fuzzy lowpass filter; (i) Image after convolution lowpass filter with NMSE of 66.2% and SNR improvement of 1.65 dB; (j) Image after fuzzy lowpass filter with NMSE of 36.59% and SNR improvement of 4.03 dB.

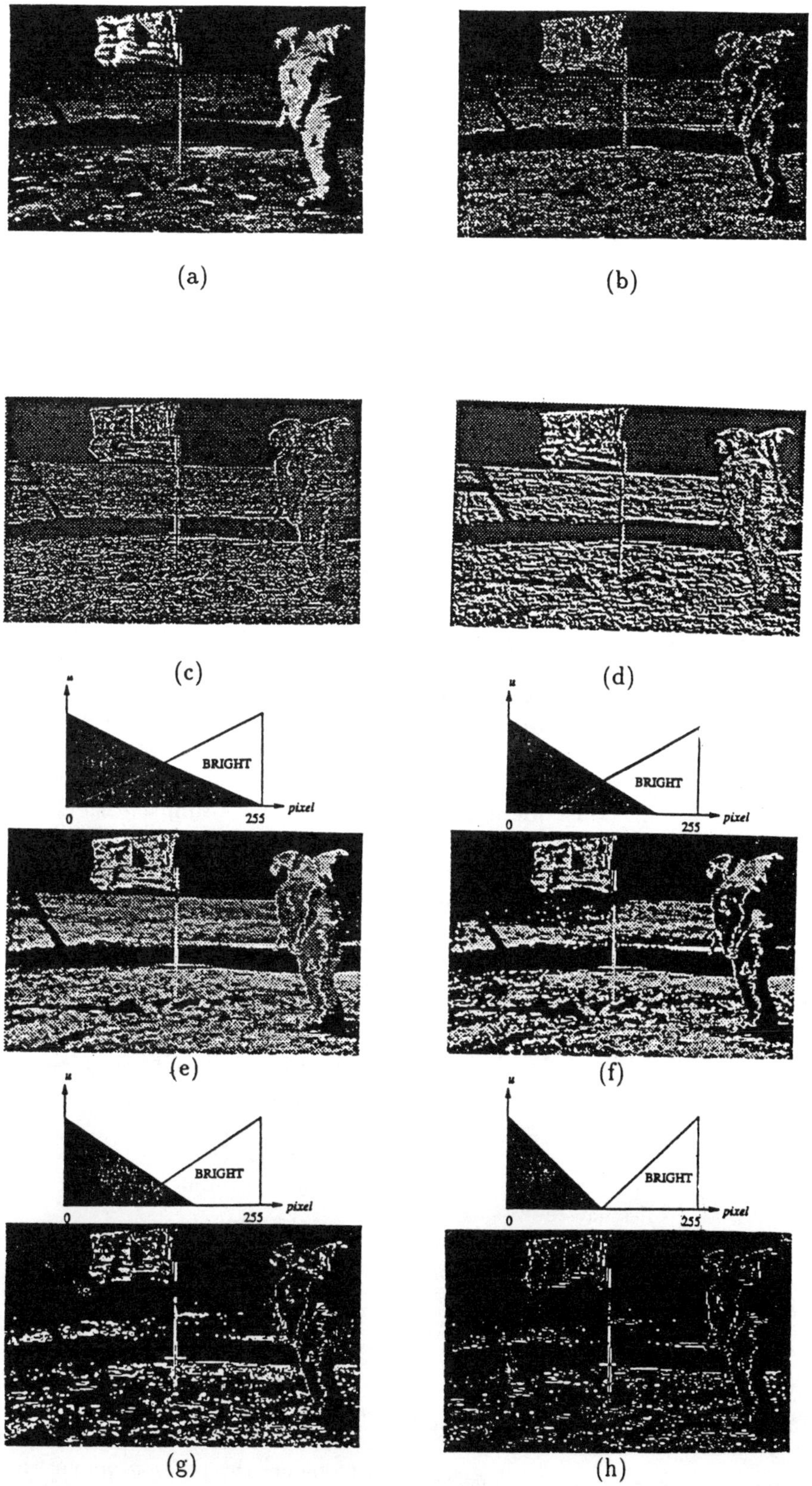


Figure-5. Image Edge Detection: (a) Original image; (b) Edge detection image by *gradient* algorithm; (c) Edge detection image by *Laplacian* algorithm; (d) Edge detection image by *Sobel's* algorithm; (e)-(h)Fuzzy edge images with different membership functions.

# Empirical Study on Learning in Fuzzy Systems

Hisao Ishibuchi, Ken Nozaki, Hideo Tanaka
Department of Industrial Engineering, University of Osaka Prefecture
Gakuencho 1-1, Sakai, Osaka 593, JAPAN

Yukio Hosaka and Masanori Matsuda
Satake Corporation
Nishihonmachi 2-30, Saijo, Higashihiroshima, Hiroshima 724, JAPAN

***Abstract*— The aim of this paper is to examine the ability of fuzzy systems as approximators of non-linear mappings by computer simulations on real-life data. The relation among six factors in the sensory test on rice taste is modelled by fuzzy systems with five input variables (flavor, appearance, taste, stickiness, toughness) and a single output variable (overall evaluation). Fuzzy if-then rules with non-fuzzy singletons in the consequent part are employed in fuzzy systems. A learning rule based on a descent method is applied to the consequent part of each fuzzy if-then rule. By a random subsampling technique, the performance of fuzzy systems for test data and training data is compared with that of multi-layer neural networks. A simple method for specifying initial fuzzy if-then rules is proposed in order to improve the performance of fuzzy systems.**

## I. INTRODUCTION

Fuzzy systems with fuzzy if-then rules have been successfully applied to control problems[6,11]. In many applications, fuzzy if-then rules were generally derived from human experts. Recently several approaches have been proposed for automatically generating fuzzy if-then rules from numerical data (for example, see[1-4,8,12]). Since fuzzy systems are universal approximators of non-linear mappings[5,13], they are suitable for the modelling of complex non-linear systems.

The aim of this paper is to empirically examine the ability of fuzzy systems as approximators of non-linear mappings by computer simulations on real-life data. Both the fitting ability to training data and the generalization ability to test data are examined by a random subsampling technique. The performance of fuzzy systems is compared with that of multi-layer neural networks[10].

In computer simulations, fuzzy if-then rules with real numbers (i.e., non-fuzzy singletons) in the consequent part are employed for modelling the relation among six factors in the sensory test on rice taste. Real numbers in the consequent part are updated by a learning rule based on a descent method. In order to improve the performance of fuzzy systems, a simple method is proposed for specifying initial values of real numbers in the consequent part.

## II. SENSORY TEST DATA ON RICE TASTE

Subjective qualification of the taste of foods is very important but difficult. The qualification of rice taste is usually performed by a kind of subjective evaluation called the sensory test[9]. Plural panelists (usually 24 persons) are involved in this test to evaluate the following six factors of each kind of rice:

$x_1$: flavor,
$x_2$: appearance,
$x_3$: taste,
$x_4$: stickiness,
$x_5$: toughness,
$y$: overall evaluation.

The sensory test data shown in Table I have been obtained by such a subjective evaluation for 105 kinds of rice (see Matsuda & Kameoka[7]). The task of fuzzy systems is to model the relation among the six factors. The unknown non-linear mapping to be realized by fuzzy systems can be represented as

$$y_p = f(x_{p1}, x_{p2}, x_{p3}, x_{p4}, x_{p5}),\ p = 1, 2, ..., 105. \qquad (1)$$

0-7803-0614-7/93$03.00 ©1993IEEE

**TABLE I.** SENSORY TEST DATA

| No. | Input variables | | | | | Output |
|---|---|---|---|---|---|---|
| $p$ | $x_{p1}$ | $x_{p2}$ | $x_{p3}$ | $x_{p4}$ | $x_{p5}$ | $y_p$ |
| 1 | 0.831 | 0.600 | 0.953 | 0.953 | -0.163 | 1.141 |
| 2 | 0.221 | 0.708 | 0.737 | 0.923 | -0.495 | 0.899 |
| ... | ... | ... | ... | ... | ... | ... |
| 105 | -1.212 | -0.074 | -0.317 | -0.144 | 0.180 | -0.658 |

## III. FUZZY SYSTEMS

### *A. Fuzzy Inference and Fuzzy If-Then Rules*

For the modelling of the unknown mapping in (1), fuzzy if-then rules of the following type are employed in a fuzzy system:

$$\begin{aligned} \textit{Rule } j \ : \ & \textit{If } x_1 \textit{ is } A_{1j} \textit{ and} \cdots \textit{and } x_5 \textit{ is } A_{5j} \\ & \textit{then } y \textit{ is } w_j, \ j = 1, 2, ..., n, \end{aligned} \tag{2}$$

where $j$ is the index of rule, $A_{ij}$ is a fuzzy subset on the $i$-th axis of the input space and $w_j$ is a real number. This type of fuzzy if-then rules was employed in [3,8] and can be viewed as a special kind of fuzzy if-then rules in Takagi & Sugeno[12].

Let $o(\boldsymbol{x})$ be the output from the fuzzy system corresponding to the input vector $\boldsymbol{x} = (x_1, x_2, x_3, x_4, x_5)$. In this paper, the following fuzzy inference method[3,8] is employed to calculate $o(\boldsymbol{x})$:

$$o(\boldsymbol{x}) = \sum_{j=1}^{n} \mu_j(\boldsymbol{x}) \cdot w_j / \sum_{j=1}^{n} \mu_j(\boldsymbol{x}), \tag{3}$$

$$\mu_j(\boldsymbol{x}) = A_{1j}(x_1) \times \cdots \times A_{5j}(x_5), \tag{4}$$

where $A_{ij}(x_i)$ denotes the membership function of $A_{ij}$. It should be noted that the product operator is used instead of the minimum operator. When the fuzzy partition is a simple fuzzy grid with triangular fuzzy subsets which intersect each other at the level of 0.5 as shown in Fig.1, the denominator in (3), i.e., the sum of $\mu_j(\boldsymbol{x})$, is 1 for any $\boldsymbol{x}$. In this case, (3) is rewritten as

$$o(\boldsymbol{x}) = \sum_{j=1}^{n} \mu_j(\boldsymbol{x}) \cdot w_j. \tag{5}$$

In computer simulations, we employed four different fuzzy partitions shown in Fig.2 where $K$ is the number of fuzzy subsets on each axis of the input space.

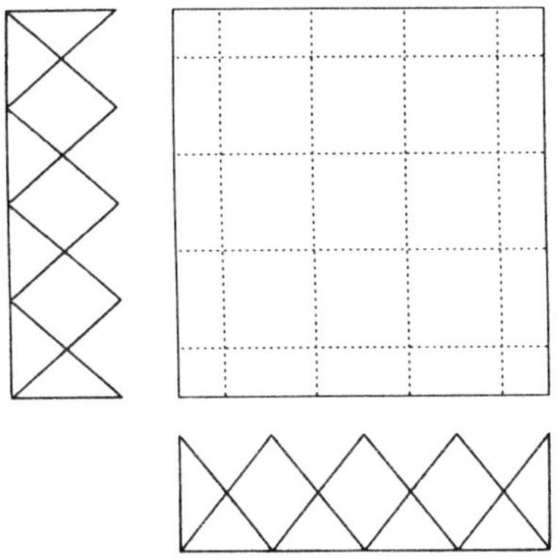

**Fig.1.** An example of fuzzy partition by a simple fuzzy grid ($K = 5$).

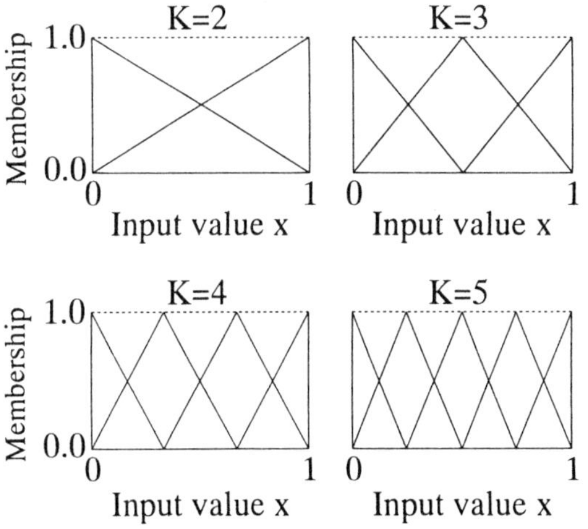


**Fig.2.** Fuzzy partitions of each axis of the input space.

### *B. Learning of Fuzzy If-Then Rules*

Let us define an error measure for the $p$-th input-output pair $(\boldsymbol{x}_p, y_p)$ as

$$E_p = \{o(\boldsymbol{x}_p) - y_p\}^2/2, \tag{6}$$

where $o(\boldsymbol{x}_p)$ is the output from the fuzzy system corresponding to the input vector $\boldsymbol{x}_p$ (see (3)-(5)). The total error is defined as

$$\begin{aligned} E &= \textstyle\sum_p E_p \\ &= \textstyle\sum_p \{o(\boldsymbol{x}_p) - y_p\}^2/2. \end{aligned} \tag{7}$$

The learning of $w_j$ in the consequent part of each fuzzy if-then rule in (2) is performed based on the descent method as

$$w_j(t+1) = w_j(t) + \beta(-\partial E_p / \partial w_j)$$

$$= w_j(t) - \beta \cdot \mu_j(\boldsymbol{x}_p) \cdot \{o(\boldsymbol{x}_p) - y_p\},$$
$$j = 1, 2, \cdots, n, \quad (8)$$

where $\beta$ is a learning constant and $t$ indexes the number of the adjustments of $w_j$. In computer simulations, $\beta$ was specified as $\beta = 1.0$. This learning rule was employed in Ichihashi & Watanabe[3] and Nomura et al.[8].

## IV. SIMULATION RESULTS

### *A. Computer Simulations*

In computer simulations, the data in Table I were normalized as

$$x_{pi} := (x_{pi} - \min\{x_{pi}\})/(\max\{x_{pi}\} - \min\{x_{pi}\}),$$
$$i = 1, 2, 3, 4, 5;\ p = 1, 2, \cdots, 105, \quad (9)$$

$$y_p := (y_p - \min\{y_p\})/(\max\{y_p\} - \min\{y_p\}),$$
$$p = 1, 2, \cdots, 105, \quad (10)$$

where

$$\min\{x_{pi}\} = \min\{x_{pi} : p = 1, 2, \cdots, 105\}, \quad (11)$$
$$\max\{x_{pi}\} = \max\{x_{pi} : p = 1, 2, \cdots, 105\}, \quad (12)$$
$$\min\{y_p\} = \min\{y_p : p = 1, 2, \cdots, 105\}, \quad (13)$$
$$\max\{y_p\} = \max\{y_p : p = 1, 2, \cdots, 105\}. \quad (14)$$

Fuzzy systems based on four different fuzzy partitions in Fig.2 were applied to the modelling of the non-linear mapping in (1) from the normalized data. Since each fuzzy system has five inputs, the number of total fuzzy if-then rules is $K^5$ where $K$ is the number of fuzzy subsets on each axis (see Fig.2).

Three-layer feedforward neural networks with various numbers of hidden units (i.e., $5, 10, 15, 20, 30, 40, 50$) were also applied to the same task. The same architecture and the same learning rule as in Rumelhart et al.[10] were employed. The learning constant and the momentum constant were specified as $\eta = 0.2$ and $\alpha = 0.9$ in computer simulations.

The following random subsampling procedure was iterated 10 times for each method with each parameter setting:

**[Random subsampling procedure ]**

**Step 1 :** Randomly divide the 105 samples into 75 samples (training data) and 30 samples (test data).

**Step 2 :** Train a fuzzy system (or a neural network) using the training data. Calculate the total error $E$ in (7) for the training data and for the test data every 10 iterations of the learning rule (or every 1000 iterations of the BP algorithm[10]). Continue the learning until 500 iterations of the learning rule (or 50000 iterations of the BP algorithm).

Average total errors over 10 runs were calculated for the training data and for the test data after computer simulations. In computer simulations, the initial value of $w_j$ in fuzzy systems was specified as $w_j(0) = 0$ for $j = 1, 2, ..., n$. The initial values of weights and biases in neural networks were random real numbers in the interval $[-0.2, 0.2]$.

### *B. Performance for Training Data*

Average total errors for the training data by fuzzy systems are shown in Fig.3. Fig.3 is the simulation results after 500 iterations of the learning rule in (8). Fig.4 shows how the average errors decreased during the learning phase of fuzzy systems. From these figures, we can see that the fitting ability to the training data of fuzzy systems depends on the fuzzy partition (i.e., the value of $K$) and fine fuzzy partitions (i.e., large values of $K$) lead to the good fitting to the training data. Thc total error $E = 0.0055$ after 500 iterations of the learning rule for the fuzzy system with $K = 5$ can be viewed as almost zero in comparison with the initial total error $E = 13.9817$ before the learning.

Fig.5 and Fig.6 show the simulation results by neural networks. Fig.5 shows average total errors after 50000 iterations of the BP algorithm[10] and Fig.6 shows the results during the learning phase. From Fig.5, we can see that the fitting ability to the training data of neural networks is not sensitive to the choice of the number of hidden units.

From the comparison of Fig.4 with Fig.6, we can see that the learning of fuzzy systems based on fine fuzzy partitions ($K = 4$ and $K = 5$) is much faster than that of neural networks (Nomura et al.[8] obtained the same result by computer simulations on artificial numerical data.)

### *C. Performance for Test Data*

Average total errors for the test data by fuzzy systems are shown in Fig.7 and Fig.8. Fig.7 shows the best result for each fuzzy system during the learning phase (e.g., 0.044 for $K = 2$ is the result after 20 iterations of the learning rule.) Fig.8 shows all the results during learning phase for fuzzy systems with $K = 2$ and $K = 3$ (Those for $K = 4$ and $K = 5$ are out of the range of the vertical axis.) From Fig.7, we can see that the generalization ability to the test data of fuzzy systems depends on the choice of a fuzzy partition (i.e., the value of $K$) and fine fuzzy partitions (i.e., large values of $K$) lead to poor performance. In

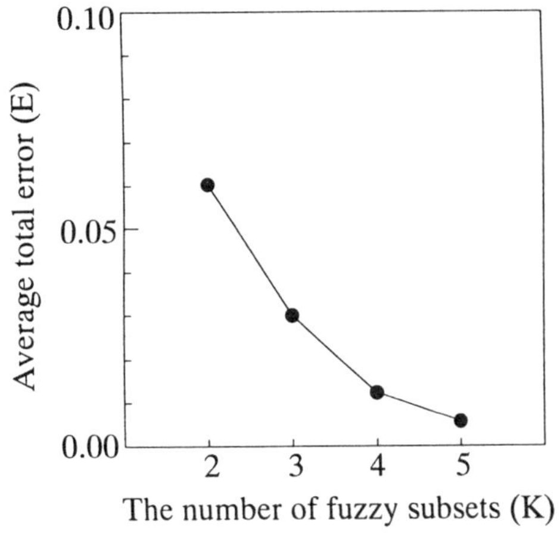


**Fig.3.** Average total errors for training data by fuzzy systems.

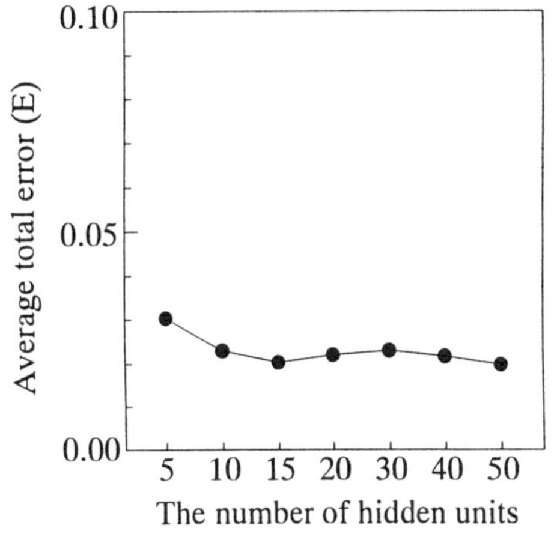


**Fig.5.** Average total errors for training data by neural networks.

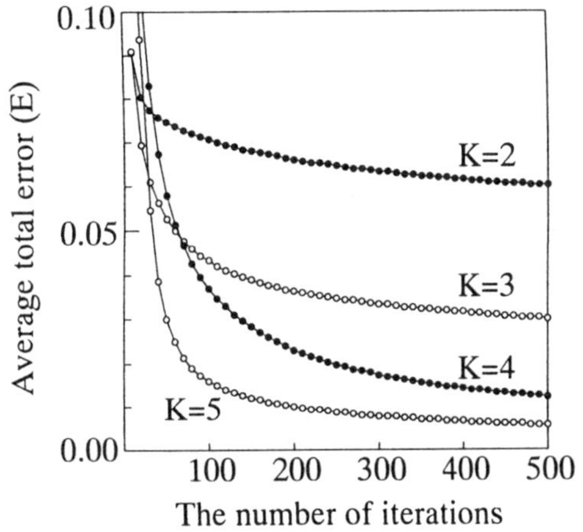


**Fig.4.** Average total errors for training data during learning phase for fuzzy systems.

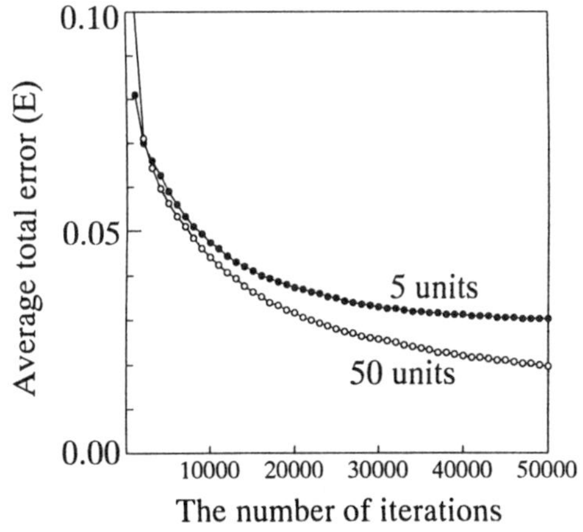


**Fig.6.** Average total errors for training data during learning phase for neural networks.

Fig.8, we can observe the over-fitting of fuzzy systems to the training data during the learning phase. The average total errors for the test data increased after attaining the minimum points: 20 iterations for $K = 2$ and 60 iterations for $K = 3$. Fig.7 and Fig.8 indicate that the choice of both the value of $K$ and the number of iterations is very important in the modelling by fuzzy systems.

Average total errors for the test data by neural networks are shown in Fig.9 and Fig.10. Fig.9 shows the best result during the learning phase. Fig.10 shows all the results during the learning phase for the neural networks with 5 and 50 hidden units. From Fig.9, we can see that the generalization ability to the test data of neural networks is not sensitive to the choice of the number of hidden units. In Fig.10, we can observe the over-fitting of neural networks to the training data during the learning phase. Fig.9 and Fig.10 indicate that the number of iterations of the BP algorithm has larger influence on the generalization ability to the test data than that of hidden units.

From the comparison of Fig.8 with Fig.10, we can see that the generalization ability to the test data of fuzzy systems with appropriate fuzzy partitions (i.e., appropriate values of $K$) is similar to that of neural networks.

## V. SIMPLE METHOD FOR SPECIFYING INITIAL FUZZY IF-THEN RULES

The simulation results in Subsection IV.C indicated that the generalization ability to the test data of fuzzy systems was sensitive to the choice of a fuzzy partition (i.e., the value of $K$) and fuzzy systems with fine fuzzy partitions (i.e., large values of $K$) led to poor performance. In computer simulations in the last section, the initial value of $w_j$

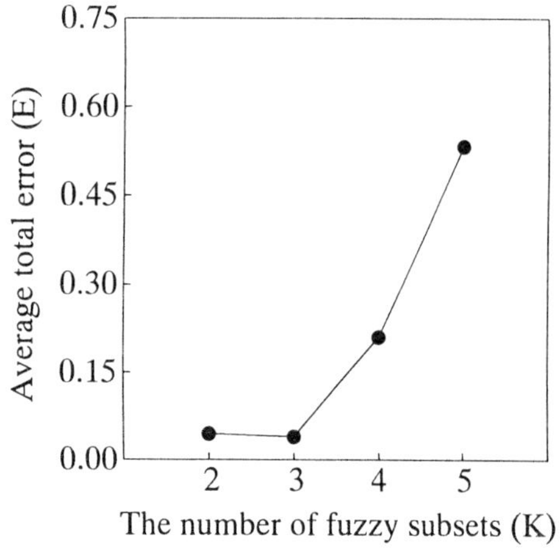


**Fig.7.** Average total errors for test data by fuzzy systems.

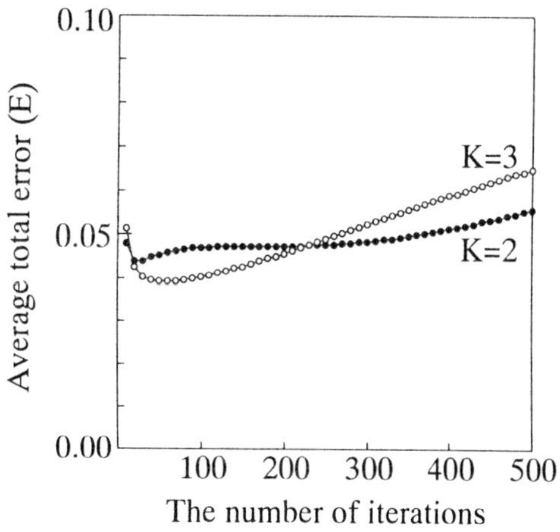


**Fig.8.** Average total errors for test data during learning phase for fuzzy systems.

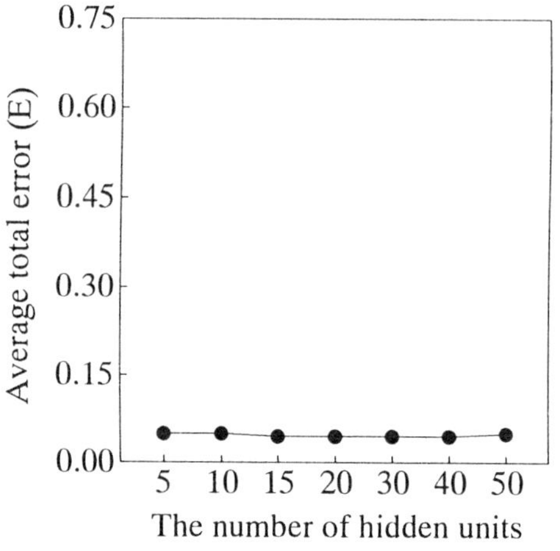


**Fig.9.** Average total errors for test data by neural networks.

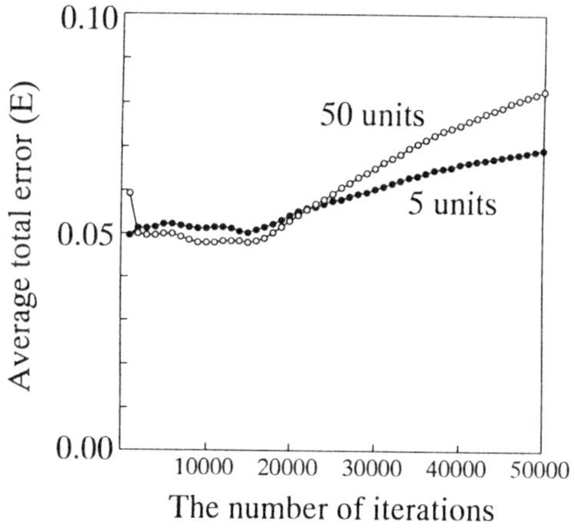


**Fig.10.** Average total errors for test data during learning phase for neural networks.

was specified as $w_j(0) = 0$ for $j = 1, 2, \cdots, n$. This specification is a possible cause of poor performance for test data of fuzzy systems. We propose the following heuristic method for specifying the initial value of $w_j$:

$$w_j(0) = \sum_{p=1}^{105} \mu_j(\boldsymbol{x}_p) \cdot y_p / \sum_{p=1}^{105} \mu_j(\boldsymbol{x}_p), \; j = 1, 2, \cdots, n. \quad (15)$$

That is, $w_j(0)$ is the average of $y_p$ weighted by $\mu_j(\boldsymbol{x}_p)$.

The simulation results for the test data by fuzzy systems with this method are shown in Fig.11 and Fig.12. Fig.11 and Fig.12 are the simulation results on the same conditions as in Fig.7 and Fig.8, respectively. From the comparison of Fig.11 with Fig.7, we can see that the performance of fuzzy systems was improved by the proposed method for $K = 4$ and $K = 5$. This indicates that the choice of the initial value of $w_j$ has large influence on the performance for the test data of fuzzy systems. For the training data, simulation results similar to Fig.3 and Fig.4 were obtained (The details are omitted in this paper.)

## VI. CONCLUSION

In this paper, the ability of fuzzy systems as approximators of non-linear mappings was empirically examined by computer simulations on real-life data. Both the fitting ability to training data and the generalization ability to test data were examined. By the simulation results, we can conclude the followings:

(i) Fuzzy systems attained better fitting to training data in a much smaller number of iterations of the learning rule than neural networks.

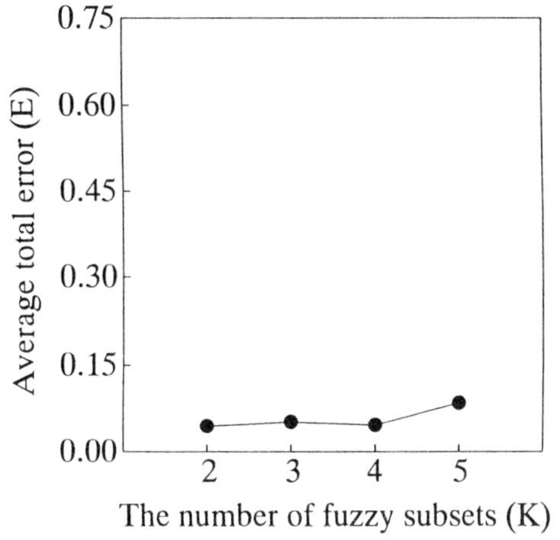


**Fig.11.** Average total errors for test data by fuzzy systems with the proposed method.

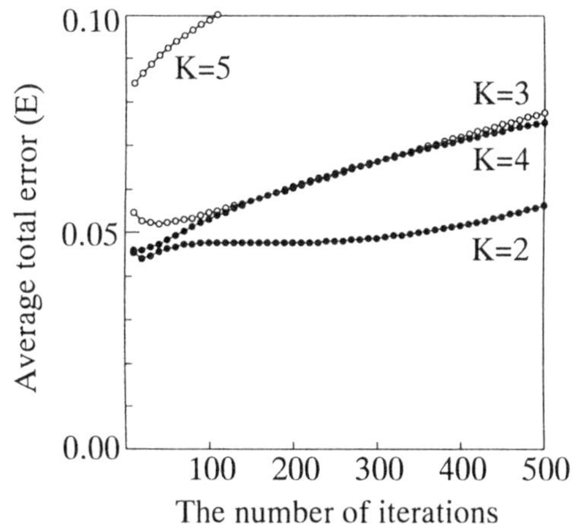


**Fig.12.** Average total errors for test data during learning phase for fuzzy systems with the proposed method.

(ii) The performance for test data of fuzzy systems with appropriate fuzzy partitions is similar to that of neural networks.

(iii) Fine fuzzy partitions led to poor performance for test data. The performance for test data of fuzzy systems with fine fuzzy partitions was improved by specifying initial fuzzy if-then rules by the proposed heuristic method.

These results suggest that comparable performance to neural networks can be obtained by trainable fuzzy systems within a small number of iterations of a simple learning rule if the choice of fuzzy partition is appropriate.

## References


[1] D.G.Burkhardt and P.P.Bonissone, "Automated fuzzy knowledge base generation and tuning," Proc. FUZZ-IEEE'92 (San Diego, California, March 8-12, 1992) pp.179-188.

[2] I.Hayashi, H.Nomura, H.Yamasaki and N.Wakami, "Construction of fuzzy inference rules by NDF and NDFL," International Journal of Approximate Reasoning, vol.6, pp.241-266, 1992.

[3] H.Ichihashi and T.Watanabe, "Learning control by fuzzy models using a simplified fuzzy reasoning," Journal of Japan Society for Fuzzy Theory and Systems, vol.2, pp.429-437, 1990.

[4] J.S.R.Jang, "Fuzzy controller design without domain experts," Proc. FUZZ- IEEE'92 (San Diego, California, March 8-12, 1992) pp.289-297.

[5] B.Kosko, "Fuzzy systems as universal approximators," Proc. FUZZ-IEEE'92 (San Diego, California, March 8-12, 1992) pp.1153-1162.

[6] C.C.Lee, "Fuzzy logic in control systems: Fuzzy logic controller-Part I and Part II," IEEE Trans. on Systems, Man and Cybernetics, vol.20, pp.404-435, 1990.

[7] M.Matsuda and T.Kameoka, "Application of fuzzy measure, fuzzy integral and neural network to the system which estimate taste by using industrial analysis," Proc. IIZUKA'92 (Iizuka, Japan, July 17-22, 1992) pp.601-606.

[8] H.Nomura, I.Hayashi and N.Wakami, "A learning method of fuzzy inference rules by decent method," Proc. FUZZ-IEEE'92 (San Diego, California, March 8-12, 1992) pp.203-210.

[9] Research Committee of Sensory Evaluation, JUSE: Sensory Evaluation Handbook. JUSE Press, 1973.

[10] D.E.Rumelhart, J.L.McClelland and the PDP Research Group, Parallel Distributed Processing, vol.1. MIT Press, Cambridge, 1986.

[11] M.Sugeno, An introductory survey of fuzzy control, Information Sciences, vol.36, pp.59-83, 1985.

[12] T.Takagi and M.Sugeno, Fuzzy identification of systems and its applications to modeling and control, IEEE Trans. on Systems, Man and Cybernetics, vol.15, pp.116-132, 1985.

[13] L.X.Wang, Fuzzy systems are universal approximators, Proc. FUZZ-IEEE'92 (San Diego, California, March 8-12, 1992) pp.1163-1170.

# Integrating Design Stages of Fuzzy Systems using Genetic Algorithms

Michael A. LEE and Hideyuki TAKAGI*

Computer Science Division, University of California, Berkeley, CA 94720 USA
lee@cnmat.berkeley.edu, takagi@diva.berkeley.edu, *FAX (510)642-5775

*Abstract*— **This paper proposes an automatic fuzzy system design method that uses a Genetic Algorithm and integrates three design stages; our method determines membership functions, the number of fuzzy rules, and the rule-consequent parameters at the same time. Because these design stages may not be independent, it is important to consider them simultaneously to obtain optimal fuzzy systems. The method includes a genetic algorithm and a penalty strategy that favors systems with fewer rules. The proposed method is applied to the classic inverted pendulum control problem and has been shown to be practical through a comparison with another method.**

## 1 INTRODUCTION

Fuzzy systems have become popular components of consumer products because they are inexpensive to implement, able to solve difficult non-linear control problems, and exhibit robust behavior. Designers are especially attracted to fuzzy systems because fuzzy systems allow them to capture domain knowledge quickly using rules that contain fuzzy linguistic terms. These attributes allow products with embedded fuzzy systems to be both cost effective and high performance.

While it is easy to describe human knowledge with fuzzy linguistic terms, it is not easy to define the terms by membership functions. In addition, fuzzy system design requires two other stages: determining the number of rules and determining the rule-consequent parameters. Some papers propose automatic methods using neural networks [5, 6, 7, 18], fuzzy clustering [4], genetic algorithms [8, 9, 10, 11, 12, 14, 16, 17, 22], or gradient methods[1, 13, 15]. Although these methods produce systems that perform better than systems designed by humans, they may be suboptimal because they treat only one or two of the three design stages.

This paper proposes an automatic fuzzy system design method that uses a Genetic Algorithm and integrates three design stages; our method determines membership functions, the number of fuzzy rules, and the rule-consequent parameters at the same time. As a sample fuzzy system, we use the Takagi-Sugeno-Kang (TSK) fuzzy model [19]. Rules in a TSK fuzzy model use traditional fuzzy variables for antecedents. However, the consequent values are computed by summing weighted combinations of the input values. We have formulated a TSK fuzzy model representation that parameterizes membership function shape and position and rule-consequent parameters. By combining our representation with the target application's boundary conditions, we can represent fuzzy systems with different numbers of rules. A genetic algorithm operates on this representation and optimizes the fuzzy system parameters with respect to performance and resource requirements. We chose a genetic algorithm optimization technique because genetic algorithms because genetic algorithms are robust, search many points simultaneously, and able to avoid local minima.

In the following sections we briefly review genetic algorithms and automatic fuzzy system design research. Next we discuss our TSK fuzzy model, how we incorporated genetic algorithms into the design process, and parameters of our design method. We demonstrate our method by deriving a four rule fuzzy system that balances an inverted pendulum. We conclude by comparing the performance of our controller with a controller derived by Another method.

## 2 REVIEW

### *2.1 Genetic algorithms*

A genetic algorithm is a probabilistically guided optimization technique modeled after the mechanics of genetic evolution. Unlike many classical optimization techniques, genetic algorithms do not rely on computing local derivatives to guide the search process. Genetic algorithms also include random elements, which helps avoid getting trapped in local minima.

Genetic algorithms explore a population of solutions in parallel. The size of the population is a free parameter, which trades off coverage of the search space against the time required to compute the next generation. Each solution in the population is coded as a binary string or gene, and a collection of genes forms a generation. A new generation evolves by performing genetic operations, such as reproduction, crossover, and mutation, on

*The author is a Visiting Industrial Fellow at the University of California at Berkeley and a Senior Researcher for Central Research Laboratories, Matsushita Electric Industrial Co., Ltd.

0-7803-0614-7/93/$03.00 ©1993 IEEE

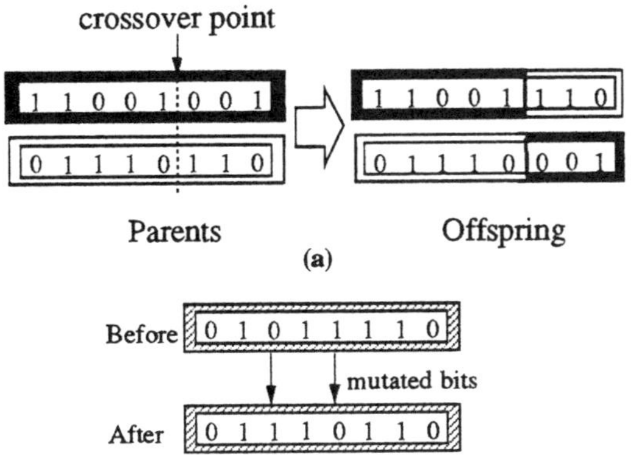


Figure 1: Genetic operations: (a) cross over (b) mutation

genes in the current population and then placing the products into the new generation.

In a simple genetic algorithm, operations are performed in the following order; reproduction, crossover, and then mutation. Reproduction involves selecting two parent genes from the current population. Selection is based probabilistically on a gene's fitness value; the higher the fitness of a gene, the more likely it can reproduce. After selecting two parents, crossover is performed according to a crossover probability. If crossover is to be performed, offspring are constructed by copying portions of parent genes designated by random crossover points (single point crossover shown in Figure 1). Otherwise, an offspring copies its entire gene from one of the parents. As each bit is copied from parent to offspring, the bit has the probability of flipping, or mutating. Mutation is believed to help reinject any information that may have been lost in previous generations [3]. Variations of these operators are discussed in [2].

A gene's fitness is evaluated by decoding the gene's binary representation and then passing it through a fitness function. The fitness function is a means of ranking solutions in the population and can include penalty terms in addition to raw performance measures. In our method, we included a penalty strategy that favors production of systems with fewer membership functions and rules. More details are given in section 3. (See [3] for extensive discussions on constructing genetic objective functions.)

### 2.2 *Automatic design of fuzzy systems*

Several papers, mentioned in section 1 , have proposed automatic design methods. Much of the work has focused on tuning membership functions. For example, [18] uses neural networks as a membership values generator and [15] treats fuzzy systems as networks and use back-propagation techniques to adjust membership functions. Nodes in these networks perform parameterized functions. These parameters are tuned by computing derivatives of the network, with respect to these parameters, and then back-propagating the error as in traditional neural networks.

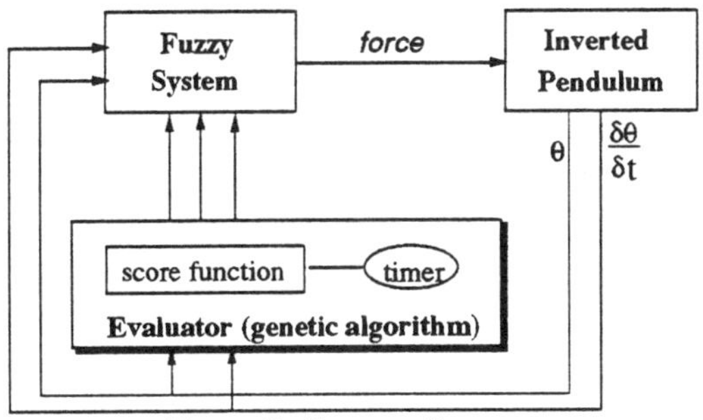


Figure 2: System diagram

Other methods use genetic algorithms to determine the number of fuzzy rules [9, 20]. Karr has developed a method for determining membership functions and number of fuzzy rules using genetic algorithms [9]. In this paper, Karr's method first uses a genetic algorithm to determine the number of rules according to a predefined rule base. Following this stage, the method uses a genetic algorithm to tune the membership functions. Our method differs from Karr's in that we perform both operations simultaneously, as opposed to sequentially, and we include a penalty strategy that favors systems with fewer rules.

Although many of the proposed methods offer an improvement on human designed fuzzy systems, they usually combine only one or two of the design stages. Because these design stages may not be independent, it is important to consider them simultaneously to find the optimal solution.

## 3 Experimental Environment and Representations

The goal of our work is to develop an automatic fuzzy system design that uses minimal knowledge of the system to be controlled. As a sample fuzzy system, we chose the TSK fuzzy model, which is widely used in actual applications. In this section we first introduce our TSK fuzzy model representation used in our experiments. Second we present the inverted pendulum application used to illustrate our technique. Lastly we present our method for evaluating a fuzzy system's performance in our application context. A system diagram showing the relationship between the components discussed in this section is given in Figure 2.

### 3.1 *Fuzzy system representation for automatic design*

We based our fuzzy system on the TSK fuzzy model. In this model the input variables are traditional fuzzy sets, however output variables are computed from linear combinations of the input values. For example, a typical rule in a TSK fuzzy system might be:

$$\text{IF X is A and Y is B THEN C} = w_1\text{X} + w_2\text{Y} + w_3$$

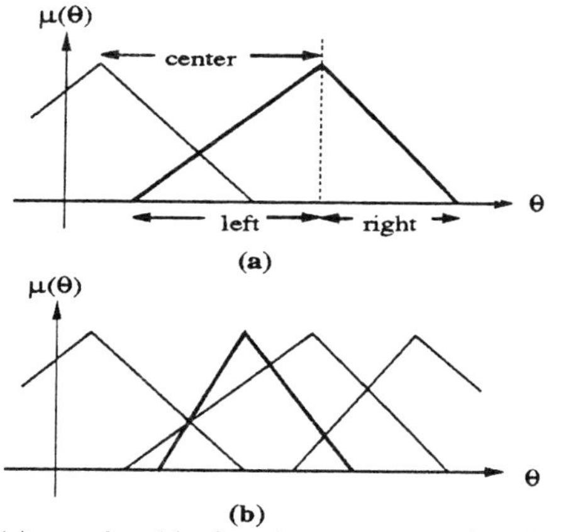

Figure 3: (a) membership function representation (b) possible member functions

where $w_n$ are rule-consequent parameters.

The representation we formulated provides the flexibility to parameterize membership functions, number of fuzzy rules, and consequent parameters. Each membership function is triangular and parameterized by left base, right base, and distance from the previous center point (see Figure 3(a)). By encoding the centers as a distance from the previous center (the first center was given as an absolute position) and the base values as the distance from the corresponding center point (see Figure 3(a)), we can use the boundary conditions of the application to eliminate necessary membership functions, which has the direct effect of eliminating rules. For example, all $\theta$ membership functions sets with center positions greater than 90° can be eliminated.

Each membership function requires three parameters, and the consequent part of each fuzzy rule requires three parameters. Consider a two dimensional input space: one input dimension fuzzy-partitioned into $m$ and the other into $n$. The number of membership functions are $m + n$ and the number of rules are $nm$. The total number of system parameters is $3(m+n)+3mn$.

Unlike most other methods, overlap restrictions were not placed on the sets in our system and the possibility of complete overlap existed (see Figure 3(b)). The final output values are computed by summing the weighted output of each rule according to its firing strength.

### 3.2 *Inverted pendulum*

For experimental purposes, we applied our method to the inverted pendulum problem. The inverted pendulum represents a classic non-linear control problem that can be described as the task of balancing a pole on a movable cart. In this simulation, the movement of both the pole and the cart is restricted to the vertical plane. The state of the system is described by the pole's angle and angular velocity (in this simulation, the cart is allowed to move infinitely in the left or right direction). The controller can exert only a constant force on the cart in either the left or right direction. The objective of the controller is to balance the pole as quickly as possible (see Figure 2 for system diagram).

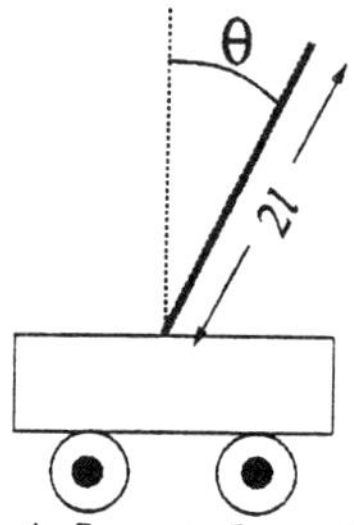

Figure 4: Inverted pendulum

The state equations for the inverted pendulum can be expressed as [6]:

$$\dot{\theta} = \frac{\delta\theta}{\delta t}$$

$$\ddot{\theta} = H_2(\theta, \dot{\theta}, force) = \frac{g\sin(\theta) + \cos(\theta)\left(\frac{-force - ml\dot{\theta}^2\sin(\theta)}{m_c + m}\right)}{l\left(\frac{4}{3} - \frac{m\cos^2(\theta)}{m_c + m}\right)}$$

where $g$ is 9.8 meters/sec$^2$, $m_c$ is the mass of the cart, $m$ is the mass of the pole, $l$ is half the length of the pole, and $force$ is the applied force in Newtons.

In this simulation, the equations of motion are defined by differential equations. A two-step forward Euler integration can be used to approximate its state at $t + h$:

$$\theta(t + \frac{h}{2}) = \frac{h}{2}\dot{\theta}(t) + \theta(t)$$

$$\dot{\theta}(t + \frac{h}{2}) = \frac{1}{2}H_2(\theta(t), \dot{\theta}(t), force) + \dot{\theta}(t)$$

$$\theta(t + h) = \frac{h}{2}\dot{\theta}(t + \frac{h}{2}) + \theta(t + \frac{h}{2})$$

$$\dot{\theta}(t) = \frac{1}{2}H_2(\theta(t + \frac{h}{2}), \dot{\theta}(t + \frac{h}{2}), force) + \dot{\theta}(t + \frac{h}{2})$$

where $h$ is the time step.

### 3.3 *Evaluating fuzzy system performance*

Unlike the fuzzy system representation, the evaluation function relies directly on the application. In this experiment, the function must be capable of ranking the fuzzy systems in the context of the inverted pendulum task. The objective of controlling an inverted pendulum is to balance it in the shortest amount of time for a wide range of initial conditions. To evaluate our fuzzy systems, we tried the fuzzy system on the inverted pendulum starting with eight different initial conditions (positions of Table 1 and their symmetric positions). Each trial terminated under one of the following three conditions: either the pole fell over, time expired, or the system balanced the pole (using some

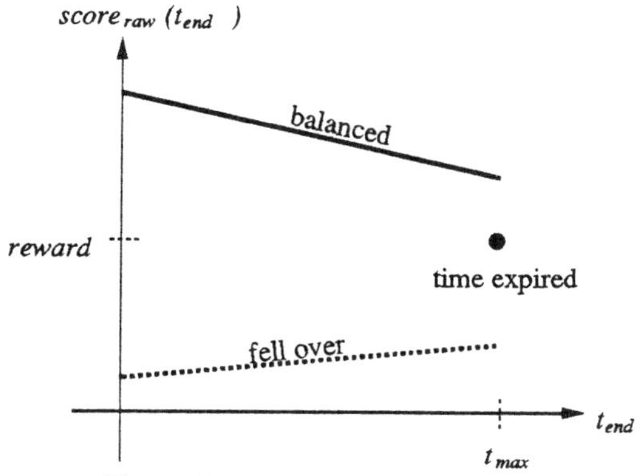


Figure 5: Raw scoring functions

epsilon criteria). Depending on the termination condition, we scored a trial in the following manner (see Figure 5):

$$score(t_{end}) = \begin{cases} a_1 (t_{max} - t_{end}) + a_2 reward & (1) \\ reward & (2) \\ b \cdot t_{end} & (3) \end{cases}$$

where $a_1$, $a_2$, $b$ and $reward$ are constants, and (1) pole balanced, (2) $t_{max} = t_{end}$, and (3) pole fell over ($|\theta| \geq 90°$).

The general idea is that if the system balances the pole, a shorter time is better than a longer time. If the pole falls over, we recognize potential success and credit the system according to the time it kept it from falling. We then added additional terms to consider steady state error and to penalize systems according to the number of rules in the system. The resulting fitness score for one trial was computed as follows:

$$score(t_{end}) = \frac{\left(score_{raw}(t_{end}) + c\sum_0^{t_{end}} |\theta_t|\right)}{\text{number of rules} + \text{offset}_{\text{rules}}}$$

The steady state error was a simple summation of the pole angle displacement and weighted with constant $c$. The $\text{offset}_{\text{rules}}$ parameter controls the degree of penalty for number of rules. The scores from the eight trials were accumulated to form a composite score, which was used as a controller's overall fitness. We varied the $\text{offset}_{\text{rules}}$ parameter in our experiments and results are discussed in section 6.

## 4 Incorporating Genetic Algorithms into Fuzzy System Design

To incorporate genetic algorithms into fuzzy system design, we must find a suitable genetic coding and determine a method to evaluate its fitness. An evaluation method was discussed in section 3.3 and can be used directly to determine fitness values. To address the coding problem, we first define a chromosome as a set of parameters that represent a higher level entity, such as a membership function or rule-consequent parameters set

| center | left base | right base |
|---|---|---|
| 10100110 | 10011000 | 01011000 |

membership function chromosome (MFC)

| $w_1$ | $w_2$ | $w_3$ |
|---|---|---|
| 10100110 | 10011000 | 01011000 |

rule-consequent parameters chromosome (RPC)

Figure 6: Composite chromosomes

(see Figure 6). By linking these chromosomes together, we can form the entire fuzzy system representation (see Figure 7). In our experiments, all parameter values are encoded as 8 bit numbers.

The maximum number of fuzzy sets per input variable was set to ten in our experiments. This limit was set through experience with the inverted pendulum application. Because we included a penalty strategy that involves the number of rules, setting this number is not so critical. There is no danger in setting this to an arbitrarily large number (with the exception of consuming excessive computing resources). The resulting genetic representation for fuzzy systems used in our experiments consisted of 360 parameters or 2880 bits (see Figure 6 and 7).

## 5 Experimental Results

Our method combines a genetic algorithm, a penalty strategy, and unconstrained membership function overlap to automatically design fuzzy systems. In this section, we present results of our method applied to the inverted pendulum problem.

In our experiments, we used a genetic algorithm with two-point crossover and mutation operators. Population size was set to 10 and crossover and mutation probabilities were 0.6 and 0.0333 respectively. We also used an elitist strategy in which the member with the highest fitness value automatically advanced to the next generation. All members were initialized with random values in most experiments. In some experiments we used apriori knowledge to initialize one member of the population with seven uniformly spaced fuzzy sets for both $\theta$ and $\delta\theta/\delta t$. Each of its 49 rules were initialized with $w_1 = w_2 = 0$ and $w_3$ equal to a value computed using the center points of the antecedent membership functions and the control law: $\text{force} = c_1 \sin(\theta) + c_2 \frac{\delta\theta}{\delta t}$.

The automatic design process was initiated by first setting the $\text{offset}_{\text{rules}}$ parameter and then letting the genetic algorithm generate 5000 generations of solutions. The best solution was kept and the rest were discarded. In one experiment with the $\text{offset}_{\text{rules}}$ parameter set to 100, the method produced a symmetric system with only four rules. Figure 8 shows the fitness level as a function of generation for this experiment comparing combinations of heuristic initialization and structural symmetry constraints. Figures 9 and 10 show trajectory plots for several

| fuzzy variable $\theta$ | | | fuzzy variable $\delta\theta/\delta\tau$ | | | rule-consequent parameters | | |
|---|---|---|---|---|---|---|---|---|
| $MFC_1$ | ··· | $MFC_{10}$ | $MFC_1$ | ··· | $MFC_{10}$ | $RPC_1$ | ··· | $RCP_{100}$ |

Figure 7: Gene map

Table 1: Initial conditions of pendulum

| | | | | |
|---|---|---|---|---|
| $\theta$ | 5.22 | 5.11 | -8.41 | 6.22 |
| $\delta\theta/\delta t$ | 6.93 | 6.97 | -1.37 | -7.14 |

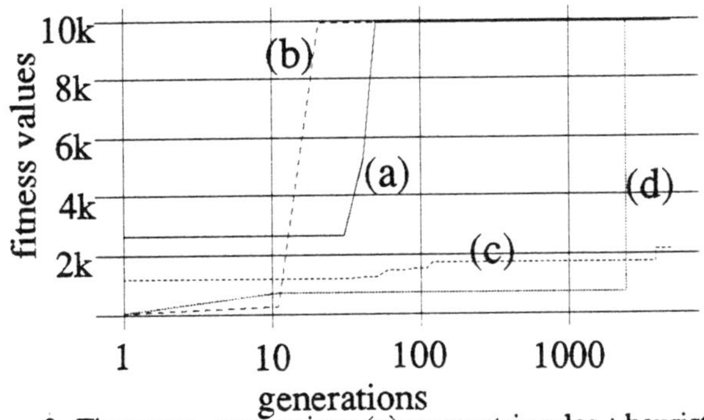


Figure 8: Fitness vs. generation: (a) symmetric rules + heuristic initialization, (b) symmetric rules + random initialization, (c) asymmetric rules + heuristic initialization, and (d) asymmetric rules + random initialization.

initial conditions using this system. The symmetric rules are:

IF $\theta$ is $A_i$ and $\frac{\delta\theta}{\delta t}$ is $B_i$ THEN $y = w_{1i}\theta + w_{2i}\frac{\delta\theta}{\delta t} + w_{3i}$, where $i = 1 \sim 4$. The obtained parameters in four consequent parts were $(w_{1i}, w_{2i}, w_{3i})$ = (0.44, 1.02, -31.65), (1.54, -0.61, -30.14), (1.54, -0.61, 30.14), and (0.44, 1.02, 31.65). The obtained triangular membership functions, $A_i$ and $B_i$ were $A_1 = A_3 = \{-119.65, -62.12, 4.59\}$, $A_2 = A_4 = \{-4.59, 62.12, 119.65\}$, $B_1 = B_3 = \{-219, -1.99, 238.56\}$, and $B_2 = B_4 = \{-238.56, 1.99, 219.64\}$.

## 6 Discussion

Our results show that our proposed fuzzy system design method can automatically determine fuzzy system membership functions, number of rules, and consequent parameters simultaneously. In one experiment, the resulting system balanced a pendulum for all initial angles in about 0.4 seconds. This performance compares favorably with a system produced by Jang [6, 7]. Both methods produce fuzzy systems that have four rules, however our system automatically determines the number of rules while Jang sets the rule number by hand.

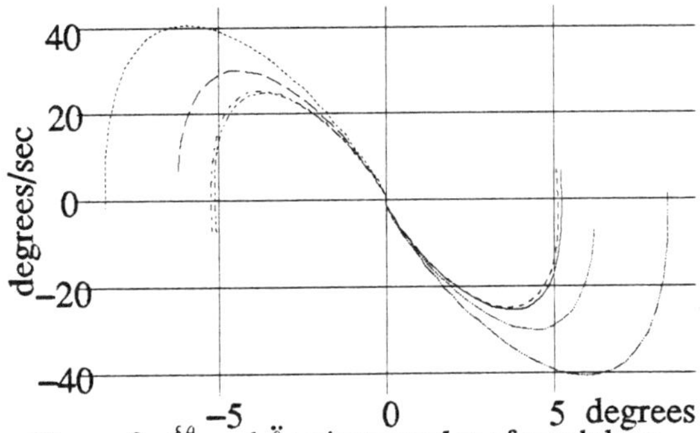


Figure 9: $\frac{\delta\theta}{\delta t}$ and $\ddot{\theta}$ trajectory plot of pendulum

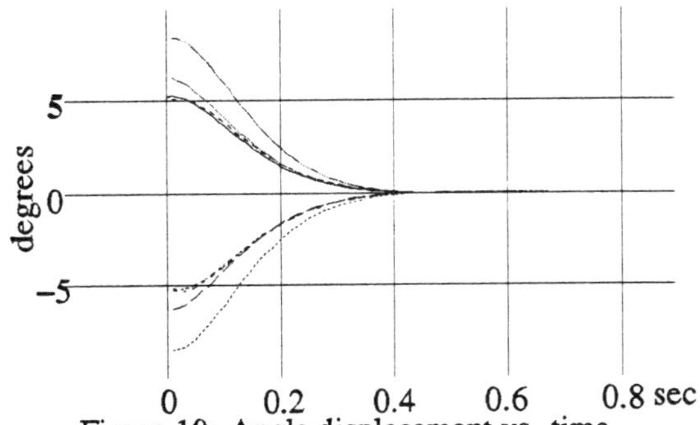


Figure 10: Angle displacement vs. time

Although the inverted pendulum involved only two input and one output variable, our design method can be extended to handle problems of higher dimension. The mechanics of the genetic algorithm remains unchanged, only the fitness function evaluation changes.

In our experiments we also found that the time to find a solution that balances all initial conditions was not correlated with the rule penalty weight. In some cases, the experiments with higher rule penalty settings found solutions before experiments with lower penalties settings. This may be due to initial condition effects due to the small population or due to the nature of the application. The sensitivity of our method to penalty settings must be studied in more detail.

A large percentage of the computing resources is devoted to evaluating population members simulating the pendulum. The time to perform the genetic operations is small compared to time to simulate the pendulum under the eight different initial conditions. If the epsilon criteria for determining whether the pendulum was balanced were relaxed, some trials might require fewer time steps.

## 7 Conclusions and Further Research

We have proposed a method for automatically designing complete fuzzy systems. Our method uses a genetic algorithm and a penalty strategy to determine membership function shape and position, number of fuzzy rules, and consequent parameters simultaneously. Our experimental results demonstrate the practicality of our method, by producing systems that perform comparably to a system produced by another method.

Other extensions to this work that need to be explored include applying this method to more complex tasks, directly comparing results with a sequential method, applying this method to other types of fuzzy systems, and eliminating unnecessary rules by considering overlap.

*Acknowledgments*

This research is supported in part by NASA Grant NCC-2-275, MICRO State Program Award No.90-191, and EPRI Agreement RP8010-34. We also would like to thank Prof. David Wessel and the Center for New Music and Audio Technologies at UC Berkeley for use of computing resources.

## References

[1] Araki, S., Nomura, H., Hayashi, I., and Wakami, N., "Self-generating method of fuzzy inference rules", Int'l Fuzzy Engineering Symposium (IFES'92), 1992, pp.1047-1058

[2] Davis, L., ed. Handbook of Genetic Algorithms, Van Nostrand, Reinhold, 1991

[3] Goldberg, D., Genetic Algorithms in Search, Optimization, and Machine Learning, Addison-Wesley, 1989

[4] Hirota, K. and Yoshinari, Y., "Identification of Fuzzy Control Rules Based on Fuzzy Clustering Method", 5th Fuzzy System Symposium, June, 1989, pp.253-258 (in Japanese)

[5] Ichikawa, R., Nishimura, K., Kunugi, M., Shimada, K., "Auto-Tuning Method of Fuzzy Membership Functions Using Neural Network Learning Algorithm," Proc. of the 2nd Int. Conf. on Fuzzy Logic and Neural Networks (IIZUKA'92), 1992, pp.345-348

[6] Jang, R, "Fuzzy Controller Design without Domain Experts," Proc. IEEE Int. Conf. on Fuzzy Systems (FUZZ-IEEE'92), 1992, pp. 289-296

[7] Jang, R, "Self-Learning Fuzzy Controllers Based on Temporal Back Propagation," IEEE Trans. on Neural Networks, Vol.3, No.5, 1992, pp.714-723

[8] Karr, C., Freeman, L., Meredith, D., "Improved Fuzzy Process Control of Spacecraft Autonomous Rendezvous Using a Genetic Algorithm," Proc. of the SPIE Conf. on Intelligent Control and Adaptive Systems, Orlando, FL, 1989, pp.274-283

[9] Karr, C., "Applying Genetics to Fuzzy Logic," AI Expert, Vol.6, No.2, 1991, pp.26-33

[10] Karr, C., "Design of an Adaptive Fuzzy Logic Controller using a Genetic Algorithm," Proc. of the Int. Conf. of Genetic Algorithms (ICGA'92), 1992, pp. 450-457

[11] Karr, C., Gentry, E., "A Genetics-Based Adaptive pH Fuzzy Logic Controller," Proc. of the Int. Fuzzy Systems and Intelligent Control Conf. (IFSICC'92), Louisville, KY, 1992, pp.255-264

[12] Karr, C., Sharma, S., Hatcher, W., Harper, T., "Control of an Exothermic Chemical Reaction using Fuzzy Logic and Genetic Algorithms," Proc. of the Int. Fuzzy Systems and Intelligent Control Conf. (IFSICC'92), Louisville, KY, 1992, pp.246-254

[13] Katayama, R., Kajitani, Y., Nishida, Y., "A Self Generating and Tuning Method for Fuzzy Modeling using Interior Penalty Method," Proc. of the 2nd Int. Conf. on Fuzzy Logic and Neural Networks (IIZUKA'92), 1992, pp.349-352

[14] Nishiyama, T., Takagi, T., Yager, R., and Nakanishi, S., "Automatic Generation of Fuzzy Inference Rules by Genetic Algorithm", 8th Fuzzy System Symposium, 1992, pp.237-240 (in Japanese)

[15] Nomura, H., Hayashi, I., and Wakami, N., "A self-tuning method of fuzzy control by descent method", 4th IFSA Congress, Vol. Engineering, July, 1991, pp.155-158

[16] Nomura, H., Hayashi, I., Wakami, N., "A Self-Tuning Method of Fuzzy Reasoning By Genetic Algorithm," Proc. of the Int. Fuzzy Systems and Intelligent Control Conf. (IFSICC'92), Louisville, KY, 1992, pp.236-245

[17] Qian, Y., Tessier, P., Dumont, G., "Fuzzy Logic Based Modeling and Optimization," Proc. of the 2nd Int. Conf. on Fuzzy Logic and Neural Networks (IIZUKA'92), 1992, pp.349-352

[18] Takagi, H. and Hayashi, I., "NN-driven Fuzzy Reasoning", Int'l J. Approximate Reasoning (Special Issue of IIZUKA'88), Vol.5, No.3, 1991, pp.191-212

[19] Takagi, T. and Sugeno, M., "Fuzzy Identification of Systems and Its Applications to Modeling and Control", IEEE Trans. SMC-15-1, 1985, pp.116-132

[20] Takahama, T., Miyamoto, S., Ogura, H., and Nakamura, M., "Acquisition of Fuzzy Control Rules by Genetic Algorithm", 8th Fuzzy System Symposium, 1992, pp.241-244 (in Japanese)

[21] Thrift, P., "Fuzzy Logic Synthesis with Genetic Algorithms," Proc. of the Int. Conf. of Genetic Algorithms (ICGA'92), 1992, pp.509-513

[22] Tsuchiya, T., Matsubara, Y., and Nagamachi, M., "A Learning Fuzzy Rule Parameters Using Genetic Algorithm", 8th Fuzzy System Symposium, 1992, pp.245-248 (in Japanese)

# Application of Neuro-Fuzzy Hybrid Control System to Tank Level Control

Tetsuji Tani
Maintenance & System Development Section
Manufacturing Department
Idemitsu Kosan Co., Ltd., Japan

Shunji Murakoshi and Tsutomu Sato
Institute of Information Technology
Information Systems Department
Idemitsu Kosan Co., Ltd., Japan

Motohide Umano
Department of
Precision Engineering
Osaka University, Japan

Kazuo Tanaka
Department of
Mechanical Systems Engineering
Kanazawa University, Japan

**Abstract : This paper proposes a practical control method using neural networks and fuzzy control techniques, where neural networks estimate the target of fuzzy control. Neural networks estimate the transient state of the plant which has non-linear process such as refrigerating and filtering. Based on the estimation, the suitable control target pattern for fuzzy control is selected.**

**This method is applied to the tank level control of the solvent dewaxing plant. And it is shown that this proposed system can control the tank level effectively not only in steady state but also in transient state.**

## 1. Introduction

Since phenomena in the real plant are too complicated to build a theoretical model, it is very difficult to design a control system of such a plant. The operator, however, can control such a plant using his experience. Recently, fuzzy logic, neural network, or both are applied to the real process rather than mathematical models[1, 2, 3]. Fuzzy logic deals with the linguistic and imprecise rules by expert's knowledge. Neural network is also applied to control plants.

This paper deals with a tank level control including non-linear process such as refrigerating and filtering.

A real process often has more than one purpose of control. Our purposes of control are,

(1) to change the flow rate from the tank smoothly,

(2) to keep the tank level stable.

These are contrary to each other.

To overcome these problems, we observe an experienced operator's procedure. He can estimate the suitable target of the tank level and keep the tank level stable. The aim of this paper is to design a neuro-fuzzy hybrid control system which replaces expert's operation. Neural networks estimate the transient state of the plant and based on this estimation the control target pattern of fuzzy control for the smooth change of the flow rate is selected.

This neuro-fuzzy hybrid method is applied to the real tank level control of the solvent dewaxing plant, and good results are obtained not only in steady state but also in transient state.

## 2. Description of Process

In the process of the vacuum distillation for producing lubricant oil, distillate oil and reduced oil are produced, including wax. Such oils including wax generally have a high

0-7803-0614-7/93$03.00 ©1993IEEE

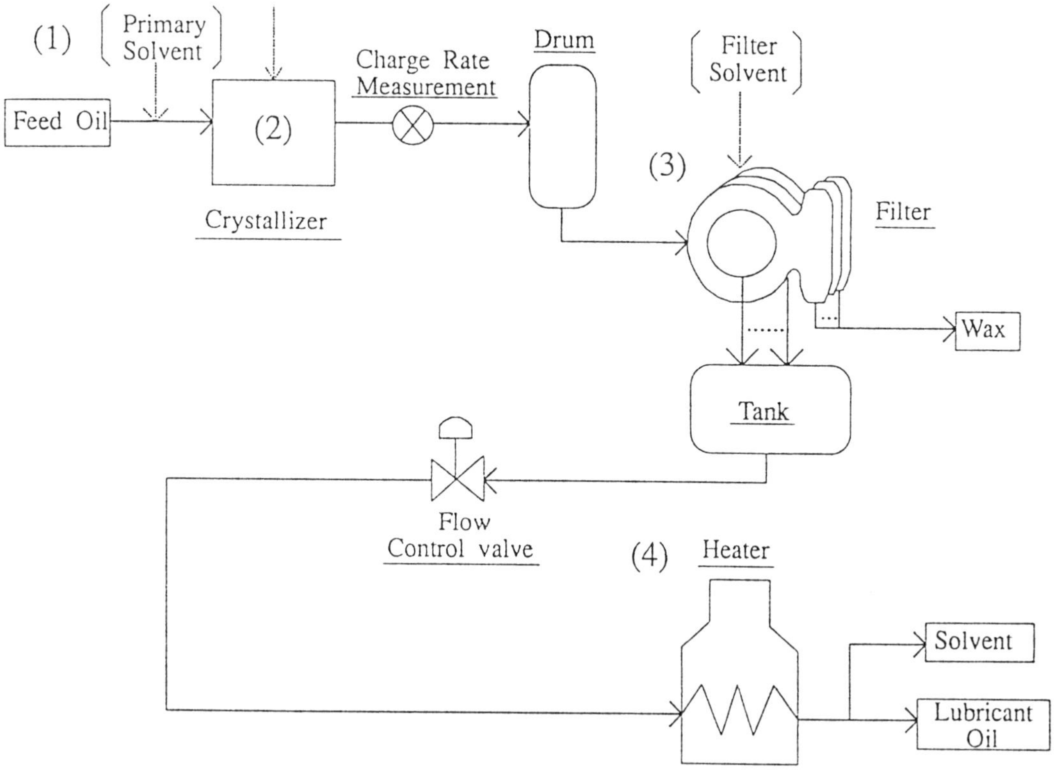


Fig.1. Process flow of the solvent dewaxing plant

solidifying point, so we have to remove the wax not to freeze at low temperature.

The solvent dewaxing plant for removing the wax is outlined in Fig.1. This plant uses solvent to remove the freezed wax easily. We have several steps as follows:

(1) The primary solvent is added to the feed oil.

(2) The feed oil is refrigerated in crystallizer on adding secondary solvent.

(3) The congealed wax is removed by the filter on spraying the filter solvent. The filter is composed of a vacuumed rotating drum which separates the congealed wax from the feed oil. The dewaxing oil, which is a mixture of lubricant oil and solvent, is sent to the tank.

(4) The heater makes solvent evaporate. As a result, low fluid point lubricant oil is produced.

Our control purpose is to keep the tank level constant in all conditions. However, we have the following difficulties for keeping the tank level constant.

(1) The inflow rate to the tank varies with the filter plugging. Since the response of the filter plugging has a long delay time when the feed oil is switched (the delay time also depends on kinds of feed oil ), it is difficult to keep the tank level constant using a feed-forward controller.

(2) The heater has a limit in the changing of the flow rate.

(3) We have the feed oil switching frequently (every three or four days).

These factors are combined in a complicated fashion, where an experienced operator used to control the flow rate manually.

## 3. Operator's Procedure

We have two operation states. One is steady state for everyday operation and the other is transient state of the feed oil switching.

*(1) Steady State*

Several filters are in operation and one is stopped periodically for washing. The tank level, therefore, goes down and up periodically. An experienced operator,

(a) estimates the flow rate roughly by observing the tank level over several hours,

(b) compensates the flow rate by observing the tank level and the time of filter washing.

*(2) Transient State*

When the feed oil will be switched, an operator controls the tank level beforehand to make the change of the flow rate more smoothly. For example, if the inflow rate of the tank is expected to be lower after the feed oil switching, the flow rate is beforehand decreased to keep the tank level constant. This prevents the tank from becoming empty and the flow rate from changing rapidly.

We model such experienced operator's procedures to control the tank level as follows:

Step1 : To obtain long-time tendencies of the flow rate, we calculate the average from operation data. This is equivalent to the expert's estimation of rough flow rate.

Step2 : To compensate the flow rate, we use a fuzzy control system based on expert's knowledge. The input variables of this fuzzy control system are the tank level and the time of filter washing.

Step3 : To control the tank level beforehand, we find transient state by using neural networks. This is equivalent to the expert's predictions. The expert predicts the transient state and changes the control target of the tank level.

## 4. Structure of Neuro-Fuzzy Hybrid Controller

We design neuro-fuzzy hybrid control system which

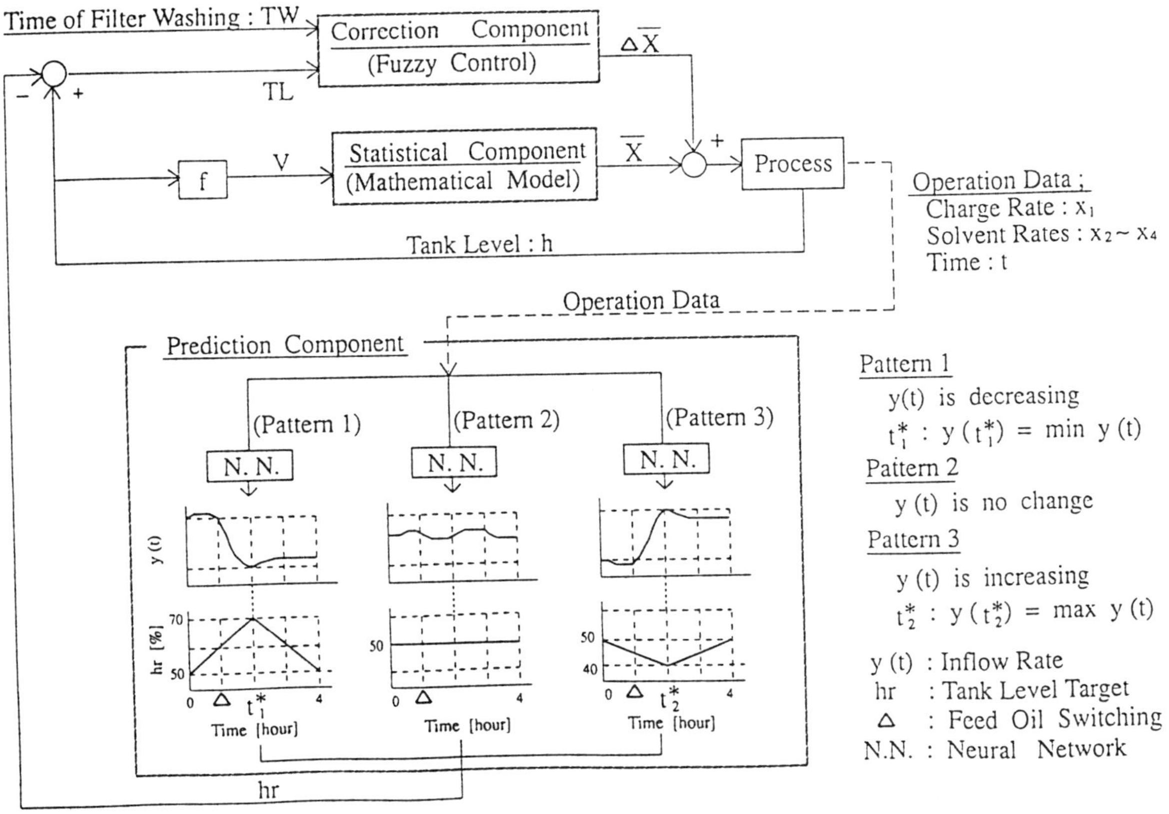


Fig.2. Outline of neuro-fuzzy control system

replaces expert's operation[4, 5, 6]. The controller consists of three components, (1) a statistical component, (2) a correction component (fuzzy controller) and (3) a prediction component (neural networks) in Fig.2.

### *4.1 Statistical Component*

The statistical component is a statistical model for calculating long-time tendencies of the flow rate from operation data. An experienced operator sets the flow rate based on long-time tendencies for the tank level. For example, if the level has a tendency to increase, the flow rate gradually increases.

The difference of the tank amount $\Delta V$ is defined as

$$\Delta V(t) = f(h(t)) - f(h(t-1)) \qquad (1)$$

where h(t) and h(t-1) are the tank level and f converts its level to the corresponding amount. And the average flow rate $\overline{X}$ is defined as

$$\overline{X}(t) = \overline{X}(t-1) + \alpha \Delta V(t) \qquad (2)$$

where $\alpha$ is a real number ( $0 < \alpha \leqq 1$ ) and determined by experience. And $\alpha = \alpha_1$ in the steady state and $\alpha = \alpha_2$ is in the transient state, where $\alpha_1 < \alpha_2$.

### *4.2 Correction Component*

The correction component is a fuzzy controller for compensating the flow rate from the statistical component to stabilize the tank level. We use a simplified method of fuzzy reasoning[7].

The control rules of experienced operators to stabilize the tank level are shown in Table 1. These rules mean that,

- when the tank level is near the target, operators focus on the rate of level changing,
- when the tank level is far from the target, operators focus on the time until the next washing.

As a example : If TL is PS and $\Delta$TL is PS then $\Delta\overline{X}$ is PS, where $\Delta\overline{X}$ is compensation of $\overline{X}$.

The tank level target for fuzzy control is 50% of the tank capacity in a steady state. In a transient state, it is set by neural networks which will be described in the next section.

### *4.3 Prediction Component*

When the feed oil will be switched, we have to predict the inflow rate of the tank. But it is too complicated process to build a mathematical model. We use a neural network approach to predict the inflow rate.

The prediction component is neural networks for predicting the inflow rate to estimate the target of fuzzy controller. We use a three layers model whose learning method is back propagation algorithm[8]. Our neural network is shown in Fig.3. The input layer has five units, the hidden layer has ten and output layer has one. We had an interview with experienced operator to decide the input variables. Input are charge rate of

Table 1. Control rule table

| | | ΔTL | | | | | TW | | |
|---|---|---|---|---|---|---|---|---|---|
| | | PB | PS | ZE | NS | NB | PB | PS | ZE |
| TL | PB | — | — | — | — | — | PB | PB | PS |
| | PS | PB | PS | — | ZE | NB | — | — | — |
| | ZE | ZE | ZE | ZE | ZE | ZE | ZE | ZE | ZE |
| | NS | PB | ZE | — | NS | NB | — | — | — |
| | NB | — | — | — | — | — | NS | NB | NB |

[Input of fuzzy control]

- TL : h - hr , where h is tank level and hr is tank level target
- ΔTL : rate of the changing level
- TW : time until next filter washing

[Compensation of the average flow rate ]

- PB : Positive Big
- PS : Positive Small
- ZE : Zero
- NS : Negative Small
- NB : Negative Big

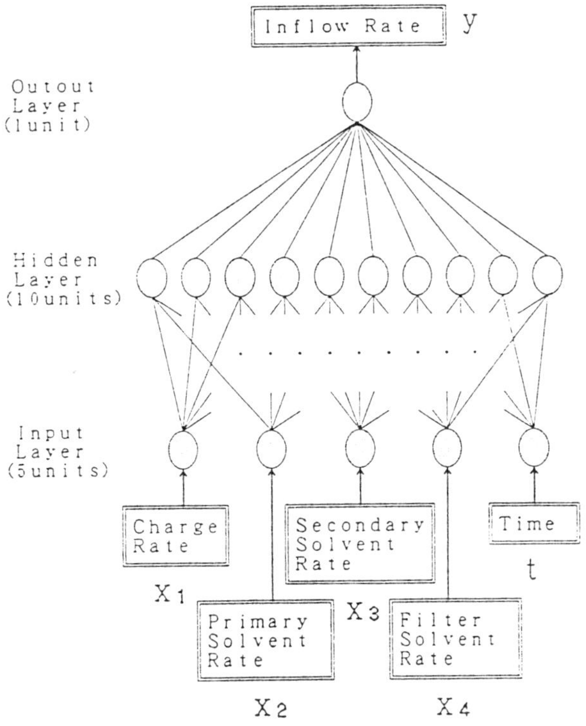


Fig.3. Outline of neural network

the feed oil, primary solvent rate, secondary solvent rate, filter solvent rate and time. Output is the inflow rate. We use several neural networks for the different feed oil switching patterns, e.g., oil A to oil B and oil B to oil C.

From the prediction of the inflow rate by the trained neural networks, we can find the followings:

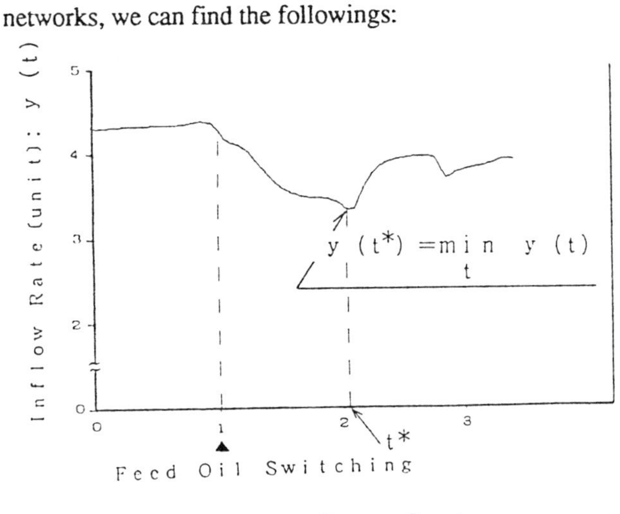


Fig.4. Example of inflow rate pattern

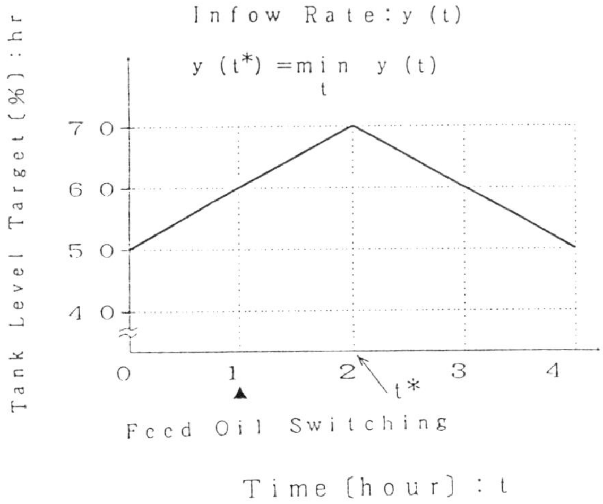


Fig.5. Example of tank level target patten

(1) The inflow rate changing pattern after the feed oil switching: increasing, decreasing and no change.

(2) The time of the highest stage or the lowest stage for the inflow rate.

For example, Fig.4 shows a prediction of a inflow rate pattern by using the trained neural network when the oil is switched oil A to oil B. We can find the followings:

(1) The inflow rate of the tank is decreasing.

(2) The inflow rate of the tank becomes the lowest stage about an hour after the feed oil switching.

Fig.5 shows the target pattern when the oil is switched from oil A to oil B. This means that it takes 2 hours that the tank level must be increased for compensation to the lowest stage of the inflow rate, and then it is decreased to 50% of the tank capacity. This is equivalent to the expert's action. He makes the tank level the highest at the lowest stage of the inflow rate.

## 5. Results

We applied the proposed method to the real plant of the solvent dewaxing plant at Idemitsu Chiba refinery. As results of on-line test, we got the stability for the tank level, and

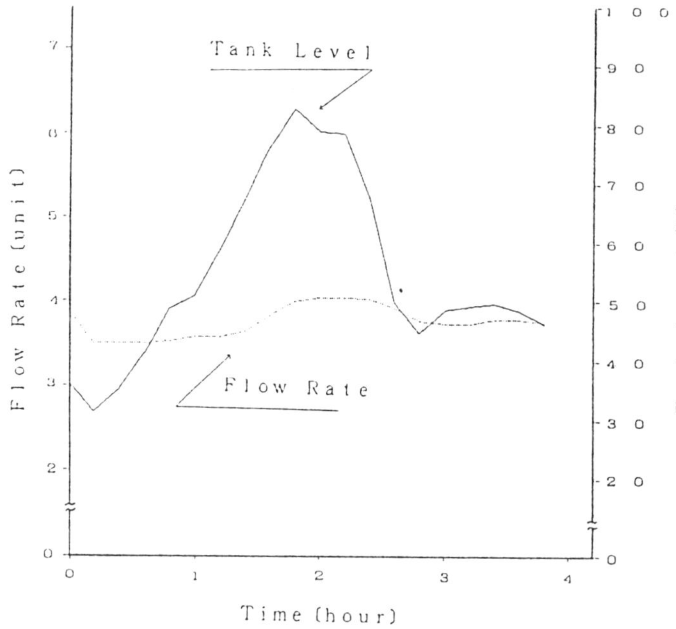


Fig.6. Example of manual operation

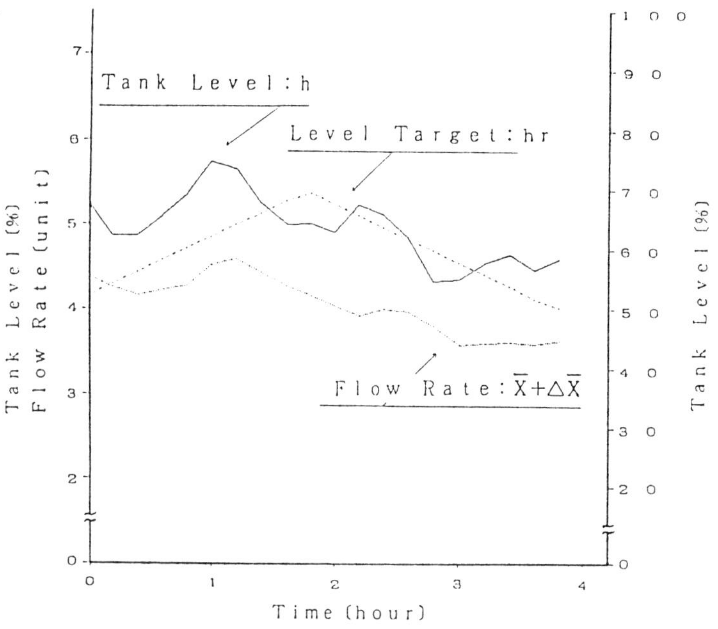


Fig.7. Example of neuro-fuzzy control system

smoothness for the flow rate not only in the steady state but also in the transient state.

Fig.6 shows a result of the manual control when oil A is switched to oil B. The experienced operator has raised the tank level before oil switching. On the other hand, Fig.7 shows a result of the proposed method for the same oil switching. The tank level ranges from 35% to 75% by the operation of proposed method, although it raises from 30% to 80% by manual operation. And the flow rate by the proposed method is as smooth as that by the well-experienced operator.

## 6. Conclusion

A practical control method of neuro-fuzzy hybrid control is proposed. Neural networks are used for estimating the target of fuzzy control.

This method is applied to the tank level control of the real process. The process has two purposes that are contrary each other. The hybrid controller shows its usefulness not only in steady state but also in transient state.

### Reference

[1] M. Sugeno, Ed. : Industrial Applications of Fuzzy Control, North-Holland(1985).

[2] A.Guez, J.L.Elibert and M.Kam : Neural Network Architecture for Control, IEEE Control Systems Magazine, Vol.8, No.2, pp.22-25(1988).

[3] K. Suzuki and Y. Nakamori : Model Predictive Based on Fuzzy Dynamic Models, 4th Inter. Conf. on Process Sys. Engg, Vol.II, pp.18.1-18.15(1991).

[4] T. Tani and K. Tanaka : A Design of Fuzzy-PID Combination Control System and Application to Heater Output Temperature Control, Trans.Society of Instrument and Control Engineers, Vol.27,No.11, pp.1274-1280(1991) (in Japanese).

[5] T. Takagi and M. Sugeno : Fuzzy Identification of Systems and Its Application to Modeling and Control, IEEE Trans. on Sys. Man and Cybernetics, Vol.SMC-15, No.1, pp.116-132(1985).

[6] G. T. Kang and M. Sugeno : Fuzzy Modeling and Control of Multilayer Incinerator, Fuzzy Sets and Systems, Vol.18, pp.329-346(1986).

[7] M. Mizumoto : Fuzzy Controls by Product-Sum-Gravity Method, Advancement of Fuzzy Theory and Systems in China and Japan (ed. by X.H.Liu and M.Mizumoto), International Academic Publishers, pp.c1.1-c1.4(1990)

[8] D. E. Rumelhart, G. E. Hinton, and R. J. Williams : Learning internal representations by error propagation, Parallel Distributed Processing, Vol.1, MIT press, Cambridge, pp.318-362(1986)

# A SIMPLE NEURAL MODEL FOR FUZZY REASONING

José Alberto Baptista Tomé
Instituto Superior Técnico-Universidade Técnica de Lisboa
INESC
R. Alves Redol nº9 , 1000 LISBOA, PORTUGAL

*ABSTRACT*-**A very simple neural architecture for fuzzy reasoning is presented. It is shown that fuzzy rules may be implemented with such nets. The net is layered and the concept of variables and predicates may be associated with areas in those layers. It is the density of activated neurones which defines the membership grades. Fuzzy logic operations are induced in a natural way by the random connections of the neurones from layer to layer. The layered structure of the model, its simplicity and the randomness of the connections makes this model adequate for representing natural systems.**

## I. INTRODUCTION

Binary neural nets have been used for different purposes such as pattern recognition (1,2), image processing(3) and universal logic function construction (4).

Here a very simple binary neural model (that is, where the nodes are Boolean functions) for fuzzy reasoning is presented. It is shown that fuzzy production rules may be implemented by a layered structure of binary neurones, where the neurones are of two types only : those which implement logic AND's and those which implement logic OR's. These processing levels (where the rules are implemented) may interact with an input layer at one end (with sensor neurones) and with an output layer at the other end. It is shown that, if neurones are aggregated in areas associated with the different variable predicates and if the density of active neurones in those areas represent the membership grade of those predicates, the fuzzy logic operations are naturally induced by the atomic logic operations of the neurones. It is the topology of the net (which is organised at a macro level but random at the level of neurone connections) which defines the membership functions of the fuzzy sets associated with the different predicates.

## II. MODEL DEFINITION

### A *Definition of Fuzzy Logic operations*

As will follow in the text, membership grades of fuzzy variables will be associated with densities of activated neurones in given areas or with the probability of finding activated a randomly chosen neurone at that area.

The fuzzy operations defined are then:

The Fuzzy complement is:

$$c(p) = 1-p$$

The axioms for the complement are obeyed:

1. c(o)=1 ; c(1)=0 since 1-0=1 and 1-1=0

2. if pa<pb then c(pa)>c(pb) since 1-pa>1-pb

3. c(c(p))=p since 1-(1-p)=p

The fuzzy AND (t norm) is defined by the probability product pa.pb . This also obeys the axioms for the fuzzy intersection:

1. pa t pb = pb t pa since pa.pb=pb.pa

2. (pa t pb) t pc = pa t (pb t pc) since (pa.pb).pc=pa.(pb.pc)

3. if pa<=pb ; pc<=pd, pa t pc <= pb t pd since pa.pc<=pb.pd

4. p t 0 = 0 , p t 1 = p since p.0=0 and p.1=p

This operation is also Archimedean and strict.

For the fuzzy union (s norm) the definition is :
pa+pb-pa.pb , known as the probability union which obeys also the axioms for fuzzy union:

1. pa<=pb ; pc<=pd, pa s pc <= pb s pd since pa+pc-pa.pc<=pb+pd-pb.pd

2. pa s pb = pb s pa since pa+pb-pa.pb=pb+pa-pb.pa

3. (pa s pb) s pc=pa s (pb s pc) since (pa+pb-pa.pb)+pc-(pa+pb-pa.pb).pc= =pa+(pb+pc-pb.pc)-pa.(pb+pc-pb.pc)

0-7803-0614-7/93$03.00 

4. $p \text{ s } 0 = p$ since $p+0-0=p$

$p \text{ s } 1 = 1$ since $p+1-p=1$

B *Variables*

In our model it is considered that different areas (or regions) are allocated to different variables. These areas are dense regions of binary neurones. Moreover sub-areas are considered, one for each possible fuzzy set associated with that variable (including fuzzy predicates).

It is the average density of activated neurones in each of these areas which defines the membership grade of the variable to the fuzzy set associated with that area.

To visualise this concept, in Fig. 1 three areas are shown with different densities to define the membership grades of the variable "age" to the fuzzy sets "child", "young" and "old" for a person with 20 years of age.

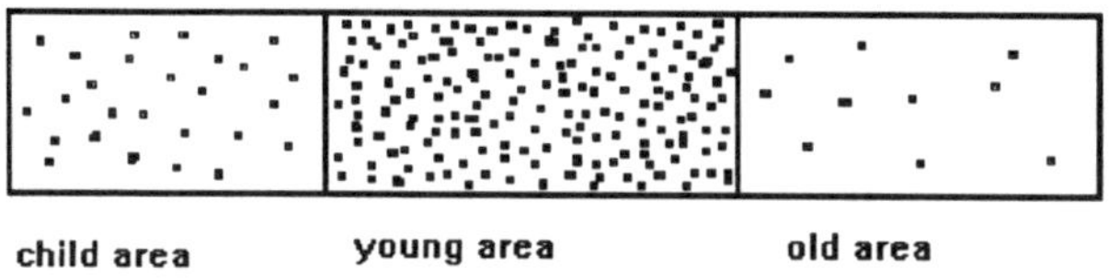


Fig. 1 Neurone activation in different areas

C *Membership-Functions*

Membership functions for each of the fuzzy sets are defined entirely by the topology of connections. In particular one possible model for defining piece wise linear membership functions is presented in the section E.

D *Production Rules*

The implementation of fuzzy rules in this model is a very simple operation. Neurone outputs of the areas associated with the antecedents are linked to the area associated with the consequent. Every neurone output is randomly mapped to an input of a neurone at the reception area. The operations performed in the neurones of the consequent area are simply AND/OR Boolean operations. More specifically, for each neurone in the consequent area each one of its inputs is taken as the output of a randomly chosen neurone from each one of the areas associated with the antecedents.

See Fig. 2 to visualise the following rules (only two neurones per area are shown):

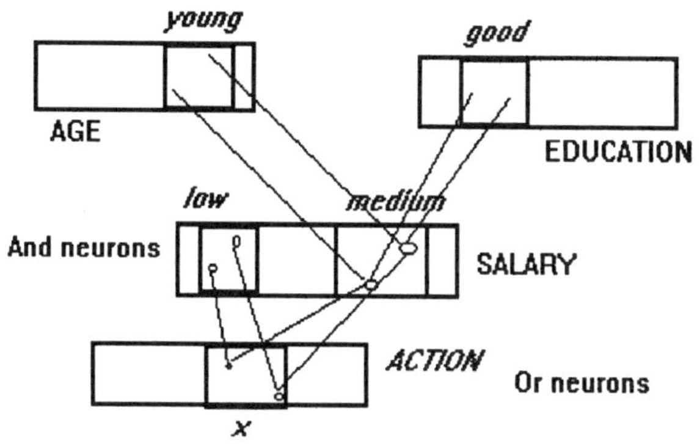


Fig 2 Connections for rule implementation

IF Age is Young AND Education Good THEN Salary Medium

IF Salary Medium OR Salary Low THEN Action X

These type of connections and logic operations induce fuzzy logic AND and fuzzy logic OR operations, (as defined in a previous section) as follows:

The average density of active neurones in a region **i** is defined by $Ni/N$, where $Ni$ is the number of active (output 1) neurones in that area and N is the total number of neurones in it. This is the same as the probability **pi** that a randomly chosen neurone in the area is activated.

The probability that a given neurone in the consequent area is activated (also the activated neurone density in that area) is:

$px = \Pi pi$ for all $i \varepsilon I$, and I is the enumeration set of the antecedent areas.

This is so because the atomic operation in each consequent neurone is a logic AND, and only when every input is a "1" will the AND give a "1".

Then, the above defined (section A) fuzzy logic AND is obtained by the model implementation of antecedent ANDING.

Also, if the atomic operation realised in the consequent area is taken as the logic OR, the probability that one given neurone in that area is activated is ( for antecedent areas i and j):

$$px = pi+pj-pi.pj$$

This is obvious since a neurone in the consequent area is 1 if one one or both of the two inputs are 1.

This operation may be extensible for any number of input neurones (pz = px+pk-px.pk, for antecedent areas i,j and k).

Again, this implements the fuzzy logical OR defined in the section A.

E *Input-Output Layers*

One of the basic assumptions of the model is that neurones are aggregated in areas (or regions) according to predicates of variables (e.g. low, medium, high for the variable temperature), which seems a good representation for fuzzy reasoning but does not seem to agree with the peripheral levels of neural processing. At these levels (input/output) it appears more natural that the intensity of a given variable may be represented by the number of activated neurones in the area associated with the variable (for example, the number of activated neurones proportional to the value or amplitude of the variable).

For this purpose possible input/output layers and interaction with the processing levels (rule implementation levels) are presented.

1) *Input-Level*

At this level it is supposed that there are sets of sensor neurones, with each set responding to different amplitude input values. It is also supposed that the output of these neurones saturate at "1" with those input amplitudes (see Fig. 3 ).

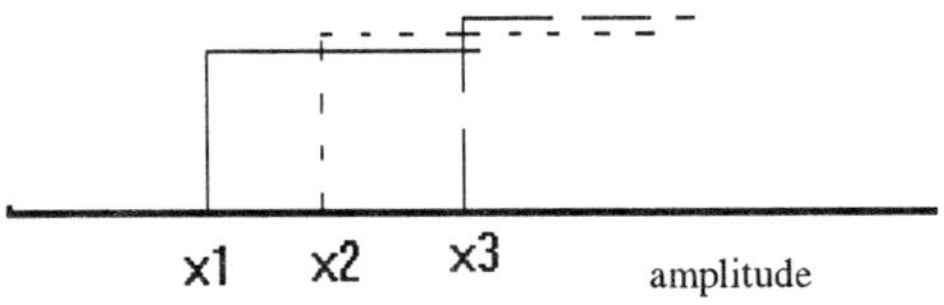


Fig. Neurone outputs with different input levels

If the number of sensor neurones which fire in every Δx ( Δx--> 0 ) is the same, the total number of activated sensor neurones is proportional to the input value of x.

Linear piece wise membership functions may be obtained simply by connecting the outputs of these sensor neurones to the first level of processing neurones - to the area associated with the membership function . This is made trough AND neurones at this processing level. Each neurone at this level has two randomly chosen inputs, one from a neurone in the area associated with the amplitude interval corresponding to the increasing part of the membership function and the other from the decreasing part.

Fig. 4 illustrates this concept. Here sensor neurones which fire at different xi are shown, for the sake of simplicity, in different strip slices.

If Δx is a discrete value, the membership functions are approximated by stairs instead of a continuous slope, which presents no problem if Δx is sufficient low.

2) *Output-Layer*

Similarly, it is considered that the last processing level (always corresponding to a consequent area) is connected to an output level, where the number of activated neurones is proportional to the desired output (this corresponds to the defuzzification process).

A simple mapping is enough, as follows :

Taking K as the number of neurones of each of the areas of the processing level (and associated with a given predicate of one variable), R the number of such areas associated with a given variable ( there is one area per label of the variable ), and fi the number of activated neurones in area i, one randomly maps the neurones of these R areas to different output layer neurones of a total of K.R .

The neurones of area i are mapped in a fraction Xi (Xi<1). Thus the probability of a given output layer neurone to be activated (thus the density of active neurones at that level) is:

$$\Sigma \text{ Xi.fi/K.R}$$

This is true if the mapping is one to one and the output ırones without input (since Xi<1) will remain at 0.

If Xi represents a weight associated with strength of that area i this is a known defuzzification process. Fig. 5 tries to represent this mapping.

At this output level there is , thus, an area per output variable where the number of activated neurones represents the amplitude of that variable, resulting from the processing of the fuzzy rules.

## III Conclusions

The main conclusion of this model is that fuzzy reasoning seems to be according simple neural topologies which have no need of complex computations to achieve their objectives. In fact, all it is needed to achieve the rules implementation is a structured layered net of binary neurones where some organisation is detected (neurone areas associated to the rule antecedents are linked together at an area associated to the consequent of the rule), but

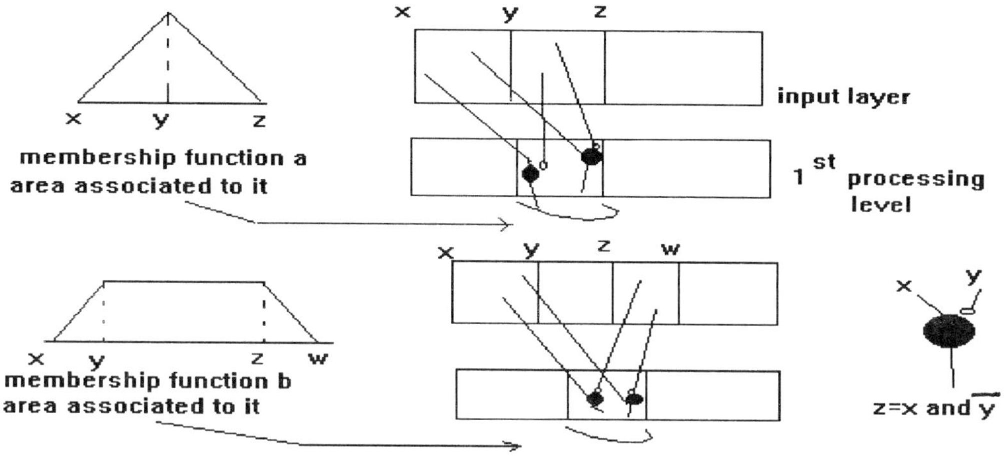


Fig 4 Linear membership functions implementation

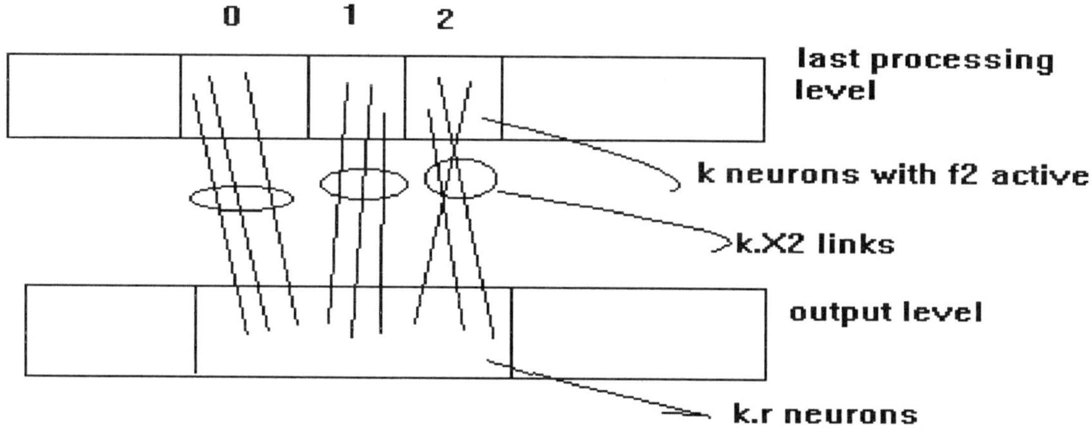


Fig. 5 Connection to output level

where randomness plays the most important place - in fact all neurone connections from area to area are random.

Therefore the model presents very attractive properties in order to represent real natural systems reasoning, among which the simplicity, layered structure and randomness of connections are probably the more important.

REFERENCES

[1] ALEKSANDER,I, "The logic of Connectionist Systems" personal communication

[2] .TOME,J., "Low order arrays for geometric pattern detection"Proceedings of Computer Architecture for Pattern Analysis and Database Management" , Miami, November 1985

[3] .DUFF,M.,FOUNTAIN,T.,"Cellular Logic image Processing", Academic Press, 1986

[4] VIDAL,J.J., "Implementing Neural Nets with Programmable Logic", IEEE Transactions on Acoustics, Speech and Signal Processing; July 1988

[5] .ZADEH,L.A.,"Fuzzy Sets", Information and Control,8,1965

[6] .ZADEH,L.A.,"Fuzzy Logic and Approximate Reasoning",Synthese, 30,1975

[7] .KAUFMANN,A.,Gupta,M., Introduction to Fuzzy Arithmetic : Theory and Applications, New York: Van Nostrand, 1985

[8] .KLIR,G.,FOLGER,T., Fuzzy Sets,Uncertainty and Information, Englewood Cliffs, Prentice Hall, 1988

[9] .MAMDANI,E.H.,GAINES,B.R., Fuzzy Reasoning and Applications, Academic Press, 1981

[10] WITOLD,P., Fuzzy Control and Fuzzy Systems, Research Studies Press Ltd., 1989

[11] TERANO, T., SUGENO, M., MUKAIDONO, M.SHIGEMASU,K., Fuzzy Engineering toward Human Friendly Systems, Proceedings of the International Fuzzy Enginnering Symposium '91, Yokohama,1991

# Fuzzy RCE Neural Network

Sing-Ming Roan, Cheng-Chin Chiang and Hsin-Chia Fu
Department of Computer Science and Information Engineering
National Chiao-Tung University
Hsinchu, Taiwan 300, R.O.C.
E-mail: hcfu@hsinchu.csie.nctu.edu.tw

***Abstract*— This paper proposes a supervised fuzzy neural network called *Fuzzy Restricted Columb Energy* (Fuzzy RCE) network for classification problems. In Fuzzy RCE, each hidden neuron is a fuzzy prototype which can be used to represent one or many training patterns. At the learning stage, the fuzzy membership functions of prototype neurons can be automatically adjusted according to the training data. The simulation results on the handwritten alphanumeric characters show that the proposed model learns very fast and has good recognition performance.**

## 1 Introduction

In these years, neural networks have been more and more attractive for various application domains. In general, a neural network involves only simple arithmetic operations with a simple control structure. They usually can be either implemented on standard sequential machines, or can be easily parallelized on connection machine. Besides, the self-adjusting property of neural networks also makes neural networks more suitable for modeling of unknown systems. Therefore, neural networks have good potential to become feasible tools for information processing.

Simutaneously, the study on fuzzy set theory also gets much attention from researchers [1, 3]. Particularly, the theories of fuzzy sets have a strong impact on techniques of pattern recognition [4]. In comparison with the conventional crisp sets, the concepts of fuzzy sets provides more robust representations to model real-world objects. Therefore, this paper is to integrate the techniques of neural networks with the concepts of fuzzy pattern recognition such that we can develop a more feasible system for classification problems.

Following this introduction is a brief descriptions on the concepts of fuzzy recognition. In Section 3, a neural network called *Fuzzy Restricted Columb Energy* (Fuzzy RCE) network, which combines the fuzzy

0-7803-0614-7/93$03.00 ©1993IEEE

recognition concepts with the RCE neural network model [2], together with its learning procedure and retrieving procedure are proposed. In Section 4, experimental results of the application on handwritten alphanumberic character recognition are presented. The final section provides some concluding remarks on this research.

## 2 Concepts of Fuzzy Recognition

Using the formal definition of fuzzy sets presented by Klir [3], let $U$ denote a universal set. Then, the membership function $\mu_A$ of a fuzzy set A is usually defined as

$$\mu_A : U \rightarrow [0,1], \qquad (1)$$

where $[0,1]$ denotes the interval of real numbers from 0 to 1. In fuzzy pattern recognition, an input is composed of one or many features. The space of each feature is assumed to be an universal set $U$. Within each feature, many fuzzy sets may be defined. For example, each tree has the height feature. The space of a tree's height may be any value in the interval (10cm,1000cm). Thus, the universal set $U$ is the interval (10cm,1000cm). For this feature, we can define four fuzzy sets such as "very high", "high", "short", and "very short". For each fuzzy set, we need to define its membership function over $U$. Thus, given a height, we can compute its corresponding degrees of membership for these four fuzzy sets.

The process of pattern classification is to assign an input to a category. Let $\mathbf{f} = [f_1, f_2, \ldots, f_n]$ be an input vector, where $f_i$'s are its features. In general, a category of objects may contains several representative prototypes. For example, the category of coconut trees generally contains a prototype with properties "*very high*", "*with large fruits*", etc. On the other hand, the category of orange trees may contain a prototype with properties "*short*", "*with small fruits*", etc. For the above two example prototypes, the words 'very', 'large', 'short' have introduce the fuzzy concepts into its features. Thus, we can use fuzzy sets to represent the prototypes of a categories of objects. Now, consider a prototype $P_i$ of a category $C$. Assume that $P_i$ has fuzzy properties in each feature. Let $A_i$'s $(1 \leq i \leq n)$ denotes the fuzzy sets which represents the fuzzy properties. Thus, we need $n$ membership functions, which is denoted as $\mu_{A_i}$, for $P_i$. When an input vector $\mathbf{f}$ is given, all these $n$ membership functions are evaluated according to $f_i$'s. Then, we can define a "goodness of fit" (say $\mathcal{G}_i(\mathbf{f})$) for $P_i$ with respect to $\mathbf{f}$ as

$$\mathcal{G}_i(\mathbf{f}) \;=\; \sum_{j=1}^{n} \mu_{A_j}(f_j). \qquad (2)$$

With the goodness of fit defined on each prototype of each category, we can assign an input vector $\mathbf{f}$ to category $C$, if there exist a prototype $P_k$ of category $C$ such that $\mathcal{G}_k(\mathbf{f}) = \max_i\{\mathcal{G}_i(\mathbf{f})\}$. Of course, we can also reject $\mathbf{y}$ if the maximum goodness of fit does not exceed a prespecified threshold.

The process of fuzzy pattern recognition stated above needs an important step to define the membership functions for each fuzzy properties of prototypes in each category.

Since it may be impossible to known what fuzzy properties a prototype should have, we cannot directly define the membership functions by observing the feature vector of training patterns. Thus, the self-adjusting capability of neural network can be utilized to handle this problem. Through the learning process of neural networks, the prototypes together with their membership functions can be generated automatically for all categories. In the next section, a fuzzy neural network together with its learning and retrieving procedure is proposed.

## 3 Fuzzy RCE Neural Network

RCE [2] network is one of the neural network models which had been used as a pattern classifier for forming nonlinear boundaries of pattern classes. Using the concepts of RCE, each hidden neuron can be used to represent a prototype. The network architecture of the model is shown in Fig. 1. In this network, there are three layers, input layer (I), prototype layer (P) and output layer (O). The neurons of input layer present the feature vector of input pattern. Each neuron in output layer represents one category. As stated earlier, each prototype neuron must define its fuzzy sets together with their corresponding membership functions for fuzzy properties. In our model, all membership functions are triangular functions which are of the form shown in Fig. 2. Each triangular membership function has two parameters $a$ and $b$ which are used to determines the center and the width of this function, respectively.

The triangular membership function can be formally defined as:

$$\mu(x) = \begin{cases} \frac{1}{b}(-|x-a|+b), \\ \text{if } (a-b) < x < (a+b) \\ 0, & \text{otherwise.} \end{cases} \quad (3)$$

Given that the dimension of input feature vectors is $n$, then $n$ fuzzy membership functions ($\mu_{i,j}(x_j)$) should be defined within each prototype. In other words, we need parameters $\{a_{i,j}|1 \le j \le n\}$ and $\{b_{i,j}|1 \le j \le n\}$ for those membership functions in each hidden neuron $i$. However, to simplify the network model, we use only one common $b_i$, instead of $\{b_{i,j}|1 \le ji \le n\}$, in each neuron. Thsu, the internal structure of each prototype neuron is shown as Fig. 3.

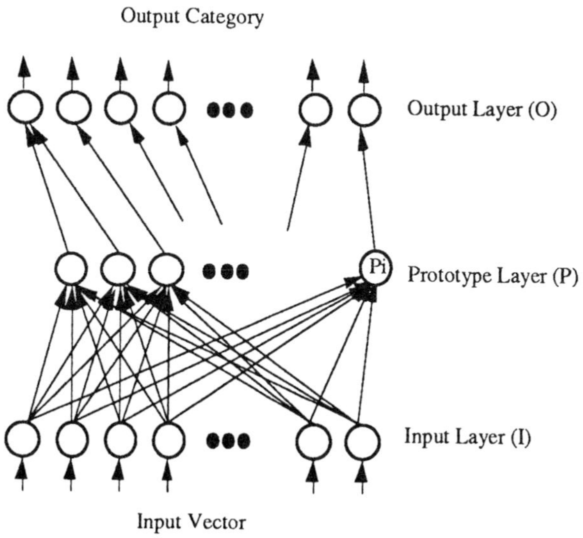


Figure 1: Fuzzy RCE net topology.

### A. Fuzzy RCE Learning

Fuzzy RCE net is a supersived learning neural model. Each pattern presented to the net-

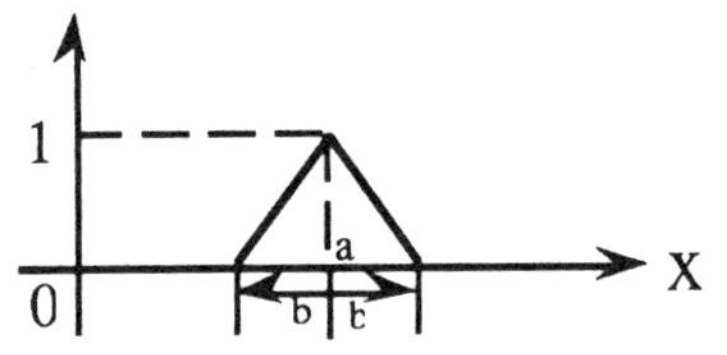


Figure 2: A fuzzy set with triangular membership function.

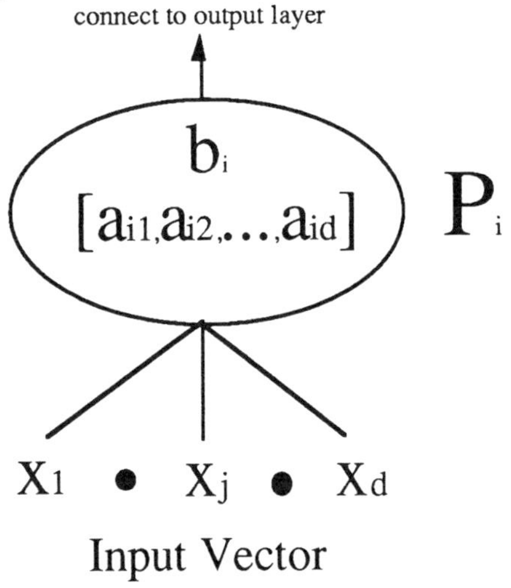


Figure 3: The structure of one prototype neuron in Fuzzy RCE.

work must contain the input feature vector as well as its associated class output. Given a training input vector $\mathbf{f} = [x_1, x_2, \ldots, x_n]$ belonging to the $K^{th}$ class, the learning procedure of fuzzy RCE network is described as follows:

1. calculate the output of all prototype neurons by

$$OUT_{P_i} = \Phi(\frac{1}{n}\sum_{j=1}^{n} \mu_{i,j}(x_j)), \qquad (4)$$

where

$$\mu_{i,j}(x_j) =$$

$$\begin{cases} \frac{1}{b_i}(-|x_j - a_{i,j}| + b_i), \\ if (a_{i,j} - b_i) < x_j < (a_{i,j} + b_i) \\ 0, otherwise \end{cases}$$

$$\Phi(x) = \begin{cases} 0, & \text{if } x \leq TH \\ x, & \text{if } x > TH \end{cases} \qquad (5)$$

The "TH" is a prespecified threshold. If $OUT_{P_i} > 0$, then we say that prototype $P_i$ is "fired".

2. Modify the parameters of prototype neurons according to the following steps.

   (a) *New Classification*
   If none of the prototype neurons in the $K^{th}$ category is fired, then a new prototype $P_j$ which connects to the $K^{th}$ output neuron is created in prototype layer. The parameters of fuzzy set in prototype

$P_j$ are set as follows:

$$\begin{bmatrix} a_{j,1} \\ a_{j,2} \\ \vdots \\ a_{j,n} \end{bmatrix} = \begin{bmatrix} x_1 \\ x_2 \\ \vdots \\ x_n \end{bmatrix} \quad (6)$$

$$b_j = b_0 \quad (7)$$

The $b_0$ is a prespecified initial value for all membership functions.

(b) *Confusion*

If any prototype $P_i$ which does not connect to the $K^{th}$ output neuron is fired, then the parameter $b_i$ is reduced to $b_i'$ such that the prototype $P_i$ cannot be fired. The value of $b_i'$ can be derived by solving the equation $OUT_{P_i} = TH$. Thus

$$b_i' = \frac{\sum_{j=1}^{n} f(x_j)}{(n-k) - n * TH} \quad (8)$$

where

$$f(x_j) = \begin{cases} |x_j - a_{i,j}|, \\ \text{if } (a_{i,j} - b_i) < x_j < (a_{i,j} + b_i) \\ 0, \text{otherwise} \end{cases}$$

$k$ is the number of $f(x_j)$'s such that $f(x_j) \neq 0$, $j = 1, 2, \ldots, n$.

These two rules for prototype acquisition and modification will guarantee the network to learn the geography of the pattern categories.

### B. *Fuzzy RCE Retrieving*

The network retrieving procedure for an input feature vector **f** is described as follows:

1. Calculate the output of all prototype neurons in prototype layer according to Eq.( 4).

2. Find the prototype neuron with the maximum output value. If this found prototype neuron connects to the $K^{th}$ output neuron, then assigned **f** to the $K^{th}$ category. If no prototype neuron is fired, then **f** is rejected.

## 4 Experimental Fuzzy RCE Recognizer and Testing Results

To applied the Fuzzy RCE network to a practical application, an experimental recognizer is built on a PC-386 AT based on the Fuzzy RCE network. This recognizer is used to recognize the handwritten alphanumberic characters. The characters input to this recognizer is through a digitizer. All sampled characters are transformed to their bit array representations. After normalizing the input bit array, the features are extracted from this bit array. The main features utilized for each character are the strokes in eight orientations. The technique for stroke extraction is based on the pixel density in each orientation [5] instead of based on tracing each stroke in the bit array. This approach reduces much computation complexity in feature extraction.

In order to demonstrate the performance of this recognizer, we present the testing results in the following. The experimental is to recognize on-line handwritten alphanumeric characters. We prepared 8 data sets each of which contains $36 \times 5(= 180)$ characters as the database. In this experiment, we try to train the recognizer with different sizes of training sets and evaluate the recognition performance on the remained untrained training set. Table 1 shows the testing results. In Table 1, the training set s1-$i$ denotes a training set which containing $i$ training data sets chosen from the eight training data sets. The second row of this table shows the number of prototypes generated after the training. The third row is the number of iterations used for the training. The fourth row shows the rejection rates. The fifth and sixth rows denote the recognition rate with rejection rate and without rejection rate, respectively.

Table 1: The testing results of the Fuzzy RCE recognizer.

| | s1 | s1-2 | s1-3 | s1-4 | s1-5 | s1-6 |
|---|---|---|---|---|---|---|
| Prototype | 116 | 179 | 216 | 259 | 317 | 364 |
| Iteration | 5 | 5 | 5 | 5 | 6 | 5 |
| Reject | 10.16 | 6.67 | 5.84 | 5.56 | 3.43 | 2.23 |
| Correct_1 | 92.54 | 94.03 | 94.37 | 95.64 | 96.55 | 98.16 |
| Correct_2 | 83.18 | 87.83 | 88.89 | 90.35 | 93.24 | 95.97 |

## 5 Conclusions

A new neural network model, called fuzzy RCE is presented. The advantages of fuzzy RCE include

1. The training time of fuzzy RCE is very short and limited.
2. The membership degree measures has good tolerance for input pattern variety.
3. The calculation of outputs for prototype neurons involves no exponential operations, thus the compuation complexity is reduced.
4. The parameters of network have their physical meaning and can be easily interpreted.

## References


[1] L. Zadeh, "Fuzzy sets," Information and control, Vol.8, pp.338-353, 1965.

[2] D.L. Reilly, L.N. Cooper, C. Elbaum, "A Neural Model for Category Learning," Biological Cybernetics v.45,pp. 35-41,1982.

[3] G.J. Klir, T.A. Folger, "Fuzzy sets, Uncertainty, and Information," Prentice-Hall International,Inc. 1988.

[4] W. Pedrycz, "Fuzzy sets in pattern recognition: Methodology and Methods," Pattern Recog. Vol.23,pp.121-146, 1990.

[5] S.M. Roan, "The study of neural network for on-line handwritten alphanumeric character recognition," Master Thesis, NCTU Taiwan ROC, 1992.

# Fuzzy Artificial Network and its Application to a Command Spelling Corrector

N. Imasaki,* T. Yamaguchi,** D. Montgomery,*** and T. Endo**

*,** Systems & Software Engineering Laboratory, Research & Development Center, TOSHIBA Corporation
70, Yanagi-cho, Saiwai-ku, Kawasaki-shi 210, JAPAN
*** Faculty of Engineering, University of Victoria, CANADA
(* Currently a visiting researcher at University of California, Berkeley, U.S.A.)

***Abstract* - This paper proposes a Fuzzy Artificial Network (FAN) which utilizes associative memories and is constructed by a method which makes it easy to represent and to modify fuzzy rule sets. While conventional fuzzy inference methods induce much fuzziness on multi-layered fuzzy rule sets, the associative memory based FAN results in inferences which fit human sense better. We call this type of fuzzy inference "associative inference." For memorizing fuzzy rule sets, the proposed FAN system employs a correlation matrix which is constructed from a nominal correlation matrix, a bias matrix, and a scale parameter, so that it is easy to carry out refinement and cut-and-paste operations for rule sets. Using a FAN development system, we compose a command spelling corrector which uses a multi-layered fuzzy rule set. The spelling corrector application shows the eligibility of associative inference for multi-layered fuzzy rule sets.**

## 1. INTRODUCTION

In order to compose intelligent systems, not only linguistic (conceptual) value processing but also physical (numerical) value processing is necessary [1],[2]. Fuzzy set theory [3],[4] is known as a suitable theory for an interface between linguistic value and physical value, and has been applied to many actual intelligent systems [5],[6]. In this movement, we are aware that fuzzy systems should advance so that they achieve the following two points: 1) inference which fits human sense better, and 2) easier refinement of rule sets. Toward the realization of these points, many trials to combine fuzzy set theory and neural networks have been performed. We call, in this paper, neural network systems which represent fuzzy rules "Fuzzy Artificial Network (FAN)." FAN systems using feed forward neural network learning techniques are known to be able to refine fuzzy rules automatically (one area of "inductive learning"). On the other hand, FAN systems which employ associative memory networks, i.e., Bidirectional Associative Memories (BAM) [7], result in inferences which fit human sense reasonably well [8],[9], whereas conventional fuzzy inference methods induce results with much fuzziness when the inference is made on a multi-layered fuzzy rule set. We call such an associative memory based fuzzy inference "associative inference" [10].

This paper proposes a FAN system which inherits both the inductive learning and the associative inference. For memorizing fuzzy rule sets, the FAN system employs a correlation matrix which is constructed from a nominal correlation matrix, a bias matrix and a scale parameter. This employment makes it easy to carry out both refinement and cut-and-paste operations for fuzzy rule sets, which is difficult for conventional FAN systems. The FAN system stores the nominal correlation matrix (given by an actual relationship on the fuzzy rule set) instead of a conventional correlation matrix (calculated from input-output data converted to bipolar elements on the fuzzy rule set). As the nominal correlation matrix shows the actual relationships on the fuzzy rule set, it is easy to carry out both refinement and cut-and-paste operations for the fuzzy rule set, and easy to check the memorizing ability of associative memories.

For the purpose to apply the FAN system to various fields, we have developed a design tool called "FAN Development System (FANDeS)." Using the FANDeS package, we composed a command spelling corrector which employs a multi-layered fuzzy rule set. In this application, we mainly show the eligibility of the associative inference for multi-layered fuzzy rule sets. The FAN based spelling corrector introduces two types of recognition: 1) recognizing a correct character in the neighborhood of a keyboard-input character using fuzzy membership functions [11], and 2) recognizing a command (word) as an element of the command line (sentence). The FAN system combines these two types of recognition by simultaneous bottom-up and top-down associative inference processing.

## 2. ASSOCIATIVE MEMORY BASED FUZZY ARTIFICIAL NETWORK

The FAN system which this paper proposes uses associative memories so that it represents a fuzzy rule set as shown in Fig.1. It consists of three elements: 1) if-part, 2) if-then-rule-part, and 3) then-part. The fuzzy rule set represents expert knowledge describing the relationships between "conditions" and "operation models." The if-part has membership functions which abstract and characterize the conditions. The then-part has membership functions or input-output functions which represent the operation models. Associative memories (a kind of BAM) store the relationships between the if-part's conditions and the then-part's operation models.

BAM, which has two layers, memorizes pattern (vector) pairs in terms of a correlation matrix and its transpose matrix. Assume that a BAM memorizes $x$-$y$ pair (each element $\in [0,1]$) and a noisy $x$ is given to the $x$ layer, the BAM recalls

0-7803-0614-7/93$03.00 ©1993IEEE

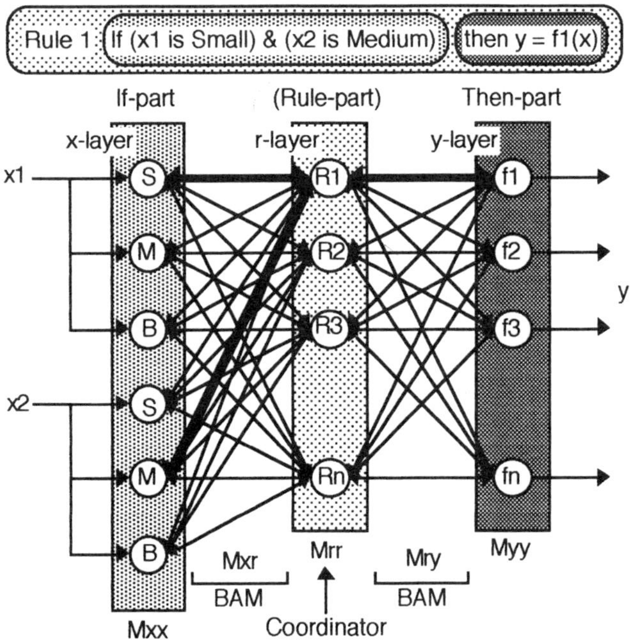


Fig.1 FUZZY RULES REPRESENTATION BY FAN

the $x$-$y$ pair on its two layers. BAM recalls a pair from its memory by reverberation which is performed as follows:

$$Y_t = \phi(M \cdot X_t), \quad X_{t+1} = \phi(M^T \cdot Y_t). \tag{1}$$

In (1), $X_t=[ax_1, ax_2, ..., ax_m]^T$, $Y_t=[ay_1, ay_2, ..., ay_n]^T$ are activation vectors on the $x$, $y$ layer respectively at reverberation step $t$, and $\phi(\cdot)$ is a sigmoidal function of each unit. The correlation matrix $M$ and $M^T$ are given as follows:

$$M = \sum_{i=1}^{N} \beta_i \cdot y_i \cdot x_i^T, \quad M^T = \sum_{i=1}^{N} \beta_i \cdot x_i \cdot y_i^T. \tag{2}$$

In (2), $x_i$, $y_i$ $(i=1 \sim N)$ are memorized pairs, $\beta_i$ is an association parameter.

In conventional associative memory systems, each element of vector $x_i$, $y_i$ $(\in\{0,1\})$ is often converted to bipolar form $\{-1,1\}$ before the calculation of (2) in consideration of an energy function [7]. We denote this conventional correlation matrix using bipolar elements "$M_b$."

BAM recollection by means of reverberation has a feature that the most suitable pair for input conditions appears as a result of decreasing fuzzy entropy which shows fuzziness of inference output.

The FAN system introduces this BAM's feature to control fuzziness at fuzzy inference. A single BAM has not so high memorizing capacity that the system uses two BAM-like associative memories as shown in Fig.1 to represent fuzzy rules. Each unit in $x$ layer represents the if-part membership function given by the fuzzy rule set as conditions. Each unit in $y$ layer represents the then-part operation model. Each unit in $r$ layer represents a fuzzy rule. Two BAMs connect the $x$ layer to the $r$ layer and the $r$ layer to the $y$ layer. The correlation matrices are $M_{xr}$ and $M_{ry}$. The $r$ layer has mutual negative weighted connections between the units in order to recall the most suitable rule for input conditions, and minor

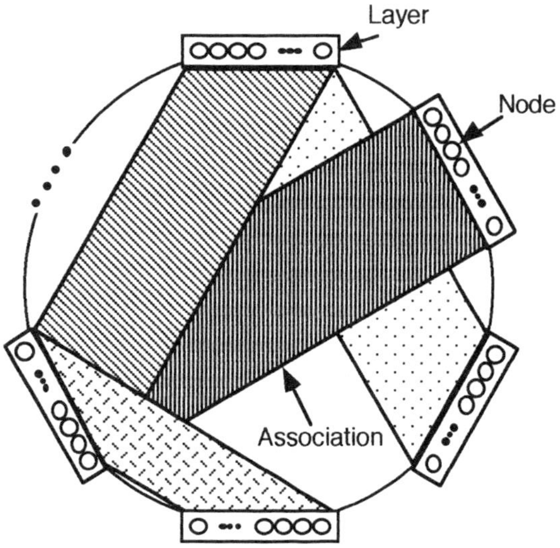


Fig.2 MULTI-LAYERED FAN

positive weighted recessive connections on each unit in order to assist its activation. These connections are represented by the $r$ layer's correlation matrix $M_{rr}$, called the coordinator, which stores fuzzy rules. There are also minor positive weighted recessive connections at $x$ and $y$ layers in order to assist activation. The $x$ and $y$ layers' self correlation matrices are denoted as $M_{xx}$ and $M_{yy}$, respectively.

A reverberation with those correlation matrices($M_{xx}$, $M_{xr}$, $M_{rr}$, $M_{ry}$, $M_{yy}$) and their transpose matrices carries out the associative inference. The associative inference comes up with a weak point of the conventional fuzzy inference, that is fuzziness increasing with respect to the inference output, because the association can regulate the fuzziness. Extraction of activation values before the sufficient regulation provides some imitated fuzzy inference methods, free from worrying about excessive induced fuzziness. We use some activated fuzzy rules to synthesize each output. Suppose that $\gamma_i$ is the normalized activation value of the $i$-th unit in the $y$ layer, the output of fuzzy rule set (which has $n$ output nodes) is given as follows:

$$y^* = \sum_{i=1}^{n} \gamma_i \cdot f_i(x). \tag{3}$$

FAN systems using the conventional correlation matrix $M_b$, are weak in that it is difficult to carry out fuzzy rule cut-and-paste operations and difficult to check the memorizing ability after setting $M_b$. In order to improve this point, we introduce an another correlation matrix $M_e$ which is composed from a normalized version of the nominal correlation matrix $M$, a bias matrix $B$, and a scale parameter $a$ as follows:

$$M_e = a\,(M + B), \tag{4}$$

where

$$a = 1 / (3c), \quad B = \begin{bmatrix} -c & -c & \cdots & -c \\ & & \cdots & \\ -c & -c & \cdots & -c \end{bmatrix}, \quad c = \frac{1}{2} max\ (\ element\ of\ M\ ). \tag{5}$$

The usage of the nominal correlation matrix $M$ which represents actual relationships in the rule set, makes it easy to carry out cut-and-paste operations for fuzzy rules and to check the memorizing ability (which is calculated from the Hamming distance between the memorized pairs) even after setting the correlation matrix $M_e$. The cut-and-paste operations change the dimension of the matrix $M$ according to the number of rules.

If we need a multi-layered fuzzy rule set, the FAN system connects necessary number of layers as shown in Fig.2 [10]. For example, the 1st rule output layer connects to the 2nd rule input layer. The usage of the nominal correlation matrix $M$ also makes it easy to build a multi-layered rule system.

## 3. FAN DEVELOPMENT SYSTEM

A software package which allows us to design associative memories including the FAN system, add as many associations (or layers) as desired, and then load the network into an application program has been developed.

The system, called Fuzzy Artificial Network Development System (FANDeS), treats an associative memory network as a group of connected "layers," each containing neural nodes. The following declaration defines the data structure used for each layer.

```
LAYER {
  char layername[LAYERNAME_LENGTH];
  char numofnodes;
  MATRIX *oldactivations;
  MATRIX *newactivations;
  double inputweight;
  double lambda;
  double coordinator;
  NODE *nodes;
  CONNECTION *connectedlayers;
  LAYER *nextlayer;
};
```

The parameter `layername` is simply the name of the layer in string form. Similarly `numofnodes` describes the number of nodes the layer contains. The pointers `*oldactivations` and `*newactivations` point to matrices (pointer based matrix implementations) of size (1, `numofnodes`) which store the activations of each node in the layer. These pointers are updated during each iteration of the reverberation calculations. The parameter `inputweight` simply defines the weight assigned to the input vector (which is stored as part of the structure `NODE`). This parameter comes into play during the reverberation calculations. The next parameter is `lambda`, which is a parameter used in the sigmoidal function during reverberation calculations. Next is `coordinator`, which defines the layer's self-coordinator matrix. The coordinator can be either positive-negative (positive weighted connections from each node to itself and equal magnitude connections from each node to each other node in the layer), positive-zero (just positive weighted connections from each node to itself), or zero (no coordinator connections). The next parameter is the pointer `*nodes`, of type `NODE`, which is defined as follows.

```
NODE {
  char nodename[NODENAME_LENGTH];
  double input;
  NODE *nextnode;
};
```

As can be seen, the structure `NODE` contains each each node's name, its input value (which is used when the layer's inputweight is nonzero), and a pointer to the next node in the linked list.

The next parameter in the `LAYER` structure is the pointer `*connectedlayers`, which describes the connections between layers. This parameter is also a linked list, so each layer can be connected to as many other layers as the user wants. The final parameter is a pointer to the next layer in the linked list.

The structure `CONNECTION` is defined as follows.

```
CONNECTION {
  LAYER *connectedlayer;
  MATRIX *connectionweights;
  CONNECTION *nextconnection;
};
```

This structure is used as a linked list describing the connections of each layer. When two layers are connected, each layer has a "connection" added to its connection list. Each connection points to the other layer, and they both point to the same connection matrix. The matrix is defined by adding associations between the two layers and then is normalized by (4) and (5) to improve network energy function, thus improving recollections of the memorized pair.

The connection associations are defined by another linked list structure.

```
ASSOCIATION {
  CONNECTION *C1, *C2;
  MATRIX  *V1, *V2;
  double beta;
  char flag;
  ASSOCIATION *nextassoc;
};
```

This structure describes everything about an individual association. The user can define as many associations as desired.

After the network has been built, the user can save it to a file and then write a simple C program to use it. There are functions in the software module which can be linked into a program which allow one to load the network, set input vectors, and make reverberation calculations. The reverberation algorithm is as follows:

```
if first-iteration {
   set all old activations based on input vectors;
}
for (all layers) {
   if (input weight) {
      set new activations based on input;
   } else clear new activations;
   if (coordinator) {
      add (old activations) * (coordinator matrix)
      to new activations;
   }
   for (all connected layers) {
      add (connected layer old activations)
      * (connection matrix) to new activations;
   }
   new activations = φ( new activations );
}
for (all layers) {
   old activations = new activations;
}
```

Essentially this is the algorithm described in (1).

*EXAMPLE:*

Consider the following fuzzy rules.

R1: if (X1 is BIG) and (X2 is BIG) then (Y is NEGATIVE)
R2: if (X1 is BIG) and (X2 is SMALL) then (Y is ZERO)
R3: if (X1 is SMALL) and (X2 is SMALL) then (Y is POSITIVE)

The network in Fig.3 is created using FANDeS. The associations are made with the following vectors:

R1: [1 0 1 0] ↔ [1 0 0] ↔ [0 0 1]
R2: [1 0 0 1] ↔ [0 1 0] ↔ [0 1 0]
R3: [0 1 0 1] ↔ [0 0 1] ↔ [1 0 0].

A simple C program is written to test this network. After loading the network, the program executes a loop which takes as input two real numbers, X1 and X2, assigns layer X input based on these numbers, and then lets the network relax. The layer X input is set by the fuzzy membership functions shown in Fig.4.

Table 1 shows the result of this experiment (after 4 iterations) with various input values.

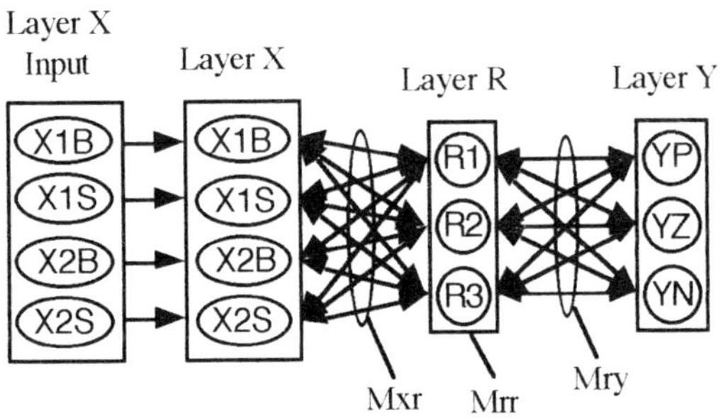


Fig.3. FAN EXAMPLE

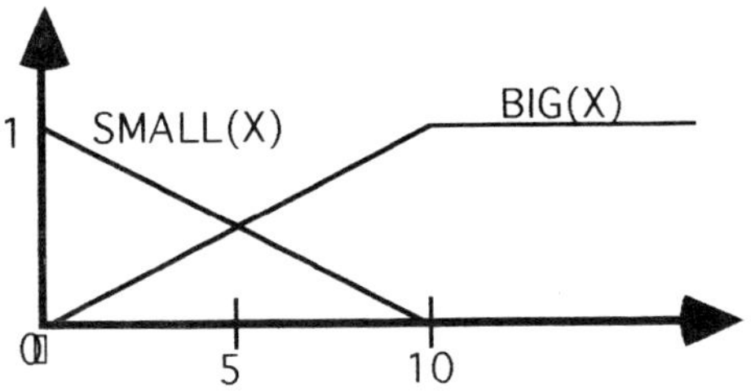


Fig.4. MEMBERSHIP FUNCTIONS FOR SMALL AND BIG

Table 1 EXAMPLE RESULT

| X1 | X2 | X1B-in | X1S-in | X2B-in | X2S-in | YP | YZ | YN |
|---|---|---|---|---|---|---|---|---|
| 10 | 10 | 1.0 | 0.0 | 1.0 | 0.0 | 0.170 | 0.224 | 0.988 |
| 10 | 0 | 1.0 | 0.0 | 0.0 | 1.0 | 0.182 | 0.989 | 0.182 |
| 7 | 3 | 0.7 | 0.3 | 0.3 | 0.7 | 0.183 | 0.989 | 0.183 |
| 0 | 0 | 0.0 | 1.0 | 0.0 | 1.0 | 0.988 | 0.224 | 0.170 |

## 4. AN APPLICATION: COMMAND SPELLING CORRECTOR

A common problem that occurs when typing commands on computers is that when typing quickly, one often makes simple errors that go unnoticed. A common source of these errors is the proximity of keys on the keyboard. For example, the key 's' is very close (i.e., next) to 'd' and 'a.' When typing quickly, sometimes one will press 'd' instead of the intended 's.'

We propose a command spelling corrector based on the fuzzy "closeness" of keys. This concept stems from Araki and Kaguei [11], where a command corrector was constructed based on fuzzy Hamming distance. In this paper, inputted commands are recognized by two factors: 1) the actual inputted character sequence, and 2) the number of arguments (or "words") in the command line (or "sentence").

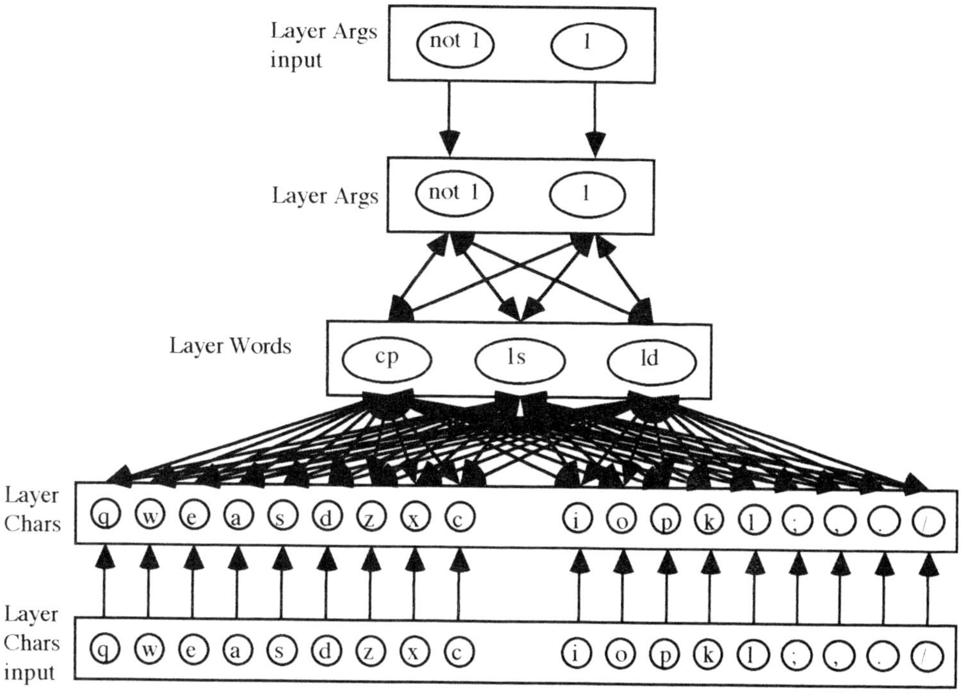


Fig.5 FAN BASED COMMAND SPELLING CORRECTOR

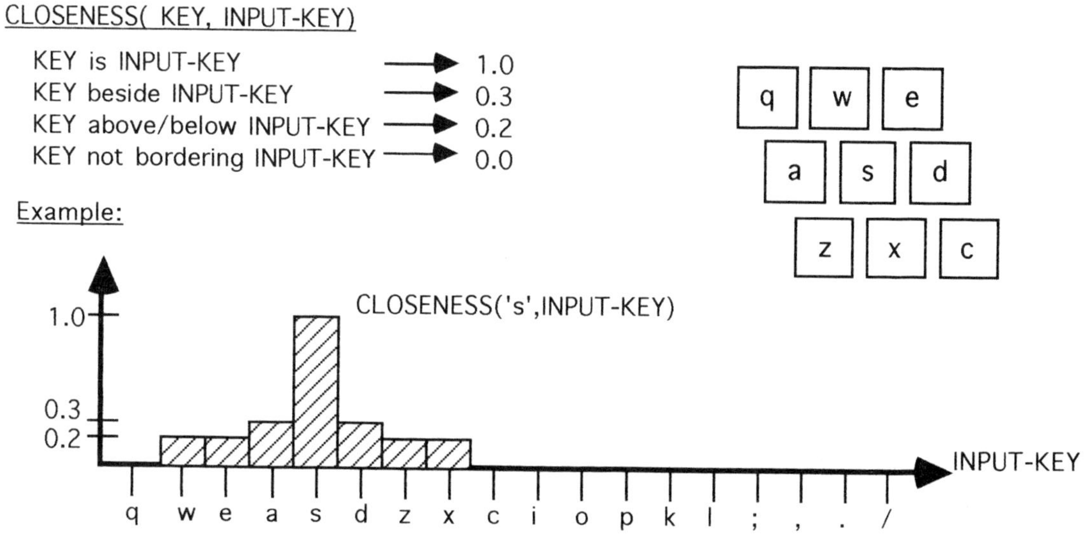

CLOSENESS( KEY, INPUT-KEY)
KEY is INPUT-KEY → 1.0
KEY beside INPUT-KEY → 0.3
KEY above/below INPUT-KEY → 0.2
KEY not bordering INPUT-KEY → 0.0
Example:
q w e
a s d
z x c
CLOSENESS('s',INPUT-KEY)
1.0
0.3
0.2
INPUT-KEY
q w e a s d z x c i o p k l ; , . /


Fig.6 CLOSENESS

A FAN system for this application has two input layer so that rules are multi-layered. The system employs a simultaneous top-down and bottom-up inference. Figure 5 shows the FAN (constructed using FANDeS) for a simple (18 keys, 3 commands) prototype version of the system. A program was written which loads the corrector FAN, sets the input vectors according to keyboard input, and then reverberates the network to make an associative inference as to what the input probably was intended to be. The program sets the input to layer "Args" by counting the "words" in the command line. The input to layer "Chars" is derived from the fuzzy "closeness" of each key with each inputted character in the first "word" of the command line. The membership function for "closeness" is shown in Fig.6. The program was executed for the input strings shown in Table 2. Table 3 shows the results. We built the networks to recognize some UNIX commands. The input "xp a b" is clearly incorrect input - there is no command "xp." The command "ld" exists, but has to have arguments, so one can assume that "ld" by itself was intended to be "ls." The command "la" does not exist, so the input "la a" was probably meant to be "ls a."

As can be seen, the association network successfully recalls the intended command in each case. Successful recollection is assisted by the normalized correlation matrix ***Me***, which improves the energy function of associative memory networks.

Some future enhancements are recommended and are currently being worked on: 1) expansion of the network to include more commands and the full keyboard, and 2) introduction of a Hebbian learning algorithm for the fuzzy membership functions, such that a commonly made error will make a stronger association to the intended key.

It is of note that the network architecture used for this application (simultaneous top-down and bottom-up inference) is also useful in many other applications including image processing (facial pattern recognition) and conceptual fuzzy sets [12]. The multi-layered FAN has shown itself to be a useful and versatile inference technique.

Table 2 TRIAL INPUT STRINGS

| Correct Input | Incorrect Input |
|---|---|
| cp a b | xp a b |
| ls | ld |
| ls a | la |
| ld a | la a |

Table 3 SPELLING CORRECTOR RESULTS

| Input line | cp | ls | ld | Probable command |
|---|---|---|---|---|
| cp a b | 1.000 | 0.025 | 0.034 | cp a b |
| xp a b | 0.999 | 0.077 | 0.117 | cp a b |
| ls | 0.001 | 1.000 | 0.028 | ls |
| ld | 0.000 | 0.997 | 0.754 | ls |
| ls a | 0.002 | 0.995 | 0.861 | ls a |
| la | 0.002 | 1.000 | 0.068 | ls |
| ld a | 0.003 | 0.476 | 0.997 | ld a |
| la a | 0.005 | 0.982 | 0.948 | ls a |

## 5. CONCLUSION

This paper proposed a Fuzzy Artificial Network (FAN) which uses associative memory networks and is constructed by a method which allows easy representation and modification of fuzzy rules. The proposed FAN system brought inference results which fit human sense reasonably well compared to conventional fuzzy inference, which induces much fuzziness when making inference on a multi-layered fuzzy rule set. The FAN system uses a correlation matrix which is derived from the normalization of a nominal correlation matrix and thus it is easy to carry out not only refinement but also cut-and-paste operations on fuzzy rules. These operations are difficult for the conventional FAN systems. Using a FAN development system, we composed a command spelling corrector which used a multi-layered fuzzy rule set. In this spelling corrector application, we mainly explained the usefulness of associative inference for the multi-layered fuzzy rule set using simultaneous bottom-up and top-down inference processing. The corrector corrected each command using information on keyboard input closeness and information on the entire command line (number of arguments).

## REFERENCES

[1] J. Russmussen: Skills, Rules, and Knowledge, "Signal, signs, and symbols, and other distinctions in human performance models," *IEEE Trans. on System, Man and Cybernetics*, vol.13, no.3, 1983, pp. 257-266.

[2] T. Sawaragi, T. Norita and T. Takagi, "Fuzzy theory," *Journal of the Robotics Society of Japan*, vol.9, no.2, 1991, pp. 238-255 (*in Japanese*).

[3] L. A. Zadeh, "Fuzzy algorithms," *Information and Control*, vol.12, 1968, pp. 94-102.

[4] T. Terano, *Fuzzy System Theory and Its Applications*, Ohmu-sya, 1987 (*in Japanese*).

[5] T. Yamaguchi, K. Goto, T, Takagi, K. Doya and T. Mita, "Intelligent control of a flying vehicle using fuzzy associative memory system," *IEEE Int. Conf. on Fuzzy Systems 1992*, vol.2, pp. 1139-1149.

[6] T. Yamaguchi, T. Takagi and T. Mita, "Self-organizing control using fuzzy neural networks," *Int. J. Control*, 1992, *in press*.

[7] B. Kosko, "Adaptive bidirectional associative memories," *Applied Optics*, vol.26, no.23, 1987, pp. 4947-4960.

[8] T. Yamaguchi, N. Imasaki and K. Haruki, "Fuzzy rule realization on associative memory system," *IJCNN-90-WASH-DC Conf. Proc.*, vol.2, 1990, pp. 720-723.

[9] T. Takagi, T. Yamaguchi and M. Sugeno, "Conceptual fuzzy sets," *International Fuzzy Engineering Symposium: IFES'91 Conf. Proc.*, vol.2, 1991, pp. 261-272

[10] S. Yamamoto, T. Yamaguchi and T. Takagi, "Fuzzy associative inference system and its features," *IEEE Int. Conf. on System Engineering*, 1992, *in press*.

[11] H. Araki and S. Kaguei, "Error correction of key-inputted instructions using fuzzy Hamming distance," *Proc. of the 2nd Int. Conf. on Fuzzy Logic & Neural Networks*, 1992, pp. 701-704.

[12] T. Takagi, A. Imura, H. Ushida and T. Yamaguchi, "Multilayered reasoning by means of conceptual fuzzy sets,"*Third International Workshop on Neural Networks and Fuzzy Logic'92*, NASA Johnson Space Center, 1992, *in press*.

# Operator Tuning in Fuzzy Production Rules Using Neural Networks

MIYOSHI Tsutomu, TANO Shun'ichi, KATO Yasunori, Thierry ARNOULD
Laboratory for International Fuzzy Engineering Research
89-1 Yamashita-cho, Naka-ku, Yokohama, 231, JAPAN

*Abstract* — **In production rules, the total matching degree of the condition part is calculated from the matching degree of each condition by aggregation operators. In ordinal production systems, simple logical "and" and "or" functions are used as aggregation operators because the matching degrees are crisp values.**

**In the case of fuzzy production rules, there are several promising approaches for handling the uncertainties of matching degrees: 1) fuzzification of the matching degree of each condition, 2) expansion of aggregation operators, etc. Many studies have been conducted on tuning 1), that use membership functions, but few have been done on tuning 2). Most of the systems use "min" and "max" functions as the aggregation operators in spite of their strong influence.**

**In this paper, we consider automatic operator tuning of parametric T-norms and T-conorms whose characteristics can be modified by parameters.**

## I. Introduction

In production rules, the total matching degree of the condition part is calculated from the matching degree of each condition by aggregation operators. In ordinal production systems, simple logical "and" and "or" functions are used as aggregation operators because matching degrees are crisp values.

In the case of fuzzy production rules, there are several promising approaches for handling the uncertainties of matching degrees: 1) fuzzification of the matching degree of each condition, 2) expansion of aggregation operators, 3) expansion of implication, 4) expansion of combination, etc.

At the Laboratory for International Fuzzy Engineering Research, we have developed an expert system called FOREX (FOReign exchange trade support EXpert system), where expert knowledge are represented by more than five thousand fuzzy production rules [1] . We found it very difficult and important to tune aggregation, implication, and combination functions as well as membership functions.

Many studies have been conducted on tuning 1) which use membership functions, but few have been on tuning 2). Most systems use "min" and "max" functions as aggregation operators in spite of their strong influence.

In this paper, as one step, we consider automatic operator tuning for the parametric T-norms and T-conorms whose characteristics can be modified by parameters.

## II. Production Rule

Let's think about the expansion of production rules using fuzzy technology. By using the ability to handle uncertainty, we can obtain advantages such as fewer rules and easy maintenance. Furthermore, by using approximate reasoning, it becomes possible to get approximate answers to the inputs that are not assumed.

To handle fuzzy production rules, there are several promising approaches to expand the production system: 1) fuzzification of matching degree of each condition, 2) expansion of aggregation operators, 3) expansion of implications, 4) expansion of combinations, etc.

### *A. Membership Functions*

By using membership functions, we can fuzzify the matching degree of each condition which will become a real number or an interval of [0, 1] instead of the crisp values of 0 or 1.

### *B. Aggregation Operators*

The total matching degree of the condition part is calculated from the matching degree of each condition by aggregation operators. Most systems use simple logical "and" and "or", or "min" and "max" functions as the aggregation operators. To choose aggregation operators, we can adjust the matching degree to match human intuition.

### *C. Implications*

Implication functions express the relationship between a condition part and an action part. To choose implication functions, we can adjust the matching degree to match human intuition.

### *D. Combinations*

Combination functions express the relationship between rules that are candidates for firing. To choose combination functions, we can adjust a conclusion to match human intuition.

0-7803-0614-7/93$03.00 ©1993IEEE

### III. TUNING

Investigations have been conducted on the automatic tuning of membership functions using neural networks [2, 3] . However, if a complex relationship exists between the conditions, tuning methods are not able to adjust for errors by tuning the membership functions. If similar characteristics exist in the relationships in some rule blocks, we must be able to efficiently tune the rule blocks by tuning the aggregation operators. But, there are few studies on tuning of aggregation operators or tuning of implications.

From this standpoint, as one step, we considered tuning the aggregation operators. In this paper, we consider automatic operator tuning of parametric T-norms and T-conorms whose characteristics can be modified parametrically.

Using typical T-operators, Gupta and Qi showed that the performance of the fuzzy controller for a given plant depends on the choice of the T-operators [4] . There are various parametric T-operators whose characteristics can be modified parametrically. So we will consider automatic tuning of the fuzzy system by parametric T-operators using back-propagation.

### IV. SYSTEM STRUCTURE

The system consists of two networks: the fuzzy controller network and the plant network. The system structure is shown in Fig. 1.

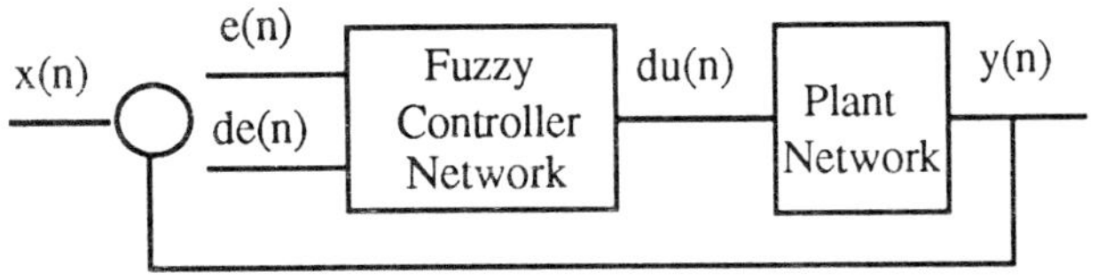


Fig. 1. System Structure.

There have been many investigations on neural networks learning about real plants. First, the plant network learns about target plant, then the fuzzy controller network learns using the plant network. It is assumed that the parameters of plant network are stable while the fuzzy controller network is learning.

#### A. Fuzzy Controller

In general, a fuzzy controller has N rules. The i-th rule can be represented as the fuzzy relation $R_i$ and its membership function is given by

$$\mu_{Ri}(e,de,du) = f(\mu_{Ei}(e), \mu_{dEi}(de), \mu_{dUi}(du)) \quad (1)$$

where e, de and du are linguistic variables for the process error, the change in error and the change in the process input, and their values are fuzzy sets such as large, small or zero, denoted by $E_i$, $dE_i$, and $dU_i$. $f(\bullet,\bullet,\bullet)$ is a general representation of the implication function. Now we use Mamdani's implication function, so the overall fuzzy relation R is given by

$$\mu_R(e,de,du) = \overset{N}{\underset{i=1}{T^*}}\left[\mu_{Ri}(e,de,du)\right] \quad (2)$$

where T[ ] represents a T-norm and $T^*$[ ] represents a T-conorm.

If the actual process error, the actual change in error, and the actual change in the process input take on the fuzzy values E', dE', and dU', then the actual change in the process input dU' can be obtained by:

$$\mu_{dU'}(du) = \sup_{(e,de)} T\left[\mu_{E'}(e), \mu_{dE'}(de), \mu_{dU'}(du)\right] \quad (3)$$

We use a simple fuzzy controller whose response is a function of the T-operator. The following set of control rules is used for the controller.

If e is negative and de is negative, then du is negative.
If e is negative and de is positive, then du is zero.
If e is positive and de is negative, then du is zero.
If e is positive and de is positive, then du is positive.

The following fuzzy sets and membership functions (see Fig. 2 through Fig. 4) are used to represent fuzzy rules.

| | |
|---|---|
| NE: | negative error |
| PE: | positive error |
| NdE: | negative change in error |
| PdE: | positive change in error |
| NdU: | negative change in process input |
| ZdU: | zero change in process input |
| PdU: | positive change in process input |

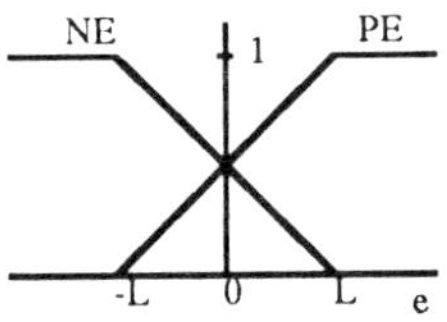


Fig. 2. Membership Function for E.

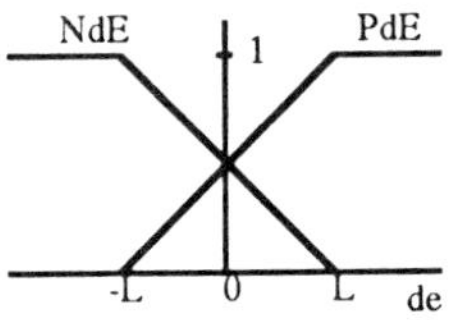


Fig. 3. Membership Function for dE.

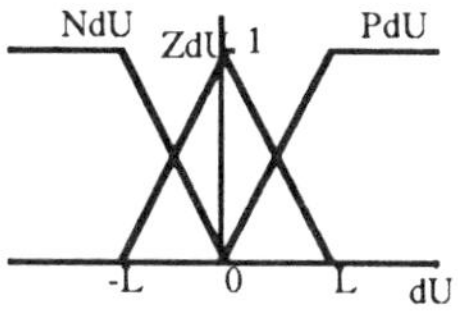


Fig. 4. Membership Function for dU.

The actual error and the actual change in error at any sampling instant are calculated as follows:

$$e(n) = [x(n) - y(n)] \times G_E, \quad de(n) = [y(n-1) - y(n)] \times G_{dE}, \tag{4}$$

where y(n) is the process output at the n-th sampling, $G_E$ and GdE are scaling factors, and x(n) is the reference input.

e(n) and de(n) are always crisp values, therefore, we obtain the following fuzzy process input dU'.

$$\begin{aligned}\mu_{dU}(du) = T^*\big[&T[\mu_{NE}(e), \mu_{NdE}(de), \mu_{NdU}(du),],\\ &T[\mu_{NE}(e), \mu_{PdE}(de), \mu_{ZdU}(du),],\\ &T[\mu_{PE}(e), \mu_{NdE}(de), \mu_{ZdU}(du),],\\ &T[\mu_{PE}(e), \mu_{PdE}(de), \mu_{PdU}(du),]\big].\end{aligned} \tag{5}$$

The crisp change in the process input du(n) is calculated by the Center of Gravity method from dU'.

*B. Error Transmission*

The network representing the fuzzy controller is shown in Fig. 5. The normal lines indicate crisp signals and the bold lines indicate fuzzy sets in Fig. 5. Refer [3] for more information.

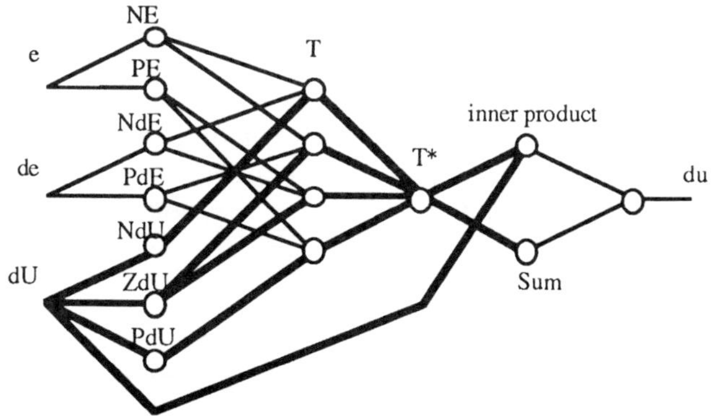


Fig. 5. Network of Fuzzy Controller.

We will show how to tune T and $T^*$ in this network using the back-propagation algorithm.

The system is evaluated by the following error. $G_U$ is scaling factor.

$$\begin{aligned}E_e &= \sum_n \frac{1}{2}\left(e(n)^2 + G_U \times du(n)^2\right)\\ &= \sum_n \frac{1}{2}\left((x(n) - y(n))^2 + G_U \times du(n)^2\right)\end{aligned} \tag{6}$$

The minimization of this error represents the quicker response with a minimum input of energy. We can transmit the error through the plant network as it is done in neural network learning.

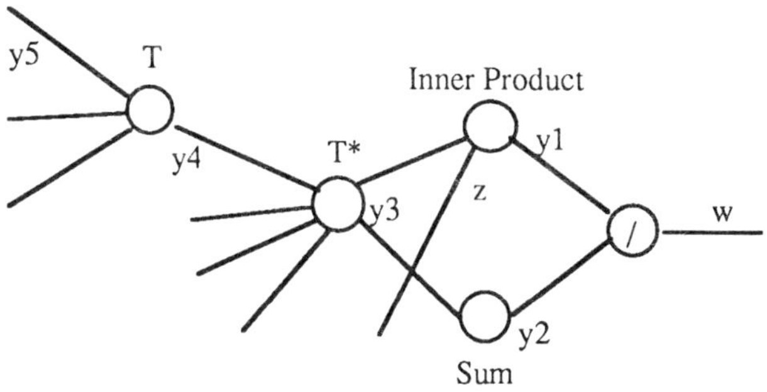


Fig. 6. Output Value of Each Node.

Let's assume that the outputs of each node of the network are $y_1$ through $y_5$, z and w (see Fig. 6). The values of $y_0$ through $y_5$ and w are:

$$w = \frac{y_1}{y_2}, \tag{7}$$

$$y_1 = \sum_z z \cdot y_3, \tag{8}$$

$$y_2 = \sum_z y_3, \tag{9}$$

$$y_3 = T^*(p_{T^*}, y_4, \cdots), \tag{10}$$

$$y_4 = T(p_T, y_5, \cdots). \tag{11}$$

Note that the values of $y_0$ through $y_5$ and w can be differential. Then we can obtain the following from differential equations for each output.

$$\frac{dE_e}{dp_{T^*}} = \frac{dE_e}{dy}\frac{dy}{du}\frac{du}{dw}\left(\frac{dw}{dy_1}\frac{dy_1}{dy_3} + \frac{dw}{dy_2}\frac{dy_2}{dy_3}\right)\frac{dy_3}{dp_{T^*}}, \tag{12}$$

$$\frac{dE_e}{dp_T} = \frac{dE_e}{dy}\frac{dy}{du}\frac{du}{dw}\left(\frac{dw}{dy_1}\frac{dy_1}{dy_3} + \frac{dw}{dy_2}\frac{dy_2}{dy_3}\right)\frac{dy_3}{dy_4}\frac{dy_4}{dp_T} \tag{13}$$

Thus, the automatic tuning of the system becomes possible by using these differential equations.

## V. SIMULATION

Gupta and Qi have produced simulation examples using typical T-operators [4] . Based on their paper, we showed in this paper, that the performance of the fuzzy controller for a given plant depends on the choice of the T-operators. We used two kinds of plants, four types of typical T-operators, and two types of parametric T-operators. First, we will introduce the plants and operators, then present the results of automatic tuning.

### *A. Plants*

The target plants used for tuning are:

- First-order plant: y' + ay = bu
- Second-order plant: y" + ay' + by = cu

where a, b, and c are constants.

### *B. T-operators*

Typical T-operators are shown below:

TABLE 1
TYPICAL T-OPERATORS

| N | $T_N(x, y)$ | $T^*_N(x, y)$ |
|---|---|---|
| 1 | min(x, y) | max(x, y) |
| 2 | xy | x + y - xy |
| 3 | max(x + y -1, 0) | min(x + y, 1) |
| 4 | xy / (x + y - xy) | (x + y - 2xy) / (x - xy) |
| 5 | x, if y = 1<br>y, if x = 1<br>0, otherwise | x, if y = 0<br>y, if x = 0<br>1, otherwise |

Those typical T-operators are ordered as follows:

$$T_5 < T_3 < T_2 < T_4 < T_1, T_1^* < T_4^* < T_2^* < T_3^* < T_5^*. \quad (14)$$

Various kinds of parametric T-operators are introduced in [4] , we selected the Dombi T-operator, that covers a wide range of strengths, and the Hamachar T-operator, that requires less computer power.

- Dombi T-operator

$$T_D(p,x,y) = \frac{1}{1+\sqrt[p]{\left(\frac{1-x}{x}\right)^p + \left(\frac{1-y}{y}\right)^p}} \quad (15)$$

$$T^*{}_D(p,x,y) = \frac{1}{1+\sqrt[p]{\left(\frac{x}{1-x}\right)^p + \left(\frac{y}{1-y}\right)^p}} \quad (16)$$

p > 0,
if p --> 0, then T = $T_5$ and $T^* = T^*_5$
if p = 1, then T = $T_4$ and $T^* = T^*_4$
if p --> ∞, then T = $T_1$ and $T^* = T^*_1$

Note that the Dombi T-operator encompasses all the operators from the logical operator to the drastic operator.

- Hamachar T-operator

$$T_H(p,x,y) = \frac{xy}{p+(1-p)(x+y-xy)} \quad (17)$$

$$T^*{}_H(p,x,y) = \frac{x+y-xy-(1-p)xy}{1-(1-p)xy} \quad (18)$$

p ≥ 0
if p = 0, then T = $T_4$ and $T^* = T^*_4$
if p = 1, then T = $T_2$ and $T^* = T^*_2$
if p --> ∞, then T = $T_5$ and $T^* = T^*_5$

Note that Hamachar T-operator encompasses all the operators from the non-parametric Hamachar operator to the drastic operator.

### *C. Result for the First-order plant*

Fig. 7 shows the results for a fuzzy controller that uses typical T-operators applied to the first-order plant. In Fig. 7, tn1 represents $T_1(x, y)$ and tcn1 represents $T^*_1(x, y)$ of Table 1. This result is the same as the result presented in [4] .

Fig. 8 shows the result of automatic tuning. In Fig. 8, tcnD represents the Dombi T-conorm and tcnH represents the Hamachar T-conorm. Fig. 9 is a close-up view of Fig. 8. To make the figure easy to read, a part of the tn3-tcn3 line is reduced.

The Dombi T-conorm is better than the Hamachar T-conorm because of the range it covers. The asymptotic limit of the parameter for the Dombi T-conorm is 3.063. This means that the strength of the Dombi T-conorm is situated between $T^*_1$ and $T^*_4$. The parameter for the Hamachar T-conorm under flows to 0.0. This means that the strength of the Hamachar T-conorm is $T^*_4$ and the suitable strength is weaker than $T^*_4$.

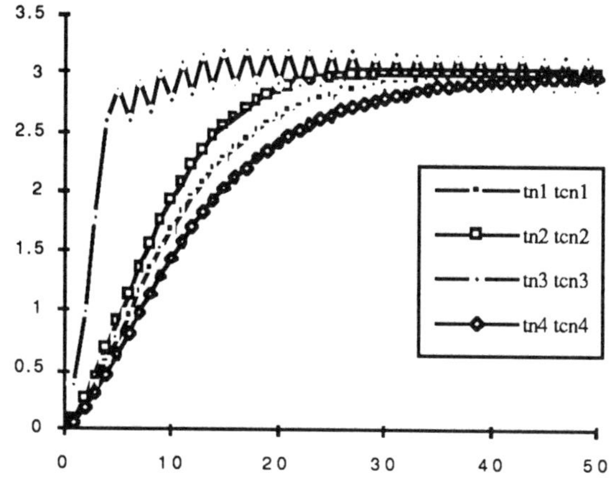


Fig. 7. Results for Typical T-operators.

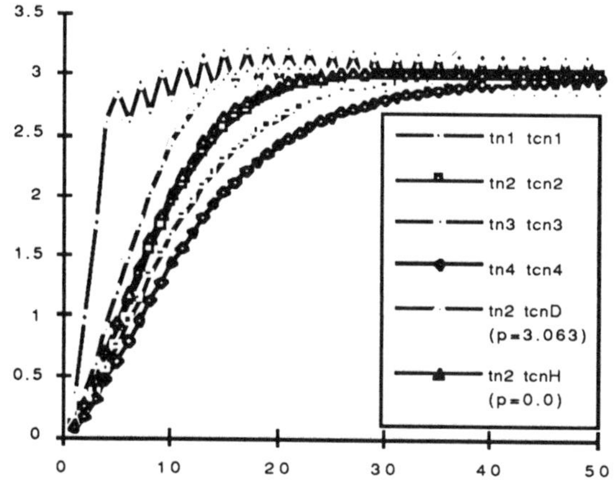


Fig. 8. Results for Parametric T-operators.

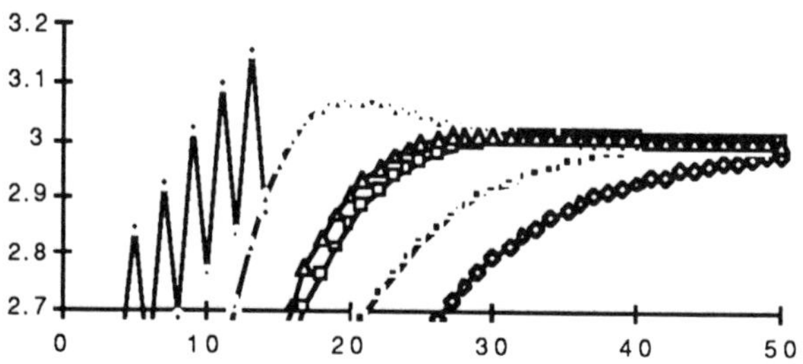


Fig. 9. Close-up view of Fig. 8.

*D. Results for the Second-order plant*

Fig. 10, 11, and 12 illustrate the results for the second-order plant. The asymptotic limit of the parameter of the Dombi T-conorm is 1.3345 and the Hamachar T-conorm under flows to 0.0. The same interpretations for the result of the first-order plant apply here.

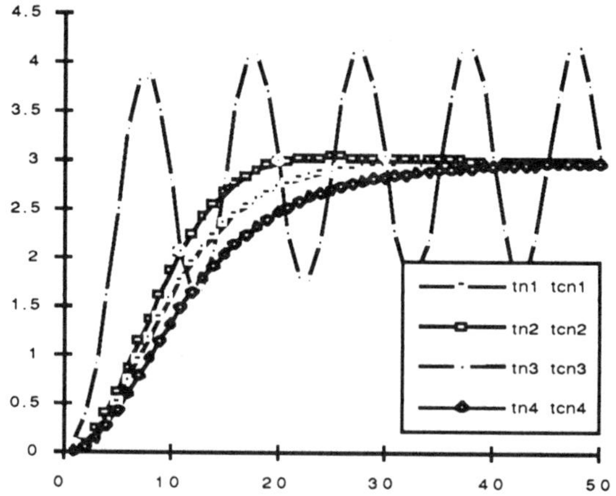


Fig. 10. Results for Typical T-operators.

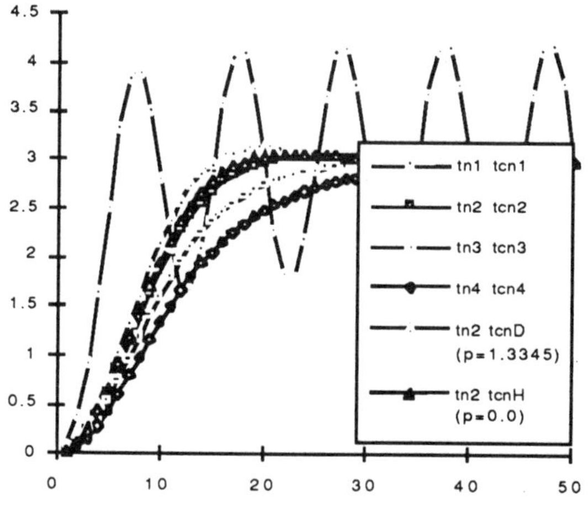


Fig. 11. Results for Parametric T-operators.

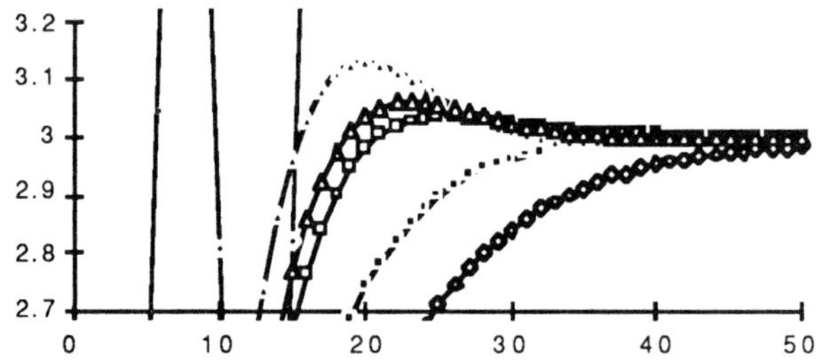


Fig. 12. Close-up view of Fig. 11.

## VI. CONCLUSION

To get the optimum strength for the aggregation operator, it is not sufficient to use only typical T-operators, because the improved strength for the target plant often falls between typical T-operators. Some of the parametric T-operators are advantageous to improve the strength of the operator.

We proposed a system that consists of a fuzzy controller network and a plant network and presented a method for automatic tuning of parametric T-operators.

Through computer simulations, we obtained the optimum strength of T-operators for each kind of plant through automatic tuning of the parameter of the T-operator.

We already presented the results of two classes of parametric T-operators, and will investigate the possibility and the sensitivity of all known kinds of parametric T-operators.

Through the simulations, we found that not only the characteristics of the T-operators, but also the characteristics of differential equations of the T-operators are important in automatic tuning.

We will study the preferred characteristics for parametric T-operators and their differential equations, and the relationship between membership functions and aggregation operators from the perspective of automatic tuning.

We plan to expand the system proposed here to a total tuning environment that can manually and automatically tune membership functions, aggregation operators, and implications.

## REFERENCES

[1] H. Yuize, T. Yagyu, M. Yoneda, Y. Katoh, S. Tano, M. Grabish, and S. Fukami, "Decision Support System for Foreign Exchange Trading - Practical Implementation -", *IFES'91*, Yokohama, JAPAN, pp. 971-982, 1991.

[2] S. Horikawa, T. Furuhashi, S. Okuma, and Y. Uchikawa, "A Fuzzy Controler Using A Neural Network and Its Capability to Learn Expert's Control Rules", *International Conference on Fuzzy Logic & Neural Networks (IIZUKA'90)*, Iizuka, JAPAN, Vol. 1, No. 2, pp. 103-106, 1990.

[3] A. Maeda, R. Someya, S. Yasunobu, and M. Funabashi, "A Fuzzy-based Expert System Building Tool with Self Tuning Capability or Membership Functions", *World Congress on Expert Systems*, USA, pp. 639-647, 1991.

[4] M.M. Gupta and J. Qi, "Design of fuzzy logic controllers based on generalized T-operators", *Fuzzy Sets and Systems*, Vol. 40, pp. 473-489, 1991.

# Neufuz: Neural Network Based Fuzzy Logic Design Algorithms

Emdad Khan Prahlad Venkatapuram

Embedded Systems Division, National Semiconductor, Santa Clara, CA 95052

## Abstract

*In this paper, a novel design of fuzzy logic - Neufuz, using neural net learning is proposed. Artificial neural net algorithms are used to generate fuzzy rules and membership functions. Combination of learned fuzzy rules, membership functions and a new fuzzy design technique based on a new fuzzy inferencing and defuzzification, significantly improves performance, accuracy, reliability and reduces design time. Neufuz also minimizes system cost by optimizing number of rules and membership functions. Simulations show very encouraging results.*

## 1.0 Introduction

Fuzzy logic (one type of approximate reasoning) recently received growing attention from many areas in industry, university and others. Developed by Zadeh [Zade86] in 1965, fuzzy logic has been proven very successful in solving problems in many areas where conventional model based (mathematical modeling of the system) approach is either very difficult or inefficient/costly to implement. Fuzzy logic based design has several advantages including simplicity & ease in design. However, fuzzy logic design is associated with some critical problems as well. As the system complexity increases, it becomes difficult to determine right set of rules and membership functions to describe the system behavior. A significant amount of time is needed to properly tune the membership functions and adjust rules before a solution is obtained. For more complex systems, it may be even impossible to come up with a working set of rules and membership functions. Besides, once the rules are determined, they remained fixed in the fuzzy logic controller i.e controller cannot learn from experience.

Use of neural nets to learn system behavior seems to be a good way to solve above mentioned problems associated with fuzzy logic based designs. Using system's input-output data, neural nets can learn systems behavior and accordingly can generate fuzzy rules [Kosk92] and membership functions. Neural net tries to mimic human brain instead of using approximate reasoning (using rules). Like fuzzy logic, it also uses numerical techniques. In a neural net, many simple processing elements are interconnected together by variable connection strengths. The net learns by appropriately varying the connection strengths. It is a data driven system and does not use programming. By proper learning, neural net can develop good generalization capabilities and thus, can solve many problems that are either unsolved or inefficiently solved by existing techniques.

Clearly, neural nets can solve many complex problems. But it may not be the most effective way to implement it, as today, implementation of neural network is more costly compared to fuzzy logic implementation. A conventional embedded controller can be easily used to implement fuzzy logic by proper programming. Neural implementation by programming is also possible but will be slower. A dedicated hardware implementation is also more common today for fuzzy logic than neural net, especially considering cost. The other problem with neural net solution is its "Black Box" nature - i.e the relation of the changes in weights with the input-output behavior. We do not have a good understanding of the "Black Box" compared to a fuzzy rule based description of the system. Thus, generating fuzzy rules and membership functions based on neural net learning and then a fuzzy based solution using those generated rules and membership functions seems to be the most logical and optimum way to solve complex problems. Expressing the weights of the neural net by fuzzy rules also provides better understanding of the "Black Box" and thus help better design of the neural net itself.

0-7803-0614-7/93$03.00 ©1993IEEE

In this paper, we have presented an elegant scheme to combine the neural net and fuzzy logic. We also presented a new scheme of fuzzy logic design that uses neural net based fuzzy antecedent processing, fuzzy inferencing and defuzzification. The neural net based fuzzy inferencing (or rule evaluation) and defuzzification are described in detail in [Khan92c]. We used the proposed techniques to several applications (including home appliances like vacuum cleaner, coffee maker & shaver) and obtained encouraging results.

## 2.0 Generating Fuzzy Rules and Membership Functions

While for a relatively simple system an expert can easily come up with a working set of rules and membership functions, it is very difficult to do so for a reasonably complex system. As already mentioned, use of neural networks to learn the appropriate set of fuzzy rules and membership functions is very crucial for such complex systems. Fig.1 shows a top level description of a neuro-fuzzy system to generate fuzzy rules and membership functions. The box represent a multilayered feedforward neural network to learn system's input-output behavior by using system's input-output data. We have used a *modified* version of the original back propagation neural net proposed by Werbos and Rummelhart ([Werb74], [Rumm86]). The multilayered neural nets, which we call Fuzzy Rule Generator (FRG), directly maps weights of different layers into fuzzy rules and membership functions. For simplicity, we are using only a 3-layered neural nets (Fig.2) to represent learning of fuzzy rules and membership functions (of a 2-input, one output system) in more detail.

The 1st layer neurons in fig.2 includes the fuzzification whose task is to match the values of the input variables against the labels used in the fuzzy control rule. The 1st layer neurons and the weights between layer 1 and layer 2 are also used to define the input membership functions. The middle layer neurons represent the rule base. We have used multiplication, rather than summation, based neurons in the middle layer. Also, linear neurons with a slope of unity are used for the middle and output layer neurons. Thus, the equivalent error at the output layer is

$$d_k^{out} = (t_k - o_k)\, f'\,(net_k) \quad \ldots\ldots\ldots\ldots\ldots\ldots\ldots\ldots\ldots\ldots\ldots\ldots\ldots\ldots \quad (1)$$

where $o_k$ is the output of the output neuron k
$t_k$ is the desired output of the output neuron k
$f'\,(net_k)$ is the derivative which is unity for the hidden & output layer neurons as mentioned above.

The weight modification equation for the weights between the hidden layer and output layer, $W_{jk}$ is

$$W_{jk}^{\ new} = W_{jk}^{\ old} + \varepsilon . d_k^{\ out} . o_j \ \ldots\ldots\ldots\ldots\ldots\ldots\ldots\ldots\ldots\ldots\ldots\ldots \ (2)$$

where $\varepsilon$ is the learning rate
$W_{jk}$ is the weight between hidden layer neuron j and output layer neuron k
$o_j$ is the output from the hidden layer neuron j

The general equation for the equivalent error at the hidden layer neurons using Back Propagation model is

$$d_j^{hidden} = f'\,(net_j) \sum_{k} d_k^{out} . W_{jk} \ldots\ldots\ldots\ldots\ldots\ldots\ldots\ldots\ldots\ldots\ldots\ldots (3)$$

However, for the fuzzification layer in fig. 2, the equivalent error is different as for the middle layer, the $netp_j$ is

$$netp_j^{hidden} = \Pi\, W_{ij} . o_i \ \ldots\ldots\ldots\ldots\ldots\ldots\ldots\ldots\ldots\ldots\ldots\ldots\ldots\ldots\ldots\ldots (4)$$

where $o_i$ is the output of the input layer neuron i. In the summation type neuron, the derivative only contains the j-th term as other terms are constants yielding zero derivatives. But here, because of the multiplication type neurons, all other terms except the j-th term remain in the derivative as a multiplier.

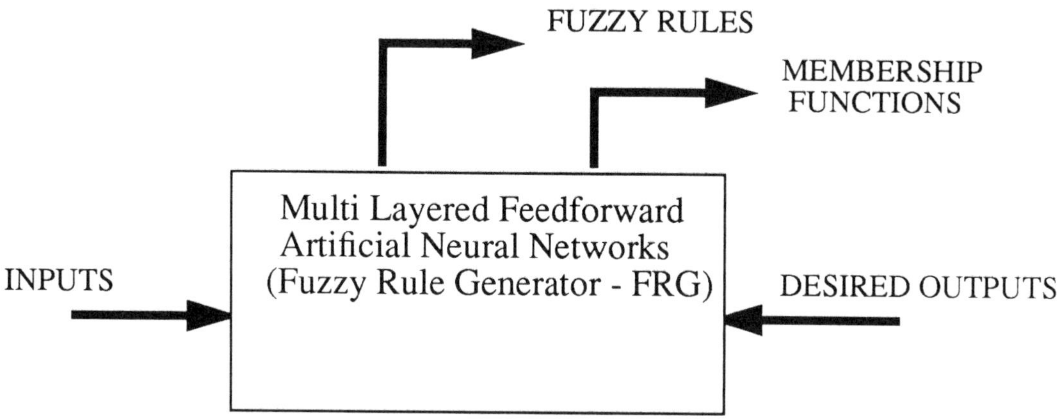


Fig. 1: Generating fuzzy rules and membership functions based on neural network learning. Neural network just uses system input-output data to learn systems behavior and accordingly generates corresponding fuzzy rules and membership functions.

Thus, for the input layer (fig.2) the equivalent error expression becomes

$$d_i^{input} = f'\ (netp_i)\ \sum_j d_j^{hidden}\ .\ W_{ij}\ .\ (\prod_{i \neq k} W_{kj}\ .\ o_k)\ ..............(5)$$

where both i & k are indices in the input layer and j is the index in the hidden layer.

The weight modification equations are equivalent to equation (2). To generate membership functions of different shapes, the input layer in fig.2 is normally replaced with multiple layer of suitable type (including sigmoid) neurons. The weight modification equations accordingly get changed for these additional layers of neurons that learn the membership functions.

As shown in fig.2, the inputs to the middle layer neurons are the preconditions or antecedents of the rules and the output is the conclusion or consequent. Thus, N1,2 can be interpreted as --

"if the input 1 is Low and input 2 is Low then the output is X"

where X can be used as the fuzzy conclusion from rule 1.

From equation (4) above, it is clear that our antecedent processing uses **multiplication** as opposed to **minimum** operation in conventional fuzzy logic design. Clearly for 2-input and one output system we need maximum $3^2 = 9$ rules if 3 membership functions are used. Thus, we need 9 middle layer neurons. The output neuron finally do the rule evaluation and defuzzification.

The neural net (i.e the fuzzy rule generator, FRG) is first initialized with some suitable values of weights that help expedite learning and convergence. After applying a good set of system input-output data for several cycles, the net converges. At this point, we can extract the generated fuzzy rules and membership functions. The inputs should be selected so that it covers the whole input range very well. This will help better learning and hence better rules and membership functions. Also, proper attention should be paid to the learning rate so that the net does not oscillate and converges to a good local minima. The neural net also optimizes the number of rules by proper learning. We can also approximate membership functions to any conventional shape (e.g trapezoid, triangle etc.) for suitable implementation in an embedded processor.Thus, this solution usually costs less compared to conventional method by using less number of rules and better shapes for the membership functions.

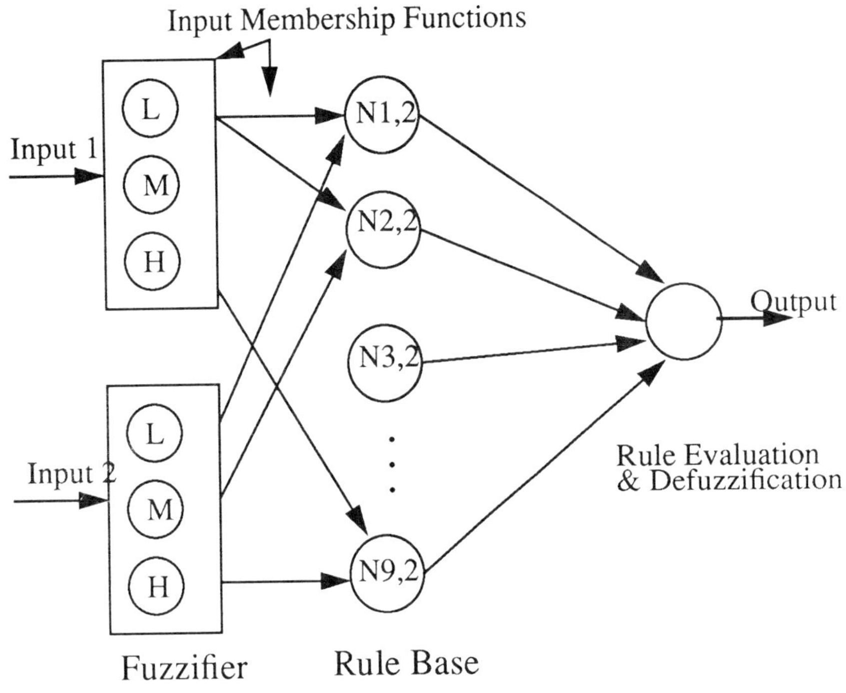


Fig. 2: Learning mechanism of Fig. 1 is shown in more detail. The net is first trained with system input-output data. Learning takes place by appropriately changing the weights between the layers. After learning is completed, the final weights represents the rules and membership functions. The learned neural net, as shown above, can generate output very close to the desired outputs. Equivalent fuzzy design can be obtained by using generated fuzzy rules and membership functions as described in section 4.

## 3.0 Fuzzy System Design Issues in Neufuz

The learned (generated) fuzzy rules and membership functions are properly used to complete fuzzy system design. Since, our fuzzy design is different than conventional, some terms needs further explanation.

### 3.1 Membership Functions

Since we use neural net learning, the learned membership function can have different shapes and size. The maximum value of the membership function may be less than unity. Number of membership functions need to be learned should be specified before learning. If necessary, the shape of the learned membership functions needs to be approximated to fit well with the implementing processor.

### 3.2 Fuzzy Rules

The format of the rules is also different as this also is dictated by neural net. For example, the format of a rule is

"if input 1 is LOW and input 2 is HIGH then output is X" where X is a number.

Thus, instead of using an output membership function, we are using singletons. These singletons are also learned by the neural net. Also, the AND operation of the antecedents is actually a MULTIPLICATION operation as mentioned above. Thus the equation to combine two antecedent is

$$\upsilon_c = \upsilon_a \cdot \upsilon_b \quad \text{......(6)}$$

where $\upsilon_c$ is the membership function of the combinations of the membership functions $\upsilon_a$ & $\upsilon_b$.

### 3.3 Rule Evaluation & Defuzzification

Our rule evaluation (fuzzy inferencing) and defuzzification methods are different than conventional and have been developed based on the equations dictated by neural net learning. To properly match with the neural net, these two steps are combined. This combination does not use any division as opposed to conventional COG defuzzification which uses a division. Thus, our proposed defuzzification saves number of cycles. Also, it is more accurate as it is based on neural net learning where accuracy can be controlled during learning by selecting the desired error criterion. Please refer to [Khan92c] for details.

## 4.0 Verification of Generated Fuzzy Rules and Membership Functions

A fuzzy rule verifier (FRV) has been developed to evaluate the rules and membership functions generated by the neural net (i.e the FRG) for their correctness. It basically uses the methods described in section 3.0, thus making automatic implementation of **Neufuz**. The inputs to this FRV are the Fuzzy Rules set generated by the FRG and the system inputs for which the output needs to be calculated from the above fuzzy rules (Fig. 3). To verify the rules and membership functions, we need to compare the outputs of the FRV with the desired outputs. Some of the desired outputs were used during the neural net training phase. The other desired outputs may be obtained either by more measurements or performing the forward calculation on the Learned neural net which is termed as a Recall operation on the neural net. The FRV includes fuzzification, rule evaluation and defuzzification.

Apart from the rule verification, the FRV provides other important functionalities. All different functionalities/capabilities of FRV are given below:

1. Verify whether the computed output for a set of inputs from the defuzzification process is same as that obtained from the Recall operation. This directly checks for the accuracy of the fuzzy rules and membership functions generated by the neural net.

2. Since the neural net based fuzzy rule generation algorithm has the capability to reduce the number of rules methodically, the FRV facilitates in verifying whether the defuzzified output using lesser number of rules is close to the value obtained from the Recall operation. The lesser the number of rules, the better is the response time of the processor that implements the fuzzy logic design (i.e fuzzification, rule evaluation and defuzzification).

3. Depending on the type of processor that is being used to implement the fuzzy design, it may not be possible to process the complexity in the shape of the input membership function dictated by the neural net. In such cases it is possible to approximate the shape to a more conve nient geometric shape and use FRV to compute the defuzzified output and compare it with the output generated from the Recall operation. In most cases, a good solution (with tolerable results) is still obtained when approximated membership functions are used.

4. User may also like to adjust the generated rules in his particular domain of interest. FRV also let user to perform such adjustments.

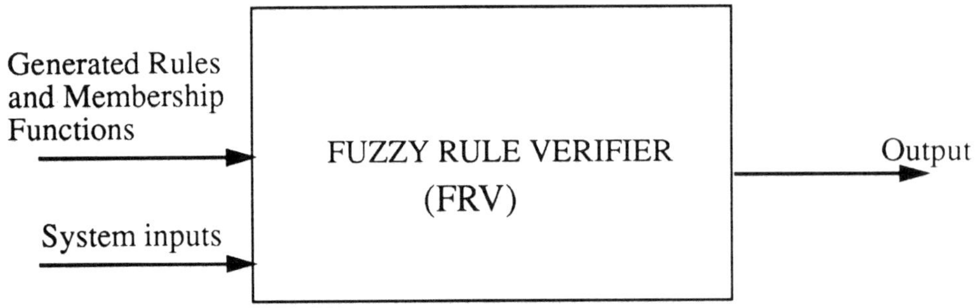


Fig. 3: Fuzzy Rule Verifier (FRV). Mainly verifies generated fuzzy rules and membership functions along with some other functionalities.

## 5.0 Automatic Code Conversion: Towards a Complete Solution

After detrmining the optimum number of rules and membership functions (including their shapes) using FRV, the fuzzy system design can be completed. The completed design is then implemented on a chosen processor. This step can be done manually or automatically. Manual implementation is time consuming. To ease this final step of implementation, we have developed an automatic code converter. This code converter module takes the fuzzy rules and the shape of the input membership functions and generates the appropriate assembly code that can be executed by the chosen embedded processor.

Thus, the Neufuz software essentially takes the user Input-Output data, learns the system behavior, generates the fuzzy rules and shape of the membership functions, allows the user to modify the shape of the membership functions, optimize the number of rules and provides a platform to verify the fuzzy rule set and approximated membership functions, and finally generates an assembly code that can be executed on a chosen embedded processor. This type of system solution will reduce the development cycle for the user for any complex control problems.

## 6.0 Simulations & Results

We have applied our proposed methods to several applications including complex digital & analog functions. The simulations showed very encouraging results. For brevity, we presented here only a highly nonlinear analog data using the function below [ equation (7)].

$$Y = 2.0*X1^3 + 3.0*X2^2 - 1.0 \quad (7)$$

where X2 is assumed to be 2.0*X1 for simplicity.

The corresponding input data provided to the rule generator software (FRG) for learning is shown in Table 1. Table 2 shows important learning parameters used. The learning factor determines learning rates in the inner layers and learning rate corresponds to the output layer only. After the neural net learned the above data to an accuracy factor of 0.008 using 7 input membership functions for each input, the fuzzy rules were generated (Table 3). As we have explained before, the data in Table 3 are numbers (singletons) as opposed to a fuzzy numbers in conventional fuzzy logic. The shape of the Input membership functions for input 1 is shown in fig.4 as generated by the Fuzzy Rule Generator (the shape of membership functions for input 2 is essentially similar as inputs 1 and 2 are related by the equation given above). The number of rules generated were $7^2 = 49$ . The number of rules were optimized to 25 using FRV. The optimization was done by deleting unimportant rules as indicated by small values in Table 3. A set of recall inputs was used along with the generated rules and membership functions and the result is shown in Table 4.

| Learning Rate | Learning Factor | Accuracy Factor |
|---|---|---|
| 0.01 | 0.20 | 0.008 |

Table 2: Critical learning parameters used by the neural net.

| Input 1 X1 | Input 2 X2 | Output Y |
|---|---|---|
| -2.10 | -4.20 | 33.40 |
| -1.70 | -3.40 | 23.85 |
| -1.30 | -2.60 | 14.89 |
| -0.90 | -1.80 | 7.26 |
| -0.50 | -1.00 | 1.75 |
| -0.10 | -0.20 | -0.88 |
| 0.30 | 0.60 | 0.13 |
| 0.70 | 1.40 | 5.57 |
| 1.10 | 2.20 | 16.18 |
| 1.50 | 3.00 | 32.75 |
| 1.90 | 3.80 | 56.04 |
| 2.10 | 4.20 | 70.44 |

Table 1: Input-output data generated by using equation (7) for learning

| Input1 X1 | Input 2 X2 | Comp. Out (49 Rules) | Comp. Out (25 Rules) |
|---|---|---|---|
| -2.100 | -4.200 | 33.321 | 33.321 |
| -1.900 | -3.800 | 28.594 | 28.594 |
| -1.500 | -3.000 | 19.245 | 19.245 |
| -1.100 | -2.200 | 10.805 | 10.805 |
| -0.700 | -1.400 | 4.212 | 4.212 |
| -0.300 | -0.600 | 0.016 | 0.016 |
| 0.100 | 0.200 | -0.836 | -0.836 |
| 0.500 | 1.000 | 2.199 | 2.197 |
| 0.900 | 1.800 | 10.063 | 10.059 |
| 1.300 | 2.600 | 23.697 | 23.694 |
| 1.700 | 3.400 | 43.508 | 43.507 |
| 2.100 | 4.200 | 70.397 | 70.397 |

Table 4: Computed Output from the Fuzzy Rules and Membership Functions using FRV

| | LH | MH | SH | MD | SL | ML | LL |
|---|---|---|---|---|---|---|---|
| LH | 8.79 | 14.58 | 11.36 | 0.04 | 0.01 | 0.01 | 0.01 |
| MH | 14.58 | 3.62 | 9.86 | 0.20 | 0.01 | 0.01 | 0.01 |
| SH | 11.36 | 9.86 | 22.93 | 19.68 | 0.07 | 0.01 | 0.01 |
| MD | 0.04 | 0.20 | 19.68 | -1.63 | 4.46 | 0.66 | 0.07 |
| SL | 0.01 | 0.01 | 0.07 | 4.46 | 9.26 | 1.06 | 0.89 |
| ML | 0.01 | 0.01 | 0.01 | 0.66 | 1.06 | 5.89 | 11.53 |
| LL | 0.01 | 0.01 | 0.01 | 0.07 | 0.89 | 11.53 | 13.47 |

Table 3: Fuzzy Rules - Horizontal is Input2 and Vertical is Input1.e.g. for 8.79 fuzzy output value, rule is: If X1 is LH and X2 is LH then output is 8.79.

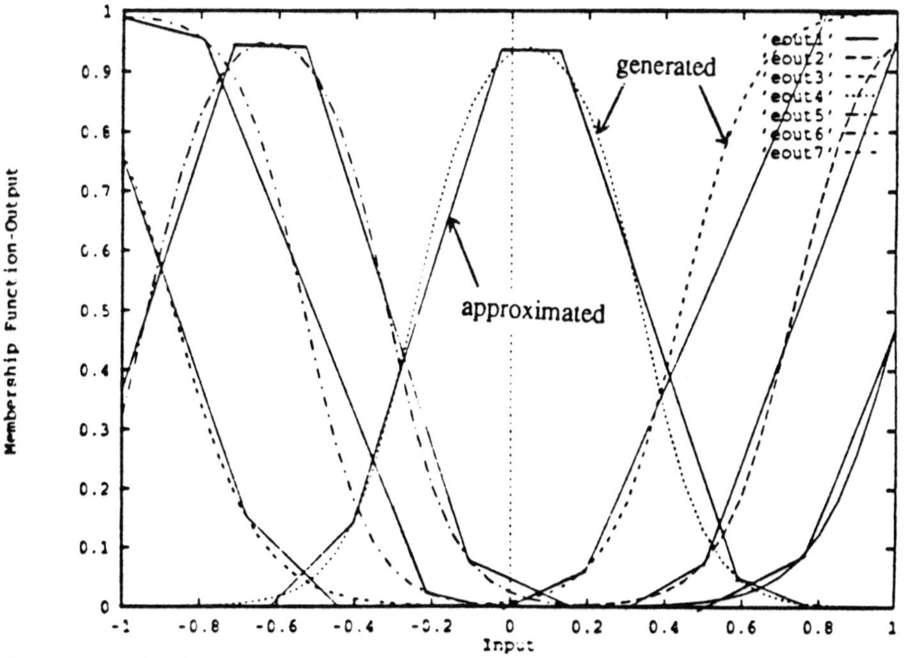


Fig. 4: Input membership function generated by the Fuzzy Rule Generator and approximated shapes.

The following critical points can be derived from the above results:

1. Using all the possible generated rules of 49, the output of FRV (i.e computed by going thru the fuzzy design) is basically equal to the neural net recall values.

2. The computed output values using lesser number of rules of 25 produced almost similar results. This is extremely advantageous as it will take less time (as well as less memory) to compute the output.

3. Further reduction in number of rules is possible but with noticeable errors.

4. By looking at the shape of the membership functions, it is obvious that implementing such shapes on high end processors is not difficult. However, for low end processors, implementation of exact membership function shapes may not be feasible. In such cases, reasonable approximation of the shape of the membership functions to a convenient geometric shape (e.g triangle, trapezoid or combination etc. as shown in Fig.4) would yield reasonably accurate defuzzified outputs. Simpler approximation (like just a triangle or a trapezoid) might result increased error in the defuzzified output.

## 7.0 Conclusion

The need of generating fuzzy rules and membership functions in fuzzy logic based design is emphasized. A method to generate the fuzzy rules and membership functions is proposed based on the concept of neural networks learning. Also presented is an elegant fuzzy logic design based on a new antecedent processing, fuzzy inferencing and defuzzification - also based on neural net learning. The generated rules and membership functions have also been verified for several real world systems input-output data and obtained very encouraging results.

## References


[Kosk92], B. Kosko, " Neural nets & Fuzzy Systems", Prentice Hall, 1992

[Klir88], G. Klir et al , "Fuzzy systems, Uncertainty and Information:, Prentice Hall, 1988

[Zade86], L. Zadeh et al, " Fuzzy Theory and Applications", Collection of Zadeh's good papers,1986

[Bere91] Hamid R. Berenji, "Refinement of Approximate Reasoning Based Controllers by Reinforcement Learning", Machine learning: proceeding of the 8th international workshop, Jun2 27-29, 1991

[Khan92a] E. Khan, "Reinforcement & Unsupervised Learning in Fuzzy-Neuro Controllers", Proceeding of SPIE conference, April, 92

[Khan92c] E. Khan, "Neural net Based Algorithms for Rule Evaluation & Defuzzification in Fuzzy Logic Design", Submitted to the WCNN93.

[Nie92] J. Nie et al , "Fuzzy Reasoning Implemented by Neural Nets", Proceeding of IJCNN92, pp II702-707

[Buck92] J. Buckley et al, "On the Equivalent of Neural Nets and Fuzzy Logic", Proceeding of IJCNN92, pp II691-695

[Haya92] Y. Hayashi et al, "Fuzzy Neural Network with Fuzzy Signals and Weights", Proceeding of IJCNN92, Vol II (Baltimore)

[Werb74] P.J Werbos, "Beyond Regressions: New Tools for Prediction and Analysis in the Behav ioral Sciences", PhD thesis. Harvard University, Cambridge, MA.

[Rumm86] D.E Rummelhart et al, "Learning Internal Representations by Error Back Propagation", Edited by James Anderson and Edward Rosenfield, MIT press, 1988.


----------------------------

# A Neural Network for Fuzzy Decision Making Problems

Roseli Ap. Francelin
USP - ICMSC - SCE
13560 - São Carlos - SP - Brasil

Fernando A. C. Gomide *
UNICAMP - FEE - DCA
13081 - Campinas - SP - Brasil

***Abstract*— Many decision problems can be formulated as fuzzy decision models. In recent years a number of different methods have been proposed to solve fuzzy decision-making problems. This paper proposes an artificial neural network to solve fuzzy decision problems that can be solved by the dynamic programming approach. This approach presents some advantages with regards to alternative approaches because of the inheriting parallelism of neural networks. Some applications are addressed to illustrate the usefulness of the approach proposed.**

## I. Introduction

This paper proposes an artificial neural network to solve fuzzy decision making problems which is a static network that designs his own interconnections and weights. It is organized as follows. First, the decision-making problem type to be solved by the proposed approach is stated in Section II. Next, the neural network's characteristics are presented in Section III and the use of the proposed neural network is illustrated by an application example in Section IV. Finally conclusions and future work are addressed in Section V.

## II. DP and Fuzzy Decision Making Problems

Decision-making problems, particularly in case of problems with non-stochastic uncertainty or ill-formed vagueness, can be formulated as fuzzy decision-making models. Bellman and Zadeh [4] has suggested a different framework to solve problems that can be modeled as fuzzy decision problems. In this framework goals and constraints of a decision problem are treated exactly in the same way. Both are defined as fuzzy sets in the space the alternatives and thus, they can be treated identically in the formulation of a decision.

Suppose that we have $n$ fuzzy goals $G_1, \ldots, G_n$ and $m$ constraints $C_1, \ldots, C_m$. The resultant decision is the intersection of the given goals $G_1, \ldots, G_n$ and the given constraints $C_1, \ldots, C_m$. That is,

$$D = G_1 \cap G_2 \cap \cdots \cap G_n \cap C_1 \cap C_2 \cap \cdots \cap C_m \quad (1)$$

and correspondingly

$$\mu_D = \mu_{G_1} \wedge \mu_{G_2} \wedge \cdots \wedge \mu_{G_n} \wedge \mu_{C_1} \wedge \mu_{C_2} \wedge \cdots \wedge \mu_{C_m} \quad (2)$$

where $\wedge$ symbolizes the min operator.

Let $f$ be a mapping from $X = \{x\}$ to $Y = \{y\}$ and $G_1, G_2, ..., G_n$ fuzzy sets in Y. Given a fuzzy set $G_i$ in Y, one find a fuzzy set $\tilde{G}_i$ in X which induces $G_i$ in Y. The membership function of $\tilde{G}_i$ is given by the equality

$$\mu_{\tilde{G}_i} = \mu_{G_i}(f(x)), \quad i = 1, \ldots, n \quad (3)$$

If $\gamma$ is a constant and $f$ is any function of $x$ we have the identity

$$\max_x(\gamma \wedge f(x)) = \gamma \wedge \max f(x) \quad (4)$$

As an application of problems involving multistage decision making in a fuzzy environment consider the discrete optimal control problem described as follows. Assume that the system under control is a time invariant finite deterministic system whose dynamic satisfy

$$x_{t+1} = f(x_t, u_t) \;\; t = 0, 1, 2, \ldots \quad (5)$$

where the state, $x_t$, at time $t$, $t = 0, 1, 2, \ldots$ ranges a finite set $X = \{ \sigma_1, \sigma_2, \ldots, \sigma_n \}$; the input, $u_t$, ranges over a finite set $U = \{\alpha_1, \alpha_2, \ldots, \alpha_m \}$ and $f : XxU \mapsto X$ is a nonfuzzy and nonrandom function.

Assume also that at each time $t$, the input is subjected to a fuzzy constraint $C_t$, which is a fuzzy set in U characterized by a membership function $\mu_t(u_t)$. Furthermore, assume that the goal is a fuzzy set $G^N$ in X, which is characterized by a membership function $\mu_{G^N}(x^N)$, where $N$ is the time of termination of the process. The decision $D : U\mathbf{x}U\mathbf{x} \ldots \mathbf{x}U \mapsto R$ may be expressed, applying (2), at once as $D = C^0 \cap C^1 \cap \ldots \cap C^{N-1} \cap \tilde{G}^N$ where $\tilde{G}^N$ is the fuzzy set in U x U x ... x U which induces $G^N$ in X. More

*This work was partially supported by CNPq, the Brazilian National Research Council for grant 300729/86-3.

0-7803-0614-7/93$03.00 ©1993IEEE

explicitly, in terms of membership functions, the decision can be expressed by

$$\mu_D(u^0, u^1, \ldots, u^{N-1}) = \mu_0(u^0) \wedge \mu_1(u^1) \wedge \ldots \wedge \mu_{G^N}(x^N)$$

where $x^N$ is expressible as a function of $x^0, u^0, \ldots, u^{N-1}$ through the iteration of (5). The main purpose is to find a sequence of inputs $u^0, \ldots, u^{N-1}$ which $\max_{\mu_D}(u^0, u^1, \ldots, u^{N-1})$.

To obtain the decision, $u_M^0, u_M^1, \ldots, u_M^{N-1}$, the dynamic programming approach can be used since the problem requires a sequence of interrelated decisions. Hence, adopting backward recursion,

$$\begin{aligned} \mu_D(u_M^0, u_M^1, \ldots, u_M^{N-1}) &= \max_{u^0,\ldots,u^{N-2}} \max_{u^{N-1}} \{\mu_0(u^0) \wedge \\ \ldots &\quad \wedge \mu_{N-2}(u^{N-2}) \wedge \mu_{N-1}(u^{N-1}) \\ \wedge &\quad \mu_{G^N}(f(x^{N-1}, u^{N-1}))\} \end{aligned} \quad (6)$$

from the result provided in (4), this may be written as

$$\begin{aligned} \mu_D(u_M^0, u_M^1, \ldots, u_M^{N-1}) &= \max_{u^0,\ldots,u^{N-1}} \{\mu_0(u^0) \wedge \ldots \\ \wedge &\quad \mu_{N-2}(u^{N-2}) \wedge \mu_{G^{N-1}}(x^{N-1})\} \end{aligned} \quad (7)$$

where

$$\begin{aligned} \mu_{G^{N-1}}(x_{N-1}) &= \max_{u_{N-1}} \{\mu_{N-1}(u^{N-1}) \\ &\wedge \mu_{G^N}(f(x^{N-1}, u^{N-1}))\} \end{aligned}$$

may be regarded as the membership function of a fuzzy goal at time $t = N-1$ which is induced by the given goal $G^N$ at time $t = N$. Thus, the following set of recurrence equations is obtained:

$$\begin{aligned} \mu_{G^{N-v}}(x^{N-v}) &= \max_{u^{N-v}} \{\mu(u^{N-v}) \wedge \mu_{G^{N-v+1}}(x^{N-v+1})\}; \\ x^{N-v+1} &= f(x^{N-v}, u^{N-v}) \end{aligned} \quad (8)$$

$v = 1, \ldots, N$ which yields the solution to the problem.

## III. A Neural Network for Fuzzy Decision Making Problems

Artificial neuron models are conventionally nonlinear units with a number of inputs and a single output response. These neurons possess a single nonlinear function to provide an output. The signum, threshold output and sigmoid functions can be used as the nonlinear function. The approach developed, as oposed to the usual nonlinear neuron, is based on a generalized neuron model, where the max and min neurons are particular instances. The max and min neurons are reviewed below, because they are essential in building the FDPNN - fuzzy dynamic programming neural network, proposed in this paper. Generalized neuron models are introduced in [7, 8, 9].

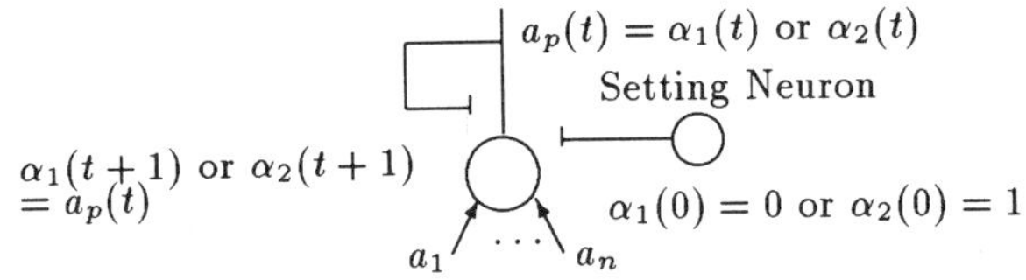


Figure 1: Recurrent Neuron

### A. *Max and Min Neural Processing*

The artificial neuron is usually assumed to be a computational device which: power averages its $n$ inputs $a_i$, $v = \sum_{i=1}^{n} w_i a_i$, according to the weights of the synapses linking the input (pre-synaptic) neurons $n_i$ the post-synaptic neuron $n_p$, and recodes $v \in V$ into the axonic activation $a_p$ of the post-synaptic neuron:

$$a_p = \begin{cases} f(v) & \text{if } v \geq \alpha \\ 0 & \text{otherwise} \end{cases}$$

where $\alpha$ is the axonic threshold and $f$ is the enconding function or transfer function. If $f(v) = 1$ for all $v \geq \alpha$ the neuron is a **binary neuron**, or a McCulloch and Pitts neuron. Otherwise, $f : V \mapsto [0, 1]$ two axonic thresholds $\alpha_1$ and $\alpha_2$ may also be defined such that:

$$a_p = \begin{cases} 1 & \text{if } v \geq \alpha_2 \\ f(v) & \text{if } \alpha_1 \leq v \leq \alpha_2 \\ 0 & \text{otherwise} \end{cases} . \quad (9)$$

The values of $\alpha_1$ and $\alpha_2$ in many applications are furnished by a special type of neuron, called bias cell.

A modification of the McCulloch-Pitts neuron related with the axonic threshold control has been proposed by Gomide and Rocha [7, 8] to define a special type of neuron, called recurrent neuron. The recurrent synapsis is established if the axon of the neuron $n_j$ makes contacts with dendrites or the cell body of $n_j$ itself (Figure 1). If the recurrent synapsis is located near the axon, then it may control **the axonic threshold** as a function of the $n_j$'s activity itself.

**Max-Neuron** (Figure 2): is defined if the axonic threshold $\alpha_1(t)$ at $t = 0$ in (9) is set as 0 by the setting neuron, and then at $t$ it is set equal to the firing level $a_p(t-1)$ at $t-1$:

$$\alpha_1(t) = a_p(t-1); a_p(t) = \begin{cases} \alpha_1(t) & \text{if } v(t) \leq \alpha_1(t) \\ v(t) & \text{otherwise} \end{cases} \quad (10)$$

where $v(t)$ is the post-synaptic activation at t. In this condition, the output $a_p(t)$ of the neuron at t encodes

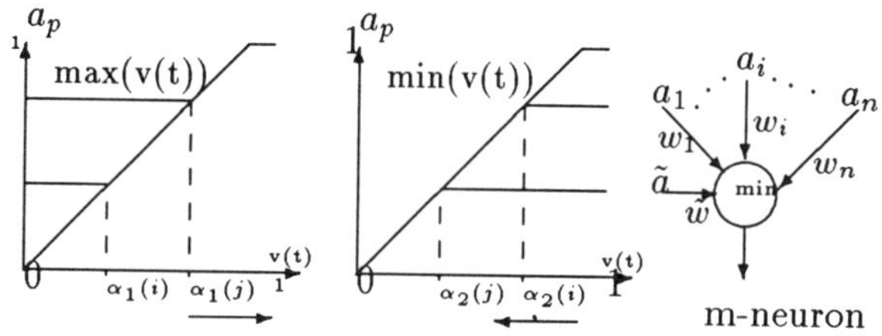


Figure 2: Max, Min Processing and m-neuron

(Figure 2): $a_p(t) = \bigvee_{k=1}^{t}(w_k a_k)$. If $w_k = 1$ for all $k$:

$$a_p(t) = \bigvee_{k=1}^{t} a_k \tag{11}$$

where $\bigvee$ is the maximum operator.

**Min-Neuron** (Figure 2) can be defined analogously if an axonic threshold $\alpha_2(t)$ at $t = 0$ is set as equal to 1 by the setting neuron, and then at $t$ is set equal to the firing level $a_p(t-1)$ at $t-1$.

A **m-neuron** (Figure 2) is a neuron that receives input signals $a_i$ which are synchronized by the same mecanism adopted by Min-Neuron (or Max-Neuron) to provide an output $s$. The m-neurons's output $s$ is given by:

$$s = \mu * w \ \wedge \ w_i * a_i$$

where $\mu$ is a bias input (previously fixed value), $a_i$ are values received from other neurons, $w$ and $w_i$ are the correspondly weights. In the FDPNN, $s = \mu \wedge a_i$ because $w = w_i = 1$.

A **M-neuron** is a Max-Neuron previously defined. In the FDPNN to be derived in the next section, the m-neuron's inputs will have the following meaning, assuming without loss of generality, a backward scheme. Let: $X_k = \{\alpha_k^1, \alpha_k^2, \ldots, \alpha_k^p\}$ be the set of the discrete values that the decision variable $x_{N-k}$ can assume at stage $N-k$, $k = 1, 2, \ldots, N$; $Y_k = \{\sigma_k^1, \sigma_k^2, \ldots, \sigma_k^r\}$ be the set of the discrete values that the state variable $y_{N-k}$ can assume at stage $N-k$, $k = 1, 2, \ldots, N$;

$$\begin{aligned}(XY)_k^j &= \{(\alpha_k^i, \sigma_k^j), \alpha_k^i \in X_k, \sigma_k^j \in Y_k \\ &\quad / \ f_{N-k}(\alpha_k^i, \sigma_k^j) \text{ exists}\},\end{aligned}$$

where $f_{N-k}$ is the fuzzy function given by (5); $n$ be the number of elements in the set $(XY)_k^j$. Thus, considering $n$ input pairs in the $(t^{th})$ m-neuron in a given layer:

$$(\mu^t, a_1), (\mu^t, a_2), \ldots, (\mu^t, a_n)$$

we have that: $\mu^t$: corresponds to the value provided by the membership function $\mu$, calculated for the $(t^{th})$ value of variable $x_{N-k}$ and $a_i$ are the inputs received from other neurons.

*B. Connections of the DPNN Neurons*

The FDPNN is composed by two types of neurons: m-neurons and M-neurons, both defined in the previous section. The layers of the DPNN alternate between m-neurons layers and M-neurons layers, denoted here by m-layers and M-layers, respectively. The first network layer is a M-layer denoted by M-layer-0. The $(i^{th})$ layer of the FDPNN is either m-neurons layer, denoted by m-layer-i or a M-neuron layer, denoted M-neuron-i.

The concepts of transmitters, receptors and controllers are fundamental to describe biological neurons learning and dynamics. A detailed study of these concepts is provided in [9]. The FDPNN is constructed based on those concepts, but in what follows the amounts of transmitters, receptors and controllers are interpreted as quantities defined by functions $q(.)$ associated with the return functions components and the constraints of the optimization problem (see [10] for a detailed description).

The connections (weights) from layers of m-neurons to layers of M-neurons, are assigned, a priori, based on the state transition table, as follows:

$$W(m_k^i, M_k^j) = 1 \text{ if } f_{N-k}(q(R(m_k^i)), q(T(M_k^j))) \text{exists.} \tag{12}$$

where $W(m_k^i, M_k^j) = 1$ denotes the connection between the $(i^{th})$ m-neuron of the m-layer-k and the $(j^{th})$M-neuron of the M-layer-k, whereas the connections from layers of M-neurons to layers of m-neurons are dynamically determined by the own network.

*Designing the connections between M-layer-(k-1) and m-layer-k neurons (Figure 3)* Consider the $(i^{th})$m-neuron in the m-layer-k and the amount of its receptor named $q(R(m_k^i))$. In the FDPNN, this neuron receives a amount $q(T(M_k^j))$ transmitted by $(j^{th})$M-neuron of the M-layer-(k). The combination of these quantities, activates the controller of this m-neuron. The controller function, in the FDPNN proposed here, is defined to be the transformation equation in (8) or state transition function $f_{N-k}$ in (5), at the stage $N-k$, i.e.,

$$\begin{aligned} q(C(m_k^i)) &= y_{N-k} \\ &= f_{N-k}(q(R(m_k^i)), q(T(M_k^j))) \end{aligned}$$

where the pair $(q(R(m_k^i)), q(T(M_k^j)))$ corresponds to $(\alpha_k^i, y_k^j) \in (XY)_k^j$. Since the controller may participate in specification and regulation of the transmitter release at pre-synaptic terminal, it exerts an action in the m-neuron which transmits a amount $q(C(m_k^i))$ to all M-neurons of the M-layer-(k-1). This action will activate M-neurons of the M-layer-(k-1) (pre-synaptic cell), i.e., the $(t^{th})$ Y-neuron of the Y-layer-(k-1), for which

$$q(R(M_{k-1}^t)) = q(T(m_k^i)) = q(C(m_k^i))$$

Thus, the degree of matching $\mu(T(m_k^i), R(M_{k-1}^t))$ between the transmitter $T(m_k^i)$ and its $R(M_{k-1}^t)$ is a function of the amount of the controller. The connections (binding) between neurons of the M-layer-(k-1) and the m-neurons of the m-layer-k, are established as follows:

$$W(M_{k-1}^t, m_k^i) = \begin{cases} 1 & \text{if } q(T(m_k^i)) = q(C(m_k^i)) \\ & = q(R(M_{k-1}^t)) \\ 0 & \text{otherwise} \end{cases} \tag{13}$$

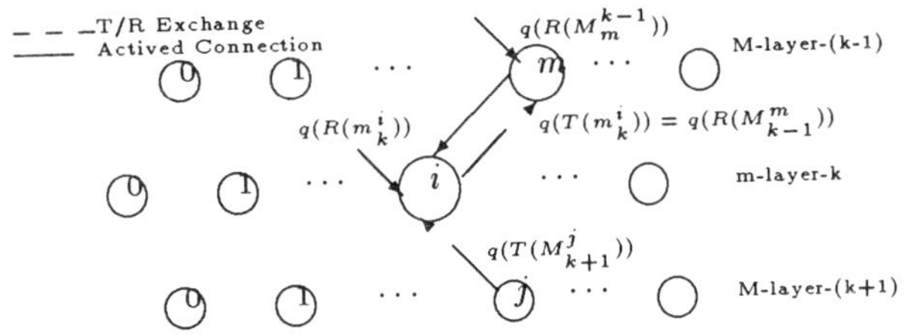


Figure 3: A connection between M-neuron to m-neuron

where $W(M_{k-1}^t, m_k^i) = 1$ denotes a connection between the $(t^{th})$M-neuron of the M-layer-(k-1) and the $(i^{th})$m-neuron of the m-layer-k.

*C. DPNN Construction and Processing Algorithm*

Denote: $X_k$ and $Y_k$ be the sets as defined in the previous section;

$a_i(.)$'s be the inputs to the M-neuron of the M-layer-k;

N be the number of stages; $s$ be the output of a neuron in the network;

$\mu_k = \mu_k(.)$ as the membership function of the constraint at stage $N-k$; $k = 1, \ldots, N-1$;

$\mu_N = \mu_N(\sigma_N^i)$ as the membership function of the goal at stage $N$;

$f_{N-k}$ as the state transition function at the stage $N-k$; $k = 1, \ldots, N$.

**Initialization Pass**
**For** each $(i^{th})$ Neuron in the layer-0 **do**
  read$(q(R(M_0^i))) = y_N^i \in Y_0$
  $s(M_0^i) = \mu_N(q(R(M_0^i)))$
**For** each m-layer-k **do**
  **For** each $(i^{th})$m-neuron
    read( $q(R(m_k^i))) = x_{N-k}^i \in X_k$
**For** each M-layer-k **do**
  **For** each $(i^{th})$M-neuron
    **Begin**
      read( $q(R(M_k^i))) = y_{N-k}^i \in Y_k$
      $q(T(M_k^i)) = q(R(M_k^i))$
    **End**

**Connections Pass**
**For** each $(j^{th})$M-neuron in the M-layer-(k) **do**
  **Begin**
    $p = 1$
    **For** each $(i^{th})$m-neuron in the m-layer-k **do**
      **Begin**
        active $q(C(m_k^i)) = f_{N-k}(q(T(M_k^j)), q(R(m_k^i)))$
        $q(T(m_k^i)) = q(C(m_k^i))$
        $W(m_k^i, M_k^j) = 1$
        *Creating connections: M-layer-(k-1) and*
        **For** $(t^{th})$ M-neuron of the M-layer-(k-1) **do**
          **If** $q(R(M_{k-1}^t)) = q(T(m_k^i))$ **then**
            $W(M_{k-1}^t, m_k^i) = 1$
            Receive the output $s$ from the $(t^{th})$ M-neuron, $s(M_{k-1}^t)$
            Compute the output signal
              $s(m_k^i) = \min\{\mu(q(R(m_k^i))), s(M_{k-1}^t))\}$
            Compute $a_p(M_k^j) = s(m_k^i)$
            $p = p + 1$
      **End**
  **End**

**Output Pass-$l$**
**For** each $(t^{th})$M-neuron of the M-layer-l
  Compute $s(M_l^t) = \max_p\{a_p\}$

**Main Algorithm**
**Begin**
  Perform Initialization Pass
  **For** $k = 1, \ldots, N$ **do**
    **Begin**
      Perform Connections Pass
      Perform Output Pass-$k$
    **End**
**End**

**Equivalence Theorem** The solution provided by FDPNN is equivalent that provided by the DP. *The proof can be found in* [10].

## IV. Application Example

Several detailed illustrative and tutorial examples are presented in [10]. In this section we present an important problem which has been solved by the network proposed here.

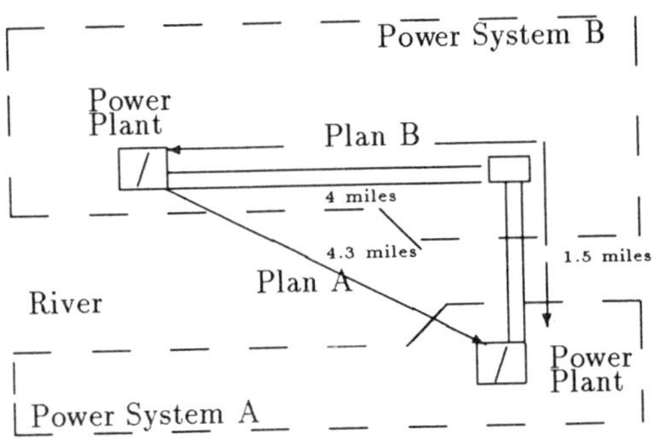


Figure 4: Interconnection Plans

Effective decision analysis is particularly important in power system long-range planning because of the requirements of large financial commitments and resource allocation. Most of the decisions in long-range system planning take place in an environment in which the objectives, the constraints and the consequences of possible decisions are not known precisely.

A feasibility study of an interconnection project between two power systems is considered [12]. The power transfer capacity is assumed to be in the 2500-3000 MW range. The main considerations in the choice of HVDC or HVAC for an interconnection are cost, availability of the cable, benefits and implications for the reinforcement of the two systems arising from the installation of the proposed interconnection.

Figure 4 illustrates two possible plans for the interconnections between System A and System B. Power Plant #1 and #2 are the selected bulk-power generation expansion sites for System A and B, respectively. The proposed interconnection project may be conceived fifteen or twenty years later. A long-range feasibility study is necessary to determine whether it will allow substantially less generation installation in both systems.

The different feasible alternatives for interconnection are showed in Figure 5.

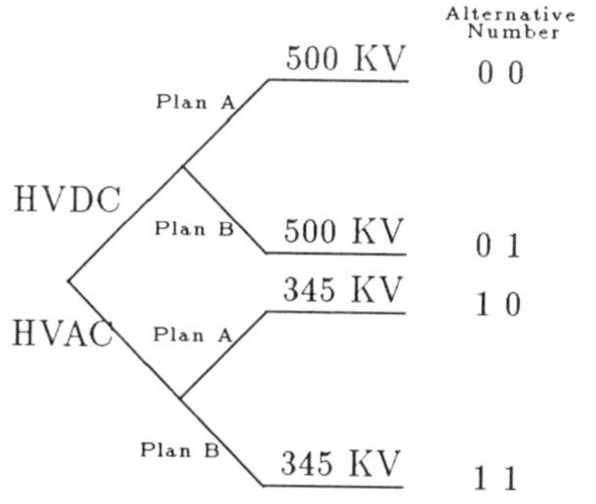


Figure 5: Different feasible alternatives

The merits of all alternatives, that will be considered here, for analyzing and selecting an alternative are capital investment costs, annual operating costs and realiability. The power system in the horizon year may be in any one of the 11 states whose membership grades are presented in Table I. The system states represent all possible system conditions in the horizon year which may have either positive or negative impact on the proposed project.

The state transition tables that define the functions $f_1, f_2, f_3$ relatively to the merits cited above are showed in Table II, Table III and Table IV, respectively.

To construct the corresponding FDPNN, the decision problem is formulated as follows. Let: $A = \{a_1, a_2, a_3, a_4\}$ be the alternatives set, Fuzzy goal at time $t = 3$ be the fuzzy set whose membership function is given by Table I, Fuzzy constraints at time $t = 2$, $t = 1$, $t = 0$ be the different merits, capital investment costs, annual operating costs and realibility whose membership functions are given by:

$$\mu_2(a_1) = 0.6, \mu_2(a_2) = 0.6, \mu_2(a_3) = 0.79, \mu_2(a_4) = 0.89$$

$$\mu_1(a_1) = 0.5, \mu_1(a_2) = 0.5, \mu_1(a_3) = 0.84, \mu_1(a_4) = 0.84$$

$$\mu_0(a_1) = 0.7, \mu_0(a_2) = 0.6, \mu_0(a_3) = 0.88, \ \mu_0(a_4) = 0.8$$

respectively.

Through the FDPNN the membership function of the fuzzy goal induced at $t = 2$, $t = 1$, $t = 0$ and the corresponding maximizing decision can be found as well. The optimal solution can be recovered through connections, providing the optimum decision :

$$a_3 \text{ or } a_4, a_2, a_1$$

with the corresponding value $\mu_2$ being 0.8. For detailed description and neural network topology see[10].

| $t$ | 1 | 2 | 3 | 4 | 5 | 6 | 7 | 8 | 9 | 10 | 11 |
|---|---|---|---|---|---|---|---|---|---|---|---|
| $\mu$ | .25 | .5 | .8 | .5 | .25 | .3 | .4 | .7 | .3 | .7 | .3 |

Table I: Horizon year system states

| Alt. | System States | | | | | | | | | | |
|---|---|---|---|---|---|---|---|---|---|---|---|
| | 1 | 2 | 3 | 4 | 5 | 6 | 7 | 8 | 9 | 10 | 11 |
| 1 | 5 | 4 | 2 | 1 | 1 | 2 | 3 | 6 | 1 | 1 | 7 |
| 2 | 5 | 4 | 2 | 1 | 1 | 2 | 3 | 6 | 1 | 1 | 7 |
| 3 | 9 | 8 | 7 | 2 | 1 | 9 | 8 | 10 | 2 | 3 | 10 |
| 4 | 10 | 9 | 8 | 3 | 2 | 10 | 9 | 10 | 2 | 4 | 10 |

Table II: Capital Investment Cost

| Alt. | System States | | | | | | | | | | |
|---|---|---|---|---|---|---|---|---|---|---|---|
| | 1 | 2 | 3 | 4 | 5 | 6 | 7 | 8 | 9 | 10 | 11 |
| 1 | 5 | 4 | 2 | 1 | 1 | 4 | 2 | 5 | 1 | 1 | 5 |
| 2 | 5 | 4 | 2 | 1 | 1 | 4 | 2 | 5 | 1 | 1 | 5 |
| 3 | 10 | 9 | 3 | 2 | 1 | 9 | 8 | 10 | 2 | 3 | 10 |
| 4 | 10 | 9 | 3 | 2 | 1 | 9 | 8 | 10 | 2 | 3 | 10 |

Table III: Annual Operating Costs

| Alt. | System States | | | | | | | | | | |
|---|---|---|---|---|---|---|---|---|---|---|---|
| | 1 | 2 | 3 | 4 | 5 | 6 | 7 | 8 | 9 | 10 | 11 |
| 1 | 4 | 3 | 2 | 1 | 6 | 4 | 5 | 7 | 1 | 1 | 8 |
| 2 | 2 | 2 | 2 | 1 | 5 | 3 | 4 | 6 | 1 | 1 | 7 |
| 3 | 10 | 8 | 6 | 4 | 10 | 9 | 9 | 10 | 2 | 2 | 10 |
| 4 | 8 | 7 | 4 | 3 | 8 | 7 | 8 | 8 | 2 | 2 | 8 |

Table IV: Reliability

## V. Conclusions and Future Work

This paper has proposed a neural network to solve fuzzy decision-making problems. The theoretical basis, an algorithmfor FDPNN implementation and an equivalence theorem between the FDPNN and the dynamic programming procedure have also been addressed. The FDPNN was tested for some applications reported in the literature. It presents some advantages because of the inheriting parallelism of the neural networks. Most of the neural networks developed for optimization purposes, the interconnections are designed a priori and defined in the initialization phase. In the approach proposed here, the M-layers neurons establish his own interconnections and weights with m-layers neurons, provided that the discrete variables values, transformation functions and return functions are given. The discrete variables values are represented by the amounts of receptors and transmitters of each neuron in the FDPNN. It is believed that many large scale optimization problems, that requires a sequence of interrelated fuzzy decisions, can be faster solved by this new approach as long as neural fuzzy computers are available. Thus, our objective is to investigate the potential of this new approach in neurocomputers implementation in the future.

## REFERENCES


[1] Francelin, R.A., Gomide, F., and Loparo, K., "System optimization with artificial neural networks", *Proc. of the Int. Joint Conf. on Neural Neworks*, IJCNN'91, vol. 3, pp. 2639-2644, Nov. 1991.

[2] Francelin, R. A., Ricarte, I., and Gomide, F., "System optimization with artificial neural networks: parallel implementation using transputers", *Proc. of the Int. Joint Conf. on Neural Neworks*, IJCNN'92,vol. 4, pp. 630-635, June 1992.

[3] Bellman, R., "Dynamic programming", *Princeton N. J. :Princeton University Press*, 1957.

[4] Bellman, R. E. and Zadeh, L. A., "Decision-making in a fuzzy environment", *Management Science*, vol 17, no. 4, pp. 141- 164, Dec. 1970.

[5] Zadeh, L. A. , "Fuzzy Sets", *Information on Control*, no. 8, pp 338-353, 1965.

[6] Negoita, C. V., *Expert systems and fuzzy systems*, The Benjamin / Cunnings Publishing Company, Inc., 1985.

[7] Gomide, F. and Rocha, A. F., "Neurofuzzy controllers", *Proc. of 2nd Int. Conf. on Fuzzy Logic and Neural Networks*, IIZUKA, JAPAN, July 1992.

[8] Gomide, F. A. and Rocha, A. F., "Neurofuzzy components based on threshold", *Proc. of IFAC Symposium on Intelligent Components and Instruments for Control Applications*, SICICA'92, Málaga, Spain, May 1992.

[9] Rocha, A. F., *Neural Nets: A theory for brains and machines*, Lecture Notes in Artificial Intelligence, Springer-Verlag, 1993.

[10] Francelin, R. A. and Gomide, F. A. C, *A Neural network to solve fuzzy discrete dynamic programming problems*, DCA/FEE/ UNICAMP Technical Report no. 018/92, sept. 1992.

[11] Nemhauser, G., *Introduction to Dynamic Programming*, John Wiley Sons, Inc., 1966.

[12] Dhar, S. B., "Power system lon-range decision analysis under fuzzy environment", *IEEE Trans. on Power Apparatus and Systems*, Vol. PAS-98, no. 2, March/April, 1979.

# On Identification of Structures in Premises of a Fuzzy Model Using a Fuzzy Neural Network

Shin-ichi Horikawa, Takeshi Furuhashi, and Yoshiki Uchikawa
Department of Electronic-Mechanical Engineering, Faculty of Engineering, Nagoya University
Furo-cho, Chikusa-ku, Nagoya, 464-01 Japan
Tel:+81-52-781-5111 ext.2792 Fax:+81-52-781-9263
E-mail:horikawa@uchikawa.nuem.nagoya-u.ac.jp

***Abstract*—Fuzzy modeling is a method of describing characteristics of systems using fuzzy reasonings. The method has a distinguishing feature in that it can express complex nonlinear systems linguistically. A lot of researches on automatic identification of fuzzy models have been done. The authors have proposed a fuzzy modeling method using fuzzy neural networks (FNNs). The FNNs have a high performance in representing complex nonlinear input-output relationships. The problem of the FNNs is that it is not known whether the tuning of the membership functions in premises with the back-propagation (BP) algorithm are proper or not.**

**This paper studies the tunings of the membership functions in the premises of an FNN and shows that the BP algorithm realizes appropriate tunings. Based on the results of this study, this paper also presents a method to identify the fuzzy models with the minimal number of the membership functions in the premises.**

## I. Introduction

Fuzzy modeling is a method of describing characteristics of systems using fuzzy reasonings [1][2]. The method has a distinguishing feature in that it can express complex nonlinear systems linguistically. The process of identifying a fuzzy model is generally divided into the identification of the premises and the identification of the consequences. The identification of the consequences is the same as the conventional linear programming problem. The identification of the premises is a particular problem of fuzzy modeling.

The identification of the premises consists of the identification of the structures and the identification of the parameters. In the premises, the structures mean the way of fuzzy partition of the input space and the parameters determine the shapes of the membership functions. Both of them are important to determine the performance of the fuzzy model in representing the nonlinear input-output relationships. A lot of researches on automatic identification of the premises have been done [1]–[10]. Most of the proposed methods in the literature search for the best structure by increasing the number of the fuzzy partitions of the input space step by step [1]–[5]. The method in [6] finds out the best structure with the genetic algorithm. To identify the parameters, various methods such as complex method [1][2], steepest descent method [7]–[10] and penalty method [5] are used.

The authors have proposed a fuzzy modeling method using fuzzy neural networks (FNNs) [11][12]. The FNNs can identify the fuzzy rules and tune the membership functions automatically using the back-propagation (BP) algorithm [13]. The modeling method in [11][12] identify the structures in the premises by increasing the number of the input variables one by one and identify the parameters with the learning capabilities of the FNNs. Although the method has a high performance in obtaining fuzzy models of complex nonlinear systems, it is not known whether the tunings of the membership functions with the BP algorithm are proper or not.

This paper studies the tunings of the membership functions in the premises of an FNN using the input-output data of which the characteristics are known and shows that the BP algorithm realizes the appropriate tunings for representing the characteristics of teaching signals. Based on the results of this study, this paper also presents a method to identify the fuzzy models with the minimal number of the membership functions in the premises.

## II. Fuzzy Neural Network

The FNNs proposed by the authors are multi-layered BP models of which the structures are designed to realize the fuzzy reasoning and to make the connection weights of the networks correspond to the parameters of the fuzzy reasoning. By modifying the connection weights of the network through the learning with the BP algorithm, the FNNs can identify the fuzzy rules and tune the membership functions of the fuzzy reasoning automatically.

The FNN used for the fuzzy modeling in this paper realizes simplified fuzzy reasoning whose consequences are expressed by constants. The inference method with two inputs $(x_1, x_2)$ and one output $(y)$ is given by

$$R^i : \text{If } x_1 \text{ is } A_{i_1 1} \text{ and } x_2 \text{ is } A_{i_2 2} \text{ then } y = f_i \quad (i = 1, 2, \cdots, n_R;\ 1 \le i_1 \le m_1, 1 \le i_2 \le m_2) \tag{1}$$

$$\mu_i = A_{i_1 1}(x_1) A_{i_2 2}(x_2), \quad \widehat{\mu}_i = \mu_i \Big/ \sum_{k=1}^{n_R} \mu_k \tag{2}$$

0-7803-0614-7/93$03.00 ©1993IEEE

$$y^* = \sum_{i=1}^{n_R} \mu_i f_i \Bigg/ \sum_{i=1}^{n_R} \mu_i = \sum_{i=1}^{n_R} \widehat{\mu}_i f_i \qquad (3)$$

where $R^i$ is the $i$-th fuzzy rule, $A_{i_1 1}$ and $A_{i_2 2}$ are fuzzy variables in the premises, $f_i$ is a constant, $n_R$ is the number of fuzzy rules, $m_j$ $(j = 1, 2)$ is the number of membership functions in the premises for the input variable $x_j$, $\mu_i$ is the truth value of the premise of $R^i$, $\widehat{\mu}_i$ is the normalized value of $\mu_i$, $y^*$ is the inferred value.

Fig. 1 shows the configuration of the FNN. The figure shows the case where the FNN has two inputs, one-output and three membership functions in each premise. The circles and the squares in the figure represent the units of the network. The denotations $w_c, w_g, w_f$ and $1, -1$ between the units mean connection weights.

The grades of the membership functions in the premises are calculated in layers A through D. The square units with a symbol of 1 are the bias units with outputs of unity. The outputs of the units with symbols of $\Sigma$ are the sums of their

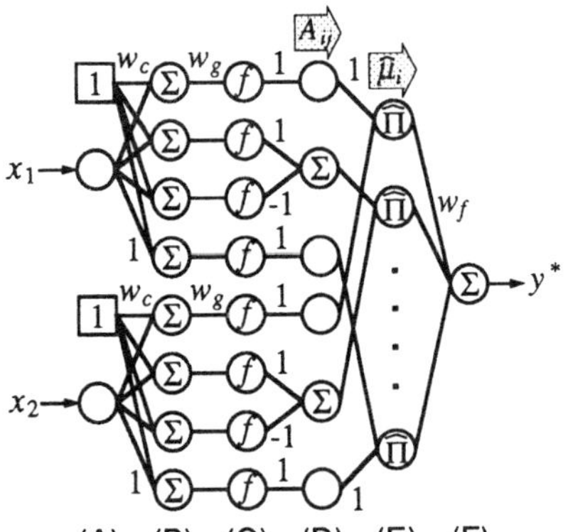


Fig. 1. Fuzzy Neural Network

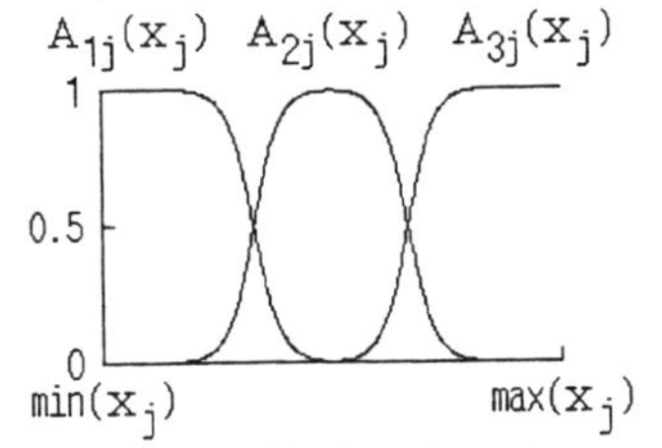


Fig. 2. Membership functions in premise

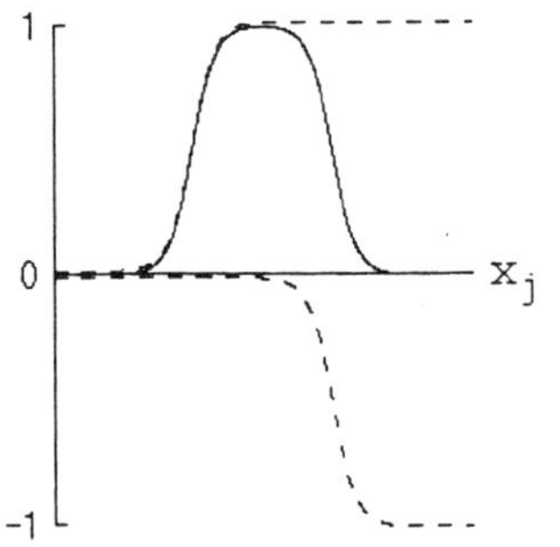


Fig. 3. Composition of membership function $A_{2j}(x_j)$

inputs and the units with symbols of $f$ have sigmoid functions being their inner functions. The units without any symbol just deliver their inputs to the succeeding layers. The inputs of the sigmoid functions in layer C are given by $w_g(x_j + w_c)$. The outputs of the units in layer C are thus expressed as

$$O^{(C)} = 1/[1 + \exp\{-w_g(x_j + w_c)\}]. \qquad (4)$$

The connection weights $w_c$ and $w_g$ are the parameters which determine the central positions and the gradients of the sigmoid functions, respectively. By appropriately initializing the weights, the membership functions in the premises $A_{1j}(x_j), A_{2j}(x_j)$ and $A_{3j}(x_j)$ can be allocated on the universe of discourse as shown in Fig. 2. The pseudo-trapezoidal membership function $A_{2j}(x_j)$ is composed by using two sigmoid functions as illustrated in Fig. 3. The sigmoid functions are shown by the dotted lines and the two functions have opposite signs. Then the truth values of the premises are calculated in layer E. The outputs of the units with a symbol $\widehat{\Pi}$ is the normalized products of their inputs. Since the inputs of the units in layer E are the grades of the membership functions in the premises, the normalized truth values of the premises $\widehat{\mu}_i$ in (2) are calculated in layer E. The inferred value $y^*$ in (3) is obtained as the output of the unit in layer F which is the sum of the products of $\widehat{\mu}_i$ and the connection weights $w_f$ corresponding to $f_i$ in (1). Thus the FNN shown in Fig. 1 realizes the fuzzy reasoning in (1)~(3).

The membership functions in the premises are initialized so that the universes of discourse are divided at the same interval as shown in Fig. 2 and the membership functions are tuned by modifying their parameters $w_c$ and $w_g$ through the learning of the FNN. The connection weights $w_f$ are initialized to be zero and the fuzzy rules are identified by modifying $w_f$ through the learning from the condition with no fuzzy rules.

## III. Tunings of Membership Functions in Premises of Fuzzy Neural Network

The fuzzy models express the input-output relationships of systems with a linear function in each fuzzy subspace of the input space and represent the nonlinear characteristics of the systems by connecting these linear functions smoothly [1]. The performance of the fuzzy models in representing the nonlinear input-output relationships is determined by the way of fuzzy partition of the input space. This paper defines the position of the fuzzy partition as the intersecting point of membership functions and studies the changes of the positions/shapes of the membership functions around the intersecting points for verifying the learning algorithm. The universes of discourse for all input variables are normalized to be in $[-1, 1]$ and the input-output data of a system are distributed in the input space uniformly.

This paper first studies the fuzzy models with one input and one output expressed as

$$R^i : \text{If } x \text{ is } A_i \text{ then } y \text{ is } f_i \qquad (1 \le i \le n_R). \qquad (5)$$

The fuzzy partitions are made by the two kinds of sigmoid functions of the FNN as shown in Fig. 2. The shapes of the fuzzy partitions are determined by the connection weights $w_c$ and $w_g$ which are the parameters of the sigmoid functions. The variations of the connection weights $dw_c$ and $dw_g$ for the $p$-th teaching signal $t_p$ $(1 \leq p \leq N)$ using the BP algorithm are given as follows [11] :

$$dw_c^{(kp)} \propto E(t_p)E(f_{i(k)})F_k'(x)C_k \quad (6)$$

$$dw_g^{(kp)} \propto E(t_p)E(f_{i(k)})F_k'(x)C_kI_k(x) \quad (7)$$

$$E(t_p) = t_p - y^* \quad (8)$$

$$E(f_{i(k)}) = f_{i(k)} - y^* \quad (9)$$

$$F_k'(x) = F'\{w_g^{(k)}(x + w_c^{(k)})\} \quad (10)$$

$$C_k = w_g^{(k)}w_{CD}^{(k)} \Big/ \sum_{l=1}^{n_R} \mu_l \quad (11)$$

$$I_k(x) = |w_g^{(k)}|(x + w_c^{(k)}) \quad (12)$$

where $k$ means the $k$-th sigmoid function $F_k$ numbered from the left hand side on the universe of discourse, $i(k)$ is the number of the fuzzy rule whose premise has the membership function (fuzzy variable) composed by the sigmoid function $F_k$, $F'(\cdot)$ is the derivative of sigmoid function, $w_{CD}^{(k)}$ is the connection weights for $F_k$ between layers C and D. In this paper, the connection weights are modified with the sums of their variations for all teaching signals.

The changes of the connection weights $dw_c$ and $dw_g$ are calculated from the terms in (8)~(12). In the case of the FNN with the number of fuzzy rules $n_R = 2$, these terms have such characteristics on the universe of discourse as shown in Fig. 4. The FNN expresses an increasing function. In the figure, only the connection weights $w_f$ which correspond to the singletons in the consequences of the fuzzy rules are modified. Fig. 4 (a) shows the constants in the consequences $f_1, f_2$ and the inferred value $y^*$ of the fuzzy model. The terms in (9)~(12) for each sigmoid function of the model are calculated as shown in Fig. 4 (b)~(e), respectively. Fig. 4 (f) shows the characteristics of the products of the terms in (9)~(11) $\left(dw_c^{(kp)}/E(t_p)\right)$ and (g) shows those of the products of the terms in (9)~(12) $\left(dw_g^{(kp)}/E(t_p)\right)$. In Fig. 4 (b)~(g), the figures on the left hand side show the characteristics given by the sigmoid function $F_1$ which composes the membership function in the premise $A_1(x)$. The figures on the right hand side show those of the sigmoid function $F_2$ for $A_2(x)$. The dotted lines in the figures express the central positions of each sigmoid function $\left(x = -w_c^{(k)}\right)$. Since the learning data are distributed in the input space uniformly, the changes of the sigmoid functions are determined by the integral values of $dw_c^{(kp)}$ and $dw_g^{(kp)}$. $E(f_{i(k)})$ in (9) shown in Fig. 4 (b) is the difference between the singleton in the consequence of the fuzzy rule $R^{i(k)}$ and the inferred value. From the characteristics of the fuzzy reasoning, the absolute value of $E(f_{i(k)})$ is minimal at the point where the grade of the membership function in the premise is maximal and becomes larger with the smaller grade. $F_k'(x)$ in (10) shown in Fig. 4 (c) is the derivative of the sigmoid function $F_k$. $F_k'(x)$ is maximal at the central position of $F_k$. $C_k$ in (11) shown in Fig. 4 (d) is the product of the connection weights $w_g, w_{CD}$ for $F_k$ and the inverse of the sum of the truth values in the premises. $C_k$ is almost constant and independent of $x$. $I_k(x)$ in (12) shown in Fig. 4 (e) is the linear function which crosses the $x$-axis at the central position of $F_k$. These characteristics of the terms in (9)~(12) will not be changed through the learning. From Fig. 4 (f) and (g), the change of the sigmoid function on the left hand side are determined by $E(t_p)$ around the right hand side of its central position and the change of the sigmoid function on the right hand side are determined by $E(t_p)$ around the left hand side of its central position. In the case of the fuzzy models with the number of fuzzy rules more than $n_R = 2$, the two sigmoid functions at each fuzzy partition are changed in the same way as those of the model with $n_R = 2$ since the characteristics of the terms in (9)~(12) is the same around each fuzzy partition.

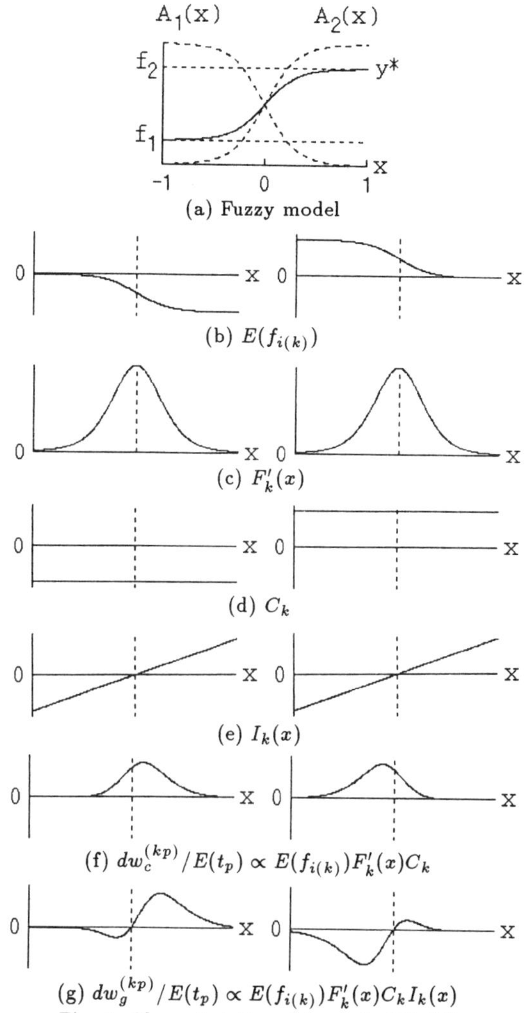

Fig. 4. Characteristics of terms in (6), (7)

This paper studies the tunings of the membership functions in the premises of the FNN using the functions shown in Fig. 5. Fig. 5 (a) is a nonlinear function with $dt_p/dx > 0$ and $d^2t_p/dx^2 > 0$, (b) is a linear function with $dt_p/dx > 0$ and $d^2t_p/dx^2 = 0$, and (c) is a step function. Local portions of

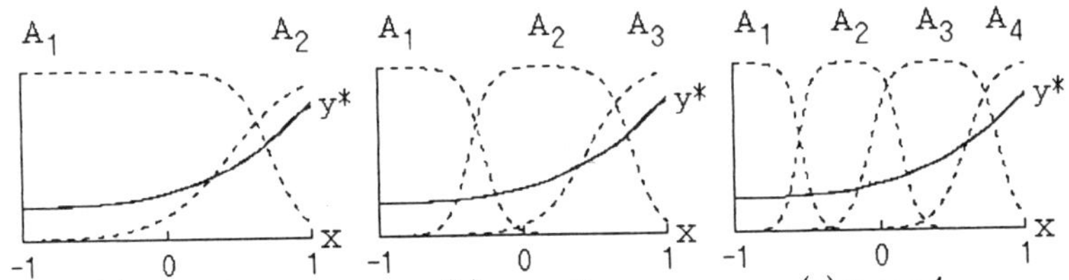


(a) $n_R = 2$ (b) $n_R = 3$ (c) $n_R = 4$

Fig. 6. Tuning results of membership functions in premise for nonlinear function

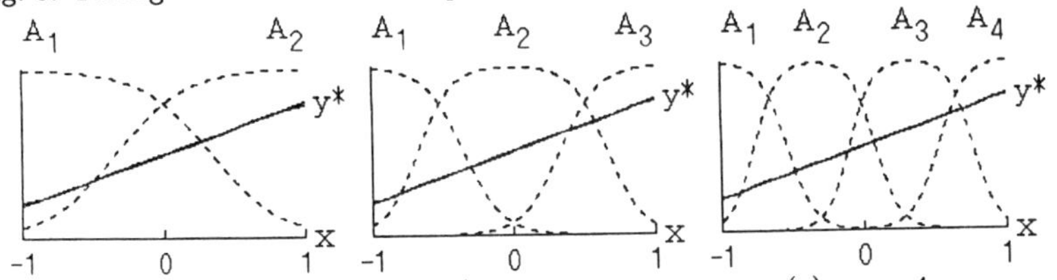


(a) $n_R = 2$ (b) $n_R = 3$ (c) $n_R = 4$

Fig. 7. Tuning results of membership functions in premise for linear function

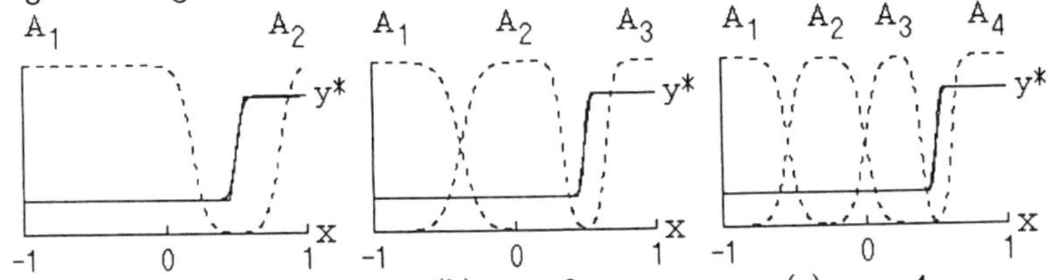


(a) $n_R = 2$ (b) $n_R = 3$ (c) $n_R = 4$

Fig. 8. Tuning results of membership functions in premise for step function

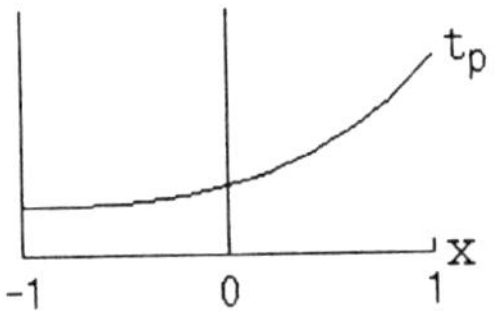


(a) Nonlinear function with $\frac{dt_p}{dx} > 0$ and $\frac{d^2t_p}{dx^2} > 0$

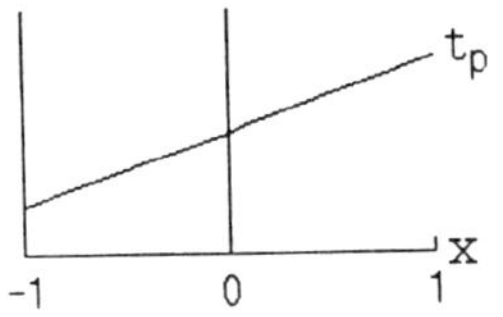


(b) Linear function

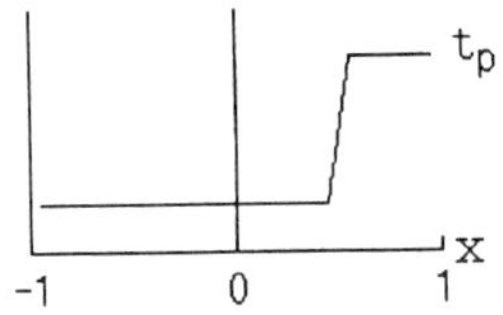


(c) Step function

Fig.5 Input-output relationships of teaching signals

many nonlinear functions can be classified into one of the three functions. In the case where the input-output data shown in Fig. 5 (a) are given, the fuzzy models with the number of fuzzy rules $n_R = 2, 3$ and 4 are tuned by the learning of the FNN as shown in Fig. 6 (a), (b) and (c), respectively. The solid lines are the inferred values $y^*$ of the fuzzy models and the dotted lines are the shapes of the membership functions in the premises of the models. In the learning of the FNN in Fig. 6, the connection weights $w_f$ which mean the singletons in the consequences of the fuzzy rules are first modified. Then $w_c$ which determine the central positions of the sigmoid functions are modified together with $w_f$ and finally all connection weights including $w_g$ which represent the gradients of the sigmoid functions are modified. This learning process makes the characteristic of $E(t_p)$ independent of the values of the teaching signals when $w_c$ are modified and of the initial fuzzy partitions when $w_g$ are modified. The membership functions in the premises of the FNN are always tuned as shown in Fig. 6 in the case where the input-output data have such characteristics as shown in Fig. 5 (a).

In Fig. 6 (a), the position of the fuzzy partition is shifted to the right hand side of the universe of discourse and the gradient of the sigmoid function on the left hand side is steeper than that on the right hand side. The sigmoid function on the left hand side works as an offset. The function on the right mainly forms the nonlinearity of the teaching signals. The fuzzy model shown in Fig. 6 (a) realizes the nonlinear function shown in Fig. 5 (a) with appropriately tuned fuzzy partition. In the case where the linear input-output data shown in Fig. 5 (b) are given, the fuzzy models with $n_R = 2, 3$ and 4 are tuned as shown in Fig. 7 (a), (b) and (c), respectively. In Fig. 7 (a), the positions of the fuzzy partitions have no change through the learning and both the two sigmoid functions are tuned to have gentle gradients to express the linear input-output relationships. Fig. 8 shows the learning results of the fuzzy models for the step function shown in Fig. 5 (c). The position of the fuzzy partition of the fuzzy model with $n_R = 2$ shown in Fig. 8 (a) moves to the stepped-up point of the teaching signals with $w_c$ being modified. The two sigmoid functions of the fuzzy model composes the crisp partition with their steep gradients. In Fig. 6~8 (b) and (c), the fuzzy partitions of the fuzzy models with $n_R = 3$ and 4 are also tuned in the same way as those of the models with $n_R = 2$. In Fig. 6, the sigmoid functions near the point where the data have strong nonlinear characteristics are modified well to be the asymmetric shapes. The shapes of the sigmoid functions of the fuzzy models shown in Fig. 7 are the same for expressing the linear equation. In Fig. 8, only the fuzzy partitions near the dividing points of the teaching signals are tuned.

The FNN can appropriately tune the fuzzy partitions of fuzzy models with the BP algorithm for expressing the characteristics of the input-output data of systems. The above study uses only the increasing functions as shown in Fig. 5. For the decreasing functions which are symmetric to the functions in Fig. 5 at any $y$-axes, the fuzzy partitions are also tuned symmetrically at the axes to those in Fig. 6, 7 and 8. For the functions symmetric to those in Fig. 5 at any $x$-axes, the same tunings of the fuzzy partitions as in Fig. 6~8 will be done since the signs of the products of the terms in (9)~(12) which determine the modifications of the fuzzy partitions will not be changed. In the case of the multi-inputs, $E(f_{i(k)})$ in (9) of the learning algorithm is different from that with one input. The fuzzy partitions will be tuned for expressing the average characteristics.

## IV. Identification of Structures in Premises of Fuzzy Model Using Fuzzy Neural Network

The input-output relationships of the teaching signals used in the previous chapter can be realized by the fuzzy models with the number of fuzzy rules $n_R = 2$. The fuzzy models with $n_R = 3$ and 4 have redundant membership functions in the premises. The number of the fuzzy rules of a fuzzy model used in this paper is equal to the product of the numbers of the membership functions in the premises for the input variables. This means that the larger the number of the membership functions is, the larger the number of fuzzy rules is. The fuzzy model with a large number of fuzzy rules is difficult to understand.

This paper presents a method to identify the fuzzy model with the minimal number of the membership functions in the premises. Based on the results of the study in the previous chapter, the method judges the necessities of the membership functions for realizing the input-output relationships of a system from the shapes of the sigmoid functions after the learning. The algorithm of the method is expressed as follows :

(i) The number of initial fuzzy partitions for the input variables are set to be more than that supposed to be required by the teaching function.

(ii) For the two sigmoid functions $F_{2m-1}$ and $F_{2m}$ which compose the $m$-th fuzzy partition with the intersecting point $P_d^{(m)}$, the connection weights $w_c$ are initialized as $w_c^{(2m-1)} = w_c^{(2m)} = -P_d^{(m)}$ and $w_g$ are initialized so that the membership function in the premise with the most narrow width has an appropriate shape. $w_f$ are initialized to be zero.

(iii) In the same way as in the previous chapter, the learning of the FNN is done by increasing the number of connection weights to be modified in the order of ($w_f$), ($w_f$, $w_c$) and ($w_f$, $w_c$, $w_g$). The FNN is trained until the error of its output converges to a preset level in each stage with the above combination of variable weights.

(iv) In the identified fuzzy model, the ratio of the gradients $w_g^{(2m-1)}$ and $w_g^{(2m)}$ at $P_d^{(m)}$ $\left(r_g^{(m)} = w_g^{(2m-1)}/w_g^{(2m)}\right)$ is calculated.

(v) Whether the input-output equation at the $m$-th fuzzy partition is an increasing one or a decreasing one is judged from the constants in the consequences of the fuzzy rules. The characteristics of the teaching function at the $m$-th fuzzy partition are classified using $r_g^{(m)}$ and the constant $\theta$ ($\theta > 1$ and $\theta \simeq 1$) as follows :

(a) $r_g^{(m)} > \theta$
$\longrightarrow$ Nonlinear with $(dt_p/dx) \cdot (d^2t_p/dx^2) > 0$

(b) $r_g^{(m)} < 1/\theta$
$\longrightarrow$ Nonlinear with $(dt_p/dx) \cdot (d^2t_p/dx^2) < 0$

(c) $1/\theta \le r_g^{(m)} \le \theta$
$\longrightarrow$ Linear or Constant.

(vi) If a series of the fuzzy partitions $P_d^{(m)}$ ($k_1 \le m \le k2$) having the same characteristics, increasing or decreasing, consists of either of the combinations of {(a), (c)} or {(b), (c)} in (v), these fuzzy partitions are put together in a fuzzy partition $(P_d^{(k1)} + P_d^{(k2)})/2$.

(vii) If the number of the fuzzy partitions cannot be reduced further and all the deviations of $P_d^{(m)}$ through (ii) to (vi) are less than a constant $\varepsilon$, the fuzzy modeling is ended. Otherwise, return to (ii).

This algorithm is to make the structure in the premises of a fuzzy model as simple as possible by putting the fuzzy partitions with the same characteristics together. Using the above algorithm, the fuzzy models with $n_R = 3$ and 4 in Fig. 6~8 can be reduced to the models with $n_R = 2$. This paper also examines the feasibility of the algorithm using the following nonlinear functions [4] :

$$y_1 = \begin{cases} -x & ; -1 \le x < 0 \\ x^2 & ; 0 \le x \le 1 \end{cases} \tag{13}$$

$$y_2 = 2^{-2/3}(1.27x + 0.27)^2\{1 - (1.27x + 0.27)\}^{2/3} \quad ; -1 \le x \le 1. \tag{14}$$

Fig. 9 shows the input-output relationships of these functions. $y_1$ consists of a linear function and a quadratic function. Judging from its characteristics, the function needs at least two fuzzy partitions to be expressed by a fuzzy model. $y_2$ is a fractional function. The nonlinear characteristics of the function is not uniform on the universe of discourse. The fuzzy model with at least four fuzzy partitions seems to be able to realize the function. The learning of the FNN is carried out with 43 input-output data at the same interval on the universe of discourse in $[-1, 1]$ for each equation in the same way as in [4][5]. The function with two inputs expressed as $y_3 = y_1(x_1) + y_2(x_2)$ is also identified using 961 data from the universes of discourse divided into 31 points at the same interval for each input variable $x_j$ ($j = 1, 2$).

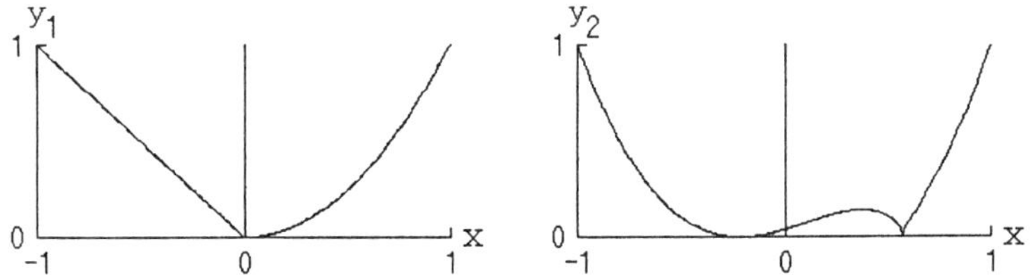


(a) $y_1$ (b) $y_2$

Fig. 9. Complex nonlinear functions

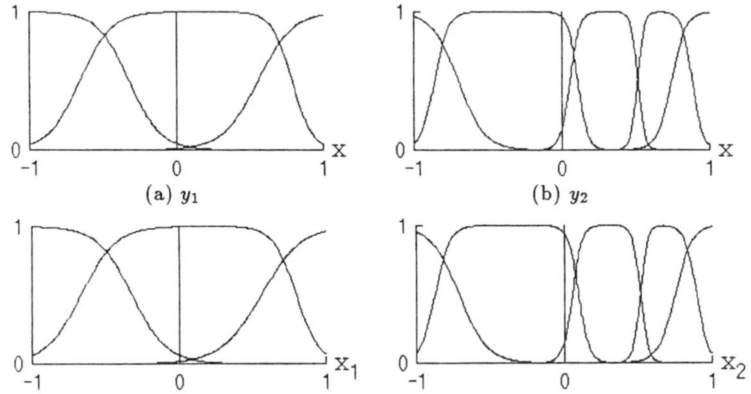


(c) $y_3 = y_1(x_1) + y_2(x_2)$

Fig. 10. Identified structures in premises

Fig. 10 shows the identified structures in the premises of the fuzzy models. The results are obtained from the condition with the initial number of the membership functions in the premises being 11. $\theta$ in (v) is set to be 1.2. This value is determined through the study in the previous chapter. $\varepsilon$ in (vii) is set to be 0.02 corresponding to 1% of the width of the universe of discourse. The identified fuzzy models with both single input and two inputs have the same numbers of the fuzzy partitions as expected before. The errors of the models identified by the FNN for $y_1$ and $y_2$ are less than the half of the errors of the models with the same numbers of fuzzy rules identified by the methods in [4][5].

## V. Conclusions

This paper studied the tunings of the membership functions in the premises using the FNN with the BP algorithm. The results showed that the fuzzy partitions of the input space of the FNN are tuned appropriately according to the characteristics of the input-output data of systems. This paper also presented a method to identify the fuzzy models with the minimal number of the membership functions in the premises. The feasibility of the modeling method was examined using the nonlinear functions with single input and two inputs. The method can identify the fuzzy models with the simple structures in the premises. The fuzzy rules of the obtained models are easy to understand.

## References


[1] T. Takagi and M. Sugeno, "Fuzzy identification of systems and its applications to modeling and control," *IEEE Trans. Syst., Man, Cybern.*, Vol. 15, No. 1, pp. 116-132, 1985.

[2] G. Kang and M. Sugeno, "Fuzzy modeling," *Trans. SICE*, Vol. 23, No. 6, pp. 650-652, 1987 (in Japanese).

[3] H. Ichihashi and T. Watanabe, "Learning control by fuzzy models using a simplified fuzzy reasoning," *J. SOFT*, Vol. 2, No. 3, pp. 429-437, 1990 (in Japanese).

[4] S. Araki, H. Nomura, I. Hayashi, and N. Wakami, "A self-generating method of fuzzy inference rules," in *Proc. IFES'91*, 1991, pp. 1047-1058.

[5] Y. Kajitani, R. Katayama, and Y .Nishida, "Acquisition of fuzzy rules using penalty method," in *Proc. 8th Fuzzy System Symposium*, 1992, pp. 257-260 (in Japanese).

[6] H. Nomura, S. Araki, I. Hayashi, and N. Wakami, "A learning method of fuzzy reasoning by delta rule –a structure identification of antecedent part by genetic algorithm–," in *Proc. 1st FAN Symposium*, 1991, pp. 25-30 (in Japanese).

[7] H. Nomura, I. Hayashi, and N. Wakami, "A learning method of fuzzy inference rules by descent method," in *Proc. FUZZ-IEEE'92*, 1992, pp. 203-210.

[8] T. Watanabe and H. Ichihashi, "Fuzzy control of a robotic manipulator learning inverse kinematics and dynamics models," in *Proc. 6th Fuzzy System Symposium*, 1990, pp. 535-538 (in Japanese).

[9] T. Watanabe and H. Ichihashi, "Iterative fuzzy modeling using membership functions of degree $n$ and its application to a crane control," *J. SOFT*, Vol. 3, No. 2, pp. 347-356, 1991 (in Japanese).

[10] H. Ichihashi, "Hierarchical fuzzy model of class $C^{\infty}$," in *Proc. 7th Fuzzy System Symposium*, 1991, pp. 505-508 (in Japanese).

[11] S. Horikawa, T. Furuhashi, Y. Uchikawa, and T. Tagawa, "A study on fuzzy modeling using fuzzy neural networks," in *Proc. IFES'91*, 1991, pp. 562-573.

[12] S. Horikawa, T. Furuhashi, and Y. Uchikawa, "On fuzzy modeling using fuzzy neural networks with the back-propagation algorithm," *IEEE Trans. Neural Networks*, Vol. 3, No. 5, pp. 801-806, 1992.

[13] D. E. Rumelhart, J. L. McClelland, and the PDP Research Group, *Parallel Distributed Processing*. Cambridge, MA: MIT Press, 1986.

# Conference Author Index